ACCIDENT PREVENTION MANUAL FOR BUSINESS & INDUSTRY

Engineering & Technology
14th EDITION

EDITORS:

PHILIP E. HAGAN, JD, MBA, MPH, ARM, CIH, CET, CHMM, CHCM, CHSP, CEM

JOHN F. MONTGOMERY, PHD, CSP, CHMM

JAMES T. O'REILLY, JD

National Safety Council
Itasca, IL

NSC Press Editor: Deborah Meyer
Cover Design, Interior Design, and Composition: Jennifer Villarreal
Executive Director Publications: Suzanne Powills
Cover Photo: SafakOguz/iStock/Thinkstock

Copyright, Waiver of First Sale Doctrine

The National Safety Council's materials are fully protected by the United States copyright laws and are solely for the noncommercial, internal use of the purchaser. Without the prior written consent of the National Safety Council, purchaser agrees that such materials shall not be rented, leased, loaned, sold, transferred, assigned, broadcast in any media form, publicly exhibited or used outside the organization of the purchaser, reproduced, stored in a retrieval system or transmitted in any form or by any means, electronic, mechanical, photocopying, recording, or otherwise. Use of these materials for training for which compensation is received is prohibited, unless authorized by the National Safety Council in writing.

Disclaimer

Although the information and recommendations contained in this publication have been compiled from sources believed to be reliable, the National Safety Council makes no guarantee as to, and assumes no responsibility for, the correctness, sufficiency, or completeness of such information or recommendations. Other or additional safety measures may be required under particular circumstances.

Copyright © 1946, 1951, 1955, 1959, 1964, 1969, 1974, 1981, 1988, 1992, 1997, 2001, 2009, 2015
by the National Safety Council
All Rights Reserved
Printed in the United States of America
24 23 22   10 9 8 7 6 5 4 3

Library of Congress Cataloging-in-Publication Data
Accident prevention manual for business & industry. Engineering & technology/[edited by] Philip E. Hagan, John F. Montgomery, James T. O'Reilly. — 14th edition.
    pages cm
  Includes index.
  ISBN 978-0-87912-322-2
  1. Industrial safety—United States—Handbooks, manuals, etc. 2. Accidents—United States—Prevention—Handbooks, manuals, etc. I. Hagan, Philip (Philip E.), editor. II. Montgomery, John F. (Johnny Franklin), 1944– editor. III. O'Reilly, James T., 1947– editor. IV. Title: Engineering & technology. V. Title: Accident prevention manual for business and industry. Engineering and technology.
  T55.A3333 2015
  658.3'82—dc23
                    2014045956
Product Number: 121590000

CONTENTS

Preface v

New and Revised Material v

Definitions of Terms vi

Acknowledgments vi

Contributors vi

PART 1   FACILITIES 1

1   Safety through Design 3

2   Buildings and Facility Layout 27

3   Construction of Facilities 59

4   Maintenance of Facilities 101

5   Fired Pressure Vessels (Boilers) and Unfired Pressure Vessels 129

PART 2   WORKPLACE EXPOSURES AND PROTECTIONS 151

6   Safeguarding 153

7   Personal Protective Equipment 179

8   Electrical Safety 221

9   Fire Protection 267

10   Flammable and Combustible Liquids 319

PART 3   MATERIALS HANDLING 347

11   Nanomaterials in the Workplace 349

12   Materials Handling and Storage 353

13   Hoisting and Conveying Equipment 389

14   Ropes, Chains, and Slings 447

15   Powered Industrial Trucks 477

16   Haulage and Off-Road Equipment 501

PART 4 PRODUCTION OPERATIONS 515

17 Hand and Portable Power Tools 517

18 Woodworking Machinery 551

19 Welding and Cutting 573

20 Metalworking Machinery 601

21 Working with Hot and Cold Metals 621

22 Automated Lines, Systems, or Processes 679

23 The Computer as a Safety Information Tool 705

PART 5 INDUSTRY-SPECIFIC SAFETY ISSUES 715

24 Process Safety Management 717

25 Aviation Safety 735

26 Oil and Gas Safety 751

27 Waste and Recycling Safety 763

APPENDIX 1 SAFETY AND HEALTH TABLES 773

APPENDIX 2 CONVERSION OF UNITS 783

APPENDIX 3 GLOSSARY 791

Index 811

PREFACE

The 14th edition of the *Accident Prevention Manual for Business & Industry: Engineering & Technology* continues a tradition begun in 1946 with the publication of the first *Accident Prevention Manual*. This Manual brings to the safety/health/environmental professional the broad spectrum of topics, specific hazards, best practices, control procedures, resources, and sources of help known in the field today. To accommodate the expansion of knowledge and topics, the Manuals are now printed in five volumes: *Administration & Programs* (14th edition), *Engineering & Technology* (14th edition), *Environmental Management* (3rd edition), *Security Management* (2nd edition), and *Accident Prevention Manual Essentials* (1st edition).

This 14th edition builds on the excellent work of previous contributors to the National Safety Council's flagship series. Volunteer experts from many different subject areas have come together to make this book an important resource to be used in support of safety programs and related education. In addition to the expertise of National Safety Council volunteers and staff, we have received expert assistance in developing, writing, and reviewing from contributors representing various disciplines and from the editors, Philip E. Hagan, John F. Montgomery, and James T. O'Reilly. If you have different ideas and want them to be considered for the 15th edition, your suggestions are welcome and can be sent to the National Safety Council, 1121 Spring Lake Drive, Itasca, IL 60143, attn. Deborah Meyer or deborah.meyer@nsc.org.

The audience served by this textbook is widespread. Safety professionals with years of experience, individuals new to the field, managers tasked with safety responsibilities, and educators preparing students for careers in the field of safety will find that these volumes are a valuable source of information. Those who work in the fields of risk management and loss control, human resources, and engineering will also find programs and information that can be incorporated successfully into working goals and objectives that will add value to any organization's safety program.

NEW AND REVISED MATERIAL

The Accident Prevention Manuals are intended for a wide range of users: for students using them as textbooks, for corporate or company managers searching for solutions to safety and health problems, for new safety specialists who must plan and organize a safety and health program within a company, or for experienced safety professionals seeking to improve an operating program and to learn more about advances in the field of safety and health. To increase their usefulness, the 14th editions of the *Administration & Programs* volume and the *Engineering & Technology* volume contain new chapters as well as completely revised material in all chapters.

All chapters in both volumes were reviewed, revised, and updated by safety professionals with expertise in the specific subject area. In addition to a new layout and two-column design of the textbook, major changes include:

Administration & Programs volume:

- Chapter 3: Safety Culture—updated and expanded
- Chapter 5: Legal and Regulatory Issues for the Safety Manager—extensively rewritten
- Chapter 11: Injury and Illness Record Keeping, Incidence Rates, and Analysis—new data and analysis added
- Chapter 14: Environmental Management—extensively rewritten
- Chapter 19: Workplace Violence—updated and expanded
- Chapter 21: Industrial Sanitation and Personnel Facilities—updated and moved from *Engineering & Technology* volume
- Chapter 22: Occupational Medical Surveillance—updated and moved from *Engineering & Technology* volume
- Chapter 23: Workers with Disabilities—updated and moved from *Engineering & Technology* volume
- Chapter 25: Transportation Safety—extensively rewritten
- Chapter 29: Homeland Security Compliance in the Workplace—updated and expanded
- Chapter 30: Motivation—extensively rewritten
- Chapter 31: Safety and Health Training—extensively rewritten
- Chapter 32: Media—extensively rewritten

Engineering & Technology volume:

- Chapter 6: Safeguarding—updated and expanded
- Chapter 8: Electrical Safety—extensively rewritten
- Chapter 11: Nanomaterials in the Workplace—updated and expanded
- Chapter 19: Welding and Cutting—extensively rewritten
- Chapter 21: Working with Hot and Cold Metals—updated and combines two metals chapters into one

- Chapter 24: Process Safety Management—updated and moved from *Administration & Programs* volume
- Chapter 25: Aviation Safety—new chapter
- Chapter 26: Oil and Gas Safety—new chapter
- Chapter 27: Waste and Recycling Safety—new chapter

DEFINITIONS OF TERMS

As the concerns and responsibilities of safety/health/environmental professionals expand, so must their ability to communicate and educate. Technical terms are defined in the text where they are used and also in Appendix 3, Glossary, in the *Engineering & Technology* volume. However, the terms *incident* and *accident* deserve a special note. In the years since the original publication of this manual, many theories of accident causation and definitions of the term *accident* have been advanced. The National Safety Council continues to work to increase awareness that an *incident* is a near-accident and that so-called *accidents* are not random events but rather preventable events. To that end, the term *incident* is used in its broadest sense to include incidents that may lead to property damage, work injuries, or both. The following definitions are generally used in this manual:

- **Accident:** That occurrence in a sequence of events that produces unintended injury, death, or property damage. Accident refers to the event, not the result of the event (see unintentional injury).
- **Incident:** An unintentional event that may cause personal harm or other damage. In the United States, OSHA specifies that incidents of a certain severity be recorded.
- **Near-miss incident:** For purposes of internal reporting, some employers choose to classify as "incidents" the near-miss incident; an injury requiring first aid; the newly discovered unsafe condition; fires of any size; or nontrivial incidents of damage to equipment, building, property, or product.
- **Unintentional injury:** The preferred term for accidental injury in the public health community. It refers to the result of an accident.

With proper hazard identification and evaluation, management commitment and support, preventive and corrective procedures, monitoring, evaluation, and training, unwanted events can be prevented.

ACKNOWLEDGMENTS

General Editor Phil Hagan thanks his many friends who still remain friends even when he disappears for long periods of time to work on book projects and such—Bil and Sharon, Paul and Shenfen, Fannie and Michael, Enrique and Michael, Dale and Miyoko, Larry, Don, and most of all, Paxton. He also thanks Deborah Meyer for keeping the team on the straight and narrow—always moving forward and keeping the team focused on getting to press.

General Editor John Montgomery thanks his wife, Karen, and the memory of Christopher for their support and encouragement during the editing and rewriting process. He also wishes to thank all chapter contributors and reviewers for their diligent review and manual updates, with special thanks to Teddy Gil, Curt Lewis, and JB Gregory for their willingness to take on the task of writing new chapters. He also wishes to thank Air Serv Corp/ABM for its support and encouragement during the review of the manuals.

General Editor Prof. O'Reilly thanks Richard Hackman, Jack Hulon, Charles Geraci, and Jack McAneney for their insights on these complex issues. His research assistant, Marina Schemmel, University of Cincinnati Law Class of 2015, provided exceptional aid in locating and digesting source materials. He also thanks his family—Carol, Jessie, and CB—for their support and encouragement.

Thank you for taking the time to consider and utilize some of these ideas. Of course, none of the book's comments take the place of legal advice, medical advice, or professional advice. Please be certain to discuss the contents of this text with the appropriate professional advisers. We do not offer this as a substitute for the timely, prudent expertise of your organization's regular advisers.

CONTRIBUTORS

The following safety, health, and environmental professionals have contributed to the 14th editions as editors, writers, and/or reviewers of chapters or sections. The National Safety Council very much appreciates the dedication and professional expertise they have contributed to the cause of safety, health, and environmental education.

Alan Barr, co-founder of ErgoMek, LLC, has been a Development Engineer for the UCB/UCSF Ergonomics Program since 2001. He has extensive knowledge in design, CAD modeling, and fabrication of mechanical interventions designed to reduce injury in the workplace. He received a BS in Biomechanics and an MS in Biomedical Engineering from the University of California at Davis in 2000.

Stephen Bennett, ARM, is co-founder of a web-based membership cooperative that brings together a network of independent safety and health consultants, providing a wide range of safety, loss prevention, regulatory management, industrial hygiene, workers' compensation, and return-to-work solutions to public and private employers. Previously, Bennett has held regional to global consulting leadership positions at Sedgwick CMS; Marsh, Johnson & Higgins; and The Travelers. E-mail: smbennett54@sbcglobal.net.

Jairo Betancourt, has more than 25 years of experience in biomedical research and laboratory safety. His experiences include designing, implementing and managing laboratory safety programs for universities and research institutions both domestically and internationally.

He is an active member of the American Biological Safety Association (ABSA). Currently he is involved in the International Working Biosafety Group (IBWG, www.internationalbiosafety.org), the ABSA Philanthropy task force, and co-editor of the Biosafety Compendium. He frequently conducts education and training programs on biosafety and biosecurity in Spain and Latin American Countries throughout Central and South America (Colombia, Mexico, Dominican Republic, Venezuela, and Argentina).

Steve Blackwell, MS, REHS, is an environmental health consultant retired from the U.S. Public Health Service Commissioned Corps. He has extensive experience in the environmental health field with the cruise ship industry, Centers for Disease Control, the Indian Health Service, the U.S. Coast Guard, and the Agency for Toxic Substances and Disease Registry, a federal public health agency of the U.S. Department of Health and Human Services. He has a BS from East Carolina University and an MPH from Florida International University. E-mail: srb0@comcast.net.

Janice Comer Bradley, MS, CSP, is Senior Vice President, National Waste & Recycling Association, where she manages the program group addressing safety, ANSI standards, technical issues, statistics, and education for the waste and recycling industry. She is also the past Vice President of the International Safety Equipment Association, where she managed the activities of Fortune 500 companies involved in supplying safety and health equipment and services. Prior to her work at ISEA, she was the director of environmental health and safety for the Rockefeller University in New York City, the university health and safety officer for Brown University, and the safety specialist for the Department of Veterans Affairs Medical Center in Dayton, Ohio. Bradley is a member of the National Academy of Science, Institute of Medicine, and Research Programs Board on Health Sciences Policy, and she lectures at Georgetown University. She earned a BS from the University of Dayton and a master's degree in environmental studies from Brown University. E-mail: jcbmarch@yahoo.com.

Thomas Bush is a safety professional with close to 30 years of extensive experience in the transportation, manufacturing, insurance, training, aviation, ground handling, and construction industries. He has worked as an Incident Commander on numerous emergency response projects and as a Project Manager for large environmental remediation projects in North America. Bush has been a Professional Member of the American Society of Safety Engineers for the past 28 years. Bush is an instructor for the Texas A&M University Engineering Extension Service.

Salvatore Caccavale, CHMM, CPEA, is Corporate Senior Manager Environmental, Health and Safety at A.M. Castle & Co., 3400 North Wolf Road, Franklin Park, Illinois 60131, (847) 349-2601.

Denis Clark, MS, PE, is a consulting Welding Engineer. He holds a BS from Cornell University and an MS from The Ohio State University, has a PE license in Metallurgical and Materials Engineering from the State of Idaho, and is an AWS Certified Welding Inspector. He has worked in the welding area for 40 years—most of that time at the Idaho National Laboratory. His areas of research have included welding process sensing and control, spray forming, cupola furnace control, the weldability of new alloys, and the application of standard AWS and ASME codes to the welding of nuclear components. Clark is an adjunct professor at Montana Tech University in Butte and has also taught as an adjunct professor at the University of Idaho. He is currently Chair of the AWS Safety and Health Committee. E-mail: denis.clark.51@gmail.com or www.declark-engineering.com.

Robert Wayne Clifton, CSP, PE, ALCM, CPCU, CIE, is presently working as the Global AVP for ESIS Inc. In his current role, he manages a staff of safety professionals in the United States, United Kingdom, Singapore, China, and Malaysia. Clifton has more than 37 years of experience as a safety consultant. He holds a master's degree in safety and health. E-mail: wayne.clifton@esis.com.

Patrick J. Conroy, OHST, CHST, is Special Assistant to the Board of Directors Council on Certification of Health, Environmental and Safety Technologists, 208 Burwash Avenue, Savoy, IL 61874-9571, Office (239) 599-8907. E-mail: pat@cchest.org.

John DeLaHunt, MBA, ARM, has managed environmental, health, safety, and risk issues in higher education since 1989. At Colorado College, he launched a comprehensive EHS program. At the University of Texas at San Antonio, he manages property insurance, workers' compensation, property conservation, and fire protection while serving as the university's Fire Marshal and Risk Manager.

DeLaHunt writes a bimonthly column for the *Journal of Chemical Health & Safety*, edited both editions of the *Environmental Compliance Assistance Guide for Colleges & Universities*, and has presented on diverse topics at dozens of conferences. He has been published in the *Journal of Chemical Health and Safety* and the Association of Higher Education Facilities Officer's Body of Knowledge. He was the charter President of the College and University Hazardous Waste Conference. He serves on the University Risk and Insurance Management Association's government relations and affairs committee and the editorial board for the *Journal of Chemical Health and Safety*.

DeLaHunt holds a bachelor's degree in Chemistry from Colorado College and an MBA in Finance and Management from the University of Colorado–Colorado Springs. E-mail: john_delahunt@msn.com.

Jane Dolezal is the Safety and Compliance Manager for Homewood Disposal Service. She has more than 20 years of combined experience in both the regulatory and the waste and recycling sectors involving the implementation and management of DOT, EPA, and OSHA compliance programs. She is a member of several ANSI committees and has a BS in Environmental and Natural Resources Policy and Studies from Michigan State University.

J. Nigel Ellis, PhD, PE, CSP, CPE, can be contacted at Ellis Fall Safety Solutions, 306 Country Club Drive, Wilmington, DE 19803. E-mail: www.Fall-Safety.com or efss@fallsafety.com.

Michael J. Fagel, PhD, CEM, has more than three decades of public service. He has been in fire service, emergency medical service, public health, law enforcement, and emergency management, as well as corporate safety, security, and threat risk management. Fagel is currently an instructor at the University of Chicago in its new Masters of Threat Risk Management program. He is also an instructor for Benedictine University's Masters in Public Health Program, as well as an instructor at Eastern Kentucky University in its Loss Prevention Masters Program. He was a team leader at the Louisiana State University's National Center for Bio Medical Research and training in its Response to Agricultural Terrorism Training program, as well as its Public Health Programs in Response to High Consequence Events. Fagel has delivered more than 350 lectures and has written more than 100 articles on safety and disaster planning. He has published two textbooks on safety and disaster management and was on the National Domestic Preparedness Office SLAG team at the FBI.

Fagel spent 10 years with the Federal Emergency Management Association in its Occupational Safety & Health Cadre, responding to incidents and disasters such as the Oklahoma City bombing, where he worked as a safety officer and CISD Debriefer. He has also consulted on terrorism-related issues at home and abroad. Fagel can be reached at PO Box 211, Sugar Grove, IL 60554, (630) 907-2020. E-mail: mjfagel@aol.com.

Cristine Z. Fargo has held the position of Manager, Standards Programs for the International Safety Equipment Association for close to 20 years. She manages the voluntary standards setting activities of 13 product groups representing manufacturers and suppliers of safety and health equipment. She represents the ISEA on numerous industry standards committees and works with federal regulatory agencies and outside standards bodies to influence activities that affect the manufacture, use, and distribution of safety equipment. In addition, Fargo speaks at various industry functions about ISEA and its role in the safety equipment business. She holds a degree in Political Science from West Virginia University and can be reached at the ISEA, (703) 525-1695. E-mail: cfargo@safetyequipment.org.

Dave Felinski, Safety Director, Association of Manufacturing Technology, 7901 Westpark Drive, McLean, VA 22102, (703) 893-2900. E-mail: www.amtonline.org.

Anne M. Germain, PE, BCEE, has been the Director of Waste & Recycling Technology for the National Waste & Recycling Association since September 2013. Prior to that, she was the Chief of Engineering and Technology for the Delaware Solid Waste Authority. In addition, she is a Past President of the Solid Waste Association of North America. She has written more than 20 papers and has presented nationally and internationally on solid waste matters. Germain is a Professional Engineer and a Board Certified Environmental Engineer. She has been active in ABET, evaluating environmental engineering programs for accreditation. She graduated from Virginia Tech with a BS in Civil Engineering and received her master's, also in Civil Engineering, from the University of Delaware. E-mail: amgermain@wasterecycling.org.

Teddy Gil, BME, MBA, ASNT Level III ET, FAA A&P License, has held varying positions throughout his career. Most recently, he has served as Process Improvement & Engineering Consultant for Delta Air Lines, implementing executive leadership for short fuse interior configuration upgrades and fabrication; Vice President of Business Development & Quality for Global Integrated Security Services, providing risk assessment consulting and training; and Vice President of People Development and Training for Air Serv Corporation. He also led the Delta Air Lines Technical Operations Employee Council responsible for implementing positive improvement and positive cultural changes throughout the organization for the more than 12,000 Technical Operations employees. Gil was also the presiding officer of the Conflict Resolution Program from 2002 to 2004 for Delta Air Lines Technical Operations. In addition, Gil has published extensively, including several book chapters for the National Safety Council.

Allen Gilley, CIH, CSP, ARM, ALCM, is a Managing Consultant in the Marsh Insurance Company's Atlanta, Georgia, office. He is responsible for the coordination and delivery of professional loss-control services to his clients, which include a wide variety of manufacturers, broadcast and print media, public entities, health care providers, contractors, restaurants, distributors, property management firms, and service organizations. Gilley is manager of the Atlanta Workforce Strategies Group, practice leader of the Southern Regional Workforce Strategies Group, and national Safety and Health practice leader. He also serves as the chairman of the Marsh Atlanta Restaurant Roundtable and chairman of the Marsh Newspaper/Media Risk Control Roundtable. He is also the MRC Practice Leader for Hospitality/Restaurants. E-mail: allen.m.gilley@marsh.com.

JB Gregory, MEd, is an experienced trainer of OSHA regulations for both general industry and construction. He has more than 28 years of experience working with federal, state, and local agencies as well as private employers in the field of regulatory compliance. Gregory has worked with both national and international clients in companies ranging from 2 to more than 800,000 employees in both union and nonunion environments. Services provided to past clients/employers include EHS project management and oversight; EHS program evaluation, development, and implementation; EHS audits; and EHS training.

Gregory has worked with various oil and gas companies in both upstream and downstream EHS operations and has conducted Hazardous Waste Operations Training for federal, state, and local agencies, as well as private employers. He has been on staff at the Texas A&M Engineering Extension Service, OSHA Southwest Education Center, for 15 years. He received a BA and an MEd with an emphasis in Training and Development from the University of Oklahoma.

Von M. Griggs-Laws has 28 years of experience as a safety professional with practical, hands-on occupational safety and health experience in such areas as aircraft maintenance, medical facilities, transportation, civil engineering, communications, supply/warehousing, petroleum distribu-

tion, food services, retail, and other industrial and administrative functions. Griggs-Laws is certified as an OSHA Instructor—General Industry & Construction Safety, Oil & Gas, and is a Certified Safety & Health Official/CSHO Texas A&M; a Certified Government Environmental Specialist/World Safety Organization; a Safety Trained Supervisor/SCHEST; a Drug & Alcohol Defensive Driving Instructor/National Safety Council; a Construction Engineering Concepts/Turner Construction Instructor; and a Work Zone Trainer/MUTCD and TxDOT. She is an Adjunct Instructor in Health & Safety for Texas A&M University, Texas Engineer Extension Service (TEEX).

William Grimes, CSP, is currently the Vice President of Safety and Security for CitationShares, a division of Cessna Aircraft Company that provides fractional jet service. Grimes is a CSP with more than 25 years of safety and health experience and also holds an FAA Airline Transport Pilot rating and type ratings in the Citation 550 and Citation 560 Excel aircraft. Prior to CitationShares, Grimes worked for TAG Aviation and Marsh Risk Consulting as a managing director. He was recognized by the Flight Safety Foundation with the President's Safety Citation award in 2007. E-mail: wgrimes@citationshares.com.

Richard J. Hackman, CIH, QEP, is Associate Director at The Procter & Gamble Co. Hackman has been with P&G for 28 years. He directs the P&G North America Regulatory and Technical Relations organization, which is responsible for regulatory compliance, regulatory/legislative influence, and technical external relations activities across all products and operations in the region. Hackman has a BS in Biology from the University of Cincinnati, an MS in Environmental Sciences from the University of Cincinnati, and an MBA in Finance from Xavier University. He is certified by the American Board of Industrial Hygiene in Comprehensive Practice, as well as by the Institute of Professional Environmental Practice as a Qualified Environmental Professional. E-mail: hackman.rj@pg.com.

Philip E. Hagan, JD, MBA, MPH, ARM, CIH, CHMM, CHCM, CET, CHSP, CEM, is Assistant Professor in the Department of Human Sciences at Georgetown University, Washington DC, and Principal with International Risk Management, LLC. In addition, he is a practicing attorney specializing in tort, safety, environmental, and business-related law issues. He is the former Director of Safety & Environmental Management at Georgetown University. He has consulted internationally on risk, safety, environmental, and emergency management issues in Italy, Qatar, India, Turkey, and China. Hagan has co-authored texts on environmental and workplace safety, legal liability, training, and indoor environmental quality; and he has been a general editor for the *Accident Prevention Manual for Business & Industry* for three editions. He was also the lead editor on the new *Accident Prevention Manual Essentials* (2014). In addition, he has presented to a diverse group of audiences on subjects ranging from hazardous waste disposal to business continuity and emergency management. He is a member of several American Bar Association committees dealing with environmental issues in business transactions, toxic torts, workers' compensation, and international environmental law. He has been a peer reviewer for various safety-related ANSI consensus standards and guidelines. He holds a BS from East Carolina University in Environmental Health, an MPH from George Washington University, an MBA from Georgetown University, and a JD from the George Mason University School of Law. E-mail: haganp@georgetown.edu.

Valienti Antonio Henry, MBA, is board certified in safety and a Certified Lean-Six Sigma Black Belt. Currently, he is a senior manager of Loss Prevention and Reduction at the University of Miami. In addition, Henry also maintains a consulting practice in both the private and public sectors, which includes the development of loss control and risk management programs using specialized Lean-Six Sigma methodologies to maximize efficiency and decrease costs for diverse clients in the health care and financial sectors. He has an MBA in International Business and Finance from the University of Miami. E-mail: vhenry@miami.edu.

David Hibbard, MPH, CIH, is the Director of Environmental Health & Safety at the University of Kentucky. He has more than 25 years of combined experience in the military, regulatory, chemical manufacturing, and higher-education sectors involving the implementation and management of health and safety and industrial hygiene programs. He holds a BS from East Carolina University and a master's in Public Health from Eastern Kentucky University. E-mail: dwhibb0@uky.edu.

Gary A. Higbee, EMBA, CSP, is President/CEO of Higbee & Associates Inc., a full-service international consulting firm. Higbee has more than 40 years of experience in the safety field and is an international expert in industrial and construction safety. He can be reached at (515) 270-6623. E-mail: g.higbee@mchsi.com.

Richard Hislop, PE, CSP, is a Chicago resident and Management Consultant focusing on safety program development and implementation of work planning and control processes for construction projects and operating facilities. E-mail: richard.hislop@gmail.com.

Dwight Hyche, Vice President Senior Boiler, Machinery & Equipment, of the Marsh Risk Consulting Practice Group, serves as a Boiler, Machinery & Equipment Risk Consultant for the Southeast Region of Marsh USA. In this capacity, Hyche specializes in servicing pulp and paper clients, electric utilities, food processors, and other clients with large investments in machinery and equipment assets. He also reviews and coordinates insurance carrier inspection services, provides guidance on recommended risk improvements and plant ratings, and assists in determining marketing strategies. In addition, Hyche performs boiler and machinery risk assessments, loss control inspections, and marketing reports where these services are needed. E-mail: dwight.hyche@marsh.com.

Wendy R. Keys, BS, MS, CS, is certified by the Board of Certification in Professional Ergonomics. Fifteen years ago, she established Ergonomics Engineering Consultants, providing ergonomic and safety services to the chemical, food processing, and manufacturing industries. Keys has been an Instructor for the Texas A&M University Engineering Extension Service since 2007, specializing in Ergonomics and Process Safety Management. Prior to EEC, Keys was employed by Neutral Posture Ergonomics, providing ergonomic consultation to NPE customers nationwide. She also spent 5 years at the corporate headquarters of El Paso Energy as a Senior Safety Engineer. Her primary responsibility with El Paso Energy included office and field ergonomics, process safety management, hearing conservation, asbestos and lead management, OSHA injury reporting, case management, incident statistics, and corporate safety auditing for construction and operational sites. She earned a BS in Industrial Engineering in 1992 and an MS in Industrial Engineering in 1994 at Texas A&M University.

Ken Kolosh directs the National Safety Council statistical reporting and statistical estimating systems. Kolosh leads the development of *Injury Facts*, an annual NSC statistical report on unintentional injuries, their characteristics, and costs. Prior to joining NSC, Kolosh worked in the corporate e-learning industry. He served as advanced strategy and systems consultant with Element K and managing consultant with NETg's strategic services team. He was also senior researcher for NETg's research and development group. Kolosh holds a BA from DePauw University and an MA from Western Kentucky University. He is on the editorial board of the NSC's *Journal of Safety Research*. E-mail: ken.kolosh@nsc.org.

John Kurtz, International Staple, Nail and Tool Association, 512 West Burlington Avenue, Suite 203, La Grange, IL 60525-2245. E-mail: isanta@ameritech.net.

Jim E. Lapping, MS, PE, has more than 30 years of construction industry experience with 20 years as an administrator of the National Safety and Health Education and Training Program. His responsibilities included representation before congressional committees and regulatory agencies on safety, health, and environmental issues. Other accomplishments include development, implementation, and evaluation of more than 40 cooperative safety programs for major construction projects. Lapping has also held positions as Senior Advisor to Assistant Secretary of OSHA, Washington DC, and vice president and safety director of a major construction company.

James Larson, JD, MBA, PE, has a long history of providing consulting services to both internal and external customers. As a student of the Theory of Constraints, Larson has used root cause problem solving to tackle many substantial and pressing issues in the business, construction, manufacturing, educational, and energy industries. He has worked for such notable names as Booz Allen Hamilton, Intel Corporation, Georgetown University, and the U.S. Department of Veterans Affairs. Larson is a licensed professional engineer and attorney, and holds an MBA from Georgetown University.

Curt L. Lewis, PE, CSP, is the President/Owner of Curt Lewis & Associates LLC, a consulting firm specializing in aviation/airline safety, accident investigation and reconstruction, industrial safety, forensic investigation, product safety, system safety, automotive crash worthiness, railroad crossing collision investigations and reconstruction, and airport and aircraft security. He worked for American Airlines/AMR Corporation for 17 years as the Corporate Manager System Safety and previously as the Corporate Manager of Flight Safety and Flight Operational Quality Assurance for American Airlines. Currently, he is an Adjunct Assistant Professor of Occupational Safety & Health at Southeastern Oklahoma State University and an Adjunct Assistant Professor with Embry-Riddle Aeronautical University, teaching aviation, system, human factors, and industrial safety courses (Outstanding Faculty Award—2003). E-mail: curt@curt-lewis.com.

Bob LoMastro is a former Army Green Beret, Navy Hospital Corpsman, and Supervisor of the National Safety Council's Safety Training Institute. He holds a master's degree in Safety Management & Engineering and several bachelor's degrees. LoMastro developed his unique, interactive teaching style as an instructor at the Naval School of Health Sciences. As President of LoMastro & Associates Inc., he draws on a career spanning more than 30 years of teaching health- and safety-related topics worldwide for the military, general industry, and construction companies.

Patrick Lorimer, BS, MPH, has more than 20 years of experience in the Health and Safety industry and is currently the Director of Operations for the EH&S practice Partner Engineering & Science, a large, national environmental engineering firm. Lorimer has worked as a Corporate Risk Manager for a large regional health system and has presented many professional seminars in the field of health safety and environmental hygiene on topics ranging from indoor air quality and risk management to environmental management systems. He holds a BS from East Carolina University and an MPH from George Washington University. E-mail: plorimer@partneresi.com.

Steven L. Lubetkin, APR, Fellow, PRSA, is managing partner of Lubetkin Communications LLC, a strategic communications consulting firm in Cherry Hill, New Jersey. His background includes nearly 30 years in senior corporate communications positions with Fortune 100 companies in transportation, technology, and financial services. E-mail: steve@lubetkin.net.

Brian Maitland, CSP, CET, is a Technical Director in Partner Engineering and Science Inc.'s Health, Safety & Environmental Hygiene discipline. He currently performs regulatory compliance services for public and private entities. His areas of expertise include loss prevention safety audits, accident investigations, job safety observations, and education and training on regulatory compliance issues standards. He has a BS in Environmental Science and Public Health from East Carolina University.

Fred A. Manuele, CSP, PE, is President of Hazards, Limited, the company he formed after retiring from Marsh & McLennan, where he was a Managing Director and Manager of M&M Protection Consultants. His experience in safety spans decades. He is a Certified Safety Professional and a Registered Professional Engineer. Manuele's book, *Advanced Safety Management: Focusing on Z10 and Serious Injury Prevention*, was published in 2008. His book, *On the Practice of Safety*, is in its fourth edition and has been adopted by several professors for safety science degree programs. Other books include Innovations in *Safety Management: Addressing Career Knowledge Needs* and *Heinrich Revisited: Truisms or Myths*. He is also co-editor of *Safety Through Design* and has published a number of papers on safety management. Manuele was awarded the honor of Fellow by the American Society of Safety Engineers and was given the Distinguished Service to Safety Award by the National Safety Council. He is a former board member of the American Society of Safety Engineers, the Board of Certified Safety Professionals, and the National Safety Council.

David E. Marquette, CSP, is the owner of SafeMarq Risk Advisors LLC. He has 40 years of consulting experience in a broad spectrum of risk control topics, including occupational safety, ergonomics, product safety and liability prevention, business continuity plans, and property loss prevention. He has consulted with industrial manufacturers of all types, major airlines, railroads, wholesalers and retailers, and heavy-highway and commercial building contractors. He specializes in aviation ground safety, behavior-based safety, ergonomics, and safety management systems. Marquette is a respected lecturer and regularly presents papers at national and international safety conferences. He was co-editor and contributor to the *Aviation Ground Operations Safety Handbook*, 6th edition, published by the National Safety Council in 2007.

James D. Mayers II, JD, is a Georgia- and DC-barred attorney who earned his JD from Mercer University. His undergraduate concentration was in biology while at Harvard. He has worked as a legal editor as well as clerked for the Macon Circuit Public Defenders' Office.

John J. McAneny, CIH, The Procter & Gamble Company, 2 Procter & Gamble Plaza, Cincinnati, OH 45202.

Bradley A. McPherson, MS, CSP, is a Certified Safety Professional and has worked for Allegheny Energy as a Safety and Health Consultant for 9 years. He has a master's degree in Occupational Hygiene and Safety from West Virginia University and a bachelor's degree in Safety Engineering Technology from Fairmont State University. He can be reached at Route 4 Box 695C, Fairmont, WV 26554.

Bill Montante, CSP, is Vice President and Senior Consultant in casualty hazard control, serving Marsh US and global clients. He has more than 25 years of manufacturing, safety management, ergonomics, and consulting experience. He has written numerous articles and received awards for technical writing excellence. He served as Ergonomics Practice Leader and continues active involvement with this practice group as well as other specialty technical task groups within and external to Marsh Risk Consulting. He has a BSE in Industrial Engineering and Human Factors from the State University of New York and holds several certifications.

Christopher Montgomery is currently a Risk Management and Workers' Compensation Associate with Curt L. Lewis and Associates and works in logistics for the Landstar Transportation Company. Prior to his current position, he was the East Coast Workers' Compensation Manager for Menzies Aviation, with responsibilities for all East Coast operations and the states of Texas and Illinois. He also served on special assignments at the Fort Lauderdale, Florida, and the Seattle, Washington, airports for Menzies Aviation. E-mail: cmonty75@hotmail.com.

John F. Montgomery, PhD, CSP, CHMM, is the Senior Vice President of ES&H for Air Serv Corp, with responsibility for 54 domestic and international stations, and is an Instructor for the Texas A&M University Engineering Extension Program. Prior to joining Air Serv, he was with American Airlines for 18 years, where he served as the Corporate Manager of Ground Safety, Corporate Manager/Acting Managing Director of the Environmental Department, and the Manager of the Noise and Emissions Regulatory Program. Prior to joining American, he was the Corporate Manager of Safety and Lost Time at Sky Chefs, spent time in the industrial sector, and was an Assistant Professor/Lecturer at several universities, including Texas A&M, University of Central Missouri, Central Oklahoma University, and Lamar University.

Montgomery holds three advanced degrees, including a PhD in Philosophy from Texas A&M University, and has two Professional Certifications: Certified Safety Professional (Safety) and Certified Hazardous Material Manager (Environmental). He is a frequent speaker at industry meetings and seminars, was an editorial advisor to *Safety + Health Magazine*, and was the General Editor for the 1997, 2001, and 2009 two-volume *Accident Prevention Manual for Business & Industry*, published by the National Safety Council. He was also a contributor to the *Accident Prevention Manual for Environmental Management* (1995 and 2000) and the first edition of *Accident Prevention Manual Essentials* (2014). He was named Rapporteur of the International Air Transportation Association Emissions Sub Group and served as a delegate to the United Nations' International Civil Aviation Organization. He was a contributor/reviewer for the United Nations' Intergovernmental Panel on Climate Change's review of the effect of aircraft emissions on the environment. E-mail: DrJFMonty@sbcglobal.net.

Patrick Moylan, MBA, CSP, is the Senior Director of Safety and Security at BBA Aviation's flight support businesses, Signature Flight Support and Aircraft Service International Group—two leading ground service providers to the global commercial airline and business aviation sectors. In this capacity, he is responsible for the design and implementation of the company's Safety Management System and for nurturing a positive safety culture in its 200+ airport locations around the world.

Prior to his current position, Moylan served as a safety management consultant at the Federal Aviation Administration's Office of System Safety and the Global

Aviation Information Network, an international aviation safety program promoting the collection, analysis, and sharing of aviation safety information. His professional experience also includes positions in airline flight operations, in general aviation operations management, and as an air traffic controller in the U.S. Marine Corps. He has served as an adjunct faculty member at the University of Maryland Eastern Shore and the Community College of Baltimore County, teaching courses in aviation safety, airport management, and air traffic control. Moylan earned his BA and MBA from Michigan State. E-mail: patrick. moylan@bbaaviation.com.

Michael O'Berry, MEd, has 35 years of industrial, fire, and oil field safety experience and is currently the Program Chair of Occupational Safety Engineering and Environmental Management at Eastern New Mexico University. He also is a Director at the OSHA Training Institute Southwest Education Center and is on staff at the Texas Engineering Extension Service within the Texas A&M University System and is an Adjunct Instructor in the New Mexico Junior College System. His expertise includes oil field and natural gas training and emerging alternative energy field sources.

O'Berry hold several degrees, including an AAS from Trinidad State Junior College, Colorado; a BS in Safety and Occupational Education from Wayland Baptist University, Plainview, Texas; and an MEd, in Curriculum Development and Instruction: Technology Integration, from Grand Canyon University, Phoenix, Arizona.

James T. O'Reilly, JD, College of Law and College of Medicine, University of Cincinnati, Cincinnati, Ohio, has authored 45 texts and 200 articles. His scholarly work was acknowledged in a March 2000 decision of the U.S. Supreme Court, quoting one of his textbooks as the "expert" in its field. He was formerly Associate General Counsel of The Procter & Gamble Company and Chair of the Local Emergency Planning Committee for Cincinnati and has served as chair of the American Bar Association's Section of Administrative Law, 1996–1997. He also served as vice mayor of an Ohio city and as a member of the regional council of governments. He has acted as a general editor for both volumes of the *Accident Prevention Manual for Business & Industry* for three editions in addition to the new *Essentials* text produced in 2014. E-mail: joreilly@fuse.net.

Richard Payant, DBA, CFM, CPE, CHS, has more than 20 years of experience as Director of Facilities Management at Georgetown University and more than 23 years of experience with the Army Corps of Engineers. He is a Certified Facility Manager and Plant Engineer and holds a certification in Homeland Security. Payant is also an Adjunct Professor teaching Facilities Management at George Mason University. He is the author of several professional publications and co-authored the *Facility Inspection Field Manual*, the *Facility Manager's Emergency Preparedness Handbook*, and the *Facility Manager's Maintenance Handbook*, 3rd and 4th editions. He holds a BS from Norwich University, an MA from Central Michigan University, and a PhD in Business Administration from Northcentral University. E-mail: rich.payant@gmail.com.

Richard Pifer, BS, MS, is currently a consultant in facilities management. He is the past Associate Vice President for Facilities and Services at the University of Rochester and was in that position from 1999 to 2014. Prior to assuming his role at the University of Rochester, he worked for more than 9 years in the Georgetown University Facilities Department, holding a variety of positions in Facilities Management Administration. He was a career military officer, has a graduate degree in management, and is active on a variety of university and community boards and committees.

Antonello Pileggi, MD, PhD, is a Research Professor at the Division of Cellular Transplantation of the DeWitt-Daughtry Family Department of Surgery and at the Departments of Microbiology and Immunology and Biomedical Engineering at the University of Miami. Since 2003, he directs the Preclinical Cell Processing and Translational Models Program at the Cell Transplant Center of the Diabetes Research Institute. His research has been funded through the National Institutes of Health, Juvenile Diabetes Research Foundation, the Diabetes Research Institute Foundation, and the University of Miami, as well as by other industries. He has lectured at national and international institutions and professional meetings.

Pileggi has served as ad hoc Reviewer and/or Study Section Member for the National Institutes of Health; the Italian Republic's Ministry of Health; American Diabetes Association; Czech Science Foundation; Regenerative Medicine Research Committee, Medical Research Council, United Kingdom Diabetes UK; and the Biomedical Research Council & National Medical Research Council, Singapore; among other national and international agencies. Pileggi has authored 20 scientific book chapters and more than 150 peer-reviewed publications in the fields of organ and cellular transplantation, immunobiology, and regenerative medicine.

Joy Prescott, MS, is Manager of Training at Texas A&M Engineering Extension Service, Infrastructure Training and Safety Institute and OSHA Training Institute Education

Center. She has more than 30 years of experience in the Safety and Health profession. Prescott has served several large corporations directing both national and global safety, health, environmental, and medical programs over her tenure as well as actively participating in many professional organizations. For the past 2 years, her efforts have centered on giving back her industry knowledge through her work at the OSHA Training Institute Education Center.

Cynthia Roth, RN, has been a professional in the ergonomics, safety, and health industry since 1987. In 1993, she co-founded Ergonomic Technologies Corp. Prior to ETC, Roth was Executive Vice President of Biomechanics Corporation of America and the Senior Vice President and Business Manager for the Langer Biomechanics Group. She has lectured on Ergonomics/Biomechanics/Safety and Health to Fortune 500 companies, to Fortune 200 international companies, and at universities and colleges around the world. She is a Trustee and past Chairperson of the American Society of Safety Engineers Foundation Board. With extensive international experience, she has also been appointed a permanent member of New York State's Commission on International Trade and is on the Advisory Boards of the NYC Department of Mental Health and Hygiene, the Ergonomics Exposition, and the publications of *Occupational Hazards* and *CTDNews*. Roth serves on many national committees and is very well published, with articles appearing worldwide. She received a degree from the University of Pittsburgh as a professional registered nurse with specialties in Occupational Nursing and Biomechanics. E-mail: Croth@ergoworld.com.

Steven G. Schoolcraft, PE, CSP, MBA, is the Examination Director for the Board of Certified Safety Professionals and has served in this role since 2002. Prior to joining the board staff, he worked for the American Bureau of Shipping, where he worked closely with the Coast Guard and other components of the U.S. government and government contractors in areas such as risk management and safety engineering. His undergraduate degree is in chemical engineering from Texas A&M University, and he joined NASA's Goddard Space Flight Center after college. Schoolcraft left NASA in 1997 and joined the American Bureau of Shipping. He is a Certified Safety Professional and a licensed professional engineer in several states. He has also earned an MBA and has been in a management role in the safety profession for more than 15 years.

Bonnie Martin Steward is the founder and owner of Martin Safety Consulting, a 10-year-old ES&H consultant company. She has been an Environmental Health and Safety Professional for 30 years and has worked in a number of companies: Adjunct Instructor in Health & Safety faculty for Texas A&M University, Texas Engineer Extension Service; External Reviewer for ISNetworld; Environmental Health & Safety Engineer, Reckitt Benckiser Pharmaceutical Manufacturer, Fort Worth, Texas; Campus Safety Trainer and Training Coordinator, The University of Texas at Arlington; Safety/Environmental Manager, Paragon Trade Brands; Instructor—Occupational Safety & Health Program, Texas State Technical College; Safety Engineer, LTV Aerospace and Defense; Safety Technician, United Technologies; and Safety Technician, M&M Mars Candy. She holds certifications as an OSHA 500 Trainer; Medic First Aid Certified—Train the Trainer; and First Aid/AED/CPR for Adult, Youth, and Infants; and is a Certified NIOSH Defensive Driving Instructor.

Ralph Stuart, MS, CIH, CCHO, is Chemical Hygiene Officer at Keene State College in Keene, NH. He has an MS from the University of Vermont in Environmental Engineering and has been active at the national level in laboratory safety innovations since 1989. These innovations include development of professional health and safety Internet information resources, as well as the EPA Project XL regulatory reinvention project for laboratory chemical waste management. He is currently secretary of the Division of Chemical Health and Safety of the American Chemical Society and chair of the ACS Safety Advisory Panel. E-mail: rstuartcih@me.com.

Patricia L. Thomas, CSP, CHMM, CET, is a Certified Safety Professional, Certified Hazmat Manager, Certified Environmental Safety and Health Trainer, and the founder and CEO of a small specialty consulting firm which provides safety, industrial hygiene, fire protection, and emergency response services, training, and consulting to a variety of industrial, municipal and construction clients. Thomas has responded as the Site Safety Officer to large hazardous material spills on the waterways including the BP Gulf Oil Spill. She has developed and presented a number of specialty courses on loss control including regulatory industrial training classes and safety and health manuals for clients.

Thomas has served at the corporate level as a consultant to five refining facilities in addition to working in several refineries and chemical plants, has provided technical expertise and consultation on refinery/chemical plant safety issues, writing safety procedures, conducting training, and emergency response. She has provided liaison services to OSHA and insurance inspectors/auditors, conducted design reviews, and has consulted on new construction projects. She also has experience directing activities for implementation of the Process Safety Management Standard.

Thomas has dual Bachelor of Science degrees from Oklahoma State University in Fire Protection and Safety Engineering Technology and Business Administration. She has spent 30 years as a master instructor at Texas A&M Industrial Fire School and has been a part-time instructor for the Southwest OSHA Education Training Center for 14 years. She is a member of NFPA, a Professional Member of ASSE, and an Associate Member of the Ft. Worth and Dallas County IEC chapters. She is also an advisory board member to the Environmental Safety and Health Program at Tarrant County Community College, and a member of the National Safety Committee for Independent Electrical Contractors.

Treasa Turnbeaugh, PhD, MBA, CSP, CET, is the Chief Executive Officer for the Board of Certified Safety Professionals. She is responsible for the overall operations of the BCSP as well as its contribution to the safety, health, and environmental profession. Turnbeaugh is experienced in the safety, health, and environmental field and in the field of professional certification. Additionally, she brings experience and leadership in the business arena of both for-profit and not-for-profit organizations.

Turnbeaugh has more than 25 years of experience in the safety profession, with experience in workers' compensation cost reduction, ergonomics, industrial hygiene, indoor air quality, behavior-based safety, cultural assessments, diagnostics and metrics, injury management, and safety process improvement. She is experienced in servicing a variety of industries, including manufacturing, health care, gaming, higher education, agribusiness, and municipalities.

Turnbeaugh holds a PhD in Health Services Research, with a minor in Epidemiology, and an MPH from Saint Louis University; an MBA from Lindenwood University; and both an MS and BS in Occupational Safety and Health, with a specialization in Industrial Hygiene, from Murray State University. She is a member of the American Society of Safety Engineers, the American Industrial Hygiene Association, and the American Society of Association Executives. She has held her CSP certification more than 20 years and is a Certified Environmental, Safety & Health Trainer.

Nicholas Valter, JD, MBA, has more than 20 years of international experience in monitoring safety in work environments. He is currently a practicing attorney and an in-flight safety coordinator for a major airline. He has experience as a union negotiator for airline safety. He has worked internationally with his own companies in both the retail and materials distribution/warehouse export sectors. His undergraduate degree is from Chaminade University, his MBA is from Georgetown University, and he holds a JD from the University of Hawaii. E-mail: nvalter@gmail.com.

Sherrie Wilson was the first female fire fighter–paramedic with the Dallas Fire Rescue Department. She currently serves as the Founder, President, and CEO of Emergency Management Resources LLC and FireHouseCommunications.com. She served on editorial and advisory boards for *Industrial Fire World Magazine*, *Emergency Medical Services Magazine*, and *Texas EMS Magazine* and as a reviewer of EMS text for both Mosby Lifeline and Brady Publishing. She has more than 50 published fire-, EMS-, and emergency-incident-related articles, publications, and conference presentations. She has authored two manuals, *Rescue Team Training* and *Medical Terminology*.

Wilson received a BS in Public Administration from Hawthorne University and completed Paramedic training at the University of Texas Health Science Center, Dallas, Texas. She serves as adjunct Health & Safety faculty for Texas A&M University, Texas Engineer Extension Service.

James A. Wolf, CFPS, CXLT, CMGT, retired in 2007 after serving the last 20 years of his career as a risk control consultant with ESIS Global Risk Control Services. He dedicated many years of service to risk management and safety engineering, most particularly in the manufacturing and aerospace industries. E-mail: wolfhamm@comcast.net.

Lynne Zarate, MSE, CIH, CHMM, is Director of the Division of Maintenance and formerly an Environmental Safety Coordinator with the state of Maryland, Montgomery County Public Schools System, which is the 17th largest public school system in the United States. She has been with the school system since 2003, managing a variety of systemwide facilities-related programs. Prior to joining the school system, she worked as the Safety Manager at Georgetown University. Other experience includes biomechanics research in automobile safety, research and operations in the paper manufacturing industry, and space shuttle development and testing at Kennedy Space Center. Zarate holds a master's degree in Environmental Engineering and a bachelor's in Chemical Engineering. E-mail: zaratelm@hotmail.com.

PART

1

Facilities

To achieve continuous improvement in safety and health, companies must examine the interactions between people and the physical structures in which they work. Part 1, Facilities, examines the design, layout, construction, and maintenance of facilities where work is done and describes how these factors can help create safer working conditions and greater health and safety for workers. The goal is to consider and to design safety features as facilities are being planned. However, even when the design was done decades ago, the facility can be laid out, retrofitted, or rehabilitated to offer a safer working environment today.

Safety through Design

Fred A. Manuele, CSP, PE
John F. Montgomery, PhD, CSP, CHMM

Introduction
Defining Safety through Design ▶ The Safety through Design Model ▶ Benefits of Safety through Design ▶ Relating Safety through Design to Quality Management ▶ General Principles and Definitions

Making Hazards Analyses and Risk Assessments
The Hazard Analysis/Risk Assessment Process ▶ Risk Assessment Matrices

The Hierarchy of Controls
The Logic of Taking Action in the Descending Order Given

Role of the Safety Professional
Behavior Modification versus Workplace Redesign ▶ Ergonomics and Human Factors in Safety through Design ▶ Haddon's Unwanted Energy Release Concept: General Design Requirements

General Safety Design Checklist

Management of Change

Including Safety Specifications in Purchasing Documents

Summary

References

Review Questions

Addendum A: designsafe Report

Addendum B: Potential Failure Mode and Effects Analysis Sequence

Addendum C: Pre-Job Planning and Safety Analysis Outline

INTRODUCTION

Identifying and analyzing hazards and making risk assessments in the design and redesign processes are the core of safety through design. In the recent past, awareness has grown that as safety through design methods are included in safety and health management systems, the frequency of injuries, illnesses, and fatalities will be reduced. Consider the following.

- On July 25, 2005, the American National Standards Institute (ANSI) approved a new standard: ANSI/AIHA Z10–2005, Occupational Health and Safety Management Systems. Thus, for the first time in the United States, a national consensus standard was issued for safety and health management systems applicable to organizations of all sizes and types.

  This is a major development. Z10 is state of the art. This standard will have a significant and favorable impact on the content of the practice of safety. It will become the benchmark against which the adequacy of safety and health management systems is measured. The standard includes several safety through design provisions that require processes to be in place so that:
  - Safety and health needs are addressed in the design and redesign processes.
  - Hazards are identified and analyzed and risks are assessed and prioritized.
  - A prescribed hierarchy of controls is used to reduce risks to an acceptable level.
  - A management of change procedure is implemented so that hazards and risks are properly considered when changes are made.
  - Safety and health specifications are included in purchasing documents and contracts to avoid bringing hazards and risks into the workplace.

- In July 2007, the National Institute for Occupational Safety and Health (NIOSH) held a Workshop on Prevention through Design (PtD). In an invitation letter, Dr. John Howard, director at NIOSH, said, "This event will mark an important starting point for a major national initiative to move design considerations into daily practices to make workplaces safer."

  The potential impact of this initiative on the practice of safety is immense. Several speakers at the workshop described the undertaking as "transformational," meaning that the focus in the practice of safety will be moved upward in applying the hierarchy of controls to give greater emphasis to design and redesign considerations. NIOSH is exclusively an occupational safety and health entity, and its definition of PtD is thus limited:

PtD: Addressing occupational safety and health needs in the design and redesign processes to prevent or minimize the work-related hazards and risks associated with the construction, manufacture, use, maintenance, and disposal of facilities, materials, and equipment.

- A trend has developed in which provisions requiring that hazards be identified and analyzed and risk assessments be made are being included in safety standards and guidelines, a few of which are listed here:
  - *Aviation Ground Operation Safety Handbook,* issued by the International Air Transport Section of the National Safety Council (6th edition, 2007)
  - *Guidance Document for Incorporating Risk Concepts into NFPA Codes and Standards,* issued by the Fire Protection Research Foundation, 2007
  - ANSI/PMMI B155.1–2006, American National Standard for Safety Requirements for Packaging Machinery and Packaging-Related Converting Machinery
  - *SFPE Engineering Guide to Fire Risk Assessment,* issued by the Society of Fire Protection Engineers, November 2006 (A course titled "Introduction to Fire Risk Assessment" related to the text is available on the Internet.)
  - SEMI S2–0706, Environmental, Health, and Safety Guideline for Semiconductor Manufacturing Equipment
  - CSA Z1000–2006, Occupational Health and Safety Management, issued by the Canadian Standards Association
  - ANSI/AIHA Z10–2005, American National Standard for Occupational Health and Safety Management Systems
  - ANSI/ASSE Z244.1–2003, American National Standard for Control of Hazardous Energy, Lockout/Tagout and Alternative Methods.

Prudent safety professionals will recognize these recent developments and acquire the knowledge and skill necessary to give counsel on safety through design concepts. To assist in these endeavors, this chapter discusses the rationale for and gives guidelines on incorporating safety into the design processes. Topics covered include:

1. safety through design concept—and its model
2. benefits of safety through design
3. relating safety through design to quality management
4. general principles and definitions
5. making hazards analyses and risk assessments—and the process
6. risk assessment matrices
7. hierarchy of controls and the logic in support of its application

8. role of the safety professional
9. behavior modification versus workplace redesign
10. ergonomics and human factors in safety through design
11. objectives/principles for safety through design
12. Haddon's energy release concept; general design requirements
13. safety design checklist
14. management of change
15. including safety specifications in purchasing documents.

Defining Safety through Design

Safety through design is defined as integrating hazard analysis and risk assessment methods early in the design and redesign processes and taking the actions necessary so that the risks of injury or damage are at an acceptable level. This concept encompasses facilities, hardware, equipment, tooling, materials, layout and configuration, energy controls, environmental concerns, and products.

Safety through design is not a program with attendant whistles, bells, slogans, and banners running to a sputtering end and forgotten, as many programs are. What is needed is an agreed-upon and well-understood concept—a way of thinking—that is translated into a process that effectively addresses hazards and risks in the design and redesign processes.

The Safety through Design Model

In Figure 1–1, emphasis is given to moving safety from an afterthought to a forethought in the design and redesign of facilities, processes, and products. Thus, considerations of hazards and risks would be moved as far "upstream" as possible in the design process. "Upstream" includes early decision making at the concept stage, when new ideas are considered and when changes are to be made in operations.

As Figure 1–1 indicates, integrating consideration of hazards and risks early in the concept and design stages results in easier and less costly safety implementation and avoids expensive retrofitting in the building, operation, maintenance, and decommissioning periods.

Benefits of Safety through Design

If decisions affecting safety, health, and the environment are integrated into the early stages of the design processes, the following benefits will be derived:

- Significant reductions will be achieved in injuries, illnesses, damage to the environment, and their attendant costs.
- Productivity will be improved.
- Operating costs will be reduced.
- Expensive retrofitting to correct design shortcomings will be avoided.

A developing awareness that application of safety through design concepts has a favorable impact on productivity, unit costs, and avoiding the expensive retrofitting costs has given impetus to the growth of the safety through design movement. Astute safety practitioners recognize the opportunity provided by the expansion of the adoption of safety through design concepts to further incorporate related safety elements into overall management systems. Applying the concepts brings safety practitioners into the mainstream of an entity's business.

Relating Safety through Design to Quality Management

There is a remarkable correlation between quality management and safety through design principles. The same system design and continuous improvement processes that ensure that a product meets quality expectations will also ensure that safety expectations will be met. This provides a vital connection between safety through design and entity goals.

We borrow from W. Edwards Deming, who was world renowned in quality management, to support that premise. As Deming stressed again and again,

> Processes must be designed to achieve superior quality if that is the quality level desired, and … superior quality can not be attained otherwise.

The same principle applies to safety. For an example of what Deming intended, we quote the fifth premise in what Deming called a "Condensation of the 14 Points of Management," as listed in his book *Out of the Crisis* (Deming 1986):

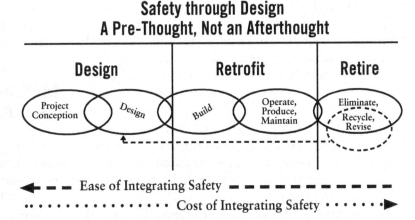

Safety includes: fire, environment, ergonomics, health, vehicle, construction workers.
Projects include: facilities, processes, equipment, products.

Figure 1–1. The model for safety through design.

Improve constantly and forever the system of production and service, to improve quality and productivity, and thus constantly decrease costs.

If you want superior quality, or superior safety, you must design it into new systems, and you must also maintain a continuous improvement program for the redesign of existing workplaces and work methods.

General Principles and Definitions

For the purposes of this chapter, the term *safety design processes* applies to:

- facilities, hardware, equipment, tooling, material selection, operations layout and configuration, energy control, and environmental concerns
- work methods and procedures, personnel selection standards, training content, work scheduling, management of change procedures, maintenance requirements, and personal protective equipment needs
- industrial, commercial, and consumer products for human use.

In applying the safety through design concept, the following definitions and principles govern:

- *Acceptable risk:* The risk for which the probability of a hazard-related incident or exposure occurring and the severity of harm or damage that may result are as low as reasonably practicable (ALARP) and tolerable in the setting being considered.
- *ALARP:* The level of risk that can be further lowered only by an increment in resource expenditure that cannot be justified by the resulting decrement of risk.
- Acceptable risk levels are to be sought with respect to new technology, facilities, materials, and designs; designing new production methods; designing products for human use; and all redesign endeavors.
- Acceptable risk does not mean zero risk, which is unattainable.
- *Hazards:* The potential for harm. Hazards include all aspects of technology and activity that produce risk. Hazards include the characteristics of things (equipment, dusts) and the actions or inactions of people.
- All risks to which the concept of safety through design applies derive from hazards. There are no exceptions.
- Thus, hazards and the risks deriving from hazards must be the focus of design efforts to achieve safety.
- *Safety* is defined as that state for which the risks are judged to be acceptable.
- Hazards are most effectively and economically avoided, eliminated, or controlled if they are considered early in the design process and, where necessary, as the design progresses.

- Both the technology and human activity aspects of hazards must be considered in design decision making.
- If a hazard is not avoided, eliminated, or controlled, its potential may be realized, and a hazards-related incident or exposure may occur that has the potential to, but may or may not, result in harm or damage, depending on exposure.
- Hazards analyses and risk assessments must be integral parts of the design process.
- *Hierarchy of controls:* A systematic way of thinking, considering steps in a ranked and sequential order, to choose the most effective means of eliminating or reducing hazards and the risks that derive from them.
- *Probability:* The likelihood of a hazard being realized and initiating an incident or exposure that could result in harm or damage—for a selected unit of time, events, population, items, or activity being considered.
- *Residual risk:* The risk remaining after preventive measures have been taken. No matter how effective the preventive actions, there will always be residual risk if a facility or operation continues to exist.
- *Risk:* An estimate of the probability of a hazards-related incident or exposure occurring and the severity of harm or damage that could result.
- In the design and redesign processes, the two distinct aspects of risk must be considered:
  ○ avoiding, eliminating, or reducing the *probability* of a hazards-related incident or exposure occurring
  ○ minimizing the *severity* of harm or damage, if an incident or exposure occurs.
- *Risk assessment:* A process that commences with hazard identification and analysis, through which the probable severity of harm or damage is established; it concludes with an estimate of the probability of the incident or exposure occurring.
- *Severity:* The extent of harm or damage that could result from a hazard-related incident or exposure.

As a matter of principle, for an operation to proceed, its risks must be acceptable.

MAKING HAZARDS ANALYSES AND RISK ASSESSMENTS

For many hazards and the risks that derive from them, knowledge gained by management personnel, design engineers, and safety professionals through education and experience will lead to proper conclusions on how to attain an acceptable risk level without bringing teams of people together for discussion. For more complex situations, it is vital to seek the counsel of experienced personnel, at all levels, who are close to the work or process.

Reaching group consensus is a highly desirable goal. Sometimes, for what a safety professional considers obvious, achieving consensus is still desirable so that buy-in is obtained for the actions to be taken. The goal of the risk assessment process is to achieve acceptable risk levels. The risk assessment process is not complete until acceptable risk levels are achieved.

A general guide follows on how to make a hazard analysis and how to extend the analysis into a risk assessment. There are many risk assessment methods. For example, in the *System Safety Analysis Handbook*, 101 methods are described. As a practical matter, having knowledge of three risk assessment concepts will be sufficient to address most occupational safety and health risk situations: initial hazard analysis and risk assessment, the what-if–checklist analysis methods, and failure modes and effects analysis. Addendum A in this chapter is an example of an initial hazard analysis and risk assessment. Addendum B displays the failure modes and analysis method. Checklists should be developed to address the hazards and risks in the locations in which they will be used.

Discussions of those techniques, and many others, can be found in books such as *System Safety Engineering and Management,* by H. E. Roland and Brian Moriarty; *System Safety for the 21st Century,* by Richard A. Stephans; *Basic Guide to System Safety,* by Jeffrey W. Vincoli; *System Safety Analysis Handbook, A Sourcebook for Safety Practitioners,* by R. Stephans and W. W. Talso; and *On the Practice of Safety* and *Advanced Safety Management: Focusing on Z10 and Serious Injury Prevention,* both by Fred A. Manuele (see References).

Whatever the simplicity or complexity of the hazard/ risk situation, and whatever analysis method is used, the following thought and action process is applicable. (This hazard analysis/risk assessment process and the following risk assessment matrices are adaptations of material in *On the Practice of Safety* [Manuele 2003].)

The Hazard Analysis/Risk Assessment Process

1. *Establish the analysis parameters.* Select a manageable task, system, process, or product to be analyzed; establish its boundaries and operating phase (standard operation, maintenance, start-up); and define its interface with other tasks or systems, if appropriate. Determine the scope of the analysis in terms of what can be harmed or damaged: persons (the public, employees), property, equipment, productivity, the environment.

2. *Identify the hazards.* A frame of thinking should be adopted that gets to the bases of causal factors, which are hazards. Ask: What are the aspects of technology or activity that produce risk? What are the characteristics of things (equipment, dusts) or the actions or inactions of people that present a potential for harm? Depending on the complexity of the hazardous situation, some or all of the following may apply:

   ○ Use intuitive engineering and operational sense: this is paramount, throughout.
   ○ Examine system specifications and expectations.
   ○ Review codes, regulations, and consensus standards.
   ○ Interview current or intended system users or operators.
   ○ Consult checklists.
   ○ Review studies from other, similar systems.
   ○ Consider the potential for unwanted energy releases.
   ○ Take into account possible exposures to hazardous environments.
   ○ Review historical data—industry experience, incident investigation reports, the Occupational Safety and Health Administration (OSHA) and National Safety Council data, manufacturer's literature.
   ○ Brainstorm.

3. *Consider the failure modes.* Define the possible failure modes that would result in realization of the potentials of hazards. Ask: What circumstances can arise that would result in the occurrence of an undesirable event? What controls are in place that mitigate against the occurrence of such an event or exposure?

4. *Determine the frequency and duration of exposure.* For each harm or damage category selected for the scope of the analysis (persons, property, business interruption, etc.), estimate the frequency and duration of exposure to the hazard. This is a very important part of this exercise. For instance, in a workplace situation, ask: How often is a task performed? How long is the exposure period? How many people are exposed? More judgments than one might realize will be made in this process.

5. *Assess the severity of consequences.* The purpose is to determine the magnitude of harm or damage that could result. Informed speculations are made to establish the consequences of an incident or exposure: the number of injuries or illnesses and their severity, and the number of fatalities; the value of property or equipment damaged; the time for which productivity will be lost; and the extent of environmental damage. Historical data can be of great value as a baseline. On a subjective basis, the goal is to decide on the worst credible consequences should an incident occur, not the worst conceivable consequence. *When the severity of the outcome of a hazards-related incident or exposure is determined, a hazard analysis has been completed.*

6. *Determine occurrence probability.* Extending the hazard analysis into a risk assessment requires one additional step of estimating the likelihood, the probability, of a hazardous event or exposure occurring. Unless

empirical data are available, which would be a rarity, the process of selecting incident or exposure probability is subjective.

For the more complex hazardous situation, brainstorming with knowledgeable people is necessary. To be meaningful, probability has to be related to an interval base of some sort, such as a unit of time or activity; events; units produced; or the life cycle of a facility, equipment, process, or product.

7. *Define the risk.* Conclude with a statement that addresses the probability of a hazards-related incident or exposure occurring, the expected severity of adverse results, and a risk category (e.g., high, serious, moderate, or low). Using a risk assessment matrix for that purpose assists in communicating the risk level.

8. *Rank risks in priority order.* A risk ranking system should be adopted so that priorities can be established. Because the risk assessment exercise is subjective, the risk ranking system would also be subjective. Prioritizing risks gives management the knowledge needed on the potentials risks have for harm or damage so that intelligent resource allocations can be made for their elimination or reduction.

9. *Develop remediation proposals.* When the results of the risk assessment indicate that risk elimination or reduction measures are to be taken, alternate proposals for the design and operational changes necessary to achieve an acceptable risk level would be recommended. In their order of effectiveness, the actions as shown in the Hierarchy of Controls section (discussed later) would be the base upon which remedial proposals are made.

For each proposal, remediation cost would be determined and an estimate would be given of its effectiveness in achieving risk reduction. Risk elimination or reduction methods would be selected and implemented to achieve an acceptable risk level.

10. *Follow up on actions taken.* Although a hazard analysis and a risk assessment result from applying the steps in the preceding outline, good management requires that the effectiveness of the actions taken be determined. Follow-up activity would determine that the:
    ○ problem was resolved, only partially resolved, or not resolved
    ○ actions taken did or did not create new hazards.

If the problem is not resolved or if new hazards are introduced and the residual risk is not acceptable, the risk is to be reevaluated and other countermeasures are to be proposed. *Residual risk* is the risk remaining after preventive measures have been taken. No matter how effective the preventive actions, there will always be residual risk if an activity continues.

Attaining zero risk is not possible. If the residual risk is not acceptable, the action outline set forth in the foregoing hazard analysis and risk assessment process would be applied again.

11. *Document the results.* Documentation, whether compiled under the direction of location management or by the provider of equipment or services, should include comments on the risk assessment method used, the hazards identified and the risks deriving from them, and the risk reduction measures taken to attain acceptable risk levels.

Risk Assessment Matrices

A risk assessment matrix provides a method to categorize combinations of probability of occurrence and severity of harm, thus establishing risk levels. A matrix helps in communicating with decision makers and influencing their decisions on risks and the actions to be taken to ameliorate them. Also, risk assessment matrices can be used to compare and prioritize risks and to effectively allocate mitigation resources. Safety professionals must understand that definitions of the terms used for incident probability and severity and for risk levels vary greatly in the many risk assessment matrices in use. *Thus, safety professionals should create and obtain broad approval for a risk assessment matrix that is suitable to the hazards and risks with which they deal.*

Several variations of risk assessment matrices are shown in *Advanced Safety Management: Focusing on Z10 and Serious Injury Prevention* (Manuele 2008). Three examples of risk assessment matrices are provided here to serve as references. Table 1–A is an adaptation from a matrix in MIL-STD-882-D, the Department of Defense's Standard Practice for System Safety.

TABLE 1–A. Risk Assessment Matrix

Occurrence Probability	Severity of Consequence			
	Catastrophic	Critical	Marginal	Negligible
Frequent	High	High	Serious	Medium
Probable	High	High	Serious	Medium
Occasional	High	Serious	Medium	Low
Remote	Serious	Medium	Medium	Low
Improbable	Medium	Medium	Medium	Low

Table 1–B, the second exhibit of a risk assessment matrix, is a composite of matrices that include numerical values for probability and severity levels that are transposed into risk scorings. It is presented here for people who prefer

TABLE 1–B. Risk Assessment Matrix: Numerical Gradings

Severity Levels and Values		Occurrence Probabilities and Values				
		Frequent (5)	Likely (4)	Occasional (3)	Seldom (2)	Unlikely (1)
Catastrophic	(5)	25	20	15	10	5
Critical	(4)	20	16	12	8	4
Marginal	(3)	15	12	9	6	3
Negligible	(2)	10	8	6	4	2
Insignificant	(1)	5	4	3	2	1

Very high risk: 15 or greater; high risk: 9 to 14; moderate risk: 4 to 8; low risk: under 4

to deal with numbers rather than qualitative indicators. (Take care, though: the numbers are arrived at judgmentally and are qualitative.)

Table 1–C is an adaptation from the Risk Estimation Matrix in ANSI B11.TR3–2000, a technical report titled Risk Assessment and Reduction—A Guide to Estimate, Evaluate and Reduce Risks Associated with Machine Tools. This matrix is referenced in the example of an initial hazard analysis and risk assessment shown in Addendum A, provided by Design Safety Engineering.

TABLE 1–C. Risk Estimation Matrix—TR3

Probability of Occurrence	Severity of Harm			
	Catastrophic	Serious	Moderate	Minor
Very likely	High	High	High	Medium
Likely	High	High	Medium	Low
Unlikely	Medium	Medium	Low	Negligible
Remote	Low	Low	Negligible	Negligible

THE HIERARCHY OF CONTROLS

A *hierarchy* is a system of persons or things ranked one above the other. A hierarchy of controls provides a systematic way of thinking, considering steps in a ranked and sequential order, to choose the most effective means of eliminating or reducing hazards and the risks that derive from them. Acknowledging that premise—that risk reduction measures should be considered and taken in a prescribed order—represents an important step in the evolution of the practice of safety.

A major premise to be considered in applying a hierarchy of controls is that the outcome of the actions taken is to be at an acceptable risk level. The definition of acceptable risk is repeated here for emphasis.

Acceptable risk is that risk for which the probability of a hazard-related incident or exposure occurring and the severity of harm or damage that could result are as low as reasonably practicable and tolerable in the situation being considered.

That definition requires taking into consideration the:
- practicable minimization of each of the two distinct aspects of risk as risk reduction actions are decided upon:
  - avoiding, eliminating, or reducing the *probability* of a hazard-related incident or exposure occurring
  - reducing the *severity* of harm or damage that may result, if an incident or exposure occurs
- feasibility and effectiveness of the risk reduction measures to be taken, and their costs, in relation to the amount of risk reduction to be achieved.

Decision makers should understand that, with respect to the six levels of action shown in the following hierarchy of controls:
- The ameliorating actions described in the first, second, and third levels are more effective because they:
  - are *preventive* actions that eliminate or reduce risk by design, substitution, and engineering measures
  - rely the least on personnel performance
  - are less defeatable by supervisors or workers.

- Actions described in the fourth, fifth, and sixth levels are *contingent* actions and rely greatly on the performance of personnel.

The literature contains many variations of hierarchies of control. Some have as few as three elements; others have four or five. The following hierarchy has six elements and is considered state of the art. It is compatible with the hierarchy of controls in ANSI/AIHA Z10–2005. But, in one respect it differs. In the Z10 model, the first element reads "Elimination." This author believes that "Elimination" used alone is inappropriately limiting. The following is a list of the hierarchy of controls:

1. Eliminate or reduce risks in the design and redesign processes.
2. Reduce risks by substituting less hazardous methods or materials.
3. Incorporate safety devices.
4. Provide warning systems.
5. Apply administrative controls (work methods, training, work scheduling, etc.).
6. Provide personal protective equipment.

The Logic of Taking Action in the Descending Order Given

Comments follow on each of the action elements listed in the preceding hierarchy of controls, including the rationale for listing actions to be taken in the order given. Taking actions in the prescribed order, as *feasible and practicable*, is the most effective means to achieve risk reduction.

Eliminate or Reduce Risks in the Design and Redesign Processes

The theory is plainly stated. If hazards are eliminated in the design and redesign processes, risks that derive from those hazards are also eliminated. But, elimination of hazards completely by modifying the design may not always be practicable. Then, the goal is to modify the design, within practicable limits, so that the:

- probability of personnel making human errors because of design inadequacies is at a minimum
- ability of personnel to defeat the work system and the work methods prescribed, as designed, is at a minimum.

Examples include designing to eliminate or reduce the risk from:

- fall hazards
- ergonomic hazards
- confined-space hazards
- noise hazards
- chemical hazards.

Reduce Risks by Substituting Less Hazardous Methods or Materials

Methods that illustrate substituting less hazardous methods, materials, or processes for that which is more hazardous include:

- using automated material-handling equipment rather than manual material handling
- providing an automatic feed system to reduce machine hazards
- using a less hazardous cleaning material
- reducing speed, force, amperage

- reducing pressure, temperature
- replacing an ancient steam heating system and its boiler explosion hazards with a hot air system.

Substitution of a less hazardous method or material may or may not result in equivalent risk reduction in relation to what might be the case if the hazards and risks were reduced to a minimum through system design or redesign. Consider this example.

Considerable manual material handling is often necessary in a mixing process for chemicals. A reaction takes place, and an employee sustains serious chemical burns. There are identical operations at two of the company's locations. At one, the decision is made to redesign the operation so that it is completely enclosed, automatically fed, and operated by computer from a control panel, thus greatly eliminating operator exposure.

At the other location, funds for doing the same were not available. To reduce the risk, a substitution took place. It was arranged for the supplier to premix the chemicals before shipment. Some mechanical feed equipment for the chemicals was also installed. The risk reduction achieved by substitution was not equivalent to that attained by redesigning the operation.

Incorporate Safety Devices

When safety devices are incorporated in the system in the form of engineering controls, substantial risk reduction can be achieved. Engineered safety devices are intended to prevent access by workers to the hazard. They are used to separate hazardous energy from the worker and deter worker error. They include devices such as:

- machine guards
- interlock systems
- circuit breakers
- start-up alarms
- presence-sensing devices
- safety nets
- ventilation systems
- sound enclosures
- fall-prevention systems
- lift tables, conveyors, and balancers.

Provide Warning Systems

Warning system effectiveness—and the effectiveness of instructions, signs, and warning labels—rely considerably on administrative controls, such as training, drills, the quality of maintenance, and the reactions of people. Further, although vital in many situations, warning systems may be reactionary in that they alert persons only after a hazard's potential is in the process of being realized (e.g., a smoke alarm). Examples are:

- smoke detectors
- alarm systems
- backup alarms
- chemical detection systems
- signs
- alerts in operating procedures or manuals.

Apply Administrative Controls

Administrative controls rely on the methods chosen as appropriate in relation to the needs, the capabilities of people responsible for their delivery and application, the quality of supervision, and the expected performance of the workers. Some administrative controls are:

- personnel selection
- developing appropriate work methods and procedures
- training
- supervision
- motivation, behavior modification
- work scheduling
- job rotation
- scheduled rest periods
- maintenance
- management of change
- investigations
- inspections.

Achieving a superior level of effectiveness in all of these administrative methods is difficult, and not often attained.

Provide Personal Protective Equipment

The proper use of personal protective equipment relies on an extensive series of supervisory and personnel actions, such as the identification of the type of equipment needed and its selection, fitting, training, inspection, maintenance, and so forth. Examples include:

- safety glasses
- face shields
- respirators
- welding screens
- safety shoes
- gloves
- hearing protection.

Although the use of personal protective equipment is common and is necessary in many occupational situations, it is the least effective method when dealing with hazards and risks. Systems put in place for their use can easily be defeated. In the design processes, one of the goals should be to reduce reliance on personal protective equipment to a practical minimum.

For many risk situations, a combination of the risk management methods shown in the hierarchy of controls is necessary to achieve acceptable risk levels. But the expectation is that consideration will be given to each of the steps in a descending order and that reasonable attempts will be made to eliminate or reduce hazards and their associated risks through steps higher in the hierarchy before lower steps are considered. A lower step in the hierarchy of controls is not to be chosen until practical applications of the preceding level or levels are exhausted.

A yet unpublished document, MIL-STD-882E, includes provisions that further explain the thought process that should govern when the hierarchy of controls is applied. Excerpts follow from the "System safety mitigation order of precedence" section:*

> In reducing risk, the cost, feasibility, and effectiveness of candidate mitigation methods should be considered. In evaluating mitigation effectiveness, an order of precedence generally applies as follows.
>
> a. Eliminate hazard through design selection.
>   Ideally, the risk of a hazard should be eliminated. This is often done by selecting a design alternative that removes the hazard altogether.
> b. Reduce mishap risk through design alteration.
>   If the risk of a hazard cannot be eliminated by adopting an alternative design, design changes should be considered that reduce the severity and/or the probability of a harmful outcome.
> c. Incorporate engineered safety features (ESFs).
>   If unable to eliminate or adequately mitigate the risk of a hazard through a design alteration, reduce the risk using an ESF that actively interrupts the mishap sequence.
> d. Incorporate safety devices.
>   If unable to eliminate or adequately mitigate the hazard through design or ESFs, reduce mishap risk by using protective safety features or devices.
> e. Provide warning devices.
>   If design selection, ESFs, or safety devices do not adequately mitigate the risk of a hazard, include a detection and warning system to alert personnel to the presence of a hazardous condition or occurrence of a hazardous event.
> f. Develop procedures and training.
>   Where other risk reduction methods cannot adequately mitigate the risk from a hazard, incorporate special procedures and training. Procedures may prescribe the use of personal protective equipment.

---

* The dissertation here on the hierarchy of controls is an adaptation of materials in *Advanced Safety Management: Focusing on Z10 and Serious Injury Prevention* (Manuele 2008).

ROLE OF THE SAFETY PROFESSIONAL

The safety professional is often the driving force in a company for including safety decisions during design stages. Although there are numerous standards, regulations, specifications, design handbooks, and checklists that establish the minimums for specific design subjects, no specific standard clearly describes the principles to be applied in designing for safety and the goals to be achieved. The safety professional must work to make safety through design a part of the organization's philosophy and standard operating procedure.

The safety professional can influence the design of the workplace and work methods at three critical points:

- preoperational, in the design process—where the opportunities are greatest and the costs are lower for hazard and risk avoidance, elimination, or control
- in the operational mode—where hazards are to be eliminated or controlled and risks reduced, before their potentials are realized and hazards-related incidents or exposures occur
- postincident—as investigations are made of hazards-related incidents and exposures for causal factor determination and risk reduction.

Behavior Modification versus Workplace Redesign

Because many organizations still adopt a reactive mode in designing for safety, management and safety professionals tend to focus on behavior modification or training as solutions when the problem is workplace or work methods design. Although behavior modification and training are important elements of a safety and health initiative, such measures are misdirected when applied to solving workplace or work methods design problems.

If the design of the work is overly stressful or if the work situation encourages employees to take risks, then the causal factors are principally systemic. To label the causal factors as "employee error" or "unsafe act" would be inappropriate and ineffective, as the following actual case histories illustrate.

- Bags weighing 100 lb (45 kg) were delivered to workstations on pallets. Workers slit open the bags and lifted them to shoulder height to pour the contents into hoppers. The job required a fast work pace, with workers stooping and twisting to lift the bags. Back injuries were frequent. Investigative reports always listed the causal factors as improper lifting. The corrective action was always "reinstructed the worker in proper lifting techniques."

  Obviously, this was a work methods design problem; most of the population would be overstressed in this work situation. As a result, no amount of training, rein-

struction, or behavior modification would correct what was an inherently unsafe act. Yet, in a similar situation, many safety personnel would simply look to past investigation and analysis practices and recommend another employee training program on "how to lift safely."

- Because of a glitch in production scheduling, delivery of parts by a conveyor to a workstation ceased. The design of the conveyor allowed parts to fall off and accumulate beneath the belt. An employee, wanting to keep up with production needs, went beneath the conveyor to retrieve the parts that had collected there. Her hair got caught in a drive belt. When the conveyor started up again, part of her scalp was torn away.

  At first, the causal factor for this incident was recorded as the unsafe act of the employee. Line workers were cautioned not to enter the space beneath the conveyor. Later, however, investigators examined the contributing factors of the production scheduling glitch and the design of the conveyor. If parts had not fallen off the conveyor, the worker would not have been tempted to retrieve them. As a result of this investigation, the design of the conveyor was modified.

- A worker failed to follow the established procedure to lock out and tag out the electrical power during a maintenance operation and was electrocuted.

  The incident investigation report recorded the causal factor as "employee failed to ... ." However, investigators also determined that the distance to the power shutoff was 216 ft (66 m). They showed that the design of the energy system, which made the power shutoff so inconvenient, "encouraged" the employee's risky behavior. Other employees confirmed the findings by expressing their own dissatisfaction with the energy system layout, which promoted "employee error."

Had these organizations studied their work processes as they were being designed—the weight of the bags, operation of the conveyor belt, and the energy system layout—the studies would have raised questions of safety. Management could have foreseen the hazards represented by these work practices or procedures and could have avoided them by designing safer methods. In a reactive mode, the companies looked at design issues only after an incident had occurred.

Organizations might find that for many incidents, the causal factor labeled "employee error"—an unsafe act—is actually "programmed" into the prescribed work method. This is particularly the case when the design of the work is overly stressful, provokes errors, or encourages riskier actions than desired. If the work is so designed, it is reasonable to assume that the "performance deviation" is principally a systemic problem rather than a personal action problem.

Alan D. Swain, in a paper titled "Work Situation Approach to Improving Job Safety," spoke of the workers being in work situations created by management and suggested that management "forego the temptation to place the burden of accident prevention on the individual worker." As he states in his paper:

[A] means of increasing occupational safety is one which recognizes that most human initiated accidents are due to the features in a work situation which define what the worker must do and how he must do it ... . [T]he situation approach, emphasizes structuring or restructuring the work situation to prevent accidents from occurring. Use of this approach requires that management recognize its responsibility (1) to provide the worker with a *safety-prone* work situation and (2) to forego the temptation to place the burden of accident prevention on the individual worker.

One of the ways to create safer workplace designs is to recognize the capabilities and limitations of workers. Designers and safety professionals can use the principles of ergonomics to help them accomplish this goal.

Ergonomics and Human Factors in Safety through Design

Alphonse Chapanis, a leading authority on designing work to fit the capabilities and limitations of people, often stated that companies can benefit greatly by designing work that is not error-provocative. The following excerpt is from his chapter, titled "The Error-Provocative Situation," in *The Measurement of Safety Performance* (1980):

Many work situations and equipment setups are error-provocative. The evidence is clear that people make more errors with some devices than they do with others.

The improvement in system performance that can be realized from the redesign of equipment is usually greater than the gains that can be realized from the selection and training of personnel.

Design characteristics that increase the probability of error include a job, situation, or system which:

- violates operator expectations
- requires performance beyond what an operator can deliver
- induces fatigue
- provides inadequate facilities or information for the operator
- is unnecessarily difficult or unpleasant
- is unnecessarily dangerous.

A central point in Dr. Chapanis's work is that "The improvement in system performance that can be realized from the redesign of equipment is usually greater than the gains that can be realized from the selection and training of personnel."

If the environment constructed by management—which includes the design of both the workplace and work methods—requires behavior that is considered unsafe, then management's focus should be mainly on altering the work environment.

Haddon's Unwanted Energy Release Concept: General Design Requirements

Dr. William Haddon was the first director of the National Highway Safety Bureau. Haddon espoused the concept that unwanted transfers of energy can be harmful (and wasteful) and that a systematic approach to limiting such possibility should be taken. His work is considered seminal. Modified and extended for the workplace, the theory—practicably applied—is this: for all injuries and illnesses, an unwanted and harmful transfer of energy or exposure to an injurious environment is a factor.

Although Haddon stated in his paper titled "On the Escape of Tigers: An Ecological Note" that "the concern here is the reduction of damage produced by energy transfer," he also said that "the type of categorization here is similar to those used for dealing systematically with other environmental problems and their ecology." This excerpt is from Haddon's breakthrough paper:

A major class of ecologic phenomena involves the transfer of energy in such ways and amounts, and at such rapid rates, that inanimate or animate structures are damaged. Several strategies, in one mix or another, are available for reducing the human and economic losses that make this class of phenomena of social concern. In their logical sequence, they are as follows:

- prevent the marshalling of the form of energy
- reduce the amount of energy marshaled
- prevent the release of the energy
- modify the rate or spatial distribution of release of the energy from its source
- separate, in space or time, the energy being released from that which is susceptible to harm or damage
- separate, by interposing a material barrier, the energy released from that which is susceptible to harm or damage
- modify appropriately the contact surface, subsurface, or basic structure, as in eliminating, rounding, and softening corners, edges, and points with which people can, and therefore sooner or later do, come in contact
- strengthen the structure, living or nonliving, that might otherwise be damaged by the energy transfer
- move rapidly in detection and evaluation of damage that has occurred or is occurring, and counter its continuation or extension
- after the emergency period following the damaging energy exchange, stabilize the process.

All hazards are not addressed by the unwanted energy release concept. Such examples are the potential for asphyxiation from entering a confined space filled with inert gas or inhalation of asbestos fibers. But all hazards are encompassed within a goal that is to avoid both unwanted energy releases and exposures to hazardous environments.

Keeping Haddon's unwanted energy release concept in mind will be particularly beneficial as managements, supervisors, engineers, designers, and safety professionals consider adopting safety through design methods. To provide guidance to those who apply the hierarchy of controls, a duplication appears on page 15 of "General Design Requirements: A Thought Process for Hazard Avoidance, Elimination, or Control," as it appears in *On the Practice of Safety* (Manuele 2003). This guideline is the author's extension of the incident and exposure prevention aspects of Haddon's work. The guideline gives advice on designing the workplace and the work methods. It addresses nine major subjects. Haddon listed 10 strategies, one of which is divided here into two parts, becoming items 2 and 3. Haddon's last two subjects pertain to recovery actions to be taken after an incident occurs. They relate to emergency preparedness, for which a good reference is Chapter 18, Emergency Preparedness, in the *Administration & Programs* volume. In no way is it suggested that the following guideline addresses all hazard and risk elimination or amelioration possibilities. It can be helpful as a reference, and as a teaching tool.

GENERAL SAFETY DESIGN CHECKLIST

While the previously given general design requirements presented a thought process for hazard avoidance, elimination, or control, a checklist now follows that is in much more detail. It also gives recognition to Haddon's unwanted energy release concept. The questions in "A. Introduction: Basic Considerations" relate to Haddon's theory and are presented as general concepts to be considered when using the checklist; yes or no answers are to be obtained for these questions. They emphasize that the two distinct aspects of risk are to be considered in the design process:
- avoiding, eliminating, or reducing the *probability* of a hazard-related incident or exposure occurring
- minimizing the *severity* of harm or damage if an incident occurs.

It must be understood that no checklist can be drafted to cover all possible hazards and risks. The following is presented as a basis from which safety professionals can choose and add as they develop checklists suitable to the operations to which they give counsel. This checklist is comparable to that contained in *Advanced Safety Management: Focusing on Z10 and Serious Injury Prevention* (Manuele 2008).

A. Introduction: Basic Considerations
1. Can production of hazardous materials or energy be eliminated?
2. Will the amount of the hazardous materials or energy be limited?
3. Can less hazardous materials be substituted?
4. Can hazardous material or energy buildup be prevented?
5. Can release of hazardous materials or energy be slowed down?
6. Can unwanted energy release be separated in space or time from that which is susceptible to harm or damage?
7. Can barriers be interposed to separate the unwanted energy release from that which is susceptible to harm or damage?
8. Will surfaces with which people come in contact be modified to reduce the risk of injury?

B. Designing for Those with Disabilities
1. Do the designs take into consideration the requirements of the Americans with Disabilities Act (ADA)?
2. Are reasonable accommodations made for the disabled?

C. Confined Spaces
1. Have confined spaces been eliminated by design where possible?
2. Are any confined spaces to be permit required? [OSHA Standard 29 CFR 1910.146(c)(1)]
3. Have confined spaces been designed for ease of ingress, prompt egress, and, where possible, elimination of hazardous atmospheres?
4. Can confined spaces be designed with multiple, large accesses?
5. Are accesses provided with platforms that will support all required personnel and equipment?
6. Will access ports be large enough to permit entry when personnel are using personal protective equipment?
7. Will pipes or ducts limit entry to access ports?
8. Are locations of ladders and scaffolds in the space identified?
9. Are fall-protection needs fulfilled (such as anchorage points)?
10. Can the necessary equipment be moved through accesses?
11. Does the design provide for isolation of the confined space from hazardous energy (i.e., electrical, chemical, etc.)?
12. Does the design provide for isolation by valve blocking, spools, double blocks and bleeds, flanges, and flushing connections?

General Design Requirements: A Thought Process for Hazard Avoidance, Elimination, or Control

1. Avoid introduction of the hazard: prevent buildup of the form of energy or hazardous materials.
   - Avoid producing or manufacturing the energy or the hazardous material.
   - Use material handling equipment rather than manual means.
   - Don't elevate persons or objects.

2. Limit the amount of energy or hazardous material.
   - Seek ways to reduce actual or potential energy input.
   - Use the minimum energy or material for the task (voltage, pressure, chemicals, fuel storage, heights).
   - Consider smaller weights in material handling.
   - Store hazardous materials in smaller containers.
   - Remove unneeded objects from overhead surfaces.

3. Substitute, using the less hazardous.
   - Substitute a safer substance for a more hazardous one: when hazardous materials must be used, select those with the least risk throughout the life cycle of the system.
   - Replace hazardous operations with less hazardous operations.
   - Use designs needing less maintenance.
   - Use designs that are easier to maintain, considering human factors.

4. Prevent unwanted energy or hazardous material buildup.
   - Provide appropriate signals and controls.
   - Use regulators, governors, and limit controls.
   - Provide the required redundancy.
   - Control accumulation of dusts, vapors, mists, etc.
   - Minimize storage to prevent excessive energy or hazardous material buildup.
   - Reduce operating speed (processes, equipment, vehicles).

5. Prevent unwanted energy or hazardous material release.
   - Design containment vessels, structures, elevators, materials handling equipment to appropriate safety factors.
   - Consider the unexpected in the design process, to include avoiding the wrong input.
   - Protect stored energy and hazardous material from possible shock.
   - Provide fail-safe interlocks on equipment, doors, valves.

   - Install railings on elevations.
   - Provide nonslip working surfaces.
   - Control traffic to avoid collisions.

6. Slow down the release of energy or hazardous material.
   - Provide safety and bleed-off valves.
   - Reduce the burning rate (using an inhibitor).
   - Reduce road grade.
   - Provide error-forgiving road margins.

7. Separate in space or time, or both, the release of energy or hazardous materials from that which is exposed to harm.
   - Isolate hazardous substances, components, and operations from other activities, areas, and incompatible materials, as well as from personnel.
   - Locate equipment so that access during operations, maintenance, repair, or adjustment minimizes personnel exposure (e.g., hazardous chemicals, high voltage, electromagnetic radiation, cutting edges).
   - Arrange remote controls for hazardous operations.
   - Eliminate two-way traffic.
   - Separate vehicle from pedestrian traffic.
   - Provide warning systems and time delays.

8. Interpose barriers to protect persons, property, or the environment exposed to an unwanted energy or hazardous material release:
   - insulation on electrical wiring
   - guards on machines, enclosures, fences
   - shock absorbers
   - personal protective equipment
   - directed venting
   - walls and shields
   - noise controls
   - safety nets.

9. Modify the shock concentrating surfaces:
   - padding on low overheads
   - rounded corners
   - ergonomically designed tools
   - "soft" areas under playground equipment.

13. Can spaces be designed so that maintenance and inspection can be performed from outside or by self-cleaning systems?

D. Electrical Safety
1. Overall, will the electrical system meet OSHA/ National Electrical Code requirements?

2. Will the system be sufficiently flexible to allow for future expansion?
3. Will emergency power be provided for critical systems?
4. Is grounding adequate?
5. Are ground-fault interrupter circuits to be installed where needed? [29 CFR 1910.304(f)(7)]
6. Are grounding connections to piping and conduits

eliminated to prevent accumulation of static electricity?

7. Is grounding provided for lightning protection on all structures?

8. Are accommodations made for special-purpose or hazardous locations? [29 CFR 1910.307]

9. Is the design adequate where there may be combustible gases or vapors?

10. Is high-voltage equipment isolated by enclosures such as vaults, security fences, and lockable doors and gates?

11. Are nonisolated conductors such as bus bars on switchboards or high-voltage equipment connections that are located in accessible areas protected to minimize hazards for maintenance and inspection personnel?

12. Where injury to an operator may occur if motors were to restart after power failure, are provisions made to prevent automatic restarting upon restoration of power? [29 CFR 1910.262(c)(1)]

13. Are electrical disconnect switches lockable, readily accessible, and labcled? [29 CFR 1910.303(f)]

14. Are breakers/fuses properly sized? [29 CFR 1910.303(b)]

15. Has the polarity of all circuits been checked? [29 CFR 1910.403(a)(2)]

16. Do electrical cabinets and boxes have appropriate clearances? [29 CFR 1910.303(g) & (h)]

17. Are exposed live electrical parts operating at 50 volts or more guarded against accidental contact by approved cabinets or enclosures, by location, or by limiting access to qualified persons? [29 CFR 1910.303(g)(2)(i)]

18. Are rooms or enclosures containing live parts or conductors operating at more than 600 volts nominal or designed to be kept locked, or have provisions been made to be under the observation of a qualified person at all times? [29 CFR 1910.303(h)(2)]

19. Are the electrical wiring and equipment located in hazardous (classified) locations intrinsically safe, approved for the hazardous location, or safe for the hazardous location? [29 CFR 1910.307(b)(1–3)]

E. Emergency Safety Systems—Means of Egress

1. Are means of egress adequate in number, remote from each other, and properly designated, marked, lighted, and easily recognized?

2. Has emergency lighting been provided for means of egress and elsewhere where needed?

3. Does the design contemplate emergency lighting where workers may have to remain to shut down equipment?

4. Do means of egress exit directly to the street or open space?

5. Are doors, passageways, or stairways that do not lead to an exit marked by signs reading "not an exit" or by

a sign indicating actual use?

6. Does the design provide internal refuge areas for workers who cannot escape?

7. Will reliable emergency power be provided for critical and life-support systems?

8. Will emergency safety showers and eyewash stations be adequate and be properly placed?

9. Will adequate first-aid stations, spill carts, and emergency stations be provided?

F. Environmental Considerations
(Some are operational, beyond design)

1. Have waste products been identified and a means of disposal established?

2. Will provisions be made for responding to chemical spills (containment, cleanup, disposal)?

3. Is there an existing spill control plan for chemicals?

4. Have all waste streams been identified?

5. Are adequate pretreatment facilities provided for process waste streams?

6. Will an adequate storage area be available for wastes held prior to treatment or disposal?

7. Will waste storage areas have adequate isolation or containment for spills?

8. Will hazardous wastes be disposed of at approved treatment, storage, and disposal facilities?

9. Has special equipment or specially trained personnel been provided for treatment operations?

10. Has the acquisition of permits been addressed for the treatment or disposal of waste streams?

11. Have state or local requirements for permitting been evaluated and factored into the project?

12. Can the facility meet regulations for reporting spills or the storage of chemicals?

13. Have adequate provisions been made for cleaning the process equipment?

14. Have provisions been made for a catastrophic release of chemicals?

15. Have provisions been made for any necessary demolition and the resulting waste?

16. Have requirements for remediation at the site prior to construction been addressed?

17. Will all feasible measures for waste minimization be implemented?

18. Have the processes that generate air pollution been evaluated for minimization potential?

19. Will adequate air pollution controls be installed (scrubbers, fume hoods, dust collectors)?

20. Have handling and cleaning of air pollution control systems been addressed?

21. Have the processes that generate wastewater been evaluated for minimization potential?

22. Will indoor spills be protected from reaching drains?
23. Will outdoor spills be protected from reaching storm water drains and sewer manholes?
24. Are adequate water disposal systems available?
25. Will pretreatment methods be necessary and provided?
26. Will the discharges of domestic and industrial waste-water be in accord with regulations?

G. Ergonomics—Workstation and Work Methods Design

1. Generally, have material-handling designs considered worker capabilities and limitations, to accommodate the employee population at the 95% level?
2. Do material-handling designs promote the use of mechanical material-handling equipment, such as conveyors, cranes, hoists, scissor jacks, and drum carts?
3. Do design layouts minimize:
   a. constant lifting
   b. twisting and turning of the back when moving an object
   c. crouching, crawling, and kneeling
   d. lifting objects from floor level
   e. static muscle loading
   f. finger pinch grips
   g. work with elbows raised above waist level
   h. twisting motions of hands, wrists, or elbows
   i. hyperextension or hyperflexion of wrists
   j. repetitive motion
   k. awkward postures?
4. Are workstations designed to provide:
   a. adequate support for the back and legs
   b. adjustable work surfaces that are easily manipulated
   c. delivery bins and tables to accommodate height and reach limitations
   d. work platforms that elevate and descend, as needed
   e. powered assists and suspension devices to reduce the use of force?
5. Has adequate attention been given to:
   a. lighting (to Illuminating Engineering Society requirements)
   b. heat
   c. cold
   d. noise
   e. vibration?
6. Does the design accommodate the hazards inherent in servicing, maintenance, and inspection?
7. Will there be adequate clearance and ready access to equipment for servicing?
8. Will controls be efficiently located in a logical and sequential order?
9. Will indicators be easy to read, either by themselves or in combination with others?

H. Fall Avoidance

1. Overall, has the design minimized the need for ladders and stairs?
2. Where work at heights is to be done, has adequate consideration been given to providing work platforms or fixed ladders?
3. Are parapets or guardrails provided at roof edges?
4. Is equipment designed to minimize fall hazards during maintenance, inspection, and cleaning?
5. Does the design provide for fall-arrest measures, such as anchorage points and fall restraining systems?

I. Fire Protection

1. Overall, in the design, will national and local fire codes and insurance requirements be met?
2. Will fire pumps, water tanks/ponds, and fire hydrants be adequate?
3. Will risers and post valves be accessible and protected from damage?
4. Will small hose standpipes be adequate?
5. Will sufficient hose racks be provided?
6. Will special fire suppression systems be provided?
7. Has containment of fire suppression water been addressed?
8. Will there be adequate external fire zones?
9. Will emergency vehicle access be adequate?
10. Will flame arresters be installed where needed on equipment vents?
11. Will fire extinguishers be of appropriate types, adequate, and mounted for easy access?
12. Will the design for location of flammables be appropriate?
13. For flammables, will storage rooms and cabinets meet national fire codes and insurance requirements?
14. For flammable liquid dispensing, will grounding, bonding, and ventilation be adequate?
15. Will fire sensors, pull stations, and alarms be adequate?
16. Are flooding systems designed to provide a predischarge alarm that can be perceived above ambient light or noise levels before the system discharges, giving workers time to exit from the discharge area?
17. Has the project been reviewed by insurance personnel?

J. Hazardous and Toxic Materials

1. In the design process, have all materials in this category been identified?
2. Have the physical properties of the individual chemicals been identified?
3. Have the most conservative exposure limits been established as the design criteria?
4. Has a determination been made to use intrinsically safe equipment?
5. Have safety data sheets been obtained for all materials?

6. Are the reactive properties known for chemicals that will be combined or mixed?

7. Have measures been taken to eliminate, substitute for, or minimize the quantities of hazardous chemicals?

8. Does the design emphasize closed process systems?

9. Will the design properly address all occupational illness potentials and minimize the need for monitoring, testing, and personal protective equipment?

10. Are storage facilities designed to separate hazardous from nonhazardous substances?

11. Does the design consider the chemical compatibility issues?

12. Have adequate provisions been made for chemical release, fire, explosion, or reaction?

13. Have provisions been made to contain water used in hazardous release control?

14. Are ventilation systems adequate to handle an emergency release?

15. Is the storage of hazardous chemicals below ground avoided?

16. Are storage tanks located so as to minimize facility damage or damage to the public in a catastrophic event?

17. Are adequate storage tank dikes provided?

18. Will emergency ventilation be provided for accidental releases?

19. For extraordinary releases, will special ventilation, relief, and deluge systems be provided?

20. Will the normal use of chemicals allow operating without personal protective equipment?

21. Will the design of bulk loading/unloading facilities contain anticipated leaks and spills?

K. Lockout/Tagout—Energy Controls

1. In the design process, has adequate attention been given to lockout/tagout requirements to prevent hazardous releases from these energy sources:
   a. electrical
   b. mechanical
   c. hydraulic
   d. pneumatic
   e. chemical
   f. thermal
   g. nonionizing radiation
   h. ionizing radiation?

2. Are lockout/tagout devices adequate in design and number, readily accessible, and operable?

3. Are lockout/tagout devices standard throughout the facility?

L. Machine Guarding

1. Overall, do the designs prevent workers' hands, arms, and other body parts from making contact with dangerous moving parts? [29 CFR 1910.212(a)(3)]

2. Have the requirements of all applicable machine-guarding standards of the American National Standards Institute been identified and met?

3. Are safeguards firmly secured and not easily removed? [29 CFR 1910.212(a)(2)]

4. Do safeguards ensure that no object will fall into moving parts? [29 CFR 1910(a)(1)]

5. Do safeguards permit safe, comfortable, and relatively easy operation of the machine? [29 CFR 1910.212(a)(2)]

6. Can machines be oiled without removing safeguards? [29 CFR 1910.212(a)(2)]

7. Does the design include a system that requires shutting down machinery before safeguards are removed?

8. Are fixed machines soundly anchored?

9. Are in-running nip points properly guarded? [29 CFR 1910.212(a)(1)]

10. Will the design properly address point-of-operation exposure? [29 CFR 1910.212(a)(3)]

11. Are all reciprocating parts properly guarded? [29 CFR 1910.212(a)(3)(iv)]

12. Are all rotating parts properly guarded? [29 CFR 1910.212(a)(3)(iv)]

13. Are all shear points properly guarded? [29 CFR 1910.212(a)(3)(iv)]

14. Are exposed set screws, keyways, collars, and so forth properly guarded? [29 CFR 1910.212(a)(3)(iv)]

15. Does the design eliminate the potential for flying chips? [29 CFR 1910.212(a)(i)]

16. Has the potential for any sparking been eliminated? [29 CFR 1910.212(a)(i)]

17. If robots are to be used, are they designed to ANSI/RIA R15.06–1999 (American National Standard for Industrial Robots and Robot Systems—Safety Requirements)?

M. Noise Control

1. In the design process, have maximum noise levels been established that are to be stipulated in specifications for new equipment?

2. Is emphasis given to controlling noise levels through engineering measures?

3. Are the size or shape of rooms and proposed layout of equipment, workstations, and break areas to be evaluated for noise levels?

4. Will workers be separated from noise by the greatest feasible distance?

5. Will barriers be installed between noise sources and workers?

6. Are enclosed control rooms to be provided for operators in areas where the noise is above trigger levels?

7. Are lower noise level processes to be selected, where feasible?
8. Have equipment and workstations been located so that the greatest sources of noise are not facing operators?

N. Pressure Vessels
1. Will all pressure vessels be designed to American Society of Mechanical Engineers and insurance company requirements?
2. Will pressure vessels containing flammables or combustibles meet OSHA 29 CFR 1910.106 and NEC standards?
3. Will pressure-relief valves be:
   a. correctly sized and set
   b. suitable for intended use
   c. directed to discharge safely?

O. Ventilation
1. Have all sources of emission been identified and their hazards characterized?
2. Have ways to reduce personnel interaction with the emission sources (location, work practices) been considered in the design process?
3. Have assessments been made with respect to incompatible emission streams (cyanides and acids, etc.)?
4. Has consideration been given to weather conditions and seasonal variations?
5. Will the design requirements of the ANSI Z9 series, the ACGIH ventilation manual, the ASHRAE guidelines, and NFPA 45 and 90 be met?
6. Will local ventilation effectively capture contaminants at the point of discharge?
7. Will room static pressures be progressively more negative as the operation becomes "dirtier"?
8. Will ventilation systems provide a margin of safety if a system fails?
9. Will emergency power and lighting be provided on critical units?
10. Will the ventilation equipment be remote and/or "quiet"?
11. Will spray booths and degreasers meet OSHA standards?
12. Will laboratory or contaminated air be totally exhausted?
13. If contaminated air is cleaned and reused, will it meet good safety requirements?
14. Will the makeup air to hoods be clean and adequate?
15. Have flow patterns been established to prevent exposure to personnel?
16. Does the design provide for proper gauging and alarm systems with respect to a sudden pressure drop?
17. Are ventilation controls easily accessible to operators?

P. Walking and Working Surfaces, Floor and Wall Openings, Fixed Stairs and Ladders
1. Will aisles, loading docks, and through doorways have enough clearance to allow safe turns where material-handling equipment is used? [29 CFR 1910.22(b)(1)]
2. In the aisles, are persons and vehicles adequately separated?
3. Are permanent aisles to be marked with lines on the floor? [29 CFR 1910.22(b)(2)]
4. Does the design provide for floors, aisles, and passageways being free from obstruction? [29 CFR 1910.22(b)(1)]
5. Has a logistics study been made to provide safe and efficient flow of persons and materials?
6. Will the construction texture of walking surfaces be nonslip?
7. Will the floors be designed to stay dry?
8. Will water and process flows be designed to keep off the walkway?
9. Will the floors be sloped and drained?
10. Will utilities and other obstructions be routed off the walking surfaces?
11. Will the design allow future utility expansion, with added facilities not having to be above, and thereby crossing floors, and obstructive?
12. Will designs for floor and wall openings meet the requirements of OSHA 29 CFR 1910.23?
13. Do the designs for fixed stairs and ladders meet the requirements of OSHA 29 CFR 1910.23, 1910.24, and 1910.27?

MANAGEMENT OF CHANGE

The objective of a management of change process is to prevent the introduction of new hazards and risks into the work environment when changes are made in technology, equipment, facilities, work practices and procedures, design specifications, raw materials, organizational or staffing changes affecting skill capabilities, and standards or regulations. In this safety through design application, applying the change analysis concept is essential in a management of change process. A change analysis is to ensure that:
- the hazards and risks that may arise when a change is to be made have been identified and assessed and that appropriate control measures are taken
- new hazards are not created by the change
- the change does not negatively affect previously resolved hazards
- the change does not make the potential for harm of an existing hazard more severe.

Having management of change processes in place has been required by the OSH Act's Rule for Process Safety Management of Highly Hazardous Chemicals (29 CFR 1910.119) since it was promulgated in 1992. ANSI/AIHA Z10, the Occupational Health and Safety Management Systems standard adopted in 2005, also requires that processes be implemented and maintained to control the risks when changes are made. Unfortunately, little literature exists on including management of change provisions in safety management systems, other than for the chemical industry. A chapter in *Advanced Safety Management: Focusing on Z10 and Serious Injury Prevention* (Manuele 2008) is devoted to management of change processes, with an emphasis on other-than-chemical operations.

For moderate-sized locations, safety professionals should consider drafting and proposing the implementation of a pre-job planning and safety analysis system to fulfill management of change needs. The purpose of a pre-job planning and safety analysis system is to provide a means for supervisors and their staffs to review how the work is to be done and the hazards and risks that may be encountered—before the work is commenced.

Addendum C in this chapter provides a framework from which a pre-job review system can be developed. It should not be adopted as presented. For example, revisions will almost always be necessary in item 9, which is purposely an extensive list of hazards.

INCLUDING SAFETY SPECIFICATIONS IN PURCHASING DOCUMENTS

This author places great emphasis on having safety specifications included in purchasing documents because doing so prevents introducing hazards and risks into the workplace. This emphasis derives from the following sequence of thoughts. Risks of injury derive from hazards. If hazards are properly addressed and eliminated or brought under control in the design process so that the risks deriving from them are at an acceptable level, the potential for harm or damage and operational waste is minimized. The logical extension of addressing hazards and risks in the design process is to have the design specifications the organization decides upon included in purchase orders and contracts so that suppliers and vendors know what safety specifications are to be met. That reduces the possibility of bringing hazards into the workplace.

Although having safety specifications included in purchase orders or contracts is not a broadly applied practice, safety professionals are encouraged to consider the benefit to be achieved if they are. If the ideal is attained in the purchasing process and hazards and risks brought into the workplace are at a practical minimum, significant risk reduction results, and the outcome will be fewer injuries and illnesses.

Applications of safety-related design standards that become purchasing specifications are not easily acquired. (Examples are included in *Advanced Safety Management: Focusing on Z10 and Serious Injury Prevention* [Manuele 2008].) Most companies consider their specifications proprietary and don't make them available to others freely.

For the moderate-sized company with a limited engineering staff, writing design and purchasing specifications will not be easy to do. It seems appropriate to suggest that organizations prevail upon the business associations of which they are members to undertake writing generic design specifications and purchasing specifications that relate to the hazards and risks inherent in their operations.

SUMMARY

- Over time, the level of safety achieved will relate directly to the caliber of the initial design of facilities, hardware, equipment, tooling, operations layout, the work environment, and the work methods—and their redesign as continuous improvement is sought.
- Safety through design is defined as the integration of hazard analysis and risk assessment methods early in the design and redesign processes and taking the actions necessary so that the risks of injury or damage are at an acceptable level.
- In the design and redesign processes, management seeks to avoid, reduce, or eliminate the probability and severity of a hazard potential being realized and causing an incident.
- The design and redesign stages offer the greatest opportunity to anticipate, analyze, eliminate, or control hazards and the risks that derive from them.
- Organizations should apply the following priorities to design and redesign processes: eliminate or reduce risks in the design and redesign processes, reduce risks by substituting less hazardous methods or materials, incorporate safety devices, provide warning systems, apply administrative controls, and provide personal protective equipment.
- The safety professional can influence the design of the workplace and work methods at three critical points: the preoperational, operational, and postincident stages.
- Organizations should take a proactive stance regarding safety through design. They should also examine work procedures and systems more closely for the causal factors they may contain, rather than assuming worker behavior as the principal causal factor for an incident.

- Organizations should consider the strengths and limitations of workers when designing the workplace and the work methods.
- In implementing a safety through design process, organizations should establish clear-cut objectives, assess hazard probability/severity, conduct hazards analysis and risk assessments, establish design review procedures, and use project checklists.
- Management of change procedures should be implemented so that hazards and risks are properly addressed when changes are made.
- Safety specifications should be written into purchasing documents so as to avoid bringing hazards and risks into the workplace.

REFERENCES

Accident Prevention Manual: Administration & Programs. Itasca, IL: National Safety Council, 2015.

ANSI/AIHA Z10–2005. Occupational Health and Safety Management Systems. Fairfax, VA: American Industrial Hygiene Association, 2005. aiha.org/marketplace.htm.

ANSI/ASSE Z2441–2003. Control of Hazardous Energy: Lockout/Tagout and Alternative Methods. Des Plaines, IL: American Society of Safety Engineers, 2003.

ANSI/PMMI B155.1–2006. Safety Requirements for Packaging Machinery and Packaging-Related Converting Machinery. Arlington, VA: Packaging Machinery Manufacturers Institute, 2006.

ANSI B11.TR3–2000. Risk Assessment and Reduction—A Guide to Estimate, Evaluate and Reduce Risks Associated with Machine Tools. McLean, VA: Association for Manufacturing Technology, 2000.

Aviation Ground Operation Safety Handbook. 6th ed. Itasca, IL: National Safety Council, 2007.

Chapanis, A. "The Error-Provocative Situation." In *The Measurement of Safety Performance*, edited by W. E. Tarrants. New York: Garland Publishing, 1980.

Christensen, W. C. "Retrofitting for Safety: Career Implications for SH&E Personnel." *Professional Safety* (May 2007).

———. "Safety through Design: Helping Design Engineers Answer 10 Key Questions." *Professional Safety* (March 2003).

Christensen, W., and F. A. Manuele, eds. *Safety through Design.* Itasca, IL: National Safety Council, 1999.

CSA Z1000–2006. Occupational Health and Safety Management. Mississauga, Ontario, Canada: Canadian Standards Association, 2006.

Deming, W. E. *Out of the Crisis.* Cambridge, MA: Center for Advanced Engineering Study, Massachusetts Institute of Technology, 1986.

Guidance Document for Incorporating Risk Concepts into NFPA Codes and Standards. Quincy, MA: The Fire Protection Research Foundation, 2007.

Haddon, W. J., Jr. "On the Escape of Tigers: An Ecological Note." *Technology Review* (May 1970).

———. "The Prevention of Accidents." *Preventive Medicine* (1966).

Lowrance, W. W. *Of Acceptable Risk: Science and the Determination of Safety.* Los Altos, CA: William Kaufman, 1976.

Manuele, F. A. *Advanced Safety Management: Focusing on Z10 and Serious Injury Prevention.* Hoboken, NJ: John Wiley & Sons, 2008.

———. ANSI/AIHA Z10–2005. The New Benchmark for Safety Management Systems. *Professional Safety* (February 2006).

———. *On the Practice of Safety.* 3rd ed. Hoboken, NJ: John Wiley & Sons, 2003.

———. "Risk Assessments and Hierarchies of Control." *Professional Safety* (May 2005).

MIL-STD-882-D. *Military Standard System Safety Program Requirements.* Washington DC: U.S. Department of Defense, 2000.

OSHA's Rule for Process Safety Management of Highly Hazardous Chemicals. 29 CFR 1910.119. Washington DC: U.S. Department of Labor, 1992.

Roland, H. E., and B. Moriarty. *System Safety Engineering and Management.* 2nd ed. New York: John Wiley, 1990.

SEMI S2-0706. Environmental, Health, and Safety Guideline for Semiconductor Manufacturing Equipment. San Jose, CA: Semiconductor Equipment and Materials International, 2006.

SFPE Engineering Guide to Fire Risk Assessment. Bethesda, MD: The Society of Fire Protection Engineers, 2006.

Stephans, R. A. *System Safety for the 21st Century.* Hoboken, NJ: John Wiley & Sons, 2004.

Stephans, R., and W. W. Talso, eds. *System Safety Analysis Handbook, A Sourcebook for Safety Practitioners.* Albuquerque, NM: System Safety Society, 1999.

Swain, A. D. "Work Situation Approach to Improving Job Safety." Albuquerque, NM: Sandia Laboratories, 1962.

Swartz, G., ed. *Safety Culture and Effective Safety Management.* Itasca, IL: National Safety Council, 2000.

Vincoli, J. W. *Basic Guide to System Safety.* Hoboken, NJ: John Wiley & Sons, 1993.

REVIEW QUESTIONS

1. List the safety through design provisions in ANSI/AIHA Z10.
2. Define safety through design.
3. What benefits are obtained by applying safety through design concepts?
4. How do safety through design concepts relate to quality management?
5. Define the following terms.
   a. acceptable risk
   b. safety
   c. hazards
   d. risk
   e. probability
   f. severity
   g. residual risk
6. Outline the hazard analysis/risk assessment process.
7. What would a risk assessment matrix include and what benefit derives from its use?
8. Outline the hierarchy of controls.
9. Why do the ameliorating elements in the first, second, and third levels of the hierarchy achieve more effective control of risk?
10. What is the desired outcome in applying the hierarchy of controls?
11. What concepts must be taken into consideration in achieving an acceptable risk level?
12. List the three critical points during which a safety practitioner can influence the design of the workplace and work methods.
    a.
    b.
    c.
13. Why may behavioral modification techniques be inadequate to resolve occupational risk design issues?
14. Alan D. Swain suggested that management forego the temptation to _____ .
15. What is the central point of Dr. Chapanis's work?
16. What is the central point of Dr. Haddon's unwanted energy release theory?
17. Why are the "General Design Requirements: A Thought Process for Hazard Avoidance, Elimination, or Control" included in this text?
18. List the nine major requirements in the "General Design Requirements."
    a.
    b.
    c.
    d.
    e.
    f.
    g.
    h.
    i.
19. What is the objective of a management of change process?
20. What benefits are obtained by including safety specifications in purchasing documents?

Addendum A

designsafe Report **Provided by design safety engineering, inc. www.designsafe.com**

Application:	Transfer Line, Machine #334185
Description:	Sample assessment for demonstration
Analyst Name(s):	Steve, Rick, Rob plant operators, Bruce Jones, safety, Tom Woods, engineering
Company:	ABC Company
Facility Location:	Washington, DC
Product Identifier:	Model 89RX-1
Assessment Type:	Detailed
Limits:	This initial risk assessment is for certain Operator tasks
Sources:	on site investigations, discussions w/ plant personnel
Risk Scoring System:	ANSI B11 TR3 Two Factor

Guide sentence: When doing [task], the [user] could be injured by the [hazard] due to the [failure mode].

Item Id	User / Task	Hazard / Failure Mode	Initial Assessment Severity Probability	Risk Level	Risk Reduction / Comments	Final Assessment Severity Probability	Risk Level	Status / Responsible / Reference
1-1-1	operator(s) tool change	mechanical: cutting / severing	Moderate Remote	Negligible	gloves / issue to all new hires	Minor Remote	Negligible	Complete Joe
1-1-2	operator(s) tool change	mechanical: impact dropping heavy tool	Moderate Unlikely	Low	lift assist, standard procedures	Minor Remote	Negligible	Complete
1-1-3	operator(s) tool change	mechanical: pinch points	Minor Remote	Negligible	standard procedures	Minor Remote	Negligible	Complete
1-1-4	operator(s) tool change	mechanical: head bump on overhead objects	Minor Remote	Negligible	other	Minor Remote	Negligible	Complete
1-1-5	operator(s) tool change	ergonomics / human factors: lifting / bending / twisting	Minor Remote	Negligible	look into lift assists or quick release fasteners	Minor Remote	Negligible	Complete Jane
1-1-6	operator(s) tool change	slips / trips / falls: slips	Serious Likely	High	graded floors non-slip flooring, contain coolant, footwear	Minor Remote	Negligible	In-process Jane
1-2-1	operator(s) remove reject parts	mechanical: cutting / severing	Moderate Remote	Negligible		Minor Remote	Negligible	Complete
1-2-2	operator(s) remove reject parts	mechanical: drawing-in / trapping	Catastrophic Likely	High	interlocked barriers, presence sensing devices, stop line to pull part / requisition submitted	Minor Remote	Negligible	In-process John
1-2-3	operator(s) remove reject parts	mechanical: impact by dropped parts	Moderate Unlikely	Low		Minor Remote	Negligible	Complete
1-2-4	operator(s) remove reject parts	ergonomics / human factors: lifting / bending / twisting	Moderate Remote	Negligible		Minor Remote	Negligible	Complete
1-3-1	operator(s) probe check	Other: None, no hazards						Complete

Addendum B

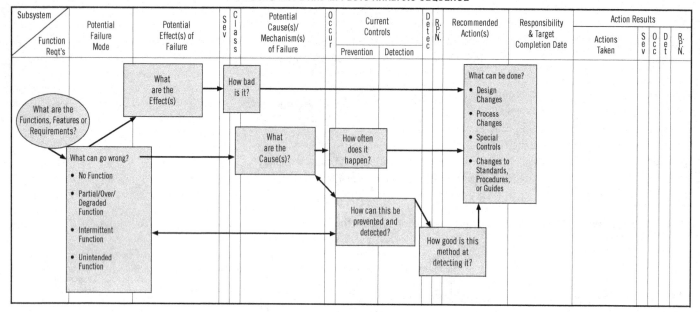

Addendum C

Pre-Job Planning and Safety Analysis Outline

1. Review the work to be done. Consider both productivity and safety:
   a. Break the job down into manageable tasks.
   b. How is each task to be done?
   c. In what order are tasks to be done?
   d. What equipment or materials are needed?
   e. Are any particular skills required?
2. Clearly assign responsibilities.
3. Who is to perform the pre-use of equipment tests?
4. Will the work require a hot work permit, a confined entry permit, lockout/tagout (of what equipment or machinery)?
5. Will it be necessary to barricade for clear work zones?
6. Will aerial lifts be required?
7. What personal protective equipment will be needed?
8. Will fall protection be required?
9. What are the hazards in each task? Consider:

Access	Work at heights	Work at depths	Fall hazards
Worker position	Worker posture	Twisting, bending	Weight of objects
Elevated loads	Welding	Fire	Explosion
Electricity	Chemicals	Dusts	Noise
Weather	Sharp objects	Steam	Vibration
Stored energy	Dropping tools	Pressure	Hot objects
Forklift trucks	Conveyors	Moving equipment	Machine guarding

10. Of the hazards identified, do any present severe risk of injury?
11. Develop hazard control measures, applying the hierarchy of controls.
    - Eliminate hazards and risks through system and work methods design and redesign.
    - Reduce risks by substituting less hazardous methods or materials.
    - Incorporate safety devices (fixed guards, interlocks).
    - Provide warning systems.
    - Apply administrative controls (work methods, training, etc.).
    - Provide personal protective equipment.

12. Is any special contingency planning necessary (persons, procedures)?
13. What communication devices will be needed (two-way, hand signals)?
14. Review and test the communication system to notify the emergency team (phone number, responsibilities).
15. What are the workers to do if the work doesn't go as planned?
16. Considering all of the foregoing, are the risks acceptable? If not, what action should be taken?

Upon Job Completion
17. Account for all personnel.
18. Replace guards.
19. Remove safety locks.
20. Restore energy as appropriate.
21. Remove barriers/devices to secure area.
22. Account for tools.
23. Turn in permits.
24. Clean the area.
25. Communicate to others affected that the job is done.
26. Document all modifications to prints and appropriate file.

Buildings and Facility Layout

2

Patricia Thomas
John F. Montgomery, PhD, CSP, CHMM

Design for Safety
General Considerations ▶ Design Considerations ▶ Buildings, Processes, and Personnel Facilities ▶ Codes and Standards

Site Selection
Location, Climate, and Terrain ▶ Space Requirements

Outside Facilities
Enclosures and Entrances ▶ Shipping and Receiving ▶ Roadways and Walkways ▶ Trestles ▶ Parking Lots ▶ Landscaping ▶ Waste Disposal ▶ Air Pollution ▶ Confined Spaces ▶ Outside Lighting ▶ Docks and Wharves

Facility Railways
Clearances and Warning Methods ▶ Track ▶ Loading and Unloading ▶ Overhead-Crane Runways ▶ Types of Motive Power ▶ Tools and Appliances ▶ Car Movers ▶ Safe Practices

Facility Layout
Location of Buildings and Structures ▶ Layout of Equipment ▶ Aisles ▶ Parking and Ramps ▶ Electrical Equipment ▶ Heating, Ventilation, and Air Conditioning (HVAC) ▶ Inside Storage

Lighting
Daylight ▶ Electric Lighting ▶ Quality of Illumination ▶ Quantity of Illumination ▶ Glare ▶ Lighting Management ▶ Safety ▶ Hazardous (Classified) Locations ▶ Wet Locations ▶ Protective Lighting ▶ Security of Facilities

Use of Color
Color and Light in the Workplace ▶ Human Response to Color ▶ Color-Coding ▶ Accident Prevention Signs

Building Structures
Runways, Platforms, and Ramps ▶ Aisles and Corridors ▶ Stairways ▶ Walkways ▶ Exits ▶ Flooring Materials ▶ Floor Loads ▶ Workstation Design

Summary

References

Review Questions

DESIGN FOR SAFETY

By carefully planning the design, location, and layout of a new facility or of an existing facility that needs major alterations, safety and health professionals can greatly improve the safety and productivity of a facility's operations. Numerous accidents, occupational diseases, explosions, and fires might be prevented if safety measures are incorporated during the early planning stages of a facility. The topics covered in this chapter include:

- general considerations in designing for safety and some of the significant codes and standards involved
- safety factors to consider when selecting a facility site
- hazards and safety factors to consider in outside facilities
- safety concerns regarding facility railways
- safety design decisions to make when developing facility layouts
- use of lighting and color to enhance safety in the workplace
- how to make building structures safe.

Ideally, safety and health professionals conduct a safety and health study of a proposed facility while the designing and engineering are in the developmental stages. Safety and health professionals should approach this study from the viewpoint of removing hazards rather than adding protective equipment. For example, they could suggest eliminating the storage of hazardous materials or substituting a less hazardous product. Also, they could suggest ways to reduce risks, such as intensifying a rate of mixing, storing a gas at a lower temperature and pressure, or simplifying the facility's design. (For more detailed information on these approaches, see the References at the end of this chapter as well as Chapter 9, Fire Protection, and Chapter 10, Flammable and Combustible Liquids, both in this volume.)

General Considerations

Effective human performance is a key factor in efficient production. Therefore, plan industrial systems with workers in mind, and ask the following questions when changes in the workplace are required:

- What will workers do?
- How should workers do it?
- Where should workers do it?
- Why should workers do it?
- What can happen to workers who do it?

Machinery is the second important factor when planning facilities. A third factor is the flow of raw materials, in-process materials, component parts, and the final product.

Design Considerations

Factors to consider in the general design of a workplace include the following:

- illumination
- noise and vibration control
- product flow
- ventilation (particularly around dust, vapors, and fumes)
- control of temperature and humidity
- workstations and movements of employees
- supervision and communication
- support requirements for vehicles, portable ladders, material-handling devices, monitoring and controlling systems, and cleaning and maintenance.

Factors to consider when designing machine tools and equipment include the following:

- construction and procedures
- visual displays, signs, and labels
- protective features and guards
- controls and handles
- maintenance and service needs
- accident prevention signs.

During the design planning stage of a facility, safety and health professionals should also consider the human interaction factor and help devise safe and efficient ways for employees and supervisors to communicate.

Buildings, Processes, and Personnel Facilities

Important factors to consider when determining the appropriate sizes, shapes, and types of buildings and structures include the following:

- the nature of the business and its processes
- the nature of the production materials
- maintenance
- heating, ventilation, and air conditioning equipment
- working conditions
- shipping and receiving materials
- economic circumstances.

Facilities for employee-oriented activities (e.g., lunchrooms and medical, safety, and disaster services, etc.) should be planned and situated for workers' convenient and efficient use.

Codes and Standards

Many companies have the policy that their safety and health specialists, as well as their insurance companies, must review plans and specifications for new facilities or for facilities that need remodeling. Safety and health professionals should also ensure that the plans include provisions for fire prevention, safe work practices for all workers, and periodic inspections to verify the integrity and strength of the building's structure throughout the construction process. Ideally, safety and health professionals are allowed to review and approve plans before the plans are released for bids. If a company adheres to this policy, it will not have the costs of alterations or new installations that would be necessary because the facility fails to satisfy local and state fire, safety, and health regulations.

Most local ordinances and state or provincial laws require governmental authorities to review and approve building plans for emergency and nonemergency exits. In some states, the appropriate authorities must approve plans for the installation of emergency lighting, fire alarms, and automatic sprinkler systems. In addition, exhaust and ventilating installations must be approved in certain states or provinces.

Many national and local codes require companies to have a means of controlling air-polluting industrial contaminants. In the United States, the Environmental Protection Agency has specific emission standards for raw waste disposal and industrial by-products. Be sure to check these codes and standards during the planning stage.

Several organizations have developed voluntary safety codes that establish standards for structures and equipment. Specifications for the construction of floor and wall openings and of railings, for example, are given in the American National Standards Institute's (ANSI) A1264.1, Safety Requirements for Workplace Floor and Wall Openings, Stairs, and Railing Systems, and in OSHA 29 CFR 1910 Subpart D. Proper electrical wiring and electrical installations are covered in NFPA 70, National Electrical Code, issued by the National Fire Protection Association (NFPA).

Fire-extinguishing equipment requirements and fire protection standards and codes for flammable liquids and gases, combustible solids, dusts, chemicals, and explosives are provided in the 12 volumes of the National Fire Codes, which were developed by the NFPA and the International Code Council. Be sure to consult the latest edition of the standards and codes. See also Chapter 9, Fire Protection, in this volume. For the addresses of the appropriate agencies in countries other than the United States, consult the Sources of Help appendix in the *Administration & Programs* volume.

Remember that "designing by the code" is no substitute for intelligent engineering. Codes merely establish a minimum standard that, in many situations, must be exceeded.

SITE SELECTION

By carefully selecting the site of a new facility, safety and health professionals can help ensure that the facility complies with the local, state, and federal health and safety codes for buildings. When selecting a site for a facility, consider the following factors:

- the relationship of the new structure to climate and terrain
- the space requirements
- the type and size of the building
- the locations of the necessary disposal facilities
- the transportation to and from the facility
- the market
- the labor supply
- the facility's hazards to the community.

In the predesign stage, safety and health professionals can also help planners detect potential safety problems. Relief models of the site, made to scale, along with maps can help planners design safety features into the facility before construction begins.

Location, Climate, and Terrain

Planners should study the climate and terrain of the site on which the facility will be located. The prevailing winds, for example, may influence the decision of the best place for the processing equipment in relation to the administrative offices and the population in the area. In areas prone to hurricanes, tornadoes, earthquakes, or floods, plans and specifications should include protective measures for personnel, and such safety factors must be designed into the facility. (See the *Administration & Programs* volume, Chapter 18, Emergency Preparedness. Also consult the Environmental Protection Agency's hazardous-waste conformity issues in containment areas discussed in the *Administration & Programs* volume, Chapter 14, Environmental Management.)

Space Requirements

Fire protection codes specify minimum distances between buildings according to their sizes, types, and occupancies. Laws governing the storage of explosives and other highly flammable materials specify minimum distances between manufacturing areas and storage facilities for such materials (15 ft to 200 ft, depending on the materials and their amounts; see NFPA Table 2.51.d). Minimum distances between both toxic and flammable materials and adjoining property are also specified.

The necessary size of a site may be determined by both current space requirements and possible future expansion. For example, some companies, anticipating an increased use of air transportation, allocate space for landing fields or heliports. Plans for such expansion should include all necessary safety precautions as well as ample space for outdoor storage areas. When the areas for storing materials adjacent to the facility become insufficient, space needs to be provided elsewhere. Storing materials away from the facility, however, requires additional handling and transportation, thus increasing both costs and the possibility of accidents.

Parking lots are best situated inside a facility's fence for the convenience, protection, and safety of employees and visitors. Because a substantial area may be necessary for employees' and visitors' vehicles, parking needs must be considered during site planning.

Well-situated disposal areas for solid and liquid wastes must also be included when a site is being laid out. Plan drainage and waste disposal in relation to space, terrain, and facility needs as well as in relation to their effects upon the surrounding municipal systems.

OUTSIDE FACILITIES

When planning outside facilities, safety and health professionals should keep safety precautions in mind in order to reduce the chance of accidents. Several outside facilities and their safety parameters are discussed in this section.

Enclosures and Entrances

A fence around yards and grounds serves many purposes. For example, fencing protects employees and visitors from transformer stations, pits, sumps, stream banks (under certain circumstances), and similar dangerous places. Fencing also keeps out trespassers, who may interfere with the work taking place or be injured on the property. A galvanized, woven-wire fence is a good enclosure.

Enough entrances should be planned to accommodate the facility's traffic volume. Entrances should provide clearance for the largest expected delivery vehicles, and good visibility in all directions is essential at all entrances. Entrances and exits should also be wide enough and numerous enough for building occupants to be able to evacuate quickly in the case of an emergency. Requirements for exit widths can be found in NFPA 101, Life Safety Code.

Because it is unsafe for pedestrians to use the same entrances that railroad cars and motor vehicles use, designate gates that are convenient for pedestrians to use. If a pedestrian entrance must be located near railroad tracks, fence off part of the right-of-way, which will prevent employees from taking shortcuts along the tracks. If pedestrian entrances must be located near busy thoroughfares or if workers cross the railroad tracks on which trains frequently run, install traffic signals and build subways or pedestrian bridges (Figure 2–1). Such precautions are especially important when parking lots are located at a significant distance from the facility.

Figure 2–1. A bridge provides the safe crossover of a freight yard. Be sure that the bridge's construction and personnel protection are adequate. *(Courtesy Guardian Engineering & Development Company)*

Shipping and Receiving

Shipping and receiving facilities should coordinate with the overall flow of materials within the company or facility. They should also contribute to the efficient flow of materials into and out of production areas. Design shipping and receiving areas to minimize heat and cooling losses from the building. The use of self-leveling dock boards, truck levelers, and cranes speeds up the loading and unloading processes. In addition, dock locks increase the safety of the loading and unloading processes.

Railroad sidings—commonly used as shipping and receiving facilities—require planning, especially if it is advantageous for a company to use bulk raw, process, and maintenance materials. Tank-car lots of hazardous materials require special considerations for pressure piping, breakaway piping, valves, pumps, derails, excess-flow valves, and vapor return lines. Each sidetrack should be protected from main line and public thoroughfares. Proper clearance between the main facilities and the cars should also be observed. (See Chapter 10, Flammable and Combustible Liquids.)

Roadways and Walkways

The safety and health professional should help the civil engineer design for optimum safety. Roadways in facility yards and grounds are locations of frequent accidents unless these roadways are carefully laid out, well constructed, well surfaced and drained, and kept in good condition. Lighting for nighttime use must be included to ensure personnel safety and security.

Roadways

Hauling by heavy-duty trucks requires roadways up to 50 ft (15 m) wide for two-way traffic, with ample radii at curves. Grades, in general, should be limited to a maximum of 8%. A slight crown is necessary for drainage, as are ditches to carry off water.

Locate roadways at least 35 ft (11 m) from buildings, especially building entrances. At loading docks, allow one truck length to make backing up easier.

The regulation and control of traffic signs, road layout, and markings should conform to federal and state or provincial practices. The U.S. Department of Transportation's *Manual on Uniform Traffic Control Devices for Streets and Highways* provides guidelines on these matters.

Traffic signs and signals are essential for regulating speed and movement at hazardous locations. Stop signs are specific for railroad crossings and at entrances to main thoroughfares. SOUND YOUR HORN signs are necessary at sharp curves (blind corners)—where view is obstructed—and at entrances to buildings. Convex mirrors mounted on the sides of buildings provide visibility around sharp turns or around building corners and help prevent accidents if roadways must be built close to buildings. Use barricades and MEN WORKING signs at construction and repair sites. Traffic signs for roadways used at night should be made of reflective or luminous materials.

Walkways

Good walkways between outside facilities keep employees from being injured by helping them avoid stepping into holes and ruts in cracked ground. Concrete is preferred for sidewalks, especially in often-used areas like entrances and between main buildings. To discourage shortcutting, walkways should be the shortest distance from one building to another. A fence or railing should separate a walkway that must be next to railroad tracks. Install warning signs at railroad crossings and other hazardous places. Situating walkways clear of the eaves of buildings reduces the danger of falling icicles. In some areas, covered walkways increase comfort and protect personnel from the elements and slippery walkways. Keep walkways in good condition, especially where they cross railroad tracks, and clear of ice and snow. If site plans call for bridges over streams, ditches, or other hazards, segregate pedestrian traffic with a fence or handrails that are 42 in. (1.1 m) high and with intermediate rails (Figure 2–1). On a larger site, if employee wellness is a core value of the company, the company should consider providing walkways for employees to use when walking or jogging for exercise.

Trestles

If employees are required to perform duties on trestles, provide a footwalk that is 5 ft 1 in. (1.5 m) wide, measured from the nearest rail, on at least one side. The foot-walk should have a railing with the top-edge height of the toprails, or equivalent guardrail system parts, 42 in. (1.1 m) above walking/working levels. Midrails should be installed at a height that is midway between the top edge of the guardrail system and the walking/working surface, and on the exposed side, toeboards should be 4 in. (10 cm) high. [See 29 CFR 1910.23(e)(2) and (e)(4).]

If employees travel on both sides of the track, place crosswalks at frequent and convenient locations. Metal gratings or screens installed over walkways or passages under trestles protect employees from falling materials. Openings for conveyors or hoppers require gratings, or a cover with bars spaced not more than 12 in. (30.5 cm) apart, that prevent employees from falling into the openings.

Parking Lots

To reduce traversing across the facility's grounds, the parking lot should be located so that employees do not need to cross a roadway to go from the parking lot to the facility. To keep the parking lot secure, fence the entire parking area and separate it from other areas of the facility. The surface of the parking lot should be smooth and solid so that employees do not injure themselves by falling on its uneven surface. Lots should be as level as possible and have an adequate slope for drainage.

Using white lines, 4 to 6 in. (10 to 15 cm) wide, to designate parking stalls reduces confusion as well as accidents that might result when drivers back out of the spaces. Standard stalls are 9 ft (2.7 m) wide and 20 ft (6.1 m) long. The appropriate center-to-center distance between parked vehicles depends on the method of parking.

Angle parking has both advantages and disadvantages. For example, the smaller the angle, the fewer the number of vehicles that can be parked in the area. Although aisle widths can be narrower, traffic is usually restricted to going in only one direction. On the other hand, angle parking is easier for drivers and does not require a lot of space to make sharp turns.

The area that should be provided per vehicle in parking lots varies from 200 ft (19 m) to more than 300 ft (28 m) if aisles are included. Large, economically laid-out lots might approach the 200-ft (19-m) figure; small or poorly configured lots might have a larger amount of aisle space and approach 300 ft (28 m) per vehicle. A large commercial parking lot with an attendant is considered efficient if the layout restricts the parking space to 240 ft (22 m) per vehicle.

The Americans with Disabilities Act (ADA) specifies the size, location, and number of parking stalls reserved for the vehicles of disabled persons. Stalls must be 8 ft wide and must be next to a 5-ft-wide access aisle. Two

8-ft-wide stalls may share an access aisle. The required number of accessible stalls is based on the total number of stalls in the parking lot. For more details, consult Access to Buildings in Chapter 23, Workers with Disabilities, in the *Administration & Programs* volume, and the ADA.

For orderly traffic movement, parking lots should have separate entrances and exits for incoming and outgoing vehicles. Designate such entrances with appropriate signs. Control the traffic at exits to heavily traveled streets with either a traffic light or an acceleration or merging lane. Be sure that the parking area does not encroach on fire hydrant zones, approaches to corners, bus stops, loading zones, or clearance spaces for island exits. In addition, driveways should not be obstructed.

Install speed limit signs and signs for visitors where needed. These signs should conform to recommended standards and should be similar to nearby street and highway signs. The U.S. Department of Transportation's *Manual on Uniform Traffic Control Devices for Streets and Highways* gives details on signs and pavement markings.

If the parking lot will be used at night, provide adequate lighting for the safety and security of parking lot users. About 1 to 5 footcandles per square foot (fc/ft²; 11 to 54 lux per m²) at a height of 36 in. (90 cm) should be adequate. During the planning stage, include provisions for removing ice and snow from the lot.

Landscaping

Many companies landscape the grounds of both old and new facilities. Design landscaping so that trees and shrubbery do not create blind spots at roadway or walkway intersections. Proper maintenance is required to prevent bushes from creating blind spots. Bushes should also be kept trimmed to eliminate hiding places, especially near windows and entrances. (See Chapter 3, Construction of Facilities, in this volume, for details on grounds maintenance.)

Waste Disposal

Unsafe methods of waste disposal may injure workers and the public and damage property. Knowing the characteristics of wastes is essential to planning suitable disposal methods. These disposal methods must also conform to applicable municipal and state or provincial regulations. Because treating and disposing of wastes require specialized knowledge and training, safety and health professionals should consult qualified personnel.

If the use of the city or district sewage system is planned, the officials in charge of the system should be informed of the kinds and amounts of the wastes. Note that if the wastes' properties will interfere with the operation of the sewage disposal facility, the officials might refuse to accept the wastes or might require provisions such as pretreating the wastes prior to their discharge into the sewer system.

Under no circumstances should toxic, corrosive, flammable, volatile, or radioactive wastes be discharged into a public sewage system. The facilities that handle or process these types of materials—even in small quantities—must conform precisely to local and state or provincial regulations for the disposal of such wastes.

Many wastes can be disposed in landfills if state or provincial and local laws permit. Chemical wastes should be rendered harmless before disposal; potent acids, for example, should be neutralized. Poisonous materials, magnesium chips, explosives, and similar substances require special procedures to be disposed of safely. Combustible materials such as wood, scraps, and paper may be burned in an incinerator that conforms to applicable laws, is allowable for the area, is safely situated, and is properly attended.

In some situations, a safety and health professional might contract a private disposal service to dispose of waste materials. If so, the safety and health professional must inform the service of any hazardous materials.

Air Pollution

Smoke and inert dust may be nuisances or even hazards to the public. Therefore, prior to the construction of a new facility, check federal and local emissions standards to be certain that the facility's emissions will comply with and not violate these standards.

Toxic smoke, fumes, and dust are hazardous emissions in some industries. Although tall stacks and/or scrubbers are often used to diffuse gases into the atmosphere, their effectiveness depends on the nature and volume of the gases, the location of the facility, the direction of the prevailing wind, and the conditions of the atmosphere. For example, rain may absorb harmful gases and, upon falling, extensively damage crops and the environment. Atmospheric conditions resulting in poor diffusion occasionally lead to temporary shutdowns of facilities that have high stacks.

Investigate the possibilities of minimizing wastes, preventing pollution, and recovering usable or marketable materials from wastes. Filters, cyclones, electronic precipitators, and similar equipment recover dusts and fumes for their intrinsic value and to prevent them from polluting the atmosphere. Spray towers and similar equipment can often economically recover useful gases and vapors.

Confined Spaces

Confined spaces are the locations of many serious and fatal injuries to workers and would-be rescuers. The most common underlying cause of these injuries is that workers are not prepared for unexpected hazards.

A confined space is any location that is not designed for continuous human occupancy. It has limited access and egress and may also be vulnerable to hazards such as the inundation of water, gas, or solid particulates. The confined space might have sloping sides because of a bin or hopper that leads to a crusher, auger, or other obstruction. Other dangers include electrical hazards, oxygen deficiency, possibilities of falls, radiation, toxic gas or vapor, and fire or explosion.

For newly designed or constructed areas, look for ways to engineer the project so that confined spaces are not created. If doing so is not possible, investigate ways, such as installing remote control valves, valve-monitoring equipment, etc., to keep people from having to enter the area. If even that is not possible, then the requirements of ANSI Z117.1–2009 and OSHA regulations must be followed.

To address the problem of confined-space entry, confined spaces must first be identified. Carefully evaluate all operations at the facility. If the company uses tanks, silos, boilers, pits, manholes, trenches, chemical storerooms, subfloors, or any location with an access problem, the company should post warnings at these locations to alert employees of the danger.

Second, identify all the potential hazards at each location and the methods that can eliminate or mitigate those hazards when work must be performed at each location.

Third, create a form for confined-space work that documents that a confined-space entrant and attendant know how to function safely in that confined space. Consult OSHA (29 CFR 1910.146) and ANSI (Z117.1–2009) for the information required on this form and for confined-space entry requirements. In some cases, it will be necessary to complete a confined-space work permit each time an entry will be made. Other permits may authorize entry over longer time periods. Each permit must also be site specific.

Fourth, train personnel in the hazards of and safe behavior in confined spaces. Also include training in the methods of safe entry for the personnel who must work in these confined spaces. This training should include ventilation techniques and systems, respiratory protection (Figure 2–2), atmosphere testing (Figure 2–3), lockout/tagout procedures, using protective equipment, and evacuation procedures (Figure 2–4). These personnel should be cross-trained in the duties and responsibilities of the outside attendant. The outside attendant should maintain continuous communication with the entrant and, under no circumstances, should enter the confined space to assist or rescue the entrant because the attendant is the entrant's only link to the outside world and to assistance.

Figure 2–2. The operator is wearing a coverall, gloves, and a supplied-air respirator while spraying the inside of a tank. *(Courtesy Bridgeport Chemical Corporation)*

Figure 2–3. Before entering a confined space, test for toxic gas or vapor and oxygen deficiency. *(Courtesy Bridgeport Chemical Corporation)*

Finally, ensure that a trained and equipped rescue team is available to respond to emergencies. This team may consist of company personnel, a contracted service, or municipal emergency responders. No matter who comprises the team, team members must be knowledgeable about the potential hazards at the facility and be able to respond quickly. Practicing rescue simulations using mannequins provides the necessary training for a rescue team but must be done carefully so that even this activity does not lead to injuries (Figure 2–4).

Figure 2-4. Rescue personnel should practice simulated rescues so that they become proficient with the equipment and the procedures. This rescuer is using a hoist and winder tripod to assist in rescuing a disabled worker from a manhole. *(Courtesy Miller Equipment)*

Outside Lighting

Outside lighting should not only aid production but also function as part of a facility's security system. Maintenance personnel should adjust the timing devices on lighting as daylight hours shorten. Lighting units with different types of light sources are available for special applications. Consider lamp life and ease of maintenance when selecting lighting units. Keep in mind that outside lighting must be able to withstand exposure to the elements and not become dimmer. (See the Lighting section later in this chapter.)

Docks and Wharves

The condition of a sea, lake, or river is an important consideration when designing and constructing docks and wharves. A soft, deep bottom restricts the use of concrete and bulky fire-resistant materials. Wood pilings must be protected if marine borers are present. Flexibility and elasticity of docks and wharves are essential in tidal waters and where waves force vessels against piers. Ice buildup from tidal movement is another consideration.

Safety requirements on docks and wharves include good illumination for nighttime work, a floor that will withstand heavy trucks, and traffic control equipment. The sizes and speeds of the vehicles that will travel on the piers must also be considered during dock and wharf design.

FACILITY RAILWAYS

The prevention of facility railway hazards must be incorporated into the design of a new facility. Horizontal and vertical track clearances and the proper installations of tracks, fittings, and structures are primary considerations.

For all phases of track construction, the American Railway Engineering Association's (AREA) recommended practices are an excellent resource. However, some state or local regulations for overhead clearances may differ from the AREA recommendations.

Clearances and Warning Methods

Where platforms, building entrances, or structures are located along curved tracks, allow for additional clearance on both sides of the curve for the sideways movement of railcars. Also allow for additional track clearance when awnings, jalousie windows, or louvered windows are installed in buildings adjacent to the tracks. Install a stop sign along with a warning sign far enough away from a tight area to permit personnel riding on the sides of cars to dismount before reaching the area that has narrow clearance.

Standard clearances may not be sufficient if tracks pass doorways, corners of buildings, or other places where workers could walk directly onto tracks and in front of moving cars. Safeguard these locations by installing fixed railings that force workers to take a short detour before stepping onto the track. If a barrier railing is impractical, use hinged bars or gates that swing horizontally through an angle of not more than 90 degrees.

Another way to protect workers is to install convex mirrors at a 45-degree angle at the intersections of passageways and the track. This way, workers can see approaching trains before reaching the intersection.

Various methods are used inside facilities to alert workers to railway crossings. Inside buildings, automatic blinking lights and gongs or bells are more effective than signs. To safeguard workers on heavily traveled passageways, use both gates equipped with red lights and crossing guards furnished with whistles and shielded red lanterns. Increase the visibility of gates by painting them with alternating red and white stripes at 45-degree angles. Another means of protecting workers is the bell warning system, shown in Figure 2-5.

In locations where facility or railroad personnel may be required to switch cars or perform other tasks at night, provide adequate lighting by installing high-intensity,

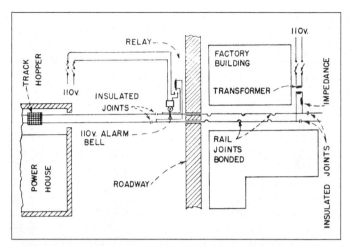

Figure 2–5. A layout of a bell warning system for blind crossings inside a plant.

high-pole lights that are arranged to cast as few shadows as possible.

Track

Although the regulations of the Federal Railroad Administration (FRA) are not mandatory for most industrial facility railways, FRA standards and AREA-recommended practices are excellent resources. However, these recommended practices usually apply only to standard-gauge track. Therefore, inspect rails and fittings periodically, and repair them as needed. Significant accidents might result if defects remain unrepaired.

Tracks

Where possible, arrange for tracks to be level at loading points. Workers have been killed by railcars rolling down even slight slopes. If it is necessary to leave cars on grades, set the brakes firmly and use car blockers, rail clamps, or track skates. Where tracks end, install a standard bumping post or earth mound.

Derailers

Install derailers at the bottoms of steep slopes and where sloping switch tracks connect to main lines. Also, install derailers on the approach to permanent shipping and receiving areas, whether they are located within or outside a building. In areas where tracks are open at both ends, place derailers beyond both ends of the shipping and receiving areas. Do not place derailers in hard-paved areas because this type of surface defeats the purpose of derailers.

Trestles

Provide trestles with a footwalk not less than 5 ft 1 in. (1.5 m) wide. The footwalk's railing, or equivalent guardrail system parts, should have a top-edge height that is 42 in. (1.1 m) above walking/working levels. Midrails shall be installed at a height midway between the top edge of the guardrail system and the walking/working surface; and on the exposed side, toeboards should be 4 in. (10 cm) high. [See 29 CFR 1910.23(e)(1) and (e)(4).] Railings should have the standard side clearance required by the railroad or the state or province. If footwalks are necessary on both sides of the track, build crosswalks to connect them. Design and build trestles so that they can support anticipated loads and also withstand the vibrations and jolts from the anticipated loads.

Switches

Switches should have rounded corners to lessen the possibility of cuts, scratches, and torn clothing. Provide switch lamps or reflectorized areas if tracks are to be used at night. Install blocking in switch points, and use covers to prevent employees from getting their feet caught.

Covers in Open Areas

Cover hoppers or trackside bins into which materials are dumped by placing thick steel bars over them. Openings at ground level for conveyors and similar equipment used to unload cars should have covers that remain in place even when the equipment is not in use. This will prevent workers from falling or being carried through the openings.

Cover walkways that are under trestles in order to prevent workers who use them from being hit by falling materials.

Loading and Unloading

Tracks at loading and unloading areas warrant special attention because many accidents occur when loads are moved into and out of railcars. Sufficient clearance for trucks requires large bridge plates or dock boards. If the level of the dock is considerably higher or lower than the level of railcar door openings, truck movement is more dangerous.

The dock should be wide enough to provide a temporary storage area without interfering with lift-truck loading and unloading movements. A narrow dock may compel lift trucks or pallet jacks to turn onto dock boards at dangerous angles. This hazard can be reduced by making dock boards wider at the dock side, with flanges on the plates that turn wheels away from the edge. Portable dock boards must be securely anchored and strong enough to support the load imposed on them. Provide handholds or other equipment that permit safe loading and unloading.

Prevent cars designated for loading or unloading from being moved by switching crews. Standard blue flags for daytime and blue lights for nighttime warn train crews not to move these cars. Place signals between the rails at both ends of a car that is accessible from either direction. Firmly prohibit train crews from coupling engines or cars to any cars thus protected. Only the employees engaged in the loading or unloading operations should remove the blue signals and should do so only when they are ready to release the cars. (See 49 CFR 218.21d.)

To warn personnel that switching operations are taking place, install bells and oscillating warning lights along the tracks in work areas. Facility supervisors should turn on the warning lights and bells before switching operations begin. Facility management is responsible for removing employees from railroad cars before releasing the track to railroad employees for switching. Therefore, before derailers and blue flags are removed, facility supervisors must make sure that:

- building doors are opened and other obstructions are removed to provide the standard side clearance from the track area
- all overhead building cranes in the area being switched have stopped operations and are clear of the tracks
- all dock boards (bridge plates) are removed from railroad cars
- all counterweighted, retractable service platforms are retracted and secured
- all equipment for moving railroad cars (cables, hooks, etc.) is removed from the cars
- all car doors, hopper doors, and so forth, are closed and properly secured.

Use tracks intended for loading or unloading flammable liquids or other dangerous materials for those purposes only. For additional protection when loading or unloading dangerous materials, provide locks for switches. Specific recommendations for tank cars, grounding, rail bonding, and so on, are given in Chapter 10, Flammable and Combustible Liquids, in this volume.

After unloading the car, the consignee is responsible for cleaning the car before releasing it. When pieces of crating and dunnage, nails, and strapping are left scattered in the car, they become hazards to railroad employees and others who enter the car later. When railcars being loaded or unloaded are damaged or otherwise in need of repair, notify the rail carrier or switching crew so that proper safety precautions can be taken.

In addition, it is the consignee's responsibility to make sure that the doors are properly closed and secured. Plug doors on boxcars are a significant hazard when the doors are not properly secured (Figure 2–6). Most rail carriers instruct switching crews not to move such cars until the doors have been secured.

Overhead-Crane Runways

A significant hazard exists where an overhead-crane run-

Figure 2–6. Plug doors should be locked before a railroad car is moved. *(Courtesy Inland Steel Company)*

way crosses above a railroad track inside or outside a building. The crane's load or hook blocks may strike locomotives or railcars while switching takes place.

To prevent movement of the crane near an occupied track, install a set of interlocked signal lights, with one set visible to the crane operator and the other set visible to the switching crew. All personnel involved, especially crane operators, must be instructed to heed the signals without exception. The signals can be started manually with a key switch; the area supervisor should keep the key. The signals can also be interlocked using a derailer.

Another way to control a crane's movement is with a zone power cutoff for the runways near the vicinity of the track. The cutoff is engaged by a key switch that is under the control of the area supervisor.

Require switch crews to get clearance from the area supervisor before moving into the overhead-crane area. Make supervisors responsible for keeping cranes out of the way until the switch engine and cars move out. (See also Chapter 13, Hoisting and Conveying Equipment.)

Types of Motive Power

Knowing the type of motive power necessary for a facility railway is important to prevent accidents and fire. Explosive gases are easily ignited by flames or sparks from fuel-fired locomotives. Where such gases may be present, locomotives powered by electricity, compressed air, or storage batteries should be used.

In addition, where ventilation is insufficient to keep the concentrations of noxious and toxic exhaust gases at a safe level, such as in mines, prohibit the use of fuel-fired locomotives. Diesel engines, however, can be equipped with devices that eliminate toxic gases from the exhaust. The diesel engine locomotive is thus used in some adequately ventilated mines but not in coal mines.

In addition to sparking in explosive atmospheres, a major hazard of electric locomotives is employees touching the overhead wires or third rail. Place guards at all points where employees can contact electrified equipment or have the wire high enough that employees cannot come in contact with it.

Boilers in steam locomotives should be constructed in accordance with the American Society of Mechanical Engineers' (ASME's) Boiler and Pressure Vessel Code and should be inspected in accordance with ASME's Laws, Rules and Instructions for Inspection and Testing in Steam Locomotives and Tenders and Their Appurtenances. (See the References at the end of this chapter.) Although the ASME regulations are not mandatory for most industrial facility railways, the standards are an excellent guide.

Equip diesel locomotives, which function more quietly than steam locomotives, with bells. Some companies paint stripes on the front and rear of diesel locomotives in contrasting colors such as yellow and black. Install handrails around the outside of deck walks, and provide each locomotive with a fire extinguisher for oil fires.

The tanks of compressed-air locomotives should also be constructed in accordance with ASME's Boiler and Pressure Vessel Code. Each air receiver needs an air pressure gauge, a pop safety valve, and a drainpipe with a valve at the bottom.

A battery-powered locomotive should have a deadman's switch so that the operating lever returns automatically to the OFF position when released or when the operator leaves the controls and breaks the circuit. Safety provisions for battery-charging rooms are covered in Chapter 15, Powered Industrial Trucks, in this volume.

Equip all facility locomotives and railcars with safety appliances and standard automatic locomotives, couplers, and air brakes, as required by federal law for common-carrier railroads. Maintain all equipment and appliances in sound, secure operating conditions.

Tools and Appliances

Using the correct tools and appliances is important for safely operating a facility's railway. For example, automatic couplers eliminate the danger of workers needing to go between standard- or narrow-gauge cars to manually insert or withdraw pins.

Use only standard car wrenches to open cars with hopper bottoms. Using the specific tool to close the latches on bulk cars eliminates the need for a person to go on top of the car and risk falling from it. Provide rerailers for narrow-gauge cars; other methods of rerailing these cars are generally dangerous.

Car Movers

Using manual methods to move cars often results in accidents. A much safer procedure uses a switch engine. If a standard handcar mover is used, however, place a shield around the towbar. In that way, employees will not strike their hands or otherwise injure themselves if the pry tool slips. Crowbars, push-poles, and other makeshift tools should never be used to move cars.

When a car is on a grade, test the hand brake to make sure that it engages and that excess slack in the brake's chain is taken up. A worker should remain on the brake platform to use the brake to stop the car at the required point.

Although winch-type car pullers are used extensively, the operator could be killed if the rope breaks. Because the operator is in line with the rope while operating the equipment, install a shield of steel plate or expanded metal as protection. A forged-steel hook should fasten the rope to the car.

A self-propelled, rider-operated car mover is used in many industries (Figure 2–7). This vehicle has both rubber-tired wheels and steel rail wheels. Because both sets of wheels are retractable, the car mover can run either on the facility's grounds or on rails. This type of car mover is fitted with standard couplers and, depending on the model, can have a drawbar pull of 8,400 to 18,000 lb (38 to 80 kN). Its use eliminates the need for handcar movers, winches, capstans, or powered equipment, none of which is designed for the purpose and all of which are often dangerous.

Figure 2–7. Self-propelled, rider-operated car mover. *(Courtesy Whiting Corp.)*

Safe Practices

Transportation personnel as well as all other employees should observe the safety rules observed by all large railroad systems. When safety problems arise in the facility's railway, seek help and suggestions from safety professionals and from operating officers of the facility's connecting railway line.

Hold safety meetings and other educational activities to inform the facility's personnel about railway hazards.

Identifying every practice required to safely operate railway equipment would fill an entire book. However, the following are a few of the most important safety factors:

1. Stop and look both ways before crossing any track.
2. Expect trains or cars to move at any time, on any track, and in either direction.
3. Step over rails when crossing tracks. Never step, walk, or sit on any rail.
4. Never go between moving cars, or cars that may move, to adjust couplers or for any other purpose. (The once-common practice of kicking couplers to align them is especially dangerous.)
5. Make a hand or lamp signal to stop. Be sure to receive an acknowledgment of the signal before moving between standing engines or cars.
6. To close a boxcar door, place one hand on the door handle and the other hand on the back end of the door.
7. Step—do not jump—down from cars.

Railroads often have their own standards for operations on their properties. These standards may be available from the relevant railroad's safety department.

FACILITY LAYOUT

Size, shape, location, construction, and layout of buildings and facilities should lead to the most efficient use of materials, processes, and methods. Safety and health professionals should strive for efficient production and maximum employee safety. Some principles to consider when planning a facility's internal layout follow:

- Employees should be able to recognize how materials, people, and products flow through the facility.
- Employees should learn where tools and equipment are located.
- Employees should be able to move easily within, to, and from the facility.
- Amenities for employees such as the lunchroom, locker room, and personnel office should be conveniently located.
- Supervisor offices should be situated near the work area so that communication with employees is as convenient as possible.

- Physical separation from areas that are extremely noisy or hazardous should be provided.

Location of Buildings and Structures

To minimize the hazards of fires and explosions, store both raw materials and finished products—as well as volatile, flammable liquids and liquefied petroleum (LP) gas—far from processing buildings. By doing so, fires and explosions are more easily controlled. The costs of separate storage facilities eventually may be less than the costs of the special systems and controls needed to store these materials in processing buildings. Provide ample space between storage units and flame sources such as boilers, shops, streets, and adjoining property. Follow the codes of the local and state or provincial authorities and of NFPA 30, Flammable and Combustible Liquids Code, when planning the locations of a facility's buildings.

Federal, state, and local laws govern the storage of explosives. Situate and manufacture magazines according to the recommendations of the Institute of Makers of Explosives. The type of retardant required in relation to the distance between buildings of frame, brick, and fire-resistant construction, as well as the minimum necessary distances between buildings, are given in NFPA 80A, Recommended Practice for Protection of Buildings from Exterior Fire Exposures. Factory Mutual data sheets and DOD guidelines are also useful references.

Many facilities use and store flammable liquids having flash points below 100°F (37.8°C). Plans and layouts for such facilities should conform strictly to the requirements of local fire prevention authorities and to the flammable-liquids handling and storage specifications in NFPA 30, Flammable and Combustible Liquids Code. NFPA 30 identifies the conditions under which flammable and combustible liquids of various classes can be stored in and around buildings. (See also Chapter 10, Flammable and Combustible Liquids, in this volume.)

Meticulous specifications must be complied with when storing flammable liquids outside of buildings in underground or aboveground tanks. Refer to current federal and state or provincial regulations regarding the amounts and kinds of liquids allowed in underground tanks. Standards for tanks specify proper types and thicknesses of materials; provisions for relieving excessive internal pressure; and details about proper grounding, insulation, and piping. The appropriate material to construct storage tanks depends on the type, corrosive properties, and processing requirements of the liquid.

The necessary distance between buildings and the location of an aboveground tank on adjoining property depend on the content, construction, fire-extinguishing equipment, and greater dimension (diameter or height) of the tank. A secondary containment vault, properly lined with monitors, may be required. NFPA 30 describes four types of tanks and each one's minimum distance from buildings. The American Petroleum Institute (API) also has guidelines for tank storage.

An automatic extinguishing system is considered fundamental in hazardous locations. Other safety features should include spark-resistant, conductive flooring; grounding of equipment and structures; and specialized electrical equipment and wiring. All tools, trucks, and similar equipment should be made from spark-resistant metal. Emergency exit doors should also be provided. (See Chapter 8, Electrical Safety, in this volume, and NFPA 30 and 70 for discussions of equipment designed for use in hazardous locations.)

Layout of Equipment

A detailed flow sheet is a useful guide for laying out facilities, particularly those incorporating dangerous materials and complex processes. By using a flow sheet, the nature of the materials and the process in each manufacturing stage can be studied and provisions can be made to eliminate or control any attendant hazards.

Other methods used to determine the safest and most efficient layout of machines and equipment include the two-dimensional method and the three-dimensional model. The two-dimensional method consists of making templates to scale and fitting them into a plan of the site or floor area. A more effective method uses three-dimensional models made to scale and set up on a scaled floor plan. The models can be rearranged until the safest and most efficient layout has been discovered. Computer simulations of arrangements, which are rapidly replacing model technology, should also be considered.

Layout studies indicate the most suitable locations for procedures such as spray painting, welding, and other work that generally requires segregated areas. Use the three-dimensional model to anticipate and prevent congested areas. The frequent handling of materials in these areas often leads to many unnecessary movements and to poor housekeeping—itself a source of accidents.

Insufficient headroom at aisles, platforms, pipelines, overhead conveyors, and other installations can also be determined from studying the models. A vertical distance of at least 7 ft (2.1 m) is generally stipulated to provide ample clearance between passageways and stairways and overhead structures. Overhead cranes and conveyors need at least 24 in. (61 cm) of vertical and horizontal clearances.

Integrated computer systems require a location with controlled temperatures, good ventilation, and sufficient electrical power. Layout studies also reveal the best locations for these systems.

Figure 2–8. This aisle is wide enough for small industrial trucks and is well marked.

Aisles

Forklifts and other powered industrial vehicles require ample room for movement—backing up and turning—that does not endanger workers or equipment. Aisles should also be wide enough for vehicles to be able to pass each other without colliding. For one-way traffic, aisles should not be less than 3 ft (0.9 m) wider than the widest vehicle. Aisles for two-way traffic should not be less than 3 ft (0.9 m) wider than twice the width of the widest vehicle. These minimum widths are exceeded considerably in new buildings where, because of anticipated heavy traffic, aisles from 12 to 20 ft (3.7 to 6.1 m) wide have been specified. In addition, a safe layout of aisles requires a lack of blind corners and an adequate turn radius for vehicles. A 6-ft (1.8-m) radius is sufficient for vehicles and small industrial trucks (Figure 2–8).

Clearly delineate aisles with approved markings in either traffic paint or striping material. ANSI Z535.1, Safety Color Code for Marking Physical Hazards, and OSHA 29 CFR 1910.22(b)(2) provide requirements and guidelines on markings. Selecting the materials for markings should be based on floor surface and use patterns. Many companies use plastic buttons that are glued or anchored to the floor with metal fasteners because of their durability.

Parking and Ramps

Designate parking areas for both manual and powered lift trucks. Large facilities often include garages with room for both storage and maintenance of these trucks. Also provide battery-charging rooms with good ventilation to prevent the buildup of hydrogen gas during the recharging process. (See Chapter 15, Powered Industrial Trucks, in this volume.)

If building plans include ramps for use by pedestrians and vehicles, earmark a 3-ft- (0.9-m-) wide section as a walkway. Sharp turns into aisles at the tops and bottoms of ramps are hazardous and should be prevented. Provide an abrasive coating where floors could become slippery. Ramp inclines should be minimized to allow for forklift ease of use.

Electrical Equipment

Completely metal-enclosed and grounded unit substations have been developed for industrial facilities. However, if transformers must be installed in confined areas or near flammable materials, be sure that they are noncombustible transformers. Furthermore, if these substations are in enclosed areas, provide ventilation to reduce the buildup of gases and encourage the release of heat.

Short-circuit-protective devices should be able to carry the load and should be designed to function despite any damage from a maximum short-circuit current. Circuit breakers, fuses, and safety switches that fail because of a short-circuit current may explode and cause severe damage and injuries. Therefore, use motor controls capable of lockout/tagout and other electrical safeguards.

Design safety measures for grounding systems and battery-charging rooms. Evaluate every grounding system to determine whether the system is capable of conducting the necessary amount of voltage. When direct-current voltage is supplied from batteries, isolate the battery-charging room from the work area. A battery-charging room should be well ventilated, and smoking should be prohibited.

In an up-to-date industrial electrical system, sections must be able to be de-energized for maintenance and other work without the entire system shutting down.

In addition, all electrical installations must conform to NFPA 70, National Electrical Code, as well as to local ordinances. (See also Chapter 8, Electrical Safety, in this volume.)

Heating, Ventilation, and Air Conditioning (HVAC)

Heating, ventilation, and air conditioning not only provide worker comfort but also are often needed for secure processing conditions. Personal comfort affects workers' efficiency. Make every effort, therefore, not only to have typical office and facility conditions comfortable but also to eliminate—or at least reduce—inadequate conditions that can contribute to excessive employee fatigue and discomfort. In buildings where flammable liquids or vapors are handled, provide adequate ventilation to prevent explosive concentrations from forming.

Because of the noise and vibrations from boilers, fans, and air conditioning equipment, this equipment should be separated from general work areas. Boilers should receive adequate air, and combustion by-products should be removed safely. Also, when situating incinerators, be sure that a negative-pressure differential in a building does not result in an incinerator stack serving as an air source.

For maintenance, authorized employees should have easy access to this machinery. That is, there must be sufficient space around the equipment for employees to be able to replace parts; for example, there must be room to pull tubes if necessary. For laboratory settings, consider ventilation criteria. See the *Biosafety in Microbiological and Biomedical Laboratories* (5th ed.) manual, cdc.gov/OD/OHS/biosfty/bmbl5/BMBL_5th_Edition.pdf. The National Safety Council's (NSC's) *Fundamentals of Industrial Hygiene*, 6th edition, provides more specific suggestions, as does NFPA 90A, Installation of Air Conditioning and Ventilating Systems.

Inside Storage

The storage space needed for raw materials and finished products may be estimated based on maximum production requirements. Make allowances for shortages, seasonal shipping, and quantity purchases.

Today's mechanical handling and stacking equipment permits the considerable use of vertical space through multiple decking. If this method is anticipated to be used, be sure that the flooring can support the maximum projected load. Also, ensure that new shelving/racking installations meet or exceed seismic code structural requirements and are secured to the floor and any walls that may be behind shelving.

Employees should easily be able to reach storage areas for supplies, finished products, and empty or full pallets. Additional features of storage areas should be stable piles or stacks as well as a properly functioning fire-extinguishing system. The fire-extinguishing system's density and area of application depend on the types of materials stored and the heights of the piles or stacks.

Room near working areas to store supplies, tools, flammable liquids, and infrequently used equipment is seldom included in layouts. The result is that such items often are left in unsafe positions and locations. To discourage employees from leaning heavy materials or equipment against walls, where they can fall, designate storage places for these materials or equipment.

Also, include space for racks, bins, and shelves. Store, above eye level, materials that project into aisles or walkways from racks, bins, and shelves. Use metal baskets or special racks provided with drip pans for storing machine parts covered with cutting oils. To store liquids, use confinement dikes, curbs, or drains. When spills occur, they will thus be prevented from affecting other areas of the facility.

Include closets for the storage of janitorial supplies such

as waxes, soaps, and other cleaning supplies. Plan for these closets to have floor sinks for filling pails, thus preventing personnel from having to lift heavy pails.

Storage of waste material may take considerable room, especially if the waste is bulky or is produced in large quantities. In buildings with storage on lower levels, chutes leading to waste bins may help prevent accumulations of waste in work areas. To dispose of small quantities of sharp-edged waste, designate boxes with handles that are shielded from sharp materials or similar containers. (See Chapter 12, Materials Handling and Storage, in this volume.) Specially designated areas meeting EPA regulations on hazardous waste accumulation must be included in a facility's layout.

LIGHTING

Electrical lighting that is properly complemented by daylight can satisfy a facility's lighting needs and also provide energy savings from processes that operate during daylight hours only.

Daylight

The amount of daylight can be predicted using the procedures in Calculation of Daylight Availability, Illuminating Engineering Society (IES) publication RP-21. To use it advantageously, take into account the following design factors:

- variations in the amount and direction of incidental daylight
- brightness distributions from clear, cloudy, partly cloudy, and overcast skies
- variations in the intensity of sunlight
- effects of local terrain, landscaping, snow reflectance, and nearby buildings on available light
- geographical location and orientation of the building.

The amount of natural light that enters a building and is available for use depends on the design of skylights and openings in the walls of the structure as well as on interior design and furnishings. Skylights and fenestration have at least three purposes in industrial buildings: (1) admitting, controlling, and distributing daylight for vision; (2) providing a distant focus for eye muscles to relax; and (3) reducing the claustrophobia that some people experience in completely closed-in structures. The design and installation of windows should also make washing them easy and safe.

Skylights may be installed to increase the amount of natural light. When considering skylights, a cost/benefit analysis must be undertaken to evaluate the effects of the skylights on the structural, heating, ventilating, and air conditioning systems as well as how well or poorly the skylights illuminate task performance. The skylights must also be specially reinforced so that employees who work on the roof do not fall through them.

Because the amount of daylight varies with the time of day, the day of the year, and weather conditions, it is necessary to install an electric lighting system to satisfy the total lighting requirements of a facility.

The primary advantage of daylight is the reduction in the amount of electric energy needed. To be effective, the daylight and electric light components must be coordinated so that they consistently provide the amount of lighting needed for employees' visual comfort and correct task performance. Consider using a lighting management control system to regulate the quantity and quality of illumination.

Such a control system, either manual or automatic, can vary the output of the electric lighting system and supplement the available daylight with the amount of artificial light needed to maintain the appropriate levels of lighting to complete particular tasks.

However, because of the critical nature of many production tasks, daylight cannot always be relied on. An example of a good use of daylight is in warehouse and storage facilities where skylights and simple photosensitive controls are effective. For a more comprehensive discussion of this subject, see the IES Recommended Practice for Daylighting.

Electric Lighting

The most important requirement of industrial lighting is that it provides a safe working environment and suitable visibility for all types of industrial activities. Under appropriate lighting conditions, workers can correctly observe and effectively control the operation and maintenance of all machines and processes.

In most industrial work areas, a sufficient quantity of natural light is not available, even during optimal daylight conditions. Therefore, electric lighting is needed to maintain proper visibility. The electric lighting system must be designed and installed so that it provides the necessary levels of illumination in areas adjacent to skylights, windows, or walls and thus ensures that good lighting is available throughout the work area.

The distribution of light from a lighting source is also important. Highly concentrated lighting makes high mounting heights economically feasible, whereas low mounting heights necessitate a widespread light distribution.

Four forms of electric lighting are used in industrial areas:

1. *General lighting* produces relatively uniform illumination throughout the area. According to the Guide on Interior Lighting (publication 29.2 of the International Commission on Illumination), general lighting should be designed so that the ratio of the minimum illuminance

to the average illuminance is not lower than 0.8 so that equivalent amounts of lighting are provided throughout the interior of a building. The average illuminance of the general area should not be lower than one-third of the average illuminance of the task areas. The average illuminance of adjacent interiors should not vary from each other by a ratio exceeding 5:1. Care must be taken not to exceed the recommended spacing criterion (SC) or the spacing-to-mounting-height (S/MH) ratios for the lighting equipment used. "Point-by-point" illuminance calculations are often used to examine the uniformity of general lighting levels.

2. *Localized general lighting* is an alternative to uniformly spaced illumination throughout an area and provides light at work zones.

3. *Supplementary lighting* is used to provide intense illuminances in small or restricted areas where general lighting cannot readily or economically provide such illuminances. Supplementary lighting is also used to furnish a specific level of brightness or a specific color or to provide a specific focusing or positioning of light sources.

4. *Emergency lighting* must be planned carefully for every facility in accordance with NFPA 101, Life Safety Code. Lighting for means of egress, exits, and stairwells may be provided by battery-powered lighting equipment or by a standby generating system that powers only selected lights.

Quality of Illumination

The quality of illumination depends on the distribution of brightness in the visible environment. Glare, diffusion, direction, uniformity, color, brightness, and brightness ratios all have significant effects on the ability to see easily, accurately, and quickly. Environments with poor-quality lighting are difficult and possibly hazardous to work in. Although moderately deficient lighting is not readily noticeable, the cumulative results of even slight glare can be significant fatigue, loss of visual acuity, and reduced productivity.

Quantity of Illumination

The quantity of light necessary in any particular environment depends primarily on the work that needs to be done in that environment. Studies show that as the illumination of a task increases, the ease, speed, and accuracy of accomplishing that task also increase. The quantity of illumination is expressed in footcandles (1 fc = 10.8 lux) and measured with an illuminance meter, which indicates the number of footcandles reaching the work area. Currently, the IES-recommended illuminances of industrial areas are given in ANSI/IES RP-7, Practice for Industrial Lighting. (See also Appendix 1, Safety and Health Tables.)

Glare

Glare may be defined as any brightness within the field of vision that is of such magnitude that it causes discomfort, fatigue, aggravation, or loss of visual performance and visibility. Glare basically reduces the efficiency of the eyes and thus increases of the risks of accidents. There are four types of glare:

1. direct glare
2. reflected glare
3. discomfort glare
4. disability glare.

Direct glare results from a source of light (whether daylight or artificial light) within the normal field of vision. Direct glare can be reduced by (1) decreasing or shielding the brightness of the light source, (2) repositioning the light source so that it no longer falls within the field of vision, or (3) increasing the brightness of the area surrounding the source of glare and against which the glare is seen (thus reducing the difference between the brightness of the source and the brightness of the background).

Reflected glare results from intensely bright images or from the differences in the brightness reflected by shiny ceilings, walls, desktops, materials, and other surfaces within the field of vision. Reflected glare may be more bothersome than direct glare if the reflected glare appears near the task or is superimposed over the task. For example, some tasks involve working on objects that reflect glare from the light source. Furthermore, reductions in contrast resulting from "veiled" reflections often take place and can significantly reduce the contrast necessary to be able to accurately see and discern details while completing a task.

Reflected glare may be reduced by (1) decreasing or shielding the brightness of the light source, (2) repositioning the light source or the task so that the reflected image is directed away from the worker's eyes, (3) increasing the level of illumination by increasing the number of light sources and thus reducing the relative brightness of the glare, or (4) in special cases, changing the characteristics of the surface to eliminate the temporary reflection and the resulting reflected glare.

Discomfort glare, although bothersome does not necessarily interfere with visual performance or visibility. Discomfort Glare in the Interior Working Environment, CIE No. 55, which is available from the IES, describes methods of controlling discomfort glare.

Disability glare, which reduces a worker's ability to see and thus the worker's performance, is accompanied by discomfort.

Lighting Management

Effectively monitoring and controlling illumination may be referred to as "lighting management." Lighting management provides the appropriate illumination when and where it is needed and minimizes electricity use. Monitoring uses sensors to register illuminances and worker presence, and its controls are either ON-OFF or adjustable (and may be manual or completely automatic). In lighting management's simplest form, the worker uses a control to switch on the lighting circuit and to adjust the illumination. More advanced systems are microprocessor based and receive sensor inputs and respond with control outputs that engage and adjust the illumination using low-voltage control relays, controllable circuit breakers, and adjustable dimming systems.

Caution must be exercised when selecting and applying automatic lighting controls within industrial facilities. When appropriate task performance is the primary objective, the worker must be able to readily adjust or override the lighting controls whenever necessary and at the worker's discretion.

Safety

Because safe working conditions are essential in any industrial facility, illumination's effects on safety must be considered. Design a facility's environment to compensate for the limitations of human ability. For example, any factor that aids vision increases the possibility that a worker will notice a potential hazard and be able to prevent it.

Usually when accidents are attributed to poor illumination, the cause is recorded as "very noticeable poor quality of illumination" or "practically no illumination at all." However, many less tangible factors associated with poor illumination are often important contributing causes of industrial accidents. Some of these factors are direct glare, reflected glare from the work, and excessive shadows—all of which hamper vision and increase visual fatigue. The following are other hazards associated with incorrect illumination design:

1. How regulatory safety colors appear under different light sources must be considered. High-pressure sodium lamps and low-pressure sodium lamps, as well as certain mercury lamps, tend to change the appearance of safety colors when these colors are painted using standard paints or protective coatings. Carefully investigate the safety colors and paints or protective coatings to obtain the appropriate visual result. Paint or protective coating manufacturers should also be consulted to find out the proper paint or protective coating formulations that will allow the safety colors to appear as their true colors under these light sources.

2. "Flicker," or the "stroboscopic effect," is the variation in the light output of a lamp run on an alternating-current circuit. The stroboscopic effect is most noticeable from rotating or oscillating machinery and is especially noticeable from high-pressure sodium lamps. Note that all sources can be operated on three-phase power, which helps overcome the stroboscopic effect.

Although visual tasks tend to be difficult, they are vital to profitable operation. Careful observations of equipment and instruments and quick physical responses to minute changes in equipment and instruments require high-quality illumination. Proper lighting thus safeguards a company's investments in both its machines and its skilled personnel.

Although the facility safety and health professional may be generally familiar with the lamps, reflectors, and lighting requirements in industrial environments, for extensive or difficult lighting installations, a qualified illuminating engineer should always be consulted. For a more extensive discussion of industrial lighting, see Appendix 1, Safety and Health Tables, in this volume, and the following references: ANSI/IES RP-7, Practice for Industrial Lighting, and the IES *Lighting Handbook*.

Hazardous (Classified) Locations

Areas in industrial facilities may be classified as hazardous locations under ANSI/NFPA 70, National Electrical Code, Chapter 5, Special Occupancies, because of the presence or potential presence of flammable or combustible liquids and vapors, flammable dusts, or flammable fibers. Such areas require the specialized lighting equipment and motorized equipment such as forklifts listed in the publications Electric Lighting Fixtures for Use in Hazardous (Classified) Locations, ANSI/UL 844, and ANSI B56.1. Equipment listed as "UL 844" provides the required illumination without introducing hazards to life and property. Because each type of lighting fixture is designed to meet specific requirements, these lighting fixtures are usually not interchangeable. (Consult Chapter 5 of the National Electrical Code to determine the requirements for lighting equipment and circuits in hazardous [classified] locations.) When any questions or doubts arise, consult the local electrical code inspector. (See Chapter 8, Electrical Safety, in this volume.)

Wet Locations

In normal circumstances, electrical equipment should not come in contact with water or other liquids. If unrated electrical equipment comes in contact with water, the equipment must be repaired immediately to prevent an arc flash event. When equipment could come in contact with a liquid, such as at outdoor locations or during wet processes, appropriate electrical fittings, fixtures, and cords should be used. The electrical equipment used at these times must be listed by Underwriters Laboratories (UL). Applicable standards are ANSI/UL 1570, 1571, and 1572 and OSHA 29

CFR 1910.303(b)(1)(viii) and (b)(6). Requirements for the use of ground-fault circuit interrupters (GFCIs) to protect personnel are found in ANSI/NFPA 70, National Electrical Code, and in 29 CFR 1910.304(b)(3).

Protective Lighting

Protective lighting is necessary for the nighttime monitoring of outdoor areas. Such lighting discourages would-be intruders or renders them visible to facility guards. It may also reduce the chances of a fire spreading. Because protective lighting is usually not adequate for efficient facility operation, this type of lighting is generally considered an auxiliary to productive lighting.

Protective lighting takes the form of positioning adequate light in the outside areas of buildings or, in some cases, by setting up equipment that produces blinding light in the eyes of an intruder but with no light appearing on the guard. Lighting units should be arranged to eliminate concealing shadows.

In general, four types of lighting units are used in protective lighting systems: floodlights, streetlights, Fresnel lens units, and searchlights. Infrared lamps, which are invisible to trespassers, are also used along with infrared TV monitors. Refer to IES publication RP-10, Protective Lighting.

The amount of time required for an emergency evacuation of personnel should influence the choice of lighting units. Correlate units' battery capacities with the number of needed lamps and their wattages to determine the lighting needed for the length of time required for the complete evacuation of personnel: 90 minutes, if using the recommendations of NFPA 101, Life Safety Code, or NFPA 70, National Electrical Code.

When longer durations of emergency lighting are required, engine-powered generators should be the power source. Engine-generator sets for emergency lighting must start up and provide light automatically upon failure of the usual power supply. An automatic switch transfers power from the usual power supply to the emergency power supply and reverses the procedure upon the reestablishment of the normal power supply.

Outdoor lighting (on the roofs or walls of buildings, in parking lots, on roadways, etc.) of any design should be compatible with the lighting in the surrounding neighborhood and prevent light "spill" and glare directed onto passing motorists. Refer to IES publication CP-46, A Statement on Astronomical Light Pollution and Light Trespass.

Security of Facilities

A security system is an essential consideration when planning or remodeling an industrial facility. The system should be both functional and cost effective. Beginning with site selection, design the facility's environment for minimum loss and maximum security. Typical security considerations include the following:

- Keep the number of openings to a minimum.
- Secure all windows.
- Use protective lighting.
- Make sure that entrances and service doors lead to a reception area.
- Install alarm systems that detect fire, fumes, vapors, and intruders.
- Install automatic extinguishing systems.
- Limit access to docks and other receiving areas.

See Chapter 23, Workers with Disabilities, in the *Administration & Programs* volume for details about the necessary security for workers with disabilities.

USE OF COLOR

The following excerpt, written by Linda Trent, is used with permission from The Sherwin-Williams Company, Cleveland, Ohio:

> At work or at play, consciously or subconsciously, people respond to the colors around them. And, in growing numbers, industrial designers and managers are paying more attention to the interactions between color, lighting, and human behavior. Managers are becoming more attuned to the industrial psychologists' message that the quality and appearance of work areas can stimulate interest or create boredom. As such, they are receptive to the idea that the proper use of color can generate a positive response to the work environment—favorably affecting workers' housekeeping efforts, safety, and overall productivity. Moreover, since it doesn't cost any more to paint work areas in scientifically chosen colors than in colors chosen entirely at random, many managers are willing to consult a professional color stylist.

Color and Light in the Workplace

The function of paint, in addition to protecting surfaces, is to absorb or remove some parts of the color spectrum and to transmit or reflect other parts of that spectrum. Color is determined by which parts of the visible spectrum are reflected. Consequently, color is visibly modified by different light sources. For that reason, the effects of a facility's lighting system should be a major consideration when selecting colors for the facility.

Typically, a facility's lighting system should be selected based on characteristics such as the amount of light produced per watt used, ease of maintenance, ease of shielding, ease of directional control, and overall cost. The light system's effects on color and color-rendering properties are usually secondary considerations. Lighting systems currently used in industrial settings include a range of fluorescent and high-intensity discharge lamps and combinations of illuminants. These various light sources' effects on color are listed in Table 2–A.

Interestingly, light affects color, and color affects the quality of light. When referring to surface colors, this effect on light is called the light-reflectance value (LRV) of color. It is an important property because reflections from painted surfaces—ceilings, walls, machinery, and floors—act as secondary light sources. With proper color styling and recommended reflectance, work area surfaces maximize the available light and reduce shadows.

In general, pale colors reflect light whereas dark colors absorb light. White, and then the pastels, have the highest light-reflectance values; black affords no reflectance. Often, paint suppliers refer to an LRV that is equivalent to the reflectance of a material's surface as defined in the IES *Lighting Handbook*.

In a work environment, the LRVs of colors can contribute significantly to workers' being able to see a task clearly. Objects are discerned only in contrast with their surroundings; the most effective contrasts can be obtained by selecting colors based on their light-reflectance characteristics. In industrial settings, high-contrast conditions are provided for tasks such as inspecting products but are deemed unnecessary for tasks such as retrieving goods from seldom-used storage areas.

As a rule of thumb, surfaces in industrial facilities should provide light-reflectance values within the ranges given in Table 2–B. Color more than types of materials affects light reflectance. Keep wall reflectance within the recommended ranges except when atypical conditions make alterations necessary. For example, walls in areas with high exposed ceilings might have a high-reflectance ceiling that is brought down to the level of suspended light fixtures, which direct some light upward. This upper wall surface can increase the light reflectance in a room by as much as 10% (Edmonds 1977).

TABLE 2–A. Color Effects of Light Sources*

Lamp Type	Appearance on Neutral Surface	Effect on "Atmosphere"	Colors Strengthened	Colors Grayed	Effect on Complexions	Remarks
FLUORESCENT						
Cool white°	white	neutral to moderately cool	orange, yellow, blue	red	pale pink	Blends with natural daylight; good color acceptance
Deluxe cool white°	white	neutral to moderately cool	all nearly equal	none appreciably	most natural	Best overall color rendition; simulated natural daylight
Warm white⁺	yellowish white	warm	orange, yellow	red, green	sallow	Blends with incandescent light
Deluxe warm white⁺	yellowish white	warm	red, orange, yellow, green	blue	ruddy	Good color rendition; simulates incandescent light
INCANDESCENT						
Filament⁺	yellowish white	warm	red, orange, yellow	blue	ruddiest	Good color rendering
HIGH-INTENSITY DISCHARGE LAMPS						
Deluxe white mercury°	purplish white	warm, purplish	red, yellow, blue	green	ruddy	Color acceptance similar to cool-white fluorescent
Metal halide multi-vapor°	greenish to pinkish white	moderately cool, greenish	yellow, green, blue	red	grayed	Color acceptance similar to cool-white fluorescent
High-pressure sodium°	golden white	warm, yellowish	yellow, orange, green	red, blue	golden	Color acceptance approaches that of warm-white fluorescent

*Table based on information from The General Electric Co.
°Greater preference at higher levels.
⁺Greater preference at lower levels.

Another exception to the recommended values in Table 2–B involves the peripheral areas of a room or work space. If these areas are not in the direct line of vision of a task and are restricted to about 10% of a worker's visual area, they usually do not compromise the efficiency of the lighting system. In fact, intense tones with low-reflectance characteristics can be used as accents and focal points to make the workplace more pleasant. In addition to reflectance, several other factors affect color in the workplace.

TABLE 2–B. Reflectance Values Recommended for Facility Surfaces

Surface	Reflectance Values	
	Manufacturing Areas (%)	Office Areas (%)
Ceilings	80–90	80–90
Walls	50–65	60–70
Floors	15–30	25–40
Machinery	30–50	—
Desktops	—	40-50

Geographical Location and Exposure

Regional color preferences can be incorporated into a color plan (e.g., desert colors in the Southwest or patriotic colors in the East). In a facility with windows, a particular exposure can be contrasted with color (e.g., warm colors [reds, yellows, or oranges] for a northern exposure).

Age of the Facility

Because newer buildings don't have many windows, pale colors are used to improve overall lighting and employee morale. In some older buildings, windows no longer lead to the outdoors, making an area look gloomy and the windowed wall look disordered. In these cases, paint the windows to increase light-reflectance values and eliminate distractions.

Demographics

Consider the number of female versus male employees in the facility. Color-preference studies indicate that men prefer blue, followed by red. Women, on the other hand, tend to prefer lighter colors such as peach and mauve.

Noise Level

No color can diminish the noise of machinery. The use of blues, greens, and neutral colors, however, tends to lessen workers' noise-induced tension.

Types of Equipment

Most types of equipment are painted in a neutral color to reduce glare and provide good contrast with safety colors. Very large machinery, in some instances, can accommodate a two-tone combination that emphasizes operating parts and provides visual contrast.

Psychological Factors

Dark, intense colors make surroundings appear confined and can sadden workers. On the other hand, lighter colors such as pastels lift workers' spirits and create the illusion of spaciousness and tranquillity. To reduce visual clutter, paint structural work (pipes, I-beams, etc.) the same color as the adjacent wall or ceiling.

Safety

The Occupational Safety and Health Administration (OSHA) and ANSI have set standards for safety colors throughout industry. Table 2–C summarizes the OSHA and ANSI safety color code.

Some companies are very faithful to the colors used in their corporate logos and in the packaging of their products and incorporate those colors into their facilities' interiors, either in graphics or as understated embellishments on doors or trims. Consider the effects of using patterns in addition to color. Certain color and pattern combinations on carpets, walls, and other areas can cause dizziness, especially in elderly individuals, which increases the risks of falling.

Human Response to Color

When Sir Isaac Newton defined the visual process that takes place in optics, he described the three stages of color perception in terms of the physics, physiology, and psychology of the process:

1. the light entering the eye and the factors that determine the spectral composition of the light (physics)
2. the response in the retina and visual pathways to the light (physiology)
3. the color actually being perceived and the mind's response to the color's appearance (psychology).

Scientists who study color today incorporate Newton's observations but use a slightly different vocabulary. They say that reflexive responses to color, the most observable of the three types of responses, result from the physical structure of the eye. The most apparent reflexive mechanism can be seen when certain colors cause an advancing/retreating effect, which results from the eye not focusing all wavelengths of light in the same place. Colors with longer wavelengths—red, yellow, and orange—seem to move toward the observer. Cool, dark colors—blue and green—on the

TABLE 2–C. Summary of OSHA and ANSI Safety Color Code Corporate Colors*

Color	Designation
Red	Fire: Protection equipment and apparatus, including fire-alarm boxes, fire-blanket boxes, fire extinguishers, fire-exit signs, fire-hose locations, fire hydrants, and fire pumps. Danger: Safety cans or other portable containers of flammable liquids, lights at barricades and at temporary obstructions, and danger signs. Stop: Stop buttons and emergency stop bars on hazardous machines.
Orange	Dangerous Equipment: Parts of machines and equipment that may cut, crush, shock, or otherwise injure.
Yellow	Caution: Physical hazards such as stumbling, falling, tripping, striking against, and being caught in between.
Green	Safety: First-aid equipment.
Blue	Warning: Caution limited to warning against starting, using, or moving equipment under repair.
Black on yellow	Radiation: x-ray, alpha, beta, gamma, neutron, proton radiation.
Black and white	Boundaries of traffic aisles, stairways (risers, direction, and border limit lines), and directional signs.

* See full text under Section 1910.144 of Occupational Safety and Health standards. For piping colors, see ANSI Standard Scheme for the Identification of Piping Systems, A13.1981.

other hand, have shorter wavelengths and appear to move away from the observer. This reflexive response also creates the illusion of a larger or a smaller space. Moreover, safety engineers count on this response and use warm, bright colors to call attention to dangerous machine parts, fire hazards, and physical hazards that might cause workers to lose their footing.

Standard physiological responses include all of color's effects that cannot be considered reflexive or conditioned responses. Of these standard responses, the photoreactive effects are the easiest to describe and observe. For example, physicians use blue light to treat hyperbilirubinemia, a dangerous condition in newborn babies in which their blood contains potentially fatal levels of the hemoglobin by-product bilirubin and gives them a yellow hue (jaundice). Because bilirubin is photoreactive, it breaks down when exposed to certain wavelengths of radiation; the "bililight" is thus a treatment used for jaundiced babies.

Other physiological responses to color are (1) the relationships between hue and perceived heat and (2) the relationships between certain hues and metabolic activities. The basic idea behind the former is that "warm" colors (reds, yellows, and oranges) tend to make people feel warmer, whereas "cool" colors (blues and greens) tend to make people feel cooler. Early studies suggested that the perceived temperature differences of these two groups of colors might be as great as 7°F. More carefully controlled, recent studies, however, suggest that the differences might actually be less than 1°F or even completely indistinguishable (Faber Birren and Company). Despite these studies' results, researchers conclude that, in general, cool colors calm people, whereas warm colors excite them.

These hue/activity correlations have practical, psychological applications in selecting colors for a facility's interiors: use cool colors to ease tensions resulting from highly detailed work or noisy machinery; use warm colors in a lunchroom to dispel boredom and stimulate conversation. The following list and Table 2–D give more suggestions for selecting colors to enhance productivity in the workplace.

- Use color schemes to differentiate and consolidate work areas that would otherwise blend into large facilities that have a variety of production activities.
- Use neutral colors, which have low LRVs, in laboratories, where reflected color might hinder accurately seeing the materials being tested and analyzed. However, do make the area interesting by using more intense colors on furniture and doors, where visual observation is not critical.
- Use intense, bright colors in time clock and locker room areas. In such non–work areas, color can provide a cheerful atmosphere for the employees at the beginning and at the end of the workday.
- Use colors with high reflectance values in warehouses to offset the typically low light levels there and to maximize the availability of light. Color-code storage areas to facilitate locating materials.
- Use contrasting colors in production areas to focus workers' attention on the tasks and to increase visibility.
- Use intense shades of warm colors sparingly to avoid confusing workers or making them anxious.
- Use high-reflectance colors on stairways and intense accent colors on rails and doors to demarcate points of orientation.

Color-Coding

Color is used extensively for safety purposes. Although color should never be used as a substitute for good safety measures or mechanical safeguards, specific colors are used to identify specific hazards. Be sure to check the latest regulations for in-facility use, shipping, or consumer protection. Color-coding is generally as follows:

- Red designates fire protection equipment, danger, and emergency stops on machines.
- Yellow is the standard color for (1) denoting hazardous

TABLE 2–D. Characteristics and Suggested Uses for Various Colors

Color	Impression	Suggested Use for Interiors
WARM COLORS		
Red, orange, yellow	Attract attention, create excitement, promote cheerfulness, stimulate action.	Nonproductive areas, including employee entrances, corridors, lunchrooms, break areas, locker rooms, etc.
COOL COLORS		
Blue, turquoise, green	Cool, relaxing, refreshing, peaceful, quieting. Encourage concentration.	Production areas, maintenance shops, boiler rooms, etc.
LIGHT COLORS		
Off-whites and pastel tints	Make objects seem lighter in weight; areas seem more spacious. Will usually give people a psychological lift. Reflect more light than darker tones.	Most production areas, especially small rooms, hallways, and warehouse and storage areas. Poorly illuminated rooms.
DARK COLORS		
Deep tones, gray, black	Make objects seem heavier; absorb light. Will make rooms appear smaller and surroundings cramped. Long exposure will create monotony and depression.	Not normally recommended for large areas because of light absorption qualities. Use should be confined to small background areas where contrast is needed.
BRIGHT COLORS		
Notably yellow, as yellow-green, orange, red-orange, red	The purer these colors are, the more compellingly they attract the eye. Make objects appear larger and create excitement.	Complement to basic wall colors; small objects such as doors, columns, graphics, time clocks, time-card racks, bulletin boards, tote boxes, dollies, etc.
White	Pure, denotes cleanliness, reflects more light than any other color.	All ceilings and overhead structures, and rooms where maximum light reflection is needed. Can also be used on small objects for greater contrast.

areas that may result in accidents from slipping, falling, striking against something, and so forth; (2) identifying flammable liquid-storage cabinets; (3) marking a band on red safety cans; (4) denoting materials-handling equipment such as lift trucks and gantry cranes; and (5) designating radiation hazard areas or containers (Safety Black on Safety Yellow). Black stripes or "checkerboard" patterns are also often used with yellow.

- Green designates the locations of first-aid and safety equipment (other than fire-fighting equipment). (See also the Blue entry.)

- Black and white, and combinations of them in stripes or checks, are used for housekeeping and traffic markings. They are also permitted to be used as contrast colors.

- Orange is the standard color for pointing out the dangerous parts of machines or energized equipment, such as the exposed edges of cutting devices, and the insides of (1) movable guards and enclosure doors and (2) transmission guards.

- Blue is used on informational signs and bulletin boards that are not of a safety nature. If they are of a safety nature, use green except to flag railroad cars. A blue flag is used to designate chocked cars that are being unloaded.

- According to ANSI Z535.1, "The radiation hazard symbol

colors shall be Safety Black on Safety Yellow. At present Safety Purple on Safety Yellow or Safety Black on Safety radiation hazard symbols may be used until replaced."

The piping in a facility may carry harmless, valuable, or dangerous contents. Therefore, it is highly desirable to designate different piping systems. ANSI A13.1, Scheme for the Identification of Piping Systems, specifies standard colors for identifying pipelines and describes methods of applying these colors to the lines. The contents of pipelines are classified in the following way:

Classification	Color
Fire protection	Safety Red (7.5R; LRV 12)
Dangerous	Safety Yellow (5.0Y; LRV 69)
Safe	Safety Green (7.5G; LRV 6)
Protective materials (e.g., inert gases)	Safety Blue (2.5PB; LRV 5)

(Standard Munsell hue and light-reflectance values are given for each safety color.)

Apply the appropriate color to the entire length of the pipe or in bands that are 8 to 10 in. (20 to 25 cm) wide near valves and pumps and at repeated intervals along the pipeline. Stencil in black the name of the specific material at easily seen locations such as valves and pumps. Designate piping that is less than ¾ in. (1.9 cm) in diameter by using enamel-on-metal tags. Other methods may be equally effective in identifying piping networks. Note ANSI A13.1's recommendation that highly resistant colored materials instead be used where acids and other chemicals may affect paints.

Situations may arise in which pipes cannot be color-coded or labeled. For example, some pipes may not be able to be designated as carrying a single material. In these situations, personnel must be trained to be aware of the hazardous materials in unlabeled piping.

Accident Prevention Signs

Accident prevention signs are among the most widely used safety measures in industry. Therefore, uniformity in the color and design of these signs is essential. Even if employees cannot read English or are color-blind, they will know how to react correctly to accident prevention signs.

The following summarizes the requirements for signs:
DANGER—Immediate and grave danger or peril. Red oval in top panel; black or red lettering in lower panel.
CAUTION—Lesser hazards. Yellow background color; black lettering.
GENERAL SAFETY—Green background in upper panel; black or green lettering on white background in lower panel.
FIRE AND EMERGENCY—White letters on red background. Optional for lower panel: red on white background.
INFORMATION—Informational signs, bulletin boards, and railroad flags for chocked cars. Blue letters on white background.
IN-FACILITY VEHICLE TRAFFIC—Standard highway signs. See the U.S. Department of Transportation's *Manual on Uniform Traffic Control Devices for Streets and Highways.*
EXIT MARKING. See NFPA 101, Life Safety Code, Section 5–11.

BUILDING STRUCTURES

Building structures include access areas, aisles and corridors, stairways, walkways, exits, flooring, and workstations. Of these, stairways, runways, ramps, and other access structures are the principal locations of accidents. One-fifth of all industrial injuries result from falls. Of those that take place as a worker moves from one level to another, most injuries result from falls on stairs and ladders. (See Chapter 3, Construction of Facilities). Diligent design and construction, however, can prevent many serious injuries. Also be sure to heed the standards of state and municipal governments and those of ANSI.

Runways, Platforms, and Ramps

Widths of runways and ramps should be appropriate for the anticipated amount of traffic. Use standard railings to guard open sides. Platforms 4 ft (1.2 m) or higher above floor or ground level should also be guarded by a standard railing. Walkways should have railings with a top-edge height equal to that of the toprails, or equivalent guardrail system parts: 42 in. (1.1 m) above walking/working levels. Midrails should be installed at a height midway between the top edge of the guardrails and the walking/working surface, and, on the exposed side, toeboards should be 4 in. (10 cm) high (Figure 2–9). [See 29 CFR 1910.23(e)(1) and (e)(4).] A convenient access should be provided to locations that are more than 4 ft (1.2 m) above the floor.

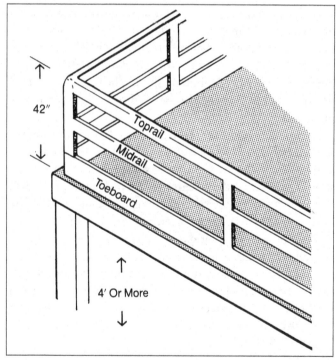

Figure 2–9. Open-sided floors or platforms that are more than 4 ft (1.2 m) above floor or ground level and scaffolds that are more than 10 ft (3.1 m) above floor or ground level should have a railing with a top-edge height of the toprails, or equivalent guardrail system parts, that is 42 in. (1.1 m) ±3 in. (8 cm) above walking/working levels. Midrails should be installed at a height midway between the top edge of the guardrail system and the walking/working surface, and, on the exposed side, toeboards should be 4 in. (10 cm) high. [See 29 CFR 1926.502(b)(1) and (b)(2)(i).] If persons can pass beneath or if there is moving machinery or other equipment from which falling materials could create a hazard, the guardrail should have a 4-in.- (10-cm-) high toeboard. Screening can also be added.

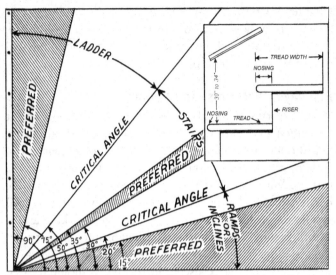

Figure 2–10. Preferred angles for fixed ladders, stairs, and ramps.

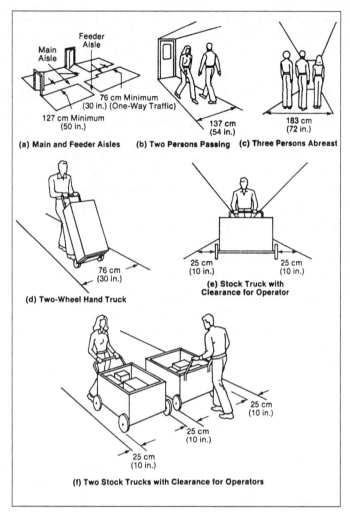

Figure 2–11. Minimum clearances for aisles and corridors. (Reprinted with permission from *Ergonomic Design for People at Work*, Eastman Kodak Company.)

Wire screen enclosures are necessary when materials must be stored on platforms and when these materials or fragments of materials could fall off the platform and injure anyone present beneath the platform. Use wire netting of No. 16 U.S. gauge wire with 1-in. mesh (1.5-mm-diameter wire with 38-mm mesh) for the screen. Plywood or other materials may also be used to enclose platforms on all sides.

Construct ramps with the smallest slope practicable—some states require a slope of 1 in 10 (5°43'). Fifteen degrees (a slope of 2.68 in 10) is a recommended maximum; a slope should never exceed 20 degrees (3.64 in 10) (Figure 2–10). Consult the ADA for specifications for ramp slope if the ramp might be used as a wheelchair access ramp.

Except where dislodged abrasives would damage equipment or process materials, apply abrasive coatings or pressure-sensitive adhesive strips to ramps to provide safe footing. Install toeboards when a ramp extends over a workplace or a passageway. On steep inclines, place cleats 16 in. (41 cm) apart. Make sure that planks run along the length of the ramp and do not overlap.

Aisles and Corridors

Design aisles and corridors so that they provide at least minimum clearances for equipment (Figure 2–11). Keep the following suggestions in mind when designing aisles and corridors:

- Keep aisles clear of structural supports, columns, and so forth.
- Incorporate traffic guides.
- Situate pathways (aisles) at the shortest possible distance between work areas.
- Design aisles so that as users make their way down the aisles, the users do not inadvertently hit machines and equipment.
- Try not to design blind corners; use mirrors if necessary.
- Make sure that doors do not open into corridors.
- Make sure that traffic does not have to go down the aisle in only one direction.
- Do not design aisles to be against walls.

Stairways

Spiral stairs are generally not acceptable in industrial locations, according to OSHA 29 CFR 1910.24(b) and NFPA 101, Life Safety Code. Fixed ladders with a slope of 75 to 90 degrees may be alternatives to spiral stairs. Refer to OSHA 29 CFR 1910.27(e)(1). For industrial stairways, be sure to cover treads with a durable, slip-resistant material.

The preferred slope of a stairway is between 30 and 35 degrees from the horizontal (see Figure 2–10 and Table 2–E). The most suitable slope of a fixed ladder is from 75

TABLE 2–E. Slope and Dimensions of Treads and Risers

Angle of Stairway with Horizontal	Riser (inches)	Tread and Nosing (inches)
30°35′	6½	11
32°08′	6¾	10¾
33°41′	7	10½
35°16′	7¼	10¼
36°52′	7½	10
38°29′	7¾	9¾
40°08′	8	9½
41°44′	8¼	9¼
43°22′	8½	9
45°00′	8¾	8¾
46°38′	9	8½
48°16′	9¼	8¼
49°54′	9½	8

Reprinted with permission from OSHA Standards §1910.24.
(1 in. = 2.54 cm)

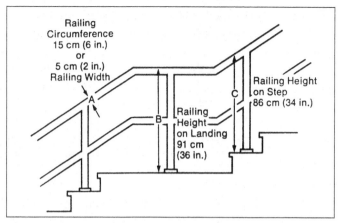

Figure 2–12. Handrail design guidelines. *(Reprinted with permission from Ergonomic Design for People at Work, Eastman Kodak Company.)*

to 90 degrees. A tread width of not less than 11 in. (28 cm), including a nosing of 1 in. (2.5 cm), is recommended. Riser height should be no more than 8 in. (20 cm) or less than 5 in. (12.5 cm) and should be uniform on each flight. Stairs and landings should also be able to support an active load of at least 100 lb/ft. (48.82 kg/m), with a safety factor of 4.

A flight of stairs with four or more risers should have a handrail as specified in ANSI A1264.1–1995, Safety Requirements for Workplace Floor and Wall Openings, Stairs, and Railing Systems. Rails should be 30 to 34 in. (76.2 to 86.4 cm) from the top surface of the stair tread, measured in line with the face of the riser. (Up to 42 in. [1.1 m] is permitted on steep angles.) An intermediate handrail is recommended for stairways that are more than 88 in. (2.2 m) wide. Hardwood handrails should have at least a 2-in. (5-cm) diameter (Figure 2–12). If metal pipe is used, the pipe should be at least 1 in. (2.54 cm) in outside diameter. Design for clearance between the handrail and the wall of at least 3 in. (7.62 cm). To provide a level surface along the top and both sides of a handrail, mount the handrail directly onto a wall or partition using brackets that are attached to the lower part of the handrail. Space brackets no more than 8 ft (2.5 m) apart. The top of the handrail's post should be able to withstand 200 lb (91 kg) applied from any direction. Also be sure to consult applicable local and state codes.

Because of space limitations, a permanent stairway sometimes has to be installed at an angle greater than 50 degrees. Such a stairway (commonly called an inclined ladder or "ship's ladder") should have handrails on both of its sides and open steps.

Other safety precautions to keep in mind when designing stairways are the following:
- Provide adequate lighting in stairways.
- Locate the lighting so that it does not cause glare.
- Enclose outside stairways to block out rain, snow, and ice.
- Enclose all inside stairways in partitions of fireproof or fire-resistant material.
- Install approved fire doors to prevent the spread of smoke or flames from one floor to another. Check NFPA, federal, state or provincial, or local codes regarding fire door requirements.

Walkways

Building plans should include elevated walkways and platforms for tanks, bins, and machinery as well as for locations where workers must go during a normal workday. Provide conveyors with slip-resistant surfaces, crossovers, railings, guarding, and toeboards.

Equip walkways and ramps with a standard railing, which should consist of a toprail, an intermediate rail, and posts and have a vertical height of 42 in. (1.1 m) from the upper surface of the toprail to the floor, platform, runway, or ramp level. A standard toeboard should have a vertical height of 4 in. (10 cm) from its top edge to the level of the floor, platform, runway, or ramp (Figure 2–9).

Exits

Situate exits so that in the case of fire or other emergencies, all persons can quickly evacuate the building. A building needs to have the appropriate number and size of exits from the outset because altering or adding exits after a building has been constructed is expensive. Also be sure that the exits conform to NFPA, federal, state or provincial, and local requirements.

NFPA 101, Life Safety Code, generally requires two exits on each level, including basements. Depending on a floor's occupancy, one exit may be an inside stairway or smokeproof tower, and the other exit may be a movable stairway or horizontal exit. Some government regulations require two or more exits that are distant from each other to be provided on all floors of industrial buildings that are two or more stories high. One exit must be a stairway made of fire-resistant material and lead directly to the outside at grade (ground) level.

To a considerable extent, the appropriate number and size of exits depend on the processes that take place in a building. In buildings where high-hazard processes take place, no part of the building should be more than 75 ft (23 m) from an exit. For buildings where medium- and low-hazard processes occur, areas of the building may be 100 to 150 ft (30 to 45 m) away from an exit. The Life Safety Code also stipulates that aisles, passageways, and corridors leading to exits must be accessible to all employees and that the total width of passageways and aisles must be at least the width of the exit. Exit doors should be clearly visible, illuminated, and denoted by signs and should open in the direction of departure. Exits should be kept unobstructed and free of any objects. They should also be secure areas that allow personnel the chance to leave the facility in an emergency. Doors and passageways that may be mistaken for exits must be denoted by NOT AN EXIT signs.

Flooring Materials

Comfort, health, and safety are closely related to the design and specifications of floors. Because flooring requirements often vary tremendously from department to department, carefully consider the following factors to determine the best type of flooring for a particular location:

- load
- durability
- maintenance
- noise
- dustiness
- drainage
- heat conductivity
- resilience
- electrical conductivity
- appearance
- chemical composition
- slip resistance.

Because a principal cause of accidents is floor slipperiness, be aware of the slip-resistant qualities of various types of floor surfaces. (See Chapter 28, Contractor and Customer Safety, in the *Administration & Programs* volume.) Different types of materials can reduce slipperiness in different areas and can minimize the factors that cause rapid deterioration of flooring. For example, use cast-metal inserts around woodworking machines. Where acid is spilled occasionally, cover the floor with materials that are acid resistant and provide drainage so that the acid can be washed away. Reduce the slipping hazard at the doorsills of elevators by installing slip-resistant surfaces.

Inserts should be installed flush with the surface of the floor. However, if an insert must be placed on top of the floor, a bevel must be on every side from which the insert can be approached. The bevel should also be at such an angle and height that a person will not trip or lose his or her balance.

Abrasive-coated fabric strips, which are secured to the floor by an adhesive, or abrasive floor coatings, also reduce slipperiness. They can be used on metal floor plates, the foot of stairs, stair nosing, and other hazardous locations.

Drainage is essential where wet processes take place. In some locations, especially along passageways, slip-resistant floor gratings are installed to reduce slipperiness, or raised floor mats are installed to raise the walking surface above the wet area. Also consider the following safety and health factors:

- Install moisture-absorbing mats or runners at entrances to reduce the amount of mud and dirt tracked into a building.
- Install noncombustible flooring where welding is performed regularly and in oven, furnace, and boiler areas.
- Indicate openings in floors with railings or barriers. Install a toprail, intermediate rail, and toeboard according to ANSI A1264.1–1995, Safety Requirements for Workplace Floor and Wall Openings, Stairs, and Railing Systems. Also install sheeting or woven wire around the opening to prevent material from falling in.
- Install adequate electrical outlets, electrical drops, or compressed-air stations to keep cords and hoses off the floor.

Several types of flooring materials are appropriate for industrial facilities. The characteristics and uses of some of these materials are discussed next.

Asphalt

Asphalt is used in various flooring materials. Dustless, malleable, odorless, and warm to the feet, asphalt provides especially good flooring in some facilities. If the asphalt mixture contains silica, the flooring will be resistant to acid. However, asphalt tile, which is often used in offices, may become slippery when washed or improperly waxed. Asphalt is quickly being replaced by vinyl-based sheet goods, which are more attractive and also less trouble to keep clean. Ordinary grades of asphalt are not recommended for facility roads because these grades soften in hot weather and do not hold up under heavy industrial trucks.

Paving Brick

If laid on top of a firm foundation like concrete, paving brick is suitable for heavy traffic areas. Cement mortar joints make a level surface. Foundry floors are made from hard-burned bricks that are laid face up on a concrete base with sand-filled joints.

Concrete

Concrete floors appear widely in warehouses and factories. Although smooth concrete is slippery when wet, floor coatings that provide slip resistance or chemical resistance can be used; concrete cannot withstand acids, however. In some types of work, concrete is too hard and too cold to stand on for long periods. Resilient, slip-resistant mats with low heat conductivity can be placed on concrete when workers must stand in one position for substantial periods.

Cork Tile

Cork tile is beneficial because of its insulation, resiliency, noiselessness, high slip resistance, and ability to tolerate lightweight traffic for long periods. However, it is not suitable for wet locations or where there are heavy loadings. It is also expensive.

Asphalt-Based or Vinyl-Based Tile

Use asphalt-based or vinyl-based tile or sheet goods where cleanliness and attractiveness are important, such as in offices, laboratories, and workrooms. This material is easy to clean, noiseless, and a poor conductor of heat.

Magnesite

For areas with little traffic or where weak oils are used, install magnesite flooring. It must be laid on a rigid base and should not be used where there is excessive moisture or hydrostatic pressure, such as basements. Because magnesite corrodes some metals, a coating of bituminous paint is necessary to protect metal objects such as pipes.

Cast-Iron Plates

Cast-iron plates with checkered or otherwise roughened surfaces that are laid in cement or asphalt are suitable for rigorous wear, such as in warehouses. They are relatively slip resistant unless wet or worn smooth. However, these plates are noisy and greatly conduct heat and electricity.

Metal Grates

Floors and gratings of metal grille do not collect dust, dirt, or liquids. Because this flooring is noisy, do not use it where hand trucking takes place regularly. However, this type of flooring is particularly appropriate in boiler rooms and over openings.

Parquet

Parquet is laid on an underfloor. If the material is sealed properly, little maintenance is required. Suitable for office floors, parquet withstands use but can be noisy.

Rubber

Rubber flooring is resilient. It is also highly dielectric, which is unsuitable where static electricity is common. However, conductive types of rubber flooring are available. In addition, abrasive rubber flooring can be used to eliminate slipperiness.

Terrazzo

Because terrazzo flooring has no joints in its surface, its use eliminates some of the problems encountered with some other types of flooring. For example, terrazzo can be made electrically conductive by grounded grilles, and it conducts heat. However, appropriate sealers are necessary to make such a floor impervious to most acids, and the terrazzo mixture is slippery unless it includes abrasive aggregates.

Ceramic Glazed Tile

Laboratories and other facilities such as dairies that require considerable cleanliness and sanitary conditions use ceramic glazed tile. This tile should be laid in Trinidad asphalt or in cement that has a low lime content. Two or three layers of asphalt roofing felt may be laid beneath the tile.

Wood

With the appropriate wood, properly constructed wood floors function well in many different types of work areas. The main drawback of wood floors is that they pose a fire hazard. Plank or board floors of the softer woods are generally inappropriate because they lead to many injuries from slipping and falling as well as from splinters. However, matched or jointed hardwood flooring that is nailed to a subfloor structure or to sleepers in concrete is a good floor for facilities that have light manufacturing. The necessary thickness of hardwood flooring depends on the wear to which it will be subjected. Wood should be laid with the grain parallel to truck travel.

Wood Blocks

Wood blocks satisfy many of the requirements for a good floor. Properly made, a wood-block floor produces little noise and does not become slippery or cause worker fatigue. If the blocks are laid on a level, rigid base, the floor will likely not crack and will accommodate substantial use. Wood blocks filled with creosote are necessary for floors that encounter liquids or other moisture. Expansion joints are required along walls, columns, and similar locations. If wood blocks are laid with a high-melting-point pitch, hot weather will not warp the floor. Using oils and organic solvents on the blocks, however, is problematic because they dissolve bituminous fillers and coatings.

Floor Loads

Design floors so that they can dependably support anticipated loads. Consult a registered structural engineer, and refer to ANSI A58.1, Minimum Design Loads for Buildings and Other Structures, to determine appropriate floor loads.

Figure 2–13 presents some floor-loading fundamentals. Ideally, a load is uniformly distributed over a floor area (Figure 2–13a). If the same load is concentrated at the center of the span (Figure 2–13b), twice the structural strength will be required. Conversely, if a floor is designed to support a given uniform load, only one-half of this uniform load can be concentrated at the center of the span. Figure 2–13c shows the ideal distribution of aisles and loads.

Design floors for dynamic loads and anticipated future loads. In estimating floor loads, the weight of a man is calculated to be 160 lb (73 kg); the weight of a woman is calculated to be 138 lb (63 kg). Manufacturers can supply the weight of equipment, and handbooks can supply the weight of bulk materials. (When piling bulk materials, remember that air space often develops, which reduces the overall density of the material.) Fully loaded industrial trucks may weigh as much as 60,000 lb (27.2 tons). Foundations that distribute loads and vibrations over large areas or that use springs or vibration mounting to cushion impacts help increase structural strength.

Workstation Design

When designing employee workstations, keep the following suggestions in mind:

- Place controls where employees can reach them using the least amount of movement.
- In addition to general illumination, provide lighting that is suitable to the task.
- Provide jigs and fixtures that relieve pressure on employee body.
- Provide a workbench on which each worker can sit or stand as needed.
- Provide cushioned mats for employees to stand on.
- Provide a foot bar at each workbench so that each employee can elevate his or her foot while standing at the workbench.
- Determine workflow patterns that are appropriate for and accommodating of each worker.
- Provide the opportunities for audio and/or visual signals to come from machine operators.
- Pre-position materials, equipment, products, and tools so that the worker's body can remain stationary.

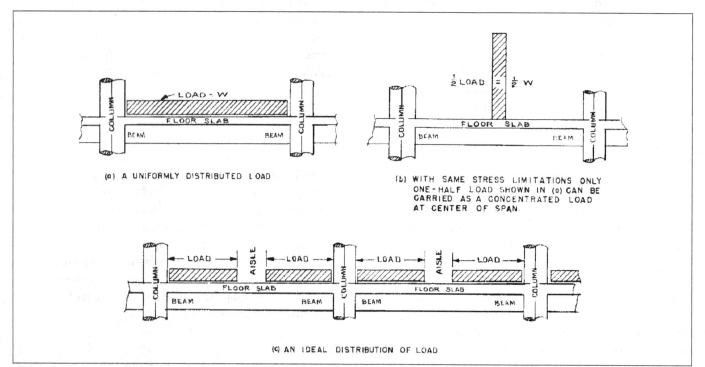

Figure 2–13. Various types of loadings. Changing the location of a load changes the total load that a floor can support.

- Place tools, controls, and materials in the employee's direct line of vision.
- If an employee must use both hands at the same time, the necessary motions should (1) begin at about the same time, (2) be somewhat alike for each hand, and (3) end at the same time for each hand. Both hands should not be idle at the same time.
- Workers should use arm motions that are steady and continuous, not irregular and abrupt. Wrists should remain straight, and forearms should be kept level with the elbow.
- Design the workflow so that the work proceeds steadily.
- Facilitate the subsequent step in the work process by providing storage of unfinished products.
- Design circular layouts for street production. (See also Chapter 16, Ergonomics Yesterday, Today, and Tomorrow, in the *Administration & Programs* volume.)

SUMMARY

- Many accidents, occupational diseases, explosions, and fires can be prevented by carefully planning the design, location, and layout of a new facility or the alterations of an existing facility.
- Ideally, safety and health professionals conduct a safety and health study of a proposed facility in the developmental stages to remove hazards and reduce risks. They should also consider the roles of workers, the functions of machines, and the flow of materials.
- Four factors should be considered in facility design and layout: (1) the general design of a workplace; (2) the workplace's compliance with appropriate codes and standards; (3) the sizes, shapes, and types of buildings, processes, and personnel facilities needed; and (4) the safety procedures and fire protection standards required.
- Site selection involves considering safety issues and possible hazards to the community; studying location, climate, and terrain; and understanding the space requirements of the company.
- When planning outside facilities, safety and health professionals should ensure that worker safety is incorporated into the designs of company grounds, shipping and receiving facilities, and all roadways, walkways, trestles, and parking lots.
- The company must develop air pollution controls and waste-disposal methods that conform to municipal and state or provincial regulations.
- As much as possible, the prevention of facility railway hazards should be part of a new facility's design. Employees must know the regulations and practices for operating and maintaining fuel-fired, electric, diesel, compressed-air, and battery-powered locomotives.
- The layout of buildings and facilities should incorporate the most efficient uses of materials, processes, and methods and minimize the possibilities of fire and explosions.
- Equipment can be laid out using detailed flow sheets and must be laid out to ensure maximum efficiency of and safety for workers.
- Both daylight and electric lighting can satisfy a facility's lighting needs. Proper illumination can help reduce accidents, minimize the number of hazardous areas, and make buildings and grounds more secure.
- Security considerations for facilities should include reducing the number of openings, securing all windows, providing protective lighting, and installing alarm systems.
- Industrial designers and managers are paying more attention to how color, lighting, and human behavior interact. The light-reflectance value (LRV) of color is color's effect on light, which can affect workers' ability to see a task and distinguish among color-coded materials.
- Colors also have regional and gender preferences, affect employees' morale, alter workers' perceptions of their surroundings, and provide an effective way to focus attention on hazardous items and safety signs.
- Access structures (stairs, ramps, etc.) must be designed for effortless use, clearly marked, kept unobstructed and well maintained, and safeguarded with rails, banisters, or other safety equipment.
- Conduct a careful study of each department in order to choose the most appropriate flooring material. In general, floors should be slip, scuff, and scratch resistant; easily cleaned; and provided with adequate drainage. Floors should also be designed to support both current and anticipated loads safely.
- To reduce strain and injuries, workstations should be well lighted, should be ergonomically sound, and should facilitate workflow.

REFERENCES

A Statement on Astronomical Light Pollution and Light Trespass, CP-46.

American National Standards Institute, 11 West 42nd Street, New York, NY 10036.

Electric Lighting Fixtures for Use in Hazardous (Classified) Locations, ANSI/UL 844–1996.

Fluorescent Lighting Fixtures, ANSI/UL 1570–1988.

High Intensity Discharge Lighting Fixtures, ANSI/ UL 1572–1990.

Incandescent Lighting Fixtures, ANSI/UL 1571–1990.

National Electrical Code, ANSI/NFPA 70–1999.

Practice for Industrial Lighting, ANSI/IES RP-7–1990.

Safety Color Code for Marking Physical Hazards, ANSI Z535.1–1998.

Safety Requirements for Building Construction, ANSI A10 Series.

Safety Requirements for Confined Spaces, ANSI Z117.1–2009

Safety Requirements for Workplace Floor and Wall Openings, Stairs, and Railing Systems, ANSI A1264.1–1995.

Scheme for the Identification of Piping Systems, ANSI/ASME A13.1–1998.

Safety Standards for Powered Industrial Trucks, ANSI B56.1.

Minimum Design Loads for Buildings and Other Structures, ANSI A58.1.

American Petroleum Institute (API), Guidelines for tank storage, 2951 N Great Southwest Parkway, Grand Prairie, Texas 75050.

American Railway Engineering Association (AREA), 4501 Forbes Blvd., Suite 130, Lanham, Maryland 20706

American Society of Mechanical Engineers, 345 East 47th Street, New York, NY 10017.

Boiler and Pressure Vessel Code.

ASME's Laws, Rules and Instructions for Inspection and Testing in Steam Locomotives and Tenders and Their Appurtenances.

Biosafety in Microbiological and Biomedical Laboratories (5th ed.) manual, cdc.gov/OD/OHS/biosfty/bmbl5/BMBL_5th_Edition.pdf.

Faber Birren and Company, 500 Fifth Avenue, New York, NY 10110. Specifications of Illumination and Color in Industry. Reprinted from *Transactions*, American Academy of Ophthalmology and Otolaryngology.

Factory Mutual System, Engineering Division, FM Data Sheets.

Federal Railroad Administration (FRA), 1200 New Jersey Avenue, SE Washington, DC 20590.

Illuminating Engineering Society of North America, 345 East 47th Street, New York, NY 10017.

Calculation of Daylight Availability, RP-21.

Discomfort Glare in the Interior Working Environment, CIE No. 55.

Glare and Lighting Design.

Guide on Interior Lighting (publication 29.2 of the International Commission on Illumination).

Institute of Makers of Explosives, 1101 14th St., NW Suite 1030, Washington, DC 20005-5635.

International Commission on Illumination, *Guide on Interior Lighting*. Babenbergerstraße 9/9A, 1010 Vienna, Austria.

National Fire Protection Association, 1 Batterymarch Park, Quincy, MA 02269.

Air Conditioning and Ventilating Systems, NFPA 90A, 1993.

Flammable and Combustible Liquids Code, NFPA 30, 1993.

Life Safety Code, NFPA 101, 1994.

Marine Terminals, Piers, and Wharves, NFPA 307, 1995.

National Electrical Code, NFPA 70, 1993.

National Fire Codes, 12 volumes.

NFPA Table 2.51.d

National Safety Council, 1121 Spring Lake Drive, Itasca, IL 60143.

Fundamentals of Industrial Hygiene, 6th edition.

Occupational Safety and Health Administration, 200 Constitution Avenue NW, Washington DC 20210.

Confined Spaces, 29 CFR 1910.146.

Electrical, OSHA 29 CFR 1910.303(b)(1)(viii) and (b)(6).

Fixed Industrial Stairs, OSHA Standards §1910.24.

Guardrail 29 CFR 1926.502(b)(1) and (b)(2)(i).

Guarding Floor and Wall Openings and Holes, 29 CFR 1910.23(e)(1) and (e)(4).

Guarding Floor and Wall Openings and Holes, 29 CFR 1910.23(e)(2) and (e)(4).

Railroad Operating Practices, 49 CFR 218.21d.

Walking and Work Surfaces, 29 CFR 1910 Subpart D.

Walking Surfaces Checklist, 29 CFR 1910.22(b)(2).

The Edmund Unique Lighting Handbook, Edmund Scientific Co., Barrington, NJ: 1977.

Protective Lighting, RP-10.

Recommended Practice for Daylighting.

U.S. Department of Justice, Civil Rights Division, Washington D.C.

Americans with Disabilities Act (ADA).

U.S. Department of Transportation, *Manual on Uniform Traffic Control Devices for Streets and Highways*.

U.S. Environmental Protection Agency, 401 M Street SW, Washington DC 20460.

Large Quantity Generators, 40 CFR 262.34(b).

Small Quantity Generators, 40 CFR 262.34(d)(2).

REVIEW QUESTIONS

1. List five of the eight factors to consider in the general design of a workplace.
   a.
   b.
   c.
   d.
   e.

2. Companies should ensure that which of the following groups review and approve their plans and specifications for new facilities or facilities that need remodeling?
   a. their safety and health specialists
   b. their insurance companies
   c. governmental authorities
   d. all of the above

3. Name the specific safety code for electric wiring and electrical installations and the organization that established it.

4. List four of the six factors to consider when designing machine tools and equipment.
   a.
   b.
   c.
   d.

5. What should be done to protect pedestrians if pedestrian entrances must be located near railroad tracks or busy thoroughfares?

6. Knowing the nature of wastes is essential for knowing the appropriate disposal methods. Which of the following wastes can be disposed of by burning in an incinerator?
   a. poisonous materials
   b. wood and paper
   c. magnesium chips
   d. all of the above
   e. none of the above

7. Describe the steps that should be taken to address the problem of confined spaces.
   a.
   b.
   c.
   d.
   e.

8. Which type of lighting can provide better illumination levels in small or restricted areas?
   a. supplementary lighting
   b. daylight
   c. general lighting
   d. emergency lighting

9. What are the six security factors to consider when designing a facility's environment?
   a.
   b.
   c.
   d.
   e.
   f.

10. Why do safety engineers use warm colors to call attention to dangerous machine parts, fire hazards, and physical hazards?

11. Neutral colors of low light-reflectance values should be used in what type of working environment?

12. Red is the standard color for:
    a. designating dangerous parts of machines, such as exposed cutting edges.
    b. marking flammable liquid storage cabinets.
    c. the radiation hazard symbol.
    d. identifying fire protection equipment, danger, and emergency stops on machines.

13. List seven factors to consider when determining the best type of floor for a particular location.

Construction of Facilities

3

Michael O'Berry, MEd
Richard Hislop, PE, CSP
Jim E. Lapping, MS, PE, CSP
John F. Montgomery, PhD, CSP, CHMM

Safety in Construction
Introduction ▶ Safety Programs ▶ Establishing a Safety Program ▶ Risks of Using Contracted Services ▶ Managing Safety on a Project

Elements of a Safety Plan
Management Commitment and Expectations ▶ Responsibility and Accountability ▶ Designated Safety Representative ▶ Reinforcement (Discipline) ▶ Inspections ▶ Training ▶ Incident Reporting and Investigations ▶ Housekeeping ▶ Substance Abuse Program ▶ Emergency Procedures

Integrating Safety into the Construction Process
Engineering and Design Phase ▶ Procurement— Contracting Phase ▶ Work—Construction Phase

Roles and Responsibilities
The Client ▶ Architects and Design Engineers ▶ Project and Field Engineers ▶ Construction Manager ▶ General Contractor ▶ Subcontractors ▶ Craftspeople

Insurance
What Is Insurance? ▶ Comprehensive General Liability ▶ Workers' Compensation ▶ Wrap-Up Insurance

Safety in Contracts
The Contract ▶ Safety Clauses

Contractor Selection
Contractor Screening ▶ Safety as a Technical Evaluation Criterion ▶ Request for Proposal and the Pre-Bid Meeting ▶ Evaluation Criteria for Contractor Selection

Role of the Field Engineer
Roles and Responsibilities of the Field Engineer ▶ Work Release Meeting ▶ Equipment Inspections ▶ Job Site Monitoring ▶ Inspections ▶ Enforcement ▶ Accident Investigation and Reporting ▶ Progress Meetings ▶ Work Safety File

Contract Close-Out
Construction Contract Environment, Safety, and Health (ES&H) Requirements ▶ Contract Compliance ▶ Acceptance Certifications ▶ Loaned Equipment and Permits ▶ Safety Program Documentation and Retention ▶ Safety Statistics ▶ Accident Reports and Personal Injury Information ▶ Other Safety-Related Information ▶ Work and Safety Planning Information ▶ Monitoring Records

Summary

References

Review Questions

SAFETY IN CONSTRUCTION

Introduction

This chapter addresses the management of construction and demolition operations carried out in an industrial environment. It has been developed for those looking for an overview of the safety and health process that should be in place for construction operations performed on their property. The term *construction* as used throughout this chapter includes demolition operations.

Safety Programs

Given the technical resources and safety information available today, the goal of every project should be zero incidences resulting in injury, illness, or damage to equipment or property. Unfortunately, high-profile stories such as the collapse of tower cranes in New York continue to provide headline news. Less newsworthy, but just as devastating, are the avoidable construction-related injuries and fatalities that continue to occur with alarming frequency.

Responsibility for construction site safety has been the focus of heated debate for some time. Plaintiff and defense attorneys argue the subject regularly. Architects, engineers, and clients often contend safety is the responsibility of the construction manager or general contractor.[1] Construction managers attempt to skirt the issue of responsibility for safety by arguing limited contractual authority. General contractors point out that their contracts place responsibility on the subcontractors to follow OSHA regulations and safe work practices. In turn, subcontractors and craftspeople point back up the line to both the owner and the general contractor as being the ones controlling—and, therefore, responsible for providing—a safe job site. To further complicate the issue of safety responsibility, many contracts contain indemnification clauses requiring the contractor doing the work to defend the party letting the contract against third-party claims.

Responsibility for safety on construction sites tends to be confusing because lines of responsibility are often blurred by the attempts of those involved to transfer responsibility and accountability for safety to others. The issue of who is responsible for safety is becoming more contentious as the costs of accidents, insurance, litigation, workers' compensation, and other associated expenditures escalate.

The primary factor in the success of an effective safety program is management and its involvement in the safety program. Although a dollar value cannot be placed on the humanitarian aspects of a safety program, it is also impossible to place a dollar value on the negative effects personal injuries and fatalities have on labor relations and publicity. Merely incorporating safety clauses in contracts and printing a safety program will not yield the desired results without a serious and persistent management commitment to make the program work. It is human nature to place emphasis on the program (or programs) that will be evaluated. Thus, safety will receive attention proportional to the importance placed on it by management.

The safety of all employees, engineers, managers, subcontractors, visitors, and bystanders in the vicinity of contracted work should be of significant concern to everyone involved in the construction process. Only through the clear definition of responsibility and accountability for safety can personal injuries and other related losses be minimized in a continually changing environment such as that on a construction site.

Owners ultimately pay the cost of the safety program on a job site. The effectiveness of the implementation of a safety program will have a direct bearing on the losses sustained by the project. Therefore, it is in the interest of owners to establish an environment within which contractors are required to follow safe work practices. Fear of incurring liability by becoming involved in establishing safety requirements should not be a reason for failing to take a proactive role in defining expectations of an effective construction safety program. All incidents and personal injuries can be prevented. This has been demonstrated by the organizations that have implemented world-class safety programs. A useful reference in this regard is *Construction Site Safety: A Guide for Managing Contractors* (Hislop 1999).

Establishing a Safety Program

A safe work environment does not just materialize. To establish a safe worksite, roles and responsibilities of each project participant must be clearly and unambiguously defined. This is particularly the case where one party assumes multiple roles in the construction process, such as an owner who designs and chooses to oversee the execution of the work by multiple prime contractors. What is the role of the owner? How much of the safety program is the owner responsible for? Who will be responsible for planning and coordinating the project safety plan? Who is responsible for safety audits and inspections? Whose job is it to mitigate identified safety hazards? Frequently, the task of assigning responsibilities for safety is overlooked by the owner in the initial enthusiasm and accelerating momentum of a new project.

Construction projects that had extended periods with no lost-time injuries and projects without OSHA-recordable

---

[1] *Construction manager*—The single entity that will be responsible for providing controls over contracts awarded, seeing that they are within the estimated budget, and administering construction without having to engage any field employees or tradespeople. *General contractor*—Contractor who has subcontractors to do some part (or all) of the work that the general contractor has undertaken to do for the owner. Previously, the term referred to the contractor who employed workers of different trades and who undertook to do most (or all) parts of the work as directed by a specialist (trade) contractor, who normally undertook the work of only one trade.

injuries have the following common safety management approaches and practices:

1. demonstrated management commitment to safety, as shown by regular field presence and tracking project safety indices
2. management and supervisor accountability for safety performance
3. adequate safety staff, who report directly to senior management
4. project-site-specific safety plan and a requirement that subcontractors submit project-specific safety plans
5. worker involvement and participation in work planning
6. task preplanning meetings held by foreman with his/her crew before starting each task
7. structured safety observation program
8. accident/incident reporting and investigation, which includes top management participation
9. safety orientations and safety training
10. drug and alcohol testing
11. safety recognition and rewards
12. imposition of sanctions on work crews that do not comply with safety requirements.

Risks of Using Contracted Services

An owner has two options with regard to establishing a construction safety program: (1) develop and manage the safety program directly or (2) have it developed and managed by a second party such as a construction manager (CM). In the first case, where the owner has the technical expertise and resources to do so, the owner might manage the safety program directly. Typically, owners do not have professionals on staff with the technical expertise to direct or manage construction safety programs in a contracted work environment. Where owners do not have the technical resources in-house, they generally opt to retain a CM. The CM acts as the owners' agent and manages the work, including safety. In this scenario, the owner must clearly communicate its expectations regarding safety responsibilities and how performance will be measured. The owner might go so far as to define specific program elements that are important to its particular culture or require that technical criteria be applied to the selection of contractors.

Regardless of the type of approach chosen, the owner/client must adopt and support certain basic tenets, such as:

- supporting the safety program, including adequate budget
- defining minimum expectations of the project safety program
- recognizing that construction safety is part of a dynamic process where decisions must be made in a timely manner
- addressing problems as they occur
- monitoring and verifying that the safety program and site-specific plan are being effectively implemented.

The establishment of a safe worksite requires that those in control commit to the importance of having a well-defined safety program during early project conceptual development. It requires them to carry and communicate this commitment through contractual negotiations, work implementation, and finally project completion, never wavering in their commitment to safe work planning and practices.

Managing Safety on a Project

Monitoring day-to-day activities of construction trades and contract compliance is an essential function of the owner. Usually, it is only after a tragic, personal injury or a serious, safety-related loss that the employer begins to take an interest in safety. At this stage, it is difficult and often very expensive to implement an effective safety program.

It is easy to assume that individuals who have been assigned supervisory and management positions, and contractors who bid to perform work, have all been anointed with safety awareness and insight. This is not so. Safety awareness, the ability to recognize hazards, and the technical background to be able to eliminate or control identified hazards are learned. Further, safety often seems to be practiced only when it is clear that the project owner considers safety to be important. Perhaps it boils down to the fact that individuals strive to perform well in those areas in which they know they are going to be measured.

The project owner must define the expected outcome of the safety program and assign responsibility for managing it even before developing the contract language. Safety is just as important as schedule, quality, material control, or any other facet of construction. Like schedule and quality, safety does not just happen. There must be a clear and unambiguous description of the safety program criteria, as in any other aspect of business. Most important, perhaps, is the assignment of responsibility for safety; because, generally, there is not a contractual relationship between subcontractors or even between prime contractors, it is the role of the owner to clearly establish responsibility for safety down through the organizational hierarchy.

There are a number of classical organizational relationships in construction that depend on the familiarity of the client with construction practices and the complexity of the work to be performed. The construction manager is responsible for integrating the skills and performance of the participants into a cohesive project team.

ELEMENTS OF A SAFETY PLAN

Owner requirements, government regulations, and industry standards for safety contain the basic elements of the overall *construction safety program*. Extracting the provisions from the program that apply to a specific project results in the *project safety plan*, which identifies the entities and individuals responsible and accountable for implementing the safety plan. The safety program describes what must be included in the safety plan, and the safety plan describes how and by whom the program elements will be implemented. A successful safety plan must address site-specific hazards, define safety expectations with regard to safe work practices, and clearly define safety roles and responsibilities. Federal, state, and local safety regulations and standards define the minimum expectations that should be in place for a given project. Safety plan requirements and their emphasis will differ for each project. A project that involves significant work with buried utilities will have very different safety plan requirements and emphasis than those of a project with extensive work at heights. To ensure a clear understanding of the owner's safety expectations, they should be documented in the project safety plan.

Management Commitment and Expectations

To promote the integration of safety into regular work practices, an owner must define its expectation to the individuals controlling the work that "safety is important." Safety begins with the attitude that incidents are preventable through the commitment of management and the requirement that planning and safe work practices must be followed. The owner should establish that safety be included as the first item of discussion at all meetings, including those dealing with cost, schedule, sales, and training. Once again, a safety program will be only as good as the owner's commitment to safety and its visible support of it. The importance of safety over expediency must be regularly emphasized so that supervisors and the work force understand that although schedule and production are important, above all else, work must be performed safely.

Management must clearly define its commitment to safety and its expectations in that regard on a regular basis. To document its safety expectations, the owner must commit the safety requirements to writing in the contract, including all entities performing work on the project and clearly identifying responsibility and accountability for safety.

Policy Statement

The cornerstone of the safety program is the owner's affirmation of its position on safety (i.e., its *safety policy*). This is a written statement of the principles and general rules embodying the company's commitment to workplace safety and health. The policy statement can be brief, but should address:
- management's commitment to employee safety
- the organization's safety philosophy

- who is accountable for the occupational health and safety program
- acknowledgment of its responsibilities for the safety of all employees
- that safety should not be sacrificed for expediency
- that the work and the work environment must be evaluated for potential hazards before any work is initiated and that all potential hazards must be eliminated or controlled.

The policy should be stated in clear and unambiguous terms, and it should convey that management is sincerely committed to safety.

This is particularly important in an environment where the safety message is different from safety practices in the past. Each organization must determine how best to communicate its safety policy and safety rules to its contractors and work force.

Safety Procedures

Safety procedures are needed to simply and concisely define accepted work practices. Poorly defined procedures, or those procedures with little relevance to the work being performed, may be interpreted as an indication that safety is not important. The following are some guidelines for establishing safety procedures.

Procedures should be:
- available to all employees in written form
- specific to safety concerns in the workplace
- stated positively and in understandable terms
- explained
- enforceable
- reviewed periodically to evaluate their effectiveness and to make changes for improved effectiveness.

One approach to initially establish safety procedures is to address the most frequent causes of injuries on construction sites or those that have occurred with most regularity at the owner's facility.

Safe Job Work Procedures

Many job-related injuries occur because employees are not aware of existing or potential hazards and how to control the hazards related to their work. This may be because they were never trained to perform the work safely in the first place, or they may not have received training to help them identify hazards, or they do not have the background to develop appropriate controls to deal with those hazards once they have been identified. Prior to starting work, each work task should be evaluated to identify the most efficient way of performing it safely. The agreed-upon safe work practices defined as the result of this evaluation should then be communicated to the individuals required to perform the work by those responsible for supervising it at the pre-task meeting.

Job Safety Analysis

A means of systematically identifying and evaluating safety issues associated with a work task is a job safety analysis (JSA). Each step of a task is evaluated to identify the hazards associated with the work. Where issues are identified, an effective means of performing the work in a safe manner is agreed to. This process should involve the participation of several individuals, including supervisors, individuals who perform the work, and technical specialists with an awareness of safety-related considerations to help recommend safe procedures to execute the work.

A JSA should be developed for all critical tasks with priority given to addressing tasks:

- where frequent accidents and injuries occur
- where severe accidents and injuries occur
- with a potential for severe injuries
- that are new or that are a modification to a previous procedure
- that are infrequently performed.

A job safety analysis generally consists of the following steps:

1. Select the job.
2. Break down the job into a sequence of steps.
3. Identify the hazards associated with each step.
4. Define preventive measures.

Where a job consists of more than one specific task, each separate task should be analyzed.

After workers have been briefed on the manner in which work is to be performed, supervisors should monitor their subordinates' work to ensure the procedures have been understood and are being followed.

Responsibility and Accountability

For a safety program to be successful, management, supervisors, and workers must recognize that they each have a role in the safety process. Management has a responsibility to provide its employees with a safe and healthful work environment. Management must also ensure that employees have the tools, personal protective equipment (PPE), and other resources needed to execute their work safely. Supervisors have the responsibility to ensure that each of their subordinates clearly understands how to perform their work safely and that they follow safe work practices. It is the responsibility of workers, prior to starting a task, to make sure they understand how they are expected to perform the work safely and to have the tools and protec-

tive equipment they require to perform the work following approved safe work procedures.

Documented safety programs/plans should define the safety responsibilities of individuals at each level of the project organization—from management to the subcontractor's craftspeople. This information will enable everyone on the project to understand what his or her respective responsibilities are and who is responsible for what in the project safety program.

Workers' responsibilities:

- knowing and complying with safety regulations
- following safe work procedures
- using PPE appropriate for the work being performed
- correcting or reporting unsafe work practices and unsafe conditions
- helping new employees recognize job site hazards and follow proper work procedures
- reporting injuries or illnesses immediately.

First-line supervisors' responsibilities:

- instructing workers on how to identify unsafe conditions
- requiring subordinates to follow safe work practices
- ensuring that individuals assigned to operate equipment are adequately trained and authorized to operate equipment
- correcting unsafe acts and unsafe conditions
- enforcing health and safety regulations
- promoting safety awareness among workers
- ensuring required PPE is worn by workers and that workers understand the reason for its use
- inspecting their own and surrounding work areas and taking action to control or eliminate hazards
- ensuring injuries are treated and reported
- investigating all accidents/incidents.

Management's responsibilities:

- providing a safe and healthy workplace
- establishing and maintaining a health and safety program
- ensuring workers are trained or certified, as required
- providing workers with safety and health information
- ensuring PPE is available
- supporting supervisors in their safety and health activities
- providing medical and first-aid facilities
- evaluating safety performance of supervisors
- reporting accidents and occupational illnesses to appropriate authorities.

Designated Safety Representative

Although the project manager is ultimately responsible for ensuring that an effective job site safety program is in place, it may be necessary to appoint a knowledgeable individual to provide the project manager with technical support and assist in the responsibility of overseeing the implementation of the safety program. This individual is the *designated safety representative* (project safety manager). Supervisors, including foremen, must understand that even though a project safety manager is included in the project staff, the supervisor will still be held accountable for his or her own safety-related decisions and the performance of his or her respective subordinates. Workers must also be made to understand that they are expected to look out for their own safety and to adhere to the safety rules.

It is the role of the designated safety representative to bring safety issues and concerns to management's attention and to guide supervisors and workers in the implementation of their respective safety responsibilities. The designated safety representative is not the safety program but is the technical safety support for the project.

Competence Commensurate with Responsibilities

It is often mistakenly assumed that individuals who have been conferred the status of craftsman, supervisor, or manager—by virtue of the excellence of their technical knowledge—should also be able to recognize safety hazards associated with their work.

A second mistake is to assume that those individuals who are able to recognize hazards will include safety considerations to mitigate those hazards in their work planning, thereby eliminating or controlling the hazards associated with their work. Although some aspects of safety are intuitive, the ability to recognize hazards and to develop means by which to control the hazards must be learned by most individuals. They must also understand that the owner is willing to pay them for the additional time the supervisor perceives that it will take to incorporate the necessary safety controls into the performance of their work.

Therefore, make sure that supervisors to whom work is delegated understand their safety responsibility and have the technical background and experience to fulfill this responsibility. Also make sure that they make similar assurances regarding the training and safety awareness of the individuals to whom they, in turn, assign work. Even when owners contract with general contractors with national name recognition, they must make a concerted effort to ensure that the specific individuals who will be working on the project can demonstrate that they are committed to safety and that this is reflected in their past track record.

Competent Person

A *competent person* is defined as an individual who, as the result of training and/or experience:
- is capable of identifying or predicting hazardous situations
- has the authority and responsibility to take prompt corrective measures to eliminate them

- takes action to implement corrective measures that have been determined to be necessary to control a recognized hazard.

The fact that regulators have felt the need to define a requirement for a competent person classification suggests that, in many cases, individuals assigned supervisory and oversight responsibility do not effectively identify and control known hazards. *Competence* in this context is about action, not certification. The following is a listing of some activities, operations, and health areas in which 29 CFR 1926 and various ANSI standards require that competent persons be designated:

Competent person categories:
excavation
scaffolds
fall protection
cranes and derricks
hazardous chemicals
underground construction
ionizing radiation
flammable liquids
painting
lead
slings
ladders
material handling
hearing protection
powder-activated devices
helicopters
respiratory protection
shipyards
telecommunication systems
blasting agents.

Reinforcement (Discipline)

Consequences control behavior, and behavior is based on work direction and work conditions. A mechanism must be in place to address situations where the violations of safety principles occur. It is accepted practice in construction that once training, guidance, and encouragement have been exhausted, discipline is the remaining recourse to reinforce the application of safe work practices. The expectation of safety compliance should be defined in the contract, as should the repercussions of failing to do so.

Disciplinary programs in contracted work environments generally follow the same framework. Individuals identified violating safety rules are first given a verbal warning. A second violation results in a documented warning and possibly being barred from the site for a short period of, say, 2 days. (The owner cannot generally fire a contractor's employee but can deny the individual access to the project site.) The third time an individual violates a safety rule, the practice is that the individual is barred from the site for 6 months or denied access to the site for the duration of the project. When an employee is involved in a case of serious or gross misconduct, the progressive disciplinary action is not generally invoked; immediate action is taken.

However, the behavior of the employee's supervisor should be seriously evaluated as well. In the disciplinary process, the employer responsible for the work that is not being performed safely generally does not assume any blame. However, it might be helpful to question why the employee failed to follow safe work practices. Was it the result of specific direction or implied directions?

Do not resort to disciplinary action until the reason for the unsafe behavior has been clearly established as a willful safety violation. When an incident does occur, focus on identifying what must be changed in terms of work procedures, equipment, and educating the workers. Punishment may motivate workers to remain secretive about unsafe acts that could become the source of a serious incident.

Inspections

When work is in progress, a regular evaluation of the job site conditions must be conducted to verify that workplace hazards are being identified and controlled and that safe work practices are being followed. Traditionally, safety representatives and supervisors conduct safety inspections.

This approach is limited in value as it is restricted to the availability of and technical knowledge of a few individuals. In world-class safety environments, management and supervisory personnel, as well as other project team members and workers, participate in the hazard identification process. Each organizational tier in a project hierarchy—from the client to the subcontractor performing the work—has a vested interest in ensuring that safety is integrated into work practices. Each organizational element should conduct inspections that focus on issues of concern at their respective levels. The owner should verify and monitor that the construction manager is enforcing a requirement that the general contractors have site-specific safety plans, and so on down the chain of command.

Field engineers should inspect their work areas prior to the start of work and on a frequent basis to identify existing or potential hazards not anticipated and planned for during the project hazard analysis or not covered in the safety plan. They should monitor job site activities under their control to assess the effectiveness of their safety plan implementation and to identify hazardous conditions or activities that need immediate correction or require additional training or retraining.

Because the greatest risk of injury is present at the worker level, each contractor should be required to conduct and submit regular safety inspection reports of their own work to the project manager or the host employer's field engineer. This compels the contractor to participate in the process of looking out for conditions and work practices that could put his or her own employees at risk of injury. Contractor inspection reports also serve to document the contractor's involvement in the identification and correction of unsafe work conditions and work practices.

Current safety legislation requires workplace inspections be conducted on a "frequent and regular basis" by a competent person to ensure workplace safety. The frequency of planned formal inspections must be defined by each organization, based on the rate of change of physical conditions on the job site, its record of compliance with safe work practices, and the frequency of accidents and injuries. Participation in formal inspections should be part of every supervisor's and manager's job description. They should include contracted work activities during their inspections.

The safety program and plan should answer these questions:
- Who should conduct inspections?
- When should inspections be conducted?
- What is the focus of the inspections?
- Who should receive the reports?
- How should deficiencies be addressed?
- What records should be kept?

A process should be established to collect the results of all these observations, electronically if possible, as this allows for rapid analysis of the information in order to identify if there are any areas of greater-than-expected incidents.

Training

Training should be an integral part of all safety programs. Only through regular communication of hazards, control measures, and safe work practices can the incidence rate of work injuries be systematically reduced. The most frequent incident root cause and most frequently issued OSHA citation following workplace accident investigations involves deficiencies in employee training programs. To perform their work safely, workers must be trained. For the training to be effective, each task must be thoroughly analyzed and safe work procedures developed and communicated to the work force.

According to the Business Round Table, approximately 250,000 new workers join the construction work force each year. About 80% are nonunion and do not have the benefit of formal construction trade training or safety training. This is a significant concern given the increasing sophistication and complexity in construction. The unfortunate result is the very high fatality and injury rate being experienced by construction workers.

Statistics from 2013 indicated that there was a reduction in construction-related fatalities when compared to 2012, according to results from the Census of Fatal Occupational Injuries (CFOI) conducted by the U.S. Bureau of Labor Statistics. The rate of fatal work injury for U.S. workers in 2013 was 3.2 per 100,000 full-time-equivalent (FTE) workers, compared to a final rate of 3.4 per 100,000 in 2012.

Just as traumatic are the multitude of disabling injuries and the thousands of lost-work-time injuries that occur for each fatality. There must be a process in place to ensure that individuals coming to the job site are aware of project-specific safety issues and are trained to handle them.

As a point of reference, between 29 CFR 1910 and 29 CFR 1926, there are 192 nonoverlapping requirements for the provision of training. A single employer will not perform all the activities that require training. However, there are a significant number of requirements for training, and employers should be aware of this and have a process in place to address the needs of the workers. OSHA has placed the burden on employers to provide their employees with safety training. There is no guidance regarding the frequency of training other than when employees are observed *not* following established procedures; then training or retraining is required.

The host employer should ensure that individuals coming to its facility to perform work are familiar with, at the very least, site-specific safety issues and their controls. Owners should include in their contract specifications that contractors employ only trained and qualified tradespeople on the projects. The host employer should also define an expectation that each subcontractor employee be briefed by the respective supervisors on the work to be performed that day and be given the opportunity to address specific safety issues that may surface as the work evolves.

Safety Orientations

Job site safety orientations provide a forum for the host employer to convey its commitment to providing a safe working environment. The orientation is an opportunity to point out site-specific information regarding the facility safety hazards, and it is an opportunity to remind workers of their responsibility to give due consideration to safety while performing their work. Even the most seasoned workers and supervisors need to become familiar with job site layouts, project management personnel, company policies, and other information related to an unfamiliar project. This is the time to explain what is expected of them when performing work that exposes them to a fall hazard, toxic fumes, excessive dust, and so forth. It is an opportunity to remind workers to bring hazards they might create to the attention of their supervisors so other individuals who might be affected by them can be advised.

Sadly, the conduct of effective safety orientations is often the exception rather than the norm on most job sites. This step in the construction sequence is often skipped in the interest of the expediency of getting the workers onto the job site to start the work.

Safety awareness and hazard communication training should be conducted before any individual is permitted to begin work on or visit an unfamiliar site. Statistically, the majority of injuries occur to employees who are not familiar with site-specific safety expectations and those individuals who are not familiar with the job-related hazards. Approximately 50% of construction industry deaths occur to individuals who have been on a job site less than 30 days. Accident frequency decreases with increased experience and greater safety awareness. The awareness of safety issues can be accelerated through safety and health training.

Orientations must be conducted with sufficient frequency to enable contractors to get their personnel trained and onto the job site; otherwise, they may find a way to circumvent the process. Visitors and contractors delivering materials do not necessarily need the same course detail required by a specialty contractor as long as they are accompanied by a qualified person while on the job site.

The contents of site orientations should include, as a minimum:
- introduction to the job site
- site rules and regulations
- site-specific hazards
- requirements for PPE
- fire protection system and emergency procedures
- first-aid and treatment program
- immediate reporting of injuries
- permit requirements
- disciplinary program
- introduction to key project personnel.

The message that must be reinforced is the fact that working safely is the corporate objective. *No one has authority or permission to work unsafely or to take shortcuts that place him or her or the project at risk.*

One means of ensuring that every person on the job site has attended an orientation is to issue hard-hat decals on completion of the training. Another means is to issue site access picture badges to avoid attempts to bypass the orientation requirement. Issuing identification badges is not a complex or costly process, given the technology available today.

A brochure with a concise summary of the project safety program and site rules should be provided to each orientation attendee. The brochure should be a reminder of the important points that were addressed during the orientation. Generally, a single safety training session is not sufficient to convey all the requisite information needed in construction and to maintain a heightened sense of safety awareness. To reinforce what has been addressed at the site, regularly scheduled safety orientation follow-up training is required.

Pre-Work Meeting

The best opportunity to reinforce safety is at the start of each workday during the supervisor's work assignment meeting. When each supervisor assembles his or her work crew to discuss that day's activities is the opportune moment to reinforce the importance of safety and associate it with the work to be performed.

A pre-work meeting should be held at the start of each day and when there are changes in work assignments. It need only be a short assembly with the following agenda:

- review previous day accomplishments
  - what was done well
  - opportunities for improvement
- new work assignments
  - work practices (JSA)
  - safety issues
  - coordination within the crew and with adjacent activities
- questions and answers.

Work practices will improve to the extent individuals are provided with feedback on what is expected of them.

New work assignments and a review of the associated JSA for the work to be performed that day should be addressed. During this discussion, the supervisor confirms that everyone understands what is expected of him or her. The foreman should solicit feedback from the workers on the content of the JSA and what changes, if any, might be required to complete the work safely. Workers who have been on other jobs will most likely be able to contribute valuable suggestions to improve the effectiveness of the JSA. Once they have critiqued it and have received positive feedback from the supervisor regarding their observations, they will be more likely to be committed to following the JSA.

The pre-work meeting also provides an opportunity for the supervisor to speak with each employee to determine if he or she is prepared to work that day. The supervisor has the opportunity to observe each worker's behavior and to judge if alcohol, drugs, or some preoccupation impairs him or her. Workers who are unfit for work are a danger to themselves and to the rest of the work crew. The supervisor should document the main points addressed at the pre-work meeting in the daily log.

Weekly Toolbox Talks

Weekly *toolbox talks* provide the opportunity to reinforce the company's safety policy, note changes in safety proce-

dures, review lessons learned from accidents that may have occurred within the company at other sites, address safety topics of a general nature, or discuss the development of JSAs for upcoming work.

Traditionally, contractors who remain on a job site for 2 or more weeks are required to conduct weekly toolbox talks. The toolbox talks are expected to last no less than 5 minutes and to address topics relevant to the work being performed.

The downside of toolbox talks is that it is not reasonable for supervisors to be able to anticipate all the safety issues that will surface during the next week. Nor is it reasonable to expect the workers to retain this information or put it into practice several days later. It is a common practice among some companies to send purchased toolbox talks to the field that may have nothing to do with the work being performed. For this reason, pre-work meetings and onsite safety coordination are powerful safety communication tools.

Where toolbox talks are required, a field engineer should monitor them periodically to verify that they are being conducted effectively. If not monitored, toolbox talks may become a general work coordination meeting and fail to address safety issues. One means of ensuring that they are being conducted is to require that a copy of all toolbox talks' outlines, with employee signatures, be attached with labor invoices as a criterion for payment.

Incident Reporting and Investigations

Accidents and incidents must be investigated so that measures can be identified to prevent a recurrence of similar events. Although investigations represent an "after-the-fact" response, a thorough investigation may uncover hazards or problems that can be eliminated to prevent the occurrence of future incidents.

The safety program should specify:

- what is to be reported
- to whom it is to be reported
- how it is reported
- which incidents are investigated
- who will investigate them
- what forms are used
- what training investigators will receive
- what records are to be kept
- what summaries and statistics are to be developed
- how often reports are prepared.

Housekeeping

Slips, trips, and falls are the leading causes of injuries on construction worksites. Litter and debris conceal tripping hazards and increase the possibility of other injuries. As a general rule, a site with poor housekeeping practices also has a poor safety record. Debris on a site also creates extra work, as it frequently needs to be relocated for worksite access. However, as with any other objective, there must be regular reinforcement of the importance of this requirement.

Nothing sends a clearer psychological message to employees and subcontractors than management's commitment to providing a safe and healthful work environment and insistence on an orderly project work area.

Substance Abuse Program

Eleven percent of the drivers on U.S. roads today do not have a valid driver's license. These individuals account for 90% of all traffic accidents. The Center for the Protection of Worker's Rights has determined that in some regions of the United States, up to 30% of construction workers report to work under the influence of some behavior-modifying substance. So, how many human error accidents are attributable to substance abuser error? The use of behavior-altering substances affects job performance and also places the employee and his or her peers at risk. Any worker, regardless of ethnic origin, socioeconomic background, or occupation could be a substance abuser, and it is not always the substance abuser that is injured. Drug counselors estimate that a chronic substance abuser functions at between 50 and 67% of his or her capacity on the job.

Substance abuse is widespread throughout the United States. Unless an organization institutes a substance abuse program, drug and alcohol abusers probably work there. Substance abuse screening is becoming a necessity in today's workplace. A study conducted by the Institute for Health Policy at Brandeis University found substance abuse to be the foremost health problem in the United States, resulting in more deaths, illnesses, and disabilities than any other preventable health condition. Companies that screen job applicants report that a high percentage of candidates initially test positively; as word spreads of the company's requirements, this number reduces to between 5% and 8% of applicants.

Drugs and alcohol:
- cause workers to take more chances
- increase the potential for injury
- increase absenteeism and tardiness
- decrease productivity and quality of workmanship
- increase health insurance and workers' compensation costs
- increase theft of materials, tools, and equipment.

Workplace Drug Testing

To identify employees who abuse drugs, a workplace drug-testing program must be in place. This requirement is generally opposed in construction unless it is for cause. However, approximately 20% of the American work force has a drug-testing policy in their workplaces. Companies that do impose drug-testing requirements send a strong message that they support a drug-free environment. Surprisingly, most employees approve of drug-testing programs.

There are several approaches to workplace testing:
- preemployment testing (from the site owner's perspective, this occurs following the site safety orientation and before the individual begins to perform work on behalf of a subcontractor)
- post-accident or for-cause testing
- scheduled testing (during employee medicals)
- random testing.

It is imperative that organizations implementing a drug-free-workplace program have written policies and procedures in place. A lawyer experienced in labor and contract matters should review the procedures and help design a program that complies with state laws and meets the specific needs of the employer.

In the contracted work environment, the host employer should establish drug-testing requirements in the contract language. At high-risk facilities such as refineries and chemical plants, the constructor or service provider working on behalf of the host employer should be required to provide documented evidence that all direct hire and subcontracted workers have been tested. The contract may further stipulate that testing for cause will be the host employer's prerogative.

Emergency Procedures

Emergency procedures are plans for dealing with emergencies such as major injuries, fires, explosions, releases of hazardous materials, violent occurrences, or natural hazards. When such events occur, the urgent need for rapid decisions in a short time can lead to chaos if there is a lack of resources and trained personnel to deal with the situation. At a minimum, procedures for prompt medical response must be established. OSHA expects that any employee should receive medical response within 3 to 4 minutes. This requires that, at a minimum, there must be someone on each work crew or shift with first-aid and CPR training.

The objective of the emergency plan is to prevent or minimize fatalities, injuries, and damage. The organization and procedures for handling these sudden and unexpected situations must be clearly defined. The process of establishing an emergency plan is not all that complex and is strongly advocated by individuals and organizations that have been caught without one. However, few organizations have established emergency procedures or a crisis management plan.

To develop a set of emergency procedures, compile a list of the hazards (e.g., fires, explosions, earthquakes, and floods). Identify the possible major consequences of each (casualties, damage to equipment, or impact on the public). Determine the required countermeasures (which might include evacuation, rescue, fire fighting, etc.). Inventory the resources needed to carry out the planned actions (e.g., medical supplies, rescue equipment, and trained personnel). Based on these considerations, establish the necessary emergency organization and procedures. Communication, training, and periodic drills are required to ensure adequate performance when the plan must be implemented.

INTEGRATING SAFETY INTO THE CONSTRUCTION PROCESS

Successful construction projects are the result of effective planning and execution and the collective effort of the entire construction project team. However, only when project teams integrate safety into the planning and execution of routine work practices will the success of the safety program be ensured as well. Defined in this section is the sequence of events common to most construction projects and the associated safety responsibilities of the individuals involved with this process. The duration of the events contained in this process will vary depending on the complexity of the work, technical competence of the individuals involved, and their familiarity with the process. For example, the duration of the planning phase preceding the installation of a fan unit will be quite different than it will be for the construction of an office complex. However, the steps defined here, when followed, result in highly productive and incident-free construction projects.

Engineering and Design Phase

Project Conceptualization

All work begins with the *originator*, who identifies the need for a new installation or modification to an existing facility. The originator's concept is generally presented to an architect or engineering design group (*designers*) to be developed. Most originators lack the technical background to convert their concept into a format with the detail necessary to guide a constructor to complete the work as conceptualized.

The originator's contribution to safety is the identification of safety and health issues that may not be readily apparent to the designers. The originator may be aware of environmental issues, such as subsurface contamination or unusual design features in the facility infrastructure, or that there are sporadic, very high ambient noise levels in the vicinity of where the work is to be performed.

The originator/client's safety responsibility during the design and planning stage of a project is to ensure that sufficient resources, including time, are allocated to the project to enable it to be completed safely and that competent people are appointed as coordinators, designers, and contractors. Some companies are using advanced building information modeling (BIM) safety assessment and planning on projects to enhance safety efforts during the planning stages. Use of BIM allows project models to be reviewed to identify potential risks. Identification of these potential risks shows where prefabrication, safety hazards analysis, daily site coordination, and plans to optimize facilities' operations and maintenance efforts can be used to promote safety both during and after the construction project is completed.

"Project" Coordinator Identification

An architect or engineer is appointed to coordinate the development of the project design. Each organization undoubtedly has its own nomenclature for the individual responsible for the development of the work project. For the sake of simplicity, this individual is referred to as the *coordinating engineer*. The role of the coordinating engineer is to facilitate development of the project design by ensuring that relevant project planning information is communicated to the designers and that they take proper account of safety and health considerations in their design work.

The coordinating engineer's safety responsibilities at the design and planning stage include the following:
- coordinating the identification of safety and health considerations during the design and planning phase of the project
- initiating the baseline safety review
- coordinating the 90% design review.

Initial Cost and Schedule Estimate Development

The coordinating engineer should initially develop a preliminary cost and schedule estimate for the work as he or she understands it, with consideration of the safety issues associated with the work, and present this information to the originator.

Originators often have an unrealistically modest concept of the time it takes to develop a project design and complete the work. They often have an unrealistically low appreciation of the cost of construction. Although not immediately apparent, both of these aspects of projects have significant safety and health implications.

Unrealistic schedule and cost expectations may induce the originator to attempt to suggest shortcuts to the work process when faced with schedule and cost overruns. Therefore, the originator must be given a fair initial cost estimate to permit that individual to determine if adequate resources are available to fund the project or if the size of the work project must be re-scoped. Potential issues such as schedule delays should be addressed with the originator to determine if special expediting will be required to meet any particular project completion requirements.

Baseline Safety Review

The coordinating engineer must conduct a baseline safety review in order to identify safety hazards associated with use of the space in which the project will be built and considerations of the subsequent occupancy and maintenance requirements of the completed work. The baseline safety review will assist the designers to identify preexisting and potential hazards associated with the construction and operation of the completed facility. The following specialists may be included in performing the baseline safety review:

- current occupants
- safety engineering
- fire protection engineering/fire departments
- industrial hygiene
- environmental compliance
- emergency management.

Armed with the information compiled in the review, the designers can address mitigating or proposed controls for known hazards during the engineering of the work. This will help provide for a safe work environment for the contractors and avoid the cost of developing controls for these hazards during construction. It will also avoid the greater cost of retrofitting hazard controls once the work is completed.

Design and Engineering

The designers define the configuration and components of the work through the plans and specifications. The nature of the design influences the means and methods of the project construction. Therefore, designers play a critical role in the identification and control of safety hazards that may exist on the job site or may arise as the result of the facility design. Many hazards faced by the construction workers are created by the designers, as are the hazards faced by maintenance and operations personnel of the completed work. A good book on the subject of identification of hazards is *Construction Safety Engineering Principles* (MacCollum 2007).

The designer's safety responsibilities at the design and planning stage include the following:

- Structures must be designed, to the extent possible, to avoid or minimize risks to health and safety while they are being built and maintained.
- Where hazards cannot be engineered out, adequate notice must be provided to the constructor.

Design and Constructability Review

When a project's design is approximately 90% complete, an evaluation of the design should be conducted to ensure the environment, safety, and health concerns identified in the baseline safety review have been adequately addressed. Participants in this review should be representatives from organizations with a vested interest in the successful outcome of the project. These should be individuals expected to oversee the contracted work and occupy, operate, and maintain the completed project.

The types of questions to be considered in this review should include the following:

- Can the facility, as designed, be constructed safely?
- Does the design meet the needs of the occupant and National Fire Protection Association (NFPA) Life Safety Code requirements?
- Will maintenance and operations personnel responsible for keeping the facility operating be able to do so without personal risk?

The answer to the first question begins with a systematic evaluation of the probable work sequence each contractor will most likely follow—from mobilization to completion of the work. Comments and suggestions generated by this review enable the designers to reevaluate the project plans to determine where additional safeguards must be considered. With access to this information, the project manager may be able to eliminate or minimize hazards through judicious site layout and when planning the work sequencing. Where identified hazards cannot be reasonably controlled, information regarding these hazards can be brought to the constructor's attention through the contract language and at the pre-bid meeting.

Unfortunately, in the interest of expediency, the 90% design reviews are often overlooked. The risk of bypassing this step is potential for delays after construction has started, when the originator or an inspector identifies needed changes that could have been addressed during the review of the plans. Additionally, safety considerations not addressed during the initial design phase often cost significantly more to retrofit following the completion of the project.

Representation in the 90% design review for safety should include:

- originator
- maintenance
- safety
- scheduling planner
- estimators
- procurement.

The design review is a good opportunity to introduce the *Procurement* personnel and the *field engineer*, who will be assigned to oversee the construction of the project. Having the opportunity to participate in discussions associated with the project design will help them develop an insight into issues that should be addressed in the procurement documentation and coordinate involvement of safety representatives during the 90% design review.

The facility safety representative's key safety responsibilities at the design and planning stage include the following:

- Give advice, as requested, to the coordinating engineer on the adequacy of provision for safety and health features in the design.
- Assist with the conduct of the baseline safety review.

Pre-Work Planning

While the design is in progress, the coordinating engineer and the field engineer, who will manage the contracted work, have the opportunity to evaluate the anticipated means and methods required to complete the work. Through this evaluation process—and with the results of the previously completed baseline safety review—they will be able to develop an inventory of potential hazards. Those hazards that cannot be engineered out of the project or minimized through work planning must be brought to the attention of the bidders in the contract language and during the pre-bid meeting to minimize the potential that they may be overlooked. Information regarding the hazards should also be documented at this point for the benefit of the project or job site manager so that it can be flagged in the project schedule.

Statement of Work

Potential hazards identified associated with the work to be performed should be included in the statement of work, along with the description of the work to be performed and the list of special conditions. At this time, the role of the coordinating engineer draws to a close, and the work is assigned to a field engineer, who will be responsible for managing the execution of the work.

Procurement—Contracting Phase

Bid Package

Procurement assembles the bid package that consists of the scope of work, construction drawings, and specifications. Detailed reference should be made to specific safety hazards that will be the responsibility of the successful bidder to control.

The host employer's key safety responsibilities during the procurement and contracting phase include the following:

- Consider safety and health issues when preparing and presenting invitations to bid.
- Develop the safety and health program for the project.

Invitation to Bid

Procurement issues an invitation to bid to prospective contractors and sets a date for a pre-bid meeting.

Pre-Bid Meeting

The pre-bid meeting provides the forum to advise prospective bidders of the scope of work to be considered and to emphasize the importance placed on safety, along with quality assurance and other criteria specific to the work. It also offers the opportunity for contractors to raise questions and clarify any issues they might have with regard to the work and associated safety requirements.

Bid Review, Evaluation, and Contractor Selection

On receipt of the bids, a technical safety evaluation of the apparent low bidder's ability to perform the work safely is completed. This evaluation is to confirm that the low bidder is able to perform the work safely as judged by the technical selection criteria. (If contractors have been prescreened, this process would have been completed prior to the invitation to bid.)

Contract Award

The lowest bidder that meets the technical selection criteria is notified of its selection by Procurement. The chosen contractor is asked at this time to produce its project-specific safety plan and job safety analysis for the work to be performed. The contractor is also asked to identify the persons to whom he or she has delegated the specific responsibility of coordinating safety-related issues and to provide a list of the designated competent persons, as required.

Preconstruction Meeting

Once the successful contractor is selected, a meeting should be held with the apparent low bidder's supervisory personnel expected to perform the work and the facility personnel. The objective of this meeting is to establish that the contractor has a clear understanding of the contract scope, as well as the job-specific hazards and job safety requirements. The meeting should review how the contractor will conduct business and the procedures the contractor and other facility organizations will follow before the start of and during the performance of the work.

The field engineer's key safety tasks during the procurement and contracting phase include the following:

- Bring to the attention of the prospective bidders information regarding significant safety and health hazards associated with the work/project and the risks of other work that will be under way in the area of the work/project.
- Develop responses to contractor queries via formal addenda to the invitation for bids.

Safety Program and Job Safety Analysis Review

The field engineer reviews the contractor's safety program and job safety analysis for completeness and applicability to the work to be performed. The objective of this review is to ensure the contractor understands the risks associated with the work and has proposed reasonable safeguards. Where the need for changes to the contractor's safety controls is identified, these are brought to the contractor's attention for resolution. Once all documentation is in order, a pre-work release meeting is scheduled. A useful guide for the conduct of this effort is the book *Construction Safety Planning* (MacCollum 1995).

Pre-Work Release Meeting

The field engineer's safety expectations are reviewed with the contractor's supervisors. Often, this is the first time they may have heard the project safety requirements. It is not uncommon for the individuals who received the request for bid and attended the pre-bid meeting to neglect to communicate this information to the field personnel. Once all requirements for documentation have been satisfied, Procurement issues a notice to proceed to the contractor.

The facility safety representative's responsibilities during the procurement and contracting phase include the following:

- Provide guidance for the selection of contractors.
- Assist the field engineers and coordinators as they evaluate the hazard risks in the work to be performed.

Work—Construction Phase

When the contractor's employees initially report to work at the job site, they should be briefed on the site rules and regulations, the general hazards, and any special emergency response requirements specific to that worksite before they begin work. Workers need to be aware of the hazards peculiar to the site. Increasingly, site hosts are communicating this information to contractors themselves to ensure it is done to their satisfaction.

The contractor's key safety responsibilities during the work/construction phase include the following:

- Provide relevant information on safety and health risks created by the work and how such risks will be controlled.
- Develop and implement a site-specific safety and health plan.
- Ensure workers are following safe work practices.
- Ensure workers bring hazards that might affect others to the contractor's attention or the attention of the field engineer.

The field engineer's key safety responsibilities during the work phase include the following:

- Be satisfied that contractors are following recognized safe work practices.
- Ensure required training for safety and health is completed.
- Allow only safe equipment and authorized people into construction areas.

- Assist contractors to obtain safe work permits where necessary.

Specific Worksite Safety Briefing

Where appropriate, contractors should be briefed on facility-specific work practices and requirements such as lockout/tagout or work entry permit requirements. At this time, the contractor should inform the workers of the specific individuals who will approve permitted work and how they should be contacted. They should be briefed on all site-specific safety hazards and applicable emergency response requirements.

Job-Specific Training

Contractors review the job safety analysis (JSA) they prepared for their specific job with their respective employees prior to the start of work. Specific training must be provided to address hazards identified in the JSA. Contractor employees should sign the JSA to indicate they have read or been briefed on the information contained in the JSA and understand it. The JSA should be available for the field engineer's review and for the reference of the workers.

Contractor Equipment Inspection

All heavy equipment, such as man-lifts and cranes, should be inspected by qualified persons for general condition and serviceability prior to being permitted onto the job site. The inspection should ascertain that the equipment is in good operating condition, that appropriate safeguards are mounted on the equipment, and that the operators are certified to operate the equipment.

Contractor Tool Inspection

Prior to starting work, contractors should affirm that the tools and equipment to be used on site are serviceable. Tools include all manner of material being used by the contractor, such as extension cords, hand and power tools, and so forth. The field engineer and facility safety representative should periodically assess the condition of tools in use by the contractors as they perform their routine site evaluations. This is one measure of the contractor's implementation of the safety program.

Progress and Coordination Meetings

Regularly scheduled project coordination meetings should include safety as the first order of business.

The facility safety representative's key safety responsibilities during the work phase include the following:

- Be reasonably satisfied that contractors are following recognized safe work practices.
- Ensure required training for safety and health is completed.

Permits as Needed

The object of permits is to ensure that work known to contain recognized hazards is implemented within accepted safe work practices and in conformance with the owner and regulatory requirements.

Types of permits:

- weld/burn/open flame
- hot work (electrical)
- concrete coring
- excavation
- confined-space access
- operation work entry
- equipment movement.

Contractor Progress Reports

Depending on the work to be performed and the length of the contracted service, contractors may be required to submit reports as evidence of the completion of work activities.

- minutes of meetings (when held)
- toolbox meeting records (weekly)
- man-hour and injury report (weekly)
- schedule updates
- material certifications (as needed; e.g., concrete, stainless steel, cable)
- conformance to code statement (if needed)
- vendor inspection sheets (if requested).

The contractor's key safety responsibilities during the work phase include the following:

- Abide by the approved safety program and plan.
- Be reasonably satisfied that when arranging for subcontractors to carry out work, such subcontractors are competent and recognize the safety provisions they are expected to follow.
- Ensure required safety training is completed.
- Allow only authorized people in the construction area.
- Display required safety postings.
- Identify the hazards of the work, assess the risks arising from these hazards, and define methods to control them.
- Report accidents and injuries.

Inspections and Audits

Both the contractor and the host employer should conduct regular inspections of the workplace to ensure that housekeeping is being kept up and that no new hazards are being created by the work in progress. Audits to evaluate how effective the safety program and plan are being implemented should be conducted by a third party, with reports being provided to the owner and contractors. Specific actions and a timetable should be established for correction of hazards identified during inspections and audits.

Beneficial Occupancy

In some facilities, a process is in place whereby space cannot be reoccupied following contracted work prior to the approval of the fire department or fire safety engineering. Approval to reoccupy space is granted contingent upon the verification of life safety compliance requirements.

Final Acceptance Determination

Once advised by the contractor that the work is complete, it is in the interest of the field engineer to determine that the originator is satisfied with the end product. The field engineer should ensure that the following (where applicable) are completed as well.

End of work/project documentation:

- punch list closed
- as-built/shop drawings submitted
- specification changes submitted
- acceptance criteria listing completed (if required)
- equipment acceptance tests completed
- maintenance training and sign-off sheets submitted
- maintenance manuals delivered.

Contract Close-Out

A close-out meeting with the contractor should be considered to review performance and submit close-out comments.

Upon satisfactory completion of the work/project, the field engineer should inspect and verify that the contractor has left the job site in a safe condition and that all safety-related construction features have been performed in accordance with approved drawings and specifications.

The field engineer may wish to call in the facility safety representative to verify that there were no outstanding safety-related issues and obtain his or her signature on an Inspection and Acceptance Memorandum generated by Procurement to release the contractor from the project and pay the retention.

When first presented with this structure, the reaction of many organizations is, "If we have to follow all those steps, we will never complete a project." However, when all the project team members, including the originator, the project managers, field engineers, safety personnel, and so forth, are assembled and asked to help define the sequence of events necessary to complete a project, they generally develop a similar structure. In fact, it is a good idea to invite all project team members to get together before a large undertaking to agree to the sequence of events and the involvement each member would like to have at each step. In this manner, there will be a consensus of what must take place and who is responsible for its successful execution.

ROLES AND RESPONSIBILITIES

The Client

Clients must clearly define their safety expectations to the project designers and construction manager (CM), just as required quality standards and the scheduled date of completion of the project are defined. The client must regularly reinforce its commitment to safety by addressing the subject at meetings and when following up on design issues so everyone involved in the project realizes the client is serious about its commitment to safety. The client should require that the designers conduct constructability reviews of their work and that the construction manager regularly report on the safety performance of the contractors.

The client should periodically conduct its own assessment of the effectiveness of the construction manager's implementation of safety program and plan requirements and the degree to which safety is being integrated into routine work practices.

Architects and Design Engineers

Included in the architect's and design engineer's (designer's) responsibilities should be the assurance that the design they develop can be constructed, operated, and maintained safely. Designers do not generally address construction worker safety for a variety of reasons. They are not typically educated or trained to address worker safety. They claim that they do not have the tools or information to help them design work safety. The real reason is most likely that their legal advisers and insurance companies would prefer that they not address the issue to minimize the designers' liability exposure.

However, design professionals' responsibility as outlined in their code of ethics states:

> Engineers, in the fulfillment of their professional duties, shall:
> Hold paramount the safety, health, and welfare of the public.[2]

Construction workers are members of the general public and are unique facility users clearly at risk when building a structure or system developed by a designer. Therefore, designers should show the same regard for the safety of contractors performing their work as they do for the facility end-user.

Architects and design engineers often disavow responsibility for safety issues associated with the construction of their work. This will change only if clients insist that

designers address construction safety concerns. Clients should require that their designers employ a construction safety professional. This individual should review the project plans and specifications for potential safety hazards associated with the design and the anticipated means and methods required of the contractors to complete the work. In this way, measures can be taken in a timely manner to eliminate or reduce the potential for accidental losses during the construction of the work. This approach may also reduce hazards to operating and maintenance personnel occupying the completed facility.

The client's expectation that the architect include consideration for constructability and safety of the completed design should be clearly defined at the onset of the project. Unless this is done, the participants in the design and construction process will assume safety is someone else's responsibility. Costs associated with worker injuries and fatalities are ultimately borne by the client. Insisting that safety be included in design considerations will prevent the occurrence of injuries and ultimately reduce construction costs. The requirement to address safety will likely result in higher designer fees to cover added effort and responsibility. For the construction industry, the return on investment of the higher design fees will be a safer workplace and fewer litigation and injury claim costs.

Project and Field Engineers

Project and field engineers should be given specific direction regarding responsibility for ensuring that due consideration is given to safety and the protection of individuals involved in the construction work. In 1996, the New Jersey Supreme Court rendered the opinion that even when an engineer has no contractual obligation concerning safety, an engineer with actual knowledge of a dangerous condition to which a job site worker is exposed needs to act to prevent an injury.

If the engineer does not bring the hazard to the attention of the exposed party, the engineer may be held liable. Similar judgments were also rendered in the case of *Carvlho v. Toll Brothers et al.* (651 A.2d 492; Super. A.D.N.J. 1995). This case follows the similar Kansas Supreme Court case of *Balagna v. Shawnee County* (668 P.2d 157; Kan., 1983). In the latter, the court overruled the argument that the engineer would have exceeded its contractual authority by notifying a worker of an obvious unsafe condition.

The court acknowledged that the engineer did not have a contractual responsibility for safety. However, the absence of a contractual provision imposing such responsibility does not relieve an individual of exercising reasonable care to take some action when circumstances present at the job site demand such intervention to protect a fellow worker from harm.

Project engineers must understand what "stop work" means and their authority and responsibility in regard to

---

[2] The first fundamental canon of the National Society of Professional Engineers' Code of Ethics (NSPE 1996).

safety. Ignoring unsafe behavior condones that behavior. Each time a project engineer walks by an activity being performed unsafely, the workers may reason, "I did this before and no one said anything, so it must be OK here." Accepting unsafe practices undermines efforts to convince the employee that this job is different from the last one (where perhaps only schedule was a concern).

A concern that must be addressed with project engineers is the issue of liability to which they might be exposed if they become involved in a safety-related issue or identify a hazard. Often, avoidance is the stance of choice (as illustrated earlier). Such a position is based on the false assumption that any action taken to prevent an injury will be used as an example of "control" over work with regard to safety. This assumption is then extended to the concern that an injured employee may claim that anyone who acted to prevent an injury on one occasion has a duty to do so on others, particularly in the injured party's case.

Preventing an injury is the best way to avoid a lawsuit. Thus, the project engineer, who is often the client's representative, must document that a safety concern was identified and formally request, in a timely manner through the general contractor, that the subcontractor rectify the situation. After all, it is the subcontractor's contractual responsibility to do so.

Construction Manager

The construction manager retained to act on the owner's behalf to manage the development and construction of a project is also responsible for the safe execution of the work. The owner should unambiguously define the construction manager's responsibility and authority for safety. The owner should hold the construction manager accountable for developing requisite protocols and monitoring the safety program's implementation.

In the past, a single general contractor usually completed large contracted work projects. A project manager controlled the entire construction process from the development of the work packages to the execution of the work by direct-hire craftspeople. As projects became larger and more complex, project owners opted, with increasing frequency, to employ construction managers to coordinate and oversee the work being performed by several general contractors. The construction manager is expected to ensure that the project is built to the specifications provided by the engineering design or architectural firm.

Safety is a project management consideration that is sometimes overlooked or is consciously avoided. Many project managers are so busy dealing with coordination issues related to the numerous specialty contractor trades, material suppliers, and providers of rental equipment that they are hard pressed to find time to deal with safety. To complicate this situation, construction managers are occasionally brought onto projects after the contracts have been awarded and work has started. The construction manager may be so busy catching up that he or she may not have the opportunity to develop a good understanding of the potential hazards inherent in the project before safety problems begin to develop.

Another reason for not addressing safety is the concern of incurring liability and the potential of being named in safety-related litigation. "The CM can either face this challenge, or hide his head in the sand. However, no exculpatory clause will assure immunity from liability. Given this choice, the reasonable approach of many CMs is to recognize and take hold of the risks through a pro-active approach, placing the CM in control of the circumstances creating risks of project injury" (Connor 1991).

The owner is certainly within its rights to require that the construction manager explain the manner in which he or she proposes to implement the safety program. The client should also require that the construction manager periodically report on the results of the safety program and hold him or her accountable for the results.

Occasionally, clients make the mistake of being too prescriptive in defining safety program requirements and procedures, and they then attempt to micromanage the implementation of the program. The construction manager assigned responsibility for the site safety program should be given either specific project safety goals or the detailed program and procedures he or she is to follow and should then be given the authority to run the safety program.

With the responsibility for development and implementation of safety programs come very real liability considerations. Therefore, the construction manager must be given the authority to enforce the safety program requirements. Some clients make the mistake of assigning responsibility for safety to the construction manager but fail to confer the authority to stop work, withhold payment, or use other means of leverage to achieve requisite compliance.

General Contractor

General contractors are responsible for defining the safety practices for the means and methods to be implemented in the execution of the work for which they are responsible and to ensure that their subcontractors implement those practices.

The general contractor must ensure that his or her subcontractors are aware of the site safety requirements and the standards against which their performance will be measured. The general contractor should review and agree to the manner in which the subcontractors will perform their work as defined in the subcontractors' site-specific safety plan and the job safety analysis (JSA) produced by the subcontractors for each phase of the work they are expected to perform.

Neither the client nor the construction manager should delegate the decision regarding the standards to be met to the general contractors. Doing so will result in a broad disparity in the implementation of the safety program, eventual discord, and circumvention of safe work practices.

Subcontractors

Subcontractors are expected to supply the labor and tools to complete the work as scheduled and within defined specifications. They are responsible for ensuring that the individuals they bring to work are technically and physically capable of performing the work assigned to them. They are also responsible for ensuring that the individuals they bring to work have the required equipment and personal protective equipment to perform their work safely.

To be assured that this is the case, subcontractors should be required to produce a safety plan that includes JSAs for each phase of their work. Where their work creates hazards (such as the release of toxic fumes, excessive noise, radiation, etc.), these will be documented and be apparent to the general contractor reviewing the JSA. The JSA will define how employees will be protected from the hazards and the means by which the subcontractor proposes to alert others who might be exposed to those hazards.

Craftspeople

Last in the hierarchical chain are the craftspeople. They are expected to apply themselves and the tools of their trade to produce work of a defined standard. They are expected to perform their work in an informed and safe manner by complying with accepted safe work practices as defined and communicated by the owner, construction manager, general contractor, and their immediate employer.

Craftspeople are responsible for their own safety in regard to the work they are performing. They must understand that it is their responsibility to ensure that their tools are in safe working condition and that they have the knowledge to perform the work safely. The craftspeople must inspect their own equipment (such as ladders and scaffolds) regularly for obvious defects. They must be aware of site-specific requirements such as work entry permits or whether the site host insists on placing the first lock on lockouts/tagouts. This information will be included in the subcontractor safety plan and will be available through review of the JSA specific to the work that the craftspeople are to perform.

At the onset of a project's development, the owner funding the construction work must clearly establish that safety is a serious consideration. The owner must then support this position throughout the project development and implementation. When an owner assigns responsibility for safety to a construction manager, the construction manager must define the program criteria and see that the resulting process fosters a safe work environment.

Contractors are responsible for and should be held accountable for the safety of the work practices they employ and the safety of their respective employees. Contractors should be required to systematically evaluate the hazards associated with their work in order to protect their own employees and to implement precautionary measures to prevent other individuals from being affected by these hazards. The hazards that cannot be eliminated or effectively controlled should be brought to the attention of the general contractor to be communicated to other contractor employees.

Contrary to folklore, taking time to consider safety will not delay a job. In fact, extensive field experience in construction shows that each project in which safety was considered an integral part of business was completed ahead of schedule. Those work groups on projects that gave due consideration to safety frequently had to wait for groups that did not give high regard to safety. Organizations that fail to implement a good safety program spend substantial time and effort addressing minor, repeatedly occurring inconveniences.

INSURANCE

Construction insurance is a major project-related cost, following materials and labor. A study conducted by Stanford University, on behalf of the Business Round Table, concluded that insurance premiums on typical industrial projects cost 7% of direct labor costs for workers' compensation insurance and another 1% for builders' risk and liability insurance. Because labor can represent about one-third of the total project cost, insurance represents a measurable portion of total project cost (Samelson and Levitt 1982). Owners or contractors who are able to transfer responsibility for loss exposures to another and have that other entity pay for the insurance to cover that responsibility are able to substantially reduce their own loss-related costs.

While some risk managers would advocate the transfer of risk and support insurance as the sole solution to loss control, this does not of itself eliminate the hazards or moderate unsafe work practices that are the sources of the majority of losses. The transfer of risk does not eliminate the potential of being involved in litigation or having to defend a third-party lawsuit. The selection of insurance coverage and determination of limits should be a risk-based decision to protect the project and project-related assets from losses that could disrupt the project schedule and its timely completion or, more fundamentally, the operational viability of the enterprise. Insurance should not be considered to be an alternative to a safety program.

The subject of insurance can be addressed in other books. The purpose of this section of the chapter is limited to providing an insight into some of the types of coverage available to the construction industry and is certainly not intended to advocate one type of insurance over another. There are three distinct categories of accidental losses from a risk management perspective:

1. There are the numerous small losses that all projects experience whose aggregate costs have minimal financial impact. From a safety perspective, these are first-aid and minor medical treatment cases. It is generally more efficient for the project to absorb the costs related to these occurrences as part of the cost of doing business rather than recovering them through standard insurance claim processes.

2. The next group of accidental losses involves less frequent occurrences resulting in larger individual losses, with larger annual aggregate costs. These losses may take the form of incapacitating personal injuries, vandalism, or major equipment damage. The most common means of recovery from these sorts of losses is insurance.

3. The final category is catastrophic loss. These instances happen infrequently; however, their cost is so great that their occurrence could affect the liquidity and even solvency of the business enterprise. A major refinery fire resulting from a contractor error and natural causes such as an earthquake are examples of catastrophic losses. Insurance is the preferred means of recovery in these occurrences as well.

What Is Insurance?

Insurance is simply an agreement that, for some financial consideration, another organization (typically an insurance company) will cover potential, predefined losses. Insurance companies, not being philanthropic organizations, will attempt to quantify a company's potential loss claims based on the firm's historical losses, the nature of its operation, and the effectiveness of its safety program. They will then estimate the administrative cost of handling those loss claims and the probability of each type of occurrence, to which they will attribute a cost and, of course, build in a profit margin for themselves. This will be the basis for their determination of the cost of providing the company with insurance and becomes the basis for the insurance premiums it will be charged. The funds used by the insurance company to cover claimed losses come from the cumulative pool of funds collected from all its clients.

Cost of insurance = Expected losses based on average past loss experience + Administrative costs + Profit

Relying on insurance as the sole remedy to cover losses resulting from construction accidents is a very costly loss control approach. It offers a poor return on investment as compared with the decrease in loss exposure achieved by directing that same money into eliminating or controlling hazards. As seen in the preceding formula, the fact is that in the long run, insurance companies recover the entire cost of all incurred loss; the insurance company charges an administrative fee for providing insurance coverage and handling claims—and it builds in a profit.

As long as hazards are present, the potential for a loss is present. Eliminating or controlling hazards reduces the potential for losses and, therefore, the need for insurance. When an accident does occur, more often than not, one of the accident investigation recommendations will be to eliminate or more effectively control the hazards that caused the occurrence. Should there be an injury resulting from a known hazard, all parties aware of its existence run the risk of being named in an associated lawsuit, and certainly the hazard causing the injury will then be mitigated. The logical conclusion here is that it is in the best interest of every project to conduct a project risk analysis and implement a safety process through which workplace hazards are identified and eliminated.

Very little insurance coverage is required by law. The only insurance coverage typically required is workers' compensation and vehicular liability. Workers' compensation covers medical costs of employees injured while on the job, disability, and lost wages. However, there are many other types of insurance coverage that can protect an organization from irrecoverable losses. A qualified insurance broker can explain the specific details of coverage provided by each type of policy and their related costs.

Comprehensive General Liability

General liability exposures vary widely. In construction, general liability exposure arises from the fact that a worksite exists, construction operations are taking place, and independent contractors are present and performing work. Insurance for this type of operation is termed *comprehensive general liability coverage*.

One specifically defined insurance coverage—operations and premises liability (under comprehensive general liability coverage)—is for legal liability for damages resulting from bodily injury or property damage caused by defined occurrences, subject to stated exclusions. This includes the premises on which construction work is being performed or operations are being conducted by the insured party. The exposures here are bodily injury or damage to someone else's property.

Independent Contractor's Protective Liability

Independent contractor's protective liability covers the insured party's legal liability for bodily injury and property damage caused by an occurrence resulting from operations performed for him or her by an independent contractor. It also includes occurrences associated with the insured party's general supervision of that work. When a comprehensive general liability policy is issued to a contractor, automatic insurance is provided for the liability that may result from subcontracted operations. Owners may opt to require that general contractors name them, their designers, and their general contractor as additional insureds under the contractors' comprehensive general liability policy instead of writing a separate policy. This requirement should be clearly stated in the owner's contract documents, and the owner should request evidence of such coverage being in place before work is authorized to begin.

This insurance is important to cover defense-related costs in today's construction world, where it is common practice to sue anyone even remotely associated with the work. Further, in an environment in which judgments may be awarded against the owner, general contractor, or construction manager even though the work was sublet, this coverage becomes advisable.

Completed Operations Liability

A rider to a comprehensive general liability policy is *completed operations and products liability*. This coverage limits the legal liability for bodily injury or property damage caused by an occurrence that takes place in a completed or abandoned operation, or away from premises owned or rented by the named insurer. It also covers goods or manufactured products sold, handled, or distributed by the named insured or by others trading under his or her name.

Indemnity Agreements

Indemnity agreements are those in which one party to a contract says he or she will defend, pay for, and hold harmless the other party. Indemnity agreements are contract clauses that "pass the buck" and are also called waivers of subrogation. General contractors indemnify the owner, subcontractors indemnify the general contractor, subs to subcontractors indemnify the subcontractor, and so on down the "food chain." Everyone from the top of the line on a job site and down each tier wants to have the person below waive his or her legal right to seek recovery from a loss. This is the case even when the person on top messes up, accidentally or even deliberately. It is like a game of musical chairs, where there is only one winner. In the game of insurance musical chairs, it is the loser who pays. In some jurisdictions, the indemnifying party is required to cover all losses—even if the party being indemnified is negligent. In these jurisdictions, a company will want to bargain this in or out depending on whether it is the indemnitor or the indemnitee.

Some insurance carriers routinely provide waivers in liability policies. For other carriers, the waiver must be endorsed to a policy on a case-by-case basis. Each state has different statutes dealing with such waivers. Additionally, each state's courts will interpret these statutes differently. In several states, the waivers are not legal at all. Although many states have laws against these clauses, a lawyer can always be found who has an interpretation that allows clients to find some way to protect themselves. Therefore, companies and individuals should always check the contracts and know what they are signing. If there is such a clause in the documents, make sure there is a strategy to deal with it.

Case Scenario: The Joint Venture Shuffle

Assume a general contractor awards a subcontract to a joint venture made up of two separate companies to provide cast-in-place concrete, and the subcontract contains an indemnification clause. One of the joint venture partners prepares the rebar work and the other the framing, pouring, and finishing. Now assume that an employee of the joint venture contractor installing the rebar gets injured and sues the general contractor.

When the general contractor seeks to enforce the indemnity provision and tenders the defense to the rebar entity, to his great surprise he may find that there is no insurance from the joint venture—the legal entity who contracted to indemnify. The insurance carrier can say that the rebar contractor did not contract to indemnify.

Many owners and general contractors who award subcontracts to joint ventures make the mistake of accepting the insurance of each of the joint venture partners. The exposure here is that the joint venture itself does not have insurance. The joint venture is, in the eyes of the courts, the real contracting party. A joint venture is a legal entity formed for a specific purpose. The insurance of the separate parties is not related to the operation of the joint operation. In this case, the general contractor can be stuck defending the lawsuit and paying the judgment.

Builders' Risk

Property insurance to protect the project or building during the course of construction is provided by *builders' risk*. If the work is damaged or destroyed during construction and prior to acceptance by the job site manager, the builders' risk policy will pay to rebuild. Although the "work" usually belongs to the general contractor until it has been accepted, the owner generally will require that the general contractor have such a policy in place. Occasionally, the owner assumes that if the contract document requires the general contractor to obtain the policy, the policy will be purchased. The owner should request evidence that the policy has been obtained before work begins and should be sure that the owner is named in the policy as the beneficiary.

Builders' risk policies have deductibles just as in automobile insurance. Builders' risk deductibles range anywhere from a couple of thousand dollars to several million dollars on larger projects. In the area of builders' risk, it is not a bad idea for the general contractor to pass the deductible down to its subcontractors. If the subcontractor knows he or she is liable for the deductible, he or she may pay more attention to loss prevention.

If the general contractor intends to pass the policy deductible down to the party who incurs the loss (the subcontractor), the general contractor should make sure the subcontractor is aware of this fact. This fact may not be explicitly mentioned in the subcontract with the general contractor, due to one of those typical subcontract provisions that make the owner/general contractor agreement part of the agreement. Otherwise, the subcontractor may find herself or himself paying an unexpected deductible.

Offsite and Transit Limits

Offsite and transit limits are an aspect of builders' risk. They cover construction materials while stored off site or while in transit to the site. Assume the general contractor has a large number of precast concrete components. The precaster, by the nature of the product, produces panels, beams, and so forth well ahead of when they are needed. They may be stored in a nearby vacant lot, where their value exceeds the offsite limit. Other examples of this sort of material could be HVAC units for a large structure that are being trucked to a site where the value of these units might exceed the transit limit.

Review the certificate of insurance or the binder on the builders' risk. Make sure it is in place before the work begins. Consider the frustration, the cost, and the delay of not having insurance funds necessary to replace lost units. Determine the policy limits and plan operations accordingly. Insert a provision in the contract that prohibits exceeding the limits.

Workers' Compensation

There is a growing crisis in workers' compensation insurance as premiums are being driven increasingly higher by rapidly increasing losses. Plagued with runaway medical costs, widespread abuse of benefits, and rapidly escalating insurance premiums, businesses find their competitiveness is being threatened.

Increased attention on occupational health issues such as stress, mental illness, and physical impairment due to exposure to asbestos, lead, silica, workplace chemicals, and ergonomic issues is putting added pressure on an already overburdened system. For owners, who ultimately pay the bill, increased cost of capital construction means fewer facilities and higher fixed costs at a time when world competition demands lower costs.

Even though workers' compensation laws are governed by individual states and vary from state to state, the laws are essentially quite similar, even if costs and benefits are not. Each business must meet its workers' compensation liability in some way. State compliance with this requirement falls into four groups:

Open states: there are no state fund plans, and workers' compensation insurance is provided by private insurance carriers.

Monopolistic states: workers' compensation insurance can be purchased only from a state-sponsored insurance fund. There are eight states and territories where this is the case:

Nevada	Ohio	West Virginia	U.S. Virgin Islands
North Dakota	Washington	Wyoming	Puerto Rico

Competitive states: workers' compensation insurance can be purchased from either a safety fund or from a private carrier. This is the option in 12 states:

Arizona	Idaho	Montana	Oregon
California	Maryland	New York	Pennsylvania
Colorado	Michigan	Oklahoma	Utah

Self-insurance states: all states except the four listed below permit employers to self-insure. The financial requirements are very rigid and include high bonding limits. Workers' compensation costs generally must exceed $200,000 per year to consider this option.

Nevada	North Dakota	Texas	Wyoming

Fundamentally, workers' compensation insurance premiums are determined by using three elements:

- experience modification rate (EMR)
- manual rate
- payroll.

Experience Modification Rate

The National Council on Compensation Insurance (NCCI) formulates the experience modification rate. This organization compiles workers' compensation payroll and injury data from approximately 600 insurance carriers. Contractors that have reached a minimum premium threshold size in their particular state are rated and receive a new EMR each year. The calculation of this rate is based on each contractor's payroll and injury data, actuarial factors, weights, and ballast. The resulting EMR for each contractor is sent to the contractor's insurance carrier by the NCCI. This information is used by the carrier to calculate the contractor's annual workers' compensation insurance premiums. Therefore, the EMR is an unbiased means of judging the relative effectiveness of a contractor's safety program and the rate at which it experiences injuries.

The EMR is a reflection of an employer's safety performance as compared with the average contractor in a specific specialty. An experience modifier of 1.00 is average. The EMR is the value by which an insurance carrier's base rate is multiplied to determine the insurance premiums paid by its client, the contractor. An employer with above-average losses is assessed a modifier higher than 1.00; conversely, employers with fewer than average losses will have an EMR of less than 1.00.

For example, the workers' compensation insurance premiums a contractor with an EMR of 0.80 will pay will be 20% less than the premiums for the average contractor performing similar work. This fact is an important consideration when a company reaches the point in the contracting process of judging the ability of contractors to perform work safely and when assessing if, in fact, the contractor can complete the work for the bid price.

A contractor who says, "I have a bad EMR because my work is so dangerous" is either trying to pull a fast one or does not understand what an EMR is. He has a high EMR because he has a poor safety performance history as compared with his peers in his industry, in his state. His high EMR is the result of not placing appropriate emphasis on safety.

A contractor's EMR reflects his or her loss experience for the previous 3 years. (A contractor who has been in business for less than 3 years will have an EMR of 1.00.) Each year when determining an employer's EMR rating, the oldest year of experience is dropped and the most recent year is added.

The EMR formula is designed to account for statistical variations in the size of employers. An employer with a large number of employees will usually have accident claims that result in a fairly even distribution of both the number and severity of claims. In the case of small employers, frequency and severity are adjusted to minimize the unpredictability. With smaller employers, primary weight is given to frequency and secondary emphasis to severity. As the size of the employer increases, statistical variations decrease, and the experience modification formula reduces to the simple ratio of actual to expected losses.

EMRs generally vary between 0.20 and 2.60. A contractor with an EMR of 2.60 pays 2.6 times more in workers' compensation insurance premiums than does the contractor with an EMR of 1.00. The greater the contractor's EMR, the greater the proportion of the contractor's bid that is being allocated to cover insurance costs. It should be obvious who has the competitive advantage. Logically, when a contractor's EMR is higher, proportionally less is being paid for the quality of material and labor to complete the work than would be paid by a contractor with a low EMR. The additional benefit to owners of using contractors with low experience modification rates is the proportional reduction in the number of accidents that will occur on the job site and the lower probability of being involved in litigation.

Manual Rates

The manual rate reflects the cost of covering the workers' compensation costs for each trade in each state. It does not reflect a single employer's accident history. Insurance coverage provided to each employer using the same trades or crafts in the state will be based on the same manual rate. The variation in manual rates among states is based on the benefits and entitlements in the local compensation laws. The rates are then calculated by determining the total cost of all claims for each specific classification code and then dividing by the total amount of payroll for the classification code; the rate is then expressed as a dollar rate per $100.00 of payroll.

Insurance for a pipe bender will be different than that for an ironworker erecting steel. Besides the variation in class of work, manual rates also vary widely from state to state. The variation in workers' compensation rates from state to state is dramatic. For example, the manual rate for structural steel erection across the United States ranges from a low of $14.00 in Nebraska to a high of $103.85 in Minnesota. Clerical workers are not injured on the job frequently; when they are injured, their injuries are generally not as severe as steel workers'. The manual rate for a clerical worker might typically be around $0.40 per hundred dollars of payroll, while the manual rate for a roofer would typically be more like $40.00 per hundred dollars of payroll.

Payroll

In most cases, when a business starts a policy, the first year's workers' compensation premium is actually a deposit that will be verified by an audit to determine if any audit premium is owed by the business to the insurer. The audit is used to determine the true payroll and associated employee exposure. This issue can sometimes lead to conflict between the carrier and the policyholder.

A potential source of conflict is related to how "payroll" is defined by rating agencies (NCCI, WCIRB, etc.). Actual wages paid to an employee are easy to determine. However, "payroll" used to calculate workers' compensation premiums typically includes vacation pay, holiday pay, bonuses, sick pay, auto allowances, and commissions.

Example: Comparison of Contractors

Contractor A has an effective safety program and good performance history. He enjoys a favorable EMR of 0.60. Assume his direct payroll is $10 million, and he has a composite manual rate of $15.00. His workers' compensation (WC) insurance premium, modified for good performance, is $900,000.

$$\text{WC premium} = \$10,000,000 \times 0.60 \times \frac{\$15 \text{ manual rate}}{\$100 \text{ of payroll}}$$

$$\text{WC premium} = \$900,000$$

Contractor B has not implemented an effective loss control/safety program except to stay clear of OSHA. Her attitude is "we are in a dangerous business—we should expect some people to get hurt." Because of a continuing stream of costly injuries, she has developed an EMR of 1.40. She has the same payroll, in the same line of work as her competitor, Contractor A. Her workers' compensation premium, modified for poor performance, is $2,100,000.

$$\text{WC premium} = \$10,000,000 \times 1.40 \times \frac{\$15 \text{ manual rate}}{\$100 \text{ of payroll}}$$

$$\text{WC premium} = \$2,100,000$$

The difference in insurance costs between the two contractors is:

$$\$2,100,000 - \$900,000 = \$1,200,000$$

This is 12% of her direct labor costs! The magnitude of savings will, of course, vary from state to state, but even in states with moderate workers' compensation costs, the savings are significant.

Wrap-Up Insurance

The term *wrap-up* is being heard with increasing regularity in construction insurance–related discussions. Most contractors have heard of the concept—it has been available for the past 60 years. However, it has only been during the past 35 years that it has come into common use.

Wrap-up is a generic reference to an arrangement in which a single entity furnishes insurance for an entire project. Wrap-up insurance, consolidated insurance programs, or owner-controlled insurance programs (OCIPs) are those in which there is a single insurance policy covering the entire project, instead of where each project contributor— such as the owner, architect/engineer, general contractor, or construction manager and perhaps 15 to 30 prime and subcontractors—insures only its own portion of the work through many different insurance sources. The same approach has been adopted more recently by construction managers and general contractors and is referred to as a contractor-controlled insurance program (CCIP). Both programs share similar key concepts and many advantages and disadvantages. In a wrap-up, the policyholder pays the cost of the insurance directly rather than having each subcontractor include insurance costs in its bid. The wrap-up, therefore, eliminates the requirement that the owner police the contractors to ensure that certificates of insurance have been obtained and are correct. Given the economy of scale, the policyholder can provide insurance coverage to the project subcontractors at a competitive price.

A wrap-up program offers several financial and administrative advantages to the policyholder. The most significant advantage is the broader coverage and higher limits available because of volume buying power. The centralization and unified onsite loss control functions and claims handling promote uniform loss control implementation. A wrap-up policy can be either job specific and run for the duration of the construction project, or it can be a "rolling wrap-up" that covers several projects managed by the same entity over a period of time. It can cover workers' compensation and general liability under a master contract. Individual workers' compensation policies are issued to the subcontractors to maintain a record of their experience and payrolls. Wrap-up policies are written for the term of the project, with claims coverage extending as long as 10 years after the completion of the project, thus ensuring continuity of insurance policy terms, conditions, and exclusions.

Wrap-up programs provide some unique advantages. The owner/general contractor receives peace of mind knowing that there are no uninsured workers on the job and that there is no difference in insurance coverage and limits among the various participants. Budgeting becomes clear-cut as the insurance cost is assessed consistently. Subcontractors are not incorporating it into their bid costs on a subjective basis. The carrier provides loss control and claim services to the insured's specifications, ensuring a uniform result and more effective overall management of workplace issues, including application of managed-care approaches and the possible use of alternative dispute resolution (ADR), which results in additional savings. Further, because insurance is provided by a common carrier, costs and time delays when settling claims are reduced because there will be no cross-carrier disputes.

The key to management of a successful wrap-up insurance program is the control of:
- all project insurance
- subcontractor insurance in all tiers
- loss prevention programs
- claims management.

Under a fragmented insurance program, the owner cannot be assured that completed operations coverage would still be in force, nor will coverage be in force for losses that are caused by the contractor's negligence, but that occur after the work has been completed. The wrap-up provides for this coverage to continue for a specified period (which may be as long as 10 years after completion of the project).

Advantages of wrap-up insurance:
- economy of scale
- reduced administrative costs
- avoidance of double coverage
- longer coverage periods
- elimination of cross-liability suits
- wider broker selection.

A consistent approach to unique coverage issues is another advantage of the wrap-up program. Professional liability issues arising from the design/build process can be addressed, as can potential losses due to local laws. The insured has better control of potential third-party liability issues that may come from workers' compensation claims. Some exposures that may not be covered under traditional general liability contracts can be included in a wrap-up program, including pollution, asbestos, and lead abatement liabilities.

Until recently, a single project with an estimated cost of $100 million was considered the threshold for a wrap-up. This threshold has been reduced because brokers and carriers have formed dedicated teams to handle wrap-up programs. Wrap-up programs begin to become economically viable once there are $1 million of premiums plus losses per year. Wrap-up programs have been written and managed for projects as low as $50 million, with rolling wrap-up programs at $75 million over 3 years. However, some states such as Minnesota and Michigan have minimum requirements of $80 million in total project costs. The U.S. Government Accountability Office suggests $50 million as the project cost threshold for considering a federally funded wrap-up program.

Wrap-ups not only offer the security contractors are looking for but also make good economic sense. However, it is important to keep in mind that a wrap-up program will be profitable only when the job site manager controls losses through a good safety program.

Two potential disadvantages of wrap-up insurance are greater administrative costs and the possibility for high up-front premiums. Project owners are responsible for administering wrap-up insurance and must provide administrative support either internally or through outsourcing. Some insurance companies require owners to make large premium payments at the start of a construction project or to establish a special reserve (typically not required with a traditional insurance policy).

Insurance is one of those items that owners generally do not like to pay for, but sufficient insurance can be critical to the success of a project. Without proper insurance, a company could lose its entire investment as well as the time and effort put into its project.

A project manager's decision is not just to insure or self-insure an exposure. The project manager needs to know what the coverage really covers! Many construction bankruptcies are the result of the failures by the contractor to understand the coverage. Most of the errors and omissions lawsuits between the contractors and their brokers involve failure to communicate what policy coverage gaps exist. Policy coverage is limited, both in amount of money paid at the time of a loss and in the events that are covered. Read the "What Is Covered," "Exclusions," and "What We Will Pay on Your Behalf" portions of the policies, and even the "Workers' Compensation" policy. Most contractors are operating "naked" of coverage or are underinsured for what they presume to be routine operations.

The types and amounts of coverage purchased for a project must be evaluated on a cost/benefit basis like any other commodity. Enlist an accountant and insurance agent to help review the amount of coverage needed. The bottom line is that insurance is not a substitute for an effective safety program.

SAFETY IN CONTRACTS

In a contracted work environment such as construction, there is often distrust between the contracting entities. The owner feels that the general contractor is going to gouge him, cut corners, and try in every way to make as large a profit as possible. On the other hand, the general contractor is concerned that the owner is going to criticize every aspect of the work, often rejecting the work, and requiring that it be redone until the general contractor goes broke. In such an environment, the contract is the means whereby all the parties to the work to be performed define what is expected and where commitments are made as to what will be delivered at what price.

People are identified by what they do. People who build buildings, roads, bridges, and other such structures are called *contractors*. This is an appropriate label because all work in construction is based on a *contract* (Keres 1996).

In construction, if work is not called for in a contract, it is not done. If an owner wants a contractor to do something not included in the original contract, a change is made to the contract. In other words, owners tell contractors what they want, through contract documents. Contractors perform contracted work and subcontractors perform their trade specialties pursuant to their respective agreements.

Within the construction industry, everyone is a party to a contract, whether it is an owner employing contractors, a contractor hiring a subcontractor, or a contractor providing construction services. In spite of this, it is ironic how little there generally is in contracts regarding safety. Those references that are made to safety are typically vague or indefinite. To bind a contractor to a desired level of performance, the requirements must be clearly stated in the contract.

It is important to note that there is a power shift in the relationship from owner to contractor once the contract has been signed. Therefore, as an owner, make sure the contract defines what should be done. Do not assume the contractor can or is willing to interpret the owner's intentions. Define what is wanted as explicitly as possible, particularly in the area of safety.

Unfortunately, there are no magic phrases or clauses, nor is there a single key to a perfect contract. This section explores some common practices and methods and defines guidelines for the development of contracts with good safety criteria.

The Contract

The contents of contracts may change as owners add or modify clauses to close the gaps they may have identified in previous contracts. Contractors must be aware that what was in the last contract may not be contained in the next one. There are a multitude of requirements that can be included and numerous ways to include them in a contract. Yet, there are some basic characteristics that must be present in each contract and there are some common methods of including them in contracts.

The first rule of contract methodology is this:

The drafter is the master of the contract.

The most common provision in construction contracts from the perspective of safety is that the writer, the entity at the high end of the food chain, generally states that: "I am not responsible for safety, you are." To fortify this, the drafter relies on indemnity and insurance requirements through which he or she endeavors to pass on responsibility for safety and associated losses to sub-tier contract participants. This is the way it is. However, as discussed in previous sections, to effectively control losses, a proactive stance must be taken and expectations must be clearly defined in regard to the safe performance of work. Define what should be built, what material should be used to build it, and what quality standards should be met. Then, also define the safety expectations.

Where to Start

First, the general structure of contracts must be understood. The most widely used contract form in construction is that developed by the American Institute of Architects, commonly referred to as an AIA. Within the AIA documents, the A201-2007, general conditions of the contract for construction (or hybrids thereof), consists of approximately 38 pages. In this document only three pages deal specifically with safety: Article 10—Protection of Personnel and Property.

Priorities of Contract Documents

To determine what is included in a contract with regard to safety, it is important to know how contract documents relate and affect one another and to know what to do or which of the multitude of contract requirements will prevail.

For every contractor on a project, there has to be a contract. The owner has a contract with the construction manager or the general contractor. The construction manager has a contract with the general contractor or, on large projects, the generals. Then, the general has subcontracts with its subcontractors, and the subcontractors have contracts with their respective subcontractors and service providers.

Because the owner is typically the first drafter, the owner's contract has precedence. However, there are numerous phrases and insertions that can change this. The general contractor will, in most cases, have within his or her subcontract agreement language to the effect that: "The Subcontractor agrees to be bound to the General as the General is bound to the Owner." This means that whatever is required of the general by the owner, the general will require the same of the subcontractor. Additionally, there will be language that states, "The terms of the Contract Documents between the Owner and the General Contractor are incorporated herein and the Subcontractor is bound thereby." These terms may not be explicitly defined in the general contractor's contract document. The subcontractor is often expected to ask to see those documents to learn what those requirements might be.

Of course, the subcontractor puts language in its subcontract that the terms of its agreement with the general contractor, which include the general contractor's contract with the owner, are incorporated therein and the sub-sub is bound thereby, and so forth down the organizational chain. This results in a contract maze, and because the contracts themselves are made up of numerous parts, there is no single contract—there are *contract documents*.

The first thing to establish when reviewing a set of contract documents is which of the numerous documents prevails. There will be a clause or section somewhere that will prioritize the contract documents. This will typically be in the owner/contractor contract or the instructions to bidders. The subcontractor is bound by all of the documents between the general contractor and the owner.

Typically, the priority of contract documents in the owner/general contractor format is:

1. the actual written contract
2. addenda/change orders to the written contract
3. the general conditions of the contract
4. the supplementary general conditions
5. the plans and specifications (although some contracts will say the plans prevail over the specifications)
6. instructions to bidders.

What Is the Contractor Really Bound By?

Some subcontracts say, "The Subcontractor is bound as the General is bound"—which is not much help, nor is it informative, to either the general or the subcontractor. Other subcontracts will say that in case of conflict between the subcontract and the other documents, "The most stringent requirements will apply." This means that the contract provision enforcing the greatest duty or penalty on the subcontractor will prevail. Still other subcontracts will say that in case of conflict, "The terms of the Subcontract will prevail." This is better for the subcontractor, but if the subcontract is silent, the original priority of documents holds.

Read *all* the contract documents! Owners and contractors take a significant risk if they sign a contract that they have not read and do not completely understand.

So, how should this complex system of documents incorporating other documents and one document changing the terms of another be handled? First, get a complete picture of the entire set of contract documents, including those documents that are incorporated by reference. Then, chart out what is really said. Nothing is more frustrating than reading one document and planning accordingly only to read later on that another document changes the first document. It is more productive to read the entire set *twice:* the first time to know what it says and the second time to unravel the changes and interrelationships.

Safety Clauses

10.2.1 "The Contractor shall take *reasonable* precautions for safety of, and shall provide *reasonable* protection to prevent damage, injury or loss to:"

What does this paragraph in Article 10 of the AIA 201 contract form mean? The word *reasonable* is subject to interpretation and is not definitive. What the owner thinks is reasonable, the contractor might not agree with. As the owner, it might be beneficial to replace the word *reasonable* with the word *all* or the word *necessary*. This puts a more stringent burden on the contractor. Even the contractor may wish to remove the word *reasonable*. A judge can interpret the word *reasonable* to fit the situation. Apply reverse legal logic here. If an accident happened, and it is assumed that all accidents are preventable, the logical conclusion would

be that "reasonable" precaution was not taken or protection provided.

Other adjectives to look out for are *routine, normally accepted,* or *as common in the industry*. These types of words and phrases are qualifiers without definitiveness. It is common to think nothing of this type of language when reading through a contract, but after more careful consideration, they leave a company or individual exposed to someone else's interpretation of how much effort should have been expended in providing a safe work environment. With this type of language in a contract, there is no objectivity with regard to safety. Signing a contract with words or phrases like this subjects the signer to someone else's interpretation.

Objectivity and Specificity

Although many owners and general contractors have detailed written safety programs, few of them impose these standards on their subcontractors directly. Consider including the written safety program as a specific contract document. This specificity will increase the owner's or contractor's ability to ensure work is conducted in a safe manner. Safety articles or clauses should include specificity with regard to safety expectations because the safety requirements of each job site will depend on the work to be performed.

Stating that it is a requirement to comply with OSH Act is a start. Clarify those areas in which there is potential for argument or vagueness. Insert site-specific safety requirements in the contract language, such as "100% fall protection is required for all employees working above 6 feet." To require that the contractor have a drug-testing program in place, define the requirement in the contract. Insert safety requirements item by item into all agreements as a separate addendum to the contract.

Comply with All Laws

As in Article 10.2.2, there is typically a requirement that states, "comply with all laws, etc." This is an example of an incorporating provision. The "comply-with-all-laws" clauses suggest that sub-tier contractors are expected to comply with all federal, state, and local safety rules and regulations. This is not bad, but is it enough?

Until 1997, what did OSHA really say regarding fall protection for ironworkers? In fact, the interpretation of the requirement was such a problem that the Steel Erection Negotiated Rule Advisory Committee (SENRAC) was established to clarify the ambiguities in 29 CFR 1926.750 regarding fall protection requirements for ironworkers. After protracted discussions and a considerable period of time, this committee, consisting of experts from across the United States, was able to reach an agreement. They have since defined the requirements for fall protection for work involving steel erection. To say in a document "to comply with OSHA" is often open to interpretation. Putting this phrase in a subcontract does not ensure that safe work practices are followed or eliminate OSHA citations, the general duty clause interpretation, or perhaps potential litigation.

Laws of the Jurisdiction

Contractors must know the laws of the jurisdictions in which they work. For example, in Chicago there is a section of the building code that deals with the structural support of tower cranes. Have an accident with a crane in Chicago, and the lawyers will have a field day. A jury will not want to hear that there was no compliance with the ordinance because the contractor was unaware of it, or did not read it, or disregarded the general condition requiring compliance with it.

One specific statute that appears regularly is the duty to provide lateral and subjacent support to adjoining property. Owners must be alert to their duty to give notice to adjoining landowners and to support those buildings. If this is the case, make sure to comply, or specifically transfer this obligation to the general contractor. If it is the owner's intent to transfer this responsibility to the general contractor, ensure that he or she is explicitly aware that this is included in the requirements that he or she will "comply with all laws."

Notice of Injuries

Many host employers include the clause that requires the contractor or subcontractor to send all notices of injuries to the host employer or general contractor. This is not the means by which employers or contractors should expect to learn of occurrences within their own operations. Field engineers should be advised of the occurrence of injuries as soon as they happen. However, this requirement is a useful clause to eliminate cracks in communication.

Notices of injury clauses are seldom complied with. So, put some teeth into the requirement. The host employer wants to know of the occurrence of incidents and significant "near-hits." Consider inserting the requirement that to get paid, the payee must first submit reports of injury or a statement to the effect that there were no occurrences. Include this in the contract clause that deals with documents necessary for payment, such as lien waivers and the like. In this way, someone in the office responsible for submitting documents to get money will look up those reports and send them in.

To add even more teeth, consider implementing a $100 or $250 penalty on each report not filed with a pay request. Make it known that the construction site safety manager's budget will get credited with the penalty fee money if there are accidents that subcontractors did not report. The safety manager will be motivated to ensure that his or her records are in order and that he or she is aware of all occurrences and near-misses. It will also motivate subcontractors to report injuries. Having notice of all accidents can prevent that sinking feeling when served with a lawsuit several years later and there is no documentation for a defense.

Contractors are required to have workers' compensation coverage and are required to provide their accident information to their insurance carrier. Because they must provide a certificate of insurance prior to starting work, owners have a contractor's insurance carrier's address. Approach the insurance carrier and ask it to give notice of injury reports to confirm the company's records are complete, and ask whatever other information the carrier is able to share within the confines of dealing with its customer. This will provide one more means by which to capture this information.

Hidden Provisions

Hidden provisions are typically important provisions placed in a subsection of the general conditions, where they might not be seen; they are usually one of those documents never seen, but that were "incorporated."

Take, for example, the host employer who has a clause titled "Watchmen" and basically says in five long sentences that the general contractor will provide a watchman for any and all hours that the general contractor does not have staff present on the job site. The last sentence then says, "The General Contractor assumes all responsibility for and indemnifies the host employer for all fire, vandalism, theft, or any occurrence caused by a third party or an outside source."

This is certainly not an ethical practice. This observation is included as a note of caution to those individuals who do not always read all the contract language or make the effort to locate all contract-related documentation.

What Is Not There

A word of warning: be sure to do more than read. In every contract, what is *not* there must also be determined. Do not become so caught up in reading and analyzing that missing provisions are overlooked. The contract documents might *not* include a term or provision that is very important.

Therefore, always know what is wanted and needed in the contract documents. If there is no reference to safety or there is no reference to something that should be assumed to be normal, then insert it. Make sure that what is wanted is really there.

Take the case of the absence of a clause that the work completion date is contingent on the arrival date of critical equipment. The contractor without a clause that defines this contingency may be bound to complete work by a specific date without regard to the actual arrival date of a critical component, over which the contractor may not have any control. Beware what is not in the contract language.

Safety should be part of every contract. The most common excuse for not including it is that, "Well, if I define safety requirements in a contract, I am accepting responsibility to enforce them, and I might get sued."

If there is no effective safety program, the probabilities of being sued are even higher, and there will not be an effective defense. Defining safety requirements in contracts provides more leverage and gives the contractor no excuse for not having included safety requirements in the cost estimates and work planning considerations.

CONTRACTOR SELECTION

Construction is an industry that, by its very nature, contains danger if conditions are not aggressively addressed. Owners can protect workers and themselves—if they are willing to lead the way and confront the risks head-on. Dynamic leadership by individual owners can help reduce injuries, disabilities, and deaths caused by construction accidents. Project safety will be only as good as that of the poorest performer on the job site. Selecting contractors with a demonstrated record of safe performance is perhaps the most effective means of improving the odds of having a safe job site. Both project managers and owners are becoming increasingly aware, as they experience the ever-increasing rise in injury claims and accident costs, of the need to be more selective in their choice of contractors. Safety practitioners, owners, and general contractors who regularly use contracted services now advocate the consideration of contractors' past safety performance in their selection criteria.

It has generally been the practice in construction to base the selection of contractors solely on the lowest bid. Consideration of safety in the selection process tends to be the exception rather than the rule. Why this continues to be so is difficult to understand.

Rationalizations such as, "I am compelled to accept the contractor with the lowest bid" or "I am not permitted to pre-select contractors, so I can't reject a contractor based on past accident experience," can be overcome and should not be impediments to the selection of safe contractors. Technical selection criteria that include contractors' past safety performance can be used in order to evaluate safety suitability as well as quality and technical competency.

In its A-3 report published in 1989, the Business Round Table (an organization whose membership includes chief executive officers and chief operating officers of 200 of the leading Fortune 500 businesses) states that consideration of safety in the bidding process measurably improves safety performance. Further, the report indicates that contractors with good safety performance are more efficient in the execution of their other functions.

Construction Users Round Table studies have shown that owners who proactively seek out and hire safe contractors experience better safety results on their projects. Contractor prequalification requires significant owner effort and project calendar time, but it ultimately pays off in reduced numbers of injuries, increased productivity, and lower project costs.

Consideration of safety is successfully included in the contractor selection process by enlightened organizations. The question that begs to be answered is why safety is not given more weight in the contractor selection process by more organizations. Surely it makes sense to choose to work with contractors who have a demonstrated track record of good safety performance, thereby decreasing the odds of having accidents on the job site.

The rising incidence of accident litigation and escalating workers' compensation costs are getting corporate attention, and corporations are starting to hold job site managers accountable. Those host employers who have not yet incorporated safety into their selection criteria are beginning to look for solutions. The following example demonstrates the effect that accidents have on contractors' bottom line.

Example

This scenario compares the net profit for three carpentry contractors with different accident experience rates. Consider that each contractor is retained for a job requiring 400,000 man-hours of effort ($8 million in labor costs).

Example Company	Lost-Time Rate	Lost-Time Cases	Average Claims plus Indirect Costs (@ $25,000/Case[3])	Profit Margin 5%	Net Profit
A	3	6	$125,000	$400,000	$275,000
B	5.4	11	$275,000	$400,000	$125,000
C	9	18	$600,000	$400,000	($200,000)

Let us assume that the industry-average lost-time rate for the carpentry trade is 5.4 cases per 200,000 man-hours worked. At a 5% profit margin on $8 million of labor costs, this would provide each contractor with a potential profit of $400,000. The medical expenses in this example reflect the average cost of lost-time accidents on net profits, as reported by the National Safety Council.

In this case, Contractor C will have to allocate more financial resources to deal with job-related accident costs. As a result, fewer funds will be available to cover material and labor costs. Not generally known as philanthropic organizations, contractors have little choice but to pass on costs resulting from past losses to future customers. Contractors whose work force experiences high rates of injuries also bear the burden of higher insurance premiums than do their safer competitors.

---

[3] National Safety Council 1997 Estimate of Lost-Time Case Total Costs.

If a contractor has a high overhead resulting from past accidents, where are the savings to be achieved in a competitive bid environment? There are really few areas where savings might be realized:

- cheap (inexperienced) labor
- distributed (frequently absent) supervision
- inexpensive (substandard) material
- anticipation of numerous change orders and filing loss claims.

No one wants to see others lose money, but neither do they want to pay for losses contractors sustained on previous projects. Occurrences resulting in injuries potentially increase the risk of schedule delays and the additional administrative burden necessary to deal with litigation that often follows personal injury occurrences.

Job site managers generally concur that contractors who effectively meet technical specifications typically require less oversight than do those who are known to have failed to meet requirements on past jobs. Similarly, contractors with a high incidence of injuries can be expected to continue to experience losses at a similar rate in future work. Consequently, contractors with poor safety records require additional scrutiny and supervision. Further, they may put owners and their employees at risk of injury as well. Which contractor would an owner want on its job site? Which is the least expensive contractor in the long run? Most likely, owners would prefer to work with the contractor with a low rate of accidents.

Contractor Screening

To ensure a safe job site, establish a process to screen out contractors with unacceptable safety practices. Contractors providing low-risk services such as restocking food and drink machines, delivery of laundry, copy machine maintenance, and the like generally do not need to be subjected to the rigorous detail that would be applied to the screening of a piping contractor expected to work in a refinery environment. There are two approaches to contractor screening. It is particularly important to select safe contractors for high-risk jobs such as construction, where loss exposures are the greatest.

Safety as a Technical Evaluation Criterion

Just as selection criteria are established to evaluate contractors' capability to produce a product to defined technical specifications or to evaluate whether they can demonstrate the financial robustness to remain in business throughout the duration of the contract, so should criteria be established to judge the contractor's ability to complete the work without personal injuries that would expose the host employer to subsequent litigation.

The effectiveness of the contractor's risk reduction and loss control practices should be the basis for contractor safety selection criteria. The principle being applied in this selection process is that organizations, and the individuals in them, will continue to behave as they have in the past. They are unlikely to change their work practices unless something significant forces them to do so. It is unlikely that a contractor with a history of high injury rates will suddenly stop having accidents simply because the contract states that a high degree of importance is placed on safety.

Commonly used measures of contractor safety performance effectiveness include the following:

1. experience modification rating
2. injury frequency and severity rates
3. safety program evaluations and evaluation of key personnel
4. OSHA citation history
5. references from others who have employed the contractor previously
6. evaluation of the contractor's integration of safety into his or her work practices.

Experience Modification Rating

Criteria: Experience modification rate (EMR) of 1.0 or less.

EMRs are calculated annually by each policyholder's insurance carrier. The calculation is based on past claims experience as compared with the average claims submitted by other policyholders in the industry in their respective states. The higher the policyholder's accidents experience and number of claims filed, the higher its EMR. The higher its EMR, the greater its premium paid to its insurance carrier.

The use of EMR rates as an evaluation criterion is unbiased because the contractor's own insurance company calculates the rate. An EMR of 1.0 indicates the company submits injury claims at a rate that is considered average for its industry group. A rating greater than 1.0 indicates the company files more injury claims than other, similar organizations. A rating of less than 1.0 means the company has filed fewer injury claims than the average contractor in its industry.

It is becoming common practice to request that contractors submit their EMR rates for the past 3 years as a part of the evaluation criteria to assess the contractor's relative accident experience and to determine his or her accident experience trend. Because the objective is to work with contractors whose accident experience is as least as good as the average in their industry, the threshold acceptance criterion is generally an EMR of 1.0 or less. High-risk environments such as petrochemical plants generally set more demanding criteria and may accept only contractors with EMR rates of 0.8 or lower.

The comparison of contractor EMRs should be limited to those companies within the same industry group (Standard Industrial Classification [SIC] code) and only those from the same state. EMR rates for contractors with the same relative injury and claims experience may vary among states based on their interpretations of reporting requirements and the general work culture within each state. A contractor from Indiana with an EMR of 1.0 may have a different lost-time accident rate than a contractor from Michigan with an EMR of 1.0. The insurance industry recognizes this fact by varying insurance rates from state to state.

An important fact to bear in mind when evaluating EMR rates is that they can also be manipulated to some degree. An organization with a high deductible will file fewer claims than one with a lower deductible. In this situation, it will appear that the organization has experienced fewer injuries. A tactic of devious firms with high EMRs is to dissolve and then re-incorporate under a new name. They then resume business with an EMR of 1.0 and continue with their unsafe work practices for 3 years.

Injury Frequency and Severity Rates

Criteria: Accident experience and rates equal to or less than industry experience for similar industries as reported by the Bureau of Labor Statistics (BLS) or the National Safety Council.

Historical injury information provides a retrospective view of the level of safety practiced by a contractor. The evaluation criteria commonly used by most organizations in the United States include the following:
- fatalities
- injury cases resulting in lost workdays
- workdays lost as the result of injuries
- injuries resulting in restricted work
- number of workdays of restricted work
- cases requiring medical attention.

Injury Case Rates. To equitably compare injury and illness experience, injury cases are converted into incidence rates that are based on man-hours worked. Although contractors are usually asked to provide their incident rates, it is often the case that they do not know how to calculate them.

To verify incident rates, it is common practice to request contractors to provide raw accident information numbers and the number of man-hours worked for each of the past 3 years to verify the rates quoted.

Incident Rate Calculation. Incidence rates are calculated using the following generally accepted formula:

$$\text{Incidence rate} = \frac{\text{Number of incidents} \times 200{,}000}{\text{Number of hours worked}}$$

Contractors should be asked to provide injury and illness data from all sites where work has been performed. If the contractors have affiliated businesses (e.g., hourly workers in one and white-collar/management in another), that information should be provided as well. Contractors should have this information on hand because all companies with 10 or more employees are required to maintain an OSHA 300 log. This log is a record of illnesses and occupational injuries that have occurred during the past calendar year.

Clarification of Definitions. When asking contractors to provide their injury incident rates, there must be a clear definition of terms and equations used to calculate them.

Corporate Safety Philosophy. Another relevant inquiry is to determine the corporate philosophy of assigning injured employees to restricted or light-duty work. Some organizations return injured employees to work as quickly as possible after accidents, even if the employee's productivity is limited. Others prefer that injured employees take as much time as is legitimately needed to recuperate before returning to work. These two very different approaches have significant impacts on frequency and severity rates.

Both physical absence from work and restricted work following an injury are considered lost time by OSHA. Organizations with aggressive injury management programs are increasingly encouraging employees to return to work as soon as they are able, even if in a modified duty capacity. It is their experience that such employees return to full productivity sooner than employees who delay their return to work.

Further, the definition of time away from work must be understood, particularly in international construction environments. In some locations, counting lost workdays starts with the next regularly scheduled workday following the injury. Other locations do not begin measuring "lost time" until the employee has been off work for at least 3 workdays. Adding to the potential confusion is the practice of some companies of switching employees' days off to minimize lost-time reporting.

Head Count. The method by which contractors determine the number of "employee hours worked" used in loss-rate equations is also a relevant subject for inquiry. Does the number of man-hours used represent all employee hours worked by subcontractors, or does it represent only office workers? Office workers are exposed to fewer hazards and, therefore, have fewer accidents. Including home-office work hours, while perfectly legitimate, will reduce loss rates. It is not unreasonable to include home-office hours as long as the basis of comparison is consistent among organizations. Consider requesting contractors to provide their man-hours in two segments—office and trade hours—to provide the information for an equitable basis of comparison.

Types of Accidents. Ask contractors to provide a description of what caused their accidents and what they did to prevent the reoccurrence of those incidents on future jobs. Do they have a proactive program in place to continuously improve their safety program?

Is the contractor experiencing the same type of occurrences repeatedly? What is causing the accidents? Are injuries occurring repeatedly to the same work crew supervisors? If so, ownership certainly would not want those particular supervisors on the job site even if the contractor's experience as a whole meets the established evaluation criteria.

Safety Program Evaluations

Criteria: Integration of safety into work practices and guidelines to do so.

A third evaluation measure is to analyze each prospective bidder's safety program and his or her compliance with safety regulations in the field. The evaluation of a contractor's safety program documentation provides useful insight into contractor awareness of hazards inherent in the work and in the environment in which the work is performed. The safety program should provide evidence of the necessary controls and accepted work practices to perform the work safely. If the program does not address this effectively, the organization may not have effectively analyzed its work or appropriately educated its employees.

Although the absence of a documented safety program does not necessarily mean that safety is not being integrated into work planning and execution, it may be an indication that there is no uniform approach or standard by which safety is managed. This is not to suggest that every line manager must manage safety exactly the same way. However, for a safety program to be effective, there must generally be a common philosophy regarding safety implementation. Otherwise, there exists the potential of conflict and error.

The evaluation of safety program documentation should verify that contractors are conducting regular worksite inspections and are scheduling specialized inspections as required. Inspection reports should be evaluated for consistency and frequency of inspections, thoroughness of the inspection, clarity of the report, appropriate classification of identified hazards, and timely and appropriate remedial actions. Specialized inspection records such as preventive maintenance and pre-use inspections should be checked to ensure their regular application.

Benchmarks against which to evaluate general corporate safety documentation are 29 CFR 1926 and the American National Standards Institute's construction and demolition standards (ANSI A10.33 and ANSI A10.38).

These references contain direction regarding what industry consensus has defined to be minimum criteria for effective safety programs.

Some organizations hire consultants to conduct annual program reviews. This brings in an unbiased set of eyes to evaluate the program. Another alternative is to invite individuals from other worksites to conduct mutual inspections. If auditors identify problems that routine inspection reports have not addressed, the problem may be that the individuals conducting the routine inspections need more training in recognizing hazards and unsafe work practices. If the same observations are made repeatedly, there may be a problem with the process of controlling hazards that results in the hazard recurring.

A safety program evaluation is clearly more time-consuming than analyzing injury statistics, but it is ultimately more revealing, as it enables the reviewer to focus on the contractors' safety practices in a work environment relevant to the work to be done. The object here is to determine if contractors *manage safety and if it is part of their routine work practices* as opposed to *managing safety as an afterthought of their work process*. Remember: the chosen contractor is going to be a partner in the business for some period of time, and no owner wants to work with an organization that takes safety less seriously than it does itself, thereby potentially disrupting the project's execution.

OSHA Citation History

Criteria: Absence of a negative trend of OSHA citations.

Another source of information regarding a contractor's past safety performance is the evaluation of his or her OSHA citation history. This information is often neglected out of concern about requesting confidential information. In fact, OSHA inspection records are public record and can be obtained directly from OSHA through the Freedom of Information Act if the prospective bidder is reluctant to provide this information.

Just as the absence of citations does not indicate outstanding safety performance, the presence of citations should not be an automatic basis for rejection. Where there are repeated citations for the same type of violation or regularly occurring citations that span some extended period of time, there may be cause for concern.

If a contractor has been the subject of regular OSHA scrutiny and citations, he or she may not be a desirable contractor for the job site. The presence of that contractor may motivate OSHA representatives to consider visiting the site more frequently than they might have in the past.

References from Previous Employers

Criteria: Were previous employers satisfied with the contractor's safety performance?

Ask for references. Was the contractor responsive? What problems did employers experience that might occur if that contractor should come to work on the new project?

Integration of Safety on Current Jobs

Criteria: Demonstrated effectiveness of integrating safety into current work practices on current jobs.

The most effective means of evaluating a contractor's ability to work safely is to visit a job site where he or she is working to see how he or she performs work. If the new project requires the use of cranes, take a look at the cranes the contractor is using. What condition are they in? How are they being managed? Are they equipped with load moment and anti-two-blocking devices? The condition of the work area and how the contractor manages the work will provide some insight into the corporate culture regarding the standard of care a contractor applies to his or her work.

Interview the prospective contractor. Does he or she have a documented safety program? If so, does the program address the hazards to which workers will be exposed, and does the program comply with applicable regulations? Can the contractor/manager produce his or her documented safety program? The attitude toward safety of the prospective project manager and superintendent is perhaps the most critical factor in the evaluation of a construction manager or general contractor. There are only a limited number of "A" players available, and it is important to ensure that the individuals an approved, selected organization's office sends to a job site have the right attitude as well.

Therefore, these criteria should be applied to both the contractor and the performance of the contractor's key personnel, project manager, superintendent, and designated safety representative, who will directly influence the work on the job site. What better means of improving the odds of good safety performance on a project than the selection of organizations and individuals with proven track records?

Request for Proposal and the Pre-Bid Meeting

Having established selection criteria and developed the request for bid (RFB) documentation, prospective bidders should be invited to a pre-bid meeting. Prospective bidders should be provided with an opportunity to review the RFB requirements, ask questions, and obtain clarifications as directly as possible.

A clear and concise definition of safety requirements in bid invitation documents is an important element in effective contractor safety management. Prospective contractors must understand the project safety requirements before they can develop a realistic bid. Implied requirements or requirements that are not clearly stated in the bid documents may become the source of conflict later. Many conflicts that develop during the contracted work process can be traced to misunderstandings and miscommunications regarding requirements and standards that the job site manager considers relevant.

Where a preferred bidders list has not been developed and the lowest apparent bidder is to be evaluated, the safety evaluation criteria to be used should be defined prior to the start of the contract bidding process. A statement to the effect that safety will be included as one of the technical evaluation criteria should be included in the RFB. It is only fair that prospective contractors be made aware, prior to investing time and effort in bid preparation, that their past safety performance will be evaluated. Contractors who recognize that they have a poor safety record or a weak safety program may opt not to participate in the bidding process with the knowledge that there exists a real possibility that they may be rejected based on poor safety experience. With this knowledge, if, in the event the lowest apparent bidder is not awarded the contract, there will be no basis for the low bidder to contest the bid evaluation by claiming he or she was unaware of the safety selection criteria.

To ensure that prospective contractors have the information they require to develop estimates to address job site hazards, some organizations go so far as to include phase analysis evaluation as an addendum to their RFB. It may not be necessary to include the entire analysis, but the RFB should address at least the following items:

1. work to be done
2. identified hazards and work restrictions
3. work permits and license requirements
4. contractor safety qualification requirements
5. special job site safety program requirements
6. contract management arrangements
7. orientation and training requirements
8. audits of contractor performance
9. onsite control of work
10. code compliance requirements
11. record-keeping and reporting requirements
12. contract termination/completion criteria.

Work to Be Done

The work to be done should be described in sufficient detail to enable prospective contractors to understand the content of the work required as well as the work environment.

Identified Hazards and Work Restrictions

Unusual hazards and potential restrictions should be brought to the contractor's attention in the RFB and repeated during the pre-bid meeting. It may sometimes be difficult for prospective contractors to get an accurate picture of the scope of work in a particular project by just reading the contract documents. A good way of communicating this information is to include a worksite visit in the "pre-bid meeting" agenda. This provides the prospective bidders with an opportunity to see the job site and ask questions relevant to their bid development. Make sure that all requests for clarification are also submitted in writing. Then provide written answers to all the prospective bidders.

Work Permits and License Requirements

Work permits and licenses required by the site or by local code should be identified. The methods by which they will be issued and controlled should also be addressed. Some areas to be considered are:
- onsite entry and security
- vehicle passes
- confined-space entry
- excavation
- lockout/tagout requirements
- working hot (electrical)
- explosive devices
- hazardous materials handling and transportation
- hazardous-waste disposal
- radiation source management
- environmental permits.

The RFB document should indicate that the successful bidder and his or her subcontractors will be responsible for the enforcement of safe work practices outlined in the permits and licenses, as well as all other site or local code requirements that apply to the work under his or her control.

Contractor Safety Qualification Requirements

Contractor qualifications are those criteria the selection team will use to evaluate the prospective contractors and their bids.

Special Job Site Safety Program Requirements

The job site manager must define all site-specific safety requirements, which might include specific types of tools, restrictions on certain work practices, special lockout procedures, or site-specific fall protection requirements. The contractor and his or her subcontractors should provide detailed documentation that demonstrates that their safety programs meet the job site manager's requirements.

Contract Management Arrangements

The RFB document should define the job site organization hierarchy and the key contact points. Critical questions that should be addressed include the following:
- What is the job site organization structure?
- Who within the host's organization is responsible for contract coordination and administration?
- Who will address safety and health issues?
- What technical support, if any, will be provided?

- How will work delays, challenges, accidents, and similar events be reported to the job site coordinator?
- What information must the contractor report to the job site coordinator during the life of the contract, and to whom should the reports be made?
- What anticipated hazards will be present on the job site, and how will the host control them?

Orientation and Training Requirements

The RFB should identify specific training requirements that the job site manager feels are necessary for successful completion of the work. The contractor should be required to demonstrate, to the satisfaction of the job site manager, that his or her employees and subcontractors have completed the site orientation and job-specific training, as well as any other qualifications specified by the job site manager.

Some host organizations prefer to provide and manage certain orientation and training requirements they feel are particularly important prior to granting contractor employees permission to be on the worksite, rather than relying on the contractor to communicate this information.

Audits of Contractor Performance

The RFB should specify that the job site manager will conduct periodic assessments of the contractor's safety and health program implementation during the contract period. A copy of the audit protocol and performance requirements should be available if the contractor would like to see what would be evaluated. The job site manager should carefully consider the practicality of each type of measurement in consideration of the length and risk level of each individual contract.

Onsite Control of Work

Contractors should be expected to have a process in place to monitor for unsafe work practices. Appropriately trained individuals must do this. Specific requirements are now in place to recognize the individuals who are able to competently inspect specialized work such as excavations and scaffolding. The contractor should be asked to explain the basis upon which these individuals have been determined to be competent to inspect the work. The contractors should also be aware that they will be expected to provide for direct supervision of employees at all times. Ideally, the contractor should provide senior managers, supervisors, and onsite safety coordinators with training in hazard recognition and safety and health management.

Code Compliance Requirements

The RFB documents should identify specific code requirements and relevant regulations, including licensing of individuals and certification of equipment.

Record-Keeping and Reporting Requirements

All records and reports required by the job site manager pertaining to the contractor's safety program should be identified. Such records generally should include:

- weekly personal injury reports and man-hour reports
- records of personnel training
- documentation and follow-up of identified safety problems
- inspection reports
- accident/incident investigation reports
- equipment inspections
- safety meetings.

Standard forms to be used for records and reports may be specified in the RFB; however, the major concern of the job site manager should not be whose form is to be used, but that the critical information on each issue is provided.

Contract Termination/Completion Criteria

The RFB documents should clearly define contract penalties and termination procedures in case of substandard safety performance. The documents should also specify when and how disciplinary measures will be imposed. Consideration should also be given to defining what will be required to make the determination that the contract has been satisfactorily completed.

Preference for contractors with good safety experience will improve the probability of completing a project with a low injury incidence rate and the absence of major mishaps. If the selection of preferred bidders is not an option, then safety should be an integral aspect of the contract technical selection criteria.

Because construction is considered to be a potentially high-risk environment, contractor selection is being increasingly advocated in consensus standards such as ANSI A10.33 and A10.38 and the OSHA Process Safety Management (PSM) standard (29 CFR 1926.64; see also 29 CFR 1910.119).

A detailed job plan, which identifies risks inherent in a project, is critical to the successful management of contractor safety and health. If the "work to be done" or the expected performance standards are not set out clearly and completely prior to bid request, prospective contractors may submit inadequate bids.

Legal counsel should become involved in the development of the contract language as it relates to safety requirements to make sure they are clearly defined. The true test of senior management's understanding of the value of safety—and supporting it—is being able to pass over the low bidder with a poor safety record in favor of the second or even third higher bidder able to meet the safety criteria requirements.

Thus, the value of selecting safety-conscious contractors includes less exposure to injuries and insurance claims, greater control of hazardous conditions in the workplace, increased productivity, improved morale, and decreased liability to third-party lawsuits and contractual disputes.

Enlightened owners and project constructors will soon no longer accept generic safety manuals as evidence of a corporate safety program. The more proactive organizations are insisting on job-specific safety documentation. Project managers want to see evidence of safety documentation directly related to the work to be performed on the job at hand.

Evaluation Criteria for Contractor Selection

The following is an example of technical selection criteria established for the selection of a contractor for the fabrication and installation of a nitrogen distribution system. The job site manager would be expected to conduct technical evaluations based on the criteria listed in the section entitled "Technical Criteria." Criteria 1 through 5 are of equal importance and will be scored on a "go/no go" basis. Only those offers scored "go" for each criterion will be considered *technically acceptable* and therefore eligible for award. Price will be evaluated in accordance with the section entitled "Price and Price-Related Criteria."

Technical Criteria

Criterion 1—Safety. The offeror must demonstrate the existence, within the offeror's organization and that of any selected subcontractor, of a satisfactory safety record that, in the job site manager's view, is sufficient to successfully complete this project. Note: To satisfy this requirement, the offeror must provide documentation from the offeror's and any subcontractor's workers' compensation carrier that should reflect a workers' compensation experience modification rate (EMR) of 1.0 or less and meet defined selection criteria.

Criterion 2—Past Performance. The offeror must demonstrate the successful performance of at least two projects that, in the job site manager's view, are substantially similar to the scope of this requirement within the past 3 years. Note: To satisfy this requirement, the offeror must provide details of each successfully completed project or individual product grouping, valued in excess of $100,000 (or a value commensurate with the size of the project contemplated), to include the following information:

1. project description
2. dollar value
3. inclusive dates of performance
4. clients who may be contacted as references (including name, address, and telephone number)
5. performance records with regard to delivery within schedules and quality of performance
6. explanation of the relevance of cited projects to this effort.

Criterion 3—Welding Capabilities. The offeror must demonstrate the availability, within the offeror's organization or that of any selected subcontractor, of a welding program that, in the job site manager's view, is sufficient to successfully complete this project. Note: To satisfy this requirement, the offeror must provide copies of the offeror's or subcontractor's welder qualification program and transfer line construction welding procedures.

Criterion 4—Cryogenic Cleaning Capabilities. The offeror must demonstrate the availability, within the offeror's organization or that of any selected subcontractor, of a cryogenic cleaning program that, in the job site manager's view, is sufficient to successfully complete this project. Note: To satisfy this requirement, the offeror must provide copies of the offeror's or subcontractor's cryogenic cleaning procedures.

Criterion 5—Quality Control. The offeror must demonstrate the availability, within the offeror's organization or that of any selected subcontractor, of a quality control program that, in the job site manager's view, is sufficient to successfully complete this project. Note: To satisfy this requirement, the offeror must provide copies of the offeror's or subcontractor's quality assurance manuals and/or plans that should detail quality assurance procedures to be applied to this project and that substantially incorporate ANSI/ASQC-C1, Specification of General Requirements for a Quality Program.

Price and Price-Related Criteria

Price will not be numerically scored or adjectivally rated. However, price and price-related factors specified in the solicitation are considered more important than the technical criteria listed previously in the job site manager's overall evaluation and will be the controlling factor for award *when considering all technically acceptable offers.* Technical acceptability is defined as any offer that was rated "go" on all technical criteria listed in the previous sections. (Note: This criterion was actually used for the installation of a nitrogen supply line for a laboratory facility.)

ROLE OF THE FIELD ENGINEER

It is the field engineer's job to make sure the subcontractor clearly understands the technical requirements of the work and to monitor the contractor's integration of safety into the work execution. The field engineer is the contractor's "go-to guy" for information regarding where to go and when to be there to perform work. Without this support, the contractor's efforts will not be applied efficiently during the often-short period workers are on the job site. In some environments, contractors can unknowingly place themselves in potentially dangerous situations without the guidance of a facility representative.

Even with the prescreening process, site-specific safety program reviews, and safety orientations, the host employer cannot be certain that the contractor will perform within defined job site safety standards. At this point, the human element must be introduced in the form of the field engineer. Contrary to the fact that many contractors believe, or would like to believe, that their host employer is solely responsible for all work-related safety and health matters, it is each contractor's (employer's) responsibility to see to the safety of his or her respective employees. The role of the field engineer is to ensure that the contractors understand the job site hazards, are clear as to what is expected of them, coordinate intercontractor efforts, and, finally, oversee compliance with the contract-specified safety requirements.

Roles and Responsibilities of the Field Engineer

The role of the field engineer can be boiled down to the following functions:
- Identify site-specific safety hazards to contractors.
- Establish that the contractors recognize the hazards and are prepared to deal with them.
- Coordinate the interfaces between contractors.
- Coordinate the interfaces between contractors and operating facilities.
- Verify that the contractor is performing to agreed-upon contract requirements.

Work Release Meeting

Before approving the contractor's *notice to proceed,* the field engineer must be confident that all required planning has been completed and prerequisite documentation is in place. The field engineer responsible for overseeing the work should hold a meeting with the contractor's supervisors as soon as the following actions have been completed (but before the start of any contracted work):
- Contract is signed.
- Performance, payment bonds, and insurance have been secured.
- Safety and health program is approved.
- Job-specific safety analysis is reviewed.
- Schedule is approved.

At this pre–work release meeting, the field engineer should confirm that the key contractor and facility personnel have a clear understanding of the contract scope, as well as job-specific hazards and safety requirements. The meeting agenda should review how the contractor and facility will do business with each other, the procedures the contractor will follow, and any facility-specific requirements that must be met, before the start of and during the performance of the work. A meeting agenda should be distributed to the participants before the meeting.

Meeting minutes with a summary of items discussed should be distributed to all attendees along with a list of principal contacts for the project (for both contractor and facility) with workday telephone numbers and off-hour emergency contact names and numbers. The minutes should be transmitted within 5 working days following the meeting.

Equipment Inspections

All heavy equipment brought on site should be inspected to determine that it will operate throughout its expected stay on site. The field engineer responsible for the inspection should be able to call on an individual with the technical expertise to conduct the inspection if he or she is not familiar with a specific piece of equipment. The purpose of this inspection is not to certify the equipment, but to verify that it has been certified and appears to be in good working condition.

For example, an equipment inspection might include verification of:

- seat belts
- lights
- horns
- backup alarm
- windshield in place and free of defects
- absence of leaks
- fire extinguisher
- rollover protection (where necessary)
- cranes with anti-two-blocking and cables in good condition
- inspection certification
- operator medical exam
- operator certification
- load chart
- angle indicator
- motion alarm
- proximity alarm.

Invariably, very sorry-looking equipment shows up on new projects. However, once a few cranes and heavy pieces of equipment have been denied access to the job site based on the justified determination that the equipment does not meet job site standards, the word will get around and serviceable equipment will arrive at the gate. The other equipment will be sent to less-demanding clients. The last thing a job site manager wants on site is a piece of equipment that may fail in the midst of a work activity, thus disrupting the workflow of the project. It is better to invest time in work preparation and ensure that everything is in place than to have to take remedial steps to correct a failure.

Job Site Monitoring

The degree and quality of coordination provided by the field engineer have a direct effect on contractor safety performance. The greater the support provided to the contractors, the better the contractors will understand the job site requirements and the more productive they will be. The degree to which the field engineer emphasizes safety will directly affect the emphasis the contractor places on safety performance. Field engineers must emphasize safety in their daily communications with contractors and supervisors and when they have the opportunity to interface with the workers.

Inspections

The field engineer should maintain a field notebook to record daily observations. In addition to quality, schedule, performance observations, and records of job site conditions, the field engineer should monitor safety compliance. Job site hazards that cannot be corrected immediately should be noted. The corrective action taken to address safety issues should be recorded.

Deficiencies such as masons overloading their scaffolding or missing toeboards that should be in place should be brought to the contractor's or supervisor's attention for correction. This observation should be entered in the field engineer's notebook as a record of the conversation. It puts the contractor on notice and provides the field engineer with a record of the event. If the situation surfaces again, the field engineer has the documented basis for beginning formal corrective action.

In a contracted work environment, it is not generally good form to give work direction to subcontractor employees, so it is a good idea to include the contractor or supervisor in routine job site inspections so that observed deficiencies can be identified and corrected on the spot.

Job site progress photos and videos provide a good visual record of job site conditions. Photographs are indisputable evidence in the event of an argument or can be used at job progress meetings to clearly point out housekeeping, worksite conditions, or specific work practices that need to be addressed. The availability of electronic cameras and computers enables photographs and overhead viewgraphs to be generated with little difficulty.

Enforcement

Adherence to established safety standards is a significant factor in the level of safety in the working environment. An important role of the field engineer is to monitor general work practices and bring safety discrepancies to the attention of the respective employer and its employees.

Any observations of undesired safety behavior should be noted in the field engineer's notebook. This is not the written notice issued the second time a substandard behavior is noted. The field engineer's record of the first approach will diffuse any argument regarding whether the worker was previously corrected for a similar infraction.

> "Bill, this is the second time I have seen you this week not wearing your safety glasses; we spoke about it before. You know it's a project requirement to wear safety glasses. Look, I don't want to make a federal case out of this, but I am going to make a note that I am giving you a formal notice. Would you please initial in my book that we talked about this? Bill, please wear your safety glasses from now on."

Document all safety conversations.

The field engineer should not feel he or she is being a "nice guy" by walking away from a safety issue that could hurt a worker later. Remember: what is condoned today will have to be accepted tomorrow. The objective of a proactive safety program is not to threaten the workers or their jobs, but to encourage them to work safely so they will be there the next day.

Accident Investigation and Reporting

Investigations are generally relegated to the contractor to perform. However, the field engineer should participate in the investigation process, even if it is in the role of an observer. This will provide insight into why the incident occurred, help prevent a similar incident from happening again, and ensure that proper documentation is collected for the safety file. The investigation of occurrences (accidents) is the only way the source of the incident can be determined. Investigations should not be a hunt for the guilty perpetrator, but a determination of why the incident occurred in the first place.

Progress Meetings

People generally agree that most meetings are poorly organized and run. Planning and coordination are key to the success of meetings.

Produce an agenda for each meeting. This will provide focus and demonstrate preparation for the meeting. An agenda will also help maintain the focus of the meeting.

Schedule the meeting for a specific duration, and let the attendees know what that time frame will be. Then stick to the timetable. If everything is not addressed in the allocated period of time, identify what issues were not covered and propose to address them at the next meeting. Do not go over the meeting time unless it is absolutely essential—and then only with the permission of the participants.

Meetings should never last more than 1 hour. Participants may not come back if meetings are longer than that or if the meetings consistently run over the scheduled time. Remember: the participants may have something else scheduled following the meeting. If the subject can be addressed in less time than what was scheduled, state that the topic has been sufficiently covered and then adjourn. Attendees will appreciate this and will be more willing to attend when another meeting is scheduled.

Post the objective of the meeting in a prominent location in the meeting room. It helps everyone maintain his or her focus on the meeting objective. If the conversation strays, the objective helps draw back the focus.

Work Safety File

The field engineer should maintain a job safety file. This should contain the records that will be transferred to the client or originator when the work is done. It should contain all the information the facility operator needs to know regarding the construction of the facility:

- as-built specifications and drawings
- design criteria
- general details of construction methods and materials used
- potential hazards such as locations of post-tensioned tendons
- manuals produced by equipment suppliers and specialist contractors that outline operating procedures and maintenance schedules for installed equipment
- requirements for cleaning and repair of components
- locations of utilities and service connections
- sources of potential risk to operations and maintenance personnel that were identified during the contracted work process and that should be brought to the client's attention and highlighted in the file (e.g., confined spaces).

The field engineer should discuss the proposed contents of this file with the originator/client before the work begins to identify if there is any additional information to that planned that should be included in the file. The field engineer should collect this information as the work evolves.

Prudent host employers recognize that it is not enough to simply exercise reasonable care in the initial selection and employment of specialty contractors to perform contracted work. To effectively prevent accidents when working with contractors, the host employer undertaking the project must monitor the work for the duration of the contract. The individual generally assigned the responsibility to monitor the contractors is the field engineer.

CONTRACT CLOSE-OUT

A job is not done until the paperwork is complete. This is also true with regard to managing construction safety. Imagine the following scenario: 2 years after the project completion party ended and the keys to the facility were handed to the company president, a company is served with papers to appear for a discovery deposition regarding an accident that occurred to a contractor employee at the site during construction. What is known about that occurrence? What information is available to deal with that inquiry? Is there access to the following information to prepare for the deposition or for testimony if the case goes to trial?

Construction Contract Environment, Safety, and Health (ES&H) Requirements

Before ending a construction project, each project manager should ensure that environment, safety, and health requirements have been addressed. Frequently, use of a list would minimize the chances of missing an important action item during the close-out stages:
- personnel injury reports
- project safety statistics
- site-wide and subcontractor safety meeting minutes
- inspection reports and progress photos
- job safety assessments (JSAs)
- Safety Data Sheets (SDSs)
- employee orientation records
- toolbox talks and sign-off sheets
- weekly man-hour reports
- engineers' field notes and project notebooks
- notices of unsafe acts
- work permits
- disciplinary notices.

The orderly closeout of contracts and gathering relevant documentation will minimize administrative difficulties down the road. Having access to the right documentation will certainly minimize future headaches. Once a contract is closed and any remaining retention is released, leverage to get information from contractors is gone.

Contract Compliance

What deliverables have been specified in the contract language that relate to safety?
- man-hour reports
- personnel injury reports.

Be sure project safety statistics have been maintained and there is a complete file of all accident reports by date so that a contractor does not slip out, leaving gaps in the records. The absence of man-hour information will lead to inaccuracies in the project's safety statistics.

Acceptance Certifications

Has all building and system-related documentation been completed and delivered? Are the following tests and documents on file?

- Have all operating system tests been completed?
- Have all acceptance tests been witnessed and documented?
- Have all operations manuals and warranty information been delivered?
- Are system training sessions completed?
- Have the as-built drawings been completed and submitted?

After completion of the facility, operation manuals can be a useful information source for operations personnel to develop their plant operations safety program. Operation manuals are also useful when developing the facility safety inspection checklist.

Loaned Equipment and Permits

What equipment does the contractor have on loan that should be returned? Although it is not generally the custom to loan equipment to contractors, there are certainly situations in which it is expedient to do so, such as the loan of two-way radios. Pagers are another means of locating specific personnel. An owner may have an iron-clad policy that lockout locks may not be removed other than by the person who installed the lock. In such a case, make sure that a contractor has transferred responsibility for any remaining locks to a facility person. Be sure the following items have been returned or addressed:
- site access identification cards
- radios and pagers
- ladders
- keys
- open permits have been closed
- lockout locks have been removed.

Safety Program Documentation and Retention

In today's litigious society, there is little need of a discussion regarding why project-related safety records should be retained. The challenge is to determine what should be retained. A corporation may already have guidelines in this regard; otherwise, the corporate legal counsel will surely have an opinion. Ask the following questions:
- What information should be retained?
- Where should it be stored?
- How long should it be retained?
- Who will know where to find it?

Information that will be requested first, in the event that an injury claim is filed against the firm, will be a request for documentation of the safety program and its structure (i.e., program documentation, permit and inspection forms, training syllabuses, safety posters, etc.).

The historical information that relates to the implementation of the program should include information such as the following, which are evidence of the execution of the safety program:

- inspection reports
- weekly site safety audits and progress photos
- safety meeting minutes
- job safety assessments (JSAs)
- toolbox talks and sign-off sheets
- employee orientation outline and attendance records
- approved safety variances
- enforcement/reinforcement documentation
- safety notices.

Safety Statistics

At the conclusion of a project, there are always close-out meetings and presentations to advise the client how its money was spent and the virtues of the product it is receiving. Safety statistics and related project experience should be an integral part of that information:

- man-hour reports and summaries
- log of summary and occupational illnesses and injuries
- commendations and citations
- copies of articles that reported on the project's successes.

Accident Reports and Personal Injury Information

Is there a complete file of all accidents reported and investigated? These can be filed either by contract or chronologically:

- employer's first report of injury
- supervisor's accident investigation report
- project accident investigation reports
- all other claim-supporting documentation.

Other Safety-Related Information

Did the project engineers keep daily logs of their field activities? Where are they? These sorts of documents have been quite useful in mounting an effective defense to demonstrate the active involvement of field engineers in overseeing safety on the construction site:

- field engineer's daily logs
- job progress photos/videotapes
- field inspection reports and related hazard abatement information.

Work and Safety Planning Information

It is important to maintain records of safety-related efforts from the project.

- safety meetings: site wide and contractor specific
- safety-related planning documentation and correspondence
- job safety assessments and sign-off sheets
- audit results and documentation
- resolution of audit results.

Monitoring Records

Any records of any type of monitoring related to the construction project should be cataloged and stored in an accessible, secure place at the end of the project.

- exposure and medical records (should be maintained 30 years)
- environmental test results (should be maintained 30 years)
- noise-exposure records (should be maintained 3 years).

This may be quite a bit of material, depending on the size of the project. However, if this information is collected as the project evolves, the files will then contain the records needed when the project is complete. For projects not yet started, this is an outline for a good records index.

Where Should It Be Stored?

The safest place for this information is in long-term retention. It is not underfoot and is, therefore, less likely to be dumped in a housekeeping initiative. Whatever place is selected, be sure the information will not be thrown out inadvertently. Perhaps marking the boxes with prominent labels that direct the handlers to the Legal Department before the files are eventually discarded might be useful.

How Long Should It Be Retained?

There are numerous requirements for the retention of employee-related safety records. OSHA records are required to be kept for 5 years per federal guidelines; however, individual state statutes and regulations do vary. Other routine files must be retained under 29 CFR 516 and state requirements. These retention periods are generally short—just a few years. Workers' compensation requirements for retention periods vary by state but are usually between 1 and 3 years.

The aforementioned requirements guide the minimum retention of employee-related documents, but from the perspective of developing a defense in the case of litigation for construction-type injuries, the statute of limitation is generally 6 years. Occupational illnesses may take more time to evolve. Therefore, records of work environmental condition monitoring, such as containing air pollutants, noise, radiation, and the like, should be retained for 30 years.

Who Will Know Where to Find It?

This is perhaps the most challenging issue. Copies of the document inventory listing should be forwarded to the Legal Department and to Procurement to be filed with their contract-related records. Copies should also be forwarded to the attention of the individuals who are most likely to be presented with a potential claim. In that way, the appropriate individuals and departments will be aware of where the relevant safety-related information can be found.

SUMMARY

- This chapter addresses how to plan and manage contractor safety in order to improve the odds of having an injury-free workplace. Managing safety can be learned. While some people seem to intuitively know how to motivate people to work safely, most effective managers learn this skill. Skillful safety management involves knowing what is to be done, who is to do it, and when and how it should be done. This chapter addresses many of these factors and provides a description of how to implement an effective safety program.

- The program elements addressed in this chapter apply to both large and small projects, from conceptual design to completion of the work. Large or complicated projects offer more opportunity to overlook critical aspects than do small projects, but small projects still require the application of the same fundamental safety tenets. In either case the individuals planning and coordinating contracted work must pay attention to the factors discussed in this chapter, if they are to have a project free of injuries.

- In many cases, the consequences of inept safety management are not terribly significant. The work force on a small project may experience scraped knuckles or perhaps a lost day of work. It may be the assumption of the project manager that things just happen and "after all, people do get hurt in construction." But if the circumstances leading up to a potentially minor incident are complicated with a few more factors, the occurrence may result in significant consequences. Instead of a scraped knuckle, the worker could lose fingers. Instead of a simple slip and fall near a leading edge, the worker might go over the edge if adequate physical controls and safe work procedures are not in place.

REFERENCES

American National Standards Institute, 11 West 42nd Street, New York 10036.
  Basic Elements of a Program to Provide a Safe and Healthful Work Environment, A10.38–1991.
  Safety and Health Program Requirements for Multi-Employer Projects, A10.33–1992.
The Business Round Table. *Improving Construction Safety Performance.* Report A-3, September 1989.
Code of Federal Regulations.
  29 CFR 1910.119, Process Safety Management of Highly Hazardous Chemicals.
  29 CFR 1926.64, Process Safety Management of Highly Hazardous Chemicals.
Connor, R. D. *The Agent Construction Manager's Liability for Safety Using a Pro-Active Approach to Manage Liability Exposure.* National Construction Management Conference, 1991.
Hislop, R. *Construction Site Safety: A Guide for Managing Contractors.* Boca Raton, FL: CRC Press, 1999.
Keres, F. *Safety in Construction Contracts.* National Safety Council Congress and Exposition, 1996.
MacCollum, D. *Construction Safety Engineering Principles.* New York: McGraw-Hill, 2007.
———. *Construction Safety Planning.* New York: Van Nostrand Reinhold, 1995.
National Safety Council, 1121 Spring Lake Dr., Itasca, IL 60143-3201.
  1997 Estimate of Lost-Time Case Total Costs.
National Society of Professional Engineers, 1420 King Street, Alexandria, VA 22314.
  Code of Ethics. 1996.
Samelson, N., and R. E. Levitt. *Owner's Guidelines for Selecting Safe Contractors.* December 1982.

REVIEW QUESTIONS

1. The primary factor in the success of an effective safety program depends on _____ and its involvement in the safety program.
   a. unions
   b. insurance agent
   c. management
   d. craft steward

2. _____ ultimately pay(s) the cost of the safety program on a job site.
   a. Owners
   b. Insurance companies
   c. Government
   d. Employees

3. Monitoring day-to-day activities of construction trades and contract compliance is an essential function of the
   a. union.
   b. government.
   c. owner.
   d. insurance company.

4. It is the role of the _____ to clearly establish responsibility for safety down through the organizational hierarchy.
   a. owner
   b. union
   c. contractor
   d. government

5. A successful safety plan must address the following:
   a. site-specific hazards.
   b. safety expectations.
   c. safety roles and responsibilities.
   d. all of the above
6. To ensure a clear understanding of the owner's safety expectations, they should be documented in the
   a. project safety plan.
   b. insurance policy.
   c. union contract.
   d. subcontracts.
7. A means of systematically identifying and evaluating safety issues associated with a work task is a
   a. site audit.
   b. daily inspection.
   c. job safety analysis.
   d. crew briefing.
8. A JSA should be developed for all critical tasks, with priority given to addressing tasks
   a. where frequent accidents and injuries occur.
   b. where severe accidents and injuries occur.
   c. that are infrequently performed.
   d. all of the above
9. For a safety program to be successful, who must recognize that they each have a role in the safety process?
   a. management
   b. supervisors
   c. workers
   d. all of the above
10. Workers are responsible for which of the following?
    a. knowing and complying with safety regulations
    b. following safe work procedures
    c. correcting or reporting unsafe work practices and unsafe conditions
    d. all of the above
11. It is the role of the designated safety representative to
    a. bring safety issues and concerns to management's attention.
    b. guide workers in the implementation of their respective safety responsibilities.
    c. guide supervisors in the implementation of their respective safety responsibilities.
    d. all of the above
12. A *competent person* is defined as an individual who, as the result of training and/or experience,
    a. is capable of identifying or predicting hazardous situations.
    b. has the authority and responsibility to take prompt corrective measures to eliminate hazardous situations.
    c. takes action to implement corrective measures that have been determined to be necessary to control a recognized hazard.
    d. all of the above

Maintenance of Facilities 4

Richard Pifer, MSM
Rich Payant, DBA, CFM, CPE, CHS

Facility Maintenance
Foundations ▸ Structural Members ▸ Walls ▸ Floors ▸ Roofs ▸ Tanks and Towers ▸ Stacks and Chimneys ▸ Fixed Ladders ▸ Platforms and Loading Docks ▸ Canopies ▸ Sidewalks and Driveways ▸ Underground Utilities ▸ Lighting Systems ▸ Stairs and Exits ▸ Heating Equipment

Indoor Environmental Quality Issues
Cleanup and Prevention ▸ Preventive Maintenance

Grounds Maintenance
Electric-Powered Hand Tools ▸ Gasoline-Powered Equipment ▸ Snow Shoveling ▸ Chemicals and Pesticides

Computerized Predictive Maintenance
Computerized Assistance ▸ Maintenance Diagnostic Technology

Maintenance Crews
Training ▸ Preventive Maintenance Plans ▸ Inspection of Equipment ▸ Personal Protective Equipment ▸ Lockout/Tagout ▸ Piping ▸ Crane Runways ▸ Final Check: Tools and Guards ▸ Lubrication ▸ Shop Equipment Maintenance ▸ Special Tools ▸ Keeping Up-to-Date

Summary

References

Review Questions

A sound, efficient maintenance program is essential in any industrial establishment. Such a program keeps the physical facility in good condition and prevents or removes safety and health hazards. This chapter covers the following topics:

- maintaining a building to promote worker health and safety
- addressing the issue of indoor environmental quality
- dealing with safety issues in grounds maintenance
- using computer technology to help in maintenance work
- promoting safe work practices and training for maintenance crews.

FACILITY MAINTENANCE

Maintenance includes (1) care to ensure long-term life of company assets, (2) routine care to maintain uninterrupted service and appearance, and (3) the repair work required to restore or improve service and appearance. Maintenance is recognized as one of the keys leading to improved facility productivity, safety, and good public relations. Appearance—both internal and external—affects employees, customers, and the public.

Too often, maintenance is interpreted as nothing more than making repairs. However, maintenance programs should place more emphasis on preventive maintenance and on the type of inspection and monitoring that discovers adverse conditions before they result in equipment failure and accidents. The use of computers and space-age technology has provided new tools for preventive maintenance. For example, computer programs that monitor equipment can alert personnel regarding maintenance needs before the equipment breaks down.

The facility engineer or maintenance superintendent often supervises the maintenance program. However, because the safety of employees is closely tied to the condition of buildings and equipment, safety and health professionals will find that the maintenance program has an important bearing on the safety program. Therefore, do not hesitate to point out to management that equipment and/or structures need repairs, modification, or replacement.

Foundations

Starting from the bottom, inspect and maintain footings, column bases, foundation walls, and pits. The integrity and safety of the rest of the building, as well as that of employees and equipment, depends on a firm foundation.

Footings and Columns

Although it is often hard to detect flaws in footings, it is possible to check for cracks and unusual settlement of a building's columns and footings. For example, placing level marks at known elevations above the basement floor will provide a reference point to identify changes due to settling of the building. Check these marks periodically for signs of settlement.

Excessive settlement may threaten the stability of a building as well as the effective functioning of the machines and equipment in that building. Inspectors should at once report excessive settlement to management for immediate action.

Inspectors should also check the bases of columns for dry rot and rust. Dry rot around the bottom of wood columns at the basement floor level can result if the basement floor is damp, subject to water seepage, or alternately dry or wet. Maintenance personnel should scrape away rust at the bases of steel columns, inspect the bases for notable damage, and give the bases a coating of a preservative. (See also Structural Members.)

Foundation Walls

Inspect both the inside and the outside of foundation walls for cracks, which may result from settlement of the building and shrinkage of concrete. Because these cracks are below grade, they can serve as a conduit for water intruding into the basement area. If enough water comes in through large cracks, settlement of the backfilled earth around the outside of foundation walls can result. Sidewalks and adjacent roadways can also be damaged from such settlement.

Small cracks in foundation walls can often be repaired from inside the building by applying a waterproofing material. However, if the cracks are relatively large, maintenance workers may have to dig outside the building down to the bottom of the wall, clean off all earth and other foreign materials, and apply an appropriate waterproofing compound. This should be followed by applying a membrane covering directly over the compound and then covering that membrane with another coating of compound. When unusual settlement is noted, inspectors should take settlement readings like those for footings and columns. If the crack is excessively large or the maintenance person is unsure of the integrity of the wall, a qualified engineer may need to be contacted before making any repairs.

Pits

Inspectors should also examine pits, noting cracks and having them repaired. Do not allow any debris or rubbish to collect in the pits; install guardrails or covering where needed. Because pits are usually at the low points of a building, they often collect liquids that have leaked or intruded into the structure. When a building has liquid intrusion problems, the pits should be examined whenever an event (rain, floods, etc.) occurs that could result in flooding.

Structural Members

Structural members—such as joists, beams, girders, columns, and flooring—require periodic inspection and maintenance. Steel, concrete, and wood parts each need specific inspections and care. When inspecting structural members, you should seek a professional opinion if you are unsure of the integrity of a member.

Joists, Beams, and Girders

Check joists, beams, and girders, and correct them for deflection, twisting, tipping, or other unusual positions. In many instances, joists, beams, and girders are covered by suspended or sealed ceilings, thus making them hard to access. In such cases, excessive deflection may be indicated only by a sagging floor. At least once a year, therefore, examine the entire floor system on each floor level. If major repairs are necessary, clear the floor of stored materials immediately. Before making repairs, consult a qualified engineer.

Columns

Examine building columns for unusual distortion or buckling (being out of plumb). Avoid excessive or unusual column loadings, and check for holes cut in or through columns. If holes exist, contact a qualified engineer for an evaluation.

Steel Parts

At least once a year—more often in corrosive atmospheres—check steel I-beams, channels, columns, angles, girders, and other structural parts for rust or signs of corrosion. Where rust exists, scrape the steel part and paint it.

Concrete Parts

Check floor slabs, beams, girders, and columns regularly for cracks, spalling, and chipping of concrete from the reinforcing steel. Because rust may form on exposed steel, make repairs at once. Gunite, for example, provides a protective coating for exposed steel. Any visible damage to concrete parts may indicate serious problems. Consult a qualified engineer for further investigation.

Wood

Inspect a wood floor system for shakes, checks, and splits in joists, planks, beams, stringers, posts, and columns. Look for decay or dry rot in wood columns, joists, and other parts. Because it is important that beams, joists, and girders provide full support, thoroughly investigate any evidence of movement or slippage.

Walls

Masonry buildings require periodic, minor repairs of walls and windows. Because mortar joints loosen and disintegrate from settlement of a building and from weathering, rake and point such joints. If these joints are not repaired, moisture will eventually seep into interior wall surfaces and cause further damage.

Exterior Walls

Inspect exterior walls made of brick, concrete, terra cotta, stone, cement, cinder block, or stucco for cracks or joint separation. When cracks or joint separation is identified during an inspection, it needs to be evaluated for potential impacts on the structure. Cracks can result from expansion, contraction, vibration, or settlement. Repair them immediately; otherwise, water intrusion could result in freezing and additional damage.

If brick walls are painted on a building's exterior and if high humidity exists inside the building, excessive moisture can penetrate the brick and condense under the paint film. Should the moisture freeze, it could cause spalling and joint disintegration. Look for telltale signs of paint bubbling or separating from the wall.

Windows

Because of settlement of a building, drying out of the wood, or improper setting of metal frames, caulked window joints may crack open, allowing moisture to enter. Before recaulking, remove all loose material, and cut out all material from cracks so that the new compound can bond well. Apply a suitable caulking compound with a device that forces the compound deep into the openings rather than covers just the surface joints.

Parapet and Stone Cap Repairs

Maintain and make necessary repairs to masonry, metal, and wood parapets; stone caps; and other stonework. Avoid having to make significant repairs on masonry walls by checking them carefully once a year for cracks or spalling.

Never coat brick parapets and walls above grade, on either side, with materials such as pitch, roof paper, or asphalt roof coating. These materials do not allow walls to breathe. Such coatings, which are commonly misapplied, cause spalling of the brick and disintegration of the mortar joints, especially in areas of the country where freezing temperatures occur.

Check stone caps and other stonework on brick walls for cracks at all mortar joints. Cracks might allow moisture to enter and eventually loosen the stone. Fill these cracks with a cement grout or mastic filler to prevent stone caps from falling off.

Interior Walls and Ceilings

Inspect partitions, cross walls, interior sides of main walls, and ceilings as rigorously as you inspect exterior walls. Look for such defects as cracks in interior walls, holes, loose mortar in joints, broken or missing brick, and spalled or worn areas on tile or brick walls where powered industrial trucks may have frequently scraped them. To prevent damage from trucks, install standard 3-in. (8-cm) or 4-in. (10-cm) railings near the floor level as barriers.

Ceilings require periodic painting, cleaning, and repair. Investigate unusual sags immediately and correct them. If a sag exists in a suspended ceiling, check the hangers and fastenings. However, if the sag has resulted from excessive loading of the floor above, correct the situation based on an assessment by a qualified engineer.

Floors

Accidents resulting from inadequate maintenance of floors are a major source of injuries in many facilities. In some cases, building owners and employers can be liable for injuries resulting from slips, trips, and falls if there was awareness of an unsafe condition and nothing was done to remedy the situation. There are also potential liability issues if a simple, routine inspection would have identified the issue and resulted in corrective actions. Frequencies of inspections depend on the factual circumstances and on the types of conditions likely to develop in any given situation.

Slippery conditions account for many slips and falls that workers experience. Holes and other irregularities in wood and concrete floors, both inside and outside facility buildings, also result in frequent injuries from trips, slips, and falls. Building consensus standards dictate the allowable height deviations within walking and working surfaces as well as the conditions of those surfaces.

Housekeeping and Maintenance Procedures

Often, in-house employees conduct maintenance and housekeeping duties. However, many businesses and commercial facilities have maintenance or cleaning services that are responsible for the care of facility floors. It is important to ensure that both in-house and contracted services follow appropriate safety rules.

Using the wrong cleaning materials or methods or installing the wrong surfacing can cause even the most suitable types of flooring to deteriorate and become slippery. For example, although alkaline cleaners should not be used on terrazzo, mild alkaline cleaners may be used on asphalt tile. Oils are unsuitable for rubber tile and, when applied to wood floors, increase the fire hazard. To keep floors safe and sanitary, follow the recommendations of the flooring manufacturer. Standardize and spell out in detail the maintenance procedures for floors (Figure 4–1).

In general, the routine housekeeping procedures for linoleum, marble, terrazzo, asphalt tile, and similar types of flooring used in offices, facilities, and the like, are to (1) clean the floors with a soft brush or vacuum cleaner and (2) when necessary to wipe them with a mop dampened with clean, cold water. Clean one section of floor at a time. If traffic in the area is heavy, rope off that section. When soap is used, remove the soap residue by thoroughly rinsing the area to avoid creating a slippery surface.

Wax for polishing wood, tile, and similar floor surfaces is unsuitable because of its inherently slippery nature. However, based on manufacturers' recommendations, floor oils and waxes can be used on various types of floors without adding unduly to their slipperiness, provided that users apply these oils and waxes as instructed. Soft floors such as

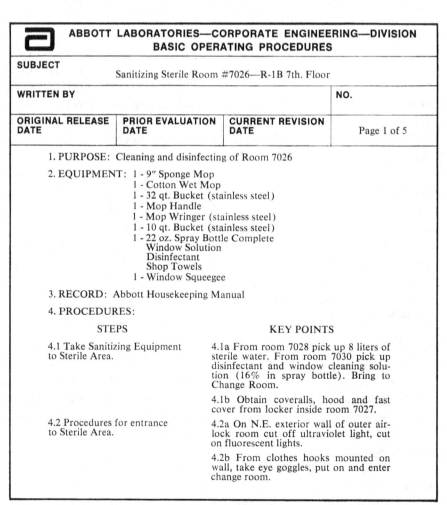

Figure 4–1. Standard operating procedures should be detailed so that each job is completed efficiently, thoroughly, and safely. Shown here is the first page of a detailed procedure for sanitizing a sterile room. *(Printed with permission from Abbott Laboratories.)*

asphalt, vinyl, and linoleum are often refinished as many as four times a year. Hard flooring such as concrete and terrazzo is usually cleaned and sealed once a year.

Oil and grease, water, paper, sawdust, and other foreign materials create slipping hazards on floors. Eliminate or contain leaks of oil from machines, leaks of water or other liquids from pipelines, and spillage from processing equipment by maintaining the equipment according to the manufacturer's recommendations. In addition, promptly tighten loose connections to minimize leaks.

When leakage cannot be readily eliminated at the source, use pans and absorbent materials on floors to minimize slipping hazards. Note that investigating such sources often reveals ways to keep slippery materials from getting on floors, such as installing splash guards on machines using cutting oils.

Promptly clean up slippery materials spilled onto floors. To remove grease and oils, cover the area with slaked lime to a depth of about ¼ in. (5 mm). After 2 or 3 hours, remove the lime with a scraper or stiff brush. Sand and various commercial cleaners also can be used.

Even an innocuous substance such as coffee can cause an accident if spilled in a high-traffic area and not cleaned up immediately. Cups containing beverages should be either covered or placed on a tray if carried to an employee's desk.

Aisles

Keep aisles clear of machinery, equipment, and raw and manufactured materials. Try to maintain at least a 28-in. (70-cm) clearance to allow emergency egress. In many cases, the allowable floor loading was calculated based on clear aisles, with no allowance made for powered industrial trucks using the aisles. For efficient and safe operations, determine whether floors and aisles are capable of sustaining the weight loads of these trucks. Provide lines on the ground that indicate the minimum aisle widths for safe passage.

Floor Load Capacity and Load Distribution

A survey of a physical facility requires accurate data on floor load capacity (Figure 4–2).

If these data are not already known or readily obtainable from building plans, have a qualified engineer conduct a structural analysis. Rough estimates based on experience, or conclusions reached by a casual glance at a handbook, are often inappropriate and can be dangerous. Use accurate weight data.

Load distribution is often a complex issue. Most buildings are designed to carry uniform weight loads. A concentrated load usually places greater stress on supporting members than does a uniform load of equal weight. Therefore, most heavily concentrated loads such as heavy

Figure 4–2. Communication is an important safety component in the workplace. Note the floor load capacity sign and the labels on the pipes. *(Courtesy International Stamping Co./Midas International Corp.)*

machines should be placed directly over beams or girders rather than over slabs or joists. As in the case of determining floor load capacity, a safety professional should consult a structural engineering specialist to accurately determine the locations of concentrated loads.

Overloading of Floors

Installation of heavy equipment, excessive weight, unequal distribution of stored raw and finished materials, and heavy truck transportation may cause overloading of floors. Post signs identifying allowable floor loads, and paint horizontal lines on the walls showing the maximum height to which materials may be piled (Figure 4–3).

Whether a floor is safely loaded or is overloaded depends on how closely the designated load capacity and the actual load capacity correspond.

Although evidence of overloading is not always readily discernible, inspectors should look for it. Deflection of flooring is the most common evidence of overloading in wood and steel beams. A sag or deflection greater than $1/360$ of a span's length suggests that a floor may be overloaded.

To measure deflection, stretch between two columns of the span a cord that is 5 ft (1.5 m) below the underside of the beams or girders at ceiling level. Measure the distance from the center of the taut cord to the bottom of the beam or girder. The difference between this measurement and 5 ft (1.5 m) is the deflection.

Floors and other structural parts show signs of overloading in several ways:

Figure 4-3. When the same type of material is stored regularly, a line can be marked on the wall to indicate the height to which the material can be piled without exceeding the allowable floor load. In addition to safe floor load signs, a warning sign reading DO NOT PILE ABOVE LINE can be placed on the wall.

- Overloaded wood beams check and crack.
- In reinforced concrete beams and girders, concrete spalls and falls away from the tensile side.
- Wood floors are punctured, and flat concrete slabs crack and spall.
- Timber columns split and crack.
- Concrete columns spall, the concrete falling away from the reinforcing rod.
- On steel columns, flanges are twisted.
- Bearing walls of masonry show extensive cracking, disintegration of the bricks, bulges, and pulling away of floor joists.

Whenever any doubt exists as to the amount of deterioration of a floor's supporting parts and, therefore, of its load-carrying capacity, have a qualified structural engineer inspect the floor and make a determination. Post the floor loads of storage areas not originally designed for storage.

Repair Procedures

Inspect wood flooring for rot, wear, and unusual stress, all of which are indicated by sag. In many cases, a section of wood flooring may need to be replaced. If so, install new flooring flush with the existing flooring. Replace badly worn or loose wood-block flooring with anchored wood blocks.

To anchor loose finished wood flooring, drill holes at an angle through the finished floor into the subfloor. Then drive flooring nails larger than the drilled holes into the subfloor.

For concrete floors, chip out the damaged area, square it off as much as possible, thoroughly clean it, and wet it down. Trowel in the appropriate cement mortar at a minimum patch thickness of 1 in. (2.5 cm). Patches 2 in. (5 cm) thick or more may require wire mesh or reinforcing steel. For finishing, a wood float gives a less slippery surface than does a steel trowel. For proper curing, keep traffic off the patch at least 3 days, unless a quick-setting cement was used.

If mixed and applied properly, epoxy resin repair materials give excellent results. A thickness of as little as $1/8$ to $3/16$ in. (2 to 3 mm) gives an extremely tough-wearing surface. Be sure to take adequate ventilation precautions when using these materials.

Roofs

Inspect and maintain roofs and roof-mounted structures on a regular basis. Roof damage can quickly lead to structural damage of other parts of the building and of equipment. Roof damage can also lead to water and moisture intrusion into interior walls and spaces, providing a potential habitat for mold to grow.

Inspection

Inspect all roofs periodically, once every 6 months or so. Check roof flashings for cracks at the parapet wall. Check roof gutters and drain connections for cracks at the roof line. Also check areas around dormers, chimneys, and valleys—all places where metal is used—to see whether the metal is tight with the roof and the drain. Because leaks will result if water rises above the flashings, keep roof drains and overflow gutters through parapet walls open. Keep gutters clean, and check gutter areas in early spring for ice damage. In some cases, it will be a good idea to put screens on gutters to keep out leaves and other debris.

It is often difficult to find the source of a leak because the fault may be distant from the place where the leak appears inside the building. On sloping roofs, check above and to the sides of the place where the leak appears. Note that flat decks overlaid with concrete are difficult to check because water tends to follow a crack's path and appear at a point distant from the source.

Abnormal Loading

Roofs are usually designed to support the maximum snow load expected for the locale. Roof configuration (especially multilevel) and wind frequently combine to deposit heavy snowdrifts over portions of a roof. Ice can also cause overloading. Where practical, remove these accumulations as quickly as possible to prevent roof collapse, major leaks, or other damage.

Observe the following precautions to prevent roof damage:

- To prevent ice and snow buildup near drain areas, clear a path from the center of the roof to the drains.
- To provide for drainage on a pitched roof with no drains, open paths that lead to the roof's edge.
- Never use blowtorches or similar devices to melt ice from drains or roof surfaces.
- To prevent the puncturing of the roof, instruct workers to use care when removing ice and snow.

Roof Anchorage

Check roof anchorage during inspections. The lifting of unanchored roofs accounts for a large percentage of wind-caused losses in U.S. industry. Securely anchor the roofs of all buildings. Anchors can be readily installed, and their cost is reasonably low.

Roof-Mounted Structures

Roof inspections should include examining penthouses, stacks, vents, air-handling units, and the supports for water tanks when these structures are flashed at the main roof level. It is also important to check the roofs of penthouses because most penthouses contain elevator machinery, which can be damaged by water seepage.

The windows, skylights, and monitor sashings of a penthouse should have necessary reglazing, reputtying, frame caulking, or painting. When workers repair or repaint operating mechanisms, they should work on a safe scaffold and wear a safety belt or harness and a lanyard tied to a lifeline. They should never work on or over unprotected skylights. Ensure that appropriate fall protection is provided for roof-top workers.

Repairs

Patch a leaky roof as soon as possible after a leak is discovered. When maintenance needs become excessive and leaks occur after each rain, replace the roof. Cold mastic applications may last about 5 years, whereas an application of hot pitch covered with pea gravel may last from 5 to 10 years. Reputable roofing manufacturers and contractors provide a guaranteed-performance bond for a number of years commensurate with the quality of the job required.

During repair, a roof can be punctured by tools, boards with nails, stones, and other sharp objects. To prevent such puncturing, use runways and protective run-board covering. If hot pitch or roofing compound is used, workers should wear gloves, goggles, and leather or fire-resistant duck leggings.

Tanks and Towers

Maintenance of tanks and towers is important for fire protection and also because structural failures may cause serious accidents. When tanks are to be cleaned or painted on the inside, strictly observe all precautions related to entering tanks. (For a detailed discussion of the required precautions, see Chapter 10, Flammable and Combustible Liquids, in this volume.)

If a tank is more than 20 ft (6 m) above the ground or the building's roof, place a wood or steel balcony around the base of the tank. The balcony should support a weight of at least 100 lb/ft² (486 kg/m²). For a tank not more than 15 ft (4.6 m) in diameter, the width of the balcony should be at least 18 in. (45 cm); for tank diameters greater than 15 ft (4.6 m), the balcony width should be 24 in. (61 cm), including railings. (See ANSI A1264.1, Safety Requirements for Workplace Walking/Working Surfaces and Their Access; Workplace Floor, Wall and Roof Openings; Stairs and Guardrails Systems.)

Stacks and Chimneys

Inspect stacks at least once every 6 months. Stacks are subject to deterioration, both inside and out, from weathering, strong winds, lightning, settlement of the foundation, and corrosive flue gases. Check a stack's ground wires to ensure that the stack has effective grounding.

A brick or concrete chimney can be protected by lightning rods of low resistance and ample current-carrying capacity. These rods must be installed from the top of the chimney and have good electrical ground. Lightning rods can be conveniently installed on chimneys when repairs call for the erection of scaffolding. Underground water pipes or buried copper plates afford good ground connections.

Fixed Ladders

The safety standard for fixed ladders (ANSI A14.3) permits the use of safety systems in lieu of cage guards on tower, water tank, and chimney ladders exceeding 24 ft (7 m) in unbroken length. A landing platform should be provided at least every 50 ft (15 m) within the length of climb. A rest platform at not more than 150 ft (46 m) with a ladder safety device should be used.

Ladder safety devices allow a climber to attach a harness to a sleeve that travels along a carrier rail or cable anchored to the ladder. The sleeve is designed to lock and suspend a person who slips and starts to fall. Many safety professionals prefer such devices to cage guards.

According to ANSI (American National Standards Institute) A14.3, fixed ladders must meet the following general requirements:
- be designed to withstand a single concentrated load of at least 250 lb (113 kg)
- have rungs with a minimum diameter of ¾ in. (19 mm) if a metal ladder or 1⅛ in. (28 mm) if a wood ladder
- have rungs at least 16 in. (40 cm) wide, uniformly spaced, and no more than 12 in. (30 cm) apart vertically

- be painted, if metal, or otherwise treated to resist deterioration
- have a pitch of 75 to 90 degrees for safe use
- have a 30-in. (75-cm) clearance with at least 24 in. (61 cm) on the climbing side of the ladder, unless caged
- have at least a 7-in. (18-cm) clearance behind the ladder to provide for adequate toe space
- have side rails that extend 3½ ft (1 m) above landings
- have a clearance width of 15 in. (38 cm) on every side of the centerline of the ladder, unless the ladder is used with cages or wells.

In addition, the provisions of OSHA's 29 CFR 1910.27, Fixed Ladders, should be followed when OSHA's General Industry Standards are applicable. These provisions address design, specific features, clearance, special requirements, and pitch.

Platforms and Loading Docks

Mechanized traffic on platforms and docks often damages platform surfaces. Check wood platforms for decay or dry rot, loose or uneven planking, and weakened or broken supporting members. Make repairs immediately.

Because they have no edge protection, concrete platforms or docks can become spalled or chipped. Ruts in the concrete's finish may cause powered industrial or hand trucks to swerve and run off a dock and into employees or material. For these reasons, provide angle-iron or channel-iron protection at the edge of the platform, and maintain it regularly. Resurface badly rutted platforms with concrete or epoxy cement.

Canopies

A canopy's roof should receive the same careful inspection as that given to the roof of the main building. Evidence of pulling away from the building should be noted and corrected. Because drainage from canopies is important, keep downspouts and gutters open and in good repair. To ensure that structural integrity is maintained, periodically scrape and paint the supporting parts of canopies if they are made of wood or steel.

Sidewalks and Driveways

Repair concrete sidewalks and driveways as soon as spalling or cracking of the concrete creates a hazardous condition. In cold-climate areas, inspect concrete sidewalks and driveways in the spring, after the ground has thoroughly thawed. Sections of sidewalks can be relaid, but bituminous driveways need to be patched with hot tar or similar material.

Whether a driveway is made of concrete, asphalt, or gravel, the driveway's drainage is important. Make repairs

as soon as possible to keep to a minimum damage resulting from poor drainage. To keep asphalt driveways in good condition, have them recoated periodically by someone who understands paving techniques.

Underground Utilities

Conducting inspection and maintenance of underground utilities is especially hazardous. Personnel should thus observe strict procedures and be closely supervised.

Sewers

At least two persons should conduct sewer maintenance. Before anyone goes into a manhole structure or sewer, he or she should obtain a confined-space entry permit and observe OHSA required confined-space entry procedures, which could include testing for oxygen deficiency, methane, hydrogen sulfide, carbon monoxide (CO), and any other suspected atmospheric contaminant. If oxygen deficiency or any contaminant is found in concentrations approaching or exceeding the OSHA permissible exposure levels, institute other characteristics of the confined-space program.

Have workers use proper respiratory equipment (see Chapter 22 on Respiratory Protection in the National Safety Council's *Fundamentals of Industrial Hygiene*, 6th ed.), or furnish ventilation by means of either blowers or suction fans to provide a safe atmosphere for entry and the tasks that need to be performed in the space. Blowers are preferable because their supply source is known, whereas suction fans may draw hazardous gases into the area from unseen pockets or crevices. Respiratory protection may be necessary for the outside attendant as well as for the inside workers.

Utility Trenches and Tunnels

When personnel must work in trenches more than 5 ft (1.5 m) deep, shoring or sloping is required. A trench has a depth greater than its width but is not wider than 15 ft.

For work in trenches, an adequately designed protective system should be installed to protect employees from cave-ins. Such a system considers soil classifications, depth of the trench cut, water content of the soil, potential impacts of weather or climatic conditions, and any operations in the immediate area. If a trench is near machinery foundations or other superimposed loading, meticulously design the bracing and shoring to be installed before the work begins. When pipe or trenches must remain open overnight, place barricades, signs, and lanterns around them to notify workers, other employees, and the public so that they don't fall into the open pipe or trench.

Check the atmospheric conditions in utility trenches, tunnel trenches, tunnels, and manholes before work begins. Proper ventilation is necessary for the safety of workers, who should work in pairs. Post signs that identify manholes, tunnels, and trenches known to be contaminated. Ensure that any employee entering follows applicable regulatory requirements. For construction involving excavations and trenches, see Subpart P, Excavations, 29 CFR 1926.650, 651, and 652. Confined-space hazards are addressed in specific standards for the general industry and shipyard employment, ranging from Section 5(a)(1) of the OSHAct (the General Duty Clause) to Shipyard Employment (29 CFR 1915 Subpart B).

Forbid smoking in or near manholes, tunnels, or trenches. Also, forbid the use of open-flame devices, such as solder pot furnaces and welding equipment, in or near manholes, tunnels, or trenches in which tests have identified the presence of combustible or flammable gas.

Waste-Disposal Facilities

The U.S. Environmental Protection Agency's (EPA's) Office of Solid Waste (OSW) regulates disposal of waste under the Resource Conservation and Recovery Act (RCRA). When on-site waste disposal is planned, obtain a permit from the EPA. RCRA Subtitle D sets criteria for municipal solid-waste landfills and other solid-waste-disposal facilities and prohibits the open dumping of solid waste. RCRA Subtitle C establishes a system for controlling hazardous waste from the time it is generated until its ultimate disposal—in effect, "from cradle to grave." 40 CFR 239 through 259 contains the regulations for solid waste, while Parts 260 through 279 contain the hazardous-waste regulations.

Because some jurisdictions enforce requirements that are more restrictive than those of the EPA, consult local regulatory bodies as well to ensure compliance with waste-disposal requirements.

Other Underground Pipelines

Before starting repairs, completely drain pipelines, block off connecting systems, and close and lock valves. Appropriate lockout and tagout procedures should be used to ensure that systems do not energize during repair and maintenance operations. Whenever workers open a line or valve, they should watch for backpressure. After a steam valve has been opened completely, it should be backed off at least one-half turn so that thermal expansion does not lock the valve in the open position. Caution workers to open valves slowly and to equalize pressure slowly. Sudden changes in pressure can wreck equipment and endanger lives.

Before people begin work on an underground pipeline, they should know its approximate location before beginning to dig. Workers cutting into or working on a pipeline should know what the line normally carries and the dangers associated with that substance. The line should be purged and/or filled with an inert gas before work begins.

Lighting Systems

All lighting systems require regular maintenance so that they provide the maximum output of light. The light output of lamps decreases as lamps age; remember this factor when replacing lamps. It is also important to ensure that lighting refits do not result in lighting levels lower than those of the original lighting.

Replacing Lamps

To reduce the number of persons who might be exposed to flying glass and dust if lamps break, replace lamp fixtures on weekends or at other times when personnel are not near the lamps. If this is not possible, ensure that there is a safety zone around the lamping operation so that others will not be injured if a lamp breaks. Those who handle the lamps should wear gloves and eye protection.

Install fluorescent fixtures in which the tubes lock in place. Equip older fixtures with shields or grids beneath the tubes to prevent the tubes from falling should vibration loosen them.

In some cases, disposal of lamps falls under the regulations for disposing universal waste in accordance with RCRA. In any case, ensure that an evaluation has been made and documented in accordance with both local and federal regulations so that lamps are disposed of in a safe manner.

To clean reflectors or glassware that cannot be taken down, shut off the current and use a cleaner that requires no rinsing. Then wipe the reflectors or glassware with a cloth.

Maintaining High Levels of Lighting

Lamp burnout and depreciation, dirt accumulation, voltage drops, and light absorption by dirty walls and ceilings reduce illumination levels. Because dark or dirty surfaces absorb as much as 80% of the light that strikes them, maintain a regular cleaning schedule. For dirty areas, this schedule might need to be every 2 weeks.

Use a light meter to measure illumination levels. Then check the readings against the minimum standards of illumination for industrial interiors (e.g., ANSI/IES RP7, Practice for Industrial Lighting). These standards, recommended by the Illuminating Engineering Society (IES) of North America, are summarized in Table 5, Levels of Illumination, in Appendix 1, Safety and Health Tables, at the end of this volume. Use these values as a guide when formulating a regular maintenance program. If cleaning lamps and replacing burnouts fail to increase the illumination to standard levels, have a qualified illuminating engineer make a complete survey of the present system.

Using Labor-Saving Devices

To simplify maintenance tasks, use labor-saving devices. When possible, use disconnecting reflectors, which permit cleaning and relamping from the floor. To reach lamps mounted low, use a stepladder, which is convenient and portable. Make sure the ladder is tall enough that workers do not need to stand on the top two steps during maintenance activities. Attach to the ladder clips and hooks that can hold spare lamps and cleaning rags. In this way, a person can perform an entire cleaning and relamping job with one trip up the ladder. When an entire installation is cleaned frequently, use special cleaning trucks with separate compartments for cleaning solutions, warm rinse water, and clean rags.

Use maintenance platforms when a great many lamps are mounted at the same height. Some platforms are made of lightweight material and are equipped with casters so that one person can maneuver them easily. Others are self-propelled aerial work platforms (Figure 4–4). This type of platform permits a person to reach several lamps safely without needing to reposition the platform.

Figure 4–4. A scissor lift with protective toeboards and railings can take the worker to the work area. The lift is stable, but mobile, and can be controlled from the platform. *(Courtesy Grove Worldwide)*

Manufacturers have designed several devices that can reach all types and styles of lamps. One of the simplest devices is a clamp grip mounted on the end of a pole and formed to fit fluorescent tubes. Many industrial facilities use this device to maintain fixtures that have open bottoms and exposed lamps. Use these clamp-grip devices between periods of regular maintenance to replace lamps in an emergency. They are also well suited for recessed reflector lamps.

Stairs and Exits

Most injuries from slip, trip, and fall incidents take place at exits or entrances to buildings. Lighting conditions in these areas are regulated by the International Building Code, the Uniform Building Code, the BOCA National Building Code, or the National Electrical Code. Exits should have a minimum of 1 footcandle of light at the floor level. Light of this magnitude is generally adequate for a worker to be able to delineate the particular physical features of the surface.

Look for the following items when inspecting the condition of stairways and exits:
- inappropriate exit signs
- improper or inadequate design, construction, or location
- lack of handrails
- handrails placed too low or having rough surfaces
- improper lighting (including emergency lighting)
- obstructions
- locked doors
- doors that open in the direction of an exit
- poor housekeeping
- wet, slippery, or damaged surfaces
- faulty treads or mats on stairs
- lack of curbing on ramps
- lack of differentiation among
    - the exit access
    - the exit
    - the exit discharge.

Whenever any of these defects is found during a maintenance inspection, repair or correct it immediately, if possible. Various types of slip-resistant materials can and should be applied directly to stairways if someone might slip on the walking surfaces.

Exits should not serve as storage areas. Keep exits well lighted and their floors smooth. Correct hazardous conditions as soon as possible after discovering them.

Check the operation of exit doors. See that they can move freely and have no obstructions and that the paths to them are clear.

Keep exit signs and lights in good repair. Test exit signs and emergency lighting that is designed to operate in the dark to see whether the lighting system fails to operate properly. (See National Fire Protection Association [NFPA] 101, Life Safety Code.)

Heating Equipment

Inspect the facility's heating equipment before winter to be sure that it is operating safely and efficiently. Breakdowns of heating equipment during cold weather are especially dangerous and costly. Observe the following precautions to reduce the possibilities of these breakdowns:

- Inspect and thoroughly clean the heating system.
- Annually inspect chimneys and vent pipes for cracks, missing mortar, and rusted holes. Repair any damage immediately.
- Keep the inside of buildings at a minimum temperature of 40°F to minimize the chances of pipes or liquid-filled systems freezing.
- Do not leave buildings unattended for long periods.
- Check at least daily that the heating system is operating properly.

INDOOR ENVIRONMENTAL QUALITY ISSUES

Indoor environmental quality (IEQ) considers all of the elements inside a building that could affect occupant well-being and health. Lighting, ventilation, chemical contaminants, biological agents, noise, vibration, air quality, ergonomics, temperature, particulates, and relative humidity are some of the elements that can undesirably affect indoor air quality. These elements could contribute to problems ranging from contamination to adverse impacts on worker comfort parameters. See Chapter 15, Indoor Air Quality, in the *Administration & Programs* volume.

Probably the most effective way to deal successfully with IEQ issues is to maintain open lines of communication with all stakeholders. Try to ensure that there is a central point of contact that coordinates communication. Communicate the results of testing in a timely fashion and listen when stakeholders have something to say regarding both their environments and their concerns.

Indoor air quality (IAQ) is a major component of IEQ. IAQ refers to the air inside a building and to its suitability for inhalation by building occupants.

IAQ problems are sometimes thought to contribute to the conditions that have been collectively referred to as either sick building syndrome or building-related illness. Sick building syndrome (also referred to as tight building syndrome) has been implicated when occupants have broad complaints of discomfort such as nausea, dizziness, dry or itchy skin, dry coughing, difficulty concentrating, muscle pain, sensitivity to odors, fatigue, and/or eye, nose, throat,

and respiratory irritation. Tying these symptoms to sick building syndrome usually requires at least 20% of the occupants to be affected. However, it should be remembered that people, not buildings, get sick, so the term *sick building syndrome* is a misnomer. In any case, affected employees should be examined by a health care practitioner who has experience dealing with such health issues. Based on the practitioner's recommendations, corrective actions should be implemented.

Building-related illnesses (BRIs) are so named when symptoms of diagnosable illnesses are identified and can be attributed directly to building contamination. Quite often, symptoms such as coughing, chest tightness, fever, chills, and muscle aches are the result of clearly identifiable causes and a clinically defined illness of known etiology. Allergic reactions, hypersensitivity pneumonitis, and humidifier fever would fall into this category. Occupants affected with BRI often recover after leaving the problematic area and receiving treatment from a health care practitioner with experience in dealing with IAQ issues.

Although sometimes difficult to identify, most IAQ scenarios have several elements in common that are key to resolving such issues:

- a susceptible population
- pollutant source(s)
- a driving force that moves the pollutant(s)
- a pathway for the pollutant(s) to travel.

It is important to treat each IAQ scenario with the goal of identifying the four elements and initiating a response plan using an interdisciplinary approach. If occupants have symptoms that could be attributed to characteristics of the indoor environment, then a susceptible population exists, and members of that population should seek health care. Although pollutant sources are sometimes difficult to identify, driving forces and pathways can always be identified.

Pollutant sources related to IEQ are usually either physical and chemical contaminants or biological agents.

Heating, ventilating, and air conditioning (HVAC) systems; differences in temperature; pistonlike effects of elevators; and outdoor weather can produce driving forces that exacerbate IAQ problems.

Pollutant pathways can be ventilation systems, doorways, electrical receptacles, process lines, elevator shafts, pipe chases, ceiling plenums, and windows.

According to the American Conference of Governmental Industrial Hygienists (ACGIH) Industrial Ventilation Manual, the need for outside air to remove indoor contaminants and to provide oxygen to people's bodies is self-evident. In most situations, enough air for these purposes enters buildings by permeating walls and infiltrating around windows. However, when 100% of the air is recirculated because of energy-conservation goals, gases and chemical solvents in the air are not diluted or removed from the work environment.

During the 1970s, when the cost of losing heated and cooled air was a major concern, companies permanently sealed and weather-stripped windows or replaced them with superinsulated units. Insulation was increased in older buildings, and new buildings were designed and built with inoperable windows and high-performance insulating materials. These features minimized the amount of outside air entering buildings and required elaborate HVAC systems that used energy-recovery components, air washes, humidifiers, and complex air-flow mechanisms. These systems kept the environment within a sealed building livable without the loss of hot or cold air.

Cleanup and Prevention

Three aspects that address indoor air-quality problems include regular preventive maintenance, preventive design, and contamination cleanup.

Remove all contamination from asbestos, PCBs, and, where possible, certain formaldehyde-containing insulation when it is found in amounts greater than government-specified limits. Removal practices must meet both OSHA and EPA standards and be conducted by a person certified to remove the particular materials.

When biological contamination has been discovered in the HVAC system or in other areas of the building, take the following steps:

- Identify and eliminate the source of water that is contributing to biological growth.
- Replace potentially contaminated air filters.
- Empty and clean condensate drainage trays.
- Use hot water instead of chemicals wherever possible to clean microbial growth from condenser coils.
- Remove contaminated materials in a manner that minimizes the potential spread of the contaminants.
- Treat contaminated carpeting and furniture with antimicrobial solutions, and then thoroughly air the carpeting and furniture.
- Check all equipment in the HVAC system after the cleanup to make sure it operates correctly.
- Keep all drains in working order.

Preventive Maintenance

Any veteran building operator who has had to deal with indoor environmental quality problems knows that a few minutes of preventive maintenance are worth hours of repairs and cleanup. The following preventive measures are recommended by experts in the field:

- Keep hot-water-supply temperatures above 120°F (40°C).

- Protect all air-handling systems from adverse environmental conditions, and provide them with a drainage system so that areas of standing water do not develop.
- Make sure that relative humidity does not exceed 70%.
- Avoid using air washes that have recirculating water systems.
- Make sure that humidifiers use steam (not boiler steam or recirculated water) as a water source.
- Avoid using spray coil systems, which are most commonly found in the southern United States.
- Find out whether the building uses fan coil and heat pump units. Such units have a system of fans that recirculate conditioned air through a secondary system. Because the secondary system has draining pans under the coils, the coils may become a source of biological pollution.
- Make sure that air filters have correctly rated dust spot efficiencies. Check pleated glass fiber filters regularly for disintegration and possible glass fiber pollution.

Preventive design starts with the following steps:

- Design and choose an HVAC system that is appropriate for the building's size and the system's anticipated uses.
- Allow for a generous number of intake and exhaust vents.
- Locate intake vents where they will receive a clean, uncontaminated supply of fresh air—that is, away from cars, buildings, and process exhausts. Try to locate intake vents so that tampering from outside sources will be minimal.
- Equip the HVAC system with regulating generators that are flexible enough to adjust to the varying air pressures of the intake and outtake vents.
- Use only steam humidifiers.
- Use prefilters to clean the air before it passes over the higher-efficiency filters.
- Institute a preventive maintenance program when any system is installed. Provide for regular inspection of drain pans, filters, and any area of the HVAC that is accessible and that might fall prey to microbial contamination.

GROUNDS MAINTENANCE

To prevent accidents and injuries resulting from tools and machines used in grounds maintenance, choose equipment for a specific purpose and use and maintain it properly. Ensure that broken equipment is repaired to its original operating condition or is replaced with new equipment. Workers should receive training specific to the tools and equipment they will be using.

Store and use fuels and chemicals, including pesticides and fertilizers, in accordance with the manufacturers' directions. The EPA defines *pesticides* to include herbicides, nematodicides, insecticides, larvicides, fungicides, and rodenticides. In addition, the EPA considers as pesticides antimicrobial compounds not regulated by the Food and Drug Administration and some swimming pool–sanitizing chemicals.

Electric-Powered Hand Tools

Apply the same rules of safe tool use outside the facility and inside the facility. The primary rules include the following:
- Use battery-powered tools outside, if possible.
- Use double-insulated tools connected to a circuit protected by a ground-fault circuit interrupter (GFCI).
- Use the correct tool for the task at hand.
- Keep landscaping tools in good condition.
- Use tools as they were meant to be used.
- Store tools in a safe place.
- Keep cutting tools sharp and clean.
- Keep tool handles smooth and strong.
- Use shovels, spades, and other digging tools with points that are smooth and properly shaped.

Electric Landscaping Tools

Read the operator's manuals carefully before using electric landscaping tools. If a tool is equipped with a three-prong plug, use a three-hole grounded receptacle and a three-wire extension cord. Use a GFCI outside. Keep visitors a safe distance from work areas, and clear the area of debris, animals, or anything else that could inflict or suffer injury or damage if encountered while the tool is operating. Do not force the tool; it will perform best and safest at the rate for which it was designed to operate. If the tool has a second handle, use both hands when operating the tool.

Never use electric-powered tools in the rain or when grass or shrubs are wet. Don't abuse the electric cord. Never carry a tool by its cord or yank the cord to disconnect the tool from the receptacle. Use tools that are Underwriters Laboratories (UL) approved or the equivalent. Always use cords that are approved for the location and that have the proper wire size to carry the current. Use the shortest cord length possible, and attach only one tool to the extension cord. Select the proper wire gauge for the extension cord. Wire gauge refers to the thickness of the copper wire, and a thick wire can carry more electricity (measured in amps) than a thinner wire. As a wire gets thicker, the gauge number gets smaller. In most cases, you can find the amp rating on the tool. In many cases, a 16-gauge extension cord up to 100 ft long will be able to handle tools up to 10 amps. For tools with 10 to 15 amps, a 14-gauge extension cord can be used for lengths up to 50 ft, and a 12-gauge extension cord should be used if the length of the cord is up to 100 ft. Be sure to avoid cutting the cord with the tool.

Always wear eye protection when using landscaping tools. Hold tools in position, ready for use, before switch-

ing on the current. Always shut off the current when resting your arms, while removing cuttings, or before changing the direction of a cut. If a tool becomes jammed or fails to start, always switch it off before trying to free the jam or before troubleshooting. When leaving the work scene, even for a few minutes, always shut off the current and disconnect the power plug.

Electric Hedge Trimmers

The usual injuries associated with using electric hedge trimmers are amputated fingers, deep cuts on fingers and hands, and cuts on knees and legs resulting from lowering the trimmer to rest the arms. These injuries result from five types of actions: (1) changing hand position with the trimmer running, (2) holding branches away from the cutting bar, (3) removing debris from the trimmer, (4) holding the trimmer with only one hand, and (5) failing to wait for the blades to stop after turning the trimmer off.

Choose hedge trimmers with the following features:
- a light enough weight to hold comfortably for a long time
- a large support handle for two hands to hold
- a switch that requires continuous finger pressure to maintain power
- a battery-powered or double-insulated mechanism
- the UL label.

When using a trimmer, workers should get into a comfortable position, use both hands, avoid working in cramped spaces, take their time, and not force the tool. They also should not overreach or lean off a stepladder. When they leave the work area, even for a coffee break, they should take the trimmer with them. Workers should also follow company regulations for securing tools if there is any chance that an unauthorized person might try to use the tools.

Lawn Trimmers and Edgers

Lawn trimmers are useful for cutting grass around tree trunks or along fences. Edgers cut borders along the edges of sidewalks, driveways, and gardens. Each year, trimmers and edgers injure several thousand people; workers should thus handle these tools carefully because the tools have metal cutting blades that can propel debris or cut a finger. Keep the guards on trimmers in place and in working order, and keep the blades sharp. Workers should not put their hands near a blade unless the machine is turned off.

Nylon-cord weed trimmers cannot cause as much injury as metal-blade trimmers or edgers can. However, getting hit by the cutter cord can sting. Operators should disconnect a trimmer's power cord when adjusting the cutter cord's length or changing the reel. Take the same precautions when using a weed trimmer as when using any other electric appliance: do not use electric edgers and trimmers in wet areas. Periodically check electric cords for cracks or breaks in the insulation.

Gasoline-Powered Equipment

Observe the following safety rules when handling gasoline:

- Never use gasoline to clean floors, tools, clothes, or hands. Gasoline should be used only as a source of energy in engines.
- Always store gasoline in an approved, closed container.
- Avoid pouring gasoline from one container to another. Doing so might generate a charge of static electricity, which could ignite the gasoline. To avoid generating static, maintain metal-to-metal contact.
- Clean up gasoline spills immediately to prevent the accumulation of vapors. Do not allow electric switches to be turned on until the gasoline vapors have dispersed.
- If gasoline is spilled, remove any saturated clothing immediately and keep yourself and the clothing away from sources of ignition. Wash the affected area of skin with soap and water to prevent a skin rash or irritation. If gasoline enters the eyes, flush them with water and get medical attention.
- When draining or dismantling gasoline tanks or equipment that likely contains gasoline, do so outdoors or in a well-ventilated area that is free from sources of ignition.
- Never smoke in fueling areas, fuel-system servicing areas, bulk-fuel delivery areas, or any other areas where fuel is present.
- Never dispense gasoline into a fuel tank while the motor is running or hot.
- Never store equipment with fuel in its tank inside a building where vapors could reach an open flame or spark. Before storing equipment in any enclosure, allow the engine to cool.
- If running an engine indoors, ensure proper ventilation to prevent a carbon monoxide hazard.

Gasoline-powered mowers and tractors should meet the standards of ANSI/OPEI B71.1–2003, Safety Specifications for Consumer Turf Care Equipment—Walk-Behind Mowers and Ride-On Machines with Mowers. Snow throwers should meet ANSI/OPEI B71.3–2005, Safety Specifications for Snow Throwers.

Power Lawn Mowers

Power lawn mowers, especially the rotary type, have proved to be a mixed blessing. Although they save time and effort and leave a lawn neatly manicured, they also take lives and cause more than 100,000 injuries a year. These injuries range in severity from minor cuts to amputations.

Reel mowers have several blades that shear the grass against a horizontal, stationary edge. Because such mowers do not need to operate as quickly as rotary mowers do, they are safer than rotary mowers. However, they are not as capable for cutting tall grass or weeds.

Safety Precautions before Mowing

Make sure the operator of a mower is well trained before using the mower. If the mower is being used for the first time that season, have the operator review the instruction manual. Before mowing, the operator should pick up rocks, glass, tree branches and twigs, and any other objects that could become lethal missiles if deflected off the mower blade. The operator also should observe the locations of fixed objects such as pipes, lawn sprinkler heads, and curbs, which could damage the mower or break apart and become missiles. The operator should make any adjustments to wheel height before starting the mower and disconnect the spark plug wire when cleaning, repairing, or inspecting the mower. Unauthorized persons should be kept out of the mowing area. Before starting the mower, the operator should quickly inspect for loose nuts and bolts, check blade condition and the engine's oil level, and fill the fuel tank using a vented can with a flex spout. The operator should wear protective footwear and safety glasses and use sunscreen, a brimmed hat, full-length trousers, and a long-sleeved shirt to protect against sunburn.

Safety Precautions during Mowing

Instruct operators to mow in daylight or good artificial light. Operators should, as much as possible, push the mower forward and avoid pulling it backward because pulling backward can injure the feet. When a slope or terrace must be mowed, have operators make a series of back-and-forth, horizontal passes along the incline. If an operator pushes the mower up the incline, the mower could drift back onto the operator's foot. If operators push the mower down the incline, they might lose their footing and fall into the mower.

Forbid operators from using the mower when the grass is wet and slippery. If the grass is damp or high, have them cut it at a slower speed, if possible, and set the cutting height higher than that for dry grass. Otherwise, the discharge chute may clog up.

Rotating blades can pick up stones, pieces of wire, nails, and other objects hiding in the grass and propel them out of the discharge chute at enormous speeds. Newer models may have a guard over the discharge chute that deflects objects downward, but the guard may have to be removed if the grass catcher is used. Some mowers have guards that automatically snap back into place when the grass bag is taken off. Others require the guard to be bolted in place any time the catcher is not used.

Before removing the grass catcher to empty it, operators should shut off the engine and wait until the blade has stopped completely. Operators should also shut off the engine when attempting to free obstructions from the discharge chute, adjusting the cutting height, or performing any operation necessitating placing hands or feet near the blade.

Riding Mowers
Riding mowers are best suited for cutting large areas of lawn. Suggested safe practices for using a riding mower include the following:

- Operators should be fully instructed on using the controls and on stopping quickly. Operators should also read the owner's manual at the beginning of each mowing season.
- Operators should clear from the work area objects that might be picked up and thrown and identify fixed objects that might damage the mower.
- Operators should disengage all attachment clutches and shift into neutral before attempting to start the engine.
- Operators should disengage the power to attachments and stop the engine before making any repairs or adjustments. They should also disengage the power to the attachments when transporting or not using the attachments.
- Before leaving the vehicle unattended, operators should disengage the power takeoff, lower the attachments, shift into neutral, set the parking brake, stop the engine, and remove the key from the ignition.
- When mowing, operators should watch for holes in the lawn and other hazards.
- Operators should use a push mower rather than a riding mower to mow sharp corners or steep slopes. If they do mow steep slopes with a riding mower, they should mow up and down rather than across. They should avoid steep slopes altogether if the slopes are wet. To prevent tipping or loss of control on slopes or sharp turns, operators should reduce speed and avoid starting or stopping suddenly.
- Operators should not mow the area between trees through which the rear wheels will not pass. Riding mowers are known to turn over backward as a result of the extreme power in the rear wheels.
- When changing direction or turning around, especially on slopes, operators should use extreme caution. They should not back up without being certain it is safe to do so. They should watch for traffic when crossing or working near roadways. When using attachments, they should direct the discharge of material away from anything that could be hurt or damaged by that material.
- Operators should keep the vehicle and its attachments in safe operating condition and keep safety devices in place. They should tighten nuts, bolts, and screws, especially the blade-mounting bolts. If the mower or its attachments strikes a solid object, the operator should stop and

inspect the mower for damage and repair that damage before restarting and operating the mower. Operators should not change the engine's governor settings or overspeed the engine.

Utility Tractors
The U.S. Consumer Product Safety Commission offers the following suggestions for the purchase, safe use, and maintenance of utility tractors.

Purchase. Specify that utility tractors, including mower attachments, have safeguards for all moving parts. This will reduce the hazard of coming into contact with belts, chains, pulleys, and gears. Buy tractors with throttles, gears, and brakes that are easy to reach and that can be operated steadily and effortlessly. Be sure that safety instructions are provided with the tractor. Warning labels should be located on the machine.

Use and Maintenance. Before using the utility tractor, operators should read the owner's manual and heed its recommendations. Operators should also observe the following precautions:

- Never allow children or unauthorized persons to operate the tractor, and keep unauthorized persons away from the cutting areas during the tractor's operation.
- Wear sturdy, rough-soled work shoes and close-fitting slacks and shirts to prevent their entanglement in the moving parts. Never operate a utility tractor in bare feet, sandals, or sneakers.
- Always turn off the machine and disconnect the spark plug wire before adjusting the machine.
- For optimum stability on slopes, drive up and down rather than across.
- Start the tractor outdoors, not in a garage, where carbon monoxide can collect.
- Do not smoke near the tractor or near gasoline storage cans. Gasoline vapors can easily ignite.
- Replace or tighten all loose or broken parts, especially blades.
- Get expert servicing regularly—doing so may prevent serious injuries.

Accident patterns involving utility tractors include the following:

- *Overturning*—This can occur when driving over uneven terrain, steep slopes, or embankments. The driver can come into contact with the tractor when it overturns or sustain injuries during a fall. Utility tractors may also overturn if they are used to pull heavy vehicles out of mud or from a ditch. In these cases, the front end of the tractor might rise and turn over on the operator.

- *Backing up*—Sometimes when a tractor backs up and when the operator is inattentive, the tractor runs over bystanders. Many of the victims are young children that the operator did not see.
- *Igniting flammable liquids*—Storing gasoline near a utility tractor can be hazardous if the gasoline spills because spilled gasoline could be ignited by a spark or heat source.

Snow Throwers

All snow throwers are potentially dangerous. Their large, exposed mechanisms, which are designed to dig into the snow, are difficult to safeguard. With proper handling, however, snow throwers are a safer method of removing snow than the backbreaking, heart-straining manual method. The safest snow throwers have guards on their drive chains, pulleys, and belts.

The auger at the front of the snow thrower presents the greatest hazard. Some snow throwers have an additional auger to produce extra throwing power. These augers, along with the moving gears, drive chains, and belts, can endanger anyone who tampers with a snow thrower while it is running. Injuries usually occur when the operator attempts to clear off debris while the motor is running. Even cleaning the machine with a stick can be extremely dangerous if the motor is left on because the spinning blades can pull the stick from the operator's hand and propel it back at the person with great force if the clutch lever is also kept on. Some models have automatic stopping devices that take effect when the handle is released.

Although snow throwers handle dry, powdered snow with little difficulty, they handle wet, sticky snow less effectively. Wet snow tends to clog the blades and vanes and often jams and sticks in the chute. Snow throwers can also pick up and throw ice, stones, and other hard objects.

The following safety instructions for snow thrower operators come from the Outdoor Power Equipment Institute:
- Read the operator's manual.
- Do not allow children to operate the machine or allow adults to operate it without proper instructions.
- Keep all bystanders a safe distance away.
- Disengage all clutches, and shift into neutral before starting the motor.
- Keep hands, feet, and clothing away from power-driven parts.
- Never place a hand inside the discharge chute or even near its outside edge while the engine is running.
- Be familiar with the controls and how to stop the engine or how to throw the unit out of gear quickly.
- Disengage the power and stop the motor before cleaning the discharge, removing obstacles, or making adjustments or when leaving the operating position.
- Adjust the machine's height to clear gravel or crushed-rock surfaces.

- Exercise caution to avoid slipping or falling, especially when operating the machine in reverse.
- Do not operate the machine on slopes or on ground where there is a risk of slipping or falling.
- Never direct discharge at bystanders or allow anyone to cross in front of the machine. Potentially dangerous debris may be hidden in the snow.
- Keep the machine in good working order and keep safety devices in place.

Snow Shoveling

If the area to be cleared of snow is small, or if no snow thrower is available, someone will have to shovel the snow by hand. Only someone in good physical condition and health should do this shoveling.

The shoveler should mentally divide the area into sections, shovel one section, and then rest before shoveling the next section. Whenever the snow begins to feel especially heavy, the shoveler should take a break. Shovelers should also keep the following information in mind:
- Wet snow is much heavier than dry snow. Adjust the rate of shoveling accordingly.
- Push or sweep as much of the snow as possible.
- If an icy crust has formed over several inches of snow, shovel the snow in layers.
- Use small amounts of rock salt or other ice-melting materials to make the job as easy as possible.
- Dress warmly for shoveling snow because cold itself can strain the body's circulation. Do not bundle up so heavily, however, that movement is difficult.

Chemicals and Pesticides

Chemicals

Thoroughly train workers in the proper handling, use, and disposal methods of the chemicals used in grounds maintenance operations. Training should be conducted in accordance with appropriate regulations: 29 CFR 1910.1200, the Hazard Communication Standard, and/or 40 CFR 170.102 to 170.260, the Federal Insecticide, Fungicide, and Rodenticide Act (FIFRA). Safety and training requirements for workers and pesticide handlers are specifically addressed in 40 CFR 170, the Worker Protection Standard. If a contractor applies pesticides, that contractor must adhere to the applicable portions of FIFRA. All pesticide handlers must receive training, and applicators must be state certified if they work with restricted-use pesticides (RUPs). Some states have pesticide regulations that are more stringent than the federal FIFRA standard. In all situations, conduct personal protective equipment (PPE) assessments to ensure that the proper PPE is available and used by workers.

PPE includes coveralls, chemical-resistant suits, chemical-resistant gloves, chemical-resistant footwear, hearing protection, respiratory protection devices, chemical-resistant aprons, chemical-resistant headgear, and protective eyewear. Train maintenance workers to recognize poisonous plants, fruits, insects, and reptiles. Workers should avoid contact with poison oak, poison ivy, and poison sumac because these plants produce the toxin urushiol, which causes an allergic contact dermatitis. However, contact with one of these plants is not necessary for exposure to occur because urushiol can be transferred to and carried on anything that touches the poisonous plant. In addition, the use of burning to dispose of some poisonous plants results in airborne toxins that can be quite hazardous when inhaled by workers. Urushiol is, however, neutralized by water. After working outdoors, workers should therefore scrub their hands thoroughly with copious amounts of soap and water.

All foreign materials, such as glass, metal, and wire, should be removed from the grounds to be maintained. Workers should treat all cuts and scratches that they receive outdoors by thoroughly cleaning the affected area and applying an antiseptic covering.

Pesticides

Under OSHA's Hazard Communication Standard, each worker who will be exposed to pesticides must receive appropriate training and information if that training and information are not covered by the EPA's Worker Protection Standard for Agricultural Pesticides. The safe use of pesticides is everyone's responsibility. The pesticide user, however, has the primary responsibility, which begins the day a pesticide is selected and purchased and continues until the empty container has been disposed of properly. A U.S. Department of Agriculture county extension agent can help with choosing the appropriate pesticide to control specific pests.

Many organizations are moving toward integrated pest management (IPM) practices. IPM is an effective and environmentally sensitive approach to pest management that relies on a combination of commonsense practices. IPM programs use current, comprehensive information on the life cycles of pests and their interactions with the environment. This information, in combination with available pest-control methods, is used to control pest damage by the most economical means and with the least amount of hazards to people, property, and the environment. The use of IPM can also result in lower costs because of the smaller amounts of pesticides used and the reduction in hazardous wastes and, thus, of hazardous-waste costs.

All pesticides sold in the United States must display an EPA registration number on the label. This number indicates that the EPA has reviewed the product and found it safe and effective when used according to directions. Every pesticide label must include a list of what the product treats, directions on how to apply the product, potential hazards, and safety measures to follow when using the pesticide.

Before using any pesticide, read the label carefully. The label describes some of the hazards involved as well as antidotes and first-aid measures. Pesticides that bear the notice DANGER—POISON on the label are highly toxic. Breathing or ingesting them, or simply allowing them to remain on the skin, can be fatal. Pesticides marked with WARNING are moderately toxic and can be quite hazardous. Pesticides that have CAUTION on the label have lower toxicity but may be harmful if ingested or misused. Follow the label's instructions for mixing, handling, and applying the pesticide. *Do not guess* when working with pesticides.

Application

Any restricted-use pesticide needed at an industrial facility must be applied by a certified handler, according to FIFRA. Use the least toxic pesticide for the job in order to reduce hazards. Manufacturers have formulated different compounds to control the same pest, so use the compound that is least harmful to the plants that you want to keep alive. For example, a severe infestation may require the use of a phosphate ester (organophosphate) insecticide, which calls for wearing protective clothing and following other precautions (Figure 4–5). If necessary, consult the supplier before using phosphate ester.

Try to purchase just enough pesticide to last one season. Doing so should cut down on storage and disposal problems. Observe the following precautions when using pesticides:

Figure 4–5. When applying insecticides, be sure personnel are well trained and wear proper equipment. Using phosphate ester (organophosphate) insecticides requires meticulous precautions. Always choose the least toxic pesticide that will do the job.

- Use a pesticide only for the purposes listed on the label.
- Keep pesticides in their original, labeled containers. Check for leaks or damage to the containers.
- Mix pesticides carefully—outdoors, if possible. Keep them off skin, and avoid breathing their dust or vapors. Use protective clothing and equipment, including respirators, when using toxic chemicals.
- Designate a set of mixing tools—measuring spoons and a graduated measuring cup—for use with sprays and dusts only. Keep these tools with the chemicals.
- Avoid spilling pesticides. Designate a level shelf or bench in a well-ventilated area, preferably outside, for mixing chemicals. A level, uncluttered surface helps prevent spills. If chemicals do spill, wash hands at once with soap and water. Then hose down the mixing area.
- Never smoke or eat while spraying or dusting. Cover food and water containers when spraying near areas where watchdogs are kept.
- During application, stay out of the spray's drift. Avoid applying pesticides outside on a windy day. If it is windy, ensure that the applicator stays upwind of the spray.
- Avoid spraying near lakes, streams, and rivers, and make every effort to keep toxic residues from entering waterways.
- Use respirators that provide effective respiratory protection.
- If a pesticide gets on skin or clothing, immediately remove the clothing and take an all-body bath or shower using plenty of soap and water. Wash clothing before wearing it again.
- When finished using pesticides, immediately wash your hands with soap and water. Do not smoke, eat, or drink without washing first.
- Never allow unauthorized personnel to enter treated areas or pesticide mixing, storage, and disposal areas.

Safe Storage of Pesticides

Store all pesticides in a well-ventilated, locked area or building. Store pesticides in their original, tightly closed containers. This way, the labels can be consulted for information in the case of accidents. Keep soap and plenty of water near storage areas. Seconds count when washing poisons off skin.

Do not store clothing, respirators, food, cigarettes, or drinks near pesticides. Such items might absorb poisonous vapors, fumes, or dusts or soak up poisonous spills.

Disposal of Pesticides

Dispose of all pesticides according to the instructions on the containers.

Emergency Information

If a dangerous situation occurs when using pesticides, obtain additional advice and information on antidotes for specific pesticides from the following agencies: the local poison control center, the state department of health, the county agricultural extension agent, and the regional office of the EPA. These agencies maintain current information on all compounds and their constituents and on recommended treatments in case of poisoning.

Another source of valuable information is a Safety Data Sheet (SDS), formerly Material Safety Data Sheet (MSDS). The Occupational Safety and Health Administration requires pesticide manufacturers to provide SDSs. Contact your supplier or the manufacturer to obtain MSDSs for the pesticides you use. An SDS contains the following information:
- company information
- identity of the chemical
- hazardous ingredients
- physical data
- fire and explosion hazard data
- health hazard data
- reactivity (instability) data
- spill or leak procedures
- special protection information
- special precautions.

COMPUTERIZED PREDICTIVE MAINTENANCE

Computerized predictive maintenance (CPM) and computerized maintenance management software (CMMS) can reduce employees' exposure to hazards, decrease equipment downtime, and optimize the effectiveness of maintenance expenditures. By reducing the costs of equipment repair and downtime, an effective CPM program reduces the total controllable maintenance costs to a minimum. These savings come from the following:
- less exposure of employees to malfunctioning equipment and thus fewer accidents
- less lost production time
- fewer unpredictable equipment failures
- efficient scheduling of equipment repairs and downtime
- fewer repairs and lower costs
- improved and safer use of labor
- longer life of equipment.

By monitoring equipment before problems start, the CPM program becomes the ounce of prevention needed to eliminate the costs incurred from breakdowns. The benefits of a CPM program, however, typically extend beyond maintenance operation and encompass other functions of the facility, including the safety function.

An effective CPM program not only alerts the proper personnel to potentially hazardous conditions, such as equipment failures, but also provides the record keeping required by state and federal safety regulations. The following sections describe how a typical CPM program works.

Computerized Assistance

The most effective and economical method of planning, scheduling, and tracking CPM tasks is to schedule them immediately after the last service date. This scheduling can be accomplished most effectively with computerized assistance. Use a computer software program that is consistent with the concepts and techniques on which maintenance system services are based. Principal features of the program might be as follows:

- CPM task definition and identification of the cycle during which the tasks are to be performed
- CPM specifications and procedures that define the craft, standard hours, work procedures, and measurements to be performed for each task
- automated scheduling based on the last service date and predetermined task cycles
- CPM preparation of work orders for all tasks according to available labor
- reports, including compliance, performance, forecast, and budget planning reports
- records of equipment history for each CPM task.

Often, the most difficult aspect of using computerized assistance is ensuring that information is entered into the database correctly and in a timely fashion.

Maintenance Diagnostic Technology

Various studies of production costs in U.S. manufacturing and processing facilities have revealed that maintenance normally accounts for 15% to 40% of the total production cost. Other studies show that maintenance costs represent an average of 28% of the total cost of goods sold.

More significantly, every dollar saved in maintenance directly lowers a company's bottom line. Although cost-effective maintenance adds to long-term profitability, top management has sought to keep costs low by refraining from investing in long-range maintenance improvements. Hence, facility engineering managers have had to struggle to improve maintenance effectiveness with minimal resources. Development of new technologies to support maintenance has lagged behind the efforts to improve and automate production, but the situation is changing.

As pressure on U.S. industries to improve quality and productivity continues, new priorities are being set, and new opportunities to improve facility maintenance by applying technology are being discovered.

Here is a review of some of the more significant technologies available.

Laser Shaft Alignment

Laser technology has now been adapted as a small, relatively inexpensive instrumentation that is useful not only in the initial alignment of machinery and shafts but also for monitoring critical shaft alignment. This new capability can help prevent many problems with rotating machinery.

Laser systems that instantly sense and display minute movements of fixed objects are now available.

Ultrasonic Testing

Instruments designed for ultrasonic testing sense the ultrasound waves produced by operating machinery as well as the turbulent flow of leakage (Figure 4–6). These instruments also provide fast, accurate diagnoses of such wasteful issues as valves in a blowy mode, faulty steam traps, and vacuum and pressure leaks. Airborne ultrasonics is extremely useful for detecting mechanical problems, especially possible bearing failure.

Figure 4–6. This portable inspection instrument uses ultrasound to detect leaks, line blockages, bearing failures, faulty valves, steam traps, and electrical problems. *(Courtesy U.E. Systems, Inc.)*

Oil Analysis

Oil analysis has become an important part of preventive maintenance. Laboratories recommend that samples for analysis be collected at scheduled intervals. Doing so is important to identify trends and detect abnormalities. The length between the sampling intervals varies depending on the types of equipment and operating conditions.

A typical oil analysis includes tests for viscosity, water or coolant, fuel dilution, solids, fuel shoot, oxidation, nitration, total acid number, total base number, particle count, and spectrographic analysis.

Wear Particle Analysis

Wear particle analysis is similar to oil analysis in that the particles to be studied are collected by drawing lubricated oil samples. But there the similarity ends. Oil analysis determines the condition of the lubricant itself, which sometimes can be used to draw conclusions about the machinery from which the sample was taken. In contrast, wear particle analysis provides specific information about wearing conditions in the machinery.

Particles in the lubricant of a machine can provide significant information about the condition of that machine. This information relates to particle shapes, compositions, sizes, and quantities.

Infrared Imaging

Infrared imaging is a visual representation of the heat energy that objects radiate in proportion to their temperatures and emissivities. At normal temperatures, most of this energy is in the infrared spectrum and is thus invisible, but it can nevertheless be measured. Most imaging systems detect infrared wavelengths in the range of the electromagnetic spectrum between 2 and 5.6 μ or between 8 and 14 μ.

A thermograph can be used directly to find and monitor thermal anomalies, or "hot spots," which indicate problems to be investigated and corrected. Successive recorded thermal images can be used to produce quantitative surveys of actual temperature trends.

For infrared surveys to be effective diagnostic tools, they should be conducted by a technician who is thoroughly trained in operating the equipment and interpreting the imagery, which requires an understanding of the facility systems being analyzed.

Vibration Analysis

The potential maintenance benefits of vibration monitoring and analysis have been recognized for decades, but only recently has vibration monitoring and analysis technology become available for widespread, practical use.

Machinery parameters that can be recorded include overall vibration amplitudes, vibration time waveforms, vibration amplitudes in specific frequency bands, phases, DC gap voltages, machine rotational speeds or other process variables, and various qualitative observations. Many data collectors provide graphic displays so that the information can be reviewed immediately. Data can be transferred between a host computer and a data collector through direct links or modems (see Figure 4–7).

MAINTENANCE CREWS

Select maintenance employees based on their experiences, alertness, and mechanical abilities. The employees should also be able to learn the essential safety principles of the machines or operations for which they will be responsible.

Training

Safety training programs for maintenance employees typically cover a wider range of subjects than do such programs for production workers.

Safety for maintenance workers involves dealing with a complex and constantly changing set of problems rather than a set pattern of activities. Furthermore, maintenance workers must know how to use not only hammers and drills but also ladders, protective equipment, chains, slings, ropes, and many other tools and equipment.

To protect themselves and others working in a maintenance area, maintenance crews must be aware of job hazards and be receptive to proper training. Their training programs should include first-aid and lifesaving techniques. In industries where irritating, toxic, or corrosive dusts, gases, vapors, or fluids are present, maintenance employees should receive training that familiarizes them with the properties of these substances and the methods to control the hazards associated with the substances. Some companies have the purchasing department notify maintenance crews when new chemicals are purchased so that the crews can institute the necessary precautions.

Before beginning nonroutine jobs, call together the maintenance crew members so that they can discuss the hazards involved and determine how to do the job safely. The crew should check the equipment and notify the supervisor before beginning work that seems unsafe. Before using the tools and tackle required to do special jobs, the crew should inspect the equipment for wear and defects. When special tools are needed to make a job safer, have the engineering department provide design and construction specifications.

For especially complicated or hazardous jobs, a safety and health professional may be called upon to help in planning. Construct scale models to determine clearances, the best methods of moving equipment and personnel, and sequences of action. After several trials have been made and the crew has agreed on a safe procedure, record the various steps as a guide for each worker (see, again, Figure 4–1).

In the course of their daily work, the maintenance crew members travel throughout the facility and become familiar with every machine and process. If properly selected and trained, each crew member can do much to locate and correct unsafe conditions in both the facility and the equipment.

In small companies, the responsibilities of maintenance personnel may include inspecting and caring for portable power tools, extension cords, and the like. If so, incorporate special procedures and personnel training.

Preventive and predictive maintenance (PPM) can reduce employee exposure to hazards, decrease equipment downtime, and optimize the effectiveness of maintenance expenditures.

As shown (right), by reducing costs due to equipment being out of service and repair costs, an effective PPM program reduces total controllable maintenance costs to a minimum.

These savings come from:
- Reduced employee exposure to malfunctioning equipment;
- Reduced production lost time;
- Fewer emergency failures of equipment;
- Scheduled equipment outages;
- Lower repair frequency and cost;
- Improved and safer labor utilization;
- Extended equipment life.

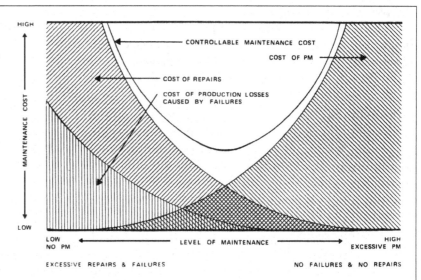

By allowing systematic monitoring of equipment and servicing it before trouble starts, a PPM program is the "ounce of prevention" needed to avert the costs incurred from unexpected breakdowns.

But the benefits of a PPM program typically extend beyond the maintenance operation to encompass other areas of the plant, including the safety function.

An effective PPM program, for example, not only alerts the proper authorities to potentially hazardous conditions—such as equipment failures—it also facilitates the recordkeeping required to conform to state and federal safety regulations.

Figure 4–7. Maintenance costs can be controlled by using computerized predictive maintenance.

Preventive Maintenance Plans

Because the function of facility maintenance is to keep equipment in top operating condition, a good maintenance system addresses equipment breakdowns before they happen—that is, crews must carry out preventive maintenance.

At minimum, supervisors should set up a preventive maintenance plan for essential equipment and for machinery that might seriously compromise the safety of workers. A good preventive maintenance program starts by listing buildings, machinery, and equipment that require periodic inspection, adjustment, cleaning, and lubrication as well as adjustments of safety guards or changes in the types of guards.

Keep detailed engineering drawings and specifications on file for each machine or building. For machinery, specifications should include dimensions, weights, sizes, locations of utility service connections, lubrication requirements, and details on bearings and on power transmission or drives. For buildings, specifications should include general layout, services available, floor load capacities, ceiling clearances, column spacing, and other features.

After analyzing its maintenance responsibilities, management should develop an organization chart that indicates where each employee is assigned. This chart gives supervisors a comprehensive overview of their work force and indicates the status and training of key employees and reserve employees. The chart should also list outside specialists who can be called on for help with unusual or hazardous jobs. This information is particularly valuable in emergencies, when competent people are needed without delay.

Inspection of Equipment

Base the schedule for inspecting equipment on the trouble points listed on maintenance records. The schedule can be determined based on the number of inspection reports turned in by full-time inspectors or by maintenance crew personnel.

Inspectors must be thoroughly familiar with the equipment to be inspected. This requirement is especially important regarding electrical equipment inspectors because electrical equipment usually gives few indications of impending trouble. Mechanical equipment, on the other hand, often explicitly signals deterioration by producing unusual noises, taking on unusual appearances, or producing substandard outputs.

Inspectors should use appropriate instruments to make their observations. Although inspectors can readily recognize burned contacts on a motor starter or hear the pounding of a worn gear, they need suitable instruments to check the insulation resistance of a motor or to measure the wear on a shaft, for example. (See Chapter 18, Woodworking Machinery, and Chapter 20, Metalworking Machinery, in this volume. See also OSHA Section 1910.132, Personal Protective Equipment, General Hazard Assessment, Training Trends.)

Personal Protective Equipment

Maintenance workers should dress appropriately for their specific jobs. They should wear snug-fitting clothes with a few small pockets. Breast pockets are often sewn closed or removed to prevent items in those pockets from dropping into machinery or hard-to-reach places when the wearer leans over. Workers should not wear neckties, wristwatches, or rings or other jewelry. Workers should also keep loose rags clear of moving machinery.

If workers carry so few tools that they do not need a toolbox, they should wear a special belt fitted with tool carriers. To prevent back and spine injuries upon landing after falls, workers should carry tools at the side instead of in the back portion of the belt. (See Chapter 17, Hand and Portable Power Tools, in this volume.)

Workers who handle rough or sharp objects should wear gloves or hand leathers. Welding gloves, rubber gloves for electrical insulation, and chemical-resistant gloves for handling acids should be worn as needed but never around moving machinery.

Every maintenance worker's tool kit should include an explosion-proof flashlight because a flashlight is often needed when working in dark places that potentially contain hazardous components. Workers should wear side-shield safety glasses while working, and every tool kit should contain goggles. Different kinds of goggles can provide additional protection from flying objects and molten metal, injurious heat and light rays, dust and wind, and acid splashes.

When workers must work in high places or must enter a manhole, bin, or tank, they should be properly trained in confined-space entry procedures and wear full personal protective equipment (PPE). (See Chapter 7, Personal Protective Equipment, for details on PPE, and Chapter 2, Buildings and Facility Layout, for more information on confined-space procedures—both chapters in this volume.)

Lockout/Tagout

The details of maintaining power presses, other types of machinery, electrical equipment, and boilers are covered in other chapters of this volume. Safe tank entry procedures are included in Chapter 5, Fired Pressure Vessels (Boilers) and Unfired Pressure Vessels; energy-isolation procedures and instructions are described in Chapter 6, Safeguarding; and electrical lockouts are discussed in Chapter 8, Electrical Safety.

Piping

Maintenance crews working on water, steam, and gas pipes must exercise extreme caution. However, when piping is properly identified and correctly handled, the hazards are greatly reduced.

Proper Identification of Piping

Accidents have resulted from improperly identified piping. This is especially true in factories where the high- and low-pressure steam lines run next to compressed-air, sprinkler-system, and sanitary lines. Supervisors can prevent maintenance crews from opening the wrong valves or disconnecting the wrong pipes by clearly identifying piping. Work out a system of color schemes, tags, and stencils so that the contents of pipes can be determined at a glance (Figure 4–8). Identification is particularly important when emergencies occur or when outside maintenance workers perform services. To prevent confusion, use a consistent color code for piping, such as the one given in the merged ASME/ANSI A13.1 2007, Scheme for the Identification of Piping Systems. This standard merged the 1981 and 1996 versions of ANSI A13.1. The ANSI/ASME A13.1 standard offers a common labeling method for pipes that can be used in all industrial, commercial, and institutional facilities and in buildings used for public assembly. The 2007 edition of ANSI/ASME A13.1 changed the color scheme requirements for the labels from four to six standard colors, basing the requirements on the characteristic hazards of the contents. Existing schemes for identification should be considered as meeting the requirements of the standard if the schemes are described in writing and employees are trained in the operation and hazards of the piping system. The requirements do not apply to pipes buried in the ground or to electrical conduits. (See also the section Use of Color in Chapter 2, Buildings and Facility Layout.) Table 4–A shows current color requirements.

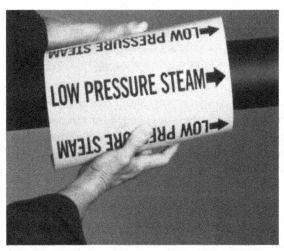

Figure 4–8. This sign indicates the direction of flow and the contents of the pipeline. *(Courtesy W. H. Brady Company, Signmark Division)*

TABLE 4–A. Designation of Colors for Piping

Fluid Service	Background Color	Letter Color
Fire-quenching fluids	Safety red	White
Toxic and corrosive fluids	Safety orange	Black
Flammable fluids	Safety yellow	Black
Combustible fluids	Safety brown	White
Potable, cooling, boiler feed, and other water	Safety green	White
Compressed air	Safety blue	White
To be defined by the user	Safety purple	White
To be defined by the user	Safety white	Black
To be defined by the user	Safety gray	White
To be defined by the user	Safety black	White

(Used with permission from the American National Standards Institute.)

Proper Isolation of Piping

Before working on a pipeline, shut off the line, lock and tag the valves, relieve pressure from the applicable section of the line, and drain the line. Shutoff valves are usually of the gate type. However, if globe valves are used, the pressure side should be under the valve seat so that the packing will not have to hold back the pressure when the valve is closed.

If pipelines that carry chemicals must run overhead, isolate or cover them so that they will not drip on workers or materials underneath. In case chemicals do drip on workers, provide emergency showers with plainly marked locations. Test the emergency showers periodically and maintain them properly. Give complete instructions to maintenance crews as well as to operating personnel about the locations and availability of emergency showers.

Take special safety precautions when maintaining pipelines, valves, and bolted flanges—especially when hazardous materials are involved. The section on Problems with Hazardous Materials in Chapter 12, Materials Handling and Storage, in this volume, contains pertinent recommendations for both operating and maintenance personnel.

When hazardous materials are encountered, the supervisor should issue special protective equipment such as chemical-protective goggles, protective suits, rubber gloves, and respiratory-protective equipment.

To prevent their hands from slipping on pipes, maintenance personnel should wipe excess oil off pipes and fittings.

Workers should wear gloves when handling pipes and fittings, especially when the ends are threaded. They should check for burrs and file them off immediately.

Industrial Gas Lines

When mechanics encounter industrial gases such as propane and butane, they should understand those gases' behaviors, storage characteristics, and pipeline arrangements. Mechanics should have a copy of the piping layout showing the location of safety features equipment such as soft heads and backfire preventers. The layout should also indicate the locations of sectional and main shutoff valves so that, in case of an emergency, maintenance personnel can find these valves quickly.

Handling Pipe

When maintenance involves a considerable amount of pipe work, arrange lengths of pipes, valves, and fittings so that the floor is not overloaded. If possible, move the materials to strategic points as the job progresses. Doing this will prevent material from accumulating in one spot and will reduce the amount of handling. Avoid touching energized electrical wires.

When lengths of pipe are transported mechanically, attach red warning flags to pipes that extend beyond the conveyance. If the load is allowed to remain near passageways, the flags should stay on the pipes, and additional warning signs or barricades should be erected to alert workers to the load's presence.

Long lengths of pipe being run overhead and continuously joined should be pulled up from the floor with overhead rigging, moved to the proper location, and secured with tie ropes, wires, or fixture straps. Using overhead rigging prevents back strains and the possibility of workers falling from ladders and also prevents the material from falling.

Crane Runways

Before mechanics work on or near an overhead crane runway, they should notify the crane operator and make sure the operator understands what will be done and what part the operator should take in the work. Mechanics should also provide a temporary rail stop between the crane and the point where the work will be done. (See Chapter 13, Hoisting and Conveying Equipment, in this volume.)

When work must be done on the crane itself, the person in charge is responsible for locking out safety switches, placing warnings indicating that people are working above, and seeing that a careful inspection is made after the job is completed. This inspection should include checking for parts or tools left on the crane, which might later drop into the mechanism or fall on workers.

When working on a crane runway, use a hand line to eliminate the need to carry material up a ladder. The line should be tied to a part of the crane, not attached to the worker's body. This will prevent the worker from becoming entangled in the line or from being pulled off the runway.

Final Check: Tools and Guards

After a machine has been repaired, mechanics should, if possible, first turn it over by hand. In doing so, they may discover mislaid tools or materials in time to prevent them from wrecking the machinery and perhaps causing injuries. To prevent tools from being left on or near machinery after a repair job, mechanics should use a toolbox with a designated spot for each tool. That way, mechanics can quickly see when a tool is missing. Mechanics could also use a tool check system to help them keep track of particular tools used for the job.

Mechanics should replace and securely fasten guards, pick up tools, and leave a work area in as good a condition as possible. Doing so helps promote good housekeeping.

Lubrication

Components that do not work because of lubrication issues often result in catastrophic failures. As a rule, lubrication should be taken care of according to its own schedule. Nevertheless, it is an important part of a preventive maintenance program. Supervisors should make a complete survey of the facility to determine lubrication requirements. This information should be entered into the record of each piece of machinery. Inspectors should then check the machinery for missing fittings, missing oil cups, and plugged oil holes. If possible, make repairs and upgrades at this time. For example, install automatic-feed oil cups, mechanical force-feed lubrication systems, and special fixtures. These features allow oiling personnel to reach parts in remote locations without being in danger of moving parts.

Provide oiling personnel with a diagram of each machine and with all the parts that require lubrication clearly marked. The diagram should include a chart that tells the oiler the kind of lubricant to use and how often to apply it. Some companies provide a color code that gives this infor-

mation. The purpose of the lubrication, the oil cups, the oil feeds, and the oil holes is indicated by the color code, and the oiler can refer to the color-coded chart.

Each oiler should have enough oils, greases, guns, fittings, and wiping cloths before starting a day's work. In large facilities, a special cart should be provided to hold all the equipment and supplies needed for the day.

Supervisors should transfer special precautions or instructions for each oiler from the master sheet to the oiler's personal schedule. These instructions should include the types of machines that must be stopped to ensure the oiler's safety and the machines that involve exposure to electrical elements and must have their power lines shut off.

Sometimes it is necessary to remove a guard from the lubrication point. Train workers to replace the guard properly in its fittings and holding devices so that it will clear all moving parts. If the guards or fittings are badly worn or damaged, workers should make a report immediately and request repairs or replacement.

If possible, shafting should be oiled while the machinery is at rest. Mechanics can use special pump oilcans with long spouts to reach overhead hangers or out-of-the-way bearings. Some of these oiling devices can be used without a ladder. If a ladder is used, it should not straddle machines that are running, and the oiler should not attempt to reach several countershafts from one position. Makeshift devices such as chairs or boxes should never be used in place of ladders.

Improper lubrication creates special problems; for example, lack of lubrication can result in overheating of bearings, shutdowns, and fires. Mechanics must be especially careful when lubricating electrical equipment. Overlubricating motor bearings causes oil to drop or to be thrown onto the insulation of the electrical windings. The oil then deteriorates the insulation, thus exposing live conductors that will arc and cause fires or cause electrically charged, ungrounded surfaces. If maintenance personnel work on overlubricated equipment, they may accidentally complete the circuit to ground with their body and be fatally shocked.

In addition, if the windings become sticky from oil and if dirt accumulates on them, defects may be covered up and go unnoticed until dangerous circumstances or a complete breakdown occurs. The oil and dirt accumulations could result in heat buildup, which could in turn cause a deterioration of the electrical windings and overheating of the bearings. The final result could be a shutdown or a fire.

Shop Equipment Maintenance

Provide repair facilities for maintenance crews. The stock of spare parts and units should be large enough to meet the probable need. Provide up-to-date machine tools, and arrange them so that they help the maintenance crew keep up an uninterrupted work flow. Mechanics should use hoists to handle heavy machinery and powered industrial or hand trucks to transport material from one department to another.

The maintenance shop should have a special area that is well ventilated, especially if welding, spray painting, or cleaning of metal machinery parts takes place in the shop. Place fire extinguishers in this area, and train the crew in using the different types of extinguishers for grease, oil, or electrical fires.

Special Tools

Use spark-resistant tools of nonferrous materials in areas where flammable gases, highly volatile liquids, and other explosive substances are stored or handled. Inspectors should check nonferrous tools before each use to be sure that they have not picked up steel particles, which could produce friction upon use and thus cause sparks.

Keeping Up to Date

The maintenance department should not overlook the potentials of new products such as cleaners, lubricants, paints, wood preservatives, insulation, floor-repair materials, protective coatings, and alloys. It should also keep up with new applications for existing products. Using new products and applications usually leads to better, safer maintenance practices.

Increased use of mechanized equipment requires a careful review of new potential hazards, especially in high-speed equipment and processes. Use of color throughout the facility, as specified in Chapter 2, Buildings and Facility Layout, will contribute to accident prevention and lead to higher operating standards.

Mechanical devices can help offset potential hazards. Centralized lubrication, centralized spray-painting equipment, floor-cleaning machines, steam cleaners, and other devices can help make a maintenance program safer. For frequently performed maintenance jobs, provide permanent accessories such as hoists, fixed ladders, and catwalks.

Reexamine maintenance procedures periodically to look for safer ways to do each job. Set up a special suggestion procedure for maintenance crews so that they can present new ideas or corrective measures.

Engineering books and service manuals should be part of the maintenance department's resource materials. Encourage supervisors and workers to become familiar with these materials and use them when necessary.

SUMMARY

- Facility maintenance includes (1) proper long-term care of buildings, grounds, and equipment; (2) routine care to maintain service and appearance; and (3) the repair work required to restore or improve service and appearance.

- The safety and health professional should be involved in maintenance and point out hazards or faulty equipment that needs attention. Maintenance inspectors should establish a regularly scheduled inspection program of the facility and grounds.

- Walls and floors must be inspected for damage, defects, and wear and regularly repaired or replaced.

- Roof-mounted structures must be regularly inspected and maintained. Roof damage can quickly lead to structural damage of other parts of the building and of the equipment inside the building.

- Tanks and towers must be maintained so that they can provide fire protection and to prevent serious accidents resulting from structural failure. Stacks and chimneys should be inspected at least once every 6 months.

- Platforms, loading docks, and concrete sidewalks and driveways should be inspected for damage or wear and repaired on a regular basis.

- Inspection and maintenance of underground utilities should be done only by trained, closely supervised personnel using proper protective equipment and other safety devices.

- Maintenance workers should replace faulty lamps, repair broken fixtures, and dispose of all lamp-related refuse in special containers. Workers are also responsible for maintaining adequate lighting levels in the building.

- Stairways and exits should be inspected for inadequate design or construction, improper handrails, poor lighting or housekeeping, and faulty treads or damaged surfaces. Exits should be kept clear and well lighted.

- To prevent accidents and injuries from grounds maintenance tools and machines, equipment specific to each job should be chosen and workers should be trained in their designated responsibilities.

- Workers must know the safe-practice rules for handling and operating electric-powered hand tools and gasoline-powered equipment.

- Workers must be carefully trained in the proper operation of utility tractors and snow throwers and must observe all safety precautions and manufacturer's instructions.

- Pesticides must be carefully selected, used, stored, and disposed of to prevent accidents and injuries. Workers should protect themselves by using PPE and decontamination procedures.

- Computerized predictive maintenance (CPM) can reduce employees' exposures to hazards, decrease equipment downtime, best use maintenance expenditures, and create efficient schedules.

- Diagnostic maintenance technology includes several methods that detect facility and grounds problems before they become serious threats to a company's operations.

- Management should select maintenance crews based on

their experiences, alertness, and mechanical abilities and train them in accident prevention.

- Supervisors should set up a preventive maintenance and inspection plan for critical equipment and machinery that could endanger worker safety and health.
- Maintenance workers must dress appropriately for each job, use proper PPE and tools, and know all safety procedures and practices applicable to the firm's operations. Maintenance supervisors and crews should keep up-to-date on the developments in their trade.

REFERENCES

American Conference of Governmental Industrial Hygienists, 6500 Glenway Avenue, Bldg D7, Cincinnati, OH 45211.
Industrial Ventilation Manual.

American National Standards Institute, 11 West 42nd Street, New York, NY 10036.
Criteria for Safety Symbols, ANSI/NEMA Z535.3–2007.
Environmental and Facility Safety Signs, ANSI/NEMA Z535.2–2007.
Inspectors' Manual for Escalators and Moving Walks, ANSI/ASME A17.2–2007.
Safety Code for Elevators and Escalators, ANSI/ASME A17.1–2007.
Safety Color Code, ANSI Z535.1–2007.
Safety Requirements for Fixed Ladders, ANSI A14.3–1992.
Safety Requirements for Material Hoists, ANSI A10.5–2006.
Safety Requirements for Portable Metal Ladders, ANSI A14.2–2000.
Safety Requirements for Portable Reinforced Plastic Ladders, ANSI A14.5–2007.
Safety Requirements for Portable Wood Ladders, ANSI A14.1–2000.
Safety Requirements for Workplace Walking/Working Surfaces and Their Access; Workplace Floor, Wall and Roof Openings; Stairs and Guardrails Systems, ANSI A1264.1–2007.
Safety Specifications for Consumer Turf Care Equipment—Walk-Behind Mowers and Ride-On Machines with Mowers, ANSI/OPEI B71.1–2003.
Safety Specifications for Snow Throwers, ANSI/OPEI B71.3–2005.
Scheme for the Identification of Piping Systems, ASME/ANSI A13.1 2007.

Claire, F. "Preventive and Predictive Maintenance—by Computer." *National Safety News*, August 1984.

Dunn, R. L. "Advanced Maintenance Technologies." *Plant Engineering* 41, nos. 80–86 (1987).

National Fire Protection Association, 1 Batterymarch Park, Quincy, MA 02269.
Flammable and Combustible Liquids Code, NFPA 30, 1993.
Life Safety Code, NFPA 101, 1994.

National Safety Council, 1121 Spring Lake Drive, Itasca, IL 60143.
Accident Prevention Manual for Business & Industry: Administration & Programs. 12th ed. 2001.
Accident Prevention Manual for Business & Industry: Environmental Management. 2nd ed. 2000.
Fundamentals of Industrial Hygiene. 6th ed. 2012.
Occupational Safety and Health Data Sheets:
Atmospheres in Subsurface Structures and Sewers, 12304–0550, 1987.
Blowtorches and Plumbers' Furnaces, 12304–0470, 1990.
Flexible Insulating Protective Equipment for Electrical Workers, 12304–0598, 1991.
Safety Hats, 12304–0561, 1992.

Outdoor Power Equipment Institute, 1901 L Street NW, Washington, DC 20036.

Payant, R., and B. Lewis. *Facility Manager's Maintenance Handbook.* 2nd ed. New York: McGraw-Hill, 2007.

Sack, T. F. *Complete Guide to Building and Plant Maintenance.* 2nd ed. New York: McGraw-Hill, 1971.

"Tight Syndrome Is a Breath of Stale Air." *Ohio Monitor,* October 1983.

Tucker, G., and D. Schneider. *The Professional Housekeeper.* 3rd ed., New York: Van Nostrand Reinhold, 1989.

Underwriters Laboratories, Inc., 333 Pfingsten Road, Northbrook, IL 60062.

U.S. Army, Corps of Engineers, Washington DC, General Safety Requirements, Manual EM385–1–1.

U.S. Department of Labor. Occupational Safety and Health Administration, 200 Constitution Avenue NW, Washington DC 20210. Code of Federal Regulations, Title 29.
29 CFR 1910.27, Fixed Ladders,

U.S. Department of Health and Human Services, National Institute for Occupational Safety and Health, Div. of Technical Services, 4676 Columbia Parkway, Cincinnati, OH 45226.

U.S. Department of the Interior, Environmental Protection Agency, Insecticide, Fungicide, and Rodenticide Act, Public Law No. 92-516, Oct. 21, 1972, published in *Code of Federal Regulations*, Title 40—Protection of Environment, Chapter 1, Environmental Protection Agency, Part 165.

REVIEW QUESTIONS

1. Facility maintenance can be described as:
   a. repair work performed to restore service and appearance.
   b. preventive care to ensure long-term life of company assets.
   c. routine care to maintain uninterrupted service and appearance.
   d. all of the above

2. List four items to check periodically for signs of excessive foundation settlement.
   a.
   b.
   c.
   d.

3. Which of the following, when inadequately maintained, is a major source of injuries?
   a. exterior walls
   b. windows
   c. floors
   d. ceilings

4. It is recommended that roofs be inspected:
   a. every year.
   b. every 6 months.
   c. every 3 months.
   d. every 2 months.

5. Describe four precautions to take to prevent roof damage from ice and snow.
   a.
   b.
   c.
   d.

6. What items should be considered when inspecting the condition of exits?
   a.
   b.
   c.
   d.
   e.

7. List five precautions to take to reduce the possibility of heating equipment breakdowns during cold weather.
   a.
   b.
   c.
   d.
   e.

8. Of the following, which have been associated with indoor environmental quality incidents?
   a. physical and chemical contaminants
   b. biological agents
   c. lack of fresh air
   d. all of the above

9. Preventive design is an important safeguard to prevent indoor environmental quality problems. List five of the seven basic elements that should be included in preventive design.
   a.
   b.
   c.
   d.
   e.

10. What are some of the hazards that can be prevented by properly training grounds maintenance workers?
    a.
    b.
    c.
    d.

11. List five worker actions that result in injuries from electric hedge trimmers.
    a.
    b.
    c.
    d.
    e.

12. Briefly explain the purpose of a computerized predictive maintenance (CPM) program.

13. The selection of maintenance employees should be based on what three criteria?
    a.
    b.
    c.

14. The use throughout the facility of which of the following contributes to overall accident prevention and leads to higher operating standards?
    a. cranes
    b. color-identification schemes
    c. goggles
    d. sprinkler system
    e. all of the above

15. What safety precautions should be taken when employees work underneath pipelines that carry chemicals?

Fired Pressure Vessels (Boilers) and Unfired Pressure Vessels

5

Teddy G. Gil, BSME, MBA, DAL Level II VT, UT, RT, PT, MT, IR

John F. Montgomery, PhD, CSP, CHMM

Introduction

Codes for Boilers and Unfired Pressure Vessels

Inspections of Boilers and Unfired Pressure Vessels

Fired Pressure Vessels (Boilers)
Design and Construction ▸ Annual Internal Inspection, Cleaning, and Maintenance ▸ Boiler Rooms ▸ Operator Training, Procedures, and Boiler Room Emergencies

Safety of High-Temperature Water
High-Temperature Water versus Steam and Cold Water ▸ Causes of Failure

Unfired Pressure Vessels
Design ▸ Internal Inspection ▸ Hydrostatic Tests ▸ Detecting Cracks and Measuring Thickness ▸ Operator Training and Supervision ▸ Safety Devices ▸ Autoclaves and Other Vessels with Quick-Opening Doors ▸ Steam-Jacketed Vessels ▸ Evaporating Pans

High-Pressure Systems

Pressure Gauges

Summary

References

Review Questions

INTRODUCTION

Fired pressure vessels (boilers) and unfired pressure vessels have many potential hazards in common as well as hazards that are unique to their specific designs and applications. Boilers and unfired pressure vessels can be found in workplaces as diverse as office buildings, hotels, hospitals, garages, warehouses, and of course manufacturing plants. These vessels can hold gases, vapors, liquids, or solids—toxic as well as benign—at pressures ranging from almost a full vacuum to thousands of pounds per square inch and at temperatures ranging from hundreds of degrees below 0 to well over 1000°F. Some examples of common household boilers are water heaters, pressure cookers, and steam heating systems. Some examples of common household and industrial unfired pressure vessels are barbeque LP gas tanks, aerosol cans, water towers, and service station fuel storage tanks. To help create a safer environment for the general public and for those who work with such equipment, this chapter covers the following topics:

- important codes governing boilers and unfired pressure vessels
- principles of inspecting this equipment, including hydrostatic testing
- boiler design, operation, and maintenance
- safety of high-temperature water systems
- unfired pressure vessel design, operation, and maintenance
- hazard control in high-pressure systems
- pressure gauges as safety controls.

Two main failure modes associated with boilers and high-temperature water heaters threaten the safety of those working around this equipment: pressure part ruptures and furnace explosions. Although very different in their causes, the results of either accident can be catastrophic. For unfired pressure vessels, pressure part rupture is the most dangerous failure mode.

Pressure parts refers to any component of a vessel, boiler, or water heater—such as tubes, drums, shells, or headers—that retains steam, hot water, or other fluids under pressure. When these parts rupture, a tremendous amount of energy can be released, often creating a violent blast that spews large amounts of steam or superheated fluid in all directions. This type of accident was very common (on steamboats and locomotives) in the 19th and early 20th centuries but has become much rarer in the last 40 years. Rupture can result from many factors, including the wasting away of pressure-retaining component thickness from corrosion or erosion, cracking of components, malfunctioning controls and inoperable pressure-relief valves, improper repairs to pressure parts, thermal shock, and operator error.

The basic cause of *furnace explosions* is the uncontrolled ignition of an excessive accumulation of fuel in the boiler furnace or its associated breeching and/or smokestack. Unlike pressure-retaining components, a boiler furnace is not designed to retain a significant amount of pressure. Rather, such a furnace is simply an enclosure in which the appropriate mixture of fuel and air is ignited and in which fuel combustion is maintained in stable condition. Many factors contribute to an unsafe accumulation of combustibles within a boiler furnace, but among the most common is the improper operation of control systems and combustion safeguards because of misapplication, faulty installation, or lack of maintenance. A common contributing factor in these situations is human error, which is itself often due to lack of understanding and failure to follow safe operating procedures, such as intentionally defeating safeguards in an effort to restore operations following plant upsets. Furnace gas explosions most often occur when lighting off a boiler, when an automatically fired boiler cycles on and off, or when a boiler was improperly isolated from fuel sources prior to being shut down for maintenance. Special care should be taken when working around a boiler while the boiler is being started, re-lit, or prepared for internal maintenance.

Fortunately, the use of up-to-date construction codes, maintenance and inspection standards, and control and instrumentation technology—combined with near-universal regulation by state, municipal, and provincial governments—have greatly reduced the frequency of boiler accidents in North America. Nevertheless, the hazards inherent in boiler and pressure-vessel operation are always present and thus require vigilance on the part of management to keep these hazards in check. To help explain the differences and provide details relating to each type of vessel, fired pressure vessels (boilers) and unfired pressure vessels will be discussed independently where applicable, with fired pressure vessels (boilers) discussed first.

CODES FOR BOILERS AND UNFIRED PRESSURE VESSELS

Design, fabrication, testing, and installation of boilers and unfired pressure vessels should comply with the applicable sections of the American Society of Mechanical Engineers' Boiler and Pressure Vessel Code (hereafter referred to as the ASME Code). The National Board of Boiler and Pressure Vessel Inspectors Code (hereafter referred to as the NBIC) governs the inspection, repair, and alteration of boilers and pressure vessels after they have been placed into service. In all 50 states, all Canadian provinces, and several U.S.

municipalities, the operations of boilers and pressure vessels are strictly regulated.

The Synopsis of Boiler and Pressure Vessel Laws, Rules and Regulations by States, Cities, Counties and Provinces, in the United States and Canada, document is available from the Uniform Boiler and Pressure Vessel Laws Society. This document indicates which governing bodies have made the ASME Code a legal requirement in their jurisdictions and what other compliances are needed. Similar information, along with contact information for the various state jurisdictional authorities, can be found on the National Board of Boiler and Pressure Vessel Inspectors' website, nationalboard.org.

For all questions regarding the installation, operation, inspection, or repair of boilers and pressure vessels, the jurisdictional authorities have the final say. Compliance with the ASME Code is determined by authorized inspectors commissioned by the National Board of Boiler and Pressure Vessel Inspectors (known as Authorized Inspectors (AIs) or Certified Individuals (CIs)) who are licensed by the state or provincial governmental authority charged with enforcing jurisdictional rules and regulations within its jurisdiction. In many cases, you can get help by contacting your property insurance carrier because many property insurance carriers employ licensed boiler inspectors who regularly provide services to the property insurance carriers' clients.

The ASME Boiler and Pressure Vessel Code contains the following 11 sections:

I. Rules for Construction of Power Boilers
II. Material Specifications
III. Nuclear Power Facility Components
IV. Rules for Construction of Heating Boilers
V. Nondestructive Examination
VI. Recommended Rules for Care and Operation of Heating Boilers
VII. Recommended Rules for Care of Power Boilers
VIII. Rules of Construction for Pressure Vessels
    Division 1
    Division 2—Alternate rules
    Division 3—Alternate rules
IX. Welding and Brazing Qualifications
X. Fiberglass-Reinforced Plastic Pressure Vessels
XI. Rules for In-Service Inspection of Nuclear Power Plant Components

These sections may be purchased directly from ASME. (See the References for ASME's address.)

The ASME Code and the NBIC pertain almost exclusively to the pressure-retaining components of a boiler and an unfired pressure vessel; the minimum requirements for the design, installation, maintenance, and operation of burners, controls, combustion safeguards, and boiler furnaces are found elsewhere. For larger high-pressure boilers (greater than 12 million Btu/hr heat input), controls and combustion safety topics are covered in the National Fire Protection Association's boiler-furnace standards, NFPA 85A, 85B, 85D, and 85E, which can be found at nfpa.org. For smaller boilers, low-pressure steam boilers, hot-water-heating boilers, and high-temperature water heaters, the applicable standard is ASME CSD-1, Controls and Safety Devices for Automatically Fired Boilers which can be downloaded from the American Society of Mechanical Engineers (ASME) at asme.org. All of these standards give guidelines for maintaining the controls and safety devices that are critical for preventing furnace gas explosions.

Before installing pressure vessels, obtain the services of a competent pressure-vessel engineering consultant or project manager. Such a professional can survey the facility or operation to determine the installation requirements, design a system that will satisfy the codes, and supervise installation and testing.

If the project manager suggests purchasing secondhand boilers or pressure vessels from another site, the permission of the jurisdictional authorities must be secured. It is best to arrange for properly commissioned inspectors to inspect used boilers and pressure vessels and indicate whether any repairs are necessary before the purchase. Again, the facility's property insurance carrier can supply inspectors licensed in the facility's state or province. In some cases, inspection by insurance company inspectors will satisfy all jurisdictional requirements, but in other cases, it may be necessary to arrange for jurisdictional inspectors to visit as well. (See the Synopsis of Boiler and Pressure Vessel Laws, Rules and Regulations by States, Cities, Counties and Provinces, in the United States and Canada, document.)

INSPECTIONS OF BOILERS AND UNFIRED PRESSURE VESSELS

As mentioned earlier, the ASME Code covers the design, fabrication, and inspection requirements only during the construction of boilers and pressure vessels. After the initial installation of the boiler or pressure vessel, the National Board of Boiler and Pressure Vessel Inspectors Code is the governing standard. The NBIC provides guidelines for inspecting, repairing, altering, rating, and rerating during the remainder of the boiler or pressure vessel's service life. Therefore, refer to the NBIC for guidance when repairing and altering boilers and pressure vessels. Also use the NBIC to supplement and elaborate on the safety and inspection procedures discussed in the rest of this chapter.

In general, install and maintain boilers and pressure vessels according to their manufacturers' instructions. Furthermore, train operating personnel not only in how to operate equipment properly but also in how to conduct routine safety checks and know when to call in qualified maintenance personnel. Any welding or welded repairs required on an ASME-constructed boiler or pressure vessel must be performed by a firm possessing certificates of authorization granted by the National Board of Boiler and Pressure Vessel Inspectors. These certificates of authorization are sometimes referred to as NB stamps because welded repairs must often be stamped with a symbol indicating that a properly authorized firm made the repair. Most such repairs are carried out under the National Board's repair authorization (i.e., R stamp).

Be sure to anticipate and prevent the following common causes of explosions in pressure vessels:
- errors in design, construction, and installation
- improper operation, human failure, and inadequate training of operators
- corrosion or erosion of heads, shells, tubes, pipes, and drums
- failure or intentional defeat of safety devices
- failure or override of automatic control devices
- failure to inspect and test thoroughly, properly, and frequently
- improper application of equipment
- overfiring
- lack of planned preventive maintenance.

In addition to explosion hazards, boilers also present fire hazards. Boilers are a significant cause of fires in hotels, stores, apartments, houses, and places of worship. Fuel leaks from oil-fired equipment are the most likely cause of boiler room fires, whereas gas leaks usually result in an explosion followed by a fire.

Because the majority of boilers in use are automatically or semi-automatically fired, they may operate unattended for long periods of time. Different state agencies have requirements about the lengths of time boilers can operate unattended. Many boilers are poorly maintained and are inspected only when they fail to operate, leaving them in less-than-perfect conditions. When fires start in unattended boiler rooms, they might gain considerable headway before being detected, unless adequate fire detection or suppression systems were installed. The means for controlling and containing fires from boilers include the following:
- Boiler rooms should be fully enclosed and built with noncombustible materials, such as ⅝-in. (1.6-cm) gypsum wallboard or thicker. Be sure to leave enough space for maintenance operations such as pulling tubes. Local building codes and some jurisdictional regulations have clearance requirements that builders must meet.

- Boiler rooms should have large door openings to allow easy access to and easy installation and removal of all boiler room equipment. Equip boiler room entryways with 1.5-hour fire-resistance-rated doors and door frames.
- A noncombustible ceiling over a boiler and automatic sprinklers over the firing end of the boiler and in areas containing gas and oil pipelines are strongly recommended. If boilers are coal fired, provide automatic sprinklers over the coal augers, feeders, chutes, and indoor coal piles. (Note: Other protection features may be required for coal piles, conveyor belts, and so on.)
- Proper clearance around the exteriors of boiler room walls prevents materials from being stored against the walls. In addition, store in that room only the materials and items that pertain to the boiler room's operation. If combustible items are needed in the boiler room, store no more than the necessary quantities of them in cabinets that are approved and designed for combustible storage.

To minimize low-pressure boiler fires and explosions caused by faulty controls and safety devices, observe the following safeguards:
- Establish a testing and servicing program in which operating controls, safety controls, gas and fuel lines and valves, and safety valves are regularly tested and maintained.
- To prevent damage to the valve seats, make sure that the safety and relief valves are always tested when the boiler is under pressure.
- Have repairs made immediately upon any discovery of malfunctioning operating controls or safety controls or leaking of safety and relief valves. Never operate a boiler that has a malfunctioning safety or relief valve.
- Have a service organization check and maintain the boiler during the heating season and also perform the normal out-of-season maintenance.
- Keep a boiler log in which operating pressures and temperatures are recorded at regular intervals and in which important safety device testing, maintenance activities, trip events, and abnormal operational incidents are recorded.
- This record can ensure that proper care is being taken of the boiler and its associated equipment.
- Retain these records because they provide a historical profile of the boiler throughout its service life and may be useful in a post-accident root cause analysis.

FIRED PRESSURE VESSELS (BOILERS)

In its simplest definition, a fired pressure vessel (hereafter referred to as a boiler) is a closed vessel in which water is heated and steam is generated or superheated under pressure by the direct application of heat. The most common heat source is the combustion of fuel, but heat sources also include electricity and waste heat from chemical processes. Heat is transferred through tubes, drums, or shells into the water and forms steam, hot water, or high-temperature water (HTW) under pressure. There are two general types of boilers, firetube and watertube, and they can be classified as high pressure, low pressure, steam boiler, or hot-water boiler. Low-pressure boilers operate at steam pressures that do not exceed 15 pounds per square inch (psi) and 160 psi for hot water, whereas high-pressure boilers operate at steam pressures that exceed 15 psi and 160 psi for hot-water boilers.

In firetube boilers, the hot products of combustion are passed through tubes that are surrounded by water contained in a cylindrical shell. In watertube boilers, water-carrying tubes are arranged so that they are surrounded by the hot products of combustion inside the boiler furnace. In general, watertube boilers are the preferred design for large or fluctuating steam loads. Boilers must be registered and certified in accordance with either state or local agencies. In some states, boilers are under the jurisdiction of the fire protection agencies. Also in some states, boiler operators must be licensed based on the amount of horsepower the boiler generates; check with your state agency or insurance provider for details.

Design and Construction

The ASME Boiler and Pressure Vessel Code provides basic guidelines for the design and construction of boilers and pressure vessels. Additional sources discussing the design and construction of boilers are *Mark's Standard Handbook for Mechanical Engineers* (Avallone and Baumeister 1996) and *Combustion: Fossil Power Systems* (Singer 1993).

Controls and Instrumentation

Requirements for the instrumentation necessary to monitor and control the combustion and steam production of boilers and water heaters (both with a heat input of 12,000,000 Btu/hr or less) and of HTW heaters can be found in ASME CSD-1, Controls and Safety Devices for Automatically Fired Boilers. For larger boilers, the requirements can be found in NFPA 85. Make sure that boiler operators are trained to check the safety devices and controls regularly. Operators should also use a checklist and fill out an inspection form (usually supplied by the insurance company or manufacturer). ASME CSD-1 and NFPA 85 furnish guidelines on daily, weekly, quarterly, and annual checks and tests that should be carried out.

The appropriate arrangement of controls relative to the location of operating personnel is given in Subsection C6 of Section VII of the ASME Code. The standard states that in general, a boiler unit should include a meter-and-control board that is situated on the operating floor. With the board in that location, the operator can see either the furnace door or the lighting ports of the burners and the water column of the boiler without needing to leave the control board. If the meter-and-control board cannot be situated in direct view of the operator, then reliable remote-indicating equipment—whose indication a second operator can confirm by making a visual check when lighting a boiler off—should be installed.

Economizers

Normally found on high-pressure steam boilers, economizers are constructed of coils or straight tubes that are arranged either inside the boiler itself near the flue gas discharge breeching or as a separate module mounted between the boiler and its smokestack or chimney. The economizer transfers heat from the exiting flue gas to the incoming boiler feedwater, thus raising the temperature of the water and reducing the amount of energy needed to convert the feedwater into steam.

Generally speaking, economizers should be equipped with at least one safety valve to protect the tubes or coils from overpressure conditions. A safety valve is especially needed if the economizer section can be closed off by isolating valves on the feedwater inlet and outlet lines.

Superheaters

Most steam heating and process loads require steam at saturation temperature—that is, the lowest temperature at which water is converted to steam at a given pressure. However, more energy can be imparted to saturated steam, and thus efficiencies can be gained by increasing the temperature of the steam already produced in the boiler. Increasing the steam's temperature is achieved through the process of superheating. The use of superheated steam is very common in the electric-power-generation and chemical-processing industries. Superheaters are sets of coils or tubes that are usually mounted in the convection section of a watertube boiler or, less commonly, are constructed as separately fired units through which saturated steam is passed to pick up additional heat. In either case, the superheater tubes, coils, and associated piping should be equipped with safety valves to prevent overpressure of these pressure-retaining components.

Air Preheaters

Another method by which boiler efficiencies are increased is the use of combustion air preheaters. Air preheaters come in several designs; however, the most common design for large boilers is the regenerative-style air heater (often referred to as Ljungström air heaters after their original designer). Regenerative-style air heaters are built like large wheels that rotate either vertically or horizontally into and out of the boiler flue gas discharge breeching. Heat is transferred from the exiting flue gas to the incoming air supplied by forced-draft fans to sustain combustion.

When a boiler is shut down for maintenance, unburned fuel may be carried through the boiler and deposited on the internal structures of the preheater. This accumulation of unburned fuel can be ignited on start-up and create a destructive fire inside, which is usually detectable only because of a sudden rise in the air heater's temperature. Fires inside air preheaters are difficult to extinguish and usually result in complete destruction of the preheater.

To prevent these fires, maintain proper combustion and use soot blowers correctly. Because an explosion could result, do not use a soot blower when a fire in the gas passages is suspected. Most importantly, prior to start-ups, thoroughly wash the internal structures of air preheaters with steam or hot water, as directed by the manufacturer.

Chimneys

Chimneys or smokestacks are used to direct out of the boiler room the products of combustion discharged from the boiler and release these gases at a safe elevation into the atmosphere. To reduce the possibility of carbon monoxide and other toxic gases being released into the boiler room and harming operators and other employees, regularly inspect chimneys for leakage. Chimneys—whether made of brick, concrete, or steel—should be equipped with grounded lightning arresters. If chimneys are not self-supporting, they should be fastened to sturdy building structures. Any ladders added to a chimney should be permanent and should be securely fastened to the chimney, protected with hoop enclosures, and designed according to applicable codes.

Ash Disposal Equipment

Solid fuel–fired boilers (i.e., coal or wood waste–burning units) must be designed with a system that removes ash. Ash removal from small boilers can be achieved by operators periodically raking the ash off the boiler furnace grates and into a disposal location. In large boilers, however, ash removal is a complex and automated system. In either case, ash coming out of the boiler is very hot and is light enough to be blown onto employees by even slight breezes. Operators working in and around ash removal areas should thus wear proper protective equipment: eye protection, gloves, and fire-resistant clothing.

Properly protect hoistways, driving machinery, conveyors, worm gears, ash sluices, and reciprocating pumps from ashes. A warning device (bell, horn, strobe light) must be hooked up to the driving machinery to indicate that the doors are about to open. Exercise special caution to prevent operating

personnel from being injured by the steam or hot water that may be present when the ash gates open. When excess carbon is present in ash pits and is not properly wetted down, a gas explosion might result when the gates open. Never store ashes near boilers or any combustible materials. Many ashes contain sulfur compounds that, on contact with water, form highly corrosive acids. These acids eat away metal surfaces and cause significant damage over time.

Water Treatment

Water treatment that is supervised by knowledgeable consultants helps extend the service life of a boiler and keep its thermal efficiencies high. Generally, water treatment should remove dissolved O_2 and CO_2 and maintain a pH that is basic enough to minimize corrosion. Removing dissolved gases from boiler water can be done mechanically using a deaerator vessel or by adding chemicals known as oxygen scavengers. The pH is usually adjusted by introducing caustic soda. The normally recommended pH for boiler water is between 8 and 11. (Note: Always check and maintain the pH recommended by either the boiler manufacturer or water treatment personnel.) Using water softeners to supplement the chemical treatment of boiler water helps reduce the buildup of scale and other deposits; such a buildup can lead to localized overheating. Because operators can sustain injuries when introducing boiler water treatment compounds into feedwater, provide operators with adequate protection against scalds and caustic burns. Using automatic feeding or softening equipment also reduces the chances of injury. Consult feedwater-treatment professionals for additional information. Storage of boiler chemicals must comply with OSHA regulations and good housekeeping practices.

Blowdown Pipes and Valves

Blowdown piping is used to remove sludge and other impurities from boiler water. If not removed, these impurities build up and could severely impede the efficiency and safety of the boiler. Generally, boilers need to be "blown down" once or twice a day, as directed by the water treatment consultant. Conduct blowdown piping and drain boilers to a discharge point that is not hazardous to operators or other personnel. All piping, operating, and discharge valves should conform to ASME Code Section I or Section IV. Galvanized piping should never be used in any steam service, including blowdown service.

Safety Valves and Fusible Plugs

Selecting, fabricating, installing, testing, and replacing safety valves and fusible plugs should comply with the ASME Code and the NBIC. Safety valves and fusible plugs (where installed) are critical to safely operating boilers and require special attention. Manufacturers, boiler inspectors from insurance companies, and other specialists can recommend the specific procedures for checking these safety devices.

Safety valves, when properly installed, relieve excess pressure or vacuum conditions (depending on the design), which could damage the equipment or injure personnel. Keep safety valves in good working order at all times, and have them inspected by qualified personnel in accordance with insurance company recommendations, rules of the local jurisdiction, and/or NBIC standards. Safety valves should be set to open at a pressure no greater than the maximum allowable working pressure of the boiler they are protecting. However, if more than one safety valve is mounted on a boiler, then only one valve needs to be set at or below the maximum allowable working pressure. The total relieving capacity of all safety valves on a boiler must exceed the boiler's maximum steam production capability. Safety valves for water heaters differ from those used for boilers in that water heater valves must sense excessive temperatures as well as overpressures. (See ANSI Z21.22, Relief Valves and Automatic Gas Shut-Off Devices for Hot Water Supply Systems.)

If a safety valve opens and fails to reseat correctly, it will continue to leak and will eventually impair the boiler's proper operation. This can occur when small bits of debris lodge between the valve's disk and its seat, thereby providing a leakage path. Sometimes using the manual lifting lever to open the valve wide a few times can blow the debris off the seat and allow the valve to reseat properly. However, if this fails after several attempts, the boiler should be scheduled out of service so that the safety valve can be repaired or replaced.

Whenever a boiler is returned to service, test the safety valves. Most safety valves are equipped with a manual test lever. Before testing, the boiler pressure must be at least as high as 75% of the set pressure on the safety valve. To manually test a safety valve, raise the test lever and hold the valve wide open long enough to blow out any dirt or debris. This kind of test ensures that the valve is not stuck in the shut position. CAUTION: Manual testing should not be conducted on boiler safety valves that operate above 400 psi.

A more meaningful test is a pressure test, which involves raising the boiler pressure above its normal operating pressure and all the way to the safety valve's set pressure. If the valve does not open when its set pressure is reached, immediately shut the boiler down and have the safety valve repaired or replaced.

If boilers are kept in continuous operation, the NBIC recommends manually lifting the valves every 6 months. The use of small chains or ropes attached to a safety valve's test lever and threaded through pulleys mounted above the boiler is a good way to facilitate this testing without needing to have an operator climb on top of a boiler, where safety valves are usually mounted.

The majority of steam and hot water released from a safety valve is carried out of the boiler room and to a safe point of release by discharge pipes. Discharge pipes should not be rigidly mounted on a safety valve's body but independently supported so that their weight does not contribute to the stress on the valve or the fitting on which the valve is mounted. Drip pans can be attached to the valve outlet to catch liquid released from the safety valve.

Safety valves are usually designed with small drainage openings in their bodies. These drain holes must remain open because water trapped in a valve body can expand and rupture the valve when exposed to freezing conditions. Drain pipes from drip pans and valve body drains should be located away from the boiler setting and discharged into an open funnel that provides a clear view of the drip.

Fusible plugs are an older technology created to indicate overheating, such as from low water levels. These plugs are small-diameter screwed fittings inserted into the boiler shell plate. They are made of a material that melts at a temperature below the temperature that would be destructive to the boiler. Thus, when excessive temperatures are reached, the fusible plug melts, and the escaping steam warns operators of the overheating. When using these plugs, make sure that they are manufactured, installed, inspected, repaired, and/or replaced according to the ASME Code and manufacturers' instructions.

Steam Pressure and Water-Level Indicators and Controls

Pressure indication and regulation inside a boiler are critical to the boiler's safe operation. Each steam boiler should have a steam gauge, with a dial range of not less than 1.5 times the maximum allowable working pressure, that is connected to the steam space or to the steam's connection to the water column. The steam gauge should be mounted on a siphon or equivalent device to ensure that the gauge tube remains filled with water. Pressure gauges installed on a multiple-boiler setup should be of the same type and graduated similarly.

On automatic and semi-automatically fired boilers, pressure control switches are used to maintain the desired steam pressure inside the boiler. These switches are designed so that the boiler cycles on when the steam pressure reaches the minimum level and cycles off when the desired steam pressure is achieved. In addition to the pressure control switch, a high-pressure cutoff switch should be included to ensure that if the pressure control switch fails and the steam pressure rises above the acceptable operating limit, the high-pressure cutoff switch will shut the boiler down. The setting on the cutoff switch should be higher than the desired boiler operating pressure but lower than the setting of the safety valves. Boiler safety valves are the last line of defense in preventing overpressure. Pressure switches should be mounted on siphon tubes in a way that ensures the tubes will not be damaged by high steam temperatures.

Water-level indicators and controls are also crucial for safe boiler operation. Each boiler must have at least one direct-reading water column or gauge glass. For boilers with a maximum allowable working pressure of 400 psi or greater, two gauge glasses must be provided. This second gauge glass can be omitted if there are two independent, remote water-level-sensing devices installed.

Boiler water level is controlled by switches, which are of various designs. The simplest and most common water-level controller is a float switch. When the water level inside a boiler falls below the predetermined limit, the water-level switch causes the feedwater pump to cycle on to supply makeup water to the boiler. When the water level reaches the desired point, the switch turns off the feedwater pump. Alternatively, some controller systems keep the feedwater pumps running continuously and use water-level switches to operate valves on the feedwater inlet lines, opening the valves when more water is needed and shutting the valves when the desired water level is reached.

When a boiler's water level falls too low, there will not be enough water to absorb the heat from the burner, and the boiler parts will overheat. If the boiler continues to fire despite the low or absent water level, the boiler's steel will begin to melt and come apart, which will not only destroy the boiler but also endanger the employees working in and around the boiler room. To prevent this from happening, steam boilers are required to have at least two low-water-level fuel-cutoff devices in addition to a water-level controller. These devices should be tested regularly when the boiler is in service and dismantled and inspected annually. Consult with insurance carrier representatives or state inspectors for advice on the type and frequency of testing recommended for your company's boiler. CAUTION: Operators should never attempt to reintroduce water into an overheated boiler. This could thermally shock the boiler, causing pressure parts to rupture violently.

Good Piping Practices

To prevent accidents, install steam lines so that they minimize the amount of maintenance work needed on them. The ANSI/ASME B31.3 Series, Pressure Piping code, (1) prescribes the minimum requirements for the design, materials, fabrication, erection, testing, and inspection of various piping systems and (2) discusses expanding, supporting, and the flexibility of lines.

Another safety consideration for good piping practice is to install valves and other boiler-operating controls so that they are easy to access. Many operators and maintenance personnel have been hurt by falling from ladders or inadequate work stands while trying to operate a hard-to-reach valve.

If it is necessary to open lines, maintenance personnel should always assume that the lines are charged and under pressure. Provide safe work standards, and secure a permit for line-breaking procedures.

When multiple boilers are connected through a main steam supply, feedwater or blowdown piping ensures that a double-block-and-bleed valve arrangement is in place. Two block valves having a free-blowing drain between them is required. This setup ensures that any boiler in the lineup can be completely isolated from sister units before being shut down and opened for maintenance.

Placing Boilers into and out of Service

This chapter cannot cover all the details involved in placing boilers into and out of service. Therefore, refer to the ASME Code and the NBIC and follow all manufacturers' recommendations.

Annual Internal Inspection, Cleaning, and Maintenance

Internal Inspection

Most state and provincial jurisdictions require boilers to be shut down annually for internal inspection, cleaning, and maintenance. When taking a boiler out of service for a prolonged period, clean it promptly and have it inspected by someone licensed by the National Board of Boiler and Pressure Vessel Inspectors and the state or local jurisdiction. In some cases, a jurisdiction may grant extensions allowing boilers to be operated up to 3 years between internal inspections. Before the inspector arrives, boilers must be cool enough so that the inspector does not have to rush through the process and must be clean enough so that the inspector can thoroughly examine metal parts for corrosion, pitting, cracking, and other defects. Also, internal parts should be readily accessible so that the inspector can thoroughly examine them. Open hand-holes and manholes, and ventilate the boiler. Provide adequate lighting and protective equipment for working inside the boiler.

Internal boiler inspections are conducted on both the fireside and the waterside of the boiler. The fireside inspection is of the burner cavity, furnace, and refractory to locate any signs of overheating, leakage, improper flame alignment, corrosion, incomplete combustion, or other deterioration. Waterside inspections are conducted to determine the effectiveness of the boiler's water treatment program in preventing scale buildup and corrosion. Thus, both the firesides and the watersides must be accessible for a proper inspection to take place. Both firetube boilers and watertube boilers are designed with openings, furnace doors, hand-holes, and manways to provide this access to inspectors.

Another very important aspect of the internal inspection is opening the low-water-level fuel-cutoff devices. Oftentimes, these devices can be rendered inoperable by scale accumulation in the piping connecting them to the boiler, float collapse, bent or broken linkages, and other adverse conditions that can be detected only by dismantling the devices during the annual internal inspection.

Cleaning

Prompt cleaning is important. The soot that accumulates on fireside surfaces absorbs moisture quickly and thus contributes to the deterioration of a boiler's metal surfaces. Remove soot and fly ash (from solid fuel–fired boilers) as soon as the boiler has cooled. Because ashes may remain hot for days, they are a hazard to anyone entering the boiler's combustion chamber. Therefore, carefully wet the ashes down with a hose. Because a jet of water driven directly into the center of a hot ash pile can cause the pile to explode, start at the outside and move toward the center when wetting down an ash pile. The operator should stay clear of any steam and dust that might fly up. When removing ashes, operators should prevent other personnel from being injured by any steam or hot water that may be present when the ash gates are opened. Operators should completely wash away the ash and dry the boiler's surfaces. Doing so prevents any ash deposits from binding to the boiler.

Annual Maintenance

Schedule shutdowns of boilers so that necessary preventive maintenance work is performed on an annual basis. Because boilers operate under harsh conditions, their mechanical structures and auxiliary equipment are subject not only to normal wear and tear but also to serious damage when adverse conditions develop during their operation and are not detected. At least once a year, have the flame-safeguard supervisory system and other safety controls inspected and tested during the scheduled shutdown. As noted previously in this section, be sure that boilers are:

- cool enough so that the inspector does not have to rush through the process
- clean enough so that the inspector can thoroughly examine metal parts for corrosion, pitting, cracking, and other defects
- readily accessible so that the inspector can thoroughly examine internal parts.

Precautions for Entering Boilers and Furnaces

General precautions for entering boilers include having proper ventilation, proper equipment, and proper protection. Observe the rules for working in confined spaces. Implement a confined-space entry procedure. To ensure that no flammable or toxic gases are present, ventilate boiler systems thoroughly, and then check the atmosphere

using a calibrated combustible gas indicator before allowing anyone to enter the boiler. This is especially important when more than one boiler is connected to a breeching or chimney—under certain circumstances, flue gases can enter a boiler from other boilers.

Because many injuries can occur when employees clean or perform other maintenance work on boilers, the need for caution cannot be overemphasized. When cleaning a boiler, employees should wear hard hats; safety goggles; approved dust masks; heavy, leather-palmed gloves; and protective footwear. Personnel working in permitted confined spaces should wear a lifeline and be under a spotter's constant observation. To maximize employee/contractor protection, a personal protective equipment (PPE) hazard assessment should be conducted to ensure that proper PPE is being used. To prevent steam, hot or high-pressure water, or hot gases from reaching employees, follow both confined-space entry and lockout/tagout procedures. (See Chapter 6, Safeguarding, in this volume.)

Because sealing off and testing for leaks may not be sufficient for fuel gases, apply positive blanking or double-block-and-bleed valve arrangements. Check all closed valves for leakage. Lines that are interconnected between boilers must be positively sealed off at both ends and locked out. Also, provide work stands for employees and protect them from ash falling overhead as they enter boilers. For ventilation, provide portable power-driven blowers that are operated outside the boiler's setting and that have canvas tubes coming in through access doors; draft fans can also be operated for short periods of time.

To prevent electrical shock, many firms permit only 6-V or 12-V lights and tools to be used inside a boiler. Such equipment is connected to small, portable power transformers outside the boiler. Battery-powered lights are even safer. In all instances, properly ground all electrically operated tools and extension cords used inside a boiler. Also, make sure that this equipment is thoroughly inspected before use. (See Chapter 8, Electrical Safety, in this volume.)

Boiler Rooms

The floors, lighting, exits, stairs, ladders, and runways of boiler rooms require special safety precautions.

Floors

Because boiler room floors can become very slippery and dirty, install a surface that can be easily cleaned. Build into the flooring ample drainage and other protection against flooding.

Lighting

In addition to standard lighting, boiler rooms should have a source of emergency lighting. Keep gauges and controls especially well lit so that they can be read easily. Provide well-maintained flashlights for employee use in power failures and other emergencies. Exits, too, should be well lit and identified.

Exits

Each boiler room should have two or more exits, which are remotely located from one another. If a boiler extends more than one story above ground level, the room should have an exit at the boiler runway or floor level of each story. These exits should lead to a fire escape on the outside of the building. If the boiler room is in a basement (or subbasement), exits should lead to outside stairways and runways and should have landings that lead to the exit doors. ASME CSD-1 currently recommends that emergency shutdown buttons be installed at each boiler room exit so that an operator evacuating the boiler room in an emergency can safely shut down the boiler on his or her way out of the room.

Stairs, Ladders, and Runways

Some regulatory codes require the installation of stairs, ladders, and runways around boilers that extend 10 ft (3.1 m) or more above floor level. Even if local, state, and/or federal codes/ regulations do not require them, provide such access so that personnel can operate and service a boiler safely and without having to step on hot-steam lines, hot-water lines, or valve stems or handles. Stairs, ladders, and runways must have standard guardrails, handrails, and toeboards. To provide a slip-resistant surface and to permit air circulation, install runways made of steel grating. Do not situate walkways near water glasses or safety-valve discharge areas, where an operator might be accidentally scalded.

Operator Training, Procedures, and Boiler Room Emergencies

Operator error is a major factor in boiler accidents. It is thus imperative that the personnel responsible for operating boilers fully understand the processes and hazards associated with boilers, including the purposes and operating principles of the control and safety devices installed. Procedures for routine start-up, shutdown, and basic troubleshooting should be developed with the help of the manufacturer or other qualified persons. In addition, emergency proce-

dures should be established to guide operators in the proper responses to upset conditions, low-water conditions, and companywide power failures.

Permanently post procedures for both routine and emergency boiler room operations. Be sure the procedures are understandable and legible (Figure 5–1). Have equipment manufacturers supply the procedures applicable to their equipment. In addition, furnish all operators and substitute operators with copies of the procedures, and formally train them in using the procedures. Supervisors should make sure that boiler room operators know the procedures and are able to perform the necessary procedures under crisis conditions.

Many facilities have only one boiler room operator. If this employee becomes sick or injured, boilers may be left unattended and an accident could occur. In facilities with isolated boiler rooms operated by a single person, it is a good idea to provide a system whereby someone periodically checks in on the boiler room to ensure that the operator is in attendance and that all conditions are normal. Facilities that have security personnel patrolling the premises during off-hours should make sure that these person-

Figure 5–1. An emergency procedure and checklist poster similar to this one should be posted permanently in the boiler room. *(Reprinted with permission from The Travelers Insurance Companies.)*

nel check the boiler room as part of their rounds. Provide an intercom, radio, or telephone system to ensure prompt response to a boiler room emergency. In addition, train one or more persons (a supervisor, a night-shift employee, or someone else) to take over the boiler room's operation in an emergency situation.

SAFETY OF HIGH-TEMPERATURE WATER

High-temperature water (HTW) is water kept in a closed system under high pressure so that it remains in liquid form rather than turns into steam. Conditions such as 400°F (200°C) and 247-psi (1,700-kPa) pressure often exist. When liquid under this pressure and temperature expands to steam at atmospheric pressure, employees can be fatally scalded.

High-Temperature Water versus Steam and Cold Water

HTW is different from steam and cold water when it discharges through a rupture in a pipe or piece of equipment. As it expands, HTW's volume increases at a very high rate, whereas its energy is released at a very low rate. Energy liberated in the expansion accelerates the particles of water and vapor and pushes air out of the way so that the steam-water mixture being formed can occupy the vacated space. Therefore, practically no energy is available to rupture equipment and impart kinetic energy to the fragments. Conversely, when steam escapes, approximately 16 times more energy is released during its expansion than when HTW escapes, and this considerable amount of leftover energy can produce an explosion. Fragments of fractured cast-iron valves on steam service have been known to penetrate a 10-in.- (25-cm-) thick brick wall. No case has been observed, however, in which parts of fractured valves on HTW service have been projected any distance.

The increase in the volume of escaping HTW continues after it leaves the pipe. The steam-water mixture does not form a long jet, unlike escaping steam or water; rather, it spreads out almost at right angles from the centerline of the jet to form a wet fog. Nevertheless, such factors should not lead to a sense of false security or negligence on the part of the design engineers and operating personnel. Although HTW is safer than steam and HTW-caused accidents are rare, such accidents still happen. Even 180°F (80°C) water can be fatal if enough of the body is exposed to it.

Causes of Failure

When equipment or piping does fail in HTW systems, it usually does so because of operating errors or mechanical forces (such as water hammer, thermal expansion, and thermal shock) and because of faulty materials. Therefore,

select and train qualified operators, and allow only experienced engineers to design HTW systems. These engineers must be able to meticulously analyze the entire design and select equipment that will prevent accidents. A good design is efficient and simple and does not overlook the essentials. Avoid overloading systems with automatic controls; if these controls malfunction, they can introduce more hazards into the system than they prevent. ASME CSD-1 provides guidance on the instrumentation required.

UNFIRED PRESSURE VESSELS

Unfired pressure vessels such as compressed-air tanks, propane tanks, deaerators, condensate tanks, steam-jacketed kettles, pulp mill digesters, and rubber vulcanizers are found throughout the commercial and industrial worlds. These vessels are designed to contain fluids under internal pressure or vacuum, but unlike boilers, they are not heated directly through the combustion of fuels or other external heat sources. Heat can be generated in an unfired pressure vessel by chemicals reacting within the vessel or by applying hot water, steam, or oil or some other heating medium either directly into the vessel or by circulating it around the vessel in a so-called jacket (or in internal coils).

Design

Unfired pressure vessels are discussed in ASME Code Section VIII, Divisions 1 and 2. The following classes of vessels, however, are exempt from this section's scope:
- vessels subject to federal regulations
- vessels with a maximum capacity of 120 gal (450 l) or less of water under pressure, in which any trapped air serves only as a cushion
- vessels having an internal or external operating pressure not exceeding 15 psi (103 kPa), with no limitation on size
- vessels with an inside diameter not exceeding 6 in. (15 cm), and no limitation on pressure
- hot-water storage tanks heated by steam or other indirect means—with a heat input of 200,000 Btu (59,000 J/s) or less, water temperature of 200°F (93°C) or less, and maximum capacity of 120 gal (450 l) or less.

This portion of the ASME Code does indicate that vessels designed for pressures greater than 3,000 psi (20,700 kPa) may be code-stamped.

Division 1

ASME Code Section VIII, Division 1, normally covers vessels with ratings of 3,000 psi or less (with the exceptions just listed). Vessels may be constructed to handle pressures greater than 3,000 psi; however, design principles and construction practices, in addition to the minimum ASME Code requirements, must be considered. Vessel designs under Division 1 rules are calculated according to the principal stress theory, and a design safety factor of 3.5 must be provided for tensile strength. Vessels built to these specifications may be used anywhere the pressure and temperature do not exceed the levels allowed by the ASME Code.

Before a pressure vessel is designed under Division 1 rules, consider the following questions:

- Will the material used in constructing the vessel affect or chemically change the material in process? For instance, carbon steel is not a good choice for vessels that will contain food products because rust may form and contaminate the process.
- Will the material in process affect or damage the material used to construct the vessel? If the material that will be processed is corrosive, then stainless steel or other high-alloy steels should be used.
- Will the filled vessel be able to safely carry the weight of its contents and withstand internal pressure? The minimum necessary thicknesses of vessel shells and heads can be calculated for given materials using formulas given in the ASME Code.
- Will the vessel be able to withstand both the pressure introduced into it and any additional pressure that a chemical reaction may cause in process? Allowances should be made in the design process for the highest possible pressure to which a vessel will be subjected.
- Will the vessel be able to withstand any intentionally or accidentally created vacuum and not collapse? If not designed for full vacuum, vacuum-breaking devices should be included.

In addition to these general requirements, specifications for construction of Division 1 pressure vessels should include the following:

- working pressure range
- working temperature range
- data as to whether or not the pressure and/or temperature range is cyclic
- description of what the vessel's contents will be
- specific information that may affect fabrication and installation of the vessel, such as stress relief, radiography, welding, and other requirements.

Division 2

Under Division 2 rules, vessel design should be based on a detailed stress analysis. A design factor of 3 must be provided for tensile strength. Although design calculations in Division 2 are more complex than those in Division 1, Division 2 calculations allow for thinner wall sections and may be used to design vessels with pressures exceed-ing 3,000 psi. The alternate rules of Division 2 apply only to vessels installed in a fixed location and subjected to a specific service. To obtain a vessel with an ASME stamp under these rules, a prospective purchaser must prepare a user's design specification and have it certified by a registered professional engineer who is experienced in pressure vessel design.

Other Codes

The ASME Code has been adopted by many governing bodies and therefore has the force of regulation. However, depending on the jurisdiction governing a vessel, other codes—such as the American Petroleum Institute's code and state and local codes—may be in force. Because these codes may impose size or service limitations that are more restrictive than those of the ASME Code, always check with local authorities before purchasing a vessel.

Secondhand Vessels

Prospective purchasers of secondhand vessels must comply with the jurisdictional requirements for secondhand vessels. Usually one of the requirements is to have the equipment inspected by a National Board (NB)–licensed inspector. Consult your property insurance carrier or your state or provincial boiler and pressure vessel safety office to locate qualified inspectors. Before purchasing a secondhand vessel, obtain a written report indicating that the equipment meets the requirements of the jurisdiction in which the vessel is to be installed. A great deal of inconvenience can arise when secondhand equipment is purchased and installed before being inspected. Also note that if secondhand equipment was condemned or rejected by one state, it should not be bought, sold, or operated in another state.

Internal Inspection

For most pressure vessels, periodic external inspections can adequately evaluate the integrity of a vessel's pressure parts. However, vessels that contain corrosive substances such as ammonia, pulp mill cooking liquor, or strong chemical reagents should be scheduled for periodic internal inspections. These inspections should be carried out by NB-licensed inspectors to ensure compliance with jurisdictional and/or insurance requirements. For certain vessels such as chemical reactors and distillation towers, additional, highly specialized inspections may be necessary.

Large chemical plants and petroleum refineries that have dozens and perhaps hundreds of unfired pressure vessels may find it advantageous to employ a full-time staff that regularly inspects all their pressure vessels. Such inspections, coupled with good preventive maintenance, will prolong the life of the vessels, piping, and associated equipment and help ensure their safe and reliable operation.

The inspection or maintenance department should keep a file or log containing original design documentation as well as the inspection and maintenance history of each pressure vessel. Include the following in each vessel's file:

- blueprints
- manufacturer's data reports and instructions
- design data, including location of dimensional checkpoints
- installation information
- records of process changes
- the vessel's historical profile, including records of all repairs and conditions found during inspections.

This log will be valuable for operating existing equipment and for designing, installing, and operating new equipment.

When inspections are carried out by state or other third-party inspectors, someone familiar with the process and service conditions of each vessel should accompany this inspector and detail each vessel's processes. When new processes are developed, inform inspectors and the operators in detail about these processes and how they may affect the pressure vessels.

Entry

To reduce the number of fatalities associated with entering pressure vessels and other dangerous, confined spaces, establish a safe procedure for entering tanks. OSHA has proposed strict and detailed regulations for confined-space entry [29 CFR 1910.146, Permit-Required Confined Spaces] and lockout/tagout systems [29 CFR 1910.147, The Control of Hazardous Energy (Lockout/Tagout)] that should be implemented; however, because these proposals were subject to change, we have refrained from listing the specific requirements they include. The hazards of working in large pressure vessels and tanks are exacerbated by the limited egress available in those environments. Compounding the problem is the difficulty of workers inside the vessel communicating with "hole-watchers" outside the vessel. (See the procedures described in Chapter 10, Flammable and Combustible Liquids). The following hazards can endanger workers in confined spaces:

- toxic materials, including inert gases like nitrogen and exhaust from portable generators or from welding machines already in—or introduced later into—the confined space
- flammable vapors, which can be ignited by welding or grinding sparks
- insufficient oxygen
- heat or smoke from a fire inside the vessel
- hot gases or other fluids entering the vessel because vessel is improperly isolated from external sources

- start-up of agitators or the confined space being set in motion.

Before anyone enters a pressure vessel, make sure that it has been properly drained, ventilated, and cleaned. Next, check the vessel's atmosphere for its oxygen content, explosiveness, and toxicity using a calibrated combustible gas indicator. Disconnect and blank all connecting pipelines, or close, lock out, and tag valves on the line. All power-driven devices such as agitators must be positively disconnected, locked out, and tagged. (See the discussion of entry-isolation procedures in Chapter 4, Maintenance of Facilities, and Chapter 6, Safeguarding, both in this volume.)

After all preparations for entry have been completed, the properly trained and authorized supervisor for the job should make a final check that the vessel is safe, that all lines are closed off, that the relevant power sources are locked out, that ventilation and personal protective equipment are adequate, and that safe work procedures have been designed. A confined-space entry permit stating that all precautions have been carried out can then be issued. Periodic rechecks of the atmosphere inside the vessel or tank during the course of the work—and upon employees returning to the confined space following work breaks—are strongly recommended.

When purging a tank, the vent should discharge outside the tank and into an area where the discharge will not create a hazard. In some instances, vessels may be purged using an inert gas such as CO_2 or nitrogen. Be aware of the government regulations covering the discharge of certain gases. Remember that an inert gas will not support life, so persons entering a vessel that has an oxygen-deficient atmosphere must wear air-supplied respirators or self-contained breathing equipment.

Using forced ventilation in confined spaces may be safer than requiring employees to wear respiratory protection. Have air blown in until tests of the exhaust and of the interior of the vessel indicate that the space is safe for entry, or have air continuously sucked out from the bottom of the vessel by venturi action. Test all areas of the vessel for inadequate oxygen and flammable and toxic gases at specific intervals (the rule of thumb is a minimum of every 2 hours) to make sure that conditions remain safe while employees are in the vessel. Introduce air to make sure that there are several exchanges of air per minute in the vessel.

Provide straight ladders or rope or chain ladders that have rigid wooden rungs. Also, do not allow employees to enter an opening that they must squeeze through. In an emergency, they may not be able to exit or be removed quickly enough to be saved.

Another necessary precaution is to have employees wear safety harnesses attached to lifelines when they enter any vessel, particularly when they must work inside vertically oriented vessels or off scaffolds constructed inside a vessel. Station an observer (or hole-watcher) outside the vessel to signal for help if needed. Although the supervisor may station outside the vessel a person who is equipped to rescue the workers, often more than one person will be needed to help with the rescue.

No matter a vessel's prior contents, the person entering the vessel should be equipped with a vapor-proof flashlight or a vapor-proof or low-voltage extension light. Chemical protective suits may be necessary. Be sure that all inspection equipment, including tools, are made of nonsparking materials (e.g., beryllium).

Cleaning and Purging

When a vessel normally contains chemicals such as petroleum products, ammonia, or other toxic or corrosive materials, the vessel's internal surfaces must be cleaned or purged before entry can be permitted. The needed method of cleaning or purging depends on the contents of the vessel. Sometimes water is the only cleaning agent necessary, but when a vessel normally contains acidic substances, a caustic solution or other neutralizing agent may be necessary to remove sludge and other vapor-containing materials. High-pressure steam is also an effective purging agent. Finally, forced ventilation is required to ensure that all harmful gases and vapors have been removed. Before permitting employee or contractor entry, the vessel's atmosphere must be tested for the presence of flammable and toxic gases as well as sufficient oxygen levels. (See Chapter 10, Flammable and Combustible Liquids.)

Hydrostatic Tests

Hydrostatic tests are conducted by completely filling a boiler or pressure vessel with water, carefully driving out all the air, and increasing the water pressure inside the vessel to the desired test point. For new construction, the ASME proscribed test pressure is 1.5 times the maximum allowable working pressure of the boiler or pressure vessel. Oftentimes, boilers and pressure vessels that have been in service a long time require periodic hydrostatic testing. This testing is most commonly performed after welding was needed to restore the pressure-retaining capability of a vessel that had deteriorated because of normal wear and tear or a service-related failure. Hydrostatic testing can also help ensure that a pressure vessel constructed without internal-inspection accessibility is safe for continued service. In these situations and at the discretion of the authorized inspector, the target test pressure can be reduced to the maximum allowable working pressure, to the set pressure of the safety valve that was installed the lowest, or to the normal operating pressure of the boiler or pressure

vessel. When performing hydrostatic tests, the temperature of the water should be between 60° and 120°F (16° and 50°C). Also consider the weight of water that will fill the vessel to be sure that the vessel's foundation or other supporting structures will be able to tolerate the extra weight without being damaged.

When pressure testing using water is not practical, a pneumatic test using air or nitrogen is permitted but requires extra caution to ensure that hazards to personnel are minimal. Follow the requirements for pneumatic testing given in the original ASME Code. (See ASME Code Section VIII for construction of new unfired pressure vessels, ASME Code Section I for boilers, and the NBIC for both boilers and pressure vessels.) Although compressed gas or compressed air can be used to test a boiler or unfired pressure vessel for leaks at pressures below the working pressure, never use compressed gas or compressed air to test a pressure vessel above its maximum allowable working pressure. Testing should follow the procedures given in the ASME Code and the NBIC and should be conducted under the supervision of qualified personnel.

As just mentioned, the required pressure for a standard hydrostatic test is normally no greater than 1.5 times the maximum allowable working pressure of the vessel being tested. Section VIII, Division 2, of the ASME Code allows the design engineer to establish upper boundaries in terms of stress-intensity limits relative to the yield strength, tensile strength, or creep-rupture strength at test temperature. Inspection of a Division 2 vessel should be made at the pressure equal to the greater of the design pressures or 75% of the test pressure.

To minimize hazards to personnel during pressure testing, isolate test areas as far as possible from other operations and provide suitable barricades to shield personnel and equipment. These procedures are especially important when conducting proof tests, which are similar to hydrostatic tests except in the former's intent to increase the internal pressure of the vessel being tested to the point of the vessel's destruction. All personnel should keep clear of a vessel under full-test pressure. Allow no one to approach the vessel until the pressure has been reduced to, or is close to, the vessel's maximum allowable working pressure.

Detecting Cracks and Measuring Thickness

Pressure vessels used to process gases or oily materials may have very small leaks that are not apparent during hydrostatic tests. To detect these leaks, nondestructive techniques (NDT) are commonly used to check a vessel in a way that will not impair the vessel's material and hence damage the future usefulness of the vessel. One example of a nondestructive technique that is sometimes used to detect cracks in vessels involves placing a small amount of ammonia inside a vessel and applying compressed air until a maximum pressure of 50% of the working pressure is attained. A swab soaked in hydrochloric (muriatic) acid is then passed over all seams and other areas with possible leaks. Leakage is indicated by a white vapor (ammonium chloride) that is formed from the mixture of the escaping ammonia and the acid. Using a burning sulfur stick is also an effective indicator because a change in the flame characteristics (various chemicals cause a color difference in the flame as it passes over them) indicates the presence of ammonia.

Nondestructive examination (NDE) techniques (also known as nondestructive testing [NDT]) such as visual testing (VT), electromagnetic testing (ET), radiography (RT), and ultrasonic examination (UT) are good technologies for locating cracks, wastage, and other undesirable conditions. Inspectors also sometimes use liquid dye penetrant testing (PT) and magnetic particle testing (MT) techniques. Cracks, pitting, and uniform thinning of vessel materials are major factors leading to leaks and

incidents involving catastrophic rupture. Because no single method will be able to detect all possible flaws, the characteristics of the flaw being evaluated must be considered. Hence, nondestructive testing must always be performed by technicians certified in the American Society for Non-Destructive Testing's SNT-TC-1A (2011) or the American Society for Non-Destructive Testing's ANSI/ASNT CP-189, according to the requirements of Section V of the ASME Code. (The newly approved American Society for Non-Destructive Testing's ANSI/ASNT CP-105, Training Outlines for Qualification of Nondestructive Personnel (2011), replaces "Recommended Training Course Outlines" in the American Society for Non-Destructive Testing's Recommended Practice No. SNT-TC-1A, Personnel Qualification and Certification in Nondestructive Testing (2001), and is included with the SNT-TC-1A.)

Table 5–A is a brief synopsis of the most common methods used in pressure vessel inspections.

TABLE 5–A. Common Pressure Vessel Inspection Methods

Method	Detection Characteristics	Advantages	Limitations
Visual testing (VT)	Surface characteristics such as finish, scratches, cracks, color, stain in transparent materials, and corrosion	Often convenient; can be accomplished by means of simple tools and magnifying equipment; low cost	Can be applied only to surfaces, through surface openings, or to transparent material; flaw size and human fatigue factors in large areas; direct visual access required
Electromagnetic testing (ET)	Changes in electrical conductivity caused by material variations, cracks, voids, inclusions, and corrosion	Readily automated; moderate cost	Limited to electricity-conducting materials; limited penetration depth; human fatigue a factor in large areas; probe size can influence size of cracks detected
Radiography (RT)	Changes in density from voids, inclusions, material variations; generally requires access to both sides	Can be used to inspect wide range of materials and thicknesses; versatile; film provides record of inspection	Radiation safety requires precautions; expensive; detection of cracks can be difficult unless perpendicular to x-ray film
Ultrasonic examination (UT)	Changes in acoustic impedance caused by cracks, nonbonded areas, inclusions, corrosion, or interfaces	Can penetrate thick materials; excellent for crack detection; can be automated; moderate cost	Normally requires coupling to material either by contact to surface or immersion in a fluid such as water; surface needs to be smooth; human fatigue a factor in large areas
Penetrant testing (PT)	Surface openings due to cracks, porosity, seams, or folds	Inexpensive; easy to use; readily portable; sensitive to small surface flaws	Flaw must be accessible from surface; not useful on porous materials or rough surfaces
Magnetic particle testing (MT)	Leakage of magnetic flux caused by surface or near-surface cracks, voids, inclusions, material/geometry changes	Inexpensive to moderate cost; sensitive to both surface and near-surface flaws	Limited to ferromagnetic material; surface preparation and post-inspection demagnetization may be required

For more information on certification requirements and levels of certifications, the American Society for Nondestructive Testing (ASNT) is the largest technical society for NDT professionals; additional resources and information can be found at asnt.org.

In many other situations, it is vital to check the thickness of an unfired pressure vessel, but such an inspection must be done without damaging the vessel. For this test, inspectors employ instruments that measure the thickness using ultrasonic testing and electromagnetic testing equipment. Using these instruments, a qualified inspector can determine the thickness of metal to within 2% or 3% of its actual thickness for very small areas or approximately 10% of the actual thickness for large inspection areas. Depending on the NDT inspection method, type of equipment, and accessories used with the equipment, cracks that extend to, below, or on the opposite side of the surface may be detected. A well-trained and -qualified inspector who knows the necessary inspection procedures or a highly qualified, Level III–certified inspector can determine crack depth with equipment appropriate for the material and thickness of the material.

Older techniques, such as the lacquer method, may still be used. This method can detect microscopic, hairline cracks when the leak's location is not readily apparent. The lacquer method consists of cleaning the area and then coating the area with clear lacquer. After the lacquer hardens, an inspector conducts a hydrostatic test. Hairline cracks or fatigue-induced stress cracks, if they extend through the entire material, will show up on the surface because of expansion and cracking of the lacquer. This method will not make subsurface cracks apparent, unlike many of the commonly used NDT methods.

At every inspection of pressure vessels such as vulcanizers, digesters, and autoclaves that have removable coverplates, heads, or doors, inspectors should check the holding bolts, coverplate bolts, slots, and retaining rings for wear and hammer-test them for soundness. (Hammer tests involve tapping gently with a hammer made of nonsparking material and listening for unusual sounds that result, which indicate cracked equipment, loose bolts, and so on.) Because these parts endure considerable wear and tear in service, they need to be periodically repaired or replaced. Nuts and bolts that can withstand the mechanical and thermal stresses of pressure vessels are relatively inexpensive and readily obtainable. Measure the width of holding slots and carefully check retaining rings and coverplates for worn areas, cracks, and other detrimental conditions resulting from the stress of improperly tightening, adjusting, or closing the doorplate. If cracks cannot be satisfactorily repaired, the vessel should be disposed of.

The supervisor of facility operations should instruct the operators of vessels with coverplates or removable doors to tighten the bolts or quick-closing lugs without damaging them or the retaining rings. A torque wrench must always be used to tighten bolts because using this wrench ensures uniform tightness, reduces wear of and damage to the bolts, and ensures that the bolts are tightened to the manufacturers' recommended torque ranges. Likewise, instruct operators to open coverplates or removable doors only after the vessel has been relieved of all pressure. Note that the improper opening of vessels can be prevented entirely by establishing a safety interlock system.

Operators should know where to look for wear on holding bolts, quick-opening lugs, and lug openings on coverplates and removable doors. They should also know when to notify the supervisor about worn bolts and when to simply replace any bolts that have questionable conditions.

To prevent operators from opening or closing the wrong valves, tag and mark valves and pipelines as described in ANSI A13.1, Scheme for the Identification of Piping Systems. Good labeling of valves, piping systems, and control equipment will help prevent accidents caused by operator error. (See also Chapter 2, Buildings and Facility Layout.)

Operator Training and Supervision

Meticulously train employees working with pressure vessels, especially those used in chemical processes, in both routine duty and emergency procedures. Supervisors, too, should be qualified and knowledgeable. Explain the entire process to a new employee being trained as an operator or an assistant. Discuss the hazards involved and exactly what part the vessel plays in the process. Detailed procedures should be written out, periodically reviewed, and made available to experienced operators as well as trainees.

Provide checklists to make certain that no step is overlooked in the start-up, shutdown, or routine processing cycle. Operators should regularly record the information from the instruments that measure pressures, temperatures, flow rates, and flow levels as well as the times and frequencies at which valve operations and other critical processing steps are performed. After each completed processing cycle, the operator should initial the checklist. Supervisors or managers might review these checklists to ensure that operators have taken care of all the duties established in the procedures.

Good communication among operators is critical, especially among the operators inside a control room and the operators working outside the control room and near the equipment. Signaling systems such as alarm sirens, flashing lights, radios, and cell phones may all be utilized to make personnel aware of equipment operations that are taking place nearby but are being controlled remotely. Note on the checklist or operations log the times that these signals are given.

Safety Devices

Pressure vessels are used to process a great variety of materials, and each vessel should be equipped with safety devices specifically designed for that vessel and for the work it is to do. Safety valves for each vessel should be ASME/NB-rated and stamped "safety valves." The vessel should have safety devices that will adequately protect it against overpressure, chemical reaction, and other abnormal conditions.

Safety Valves

The old ball-and-lever safety valve has been prohibited from use in all jurisdictions within the United States and Canada because its setting can easily be tampered with and can accidentally be reset. ASME/NB-rated and -stamped safety valves of the spring-loaded type are commonly used on pressure vessels, including vessels containing air, steam, gases, and liquids that will not solidify as they pass out through the safety valve's discharge.

Valves on pressure vessels containing air or steam should be large enough to be able to discharge the contents at a rate that prevents pressure buildup, as prescribed in the ASME Code. On vessels that contain liquids, the safety valve's seat should be constructed so that it will not collapse and thus enable the valve's contents to obstruct the discharge's opening. The safety valve's discharge line should lead to an area where it is safe and permissible to be discharged. When practical, equip valves with test levers; regular testing as set forth in the NBIC prevents the valves from sticking. However, for vessels with dangerous (e.g., toxic or flammable) contents, the safety valve should not have a test lever.

If liquid contents are heated, the safety valve should be designed to function if the vessel is overpressured as the liquid expands. For pressure vessels containing hot water or in which water is heated, calibrate the valve to relieve the contents according to the total number of Btus that can be applied in the vessel.

Rupture Disks

Rupture disks, commonly found in chemical-processing plants, are thin metallic disks designed to open and relieve pressure in a vessel or system of vessels. Various designs of rupture disks are available. Once a rupture disk opens, it must be replaced before the vessel can be returned to service. A frangible disk may not clog as easily as a spring-loaded safety valve and is easily and inexpensively replaced, whereas a rupture disk may clog or become coated with material in such processes as the manufacture of varnish and other resins. At times, this coating becomes thick enough to affect the rupturing pressure of the disk, and the disk then needs to be replaced or cleaned with a solvent. Check the condition of these disks at least once a year to make sure that they are free of any buildup of chemicals or by-products. Remember that when a disk ruptures, all pressure is relieved from the vessel, which could result in the complete loss of a product or the spoiling of in-process material.

A rupture disk must function within ±5% of its specified bursting pressure at a specific temperature. Disks may be installed between a spring-loaded safety valve (or relief valve) and the pressure vessel. This location prevents unnecessary corrosion of the valve and prevents the valve from becoming plugged up by the vessel's contents. All installations must be in accord with the ASME Code, Section VIII.

Vacuum Breakers

Just as it is important to protect a pressure vessel from bursting because of overpressure, it is also important to protect a pressure vessel from collapsing under a vacuum. Several safety devices known as vacuum breakers provide such protection. One design, the mechanical vacuum breaker, which is similar to a spring-loaded safety valve, has a spring set at a predetermined vacuum. Another design, the weight-balanced vacuum breaker, uses a weight suspended from a fulcrum attached to the vacuum breaker's gate. This vacuum breaker is generally used on pressure vessels that work intermittently on pressure and on vacuum. If the vacuum exceeds the setting of the weight on the fulcrum, the breaker opens, thus relieving the vacuum. On some vessels that ordinarily work under pressure but in which a vacuum may develop because of rapid cooling (as when steam condenses), a check valve may be installed with a flap or valve disk that faces into the vessel. Whenever a vacuum develops, the check-valve disk opens automatically. When vessels are designed for "full vacuum," no vacuum breakers are required.

Water Seal

A water seal is used on pressure vessels that operate on low pressure or under slight vacuum, such as alcohol stills and gas holders. A water seal is a U-pipe filled with water, with one end connected to the pressure side of the vessel and the other end vented into the atmosphere. Because the vessel operates under a pressure of only a few pounds, the amount of internal pressure can be regulated by the height of the water in the vent pipe. If the pressure rises above the predetermined limit, the water is forced out of the pipe, thus reducing the pressure.

Vents

In many processes, pressure must be relieved before the pressure vessel can be opened. An easy means of relieving this pressure is to vent it into the atmosphere.

Vent pipes should be large enough in diameter to reduce the contents of the vessel before excess pressure can build up. A good rule of thumb is that the diameter of the vent pipe should be at least 1.5 times the combined diameters of all inlet piping. A vent pipe, preferably with a U-bend at the atmospheric discharge, should be installed in a way that prevents dirt from clogging the pipe. Lime deposits have also been known to restrict the capacity of vents, making pressure vessels out of tanks that were not designed to retain pressure. Be careful to direct the flow of vent pipes away from the vessel so that it will not impinge on the metal and cause a fire.

Also insulate vent pipes in cold weather. Vapor may freeze as it leaves the vent, thus rendering the vent unusable as a safety device. If a vent pipe is situated where it might freeze or become clogged with dirt or lime deposits, install a relief valve on the pipe as an added safeguard.

Regulating or Reducing Valves

Some vessels operate using a steam pressure much lower than that obtained from a boiler or steam transmission line. A regulating or reducing valve lowers high-pressure steam to the pressure required for a specific function. Include a safety valve on the low-pressure side of the reducing valve to protect vessels and piping that were designed for lower pressures. If the reducing valve(s) fail(s), the relieving capability of the safety valve should be great enough to discharge the entire flow of steam without allowing the pressure to rise above the vessel's maximum allowable working pressure. To provide protection for all pressure vessels in a series of the same type of vessel, one reducing valve and one safety valve should be installed in the common steam supply line. This is the usual method for steam-jacketed kettles, in which ordinary pressures do not exceed 10 to 25 psi (70 to 170 kPa). Connect safety valves so that there is no stop valve between them and any vessel they protect.

Autoclaves and Other Vessels with Quick-Opening Doors

Pressure interlocks should be installed on all autoclaves, vulcanizers, retorts, pulp mill digesters, and other pressure vessels that contain large amounts of steam and are equipped with quick-opening doors or reclosing devices. An interlock will prevent the charging door from opening until all pressure has been relieved and will also prevent the introduction of steam or another pressurizing medium into the vessel until the charging door is fully closed. Opening an autoclave that has pressure in it will result in the door flying open with explosive force, and the contents possibly shooting out like projectiles. In addition, in reaction to the blowout, the autoclave may move back a sizable distance.

The most critical parts of these vessels are those of the closure system, such as nuts, bolts, lugs, retaining slots, and/or rings. Operators should be trained to routinely inspect these parts visually and to be able to recognize the

signs of wear or cracks. Licensed inspectors should also examine these components annually. Whether using a bolted door, rotary-lug door, shearing door, clamp door, or screwed-on door, the sealing mechanism should be maintained in good shape. (See earlier sections on Detecting Cracks and Measuring Thickness and Operator Training and Supervision.)

Additional Safety Procedures when Using Autoclaves

Management must make sure that personnel are aware of the safety procedures to follow when working with pressure vessels:

- Do not weld anything to any part of the vessel, door, locking ring, etc.
- Do not cut, drill, or fasten anything to any part of the vessel, door, locking ring, etc.
- Assume that the vessel is pressurized any time the door is closed. That is, assume any lines or pipes connected to the autoclave are at autoclave pressure. Do not remove or adjust any autoclave fittings unless the door is open.
- At no time should a person enter the autoclave unless he or she knows there is sufficient oxygen for breathing.
- Leave no flammable materials, debris, or plastics in the autoclave. If solvents are used, use quantities of less than 1 quart in approved safety containers, and ensure that no source of ignition is present (such as autoclave lights, an electric heater in the apparatus, cigarettes, or sparks).
- The operator must inspect the closure's alignment and also inspect the vessel and closure for cracks, other damage, hot spots, or any other unusual conditions. If any of these factors is observed, the operator should not operate the autoclave but instead notify maintenance or management.

Recommended Scheduled, Periodic Maintenance for Autoclaves

Experts recommend the following time table for the general maintenance of most autoclaves:

- *Doors*—lubricate wedges quarterly; remove debris and inspect door seals weekly; check hydraulic pump reservoir levels every 2 months; check entire hydraulic system annually.
- *Interior*—clean at least every 2 weeks; check thermocouple panels and screws quarterly; check all bolts, washers, and nuts for tightness annually.
- *Cooling system*—check cooling tower fan belts biannually; inspect and clean all water filter screens biannually; drain reservoir tank biannually.
- *Vacuum system*—check vacuum pump packing every 2 months; vent and drain receiver tanks annually.
- *Air regulators*—check and readjust air regulators every 2 months.

- *Transducers*—check and recalibrate transducers every 2 months.
- *Pressure gauges*—check and recalibrate analog pressure gauges biannually.

Steam-Jacketed Vessels

Steam-jacketed vessels are used to heat liquid mixtures to a moderate degree. In such a vessel, steam circulates between the outer and the inner shells of the vessel at pressures that are usually 10 to 50 psi (0.7 to 3.5 kPa). Occasionally, a process may require the vessel to be operated at pressures up to 100 psi (690 kPa). Heat is then transmitted through the inner shell to the vessel's contents. Such vessels are used principally in commercially preparing food, in candy manufacturing, and for cooking starch in laundries and textile mills. These vessels are also used in the chemical industry to create low-temperature reactions or in blending operations. On a steam-jacketed vessel with a tight cover, provide a separate safety valve for the inner kettle. If a steam-jacketed vessel can be completely valved off, keep the vessel from collapsing by using a vacuum breaker.

Observe the following precautions when operating steam-jacketed kettles:
- Thoroughly drain the steam space before releasing steam into the jacket. Because water in the steam space may cause thermal shocking and water hammer, which could lead to catastrophic failure of the trap, open drain lines even though traps are installed.
- Release steam into cold vessels slowly to allow ample time for the vessel parts to heat and expand uniformly. The importance of this step increases as vessel size and steam temperatures increase.
- Unless automatic safeguarding is provided, open vents when the steam supply is shut off. Doing so prevents damage to, and even collapse of, the inner vessel once the steam has condensed.
- Where agitators are used, paddles must not strike the interior surfaces of the inner vessel. Even a slight deformity to the inside of a vessel may require extensive repairs. Be sure that hand stirrers are also used carefully. Whether processes involve solids, semisolids, or slurries, paddles should have the appropriate tolerance or clearance of vessel walls to prevent erosion of the walls.
- Make sure that vessels with openings large enough to permit access have guardrails or other guarding systems that prevent employees from accidentally falling into the vessels.
- Be sure that vessels are filled only to the level at which undue splashing will not occur when the contents are heated or agitated. Also use splash guards or loose covers.

As a general precaution in maintenance work, never allow employees to climb above large open vats or kettles filled with hot, corrosive, or viscous fluids. Drain or cover the vats, provide safe work stands, and have employees wear lifelines. Make sure that all boards in the work stands are fastened in place and not loose. Take all other necessary precautions to make sure that people and objects do not fall into the fluids.

Evaporating Pans

Ordinarily, evaporating pans are shallow pans containing steam coils. When the pans are in operation, the steam coils are immersed in the material being processed. If the coils become exposed, the material may overheat or ignite, resulting in a fire or explosion. To prevent such occurrences, observe the following rules:
- Continuously attend to evaporating pans while they are in operation.
- After each use, thoroughly clean the pans and coils.
- After shutting off the steam in the coils, drain them to prevent the product from being drawn into lines if the steam condenses and creates a vacuum. Installing a vacuum breaker would also prevent this from happening.

HIGH-PRESSURE SYSTEMS

High-pressure-system hazards arise largely because of failures resulting from leaks, pulsation, vibration, and overpressure. The potential for injury and damage from high-pressure-system accidents is enormous. In addition to the injuries to employees from the release of high-pressure steam or gas when a vessel or piping system ruptures, serious injuries can result from fittings, flanges, gauges, and valves flying off the failed component and from the violent movement or "whiplash" of unrestrained pipe sections, tubing, or hoses.

When reciprocating pumps and compressors are utilized to generate high pressures in a piping system, the piping system's material itself is subjected to cyclical stresses from the pulsating pressures. This can lead to metal fatigue failures. Therefore, keep the piping and other system components absolutely free of internal notches or deep scratches because these kinds of defects create areas where cyclic stresses tend to congregate and lead to cracks and eventual failure. Also, prevent a pipe from having to support large amounts of stress because of ill-planned holes or cross bores. The stress-concentration factor of a radial entry to a pipe's or a cylinder's wall reduces the pressure-endurance limit by almost one-half. Because of the pulsating pressure, nothing can be considered "safe" and then just forgotten about. Instead, operators must maintain constant vigilance.

Leaks in pressurized systems can also be hazardous. Expelled liquids can easily penetrate clothing and skin.

Figure 5-2. Two methods of securing a high-pressure line: on either side of the fitting (lower right arrows) and at the bends (upper arrow). As a secondary protective measure, a channel iron was placed across the lines.

Figure 5-3. A reference gauge has been installed for each of the four separate pressure stages of this compressor. Note the plastic shield (arrow) over the gauges.

An unexpected leak might instantaneously fill an enclosure with an explosive mixture of gases. Because most leaks occur at gasket joints, minimize the number of gasket joints.

Leaks can also occur at defective welds. Welds subject to leakage can be in the main seams used to join head and shell sections as well as welds used to join nozzles, openings, or nonpressure attachments. If improperly done, welds are particularly prone to metallurgical weakening from residual stresses, porosity, lack of fusion, or incomplete penetration. When possible, radiography should be utilized on all critical welds in high-pressure systems to ensure a good-quality original weld.

If possible, limit vibration in piping with appropriate dampening. Designers use various means to dampen vibration in hydraulic and pressurized gas systems. However, because dampening can never be fully achieved, high-pressure systems require pipe support systems that allow for some movement and thermal expansion. Be sure that the supports are durable enough to resist deflection from any direction and are securely anchored to the building to prevent any severe displacement (Figure 5-2).

PRESSURE GAUGES

Pressure gauges used at 1,800 psi (12,400 kPa) or higher (except UL-listed, gas-regulator gauges) should have full-sized blowout backs; have structurally sound sides; be front-designed to withstand internal explosions; and have either a multi-ply plastic or a double-laminated safety glass cover for the gauge faces (Figure 5-3). Tests at 3,000 psi (20,700 kPa) show that gauges not constructed in this manner have a holes blown into their faces. Providing holes in the back of a gauge does not allow enough vent area for the safe clearance of gases. Furnish a solid shield for high-pressure gauges that is made of acrylic plastic at least 5/8 in. (1.6 cm) thick and meets MIL8 P5225B-Finish A Specification. Make sure the shield is free of scratches, gripper marks, tool chatter marks, and other stress increasers.

When mounted, there should be at least 2 in. (5 cm) of clearance between a gauge and the item to which it is attached. When mounting the gauge flush to a backing plate, cut a hole through the plate with a diameter at least equal to the diameter of the gauge. Of course, leave enough room to mount the face flange. Shields mounted behind gauges should be solid. Leave an unobstructed area of no less than 2 in. (5 cm) wide behind the gauge so that vented gases can emanate between the shield and the back of the gauge.

Almost all Bourdon tubes in pressure gauges eventually fail because of fatigue caused by constantly pulsing pressure. They can also fail after many cycles or even when a gauge is new. Gauges on large vessels are not subject to the large pressure oscillations that gauges are when a compressor maintains the pressure.

Restrict all but the necessary personnel from the areas where high gas-pressure systems operate. Situate behind barricades reactors, pressure vessels, and heat exchangers having substantial hazard potential, and install remote control systems and other monitoring devices to operate those vessels. Also place behind barricades vessels and systems undergoing tests.

SUMMARY

- Boilers and unfired pressure vessels have many common potential hazards that must be controlled through the use of safety devices and safe work practices.
- Management must make sure that the design, fabrication, testing, and installation of boilers and unfired pressure vessels comply strictly with all federal, state, and local codes.
- ASME and the National Board of Boiler and Pressure Vessel Inspectors have established guidelines for inspecting pressure vessels. Workers should be trained to operate this equipment according to safety standards and safe work practices.
- Boilers are closed vessels in which water is heated and steam is generated or superheated under pressure by the direct application of heat. Although the most common heat source is the combustion of fuel, heat sources also include electricity and waste heat from chemical processes.
- Operators must be carefully selected and thoroughly trained to operate boilers and pressure vessels safely and correctly, to understand emergency procedures, and to use approved procedures to inspect and maintain equipment. Supervisors must ensure that operating procedures are properly reviewed and implemented and must closely monitor all operations.
- Management should schedule regular cleaning and maintenance of boilers to minimize production interruptions. Workers must use ultimate caution when entering boilers. Management should ensure that applicable OSHA confined-space entry and lockout/tagout requirements are heeded whenever workers must enter boilers and pressure vessels.
- Safety devices on boilers and pressure vessels help prevent accidents and injuries. Workers must be sure that these devices are properly installed and maintained and are periodically tested.
- High-temperature water (HTW) must be kept in a closed system and under high pressure so that it remains in liquid form instead of turns into steam. Only trained operators and engineers should work around HTW systems.
- High-pressure-system hazards arise largely from failures caused by leaks, pulsation, vibration, and overpressure. Supervisors and workers must make sure that all elements of these systems are kept clean, are in good working order, and are frequently inspected.
- Unfired pressure vessels include air tanks, steam-jacketed kettles, digesters, and vulcanizers. Unfired pressure vessels should be inspected regularly by qualified persons. If vessels cannot be inspected internally, they should undergo periodic hydrostatic or pneumatic testing. Cleaning and purging methods should be selected according to a vessel's contents.
- Pressure gauges are critical to monitoring high-pressure systems and must be appropriately designed for the equipment. For general safety, restrict all but the necessary personnel from areas where high-pressure systems operate.

REFERENCES

American National Standards Institute, 11 West 42nd Street, New York, NY 10036.
Pressure Piping, ANSI/ASME B31.3 Series.
Relief Valves and Automated Gas Shut-Off Devices for Hot Water Supply Systems, ANSI Z21.22–1999.
Scheme for the Identification of Piping Systems, ANSI A13.1–1981 (Erratum 1998).

American Petroleum Institute, 1220 L Street NW, Washington DC 20005.

American Society of Mechanical Engineers, 3 Park Avenue, New York, NY 10016–5902.
Boiler and Pressure Vessel Code.

Avallone, E. A., and T. Baumeister, eds. *Mark's Standard Handbook for Mechanical Engineers.* 10th ed. New York: McGraw-Hill, 1996.

Combustion Institute, 5001 Baum Boulevard, Pittsburgh, PA 15213. (General.)

Compressed Gas Association, 1235 Jefferson Davis Highway, Arlington, VA 22202.
Cylinder Service Life: Seamless High-Pressure Cylinders, Pamphlet C-5.

National Board of Boiler and Pressure Vessel Inspectors, 1055 Crupper Avenue, Columbus, OH 43229.
National Board of Boiler and Pressure Vessel Inspectors Code.

National Fire Protection Association, 1 Batterymarch Park, Quincy, MA 02269.
Fire Protection Handbook, 2006.
Life Safety Code® Handbook, 2006 ed.

National Safety Council, 1112 Spring Lake Drive, Itasca, IL 60143.
Fundamentals of Industrial Hygiene. 6th ed. 2012.

Singer, J. G., ed. *Combustion: Fossil Power Systems.* 4th ed. New York: Combustion Engineering, 1993.

Technology Transfer Services, Boiler Types and Classifications © 2014 Technology Transfer Services. All Rights Reserved. techtransfer.com/resources/wiki/entry/734/

Underwriters Laboratories Inc., 333 Pfingsten Road, Northbrook, IL 60062.

Uniform Boiler and Pressure Vessel Laws Society, 2838 Long

Beach Road, PO Box 512, Oceanside, NY 11572.

Synopsis of Boiler and Pressure Vessel Laws, Rules, and Regulations by States, Cities, Counties, and Provinces, in the United States and Canada.

U.S. Department of Labor. Occupational Safety and Health Administration, 200 Constitution Avenue NW, Washington DC 20210.

29 CFR 1910.120, Process Safety Management (OSHA 3132–2000).

29 CFR 1910.146, Permit-Required Confined Spaces.

29 CFR 1910.147, The Control of Hazardous Energy (Lockout/Tagout).

REVIEW QUESTIONS

1. What is the difference between the ASME Code and the NB Code for boilers and pressure vessels?
2. List four of the seven common causes of pressure vessel explosions.
   a.
   b.
   c.
   d.
3. When inspecting safety and relief valves in a boiler, the inspection should be performed _____ to prevent damage to the valve seats.
   a. without pressure
   b. under pressure
   c. with blowdown piping
   d. without blowdown piping
4. What is a boiler?
5. What is the maximum recommended pH for water within a boiler?
   a. 3
   b. 4
   c. 7
   d. 11
6. When should a soot blower not be used?
7. What is the purpose of safety valves in boilers and pressure vessels?
8. During the cleaning process, when should soot and fly ash be removed?
9. What is the proper way to wet down an ash pile?
10. Every boiler's flame-safeguard supervisory system and other safety controls should be inspected during a scheduled shutdown period at least _____.
    a. daily
    b. weekly
    c. monthly
    d. yearly
11. What are the three general precautions for entering a boiler?
    a.
    b.
    c.
12. What is high-temperature water (HTW)?
13. What is an unfired pressure vessel?
14. List three of the six items that a vessel's history log should contain.
    a.
    b.
    c.
15. Why would you use a hydrostatic or pneumatic test to inspect a boiler or pressure vessel?
16. Under what conditions should the safety valve of a pressure vessel not have a test lever?
17. What is a water seal and when is it used?
18. The most hazardous part of an autoclave is the _____.

PART

2

Workplace Exposures and Protections

Primary prevention is the safety strategy that focuses on reducing the impact of injuries or illnesses by eliminating causative agents or sources or implementing effective engineering controls. Chapter 6, Safeguarding; Chapter 8, Electrical Safety; Chapter 9, Fire Protection; and Chapter 10, Flammable and Combustible Liquids, describe many of the primary prevention techniques appropriate to these areas. When potentially hazardous workplace exposures are identified and cannot be addressed through elimination, substitution or administrative controls, their impact can be minimized by the appropriate use of personal protective equipment as described in Chapter 7.

Safeguarding

6

Von M. Griggs-Laws
John F. Montgomery, PhD, CSP, CHMM

Definitions

Point-of-Operation Protective Devices
Openings Used for Safeguarding ▶ Guard Construction

Point-of-Operation Safeguards
Types of Safeguards ▶ Substitution as a Safeguard
▶ Matching Machine or Equipment to Operator

Guarding Power Transmissions
Methods of Guarding Actions and Motions ▶ Rotating,
Reciprocating, and Transverse Motions ▶ In-Running
Nip Points ▶ Cutting Actions ▶ Punching, Shearing,
and Bending Actions ▶ Guarding Materials

Maintenance and Servicing

Control of Hazardous Energy Sources
Sample Lockout/Tagout Procedure ▶ Purpose ▶ Scope
▶ Sequence of Lockout ▶ Restoring Equipment to
Service ▶ Other Program Requirements ▶ Group
Lockout

Robotics Safeguarding
Hazards and Hazardous Locations ▶ Safeguarding
Methods

Guards and Noise Control

**Machine Operator Required Reporting of Employee
Injuries**

Summary

References

Review Questions

153

From the earliest days of safety, using engineering controls to provide safeguards that reduce or eliminate hazards has been fundamental to occupational safety and health programs and to accident prevention. Many safety regulations address identifying and evaluating mechanical hazards and then guide actions to protect workers from the hazards. This chapter discusses the following topics:

- creating safeguards to protect employees from hazards at the points of operation and of power transmission
- safeguarding employees during maintenance and servicing operations
- controlling hazardous energy sources
- equipment and techniques for protecting workers from the hazards of robotics
- guards designed to control noise.

DEFINITIONS

Terms used in this chapter are defined as follows:

Safeguarding
Any means of preventing personnel from coming in contact with the moving parts of machinery or with equipment that could potentially cause physical harm.

Device
A mechanism or control designed for safeguarding at the point of operation. Devices include presence-sensing devices, movable-barrier devices, holdout or restraint devices, pull-back (pull-out) devices, two-hand-trip devices, and two-hand-control devices.

Guard
A barrier designed for hazard control at the point of operation. Guards include die-enclosure guards, fixed-barrier guards, interlocked-barrier guards, and adjustable-barrier guards.

Enclosure
Safeguarding with fixed physical barriers mounted on or around a machine to prevent access to the moving parts. Enclosures are most effective when designed as part of the machine or bolted or welded to the frame or the floor.

Fencing
Safeguarding by means of a locked fence or rail enclosure that restricts machine access to authorized personnel.

Location
Safeguarding by location involves making a hazard physically inaccessible under normal operating conditions or use. Both fencing and location are limited safeguarding techniques and are permitted only if other precautionary restrictions are complied with.

Nip Points or Bites
A hazardous area involving two or more mechanical parts rotating in opposite directions within the same plane and in close interaction.

Pinch Point
Any place a body part can be caught between two or more moving mechanical parts or between one fixed part and one moving part in the same plane.

Point of Operation
The area of a machine where material is positioned for processing and where processing of the material actually takes place.

Power Transmission
Power transmission involves all the mechanical parts—such as gears, cams, shafts, pulleys, belts, clutches, brakes, and rods—that transmit energy and motion from a power source to equipment or a machine.

Shear Points
A hazardous area created by the cutting movement of a mechanical part that is beyond a stationary point on a machine.

The U.S. *Code of Federal Regulations*, Title 29, Part 1910, Subpart O, states that when machinery is constructed, the risks of using that machinery must be analyzed, and where necessary, protective devices must be provided to the operator.

Technical Report ANSI B11.TR3–2000 includes proposals for identifying, analyzing, and reducing the risks from tool-making machines (Figure 6–1) (ANSI 2000).

OSHA/ANSI provides the following hierarchical procedure for risk reduction:

1. identifying and analyzing the risk
2. removing the risk using appropriate measures
3. reducing the risk by providing protective devices
4. installing warning signals and posting warning information
5. providing personal protective equipment to operating personnel
6. training operators.

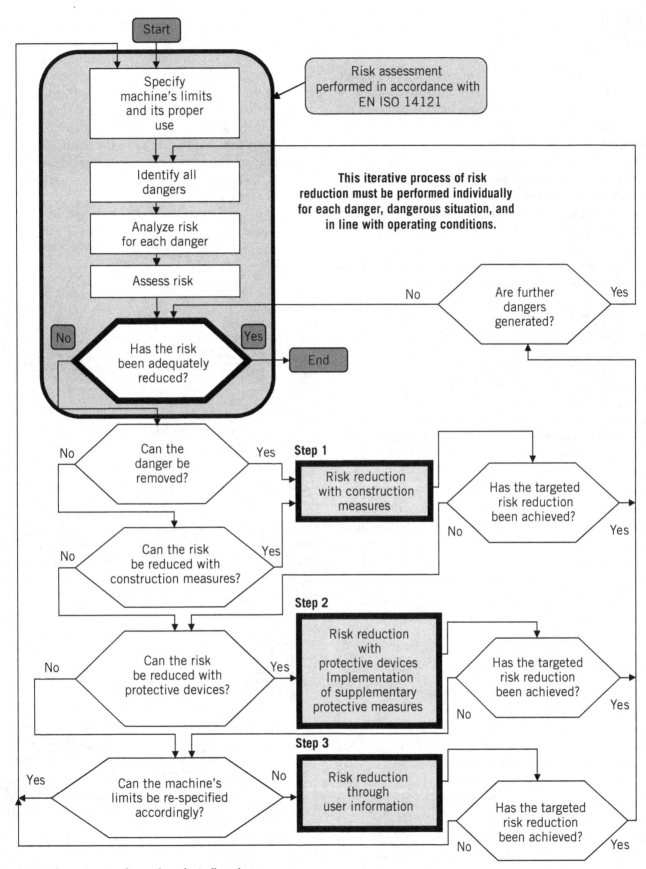

Figure 6–1. Risk assessment/hazard analysis flowchart.

POINT-OF-OPERATION PROTECTIVE DEVICES

If all machines were alike, it would be simple to design a universal point-of-operation device/guard and install it during the manufacture of each machine. Such is not the case. Further complicating the problem is that purchasers of the same machine model may use it in different ways and for different purposes. In addition, the uses of one machine may change during its lifetime. Because of these and other reasons, a machine manufacturer cannot always design and install an effective point-of-operation device or guard. In some cases, the buyer must first test a machine's functions to discover its hazards. Only from that point of awareness can an effective device/guard be designed and installed prior to using the machine.

Whenever point-of-operation protective devices/guards are needed, appropriate principles and conditions should be applied based on the type, design, construction, and location of machine safeguards. The following is a summary of these principles and important data found in OSHA 29 CFR 1910.211, 212, 213, and 217 and in various ANSI standards (B11.1, 2, 3, etc.).

For more information on point-of-operation safeguarding related to power presses, see Chapter 21, Working with Hot and Cold Metals.

Openings Used for Safeguarding

Common practice in the design of point-of-operation safeguards is to consider any opening not exceeding ⅜ in. (9 mm) to be relatively safe because such an opening would not permit any significant part of a hand to enter the guard (Figure 6–2). In many instances, however, a ⅜-in. (9-mm) opening is insufficient for processed material to pass through or under the guard. On the other hand, as the width (or height) of the opening is increased to accommodate the material, the farther inside the guard the operator is able to reach, and the closer to the point-of-operation hazard the operator becomes. Under these circumstances, it is not possible to prevent the entry of some part of the hand into the guard.

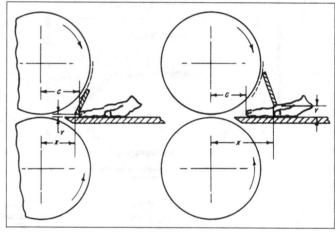

Figure 6–3. Left: A ¼-in. (6-mm) opening on the guard stops the hand. Right: A larger opening (Y) placed farther back can also stop the fingers from reaching the danger zone. Both illustrations show the ends of the fingers being stopped at approximately the same distance (C) from the danger zone. *(Printed with permission from Alliance of American Insurers; Liberty Mutual Insurance Company.)*

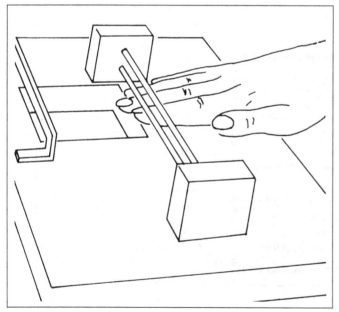

Figure 6–2. A ⅜-in. (9-mm) opening permits part of the fingers to slip past the guard. A smaller, ¼-in. (6-mm) opening prevents the fingers from crossing into the danger zone.

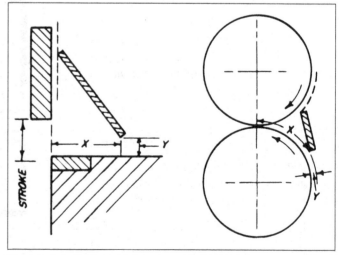

Figure 6–4. Left: Vertical shear hazard. Right: In-running roll hazard (no feed table used). *(Printed with permission from Alliance of American Insurers; Liberty Mutual Insurance Company.)*

Figure 6–3 shows sketches of an in-running roll hazard in which a feed table is used. Other types of point-of-operation hazards need an opening in the guard larger than that shown.

In Figure 6–4, the proper location of the guard's distance X for the use of required opening Y must be determined. If the dimensions of the opening and its location (distance from the hazard) are properly selected, adequate safety for the operator can be ensured. Some guard designers use this formula:

$$\text{Maximum safe opening} = \text{¼ in. (6 mm)} + \text{⅛} \times \text{Distance to guard from danger zone}$$

CAUTION: This formula is not intended for use when the distance from the guard to the danger zone exceeds 12 in. (30 cm).

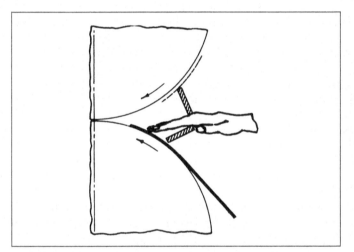

Figure 6–5. This guard protects the fingers while the hand positions materials. *(Printed with permission from Alliance of American Insurers; Liberty Mutual Insurance Company.)*

Figure 6–5 illustrates a situation in which parts of the hand and fingers must extend through the guard in order to manipulate the material inside the guard. For example, a situation can arise in which it is impossible to use hand tools or mechanical devices to manipulate material.

Test Data

For its design analysis, the Liberty Mutual Insurance Company constructed several test fixtures that approximated common guards. By testing openings and hands of different sizes, the company developed data to guide guard designers when specifying allowable openings and locations of guards. Design criteria are shown in Figure 6–6 with a feed table and in Figure 6–7 without a feed table. One set of data was established for general use. Due to variations in the sizes of hands and the inaccuracies to be expected when trying to maintain a ⅜-in. (9-mm) opening, the analysts determined that for the first 1 ½ in. (38 mm) from the danger line, no openings exceeding ¼ in. (6 mm) could be considered safe.

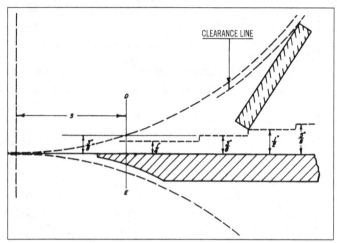

Figure 6–6. In-running roll nip when a feed table is used. DE is the stop line. S is the distance of the ⅜-in.-wide nip zone from the contact point between the rolls.

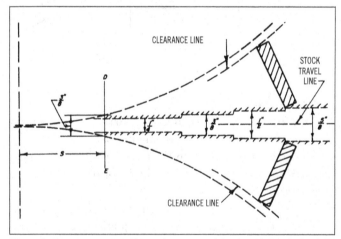

Figure 6–7. In-running rolls with a central feed and no feed table.

Most men and women have fingertips that will not travel far into a ⅜-in. (9-mm) opening. However, if a designer wishes to maintain an unquestionably safe zone beyond a ⅜-in. (9-mm) opening, the opening must be no more than ¼ in. (6 mm) wide within 1 ½ in. (38 mm) from the danger point (Figure 6–8).

Application of Test Data

A team from the Liberty Mutual Research Center applied the data from these tests to various situations that require guards.

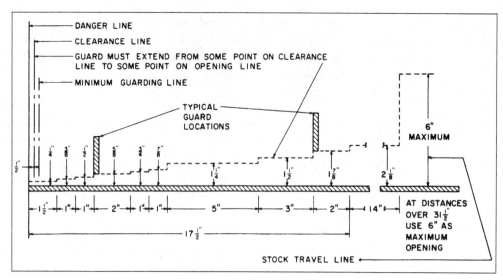

Figure 6-8. Every point-of-operation guard must prevent hands or fingers from reaching over, under, or around the guard and into the point of operation. To minimize the possibility of misuse or removal of essential parts, such a guard should have fasteners that the operator cannot easily remove. The guard should also offer maximum visibility of the point of operation. (The dimensions shown assume adult hands held flat as shown in Figure 6-3.) *(Printed with permission from Machine Tools—Mechanical Power Presses—Safety Requirements for Construction, Care, and Use, ANSI B11.1.)*

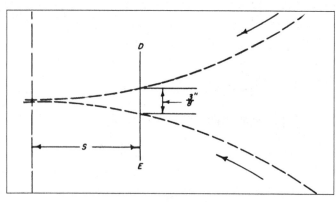

Figure 6-9. Application of test findings to designing guards for in-running rolls. See the accompanying text for an explanation of this figure.

Vertical Shear

The findings of the vertical shear portion of these tests (Figure 6-9) apply to the following items:

- all horizontal openings in the guard itself that are necessary for feeding stock into the front or sides and for ejecting finished parts or scrap
- all vertical openings in the guard itself, such as visibility slots, clearance slots for ejection devices, and stop gauges.

In-Running Rolls with Feed Table

In applying the findings of these tests to guards for in-running rolls, first consider the characteristics of a nip point. In Figure 6-8, the danger line represents a hazardous contact. For vertical shear exposures, the danger line is equivalent to the shear line. For rolls, the hazard, nip, or pinch zone is not defined by a straight line. Therefore, a ⅜-in. (9-mm) width of nip zone is considered the actual nip point through which the danger line DE is drawn (Figure 6-9). The distance of the nip zone's width from the contact points between the rolls is designated dimension S. It is recommended that rolls held less than ⅜ in. (9 mm) apart be considered rolls in contact.

Figure 6-6 shows an in-running-roll nip point in which a feed table is used. To design a properly situated barrier guard, use the following procedure:

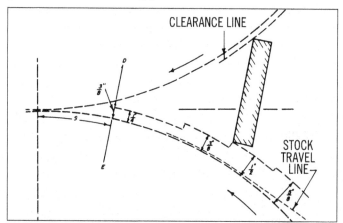

Figure 6-10. In-running rolls with stock traveling over one roll before entering the nip zone. CAUTION: When no stock is fed over the cylinder, particularly with high-speed cylinders, the space between the guard and the cylinder should be limited to ¼ in. (6 mm) to avoid creating a hazard between the moving cylinder and the guard itself. Gaps or discontinuities in the cylinder surface may also be hazards. *(Printed with permission from Alliance of American Insurers; Liberty Mutual Insurance Company.)*

1. Draw a full-scale sketch of the nip zone, with the top surface of the feed table accurately depicted. Indicate the clearance line on the top roll. If more than a 3/8-in. (9-mm) clearance is required, position the top edge of the guard in accordance with the layout for a safe opening, as shown in Figure 6–10 for a layout on a curved surface.
2. Determine the distance S from the centerline of the rolls to the point where a 3/8-in. (9-mm) space exists between the top of the feed stock and the surface of the upper roll.
3. At this distance, begin the diagram of the safe opening dimensions (Figure 6–8), up to the opening necessary for the particular guard being designed. Indicate the guard section (putting the top edge on the clearance line of the upper roll and the bottom edge at the proper point on the diagram for a safe opening), and determine the necessary dimensions for installing the guard. Also determine the appropriate width of the guard.
4. Before the guard is installed, check carefully for the possibility that hands can travel under the guard, the stability of the mounting, and the rigidity of the construction.

In-Running Rolls—Stock Traveling over One Roll before Entering Nip Zone

Figure 6–10 shows an in-running-roll nip point at which the stock travels over a portion of one roll before entering the nip zone. The stock in such an arrangement feeds either under or over a barrier.

To design a properly situated barrier guard while taking into account these conditions, use the following procedure:
1. Draw a full-scale sketch of the nip zone, with the travel line of the stock shown on the roll. Indicate the clearance line on the top roll. If more than a 3/8-in. (9-mm) clearance is required, position the top edge of the guard in accordance with the layout for a safe opening.
2. Determine the distance S from the centerline of the roll to the point where there is a 3/8-in. (9-mm) space between the rolls.
3. At this distance, begin the diagram (on the roll with the stock's travel) of the safe opening dimensions, as shown in Figure 6–8. (The diagram can be drawn on a curved surface with 1/2-in. [13-mm] divider steps.) Sketch the guard section (with one edge touching the clearance line and the other touching the diagram at the proper point for a safe opening). From this final diagram, determine the necessary dimensions for properly installing the guard as well as the appropriate width of the guard.
4. Before the guard is installed, check carefully for the possibility that hands can travel under the guard, the stability of the mounting, and the rigidity of the construction.

Figure 6–11. Safeguards should be made of sturdy materials. *(Courtesy International Stamping Co./Midas International Corp.)*

Guard Construction

To ensure that the safe opening dimensions and effectiveness of guards are maintained, the guards should be solidly constructed and secured to minimize distortion or movement (Figure 6–11). All parts of guards should be strong enough to withstand expected stresses and/or exposures. Fastenings should be durable and designed to prevent the guard from shifting and being moved or removed. Any guard with openings larger than 1/4 in. (6 mm) has an optimal construction and should be checked frequently for its alignment and condition.

Depending on the visibility and rigidity needs and on the method of feeding material through a particular guard, the designer/engineer can use the layouts shown in Figures 6–6, 6–7, or 6–10 to select the best-suited design. Figures 6–12 and 6–13 show typical designs for guards for a roll nip zone, without and with a feed table, respectively. Working from layouts similar to these, the designer/engineer can determine the appropriate location, size, and shape of the guard section. (See Chapter 18, Woodworking Machinery; Chapter 20, Metalworking Machinery; and Chapter 21, Working with Hot and Cold Metals—for more specific information on guarding.)

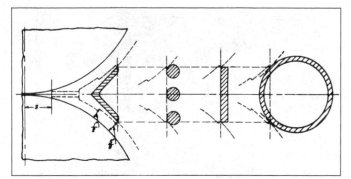

Figure 6–12. These designs for a roll nip guard are appropriate for feeding material over or under a guard and through a ⅜-in. (9-mm) opening.

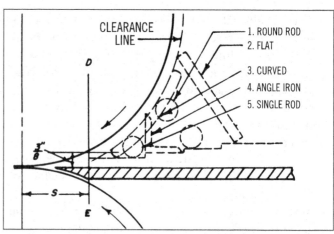

Figure 6–13. Within the diagram are different designs for an effective guard on a roll nip with a feed table.

POINT-OF-OPERATION SAFEGUARDS

Requirements for safeguarding specific machinery can be found in regulatory and related American National Standards Institute (ANSI) standards such as those listed in Table 6–A and in the References at the end of this chapter.

EN ISO 12100-1 recommends that the machine designer use the following step-by-step procedure to reduce risks:

1. Specify the limits and proper use of the machine.
2. Identify possible hazards and hazardous situations.
3. Estimate the risk of each identified hazard and hazardous situation and also consider the foreseeable malpractice or faulty operation by operating personnel.

TABLE 6–A. Selection of Important U.S. National Consensus Standards Regarding Machine Safety Guarding (this list is not complete)

Standard	Title
ANSI B11.1	Machine Tools—Mechanical Power Presses—Safety Requirements for Construction, Care, and Use
ANSI B11.2	Machine Tools—Hydraulic Power Presses—Safety Requirements for Construction, Care, and Use
ANSI B11.3	Power Press Brakes—Safety Requirements for Construction, Care, and Use
ANSI B11.4	Machine Tools—Shears—Safety Requirements for Construction, Care, and Use
ANSI B11.5	Machine Tools—Iron Workers—Safety Requirements for Construction, Care, and Use
ANSI B11.6	Lathes—Safety Requirements for Construction, Care, and Use
ANSI B11.7	Machine Tools—Cold Headers and Cold Formers—Safety Requirements for Construction, Care, and Use
ANSI B11.8	Drilling, Mining, and Boring Machines—Safety Requirements for Construction, Care, and Use
ANSI B11.9	Grinding Machines—Safety Requirements for Construction, Care, and Use
ANSI B11.10	Metal Sawing Machines—Safety Requirements for Construction, Care, and Use
ANSI B11.11	Gear Cutting Machines—Safety Requirements for Construction, Care, and Use
ANSI B11.12	Roll Forming and Roll Bending Machines—Safety Requirements for Construction, Care, and Use
ANSI B11.13	Machine Tools—Single- and Multiple-Spindle Automatic Bar and Chucking Machines—Safety Requirements for Construction, Care, and Use
ANSI B11.14	Machine Tools—Coil-Slitting Machines—Safety Requirements for Construction, Care, and Use
ANSI B11.15	Pipe, Tube, and Shape Bending Machines—Safety Requirements for Construction, Care, and Use
ANSI B11.16	Metal Powder Compacting Presses—Safety Requirements for Construction, Care, and Use (2008)
ANSI B11.17	Machine Tools—Horizontal Hydraulic Extrusion Presses—Safety Requirements for Construction, Care, and Use of
ANSI B11.18	Machine Tools—Machinery and Machine Systems for Processing Strip, Sheet, or Plate from Coiled Configuration—Safety Requirements for the Construction, Care, and Use

Standard	Title
ANSI B11.19	Performance Criteria for the Design, Construction, Care, and Operation of Safeguarding When Referenced by the Other B11 Machine Tool Safety Standards (ANSI 2010)
ANSI B11.20	Machine Tools—Manufacturing Systems/Cells—Safety Requirements for Construction, Care, and Use (ANSI 2009)
ANSI B11.21	Safety Requirements for Machine Tools—Using Lasers for Processing Materials (ANSI 2010)
ANSI B11.TR1	Ergonomic Guidelines for the Design, Installation, and Use of Machine Tools (ANSI 2004)
ANSI B11.TR2	Mist Control Considerations for the Design, Installation, and Use of Machine Tools Using Metalworking Fluids (ANSI 2000)
ANSI B151.27	Safety Requirements for the Integration, Care, and Use of Robots Used with Horizontal and Vertical Injection Molding Machines (ANSI 2003)
ANSI B56.5	Safety Standard for Guided Industrial Vehicles and Automated Functions of Manned Industrial Vehicles (ANSI 2012)
ANSI R15.06	Industrial Robots and Robot Systems—Safety Requirements
ANSI B65.1	Safety Standards for Printing Press Systems (ANSI 2011)
NFPA 70E	Electrical Safety Requirements for Employee Workplaces (NFPA 2015)
NFPA 79	Electrical Standard for Industrial Machinery (NFPA 2015)
UL 508	Industrial Control Equipment (UL 2008)
UL 614961	Electro-Sensitive Protective Equipment, Part 1: General Requirements for Design, Construction, and Testing of Electro–sensitive Protective Devices (ESPDs) (UL 2012)
UL 614962	Electro-Sensitive Protective Equipment, Part 2: Particular Requirements for Equipment Using Active Optoelectronic Protective Devices (AOPDs) (UL 2011)
29 CFR 1910.211 through 1910.222	OSHA Machinery and Machine Guarding

Alternative process for risk reduction (Source: EN ISO 121001)

4. Evaluate each individual risk and decide whether or not risk reduction is required.
5. Attempt to remove or reduce the risk by using appropriate measures. If this removal or reduction does not work, then:
6. Reduce the risk by using protective devices (separating protective devices, such as hard guards or covers; or electro-sensitive protective equipment, such as Safety Light Curtains, for example).
7. Inform and warn machine operators about the remaining risks of the machine by placing warning plates on the machine and including warning notes in the operating instructions. (EN ISO 12100-1 2011)

The first four steps deal with risk analysis and risk assessment. Risk analysis and risk assessment must be carried out methodically and be comprehensively documented.

In addition to the protective measures used by the machine designer/constructor, the operating company or machine operator may require further protective measures to reduce the remaining risk. These measures may be:
1. organizational measures (e.g., safe work processes, regular inspections, preventive maintenance, etc.)
2. personal protective devices
3. training of and instruction for operating personnel (EN ISO 121001 2013).

Proper guard design characteristics include:
1. integration with the machine
2. adequate construction, durability, and strength
3. ability to accommodate the workpiece in-feed and during ejection
4. protection from hazards
5. ease of inspection and maintenance
6. sufficiently tamper-proof or foolproof.

On the other hand, a guard should not:
1. create additional hazards
2. interfere with production
3. cause work discomfort (e.g., skin contact stress, sharp edge exposures, etc.).

Types of Safeguards

Standardized safeguards as well as improvised barriers, enclosures, and tools are designed to protect the machine operator—in particular, his or her hands—at point-of-operation areas.

Sheet metal, perforated metal, expanded metal, heavy wire mesh, or stock is commonly used to construct most types of guards. The best practice for selecting the appropriate material for new guards or barriers is to consult and follow established standards.

If parts moving through the manufacturing process

must be visible and if strength/durability of the metal is not necessary, use transparent impact plastic or safety glass. Use aluminum or other soft metals if resistance to rust is essential. Plastic and glass-fiber barriers, which are usually less expensive than metal barriers, have less strength. However, nonmetal barriers can withstand the splashes, vapors, and fumes of corrosive substances, which react with metal.

Built-in Safeguards

Because most manufacturers include safeguards as an integral part of a machine, these guards have several advantages over safeguards made by the machine user:

1. Built-in safeguards conform more accurately to the contours of the machine, thus making them superior in appearance, placement, and function.
2. Built-in safeguards, when properly designed, eliminate hazards and are able to withstand normal wear and use. Makeshift safeguards may not offer enough protection against mechanical or human failure, especially if their design and construction are not subject to hazards assessment prior to installation or do not conform to recognized standards. In addition, makeshift safeguards may give operators a false sense of security and thus increase the risk of injury. To compensate for the potential inadequacies of makeshift safeguards, operators must be extra vigilant for the presence of hazards while using the machine and its inadequately protected areas, vigilance that can increase operator stress and fatigue.
3. Built-in safeguards tend to cost less than post-installation, makeshift safeguards because the manufacturer spreads the fabrication cost over a large number of machines. Of course, besides the cost savings, the hazards presented by an inadequately guarded machine are a potent rationale for having the manufacturer design and install safeguards to conform with recognized standards.

CAUTION: Modifying a machine may invalidate the manufacturer's warranties. Before taking any such actions, end users should discuss modification plans with the manufacturer.

Barrier Guards

Barrier guards block access to the dangerous parts of a machine and are often adjustable to accommodate different types of tools or kinds of work. Once adjusted, barrier guards should not be moved or detached from the machine. Figure 6–8 shows the safe distance between a barrier guard and the point of operation, along with the permitted sizes of openings in a fixed guard.

Interlocking Barrier Guards

Interlocking barrier guards can be mechanical, electrical, pneumatic, or a combination. Until the guard has moved into a predetermined position, the interlocking guard prevents the operation of the control mechanism (actuator) that puts the machine in motion, thereby preventing the operator from reaching into the hazardous point-of-operation area. When the guard opens and exposes the operator to the point-of-operation hazard, the control mechanism is locked; that is, a locking pin or other safety device prevents the main shaft or other means of power transmission from rotating and prevents any other mechanisms from operating. Once the machine is in motion, the guard cannot be opened until the machine has come to rest or has reached a fixed position in its process of movement. When neither a fixed guard nor an interlocking guard is practical, use mechanical interlocks.

An effective interlocking guard must satisfy three requirements:

1. guard the hazardous area before the machine can be operated
2. stay closed until the rotating equipment is at rest
3. prevent operation of the machine if the interlocking device fails.

Automatic Safeguarding Devices

When neither a fixed guard nor an interlocking guard is practical, an automatic barrier guard can be used, subject to the limitations presented in Table 6–B. Such a guard must prevent the operator from coming in contact with the dangerous parts of a machine while it is in motion or must stop the machine in the event of a subnormal operating condition. Figure 6–14 shows the relationship between safeguarding devices and fixed barrier guards. (See Chapter 21, Working with Hot and Cold Metals.)

All of the types of guards just discussed are suitable for protecting operators at points of operation. Safeguarding devices can also be used in lieu of or in conjunction with barrier guards at the points of operation. Presence-sensing devices (either photoelectric or radio frequency), pull-backs, restraints, and two-hand controls are useful for specified conditions and constraints. (See Chapter 21, Working with Hot and Cold Metals.)

Guarding by Location

In some cases, the out-of-the-way location of an exposed part serves as a sufficient safeguard. OSHA standard 29 CFR 1910.219(c)(2)(i) states, "All exposed parts of horizontal shafting seven (7) feet or less from floor or working platform, excepting runways used exclusively for oiling, or running adjustments, shall be protected by a stationary casing enclosing shafting completely or by a trough

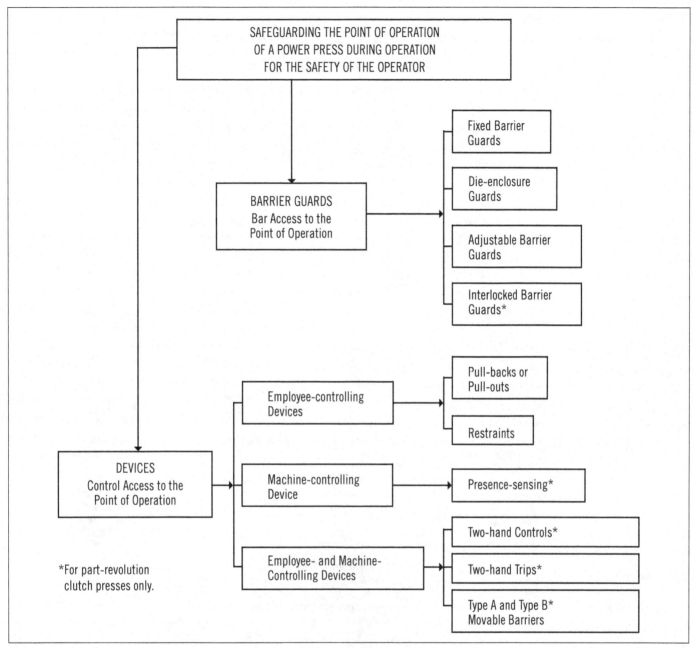

Figure 6–14. Automatic guards can protect the employee at the machine's point of operation.

enclosing sides and top or sides and bottom of shafting as location requires."

Substitution as a Safeguard

Like the installation of safeguards, substituting one type of machine function for another can sometimes eliminate or reduce machine hazards. For example, substituting direct-drive machines or individual motors for an overhead line-shaft transmission decreases the hazards inherent in transmission equipment; speed reducers can replace multi-cone pulleys; and remote-controlled, automatic lubrication can eliminate the need for workers to get near moving parts.

Matching Machine or Equipment to Operator

The safe operation of machinery involves more than just eliminating or enclosing hazardous moving parts. It is also important to assess the overall injury potential of the machine's operation, to perform a thorough machine-task demands (or ergonomic) analysis, and to ask these fundamental questions:

TABLE 6–B. Point-of-Operation Protection

Type of Safeguarding Method	Action of Safeguard	Advantages	Limitations	Typical Machines on Which Used
ENCLOSURES OR BARRIER GUARDS				
Complete, simple fixed enclosure	Barrier or enclosure that admits the stock but that will not admit hands into danger zone because of feed opening size, remote location, or unusual shape.	Provides complete enclosure if kept in place. Both hands free. Generally permits increased production. Easy to install. Ideal for blanking on power presses. Can be used with automatic or semiautomatic feeds.	Limited to specific operations. May require special tools to remove jammed stock. May interfere with visibility.	Bread slicers Embossing presses Meat grinders Metal-squaring shears Nip points of in-running rubber, paper, textile rolls Paper corner cutters Power presses
Warning enclosures (usually adjustable to stock being fed)	Barrier or enclosure admits the operator's hand but warns him/her before danger zone is reached	Makes "hard to guard" machining safer. Generally does not interfere with production. Easy to install. Admits varying sizes of stock.	Hands may enter danger zone—protection not complete at all times. Danger of operator not using guard. Often requires frequent adjustment and careful maintenance.	Band saws Circular saws Cloth cutters Dough brakes Ice crushers Jointers Leather strippers Rock crushers Wood shapers
Barrier with electric contact or mechanical stop activating mechanical or electric brake	Barrier quickly stops machine or prevents application of injurious pressure when any part of operator's body contacts it or approaches danger zone.	Makes "hard to guard" machines safer. Does not interfere with production.	Requires careful adjustment and maintenance. Possibility of minor injury before guard operates. Operator can make guard inoperative.	Calendars Dough brakes Flat roll ironers Paper box corner stayers Paper box enders Power presses Rubber mills
Enclosure with electrical or mechanical interlock	Enclosure or barrier shuts off or disengages power and prevents starting of machine when guard is open; prevents opening of the guard while machine is under power or coasting. (Interlocks should not prevent manual operation or "inching" by remote control.)	Does not interfere with production. Hands are free; operation of guard is automatic. Provides complete and positive enclosure.	Requires careful adjustment and maintenance. Operator may be able to make guard inoperative. Does not protect in event of mechanical repeat.	Dough brakes and mixers Foundry tumblers Laundry extractors, driers, and tumblers Power presses Tanning drums Textile pickers, cards
AUTOMATIC OR SEMIAUTOMATIC FEED				
Nonmanual or partly manual loading of feed mechanism, with point of operation enclosed	Stick fed by chutes, hoppers, conveyors, movable dies, dial feeds, rolls, etc. Enclosure will not admit any part of body.	Generally increases production. Operator cannot place hands in danger zone.	Excessive installation cost for short run. Requires skilled maintenance. Not adaptable to variations in stock.	Baking and candy machines Circular saws Power presses Textile pickers Wood planers Wood shapers
HAND REMOVAL/RESTRAINT DEVICES				
Hand restraints (hold-back)	A fixed bar and cord or strap with hand attachments that, when worn and adjusted, do not permit an operator to reach into the point of operation.	Operator cannot place hands in danger zone. Permits maximum hand feeding; can be used on higher-speed machines. No obstruction to feeding a variety of stock. Easy to install.	Requires frequent inspection, maintenance, and adjustment to each operator. Limits movement of operator. May obstruct space around operator. Does not permit blanking from hand-fed strip.	Embossing presses Power presses Power press brakes
Hand pull-backs or pull-outs	A cable-operated attachment on slide, connected to the operator's hands to pull the hands back only if they remain in the danger zone; otherwise, it does not interfere with normal operation.	Acts even in event of repeat. Permits maximum hand feeding; can be used on higher speed machines. No obstruction to feeding a variety of stock. Easy to install.	Requires unusually good maintenance and adjustment to each operator. Frequent inspection necessary. Limits movement of operator. May obstruct work space around operator. Does not permit blanking from hand-fed strip stock.	Embossing presses Power presses Power press brakes

6 Point-of-Operation Safeguards

Type of Safeguarding Method	Action of Safeguard	Advantages	Limitations	Typical Machines on Which Used
MISCELLANEOUS				
Limited slide travel	Slide travel limited to ¼ in. or less; fingers cannot enter between pressure points.	Provides positive protection. Requires no maintenance or adjustment.	Small opening limits size of stock.	Foot power (kick) presses Power presses
Presence-sensing device	Sensing field and brake quickly stop machine or prevent its starting if hands are in the danger zone.	Does not interfere with normal feeding or production. No obstruction on machine or around operator.	Expensive to install. Does not protect against mechanical repeat. Limited to use on machines with means to quickly stop the machine during the operating cycle.	Embossing presses Power presses Press brakes
Type A and B gate devices	Encloses danger area before machine action starts. Stays closed until hazard ceases or stops machine if opened too soon.	Interlocked with operating cycle. Allows free access to load and unload machine. Fully encloses point of operation.	Usually limited to machines on which the part or material being processed is fully within the point-of-operation area.	Power presses Plastic injection-molding machines Compression-molding machines Die-casting machines
Special tools or handles on dies	Long-handled tongs, vacuum lifters, or hand die holders avoid need for operators to put hands in the danger zone.	Inexpensive and adaptable to different types of stock. Sometimes increases protection of other guards.	Operator must keep hands out of danger zone. Requires unusually good employee training and close supervision.	Dough brakes Leather die cutter Power presses Forging hammers
Special jigs or feeding devices	Hand-operated feeding devices of metal or wood keep the operator's hands at a safe distance from the danger zone.	May speed production as well as safeguard machines. Generally economical for long jobs.	Machine itself not guarded; safe operation depends on correct use of device. Requires good employee training, close supervision. Suitable for limited types of work.	Circular saws Dough brakes Jointers Meat grinders Paper cutters Power presses Drill presses
TWO-HAND TRIP				
Electric	Simultaneous pressure of two hands on switch buttons in series actuates machine.			
Mechanical	Simultaneous pressure of two hands on air control valves, mechanical levers, controls interlocked with foot control, or the removal of solid blocks or stops permits normal operation of machine.	Can be adapted to multiple operation. Operator's hands away from danger zone. No obstruction to hand feeding. Does not require adjustment. Can be equipped with continuous-pressure remote controls to permit "inching." Generally easy to install.	Operator may try to reach into danger zone after tripping machines. Does not protect against mechanical repeat unless blocks or stops are used. Not generally suitable for blanking operations. Must be designed to prevent tying down of one button or control, which would thereby permit unsafe one-hand operation.	Dough mixers Embossing presses Paper cutters Pressing machines Power presses Power press brakes
Electric	Simultaneous pressure on two-hand switches held down until dies close.	Can be adapted for multiple operators. Operators' hands are away from hazard during die-closing portion of stroke. No obstruction to hand feeding.	Hand buttons must be spaced far enough from point of operation to stop machine upon removal before hand can reach into the hazard. Control circuit must be designed to prevent tying down of one button. Buttons must be spaced far enough apart to prevent operation with one hand and another part of the body.	Dough mixers Embossing presses Paper cutters Pressing machines Power presses Press brakes

- What task demands (physical and mental) does the machine impose on the operator?
- Do those demands exceed operator capability?
- Is there a materials-handling hazard?
- Are the limitations of a person's physical ability to lift, push, and pull recognized?
- What types, frequency, and complexity of decisions must the operator make?
- Does the design of safeguards consider physiological factors and human dimensions (anthropometry)?
- What demands (expectations) does the operator impose on the machine?
- What environmental factors or demands might affect machine operation or operator performance?

Evaluate all physical or design features of a machine and the workplace as though the machine is an extension of a person's body and can do only what that person wants it to do. To match the machine to the operator, consider the factors in the following paragraphs.

The Workplace

Arrange machines and equipment so that an operator has to do a minimal amount of strenuous lifting and moving. Provide conveyors, skids, jacks, lift tables, or other equipment to feed raw stock and chutes or gravity feeds to remove finished stock.

The Work Height

Ensure that workstations are at the optimal height for stand-up or sit-down positions of working. For the sit-down position, determine the proper height and type of chair or stool by considering various seated heights, body sizes, reach requirements, and ease of chair adjustability. Elbow height is one factor to consider when deciding on the work height that will maximize worker strength and minimize worker fatigue. In general, an appropriate work height is 41 in. (1 m) from floor to work surface, with a chair height from 25 to 31 in. (0.6 to 0.8 m). (See Chapter 16, Ergonomics Yesterday, Today, and Tomorrow, in the *Administration & Programs* volume.)

Controls and Displays

The position and design of machine controls and displays such as dials, push buttons, and levers are important to machine function and to human error reduction. Speed, ON-OFF, and emergency controls, in particular, should be readily accessible. Standardize movement controls and display functions on similar machines so that operators can easily switch from working on one machine to working on another. The layout and functional relationships of controls and displays should coordinate with process flow. (See Chapter 16, Ergonomics Yesterday, Today, and Tomorrow, in the *Administration & Programs* volume.)

Material-Handling Aids

Provide aids to minimize the manual handling of raw materials and in-process or finished parts, both to and from machines. Aids include overhead chain hoists, belt or roller conveyors, and work positioners. (See Chapter 21, Working with Hot and Cold Metals, for more on work holding and positioning tools.)

Operator Fatigue

Workers usually experience fatigue at a machine station because of the combination of physical and mental activities, not simply from expending energy. Other contributors to fatigue include excessive speed-ups; boredom from monotonous operations; awkward work motions; static muscle stresses; prolonged standing or sitting; and skin contact with hard parts of machines, guards, or work surfaces.

U.S. *Code of Federal Regulations* Title 29, Part 1910, Subpart O, defines the calculation of the minimum safety distance of a protective device for mechanical power presses (Table 6–C) (see OSHA 29 1910.217, Table 0-10). If the safety distance calculation results in a greater value, this table must be used.

TABLE 6–C. Distances Guards Should Be Positioned from Danger Line in Accordance with Required Openings

Distance of Opening From Point of Operation Hazard	Maximum Width of Opening
½ to 1 ½	¼
1 ½ to 2 ½	⅜
2 ½ to 3 ½	½
3 ½ to 5 ½	⅝
5 ½ to 6 ½	¾
6 ½ to 7 ½	⅞
7 ½ to 12 ½	1 ¼
12 ½ to 15 ½	1 ½
15 ½ to 17 ½	1 ⅞
17 ½ to 31 ½	2 ⅛

(OSHA 29 1910.217(g), Table 0-10)

GUARDING POWER TRANSMISSIONS

To varying degrees, all mechanical actions and motions are hazardous. Rotating members, reciprocating arms, moving belts, meshing gears, cutting teeth, and parts being impacted or sheared are examples of mechanical actions and motions that workers need protection from. These hazardous exposures are not unique to any one machine or industry and may be classified as follows:

1. rotating, reciprocating, and transverse motions
2. in-running nip points
3. cutting actions
4. punching, shearing, and bending actions.

Methods of Guarding Actions and Motions

Whenever hazardous machine actions or motions are present, it is essential to provide protection for operators and other workers. Multiple guarding options may be available.

Certain guarding methods may be preferable to others, but the type of operation, the size or shape of the stock to be processed, the method of handling the stock, the physical layout of the work area, the type of stock, and the production requirements or limitations may alter design priorities. How flexible operations must be may also determine the feasibility of the method used. As a general rule, fixed enclosure guards are effective for controlling power-transmission apparatus hazards.

Generally, the same principles used in point-of-operation guards apply to power-transmission guards, although openings for loading and unloading materials need to be considered. Power-transmission guards, although generally fixed, can also be hinged, sliding, or bolted coverplates. Fixed guards should be removed only for maintenance, service, and adjustments and only when the equipment is properly locked out.

Power-transmission guards generally must envelop all moving parts so that no part of an operator's body touches the hazard. In many cases, a simple flat plate or box that covers the opening is all that is necessary, especially when the parts are flush with or recessed within the frame of the machine. Where parts protrude beyond the frame, it may be necessary to build a guard that conforms to the dimensions and forms of those protruding parts (Figure 6–15). In such cases, any openings that permit shafts or other parts to go into the machine must follow the requirements for the maximum size of permitted openings relative to their distance from the moving parts (Table 6–D).

Rotating, Reciprocating, and Transverse Motions

Rotating, reciprocating, and transverse motions create hazards in two general areas—at the point of operation (i.e., where work is being done) and at the points where power or motion is trans-

Figure 6–15. These expanded metal covers safeguard employees from protruding moving parts and power transmissions.

TABLE 6–D. Standard Materials and Dimensions for Machinery Guards

Material 1 in. = 2.5 cm; 1 ft = 0.305 m	Clearance from Moving Part at All Points (inches)	Largest Mesh or Opening Allowable (inches)	Minimum Gage (U.S. Standard) or Thickness	Minimum Height of Guard from Floor or Platform Level (ft, in.)
Woven wire	Under 2 2–4 4–15	3/8 1/2 2	No. 16 3/8 in. No. 16 1/2 No. 12-2	8-0* 8-0 8-0
Expanded metal	Under 4 4–15	1/2 2	No. 18 1/2 in. No. 13-2	8-0 8-0
Perforated metal	Under 4 4–15	1/2 2	No. 20 1/2 in. No. 14-2	8-0 8-0
Sheet metal	Under 4 4–15	— —	No. 22 No. 22	8-0 8-0
Wood or metal strips, crossed	Under 4 4–15	3/8 2	3/4-in. wood or No. 16 metal	8-0
Wood or metal strips, not crossed	Under 4 4–15	1/2 the width One width		
Plywood, plastic, or equivalent	Under 4 4–15	— —	1/4 in. 1/4 in.	8-0
Standing railing	Min. 15 Max. 20	—	—	3-6

*Guards for rotating protruding objects should extend to a minimum height of 9 ft from the floor or platform.

mitted from one part of a mechanical linkage to another part. This section will deal primarily with transmitted-power situations.

Any rotating object is dangerous. Even shafts that rotate slowly and smoothly can grab onto clothing and hair. Even the briefest amount of skin contact with a rotating object can cause the rotation to force or pull a finger, hand, or arm into a hazardous situation. Incidents resulting from contact with rotating objects are infrequent, but the subsequent injury is usually severe.

Common rotating mechanisms that pose hazards include collars, couplings, cams, clutches, flywheels, shaft ends, spindles, rotating bar stock, lead screws, and horizontal or vertical shafting. Exposed bolts and oil cups and projecting keys or screw threads increase the risks of injury while a mechanism is rotating.

A rotating mechanism, which is commonly housed in a stationary case or shell, consists of a revolving cylinder, a screw, and agitator blades or paddles. Examples of this type of rotating mechanism include washing machines, extractors, raw material mixers, and screw conveyors (Figure 6–16).

Reciprocating and transverse motions are hazardous because the back-and-forth or straight-line movement may strike or ensnare a worker in a pinch or shear point between fixed or moving objects.

In-Running Nip Points

In-running nip points are a danger that is present only with the movement of rotating objects. An in-running nip point develops whenever machine parts rotate toward each other or whenever a machine part rotates toward a stationary object; when either of these rotations takes place, objects, loose clothing, and body parts can be pulled into the nip point.

Nip point hazard locations include the in-running side of rolling mills and calendars; the in-running side of rolls used for bending, printing, corrugating, embossing, or feeding and conveying stock; or the in-running side of a chain and sprocket, belt and pulley, gear rack, gear and pinion, or belt conveyor terminal (Figure 6–17).

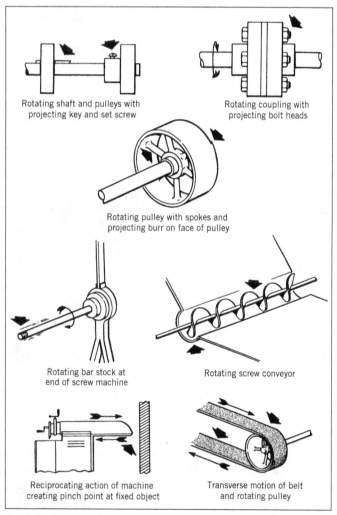

Figure 6–16. Examples of typical rotating, reciprocating, and transversing mechanisms.

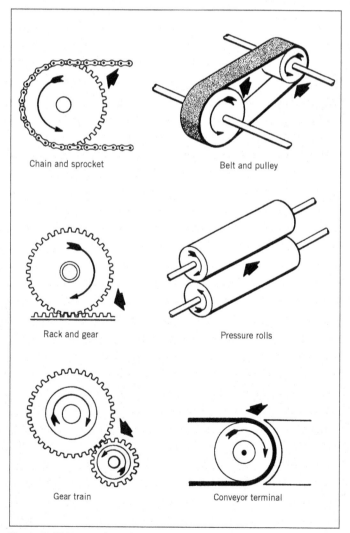

Figure 6–17. Examples of in-running nip points.

Cutting Actions

Cutting action results when a rotating, reciprocating, or transverse motion transfers to a tool, which then removes material in the form of chips. The danger of cutting action exists at the cutting edge of the machine as the edge approaches or touches the material. Such action occurs at the point of operation when cutting wood, metal, or other materials.

Mechanisms that involve cutting actions include band and circular saws, milling machines, planing or shaping machines, turning machines, boring or drilling machines, and grinding machines (Figure 6–18).

Punching, Shearing, and Bending Actions

Punching, shearing, or bending action results when power is applied to a ram (plunger) or knife to blank, trim, draw, punch, shear, or stamp metal or other materials. The danger of this type of action exists at the point where stock is inserted, maintained, shaped, or withdrawn. Equipment that involves punching, shearing, or bending actions includes power presses, foot and hand presses, bending presses, or brakes as well as squaring, guillotine, and alligator shears (Figure 6–19).

Guarding Materials

In most circumstances, metal is the preferred material for constructing guards. The framework of guards usually consists of structural shapes, pipe, strapping, bar, or rod stock. Generally, use expanded, perforated, or solid sheet metal or wire mesh for filler material (see Table 6–D). Where visibility is required, use plastic, polycarbonate, or Lexan™.

Guards made of wood have limited use because of wood's lack of durability and strength, relatively high maintenance costs, combustibility, and tendency to splinter, which can lead to injury or product contamination. In

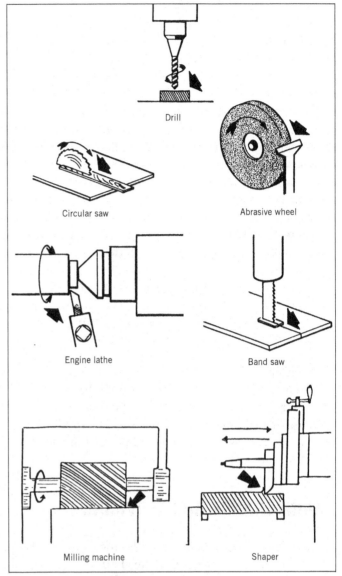

Figure 6–18. Examples of cutting actions.

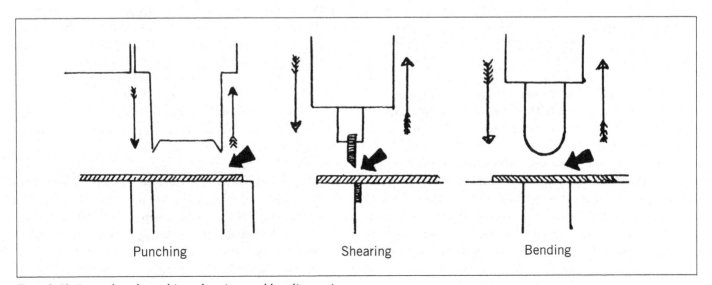

Figure 6–19. Examples of punching, shearing, and bending actions.

addition, sparks from nearby welding operations or from overheated bearings, belts rubbing together, defective wiring, and other heat sources can ignite wooden guards, especially when such guards are oil-soaked.

When resistance to rust or prevention of damage to tools and machinery is important, use guards made of aluminum, another soft metal, or plastic. When inspecting moving parts will be necessary, use plastic guards.

When it is not possible to keep out lint or fibers, provide ample ventilation through the guard. Such vents should be too small for a hand to fit through but large enough for lint or dust to fall through. Large guards should also have self-closing access doors that allow cleaning by brush or vacuum hose. Consider using latches interlocked with the power source to prevent operation of the machine while the access door is open.

The material used for a guard should be substantial enough to withstand (1) internal as well as external impacts, (2) parts under stress, (3) passing workers or vehicles, (4) environmental exposures, and (5) parts or materials that a machine is processing that can become lethal projectiles.

A machine located near heavy traffic areas is vulnerable to being damaged. If the machine cannot be relocated, be aware that an enclosure guard may be insufficient to protect both people and the machine from all possible hazards. In this case, guard railings should be used in addition to the enclosure guard.

MAINTENANCE AND SERVICING

When designing enclosure guards, incorporate ease of maintaining and servicing them. Failure to establish, reinforce, enforce, and facilitate safe maintenance and servicing procedures may be a major reason that workers do not replace guards. Especially when workers have frequent maintenance and servicing duties, it becomes easier for them to simply remove and not replace the guards, thus leading to an even more hazardous scenario (i.e., maintaining and servicing the machine while it is operating). Such an undertaking is highly dangerous for all personnel concerned. The occurrence of this hazardous scenario can be prevented in one or more of the following ways:

1. Implement engineering techniques that reduce the frequency of or eliminate a maintenance and servicing job/task. For example, situating on the outside of the guard the fittings of parts that will need servicing may eliminate the need to remove the guard, thereby permitting servicing while the machine is operating. Alternatively, by having oil or grease fittings extend through the guard, lubricating the machine is much

less hazardous. Such a design is highly suitable for machines that cannot be shut down for adjustment or maintenance.

2. Equip a machine with controls that automatically lubricate, adjust, or service the machine. Although complex equipment like this might be costly, for some machines, this cost is offset by the savings brought about by having adjustment and maintenance procedures take place habitually and by increased machine efficiency.

Also consider installing or mounting machine guards. The best way to mount guards securely is either by tack welding or by special bolting.

CONTROL OF HAZARDOUS ENERGY SOURCES

When equipment and machines need maintenance and servicing, isolate the energy sources and implement lockout/tagout procedures. Terms often used in standards and regulations to describe a machine or system with all its energy sources neutralized include *zero mechanical state*, *zero energy state*, and *energy isolation*. Types of machine energy are electrical, pneumatic, steam, hydraulic, chemical, gravity, and thermal. Energy can also be the potential energy in suspended parts and springs.

The following material is based on Control of Hazardous Energy Sources (Lockout/Tagout), 29 CFR 1910.147, and represents a minimally acceptable energy control program.

Sample Lockout/Tagout Procedure

A company must establish a procedure for controlling a hazardous energy source that includes machines being isolated from all sources of energy and potential energy. The following material is based on Control of Hazardous Energy Sources (Lockout/Tagout), 29 CFR 1910.147, and represents a minimally acceptable energy control program.

The minimally acceptable lockout procedure presented in this volume is a guide that organizations can use to develop internal lockout procedures. When energy-isolating devices are not lockable, the organization can use tagout, as long as the organization conforms to this procedure, provides training in the procedure, and performs rigorous, periodic inspections. When tagout is used and energy-isolating devices are lockable, the employer must provide full employee protection, training in the procedure, and rigorous, periodic inspections. For complex machines, more comprehensive, machine-specific procedures may need to be developed, documented, and implemented (Figure 6–20).

LOCKOUT : TAGOUT SYSTEM™

CAUTION: Servicing or maintenance is not permitted unless this equipment is isolated from all hazardous energy sources. This is the exclusive responsibility of designated "Authorized Employees" (**see listing at bottom) who must follow the ***complete*** "Lockout/Tagout Procedure" as published by the company. ***This sheet is limited and abbreviated; it must not be considered a substitute for the company's complete Procedure.***

LOCKOUT:TAGOUT DATA SHEET

For Equipment No. 172

Equipment Description: Blivet Sintering Machine with Taurus Automatic Feeder

HAZARDOUS ENERGY		ISOLATING DEVICES			CONTROL DEVICES (Check ✓)			
Type	Magnitude	Type	Location	I.D. No.	Lock	Tag	Both	Add'l Measures (See Below)*
Electric	440 V.	Switch	Right-Front Post	S819			✓	
Steam	170 psi	Valve	At wall 30 feet west	V47		✓		
Hydraulic	3000 psi	Valve	On pump 10 feet east	V16			✓	①

***ADDITIONAL SAFETY MEASURES (Refer To Table Above):**

① After closing and locking valve, drain hydraulic cylinder on machine 172.

****AUTHORIZED EMPLOYEES. Only the following are authorized to undertake the Lockout/Tagout Procedure on this equipment:**

Jerzi Brodofski; Andre Schoenzer
Paul Simpson, Nancy Mahabro
Stanley Cheng, Joe Hershko
Walter Paley
John Wilson

Prepared By: George Murdock **Date:** 8/8/90 **QUESTIONS? PHONE:** 538

© 1990 Idesco Corp., N.Y.C. Q-SIGN® V9-11

Figure 6–20. An example of one company's documentation of a lockout/tagout procedure. Note that this is only part of a comprehensive energy-isolation program. *(Courtesy Idesco Corp.)*

Purpose

This procedure establishes the minimally acceptable requirements for the lockout of a machine or equipment. Such a procedure ensures that the machine or equipment is stopped, isolated from all potentially hazardous energy sources, and locked out before employees perform any servicing or maintenance during which the unexpected energizing or start-up of the machine or equipment—or the release of stored energy—could cause injury. These minimally acceptable requirements do not encompass the full range of regulatory requirements. The complete requirements are determined by a company's particular operations, equipment, and machine-specific procedures.

Scope

All employees must comply with the restrictions and limitations imposed upon them during the use of lockout. Authorized employees are required to perform lockout in accordance with the lockout procedure. In addition, no employee, upon observing that a machine or piece of equipment is locked out for servicing or maintenance, should attempt to start, energize, or use that machine or equipment.

Sequence of Lockout

The following lockout sequence meets minimally acceptable standards for controlling a hazardous energy source.

1. Notify employees when the servicing or maintenance of a machine or piece of equipment is required. Inform them that the machine or equipment must be shut down and locked out to perform the servicing or maintenance.
2. An authorized employee should follow company procedure to determine the type and magnitude of energy that the machine or equipment utilizes, should understand the hazards of that energy, and should know the methods to control that energy.
3. If the machine or equipment is operating, shut it down using the typical stopping procedure (e.g., depress the STOP button, open a switch, or close a valve).
4. Deactivate the energy-isolating device(s) so that the machine or equipment is no longer isolated from the energy source(s).
5. Lock out the energy-isolating device(s) using the assigned individual lock(s) (Figure 6–21).
6. Stored or residual energy (such as that in capacitors, springs, elevated machine members, rotating flywheels, hydraulic systems, and air, gas, steam, or water pressure) must be dissipated or suppressed by a method such as grounding, repositioning, blocking, or bleeding down (Figure 6–22).
7. Before ensuring that the equipment is disconnected from the energy source(s), verify that no personnel are in the area. Then confirm the isolation of the equipment by working the push buttons or other normal control(s) and making certain that the equipment does not operate. CAUTION: After verifying the isolation of the equipment, be sure to return the control(s) to the neutral or the OFF position.
8. The machine or equipment is now locked out.

OSHA 29 1910.211 contains the following requirements: A control system must be constructed in such a way that

1. a fault that occurs inside the system does not prevent the normal stop process from being activated
2. another machine cycle cannot be executed before the fault has been removed
3. the fault can be revealed by a simple test, or displayed by the control system as best practice.

Subpart 3.14 of ANSI B11.19–2003 defines control reliability as follows:

> Control reliability is the capability of the machine control system, the safeguarding, other control components and related interfacing to achieve a safe state in the event of a fault within their safety related functions.

172   6 Safeguarding

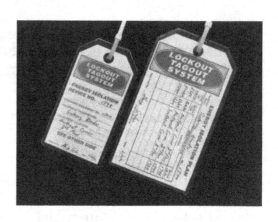

Figure 6–21. Sample assigned individual locks for energy-isolation devices. The energy-isolation plan tag lists all sources of hazardous energy and the energy-isolation device to which the tag should be attached. *(Courtesy Idesco Corp.)*

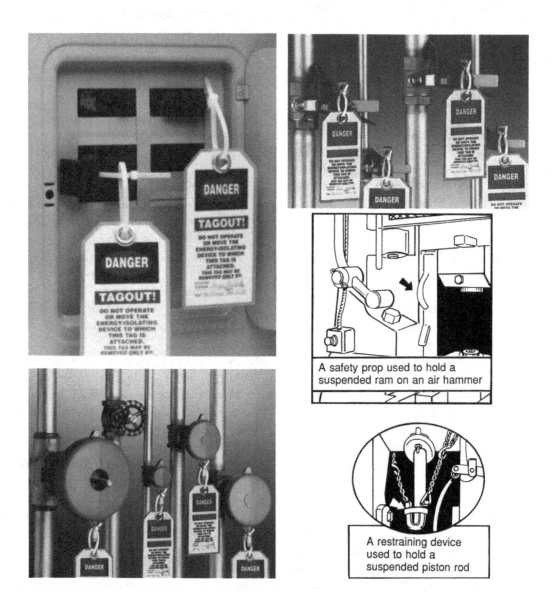

Figure 6–22. Tags accompanying sample devices for blocking out, blanking out, or locking out circuit breakers, pipes, or valves.

Restoring Equipment to Service

When the servicing or maintenance is complete and the machine or equipment is ready to return to normal operation, the following steps should be taken:
1. Check the machine or equipment and the surrounding area to ensure that nonessential items have been removed and that the machine or equipment components are operationally intact.
2. Check the work area to ensure that all employees have been safely situated in or removed from the area.
3. Verify that the controls are in neutral.
4. Remove the lockout devices and reenergize the machine or equipment. Note: Some forms of blocking may require reenergizing the machine before removing the lockout devices.
5. Notify pertinent employees that the servicing or maintenance has completed and that the machine or equipment is ready to use.

Other Program Requirements

Training
Authorized and appropriate employees must be trained and annually retrained. Provide remedial training as needed when individual actions demonstrate nonconformance with program requirements.

Lockout Equipment
Lockout and tagout devices should be durable; differentiated, color-coded, or otherwise characterized for each facility; and supplied to authorized personnel. Tags must state, at minimum, DO NOT START, DO NOT OPERATE, or DO NOT OPEN as well as who placed the tag, the date the tag was placed, and the reason the tag was placed. Tags must be able to withstand a 50-lb pull on the attachment.

Self-Audit
Conduct an annual audit of conformance to procedural requirements and the lockout program. Use the results of the audit to improve the program, the procedure, and personnel training and to provide constructive feedback to personnel.

Group Lockout

If a group of employees are locking out a piece of equipment or power source, each employee should have an individual lock and should attach the lock to the group lockout device.

ROBOTICS SAFEGUARDING

Robots are programmable, multifunctional, mechanical manipulators that have three means of power: electro- mechanical, hydraulic, and pneumatic. Robots are used for a variety of applications and processes, such as spray painting, arc and spot welding, materials handling, assembly, and machine loading and unloading.

Hazards and Hazardous Locations

The principal hazards of robots are:
1. being struck by a robot's moving parts while within the robot's operating envelope or movement zone
2. being caught between a robot's moving parts and other machinery or objects within or near the robot's movement zone
3. being struck by objects or tools the robot has dropped or ejected.

As illustrated in Figure 6–23, dividing the robotized workstation into two zones or volumes—the robot movement zone and the approach zone—helps specify sources of injury.

If technicians stand in the robot movement zone to perform a task when power is available to the robot, they will

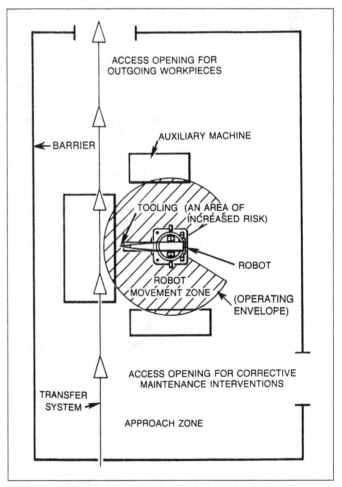

Figure 6–23. The robot movement zone (operating envelope) and the approach zone comprise the danger zone of industrial robots. Note the barrier surrounding the entire danger zone. *(Courtesy NIOSH)*

be exposed to crushing, shearing, and impact injury risks. Certain regions within the robot movement zone, such as the region around the end-effector, contain increased risk.

Just outside the robot movement zone is the approach zone. The boundaries of this zone can be specified and the limits of the protected area identified. In this zone, personnel may be exposed to thrown objects, radiation, flash, electrical hazards, and mechanical hazards of associated equipment. Furthermore, personnel in the approach zone can cross into the movement zone. However, passage from the approach zone to the movement zone can be reduced by limiting the size of the openings through which personnel or working materials must pass to reach the robot movement zone.

Effective design of the workstation control system minimizes the possibility that a robot and associated machines could move in a way that would harm an operator inside the movement zone and, therefore, satisfies the highest priority within the control logic. A powered robot's motion is initiated by the closure of a power supply switch to an actuator, such as an electric motor or a hydraulic cylinder. This closure can be accomplished by any of the following:

1. a planned step forward to an output condition in the control program
2. a person switching the robot to automatic operation
3. electromagnetic interference generating the voltage necessary for a logic switch at a microelectronic gate
4. another control circuit inputting a switching signal
5. a bug or other error in the control software
6. a hardware failure in the switching device
7. automatic restart after a power interruption.

A robot's failure to stop when commanded is also a potentially hazardous condition that should be evaluated.

If a worker is present when a motion-initiating event occurs, the ensuing robot motion can lead to injury from:

1. impact
2. puncture
3. pinch-point closure
4. dragging the person over a sharp object
5. pushing the person into another machine's point of operation.

Personnel can be struck by:

1. any part of the robot itself
2. a workpiece being handled by the robot
3. robot tooling.

Robots manipulate many kinds of end-effectors, such as grippers, welding electrodes, grinding wheels, lasers, and high-pressure water jets. Problems with these end-effectors frequently require maintenance interventions.

Contact with end-effectors because of robot movement can result in serious cuts, burns, and puncture wounds. A workpiece that a robot is handling is a source of danger when a gripper loses its hold and allows the workpiece to fall or become a projectile.

Equipment supplying power and control to a robot presents a potential electrical and pressurized fluid hazard. Ruptured hydraulic lines could become dangerous, high-pressure cutting streams or whipping hoses. A pinch point could develop if control cabinets are too close to the robot. In addition, cables on the floor are tripping hazards.

Safeguarding Methods

In order to physically limit the robot movement zone to the range of motion that a particular operation or installation requires, use some form of mechanical stops that are able to tolerate the force of momentum of a robot traveling at maximum speed and carrying a full load.

Other safeguarding methods include the following:

1. Install an amber warning light on a robot so that the robot is noticeable from all directions. This light should be on whenever the robot is energized, signifying that the robot is "live" even when it is not moving.
2. "Teach" panels or pendants should have an EMERGENCY STOP button or a deadman's switch that is hardwired into the drive-power stop circuit and not interfaced through a computer input/output register, thus allowing an operator to stop a robot's movement by interrupting the machine-drive power.
3. When a robot is in the "teach" mode of operation, limit the robot's rate of movement to 6 in. (15 cm) or fewer per second, as measured at the end of the robot's fully extended arm.
4. Program a robot so that an operator cannot place it into automatic cycle using the "teach" pendant but must instead close all interlocked gates and return to the master control panel outside the robot movement zone.
5. Place fixed guards, such as a 5-ft (1.5-m) fence, around the perimeter of the robot movement zone, and design the guards to prevent inadvertent or unauthorized entry, above or below, and also to capture objects dropped or ejected by a robot.
6. Ensure that there is sufficient clearance between each perimeter guard and the robot movement zone so that a person cannot be trapped between the two (Figure 6–23).
7. Electronically interlock perimeter access gates to interrupt main drive power if the gates are opened during the automatic cycle of a robot. Program the robot so that closing the interlock gate cannot initiate the automatic cycle; such initiation should be achievable only at the main control panel outside the perimeter guards.
8. Place warning signs at points of access to the movement

zone to make those who enter the zone aware of the customary hazards of the robot and/or any unusual hazards that could develop. Unusual hazards include overlapping movement zones of two or more robots in close proximity and of other automation or machinery whose movements extend into the work area.

9. Consider using any of the several devices that exist for sensing the presence within or intrusion of a person into the robot movement zone, such as photoelectric cells, pressure-sensitive mats, and light or sound curtains.

10. Do not allow anyone to work with a robot until he or she has received proper training in hazard awareness and operating the robot safely as it performs a particular function. Prior to giving workers hands-on experience, ensure that all applicable persons are fully aware of what will happen if the control system fails and of the safety features provided both in the workplace a with the robot.

11. Anchor a robot according to its manufacturer's recommended specifications. Situate the robot's control panel so that an operator can easily see the robot while programming or operating the master controls.

12. Provide interlocking disconnects for all sources of energy to a robot, including electricity, air, and hydraulic power.

13. Provide a means to release stored energy before servicing a robot. This energy may take the form of air and hydraulic accumulators, springs, counterweights, flywheels, or the load held by the robot.

14. Shield all solid-state electronic devices that control a robot from possible radio frequency interference, which could cause loss of control. It is important for the robot's computer to recognize frequency interference and thus prevent unexpected movements and injuries. (See Chapter 22, Automated Lines, Systems, or Processes, for more information on robotics.)

GUARDS AND NOISE CONTROL

A properly designed, constructed, and mounted guard potentially has the side benefit of reducing noise. Noise travels primarily by conduction and vibration through the air. Sound barriers can effectively stop or lessen noise by absorbing, reflecting, or confining the sound waves. Because guards are usually positioned at either the point of operation or the point of power transmission—both of which are where a machine's noise originates—they can be designed as barriers against noise as well as injury. Each situation calls for inspecting the noise source and considering the surrounding environment. Professional assistance may be required.

MACHINE OPERATOR REQUIRED REPORTING OF EMPLOYEE INJURIES

In accordance with 29 CFR 1910.217(g), employers must report within 30 days all point-of-operation injuries to machine operators or other employees. Employers should mail the following information to Directorate of Standards and Guidance (formerly Director of Safety Standards), OSHA, U.S. Department of Labor, Washington DC 20210, or to the state agency that administers a plan approved by the Assistant Secretary of Labor for Occupational Safety and Health. Employers may also e-mail the information after completing the following items:

- **1910.217(g)(2)(i):** Employer's name, address, and location of the workplace (establishment).
- **1910.217(g)(2)(ii):** Employer's name, address, and location of the workplace (establishment).
- **1910.217(g)(2)(iii):** Type of clutch used on the press (full revolution, part revolution, or direct drive).
- **1910.217(g)(2)(iv):** Type of safeguard(s) being used (two-hand control, two-hand trip, pullouts, sweeps, or other). If the safeguard is not described in this section, give a complete description.
- **1910.217(g)(2)(v):** Cause of the accident (repeat of press, safeguard failure, removing stuck part or scrap, no safeguard provided, no safeguard in use, or other).
- **1910.217(g)(2)(vi):** Type of feeding (manual with hands in dies or with hands out of dies, semiautomatic, automatic, or other).
- **1910.217(g)(2)(vii):** Means used to actuate press stroke (foot trip, foot control, hand trip, hand control, or other).
- **1910.217(g)(2)(viii):** Number of operators required for the operation and the number of operators provided with controls and safeguards.
- **1910.217(h):** Presence–sensing device initiation (PSDI), (OSHA 2015). *Courtesy of www.oshatrain.org*

SUMMARY

- Safeguarding involves identifying and evaluating all hazards—and potential sources of injuries and property and equipment loss—and then creating adequate controls or safeguards to protect employees from those hazards.
- Management should ensure that employees clearly understand safeguarding terms and procedures.
- Manufacturers cannot always install effective point-of-operation protection devices. In such cases, end users must design and install these devices according to the regulatory safeguarding requirements of each operation. Companies must also adequately construct point-of-operation safeguards and devices and inspect them frequently.

- Guards used to protect workers from hazardous areas of machines include barrier, interlocking barrier, and automatic safeguarding devices. Only fixed guards should be used around power-transmission parts. Safeguarding devices can be used instead of or in conjunction with barrier guards at points of operation.
- In some cases, "guarding by location" is an acceptable means of protecting a worker. However, this type of guarding must comply with applicable OSHA regulations.
- Power-transmission guards must envelop all moving parts to prevent an operator's body from coming in contact with those parts. In addition, the guards' material should be able to withstand internal and external blows, parts under stress, and passing workers and vehicles.
- During the maintenance and servicing of equipment and machines, prevent unexpected, injury-causing movements by using energy isolation through lockout/tagout. The maintenance program must provide full employee protection, training, and reinforcement as well as rigorous periodic inspections.
- Robotic workstations require barrier guards that surround the entire area and also require established safety procedures for operations in the robot movement zone and the approach zone.
- Guards can help reduce noise hazards by either absorbing or reflecting sound.

REFERENCES

American National Standards Institute, 11 West 42nd Street, New York, NY 10036.

Drilling, Mining, and Boring Machines—Safety Requirements for Construction, Care, and Use, ANSI B11.8–1983 (R1994).

Gear Cutting Machines—Safety Requirements for Construction, Care, and Use, ANSI B11.11–1994.

General Safety Requirements Common to ANSI B11 Machines, B11-2008, 2008.

Graphic Technology—Safety Standard—Printing Press Systems, ANSI B65.1-2005, 2011

Grinding Machines—Safety Requirements for Construction, Care, and Use, ANSI B11.9–1975 (R1997).

Industrial Robots and Robot Systems—Safety Requirements, ANSI/Robot Industry of America (RIA) R15.06–1999.

Lathes—Safety Requirements for Construction, Care, and Use, ANSI B11.6–1984 (R1994).

Machine Tools—Coil-Slitting Machines—Safety Requirements for Construction, Care, and Use, ANSI B11.14–1996.

Machine Tools—Cold Headers and Cold Formers—Safety Requirements for Construction, Care, and Use, ANSI B11.7–1995 (R2000).

Machine Tools—Horizontal Hydraulic Extrusion Presses—Safety Requirements for Construction, Care, and Use, ANSI B11.17–1996.

Machine Tools—Hydraulic Power Presses—Safety Requirements for Construction, Care, and Use, ANSI B11.2–1995 (R2000).

Machine Tools—Iron Workers—Safety Requirements for Construction, Care, and Use, ANSI B11.5–1988 (R1994).

Machine Tools—Machinery and Machine Systems for Processing Strip, Sheet, or Plate from Coiled Configuration—Safety Requirements for Construction, Care, and Use, ANSI B11.18–1997.

Machine Tools—Mechanical Power Presses—Safety Requirements for Construction, Care, and Use, ANSI B11.1–1988 (R1994).

Machine Tools—Shears—Safety Requirements for Construction, Care, and Use, ANSI B11.4–1993.

Machine Tools—Single- and Multiple-Spindle Automatic Bar and Chucking Machines—Safety Requirements for Construction, Care, and Use, ANSI B11.13–1992 (R1998).

Metal Sawing Machines—Safety Requirements for Construction, Care, and Use, ANSI B11.10–1990 (R1997).

New Standard for Automatic Guided Vehicles Released, ANSI B56.5, 2012

Performance Criteria for Safeguarding, ANSI B11, ANSI 2010.

Pipe, Tube, and Shape Bending Machines—Safety Requirements for Construction, Care, and Use, ANSI B11.15–1984 (R1994).

Power Press Brakes—Safety Requirements for Construction, Care, and Use, ANSI B11.3–1982 (R1994).

Risk Assessment & Risk Reduction—A Guideline to Estimate, Evaluate & Reduce Risks Associated with Machine Tools, ANSI B11.TR3-2000, 2000.

Robots Used with Horizontal and Vertical Injection Molding Machines—Safety Requirements for the Integration, Care, and Use, ANSI/SPI B151.27-2003, 2003.

Roll Forming and Roll Bending Machines—Safety Requirements for Construction, Care, and Use, ANSI B11.12–1996.

Safety of Machines; General Requirements and Risk Assessment, ANSI B11.0–2010, 2010

Safe Openings for Some Point of Operation Guards, Technical Guide 02–678, 2007.

Safety Requirements for Integrated Manufacturing Systems, ANSI B11-29, 2004.

Standard for Electrical Safety in the Workplace, NFPA 70E, 2015.

EN ISO 121001, www.Leuze Electronics.org., 2014.

National Fire Protection Association, 1 Batterymarch Park, Quincy, MA 02269.

Electrical Standard for Industrial Machinery, NFPA 79, 2015.

Underwriters Laboratory, 333 Pfingsten Road, Northbrook, IL, 60062.

Industrial Control Equipment, UL 508, 2008.

Standard for Electro-Sensitive Protective Equipment, UL 61496-2, 2011.

UL Standard for Safety Electro-Sensitive Protective Equipment, UL 614961, 2012.

U.S. Department of Labor, Occupational Safety and Health Administration, 200 Constitution Avenue NW, Washington DC 20210.

Code of Federal Regulations, Title 29. Sections 1910.147, Control of Hazardous Energy Sources (Lockout/Tagout), and 1910.211 through 1910.222, Subpart O–Machinery and Machine Guarding, 2014.

REVIEW QUESTIONS

1. Define safeguarding.
2. List the six characteristics of a proper guard.
   a.
   b.
   c.
   d.
   e.
   f.
3. Which of the following is a hazardous area involving two or more mechanical parts rotating in opposite directions within the same plane and in close interaction?
   a. pinch point
   b. nip point or bite
   c. point of operation
   d. power transmission
4. Name the four general types of safeguards.
   a.
   b.
   c.
   d.
5. Briefly explain the three advantages that built-in machine safeguards, which are designed and installed by the manufacturer, have over safeguards made by the machine user.
   a.
   b.
   c.
6. A point of operation is defined as:
   a. all mechanical parts—such as gears, shafts, pulleys, belts, clutches, brakes, and rods—that transmit energy and motion from a power source to a machine.
   b. a mechanism or control designed for safeguarding.
   c. the area of a machine where material is positioned for processing and where processing of the material actually takes place.
   d. a hazardous area created by the cutting movement of a mechanical part that is beyond a stationary point on a machine.
7. An effective interlocking barrier guard must satisfy what three requirements?
   a.
   b.
   c.
8. Which of the following is the preferred material to construct guards?
   a. plastic
   b. metal
   c. shatter-resistant glass
   d. wood
9. What are the benefits of nonmetal barriers?
10. What five factors should be considered when matching a machine or equipment to an operator?
    a.
    b.
    c.
    d.
    e.

11. List the five steps that should be taken after a machine or piece of equipment has been locked out for repair and is ready to return to normal operating condition.
   a.
   b.
   c.
   d.
   e.

12. Which of the following is a principal hazard when using robots?
   a. being struck by a robot's moving parts
   b. being trapped between a robot's moving parts and other machinery or objects
   c. being struck by objects or tools that the robot has dropped or ejected
   d. all of the above
   e. only a and b

13. Name three devices that can sense a person's presence within a robot's movement zone.
   a.
   b.
   c.

Personal Protective Equipment

7

Cristine Z. Fargo

Philip E. Hagan, JD, MBA, MPH, ARM, CIH, CHMM, CET, CHCM, CEM

J. Nigel Ellis, PhD, PE, CSP

A Program to Introduce PPE
Policy ▶ Selection of Proper Equipment ▶ Proper Training ▶ Use and Maintenance ▶ Enforcement ▶ Recognition Clubs ▶ Who Pays for PPE?

Head Protection
Protective Headwear ▶ Maintenance ▶ Color-Coded Protective Helmets ▶ Bump Caps ▶ Protective Hair Covering

Eye and Face Protection
Selection of Protective Eyewear ▶ Contact Lenses—Rumor versus Facts ▶ Comfort and Fit ▶ Face Protection ▶ Acid Hoods and Chemical Goggles ▶ Laser Beam Protection ▶ Eye Protection for Welding

Hearing Protection
Occupational Noise-Induced Hearing Loss ▶ Hearing Conservation Program ▶ Types of Hearing Protection Devices ▶ Hearing Protector Selection

Fall Arrest Systems
What Is Fall Arrest? ▶ When Are Fall Arrest Systems Needed? ▶ Elements of a Successful Fall Arrest Program ▶ Which Fall Arrest System to Use? ▶ Passive Fall Arrest Systems ▶ Active Fall Arrest Systems ▶ Rescue Systems ▶ Equipment Inspection and Maintenance ▶ Cleaning Fall Arrest Equipment ▶ Storage

Fall Protection Standards

Respiratory Protection
Selecting Respiratory Protection ▶ Identification of the Hazard ▶ Evaluation of the Hazard ▶ Protection Factors ▶ Types of Respirators ▶ Air-Supplying Respirators ▶ Supplied-Air Respirators ▶ Combination Supplied-Air SCBA Respirators ▶ Air-Purifying Respirators ▶ Fitting Respirators ▶ Storage of Respirators ▶ Maintenance of Respirators ▶ Cleaning and Sanitizing ▶ Inspection of Respirators ▶ Training ▶ Medical Surveillance

Hand and Arm Protection
Gloves ▶ Hand Leathers and Arm Protectors ▶ Impervious Clothing ▶ Decontamination Considerations

Protective Footwear
Metatarsal Footwear ▶ Conductive Footwear ▶ Electrical Hazard Footwear ▶ Static Dissipative Footwear ▶ Sole Puncture Resistant Footwear ▶ Foundry Footwear ▶ Other Features of Protective Footwear ▶ Cleaning Rubber Boots

Special Work Clothing
Protection against Heat and Hot Metal ▶ Flame-Retardant Work Clothes ▶ Protection against Impact and Cuts ▶ Heat Stress ▶ Cold-Weather Clothing ▶ High-Visibility Clothing ▶ Special Clothing ▶ Cleaning Work Clothing

Summary

References

Review Questions

Methods of controlling potentially harmful exposures to hazardous substances or forms of energy found in the workplace environment typically are classified into three broad, occasionally overlapping categories: engineering controls, administrative controls, and personal protective equipment (PPE). Engineering controls are passive measures designed into the work environment to prevent contact with a harmful substance or other hazard. Common examples of engineering controls are eliminating toxic materials or using less toxic substitutes, changing process design, using barriers or guards, isolating or enclosing hazards, and using local exhaust ventilation. Administrative controls include such measures as worker rotation to minimize exposure, implementing proper housekeeping practices, and devising appropriate worker training.

This chapter covers the following topics:
- developing a program to introduce PPE in a company
- types of protective head, face and eye, and hearing protection devices and how to use them
- types of fall arrest systems and their use and care
- major respiratory protection equipment, including care and maintenance
- major types of hand and arm protective gear and footwear
- protective clothing for special work situations.

Personal protective equipment refers to the use of respirators, special clothing, safety glasses, hard hats, or similar devices whose proper use reduces the risk of personal injury or illness resulting from occupational hazards. Generally speaking, use of PPE is the least desirable method of controlling exposure to harmful substances in the workplace environment.

Properly implemented engineering and administrative controls can greatly reduce or eliminate the hazard at the source. In contrast, when PPE is the primary control measure, the hazard is still present in the environment. The particular protective device provides a barrier between the hazard and the worker. Improper use or failure of the device means the worker is exposed to a direct threat to health and safety.

In some instances, PPE may be the only recourse. For example, with many physical hazards such as welding flash and sparks, dark goggles are used because no other feasible vision protection controls, short of automating the operation, are effective. On a construction site, hard hats and safety shoes are necessary to protect workers exposed to hazards from falling objects or from objects that could crush a foot. However, against such risks as chemical hazards, good industrial hygiene practice dictates that PPE be used only when the more desirable engineering and administrative controls are not feasible, when PPE is an interim control method while the "higher" controls are being implemented as a supplement, or as added protection.

Clearly, management must design a safe working environment by evaluating all hazards in the work environment, assessing the need for controls, and controlling or eliminating hazards to protect workers. Such a policy means considering the worst-case analysis of conditions. For those work environment hazards that cannot be eliminated through engineering or administrative controls, PPE becomes the best protection method. It is important for management to take a strong, positive attitude toward the proper use of PPE.

A PROGRAM TO INTRODUCE PPE

Companies should conduct an assessment of hazards in the workplace, as required by the Occupational Safety and Health Administration (OSHA), to determine their needs for PPE to protect workers. Once management decides on the use of PPE, the following steps should be taken:

1. Write a policy on usage of the PPE, and communicate it to employees and visitors as needed.
2. Select the proper type of equipment.
3. Implement a thorough training program to make certain employees know the correct use and maintenance of their equipment.
4. Enforce the use of PPE.

Policy

A written PPE program should include a policy, hazard assessment or PPE needs assessment, selection of PPE to be used, worker training and motivation in the use of PPE, and enforcement of company rules. The policy should clearly state the need for and use of PPE. It also may contain exceptions or limitations on use of PPE. Some policies or safety rules may include such details as the specific work conditions expected. Management staff must follow the same safety rules.

The following is an example of one firm's policy on wearing of PPE devices:

> For safe use of any personal protective device, it is essential the user be properly instructed in its selection, use, and maintenance. Both supervisors and workers shall be so instructed by competent persons.

Selection of Proper Equipment

After the need for PPE has been established, the next step is to select the proper type. The most important criterion is the degree of protection that a particular piece of equipment affords under various conditions.

Except for respiratory protective devices, few items of PPE available commercially are tested according to published and generally accepted performance specifications and approved by an impartial examiner. Although satisfactory performance specifications exist for certain types of PPE (notably protective helmets, devices to protect the eyes from impact and from harmful radiation, and rubber insulating gloves), there are no regulatory obligations to have such items independently tested or certified and in many cases, the performance standard does not impose third-party testing or certification.

The Safety Equipment Institute (SEI) has formulated objective policies for third-party certification of safety equipment. SEI voluntary certification programs involve both product testing and an ongoing program of quality assurance audits. Participating manufacturers are required to submit a specific number of product models to undergo demanding performance tests in SEI-authorized independent laboratories. When the laboratory has completed the test, SEI receives a pass or fail notification. For the quality assurance program, SEI conducts an audit at a manufacturer's production facilities to ensure that products coming off the assembly line are made to the same exacting specifications as the product model actually tested for certification.

The Safety Equipment Institute's existing certification programs include (1) eye and face protection, such as goggles, face shields, spectacles, and welding helmets; (2) emergency eyewash and shower equipment; (3) protective clothing; (4) protective headwear, such as hard hats; (5) protective footwear; and (6) personal fall protection. (The latest edition of the list of SEI-certified products is available on SEI's website. See References at the end of this chapter for further information.)

Proper Training

The next step is to obtain worker compliance with company requirements to wear the PPE. Several factors influence compliance, including (1) how well workers understand the need for the equipment; (2) how easy, comfortable, and convenient the equipment is to wear; (3) how effectively economic, social, and disciplinary sanctions can be used to influence the attitudes of workers; and (4) employee involvement in the decision-making process.

For organizations in which workers are accustomed to wearing PPE as a condition of employment, compliance may be only a minor problem. People are issued equipment meeting the requirements of the job and are taught how and why it must be used. Thereafter, periodic checks are made until use of the issued equipment has become a matter of habit.

However, when a group of workers are issued PPE for the first time or when new devices are introduced, compliance may be a more difficult problem. The safety and health professional or management must give workers a clear and reasonable explanation as to why the equipment must be worn. If employees are required to change their traditional work procedures, they may put up resistance, whether justifiable or not. Also, management cannot ignore the fact that workers may be reluctant to use the equipment because of bravado or vanity. Having supervisors first try out new protective equipment and devices before actual adoption and asking for their feedback usually makes them more willing to persuade workers to use the equipment.

A good deal of the resistance to change can be overcome if the persons who are going to use the PPE are allowed to choose their equipment from among several styles preselected to meet job requirements. When employees are involved in the decision about what PPE is provided to them, they are more inclined to wear it. This is especially important because the "one size or model fits all" approach may not work given the diverse workforce. The cost of providing multiple styles or sizes of PPE far outweighs the potential expense of injuries resulting from failure to use the equipment. For the convenience of their employees, some companies maintain equipment stores on the facility premises.

Employees required to use PPE should be given proper training. The training program routinely should cover the following topics:
- describing what hazard and/or condition is in the work environment
- telling what has been, can be, or cannot be done about it
- explaining why a certain type of PPE has been selected
- discussing the capabilities and/or limitations of the PPE
- demonstrating how to use, adjust, or fit the PPE
- practicing PPE use
- explaining company policy and its enforcement
- discussing how to deal with emergencies
- discussing how PPE will be paid for, maintained, repaired, cleaned, and replaced.

Use and Maintenance

All equipment must be inspected before and after each use. The company should keep records of all inspections by date, with the results tabulated. Supervisors and workers should follow the recommendations of the manufacturer for inspection, maintenance, repair, removal from service, and replacement of parts.

Enforcement

Employees need to know how the use of PPE will be enforced. Many companies have some type of progressive disciplinary action ranging from unpaid time off to termination. Management enforcement of the PPE program is critical to success.

Recognition Clubs

Several organizations sponsor recognition awards for those who avoided or minimized injury by wearing PPE. These organizations include the Wise Owl Club, sponsored by Prevent Blindness America; The Golden Shoe Club; and The Turtle Club.

Who Pays for PPE?

Employers must provide and pay for PPE required by the company for the worker to do his or her job safely and in compliance with OSHA standards. Examples would include welding gloves, respirators, hard hats, specialty glasses, specialty foot protection, and face shields. General work clothes (e.g., uniforms, pants, shirts, or blouses) not intended to function as protection against a hazard are not considered to be PPE.

For those cases in which equipment is personal in nature and usable by workers away from the job, the matter of payment may be decided by labor–management negotiations. Examples would include prescription safety glasses, safety shoes, and cold-weather outerwear of the type worn by construction workers. On the other hand, shoes or outerwear subject to contamination by hazardous substances that cannot be safely worn off site must be paid for by the employer.

OSHA's 29 CFR 1910.132 establishes the employer's obligation to provide personal protective equipment to employees as follows:

Protective equipment, including personal protective equipment for eyes, face, head and extremities, protective clothing, respiratory devices and protective shields and barriers, shall be provided, used and maintained in a sanitary and reliable condition wherever it is necessary by reasons of hazards of processes or environment, chemical hazards, radiological hazards or mechanical irritants encountered in a manner capable of causing injury or impairment in the function of any part of the body through absorption, inhalation, or physical contact.

Although there are some circumstances in which workers in a particular trade would provide their own PPE, it is still the employer's obligation to ensure that such equipment is adequate and properly maintained.

For example, the welder's helmet is almost universally supplied by management because no one could perform the job without such protection. Although work gloves are sometimes purchased by the user, welder's or other special-purpose gloves are usually considered a necessary part of the job and are issued by the employer. In some instances, safety shoes are offered at a partial reimbursement rate to encourage workers to buy them, while in other areas, a shoemobile service is available. The vendor's vehicle is completely equipped and stocked to fit and sell safety shoes.

The next few sections discuss seven major categories of PPE—protection for the head, eyes and face, hearing, respiratory system, hands and arms, feet, and body—as well as discuss fall arrest systems. Each section provides information on the standards available or proposed, some details about the equipment available, and suggestions for selecting equipment to meet the job hazard.

HEAD PROTECTION

Employees exposed to head injury hazards must be given protective headwear (see Figures 7–1 through 7–3). It is very important that workers wear hard hats as they are intended to be worn and observe any inspection or maintenance requirements of the employer or manufacturer. Some operations requiring head protection include tree trimming, construction work, shipbuilding, logging, mining, electric and communication line construction or maintenance, and basic metal or chemical production.

Protective Headwear

Protective headwear is designed to absorb the shock of a blow and shield the wearer's head from the impact and penetration of falling objects. In some cases, protective headwear is required to prevent electric shock and burns. Protective headwear can also prevent the head and hair from becoming entangled in machinery or exposed to hazardous environments. Safety and health professionals should be alert to any changes in operations that may create a need for protective headwear. For example, in a slack season, a firm might transfer some employees to duties that require protective headwear. Also, construction, maintenance, and odd jobs requiring head protection often occur in the normal operations of many companies.

OSHA recognizes the voluntary industry consensus standard ANSI/ISEA Z89.1, American National Standard for Industrial Head Protection, for purposes of complying with the obligation to provide appropriate head protection and permits products manufactured to the 1997, 2003, and 2009 editions to be used at this time. Users should be aware of the fact that while OSHA regulation makes reference to these specific versions, the most current version available and used by manufacturers is ANSI/ISEA Z89.1–2014.

Head protection devices must bear certain identifications, as prescribed by the Z89.1 standard, including the manufacturer's name, date of manufacture, and ANSI standard designation (Z89.1). They must also include the designation for both type and class:
- Type I—helmets intended to reduce the force of impact from a blow to the top of the head
- Type II—helmets intended to reduce the force of impact from a blow to the sides or top of the head

Figure 7–1. Standard protective headgear. *(Courtesy E. D. Bullard Company)*

Figure 7–2. A universal adapter with chemical goggles is added to ensure that workers will wear their goggles whenever safety conditions require protective headwear. *(Courtesy Mine Safety Appliances Company)*

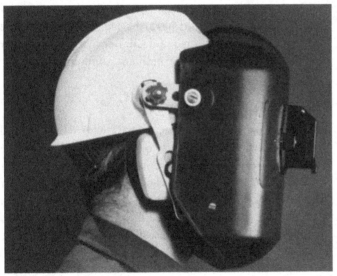

Figure 7–3. Protective welding headwear. *(Courtesy Mine Safety Appliances Company)*

- Class G (general)—limited voltage protection
- Class E (electrical)—high-voltage protection
- Class C (conductive)—no voltage protection.

Class G helmets are intended for protection against impact hazards and are typically used in heavy industrial settings, such as manufacturing or construction.

Class E helmets protect the wearer's head from impact, from penetration by falling or flying objects, and from high-voltage shock and burn. Generally, this type of helmet is constructed of insulated materials and is used by the utility services industry.

Class C helmets are used for comfort and impact protection, usually when there is a possibility of bumping a head against a fixed object and in settings where there is no danger from electrical hazards or corrosion.

It is seldom mentioned, but the use of a chin strap adds considerably to the protection offered by a helmet. A chin strap keeps the helmet from falling off in awkward positions and keeps the helmet on during impact.

Head protection devices may also include specific markings indicating certain optional features, including those that have demonstrated high-visibility properties, those that have been tested to higher or lower temperature conditions, and those that are capable of being worn in the backward (reverse-wear) position. It is important to ensure that all manufacturer instructions are followed before putting a helmet on in the backward position because it may be necessary to change the harness in order to ensure a proper fit.

Maintenance

Before each use, helmets should be inspected for cracks, no matter how small; signs of impact or rough treatment; and wear that might reduce the degree of safety originally provided. Prolonged exposure to ultraviolet (UV) radiation from sunlight or other sources like welding and chemicals can shorten the life expectancy of thermoplastic helmets. Discard all helmets that show signs of chalking, cracking, or reduced surface gloss. Any helmet that has received an impact should be removed from service. Additionally, users should consult manufacturer's product literature for guidance on product service life.

A protective helmet should not be stored or carried on the rear window shelf (cradle) of a vehicle because sunlight and extreme heat may reduce its degree of protection. Also, in case of an emergency stop or collision, the helmet could become a hazardous missile inside the truck or car.

At least every 30 days, protective helmets (in particular, their sweatbands and cradles) should be washed in warm, soapy water or in a suitable detergent solution recommended by the manufacturer, then rinsed thoroughly (Figure 7–4).

Before reissuing used helmets to other employees, make sure the helmets are scrubbed and disinfected. Solutions and powders are available that combine both cleaning and disinfecting. Helmets should be thoroughly rinsed with clean water and completely dried. Keep the wash solution and rinse water temperature at approximately 140°F

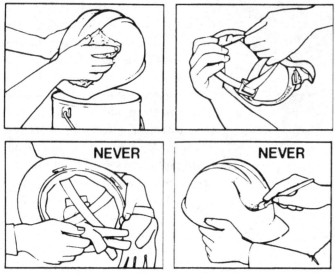

Figure 7–4. Maintenance of protective headwear. Shell should be cleaned regularly both for safety and appearance. Dirt or stains may hide hairline cracks, a reason to replace the helmet. Regular inspection of suspension system is important. Wearers should look closely for cracking, tearing, or fraying of suspension materials. Never carry anything inside the helmet. A clearance must be maintained inside the helmet for the protection system to work. Never use paint on a helmet that could affect the protective nature. Paint contains solvents that can make the shell brittle. Reflective tape is recommended for numbers or symbols.

(60°C). Do not use steam, except on aluminum helmets.

Removal of tar, paint, oil, and other materials may require the use of a solvent. Because some solvents can damage the shell, the supervisor should ask the helmet manufacturer what solvent can be used safely on the material.

Supervisors and workers should pay particular attention to the condition of the helmet's suspension webbing because it helps absorb the shock of a blow. They should look for loose or torn cradle straps, broken sewing lines, loose rivets, defective lugs, and other defects. Sweatbands are easily replaced. Disposable helmet liners made of plastic or paper are available for hats used by many people (such as visitors). The company should stock an adequate number of crowns, sweatbands, and cradles as replacement parts. Some companies replace the complete suspension webbing at least once a year.

Color-Coded Protective Helmets

Many companies use color-coded protective helmets to identify different working crews. Some colors are painted on during manufacture, and others have the color molded in. It is not recommended that paint be applied after manufacture because paint solvents may reduce the helmet's dielectric properties or affect the shell. Alterations of any sort can affect the performance of the gear. However, if painting is necessary, manufacturers should be consulted with regard to the type of paint that would be compatible with the construction of the protective helmet. Lighter-colored hats are cooler to wear in the sun or under infrared energy sources.

Bump Caps

A bump cap is not a helmet or hard hat. There is no standard that covers bump caps, except for each manufacturer's specification. Nonetheless, the bump cap has its place in some work environments. When the impact hazard is from bumping into stationary objects (such as low-slung pipes or catwalks, floor works, or well-protected machinery) or from cleaning in tight spaces, and not from overhead operations, the risk of potential injury is limited by the comparatively restricted movement of the worker's head. In these cases, the bump cap is sufficient protection.

Workers who wear the bump cap and/or other protective headgear must be trained and supervised to ensure correct usage. Bump caps should never be used where an ANSI Z89.1 protective helmet is required. Because of the danger of mistaking bump caps for helmets, some organizations prohibit their use.

Protective Hair Covering

Employees with long hair or beards who work around rotating shafts, chains, belts, or other rotating machine parts must take care to prevent their hair from contact with moving parts. Besides the danger of direct contact with the machine, which may occur when workers lean over, the hair can also be lifted into moving belts or rolls that develop heavy charges of static electricity. Because this hazard cannot be completely removed by mechanical means, workers with long hair should be required to wear protective hair coverings.

Hair nets, bandannas, and turbans are frequently unsatisfactory solutions because they do not cover the hair completely. Caps should cover the entire head of hair. If the wearer is exposed to sparks and hot metals, as in spot welding, the cap should be made of flame-resistant material. Some chemical facilities provide disposable flameproof caps. No standards have been accepted for caps, but they should be made of a durable fabric to withstand regular laundering and disinfecting, if they are not disposable.

To encourage its use, the cap should be as attractive as possible. It should have a simple design, be available either in a variety of head sizes or adjustable to fit all wearers, and be cool and lightweight. If dust protection is not required, the cap should be made of open-weave material for better ventilation. Finally, it should come with a visor and be worn with visor in front.

After a suitable cap has been chosen, management must enforce its use. For reasons of vanity, workers often wear the cap on the back of their heads so that part of the hair over the forehead is exposed. Sometimes, this practice can be discouraged by demonstrating vividly what may happen when the hair comes in contact with a revolving spindle. Management can also use vanity in the service of safety. If workers can be shown that caps preserve hair from the effects of dust, oils, and other shop conditions, employees may be more willing to wear protective hair covering.

EYE AND FACE PROTECTION

Protection of the eyes and face from injury by physical and chemical agents or by radiation is vital in any occupational health and safety program. In fact, this type of protection has the widest application and the broadest range of styles, models, and types.

The cost of acquiring and fitting eye protective devices is small when measured against the expense of eye injuries. For example, the purchase and fitting of a pair of impact-resistant spectacles may cost about $10; compensation payment for eye injury can exceed $3,600, according to Prevent Blindness America. The cost of a first-aid eye treatment may exceed $350. These numbers do not include indirect costs that an employer may incur, including lost time of the injured employee as well as those who stop work to assist the injured co-worker, lost time of management personnel to investigate an accident and complete requisite paperwork, and loss of profits on the injured employee's productivity. Some 70% of all eye injuries result from flying debris. Contact with harmful substances, chemicals, and so forth, causes more than 20% of injuries. Foreign bodies in the eye occur in about 60% of the cases.

The eye and face protection standard, ANSI Z87.1–1989, American National Standard Practice for Occupational and Educational Eye and Face Protection, sets fairly comprehensive standards to be used for protective eye and face devices purchased after July 5, 1994. (Note that while OSHA specifies this edition in its regulation, the most current version is ANSI/ISEA Z87.1–2010, Occupational and Educational Personal Eye and Face Protective Devices.)

Eye and face protective devices purchased before July 5, 1994, should comply with ANSI Z87.1–1968, USA Standard for Occupational and Educational Eye and Face Protection. In lieu of complying specifically with either version of the cited ANSI standard, the employer could demonstrate that alternative protective equipment would be equally effective, including providing products that meet the current, 2010 edition. These standards set performance standards, including detailed tests, for a broad range of eye and face protection—excluding only x-ray, gamma, and high-energy particulate radiation; lasers; and masers.

Selection of Protective Eyewear

Factors that should be considered in the selection of safety eyewear include (1) level of protection afforded, (2) comfort with which they can be worn, and (3) ease of repair. Styles now available resemble more attractive, regular eyewear. Flexible glasses are preferred by many because of their light weight and convenience, although they generally do not last as long as the sturdier frame and glass lens eyewear.

In making the selection of protective eyewear, the employer and employee should be aware of the type of hazard likely to be encountered, such as an impact hazard from flying debris, a splash hazard from liquid handling, or a possible dust or mist exposure. Once the eyewear has been selected, its use should be enforced to provide maximum protection for the degree of hazard involved. On certain jobs and in some locations, 100% eye protection is necessary.

There are a variety of eye protector configurations and styles available, and caution should be taken to ensure that consideration is given to other factors beyond just the hazard exposure. Some protectors may not be compatible with other personal protective equipment when worn together, such as goggles with face shields, goggles with respirators, and spectacles with goggles. The end user should carefully match protectors with other personal protective equipment to provide the protection intended.

Any eyewear must be designed so as not to compromise the wearer's ability to see through the lenses. Lenses must not have appreciable distortion or prism effect. ANSI Z87.1 limits the nonparallelism between the two faces to $\frac{1}{16}$ prism diopter (4 min of arc). Both the refraction in any meridian and the difference in refraction between any two meridians must be limited to $\frac{1}{16}$ diopter. Additional information on eye protection can be found in the National Safety Council's publication *Fundamentals of Industrial Hygiene*, 6th ed. (Plog 2012).

Contact Lenses—Rumor versus Facts

For more than 25 years, a rumor has persisted that welding or other electric flashes make contact lenses stick to the eyeball. This rumor has been proved false.

Incident data and studies suggest that contact lens wearers do not appear to have problems when their eyes are properly protected in the workplace. Prevent Blindness America publishes the latest research findings as a service to both business and safety and health professionals. Their purpose is to help contact lens wearers keep their eyes in good condition. The following guidelines and recommendations for contact lens use in industry are reprinted with permission from Prevent Blindness America.

Contact lenses sometimes provide a superior means of visual rehabilitation for employees who have had a cataract removed from one or both eyes, who are highly nearsighted, or who have irregular astigmatism from corneal scars or keratoconus. Except for situations in which there exist significant risks of ocular injury, individuals may be allowed to wear contact lenses in the workplace. Generally speaking, contact lens wearers who have experienced long-term success with contacts can judge for themselves whether or not they will be able to wear contact lenses in their occupational work environment. However, contact lens wearers must conform to the prerogatives and directions of management regarding contact lens use. When the work environment entails exposures to chemicals, vapors, splashes, radiant or intense heat, molten metals, or a highly particulate atmosphere, contact lens use should be "restricted" accordingly. (Contact lens use considerations should be made on a case-by-case basis in conjunction with the guidelines of the OSHA and NIOSH.)

Recommendations

Prevent Blindness America makes the following recommendations as a service to managers who must direct contact lens use and employees who must wear them:

- A specific written management policy on contact lens use should be developed with employee consultation and involvement.
- Occupational safety eyewear meeting or exceeding ANSI Z87.1 standards should be worn at all times by individuals in designated areas.
- Employees and visitors should be advised of defined areas where contacts are allowed.
- At workstations where contacts are allowed, the type of eye protection required should be specified.
- Restrictions on contact lens wear do not apply to usual office or secretarial employees (unless they must enter hazardous areas where exposure is significant).
- A directory should be developed that lists all employees who wear contacts. This list should be maintained in the medical facility for easy access by trained first-aid personnel. Foremen or supervisors should be informed of individual employees who wear contact lenses.
- Medical and first-aid personnel should be trained in the proper procedures and equipment for removing both hard and soft contacts from conscious and unconscious workers.
- Employees should be required to keep a spare pair of contacts and/or a pair of up-to-date prescription spectacles in their possession. They will then be able to continue their job functions should they damage or lose a lens while working.
- Employees who wear contact lenses should be instructed to remove their contacts immediately if redness of the eyes, blurred vision, or pain in the eyes associated with contact lens use occurs.

Guidelines for the Use of Contact Lenses in Industrial Environments

The American Optometric Association has adopted the following policy statement concerning the use of contact lenses in industrial environments (Anthony P. Cullen, MSc, OD, PhD, DSc, FCOptom, FAAO):

Contact lenses may be worn in some hazardous environments with appropriate covering safety eyewear. Contact lenses of themselves do not provide eye protection in the industrial sense.

Most successful contact lens wearers wish to wear their contact lenses in all aspects of their lives, including the workplace. This may conflict with government- or industry-imposed restrictions on the use of contact lenses in a given industrial environment. These restrictions, in turn, may be unreasonable and discriminatory.

In risk management, it is necessary to balance risk with benefits and to differentiate perceived risk from actual risk. Because both contact lenses and certain environments may produce adverse ocular effects, it is tempting to assume that there may be additive or synergistic effects when contact lenses are worn in that environment.

When considering the advisability of wearing contact lenses in a given industrial setting, a number of questions should be addressed:

- Is there an actual hazard?
- Does the wearing of contact lenses place the eye at greater risk than a naked eye?
- Does the removal of the contact lens increase the risk to the eye, the wearer, or co-workers?
- Is the risk different for various contact lens materials and designs?
- Are there other risks to the wearer or co-workers?
- Do contact lenses decrease the efficacy of other safety strategies?

Ocular hazards are greater in some occupations than others. Those who prescribe contact lenses for industrial workers should be concerned as to the advisability of wearing the lenses in a given environment. The type of work may influence the selection of lens material and design and wearing and replacement schedules. The following factors may be of value in making these decisions:

- toxic chemicals and/or physical agents that may be encountered
- raw material and by-products involved
- potential for ocular exposure
- protective equipment provided, available, and used

- hygiene facilities available
- presence or absence of safety and health personnel
- factors that may influence compliance with cleaning and wearing schedules.

An evaluation of the published material, including laboratory and human studies and well-documented case reports, indicates that contact lenses may be worn safely under a variety of environmental situations, including those that, from a superficial evaluation, might appear hazardous. Indeed, some types of contact lenses may give added protection to spectacle lens and nonspectacle lens wearers in instances of certain fume exposure, chemical splash, dust, flying particles, and optical radiation. The evidence also refutes the claims that contact lenses negate the protection provided by safety equipment or make the cornea more susceptible to damage by optical radiation—in particular, arc flashes. Thus, a universal ban of contact lenses in the workplace or other environments is unwarranted.

Regulations limiting the wearing of contact lenses in any given circumstance must be scientifically defensible and effectively enforceable. They should not be based on perceived hazards, random experience, isolated unverified case histories, or unsubstantiated personal opinions.

Conversely, it would be imprudent for a practitioner to prescribe contact lenses in order to circumvent uncorrected visual acuity standards for those occupations in which individuals may be required to function without correction on some occasions or in environments contraindicated for the type of lens prescribed.

All practitioners must stress that personal protective equipment, including safety eyewear, is not replaced by contact lenses. Where circumstances create the necessity, eye protection must be worn.

These guidelines were revised in May 1998 by the AOA Contact Lens Section.

Comfort and Fit

To be comfortable and effective, eye protective equipment must be properly fitted. Corrective spectacles should be fitted only by optometrists or ophthalmologists. An employee can be trained to adjust and maintain eye-protection equipment, however, and each employee can be taught the proper care of the device used. To give the widest possible field of vision, goggles should be fitted as close to the eyes as possible, without bringing the eyelashes in contact with the lenses.

In areas where goggles or other types of eye protection are used extensively, goggle-cleaning stations should be conveniently located. The stations should provide defogging materials and wiping tissues, along with a receptacle for discarding them. Before choosing a defogging material, test to determine the most effective type for a specific application.

Sweatbands can help prevent eye irritation, aid visibility, and eliminate work interruptions for face mopping. Sweatbands are usually made of a soft, light, highly absorbent cellulose sponge. An elastic band holds the sweatband in place on the wearer's forehead so that it does not interfere with glasses or goggles. Evaporation from the exposed surface produces a cooling effect that increases the wearer's comfort. Wearing a sweatband should not interfere with the effectiveness of other protective equipment (respirators, protective helmets, eyewear, face shields).

Face Protection

As a general rule, face shields should not be without suitable spectacles or goggles. A variety of face shields protect the face and neck from flying particles, sprays of hazardous liquids, splashes of molten metal, and hot solutions (Figure 7–5). In addition, they provide antiglare protection where required.

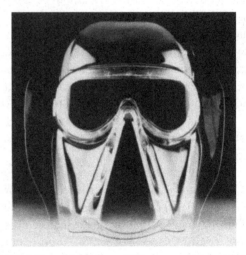

Figure 7–5. This face shield allows 160-degree peripheral vision and can be worn with prescription glasses. Its curved surfaces divert chemical splashes away from the face. *(Courtesy Millennium Safety Products, Inc.)*

Three basic styles of face shields include headgear without crown protectors, with crown protectors, and with crown and chin protectors that can accommodate a variety of replaceable window styles:
- clear transparent
- tinted transparent
- wire screen
- combination of plastic and screen.

The materials used in face shields should combine mechanical strength, light weight, nonirritation to skin, and the ability to withstand frequent disinfecting operations. The shield should be made of noncorrosive metals and slow-burning plastics. Only optical-grade (clear or tinted) plastic, which is free from flaws or distortions, should be inserted for the windows. However, plastic windows should not be used in welding operations unless they conform to the standards on transmittance of absorptive lenses, filter lenses, and plates.

On some jobs, such as pouring low-melting metals, the face shield must protect the head and face against splashes of metal. A face shield similar to an arc welder's piece but made of wire screen can be used and will provide better ventilation than a solid shield. The plain wire will not fog under high-temperature and high-humidity conditions. A metallized plastic shield that reflects a substantial percentage of heat has been developed for jobs in which the worker is exposed to high temperatures.

Acid Hoods and Chemical Goggles

The company can provide head and face protection from splashes of acids, alkalis, or other hazardous liquids or chemicals in several ways, depending on the hazard. A hood made of chemical-resistant material with a glass or plastic window can give good protection. Some manufacturers provide a hood with replaceable inner and outer windows. In all cases, there should be a secure joint between windows and hood materials.

Although hoods are extremely hot to wear, they can be made with air lines for the wearer's comfort. If so, the wearer should have a harness or belt like that on an air-line respirator to support the hose.

If protection is necessary only from limited direct splashes, the person can wear a face shield made of a material unaffected by the liquid or a flexible-fitting chemical goggle with baffled ventilation, provided the eyes are not exposed to irritating vapor. For severe exposure potential, a face shield should be worn in connection with the flexible-fitting chemical goggles.

Face shields should be shaped to cover the whole face. They should be supported by a headband or harness, so they can be tipped back and clear the face easily. Any shield should be easily removed in case it becomes contaminated with corrosive liquid.

If goggles worn under the shield are nonventilated for protection against vapor and splashing, they should also be nonfogging. If necessary, the user can use frequent applications of antifog cleaner to avoid fogging.

Laser Beam Protection

Lasers produce monochromatic, high-intensity light beams, frequently capable of causing significant eye damage. A laser beam of sufficient power can theoretically produce retinal intensities at magnitudes even larger than those produced when directly viewing the sun. Exposures to this type of laser beam have the potential for causing permanent blindness.

No one type of glass or plastic offers protection from all laser wavelengths. Consequently, most laser-using firms do not depend on safety glasses to protect an employee's eyes from laser burns. Some point out that laser goggles or glasses might give a false sense of security, tempting the wearer to unnecessary exposures.

Nevertheless, researchers and laser technicians frequently do need eye protection. Both spectacles and goggles are available—and glass or plastic for protection against nearly all the known laser wavelengths can be special-ordered from eyewear manufacturers. Typically, the eyewear will enjoy maximum attenuation at a specific laser wavelength, with protection falling off rather sharply at other wavelengths.

Laser protective goggles or spectacles, or an "antilaser eye shield," attenuate the helium-neon laser light (wavelength 6,328 angstroms [Å]) by factors of 10 (optical density [O.D.] 1), 100 (O.D. 2), 1,000 (O.D. 3), or more. An optical density of 3 or 4 still renders the beam visible in bright sunlight. Antifog-style goggles are available for use in the field, as are antifog solutions.

The American Conference of Governmental Industrial Hygienists (ACGIH) cautions that laser safety glasses or goggles should be evaluated periodically to make sure that adequate optical density is maintained at the desired laser wavelength. Laser glasses or goggles designed for protection from specific laser wavelengths must not be used with different wavelengths of laser radiation. The eyewear should clearly display the optical density values and wavelengths, which should also be marked on eyewear storage shelves.

Laser safety glasses or goggles exposed to intense energy or power density levels may lose effectiveness and should be discarded. Technical details, uses, hazards, and exposure criteria for lasers are given in the NSC's *Fundamentals of Industrial Hygiene*, 6th ed. (Plog 2012). Also see ANSI Z136.1–2014, Safe Use of Lasers.

Eye Protection for Welding

In addition to damage from physical and chemical agents, the eyes are subject to the effects of radiant energies. Ultraviolet, visible, and infrared bands of the spectrum can all damage the eyes and, therefore, require special protective measures to eliminate the hazard.

Ultraviolet radiation can produce cumulative destructive changes in the structure of the cornea and lens of the eye. Short exposures to intense UV radiation or prolonged exposures to UV radiation of low intensity will produce painful, but ordinarily self-repairing, corneal damage.

Radiation in the visible light band, if too intense, can cause eyestrain and headache and can destroy the tissue of the retina. Infrared radiation transmits large quantities of heat energy to the eye, causing discomfort, although the damage produced is superficial. Extended infrared exposure has been associated with the development of cataracts.

The protective properties of filter lenses have been established by the National Bureau of Standards. The percentage transmittance of radiant energies in the three bands—UV, visible, and infrared—is established for 15 different filter lens shades (Table 7–A). Both absorptive and filter lenses are available in polycarbonate.

Welding processes emit radiation in three spectral bands (see Chapter 19, Welding and Cutting, in this volume). Depending on the flux used and the size and temperature of the pool of melted metal, welding processes will emit more or less visible and infrared radiation; the proportion of energy emitted in the visible range increases as the temperature rises. At least one manufacturer produces an aluminized cover for the usual black welding helmet. Its purpose is to reduce infrared absorption and the resulting heat stress to the wearer.

All welding presents problems, mostly in the control of infrared and visible radiation. Heavy-gas welding and cutting operations, and arc cutting and welding exceeding 30 amps, present additional problems in control of UV radiation. Welding helmets must be used to provide head and face protection (Figure 7–3).

Welders may choose the shade of lenses they prefer within one or two shade numbers. The most commonly used shades are numbers 1.5 to 3.0, intended for glare from snow, ice, and reflecting surfaces and for stray flashes and reflected radiation from cutting and welding operations in the immediate vicinity (for goggles or spectacles with side shields worn under helmets in arc-welding operations, particularly gas-shielded arc-welding operations). Shade 4 is intended for the same uses as shades 1.5 to 3.0, but it provides more suitable protection from greater radiation intensity.

For welding, cutting, brazing, or soldering operations, use the guide for the selection of proper shade numbers of filter lenses or windows in Chapter 19, Welding and Cutting. (Recommendations are also in ANSI/ISEA Z87.1–2010, Standard for Occupational and Educational Personal Eye and Face Protective Devices.) To protect the filter lenses against pitting, they should be worn with a replaceable plastic or glass cover plate. Filter lenses that are cracked, pitted, or otherwise damaged should be discarded and replaced because these conditions may compromise eye/face impact protection and may allow UV and IR radiation to pass through to the wearer's eyes.

Eye protection having mild filter shade lenses or polarizing lenses and opaque side shields is adequate for protection against glare only. For conditions in which hot metal may spatter and in which visible glare must be reduced, management should specify a plastic face shield worn over mild filter shade spectacles with opaque side shields.

The shade of the plate in a welder's helmet can be combined with the shade of the goggle worn underneath to produce the desired total protection. This procedure has the added advantage of protecting the eyes from other welding operations or from an arc when the helmet is raised.

TABLE 7–A. Transmittances and Tolerances in Transmittance of Various Shades of Absorptive Lenses, Other Lenses, and Plates

Shade Number	Optical Density			Luminous Transmittance			Maximum Infrared Transmittance	Maximum Special Transmittance in the Ultraviolet and Violet			
	Maximum	Standard	Minimum	Maximum	Standard	Minimum		313 nm	334 nm	365 nm	405 nm
				Percent	Percent	Percent	Percent	Percent	Percent	Percent	Percent
1.5	0.26	0.214	0.17	67	61.5	55	25	0.2	0.8	25	65
1.7	0.36	0.300	0.26	55	50.1	43	20	0.2	0.7	20	50
2.0	0.54	0.429	0.36	43	37.3	29	15	0.2	0.5	14	35
2.5	0.75	0.643	0.54	29	22.8	18.0	12	0.2	0.3	5	15
3.0	1.07	0.857	0.75	18.0	13.9	8.50	9.0	0.2	0.2	0.5	6
4.0	1.50	1.286	1.07	8.50	5.18	3.16	5.0	0.2	0.2	0.5	1.0
5.0	1.93	1.714	1.50	3.16	1.93	1.18	2.5	0.2	0.2	0.2	0.5
6.0	2.36	2.143	1.93	1.18	0.72	0.44	1.5	0.1	0.1	0.1	0.5
7.0	2.79	2.571	2.36	0.44	0.27	0.164	1.3	0.1	0.1	0.1	0.5
8.0	3.21	3.000	2.79	0.164	0.100	0.061	1.0	0.1	0.1	0.1	0.5
9.0	3.64	3.429	3.21	0.061	0.037	0.023	0.8	0.1	0.1	0.1	0.5
10.0	4.07	3.854	3.64	0.023	0.0139	0.0085	0.6	0.1	0.1	0.1	0.5
11.0	4.50	4.286	4.07	0.0085	0.0052	0.0032	0.5	0.05	0.05	0.05	0.1
12.0	4.93	4.714	4.50	0.0032	0.0019	0.0012	0.5	0.05	0.05	0.05	0.1
13.0	5.36	5.143	4.93	0.0012	0.00072	0.00044	0.4	0.05	0.05	0.05	0.1
14.0	5.79	5.571	5.36	0.00044	0.00027	0.00016	0.3	0.05	0.05	0.05	0.1

Reprinted with permission from ANSI Standard Z87.1–1989.

To protect against UV and infrared radiation and against visible glare in inspection operations, protective lenses should be installed in a hand shield or welder's helmet. The shield should be made of a nonflammable material that is opaque to dangerous radiation and a poor conductor of heat. A metal shield is not desirable because it becomes hot under infrared radiation.

Some tinted lenses used in special work afford no protection against infrared and UV radiation. For instance, most melters' blue glass lenses used in open-hearth furnaces and the lenses used at Bessemer converters afford no protection against either type of harmful radiation. Short exposures while using these lenses may cause no harm. However, new personnel learning these flame-reading skills should be provided with lenses that protect in these two portions of the spectrum, and all personnel should be encouraged to use them.

The chemical composition of the lens rather than its color provides the filtering effect. This factor must be considered when selecting a filtering lens.

HEARING PROTECTION

Medical professionals have long been aware of the problem of noise-induced hearing loss (NIHL) in industry. Noise, or unwanted sound, is a by-product of many industrial processes. Sound is created by pressure changes in a medium (usually air), originating from a source of vibration or turbulence. Exposure to high levels of noise can cause hearing loss. The extent of damage depends primarily on the intensity of the noise and the duration of the exposure. NIHL can be temporary or permanent. Temporary hearing loss results from short-term noise exposures, while prolonged exposure to high noise levels over a period of time gradually causes permanent damage.

The American College of Occupational and Environmental Medicine's Noise and Hearing Conservation Committee (1987) has developed the following guidelines in response to the question "What are the distinguishing features of occupational noise-induced hearing loss?"

Occupational Noise-Induced Hearing Loss

Occupational noise-induced hearing loss, as opposed to occupational acoustic trauma, is a slowly developing hearing loss over a long period (several years) as the result of exposure to continuous or intermittent loud noise. Occupational acoustic trauma is a sudden change in hearing as a result of a single exposure to a sudden burst of sound, such as an explosive blast. The diagnosis of noise-induced hearing loss is made clinically by a physician and should include a study of the noise exposure history.

The principal characteristics of occupational noise-induced hearing loss are as follows:

- It is always sensorineural, affecting hair cells in the inner ear.
- It is almost always bilateral. Audiometric patterns are usually similar bilaterally.
- It almost never produces a profound hearing loss. Usually, low-frequency limits are about 40 dB and high-frequency limits about 75 dB.
- Once the exposure to noise is discontinued, there is no significant further progression of hearing loss as a result of the noise exposure.
- Previous noise-induced hearing loss does not make the ear more sensitive to future noise exposure. As the hearing threshold increases, the rate of loss decreases.
- The earliest damage to the inner ears reflects a loss at 3,000, 4,000, and 6,000 Hz. There is always far more loss at 3,000, 4,000, and 6,000 Hz than at 500, 1,000, and 2,000 Hz. The greatest loss usually occurs at 4,000 Hz. The higher and lower frequencies take longer to be affected than the 3,000- to 6,000-Hz range.
- At stable exposure conditions, losses at 3,000, 4,000, and 6,000 Hz will usually reach a maximal level in about 10 to 15 years.
- Continuous noise exposure over the years is more damaging than interrupted exposure to noise, which permits the ear to have a rest period.

Hearing Conservation Program

The OSHA hearing conservation standard (29 CFR 1910.95, Occupational Noise Exposure) requires a hearing conservation program for employees exposed to excessive noise. Employers must develop and maintain an audiometric testing program for all employees who are exposed to noise levels in excess of 85 dB for an 8-h time-weighted average. With the increasingly frequent use of extended-hour shifts (i.e., 10 or 12 h), the 85-dB exposures level must be recalculated to reflect the new shift length. OSHA currently enforces a 90-dBA permissible exposure limit (Table 7–B). Exposure to 115 dBA is permitted for a maximum of 15 minutes for an 8-hour workday. No exposure above 115 dBA is permitted (29 CFR 1926.52).

TABLE 7–B. Permissible Noise Exposures

Duration per Day, Hours	Sound Level dBA Slow Response
8	90
6	92
4	95
3	97
2	100
1 ½	102
1	195
½	110
¼ or less	115

Source: 29 CFR 1910.95 Table G-16.

Research demonstrates that construction workers can be exposed to noise levels of 95 to 125 dBA through daily activities (rock drilling—up to 115 dBA; chain saw—up to 125 dBA; abrasive blasting—to 112 dBA; heavy equipment operation—110 dBA; demolition—up to 117 dBA; and needle guns—up to 112 dBA.)

In fact, it is a good idea to do audiometric testing and to maintain a noise-exposure record on all employees with potential occupational exposures to noise levels of 85 dB or greater. Audiometric testing should be conducted when new employees are hired and annually thereafter. A testing program properly carried out may determine whether the hearing protective devices worn by employees are actually protecting their hearing from noise damage.

The hearing conservation program may also require the use of hearing protection devices. Before requiring any employee to wear hearing protection, management should measure and evaluate the noise in the workplace. This step serves several purposes: (1) provides the physical evidence of individual exposures; (2) identifies areas where controls need to be established; (3) helps prioritize noise-control and noise reduction efforts, including administrative controls; (4) documents exposures in the work environment for medical-legal purposes; (5) establishes documentation for state, federal, or insurance compliance requirements; (6) provides a basis for analyzing cause–effect relationships between noise exposure and hearing status; and (7) provides insights for improving education and compliance among workers, supervisors, and managers.

When translating noise measurements into exposure estimates, remember that there is no precise safe–unsafe line of differentiation. Any unprotected encounters with steady-state or intermittent noise that exceeds about 85 dB or with impulse or impact noise that exceeds about 120 dB (peak) may overtax the auditory mechanisms of the ears. (See Gasaway 1984 in the References.)

When noise measurement is completed, and other possible noise-control efforts are unsuccessful, then the need for hearing protection is clearly established. For explanation of noise measurement, evaluation, and control, see the National Safety Council's *Fundamentals of Industrial Hygiene*, 6th ed. (Plog 2012).

To develop an effective hearing protection program, companies need to have an accurate knowledge of the noise levels (and frequencies) that pose a hazard to workers. From the data obtained in the noise survey described earlier, management can select the proper hearing protection devices.

Hearing protection devices reduce (attenuate) noise levels with various degrees of success, resulting in varying levels of protection at different noise frequencies. To help management choose the right devices, firms can use the U.S. Environmental Protection Act (EPA) requirement that calls for all protectors to carry a label that indicates their noise reduction rating (NRR). The number provides an estimate of a device's degree of protection and generally can be subtracted from the decibel value of noise in the workplace. This value indicates the noise level theoretically being received in the worker's ear.

However, companies should exercise some caution in applying the full NRR when using hearing protection devices to reduce occupational exposures. Because the NRR is derived under laboratory conditions, wearing conditions of the device on the job will be less than ideal, and noise frequencies and sound levels will not be equal across the spectrum. When evaluating occupational noise exposure, OSHA derates the NRR by one-half for all types of hearing protection. On the other hand, NIOSH considers the performance of different types of hearing protectors and recommends subtracting from the NRR 25% for earmuffs, 50% for formable earplugs, and 70% for all other earplugs.

Types of Hearing Protection Devices

Hearing protectors in general can be categorized as four types: enclosure (helmets), aural (ear insert), superaural (canal caps), and circumaural (earmuffs) (Figures 7–6 and 7–7).

Before a company issues any ear insert, management should take certain measures: (1) each employee's ear canals should be examined for any abnormalities or irregularities; for example, certain otic conditions may not allow use of earplugs; (2) employees must be taught proper insertion techniques; and (3) employees must be taught proper sanitation and checking techniques.

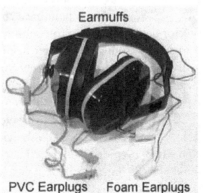

Figure 7–6. Three types of hearing protection devices.

Figure 7–7. Earmuff with voice-actuated communication system. *(Courtesy Earmark, Inc.)*

Enclosure

The enclosure hearing protector completely surrounds the head, such as an astronaut's helmet. Sound is reduced through the acoustical properties of the helmet. Additional attenuation can be achieved by wearing inserts with the enclosure helmet. Expense, temperature inside the helmet, and its bulk normally rule out general use of the enclosure hearing protector, but certain occupational tasks or industries may have specific needs for it.

Aural Insert

Commonly called inserts or earplugs, the aural insert is generally inexpensive and has a limited service life. The plug or insert falls into three broad categories: (1) formable, (2) custom molded, and (3) molded.

1. Formable aural inserts fit all ears. Many of the formable types are designed to be disposable after being worn once. Materials from which these disposable plugs are made include fine glass fiber, wax-impregnated cotton, and expandable plastic foam.

   Various models provide different degrees of noise reduction. Manufacturers will supply attenuation data for their products to help the safety and health professional evaluate their effectiveness for use in a given situation.

2. As the name indicates, custom-molded hearing protectors are made for a specific individual. A prepared mixture is carefully placed in the person's outer ear, with a small portion extending into the ear canal; as the material sets, it conforms to the shape of the individual's ear and external ear canal. Only trained personnel should attempt the process of forming these hearing protectors.

3. Molded (or premolded) aural inserts are usually made from a soft silicone rubber or plastic. The most important aspect of this protector is a snug fit to provide adequate protection. Some persons may find these inserts uncomfortable because of the irregular shape of the ear.

Superaural

The superaural, or canal cap, hearing protector depends on sealing the external edge of the ear canal to achieve sound reduction. The caps, made of soft, rubber-like material, are held in place against the edges of the ear canal by a spring band or a head suspension.

Circumaural

Cup (or earmuff) devices cover the external ear to provide an acoustical barrier between external sound and the inner ear. The attenuation provided by earmuffs varies widely due to differences in size, shape, seal material, shell mass, and type of suspension.

Hearing Protector Selection

Individual head size and shape can influence the attenuation characteristics of hearing protectors. Also, wearing other PPE such as safety helmets or safety spectacles must not compromise the efficiency of the hearing protection. Temple pieces of safety spectacles can cause noise leakage; to minimize this leakage employees can use the cable temple pieces or use aural inserts. The type of cushion used between the shell and the head also has a great influence on attenuation efficiency. Liquid- or grease-filled cushions may give better noise suppression than plastic or foam rubber types but may present leakage problems.

When selecting a hearing protection device, also consider the work area in which the employee must use it. For example, large earmuffs would not be practical for someone who works in confined areas with little headroom. For these conditions, a small or flat ear cup or insert protector would work better.

When employees must wear muff protectors in special-hazard areas (e.g., around high-voltage cables), nonconductive suspension systems may be needed in connection with muff protectors.

Another consideration when selecting a hearing protective device is how often employees are exposed to excess noise (once a day, once a week, or infrequently). For such cases, an insert or plug device may satisfy legal requirements. If the noise exposure is relatively frequent and the employee must wear the protective device for an extended time, a muff protector might be the best choice. If noise exposures are intermittent, muff protectors are probably more desirable because it is somewhat more difficult to remove and reinsert earplugs than earmuffs.

FALL ARREST SYSTEMS

Many employees in today's work force are tasked with duties that require work at heights above ground level. The tasks can be as simple as changing a light bulb or as difficult as painting a chimney. Both of these work situations require fall protection for the employee while the job is being done. At greater heights, as in construction or utility work, fall protection becomes mandatory under most safety regulations.

The impact from even a 4-ft (1.2-m) fall can be enough to cause serious injury. Companies can use many methods to prevent employees from falling. This section will deal only with fall protection systems and not with mobile elevated access equipment, ladders, aerial buckets, rescue equipment, and so on.

What Is Fall Arrest?

Fall arrest is defined as a means of preventing workers from experiencing disastrous falls from elevations. Fall arrest systems are usually classified as passive or active.

Passive Fall Arrest

This system consists of components and systems, such as nets and handrails or guardrails, that do not require any action on the worker's part. A properly designed passive fall arrest system, installed correctly, will protect the individual 100% of the time.

Active Fall Arrest

This system is made up of components and systems that require some manipulation by the workers to make the protection effective. These systems include harnesses; lanyards and their attachments; and component parts such as rope-grabbing devices, lifelines, self-retracting lanyards (SRLs), and so on. Active equipment will not work by itself and must be connected or employed by the individual to be protective.

When Are Fall Arrest Systems Needed?

The first factor to consider in selecting a fall arrest system is the height at which the worker will be performing the job. Some U.S. regulations provide limits, for example:

Height	General Regulation	U.S. OSHA Standard (29 CFR)
Over 4 ft	Guardrail	1910.23
	Midrail	1910.23
Over 6 ft	Guardrail	1926.500
Over 25 ft	Overwater	1926.105

Second, the safety and health professional should analyze the job site and specific task to be done. If the job requires working vertically, a different or modified system will be needed than if the worker must move laterally.

Third, other factors should be addressed, including rescue methods, backup systems, length of time at workstations, dry or wet conditions, number of workers needed on the job site, and environmental factors.

This complete analysis, along with a review of regulations, helps determine the fall arrest system needed. Modern fall protection encompasses a variety of technical, medical, ergonomic, and legal issues and has become a multidisciplinary science of its own. In most cases, the introduction of a fall arrest system involves much more than simply selecting and purchasing one. Quite often, a system has to be specially designed for a particular application. In such cases the design engineer, or other competent person, has to take into account not only the individual performance of every component of the system and a proper anchor for

the equipment, but also the geometry of the workplace, the environmental conditions present, and the method of post-fall rescue to be employed.

Elements of a Successful Fall Arrest Program

Some criteria need to be established for designing a fall arrest program. First, the employer must set a policy, which is clearly communicated to employees and enforced during applicable operations, that addresses these points:

- Worker qualification—Is the employee qualified to perform work at elevated conditions?
- Training—Are workers who are placed in the elevated work positions trained in the arrest system to be used?
- Selection of equipment—Is equipment being used as required to perform the job safely? Equipment purchased for the job must meet appropriate standards and, if required, be certified.
- Installation of equipment—Has equipment been installed according to acceptable standards, regulations, and manufacturer's recommendations?
- Equipment maintenance and inspection—Can equipment be maintained as recommended, and will employees inspect their personal system components daily before each use? Have employees been trained on how to properly inspect their fall protection device?
- Rescue procedures—Has a plan been developed to rescue any employee who has fallen while using a fall protection system?
- Job survey analysis—Has a job procedure been developed and implemented for every job in an elevated situation?

Which Fall Arrest System to Use?

Many different kinds of passive and active fall arrest systems are available. Choosing the one best suited to a particular task requires planning, forethought, and a thorough understanding of the systems on the market today. When falling hazards are identified and cannot be eliminated, management must adopt some means of control to minimize the risk of personal injury.

The analysis of elevated work tasks is intended to determine the most suitable match between required worker mobility and the capabilities of the fall arrest system. The company policy establishes what is to be done.

Next, the appropriate system and its components must be selected. A variety of equipment is available to help employers set up an effective fall arrest program. Generally, this includes nets, body support mechanisms, climbing arrest systems, vertical lifeline systems, horizontal lifeline systems, confined entry and retrieval systems, and controlled descent–emergency escape systems.

However, the proper selection and purchase of safety equipment alone does not constitute a fall arrest program.

The employer also has an important responsibility for choosing and using fall equipment only for the application recommended in the literature, instructions, and on the label. Commercial equipment should never be used for applications not stated by the manufacturer.

Passive Fall Arrest Systems

Passive fall arrest systems include general all-purpose nets, personnel nets, and debris nets. These devices are easy to use and have a wide range of applications.

Nets

Properly installed, nets can be a vital part of a passive fall arrest system. Nets are designed to provide protection under and around an elevated work area where fall hazards exist. The worker is not directly involved by "wearing" fall arrest equipment; rather, the net is there to catch a falling worker before she or he hits the ground or obstructions. Two major types of nets are available, one for personnel and the other for debris. Often the two types are combined to form a dual net with twofold purpose (Figure 7–8).

Figure 7–8. These nets protect workers from falls and from falling debris. *(Courtesy Pearl Weave Safety Netting Corp.)*

Personnel Nets

Personnel nets can be used for large work crews, such as those employed on bridge construction or repair or on long-term structural projects. They also provide protection where large open areas or long leading edges expose workers to height hazards (up to 25 ft [8 m]) below the work surface by current U.S. OSHA regulations), and the use of other fall arrest equipment is deemed impractical or not feasible for the work method. The advantage of nets is that individual worker training is not required. Once installed, nets are always in place and ready for use. However, other personnel fall arrest must be available during net installation and removal.

Personnel nets must be manufactured and tested in accordance with ANSI/ASSE A10.11–2010, Safety Requirements for Personnel and Debris Nets, and U.S. Code of Federal Regulations 29 CFR 1926.105 requirements. Mesh openings may not be greater than 6 in. (15 cm) × 6 in. (15 cm). Nets meeting these requirements must bear labels displaying the manufacturer's name and date of manufacture, together with testing data.

Nets should be as close to the work level as possible and no lower than 25 ft (8 m) (except with bridges, where OSHA considers the highest work level to be the lowest part of the bridge), and must extend outward 8 ft (2.4 m) from the structure. Nets must be tested in the field by dropping a 400-lb (181-kg) sandbag from a 25-ft (8-m) height, according to ANSI/ASSE A10.11–2010, repeated at successive 6-month intervals or whenever a relocation of or major repair to the net has occurred, and must not have any broken strands.

Note: The net may be certified to meet these requirements by a qualified person. Nets should be moved up regularly to avoid exceeding the 25-ft (8-m) limit as a building is constructed. For a personnel net to meet ANSI/ASSE A10.11–2010 or OSHA regulations, the manufacturer must affix a permanent label with the following information:
- name of manufacturer
- identification of the material
- date of manufacture
- date of prototype test
- name of testing agency
- serial number.

Debris Nets

Debris nets are designed to catch falling debris (i.e., tools, foreign objects, falling concrete, and other construction debris) and to protect workers and pedestrians below. The strength and size of the mesh must be sufficient to catch and contain the size, weight, and impact of the objects that are likely to fall. Popular net sizes range from ¼- to ⅓-in. (6- to 8-mm) mesh. To catch large, heavy objects as well as small, light objects, the smaller mesh nets can be used in conjunction with the larger mesh and stronger personnel nets.

These net systems can also be used to catch personnel as well as debris. In these cases, personnel nets are deployed in conjunction with debris nets. The nets must be kept clear of debris to help ensure a falling worker's safety. When these are in place, a means of rescuing a fallen worker must be available.

Active Fall Arrest Systems

Active fall arrest systems include components such as fall arresters and shock absorbers, harnesses, lifelines, and SRLs. All active systems begin with an anchorage point and have some connecting components to the worker. Harnesses and components should be used only for employee protection (as part of a personnel fall arrest system or positioning device system) and not to hoist materials.

Anchor/Anchorage Points

The critical problem in all active fall protection—the anchorage point—is the position on an independent structure to which the fall arrest device or lanyard is securely attached. Supervisors and workers must also analyze all hazards below and to the side of the anchoring point to ensure that the worker does not strike or swing into any obstacles should he or she fall. The U.S. OSHA requirement for an anchorage is a 5,000-lb (2,268-kg) minimum static load strength (needed for 6 ft [2 m] of free fall). The Canadian Standard Association's (CSA) Standard Z259.2 requires 6,000 lb (2,700 kg). The strength, location, and design must allow the worker enough mobility to perform the job.

Lanyard

A lanyard is a short, flexible rope, strap, or webbing connecting the worker to the anchor. A lanyard permits limited lateral movement on the job. Its length (and placement of the anchor) determines the amount of free fall a worker experiences before the protective device stops the fall. However, retracting lifelines require 3,000 lb (1,360.8 kg). Furthermore, OSHA has permitted a 2:1 safety factor for engineered systems.

Body Belts

Body belts are not acceptable as part of a personnel fall arrest system. However, the use of a body belt in a positioning device system is acceptable and is regulated as follows:

- Positioning devices should be rigged such that an employee cannot free-fall more than 2 ft (0.9 m).
- Positioning devices should be secured to an anchorage capable of supporting at least twice the potential impact load of an employee's fall or 3,000 lb (13.3 kN), whichever is greater.
- Positioning devices should be drop forged, pressed, or formed steel, or made of equivalent materials with a corrosion-resistant finish, and all surfaces and edges should be smooth.
- Connectors, connecting assemblies, and D-rings and snaphooks should meet requirements of 29 CFR 1926.502.

- Snaphooks should be a locking type designed and used to prevent disengagement of the snaphook by the contact of the snaphook keeper by the connected member.
- Positioning devices should be inspected prior to each use for wear, damage, and other deterioration, and defective components should be removed from service.

Harnesses

Full-body harnesses encompass the torso and are attached to other parts of the fall arrest system. A full-body harness distributes the fall arrest force over the shoulder-to-thigh body areas.

Typically made of nylon and polyester webbing straps, harnesses are the preferred worker component in fall arrest systems. By spreading the fall arrest force over the body, a harness enables the worker to avoid bodily injury. The generally accepted maximum free-fall distance is 6 ft (1.8 m). A goal of 2 ft (0.6 m) is best to enable self-rescue and avoid prolonged suspension.

Retracting Lifeline Devices

Retracting lifeline devices are portable, self-contained devices fixed to an anchorage point above the work area. They act as an automatic taut lanyard. The lifeline rope, webbing, or cable is attached directly to the worker's body harness. The line extends out of the device as distance increases and retracts as the worker moves closer. At the moment a fall occurs, a centrifugal locking mechanism is activated to arrest the movement, thereby reducing the potential free-fall distance shock load. This device is ideal for use on sloping roofs and angular structures, because the rope is never slack and does not interfere with the surface work. The extension limit varies from 6 ft (2 m) on some devices to 300 ft (91 m) on others. Unlimited lengths are possible with a counterweight variation.

Lifelines

A horizontal lifeline is an anchoring cable rigged between two fixed anchorage points on the same level (Figure 7–9). The line may serve as a mobile fixture to attach lanyards, lifelines, or retracting lifelines. The purpose is to limit swing injuries by providing a continuously overhead fixture point as the worker moves horizontally. The important factors for nonengineered systems include (1) adequate degrees of slack; (2) 500 lb (227 kg) per worker strength (where falls up to 6 ft [2 m] are anticipated and where short horizontal lifelines [25 ft (8 m)] are used); (3) supports every 20 to 50 ft (6 to 16 m) that preferably can be passed without detaching the line; and (4) for engineered systems, sufficient shock absorption and design strength at least twice the force calculated for the dynamic fall of an anticipated number of workers who may use the line. Lightweight, low-stretch synthetic cables serve as a practical alternative to steel cable.

Figure 7–9. This worker is protected by a horizontal lifeline; the lanyard extends and retracts as needed for normal operations.

Extremely careful engineering is required for all horizontal lifelines. When used with retracting lifelines, the horizontal lifeline should be arranged overhead with little sag (approximately 10 degrees to the horizontal or less, which is also usable with rope or synthetic cable). When used with lanyards, the line should be set at a maximum height of 78 in. (198 cm) at the center. The worker may not be able to travel down the horizontal lifeline slope after a fall. Slack in the line during and after a fall should be considered to determine proper clearance in relation to other obstacles and the ground.

Some organizations offer training programs to determine proper end-force and dynamic slack projections. Manufactured horizontal lifeline systems typically have an in-line shock absorber for reducing end forces.

Lifeline ("Dropline")

A dropline is a vertical lifeline that extends from an independent anchorage point and to which a lanyard is attached using a grabbing device. The dropline should be at least ⅝-in.-(16-mm-) diameter nylon, ⅝-in.-(16-mm-) diameter polyester, or ⁵⁄₁₆-in.-(24-mm-) diameter or ⅜-in.-(9-mm-) diameter steel cable with a minimum breaking strength of 5,000 lb (2,449 kg). Steel cable should be used only

in spark- or heat-producing work operations, although it is popular for use with work at extreme heights, such as chimney building and repair, or where a limit on the system elongation is important. Ropes always must be protected from abrasive or cutting edges, which may weaken the fibers.

Weather-protected nylon and polyester lifelines with neoprene jackets are available. Polypropylene ropes are popular with utilities because of the low moisture absorption and high dielectric constant. Ultraviolet-stabilized polypropylene ⅝ in. (16 mm) or ¾ in. (19 mm) in diameter is popular for many suspended scaffold operations because of its light weight and low cost; however, unstabilized ropes could be hazardous after short exposures to sunlight.

Hardware Connectors

Hardware connectors consist of bolts, shackles, D-rings, snaphooks, and metal links that connect parts of the lifeline. The OSHA regulation requires these connectors to demonstrate 5,000-lb (2,268-kg) static tensile strength for D-rings and snaphooks.

Carabiner-type snaphooks are often an alternate oval-shaped snaphook design with an automatic twist-lock spring gate to help eliminate the possibility of rollout. Snaphooks should be attached to compatible hardware and never to each other. These hooks must always be tested during inspection and maintenance to see if they fully close and lock and if they will do so on the anchorage point without stress to the gate.

Fall Arresters and Shock Absorbers

Many types of fall arresters are available in various sizes and types of lifeline. These devices slow a worker's fall or break the fall to prevent injury. Nearly all fall arresters use friction in the rope to disperse fall energy. Often there is a gradual delay action so the body does not experience a severe jolt or shock when the fall is arrested.

If a lanyard is used as the only fall protection, a shock absorber fall arrester and a full-body harness can soften the arresting force on the body. Shock absorbers may not always be required or necessary if the anchorage point height allows little or no free fall. These devices work by the unfolding and tearing of stretched woven webbing or stretching fiber.

Fall Arresting System

A fall arresting system (FAS) is engineered from components and designed for a work positioning system. The purpose of an FAS is not only to stop the fall, but also to ensure that the energy gained by the body during the fall is distributed so as to prevent the wearer from being injured. A fall arresting system is composed of an independent anchorage point, a vertical lifeline (dropline), a fall arrester, a harness, and—optionally—a lanyard and a shock absorber, equipped with all the necessary hardware (snaphooks, D-rings, and so on).

It is usually easier to slowly arrest a fall and to prevent injury to the worker during and after the fall. Injuries generally occur as a result of forces acting on the body at the instant a fall is stopped or through collision with obstacles.

Work Positioning System

This is a system that permits users of harnesses and lanyards to lean to do work. Fall arrest is a separate system that does not involve the parts of a work positioning system during fall arrest.

Restraint System

This is a system that permits users of harnesses and lifelines to move up to a fall-hazard zone but restricts movements. The system must meet fall arrest requirements because of the likelihood of a free fall under some configurations of use.

Rescue Systems

The moments following a fall can be critical in preventing worker injuries. Organizations should develop, implement, and regularly practice rescue procedures. The following sections discuss aboveground rescues and belowground or confined-spaces rescues.

Aboveground

Descent devices permanently installed or immediately available at such workstations as overhead crane cabs and grain elevator workhouses can be used effectively for lowering an injured member of the crew quickly and safely to ground level. Devices with no inherent speed control require the presence of a trained rescue team or trained co-workers to supervise the rescue operation. Automatic speed-limiting descent devices reduce or eliminate the need for trained rescue personnel; the machine itself controls the injured worker's rate of descent.

Collapsible cradles are snap-on accessories that may be used with either system. They are useful for small, cramped spaces in crane cabs and towers, but rigid stretchers should be available for victims with broken bones or internal injuries to prevent compounding the injury.

Belowground Tanks or Confined Spaces

Confined spaces are those enclosed spaces that have limited openings for entry and exit, have poor ventilation, and are not designed for continuous worker occupancy. Examples of confined spaces include storage tanks, process vessels, ship compartments, pits, silos, vats, sewers, boilers, tunnels, vaults, and pipelines. (See also Chapter 2, Buildings and Facility Layout, in this volume.)

Rescue workers can be lowered into tanks or confined spaces by means of lifelines to locate workers or to determine if they need assistance. If manholes and tanks are deeper than 10 ft (3 m) per platform or level, fall protection is required in addition to a means of rescue under OSHA standards.

A manual or air-operated winch accessory is available for use with some steel cable retracting lifeline devices; the winch helps eliminate the need for personal attention to the victim until he or she has been raised. Alternatively, a block-and-tackle or ratchet winch can provide the lifting mechanism with limited human effort after the victim has been hooked up, provided a lock or overspeed mechanism is incorporated. An anchorage point, such as that provided by a 7- or 10-ft (2.1- or 3-m) tripod, should be available before work begins at the site. (See Figure 2–6 in Chapter 2 in this volume.)

Equipment Inspection and Maintenance

The fall arrest equipment manufacturer's instructions must be incorporated into a company's inspection and preventive maintenance procedures (Table 7–C). Workers need to be trained to inspect equipment, to understand the basics of static loading of fall equipment for test purposes, and to check equipment for damage before each use. Equipment should be removed from service after exposure to the forces of arresting a fall or equivalent forces or if a built-in stress indicator or warning system has been activated.

Note: This section is not meant to be complete, but only to serve as a guide. Special situations such as ionized radiation, electrical conductivity, spark generation, chemicals, and so on, must also be considered. The recommendations should not be regarded as an industry consensus but rather a guideline for safety in developing company procedures.

Cleaning Fall Arrest Equipment

Fall arrest equipment must be cleaned regularly to ensure that it remains in good condition and top working order.

Synthetic Ropes and Harnesses

Washing this equipment in soapy water is the best way to remove loose debris. Rinse with fresh water and dry in a cool area away from UV light and excessive heat. Always make sure that labels are legible after cleaning. Industrial solvents should not be used to clean synthetic materials. These chemicals can degrade the product by leaching oils used in the manufacturing process to give greater strength to the final product.

Fall Devices

Wash fall arrest devices in soapy water. If metal parts have been soiled with caked materials or paint overspray,

TABLE 7–C. Fall Protection Equipment Maintenance and Inspection Guide

Components	Lifetime (yr)	Service Shelf	System Rating	Checking (3-6 mo)
Webbing (belts, lanyards)	2-3	7	5,000 lb 500-800 lb shock absorber	Cuts, wear, burn, pull one unit
Ropes (lifelines, lanyards)	1-2		5,000 lb	Synthetic: Pull end sample, cut in strand, worn, dirt inside Cable: kink, broken wire, terminations
Hardware hooks D-rings	>5 2-3 <5	<5 <5	5,000 lb	Cracks, distortion, wear points, corrosion
Locking Fall Devices rope grabs retracting lifeline climbing device	3-5 3-7 <5		1,000 lb before slip 3,000-lb line 350-450 lb before slip 1,000-lb proof load	Recertification/old models Distortion, wear Cleaning difficulty Comparable parts Operates manually as intended Recertification models
Safety Lowering Devices retracting lifeline escape descent-control device	3-5 5		4-9 ft/sec (3,000 lb) 3-6 ft/sec (1,800 lb) No limit (5,000 lb)	Recertification/old models Distortion, wear Controlled payout Operates manually as intended
Metal goods should have no limit on shelf-life expectation.				

Source: J. Nigel Ellis, *Introduction to Fall Protection* (Des Plaines, IL: ASSE, 1998)

consider using cleaning solvents such as wood alcohol or 1,1,1-trichloroethane to remove the contaminants. Labels must be legible after cleaning, or be replaced after calling the manufacturer.

Do not oil moving parts unless instructed by the manufacturer, as oil could interfere with a descent device's brake efficiency. For some parts, such as rollers or bearings subject to heavy use or dirt, such lubrication may be reasonable. Any lubrication used must have the manufacturer's approval. Many manufacturers offer reconditioning programs and retesting documentation to help owners maintain fall arrest devices in good working order. Employers should take advantage of the fall equipment manufacturer's expertise.

Storage

Keep synthetic materials away from bright light and UV light during storage, and maintain them in a cool, dry place. Areas where heat, moisture, oil, chemicals (or their vapors) may be present should also be avoided. When dyed synthetic color fades, it indicates UV exposure that may lead to equipment damage and failure. Equipment that is damaged or is in need of repair should not be stored in the same area as usable equipment. All fall arrest and body support devices must be inspected regularly and defective parts repaired or the equipment replaced. Manufacturer's instructions should be consulted for inspection details.

FALL PROTECTION STANDARDS

At press time, the newly approved fall protection code, ANSI/ASSE Z359, was adopted to incorporate basic fall safety principles, including hazard survey, hazard elimination and control, and education and training. These standards also offer guidance on design considerations for new buildings and facilities, recognizing that design deficiencies often increase the risk for employees who may be exposed to fall hazards. For more detailed information on the ANSI/ASSE Z359 Fall Protection Code, visit asse.org.

RESPIRATORY PROTECTION

A long-standing hierarchy of controls requires employers to use engineering and work practice controls as the primary means to protect an employee's health from contaminated or oxygen-deficient air. However, if such controls are not technologically or economically feasible (or otherwise inappropriate), an employer may rely on a respiratory protection program to protect employees.

The respiratory protection program must consist of worksite-specific procedures specifying the selection, use, and care of respirators. The program must be updated as often as necessary to reflect changes in workplace conditions and respirator use.

The respiratory protection program must cover the following basic elements, as applicable:
- a written respiratory protection program containing workplace-specific procedures necessary to protect the health of the employee from workplace contaminants or when the employer requires the use of respirators (29 CFR 1910.134)
- procedures for selecting respirators
- medical evaluations of employees required to use respirators
- fit-testing procedures for tight-fitting respirators
- use of respirators in routine and reasonably foreseeable emergency situations
- procedures and schedules for cleaning, disinfecting, storing, inspecting, repairing, and otherwise maintaining respirators
- procedures to ensure adequate air quality, quantity, and flow of breathing air for air-supplying respirators
- training of employees in the respiratory hazards and proper use of respirators, limitations of use, and applicable maintenance procedures
- procedures for regularly evaluating the effectiveness of the program.

Note: If an employee is voluntarily using a filtering facepiece, then an employer is only required to provide a copy of Appendix D, 29 CFR 1910.134, to each respective user. For cases in which employers allow the voluntary use of respirators other than filtering facepieces, the costs associated with ensuring use of the respirator itself does not create a hazard, such as medical evaluations and maintenance, must be covered at no cost to the employee.

NIOSH, under authorization of the Federal Mine Safety and Health Act of 1977 and the Occupational Safety and Health Act of 1970 (OSH Act), provides a testing, approval, and certification program to ensure that safe personal protective devices and reliable industrial hazard-measuring instruments are available commercially. Use of the terms "approved" and "certified" on the product label reflects these applicable federal regulations.

Employers can consult manufacturers for information related to applications and limitations of a respirator type, equipment needs, repairs and maintenance, and other product-specific information.

Selecting Respiratory Protection

Given the hundreds of toxic substances workers may encounter and the wide variety of respiratory protection equipment available, making the right choice of breathing equipment can be a difficult task.

The proper selection of respiratory protective equipment involves three steps: (1) identifying the hazard; (2) evaluating the hazard; and (3) selecting the appropriate, approved respiratory equipment based on the first two considerations (Figure 7–10).

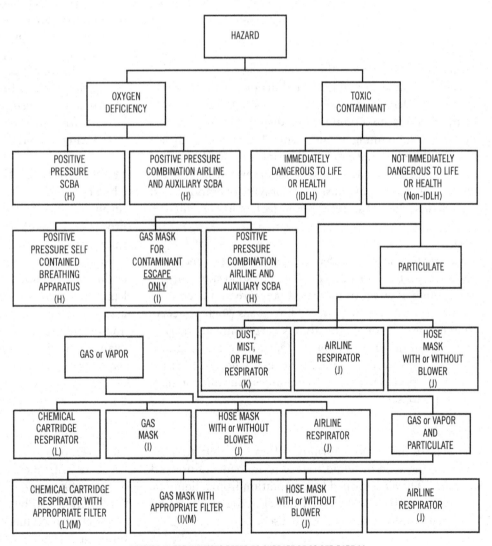

Figure 7–10. Suggested outline for selecting respiratory protective devices. (Letters in parentheses refer to Subparts of Title 30 CFR, Chapter 1, Part 11.)

Identification of the Hazard

Airborne hazards that could require respiratory protection generally fall into the following basic categories (taken from the OSHA Technical Manual, Chapter 7):

Dusts

Particles that are formed or generated from solid organic or inorganic materials by reducing their size through mechanical processes such as crushing, grinding, drilling, abrading, or blasting.

Fumes

Particles formed when a volatilized solid, such as a metal, condenses in cool air. This physical change is often accompanied by a chemical reaction, such as oxidation. Examples are lead oxide fumes from smelting and iron oxide fumes from arc welding. A fume can also be formed when a material such as magnesium metal is burned or when welding or gas cutting is done on galvanized metal.

Mists

A mist is formed when a finely divided liquid is suspended in the air. These suspended liquid droplets can be generated by condensation from the gaseous to the liquid state or by breaking up a liquid into a dispersed state, such as by splashing, foaming, or atomizing. Examples are the oil mist produced during cutting and grinding operations, acid mists from electroplating, acid or alkali mists from pickling operations, paint spray mist from spraying operations, and the condensation of water vapor to form a fog or rain.

Gases

Gases are formless fluids that occupy the space or enclosure and that can be changed to the liquid or solid state only by the combined effect of increased pressure and decreased temperature. Examples are welding gases (such as acetylene, nitrogen, helium, and argon); carbon monoxide generated from the operation of internal combustion engines; and hydrogen sulfide, which is formed wherever there is decomposition of materials containing sulfur under reducing conditions.

Vapors

Vapors are the gaseous form of substances that are normally in the solid or liquid state at room temperature and pressure. They are formed by evaporation from a liquid or solid and can be found where parts-cleaning and painting take place and where solvents are used.

Smoke

Smoke consists of carbon or soot particles resulting from the incomplete combustion of carbonaceous materials such as coal or oil. Smoke generally contains droplets as well as dry particles.

Oxygen Deficiency

An oxygen-deficient atmosphere has an oxygen content below 0.5% by volume. Oxygen deficiency may occur in confined spaces, which include, but are not limited to, storage tanks, process vessels, towers, drums, tank cars, bins, sewers, septic tanks, underground utility tunnels, manholes, and pits.

Biological Agents

Some biological agents are pathogenic microorganisms or infectious agents that could cause disease in a susceptible population. Although respiratory protection guidelines for exposure to tuberculosis (TB) are covered by a specific standard (29 CFR 1910.139), this is one example of a biological agent encountered in the workplace that has caused infection requiring medical treatment in several hundred employees. TB is spread by airborne droplets when a person coughs, sneezes, or speaks, and use of respiratory protection can be an effective response measure. Other biological agents, such as the bacterium *Chlamydia psittaci* and *Histoplasma capsulatum* spores, are examples of health hazards for which respiratory protection could be helpful in reducing potential exposures.

Evaluation of the Hazard

The next step in a respirator selection process is a walk-through survey of the facility to identify employee groups, processes, or worker environments where the use of respiratory protective equipment may be required. The physical and chemical nature of the identified hazard(s) must be evaluated, potential exposures quantified (concentrations, length and nature of exposure), the surrounding environment assessed, and other potential stressors identified (physical and psychological, if present). This information should be examined in the light of occupational exposure limits (both regulatory and consensus) and the availability and feasibility of appropriate respiratory protection. A hazard evaluation form can help the safety and health professional in conducting the survey (Figure 7–11). He or she may use instruments to determine the concentration of airborne contaminants. However, only qualified individuals should be allowed to use these instruments and interpret their results. If an organization does not have in-house qualified personnel, outside consultation will be required. Whenever processes change, a new evaluation must be conducted to quantify any new or different exposures. If hazards have worsened (or lessened), changes to the level of respiratory protection may be necessary, indicating a change in the written respiratory protection program and subsequent changes in the required respirator.

Respiratory Protective Equipment Hazard Evaluation Form

Company _____   Date _____

Division _____   By _____

Department _____   Page _____

Employee	Job Description	Limits	Respiratory Protection Required		Respiratory Equipment Type-SCP	Remarks
			Yes	No		

Figure 7–11. This sample respiratory protective equipment hazard evaluation form specifies whether respiratory protective equipment is required for a job and what type.

There will be one of two outcomes from the hazard evaluation. Either (1) respiratory protection is needed, or (2) engineering and administrative controls are sufficient to protect the health of the workers. If the hazard evaluation indicates the need for a respiratory program, then it is important to know which respirator to select for use.

NIOSH recommends that the hazard evaluation include examining the nature and extent of the hazard, work requirements and conditions, and the characteristics and limitations of the respirators available.

When the need for a respirator is ascertained, a NIOSH-certified respirator should be selected. In some cases a specific respirator for a contaminant has not been certified. When this occurs, a NIOSH-certified respirator with no limitation prohibiting its use for that contaminant can be used only if the protective features of the respirator are appropriate for the contaminant's physical form, chemical properties, and the conditions under which it will be used. And in all cases, respirators must be chosen and used according to the limitations identified on the NIOSH certification label.

A system should be in place that provides a reliable means of protecting respirator wearers from contaminant breakthrough any time an air-purifying respirator is used as protection against gases and vapors. The written respirator program should describe the change schedule, indicating how often cartridges and/or canisters should be replaced and what information was used to make this judgment regarding service life. The service life depends on many factors, including environmental conditions, breathing rates, cartridge filtering capacity, and the concentration of contaminants in the air. Odor thresholds and other warning properties cannot be used as the primary basis for determining the service life of gas and vapor cartridges and canisters. It would be a good idea for employers to apply a safety factor to the service life estimate to ensure that the change schedule is a conservative estimate.

Effective methods for determining the service life of a respirator cartridge include:

- using cartridges with end-of-service-life indicators (ESLIs)
- using an established and enforced cartridge/canister change schedule that is based on objective information or data that will ensure that canisters and cartridges are changed before the end of their service life.

For atmospheres that are immediately dangerous to life and health (IDLH), the highest level of respiratory protection and reliability is required. Only the following respirators must be provided for use in an IDLH atmosphere: either a full-facepiece pressure-demand self-contained breathing apparatus (SCBA) certified for a minimum service life of 30 min, or a combination full-facepiece, pressure-demand, supplied-air respirator (SAR) with an auxiliary self-contained air supply.

Protection Factors

Assigned protection factors (APFs) are used to indicate the level of effectiveness a respirator provides to the wearer. An employer should ensure that the APF for an assigned respirator is adequate to provide protection from the identified hazardous contaminant. In August 2006, OSHA published its final rule that mandates the use of specified APFs by employers when selecting respiratory protection for workers.

Types of Respirators

The proper selection and use of a respirator will depend on the initial determination of the concentration of the hazard or hazards present in the workplace or the presence of an oxygen-deficient atmosphere.

When using a respiratory protection program to protect the health of the employees, employers must not only provide respirators but must ensure that employees use them. Respirators provide protection either by removing airborne contaminants from the air before they are inhaled or by providing an independent supply of breathable air. There are two major classifications of respirators:

- air-supplying respirators, which provide clean breathing air
- air-purifying respirators, which remove (or filter) contaminants from the air.

Air-supplying respirators deliver clean breathing air from a source independent of the surrounding atmosphere. They do not remove contaminants from the atmosphere. These respirators are classified by the way breathing air is supplied and regulated. Air-supply respirator configurations include:

- self-contained breathing apparatus (air or oxygen is carried in a tank on the worker's back)
- supplied-air respirators (compressed air from a stationary source is supplied through a high-pressure hose connected to the respirator)
- combination self-contained and supplied-air respirators.

Air-purifying respirators remove contaminants and are grouped according to the contaminant removed:

- particulate
- vapor
- gas
- combination.

Particulates are removed by filters, while vapors and gases are removed by either chemical cartridges or canisters. Filters and canisters/cartridges are typically used up during the air-filtering (cleaning) process and, if replaceable, are exchanged for new ones when their effective life has ended. Filtering facepiece respirators (commonly referred to as "disposable respirators," "dust masks," or "single-use respirators") that cannot be cleaned, disinfected, or resupplied with a new filter are discarded after use.

Particulate-removing respirators are designed to filter nuisance dusts, fumes, mists, toxic dusts, radon daughters, asbestos-containing dusts, biological agents or fibers, or any combination of these substances. These respirators can be either single-use or fitted with replaceable filters that could be used until their effective life has expired. These respirators may be nonpowered or powered air-purifying.

Vapor- and gas-removing respirators use canisters or cartridges filled with a sorbent that cleans some portion of the vapors or gases from contaminated air before it can enter the breathing zone of the worker.

Combination cartridges and canisters are available to protect against particulates, vapors, and gases.

Respirators can be furthered classified according to pressure relationships between the air within the facepiece and the ambient air outside the facepiece. When the pressure is normally positive with respect to ambient air pressure throughout the breathing cycle, the respirator is a positive-pressure respirator. If the air pressure inside the facepiece is negative with respect to the ambient air pressure, the respirator is a negative-pressure respirator. Exposure evaluations should examine the concept of negative and positive pressure when considering potential contaminant leakage into the respirator.

In addition, respirators fall into two general categories relating to "fit": tight-fitting and loose-fitting. The tight-fitting respirator is designed to form an airtight seal with the face of the wearer; the loose-fitting respirator has a respiratory inlet covering that forms a partial seal (not airtight) with the face. Tight-fitting respirators are available in three design configurations: quarter-mask, half-mask, and full-facepiece.

Loose-fitting respirators need to supply enough air to maintain a slight positive pressure inside the respirator in relation to the outside environment. A loose-fitting respirator has a respiratory inlet covering designed to form a partial seal with the face. These include loose-fitting facepieces, as well as hoods, helmets, blouses, or full suits, all of which cover the head completely. Because the hood is not tight-fitting, it is important that sufficient air is provided to maintain a slight positive pressure inside the hood relative to the environment immediately outside the hood. In this way, an outward flow of air from the respirator will prevent contaminants from entering the hood.

Air-Supplying Respirators

Air-supplying respirators provide a breathing gas (usually air) to the worker. The different types are classified according to (1) the method used to supply breathing gas and (2) the method used to regulate the gas supply.

Self-Contained Breathing Apparatus

An SCBA provides a transportable supply of breathing air and affords protection against both toxic chemicals and oxygen deficiency (Figure 7–12). The wearer carries enough air or oxygen for up to 4 h, depending on the design. All personnel engaged in interior structural fire fighting must use SCBA. SCBAs can be classified as "closed circuit" or "open circuit."

Figure 7–12. Self-contained breathing apparatus. *(Courtesy Mine Safety Appliances Company)*

Closed Circuit

Another name for the closed-circuit SCBA is a rebreather device. After the exhaled carbon dioxide has been removed and the oxygen content has been restored by a compressed or liquid oxygen source or an oxygen-generating solid, the gas can be breathed again. The devices are designed for 1- to 4-h use in oxygen-deficient or IDLH atmospheres that might exist during mine rescues or in confined spaces.

Open Circuit

An open-circuit SCBA exhausts air to the atmosphere instead of recirculating it. Compressed air is almost always the breathing gas used; compressed oxygen cannot be used in a device designed for compressed air because minute amounts of oil or other foreign matter in the device can cause an explosion. Mine safety regulations prohibit the certification of devices that permit interchangeable use of air and oxygen.

In an open-circuit SCBA, a cylinder of high-pressure compressed air (2,000 to 4,500 psi) supplies air to a regulator that reduces the pressure for delivery to the facepiece. The regulator is usually mounted on the facepiece or is connected by a hose to the respirator inlet. Most open-circuit SCBAs have a service life of 30 to 60 min. Open-circuit SCBAs are widely used in fire fighting, for industrial emergencies, and for hazardous-waste site work. Care must be exercised in these work environments because failure of the respirator to provide the appropriate protection may result in serious injury or death. Consequently, the employer must develop and implement specific procedures for the use of respirators in IDLH atmospheres that include the following provisions:

- At least one employee ("standby employee") should be located outside the IDLH atmosphere and maintain visual, voice, or signal line communication with the employee(s) in the IDLH atmosphere.
- Standby employee(s) must be equipped with pressure-demand or other positive-pressure SCBA, or a pressure-demand or other positive-pressure supplied-air respirator with auxiliary SCBA.
- Standby employee(s) must be equipped with appropriate retrieval equipment for lifting or removing any employee in need of such assistance from the hazardous atmosphere. When such retrieval equipment cannot be used because it would increase the overall risk resulting from entry, equivalent provisions for rescue need to be readily available.

NIOSH also certifies SCBAs with less than 30-min service times, usually for escape use only. Escape SCBAs are also certified in combination with supplied-air air-line respirators.

Two types of open-circuit SCBAs are available: "demand" or "pressure demand." In a demand or negative-pressure respirator, air at approximately 2,000 psi is supplied to the regulator through the main valve. A bypass valve passes air to the facepiece in case of regulator failure. Downstream from the main valve, a two-stage regulator reduces the pressure to approximately 50 to 100 psi.

Inhalation creates negative pressure in the facepiece, opening an admission valve and allowing air into the facepiece, but only on demand by the wearer. However, a demand SCBA offers no more protection than does an air-purifying respirator with the same facepiece. Therefore, a demand open-circuit SCBA should not be used in IDLH atmospheres.

A pressure-demand or positive-pressure regulator is designed to maintain positive pressure in the facepiece at all times. All pressure-demand devices have a special exhalation valve that maintains positive backpressure in the facepiece and opens only when the pressure exceeds that preset level (1.5 to 3.0 in. water pressure).

Under certain conditions of work, a momentary negative pressure may occur in the wearer's breathing zone, although the regulator still supplies additional air on demand. Because of positive pressure, any leakage should still be outward. A pressure-demand SCBA has the same service time as a demand device, if it seals well to the wearer's face. Some open-circuit SCBAs can be switched from demand to pressure-demand mode. However, the demand mode should be used only for donning and adjusting the apparatus in order to conserve air and should be switched to pressure demand for actual use.

The escape-only SCBA, certified by NIOSH and the Mine Safety and Health Administration (MSHA), is usually for short-duration use (3, 5, or 10 min), and is small in size and weight (Figure 7–13). The user wears a container of compressed air on the back hip with a readily accessible air valve. Hood or facepiece styles are available.

Supplied-Air Respirators

Supplied-air or air-line respirators are available in demand, pressure-demand, and continuous-flow models. In the past, these have been designated type C respirators. The air line may provide air to a facepiece, helmet, hood, or complete body suit (Figure 7–14). A demand or pressure-demand air-line respirator is very similar in operation to a demand or pressure-demand SCBA. Continuous-flow air-line respirators maintain air flow at all times. Regulations specify that a flow of 115 liters per minute (lpm) for a tight-fitting facepiece and 170 lpm for a loose-fitting hood or helmet must be maintained at the lowest air pressure and the longest hose length specified.

Some special valving is available with some certified air-line respirators. These incorporate vortex tubes that allow heating or cooling of the air delivered to the worker for comfort and prevention of heat or cold stress.

Air-line respirators must be used only in non-IDLH atmospheres or those from which the wearer can escape without the use of a respirator. This limitation is necessary because the air-line respirator is entirely dependent on an air supply not carried by the wearer. If this air supply fails, the wearer might not escape immediately from a hazardous atmosphere. Another limitation of air-line respirators is that the air hose limits the wearer to a fixed distance from the air-supply source.

The air supply for air-line respirators must meet or exceed grade D or higher quality, as set forth by the Compressed Gas Association Commodity Specification for Air, G–7.1. To protect air quality, air from compressors must be continually monitored for carbon monoxide, or the equipment must have a high temperature shutoff alarm to indicate when the compressor overheats. Overheating could introduce excess car-

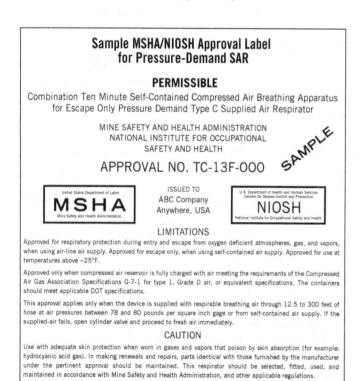

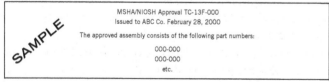

Figure 7–13. Sample MSHA/NIOSH approval label for escape-only pressure-demand self-contained compressed air breathing apparatus.

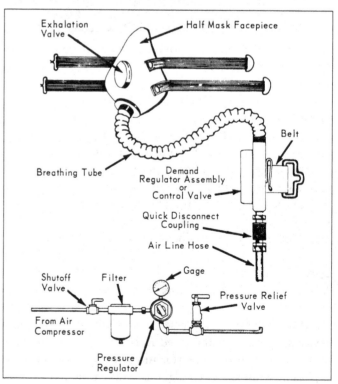

Figure 7–14. Air-line respirator parts and connections.

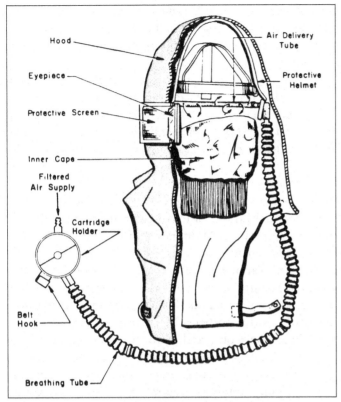

Figure 7–15. Diagram shows parts and connections for lightweight hood designs for use by persons doing abrasive blasting.

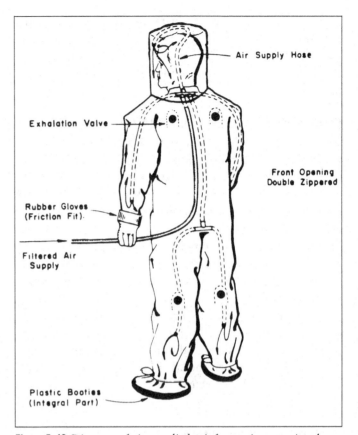

Figure 7–16. Diagram of air-supplied suit for use in corrosive chemical atmosphere.

bon monoxide into the air line. Also, where there is potential exposure to vapors or gases that poison by skin absorption, adequate skin protection must be used.

Air-Supplied Hoods

For some long-term operations that do not require a completely enclosed suit, an air-supplied hood may be used. These are particularly useful in hot, dusty environments. A vortex tube may also be used to reduce the ambient air temperature by up to 50°F (10°C). Respirable air under suitable pressure should be delivered to a hood at a volume of at least 6 ft^3/m (0.0028 m^3/s).

Abrasive Blasting Respirators

Abrasive blasting respirators are one type of air-supplied respirator. They are used to protect personnel engaged in shot, sand, or other abrasive blasting operations that involve air contaminated with high concentrations of rapidly moving abrasive particles. The requirements for abrasive blasting respirators are the same as those for an air-line respirator of the continuous-flow type, with the addition that mechanical protection from the abrasive particles is needed for the head and neck (Figure 7–15). NIOSH and MSHA staff test and certify such equipment.

Air-Supplied Suits

The most extreme condition requiring respiratory equipment is that in which rescue or emergency repair work must be done in atmospheres that are highly corrosive to the skin and mucous membranes as well as acutely poisonous and immediately hazardous to life, such as atmospheres containing vapors of ammonia, hydrofluoric acid, or hydrochloric acid. For these conditions, a complete suit of impervious clothing, with a respirable air supply, should be used (Figure 7–16).

Performance and design criteria for such products can be found in ASTM F2704-10, Standard Specification for Air-Fed Protective Ensembles. Suit material should have sufficient mechanical strength to resist rough handling and considerable abuse without tearing. The hose line supplying the air should be connected to the suit itself, as well as to the helmet. This is because wearing such a suit for a long time is not only extremely fatiguing but also dangerous—unless it is well ventilated.

Personal air-conditioning devices using a vortex tube are available for air-supplied suits or hoods. These cooling and heating devices are desirable to reduce fatigue where high ambient temperatures may be encountered (as in heat-protective clothing) or where body heat may build up (as under impermeable chemical protective clothing).

The vortex device works by taking an air stream under pressure and dividing it. One portion loses heat; the other gains heat. The cold portion passes into the suit or hood; the warm portion is vented to the atmosphere, or vice versa in cold weather.

Combination Supplied-Air SCBA Respirators

These respirators have an auxiliary air supply to protect workers against potential failure of the compressor. An air tank of 3-, 5-, or 10-min service time is typically used, mainly for emergency and escape for IDLH. The SCBA part is used only when the air line fails and the wearer must escape or when the worker disconnects the line temporarily to change locations. A combination air line and SCBA may be used for emergency entry into a hazardous atmosphere to connect the air line only if the device is rated for 15-min or longer service.

Air may be supplied to a respirator from a cylinder of compressed air or directly from an air compressor. High-pressure, low-pressure, and ambient-air supply systems are available. However, the latter cannot be used to supply pressure-demand respirators because it does not develop sufficient pressure.

Air-Purifying Respirators

Air-purifying respirators can purify the air of gases, vapors, and particulates, but do not supply clean breathing air (Figure 7–17). They must never be used in oxygen-deficient atmospheres. The useful life of the air-purifying device is limited by (1) the concentration of the air contaminant, (2) breathing demand of the wearer, and (3) removal capacity of the air-purifying medium (cartridge or filter). The air-purifying respirator has a facepiece and an attached cartridge that contains specific material needed against the contaminant. This equipment is classified as either a gas and vapor respirator or a particulate respirator.

Air-purifying respirators are available in three basic configurations. The quarter-face respirator covers the mouth and nose, with the lower edge resting between the chin and mouth. The half-face respirator fits over the nose and under the chin. It seals more reliably against the face and provides better protection against toxic materials. The full-face respirator covers the user from the hairline to below the chin and provides some eye protection. It is designed for use in higher concentrations of toxic materials than quarter- or half-face respirators. Mouthpiece-style respirators are also available for escape use only.

Gas and vapor respirators (also known as chemical-cartridge respirators) remove gases and/or vapors by passing the contaminated air through cartridges containing charcoal or other sorbents that trap these contaminants. Cartridges must be matched to the right contaminants (Table 7–D) and are used to protect against contaminants that have adequate warning properties of smell or irritation (not exceeding certain concentrations). This characteristic allows the wearer to judge when a cartridge is no longer usable. Some cartridges are dated as well and should not be used after the expiration date.

Figure 7–17. Full-face air-line respirator designed for use by employees working in the vicinity of hazardous emissions. *(Courtesy MSA)*

TABLE 7–D. Color Code for Cartridges and Gas Mask Canisters

Atmospheric Contaminants to Be Protected Against	Color Assigned
Acid gases	White
Organic vapors	Black
Ammonia gas	Green
Carbon monoxide gas	Blue
Acid gases and organic vapors	Yellow
Acid gases, ammonia, and organic vapors	Brown
Acid gases, ammonia, carbon monoxide, and organic vapors	Red
Other vapors and gases not listed above	Olive
Radioactive materials (except tritium and noble gases)	Purple
Dusts, fumes, and mists (other than radioactive materials)	Orange

Notes:
1. A purple stripe will be used to identify radioactive materials in combination with any vapor or gas.
2. An orange stripe should be used to identify dusts, fumes, and mists in combination with any vapor or gas.
3. Where labels only are colored to conform with this table, the canister or cartridge body should be gray or a metal canister or cartridge body may be left in its natural metallic color.
4. The user should refer to the wording of the label to determine the type and degree of protection the canister or cartridge will afford.

Four major rules apply to chemical-cartridge respirators:
1. They cannot be used for protection against gaseous material that is extremely toxic in very small concentrations.
2. The respirators should not be used for exposure to harmful gaseous matter that cannot clearly be detected by odor (e.g., methyl chloride and hydrogen sulfide). The former is odorless, and the latter, although foul smelling, paralyzes the olfactory nerves so quickly that detection by odor is unreliable. (Note: Odor should not be used as a primary indicator of sorbent exhaustion.)
3. Chemical-cartridge respirators should not be used against any gaseous material in concentrations highly irritating to the eyes without satisfactory eye protection (e.g., full-mask respirators).
4. Chemical-cartridge respirators do not provide protection against gaseous material that is not effectively stopped by the cartridge medium used, regardless of concentrations.

The second type of air-purifying respirator, the particulate respirator, is also known as a mechanical filter respirator. Depending on the design, the filters can screen out dust, fog, fumes, mist, spray, or smoke by passing the contaminated air through a pad or filter. These respirators consist of a facepiece with an attached mechanical filter, papers, or similar filter substance. Many types of filters are capable of trapping a range of airborne particle classes. Filters should be changed frequently, when they become clogged, or when it becomes difficult to breathe through them. Appropriate filters and cartridges can be used together where combinations of contaminants exist.

NIOSH's regulation 42 CFR 84 on nonpowered air-purifying particulate respirators identifies nine classes of filters based on levels of filter efficiency and resistance to filter efficiency degradation. Three levels of filter efficiency are 95, 99, and 99.97%. The three categories of resistance to filter efficiency degradation depend on the presence or absence of oil particles and are labeled N (not resistant to oil), R (resistant to oil), and P (oil-proof). The class of filter will be clearly marked on the filter, filter package, or respirator box. For example, a filter classified as N99 would mean an N-series filter that is at least 99% efficient. For chemical cartridges that include particulate filter elements (combination), a similar marking pertains only to the particulate filter element.

The classes of nonpowered particulate respirators require following a decision logic for selection of the proper respirator. A synopsis of the selection process for using the new particulate classification is outlined as follows:
• If no oil particles are present, use a filter of any series (i.e., N-, R-, or P-series).

• If oil particles (e.g., lubricants, cutting fluids, etc.) are present, use an R- or P-series filter.
• If oil particles are present and the filter is to be used for more than one work shift, use only a P-series filter.

Selection of filter efficiency (i.e., 95, 99, or 99.97%) depends on how much filter leakage can be accepted. Higher filter efficiency means lower filter leakage and a higher degree of protection. The choice of facepiece depends on the level of protection needed—that is, the assigned protection factor (APF) discussed earlier.

The flow chart in Figure 7–10 can be used as a guide for selection of particulate respirators.

Powered air-purifying respirators use a blower both to pass contaminated air through a filter or sorbent bed that removes the contaminant and to supply purified air to a facepiece, helmet, or hood. The purifying device may be a filter, a cartridge, or a combination of the two. A blower, usually worn on the worker's belt, is used to force the contaminated air through the element and to the respirator facepiece. The unit supplies clean air to the worker at positive pressure, so that contaminated air does not leak into the facepiece.

Gas Masks
Gas masks have been used effectively for many years for respiratory protection against certain gases, vapors, and particulate matter that otherwise might be harmful to life or health. Gas masks are air-purifying devices designed solely to remove specific contaminants from the air; therefore, it is essential that their use be restricted to atmospheres that contain sufficient oxygen to support life. Gas masks may be used only for escape from IDLH atmospheres, never for entry into such environments. Users must assess the exposure conditions carefully before selecting a specific mask for respiratory protection. If the specific exposure concentrations are suspected of exceeding established limits, only SCBA should be used.

From a practical standpoint, gas masks are generally suitable for ventilated areas not subjected to rapid change in air-contaminant levels. They should never be used in confined spaces below- or aboveground where oxygen deficiency and high gas concentrations may occur. In assessing exposure conditions, workers should remember that oxygen deficiency can occur in a confined space through the displacement of air by other gases or vapors or by means of processes (such as fire, rusting, and aerobic bacteria) that consume oxygen.

Fitting Respirators
Required fit tests must be performed before an employee uses a respirator in the workplace. Fit-testing is required for all employees fulfilling any of the following criteria:

- using negative- or positive- pressure tight-fitting respirators
- where such respirators are required by OSHA
- where the employer requires the use of such a respirator.

Fit tests must be repeated under the following guidelines:
- at least annually
- whenever a different respirator facepiece is used
- whenever a change in the employee's physical condition could affect respirator fit.

A fit test is not required for voluntary users or for escape-only respirators.

Facial hair, jewelry, corrective glasses or goggles, or other personal protective equipment must not interfere with the seal of the facepiece of tight-fitting respirators.

Air-supplied masks should also be fit-tested for appropriate size. Some facilities fit-test workers who wear supplied-air respirators with tight-fitting facepieces. Determination of facepiece fit should involve both qualitative and quantitative tests. Prior to fit-testing, the employee, together with the person administering the fit test, should check the comfort of the respirator facepiece. Make sure there is adequate time devoted to checking these items:
- proper placement on chin
- proper positioning of facepiece on nose
- comfortable strap tension
- comfortable fit across nose bridge
- ability to talk while wearing facepiece
- room for safety spectacles where required
- tendency of facepiece to slip.

Qualitative Tests
In the irritant or odorous chemical agent test, the wearer is exposed to an irritant smoke, isoamyl acetate vapor, or other suitable test agent easily detectable by irritation, odor, or taste. An air-purifying respirator must be equipped with the appropriate air-purifying element. If the wearer cannot detect any penetration of the test agent, the respirator is probably tight enough.

The advantages of a qualitative test are speed, convenience, and ease of performing the test. However, these tests rely on the wearer's subjective response, so they may not be entirely reliable.

Quantitative Tests
In quantitative testing, the employee, wearing a specially designed probed respirator, stands in a test chamber and is exposed to a test atmosphere of a nontoxic, easily detectable aerosol, vapor, or gaseous test agent. Instrumentation is used to measure the leakage into the respirator.

Protection factors can be determined from quantitative fit tests by dividing the ambient airborne concentration of the challenge contaminant by the concentration inside the facepiece. For example, if the concentration outside the facepiece is 500 parts per million (ppm), and the concentration inside the respirator is 10 ppm, the protection factor would be 50. Protection factors are used in the selection process to determine the maximum use concentration (muc), which is determined by multiplying the Threshold Limit Value (TLV) or permissible exposure level (PEL) by the protection factor.

The greatest advantage of a quantitative test is that it does not rely on a subjective response. The quantitative test is recommended when facepiece leakage must be minimized for work in highly toxic or IDLH atmospheres.

Quantitative fitting tests require expensive equipment that can be operated only by trained personnel. Because each test respirator must be equipped with a sampling probe to allow removal of a continuous air sample from the facepiece, the same facepiece cannot be worn in actual service.

Daily Fit Test
Employees who wear respirators should check the fit of their respirator each time they don it, with both negative- and positive-pressure tests as described next.

Negative-Pressure Test. The wearer can perform this test alone in the field and should use it before entering any toxic atmosphere. The test consists of closing off the inlet of the canister, cartridges, or filters by covering with the palms or replacing the seals, or of squeezing the breathing tube so that it does not pass air; inhaling gently so that the facepiece collapses slightly; and holding the breath for 10 s. If the facepiece remains slightly collapsed and no inward leakage is detected, the respirator is probably tight enough. This test, of course, can be used only on respirators with tight-fitting facepieces (Figure 7–18a).

Positive-Pressure Test. This test is similar to the negative-pressure test and has the same advantages and limitations. It is conducted by closing off the exhalation valve and exhaling gently into the facepiece (Figure 7–18b). The fit is considered satisfactory if slight positive pressure can be built up inside the facepiece without any evidence of outward leakage. The test is easy for respirators whose valve cover has a single small port that can be closed by the palm or a finger. The wearer should perform this test before entering any hazardous environment.

For some respirators, this method requires the wearer to remove the exhalation valve cover, which often disturbs the respirator fit more than the negative-pressure test does. Therefore, this test should be used cautiously if it requires removing and replacing a valve cover.

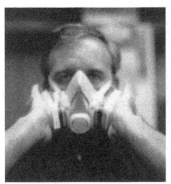 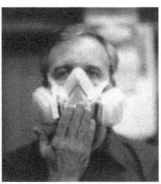

Figure 7–18. (left) A negative-pressure test; (right) a positive-pressure test.

Storage of Respirators

Respirators should be stored to protect them from dust, sunlight, heat, extreme cold, excessive moisture, and damaging chemicals. Unprotected respirators can sustain damaged parts or facepiece distortion that makes them ineffective.

Before storing the respirator, clean or wash the device according to the manufacturer's instructions. Wiping a respirator with a cloth is not acceptable practice because fibers from the cloth can be deposited on the respirator's surface. Workers should never store the respirator with folds or creases and should never hang it by the elastic headband or place it in a position that will stretch the facepiece.

Because heat, air, light, and oil cause most rubbers to deteriorate, this equipment should be kept in a cool, dry place and protected from light and air as much as possible. Today, many respirators are made of silicone rubber, which is more pliable and tends to become less hard with age. Many respirators come with their own plastic or metal cases. The equipment should not be stored in toolboxes or left on workbenches where it may be exposed to dust and damage by oil or other harmful materials.

After cleaning and inspection, place respirators in individual, sealable plastic bags. Store them in one layer with the facepiece and exhaust valve in normal position. Respirators should not be kept in lockers unless they can be protected from contamination, distortion, and damage.

Standard steel storage cabinets or steel wall-mounted cabinets with compartments are the best choice for storing air-purifying respirators. Special storage cabinets may be purchased from the manufacturer for SCBAs. The cabinets should be located in uncontaminated but readily accessible areas.

Maintenance of Respirators

The ongoing maintenance of the respirators themselves is an essential part of the respiratory protection program. If the equipment malfunctions because of poor maintenance, the employee may be exposed to a potentially fatal hazard.

The maintenance program should incorporate the manufacturer's instructions and should include provisions for disassembly, including the removal of the respirator's purifying elements, cleaning, sanitizing, inspecting for defects, repairing parts (if necessary), installing purifying elements, reassembling, packaging, and storing equipment.

The air-purifying elements (chemical cartridges or filters) should not be cleaned or exposed to excess moisture, including high humidity. Discard the elements if there is any question about their condition (Figure 7–19).

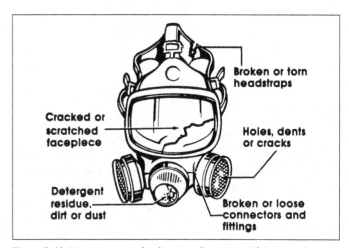

Figure 7–19. Maintenance checkpoints for air-purifying respirators.

Supervisors should be responsible for conducting daily equipment inspections, particularly of functional parts such as exhalation valves and filter elements. They should see that the edges of the valves are not curled and that valve seats are smooth and clean. Inhalation and exhalation valves should be replaced periodically. In addition, users should be trained to inspect their respirators before and after each use.

Besides the daily check, weekly inspections and inspections before and after each use, conducted by trained personnel, can ensure that respirators remain in top working order. The inspector should stretch rubber parts slightly to check for fine cracks. Users need to work the rubber every so often to prevent it from setting (becoming rigid, a cause of cracking). Inspectors need to check the headband to be sure the wearer has not stretched it in securing a snug fit. Inspectors should check emergency respirators at least monthly and keep a written record of the results.

Sometimes, in an effort to reduce resistance to breathing, workers remove exhalation valves or punch holes in the filter, the rubber facepiece, or other parts. Management should investigate the reasons for this mistreatment and correct them. For instance, it may be that, in the interest of economy, filters are allowed to become completely plugged before being replaced.

Cleaning and Sanitizing

Make sure that each time an employee uses a respirator, it is clean and sanitized. Actual cleaning may be done in a variety of ways.

The respiratory protection equipment should be dismantled and washed with whatever cleaner the manufacturer recommends in warm water using a brush, thoroughly rinsed in clean water, and then air dried in a clean place. Care should be taken to prevent damage from rough handling. This method is an accepted procedure for a small respirator program or where each worker cleans his or her own respirator.

A standard domestic clothes washer may be used if a rack is installed to hold the facepieces in a fixed position. (If the facepieces are placed loose in a washer, the agitator may damage them.) This method is especially valuable in large programs where respirators are used extensively.

Workers should follow the manufacturer's instructions regarding which cleaners to use. However, they should not use organic solvents, as these can deteriorate the elastomeric (rubber or silicone) facepiece. If bactericidal detergents are not available, use a disinfectant. Check with the manufacturer for disinfectants that will not damage the respirators.

Be sure that the cleaned and sanitized respirators are rinsed thoroughly in clean water no hotter than 120°F (50°C) to remove all traces of detergents and cleaners. Otherwise, skin irritation or dermatitis may result when the employee wears the respirator. Allow the respirators to air dry by themselves on a clean surface.

They may also be hung carefully on a line or in a specially designed drying cabinet. If management is unwilling or unable to run the maintenance program, the firm can contract with an outside service to maintain respirators in peak condition.

Inspection of Respirators

After cleaning and sanitizing, each respirator should be reassembled and inspected for proper working condition and repair or replacement of parts. The respirator should also be inspected routinely by the user immediately before each use to ensure that it is in proper working condition. An inspection checklist compiled by NIOSH recommends that management inspect disposable respirators for:
- holes in the filter (obtain new disposable respirator)
- elasticity and deterioration of straps, (replace straps, contact manufacturer)
- deterioration of metal nose clip, if applicable (obtain new disposable respirator).

Air-purifying respirators (including quarter-mask, half-mask, full-facepiece, and gas mask) should be checked for the following items:
- facepiece
  - excessive dirt (clean all dirt from facepiece)
  - cracks, tears, or holes (obtain new facepiece)
  - distortion (allow facepiece to sit free from any constraints and see if distortion disappears; if not, obtain new facepiece)
  - cracked, scratched, or loose-fitting lenses (contact respirator manufacturer to see if replacement is possible; otherwise, obtain new facepiece)
- headstraps
  - breaks or tears (replace headstraps)
  - loss of elasticity (replace headstraps)
  - broken or malfunctioning buckles or attachments (obtain new buckles)
  - excessively worn serrations on the head harness that might allow the facepiece to slip (replace headstrap)
- inhalation and exhalation valves
  - detergent residue, dust particles, or dirt on valve or valve seat (clean residue with soap and water)
  - cracks, tears, or distortion in the valve material or valve seat (contact manufacturer for instructions)
  - missing or defective valve cover (obtain valve cover from manufacturer)
- filter elements
  - proper filter for the hazard
  - missing or worn gaskets (contact manufacturer for replacement)
  - worn threads—both filter threads and facepiece threads (replace filter or facepiece, whichever is applicable)
  - cracks or dents in filter housing (replace filter)
  - deterioration of gas mask canister harness (replace harness)
  - service-life indicator, expiration date, or end-of-service date
- gas mask
  - cracks or holes (replace tube)
  - missing or loose hose clamps (obtain new connectors)
  - service-life indicator on canister (or contact manufacturer to find out what indicates the end-of-service date for the canister).

Air-supplying respirators should be checked for:
- hood, helmet, blouse, or full suit (if applicable)
  - rips and torn seams (if unable to repair the tear adequately, replace)
  - headgear suspension (adjust properly for wearer)
  - cracks or breaks in face shield (replace face shield)
  - protective screen to see that it is intact and fits cor-

rectly over the face shield, abrasive blasting hoods, and blouses (obtain new screen)
- air-supply system
  - breathing air quality
  - low-pressure alarm
  - breaks or kinks in air-supply hoses and end-fitting attachments (replace hose and/or fitting)
  - tightness of connections
  - proper setting of regulators and valves (consult manufacturer's recommendations)
  - correct operation of air-purifying elements
  - proper operation of carbon monoxide alarms or high-temperature alarms
- self-contained breathing apparatus (SCBA)
  - facepiece, headstraps, valves, and breathing tube inspection checks are same as for air-purifying respirators (consult manufacturer's literature).

In some companies, maintenance service for respirators and for other kinds of PPE can be effectively provided by traveling service carts. When a number of respirators are used regularly, the organization may set up a central station for their care and maintenance and storage, along with the care and maintenance of other items of PPE.

Some facilities have found that if two respirators are assigned to each worker, equipment lasts more than twice as long. This plan works best when users cannot clean their respirators between shifts or before the next scheduled shift.

Under such a plan, users turn in marked respirators on a set schedule—depending on use—for cleaning, disinfection, inspection, and repair. Workers wear the second until the first can be serviced and returned.

Another plan involves keeping quantities of disinfected respirators on hand for use as needed. Although this plan works well where individual needs vary, users do not readily accept the responsibility of looking after the equipment. Also, if the same respirator is worn by several persons, it must be cleaned and disinfected after each use. When not in use, the respirator must be stored in accordance with manufacturer's recommendations.

Respirators should be marked to indicate to whom they are assigned. Identification should be made in some form of permanent ink or paint so that workers cannot change the marking inadvertently or without effort. Medical evaluation must be conducted to ensure that employees are physically able to wear respirators before training commences.

Training

Once the right respirator has been selected, the wearer must be trained in its proper use and care. This step is important for every type of respirator. Each user should not only be trained when first acquiring the equipment but also be retrained periodically. Training sessions should include the following:

- reasons for respiratory protection and explanation of why other controls and methods are not being used and what efforts have been made to reduce the hazards
- explanation of the respirator selection procedure used by the safety and health professional, including identification and evaluation of specific airborne hazards
- proper fitting, donning, wearing, and removing of the respirator; importance of not modifying the respirator in any way that will impair or void its protective features
- limitations, capabilities, and operation of the respirator
- proper maintenance and storage procedures
- wearing the respirator in a safe atmosphere to allow the user to become familiar with its characteristics
- wearing the respirator in a test atmosphere under close supervision of the trainer to allow the wearer to simulate work activities and detect respirator leakage or malfunction
- recognizing and coping with emergency situations
- instructions for special use as needed
- explanation of any regulations governing the use of respirators.

The instructor should be a qualified person, such as an industrial hygienist, safety professional, nurse, or the respirator manufacturer's representative.

Medical Surveillance

Employers must provide a medical evaluation to determine each employee's fitness to wear a respirator before initial fit-testing and prior to using a respirator for the first time. Medical evaluations consist of the administration of a medical questionnaire that can be found in the mandatory Appendix C of 29 CFR 1910.134, or provision of a physical examination that elicits the same information as the questionnaire for the employee. An employer who opts to provide physical examinations to his or her employees need not also administer the medical questionnaire. These evaluations are required for all respirator users, except for employees who voluntarily use dust masks and for those individuals using escape-only respirators. SCBAs are not considered escape-only respirators. Employees who refuse to be medically evaluated cannot be assigned to work in areas where they are required to wear a respirator.

HAND AND ARM PROTECTION

No one type of PPE for the extremities is suitable for the many different work situations involved in any business or industrial operation, from the laboratory to the loading dock. Thus, management and workers must select proper protection for the hands, fingers, arms, and skin based on potential exposure to identified hazards. The specific type of protection and its material depend on the type of material being handled and the work atmosphere.

Gloves

The material to be used for gloves largely depends on what is being handled. For most light work not involving exposure to hazardous materials or microbial contamination, a cotton glove is satisfactory and inexpensive. Rough or abrasive material requires leather gloves or leather reinforced with metal stitching for safe handling (Figure 7–20). Leather reinforced by metal stitching or metal mesh or highly cut-resistant plastic gloves also provide good protection from edged tools, as in butchering and similar occupations. Double gloving affords added protection. If the outer glove starts to degrade or tears, the inner glove may offer protection until the gloves are removed and replaced. It is a good idea to check the outer glove frequently, watching for signs of deterioration (color, texture change, holes, etc.) and re-glove as necessary.

Figure 7–20. These leather or leather-reinforced work gloves protect against sharp or abrasive surfaces. *(Courtesy Ansell Edmont Industrial Inc.)*

Many plastic and plastic-coated gloves are available in materials such as neoprene, latex, and nitrile. They are designed for a variety of occupational tasks such as chemical handling and construction brick and wire carrying. Management must give careful consideration to actual permeation tests of these gloves against hazardous chemicals. Some plastic models surpass leather in durability and effective shielding. Other types have granules or rough materials incorporated into the plastic for better gripping ability, while still others are disposable. Computer software is now available to help select the appropriate glove material. Most manufacturers can provide information regarding the rate and degree of chemical permeation through their glove materials.

Exposure to proteins from the use of natural rubber latex (NRL) products may result in adverse responses in susceptible workers. These could include irritation and several types of allergic reactions. Recommended strategies for risk reduction include minimizing unnecessary exposure to NRL proteins for all workers. For example, workers in food service or landscaping industries do not need to use NRL gloves for food-handling or gardening purposes.

OSHA has made the following recommendations concerning the use of NRL gloves:

- If selecting NRL gloves for worker use, designate NRL as a choice only in those situations requiring protection from infectious agents.
- When selecting NRL gloves, choose those with a lower protein content. Selecting powder-free gloves offers the additional benefit of reducing systemic allergic responses.
- Provide alternative suitable non-NRL gloves as choices for workers who are allergic to NRL gloves.

Prudent risk reduction strategies include risk surveys, mechanisms for reporting and managing cases, safe zones (nonuse areas for products containing NRL proteins), and development of policies and procedures for reducing the risk of NRL allergies in the workplace found through an initial survey.

When workers do not need complete gloves, they can wear finger stalls or cots. These are available in combinations of one or more fingers and usually are made of rubber, duck, leather, plastic, and metal mesh. The construction of the cot depends on the degree or type of hazard to which the worker is exposed. Employees should not use gloves while working on moving machinery such as drills, saws, grinders, or other rotating and moving equipment. Machine parts might catch the glove and pull it and the worker's hand into hazardous areas.

In addition to gloves, workers also can wear mittens (including one-finger and reversible types), pads, thumb guards, finger cots, wrist and forearm protectors, elbow guards, sleeves, and capes. These protective devices are made in a wide range of materials and lengths.

Gloves or mittens having metal parts or reinforcements should never be used around electrical apparatus. Work on energized or high-voltage electric equipment requires specially made and tested rubber gloves. Workers should wear over-gloves of leather to protect the rubber gloves against wire punctures and cuts and to protect the rubber in the event of electrical flashes. Conduct frequent tests and inspections of line workers' rubber gloves and discard those that fail to meet original specifications.

Hand Leathers and Arm Protectors

For jobs requiring protection from heat or from extremely abrasive or splintery material (such as rough lumber), hand leathers or hand pads are likely to be more satisfactory than gloves. This is primarily because they can be made heavier and less flexible without discomfort.

Because hand leathers or pads are used mainly for handling heavy materials, they should not be used around moving machinery. They must always be sufficiently loose to allow workers to slip their hands and fingers out of the device if it is caught on a rough edge or nail.

For protection against heat, hand and arm protectors should be made of wool, terry, or glass fiber. Although leather can be used, it will not withstand temperatures greater than 150°F (65°C). Wristlets or arm protectors may be obtained in any of the materials of which gloves are made.

Impervious Clothing

For protection against dusts, vapors, and moisture of hazardous substances and corrosive liquids, the safety and health professional can choose from among many types of impervious or impermeable materials. These are fabricated into clothing of all descriptions, depending on the hazards involved. They range from aprons and bibs of sheet plastic to suits that enclose the body from head to foot and contain their own air supply.

Materials used include natural rubber, olefin, synthetic rubber, neoprene, vinyl, polypropylene, and polyethylene films and fabrics coated with these substances. Natural rubber is not suited for use with oils, greases, and many organic solvents and chemicals because it deteriorates over time. Make sure that the clothing selected will protect against the hazards involved and that it has been field-tested prior to actual use. For example, some synthetic fabrics used for regular work clothing in chemical facilities, where daily contact with acids and caustic solutions would cause rapid deterioration of regular cotton clothes, are not really impervious. Such clothing cannot be used where impervious materials are indicated.

Gloves coated with synthetic rubber, synthetic elastomers, polyvinyl chloride, or other plastics offer protection against all types of petroleum products, caustic soda, tannic acid, muriatic and hydrochloric acid, and even sulfuric acid (Figure 7–21). These gloves are available in varying degrees of strength to meet individual conditions.

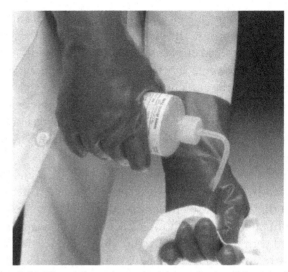

Figure 7–21. These gloves coated with synthetic rubber or plastic protect the worker from chemicals and solvents. *(Courtesy Ansell Edmont Industrial)*

Gloves should be long enough to come well above the wrists, leaving no gaps between the glove and the coat or shirtsleeve. Long, flaring gauntlets should be avoided unless they are equipped with closing snaps or straps to ensure a snug fit around the wrist.

Such gauntlets offer the best protection when acids and other chemicals are being poured. If the chemicals splash, workers can receive serious acid burns unless precautions are taken. When pouring caustic substances and harmful solvents from large to small containers, workers should wear their sleeves outside the gauntlets. This prevents any spilled chemical from running down into the protective device.

In many operations, rubber gloves with extra long cuffs have been used to advantage. The cuffs of these gloves are made with a heavy ridge near the top edge, which, when turned back, forms a trough to catch liquids and prevent them from running down the wrist or forearm. Some are made with beads near the cuff to hold inserts that form a liquid-tight seal when inserted into a sleeve.

Where acid may splash, rubber boots or rubber shoes also should be worn. Never tuck trousers into boots when working with corrosive materials. If workers wear safety shoes inside their boots, the legs of impervious trousers should cover the tops of the shoes. These precautions keep the liquid from draining off aprons or trousers into the footwear.

Decontamination Considerations

After PPE is used in a corrosive atmosphere, management must establish a strict procedure for disposing of the equipment to prevent workers from coming into contact with contaminated parts. Before the equipment is removed, it should be thoroughly washed with a hose stream whether or not it has come in contact with the corrosive chemical. Boots, coats, aprons, and hats should then be taken off, followed by removal of the gloves. This is the logical order of removal if the coat has been properly put on with the sleeves outside the cuffs of the gloves.

Workers should wash their hands thoroughly before removing their face shields and goggles, then wash their hands and face again. Ideally, a complete shower and change of clothing are far more desirable.

Ensure potentially contaminated gloves are laundered thoroughly before permitting reuse or reissue. Effective laundering can be used to sanitize the fabric and prolong glove life. Gloves used in toxic chemical service must be cleaned with special care to be sure the chemicals are thoroughly removed. Gloves that have been permeated by toxic material cannot be reused.

For protection against exposure to oil and the various other compounds that rapidly attack ordinary rubber, all the equipment discussed can be obtained in plastic and synthetic rubbers.

PROTECTIVE FOOTWEAR

Specifications for protective footwear are contained in ASTM F2413-11, Standard Specification for Performance Requirements for Protective (Safety) Toe Cap Footwear. Protective footwear is classified according to its ability to meet both the requirements for compression resistance and impact resistance (see Table 7–E). All protective footwear meeting the ASTM F2413 standard contains a protective toe box. Steel, reinforced plastic, and hard rubber are materials commonly used to make protective toe boxes. For protection in wet conditions, rubber footwear is also available with toe protection.

TABLE 7–E. Minimum Requirements for Protective Footwear
Impact
1/75 = 75 ft lbf (101.7J)
1/50 = 50 ft lbf (67.8J)
1/30 = 30 ft lbf (40.7J)
Compression
C/75 = 2,500 lb (11 121 N)
C/50 = 1,750 (7 784 N)
C/30 = 1,000 (4 4448 N)
Clearance (All Classifications)
Men—16.32 in. (12.7 mm)
Women—15.32 in. (11.9 mm)

Regulations such as OSHA standards 29 CFR 1910.132, General Requirements for Personal Protective Equipment, and 29 CFR 1910.136, Foot Protection, contain hazard assessment and foot protection requirements for employees whose work presents hazards to their feet. These hazards include objects falling onto or placed on the foot, objects rolling over the foot, sharp objects penetrating the sole of the footwear, static electricity buildup, and contact with energized electrical conductors.

Although comfort and proper fit are important factors for any footwear, it is particularly essential for protective footwear in order to encourage employee use. Management should carefully select foot protection to match the specific hazards faced by the wearer; educate employees on the need for such protection; and train workers in the proper use, care, and replacement of footwear.

Companies may arrange to have retail footwear providers supply onsite footwear sales, fitting, and training to employees through the use of portable shoemobiles.

In addition to the ASTM F2413 general requirement for impact- and compression-resistant protective footwear, the standard also describes additional protective options available, including metatarsal footwear, conductive footwear, electrical hazard footwear, static dissipative footwear, and sole puncture resistant footwear.

Metatarsal Footwear

This footwear is designed to prevent or reduce the severity of injury to the metatarsal bones and toe areas of the foot. Metatarsal footwear meeting the ASTM F2413 standard should have both a toe impact and compression rating and an integral metatarsal guard with an impact rating of 50 or 75.

Heavy-gauge, flanged, and corrugated sheet metal over the foot guards is also available for certain industrial applications. However, no standards have been developed for this equipment.

Conductive Footwear

Conductive protective toe footwear is intended to protect employees from the hazards of static electricity buildup and help equalize the electrical potential between the wearer and energized high-voltage power lines in the wearer's immediate area.

Electrical Hazard Footwear

Protective toe electrical hazard footwear is intended to protect workers against contact with exposed circuits of up to 100 V AC/750 V DC under dry conditions. This type of footwear is not intended for use in explosive or hazardous locations where conductive footwear is required. Electrical hazard footwear is intended to provide secondary electrical hazard protection on surfaces that are substantially insulated. Because the electrical insulative quality of the heel and sole provides the protection in this footwear, no metal parts should be used in these parts of the shoes.

Static Dissipative Footwear

Protective toe static dissipative footwear is designed to reduce the accumulation of excess static electricity in the body by conducting the charge to ground while maintaining a sufficiently high level of resistance to protect the wearer from electrical hazards.

Sole Puncture Resistant Footwear

Protective toe footwear with sole puncture resistance reduces the risk of puncture wounds caused by sharp objects penetrating the sole of the footwear. A protective shield inserted in the sole of the footwear covers an area from the toes to overlap the crest of the heel. This shield must be flexible yet strong enough to withstand at least a 270-lb force administered by a steel test pin in accordance with ASTM F2413 test procedures.

Foundry Footwear

Specialty types of footwear are available to protect workers in the smelting and foundry industries, where employees' feet can be exposed to molten metals. Such footwear often has no fasteners so they can be removed quickly in an emergency. In these occupations, the tops of the footwear should be covered by protective spats, leggings, and other devices that can prevent the entry of molten metal.

Other Features of Protective Footwear

Other protective footwear features not covered by standards are made available by various manufacturers. They include waterproofing, chemical resistance, and insulation against thermal extremes. Nearly all footwear manufacturers offer some variation on materials used in the soles and upper parts of the footwear for purposes of lengthening the life of the footwear and providing sole slip resistance under various field conditions. Manufacturers of protective footwear should be consulted about the products they offer that will best protect against the hazards that the wearer will encounter on the job.

Cleaning Rubber Boots

If rubber boots are reused by people on the next shift or job, great care should be exercised to disinfect boots after each shift or job. First, the boots are washed inside and outside with a hose containing water under pressure. Then they are dipped into a tub containing a solution of 1 part sodium hypochlorite and 19 parts water. The hose is used again for rinsing, after which the boots are ready for drying. Although other disinfecting agents can be used, this one has been satisfactory and is easily obtainable.

One method is to use a drying rack consisting of a tank with low-pressure steam coils and upright steel pipe boot holders that permit circulation of hot air inside the boots. After the boots are washed thoroughly and dipped in the disinfecting solution, they are completely dried in about 12 min. When several pairs of boots need cleaning, the rack with water jets can be rearranged so that many boots can be washed and rinsed at one time.

SPECIAL WORK CLOTHING

In the modern industrial environment, exposure to fire, extreme heat, molten metal, corrosive chemicals, cold temperature, body impact, cuts from handling materials, and other specialized hazards is often part of what is known as "job exposure." Special protective clothing helps to minimize the effects of these hazards, and a variety of products and materials are available to address these hazards.

Protection against Heat and Hot Metal

Leather clothing is one of the more common forms of body protection against heat and splashes of hot metal. It also provides protection against limited impact forces and infrared and UV radiation.

Garments should be made of good-quality leather, solidly constructed, and provided with fastenings to prevent gaping during body movement. Fastenings should be so designed that the wearer can remove the garment rapidly and easily. Workers should not wear turned-up cuffs or other items of clothing that can catch and hold hot metal. Garments should either have no pockets or have pockets with flaps that can be fastened shut.

For ordinary protection against hot metal, radiant heat, or flame hazards somewhat stronger than those in welding operations, wool and leather clothing is used. Specially treated clothing has been developed that is impervious to splashes of metal up to 3,000°F (1,650°C). Wool garment requirements are, in general, the same as those for leather, except that metal fastenings should be covered with flaps to keep them from becoming dangerously hot.

Asbestos substitutes, including fiberglass or other special high-temperature-resistant materials, are available. These materials are effective when made into leggings and aprons usually worn by foundry personnel working with molten metal. Such leggings should completely encircle the leg from knee to ankle, with a flare at the bottom to cover the instep. The design of the leggings should permit rapid removal in emergencies.

If the front part of the legging is reinforced, it can provide impact protection when required. Fiberboard is the most common material used for reinforcement.

When people must work in extremely high temperatures up to 2,000°F (1,090°C)—as in furnace and oven repair, coking, slagging, fire fighting, and fire rescue work—aluminized fabrics are essential. The aluminized coating reflects most of the radiant heat away, while the underlying material insulates workers against the remaining heat. Some of these suits consist of separate trousers, coats, gloves, boots, and hoods. Others are one piece from head to foot. Some suits used in industrial operations are air-fed to reduce heat and increase comfort.

Aluminized heat-resistant clothing generally falls into two classes: emergency and fire proximity suits.

Emergency suits are used when temperatures exceed 1,000°F (540°C), as in a kiln or furnace, or when workers must move through burning areas for fire-fighting or rescue operations. These suits are made of aluminized glass fiber with layers of quilted glass fibers and a wool lining on the inside.

Fire proximity suits are used in areas near high-temperature operations, such as slagging, coking, furnace repair work with hot ingots, and fire fighting where workers do not enter the flame area. These suits are seldom one-piece construction. They depend primarily on the reflective ability of an aluminized coating on a base cloth of glass fiber or synthetic fiber. Remember, never use fire proximity clothing for situations in which workers are required to enter a fire.

Flame-Retardant Work Clothes

Cotton work clothing can be protected against flame or small sparks by flame-proofing. One available commercial flame-retardant preparation can be applied to work clothing in ordinary laundry machinery after the garment is washed. Treating the material has two advantages: it makes the cloth highly flame resistant and adds little to the material's weight or stiffness.

Durable flame-retardant work clothes are readily available from many manufacturers. A high-temperature-resistant nylon fabric that chars rather than melts is available for the most severe work conditions. Mod-acrylic fabrics that resemble cotton are lightweight and have permanent fire-retardant properties. Flame-proofed clothing should be marked or otherwise distinguished to reduce the chance that workers will use untreated garments by mistake.

Protection against Impact and Cuts

A worker's body needs protection from cuts, bruises, and abrasions on most jobs where heavy, sharp, or rough material is handled. Special protectors have been developed for almost all parts of the body and are available from suppliers of safety equipment. (See also the section Hand and Arm Protection earlier in this chapter.)

For example, pads made of cushioned or padded duck will protect the shoulders and back from bruises when workers carry heavy loads or objects with rough edges. Aprons of padded leather, fabric, plastic, hard fiber, or metal can protect the abdomen against blows. Similar devices of metal, hard fiber, or leather with metal reinforcements shield the worker against sharp blows with edged tools. For jobs requiring ease of movement, workers can split their aprons or equip them with fasteners so they fit snugly around the legs.

Leg protection is required on many jobs. Guards of hard fiber or metal are widely used to protect the shins against impact. Knee pads should be worn by mold loftsmen and others whose task requires continual kneeling.

Heat Stress

When selecting PPE, management must keep in mind that some types of equipment may contribute to the potential for heat stress. (See Chapter 12, Thermal Stress, in the NSC's *Fundamentals of Industrial Hygiene*, 6th ed. [Plog 2012], for further information about heat stress.) Workers who wear SCBAs in combination with full-body impervious suits, such as those worn on hazardous-waste sites, are particularly vulnerable to heat stress. Such workers must be allowed adequate rest breaks. The employer should ensure employees are adequately hydrated and acclimatized to the heat. If necessary, cooling vortexes or vests should be supplied.

Cold-Weather Clothing

In recent years, thermal insulating underwear has become popular among outdoor workers because of its lightweight protection against the cold. Thermal knit cotton patterned after regular underwear, quilted materials, and synthetic polyester fabric quilted between layers of nylon are common types of construction.

Although polyester- and nylon-quilted material does not catch fire any more easily than does cotton, once the synthetic material starts burning, it melts, forming a hot plastic mass, not unlike hot pitch, that adheres to skin and causes serious burns. Fire-retardant quilted insulating underwear is now available to combat this danger. Other special fabrics available include a nylon material that chars at a relatively high temperature and does not melt, a glass fiber material for special uses, and a breathable fabric used with a sandwich of cotton or similar material to offer excellent cold-weather protection.

When teaching workers how to dress warmly in cold weather, make sure they check not only the thermometer but wind velocity as well. The temperature may read 35°F (1.7°C), but if there is also a wind of 45 mph (72.4 km/h), it will feel like –35°F (–37°C). (See Figure 18–4, Chapter 18, Emergency Preparedness, in the *Administration*

& Programs volume.) A high wind chill factor, as it is known, means workers must wear more layers against the cold and protect all exposed skin surfaces from frostbite and windburn.

High-Visibility Clothing

High-visibility clothing provides dramatically enhanced visibility for workers through the use of fluorescent and retroreflective materials. ANSI/ISEA 107–2010, American National Standard for High-Visibility Safety Apparel, identifies three performance classes of apparel and headwear, based on specified amounts of materials and worker hazards and tasks, complexity of the work environment or background, and vehicular traffic and speed considerations:

- Performance class 1 provides a minimum specified amount of material to differentiate the wearer from the work environment and is appropriate for occupational activities that permit full and undivided attention to approaching traffic, provide ample separation of workers from traffic, and in which the vehicle and moving equipment speeds do not exceed 25 mph.
- Performance class 2 is defined as apparel for use in activities where greater visibility is needed during inclement weather conditions or in work environments with risks that exceed those for performance class 1. As such, the specified amount of high-visibility background and retroreflective materials exceeds that of performance class 1. Workers likely to utilize these garments include airport baggage handlers, crossing guards, and survey crews.
- Performance class 3 apparel provides the highest level of visibility and is designed to provide enhanced visibility to more of the body, including arms and legs, so that the wearer can be seen through a full range of motions and be identifiable as a person. These are appropriate for employees involved in high task load projects or who are exposed to significantly higher vehicle speeds and/or reduced-sight distances, such as emergency response personnel and roadway construction workers.

When selecting the appropriate high-visibility garment, consideration should be given to the background environment to allow for contrast between the background and the wearer. For example, roadway construction workers should consider colors other than orange to differentiate themselves from barriers, traffic cones, and heavy equipment.

Special Clothing

Safety experts have developed many highly specialized types of clothing for protection against special hazards. A partial list includes such items as:

- disposable clothing made of plastic or reinforced paper for exposure to low-level nuclear radiation, for use in the drug and electronic industries, or for hazardous materials work, where contamination may be a problem
- leaded clothing made of lead glass fiber cloth, leaded rubber, or leaded plastic for laboratory workers and other personnel exposed to x-rays or gamma radiation
- electromagnetic radiation suits, which provide protection from the harmful biologic effects of electromagnetic radiation found in high-level radar fields and similar hazardous areas
- conductive clothing, made of a conductive cloth, for use by lineworkers doing bare-hand work on extra-high-voltage conductors; such clothing keeps the worker at the proper potential.

For special applications, manufacturers have a vast number of materials they can draw upon to meet specific hazards.

Cleaning Work Clothing

Manufacturers' recommendations should be followed in laundering and cleaning work clothes. Excessive water temperatures or use of certain washing preparations can deteriorate the fabric or affect its properties. Spot-cleaning with organic solvents may soften or dissolve some synthetics, while chlorine bleaches will remove most flame-retardant treatment from cotton.

Workers should not use compressed air for dusting work clothes, as it presents a potential for eye injury and may force contaminants into the skin. Instead, workers should use a vacuum system, which prevents dust from being spread into the air, where it could be inhaled or get into workers' eyes. Many industrial laundries and industrial clothing rental agencies can advise firms on cleaning and maintaining their work clothing and body.

SUMMARY

- Once a company decides on the use of PPE, it should develop a company policy on PPE usage for employees and visitors, select the proper equipment for the existing hazards, implement a training program, and enforce the use of PPE.
- Companies can encourage workers to use PPE by enlisting the aid of line supervisors and managers, letting employees have some choice in the type of equipment purchased, and establishing a sound training program with consistent enforcement of all rules and regulations.
- All workers exposed to head injury hazards must wear protective headwear to shield them from falling objects, blows, and electric shock and burns.
- Protective devices for the eyes and face include safety glasses, goggles, and face shields. Face shields alone generally do not provide adequate protection against eye

injuries and must be combined with basic eye protective glasses or goggles.

- Management must evaluate the workplace for hearing hazards and determine the need for hearing protection devices. Daily work in steady noise of more than 85 dB for 8-h shifts is considered hazardous noise exposure. Hearing protectors are categorized as enclosure, aural, superaural, and circumaural.
- Fall arrest protection, either active or passive, is defined as a means of preventing workers from experiencing disastrous falls from elevations. In selecting the right fall protection system, management should conduct a thorough job survey analysis and establish a fall protection program. Companies must also develop rescue procedures for retrieving a fallen worker from aboveground, belowground, or confined-space operations.
- To protect workers from airborne health hazards, management must provide respiratory protection equipment against gaseous, particulate, and combination contaminants and oxygen-deficient environments. Respirators are classified as air-supplying or air-purifying devices. All respirators must be routinely inspected, cleaned, and properly stored to ensure their protective effectiveness.
- Safety footwear includes steel, reinforced plastic, and hard rubber models, depending on the shoe design protective level required. Some jobs require conductive, nonconductive, foundry, or special-design safety shoes to protect workers' feet from injury.
- Special protective clothing is used to shield workers from such workplace hazards as heat, hot metal, chemical splashes, weather extremes, and electrical shock or burns.

REFERENCES

American College of Occupational and Environmental Medicine Noise and Hearing Conservation Committee. "Guidelines for the Conduct of an Occupational Hearing Conservation Program." *Journal of Occupational Medicine* 29 (1987): 981–89.

American Conference of Governmental Industrial Hygienists, Building D–7, 6500 Glenway Avenue, Cincinnati, OH 45211.
A Guide for Control of Laser Hazards. 1976.

American National Standards Institute, 11 West 42nd Street, New York, NY 10036.
Safe Use of Lasers, ANSI Z136.1–2014.

ANSI/ASSE Safety Requirements for Personnel and Debris Nets, A10.11–2010.

American Society for Testing and Materials, 1916 Race Street, Philadelphia, PA 19103.
Standard Specification for Air-Fed Protective Ensembles, ASTM F2704-10.
Standard Specification for Performance Requirements for Protective (Safety) Toe Cap Footwear, ASTM F2413-11.

American Society of Safety Engineers, 1800 E. Oakton Street, Des Plaines, IL 60018.

Canadian Standard Association's (CSA) Standard Z259.2.
CSA Z259.2.4, Fall Arrestors and Fixed Rigid Rails.
CSA Z259.2.5, Fall Arresters and Vertical Lifelines.

Committee on Respirators, P.O. Box 453, Lansing, MI 48901.
Respiratory Protective Devices Manual.

Compressed Gas Association, Inc., 14501 George Carter Way, Suite 103, Chantilly, VA 20151.
Commodity Specification for Air, G–7.1.
Oxygen, C–4.
Oxygen-Deficient Atmospheres, SB–2.

Ellis, J. N. *Introduction to Fall Protection.* Des Plaines, IL: American Society of Safety Engineers, 1988.

Gasaway, D. C. *Hearing Conservation: A Practical Manual and Guide.* Englewood Cliffs, NJ: Prentice-Hall, 1984.

International Safety Equipment Association, 1901 North Moore Street, Suite 808, Arlington, VA 22209.
High-Visibility Safety Apparel and Headwear, ANSI/ISEA 107–2010.
Industrial Head Protection, ANSI/ISEA Z89.1–2014.
Occupational and Educational Personal Eye and Face Protective Devices, ANSI/ISEA Z87.1–2010.

Mack Publishing Company, 208 Northampton Street, Easton, PA 18042.
U.S. Pharmacopoeia.

National Fire Protection Association, 1 Batterymarch Park, Quincy, MA 02269.

National Institute for Occupational Safety and Health, Division of Technical Services. Cincinnati, OH 1978.
Respiratory Protection—An Employer's Manual and Guide to Industrial Respiratory Protection.

National Safety Council, 1121 Spring Lake Drive, Itasca, IL 60143.
Hearing Conservation in the Workplace, 1991.
Occupational Safety and Health Data Sheets (available in the Council Library):
Flexible Insulating Protective Equipment for Electrical Workers, 12304–0598.

Plog, B. A, ed. *Fundamentals of Industrial Hygiene.* 6th ed. Itasca, IL: National Safety Council, 2012.

SEI Certified Products. Safety Equipment Institute, 1307 Dolley Madison Boulevard, Suite 3A, McLean, VA 22101.

U.S. Department of Health and Human Services. NIOSH Guide to Industrial Respiratory Protection, NIOSH Respiratory Protection Program in Health Care

Facilities. NIOSH Publication No. 99-143, September 1999.

————. OSHA Technical Manual, Section VIII, Chapter 2, Respiratory Protection, Appendix B-1 to 1910.134, User Seal Check Procedures, Occupational Noise Exposure, Revised Criteria 1998. NIOSH Publication No. 98-126, June 1998.

————. TB Respiratory Protection Program in Health Care Facilities, Administrator's Guide. NIOSH, CDC Publication No. 99-143, September 1999.

————. Public Health Service, Centers for Disease Control, National Institute for Occupational Safety and Health. Criteria for a Recommended Standard: Occupational Noise Exposure. Revised Criteria, June 1998.

U.S. Department of the Interior, 1849 C Street, NW, Washington DC 20240. 30 CFR Chapter 1, Subchapter B, Respiratory Protective Devices, Tests for Permissibility, Fees; Part 11. Note: The Code of Federal Regulations is available through the U.S. Government Printing Office, Washington DC 20402.

U.S. Department of Labor, Occupational Safety and Health Administration, 200 Constitution Avenue NW Washington DC 20210.

29 CFR 1910.95, Occupational Noise Exposure (general industry).

29 CFR 1910.132, Personal Protective Equipment—General Requirements (general industry).

29 CFR 1910.134, Personal Protective Equipment—Respiratory Protection (general industry).

29 CFR 1910.136, Personal Protective Equipment—Foot Protection (general industry).

29 CFR 1910.139, Respiratory Protection for M. Tuberculosis (general industry).

29 CFR 1926.52, Occupational Noise Exposure (construction industry).

29 CFR 1926.105, Personal Protective and Life Saving Equipment—Safety Nets (construction industry).

29 CFR 1926.502, Fall Protection—Fall Protection Systems Criteria and Practices (construction industry).

U.S. Public Health Service, Department of Health and Human Services.

Subchapter G—Occupational Safety and Health Research and Related Activities—42 CFR 84, Approval of Respiratory Protective Devices.

REVIEW QUESTIONS

1. What are the three broad categories of methods used to control harmful exposures to hazardous substances?
   a.
   b.
   c.
2. Define personal protective equipment (PPE).
3. Information on certified equipment is available through which of the following?
   a. Environmental Protection Agency (EPA)
   b. Safety Equipment Institute (SEI)
   c. National Institute for Occupational Safety and Health (NIOSH)
   d. all of the above
   e. only b and c
4. Which of the following adds considerably to the protection offered by a helmet?
   a. chin strap
   b. bump cap
   c. paint applied after manufacture of helmet
   d. all of the above
5. Name the standard established by the American National Standards Institute (ANSI) for eye and face protection.

6. The aspect of protective eye lenses that provides the filtering effect against infrared and UV radiation is:
   a. color.
   b. chemical composition.
   c. cost.
   d. all of the above.
7. Briefly describe the four types of hearing protection devices.
   a.
   b.
   c.
   d.
8. Name the two classifications of fall protection systems.
   a.
   b.
9. List three devices used in passive fall arrest systems.
   a.
   b.
   c.
10. How often should audiometric testing be done on employees?
    a. when new employees are hired
    b. when new employees are hired and annually thereafter
    c. annually
    d. every 2 years

11. Name five components of active fall arrest systems.
    a.
    b.
    c.
    d.
    e.
12. What are the three steps in selecting respiratory protective equipment?
    a.
    b.
    c.
13. The two main categories of respirators are:
    a.
    b.
14. Gloves or mittens having metal parts or reinforcements should never be used around:
    a. hazardous chemicals.
    b. edged tools.
    c. electrical apparatus.
15. Protective footwear is classified according to its ability to meet what two requirements?
    a.
    b.

Electrical Safety

Bob LoMastro, BA, MS, WSO-CSM
John F. Montgomery, PhD, CSP, CHMM

Introduction

Key Definitions
Arc Fault/Arc Blast ▶ Automatic External Defibrillator ▶ Bonding ▶ Current ▶ Electrical Shock and Electrocution ▶ Grounding and Grounds ▶ Qualified ▶ Resistance and Impedance ▶ Transformer ▶ Voltage ▶ Watt

Electrical Injuries
Internal Injuries ▶ Skin and Eye Injuries ▶ Injuries from Falls ▶ Cardiopulmonary Resuscitation

Electrical Equipment
Selecting Electrical Equipment ▶ Installing Electrical Equipment ▶ Wiring ▶ Guarding Electrical Parts ▶ Interlocks ▶ Barriers ▶ Warning Signs ▶ Field Marking ▶ Switches ▶ Protective Devices ▶ Arc-Fault Circuit Interrupters ▶ Ground-Fault Circuit Interrupters ▶ Control Equipment ▶ Motors ▶ Extension Cords ▶ Specialized Processes ▶ High-Voltage Equipment

Grounding
System Grounding ▶ Equipment Grounding ▶ Temporary Grounds ▶ Maintenance of Grounds ▶ Three-Wire Adapters ▶ Double-Insulated Tools ▶ Polarity Plugs

Hazardous Locations
Classification of Hazardous Locations ▶ Explosion-Proof Equipment

Maintenance
Test Equipment ▶ Lockout/Tagout (Lock-Tag-Test) ▶ Rotating and Intermittent-Start Equipment ▶ Fused Disconnects

Electrical Safe Work Practices and Protective Equipment

Employee Training

Safety Inspections

Summary

References

Review Questions

Appendices
Appendix A: Electrical Cord Codes ▶ Appendix B: Sample Electrical Work Permit

INTRODUCTION

Safe application of electricity's power has allowed for advances in science, medicine, communication, education, entertainment, and almost every aspect of modern life. But if not used properly, electricity can injure and kill.

It is not essential that the average person become an electrical expert. However, it is essential that we all learn to use electricity safely and understand the danger when used improperly. For safety professionals, the topic of electrical safety represents challenges in proper design, maintenance, equipment, and training.

Electricity is a versatile form of energy. Safety considerations must start at the design phase of any construction or remodeling project. Proper design allows us to establish safe electrical work practices and procedures for servicing and maintenance operations for electrical equipment. Electrical systems in many organizations are outdated and should be updated to incorporate changes that provide safer and more efficient systems. Safety professionals should be involved in the planning and implementation of this process.

This chapter provides an overview of basic electrical terms, components, and safety considerations to minimize employees' exposure to electrical hazards. Power-distribution systems above 600 volts (V) are not addressed here.

Before dealing with electrical equipment in any way, an understanding of a few basic electrical terms is needed. All electrical systems follow the principles of Ohm's law:

Voltage (V) = amperage (I) × resistance (R)

(See Figure 8–1.)

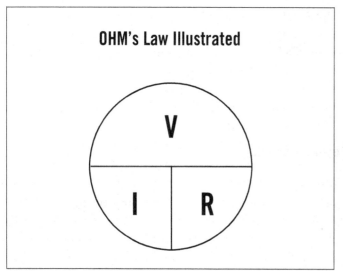

Figure 8–1. To use this symbol, you need to know two of the elements. Find the missing element by covering its symbol. Then do the math. If you need to solve for voltage, you multiply I × R. If you need to calculate the amperage (I), divide the voltage by the resistance. *(Image courtesy of LoMastro & Associates Inc.)*

Electricity flowing through a circuit is similar to the flow of water through a pipe. Although there are a few technical problems with this comparison, it serves to explain the basic function of a standard electrical system. Voltage (V) is the pressure or push, similar to water pressure forcing the water through a pipe. The amount of water that is allowed to flow is similar to amperage (I) (the number of electrons moving through the wire). The pipe will limit or restrict the amount of flow allowed; this is similar to resistance (R) in an electrical conductor (or wire). (See Figure 8–2.)

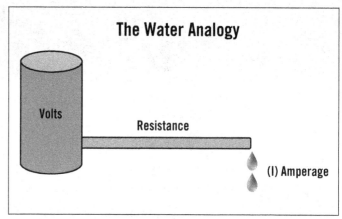

Figure 8–2. The water analogy. *(Image courtesy of LoMastro & Associates Inc.)*

KEY DEFINITIONS

Arc Flash/Arc Blast

Arcing is current leaving a conductor and traveling through a gas (usually air) to another conductor or ground path. This can result in intense light and heat to be released as the molecules in the air are superheated. This is referred to as an *arc flash* or *flashover*.

When the resulting heat reaches the point that materials (metal, plastic, etc.) vaporize, the resulting explosion is called an *arc blast*. The incident may last only a fraction of a second, but it can result in tremendous heat (up to about 35,000°F), tremendous pressure waves (copper will expand 67,000 times in volume), and intense infrared (IR) and ultraviolet (UV) light. Arc faults can be described as bolted (directed phase to phase or phase to ground) or arcing (open air). When arc faults occur in an electrical enclosure, all of the released energy is directed out of the opening. Arc flash hazard assessments must consider this additional or focused energy when establishing protective zones and clothing selection.

Automatic External Defibrillator

The *automatic external defibrillator* (AED) is a portable medical device that automatically analyzes and detects the

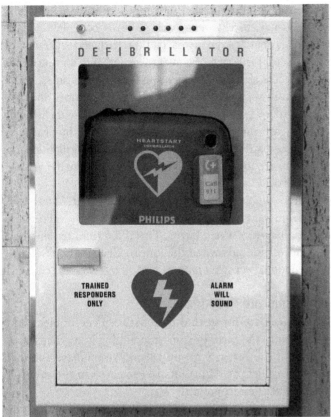

Figure 8–3. AEDs should be easily accessible for trained personnel to reach them and treat a victim within seconds. *(Roel Smart/iStock)*

cardiac arrhythmias and tachycardia of sudden cardiac arrest (SCA) in patients through electrodes; it is able to produce a measured shock, which stops the heart (contracts the heart muscles) and allows the heart to reestablish a normal rhythm. The process is called *defibrillation*. Because a person's chances of surviving an incident of SCA decrease rapidly with time, it is important that AEDs are readily available when there is a risk of electrical incidents. AED placement should allow a trained person to grab the AED and treat the victim within 90 seconds or less. (See Figure 8–3.)

Bonding

Bonding is the joining of metallic parts to establish an equal electrical charge. All metal objects have a natural electrical charge. When two metal objects are brought together, they will attempt to equalize, and it is possible that an electrical spark would result. Bonding ensures electrical continuity and balance with all of the non-current-carrying metal components in the electrical system. Because electrical current will travel from a high potential to a lower potential, bonding is designed to keep all of the conductive materials at the same potential, which will prevent arcing or sparking between conductive objects. Bonding is also important for activities such as transferring flammable liquids from one container to another because fluid in motion generates static electricity, which can also create sparks between objects. Bonding and grounding are often combined, but they are not the same thing. (See the definition for "grounding.")

Current

Current is the total volume of electrons moving (like water flowing) past a certain point, in a given length of time. Electric current is the flow of charged particles or electrons and is measured in amperes (amps). One amp (or one coulomb) is equal to 6.241×10^{18} electrons per second passing our measuring point. An amp contains 1,000 milliamps (mA), and a milliamp contains 1,000 microamps (µA). Milliamps are used when evaluating risk for electric shock or injury. It only takes a fraction of an amp, for a fraction of a second, to cause serious electrical injury or death.

If current (I) is to flow through a circuit, a complete path—from the source and back—must be available. When this path is present, the circuit is said to be complete or "closed." When the path is not complete, either intentionally or by accident, the circuit is called "open." Voltage can be present on an open circuit even though there is no current flowing.

Current can be provided as direct current (DC) or alternating current (AC). In DC systems, the current flow is traveling in one direction. In a battery, for example, direct current flows from the negative terminal through the appliance and returns to the battery on the positive terminal. The limitation with direct current is the steady loss of power through the circuit; this makes efficient transmission distances very short.

To overcome the short transmission limitations of DC, power companies use AC for electrical transmission and distribution. In alternating current, the direction of the electron flow reverses in an oscillating repetition. *Frequency*, or cycles, refers to how often the direction changes. The flow of an AC current is visualized as a wave, whereas the flow of a DC circuit is illustrated as a straight line. The alternating frequency is measured in hertz (Hz). In the United States, systems operate at 60 Hz, or 60 cycles per second. European systems operate at 50 Hz, or 50 cycles per second.

On an AC sine wave, the up curve indicates the current flowing in the positive direction, and the down curve signifies the alternate cycle, where the current moves in the negative direction. This back-and-forth is what gives AC its name. This wavelike motion allows the influencing of electrons to flow efficiently for long distances. It also makes it possible to increase or decrease the voltage using transformers. (See Figure 8–4.) Today, more electrical equipment requires the AC current to be converted to DC for final use. For example, a laptop uses both kinds of

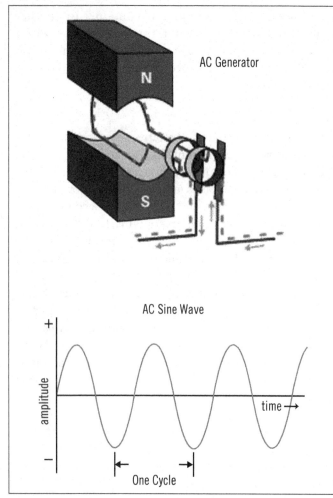

Figure 8–4. Top: As the wire loop passes the magnets in an AC generator, electrons (negatively charged) are influenced to move either toward or away from the magnetic charge. (*Reprinted from pbs.org.*) Bottom: The results can be illustrated as a sine wave. (*Image courtesy of LoMastro & Associates Inc.*)

current. The nozzle-shaped plug that goes into the computer delivers a direct current to the computer's battery, but it receives that charge from the AC receptacle on the wall. The small block that is in between the wall plug and the computer is a power adapter that transforms AC to DC and "steps down" the voltage for the computer. (See the definition for "voltage.")

Electrical Shock and Electrocution

Electric shock is caused by the passage of electric current through parts of the body. It usually involves accidental contact with exposed parts of electric circuits, but it may also result from lightning or contact with overhead wires (direct or indirect). The resulting damage depends on the intensity of the electric current, the path of the current, and the duration of current flow. Contact with alternating current, direct current, or mixed current can result in different kinds and degrees of damage. High-frequency current produces more heat than low-frequency current and may cause burns, coagulation, and/or necrosis of affected body parts. Low-frequency current can burn tissues if the area of contact is small and concentrated. Severe electric shock commonly causes unconsciousness, respiratory paralysis, muscle contractions, bone fractures, and cardiac disorders. Even passage of small electric currents through the heart can cause fibrillation. Treatment may involve such measures as cardiopulmonary resuscitation, defibrillation, and IV administration of electrolytes to help stabilize the victim. (See Electrical Injuries later in this chapter.)

Electrocution is death by electrical current. Death can result from the stopping of the heart, damage to vital organs, nerve damage, or severe burns. This is a term often misused by nonprofessionals when they mean electric shock. To be "electrocuted" means a *fatal exposure* to electrical current.

Grounding and Grounds

A *ground* is an object that connects a piece of electrical equipment to the earth or some conducting body that serves in place of the earth. A ground serves to increase and dissipate stray electrical currents to the earth in the event of an equipment malfunction.

Note: There should never be electrical current on the grounding conductor unless something is malfunctioning.

The purpose of the equipment-grounding system is to draw additional current through the grounding path, directing it into the earth, and assist the over-current devices (fuses or breakers) to open, which shuts off the flow of electricity. (See Figure 8–5.)

Many people think that equipment grounds are required to prevent human injury or death. However, this is not the

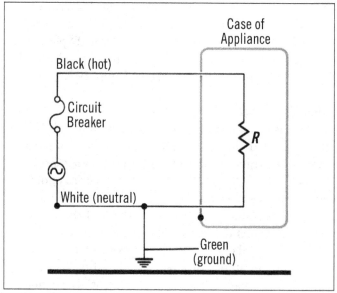

Figure 8–5. Equipment grounding. (*Image courtesy of LoMastro & Associates Inc.*)

primary purpose of electrical grounds. *Grounding* is designed to increase current flow to the earth, resulting in excess heat in the conductors, in an effort to overload the breaker or fuse (over-current device). Equipment grounding is primarily designed as fire protection. Because all of the non-current-carrying metal parts are bonded and grounded, they all become energized in the event of a short circuit. This creates a shock hazard potential for humans. But without the assistance of the grounding system, the over-current device might not respond fast enough to prevent electrical fires. When the over-current results in tripping a breaker (or fuse), it turns off the electricity to the damaged circuit. If it shuts off the power before a person contacts the energized ground, it can prevent injury while it performs its job of preventing electrical fires. This secondary effect has saved countless lives, but over-current devices are not fast enough to prevent electrical shock.

It is important for a safety professional to understand that the primary purpose of grounding is to protect the facility and equipment by assisting the flow of current to rapidly overload the over-current device. Because the earth provides an alternate path for current to return to the source (generator), it creates increased opportunities for people to become a part of that path. When protection for humans from electrical current is required (as in wet locations), additional protective devices are utilized. (See Ground-Fault Circuit Interrupters later in this chapter.)

Qualified

OSHA defines a *qualified* person as "one who has received training in and has demonstrated skills and knowledge in the construction and operation of electric equipment and installations and the hazards involved."

Some regulations and standards also specify that a qualified person "must demonstrate the skills and knowledge to work safely on energized circuits, and to be completely familiar with precautionary techniques and safety equipment to avoid the hazards involved" (National Fire Protection Association).

A person who is considered qualified for working on and around some equipment may lack the expertise to work on other equipment, making him or her unqualified in those situations. Just because a person is qualified to perform some electrical tasks does not mean he or she is automatically qualified to perform all electrical tasks. It is the employer's responsibility to determine who is or isn't qualified for any particular task.

Most safety regulations require that any electrical circuit or equipment that is energized with at least 50 V of electricity be guarded, covered, protected, or made inaccessible to everyone except qualified electrical workers. Workers exposed to energized circuits must be trained, authorized, and equipped with the correct protective clothing and tools.

When electrical panels contain energized components of different voltages, workers must be qualified for the highest voltage encountered. NFPA 70E, Electrical Safety Requirements for Employee Workplaces, requires that anyone who is *unqualified* must stay at least 42 in. from exposed circuits greater than 50 V (OSHA says 36 in.). Shock protection boundaries will be covered later in this chapter.

To become qualified, all who work with electrical wiring, transformers, and fuse boxes in both residential and commercial buildings must be licensed at some level in all 50 states at either the state, county, or local jurisdiction (Brown 2015). Members of a recognized union, such as the International Brotherhood of Electrical Workers (IBEW), are also considered qualified.

Resistance and Impedance

Resistance (R) is a blockage or friction in the conductor (like a water pipe) that impedes the movement of electrons. Resistance, which is measured in ohms (Ω), is any condition that restricts current flow. The term *impedance* is used in alternating current (AC), and *resistance* is used in direct current (DC) systems. Some metals offer more resistance or impedance to current flow than others—for example, aluminium is more resistant than copper.

Resistance results in heat. Most over-current devices respond to the levels of heat in the circuit. When too much current is flowing, the over-current device operates and stops the flow of electricity to the circuit.

Transformer

A *transformer* converts voltage from one level to another. When the change is from a higher voltage to a lower voltage, the transformer is called a "step-down" transformer. When the transformer raises the voltage, it is called a "step-up" transformer. An electrical transformer normally consists of a ferromagnetic core and two or more coils called "windings." A changing current in the primary winding creates an alternating magnetic field in the core. In turn, this influences the electrons in the wire on the secondary transformer windings. This influence (or induction) creates a voltage (or electromagnetic field) on the secondary coils. (See Figure 8–6.) Electrical transformers can be either a single-phase or a three-phase configuration. Distributing AC electrical power at high voltages allows power companies to efficiently transmit energy to service customers who live long distances away from the generation plant. As the power is distributed to varying locations, it is "stepped down" to levels usable by the customer. The electrical provider may generate power at levels as high as 230,000 V (230 kV). Commercial customers may require voltage at different levels, such as 18,000 V, 600 V, 240 V, and/or 120 V. This is accomplished by installing the necessary transformers at the facility.

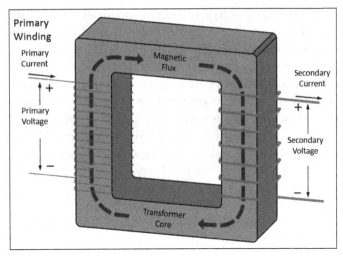

Figure 8-6. Step-down transformer. *(Image courtesy of LoMastro & Associates Inc.)*

Note: Utility-supplied transformers pose additional concerns for lockout procedures because disconnects may be located off the facility property.

Voltage

Voltage is potential difference between two points. Think of it as water pressure in the system. This electrical energy is measured in volts and may also be referred to as electromagnetic force (EMF). *Low voltage* can be an ambiguous term depending upon a person's knowledge and the application. OSHA and the NFPA use the expression "low voltage" to refer to all systems 600 V or less, and "high voltage" is defined as anything above 600 V. These terms are often misused in casual conversation. While 480 V is considered "low voltage," anyone who has come in contact with a 480-V circuit may have a different perspective. As used in this chapter, *low voltage* refers to 600 V or less. (The use of "k" means thousand, as in 15 kV or 15,000 volts.)

Power companies have slightly different definitions. They consider low voltage to be up to 600 V, medium or mid-voltage to be from 600 V to 69 kV, high voltage to be up to 230 kV, extra-high voltage to be up to 765 kV, and ultra-high voltage to be up to 1,100 kV (ANSI C84.1, 2011).

For most safety professionals and workers, the terms *high* and *low voltage* will suffice. While voltage itself does not cause injury (current and resistance do the damage), voltage provides the necessary push that forces the current to flow. According to many medical studies, potentially hazardous voltage is anything above 24 V. OSHA and NFPA regulations state that any energized parts greater than 50 V must be protected to prevent workers from contact. A car battery may be only 12 V of direct current (DC), but in a dead short, it can release very hazardous energy due to the amperage available.

Watt

A *watt* (W) is the quantity of electricity that is consumed. The watt is used to specify the rate at which electrical energy is utilized, or "the rate at which electromagnetic energy is radiated, absorbed, or dissipated." It is determined by multiplying volts (V) by amperage (I) (V × I = W). One kilowatt (1 kW) is equal to 1,000 W; 1 megawatt (1 MW) is equal to 10^6 W; 1 gigawatt (1 GW) is equal to 10^9 W.

ELECTRICAL INJURIES

Current flow, path, and time determine the severity of an electrical shock (the injury triangle). The outcome of electrical contact is determined by (1) the amount of current that actually flows through the victim, (2) the path and body parts affected as the current travels through the victim, and (3) the length of time that the tissue is exposed to the current. The preventive triangle suggests ways to eliminate one or more sides of the injury triangle. (See Figure 8-7.)

Because the amount of current that flows depends on voltage applied and the resistance of the conductors, these factors are important in planning safe work practices. Additionally, other factors that affect the extent of injury are the frequency of the current (measured in hertz), which may interrupt the heart rhythm, and the initial health of the victim. Heat is a secondary effect on body tissues. Deadly current flow, with high levels of heat, is common with contacts on low-voltage sources such as typical lighting or receptacle circuits.

A person's primary resistance to current flow is the skin's epidermal layer. Callous or dry skin has a relatively

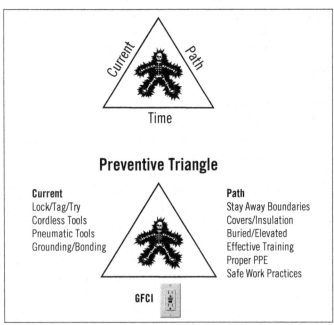

Figure 8-7. The injury triangle. *(Image courtesy of LoMastro & Associates Inc.)*

high resistance. A sharp decrease in resistance takes place, however, when the skin is moist or cut. Once the skin's resistance is broken down, current flows readily through the nerves, blood, and the other conductive body tissues.

The skin's resistance decreases rapidly with increases in voltage, time, frequency, and moisture. Just because one incidence of electrical contact doesn't result in severe injury doesn't mean that it won't in the future. A worker should always plan (and protect) for what could possibly happen in the future.

Muscle tissue responds to electrical currents by contracting. A high-voltage alternating current of 60 Hz causes extremely violent muscular contraction, which often results in throwing the victim off the circuit. Low-voltage contact also results in muscular contraction, but the effect is not so violent. The victim can lose control of the muscles of the hand, which prevents the victim from breaking the contact with the circuit. This is often referred to as "locked on" because the muscles in the hand will contract with such strength that it prevents the victim from releasing the energized circuit. As the time of exposure increases, the severity of the injury also increases.

Internal Injuries

Death or injuries from electric shock may have the following effects of current on the body:

1. The chest muscles contract, which may interfere with breathing to such an extent that death results from asphyxiation.
2. Temporary paralysis of the nerves or the nerve center occurs, which causes respiratory failure, a condition that often continues long after the victim is freed from the circuit.
3. There is interference with the normal heart rhythm, causing ventricular fibrillation (V-fib). Ventricular fibrillation is a problem that occurs when the heart beats with rapid, erratic electrical impulses. This causes the pumping chambers of the heart (the ventricles) to quiver uselessly, instead of circulating blood. The heart cannot spontaneously recover from this condition, and survival depends on effective rescue efforts. It is estimated that 50 mA (AC) is sufficient to cause ventricular fibrillation.
4. In contact with heavy current, the muscular contractions of the heart stop completely. In cases of short exposure, the heart may resume its normal rhythm when the victim is freed from the circuit. There are also numerous cases in which the heart was restarted when the victim fell and hit the ground.
5. Hemorrhage (bleeding) or destruction of tissues, nerves, and muscles occurs from heat due to current flowing through the body for extended amounts of time (measured in seconds).
6. Severe burns result from arcing contacts. This can occur even in low-voltage systems. Massive tissue damage can result from the heat generated when electricity ionizes the air to create an arc path. Clothes melting or catching fire also results in severe burn injuries.

Electrical damage is often not visible; injuries can be hidden beneath normal-appearing skin. Electricity burns from the outside in and from the inside out. The victim may not be aware of the severity of an injury due to nerve damage. Any electrical contact beyond a "tingle" should be evaluated by medical professionals. A victim's right to refuse treatment can be exercised only after the employer meets their responsibility to provide medical care.

Injuries from electrical shock are less severe when the current does not pass through nerve tissue or vital organs. Most of the electrical accidents that occur in industry, however, result from current flowing from hand to hand or from a hand to the feet. Because such a path involves both the heart and the lungs, the outcome can be severe or fatal. (See Figure 8–8.)

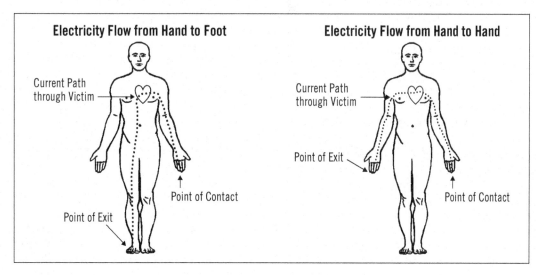

Figures 8–8a and 8–8b. Path of current. *(Images courtesy of LoMastro & Associates Inc.)*

Skin and Eye Injuries

Electricity follows the conductive minerals of the human body to complete the circuit. Tissue dies at about 300 mA. Damage to organs may not result in immediate pain. However, there have been many deaths days after an electrical shock due to undetected damage to the blood vessels, heart, kidneys, or other organs. Qualified medical personnel should evaluate electrical shock victims immediately. Clearance distances from live conductors are set by OSHA and NFPA 70E for both shock and arc flash boundaries. These protection boundaries are a critical part of any hazard assessment to prevent injuries.

Thermal burns can result from arc flashes or arc blasts. Temperatures from arc blasts may range between 10,000°F and 35,000°F. These burns are usually deep and slow to heal and may involve large areas of the body. Even people at a significant distance from the arc may receive burns or eye injuries. NFPA 70E (Electrical Safety Requirements for Employee Workplaces) establishes minimum clearance distances whenever work must be performed on energized circuits (live work). It is important to note that the flash protection boundary is the distance from an electrical point (source of arc blast) where the heat generated will be reduced to less than 1.2 cal/cm². This is considered a survivable heat level. A 1.2 cal/cm² exposure on human skin may cause second-degree burns and is never injury free.

When work must be performed on energized systems, both a shock protection boundary (which prevents contact with energized parts) and a flash boundary must be established (to prevent injury from excessive heat and pressure waves). The flash boundary is established by using either accepted engineering formulas (NFPA 70E, Annex D) or the NFPA 70E tables for arc-flash hazard PPE categories. These are discussed in Electrical Safe Work Practices and Protective Equipment later in this chapter.

While OSHA standards specify hazards for which the employer must provide protection, they do not outline specific safe work practices. However, NFPA 70E provides the necessary guidance to prevent electrical injuries and meet OSHA requirements.

Most arcing incidents are caused by the following:
- short circuits between energized bus bars or cables (loose wires, dropped tools, or conductive parts)
- improper installation or equipment design
- failure of knife switches to completely open or opening switches under load
- pulling fuses in energized circuits
- poor maintenance or housekeeping
- improper use of tools or electrical testing equipment.

Injuries from Falls

Another common source of injuries from electrical shock is falls. The worker receives a shock, which causes muscles to contract, resulting in the worker losing his or her balance. This type of accident can be caused by static electricity as well as AC or DC sources. Fall protection equipment used in energized working conditions must be rated for electrical hazards. Harnesses, lanyards, and ladders are examples of equipment that should be evaluated for suitability for electrical work.

Cardiopulmonary Resuscitation

Because electrical shock can stop the heart and lungs, be sure that workers involved in working on or near live electrical systems are trained in cardiopulmonary resuscitation (CPR), AED use, and proper rescue procedures. CPR training is provided by the National Safety Council First Aid Institute, the American Heart Association, the American Red Cross in the United States, and St. Johns Ambulance in Canada. Consult the telephone directory or the Internet for the closest provider.

Before any treatment can be administered, ensure that the victim is clear of all electrical sources. Methods of release from an energized circuit should be determined before any work begins. This may mean locating the appropriate disconnect point or providing a shepherd's hook–type rescue tool. The rescue plan should be included in the prejob briefing conducted before the work starts.

The sooner the victim is resuscitated, the better the chances of survival. Immediate application of an AED and/or CPR to a victim increases the likelihood of survival. (See Figure 8–9.) CPR should be continued until the victim is revived, or until a physician diagnoses death. Refresher training should be conducted annually. Training must be documented.

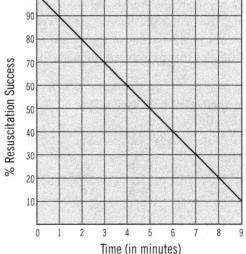

Figure 8–9. AED Survival Rates. For every minute of delayed treatment, the victim's chance of survival is decreased by 7% to 10%. *(Image courtesy of LoMastro & Associates Inc.)*

ELECTRICAL EQUIPMENT

Most items of electrical equipment are designed and built for specific types of service and environments. They operate with maximum efficiency and safety only when used for the purposes and under the conditions for which they are intended. OSHA regulations require electrical equipment to be listed or labeled by a nationally recognized testing laboratory (NRTL) and used only for the purpose for which it is approved. Most electrical equipment is designed to operate under overload conditions for limited periods of time. A continued overload, however, can result in fires, short circuits, circuit failures, or mechanical failures. Designers, installers, and operators should be thoroughly familiar with the proper use and limitations of their equipment. They should be trained to observe and report abnormal conditions. When hazardous conditions exist, the safety of the workers and the equipment should be considered more critical than production. In other words, workers and management should know when it is appropriate to shut down the equipment.

Selecting Electrical Equipment

When selecting electrical equipment, follow established codes and standards, such as the National Electrical Code (NEC), NFPA-70, and the National Electrical Safety Code, ANSI C2. In addition to these codes, check state and local codes for industrial zoning requirements. Adherence to the provisions of the NEC is required by government agencies (referred to as the authority having jurisdiction [AHJ]) but is also required by many insurance companies and local fire departments.

Engineering guidance and publications from the following groups will answer most questions concerning electrical equipment: American National Standards Institute (ANSI), Canadian Standards Association, Factory Mutual System Research Organization (FM), Illuminating Engineering Society of North America (IES), National Fire Protection Association (NFPA), and Underwriters Laboratories (UL). (See References at the end of the chapter for contact information.) Consulting services may also be available from some of these organizations.

When ordering copies of codes or standards from publishers, provide suppliers with complete information on the general type of equipment being considered, the application, and the operating conditions for best results. There is usually a charge for nongovernment codes or standards.

Installing Electrical Equipment

Where space and operating requirements permit, install electrical equipment in less congested areas of the facility or, where practical, in special rooms that only authorized persons may access. When electrical equipment is accessible by unqualified workers, OSHA requires all parts generating more than 50 V to be protected. Proper enclosures are required to shield conductors where workers may be exposed.

DANGER signs should be posted in areas with exposed conductors where unqualified workers are not permitted to enter. Field marking information must provide accurate and current information. (See Field Marking later in this chapter.)

When there is danger from vehicular traffic, install floor curbing or heavy steel barriers to prevent trucks from striking electrical equipment. Install transformers, control boards, switches, motor starters, and other electrical equipment in locations that limit the chance of accidental contact with energized parts. OSHA and NFPA standards establish clearance distances (based on voltages) for the front, top, and sides of electrical equipment in three types of conditions. The distances provide space for maintenance workers and protect them from potential ground contacts when energized work is performed. The depth of the working space should be 3 ft, 3.5 ft, or 4 ft, depending upon existing conditions. The conditions are shown in Figure 8–10.

Wiring

As a minimum standard, all electrical installations should meet the requirements of the NEC. A knowledgeable inspector should thoroughly inspect each installation prior to initial start-up or after modifications are made. The use of temporary wiring instead of permanent wiring should be avoided. Even though temporary wiring may seem rea-

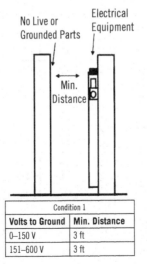

Condition 1	
Volts to Ground	Min. Distance
0–150 V	3 ft
151–600 V	3 ft

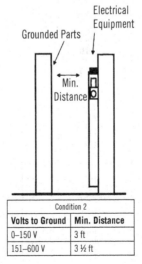

Condition 2	
Volts to Ground	Min. Distance
0–150 V	3 ft
151–600 V	3 ½ ft

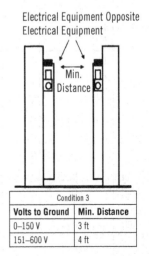

Condition 3	
Volts to Ground	Min. Distance
0–150 V	3 ft
151–600 V	4 ft

Figure 8–10. OSHA minimum clearance distances for electrical panels. Data from NEC 110.16 and OSHA 29 CFR 1910.303, Table S-1. *(Images courtesy of LoMastro & Associates Inc.)*

sonably safe when first installed, extension cords may be subjected to damage and deteriorate over time.

In the United States, the American Wire Gauge (AWG) system is used. Dimensions of the wires are given in ASTM B258-2014. The cross-sectional area of each gauge is an important factor for determining its current-carrying capacity. Increasing gauge numbers denote decreasing wire diameters, which is similar to many other nonmetric gauging systems (such as shotgun bores). The AWG tables are for single, solid, and stranded conductors. The AWG of a stranded wire is determined by the cross-sectional area of the equivalent solid conductor. Because there are also small gaps between the strands, a stranded wire will always have a slightly larger overall diameter than a solid wire with the same AWG.

When additional equipment is being added or operated under temporary conditions, avoid tapping into an existing circuit. This often results in circuit overloads. Another problem common to modifications is adding too many conductors into an existing raceway. Raceways must allow sufficient space for heat to dissipate efficiently from the conductors. Overcrowding can result in excessive heat buildup and insulation breakdown. The NEC has charts that specify the size and number of wires allowed under various conditions. Never exceed the capacities recommended by the code. (See Figure 8–11.)

Frequent inspections of wiring systems and competent supervision of maintenance activities are an important part of the overall electrical safety program. Maintenance inspection reports help prevent unsafe conditions from developing. Often in their routine work, maintenance personnel can spot hazards before they become a problem. Safety professionals should assist the maintenance department in making the inspection process as convenient as possible.

Guarding Electrical Parts

In many respects, standard machine-guarding practices can be applied to electrical equipment. Wiring systems present special hazards. Wiring must be accessible for servicing, protected from damage, and allowed to dissipate heat and service convenient control locations. Consider these hazards when developing the overall wiring plan for the facility.

Wiring must be installed in accordance with the NEC (NFPA 70) unless additional local requirements apply. It is especially important to be aware of these requirements when adding new or modifying existing circuits. For non-routine or high-hazard areas, there are special requirements for wiring and protective enclosures. Consult with an electrical engineer for nonroutine situations.

The type of wiring depends on the type of building construction, the size and distribution of the electrical loads, exposure to dampness or corrosive vapors, location of equipment, and various other factors. For typical installations, grounded metal conduit is satisfactory. Other types of wiring include armored cable, nonmetallic sheathed cable, flexible metal conduit, raceways, open wiring and insulators, and concealed knob-and-tube wiring. (See NEC Article 324

Type: THHN-THWN											
Size AWG MCM	½ in.	¾ in.	1 in.	1 ¼ in.	1 ½ in.	2 in.	2 ½ in.	3 in.	3 ½ in.	4 in.	5 in.	6 in.
14	13	24	39	69	94	154	—	—	—	—	—	—
12	10	18	29	51	70	114	164	—	—	—	—	—
10	6	11	18	32	44	73	104	160	—	—	—	—
8	3	5	9	16	22	36	51	79	106	136	—	—
6	1	4	6	11	15	26	37	57	76	98	154	—
4	1	2	4	7	9	16	22	35	47	60	94	137
3	1	1	3	6	8	13	19	29	39	51	80	116
2	1	1	3	5	7	11	16	25	33	43	67	97
1	—	1	1	3	5	8	12	18	25	32	50	72
1/0	—	1	1	3	4	7	10	15	21	27	42	61
2/0	—	1	1	2	3	6	8	13	17	22	35	51
3/0	—	1	1	1	3	5	7	11	14	18	29	42
4/0	—	1	1	1	2	4	6	9	12	15	24	35
250	—	—	1	1	1	3	4	7	10	12	20	28
300	—	—	1	1	1	3	4	6	8	11	17	24
350	—	—	1	1	1	2	3	5	7	9	15	21
400	—	—	—	1	1	2	3	5	6	8	13	19
500	—	—	—	1	1	1	2	4	5	7	11	16
600	—	—	—	1	1	1	1	3	4	5	9	13
750	—	—	—	—	1	1	1	2	3	4	7	11

Figure 8–11a. Conduit fill chart. (*Image courtesy of LoMastro & Associates Inc.*)

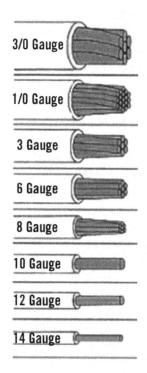

Figure 8–11b. AWG wire sizes. (*Image courtesy of LoMastro & Associates Inc.*)

for information on locations where concealed knob-and-tube wiring is not permitted.)

Wire types must be selected based on the type of service and location. The NEC lists various types of insulation used on electrical conductors, as well as how and where to use them. To prevent current from breaching the insulation, the covering must provide at least 1,000 Ω of resistance for every volt applied. A 120-V service cable that has 120,000-Ω resistance results in exposure of 1 mA.

Outside power lines are not normally made to this level of protection, and the coating is considered weatherproofing, not insulation. This is why the minimum distance for nonqualified workers to a suspended power line (with voltages between 50 V and 50,000 V) is 10 ft (per OSHA and NFPA). This is a shock protection boundary and is important because the conductor is capable of movement (swing). Suspended wire has slack. If there is a 3-ft slack, the wire can move 6 ft. Because there is insufficient insulating material, distance is used to protect people from electrical shock. Standard premise wiring is not considered insulated, either, and must be protected by pipe, raceways, or other acceptable means.

Exposed wiring is a serious hazard. All electrical panels and boxes must have proper covers or enclosures. Even the removal of a coverplate is considered an exposure to "live wires." Unused openings must be capped with approved knockout plugs or coverplates. This prevents accidental contact with live parts, contains arcs during system failures, and prevents pests from entering enclosures. Open or missing circuit breaker spaces must be filled with approved blanks.

Raceways, pipes, and cabling must be connected or attached with the properly approved fittings. Where eccentric or concentric knockouts are present, and where the pipe or raceway is part of the grounding conductor, recent codes require the use of ground bushings. These bushings provide a connection for a grounding jumper to ensure a solid connection at the fitting. (See Figure 8–12.)

Exposed wiring on power tools and cords is also dangerous and violates OSHA safety regulations. Replace damaged cords or plugs with manufacturers' approved replacement parts. Cords larger than 12 gauge can be repaired only if the repair meets the original specifications for the cord rating [29 CFR 1926.405(g)(2)(iii)].

Interlocks

An *interlock* can be defined as a device that prevents workers from making an inappropriate maneuver or that adjusts the system to a safe state if an inappropriate maneuver is used. However, interlocks for electric equipment may not be used as a substitute for lockout/tagout procedures.

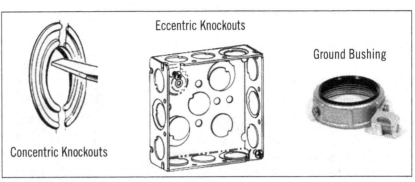

Figure 8–12. Grounding bushings. *(Image courtesy of LoMastro & Associates Inc.)*

Interlocks can prevent a user from making unsafe actions or minimize the hazard of unsafe actions by rendering the machine in a safe condition when an unsafe maneuver occurs. For example, a guard may be interlocked to prevent machine operation when a guard is removed, or a control may be interlocked to make it nonoperational if a dangerous condition will result. Only a qualified person may defeat an electrical safety interlock—and then only temporarily while he or she is working on the equipment. The interlock system must be returned to its operable condition when this work is completed. Safety interlocks may have additional or combined features to reduce hazards.

Interlock Examples

Many consumer product safety standards mandate interlocks in both industrial equipment and in everyday consumer products. Here are a few examples of consumer products that provide interlocks:

- Removal of a guard on a food processor prevents the operation of the motor and blade, thereby reducing the opportunity for spinning blade injury (example of a guard).
- Opening the filter access door on a forced-air furnace prevents operation of the blower motor and possible contact with the blower blade and/or a combustion gas recirculation hazard (example of a guard and the hazard of recirculation).
- The gear shift selector on a car allows the engine to start in the "Park" position only, which prevents the operator from starting the engine with the car in gear and prevents unexpected vehicle movement (example of limiting control operations).
- The inability to open the door on a washing machine during the high-speed extraction cycle prevents access to the spinning drum and allows the drum's rotation to stop prior to opening the door (example of a guard and a control feature).

Early interlocks were simply a safety method that relied on an electromechanical switch (like a limit or magnetic

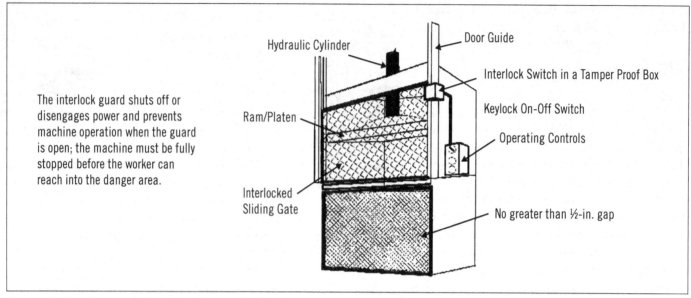

Figure 8–13. Interlock guarding of equipment. *(NIOSH)*

switch) to perform the interlocking action. However, modern interlocking mechanisms may take the form of other sensors and actuators. Interlock switch providers currently have multipole magnetic switches, unique-shaped key switches, and hidden features buried within structural components. (See Figure 8–13.)

The single-beam light curtain at the bottom of a garage door acts as an interlock to reverse the door so it cannot close on a child or animal. (See Figure 8–14.) The deadman control on a modern snow thrower (ANSI B71.3) acts as an interlock by placing the snow thrower in a safe condition (engine off or blade brake on) when the user releases the control lever to reach into the discharge chute. The thermocouple on a gas stove prevents the release of unburned gas if the gas is not ignited. The light curtain, a captured lever device, and a thermocouple are examples of interlock sensors and actuators.

Most interlocks are not designed to meet OSHA's lockout requirements. When interlocks are used as an energy control device, it is important to make sure that it is fail-safe. Interlocks should meet the following criteria:
- be equipped with fail-safe features because failure of the interlock mechanism, loss of power, short circuit, or malfunction of equipment will cause the control circuit to be interrupted
- have a visible disconnect, or opening, in the primary power circuit
- have a locking arrangement that makes attempts to circumvent the interlock impractical.

Note that interlocks for mechanical power presses must meet the ANSI B11 standard.

Figure 8–14. Light beam sensor. Upon detection of personnel or objects, the machine stops immediately. *(Photo courtesy of Omron Automation and Safety.)*

Barriers

Barriers are guards that prevent accidental contact with electrical equipment by means of a physical device. The device may be fixed, adjustable, or self-adjusting. A wide variety of materials can be used to construct barriers such as: wood, sheet metal, perforated metal, expanded metal, wire mesh, or plastic. (See Figure 8–15.)

All metal frames or guards should be grounded when electrical equipment is involved. Materials must be substantial enough to withstand whatever impact they may be subjected to and durable enough to withstand prolonged use.

8 Electrical Equipment   233

Figure 8–15. Barrier guard. *(Furchin/iStock)*

Fixed barriers have the advantage of being easily modified to work environments and of being constructed on site, and they usually require minimum maintenance. They should be designed so as not to interfere with the operator's visibility of machine operations and should not be removable without some type of tool or effort. Training for maintenance and operators must emphasize that the barrier guards must be in place during normal operations and that lockout procedures are necessary whenever they are removed.

Warning Signs

Warning signs should be posted near exposed current-carrying parts and in especially hazardous areas, such as high-voltage installations and motor control centers, and during electrical maintenance activities. Signs should be large enough to be read easily and should be visible from all possible approaches to the danger zone. The design of the warning sign should be in accordance with 29 CFR 1910.145, Specifications for Accident Prevention Signs and Tags, or ANSI Z535.4, Standard for Product Safety Signs and Labels. (See Figure 8–16.)

Field Marking

Service equipment in other than dwelling units shall be legibly marked in the field with the maximum available fault current. The field marking(s) shall include the date the fault current calculation was performed and be of sufficient durability to withstand the environment involved. (NEC 110.24)

The excerpt from the NEC indicates that common points of access—such as switchboards, panel boards, control panels, disconnects, meter sockets, or junction boxes—that are likely to require examination, adjustment, servicing, or maintenance while energized must be field marked. These requirements are intended to require owners of electrical equipment to provide labels that have enough information regarding the potential shock and arc flash hazard so that qualified electrical personnel can select procedures and personal protective equipment necessary for their safety. A variety of different-style labels are available. (See Figure 8–17.) Whichever style is selected, the label should provide the following information:

1. nominal system voltage
2. shock protection boundary
3. arc flash boundary
4. at least one of the following:
    a. available incident energy or PPE category
    b. minimum arc rating for clothing
    c. site-specific PPE requirements
5. date the hazard assessment was performed.

The label must be kept up to date. Hazard assessments should be conducted at least every 5 years or whenever changes are made that affect the information provided.

Switches

Switches, fuses, circuit breakers, ground-fault circuit interrupters, control equipment, motors, and extension cords are needed to run electrical machines safely and efficiently.

Figure 8–16. Hazard warnings. *(Images courtesy of LoMastro & Associates Inc.)*

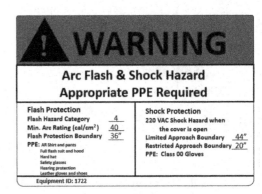

Figure 8–17. Field marking label. *(Image courtesy of LoMastro & Associates Inc.)*

Proper use of this equipment avoids hazards and prevents accidents. When switches are used to isolate power, they should be designed to accept lockout devices.

Examples of available switches include knife switches; push-button switches; snap switches; pendant switches; and enclosed, externally operated air-break switches. Many switches are designed for a specific function, such as the enclosed switch for controlling individual motors and machine tools and for lighting and power circuits. However, all switches, regardless of their function, must have approved voltage and ampere ratings that are compatible with their intended use. Manufacturers and distributors of electrical equipment can be a valuable resource in selecting and maintaining these switching devices. Circuit breakers used to switch 120- or 277-V fluorescent lighting circuits must be listed and marked switch duty rated (SWD). Circuit breakers used to switch high-intensity discharge lighting circuits must be listed and marked HID [240.83(D)].

Knife Switches

An open or exposed knife switch is extremely hazardous in systems generating more than 50 V. This is because current-carrying parts are exposed and an arc forms whenever the switch is opened or closed. Knife switches should be enclosed in grounded metal cabinets with control levers that operate outside the cabinets to prevent workers from contacting live parts. A further safeguard is the safety switch; these devices require the cover to be closed before the switch can be operated. Most modern disconnect switches provide this safety feature.

Mount a knife switch so the "blades" or moving parts are deenergized when the switch is open. Install knife switches so that gravity will not close the contacts. (See Figure 8–18.) Electrically interlocked knife switches are available for power-switching circuits. They are designed so that the switch cannot be opened when the circuit is energized, unless switches are of the load-break type, due to the arcing hazard.

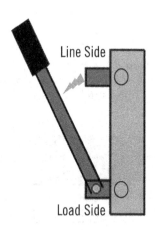

Figure 8–18. Basic knife switch. *(Image courtesy of LoMastro & Associates Inc.)*

Where it is necessary to use disconnect switches for high-voltage, heavy-current-feeder circuits for testing floors or other service installations, locate them out of reach from normal operating activities. These types of switches should be operated with (approved and tested) insulated switch sticks. Where it is impractical to locate the switch out of normal reach, protect it from accidental contact by completely enclosing it or by placing it behind a suitable fence or barricade. If it is placed behind a barrier, design the barrier to allow switches to be opened or closed remotely with an insulated hot stick.

Pendant Switches

Pendant switches are hand-operated controls used primarily as controls where wall switching is not practical. Pendant switches are also used as additional controls (in addition to the switches on walls, or in cabinets) when multiple control points are required. A common multicontrol application would be the operation of overhead cranes.

Pendants must be equipped with strain relief fittings at the point of suspension from the ceiling and at the point of entry into the device. Strain relief prevents pull or harmful tension to be exerted on the electrical wires or their terminating connections.

Other Switches

Special-use equipment may have specific regulatory requirements (fountains, outdoor lighting, cranes, etc.). Convenience may also play a factor in switch selection and location. Push-button, snap, and micro switches are often recommended because the current-carrying parts are fully enclosed, which reduces arc risks.

All switches must have indicators to show open (off) and closed (on) positions.

Protective Devices

The safe current-carrying capacity of conductors is determined by calculating the conductors' size, length, material type (e.g., aluminium versus copper), insulation, and manner of installation. If conductors are forced to carry more than the maximum safe load or if heat dissipation is limited, overheating of the conductor results. Over-current devices, such as fuses and circuit breakers, open the circuit automatically in case of excessive current flow from ground faults, short circuits, and overloads to prevent fires. All electrical circuits must have over-current protection. Cabinets or service panels are designed to hold breakers or fuses of the same type but can have different ratings. (See Figure 8–19.)

Over-current devices should interrupt the current flow whenever it exceeds the conductor's capacity. In motor control systems, the over-current device will be delayed for a short period of time to allow for the increased demand necessary for the motor to reach normal operating conditions (which may take several seconds). It takes significantly more power to start a motor than what is necessary for normal operation. Protective devices for typical electrical services should be selected to operate as fast as possible to limit fault current when a fault occurs.

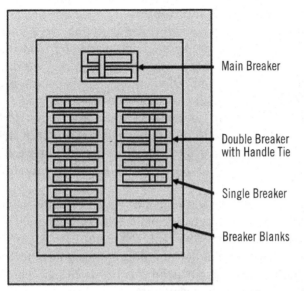

Figure 8-19. Typical service panel. *(Image courtesy of LoMastro & Associates Inc.)*

Selecting properly rated equipment does not just depend on the current-carrying capacity of the conductor. It also depends on the rating of the power supply transformer, or generator, and its potential short-circuit-producing capacities. Protection of this kind, both for personnel and for equipment, is one of the most important features of safe electrical installation. Where higher interrupting capacity is required, use specially rated high-capacity fuses or breakers.

When an over-current device "trips" or opens a circuit, an evaluation of the cause should be done before restoring power to the circuit. When the cause is unknown, a qualified electrical worker should conduct an evaluation. Repeated resetting of a circuit breaker does not fix the problem and may result in damage to the device. It can also lead to serious injury and/or property damage.

Fuses

Among the many types of fuses are link, plug, and cartridge fuses. (See Figure 8-20.) They should be used only in the type of circuit for which they were designed. Using the wrong type or the wrong-size fuse is hazardous. Over-fusing (too high a rating) is a frequent cause of overheated wiring or equipment, which can cause insulation damage and may result in fires and shock hazards.

Before replacing fuses, the system should be locked out, tagged out, and tested. Find out the cause of any short circuit or overload, and correct the problem before replacing a blown fuse. Always replace blown fuses with one of the same type and size.

Never insert fuses in a live circuit.

Link Fuses

A link fuse, as its name indicates, is a strip of fusible metal that links two terminals of a fuse block. If not enclosed, it may scatter extremely hot metal when it blows. Replaceable link fuses should be replaced only under the direction of qualified maintenance personnel or an electrician. Bypassing or "bridging" fuses must be prohibited.

Plug Fuses

Plug fuses are normally used on circuits that do not exceed 30 A at not more than 150 V to ground. In plug fuses, the fusible metal is completely enclosed. Use plug fuses that cannot be bridged inside the holder.

Cartridge Fuses

The cartridge fuse, which is widely used in industrial installations, has a fusible metal strip enclosed in a tube. If possible, use cartridge fuses that indicate when the fuse is blown. This makes identification of the blown fuse condition simple and eliminates the need for contact testing for diagnosis.

To replace or deenergize fuses, written procedures should define the locking, tagging, and testing procedures required. Only insulated fuse pullers are allowed to be used to install or remove cartridge-type fuses.

Circuit Breakers

The circuit breaker was invented in 1836 by Charles G. Page. Thomas Edison included a type of circuit breaker

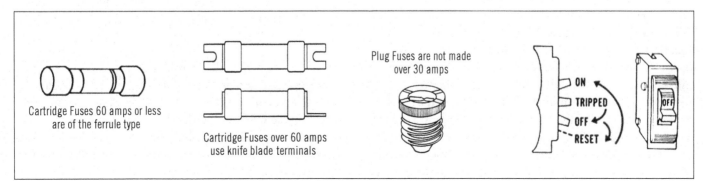

Figure 8-20. Over-current devices. *(Image courtesy of LoMastro & Associates Inc.)*

in an 1879 patent application, although his commercial power-distribution system used fuses. It was designed to protect lighting circuit wiring from accidental short circuits and overloads. The modern miniature circuit breaker, similar to the ones now in use, was patented by Brown, Boveri, and Cie in 1924. Circuit breakers were first used in high-voltage circuits, with large current capacities. Circuit breakers are now commonly used for many other kinds of circuits and provide a manual means of energizing and deenergizing a circuit and providing automatic over-current protection. They are available in a wide variety of types and sizes. Breakers can be designed to operate instantly or through a delayed timing device. Unlike fuses, which must be replaced when they open, a circuit breaker can be reset once the over-current condition has been corrected. Electrical engineers should be consulted to select the appropriate over-current devices for industrial applications. New features are being introduced every year to improve safety and reduce arcing hazards.

Circuit breakers fall into two general categories—thermal and magnetic. The thermal circuit breaker operates solely on the basis of the rise of temperature. Therefore, variations in the temperature of the room where they are installed can have an effect on the point at which it opens the circuit. Where high temperatures are a problem, consider relocating circuit breakers to locations where the temperature can be controlled or use magnetic-type circuit breakers.

A magnetic circuit breaker operates on the basis of the amount of current passing through the circuit. The current flowing creates a magnetic action to the contact arm inside the device. Excess magnetism causes the contacts to separate to an "open" position. This type of circuit breaker has the advantage of performing reliably in temperature fluctuations where heat-related tripping is a problem.

Combination breakers are also available. Breakers can provide both thermal and magnetic protection. These combination breakers are becoming very popular for use in motor control centers (MCCs). Circuit breakers can also be combined with arc- or ground-fault circuit interrupter protection (discussed later). These combinations are becoming the standard for residential applications.

Experienced maintenance personnel should regularly check and operate circuit breakers. Facility maintenance practices should include at least an annual inspection and operation of breakers. Circuit breakers must be maintained in good operating condition at all times if they are to perform properly. A common malfunction in circuit breakers is called "fusing." This occurs when the metal contacts are actually melted together and will not open manually or automatically. The breaker handle may move, but the contacts remain connected. This is why safe electrical work procedures must include locking, tagging, and testing.

Arc-Fault Circuit Interrupters

Unlike a standard circuit breaker detecting overloads and short circuits, an arc-fault circuit interrupter (AFCI) utilizes advanced electronic technology to "sense" the different arcing conditions. Often unseen, arc faults can occur anywhere in a home's electrical system, including damaged wires within walls, loose electrical connections, or electrical cords damaged by furniture or foot traffic. When arcing occurs, the AFCI analyzes the characteristics of the event and determines if it is hazardous. AFCI manufacturers test for the hundreds of possible operating conditions and then program their devices to monitor constantly for the normal and dangerous arcing conditions. AFCI standards were introduced in the 1999 NEC (National Electrical Code) and are required in new-dwelling construction and when installing new circuits in an existing dwelling. There are no industrial requirements at this time.

Ground-Fault Circuit Interrupters

Ground-fault devices measure the amount of current going to and back through a circuit. Current is constant; therefore, if more current is sent out on a circuit than returns, something is wrong in the circuit. Ground-fault circuit interrupters (GFIs) are available in a variety of response levels for different purposes. They may also be called ground-fault protectors (GFPs) and ground-fault interrupters (GFIs). In Europe, they are referred to as residual-current devices (RCDs).

GFCI (or class A) devices are designed to protect humans from exposure to electrical shock. They are fast-acting, electrical circuit-interrupting devices that are sensitive to very low levels of current loss to ground. A GFCI can sense the small amounts of current loss sufficient to cause serious personal injury (anything above 6 mA). The unit operates only on line-to-ground fault currents. It recognizes current leakage during accidental contact with an ungrounded or "hot" wire and a path to ground. GFCIs do not provide protection in the event of line-to-line contact where no current is actually lost. (See Figure 8–21.)

GFCIs are available as combination circuit breaker–GFCI devices, receptacles, and portable units. GFCIs can be an integral part of some appliances (e.g., hairdryers) or built into extension cords.

GFCIs are required when workers are using any electrical equipment in a work environment that is or may become wet or that uses a temporary power supply (e.g., generators or extension cords). NFPA 70E requires that "the employer shall provide GFCI protection where an employee is operating or using cord and plug connected tools related to maintenance and construction activities supplied by 125 volt, 15, 20, or 30 amp circuits."

For off-the-job electrical safety, GFCIs are required on

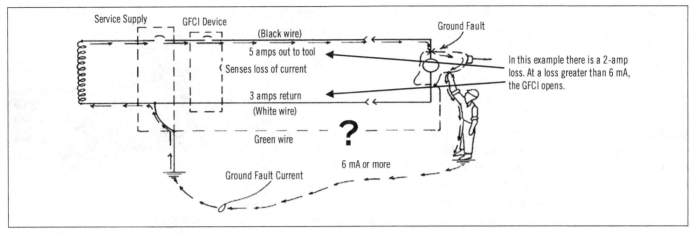

Figure 8–21. GFCI diagram. *(U.S. Department of Labor)*

15- and 20-A, 120-V outdoor receptacles, garage circuits, and bathrooms for residential locations. Consult the NEC and local codes for specific ground-fault requirements.

GFCIs must be located at the power source, such as the breaker or receptacle, to protect the entire exposed circuit. Excessive length of temporary electrical wiring or long extension cords can cause ground-fault current leaks to flow by capacitive and inductive coupling. The combined leaks of current can exceed 5 to 6 mA, thus causing the GFCI to "nuisance" trip.

GFCI fault tripping may be caused by one or more of the following items:
- wet electrical cord plug connections or wet power tools
- outdoor GFCIs unprotected from rain or water sprays
- defective electrical equipment with case-to-hot-conductor fault
- too many power tools on one GFCI branch circuit
- resistive-type heaters
- coiled extension cords (long lengths)
- poorly or improperly installed GFCIs
- defective or damaged GFCIs
- electromagnetic-induced current near high-voltage lines.

Follow the manufacturer's instructions regarding the testing of the GFCI. Testing should be conducted while the GFCI is under a load by pushing the TEST button. The GFCI should trip instantaneously. If it performs properly, remove the load and then reset it for use. Most manufacturers recommend monthly testing for permanently installed devices. Portable GFCIs should be tested prior to each use. Immediately replace any defective GFCI.

Remember that GFCIs do not replace over-current protection. GFCIs are protection against the most common form of electrical shock and electrocution: the line-to-ground fault, which may not trip an over-current device.

Proper use of GFCIs should significantly reduce the number of electrical shock incidents, which presently account for about 1,100 deaths and thousands of injuries annually.

Control Equipment

Electrical equipment should be provided with lockout capabilities for both AC and DC distribution circuits. Locate control panels to provide comfort and safety for the operator and maintenance activities. Switchboards should be away from live or moving parts of machinery. Avoid placing electrical controls in wet or damp areas, or use equipment rated for wet locations. The space around switchboards should not be used for storage and should be kept free of rubbish. Whenever possible, place the switchboard in a designated room, or use screen enclosures to keep unauthorized personnel out. The doors to the electrical enclosures should remain closed and locked to prevent unqualified workers from gaining access. This includes unqualified management personnel. Post hazard warnings as appropriate in motor control areas and electrical rooms.

Lighting in electrical service areas should be a primary consideration. Inadequate lighting may result in workers approaching energized parts too closely. Replacing blown bulbs should be considered a priority before work begins on energized equipment. Electrical control rooms should have battery backup emergency lighting to assist evacuation and rescue in emergencies.

Switch and fuse cabinets should have close-fitting doors that are kept locked to contain arcs and prevent accidental contact with conductive materials. This also prevents dirt, oil, and moisture from collecting in cabinets. The equipment framework and all metal parts must be grounded in accordance with NEC regulations.

Arrange connections, wiring, and equipment of switchboards and panel boards in an orderly manner. Plainly mark switches, fuses, and circuit breakers, and arrange them to simplify the identification of the circuits or equipment they supply. One-line diagrams must be kept up to date and available for reference.

Insulating mats provide additional protection from ground paths around switchboards, fuse panels, and control equipment. The mats should be moisture resistant, nonconductive, and able to withstand mechanical and traffic abuse.

Circuits initiated by push buttons should be low voltage (600 V or less) to prevent excessive arcing. However, push buttons may be used to control high-voltage circuits when step-down transformers or relays are provided. They can be used to limit the voltage in the control part of the circuit from exceeding 250 V for high-voltage equipment.

When voltage or current levels need to be checked often, consider installing permanent voltage and/or amp meters to avoid excessive use of portable test devices. (Misused or malfunctioning portable meters are a common cause of shock and arc blasts.)

Motors

Motors should be located so they do not interfere with the normal movement of personnel or materials. A means of disconnecting should be located within sight of the motor, if feasible. A single disconnecting means can be installed near a group of individual controllers as a main disconnect. This can simplify lockout procedures on a multimotor process. The disconnecting means must be readily accessible and must plainly indicate what it controls and whether it is in the open (off) or closed (on) position.

Exposed motors should be kept in areas free from dust, moisture, and flammable or corrosive vapors whenever possible. Motors can be isolated from personnel by mounting them on overhead supports, by installing them below floor level, or by placing them in designated motor rooms. Electrical conductors and contacts generating more than 50 V must be enclosed or guarded. (See Figure 8–22.)

Motors must be rated for the type and size required for the load and for the conditions under which they operate. There are AC and DC motors. AC motors can be synchronous or asynchronous, single-phase or three-phase, or shaded-pole or split-phase, as well as many other variances. They also vary in voltage, frequency, amperage, horsepower, and efficiency.

Safety professionals are primarily concerned with the motor area's classification and suitability for the work environment. An International Protection (IP) rating, also called the ingress protection rating, is used to provide manufacturers with a precise test that can be used to rate a product's ability to survive a particular environment. The IP code is specified using a series of two numbers and, optionally, a letter as defined by the International Electrotechnical Commission (IEC). An example of this would be an IP67 rating. The first number in this rating defines the item's solid particle protection. Thus, the number 6 indicates that the product is dust-tight. The second number in this rating defines the item's liquid ingress protection. So, here, the number 7 indicates that an IP67 product is protected for "immersion of up to 1 meter for a duration of 30 minutes." (See Figure 8–23.)

Figure 8–22. Exposed wiring. *(Eivaisla/iStock)*

There are many variations of these digits that are used to define how a product is protected against solid particles and liquids. The most commonly used variations in the motor industry are IP40, IP65, IP66, IP67, IP68, and IP69K. In addition to operating environments, consider other environmental factors such as cleaning the machine and the work area. Careful consideration of the workplace environment when selecting motors can prevent breakdowns and potential electrical hazards.

Overloads

A motor can be overloaded by (1) excessive friction within the motor itself, (2) being used for the wrong kind of job, (3) an obstruction in the driving or driven mechanism, or (4) pushing the machine to perform beyond the motor's capacity.

If one of these conditions causes the current in a motor to exceed the current rating on its nameplate, the actual heating may increase by as much as the square of the current's increase. As a result, insulation may be burned, soldered connections may be melted, or there may be excessive wear on the bearings.

To safeguard against overloads, motors have various forms of overload protection. In most cases, an additional thermal element is connected in the power circuit of the motor. This is accomplished (1) with an integral over-temperature protective device that will open the motor's circuit in the event of overheating or (2) with an external current-sensitive protective device that generates heat from an excessive current. These devices operate an overload relay and open the circuit to the motor.

Inrush Current

When started with full-line voltage, AC motors draw line currents substantially greater than their full-load-running

Value	Protection Against Ingress of Solids
0	No protection
1	Protected against solid objects over 50 mm (e.g., hands, large tools)
2	Protected against solid objects over 12 mm (e.g., fingers, large tools)
3	Protected against solid objects over 2.5 mm (e.g., wire, small tools)
4	Protected against solid objects over 1.0 mm (e.g., wire, screws)
5	Limited protection against dust ingress (no harmful deposit)
6	Totally protected against dust ingress

Figure 8–23. IP ratings. *(National Electrical Manufacturers Association)*

Value	Protection Against Ingress of Liquids	Test Configuration
0	No protection	
1	Protected against vertically falling drops of water	Test for 10 minutes, equivalent to 1 mm of rainfall per minute
2	Protected against direct sprays of water up to 15° from vertical	Test for 10 minutes, equivalent to 3 mm of rainfall per minute
3	Protected against direct sprays of water up to 60° from vertical	Test for 5 minutes, 0.7 liters/minute, 80–100 kPa pressure
4	Protected against water sprayed from any direction, limited ingress permitted	Test for 5 minutes, 10 liters/minute, 80–100 kPa pressure
5	Protected against low-pressure water jets (6.3 mm nozzle) sprayed from any direction, limited ingress permitted	Test for 15 minutes, 12.5 liters/minute, 30 kPa pressure at 3 m distance
6	Protected against high-pressure water jets (12.5 mm nozzle) sprayed from any direction, limited ingress permitted	Test for 3 minutes, 100 liters/minute, 100 kPa pressure at 3 m distance
6K	Protected against high-pressure water jets (12.5 mm nozzle) sprayed from any direction, under elevated pressure, limited ingress permitted	Test for 3 minutes, 75 liters/minute, 1000 kPa pressure at 3 m distance
7	Protected against immersion between 15 cm and 1 m	Test for 30 minutes immersed up to 1 m depth
8	Protected against long periods of immersion under pressure	Continuous immersion in water, generally up to 3 m depth
9K	Protected against close-range, high-pressure, high-temperature spray downs	

current rating. The actual magnitude of this current is called "inrush current." It is a function of the motor horsepower and design characteristics. It is also called "locked rotor current."

A letter used to indicate the design code rating on the nameplate is referred to as the code letter. The code letter of the motor is an indication of the locked-rotor kilovolt-amperes per horsepower. This is a function of the motor's design. The code letter ratings indicate the starting current a motor will draw. Code letters below F indicate a low starting current; those beyond F indicate a high starting current. (See Figure 8–24.)

Current-sensitive thermal elements of the proper capacity are listed in the NEC. Over-current devices with higher than the recommended values can result in motor damage. When over-current devices repeatedly operate, consider revising the motor output, production methods, or motor replacement.

Motor control centers often use adjustable over-current devices. Adjustments to overload sensitivity should be done only under the supervision of a qualified electrical engineer. Adjusting the response time increases the available fault current in the system and may result in extremely dangerous conditions.

Motors used in hazardous locations (areas that contain flammable vapors, combustible dusts, or fibers) require the use of equipment approved for those conditions. (See Hazardous Locations later in this chapter.)

Motor Maintenance

Regular visual inspections should be performed to identify any obvious wear, blockages to cooling fans, or environmental contamination (moisture or corrosion). Motors should be periodically disassembled to discover evidence of failed com-

Motor Code Letter			
Code	kVA/hp	Code	kVA/hp
A	0–3.14	L	9.0–9.99
B	3.15–3.54	M	10.0–11.19
C	3.55–3.99	N	11.2–12.49
D	4.0–4.49	P	12.5–13.99
E	4.5–4.99	R	14.0–15.99
F	5.0–5.59	S	16.0–17.99
G	5.6–6.29	T	18.0–19.99
H	6.3–7.09	U	20.0–22.39
J	7.1–7.99	V	≥22.4
K	8.0–8.99		

Figure 8–24. Ratings A–F indicate low starting currents. Ratings G–V are high starting currents. The motor's code letter is helpful in determining the maximum rating of the motor's electrical circuit protection. A replacement motor should have the same rating as its predecessor.

ponents, such as burnt windings, broken leads, and the like. Motor windings should be tested for insulation breakdown. Because the commutator and brush assembly are high-wear parts of a DC motor, extra time should be spent on inspecting, repairing, or replacing these vital components. Bearing inspections should be performed and worn-out or noisy bearings replaced. Motor bearings should be lubricated regularly unless sealed, nonlube bearings are used. These types of internal inspections normally require motors to be removed for bench-type examination and should be conducted only after the motor has been properly locked out. Full operational tests should be performed prior to leaving the repair shop. Common motor problems include dust, stray oil, moisture, misalignment, vibration, overload, and friction.

Friction and Wear

Friction and wear can be the result of poor motor alignment or lubrication. Repair activities should not be conducted while the motor is running. Follow the manufacturer's lubrication charts and instructions for the types and grades of lubricants, the frequency of lubrication, and other maintenance practices. Properly lubricated motors reduce friction, excessive wear, overheating of bearings, and potential fires.

Extension Cords

Extension cords are considered temporary wiring. Temporary wiring is defined as any wiring that is "not a permanent part of premise wiring." OSHA also refers to them as "cord sets." Extension cords are not a quick fix or a substitute for proper wiring of the premises. Extension cords must be listed by a nationally recognized testing laboratory. Inspect all extension cords regularly. For short-term applications, cords should be inspected prior to each use. Cord inspections should include a visual check for cracks or deformities and an examination of both ends to ensure they are free of damage or missing pins and have proper continuity. Continuity tests can be done with a volt/ohmmeter or a simple receptacle tester. Avoid kinking or excessive bending of power cords. Remove any cord with cracked or worn insulation or damaged plugs or sockets from service immediately. Typically, company policy should prohibit splicing of extension cords. If a cord must be repaired, follow procedures as outlined in NEC section 400.9 or OSHA 29 CFR 1926.403. A qualified worker may repair cords that are No. 12 AWG or larger with a suitable vulcanized or molded splice that meets the original protection design of the manufacturer.

Extension cords should not be disconnected while under an electrical load. Instruct users to turn off equipment before plugging or unplugging cords. Store extension cords neatly coiled, in a dry room, and at normal temperature. Various OSHA standards require cord sets to be inspected and documented at least quarterly. [See 29 CFR 1926.404(b)(1) for construction or 29 CFR 1910.304(b)(2) for general industry.]

Waterproof cord connectors should be used to protect plug ends from moisture in wet locations.

Extension Cords for Portable Tools

Cords for use with portable tools and equipment are made in several grades, each of which is designed for a specific type of service (see NEC section 400.4). Use heavy-duty jacketed cords with portable electric tools and extension lamps in boilers, tanks, or other grounded locations. Special types of rubber or plastic outer jackets should be used where the cord might come in contact with oils or solvents. Because the metal frames of electrical equipment should be grounded, use only power cords with a ground conductor. Double-insulated tools (nongrounded) should be protected with a GFCI.

Extension Cords for Heating Devices

Cords for heating devices, such as electric irons and water heaters, require heavy-duty construction. They are made with an insulated covering that contains a flame-retardant or thermosetting compound, such as neoprene. Such cords are designed to resist high temperatures and, in the case of neoprene, dampness.

Flexible Cords

The various types of flexible cords and cables and their approved uses and sizes are given in NEC Table 400.4. Connect flexible cords to devices and fittings so that tension will not be transmitted to joints or terminal screws. This is accomplished by special fittings or practices that provide strain relief. A variety of cord connection protection devices are available to protect against water intrusion and unintentional disconnections. All plugs that are attached to cords must have the terminal screw connections covered by suitable insulation. Avoid connecting (daisy-chaining) two or more extension cords together because this will exceed their ratings or approval listings.

Hard-service cords are marked with letters such as S, SE, SO, or ST; junior hard-service cords are marked with letters such as SJ, SJE, SJO, SJT, or SJTO. Other letters may be included after these that indicate the nature of the insulation or outer covering. (See Appendix I at the end of this chapter for an explanation of cord rating codes.) Flat-wire cords are prohibited from use on construction sites because they do not provide the puncture protection that double-insulated cords provide.

Extension Lamps

Handles of portable hand lamps must be made of nonconductive material, and there should be no metallic connection between the lamp guard and the socket's shell. For use near exposed live parts, such as the rear of switchboards, be sure that the guard itself is made of nonconductive material. Temporary lights should be placed at least 7 ft above the work surface or have guards to prevent contact with the bulb. Where flammable vapors, gases, combustible dusts, easily ignitable fibers, or flyings are present, lamps and cords must be approved for the type of hazard involved—that is, they must be intrinsically safe. Do not modify, repair, or add to these systems without approval of the manufacturer. Do not suspend temporary lights by their cords unless the manufacturer's instructions allow the practice.

In some facilities, the shock hazard of portable lamps is eliminated by using a lighting system stepped down to 12 V or less.

Specialized Processes

Among the many types of electrical equipment used in industry are the electric furnace, auxiliary heating devices, high-frequency heating equipment, electric welding equipment, x-ray laser, and ultraviolet and infrared installations. Induction heating (IH) is a common, noncontact heating process. It uses high-frequency electricity to heat materials that are electrically conductive. Because it is noncontact, the heating process does not contaminate the material being heated. It is also very efficient because the heat is actually generated inside the workpiece. This can be contrasted with other heating methods where heat is generated in a flame or heating element, which is then applied to the workpiece. For these reasons, induction heating lends itself to some unique applications in industry.

These processes or devices may introduce special operating hazards. High-frequency heating installations range in power capacity from a few hundred watts to several hundred kilowatts. Therefore, safety considerations are of prime importance. Often, protection from their electrical hazards may be secured through the same procedures recommended for use with the more common types of electrical equipment, such as complete enclosures or grounding. Following manufacturers' recommendations or industry standards is essential.

The resistance of the body to the flow of high-frequency current does not depend on the skin. At frequencies of 200 kHz to several hundred megahertz, currents flow in a very thin shell on the surface of the conductor. This tendency of high-frequency current to flow on the surface is known as "skin effect," and it increases as the frequency increases. Should the skin of a human being be punctured, the current still flows on the surface and does not penetrate to the vital organs of the body.

A person coming into contact with high-frequency electrical energy can be burned because of the natural tendency to pull away, thereby setting up an arc. High-frequency skin burns can also result from radio-frequency antennae or from waveguide exposure. Similar to serious sunburns, the burn will occur, but the person will feel nothing until after the exposure. These burns are painful and usually take longer to heal than burns from the more common thermal-heat sources.

The following methods can be used to prevent high-frequency burns to operators:
- Connect to the equipment interlocks or other devices that remove power whenever an access door is opened. This protects the operator from contacting electrical or high-frequency energy.
- In large operations, have material to be heated by induction carried by hopper feeds or conveyors in and out of the heating coil. The heating coil should then be enclosed with an effective shield so induced currents do not reach the operator.
- Locate high-frequency generators some distance from the work position. The high-frequency energy is then conveyed by a waveguide or transmission line.
- Insulate conductive coils or equipment surfaces with compatible insulation materials suitable for the application. Insulation protects the user in case of accidental contact with the coils and from the equipment's heated surfaces.

High-Voltage Equipment

In general, high-voltage equipment is more carefully guarded than low-voltage equipment because of the greater arcing and heating hazards. In addition to durability, high-voltage equipment is often more spacious to allow for heat dissipation. Changes or modifications should be done under electrical engineering supervision. (See Figure 8–25.)

Figure 8–25. High-voltage work. Only authorized and specially trained personnel should work on high-voltage equipment. Their qualifications and training should be well documented. *(wolv/iStock)*

GROUNDING

Grounding refers to both equipment grounding and system grounding. The electrical distribution system is grounded in order to prevent the occurrence of excessive voltages from such sources as lightning, line surges, or accidental contact with higher-voltage lines. Equipment grounds protect the facility electrical system and equipment. All non-current-carrying metallic parts and enclosures are grounded. This is done to assist the over-current devices to rapidly operate in the event of a ground fault. (See NFPA 70.)

Because current travels between different levels of voltage, electrical systems are grounded to keep non-current-carrying metal parts at zero voltage. According to the NEC, grounding is required for the following:
- refrigerators and similar equipment
- appliances using water, such as washers
- hand-held power tools
- motor-operated appliances
- any equipment used in damp areas
- portable hand lamps with metallic ground guards
- nonelectrical parts of equipment, such as metal equipment frames.

However, grounding is not required if the above items are:
- approved and labeled double-insulated tools
- insulated transfer tools of less than 50 V.

System Grounding

Most alternating-current systems operating at 50 V or more must be grounded. Grounding is accomplished by bonding the identified conductor (neutral) to a grounding electrode by means of an unbroken wire called a grounding electrode conductor. Identification of this grounded conductor is white or neutral gray-colored insulation or markings.

The grounded conductor (neutral) from the transformer is bonded at the service entrance panel to the grounding electrode conductor.

Bonding jumpers are then connected to a ground rod; a metal underground water pipe; a metal building framework; a bare copper conductor encircling the building; or a concrete-encased, steel-reinforcing-bar system. In residential installations, the use of water pipes for grounding is now prohibited in many local codes due to the increased use of PVC pipes for water distribution. Always check state and local codes for grounding requirements. (See Figure 8–26.)

Industrial installations are grounded in a similar manner, except that the transformer may be very large and mounted locally on a concrete pad. In this case, the grounding electrode is a metal grid carefully placed within the ground or within the concrete pad itself. As in grounding in a home, the grounded circuit conductor brought to the service equipment within the facility or commercial building must be bonded to a grounding electrode as specified in the NEC. The system ground must be capable of dispersing all potential stray electrical current available, so the design is related to the electrical service provided.

Higher resistance values can be encountered when a "made" electrode such as a rod, pipe, or plate is used. In many parts of the country, soil conditions make low-resistance values for made electrodes nearly impossible to obtain. In such cases, the NEC requires that additional electrodes

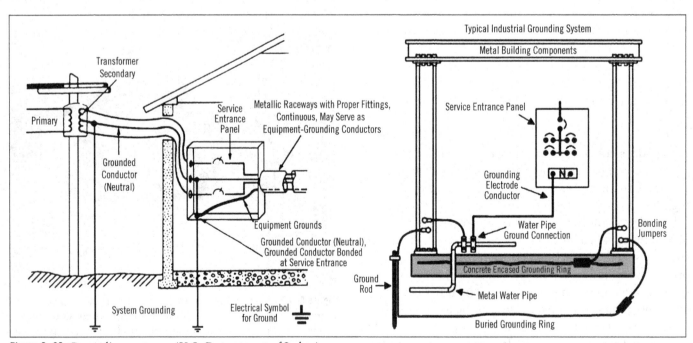

Figure 8–26. Grounding systems. *(U.S. Department of Labor)*

be used. Because the same term is applied both to system grounding and equipment grounding, it is often thought that the grounding electrode must have a very low resistance in order to dissipate ground faults. This is not true. Most of the current from a ground fault finds its way back to the source transformer via the neutral conductor, not the earth.

A grounding electrode is not intended to carry large fault currents for extended periods of time. Such electrodes provide a point of equalization so that large voltage differences both inside and outside the building are minimized. As stated earlier, some of the primary causes of such voltage differences are power surges or lightning strikes on the building, transformer, or service lines.

Once the grounded circuit conductor (neutral) is taken beyond the service entrance panel, it is not grounded again in additional subpanels or junction boxes. This establishes the service grounding point as the lowest potential and directs current back to the source. If the neutral and grounds are bonded at the subpanel neutral bar, two paths are created back to the service equipment. Both paths have low impedance (copper wire, aluminum wire, or metal conduit). Return current from the branch circuits served by the subpanel will split evenly—half on the equipment ground and half on the neutral as it returns to the source. This creates undesirable electrical exposures on the grounded equipment parts.

The major exception to this rule is the installation of additional transformers to modify voltages at various locations. In such instances, the secondary transformer may constitute a new or separately derived system. Then, the identified circuit conductor (neutral) is to be grounded to its own approved grounding electrode or system. As always, check local codes and/or with an electrical engineer before installation.

The rule prohibiting the subsequent grounding of the grounded (neutral) circuit conductor is frequently abused at the terminals of an ordinary receptacle. Many well-meaning but uninformed maintenance personnel will connect the neutral, or "grounded circuit conductors," to the equipment-grounding terminals on the receptacle. They think they have thereby doubled the grounding. However, in addition to violating the provisions of OSHA and the NEC, they have set up a dangerous condition called "bootlegging." If the neutral circuit is interrupted anywhere from the receptacle back to the point of attachment to ground, all grounded equipment enclosures will be energized at the line voltage. The equipment ground will then act as the neutral wire. Thus, anyone contacting a grounded surface would be exposed to the electrical fault energy. Generally speaking, the grounding conductor and the grounded (neutral) conductor should be connected only at the service entry point.

Not all systems are required to be grounded. If trained personnel are present, some manufacturing processes can use ungrounded systems or high-impedance grounded systems. They provide a higher degree of safety by not interrupting strategic equipment when a fault-to-ground occurs. The presence of fault-indicating equipment, together with properly trained personnel who can quickly make repairs, ensures against costly downtime of equipment and the hazardous conditions that may result. This type of system should be approved by the authority having jurisdiction (AHJ).

Equipment Grounding

When the insulation on conductors fails, the enclosures and metal parts are raised to line voltage and constitute a serious hazard for personnel. However, if the metal enclosure is attached to the main bonding jumper and to the service equipment with an equipment-grounding conductor, this voltage difference will not occur. Moreover, if the fault itself has a low resistance, and if the equipment-grounding conductor has been properly installed and maintained, a large amount of current can flow. The excess current flow should rapidly open the over-current device that protects the circuit.

The NEC and OSHA are very specific about equipment-grounding requirements because they are the key safety feature of an electrical system. (See Figure 8–27.) There are very few exceptions to grounding requirements.

Fixed equipment to be grounded includes all the exposed non-current-carrying metal parts that are likely to become energized (1) within 8 ft (2.4 m) vertically or 5 ft (1.5 m) horizontally of ground; (2) when located in a damp or wet location and not isolated; (3) when in electrical contact with metal; (4) when in a hazardous location; (5) when supplied by a metal-clad, metal-sheathed, or metal-raceway wiring method; or (6) when operated with any terminal in excess of 150 V to ground. In addition, the exposed non-current-

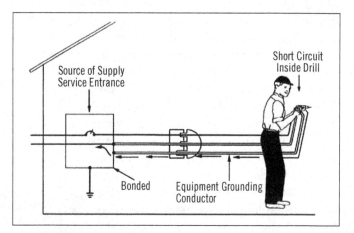

Figure 8–27. Proper equipment grounding provides a path for fault current should an insulation failure occur. In this manner, dangerous fault current will be directed back to the source—the service entrance—and will enable circuit breakers or fuses to operate, thus opening the circuit and stopping the current flow. *(U.S. Department of Labor)*

carrying metal parts of the following should be grounded, regardless of voltage: motor frames; controller cases for motors; electrical equipment in garages, theaters, and motion picture studios; accessible electric signs and associated equipment; and switchboard frames and structures.

Also, the following equipment should be grounded:
- frames and tracks of electrically operated cranes
- metal frames of non-electrically-driven elevator cars that have electrical conductors
- hand-operated metal shifting ropes or cables of electric elevators
- metal enclosures around equipment carrying voltages in excess of 750 V between conductors
- mobile homes and recreational vehicles.

Unlike the grounded circuit conductor (neutral), the equipment-grounding conductor may be grounded continuously along its length. The equipment-grounding conductor may be a bare conductor, the metal raceway surrounding the circuit conductors, or an insulated conductor. If the conductor is insulated, it must have a continuous green cover or a green cover with a yellow stripe. The equipment-grounding conductor is always attached to the green hexagonal-headed screw on receptacles, plugs, and cord connectors. When an approved metal conduit system is used as an equipment-grounding conductor, a bonding jumper is used to connect the receptacle to the box. If this is not done, the receptacle must be listed by an NRTL as being constructed to provide self-grounding.

Note: Often, the metal pipe or raceway is used as a grounding conductor instead of running a separate wire for grounds. The use of raceway grounds has resulted in many ground failures. If the connections or the pipe is broken, the grounding from that point on is compromised. Due to this problem, many local codes prohibit raceway grounds. A grounding conductor is more dependable for most situations.

For many devices—such as receptacles, switches, and lighting fixtures—an additional bonding jumper is added from the device to the grounded enclosure. This is done to prevent interruption of the grounding path if the device is removed from the junction box. The proper use of jumpers will provide flexibility when maintenance is required while still maintaining the integrity of the ground. (See Figure 8–28.)

Solder bridges are not permitted in the equipment-grounding circuit because they can easily melt, nor are switches, fuses, or any other interrupting device that could open the grounding path. Always ground the equipment-grounding conductor in the same enclosure with the conductors of the circuit that it protects.

A suitable attachment plug listed for equip-

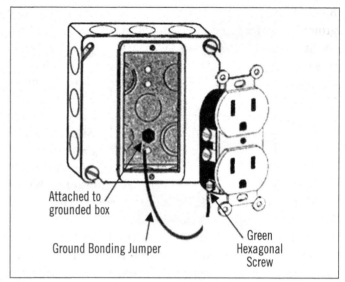

Figure 8–28. Device bonding jumper. *(Image courtesy of LoMastro & Associates Inc.)*

ment-grounding circuits is available in noninterchangeable configurations for all common voltage, current, and phase combinations. The National Electrical Manufacturing Association (NEMA) sets the standards on size, shape, and configuration for plugs and receptacles in the United States. NEMA wiring devices are made in current ratings from 15 to 60 A and voltage ratings from 125 to 600 V. Different combinations of contact blade widths, shapes, orientations, and dimensions provide noninterchangeable connectors. Each design is unique to a particular voltage, current capacity, and grounding system. Modifying plugs or receptacles to accept or connect to devices with other ratings is prohibited. (See Figure 8–29.)

General-duty devices may be used where they are not subjected to rough service or moisture. However, select the most rugged and watertight equipment if damaging exposures are anticipated. Careful selection of devices will ensure against loss of grounding continuity because of damaged components or corrosion.

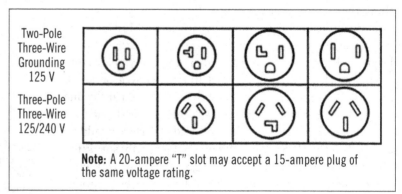

Figure 8–29. Common electrical receptacle configurations. *(Image courtesy of LoMastro & Associates Inc.)*

The size of the equipment-grounding conductor is important because it carries fault current in the event that a ground fault develops. Unless sufficient current is drawn through the over-current device to cause it to operate, the circuit will not be properly protected. The heating effects that occur at points of high resistance in a poor equipment-grounding system may result in arcs, fires, and/or explosions.

NEC Table 250.122 gives the proper sizes for equipment-grounding conductors used on various circuit ratings. Increase the conductor's size if long runs are encountered.

Workmanship is one of the most overlooked aspects of good equipment grounding. The NEC states: "Connection of conductors to terminal parts shall ensure a thoroughly good connection without damaging the conductors...." Where this provision is not observed, a point of low thermal capacity will occur. Even though the resistance of the equipment-grounding circuit is measured and determined to be quite low, if the system does not have adequate thermal capacity, high-fault currents will cause these points to overheat and melt.

Temporary Grounds

The use of temporary grounding and bonding during maintenance operations is common where voltages can be induced on a deenergized circuit. Electromagnetic induction is the production of an electromotive force across a conductor when it is exposed to varying magnetic fields. Deenergized wires can be influenced by an energized wire in the same raceway. Grounding conductors (properly sized) are attached to deenergized conductors in order to maintain a zero voltage state. This process is a part of the lockout procedure and is considered energized work activity. When these grounds are applied or removed, appropriate personal protective equipment (PPE) must be worn. Written work procedures must emphasize when temporary grounds are needed, where they are to be connected, and the danger of reenergizing the circuit with temporary grounds in place.

Maintenance of Grounds

Only personnel with knowledge of electricity should install or repair electrical equipment. They should make certain that the equipment-grounding conductor is attached to correct contact points.

Good maintenance ensures an electrically continuous equipment-grounding path from tool or electrical equipment through every attachment cord and device. This path ends at the bonding jumper at the service equipment ground or in the enclosure of a separately derived system.

Portable testing devices, such as a three-light neon receptacle tester, provide the most convenient means for checking polarity and other circuit connections. Other metered instruments are available to measure the actual impedance of the grounding circuit. Take precautions whenever using electrical testing equipment.

A receptacle-tension tester is used to measure the strength of blade contact in receptacles. The tester indicates the amount of tension on each receptacle blade. By using this tester during regular inspections, the maintenance department can replace receptacles before they result in an ineffective contact or before a fire occurs at the contact points. Tension testers can be purchased at most electrical supply houses.

Portable power tools should be checked by using an insulation-resistance tester that applies 500-V DC between the motor windings and the metal enclosure. Some insulation-resistance meters provide an additional function that permits checking and quantifying the continuity of all the conductors as well.

For the industrial user with a planned tool maintenance program, the use of a portable appliance tester will provide regular examination of tools through a series of specified tests. Tests are performed in the tool storage room. Tools should be inspected before they are issued or upon their return. Records should be maintained for each tool tested and for all repairs performed. A program for inspecting and testing power tools should be well planned and documented. Training must be provided for anyone who uses power tools, tests instruments, or repairs equipment. A well-planned testing program can predict tool failure and prevent downtime and injuries.

Three-Wire Adapters

Adapter devices that receive a three-prong plug and convert to a two-prong plug are allowed for residential use but are not allowed for commercial use. Only three-prong receptacles should be provided in commercial settings. The residential-grade adapters are often improperly connected and, therefore, fail to provide continuity to ground. When used in a home setting, the center screw on the coverplate should be attached to the ground plate on the adapter to properly complete the equipment ground. (See Figure 8–30.)

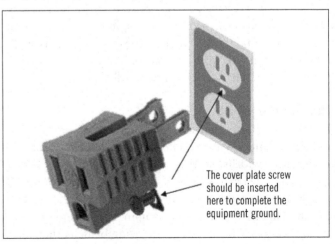

Figure 8–30. Three-wire adapter. *(Image courtesy of LoMastro & Associates Inc.)*

Double-Insulated Tools

As an alternate to grounding power tools, the NEC accepts the use of double-insulated appliances. Hand-held tools manufactured with nonmetallic cases are called double-insulated tools. If approved, they do not require grounding under the National Electrical Code. Although this design method reduces the risk of grounding deficiencies, a shock hazard can still exist. Double-insulated appliances are extremely popular due to their reduced weight and costs. They are an effective means of reducing exposure to electrical hazards. However, double insulation does not protect against defects in the cord, plug, or wet locations. Continuous inspection and maintenance for all electrical tools and devices are required. Do not use double-insulated power tools or equipment where water or a ground loop is present unless they are protected by a GFCI. (For more details, see Chapter 17, Hand and Portable Power Tools, in this manual.) Exposure to water, however, can allow a leakage path that may be of either high or low resistance. Handling equipment with wet hands, in high humidity, or outdoors after a rainstorm can be hazardous. Workers' hands must be dry before plugging or unplugging equipment.

The symbol for double insulation is the "box in a box." The words "double insulated" or the symbol must remain visible on the equipment. (See Figure 8-31.)

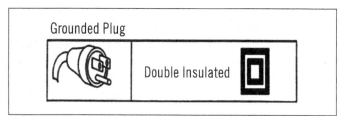

Figure 8-31. Hand-held power tools must use a three-wire plug, or the tool must show by word or symbol that the tool is double insulated. *(Image courtesy of LoMastro & Associates Inc.)*

Today, battery-powered tools are available for nearly every trade or circumstance. Cordless drills, impact tools, cutting tools, nailers, and sanders are just a few of the options available. This greatly reduces the potential for electrical injuries, but chargers and batteries must be handled properly.

Polarity Plugs

The use of polarity plugs for ungrounded equipment is an additional safety feature. The two blades on the plug are of different widths, so they will fit only one way into a receptacle. The narrow blade is the "hot" or "ungrounded" conductor. It should always be routed through the device's switch. This ensures that the equipment is energized only when the switch is operated. The wide blade is the "neutral" or return path. It is connected on the back side of the motor or appliance. (See Figure 8-32.) Reversing these con-

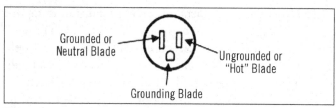

Figure 8-32. Polarity plug. *(Image courtesy of LoMastro & Associates Inc.)*

nections allows the device to be fully energized even with the switch in the OFF position. This can result in exposure to live electrical contacts even with the machine in the OFF position. If the system is bootlegged, the metal casing of the equipment can become "hot."

HAZARDOUS LOCATIONS

Hazardous (classified) locations are areas in which explosive or flammable gases or vapors (Class I), combustible dust (Class II), or ignitable fibers (Class III) are present or likely to become present. Such materials can ignite as a result of electrical arcs whenever the following two conditions exist:

1. The proportion of the flammable substance to oxygen in the air permits ignition. The percentage of oxygen and a gas/vapor creates an explosive hazard when between the lower explosive level (LEL) and the upper explosive level (UEL). When the atmosphere has at least enough fuel and oxygen (LEL) and not too much (UEL), there is an explosive atmosphere. (See Chapter 9, Fire Protection, in this manual.)
2. An electric arc, a flame escaping from an ignited substance in an enclosure, heat from an electric heater, or another source of ignition is present at a temperature equal to or greater than the flash point of the flammable mixture.

The OSHA regulations can be found in 29 CFR 1910.307 for general industry and 29 CFR 1926.407 for construction. Equipment, wiring methods, and installations of equipment in hazardous (classified) locations must be intrinsically safe, approved for the hazardous (classified) location, or safe for the hazardous (classified) location. *Standard* electrical apparatus, considered safe for ordinary applications, are not safe to use in hazardous locations. Sparks or electric arcs occurring in these locations have led to costly fires and explosions and resulted in many deaths.

Before selecting electrical equipment and wiring practices for a hazardous location, determine the exact nature of the flammable materials that will be present. An electrical fitting or device that is approved as safe for installation in an atmosphere of combustible dust may be unsafe for operation in an atmosphere containing flammable vapors or gases.

Conduct a study of the machines or devices to be used and the processes involving liquids, gases, or solid substances. Check the ratings for the equipment and flammable substance(s). When hazards have been identified and the layout of the building has been examined, decide whether only certain sections of the facility should be classified as hazardous or whether the hazardous conditions extend to all parts of the facility.

Present the results of this study to a manufacturer of explosion-proof apparatus so that electrical equipment and wiring can be matched for safe installation. Leading manufacturers of electrical equipment often maintain a staff of engineers to guide buyers in the purchase of explosion-proof fittings for nearly any situation.

Articles 500 through 503 of the NEC assign general requirements for electrical installations in hazardous locations; Articles 511 through 555 prescribe definitive requirements for specific types of occupancies.

Determining whether a hazardous situation exists in an industrial location is seldom difficult—the difficulty lies in defining the extent of the specific zones. Industries that commonly have hazardous locations often have procedures for classifying areas available through networks and associations. Defining the limits or area of the hazardous location is critical to the process. How far above, below, and outward from the source will the hazardous location extend?

For circumstances not covered in Articles 511 through 555, use this general rule: the limits of the hazardous locations are those mutually agreed upon by the owner, the owner's insurance carrier, and the authority having jurisdiction.

Classification of Hazardous Locations

Hazardous locations are classified depending on the properties of the flammable vapors, liquids, or gases; combustible dusts; or fibers that may be present. Consider each room, section, or area of the facility separately.

Hazardous locations are classified as Class I, Class II, or Class III, depending on the physical properties of the combustible substance that might be present (Table 8–A). These classes are subdivided into Division 1 and Division 2, depending on the degree of likelihood that an ignitable atmosphere might be present. Combustible substances are arranged into seven groups—A through G. The grouping depends on the reaction of the substances upon contact with an ignition source—highly explosive, moderately incendiary, and so forth. Equipment installed in hazardous locations must be approved for the applicable class, division, and group (Table 8–B). Descriptions of the three classes and two divisions in each class are:

- Class I locations are those in which flammable gases or vapors are present or likely to become present.
- Class II applies to locations in which combustible dust is likely to be present.
- Class III locations are those in which easily ignitable flyings, such as textile fibers or metal filings, are present.
- In a Division 1 location, an ignitable atmosphere could occur at any time during the course of normal operations. These areas represent the worst-case condition.
- In a Division 2 location, no ignitable atmosphere exists under normal operating conditions. However, an equipment malfunction, operator error, or other abnormal circumstance might create a hazardous environment.

TABLE 8–A. Hazardous Location Classifications

Class I Highly flammable gases or vapors		Class II Combustible dusts		Class III Combustible fibers or flyings	
Division 1	Division 2	Division 1	Division 2	Division 1	Division 2
Locations where hazardous concentrations are probable, or where accidental occurrence should be simultaneous with failure of electrical equipment	Locations where flammable concentrations are possible, but only in the event of process closures, rupture, ventilation failure, etc.	Locations where hazardous concentrations are probable, where their existence would be simultaneous with electrical equipment failure, or where electrically conducting dusts are involved	Locations where hazardous concentrations are not likely, but where deposits of the dust might interfere with heat dissipation from electrical equipment, or be ignited by electrical equipment	Locations in which easily ignitable fibers or materials producing combustible flyings are handled, manufactured, or used	Locations in which such fibers of flyings are stored or handled, except in the process of manufacture

Groups:
A—Atmospheres containing acetylene
B—Atmospheres containing hydrogen or gases or vapors of equivalent hazard
C—Atmospheres containing ethyl-ether vapors, ethylene, or cyclopropane
D—Atmospheres containing gasoline, hexane, naphtha, benzene, butane, propane, alcohol, acetone, benzol, or natural gas
E—Atmospheres containing metal dust, including aluminum, magnesium, and other metals of equally hazardous characteristics
F—Atmospheres containing carbon black, coke, or coal dust
G—Atmospheres containing flour, starch, or grain dusts

TABLE 8–B. Guidelines for Classifying Hazardous Areas

DETERMINING THE NEED FOR CLASSIFICATION
A need for classification is indicated by an affirmative answer to any of the following questions.

Class I	Class II	Class III
• Are flammable liquids, vapors, or gases likely to be present? • Are liquids having flash points at or above 100°F likely to be handled, processed, or stored at temperatures above their flash points?	• Are combustible dusts likely to be present? • Are combustible dusts likely to ignite as a result of storage, handling, or other causes?	• Are easily ignitable fibers or flyings present, but not likely to be in suspension in the air in sufficient quantities to produce an ignitable mixture in the atmosphere?

ASSIGNMENT OF CLASSIFICATION
Classification is determined as indicated by an affirmative answer to any question.

Class I, Division 1	Class II, Division 1	Class III, Division 1
• Is a flammable mixture likely to be present under normal operating conditions? • Is a flammable mixture likely to be present frequently because of repair, maintenance, or leaks? • Would a failure of process, storage, handling, or other equipment be likely to cause an electrical failure coinciding with the release of flammable gas or liquid? • Is the flammable liquid, vapor, or gas piping system in an inadequately ventilated location, and does the piping system contain valves, meters, or screwed or flanged fittings that are likely to leak? • Is the zone below the surrounding elevation or grade such that flammable liquids or vapors may accumulate?	• Is combustible dust likely to exist in suspension in air, under normal operating conditions, in sufficient quantities to produce explosive or ignitable mixtures? • Is combustible dust likely to exist in suspension in the air, because of maintenance or repair operations, in sufficient quantities to cause explosive or ignitable mixtures? • Would failure of equipment be likely to cause an electrical system failure coinciding with the release of combustible dust in the air? • Is combustible dust of an electrically conductive nature likely to be present?	• Are easily ignitable fibers or materials producing combustible flyings handled, manufactured, or used?

Class I, Division 2	Class II, Division 2	Class III, Division 2
• Is the flammable liquid, vapor, or gas piping system in an inadequately ventilated location, but not likely to leak? • Is the flammable liquid, vapor, or gas handled in an adequately ventilated location and can the flammable substance escape only in the course of some abnormality such as failure of a gasket or packing? • Is the location adjacent to a Division 1 location, or can the flammable substance be conducted to the location through trenches, pipes, or ducts? • If positive mechanical ventilation is used, could failure or improper operation of ventilating equipment permit mixtures to build up to flammable concentrations?	• Is the combustible dust likely to exist in suspension in air only under abnormal conditions but can accumulations of dust be ignited by heat developed by electrical equipment, or by arcs, sparks, or burning materials expelled from electrical equipment? • Are dangerous concentrations of ignitable dusts normally prevented by reliable dust-control equipment such as fans or filters? • Is the location adjacent to a Division 1 location, and not separated by a fire wall? • Are dust-producing materials stored or handled only in bags or containers and only stored—not used—in the area?	• Are easily ignitable fibers or flyings only handled and stored, and not processed? • Is the location adjacent to a Class III, Division 1 location?

DEFINING THE LIMITS OF THE CLASSIFIED LOCATION

The limits of the classified location—outward, upward, and downward from the source—must be determined by applying sound engineering judgment, experience gained on similar projects, and information from handbooks and other sources.

This table is based on recommendations of the National Fire Protection Association and American Petroleum Institute.

Requirements for Division 2 electrical installations are less stringent than those for Division 1 locations because, in Division 2, two possible, but improbable, circumstances must coincide for ignition. In addition, any accidental formation of an ignitable atmosphere in a Division 2 location can usually be quickly stopped and the ignitable atmosphere dispersed. As a result, any exposure to fire and explosion is usually of short duration in a Division 2 location.

The National Electrical Code, NFPA 70, contains guidelines for determining the type and design of equipment and installations that will meet this requirement. Those guidelines address electrical wiring, equipment, and systems installed in classified locations and contain specific provisions for the following: wiring methods, wiring connections, conductor insulation, flexible cords, sealing, drainage, transformers, capacitors, switches, circuit breakers, fuses, motor controllers, receptacles, attachment plugs, meters, relays, instruments, resistors, generators, motors, lighting fixtures, storage of battery-charging equipment, electric cranes, electric hoists and similar equipment, utilization equipment, signaling systems, alarm systems, remote control systems, local loud speaker and communication systems, ventilation piping, live parts, lightning surge protection, and grounding.

Class I Locations

Class I locations are areas in which flammable gases or vapors are present or are likely to become present in the air, in quantities sufficient to produce explosive or ignitable mixtures. In general, most hazardous locations in industrial facilities fall in the Class I category.

Flammable substances, such as acetylene and naphtha, have been used in industrial facilities for many years. In recent years, the use of less-common ignitable gases and liquids has increased. Hydrogen, for example, has many uses and must be given special consideration. Because of its wide explosive-mixture range, high flame-propagation velocity, low minimum-ignition energy level, and low vapor density (much lighter than air), a hazardous atmosphere can develop far above the hydrogen's source.

On the other hand, heavier-than-air vapors evolving from liquefied petroleum gases (LPGs), such as propane and butane, create special problems. LPG released as a liquid is highly volatile and has a low handling temperature; it readily picks up heat and creates large volumes of vapor. When released at or near ground level, the heavy vapors travel along the ground for long distances, thus greatly extending the horizontal plane of the hazardous location. Some flammable liquids with flash points below 100°F (38°C) may produce large volumes of vapor that may spread much farther than might normally be expected.

Lighter-than-air gases usually dissipate rapidly because of their relatively low densities. Unless released in confined, poorly ventilated spaces, low-density gases seldom produce hazardous mixtures in zones that are close to grade, where most electrical equipment is typically located.

In the case of hydrocarbons—most of which are heavier than air—the problem is not to establish the existence of a Class I location, but to define the limits of the Division 1 and Division 2 areas. Anywhere hydrocarbons are handled, used, or stored, there is a great chance that flammable liquids, gases, and vapors will be released in large enough quantities to create a hazard. Vapor can disperse in all directions as governed by the vapor's density and the air movement in the area. A very mild breeze can extend the limits of the hazardous location because the combustible mixture will not be dispersed significantly.

Division 1 Locations. Class I, Division 1 locations are areas in which one or more of the following conditions exist:

- Hazardous concentrations of flammable gases or vapors exist under normal operating conditions.
- Hazardous concentrations of flammable gases or vapors may exist frequently because of leakage, repair, or maintenance operations.
- A breakdown or faulty operation of equipment might release hazardous concentrations of flammable gases or vapors and might also cause simultaneous failure of electrical equipment.

Table 8–C gives some examples of Class I, Division 1 locations.

TABLE 8–C. Some Examples of Class I, Division 1 Locations

Locations where volatile flammable liquids or liquefied flammable gases are transferred from one container to another.
Interiors of paint spray booths and areas adjacent to paint spray booths and other spraying operations where volatile flammable solvents are used.
Locations containing open tanks or vats of volatile flammable liquids.
Drying rooms or compartments for the evaporation of flammable solvents.
Cleaning and dyeing areas where hazardous liquids are used.
Gas generator rooms and portions of gas manufacturing plants where flammable gas may escape.
Inadequately ventilated pump and compressor rooms for flammable gas or volatile flammable liquids.
All other locations where hazardous concentrations of flammable vapors or gases are likely to form in the course of normal operations.

Division 2 Locations. Class I, Division 2 locations are areas in which one or more of the following conditions prevail:

- Volatile flammable liquids or flammable gases are handled, processed, or used. However, they are confined in closed containers or systems from which they can escape only if the container is ruptured, the system breaks down, or the equipment is operated abnormally.
- Mechanical ventilation normally prevents hazardous concentrations of gases or vapors from forming. Ignitable concentrations can form only if the ventilation system fails.
- The area is adjacent to a Class I, Division 1 location from which hazardous concentrations might spread because of inadequate or unreliable ventilation supplied from a source of uncontaminated air.

Class I Groups. Maximum explosion pressures and safe operating temperatures vary widely for hazardous location electrical installations. Therefore, the electrical equipment must be approved for the specific flammable material used in the designated hazardous location (Figure 8–33 and Table 8–D).

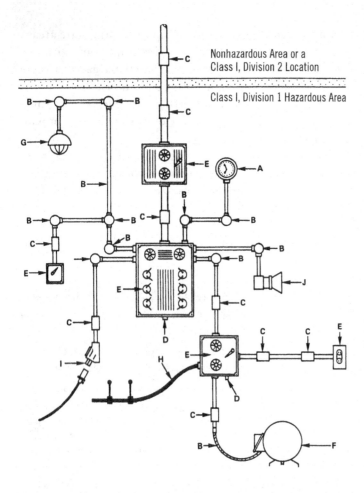

Figure 8–33. Class I, Division 1 hazardous location. (See Table 8–D.)

Substances creating Class I atmospheres fall into Groups A through D. A complete list of these substances can be found in NFPA 497M, Classification of Gases, Vapors, and Dusts for Electrical Equipment in Hazardous (Classified) Locations. Among the more common substances listed are:

- group A—acetylene
- group B—hydrogen or equivalent vapors and gases, such as manufactured gas
- group C—ethyl-ether vapors, ethylene, cyclopropane, and similar substances
- group D—gasoline, naphtha, benzene, hexane, butane, propane, alcohol, acetone, lacquer-solvent vapors, natural gas, and similar substances

Class II Locations

Class II locations are areas in which combustible dusts are present or likely to become present. A potential dust-explosion hazard exists wherever combustible dusts accumulate, are handled, or are processed. Airborne dust may be well below the concentration required for combustion; however, when poor housekeeping practices allow accumulation of dust in layers of as little as ¼ of an inch, a spark or welding slag can cause ignition and result in massive fires or explosions. Many dusts fall into the combustible category. (See Table 8–A; a complete list of these dusts is also contained in NFPA 497M.) Certain metal dusts may have characteristics that require safeguards beyond those required for atmospheres containing

TABLE 8–D. Summary of Equipment Requirements for Class I, Division 1 Hazardous Locations

A. Meters, relays, and instruments, such as voltage or current meters and pressure or temperature sensors, must be in enclosures approved for Class I, Division 1 locations. Such enclosures include explosion-proof and purged and pressurized enclosures. National Electrical Code (NEC), NFPA 70, 501-3(a)

B. Wiring methods acceptable for use in Class I, Division 1 locations include: threaded rigid metal or steel intermediate metal conduit and type MI cable. Flexible fittings, such as motor terminators, must be approved for Class I locations. All boxes and enclosures must be explosion-proof and threaded for conduit or cable terminations. All joints must be wrench tight with a minimum of five threads engaged. NEC, 501-4(a).

C. Sealing is required for conduit and cable systems to prevent the passage of gases, vapors, and flame from one part of the electrical installation to another through the conduit. Type MI cable inherently prevents this from happening by its construction; however, it must be sealed to keep moisture and other fluids from entering the cable at terminations. NEC, 501-5.
1. Seals are required where conduit passes from Division 1 to Division 2 or nonhazardous locations.
2. Seals are required within 18 in. from enclosures containing arcing devices.
3. Seals are required if conduit is 2 in. in diameter or larger entering an enclosure containing terminations, splices, or taps.

D. Drainage is required where liquid or condensed vapor may be trapped within an enclosure or raceway. An approved system of preventing accumulations or to permit periodic drainage are two methods to control condensation of vapors and liquid accumulation. NEC, 501-5(f)

E. Arcing devices, such as switches, circuit breakers, motor controllers, and fuses, must be approved for Class I locations. NEC, 501-6(a).

F. Motors should be:
1. approved for use in Class I, Division 1 locations
2. totally enclosed with positive-pressure ventilation
3. totally enclosed inert-gas-filled with a positive pressure within the enclosure, or
4. submerged in a flammable liquid or gas. NEC, 501-8(a).

G. Lighting fixtures, both fixed and portable in rigid metal conduit, must be explosion-proof and guarded against physical damage. NEC, 501-9(a).

H. Flexible cords must be designed for extra-hard usage, contain a grounding conductor, be supported so that there will be no tension on the terminal connections, and be provided with seals where they enter explosion-proof enclosures. NEC, 501-11.

the large-type dusts of aluminum, magnesium, and their commercial alloys. For example, zirconium, thorium, and uranium dusts have extremely low ignition temperatures [as low as 68°F (20°C)] and minimum-ignition energies lower than any material classified in any of the Class I or Class II groups.

Division 1. Class II, Division 1 locations are ones that meet one or more of the following criteria:
- Combustible dust is or may be in suspension in air during the course of normal operations in quantities large enough to produce explosive or ignitable mixtures.
- Mechanical failure or abnormal operation of machinery or equipment might create explosive or ignitable dust mixtures. They might also provide a source of ignition through simultaneous failure of electrical equipment, operation of protective devices, and so forth.
- Combustible dust of an electrically conductive nature may be present.

Some examples of Class II, Division 1 hazardous locations are the following:
- work areas of grain-handling and grain-storage facilities
- areas near dust-producing machinery and equipment in grain-processing, starch, sugar-pulverizing, melting, or hay-grinding facilities and in flour mills
- areas in which metal dusts and powders are produced, processed, handled, packed, or stored
- coal bunkers; coal-pulverizing facilities; and areas in which coke, carbon black, and charcoal are processed, handled, or used.

Dust that is carbonized or excessively dry is susceptible to spontaneous ignition. Therefore, electrical equipment installed in Class II locations should be able to operate at full load without developing surface temperatures high enough to cause excessive dehydration or carbonization of dust deposits that might form.

Division 2. In Class II, Division 2 locations, ignitable concentrations of combustible dusts are not usually found in the course of normal operations. However, abnormal conditions may allow combustible dusts to accumulate in large enough quantities and in large enough dust-to-air concentrations to permit ignition. Abnormal conditions could include failure of dust-control or process equipment or of containers, chutes, or other product handling equipment. The following are usually considered to be Class II, Division 2 locations:
- areas containing closed bins or hoppers and enclosed spouts and conveyors
- areas containing machines and equipment from which appreciable quantities of dust would escape only under abnormal conditions
- warehouses and shipping rooms in which dust-producing materials are handled or stored only in bags or containers
- areas adjacent to Class II, Division 1 locations.

Class II Groups. Electrical equipment installed in Class II locations must be approved for the class, the division, and the applicable group (Figure 8–34 and Table 8–E). For purposes of equipment approval, dusts are classified in the following three groups based on their conductivity:
- group E—metal dusts
- group F—carbon black, charcoal, coal, coke dust
- group G—flour, starch, grain dust.

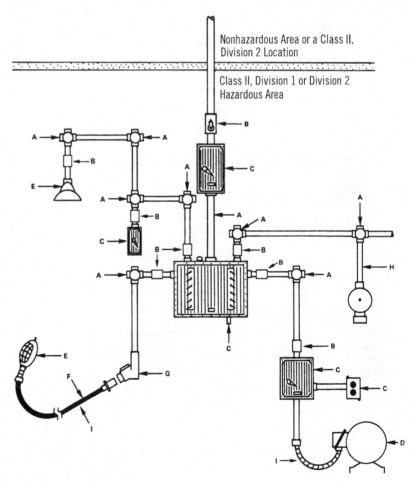

Figure 8–34. Class II hazardous locations. (See Table 8–E.)

TABLE 8–E. Summary of Class II Hazardous Locations

A. In Class II, Division 1 locations, wiring methods require that boxes and fittings containing arcing and sparking parts be in dust-ignition-proof enclosures. Threaded metal conduit or type MI cable with approved terminations is required for Class II, Division 1 locations. NEC, NFPA 70, 502-4(a).

In Class II, Division 2 locations, boxes and fittings are not required to be dust-ignition proof but must be designed to minimize the entrance of dust and prevent the escape of sparks or burning material. In addition to the wiring systems suitable for Division 1 locations, the following systems are suitable for Division 2 locations: electrical metallic tubing, dust-tight wireways, and types MC and SNM cables. NEC, 502-4(b).

B. Suitable means of preventing the entrance of dust into a dust-ignition-proof enclosure must be provided where a raceway provides a path to the dust-ignition-proof enclosure from another enclosure that could allow the entrance of dust. NEC, 502-5.

C. Switches, circuit breakers, motor controllers, and fuses installed in Class II, Division 1 locations must be dust-ignition proof. NEC, 502-6. In Class II, Division 2 areas, enclosures for fuses, switches, circuit breakers, and motor controllers must be dust-tight.

D. In Class II, Division 1 locations, motors, generators, and other rotating electrical machinery must be dust-ignition proof or totally enclosed pipe ventilated. NEC, 502-8.

In Class II, Division 2 areas, rotating equipment must be one of the following types:
1. dust-ignition proof
2. totally enclosed pipe ventilated
3. totally enclosed nonventilated, or
4. totally enclosed fan cooled.

Under certain conditions, standard open-type machines and self-cleaning squirrel-cage motors may be used. NEC, 502-8(b).

E. In Class II, Division 1 locations, fixed and portable lighting must be dust-ignition proof. NEC, 502-11.

Lighting fixtures in Class II, Division 2 locations must be designed to minimize accumulation of dust and must be enclosed to prevent the release of sparks or hot metal.

In both divisions, each fixture must be clearly marked for the maximum wattage of the lamp, so that the maximum permissible surface temperature for the fixture is not exceeded. Additionally, fixtures must be protected from damage. NEC, 502-11.

F. Flexible cords in Division 1 and 2 are required to:
1. be suitable for extra-hard usage
2. contain an equipment grounding conductor
3. be connected to terminals in an approved manner
4. be properly supported, and
5. be provided with suitable seals where necessary. NEC, 502-12.

G. Receptacles and attachment plugs used in Class II, Division 1 areas are required to be approved for Class II locations and provided with a connection for an equipment grounding conductor. NEC, 502-13.

In Division 2 areas, the receptacle must be designed so the connection to the supply circuit cannot be made or broken while the parts are exposed. This is commonly done with an interlocking arrangement between a circuit breaker and the receptacle. The plug cannot be removed until the circuit breaker is in the OFF position, and the breaker cannot be switched to the ON position unless the plug is inserted in the receptacle.

Class III Locations

Class III locations are those in which easily ignitable fibers or flyings are present. However, these fibers are not likely to be in suspension in the air in quantities large enough to produce an ignitable atmosphere. Single fibers of organic materials such as linen, cotton tufts, and fluffy fabrics, however, are quite vulnerable to a localized heat source such as an electric spark. In pure oxygen, single fibers of cotton can be ignited by a 0.02-joule (J) spark.

Textiles such as those used in clothing can be ignited and burned with repetitive or sustained high-energy electric sparks. Cotton and wool fabrics can be ignited in pure oxygen with a spark of 2.3 J. In normal air, a spark of 193 J or more is required for ignition. Silk and polyester fibers are more difficult to ignite than cotton or wool.

Fibers contaminated with oily substances can be ignited with much weaker sparks than can clean fibers. Typically, only a thousandth of the energy required to ignite a clean fabric is required for an oily sample of the same fabric. In general, the burning characteristics of fibers will be affected by the (1) specific gravity of the substance, (2) size and shape of the sample, (3) air circulation in the area, (4) oxygen concentration, and (5) relative humidity.

Division 1. Class III, Division 1 locations are ones in which easily ignitable fibers or materials, which produce combustible flyings, are manufactured, handled, or used. This classification usually includes:
- facilities that produce combustible fibers
- portions of rayon, cotton, or other textile mills
- flax-processing facilities
- clothing-manufacturing facilities
- woodworking facilities.

Division 2. Class III, Division 2 locations are ones in which easily ignitable fibers are stored or handled but not manufactured or processed. An example of a Class III, Division 2 location is a textile warehouse.

Class III—No Groups. There are no group designations associated with Class III locations. Electrical equipment installed in Class III locations need only be approved for the applicable class and division. For equipment not subject to overloading, the maximum surface temperature of equipment under normal conditions should not exceed 329°F (165°C); for equipment such as transformers and motors that are subject to overloading, maximum surface temperatures should not exceed 248°F (120°C).

Establishing the Limits of a Hazardous Location

Classifying an area as hazardous for purposes of NEC compliance is based on the possibility of flammable liquids, vapors or gases, combustible dusts, and easily ignitable fibers or flyings being present. After an area has been classified as hazardous, the next step is to determine the degree of hazard. Should the area be classified as Division 1 or Division 2? Also, the limits of the hazardous location must be defined—how far above, below, and outward from the source of the hazard.

The safest electrical installation in a hazardous location, of course, is none at all. As much as is practical, situate electrical equipment outside the area defined as hazardous. For example, lighting in a hazardous location such as a storage room may be essential, but the light switch may be located on the outside of the room to minimize the risk of arcs and sparks produced when the switch is operated. It is, however, seldom possible or practical to locate all electrical equipment outside the hazardous area. The facility engineer is responsible for ensuring that all electrical equipment in the hazardous area, and its associated wiring, conforms to the NEC and does not present a potential of explosion. (See Figure 8–35.)

Reducing Hazards. Two ways to reduce the chance of explosions from electrical sources are (1) remove or isolate the potential ignition source from the flammable material and (2) control the atmosphere at the ignition source. For an explosion to occur, the following two conditions must coexist. If either of these two conditions is eliminated, the explosion hazard is reduced to zero:
1. combustible material present in a sufficient amount and the proper concentration to provide an ignitable atmosphere
2. an ignition source powerful enough to ignite the combustible materials that are present.

Planning Electrical Installations. The hardest part of planning an electrical installation to conform to NEC requirements for hazardous locations is to define the limits of the hazardous area. How far should the hazardous location be considered to extend to ensure that safety is served, without taking unnecessary and expensive precautions? No hard-and-fast rules can be applied. Experience on comparable projects and an understanding of specific conditions at the job site provide a far better basis for defining limits than any theoretical study of flammable vapors, gases, dusts, or fibers.

The environmental aspects of an installation—prevailing winds, site topography, proximity to other structures and equipment, and climatic factors—can significantly affect the extent of a hazardous location. Among the factors that should be evaluated in establishing the limits are the following:
- size, shape, and construction features of the building
- existence and locations of doors and windows and their manner of use
- absence or presence of walls, enclosures, and other barriers
- existence and locations of ventilation and exhaust systems
- existence and locations of drainage ditches, separators, and impounding basins
- quantity of hazardous material likely to be released
- location of potential leakages
- physical properties of the hazardous material—density, volatility, chemical stability, and so forth
- frequency and type of maintenance or repair work performed on the systems containing the hazardous substance or other equipment in the area.

When establishing limits of hazardous locations, consider the area surrounding each source of hazardous material as a location and determine its individual limits. For example, consider the following rules for a Class I location:
- In the absence of walls, enclosures, or other barriers, and in the absence of air currents or other disturbing forces, a gas or vapor will be distributed in a predictable fashion.
- For heavier-than-air gases and vapors released at or near grade level, potentially hazardous concentrations are most likely to be found below grade. Heavier-than-air gases distribute themselves downward and outward. As the height above grade increases, the hazard decreases.
- For lighter-than-air gases, little or no potential exists for a hazard at or below grade. Lighter-than-air gases distribute themselves upward and outward. As the enclosed height above grade increases, the potential hazard increases. For purposes of classification, treat gases with a density greater than 75% of the density of air at standard conditions as if they were both lighter and heavier than air. Define the limits of the hazardous location accordingly.

A Sample Installation. Most hazardous locations in industrial facilities belong in Class I and involve heavier-than-air gases and vapors. Figure 8–36 provides recommendations for establishing limits for some typical situations involving such gases and vapors.

In Figure 8–36, a process pump, which handles flammable liquids at moderate pressures, is at grade elevation. Because the source of the hazard is in the open air, there are no pockets below grade level where flammable vapors can accumulate. In addition, liquid can escape only in the event

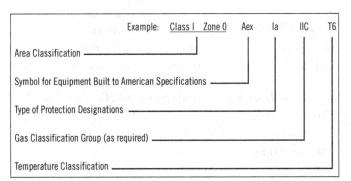

Figure 8–35. Example marking for Class I, Zone 0. (*U.S. Department of Labor*)

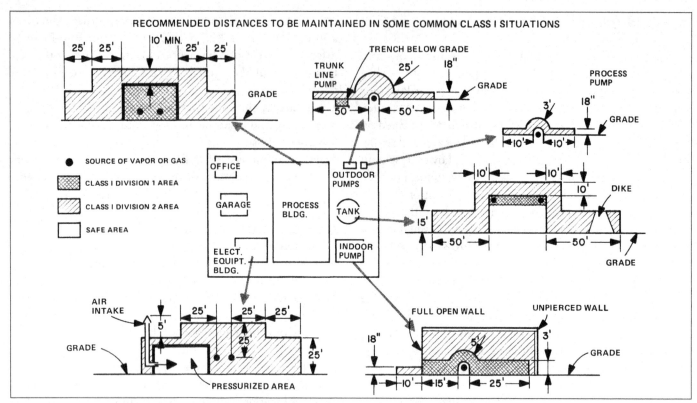

Figure 8-36. Recommended distances to be maintained in some common Class I situations, developed from standards established by NFPA and API. *(Reprinted with permission from* Plant Engineering *magazine.)*

of equipment failure—such as a leaking gasket. Therefore, the area surrounding the pump is classified as Division 2.

A different situation prevails, though, for the trunk-line pump in Figure 8–36. Even though it is located outdoors, the trunk-line pump operates at relatively high pressure. Thus, the chance of equipment failure and the volume of liquid that might be released are significantly increased. A below-grade trench is in the vicinity of the trunk-line pump. For these reasons, the limits of the Division 2 location surrounding the pump are extended, and the trench itself is classified as Division 1.

The outdoor tank in Figure 8–36 is installed with its base at grade level, and the tank is fitted with a floating roof. The space within the shell of the tank and above the roof is classified as a Division 1 location. The area surrounding the tank is classified as Division 2; it extends 10 ft (3 m) above the tank. Horizontally, the Division 2 area extends 50 ft (15.2 m) from the tank or to the dike, whichever is greater.

In Figure 8–36 the indoor pump is in a building with one fully open wall but no through ventilation. The building also contains valves, meters, and other equipment and fittings that are likely to leak. The area adjacent to the pump, and near grade, is classified as Division 1. The rest of the building and the outdoor close-to-grade area next to the open wall are classified as Division 2.

The process building has numerous sources of flammable materials and is not especially well ventilated. The entire interior is classified as Division 1, and the area surrounding the building is classified as Division 2. The Division 2 area extends horizontally from the building for 50 ft (15.2 m) in every direction and at least 10 ft (3 m) above the roof or 25 ft (7.6 m) above the source, whichever is greater. The electrical equipment room is separated, but is immediately next to the process building. It lies entirely within the Division 2 location that surrounds the process building for 50 ft (15.2 m) in all directions. No fire wall separates the electrical equipment building from the process building.

The interior of the electrical equipment building, however, is classified as a safe area because the atmosphere is controlled to prevent ignition sources from contacting an ignitable mixture. A positive-pressurization system is used to keep the building purged of any flammable mixtures. Note that the air intake for the pressurization system must be at least 5 ft (1.5 m) above the classified location. This minimum distance must not be violated under any circumstances, or this space may be invaded by a flammable mixture.

Each source of flammable material in Figure 8–36 was considered to be the focal point of a separate hazardous location. The positioning of sources of hazardous materials could cause an overlapping of Division 1 and Division 2 locations. In such cases, the more stringent classification would prevail. Thus, the area of overlap is classified as Division 1.

Explosion-Proof Equipment

Explosion-proof equipment is defined in the NEC, Article 100, as

> equipment enclosed in a case that is capable of withstanding an explosion of a specified gas or vapor, which may occur within it, and of preventing the ignition of a specified gas or vapor surrounding the enclosure by sparks, flashes, or explosion of the gas or vapor within, and which operates at such an external temperature that a surrounding flammable atmosphere will not be ignited thereby.

Explosion-proof and intrinsically safe equipment does not guarantee that vapors will not seep into electrical components. The components of the electrical system are designed to withstand the explosive force if vapors do get inside and are ignited. These fittings and enclosures have undergone exhaustive tests and are listed or labeled for use in hazardous locations (Figure 8–37).

Components must be of durable material, provide thorough protection, and be properly installed in order to protect the electrical system from interacting with hazardous conditions.

Install explosion-proof fittings on new installations and on old wiring systems where alterations are being made that create hazardous locations. Observing NEC requirements minimizes the fire and explosion risk that might result from using standard electrical components and devices in hazardous locations.

Explosion-proof components are made of durable cast material, with roomy interiors for wiring and splices. They are able to withstand high internal pressure that results from an explosion without bursting open or becoming loose. All threaded conduit must be threaded with an NPT (National Pipe Taper) standard conduit cutting die that provides a ¾-in. taper per foot. The conduit should be made wrench tight to prevent sparking when fault current flows through the conduit system and to ensure the explosion-proof or flameproof integrity of the conduit system. Equipment with metric threaded entries must be identified as being metric, or a list of adapters must be provided to permit connection to conduit of NPT-threaded fittings. Only NRTL-listed adapters can be used for connection to conduit or NPT-threaded fittings.

Heat-producing equipment for hazardous locations, such as lighting fixtures, must not only contain the explosion and vent the cooled products of combustion, but must also be designed to operate with surface temperatures below the ignition temperatures of the hazardous atmosphere.

It is impossible to prevent gases or vapors from entering the interior of either an ordinary or an explosion-resistant wiring system. Gases eventually enter the entire line through the joints and through the breathing of the conduit system caused by temperature changes. Gaseous vapors also enter whenever covers are opened or removed. For these reasons, it is impossible to (1) provide an entirely vapor-proof switch unit, (2) regulate temperatures, or (3) keep the air free from flammable gases inside the equipment. The purpose of specially designed electrical equipment for hazardous locations is to positively confine the arc, heat, and explosion within the internal limits of the explosion-proof fittings.

Conventional wiring equipment can be used instead of explosion-proof fittings when control equipment is found outside locations containing the hazardous materials. This reduces both the cost and potential hazards of the installation.

MAINTENANCE

For safe and efficient service, electric equipment must be well maintained. Motors, circuit breakers, moving parts of switches, and similar current-carrying devices wear out, break down, and need adjustment. Only trained and experienced electricians should make repairs on electrical circuits and apparatus. Refer to NFPA 70B for complete requirements for electrical maintenance.

Use only high-grade electrical equipment listed by an NRTL for electrical maintenance. Inferior or unapproved equipment can be hazardous because of defective materials, poor design, or poor workmanship.

Before beginning work on electrical systems, trained and authorized maintenance personnel should test the systems with an approved contact-type tester to confirm that it is deenergized.

Maintenance personnel involved with energized electrical

Figure 8–37. An explosion-proof and dust-tight mercury switch, listed by Underwriters Laboratories, that makes and breaks the circuit when the switch tilts the tube inside the mechanism. *(Reprinted with permission from The Appleton Electric Co.)*

work must be trained on the specific electrical testing equipment provided. They should *know and demonstrate* knowledge such as how to (1) safely access the circuit for testing, (2) select and wear the correct personal protective equipment, (3) identify testing points from the schematic diagrams, (4) properly inspect their meter (leads, clips, and probes), and (5) select the correct procedure to test circuits. Only insulated tools such as pliers, screwdrivers, testing devices, and other equipment are allowed in the restricted approach boundary.

When maintenance or repair work must be done on energized conductors, two or more employees should work together. The supervisor should conduct a pre-job briefing that details the potential hazards, procedures to be followed, and the proper protective equipment and controls to use for each potential hazard.

The kind of protective equipment needed is determined by the type of circuit, the nature of the job, and the conditions under which the work must be done. Rubber gloves with leather gauntlets, sleeves, arc-resistant clothing, mats, line hoses, insulating platforms, safety headgear, safety glasses, safety belts, fuse tongs, and insulated switch sticks are among the more common items of protective equipment used. (See Electrical Safe Work Practices and Protective Equipment later in this chapter.)

Good safety practices lie not only in using the proper protective equipment, but in taking proper care of that equipment. Protective equipment should be inspected before each use, and rubber products must be tested at frequent intervals.

To prevent accidental grounding (short circuiting) and severe injuries, maintenance personnel must constantly be aware of hazards when working around energized equipment and circuits. Grounding can easily occur if an energized wire comes loose and contacts a pipe, a conduit, a metal fixture, another wire, or anything conductive that may create a path to the earth.

Test Equipment

Various types of electrical and electronic testing equipment can detect unsafe conditions before they get out of control and cause damage. Only qualified technicians or specially trained maintenance personnel should perform electrical testing. The electrical test equipment should be listed by UL or another nationally recognized laboratory (NRTL). Testing devices are rated as Category I, II, II, or IV. (See Figure 8–38.)

It is important to note that many companies and workers claim that they do not work on live circuits. This goal is a key feature in any electrical safety program, but it is hardly 100% attainable. Often, the process of diagnosing electrical problems requires the testing of an energized circuit for data. For example, amperage cannot be tested on a deenergized circuit. Verification of an electrical lockout requires contact testing to ensure zero potential. Both of these examples are

Rated Voltage	IEC 61010-1 2nd Edition			UL 61010B-1 (UL 31111-1)		
	CAT IV	CAT III	CAT II	CAT III	CAT II	CAT I
150 V	4,000 V	2,500 V	1,500 V	2,500 V	1,500 V	800 V
300 V	6,000 V	4,000 V	2,500 V	4,000 V	2,500 V	1,500 V
600 V	8,000 V	6,000 V	4,000 V	6,000 V	4,000 V	2,500 V
1,000 V	12,000 V	8,000 V	6,000 V	8,000 V	6,000 V	4,000 V
Resistance	2 ohms	2 ohms	12 ohms	2 ohms	12 ohms	30 ohms

Figure 8–38. Categories for electrical testing devices.

common and accepted practices but are considered energized or "live" work. When it is possible to deenergize equipment before performing work, then OSHA standard 29 CFR 1910.147 (Control of Hazardous Energy) must be followed.

The following types of equipment are standard and should be considered as essential testing equipment for most industrial settings: ammeter, voltmeter, continuity tester, ground-fault indicators and locators, power analyzer, receptacle circuit tester, receptacle tension tester, and voltage detectors. Many of these functions can be performed with multimeters, and some are task specific. (See Figure 8–39.)

Workers must be trained on the specific equipment they are authorized to use and the potential hazards associated with the process. Prior to testing electrical circuits for lockout, the test instrument should be verified on a known energized circuit (live), and then the necessary testing is performed (dead); when the testing is completed, the instrument is then rechecked on the known energized circuit (live). This is referred to as the "live-dead-live" procedure and verifies that the tester works before and during the entire testing process.

Infrared thermography is a nondestructive technique for detecting temperature differences. It can identify problems such as loose connections or excessive friction in machinery or mechanical systems. A camera-like device views a large area at a time, senses infrared emissions, and converts the emissions into a visual temperature display. The equipment remains in operation, so production is not interrupted. "Hot spots" can be pinpointed quickly, saving labor and costs and focusing plant maintenance where it is needed. Specialized training and certification are available for collecting data and interpreting the image results.

Specialized testing instruments, such as volt-ohm-milliammeters, oscilloscopes, and cable testers may also be used. Instruments such as these are generally fitted for detailed engineering work or are used where ordinary recording and indicating instruments are not accurate enough.

Improper use of testing equipment may result in arc blast or serious injury. Instruments must be inspected prior to each use. The condition of the leads, fuses, displays, and batteries is commonly inspected. Defects must be corrected before the equipment can be safely used. Many manufacturers recommend that the equipment be calibrated at

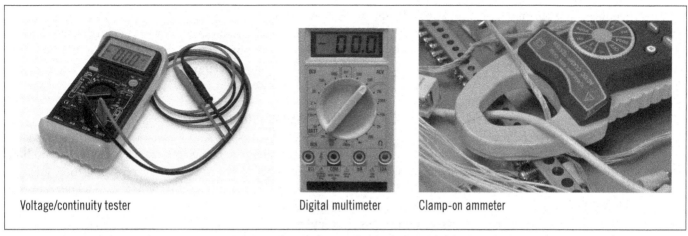

Figure 8–39. Typical electrical testers. *(From left to right: reichi/iStock; claylip/iStock; FactoryTh/iStock)*

regular intervals. Not inspecting equipment or following maintenance recommendations may violate warranties of the equipment.

Lockout/Tagout (Lock-Tag-Test)

The unexpected energization or start-up of machines and equipment by automatic or manual controls may cause injuries from electrical shock or from mechanical, chemical, or thermal radiation energies (injuries are not limited to these sources). Therefore, when electronic or electrical equipment must be serviced, maintained, repaired, or modified, open the circuit at the circuit's disconnect and lock the switch in the OFF (open) position. Apply a tag with a description of the work being done, the name of the person, the date, and the department or purpose for the lockout. Posting warning signs, or tagging alone, does not provide the positive protection that locking out equipment does.

Lockout procedures should be written and evaluated annually to ensure effectiveness. Authorized workers must also be evaluated annually to ensure they understand and follow these procedures. Energized work must be prohibited whenever feasible. Lockout is not an option, and it cannot be avoided due to inconvenience. Practices such as removing or adjusting motor brushes and cleaning or polishing commutators or slip rings should be performed on locked, tagged, and tested systems only.

Because of the risk of injuries or death, all authorized employees responsible for inspecting, testing, or maintaining the electrical system—and those affected by these activities—must be trained in lockout/tagout procedures (as authorized or affected workers) and follow safety-related work practices cited in 29 CFR 1910.331–339. (See Chapter 6, Safeguarding, in this manual, for details on an energy-isolation program.) The supervisor should see that the procedure is carried out and should provide each employee with the necessary keys, locks, and training. No two key configurations should be the same. Check to see that each key fits only one lock.

For identification, locks may be color-coded to indicate the types of craft or to differentiate shifts. Only one person has access to the key for each lock. No one should have a master key. Strict control of keys is a basic tenet in lockout safety. In emergency situations, when locks must be cut off, proper notification procedures must be followed to protect the worker whose lock is removed. When properly performed, a system locked out cannot be reenergized until everything has been inspected and each person has safely removed his or her lock.

When purchasing electrical equipment, be sure that it has lockout/tagout capabilities. To make lockout systems operable, the equipment should have built-in locking devices designed for the insertion of locks. There are devices available for locking out almost any type of equipment. Where older equipment is in use, or where explosion-proof or dust-tight equipment is installed, trained technicians or maintenance workers may construct attachments to accommodate lockout. Some cord and plug equipment may need plug control lockouts. (See Figure 8–40.) Where more than one person or group works on a piece of equipment, set up a multihasp locking device or group lockbox so that each person may apply his or her own lock.

Maintain an effective program with ongoing training in safe lockout/

Figure 8–40. Cord and plug lockout device with multihasps. *(State of California, Department of Industrial Relations)*

tagout procedures and through constant supervision. A typical lockout procedure generally follows these steps:

1. Alert the operator and other users of the system or equipment that is to be shut off.
2. Plan the shutdown to ensure that all necessary parts of the system will be controlled (off).
3. Deenergize the power supply and have each worker place his or her lock on the control switch, lever, valve, or energy control device or group lockbox.
4. Test to be sure the system is fully deenergized and that all residual energy has been drained, dissipated, released, or blocked.
5. When the work is completed, have each user remove his or her lock. Never permit workers to remove another worker's lock. Be sure workers do not expose others to danger during the restarting process. Verify that the equipment is clear, and post a watch, if necessary.
6. Reenergize the system.

A good practice to adopt is commonly called the "supervisory lockout," which requires that the area supervisor applies his or her lock first, before other workers, and is the last person to remove his or her lock before reenergizing. The process is helpful in avoiding oversights and emphasizes the supervisor's overall responsibilities in safety.

Rotating and Intermittent-Start Equipment

Before working on rotating machines or on automatic- and intermittent-start equipment, inspect all electrical controls and starting devices for proper lockout. This should be verified by the area supervisor. For example, when inspections or repairs are to be made on motors, generators, blowers, compressors, or converters, any part of which is remote-controlled or that may automatically start, the supervisor should accompany the authorized workers on a walkthrough as part of a pre-job briefing. This is especially important for any equipment that operates on timers or other automatic activators because they may "appear to be off."

Machinery connected to blowers, waterwheels, or pumps, without check valves, may start turning even when the current to the motor has been disconnected. For this reason, proper lockout would include blocking the rotor, armature, or other moving parts. (See Figure 8–41.)

Figure 8–41. A mechanical block is placed to prevent movement after the machine is deenergized. *(State of California, Department of Industrial Relations)*

Authorized employees working on or inspecting electrical equipment should not wear loose clothing because it may become entangled in couplings, coils, or other moving parts. They should remove all wristwatches, rings, metal pens, and pencils.

Fused Disconnects

An electrical fuse is a low-resistance link that is sacrificed (or melts) when an electrical circuit becomes too hot. When it is necessary to remove or replace a fuse, lockout rules apply. Testing a fused disconnect for lockout requires verification on both the line and load sides of the fuse. Employees must be trained and authorized to perform this task. Insulated fuse pullers must be provided and used to extract fuses. Always replace a fuse with one of the same type and size. Never bridge a fuse with other conductive materials. Verification of proper fusing should be included in the electrical inspection program because this is a very common hazardous condition.

High-voltage systems (more than 600 V) with fused disconnects present an elevated risk of arc blast. Fuse replacement begins with opening the controlling circuit breaker before operating the knife switches to eliminate potential arcing at the knife switches. It may be necessary to install temporary grounding conductors to each leg of the circuit before starting the work to prevent induced current from building on the conductors. Remove the grounding conductors just before the circuit is reenergized.

ELECTRICAL SAFE WORK PRACTICES AND PROTECTIVE EQUIPMENT

Persons working on electrical equipment should know that energized work is allowed only in limited and specific conditions. Years of bad or unsafe habits must be corrected in the electrical field. Electrical work practices have significantly changed over the past decade. Workers and supervisors must be trained to understand and implement these changes. NFPA 70E, Electrical Safety Requirements for Employee Workplaces, has been developed to guide employers on developing safety programs for performing work on electrical systems safely. This standard emphasizes the need to establish safe working conditions (lock-tag-test) and prohibits work on live circuits unless essential (see the following list of acceptable energized work activities) and requires proper training, planning, and safety equipment for use when live work is conducted.

Unless the following conditions exist, lockout procedures are mandatory for work on electrical systems. Failure to comply with these procedures should be considered a serious violation and should be subject to strict disciplinary actions. This includes discipline for supervisory personnel who either encourage or allow unsafe practices.

The only conditions where energized work is acceptable are the following:

1. *A system where total voltage to ground is less than 50 volts.* Unless additional hazards are present, this condition is considered reasonably safe.

2. *Additional hazards or increased risk.* This condition is created when a system is shut down. Examples of creating additional hazards might be an interruption of life-support systems, deactivation of emergency alarm systems, or shutdown of the ventilation system in hazardous locations.

3. *Infeasibility due to operational limitations or equipment design.* This condition would include performing diagnostic testing or work on a continuous process where shutdown would interrupt an integral operation to perform work on a part of that process, such as utility power distribution. (Note: There are very few industrial applications of integral work exceptions.)

Unless these circumstances exist, electrical safe work conditions (lockout) must be established and followed.

Safe work procedures should be provided in the form of written procedures. Job safety analyses or job hazard analyses (JSAs/JHAs) are often used to create and train workers in these safe work procedures. Energized work presents the risk of electrical shock and arc blasts. These hazards should be controlled using engineering and administrative means as well as personal protective equipment. Pre-work assessment should begin with an examination of the electrical system for likely hazards.

The condition of the electrical system and equipment is a key issue in establishing safe work procedures. Normal operation of electrical equipment is considered as meeting the following conditions:

1. Equipment is properly installed (complies with all applicable codes and standards).

2. Electrical equipment is properly maintained (in accordance with manufacturers' recommendations).

3. Equipment doors are closed and secured.

4. All covers and unused openings are closed.

5. There is no evidence of impending failure (such as arcing, overheating, visible damage or deterioration, loose or bound parts).

If the equipment is not in "normal" operating conditions, energized work should be avoided.

If work is permitted in an energized state (as noted earlier), an energized work permit should be completed under the following conditions:

1. *Work to be performed is inside the restricted approach boundary.* These approach boundaries represent the distance necessary to prevent electrical shock. NFPA 70E, Approach Boundaries or Shock Table [130.4(D)(a) for alternating current and 130.4(D)(b) for direct current], sets these boundaries. (See Figure 8–42.) The tables set distances for unqualified worker to approach fixed parts and movable conductors (distances normally apply to overhead power lines). Unqualified workers must remain outside of the limited approach boundary (LAB). For fixed systems 750 V and below, this distance is 42 in. (1 m). Movable conductors require a 10-ft LAB for unqualified workers. Qualified workers (trained, authorized, and properly equipped) may enter the restricted approach boundary (RAB). These boundaries represent the potential to make contact with exposed energized parts that could result in shock or electrocution while performing energized tasks.

2. *An employee interacts with electrical equipment when conductors or circuit parts are not exposed but there is a likelihood of injury from an arc flash hazard.*

Nominal System Voltage, Phase to Phase	Limited Approach Exposed Movable Conductor	Boundaries Exposed Fixed Circuit Part	Restricted Approach Boundary
< 50 V	Not Specified	Not Specified	Not Specified
50 V – 150 V	3.0 m (10 ft)	1.0 m (3.5 ft)	Avoid Contact
151 V – 750 V	3.0 m (10 ft)	1.0 m (3.5 ft)	0.3 m (1 ft)
751 V – 15 kV	3.0 m (10 ft)	1.5 m (5 ft)	0.7 m (2.17 ft)
15.1 kV – 36 kV	3.0 m (10 ft)	1.8 m (6 ft)	0.8 m (2.58 ft)
36.1 kV – 46 kV	3.0 m (10 ft)	2.5 m (8 ft)	0.8 (2.75 ft)
46.1 kV – 72.5 kV	3.0 m (10 ft)	2.5 m (8 ft)	1.0 m (3.25 ft)
72.6 kV – 121 kV	3.3 m (10.67 ft)	2.5 m (8 ft)	1.0 m (3.33 ft)

Figure 8–42. Sample of the Alternating Current Shock Approach Boundaries. See NFPA 70E Table 130.4(D)(a) for the complete table. *(Developed from NFPA 70E, 2015.)*

An energized work permit should contain the following items at a minimum:

1. description of the work to be performed, the circuit or equipment involved, and the location
2. explanation of why work must be performed on an energized system
3. description of the safe work practices that will be applied
4. result of a shock risk assessment identifying the voltage, limited approach boundary, restricted approach boundary, and necessary personal protective equipment and other protective equipment
5. result of an arc flash assessment, including the available incident energy or the arc flash PPE category, the arc-rated clothing required, and the arc flash boundary
6. methods employed to limit unqualified persons from access to the work zone
7. evidence of a pre-job briefing, including coverage of job-specific hazards
8. signature of an approving authority.

There are a few exemptions from the requirement for the work permit such as routine troubleshooting, thermography (outside the restricted boundary), and travel through work zones under a qualified worker's control. Permit exemptions do not exempt workers from the necessary planning, and all safety requirements must be followed.

A sample permit can be found in Appendix II at the end of this chapter.

When work must be performed on energized circuits, considerations for safety include selecting proper personal protective equipment. The acronym SAFE works well in the hazard assessment process.

S—for electrical shock. Principally the shock boundary for the unqualified and dielectric gloves for the qualified.

A—for arc flash and arc blast. The arc flash boundary for the unqualified and the appropriate level of arc-rated (AR) clothing for the qualified. (Note: FR clothing is not rated for electrical arc hazards.)

F—for falls, fires, and foul-ups. Proper fall protection, combustible materials, moving equipment, and confined-space issues are a few examples of additional hazards to be controlled.

E—for electrocution. Paths through humans can be deadly. In addition to PPE, energized parts and metal grounds paths should be covered or draped with insulating materials to prevent accidental contacts and paths to ground.

Workers must understand that rubber insulating gloves are not a substitute for safety devices and proper procedures. Gloves should be worn only as a supplementary measure. Two types of electrical gloves are available and designated as Type I (nonresistant to ozone) and Type II (resistant to ozone).

Voltage protection is broken down into the following classes:

- Class 00—maximum use voltage of 500 V AC/proof-tested to 2,500 V AC
- Class 0—maximum use voltage of 1,000 V AC/proof-tested to 5,000 V AC
- Class 1—maximum use voltage of 7,500 V AC/proof-tested to 10,000 V AC
- Class 2—maximum use voltage of 17,000 V AC/proof-tested to 20,000 V AC
- Class 3—maximum use voltage of 26,500 V AC/proof-tested to 30,000 V AC
- Class 4—maximum use voltage of 36,000 V AC/proof-tested to 40,000 V AC.

Before each use, gloves should be checked for punctures, tears, breakdown of rubber, or abrasions. A glove inflator should be used to perform daily checks on rubber gloves. If the glove inflator is not available, roll up the cuffs and force air into the fingers and palms of the gloves. Look, listen, and feel for air loss. If there is an air leak or breakdown of the rubber, the gloves should be destroyed and discarded.

Electrical gloves must be laboratory tested at least every 6 months. (See Chapter 7, Personal Protective Equipment, in this manual; ASTM F496; OSHA 29 CFR 1910.132 and 1910.137; or the North American Independent Laboratories for Protective Equipment Testing [NAIL for PET] for further discussion of gloves.)

Leather protectors must be worn over rubber gloves to protect the rubber from mechanical damage and from heat, oil, or grease. Electrical glove-testing services are available at approved laboratories across the country. Testing intervals should be at least every 6 months, depending on the amount of use, the type of work performed, and the conditions the gloves are exposed to. Tested gloves should be marked to indicate the test voltage and the date tested. It is also convenient to have the date of the next scheduled retest stamped on the glove. Test logs are another acceptable method of monitoring glove testing. All test results should be documented.

Where no independent laboratory is available, small companies can often have tests done at the local public utility. However, testing must comply with ASTM F496, Specifications for In-Service Care of Insulating Gloves and Sleeves.

EMPLOYEE TRAINING

Deviating from safe usage and installation practices with electrical and electronic equipment has the potential to cause severe injuries or death. Providing electrical safety training for all employees should be considered a high priority. The employer must thoroughly train all employees who work with electricity and electronic equipment. OSHA and NFPA state that both qualified and unqualified workers should receive electrical safety training. Electrical training for qualified workers must be conducted at least every 3 years.

Unqualified workers should receive awareness training on the company's lockout program. Even though they may not be authorized to perform lockout, they should understand the importance of the program and should learn to respect the process. Additionally, unqualified workers should be trained to comply with all safety boundaries and warnings, with the limits of their job in relation to electrical systems, and with activities they are prohibited from performing. Many workplace accidents occur when someone was "just trying to help."

Qualified workers need training on the specifics of the equipment they are expected to work on, the nature of electrical hazards involved, safe procedures, use of testing equipment, and lockout steps for each machine or process. Additional training for qualified employees should include CPR and AED use (with annual refreshers); the proper use of warning signs, guards, and other protective devices; and site-specific safety requirements. They must be trained to handle emergency situations and rescue procedures. It should be emphasized that energized work should never be done alone. If something goes wrong, the speed and efficiency of the emergency response may prevent death.

Recent amendments to OSHA's safety standards emphasize safety concerns when working on or near electrical equipment and energy sources. Three new paragraphs added to 29 CFR 1910.132 (General Requirements) discuss the employer's obligation to assess hazards in the workplace in order to select appropriate PPE for workers; to prohibit the use of defective or damaged PPE; and to use documented training sessions to teach workers how to use, care for, and repair PPE. The amendments underscore the need to educate workers to use proper eye and face protection (1910.133), head protection (1910.135), foot protection (1910.136), electrical protective equipment (1910.137), and hand protection (1910.138). According to OSHA officials, the new standards are performance oriented and encourage employers to continually improve protection for the workers. Officials believe that the new standards will improve worker acceptance of PPE by allowing better, more comfortable PPE and by ensuring that employers inform workers about the proper limits and use of protective equipment for the job.

Management must develop and implement safety programs to comply with OSHA's 29 CFR 1910.331–333, Electrical Safety-Related Work Practices. Both qualified and unqualified workers are covered by these standards. Sections 1910.332–333 describe the scope and content of training that employers are required to provide for workers

who handle power equipment or electrical energy sources and the selection and use of work practices to prevent injuries and fatalities to these workers. Safety programs developed by companies must meet or exceed the provisions of these standards. The OSHA standards establish the requirements for the employer regarding the "what to train." The methods to achieve these goals (the "how") are found in the NFPA 70E and 70B standards.

In developing appropriate safety training programs, management should focus on the facility's electrical system and the conditions of that electrical equipment. The program can then focus on the specific operations and changes as they occur within the systems.

Supervisors should be kept informed about existing and possible electrical hazards. Employers should require supervisors to maintain close involvement with operations that involve the use of electrical or electronic equipment. Supervisors must encourage employees to report any electrical defects or problems immediately. Unsafe tools or equipment must be repaired or replaced at once. (See Chapter 31, Safety and Health Training, in the *Administration & Programs* volume, for more information about training programs.)

Electrocutions continue to be one of the top causes of workplace fatalities. However, knowledge and protective equipment are available to prevent them.

SAFETY INSPECTIONS

Electrical safety inspections should be a regular and ongoing process in all facilities. The inspection(s) should include examination of physical conditions as well as unsafe procedures or behaviors, which may put personnel or equipment at risk. Deteriorating equipment or housekeeping conditions create increased risk of electrical incidents. Unsafe procedures must be documented and corrected before "incidents" occur. This may require a team approach and different levels of safety inspections. Safety inspectors must be trained on what constitutes correct conditions and behaviors. Inspection results can be instrumental in scheduling maintenance, repairs, system upgrades, and future training.

Safety professionals should ensure that inspections of electrical components, lockout procedures, and energized work practices are conducted on a regular basis. The frequency depends on the type of facility and workplace conditions. Supervisors should be trained to conduct continuous inspections for electrical hazards in their work areas. Qualified electrical workers and engineers should regularly examine wiring, motors, cabinets, and other electrical equipment for proper performance and maintenance.

For equipment inspections, the facilities electrician or other authorized person should deenergize equipment and perform complete lockout procedures (lock, tag, and test) to prevent exposing energized parts. Live circuits and equipment left in the operating mode are always a hazard, but during inspections, the risk of contacting energized parts or pulling wires free is significantly increased. Always assume that a circuit is live until it is proved dead by contact-type testing (such as a voltmeter). When unqualified personnel are involved in the inspection, they must observe the qualified electrician conduct tests to ensure that the parts to be exposed are deenergized. As an additional safeguard, each inspector must apply his or her own personal locks and tags to the deenergized equipment. Until the completion of all testing to verify the system is safe, the system must be considered energized.

An electrical safety inspection program should include awareness of specific items to inspect as well as commonly violated electrical standards. Common items to inspect should include the following:

1. service entrance panel—circuit identification, secure mounting, knockouts, connectors, clearances, live parts
2. system grounding—secure connections, corrosion, access, protection, proper sizing
3. wiring (general)—temporary splices protected, J box covers, insulation, knockouts, fittings, workmanship
4. electrical equipment/machinery—grounding, wiring size, over-current/disconnect devices, installation, protection
5. small power tools—attachment plugs, cords, cord clamps, leakage, grounding
6. receptacles—proper polarity, adequate number, mounting, covers, connections, protection, adapters
7. lighting—grounding, connections, attachment plugs and cords, cord clamps, live parts
8. GFCI protection—wet locations, temporary service, fountains, outdoor circuits, testing
9. switches—labeled or marked, covered as needed, lockable
10. extension cords—condition, plugs, receptacles, GFCI
11. over-current protectors—fuses, circuit breakers, link fuses.

SUMMARY

- Failure to establish or use safety practices for electrical equipment can result in property damage and serious injuries or fatalities. Workers must clearly understand the need to learn and follow safe work practices and to report electrical problems immediately.
- Severity of electrical shock is determined by (1) the amount of current, (2) the length of time the body is

exposed, and (3) the path current takes through the body. The protection triangle indicates methods to remove one or more of these elements.

- Electrical equipment should be installed in controlled areas and have fail-safe devices and guarding to protect workers and others from accidental contacts. All electrical equipment should be properly wired, grounded, and protected by over-current devices, ground-fault circuit interrupters, and control equipment installed for emergency shutoff.
- In control equipment, provide switchboards with lockout capabilities for both AC and DC circuits. Electrical motors should not interfere with the normal movement of personnel or materials.
- Extension cords must be inspected frequently to detect any wear, fraying, or breakage in the line. When service is needed for extended periods, permanent wiring should be provided.
- Make sure that workers are thoroughly trained to inspect their electrical equipment for safe operation, report any abnormal conditions, have equipment tested regularly, and observe all safety regulations and practices.
- Grounding includes system grounding, equipment grounding, and temporary grounding. Only trained personnel should install, test, and repair grounds.
- Standard electrical equipment should not be installed in locations where flammable gases, vapors, dusts, or other easily ignitable materials are present. Explosion-proof equipment is required for these locations.
- Management should determine the specific hazards in any particular location before selecting and installing electrical equipment. Management should also reduce the hazards by removing or isolating sources of possible ignition and/or by controlling the atmosphere at the ignition source.
- Before equipment inspections are performed, electrical equipment should be deenergized (whenever possible) and the systems locked, tagged, and tested to prevent accidental reenergizing while inspections are being conducted.
- Inspectors should be thoroughly trained for and properly protected from the hazards.
- All components of electrical equipment must be well maintained. Only trained and experienced electricians should make repairs on electrical circuits and apparatus.
- A facility's safety program should include thorough training for all employees who work with electrical and electronic equipment or who operate electrical systems. Supervisors should be kept informed about existing and possible hazards.
- Qualified employees should be regularly trained in work procedures, CPR/AED, rescue, and emergency procedures.

REFERENCES

American National Standards Institute, 1899 L Street, NW, 11th Floor, Washington DC 20036.

Attachment Plugs and Receptacles, ANSI/UL 498–2012.

Machine Tools Safety Package, ANSI B11.

National Electrical Safety Code, ANSI C2–2007.

Reactive Power Control, ANSI C84.1–2011.

Relays, Breakers, Switchgear Systems Associated with Electric Power Apparatus, ANSI C37.51a–2010.

Safety Requirements for Lockout/Tagout of Energy Sources, ANSI Z244.1–2003 (R2014).

Snow Throwers—Safety Specifications, ANSI/OPEI B71.3–2014.

Standard for Product Safety Signs and Labels, Includes Errata, ANSI Z535.4–2011.

American Society for Testing and Materials, 100 Bar Harbor Drive, PO Box C700, West Conshohocken, PA 19428-2959.

14 Standard Specifications for Standard Nominal Diameters and Cross-Sectional Areas of AWG Sizes of Solid Round Wires Used as Electrical Conductors, ASTM B258–2014.

Specifications for In-Service Care of Insulating Gloves and Sleeves, ASTM F496.

Specifications for Rubber Insulating Gloves, ASTM D120.

Specifications for Rubber Insulating Tape, ASTM D4325.

Brown, C. "Electrical Licensing Certification." http://www.ehow.com/info_8426839_states-require-electrician-certification.html, 2015.

Canadian Standards Association, 178 Rexdale Boulevard, Rexdale, Ontario M9W 1R3, Canada.

DHHS (NIOSH) Publication No. 97-113, *Control of Scrap Paper Baler Crushing Hazards*, Cincinnati, OH: National Institute for Occupational Safety and Health, 1997.

Factory Mutual System Research Organization, 1175 Boston-Providence Turnpike, PO Box 9102, Norwood, MA 02062.

Illuminating Engineering Society of North America, 120 Wall Street, Floor 17, New York, NY 10005-4001.

National Fire Protection Association, 1 Batterymarch Park, Quincy, MA 02269.

Classification of Gases, Vapors and Dusts for Electrical Equipment in Hazardous (Classified) Locations, NFPA 497M, 2008.

Electrical Safety Requirements for Employee Workplaces, NFPA 70E, 2015.

National Electrical Code, NFPA 70, 2014.
NEC 240.83(D).

Recommended Practices for Electrical Equipment Maintenance, NFPA 70B.

National Safety Council, 1121 Spring Lake Dr., Itasca, IL 60143-3210.

Electrical Inspection Illustrated. 3rd ed. 2008.

Occupational Safety and Health Data Sheets:

Electrical Switching Practices, 12304–0544, 1991.

Electromagnets Used with Crane Hoists, 12304–0359, 1985.

Electrostatic Paint Spraying and Detearing, 12304–0468, 1991.

Flexible Insulated Protective Equipment for Electrical Workers, 12304–0598, 1991.

Industrial Electric Substations, 12304–0559, 1991.

Portable Reamer-Drills, 12304–0497, 1989.

Power Tool Institute, 1300 Sumner Avenue, Cleveland, OH 44115-2851.

Underwriters Laboratories, Inc., 333 Pfingsten Road, Northbrook, IL 60062.

Dimensions of Attachment Plugs and Receptacles, UL 498.

Electrical Construction Materials List.

Hazardous Location Equipment List.

Insulated Wire, UL 44. "Electrical Appliance and Utilization Equipment Lists."

U.S. Department of Commerce, National Institute of Standards and Technology, Gaithersburg, MD 20899.

U.S. Department of Labor, Occupational Safety and Health Administration, 200 Constitution Avenue NW, Washington DC 20210.

29 CFR 1910.132–138, Personal Protective Equipment.

29 CFR 1910.145, Specifications for Accident Prevention Signs and Tags.

29 CFR 1910.147, Control of Hazardous Energy.

29 CFR 1910.307, Hazardous Locations.

29 CFR 1910.331–333, Electrical Safety-Related Work Practices.

29 CFR 1910.331–339, Electrical Safety Regulations Series.

29 CFR 1910.403, Wiring Design and Protection.

29 CFR 1926.403, General Requirements—Legal.

29 CFR 1926.404, Wiring Design and Protection.

29 CFR 1926.405(g)(2)(iii), Wiring Methods, Components, and Equipment for General Use.

29 CFR 1926.407, Hazardous Locations.

REVIEW QUESTIONS

1. What is a person's main resistance to current flow?
2. Why is low voltage often considered more dangerous than high voltage?
3. What is measured in ohms?
4. Name two factors that determine the outcome of an electrical shock.
   a.
   b.
5. Why is an unguarded knife switch especially hazardous?
6. How is the safe current-carrying capacity of conductors determined?
7. What is the function and primary purpose of an overcurrent device?
8. What is the limited approach boundary?
9. What are the two main categories of circuit breakers?
   a.
   b.
10. What are the three classes of hazardous locations?
    a.
    b.
    c.
11. What is the purpose of system grounds? What is the purpose of equipment grounds?
12. Where are GFCIs required, and what is their purpose?

APPENDICES

Appendix A
Electrical Cord Codes

Jacketed Wire Marking	
S	Stranded (or service wire)
J	Junior Service (300 volts). If no "J" is in the wire type, then it is hard service (600 volts).
T	Thermoplastic. If no "T" is in the wire type, then it has a rubber jacket.
O	Oil-resistant compound
W	CSA-approved for outdoor use (weather)
W-A	UL-approved for outdoor use (weather and atmosphere)
V	Vacuum—as in vacuum cleaner. This is a small O.D. jacketed wire, very flexible and initially used for vacuum cleaners. It is now used on many different types of end products. Available only in 18 AWG.

Common Codes	
HPD	Thermoset-insulated heater cord with cotton braid overall.
HPN	Neoprene-insulated heater cord.
HSJO	Rubber insulated and jacketed.
MTW	Thermoplastic-insulated machine tool wire.
NM	Nonmetallic-sheathed cable, multiconductor building wire.
R, RH	Rubber-insulated wire. Building wire used for construction.
S	Extra-hard-service flexible cord for general use at 600 volts, rated at 60°C. Rubber insulated and jacketed.
SE	Extra-hard-service flexible cord for general use at 600 volts, rated at −58°F to +221°F. TPE thermoplastic elastomer insulated and jacketed.
SEO	Type SE, but also rated oil resistant.
SEOW	Type SEO, but also rated for outdoor use at −58°F.
SEW	Type SE, but additionally rated for outdoor use at −58°F.
SJ	Hard-service cord for general use at 300 volts, rated at 60°C, rubber insulated and jacketed.
SJE	Hard-service cord for general use at 300 volts, rated at −58°F to +221°F, TPE thermoplastic elastomer insulated and jacketed.
SJEO	Type SJE, but rated as oil resistant.
SJEOW	Type SJEO, but rated for outdoor use at −58°F.
SJEW	Type SJE, but rated for outdoor use at −58°F.
SJO	Type SJ, but rated oil resistant.
SJOW	Type SJO, but rated for outdoor use at −30°F.
SJT	Hard-service cord for general use at 300 volts, rated −4°F to +140°F, PVC thermoplastic insulated and jacketed.
SJTO	Type SJT, but also rated oil resistant.
SJTOW	Type SJTO, but rated for outdoor use at −31°F.
SJTW	Type SJT, but rated for outdoor use at −31°F.
SO	Type S, but also rated oil resistant.
SPT-1	Thermoplastic parallel cord for light duty, 300 volts.
SPT-2	Type SPT-1, but heavier construction, 300 volts.
SPT-3	Type SPT-2, but heavier construction, 300 volts.
ST	Extra-hard-service cord for general use at 600 volts, rated at −4°F to +140°F, PVC thermoplastic insulated and jacketed.
STO	Type ST, but also rated oil resistant.
STOW	Type STO, but rated for outdoor use at −31°F.
STW	Type ST, but rated for outdoor use at −31°F.
SVT	Vacuum cleaner cord, PVC thermoplastic insulated and jacketed, rated 300 volts, 18 AWG and 16 AWG only, two conductors.
T, TW, THN, THHN, THWN	Thermoplastic insulated wires. Building wires used for construction.
UF	Thermoplastic underground feeder and branch circuit cable. Also used for submersible pump cable.

Appendix B
Sample Electrical Work Permit

ENERGIZED ELECTRICAL WORK PERMIT

PART I: TO BE COMPLETED BY THE REQUESTER:

Job/Work Order Number _____

(1) Description of circuit/equipment/job location: _____

(2) Description of work to be done: _____

(3) Justification of why the circuit/equipment cannot be de-energized or the work deferred until the next scheduled outage:
_____
_____

_____   _____
Requester/Title   Date

PART II: TO BE COMPLETED BY THE ELECTRICALLY QUALIFIED PERSONS *DOING* THE WORK:

Check when complete

(1) Detailed job description procedure to be used in performing the above detailed work: _____ ☐
_____

(2) Description of the safe work practices to be employed: _____ ☐

(3) Results of the shock risk assessment: _____
_____

    (a) Voltage to which personell will be exposed ☐
    (b) Limited approach boundary ☐
    (c) Restricted approach boundary ☐
    (d) Necessary shock, personal, and other protective equipment to safely perform the assigned task ☐

(4) Results of the arc flash hazard analysis: _____
_____

    (a) Available incident energy at the working distance of arc flash PPE category ☐
    (b) Necessary arc flash personal and other protective equipment to safely perform the assigned task ☐
    (c) Arc flash boundary ☐

(5) Means employed to restrict the access of unqualified persons from the work area: _____ ☐
_____

(6) Evidence of completion of a job briefing, including discussion of any job-related hazards: _____ ☐
_____

(7) Do you agree the above described work can be done safely?   ☐ Yes   ☐ No   (If *no*, return to requester)

_____   _____
Electrically Qualified Person(s)   Date

_____   _____
Electrically Qualified Person(s)   Date

PART III: APPROVAL(S) TO PERFORM THE WORK WHILE ELECTRICALLY ENERGIZED:

_____   _____
Manufacturing Manager   Maintenance/Engineering Manager

_____   _____
Safety Manager   Electrically Knowledgeable Person

_____   _____
General Manager   Date

Note: Once the work is complete, forward this form to the site Safety Department for review and retention.

© 2014 National Fire Protection Association

NFPA 70E

Fire Protection

John S. DeLaHunt, MBA, ARM
Steven G. Schoolcraft, PE, CSP
Michael J. Fagel, PhD, CEM
Philip E. Hagan, JD, MBA, MPH, ARM, CIH, CHMM, CET, CHCM, CEM

The Chemistry of Fire
Cooling a Fire ▸ Limiting Oxygen in a Fire ▸ Removing Fuel from a Fire ▸ Interrupting the Chain Reaction in a Fire ▸ Using Extinguishing Agents

Classification of Fires
Class A Fires—Solids ▸ Class B Fires—Liquids ▸ Class C Fires—Electrical ▸ Class D Fires—Metals ▸ Class K Fires—Kitchen Oils

Fire Risk
Fire Hazard Analysis ▸ Identification of Hazardous Materials ▸ Evaluating Fire Risk

Fire Prevention: Constructing Facilities
Planning for Fire Protection ▸ Site Planning ▸ Construction Materials and Interior Furnishings ▸ Fire Protection Methods and Concepts in Building Design ▸ Special Considerations for Computer Rooms

Fire Prevention: Maintaining Facilities
Inspections ▸ Hot Work ▸ Training Employees ▸ Factors Contributing to Industrial Fires

Fire Detection and Response
Human Observer ▸ Automatic Fire Detection Systems ▸ Communications ▸ Evacuation Drills ▸ Fire Brigades ▸ Alarm Systems ▸ Spacing of Detectors ▸ Location of Detectors ▸ Maintenance of Detectors

Fire Suppression and Fire Extinguishment
Portable Fire Extinguishers ▸ Sprinkler and Water-Spray Systems ▸ Fire Hydrants ▸ Fire Hose ▸ Special Systems and Agents

Facility Fire Protection Program
Objectives ▸ Preventing Fires ▸ Detecting and Responding to Fires ▸ Controlling, Suppressing, and Extinguishing Fires ▸ Recovering from Fires

Summary

References

Review Questions

Fire protection includes (1) fire prevention; (2) detecting and responding to fires; (3) controlling, suppressing, and extinguishing fires; and (4) recovering from fires to resume normal business operations. Planning for each is essential, and a facility's fire protection program is incomplete if it does not thoroughly address each of these aspects of fire protection.

The practice of fire protection engineering is a highly developed, specialized field. Some fire protection issues require the special training and experience of a licensed fire protection engineer. Implementing efficient fire protection systems requires the involvement of the building users, other engineers, architects, interior designers, building contractors, fire protection system manufacturers, the local governmental authorities having jurisdiction, and even urban planners.

Safety professionals faced with special fire protection problems should seek professional advice from qualified fire protection consultants. The National Fire Protection Association (NFPA) and the Society of Fire Protection Engineers publish excellent texts covering all aspects of fire protection, and safety professionals involved in fire protection should consider adding some of these texts to their professional libraries.

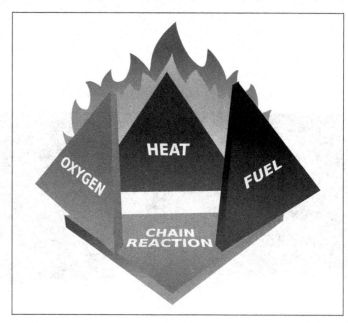

Figure 9–1. The fire tetrahedron. Fuel, air (oxygen), heat, and a sustained chain reaction are necessary components of a fire. *(Gustavb/Wikimedia Commons/Public Domain)*

THE CHEMISTRY OF FIRE

Fire, or the process of combustion, is an extraordinarily complex chemical reaction. For a fire to occur, fuel, oxygen, heat, and a chemical chain reaction must join in a symbiotic relationship (Figure 9–1), called an oxidation-reduction (redox) reaction. Analysis of the anatomy of a fire reveals that the original fuel molecules appear to combine with oxygen in a series of successive intermediate stages, called branched-chain reactions. Once this happens, combustion occurs. The intermediate stages are responsible for the evolution of flames.

As molecules break up in these branched-chain reactions, unstable intermediate products called free radicals are formed. The concentration of free radicals determines the speed of flames. The life of the free hydroxyl radical (OH^-) is very short, about 0.001 s, but long enough to be crucial in the combustion of fuel gases. The almost simultaneous formation and consumption of free radicals appear to be the lifeblood of a fire's chain reaction.

Rusting of metal is a slower, less energetic redox reaction. In combustion, a self-catalyzed reaction releases heat energy. The combustion process often involves rapid oxidation of a fuel by oxygen in the air. If the combustion process occurs in a structure or vessel such that pressure can increase, an explosion can result. Fire is a combustion process of sufficient intensity to emit heat and light.

All fires require four basic elements—fuel, oxygen, heat, and a sustained chain reaction. Common parlance for this is the "fire tetrahedron." For most fires, the chemical chain reaction is not relevant to fire control, so a simpler term—"fire triangle"—covers the relationship among fuel, oxygen, and heat.

Fires occur in one of two general forms or modes: flame fire and surface fire. Flame fires directly burn gaseous or vaporized fuel and include deflagrations. These types of fires usually involve high rates of fuel consumption and high temperatures. Premixed flame fires, like those on a gas burner or stove, are relatively easy to control. Diffusion flame fires occur when vapors or gases burn in air. These fires are more difficult to control. Surface fires occur on the surfaces of a solid fuel. The vapor given off by the flammable solid or liquid actually burns, not the solid or liquid itself. These two modes of fire are not mutually exclusive; they may occur alone or together.

Understanding how and why a fire burns suggests ways to control and extinguish it. Controlling fires requires any, or several, of the following actions.
- cooling a fire, which takes away heat
- excluding the air taken in, which limits oxygen
- removing fuel
- interrupting the chemical chain reaction by inhibiting the rapid oxidation of the fuel and the concomitant production of free radicals, which drives the fire.

Cooling a Fire

To suppress and extinguish a fire by cooling, heat must

reduce at a greater rate than the total heat the fire creates. To do this, the cooling agent must reach the burning fuel directly. The cooling action may also stop the release of combustible vapors and gases. The most common and practical extinguishing agent is water applied in a solid stream or spray or incorporated in foam. Note that when water is the extinguishing agent, steam often results. In an enclosed space, the steam displaces oxygen.

Limiting Oxygen in a Fire

Most fires burn in air; therefore, it is difficult to limit a fire's access to air in the first place. Once a fire starts, noncombustible materials such as fire blankets, dirt, sand, inert gas flooding, or foams can smother it. Smothering is ineffective for fires that involve substances containing their own oxygen supply, such as ammonium nitrate or nitrocellulose.

Removing Fuel from a Fire

Often, taking the fuel away from a fire is not only difficult, but also dangerous. However, storage tanks for flammable liquids may have piping and valves so that the fuel's supply can be isolated and pumped away, in case of fire. Shutting off the supply of flammable gas will extinguish a fire associated with that gas. Note that any other fuels in the vicinity may still be burning after the gas supply is off.

Interrupting the Chain Reaction in a Fire

Some extinguishing agents, such as dry chemicals and halogenated hydrocarbons, remove the free radicals in these branched-chain reactions from their normal function as a chain carrier. The effects that various dry chemical agents have on capturing free radicals depend on their individual molecular structures.

Using Extinguishing Agents

Some extinguishing agents help control fire by attacking more than one of its four components. For example, both plain water fog and carbon dioxide flooding can react at flame temperatures with relatively slow-burning free carbon, producing carbon monoxide and, as a result, decreasing black-smoke production. Because this reaction absorbs heat, it lowers the heat of the fire in addition to lowering the oxygen concentration.

Matching the pace at which newer and more potent fire-extinguishing agents work requires increasingly sophisticated tactics and techniques. Although any of the four basic options for fire attack will work, attacking from only one of them does not necessarily result in the most rapid extinguishing time. Attacking from more than one, by using more than one agent, however, can produce a synergistic effect that hastens extinguishing.

CLASSIFICATION OF FIRES

Fire code development agencies, including the National Fire Protection Association (NFPA) have classified types of fires. These classifications vary from country to country, or by economic regions. Fire classifications derive from the types of fuels in fires and the extinguishing agent needed to combat those fires.

Class A Fires—Solids

Class A fires occur in ordinary solid materials, such as wood, paper, rags, and rubbish. The quenching and cooling effects of water, or of solutions containing large percentages of water, are of primary importance in extinguishing these fires. Dry chemical agents (multipurpose dry chemicals) provide both rapid suppression of the flames and the formation of a coating that tends to retard further combustion. However, complete cooling with water is the most reliable means to achieve full extinguishment.

Class B Fires—Liquids

Class B fires occur in the vapor-air mixture over the surface of flammable and combustible liquids, such as gasoline, oil, grease, paints, and thinners. Limiting air (oxygen) to inhibit combustion is essential to stop Class B fires before they spread. Solid streams of water are likely to spread the fire because the liquid fuel generally does not mix with water and often floats. Spreading the fuel this way may also increase the intensity of the flame as additional vapors evolve and ignite. Under certain circumstances, water-fog nozzles are effective in the suppression of Class B fires, but not extinguishment. Generally, multipurpose dry chemicals, carbon dioxide/inert gases, and foam are effective for suppressing and extinguishing Class B fires.

Class C Fires—Electrical

Class C fires occur in or near energized electrical equipment. Responders must use nonconducting extinguishing agents when attacking Class C fires. Multipurpose dry chemicals and carbon dioxide extinguishing agents are essential for such fires. Foam and water streams are poor choices for Class C fires because both agents can expose personnel to electric shock hazards. De-ionized water mist extinguishers are effective against Class C fires. Note that after deenergizing the electricity source feeding a Class C fire, most Class C fires are either a Class A or Class B fire because the material actually burning is usually a solid or a liquid.

Class D Fires—Metals

Class D fires occur in combustible metals such as magnesium, titanium, zirconium, lithium, potassium, and sodium. These fires require specialized techniques, extinguishing agents, and extinguishing equipment. Avoid using extinguishing agents designed for Class A, Class B, and Class C fires on metal fires because of the risk of increasing the intensity of the fire due to incompatibility between extinguishing agents and the fuel. Inert dry powders are common extinguishing agents for use on Class D fires.

Class K Fires—Kitchen Oils

Class K fires typically involve cooking greases or cooking oils on commercial cooking appliances such as stoves and grills. This class of fire recognizes the special fire hazards associated with commercial cooking—large, very hot surface areas in the presence of large quantities of cooking greases and cooking oils at high working temperatures. Class K extinguishing agents saponify the oil, forming thick, long-lasting coatings over hot cooking surfaces, and penetrate the oil to cool it.

Fire Hazard Analysis

Early in the fire risk assessment process, a format and list of references for organizing a systematic, comprehensive fire hazard analysis process must be developed. The results from the fire hazard analysis will be used to identify the loss scenarios associated with the fire hazards; ultimately, they will be used to evaluate the facility fire risk. The output from the fire hazard analysis process is also an excellent tool for developing facility fire hazard inspection points.

A sample fire hazard identification structure, derived from the one presented in the NFPA *Fire Protection Handbook*, appears at the end of this chapter. It contains elements for consideration when identifying and characterizing the fire hazards present in a facility. Expanding this simple list to reflect actual conditions in a facility can result in a comprehensive report that identifies and characterizes the fire hazards present in the facility. Developing fire risk scenarios and evaluating fire risk depends on a thorough statement of the actual conditions in a facility.

FIRE RISK

Fire protection measures can be effective only if they are based on a proper analysis and evaluation of the risk of fire. A complete evaluation is important because a wide variety of methods and equipment exist to provide protection. The optimum level of fire protection is that which minimizes both the costs and the expected fire risk. Ideally, the cost of fire protection should have a predictable effect on reducing the fire risk.

Determining fire risk can be quite complicated, and in some applications, it is necessary to engage in quantitative risk assessments involving probabilistic models and other complex approaches. The purpose of this section is to provide context on performing a qualitative fire risk assessment. A far more comprehensive treatment of the subject is contained in the NFPA's *Fire Protection Handbook* (20th edition, 2008).

Figure 9–2 shows a simplified fire risk assessment process. After identifying the objectives and scope of the fire risk assessment (Figure 9–2, Step 1), the next step is to identify the fire hazards within the scope of the risk assessment (Figure 9–2, Step 2).

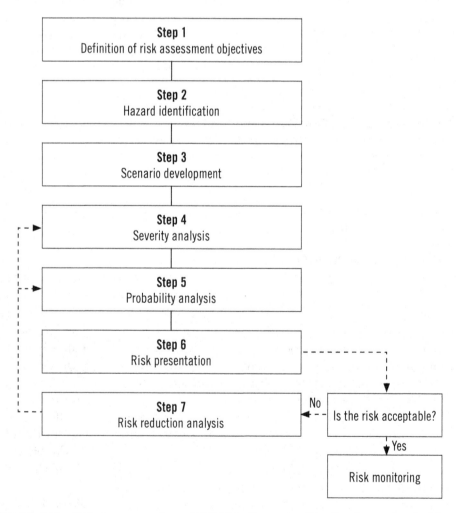

Figure 9–2. Fire Risk Assessment Steps (*NFPA Fire Protection Handbook, reprinted with permission from the National Fire Protection Association.*)

Identification of Hazardous Materials

Identifying hazardous materials used or stored in a facility and understanding the nature of hazardous materials are important parts of the fire hazard analysis. Fires often involve chemicals that have varying degrees of toxicity, flammability, and instability. Detailed information on these chemicals must be readily available to anyone in the facility as well as to personnel who may respond or who may confront emergencies involving these chemicals. Such information is conveyed via Safety Data Sheets (SDSs). When considering whether a hazardous material used in a facility is a fire hazard, consult the SDS.

In addition to consulting the SDS, the National Fire Protection Association has developed a visual system that can be used for quickly identifying the hazardous properties of chemicals. This system is sometimes called the NFPA 704 hazardous material identification system. It uses a diamond-shaped symbol and colored numerals or backgrounds to indicate the degree of hazard for each hazardous property of a hazardous material in emergency conditions (e.g., fire or spill).

Three categories of hazardous properties are identified for each hazardous material: health hazard (blue), flammability hazard (red), and instability hazard (yellow). The diamond shape also provides places at the bottom of the diamond for special hazards, such as "OX" to indicate that the material possesses oxidizing properties. The order of severity in each category is indicated numerically by five divisions ranging from 0 to 4, where higher numbers indicate higher degrees of hazard for that particular hazardous property.

Figure 9–3 shows the methods of displaying the NFPA 704 hazardous material identification system. Refer to Tables 9–A through 9–C for qualitative definitions of the meanings of each hazardous property and their related degrees of hazard. This entire hazardous materials identification system is defined in the national consensus code NFPA 704, Standard System for Identification of Hazards of Materials for Emergency Response.

Under the Global Harmonized System (GHS) of hazardous materials identification, lower numbers have more significant hazard. As a result, highly flammable chemicals carry a 4 rating in NFPA 704 and a 1 rating in GHS.

Evaluating Fire Risk

After identifying and characterizing a facility's fire hazards, it is necessary to systematically develop a set of loss scenarios associated with each fire hazard (Figure 9–2, Step 3). A loss scenario is the sequence of events leading to a destructive fire.

Developing loss scenarios requires understanding how fires might start, how they would be sustained, and the possible human and property effects of potential fire losses. Business interruptions or losses of essential business information are also relevant and important. Use the following factors along with the fire hazard analysis structure shown earlier to develop loss scenarios.

- What materials are flammable and combustible?
- What materials in a process or operation are most likely

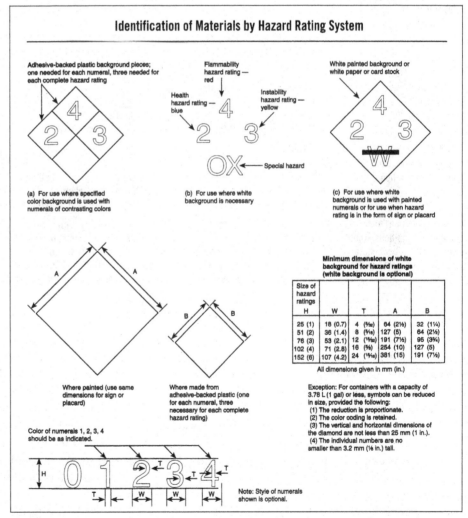

Figure 9–3. How to display the degree of hazards of hazardous materials using the NFPA 704 identification system. The meanings of the numbering systems for each hazard category are shown in Tables 9–A through 9–C. *(Reprinted with permission from the National Fire Protection Association.)*

TABLE 9–A. Degrees of Health Hazards

Degree of Hazard*	Criteria
4—Materials that, under emergency conditions, can be lethal	Gases whose LC_{50} for acute inhalation toxicity is less than or equal to 1,000 parts per million (ppm) Any liquid whose saturated vapor concentration at 20°C (68°F) is equal to or greater than 10 times its LC_{50} for acute inhalation toxicity, if its LC_{50} is less than or equal to 1,000 ppm Dusts and mists whose LC_{50} for acute inhalation toxicity is less than or equal to 0.5 milligram per liter (mg/L) Materials whose LD_{50} for acute dermal toxicity is less than or equal to 40 milligrams per kilogram (mg/kg) Materials whose LD_{50} for acute oral toxicity is less than or equal to 5 mg/kg
3—Materials that, under emergency conditions, can cause serious or permanent injury	Gases whose LC_{50} for acute inhalation toxicity is greater than 1,000 ppm but less than or equal to 3,000 ppm Any liquid whose saturated vapor concentration at 20°C (68°F) is equal to or greater than its LC_{50} for acute inhalation toxicity, if its LC_{50} is less than or equal to 3,000 ppm, and that does not meet the criteria for degree of hazard 4 Dusts and mists whose LC_{50} for acute inhalation toxicity is greater than 0.5 mg/L but less than or equal to 2 mg/L Materials whose LD_{50} for acute dermal toxicity is greater than 40 mg/kg but less than or equal to 200 mg/kg Materials that are corrosive to the respiratory tract Materials that are corrosive to the eye or cause irreversible corneal opacity Materials that are corrosive to the skin Cryogenic gases that cause frostbite and irreversible tissue damage Compressed liquefied gases with boiling points at or below −55°C (−66.5°F) that cause frostbite and irreversible tissue damage Materials whose LD_{50} for accurate oral toxicity is greater than 5 mg/kg but less than or equal to 50 mg/kg
2—Materials that, under emergency conditions, can cause temporary incapacitation or residual injury	Gases whose LC_{50} for acute inhalation toxicity is greater than 3,000 ppm but less than or equal to 5,000 ppm Any liquid whose saturated vapor concentration at 20°C (68°F) is equal to or greater than one-fifth its LC_{50} for acute inhalation toxicity, if its LC_{50} is less than or equal to 5,000 ppm, and that does not meet the criteria for either degree of hazard 3 or degree of hazard 4 Dusts and mists whose LC_{50} for acute inhalation toxicity is greater than 2 mg/L but less than or equal to 10 mg/L Materials whose LD_{50} for acute dermal toxicity is greater than 200 mg/kg but less than or equal to 1,000 mg/kg Compressed liquefied gases with boiling points between −30°C (−22°F) and −55°C (−66.5°F) that can cause severe tissue damage, depending on duration of exposure Materials that are respiratory irritants Materials that cause severe but reversible irritation to the eyes or lacrimators Materials that are primary skin irritants or sensitizers Materials whose LD_{50} for acute oral toxicity is greater than 50 mg/kg but less than or equal to 500 mg/kg
1—Materials that, under emergency conditions, can cause significant irritation	Gases and vapors whose LC_{50} for acute inhalation toxicity is greater than 5,000 ppm but less than or equal to 10,000 ppm Dusts and mists whose LC_{50} for acute inhalation toxicity is greater than 10 mg/L but less than or equal to 200 mg/L Materials whose LD_{50} for acute dermal toxicity is greater than 1,000 mg/kg but less than or equal to 2,000 mg/kg Materials that cause slight to moderate irritation to the respiratory tract, eyes, and skin Materials whose LD_{50} for acute oral toxicity is greater than 500 mg/kg but less than or equal to 2,000 mg/kg
0—Materials that, under emergency conditions, would offer no hazard beyond that of ordinary combustible materials	Gases and vapors whose LC_{50} for acute inhalation toxicity is greater than 10,000 ppm Dusts and mists whose LC_{50} for acute inhalation toxicity is greater than 200 mg/L Materials whose LD_{50} for acute dermal toxicity is greater than 2,000 mg/kg Materials whose LD_{50} for acute oral toxicity is greater than 2,000 mg/kg Materials that are essentially nonirritating to the respiratory tract, eyes, and skin

*For each degree of hazard, the criteria are listed in a priority order based on the likelihood of exposure.

to ignite, burn, or explode?

- What in the facility could be a source of ignition? Are any open sparks or flames present?
- Are high temperatures involved in any operations?
- Where are flammable and combustible materials located? Are flammable materials stored together? Do indirect connections exist? If one of the materials should burn, could the others easily ignite?
- Might any of the materials ignite because of convection or radiation?
- What toxic gases might evolve into a fire?
- How much time might it take for a fire to spread to other areas and to adjacent facilities?

- How long will recovery after the fire affect the operation?
- Are there essential, unique, or high-value processes or materials in the facility?
- Smoke and toxic gases, and sometimes heat, are largely responsible for fire deaths. What toxic gases might evolve from the burning or smothering of contents?
- How many people are likely to be involved in the facility, in adjacent facilities, or in facilities nearby?

Some fire hazard evaluations include determining the fire load, or total heat potential, of materials. Fire loads express the weight of combustible material per unit fire area. Paper and wood have a caloric value of 7,000 to 8,000

TABLE 9–B. Degrees of Flammability Hazards

Degree of Hazard	Criteria
4—Materials that rapidly or completely vaporize at atmospheric pressure and normal ambient temperature or that are readily dispersed in air and burn readily	Flammable gases Flammable cryogenic materials Any liquid or gaseous material that is liquid while under pressure and has a flash point below 22.8°C (73°F) and a boiling point below 37.8°C (100 F) (i.e., Class IA liquids) Materials that ignite spontaneously when exposed to air Solids containing greater than 0.5 percent by weight of a flammable or combustible solvent are rated by the closed-cup flash point of the solvent
3—Liquids and solids that can be ignited under almost all ambient temperature conditions. Materials in this degree produce hazardous atmospheres with air under almost all ambient temperatures, or though unaffected by ambient temperatures, are readily ignited under almost all conditions	Liquids having a flash point below 22.8°C (73°F) and a boiling point at or above 37.8°C (100°F) and those liquids having a flash point at or above 22.8°C (73°F) and below 37.8°C (100°F) (i.e., Class IB and Class IC liquids) Finely divided solids, typically less than 75 micrometers (200 mesh), that present an elevated risk of forming an ignitable dust cloud, such as finely divided sulfur, *National Electric Code* Group E dusts (e.g., aluminum, zirconium, and titanium), and bis-phenol A Materials that burn with extreme rapidity, usually by reason of self-contained oxygen (e.g., dry nitrocellulose and many organic peroxides) Solids containing greater than 0.5 percent by weight of a flammable or combustible solvent are rated by the closed-cup flash point of the solvent
2—Materials that must be moderately heated or exposed to relatively high ambient temperatures before ignition can occur. Materials in this degree would not, under normal conditions, form hazardous atmospheres with air, but under high ambient temperatures or under moderate heating could release vapor in sufficient quantities to produce hazardous atmospheres with air	Liquids having a flash point at or above 37.8°C (100°F) and below 93.4°C (200°F) (i.e., Class II and Class IIIA liquids) Finely divided solids less than 420 micrometers (40 mesh) that present an ordinary risk of forming an ignitable dust cloud Solid materials in a flake, fibrous, or shredded form that burn rapidly and create flash fire hazards, such as cotton, sisal, and hemp Solids and semisolids that readily give off flammable vapors Solids containing greater than 0.5 percent by weight of a flammable or combustible solvent are rated by the closed-cup flash point of the solvent
1—Materials that must be preheated before ignition can occur. Materials in this degree require considerable preheating, under all ambient temperature conditions, before ignition and combustion can occur	Materials that will burn in air when exposed to a temperature of 815.5°C (1500°F) for a period of 5 minutes in accordance with ASTM D 6668, *Standard Test Method for the Discrimination Between Flammability Ratings of F = 0 and F = 1* Liquids, solids, and semisolids having a flash point at or above 93.4°C (200°F) (i.e., Class IIIB liquids) Liquids with a flash point greater than 35 C (95 F) that do not sustain combustion when tested using the "Method of Testing for Sustained Combustibility," per 49 CFR 173, Appendix H, or the UN publications *Recommendations on the Transport of Dangerous Goods, Model Regulations,* and *Manual of Tests and Criteria* Liquids with a flash point greater than 35°C (95°F) in a water-miscible solution or dispersion with a water noncombustible liquid/solid content of more than 85 percent by weight Liquids that have no fire point when tested by ASTM D 92, *Standard Test Method for Flash and Fire Points by Cleveland Open Cup,* up to the boiling point of the liquid or up to a temperature at which the sample being tested shows an obvious physical change Combustible pellets, powders, or granules greater than 420 micrometers (40 mesh) Finely divided solids less than 420 micrometers that are nonexplosible in air at ambient conditions, such as low volatile carbon black and polyvinylchloride (PVC) Most ordinary combustible materials Solids containing greater than 0.5 percent by weight of a flammable or combustible solvent are rated by the closed-cup flash point of the solvent
0—Materials that will not burn under typical fire conditions, including intrinsically noncombustible materials such as concrete, stone, and sand	Materials that will not burn in air when exposed to a temperature of 816°C (1,500°F) for a period of 5 minutes in accordance with ASTM D 6668, *Standard Test Method for the Discrimination Between Flammability Ratings of F = 0 and F = 1*

Btu/lb (16 to 19 MJ/kg). A typical office is likely to have about 5 lb/ft^2 (24 kg/m^2). A flammable liquid has a heat-producing potential of 14,000 to 15,000 Btu/lb (32 to 35 MJ/kg). Fire load does not account for the rate of heat liberation or for the distribution within a structure, and it is only a guide. The fire load can be used to develop some elements of the fire protection system, although a qualified fire protection engineer should be consulted when quan-

tification of interior or exterior fire loads and related fire effects is necessary.

These loss scenarios can frame the estimation of the frequency of occurrence and reasonable consequence (Figure 9–2, Steps 4 and 5). Before beginning this process, consider defining scales of frequency and consequence that will be used to evaluate the effects of each loss scenario. These scales can be qualitative or quantitative; however, quantita-

TABLE 9–C. Degrees of Instability Hazards

Degree of Hazard	Criteria
4—Materials that in themselves are readily capable of detonation or explosive decomposition or explosive reaction at normal temperatures and pressures	Materials that are sensitive to localized thermal or mechanical shock at normal temperatures and pressures Materials that have an instantaneous power density (product of heat of reaction and reaction rate) at 250°C (482°F) of 1,000 watts per milliliter (W/mL) or greater
3—Materials that in themselves are capable of detonation or explosive decomposition or explosive reaction but that require a strong initiating source or must be heated under confinement before initiation	Materials that have an instantaneous power density (product of heat of reaction and reaction rate) at 250°C (482°F) at or above 100 W/mL and below 1,000 W/mL Materials that are sensitive to thermal or mechanical shock at elevated temperatures and pressures
2—Materials that readily undergo violent chemical change at elevated temperatures and pressures	Materials that have an instantaneous power density (product of heat of reaction and reaction rate) at 250°C (482°F) at or above 10 W/mL and below 100 W/mL
1—Materials that in themselves are normally stable but that can become unstable at elevated temperatures and pressures	Materials that have an instantaneous power density (product of heat of reaction and reaction rate) at 250°C (482°F) at or above 0.01 W/mL and below 10 W/mL
0—Materials that in themselves are normally stable, even under fire conditions	Materials that have an instantaneous power density (product of heat of reaction and reaction rate) at 250°C (482°F) below 0.01 W/mL Materials that do not exhibit an exotherm at temperatures less than or equal to 500°C (932°F) when tested by differential scanning calorimetry

Tables 9–A, B, and C. The degree of hazard in each category can be assessed quickly for any hazardous material for which the NFPA identification system is used. *(Adapted from NFPA 704, with permission from the National Fire Protection Association.)*

TABLE 9–D(a). Probability Levels

Probability	Description
Frequent	Likely to occur frequently, experienced ($p > 0.1$)
Probable	Will occur several times during system life ($p > 0.001$)
Occasional	Unlikely to occur in a given system operation ($p > 10^{-6}$)
Remote	So improbable, may be assumed this hazard will not be experienced ($p < 10^{-6}$)
Improbable	Probability of occurrence not distinguishable from zero ($p \sim 0.0$)

TABLE 9–D(b). Severity Categories

Severity	Impact
Negligible	The impact of loss will be so minor that it would have no discernible effect on the facility or its operations
Marginal	The loss will have impact on the facility, which may have to suspend some operations briefly. Some monetary investments may be necessary to restore the facility to full operations. Minor personal injury may be involved.
Critical	The loss will have a high impact on the facility, which may have to suspend operations. Significant monetary investments may be necessary to restore to full operations. Personal injury and possibly deaths may be involved.
Catastrophic	The fire will produce death or multiple deaths or injuries, or the impact on operations will be disastrous, resulting in long-term or permanent closing. The facility would cease to operate immediately after the fire occurred.

tive estimates of frequency and consequence may require significant effort. Table 9–D shows an example scale for frequency and consequence estimates.

It is appropriate to consider a range of consequences and frequencies for each loss scenario. A single loss sce-

nario reasonably may have a spectrum of consequences and related frequencies of occurrence. It may be that a particular location has a high probability of a small fire, and a low probability of a catastrophic fire.

Apply a measure of uncertainty to these estimates of

9 Fire Risk   275

		Consequence			
		Negligible	Marginal	Critical	Catastrophic
Frequency	**Frequent**	9 Scenarios	2 Scenarios		
	Probable	12 Scenarios	1 Scenario		
	Occasional	13 Scenarios	6 Scenarios	1 Scenario	1 Scenario
	Remote	65 Scenarios	21 Scenarios	7 Scenarios	3 Scenarios
	Improbable	(did not consider)	25 Scenarios	19 Scenarios	11 Scenarios

Key

Low Risk   Moderate Risk   High Risk

Figure 9–4. Example risk matrix used for presenting the risk of the loss scenarios developed from a fire risk analysis.

frequency and consequence for a loss scenario. This uncertainty factor will help prioritize risk reduction methods and identify loss scenarios that warrant additional, more detailed risk analyses.

When the estimates of frequency and consequence for the developed loss scenarios are complete, the fire risk can be presented (Figure 9–2, Step 6). Although there are numerous ways to present risk, a common method of presenting risk is using a risk matrix. An example risk matrix is shown in Figure 9–4.

There are several ways to reflect the fire scenarios and risk analysis effort, including showing the number of loss scenarios evaluated in each cell in the risk matrix or a total summation of risk in each cell. In any case, the risk matrix provides action levels and priorities for risk reduction efforts. The different cell background shades in the risk matrix represent these action levels. The darker shades represent high-risk levels; the lighter shades represent lower-risk levels. The risk tolerance of the organization should drive decisions about what risks require priority action.

The final step in the fire risk evaluation process is selecting risk reduction strategies (Figure 9–2, Step 7). These strategies should be well defined and researched, and an estimate of the effect of implementing each risk reduction method should be made. In many cases, a risk reduction method will have broad applications to several of the developed loss scenarios, and the aggregate risk reduction effect should be considered. Generally, the risk management approach to high-frequency, high-consequence events is to avoid the conditions that allow those events. Risk transfer, as with insurance, outsourcing, or other agreements, is a usual response to infrequent, high-consequence events. Systemic causes usually cause events with high frequency and a low overall consequence, so management of the processes involved is an effective response. Most organizations retain, or tolerate, the risk associated with low-frequency and low-consequence events.

This fire risk assessment process helps develop and defend fire risk reduction strategies.

FIRE PREVENTION: CONSTRUCTING FACILITIES

Much of this section is adapted from *Principles of Fire Protection*, published by the National Fire Protection Association (NFPA). Qualified fire protection engineers; local building codes and regulations; and the relevant aspects of national consensus standards such as NFPA 1 (Fire Code), NFPA 101 (Life Safety Code), NFPA 5000 (Building Construction and Safety Code), and the International Code Council's International Building Code should be consulted during all aspects of facility design, construction, and renovation.

Fire safety seeks to protect life foremost and property secondly from the ravages of fire in a building. Building design and construction must take into account a wide range of fire safety features. Not only must the interiors and contents of buildings be protected from the dangers of fire, but the building site itself must have adequate water supplies and easy accessibility by the fire department. Architects, builders, and owners may assume that national, state, and local codes provide adequate protective measures; however, these codes stipulate only minimal measures for fire safety. Planning and construction based on such codes should not reduce or limit fire-safe design efforts.

Fire codes are minimum standards, and occupants, designers, and owners should not look at them as a compliance exercise. Building and fire codes have evolved over time in response to real fires, with real loss of life and real damage to property. They are the accumulated wisdom of loss, and prudence demands that construction and use of facilities be faithful to the spirit, and the letter, of these codes.

Planning for Fire Protection

Prior to making effective decisions relating to fire safety design, a building designer must understand the specific function of the building and its general and unique conditions. Decisions regarding the fire safety design and construction of the building have the same objectives as do all fire protection measures, namely:

1. preventing fires
2. detecting and responding to fires
3. controlling, suppressing, and extinguishing fires
4. recovering from fires to resume normal business operations.

The art of probing to identify objectives is an important design function. The degree of risk that the owner and occupants will tolerate is a difficult design decision; consequently, the planning team should identify it in a clear, concise manner so that the designer can properly realize the design objectives.

Consider who will use the building and what the people using the building will be doing most of the time. The occupied building provides a great potential for fire because of the presence of large numbers of people, any one of whom could perform a careless or malicious act resulting in fire. Appliances and mechanical or electrical equipment are a potential hazard through misuse, failure, faulty construction, or substandard installation. Accumulations of combustibles, either waiting for disposal or in storage, frequently provide a ready means by which otherwise controllable fires could spread. The identification of specific functional patterns, constraints, and disabilities is vital in designing specific fire protection features that recognize occupant conditions and activities.

Consider whether there is any specific, high-value content that will need special design protection, such as intellectual property, data centers, research equipment, or artwork. The requirements with regard to protection of property within a building are often fairly easy to identify. Materials of high value that are particularly susceptible to fire and/or water damage can usually be identified in advance of building design. For example, vital records that cannot be replaced easily or quickly can be identified in advance as needing special fire protection design considerations.

Continuity of operations must take into consideration those specific functions conducted in a building that are vital to continued operation of the business and that cannot be transferred to another location. In this regard, the owner must identify for the designer the amount of time an operation can be suspended without completely compromising business success. The degree of protection required in fire-safe building design varies with the number and scope of vital operations that are nontransferable.

Two major categories of decisions should be made early in the design process of a building in order to provide effective fire-safe design. Early considerations should be given to both the interior building functions and exterior site planning. Building fire defenses, both active and passive, should be designed in such a way that the building itself assists in the manual suppression of fire.

Interior layout, circulation patterns, finish material, and building services are all important fire safety considerations in building design. Building design also has a significant influence on the efficiency of fire department operations. As a result, all fire suppression activities should be considered during the design phases.

The broad approach to the fire-safe design of a building requires a clear understanding of the building's function, the number and kinds of people who will be using it, and the kinds of things they will be doing. In addition, appropriate construction and protection features must be provided for the protection of the contents and, particularly for mercantile and industrial buildings, to ensure the continuity of operations if a fire should occur.

The majority of fire deaths and injuries in structural fires (70% to 80%) are caused by asphyxiation from exposure to smoke and the toxic gases that evolve as combustion products from fires. The carbon monoxide developed in most fires, particularly unventilated and smoldering fires, is probably the most common cause of death, although studies show that victims of smoke inhalation also exhibit signs of cyanide poisoning. Smoke also obscures visibility and thus can lead to panic situations when occupants cannot see and use escape routes. Direct exposure to flame, heat, or fire accounts for 20% to 30% of all fire-related deaths and injuries.

Site Planning

Proper building design for fire protection includes a number of factors outside the building itself. The site on which the building is located will influence the design, especially traffic and transportation conditions, fire department accessibility, water supply, and the exposure this facility has on the public. Inadequate water mains and poor spacing of hydrants have contributed to the loss of many buildings.

Traffic and Transportation

Fire department response time is a vital factor in building design considerations. Traffic access routes, traffic congestion at certain times of the day, traffic congestion from highway entrances and exits, and limited-access highways have significant effects on fire department response distances and response time and must be taken into account by building designers in selecting appropriate fire defenses.

Fire Department Access to the Site

Building designers must ask the question: Is the building easily accessible to fire apparatus? Ideal accessibility occurs where a building can be approached from all sides by existing and expected fire department apparatus. However, such ideal accessibility is not always possible. Congested areas, topography, or buildings and structures located appreciable distances away from the street make it difficult to use, or prevent effective use of, fire apparatus. When equipment such as aerial ladders, elevating platforms, and water tower apparatus cannot come close enough to the building to be used effectively, fire response suffers, possibly resulting in lost lives, injuries, property damage, pollution, and business interruption.

Bridge weight loads are a factor in fire response. Many times, businesses will place bridges near buildings for aesthetic purposes without taking into consideration the weight limits of fire equipment or construction equipment called in to handle problems that can arise at any time.

The matter of access to buildings has become far more complicated in recent years. The building designer must consider this important aspect during the planning stages. Inadequate attention to site details can place the building in an unnecessarily vulnerable position. If a building's design compromises fire defense by preventing adequate fire department access, the building itself must make up the difference in more substantial internal protection.

Fire Department Access to the Facility's Interior

One of the more important considerations in facility design is access to the fire area. This includes access to the facility itself as well as access to the facility's interior. In larger and more complex facilities, serious fires over the years have brought improvements in facility design to facilitate fire department operations. The larger the facility, the more critical it becomes to have quick access for fire fighting. In some facilities, fire fighters cannot function effectively. Where architectural, engineering, or functional requirements restrict adequate fire-fighting access and operations, alternate and additional protection should be in place. A complete automatic sprinkler system with a fire department connection is probably the best solution to this problem. Other methods that may be used in appropriate design situations include access panels in interior walls and floors, fixed nozzles in floors with fire department connections, and roof vents and access openings.

Water Supply to the Site

A facility designer must also ask: Are the water mains adequate? Are the hydrants properly located? Will water-supply volumes be adequate? The more congested the area in which the facility is to be located, the more important it is to plan what the fire department may face if a fire occurs on the property. An adequate water supply delivered with the necessary pressure is required to control a fire properly and adequately. The number, location, and spacing of hydrants and the size of the water mains are vital considerations when the facility designer plans fire defenses for the facility.

Water mains are another concern. It is of the utmost importance not to have dead-end mains. The optimum would be that all mains are on a looped water system. In addition, the system should be valved so that if there is a breakdown or a break of any type, one section can be isolated without shutting down the whole system. Breaks can occur due to freezing (in some areas), contractors cutting through mains, or almost any type of movement or impact.

Exposure Protection

Still another consideration in the design of the building is the possibility of damage from a fire in an adjoining building. The building may be exposed to heat radiated horizontally by flames from the windows of the burning neighboring building. If the exposed building is taller than the burning building, flames coming from the roof of the burning building can attack and damage the exposed building.

The damage from an exposing fire can be severe. It depends on the amount of heat produced and the time of exposure, the fuel load in the exposing building, and the construction and protection of the walls and roof of the exposed building. Other factors are the distance of separation, wind direction, and accessibility of fire fighters.

Fire severity is a description of the total energy of a fire and involves both the temperatures developed within the exposing fire and the duration of the burning. NFPA 80A, Recommended Practice for Protection of Buildings from Exterior Fire Exposures, describes estimated minimum separation distance under light, moderate, or severe exposures. The severity of the exposure is calculated on the width and the estimated fire loadings of the buildings involved. Building designers should be aware that effective separation distances between the exposing buildings can be reduced by constructing blank fire-resistant walls, closing wall openings with fire-rated construction equivalent to the fire-rated wall, eliminating combustible projections, using automatic deluge water curtains, and using fire-rated glass instead of ordinary glass.

Construction Materials and Interior Furnishings

Fire-resistive construction describes a broad range of structural systems able to resist fires of specified intensity and duration without failure. The materials are relatively noncombustible and are given a numerical rating as to their fire resistance. Nearly every building material has a fire-resistive rating. The rating is a relative term or number that indicates the extent to which it resists the effect of a fire. Ratings are usually available for most building components, such as columns, floors, walls, doors, windows, and ceilings.

Common fire-resistive components with high ratings include masonry load-bearing walls, reinforced-concrete or protected-steel columns, and poured-on-precast concrete floors and roofs. Although fire-resistive structures do not, in themselves, contribute fuel to a fire, combustible trim, ceilings, and other interior finishes and furnishings may produce an intense fire. However, because such combustible materials pose a serious threat to life safety, they must be considered in providing fire protection. Using approved building materials will lower the flame-spread ratings and limit materials that can contribute to available fuel, especially in buildings without automatic sprinkler systems.

Heavy-Timber Construction

Heavy-timber construction is characterized by masonry walls, heavy-timber columns and beams, and heavy plank floors. Although not completely immune to fire, the great bulk of the wooden members slows the rate of combustion.

Heavy-timber construction can resist fire very well. The timbers will char, and the resulting coating insulates the unburned wood. Heavy timber maintains its integrity during a fire for a relatively long time, thus providing an opportunity for extinguishment. Much of the original strength of the members is retained, and reconstruction is sometimes possible.

Noncombustible and Limited-Combustible Construction

Noncombustible and limited-combustible construction includes all types of structures in which the structure itself—exclusive of trim, interior finish, and contents—is noncombustible but not fire resistant. Exposed steel beams and columns and masonry, metal, or gypsum wallboard are the most common forms of this type of construction.

Because of the tendency of steel to warp, buckle, and collapse under moderate fire exposure, noncombustible construction is relatively vulnerable to fire damage. Therefore, it is most suitable for low-hazard occupancies or ordinary hazard occupancies. To reduce susceptibility to heat collapse, load-bearing noncombustible structural members can be protected by encasing them in concrete, covering them with gypsum wallboard, or spraying them with a protective material.

Steel. The most common building material for larger buildings is structural steel. While steel is noncombustible and contributes no fuel to a fire, it loses its strength when subjected to the high temperatures reached in a fire. At temperatures at or above 1,000°F (590°C), unprotected structural steel (e.g., ASTM A36 structural steel) loses 50% or more of its strength. Buildings built of unprotected steel will collapse relatively quickly when exposed to a significant fire.

Because unprotected structural steel loses its strength at high temperatures, it must be protected from exposure to the heat produced from fires. This protection, often referred to as fireproofing, insulates the steel from the heat. The more common methods of insulating steel are encasement of the member, application of a surface treatment, or installation of a suspended ceiling as part of a floor-ceiling assembly capable of providing fire resistance. Intumescent paints and coatings are used to increase the fire endurance of structural steel. Intumescent paints and coatings swell when heated, thus forming insulation around the steel.

Structural steel members can also be protected by sheet steel membrane shields. The sheet steel holds in place inexpensive insulation materials, thus providing greater fire endurance. In addition, polished sheet steel has been used to protect spandrel girders (the horizontal supports beneath the windows on modern high-rise buildings). The shield reflects radiated heat and protects the load-carrying spandrel.

Concrete. The resistance of reinforced concrete to fire's attack will depend on the type of aggregate used to make the concrete, the moisture content, and the anticipated fire loading. Usually, reinforced-concrete buildings resist fire very well; however, the heat of a fire will cause spalling (chipping and peeling away), some loss of strength of the concrete, and other deleterious effects.

Prestressed concrete is stronger than reinforced concrete and provides better fire resistance. However, prestressed concrete has a greater tendency to spall, with the result that the prestressed steel may become exposed. The type of steel used for prestressing is more sensitive to elevated temperatures than the type of steel that is usually used in reinforced-concrete construction. In addition, the steel used for this type of reinforced concrete construction does not regain its strength upon cooling.

Ordinary Construction

Ordinary construction consists of masonry exterior–bearing walls, or bearing portions of exterior walls, that are of noncombustible construction. Interior framing, floors, and roofs are made of wood or other combustible materials whose "bulk" is less than that required for heavy-timber construction.

If floor and roof construction and their supports have a 1-hour fire-resistance rating, and all openings through floors (including stairways) are enclosed with partitions having a 1-hour fire-resistance rating, then the construction is known as *protected ordinary construction*. Its occupancy should be limited to light or moderate hazards.

Even when sheathed, ordinary construction, unlike fire-resistive or noncombustible construction, still has combustible materials in concealed wall and ceiling spaces. Fire frequently originates in these concealed spaces or enters into them and then spreads rapidly throughout the entire room and building.

To prevent the free passage of flames through concealed spaces or openings, include the following safety features in the construction:
- Trim all combustible framing away from sources of heat.
- Provide effective fire barriers against the spread of fire between all subdivisions and all stories of the building.
- Provide adequate fire separation against exterior exposure.
- Fire-stop all vertical and horizontal draft openings to form effective barriers to stop or slow the spread of fire.

Wood-Frame Construction

Wood-frame construction consists primarily of wood exterior walls, partitions, floors, and roofs. Exterior walls may be stuccoed or sheathed with brick veneer or metal or asphalt siding. Although generally inferior to other types of construction from a fire safety standpoint, wood-frame construction can be made reasonably safe for light-hazard, low-density occupancies. Safety can be greatly increased by suitable protection against the horizontal and vertical spread of fire, provision of safe exits, and elimination of combustible interior finishes. Install enough fire detectors to alert all occupants. Automatic sprinkler protection can greatly improve the fire safety of wood-frame buildings.

The physical size of wood and its moisture content are important factors that determine whether this material will provide reasonable structural integrity. Wood is the most common material used in the construction of dwellings. If a wood-frame house is subjected to a serious fire, either from burning combustibles inside the house or from an exposure fire, it will not withstand much heat and will have little structural integrity.

Interior Finish

The way a building fire develops and spreads and the amount of damage that ensues are largely influenced by the characteristics of the interior finishes in a building. Many types of interior finishes are used in buildings, and they serve many functions. Primarily, they are used for aesthetic or acoustical purposes. However, insulation or protection against wear and abrasion are also considered major functions by building designers.

Interior finish is usually defined as those materials that make up the exposed interior surface of wall, ceiling, and floor constructions. The common interior finish materials are wood, plywood, plaster, wallboard, acoustical tile, insulating and decorative finishes, plastics, and various wall coverings.

While some building codes do not include floor coverings under their definitions of interior finishes, the current trend is to include them. Rugs and carpets are not subject to test and regulation under the Flammable Fabrics Act administered by the U.S. Consumer Products Safety Commission. Most rugs and carpets are a factor in fire spread.

Many codes exclude trim and incidental finish from the code requirements for wall and ceiling finish. Interior finishes, however, are not necessarily limited to the walls, ceilings, and floors of rooms, corridors, stairwells, and similar building spaces. Some authorities include the linings or coverings of ducts, utility chases and shafts, or plenum spaces as interior finish as well as batt and blanket insulation, if the batt faces a stud space through which fire might spread. Understanding the effects of interior finishes is essential in preventing fire, or preventing the spread of a fire, in buildings. These issues must be considered during the design phase of new building construction.

If the building on fire has combustible furniture, flames and toxic gases may spread so rapidly that occupants may not be able to escape. Poor construction practices, such as failure to protect shafts and other vertical openings, make the vertical spread of fire more rapid and the work of fire fighters more difficult.

Although the collapse of structural elements has not resulted in many deaths or injuries to building occupants, it is a particular hazard to fire fighters. A number of deaths and serious injuries to fire fighters occur each year because of structural failure. While some of these failures result from inherent weaknesses, many are the result of renovations to existing buildings that materially, though not obviously, affect the structural integrity of the support elements. A building should not contain surprises of this type for fire fighters.

Plastics. Aesthetic considerations and low cost make the use of plastic building materials desirable. All plastics are combustible, and no known treatment can make plastics noncombustible, although some have relatively low flame-spread ratings.

Cellular plastics sprayed on walls for insulation have become popular. Fire retardants can be incorporated in many of these plastics so they can meet building code requirements. However, some plastics containing polyurethane or polystyrene have been involved in serious, rapidly spreading fires.

Wood. Because untreated wood-based wallboard and paneling are highly combustible, most codes require fire-retardant treatments. Without such treatments, combustible wallboard not only enables a fire to spread so fast that people may become trapped, but also contributes fuel to the fire and creates hazardous concentrations of smoke and toxic gases.

Glass. Glass is a commonly used building material. Modern high-rise buildings contain large amounts of glass. Glass is used in three primary ways in building construction: (1) for glazing, (2) for glass-fiber insulation, and (3) for glass-fiber-reinforced plastic building products.

Most glass used for windows and doors has little resistance to fire. Fire-resistive glazing offers a rated fire separation barrier and a transparent or translucent finish, but is very expensive.

Glass-fiber insulation is widely used in modern building construction. Glass fiber is popular because it is fire resistant and is an excellent insulator. However, glass fiber is often coated with a resin binder that is combustible and that can spread flames.

Glass-fiber-reinforced plastic building products are becoming more common. The glass fiber acts as reinforcement for a thermosetting resin. Usually this combustible resin comprises about 50% or more of the material. Thus, while the glass fiber itself is noncombustible, the product can be highly combustible.

Gypsum. Gypsum, as reflected in products such as plaster and plasterboard, has excellent fire-resistive qualities. Gypsum is widely used because it has a high proportion of chemically combined water, which makes it an excellent, inexpensive, fire-resistive building material that is far superior to highly combustible fiberboard.

Masonry. Masonry (such as brick and tile) provides good resistance to heat and usually retains its integrity. Building codes describe methods for using masonry in construction.

Fire Protection Methods and Concepts in Building Design

National, state, and local fire codes provide explicit design guidance on the methods and features of fire protection in design. Adherence to at least the minimum code requirements is mandatory. The NFPA's National Fire Codes, the International Code Council's International Building Code, and other similar publications are excellent resources for understanding building design concepts, and in many cases, some or all of the codes are codified into law in local jurisdictions. The concepts presented in this section are general, and qualified fire protection engineers should be consulted before implementing any of these fire prevention methods in building construction.

Confining Fire

Using stair enclosures, fire walls, and fire doors and dividing a building into smaller units are ways to confine a fire. Plan them into the building's design. Regardless of the type of building construction, protected stair enclosures for multistory buildings are necessary to provide a safe exit path for occupants. They also retard the upward spread of fire.

Under many conditions, it is necessary to divide a floor into separate fire areas with rated fire walls. Design fire walls to rigid specifications so that they can withstand the effects of a severe fire and of building collapse on one side. To prevent the passage of heat, all openings in fire walls must be protected with approved closures at the same fire rating as the fire walls or greater. This protection includes wall penetrations for electrical conduit, pipes, cables, and HVAC ducts.

Separate Units

Traditionally, dividing buildings into separate units provided functional work areas or offered occupants some degree of privacy. From the point of view of fire safety, however, dividing a building is regarded as a way to break up the total volume of a building into small cells. In such a building, fires will remain localized and can be more easily suppressed. To prevent fire from spreading from one unit to another, building codes require (1) that units be made structurally sound enough to withstand full fire exposure without major damage and (2) that the units' boundaries be capable of acting as nonconducting heat barriers.

Fire Doors

Fire doors are the most widely used and accepted means of protecting horizontal openings. Fire doors are rated by testing laboratories as they are installed in a building and as assemblies (door frame and door as a unit). The fire doors usually have a rating of ¾ to 3 hours. They may be constructed of metal or metal-clad-treated wood materials and may be hinged, rolling (sliding), or curtain doors. Single or double doors may be specified.

When new construction is planned, select the proper types of fire doors. They will perform properly only in the uses for which they were designed. In addition, a fire door will perform properly only if it is installed with an approved frame, latching device, hardware, and closing device. The effectiveness of the entire assembly as a fire barrier may be destroyed if any component is omitted or if any component of lesser quality is substituted.

Fire doors are of value only if they will close or be closed at the time of a fire. Blocking or wedging fire doors open utterly defeats the purpose of a fire door. Prohibit this practice. Always check fire doors during fire prevention inspections. Be sure that door openings and the surrounding areas are clear of anything that might interfere with the free operation of fire doors.

NFPA 80, Standard for Fire Doors and Other Opening Protectives, requires annual inspection of fire separation doors, including the hinges, catches, latches, closers, and stay rolls of fire doors because they are especially subject to wear. This inspection also includes testing for proper operation. Immediately repair any defect that would interfere with the proper operation of fire doors. The following problems are frequently encountered in this critical test.

- Chains or wire ropes may have stretched.
- Hardware may be inoperative.
- Guides and bearings may need lubrication.
- Binders may be bent, thus obstructing the doorway.
- Stay rolls may have accumulations of paint.
- Fusible links may be painted.
- The hoods over rolling steel doors may be bent, thus interfering with the doors' operation.
- Where swinging doors are used in pairs, coordinators may need adjustment.
- Electromagnetic catches holding doors open during normal operations (and releasing doors to close during an alarm) do not work properly.
- Strike plates may have been taped or otherwise defeated for occupant convenience.

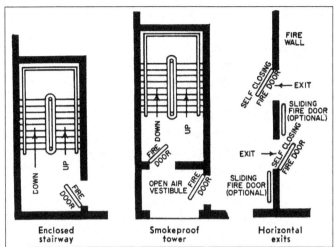

Figure 9–5. Plan views of types of exits. Stair enclosure prevents fire on any floor from trapping persons above. A smokeproof tower is better because an opening to the air at each floor largely prevents the chance of smoke on the stairway. A smokeproof tower charged with positive air pressure is more likely to prevent smoke from entering. A horizontal exit provides a quick refuge and decreases the need for a hasty flight down stairs. Horizontal sliding fire doors provided for safeguarding property values are arranged to close automatically in case of fire. Swinging doors are self-closing. Two wall openings are needed for exit in two directions. *(Reprinted with permission from the National Fire Protection Association.)*

Exits

Of the many factors involved in protecting life from fire, a building's exits are the most important. Nevertheless, exits are inadequate in many buildings. Consider the design of exits in a building's total fire safety system (Figure 9–5).

Management, architects, and others entrusted with the safety of building occupants must plan for the orderly emergency evacuation of buildings. Panic is a causal factor for loss of life during a fire or other building emergency. While fire is a common cause of panic, such dangers as boiler or air receiver explosions, fumes, or structural collapse may also threaten safe and orderly evacuation.

A building's population and degree of hazard are the major factors when designing exits. Every building or structure, and every section or area in it, should have at least two separate means of exit. Arrange them so that the possibility of any one fire blocking all exits is minimized. Designing exits involves more than a study of numbers, flow rate, and population densities. Safe exits require a safe path of escape from the fire with the least possible travel distance to the exit. The path should be arranged for ready use in case of an emergency. It should be large enough to permit all occupants to reach a place of safety before they are endangered by the fire or by smoke and toxic gases.

Evacuation

Consider the following general provisions when planning for building evacuation.

- Do not design exits and other safeguards to depend solely on any single safeguard. Provide additional safeguards in case of human or mechanical failure.
- Exit doors must withstand fire and smoke during the length of time for which they are designed to be in use. Enclose or protect vertical exits and other vertical openings to afford reasonable safety to occupants while they use the exits.
- Provide alternate exits and pathways in case one exit is blocked by fire. Also, provide exits and areas of refuge that people with disabilities can use.
- Provide alarm systems to alert occupants of fire or another emergency. Visual alarms are important for the hearing impaired.
- Provide exits and exit routes with adequate lighting, including emergency lighting.
- Mark exits with a readily visible sign. Mark access to exits with readily visible signs whenever an exit or exit route is not readily visible.
- To protect exiting personnel, safeguard equipment and areas of any unusual hazard that might spread fire and smoke.
- Practice an orderly exit drill procedure.
- Control psychological factors that can lead to panic.
- Select an interior finish and contents that prevent a fire from spreading fast and trapping occupants.
- Maintain adequate aisles in exit routes.
- Provide adequate space outside the building's exits for congregating evacuated building occupants. This should be far enough away from the building to prevent people from being hazarded by the burning structure or from fire-fighting operations.

Ventilation

Ventilation is of vital importance in removing smoke, gases, and heat so that fire fighters can reach the seat of a blaze. It is difficult, if not impossible, to ventilate a building unless appropriate skylights, roof hatches, emergency escape exits, and similar devices are provided when the building is constructed.

Ventilation of building spaces performs the following important functions:

- Protect life by removing or diverting toxic gases and smoke from locations where building occupants must find temporary shelter.
- Improve the environment in the vicinity of the fire by removal of smoke and heat. This enables fire fighters to advance close to the fire to extinguish it with a minimum of time, water, and damage.
- Control the spread or direction of a fire by setting up air currents that cause the fire to move in a desired direction. In this way, occupants or valuable property can be more readily protected.
- Provide release for unburned, combustible gases before a flammable mixture occurs.
- Install smoke dampers in air-conditioning ventilation ducts in accordance with NFPA 90A, Standard for Installation of Air-Conditioning and Ventilating Systems. This standard requires that smoke dampers operate automatically upon detection of smoke, so fire will not spread through unburned areas.

Controlling Smoke

Smoke control confines smoke, heat, and toxic gases to a limited area, dilutes them, or exhausts them, thus preventing their spread to other areas and minimizing fatalities. Smoke and hot gases generated by an uncontrolled fire, if confined within a building, can seriously impair fire-fighting operations, cause illness and death, and spread the fire under the roof for considerable distances from the point of origin. The movement of smoke within a structure is determined by many factors, including building height, ceiling heights, suspended ceilings, venting, external wind force, and the direction of the wind.

Most smoke-control systems involve a combination of methods. One method of smoke control uses a physical barrier, such as a door, wall, or damper to block the smoke's movement. Smoke-control doors may be operated manually or by some automatic detection device coupled to a door closure. An alternative method is to use a pressure differential between the smoke-filled area and the protected area.

Venting is another way of removing smoke, heat, and noxious gases from a building. Plan for smoke control during the design stage of the building by specifying vents, curtain boards, and windows. For example, use smoke- and heat-venting systems consisting of curtain boards. They protect heat-banking areas under a roof. Also, use automatic or manual roof vents to release smoke and heat through the roof.

Vents are most applicable to larger areas in one-story buildings that lack sprinklers and that do not have separate units. They are also useful in windowless and underground buildings. They are not, however, a substitute for automatic sprinkler protection. Vents and draft curtains are effective for small special housings.

So many variables affect the burning of combustible material that no exact formula can be used to compute the amount of venting required. However, vent sizes and ratios have been developed from limited experiments in test buildings without sprinklers and theories about actual fire experiences. Also, a variety of prefabricated vents is available on the market. These vents can be designed to open automatically at predetermined temperatures or smoke concentrations.

Automatic smoke-control systems often require annual testing and recertification. Because they are essentially custom designed for the facility, it is essential to collect and keep the design engineers' operations, testing, and maintenance manuals.

Connections for Sprinklers and Standpipes

Connections for sprinklers and standpipes must be carefully located and clearly marked. The larger and taller the building becomes, the greater the volume and pressure of water that will be needed to fight a potential fire. Water damage can be very costly unless adequate measures such as floor drains and scuppers have been incorporated into the building design.

Confinement of a fire in a high-rise building can only be accomplished by careful design and planning for the whole building. As buildings increase in size and complexity, more dependence on fire detection and suppression systems is necessary. Such systems are described in detail in the NFPA references listed at the end of this chapter.

Special Considerations for Computer Rooms

If a fire occurs in a computer room, the numerous electrical components and outlets involved and the quantities of paper used and stored present particular hazards. Fire suppression systems must be designed with three goals in mind: (1) extinguishing fires before they can do damage or cause injury, (2) allowing workers to escape the area unharmed, and (3) protecting vulnerable electronic hardware and software.

The following items should be considered when designing a fire protection system for computer rooms:

- Forms of equipment protection should include handheld fire extinguishers. A Class A fire extinguisher should be available for use on any paper or other ordinary, combustible-type fire, and a Class C fire extinguisher (carbon dioxide, Halotron, or de-ionized water mist) should be available for use on electrical fires. Each extinguisher should be clearly labeled with its function and class.
- An automatic fire detection system with listed smoke and heat detectors should be provided. Wire it to a central station supervisory service to ensure quick fire department response in an emergency.
- A dual system using both sprinklers and a total flood system is an effective method of automatic fire suppression. Using this method, primary protection is furnished by the total flood system; if the system fails, the sprinkler system acts as secondary protection. It is important

that the electric equipment is deenergized before water flow or system flooding. This can be done manually by operators before they leave the room or automatically by means of a single triggered activation of the system. Automatic shutdown should be provided for equipment that normally runs unattended for long periods of time. Automatic power-down programs can be added to existing software.
- The sprinkler system protecting the computer room should be valved separately from other sprinkler systems in the building. A system for drying out the equipment to reduce further damage should be developed.

The issue of using sprinklers on computer equipment has sparked considerable controversy. According to the NFPA, automatic sprinkler protection and water-spray fixed systems are essential as a means of reducing fire damage, even where electrical or electronic equipment may be exposed. Experience has shown that if a fire activates sprinklers, the sprinklers, if properly installed and maintained, provide for effective fire protection with virtually no hazard to personnel and with no measurable increase in damage to the equipment (as compared with the damage done by heat, smoke, and water from hose lines).

FIRE PREVENTION: MAINTAINING FACILITIES

The fire safety of a building depends not only on its design and construction, but also on what is continuously being done to prevent a fire from starting in the building. Good housekeeping and reducing combustible fire loads are key fire prevention factors. Keeping the fuel load down not only lessens the amount of material that can be ignited, but also provides less material that can be consumed if a fire breaks out.

Maintaining the building's active and passive fire protection features is essential in both fire prevention and control. A passive-design fire protection element is one that requires no action to function, such as a fire wall. An active design or construction element is one that requires an action in order to function, such as a fire door that must be closed. A systematic program of maintaining these fire protection elements (e.g., the integrity of fire walls, fire doors, fire detection and alarm systems, ventilation systems, and automatic sprinklers) must be developed and implemented.

Those responsible for fire prevention in an existing building are not the same as those responsible for the building's design. Decisions concerning these elements are under the control of the building owner, operators, and tenants.

The complete steps to be taken for fire prevention in existing buildings are extensive. It is essential to refer to appropriate national consensus codes and standards as well as state and local building and fire codes to understand the factors that must be considered to prevent fires in any particular building and occupancy type. Fire prevention includes activities directed specifically toward preventing a fire from starting, or if it has started, toward preventing the fire from spreading to other areas of the building.

Inspections

Set up a system of periodic fire inspections for every operation and every location in the building. The fire hazard analysis outline is an excellent tool for developing the scope and the points in inspection checklists, and the relevant national consensus codes and standards as well as state and local building and fire codes describe the specific inspection points and their related performance parameters.

Some buildings, operations, and processes require daily inspection, while others can be inspected weekly, monthly, or at other intervals. Buildings that are well designed and provided with protective devices and construction elements intended to act as fire safety features still must undergo periodic, detailed inspections.

External inspections may be performed by insurance companies, fire protection bureaus of fire departments, and the state or county/parish fire marshal's office. It is essential, however, for facility owners, operators, and tenants to conduct their own detailed fire inspections as part of their own fire safety programs. Such self-inspections must be performed by employees or contractors trained in fire inspection techniques. The individual who is in charge of fire prevention activities (such as inspections) should establish inspection schedules, determine the routing of reports, and have a complete list of all items to be inspected.

Conduct special inspections during and following any alterations in a facility or in a process. Make a complete seasonal inspection of equipment that will be or has been exposed to freezing temperatures. Do it early enough to replace or repair the equipment.

If the facility does not have its own fire protection expert, invite the local fire chief or fire marshal to inspect the fire equipment, hazardous materials used, locations for hazardous materials, and available water supply. Ask the local fire chief how the fire department can support the facility's fire protection program. Fire departments will often use preplanning forms. Fire prevention and fire safety is a communitywide effort.

Hot Work

In an effort to establish control over operations using flames or producing sparks, firms institute hot-work permitting programs. These programs require that authorization be obtained before equipment capable of igniting combustible materials is used outside the equipment's normal work areas. For processes covered by the process safety management standard (29 CFR 1910.119) administered by the Occupational Safety and Health Administration (OSHA), a hot-work permitting process is mandatory for all work on or around covered processes and the facilities in which covered processes exist. Also, NFPA 51B, Standard for Fire Prevention during Welding, Cutting, and Other Hot Work, describes methods for managing hazards induced by hot work.

When implementing a hot-work permit program, management must first develop a policy statement. The type and extent of the hot-work permit program will depend on the size of the facility, the complexity of the operations, and the degree to which hazards are present at the worksite and in the surrounding areas.

Important features of a hot-work permit program include the following:

- Inspect the area where work is to be done, and see how close combustible materials are to the work area.
- Establish dedicated fire-watch employees. Fire-watch employees should stay on duty for 30 minutes after all spark-producing equipment has been shut down.
- Equip fire-watch employees with fire-extinguishing equipment.
- Remove flammable and combustible materials from sources of ignition. If removal is not possible, these items must be protected from ignition.
- Limit unauthorized use of flame- or spark-producing equipment.
- Write a permit outlining all of the preceding hot-work features.
- Communicate with, and coordinate the activities of, all departments associated with the hot work.

A form or tag is generally used to administer the hot-work permit program. Although standard forms are available, many facilities have developed special forms that relate specifically to their operation.

If hot work is to be done by an outside contractor, establish areas of responsibility in a written contract. Ensure that the agreement answers the following questions:

- Who issues the hot-work permits? What signatures are required?
- Who provides fire-watch employees?
- Who provides standby fire-extinguishing equipment?
- Who coordinates the activities of the contractor that involve facility personnel?

During a major construction project, the general contractor often has primary and perhaps exclusive control over the job site. When that construction occurs adjacent to space controlled by the facility, establishing a system to determine which hot-work program applies is essential.

Training Employees
Employees should have periodic, recurring training in fire safety. This training should include evacuation protocols, shelter-in-place protocols, limits on storage and electrical appliances, and fire safety training specific to any operations they control. That training might include use of fire extinguishers. Training employees to extinguish incipient-phase fires may translate into a requirement to do so. In that case, OSHA regulations (29 CFR 1910.157 and 1910.38) require a written emergency management plan. This creates a record-keeping requirement and compliance burden. As a result, the decision to train employees to use fire extinguishers as a job expectation is a serious one and should take place only after careful management review.

Figure 9-6. A student in protective clothing is being trained to use a portable fire extinguisher to fight a small fire.

If the facility intends to train employees on use of fire extinguishers, the training must include recognizing incipient-phase fires, the type of fire extinguisher to use on each kind of fire, where to find the fire extinguishers, and how to use the fire extinguishers properly. Fire extinguisher training can be an essential component of fire prevention.

One of the most difficult decisions any employee faces is whether to fight a fire or to evacuate. If *all* of the following conditions are met, an employee may decide to fight a fire with an extinguisher:
1. There is a clear exit.
2. The fire brigade or fire department has been called.
3. The fire is small (incipient), such as one in a wastebasket or small tool housing.
4. The employee knows how to use the appropriate fire extinguisher.
5. The extinguisher is in working order.

An employee should *not* fight a fire (or continue to fight a fire) if any *one* of the following conditions is met:
1. The fire could block an exit.
2. The fire is spreading beyond its point of origin.
3. The employee is unsure of how to use a fire extinguisher.
4. The use of one fire extinguisher fails to suppress or extinguish the fire.

Training employees in fire extinguisher use should be periodic, and demonstrations, practice drills, and lectures should occur annually (more often if a special fire hazard exists). Employees should have printed instructions regarding the use of fire extinguishers. Printed sheets, leaflets, or cards can give both general and detailed instructions regarding the use of the fire extinguishers. In addition, instructions on using fire extinguishers should be posted near or on the fire extinguishers themselves.

Fire extinguisher manufacturers, insurance companies, local fire departments, the NFPA, and the National Safety Council offer films, posters, and cutaway displays that are useful in explaining the construction, maintenance, and operation of portable fire extinguishers. NFPA 10, Standard for Portable Fire Extinguishers, gives particularly valuable information for training employees about using portable fire extinguishers (Figure 9-6). Also, local fire departments and fire extinguisher maintenance companies may assist in fire extinguisher training.

Employees trained in use of portable fire extinguishers should know that these items are only a first means of defense. Training should reinforce to employees that, even in the hands of a trained operator, a fire extinguisher will *not* extinguish a large fire or a very fast-moving fire.

Factors Contributing to Industrial Fires
To prevent fires in existing facilities, fuel sources and sources of ignition must be identified. The fire hazard analysis outline is an excellent tool for identifying some of these sources. Electrical equipment, smoking, friction, open flames, and poor housekeeping are some of the common items that produce fuel and sources of ignition for starting fires.

Electrical Equipment
Install and maintain electrical equipment in accord with NFPA 70, National Electrical Code (NEC). Overheating electrical equipment and electrical arcs from short circuits in improperly installed or maintained electrical equipment are two leading causes of fire in buildings.

Where flammable dusts, gases, or vapors may be present, electrical equipment must be approved (listed) for use in these hazardous locations by nationally recognized testing laboratories, such as Underwriters Laboratories (UL), Intertek (ETL), or FM Global.

Hazardous locations fall into three classes, which depend on the material involved. They are further divided according to the degree or severity of the hazard. Complete definitions of the classes and divisions of hazardous locations and the types of equipment designed for each are contained in the NEC, and are discussed in Chapter 8, Electrical Safety, in this volume.

Temporary or makeshift wiring, particularly if defective or overloaded, is a very common cause of electrical fires. Do not use this type of wiring. Overloaded or partially grounded wiring may also heat up enough to ignite combustible materials without opening fuses or circuit breakers. Where flammable and combustible liquids are used, it is essential to ensure that containers are adequately bonded and grounded as described in the NEC.

Portable electrical tools and extension cords should also conform to the NEC. Inspect them at frequent intervals and repair them promptly. Use waterproof cords and sockets in damp places, and use explosion-proof fixtures and lamps in the presence of highly flammable gases and vapors.

Ground or double-insulate all electrical equipment, especially portable electrical tools. Use switches, lamps, cords, fixtures, and other electrical equipment listed by a recognized testing and certifying agency. Use such equipment only in applications for which the approval or listing was granted (see Chapter 8, Electrical Safety).

Protect lamp bulbs by using heavy lamp guards or adequately sealed transparent enclosures. Keep lamp bulbs away from sharp objects, and secure them to prevent them from falling. Never use bare bulbs if they are exposed to flammable dusts or vapors. Always consider lamp bulbs as potential hazards in such areas. Safeguard them accordingly.

Instruct employees in the correct use of electrical equipment. Prohibit them from tampering with equipment, blocking circuit breakers, using wrong fuses, bypassing fuses, or installing equipment without authorization.

Periodically inspect and test electrical installations and all electrical equipment. This ensures continued satisfactory performance and also detects deficiencies.

The NEC requires marking by Nationally Recognized Testing Laboratories for all electrical appliances. The "CE" mark used for European commerce is not equivalent.

Smoking

Carelessly discarded cigarettes, pipe embers, and cigars are a major source of fire. Prohibit smoking, especially in woodworking shops, textile mills, flour mills, grain elevators, and places where flammable liquids or combustible products are manufactured, stored, or used. It is prudent to eliminate smoking completely in a facility, and it might be desirable to eliminate it from the grounds. If this is not possible, establish special smoking areas. Using these special smoking areas must involve strict rules related to discarding cigarettes, burned pipe embers, and cigars in containers designed for this use.

Mark no-smoking areas with conspicuous signs. Everyone, including supervisors and visitors, must adhere to no-smoking regulations. It may be necessary to use more than signs to draw attention to the no-smoking areas. Lines drawn on the floor and illuminated barriers placed around areas or processes are also effective. Reinforcing the signage with disciplinary action will reduce fire risk.

Friction

Excessive heat generated by friction causes fires. A program of preventive maintenance on machinery can avert fires resulting from inadequate lubrication; misaligned bearings; broken or misaligned belts, chains, and pulleys; or broken or bent equipment—all sources of friction.

Fires frequently result from overheated power-transmission bearings and shafting in buildings such as grain elevators; cereal, textile, and woodworking mills; and plastics and metalworking facilities, where dust and lint accumulate. Make frequent inspections to see that bearings are kept well oiled and do not run hot. Keep the accumulation of flammable dust or lint to a minimum.

Provide drip pans beneath bearings, and clean them frequently to prevent oil from dripping to the floor or on combustible materials below. Keep oil holes of bearings covered to prevent dust and gritty substances from entering the bearings and causing them to overheat.

Frictional heat sufficient to cause ignition can result from the jamming of work materials during production. Another common problem is the tension adjustment on belt-driven machinery. If the belt is too tight or too loose, excessive friction may cause serious overheating.

Foreign Objects

Take every precaution to keep foreign objects from entering machines or processes. They might strike sparks where there are flammable dusts, gases, or vapors or combustible material, such as cotton lint or metal powder. For this purpose, use screens or magnetic separators such as are used in textile mills, grain elevators, and other operations where explosive mixtures of dusts are present.

Open Flames

Although open flames are probably the most obvious source of ignition for ordinary combustibles—and thought to be the most easily avoided—they still account for a large percentage of fires. Heating equipment, torches, and welding and cutting operations are principal offenders.

Air Heaters. Air heaters are commonly used on construction work and often cause fires for the following reasons.

- overheating of the air heater with resultant radiation that ignites nearby combustible materials, such as concrete form work, tarpaulins, wood structures, paper, straw, and rubbish

- failure to keep anything that will burn a minimum of 3 ft away from a portable heater
- failure to insulate air heaters from floors or other combustible bases
- failure to provide a substantial spark shield and to use fuels that do not produce high flames or sparks (see NFPA 211, Standard for Chimneys, Fireplaces, Vents, and Solid Fuel-Burning Appliances)
- failure to provide a secure base or to anchor it properly. (Vent air heaters in unventilated rooms, or enclosures to the outside, by means of an overhead hood and flue. Venting removes the toxic products of combustion and any unburned gas.)

Torches. If gasoline, kerosene, liquefied petroleum gas (LPG), acetylene, or alcohol torches are used, ensure the flames cannot ignite any surrounding combustible materials due to flame impingement or convection. Never use torches around flammable liquids, paper, and combustible packing materials.

Portable Furnaces. Provide sufficient overhead clearance for portable furnaces and blow torches so that combustible materials above them cannot ignite from flame impingement or convection. If necessary, remove or protect overhead combustible materials with noncombustible insulating board or sheet metal, preferably with a natural-draft hood and flue of noncombustible material.

Welding and Cutting. When possible, have welding or cutting done in special fire-safe areas or rooms with concrete or metal-plate floors. Flame impingement on concrete may cause concrete to spall. Consequently, keep work off the floor or protect the floor with a metal shield (see NFPA 51B).

In cases where welding and cutting operations are performed outside the special fire-safe areas, use hot-work permit programs to promote maximum fire-safe working conditions. If welding must be done over wood floors, have them swept clean and wetted down, preferably covered with flame-resistant blankets, metal, or other noncombustible covering. Keep hot metal and slag from falling through floor openings, thus igniting combustible material below.

Use sheet metal, flame-retardant tarpaulins, or flame-resistant curtains around welding operations to prevent sparks from reaching nearby combustible materials. Do not permit welding or cutting in or near rooms containing any flammable liquid, vapor, or dust unless special precautions (including a hot-work permitting process) are observed and the hazards are well understood.

Do not perform welding or cutting on a surface until combustible deposits have been removed. Also do not perform welding or cutting in or near closed tanks or other containers that have held flammable liquids. These containers must first be thoroughly cleansed and filled with water or purged with an inert gas. A combustible-gas indicator test must show that no trace of a flammable gas or vapor is present. Periodically repeat these tests to determine if any trace of flammable gas or vapor is released during the welding or cutting operation. If further tests show any trace, stop the work until all flammable gas or vapor is dispelled (see NFPA 326, Standard for Safeguarding of Tanks and Containers for Entry, Cleaning, or Repair).

Welding is sometimes necessary on equipment containing flammable material. However, such welding requires highly specialized procedures and training and should be avoided when possible. Have fire-extinguishing equipment for this type of exposure within easy reach of welding and cutting operators. If the risk is justified, and welding must be done outside of a shop or area designated for welding, station fire-watch employees, who will prevent sparks or molten slag from starting fires and who will extinguish fires should they start. Fire-watch employees should remain at the work location for at least 30 min after the welding or cutting is completed because some fires escape detection when they are starting. Install shields around spot welders to prevent sparks from reaching nearby combustible materials or from injuring employees. See Chapter 19, Welding and Cutting, in this volume for more information.

Spontaneous Ignition

Spontaneous ignition results from a chemical reaction in which there is a slow generation of heat from oxidation of organic compounds that, under certain conditions, is accelerated until the ignition temperature of the fuel is reached. This condition is reached only where there is enough air for oxidation but not enough ventilation to carry away the heat as fast as it is generated.

Spontaneous ignition usually occurs only around quantities of bulk material packed loosely enough for a large amount of surface to be exposed to oxidation yet without adequate air circulation to dissipate heat. Exposure to high temperatures increases the tendency toward spontaneous ignition.

The presence of moisture also can advance spontaneous ignition, unless the material is wet beyond a certain point. Materials such as unslaked lime promote spontaneous ignition, especially when wet. Store such chemicals in a cool, dry place away from combustible material.

At ordinary temperatures, some combustible substances oxidize slowly, and under certain conditions, they can reach their ignition point. These substances include vegetable and animal oils and fats, coal, charcoal, and some finely divided metals. Rags or wastes saturated with linseed oil or paint often cause fires, too.

The best preventives against spontaneous ignition are either total exclusion of air or good ventilation. With small quantities of material, the former method is practical. With large quantities of material, such as storage piles of bituminous coal, both methods have been used with success.

Temperatures at or above 140°F (60°C) are considered dangerous in coal piles. If temperatures rapidly approach or exceed that figure, move the pile or rearrange it to allow better circulation of air and to avoid spontaneous ignition.

Certain agricultural products are also susceptible to spontaneous ignition. Sawdust, hay, grain, and other products (e.g., jute, hemp, and sisal fibers) may ignite spontaneously, especially if exposed to external heat or to alternate wetting and drying. The best preventive measures are circulation of air, removal of external sources of heat, and storage of material in smaller quantities.

Fires in iron, nickel, aluminum, magnesium, and other finely divided metals are sometimes attributed to spontaneous ignition. These fires are thought to result from the oxidation of cutting or lubricating oils, or possibly from chemical impurities.

Iron sulfide, also called iron pyrite and pyrophoric iron, occurs as a result of lubrication of sulfur compounds in petroleum that come in contact with iron in pipes and vessels. As long as air is excluded or it is kept moist, iron sulfide does not constitute a significant ignition hazard. However, when equipment containing iron sulfide is opened to the atmosphere and dried out, it will burn. Where iron sulfide may be present, make provisions to keep the inner surface wet when opening equipment. Carefully and promptly handle the disposal of accumulated deposits of iron sulfide so it will not start burning in an area where fire would create a hazard.

Poor Housekeeping

Poor housekeeping is another factor that contributes to fires. Properly collecting and storing combustibles and disposing of rubbish, as well as maintaining locker rooms, will prevent fire hazards.

Collection and Storage of Combustibles. Many industrial fires are the direct result of accumulations of oil-soaked and paint-saturated clothing, rags, waste, packing materials, and combustible refuse. Deposit such material in noncombustible receptacles with self-closing covers that are provided for this purpose and removed daily from the work areas.

Exhaust systems of effective design will remove gases, vapors, dusts, and other airborne contaminants, many of which are fire hazards. Exhaust systems and machinery enclosures will help prevent accumulation of combustible materials on floors or machine parts. Such materials are most hazardous when airborne rather than when they have settled out.

Clean waste, although not as dangerous as oil-soaked waste, is readily combustible and should be kept in metal cans or bins with self-closing covers. Store packing materials, cotton, kapok, jute, and other highly combustible fibrous materials in covered, noncombustible containers. If large quantities of these materials are kept on hand, store them in fire-resistant rooms equipped with fire doors and automatic sprinklers. Have portable extinguishers, hose lines, or other extinguishing equipment for Class A fires available for use at such storage places.

A schedule for safe collection of all combustible waste and rubbish should be a part of the fire prevention program. Be sure that custodians or others involved in the collection of wastepaper in offices and service areas have fire-safe collection containers. Check collection practices to be sure that ash trays, which may contain smoldering material, are not emptied into combustible bags or cartons or into containers of combustibles.

At regular intervals, clean accumulations of all types of dust from overhead pipes, beams, and machines, especially from bearings and other heated surfaces. Practically anything will burn and propagate flames. Therefore, keep roofs free from sawdust, shavings, and other combustible refuse. Cleaning of such materials should be done by vacuum removal because air blowing may create dangerous clouds of dust.

Do not store such material or allow it to accumulate in air, elevator, or stair shafts; in tunnels; in out-of-the-way corners; near electric motors or machinery; against steam pipes; or within 10 ft (3 m) of any stove, furnace, or boiler.

Rubbish Disposal. To prevent fires, federal, state, and local laws forbid some methods of waste disposal, such as open burning, evaporation, and flushing to sewers. Fires are often caused by burning rubbish in yards near combustible buildings, sheds, lumber piles, fences, grass, or other combustible materials. If rubbish must be burned, the best and safest way is with a well-designed incinerator that meets the requirements of environmental pollution control laws.

If flushing or dumping waste materials into sewers is prohibited, use trap or retention tanks that can be pumped out. Thus, the waste material can be disposed of properly. Also, consider chemically altering a waste material before disposing of it.

Locker Rooms. Locker rooms in which oil-soaked clothing, waste, or newspapers are kept are always a serious fire hazard. Take every precaution to prevent such combustible materials from accumulating.

Provide lockers made of metal with solid, fire-resistant sides and backs. Doors, however, should have some open spaces for ventilation. Lockers should be large enough so that air can circulate freely around clothing hung in them. Do not permit employees to leave clothing saturated with oils or paints in lockers.

Where automatic sprinklers are used, cover locker tops with screening. Also, use locker tops made of perforated metal so that water can reach burning contents if a locker fire occurs. Heavy paper pasted over the tops will keep out dust. Lockers that have sloping tops and that stand flat on the floor will prevent the accumulation of rubbish both above and below.

Explosive Atmospheres

Dusts, gases, and vapors may create explosive atmospheres. Observe the proper precautions to prevent these atmospheres from developing.

Dusts. A dust-explosion hazard exists wherever material that will readily burn or oxidize is available in powder form. The surface, or contact, area of each dust particle is large in relation to its mass. Many synthetic resins and powders used in the plastics industry present a dust-explosion hazard comparable to that of coal. This group includes phenolic, urea, vinyl, and other types of resins and a number of molding compounds, primary ingredients, and fillers. There are two ways to prevent dust explosions:

1. Prevent the formation of explosive mixtures of dust.
2. Prevent the ignition of such mixtures if their formation cannot be prevented.

Effectively prevent an explosive mixture of dusts from forming by providing an inert atmosphere. Limit oxygen to concentrations below which flame propagation does not occur. To make this method reliable, monitor the concentration of oxygen with an oxygen analyzer. This is an effective method if enclosure of the system is possible. However, do not provide inert atmospheres if operators have to enter the atmosphere to perform work.

Take extreme precautions to prevent dust from building up to explosive proportions. Using local exhaust ventilation and performing regular cleaning will minimize this hazard. Where possible, segregate dusty operations. Also, have dust-producing equipment totally enclosed and exhausted to prevent dust from leaking into the general work area. Accumulation of dust on high surfaces, such as door lintels, window frames, and ventilation ducts, is good evidence that the ventilation system and cleaning schedules are not sufficient.

Prevent ignition of an explosive mixture of dust and air by eliminating open flames, friction sparks, static electricity, welding, and excessive heat; by increasing humidity; or by using inert gas. Take every precaution to prevent overheated bearings, smoking, friction sparks, and sparks in hand tools and in grinding or welding operations. Also control sources of static electricity.

To prevent sparks, use nonferrous material for truck wheels and bucket conveyors. Also, use pneumatic or magnetic separators to remove stones, nails, and other spark-producing foreign objects from the material being processed. Keep conveyor-belt rollers and similar moving parts properly lubricated and maintained to avoid excessive friction and heat.

As discussed earlier, use only approved dust-tight wiring, fixtures, and motors. Also, instruct employees to use only extension cords, lamps, and portable electric tools designed for protection against the hazard of dust explosions, as described in the NEC.

In buildings with high explosion hazards, such as grain elevators and plastics facilities, provide extensive dust-collection equipment. Construct these buildings so that explosive pressures will push out hinged windows or blow out wall sections or panels designed and built to fail in a predetermined area. Keep floor openings to a minimum, and seal openings for pipes and ductwork.

Have portable fire-extinguishing equipment readily available. Fog nozzles or finely divided streams of water are more effective for dust fires than are solid streams of water, which stir up dust.

Gases and Vapors. Gases and vapors that produce flammable mixtures with air or oxygen are common in industry. Such gases include hydrogen, acetylene, propane, LPG, carbon monoxide, methane, natural gas, and manufactured gases.

Highly volatile liquids that emit flammable vapors include gasoline, benzene, naphtha, and methanol. Kerosene, turpentine, stoddard solvent, and other liquids with flash points above 100°F (38°C) generally must be heated above normal room temperatures before they give off sufficient vapors to form ignitable concentrations.

Whenever using flammable liquids, especially when moving them from one container to another, bond the containers and ground them. In addition, do not transfer flammable liquids from a metal container to a plastic container because the plastic cannot be bonded and grounded effectively. Many fires have resulted from pouring gasoline from a metal can into a plastic gasoline tank.

When flammable liquids must be handled and used, allow only minimum amounts to be used in work areas, and require all such materials to be stored in approved containers.

FIRE DETECTION AND RESPONSE

Despite good construction, cleanliness, and modern fire-fighting methods, losses from fire nevertheless occur. Losses would be reduced if each developing fire were detected so it could be attacked and extinguished. Thus, fire detection is essential in every fire protection system. Means of detection could be a human observer; automatic sprinklers; smoke, flame, or heat detectors; or, more likely, a combination of these.

There are two tasks associated with detection:

1. Give an early warning to enable facility occupants and fire fighters to respond to the fire.
2. Begin fire control, suppression, and extinguishment processes.

Each fire detection system requires a sensor, which detects a physically measurable quantity of smoke, flame, or heat. For detection to occur, this quantity must undergo measurable variations when a fire begins in the vicinity of the sensor. A decision-making device coupled with the sensor then compares the measured quantity with a predetermined value. When the value is different, an alarm sounds. Thus a detector both detects and signals.

In general, there are three possible errors in any fire detection system:

1. activating a nuisance (false) alarm
2. not detecting a fire
3. detecting it too late.

The cause of nuisance alarms may be human interference, mechanical or electrical faults, or special environmental conditions.

Human Observer

A human observer is a good fire detection system for the following reason: he or she can take immediate action in a flexible way, whether calling the fire department or putting out a fire with an extinguisher. However, human observers can also respond unpredictably. For example, if a human observer detects a fire and puts it out, he or she may not report it, potentially allowing the fire to rekindle.

Early detection and notification are essential. Early notification of occupants allows time to evacuate, and early notification of the fire brigade/fire department provides a better chance of stopping the fire before extensive damage occurs.

Human observers should also report malfunctioning fire alarm systems. Fire alarm systems need to be properly maintained. Nuisance alarms have a negative effect: if an actual fire occurs, the activated alarm may be dismissed as just another "false" alarm.

Automatic Fire Detection Systems

When planning an automatic fire detection system, answer the following questions:

- What is the main purpose of the system? Typical purposes include warning the occupants of a building,

protecting irreplaceable and highly valuable goods, protecting against interruption of production, and protecting against corrosion or radioactive contamination.

- What are the possible sources of ignition?
- What kinds of material will probably be ignited first?
- What kind of building construction is used?
- What are the environmental conditions?
- What kind of detection system has been installed, and what are the reasons for choosing this system?
- How long can a fire be allowed to go undetected?

It is important to have a qualified fire protection engineer involved in the design of automatic fire detection systems. When planning the installation of fire detectors, use the following four steps:

1. Select the proper detector for each hazard area. For example, a computer area may require ionization or combination detectors. A warehouse may require infrared and ionization detectors. In low-risk areas, thermal detectors or a combination of detectors may be used.
2. Determine the spacing and locations of detectors in order to provide the earliest possible warning.
3. Select the best control system arrangement to provide fast identification of the exact source of the alarm.
4. Ensure that the system will notify the proper authorities, who can immediately respond to the alarm and take appropriate action. Every detection system must have its alarm signal transmitted to a point of human supervision.

From a systems analysis approach, a suitable detector and system must be fitted for a given environment and for detection of the presence of certain situations. Sensitivity, reliability, maintainability, and stability are the critical variables in the selection of a detection system:

- Sensitivity is established by the design. How well can the device detect what it is meant to detect?
- Reliability is the ability to perform an intended function when needed. In other words, the device should not malfunction.
- Maintainability means the unit is easy to service and keep at operating efficiency. A minimum of maintenance should be needed.
- Stability means the unit can sustain its sensitivity over time.

There are many types of fire detectors to handle various situations and to detect various fire situations. Most manufacturers and distributors offer several or all of the commonly used types. They can also design a combination of equipment into a coordinated system to meet the special needs of a facility.

Thermal Detectors

Thermal detectors detect the heat from a fire. There are several kinds of thermal detectors, each with a specific use. Thermal detectors are reliable for what they do. However, they can only detect the heat of a fire, which usually will not build up to significant levels until the last stage of a fire. Many fires start slowly, with little heat generated at the beginning, and will be well under way by the time a thermal detector comes into operation. They are generally used where no life hazard is involved and some fire loss can be tolerated.

Fixed-Temperature Detectors. These thermal detectors are based on a bimetallic element. They are made of two metals that have different coefficients of expansion. When heated, the element will bend to close a circuit, initiating the alarm. A thermal detector may also use a fusible, spring-loaded element, such as found in a sprinkler head, melting at a certain temperature and releasing an arm to close a circuit. Fixed-temperature detectors are simple, are inexpensive, and require a low-voltage draw.

Rate-Compensated Thermal Detectors. These detectors work by the expansion characteristics of a hollow tubular shell containing two curved expansion struts under compression, fitted with a pair of normally open, opposed contacts. When subjected to a rapid rise of heat, the shell expands and lengthens at a faster rate than the struts, thus permitting the contacts to close. When heated slowly, both the shell and the struts lengthen at about the same rate until the struts are fully extended, thus making contact at a preset temperature point.

Rate-of-Rise Thermal Detectors. These detectors use an enclosed, vented hemispherical chamber containing air at atmospheric pressure, with a small pressure-sensitive diaphragm on top. With a normal rise in temperature, the excess pressure is relieved through small vents. However, a rapid rise in temperature will deflect the diaphragm faster than the vents can operate, thus triggering an alarm. This unit responds quickly to a rapid rise in temperature.

Line Thermal Detectors. These detectors use a length of small-diameter tubing that can be as long as 1,000 ft (305 m). When exposed to the heat of a fire, air inside the tube expands, sending a pressure wave to expand a diaphragm at the end, which in turn triggers an alarm. This is an unobtrusive, inexpensive detector. For example, it can be run along the ceiling molding, where it is nearly invisible. No maintenance is needed. It can be painted over, and it will even work with the tubing broken, if the temperature rises fast enough.

Eutectic Salt Line Thermal Detectors. These detectors consist of pliant metal tubing containing a eutectic salt in which a wire is embedded. At a preselected temperature, the salt creates a short circuit between the internal wire and the outside tubing, thereby triggering an alarm. This detector, along with a continuous-resistance unit, has been widely applied recently to guard against fires in jet aircraft engine nacelles. The pliant tubing can be wound around and shaped to the various components of the engines to signal any increase in temperature that might be the result of a fire caused by leaking oil, hydraulic fluid, and so forth.

These line detectors are also used in conveyor-belt systems where the bearings supporting a rubberized belt may ignite because of friction and lack of lubrication. If the conveyor is transporting coal, for instance, the resultant fire could be expansive and hard to extinguish.

Bulb Detection Systems

These detectors are completely mechanical. They are especially desirable in locations in which the explosive nature of the fire hazard makes it wise or essential to avoid the use of electricity. These systems involve a number of bulbs containing air at atmospheric pressure. One or more of these bulbs are installed along the ceiling of the hazardous area, all connected to a diaphragm at the control center. When a rise in temperature strikes one or more of the bulbs, it deflects the diaphragm, and a mechanical extinguishing system can be activated. This system is used as a release mechanism for fixed carbon dioxide extinguishing systems in marine and industrial applications.

Smoke Detectors

Smoke detectors respond to the particles of combustion, both visible (smoke) and invisible. They can be triggered by either a decrease or an increase in light. When smoke enters a light beam, it either absorbs light so the receiver end of the circuit registers less light, or it scatters light so that a terminal normally bypassed by the light beam now receives part of the light.

The more sophisticated smoke detectors are responsive both to gas or products of combustion and to smoke. These products of combustion detectors are capable of detecting the beginning of a fire long before there is visible smoke or flame.

Photoelectric smoke detectors are line powered and usually include lamp supervision circuitry and an alarm in case of lamp failure. An incandescent light source may be used. Also, a high-intensity strobe lamp that generates a stronger reflection can be used so that fewer or smaller smoke particles will actuate the photocell.

Beam Photoelectric Detectors. These detectors are triggered with less light. A long beam is directed at a photocell.

Rising smoke tends to obscure the beam, decreasing light transmission and activating the circuit. These detectors are an inexpensive way to cover large spaces, such as warehouses and atria. Beam photoelectric detectors are sensitive to voltage variations; to dirt on the lamp or lens; and also to flying insects or spiders, which sometimes congregate near the lamp seeking warmth.

Reflected Beam Photoelectric Detectors. These detectors use a beam of light in a chamber, with the photocell normally in darkness. Should visible smoke particles enter the chamber, they scatter the light and reflect it onto the cell, causing a change in electric conductivity, activating the circuit.

Products of Combustion (Ionization) Detectors. These detectors sense both visible and invisible products of combustion suspended in the air. They consist of a chamber with positive and negative plates and a minute amount of radioactive material that ionizes air in the chamber. The potential between the two plates causes ions to move across the chamber, setting up a small current. When aerosols from incipient fires enter the chamber, they cling to masses of moving ions. This slows the ions' movements and increases the voltage necessary for the ions to make contact. This voltage imbalance, amplified by electrical circuitry, activates the circuit.

Ionization detectors sense fire at the earliest practical detection stage. They are the best method for detecting slow, incipient fires in commercial buildings—a cigarette in a wastepaper basket, for example, which might be in a pre-smoldering condition for 30 minutes or more. The ionization detector has the additional feature of operating in the fail-safe mode. If excessive dust is present, however, the device may activate a nuisance alarm.

Single-Chamber Ionization Detectors. These detectors are most economical. The chamber of these detectors is open to the atmosphere. Current flows between two poles and gets an increased voltage from combustion aerosols. This closes the contact and sends an alarm through the relay.

Dual-Chamber Ionization Detectors. These detectors have two identical chambers: one is a sealed chamber; the other is open to the atmosphere. The inner ionization chamber monitors the surrounding conditions and compensates for the effect on the ionization rate of barometric pressure, temperature, and relative humidity. This construction accepts a wider range of atmospheric variations without activating nuisance alarms.

Low-Voltage Ionization Detectors. These detectors require 24 V to operate, instead of operating over line power voltages.

Low-profile, low-voltage detector heads are less obtrusive, while being equally sensitive and reliable.

Aspirating Smoke Detectors. These systems use piping to convey products of combustion to a remote analyzer, which triggers an alarm signal. They often have very low thresholds of detection, resulting in very high sensitivity. These systems are useful for monitoring large areas where early detection is essential.

Flame Detectors

Flame detectors respond to the optical radiant energy of combustibles. Flame detectors sense light from the flames. Sometimes, they work at the ultraviolet end of the visible spectrum, but more often, they work at the infrared end. To avoid nuisance alarms from the effects of nearby light sources, flame detectors are set to detect the typical flicker of a flame. Some operate with delay of a few seconds to eliminate nuisance alarms from transient flickering light sources, such as flashlights or headlights.

Flame detectors have some very important applications, such as in large aircraft hangars and for guarding against fires in fuel and lubricant drips. In general, however, by the time flame is visible, a fire has a good foothold. Either an infrared or an ultraviolet detector can be used to sense the flame.

Infrared Detectors. These flame detectors sense a portion of the radiant infrared energy of flames. They are often used in operations requiring an extremely fast response—for example, where flammable liquids are stored or used.

Ultraviolet Detectors. These flame detectors react only to actual flame. They do not respond to glowing embers or incandescent radiation. Also, these units are insensitive to heat, infrared radiation, and ordinary illumination.

Combustion Gas Detectors

These detectors are closest to being general-purpose detectors. Combustion gas detectors do not rely on heat. They measure the percentage of gas present. Also, they do not wait for the dangerous condition of flames to occur before they sound an alarm. Most fires detected by combustion gas detectors can be extinguished by workers on the site.

Combustion gas detectors can usually be set to automatically sound an alarm or to set off extinguishing equipment. Some conditions may require periodic maintenance and calibration. There are areas in which combustion gas detectors cannot be used. For example, in an area where a specific level of combustion gases may be tolerated at times because a particular process emits them, flash fires from nearby flammable liquids may be anticipated. Therefore, detection must be almost instantaneous if it is to be at all effective. In such areas, a flame detector should be used. In locations where chemicals and some plastics could generate great volumes of smoke with little combustion, a smoke detector should be used.

Extinguishing System Attachments

Some automatic detection devices are often not even thought to be detectors. These devices are extinguishing system attachments. They are, nonetheless, truly detectors. Fire detection may be handled by water-flow indicators in a sprinkler system. These indicators may operate in response to a sudden increase or decrease in pressure or by detection of the flow of water by a vane inside the piping. They are designed to detect a flow of water that exceeds a set number of gallons per minute.

A second type of device indicates that the fire-extinguishing system is jeopardized. The following are a few of these devices:

- Water-level devices warn when a self-contained water storage supply is low or, in the case of a low-differential dry-pipe valve, when the level of water is too high.
- Water-temperature switches warn when the water storage supply is approaching the freezing point.
- Water-supply valve position switches signal if someone unintentionally or purposely starts to turn off the water supply. These switches should give the alarm before the handle is closed two revolutions, or one-fifth the distance from the open position. However, in practice, they are often set to signal an alarm within one-half a revolution.

Sensor Systems

In smokestacks, storage tanks, or other areas in which several variables are at work at the same time, use a linear sensor system, alone or combined with other systems, for greater accuracy in detecting fires. The common heat sensors simply provide averages. Linear sensor systems, however, give continuous point-by-point readings. They pick up and monitor trouble spots instantly over an entire area.

The concept of totally integrated fire detection brings together the facility's comprehensive fire-fighting capacity. This concept demands the continuing close cooperation of all persons involved in building design and construction, as well as in safety engineering and fire protection.

A modern fire detection system should be a combination of the various types of detecting systems in one integrated system.

Communications

Once a fire has been detected, effective methods of communications are necessary (1) as a means of alerting occupants to the emergency and (2) as a way to mobilize fire protec-tion forces, whether a facility's fire brigade, the municipal fire department, or both. The alarm system is no better than employees' training in how to respond when the alarm is activated.

Consider how effective evacuation communications will occur during power failures. Make sure that everyone can be alerted and that there is some way for everyone to get out of the building, regardless of whether the power is on.

Evacuation Drills

Planning for fire emergencies is not an exact science. Develop a comprehensive emergency plan after evaluating the particular situation at hand. Prepare an emergency guide or procedure to outline appropriate procedures and training and to assign responsibilities. Make prevention of personal injury and loss of life the prime objective of emergency planning.

Carefully plan and periodically carry out evacuation drills. Conduct them in a professional manner under rigid discipline. Training employees to leave their workplaces promptly at the proper alarm signal and to evacuate a building speedily but without confusion is largely accomplished through training, education, and practice drills. To help eliminate panic in the event of an emergency and to guarantee the smooth functioning of the emergency plan, thoroughly prepare the plan as part of an overall planning process that is comprehensive and all-inclusive.

Post up-to-date instruction sheets, including evacuation routes, and distribute them to all employees. Maps that are posted for evacuation should also show alternate routes in case the first route is closed. In addition, when performing evacuation drills, consider simulating a blocked exit to gauge how employees react.

Post fire guards in each area to make sure each area has been safely evacuated. People in places such as restrooms and high-noise areas do not always hear or see alarms. Perform a roll call at an outside location to make sure that all personnel are accounted for and that no one was left behind.

Effective emergency drills that include evacuation conducted at frequent intervals help to reinforce management's commitment and interest in all fire prevention activities. The drills should serve as a reminder to employees and supervisors that all fire prevention practices are important. Emergency drills also serve as a valuable way to check the adequacy and condition of fire exits and the alarm system. Immediately and permanently correct any deficiencies.

Fire Brigades

In some facilities, management supports a fire brigade as part of the response to an alarm. Management should not depend entirely on automatic fire protection equipment, municipal fire departments, or mutual aid agreements to minimize fire losses. Fires can get out of control before a municipal fire department arrives. One method of providing additional fire protection is to form a well-trained, -equipped, and compliant fire brigade. If an employer chooses to develop a fire brigade, note that OSHA has specific regulations and procedures outlining the requirements for the operations and establishment of fire brigades and emergency responder activities that are compliant with the appropriate regulatory agencies' guidance and rules in effect (Figures 9–7 and 9–8). Depending on the types of emergencies to which the fire brigade is expected to respond, these regulations can be very complex. Also, there are several NFPA standards applicable to fire brigades and emergency responders. These standards and regulations must be understood, along with the commit-ment necessary, before establishing a fire brigade. In addition, local fire departments should also be consulted when plans for a fire brigade are being considered.

Alarm Systems

Alarm systems can be divided into four groups: local, auxiliary, central station, and proprietary. All types of alarm systems should be equipped with a signal system that clearly communicates to all facility occupants. Whenever an alarm is sounded in any portion of the building or area, all occupants must know what the sound means. Every fire alarm system, whether it is currently in place, newly installed, or revised, should meet the following criteria:

Figure 9–7. Members of a fire brigade are being trained to use hose lines to fight a fire.

Figure 9–8. More advanced training is necessary for fire brigades expected to respond to petrochemical facilities.

- When alarms are audible, the alarm sound should be clearly and immediately distinguishable from other signals that might be used in a given building. Provision for alerting hearing impaired and visually impaired occupants is necessary.
- Locate audible and visible alarm devices so they are clearly audible and visible to all personnel. Train personnel to recognize the signal and to respond according to that location's specific response procedure.
- The fire alarm system should be composed of approved (listed) equipment and installed using methods that conform to NFPA standards.
- Maintain the alarm system in good working order. Test it at frequent intervals to ensure that it is working properly. Alarm testing protocols should conform to NFPA standards.
- All personnel should know the location of and means of contacting external fire protection sources. Conspicuously post this information in strategic areas. Each employee should also know the proper procedures for reporting a fire if he or she is the one who detects the fire.

Protected-Premises (Local) Fire Alarm Systems

A protected-premises fire alarm system consists of bells, horns, lights, sirens, or other occupant notification and warning devices located in or on the facility. Protected-premises fire alarm systems are generally used for life protection—that is, to evacuate the occupants and limit injury or loss of life from the fire. A protected-premises fire alarm system can be tied in with other fire alarm systems to summon a fire brigade or the fire department and activate fire control and suppression systems.

Some protected-premises fire alarm systems have a presignal feature. The presignal feature alerts designated remotely supervised stations on an alarm condition before an evacuation signal is activated. The presignal feature allows an initial response and investigation to occur (in less than 1 min, usually). There are three possible results from the initial response to an alarm presignal feature:

1. A determination is made that the fire detection system did not detect an actual fire condition for which the general alarm should be activated (a nuisance alarm). *Action:* evacuation signal activation is cancelled.
2. A determination is made to extinguish an incipient-stage fire. *Action:* evacuation signal activation occurs unless the incipient-stage fire is extinguished and a separate decision to cancel the ensuing activation of the evacuation signal is made.
3. A determination is made that an evacuation is warranted. *Action:* evacuation signal activation occurs.

Protected-premises fire alarm systems are available from a wide range of suppliers and are easy to install. By themselves, however, they do not provide much protection. While they alert personnel, they do not summon the fire department or a fire brigade automatically.

Auxiliary Fire Alarm Systems

Auxiliary fire alarm systems are connected to a protected-premises fire alarm system. When the protected-premises fire alarm system activates, the auxiliary fire alarm system signals the public fire service communications center.

Supervising-Station Fire Alarm Systems

Supervising-station fire alarm systems are continually staffed and operated by trained personnel. They monitor facilities' fire detection and alarm systems and make appropriate notification upon presignal alerts, alarms, and indications of system problems. There are three kinds of supervising-station fire alarm systems: central station service fire alarm systems, proprietary supervising-station fire alarm systems, and remote supervising-station fire alarm systems.

Central station service fire alarm systems are for-hire monitoring companies that continually monitor all aspects of the fire detection and alarm systems for a number of independent, unrelated organizations. Proprietary supervising-station fire alarm systems (Figure 9–9) are operated on behalf of facilities under one owner and continually monitor all aspects of the facilities' emergency systems (including fire detection systems) related to all of that one owner's facilities. Remote supervising-station fire alarm systems are similar to proprietary supervising-station fire alarm systems but are limited in scope to the alarm, supervisory, or trouble-shooting, systems of one or more specific protected-premises fire alarm systems.

Figure 9–9. This console is part of an Underwriters Laboratories–listed continuously staffed, supervising station (central station service) that has not only fire detection and alarm annunciation, but also security and emergency facility condition indicators. *(Reprinted with permission from A-1 Alarm Service Inc., Champaign, Illinois.)*

Spacing of Detectors

Because a fire-resistive structure can be expected to withstand the burnout of its contents for a longer period of time without structural collapse than a structure constructed of combustible materials, early warning of fire is of greater importance in the latter structure. Thus, combustible buildings should have more detectors than would normally be used. Such buildings should also be tied into the local fire department or central station fire alarm system. Building damage will thus be greatly minimized.

To determine the number of detectors required for a given area, consider several factors. In general, the more detectors that are installed, the greater the coverage that is provided. If the number of detectors in a given area is doubled, the distance and the time that combustion products have to travel from the farthest point in a room to a detector are proportionately reduced. The exact area to be covered by an ionization detector, for example, depends on total area, building construction, the area's contents, air movement, the value of the building and its contents, ceiling obstructions, and the cost of equipment downtime. However, the more detectors a system has, the more it costs to maintain.

According to NFPA 72, National Fire Alarm and Signalling Code, detector spacing should not exceed the linear maximum indicated by the detector's listing requirements. Closer spacing may be required due to structural characteristics of the protected area and possible drafts or other conditions affecting detector operation. Factors such as air velocity, ceiling shape and height, ceiling material, and configuration of building structure influence the proper spacing.

Location of Detectors

NFPA 72 also includes smoke detector spacing requirements. Smoke detectors should be located and adjusted to operate reliably in case of smoke at any part of the protected area. The location of detectors should be based on an engineering survey of the application of this form of protection to the area under consideration. These features include air velocity; number of detectors to provide adequate coverage of cross-sectional area of the space with respect to travel, diffusion, or stratification of smoke; and location of detectors with respect to return, supply, or circulating blowers, air-conditioning facilities, temperature variations, and the like. Such conditions vary with different installations and should be dealt with on the basis of experience.

Special consideration should be given to the storage of contents in a protected space to provide unobstructed openings for the travel of smoke to the smoke detector. Where air-conditioning or ventilating equipment serves the space to be protected by a smoke detector, particular attention should be given to the supply, return, and circulation of smoke under any condition of operation of the equipment to ensure prompt detection. There is a temptation to try to protect many thousands of square feet of building space by placing smoke detectors in the air-handling system. Because the products of combustion become diluted by air as they travel toward the detectors in the air-handling system, it follows that for detectors near the fan to detect smoke, the smoke must be very heavily concentrated in the occupied area of the building.

Maintenance of Detectors

Every fire protection system needs periodic maintenance and inspection by a knowledgeable, responsible person. Without periodic inspection and testing of each component, no fire protection system can be considered dependable. This work may be done by an organization whose specialty is installing and servicing fire detection equipment. Facility personnel also can be trained to become expert at routine inspection procedures.

Proper functioning of the system should be checked at regular intervals. Make a spot check of several detectors each month as a regular part of fire protection inspections. Actuate different detectors at each inspection so that all components of the system will have been tested in the course of a year.

Regularly inspect each detector's head screen for dust accumulation, and clean it if necessary. In a dusty location, more frequent cleaning may be necessary. On a yearly basis, check each detector on the circuit for operation and sensitivity. Also check alarm relay contacts for proper operation.

FIRE SUPPRESSION AND FIRE EXTINGUISHMENT

Portable Fire Extinguishers

Equipment used to extinguish and control fires is of two types: fixed and portable. Fixed systems include water equipment, such as automatic sprinklers, hydrants, and standpipe hoses, and special pipe systems for dry chemicals, carbon dioxide, and foam. Special pipe systems are used in areas of high fire potential where water may not be effective, such as where tanks for storage of flammable liquids and electrical equipment are located. Fixed systems, however, must be supplemented by portable fire extinguishers. These often can preclude the action of sprinkler systems. Not only can they prevent a small fire from spreading, but they can also rapidly extinguish a fire in its early stages, when such extinguishment is performed by a trained person.

Principles of Use

Even though a facility is equipped with automatic sprinklers or other means of fire protection, portable fire extinguishers should be available and ready for an emergency. The term *portable* is applied to manual equipment used on small, beginning (incipient-stage) fires immediately after discovery of a fire and before the functioning of automatic equipment or the arrival of fire fighters.

To be effective, portable extinguishers must meet all of the following criteria:

- Fire extinguishers must be approved by a recognized testing laboratory.
- Fire extinguishers must be the right type for each class of fire that may occur in the area.
- Fire extinguishers must be of sufficient size to protect against the expected exposure in the area.
- Fire extinguishers must be located where they are easy to reach for immediate use.
- Fire extinguishers must be maintained in good operating condition, inspected frequently, checked against tampering, and recharged as required.
- Fire extinguishers must be able to be operated by trained, local area personnel who can find them when needed.

Classification of Fire Extinguishers

Portable extinguishers are classified to indicate their ability to handle specific classes and sizes of fires. This classification is necessary because new and improved extinguishing agents and devices are constantly being developed and because larger portable extinguishers are available. Labels on extinguishers indicate the class and relative size of fire that they can be expected to handle. Review the section on Classification of Fires in this chapter.

Use the following paragraphs as a guide to the selection of portable fire extinguishers for given exposures (see NFPA 10, Standard for Portable Fire Extinguishers). Refer to the section on Classification of Fires, earlier in this chapter. Also, observe protection and insurance recommendations, based on fire protection requirements of the authority having jurisdiction.

- Use fire extinguishers rated for use on Class A fires for ordinary combustibles, such as wood, paper, some plastics, and textiles, where a quenching-cooling effect is required.
- Use fire extinguishers rated for use on Class B fires for flammable-liquid fires, such as oil, gasoline, paint, and grease fires, where oxygen exclusion or a flame-interrupting effect is essential.
- Use fire extinguishers rated for use on Class C fires for fires involving energized electrical wiring and equipment where the dielectric nonconductivity of the extinguishing agent is of first importance. These units are not classified by a numeral, because Class C fires are essentially either Class A or Class B but also involve energized electrical wiring and equipment. Therefore, choose the coverage of the extinguisher for the burning fuel.

- Use fire extinguishers rated for use on Class D for fires in combustible metals, such as magnesium, potassium, powdered aluminum, zinc, sodium, titanium, zirconium, and lithium. Persons working in areas where Class D fire hazards exist must be aware of the dangers in using the wrong kind of fire extinguisher on a Class D fire. These fire extinguisher units are not classified by a numerical system and are intended for special hazard protection only.
- Use fire extinguishers rated for use on Class K fires on grease and cooking oil fires in a commercial kitchen environment, where hot grills and stoves are helping to maintain combustion.

The recommendations that follow are given in NFPA 10 as a guide for marking extinguishers or extinguisher locations. They indicate which extinguisher should be used for a particular class of fire. Extinguishers suitable for more than one class of fire may be identified by multiple symbols. Apply markings by using decals, painting, or similar methods having at least equivalent legibility and durability (Figures 9–10 and 9–11).

Apply markings to the extinguisher on the front, of a size and form to be easily read at a distance of 3 ft (0.9 m). Where markings are applied to walls and panels in the vicinity of extinguishers, make them of a size and form to be easily read at a distance of 15 ft (4.6 m).

Listed fire extinguishers are rated after physical testing. These ratings, which are indicated by a numeral and a letter, define the extinguishing potential of an extinguisher because they specify the type and size or number of extinguisher that should be installed in a specific area. The numeral signifies the relative extinguishing potential, and the letter or letters signify the class or classes of fire on which the particular extinguisher is most effective for extinguishment. For example, a 4A:10B:C rating signifies that the extinguisher is recommended for both Class A and Class B fires and is safe to use if the fire is near or on energized electrical equipment.

Location of Extinguishers

Locate extinguishers close to likely hazards, but not so close that they would be damaged or cut off by the fire. Locate them along the normal path of exit from a building, preferably at the exits. Where highly combustible material is stored in small rooms or enclosed spaces, locate the extinguishers outside the door rather than inside. Locating them outside requires potential users to exit the room and then make a conscious decision to reenter the room and fight the fire.

Make the location of extinguishers as conspicuous as possible (Figure 9–12). For example, if an extinguisher is hung on a large column or post, paint a distinguishing

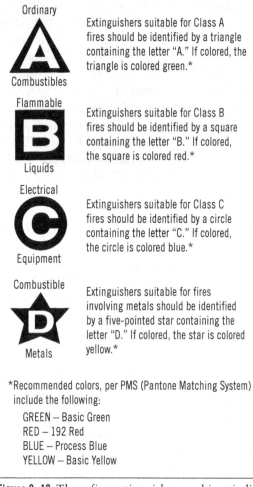

Figure 9–10. These fire extinguisher markings indicate the class(es) of fire for which the fire extinguisher is rated for use. These markings are still used, but the symbols shown in Figure 9–11 are more descriptive. *(NFPA 10, Standard for Portable Fire Extinguishers; reprinted with permission from the National Fire Protection Association.)*

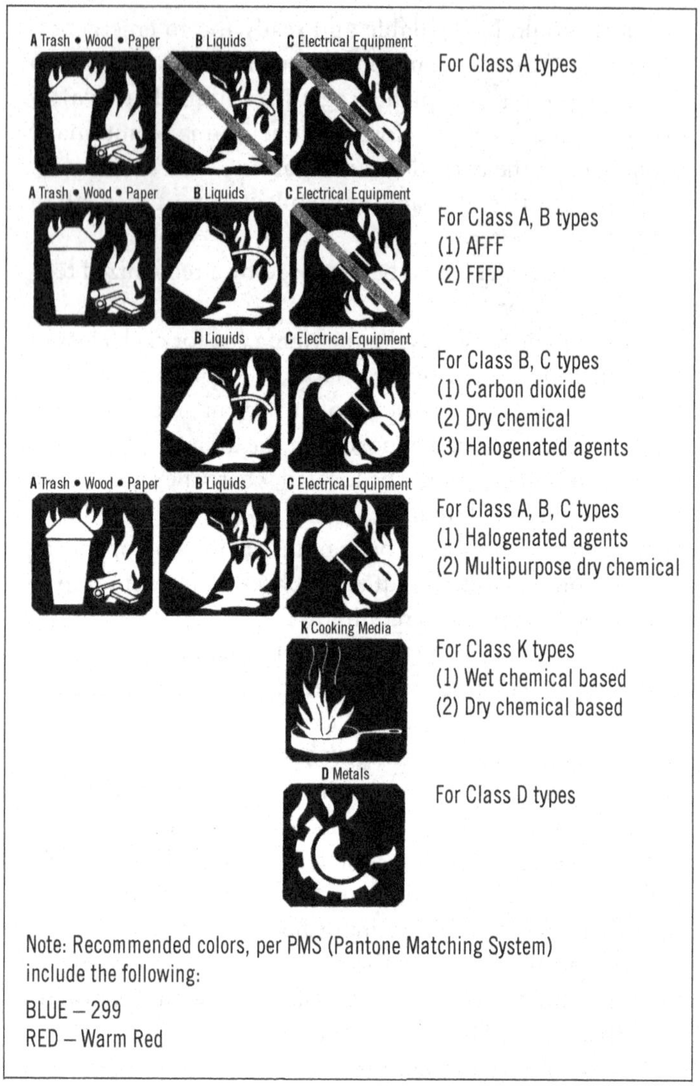

Figure 9–11. These fire extinguisher symbols indicate not only the class(es) of fire for which the fire extinguisher is rated for use, but also the class(es) of fire for which the fire extinguisher should **not** be used. *(NFPA 10, Standard for Portable Fire Extinguishers; reprinted with permission from the National Fire Protection Association.)*

red band around the post. Also, post large signs to direct attention to extinguishers. Keep the extinguishers clean. Do not paint them in any way that will camouflage them or obscure their labels and markings. If an extinguisher is not already plainly marked to indicate the classifications of fire or types of material for which it is intended, place signs or cards indicating this information on the wall close to where it hangs. Markings indicating special uses can also be stenciled on the extinguisher or on an adjacent wall. Special labels are available from manufacturers of extinguishers.

Fire extinguishers must not be blocked or hidden by stock, finished material, or machines. Hang them as directed in NFPA 10. Then they will not be damaged by trucks, cranes, and other operations or corroded by chemical processes; nor will they obstruct aisles or injure passersby. If the extinguishers are installed outdoors, protect them from the elements.

Make facility and warehouse aisles wide enough so that mobile fire protection units can be brought close to a fire and aisles can be kept free of obstructions. Mark floor spaces to allow access to fire-extinguishing equipment, and protect extinguishers with bumpers or guardrails. Extinguishers weighing more than 40 lb (18 kg) should not be more than 3 ft (0.9 m) above the floor. Maintain a clearance of at least 4 in. (10 cm) between the bottom of the extinguisher and the floor.

Figure 9–12. Fire equipment should be conspicuously located, appropriately marked, and regularly inspected.

TABLE 9–E. Fire Extinguisher Size and Placement for Class A Hazards

Criteria	Light (Low) Hazard Occupancy	Ordinary (Moderate) Hazard Occupancy	Extra (High) Hazard Occupancy
Minimum-rated single extinguisher	2-A	2-A	4-A
Maximum floor area per unit of A	3000 ft²	1500 ft²	1000 ft²
Maximum floor area for extinguisher	11,250 ft	11,250 ft	11,250 ft
Maximum travel distance to extinguisher	75 ft	75 ft	75 ft

For SI units, 1 ft = 0.305 m; 1 ft² = 0.0929 m².
Reprinted with permission from the National Fire Protection Association.

Distribution of Extinguishers

The relative hazard of the occupancy, the nature of any anticipated fires, protection for special hazards, and requirements of local codes determine the minimum number and the type of portable extinguishers to be installed for each floor or area.

Class A Extinguishers. Extinguishers suitable for Class A fire hazards are installed according to the classification of occupancy: light hazard, ordinary hazard, or extra hazard (Table 9–E).

- Light-hazard occupancies include office buildings, schools (exclusive of trade schools and shops), churches, and public buildings. Because of the relatively small amount of combustibles in such buildings, incipient fires of minimal severity may be anticipated.
- Ordinary-hazard occupancies include department stores, warehouses, and manufacturing buildings, where incipient fires of average severity in combustibles may be anticipated.
- Extra-hazard occupancies include some warehouses, woodworking shops, textile mills, and paper mills. Because the character or quantity of combustibles is more hazardous, extra-severe incipient fires may be anticipated.

Class B Extinguishers. Extinguisher requirements for Class B protection of a special hazard area, such as a laboratory or an area in which flammable liquids are stored, are necessary in addition to the requirements of extinguishers for Class A protection, except where the total area under consideration presents wholly Class B hazards. The requirements for fire extinguisher size and placement for Class B fires, other than those for flammable liquids greater than 0.25 in. (6 mm) deep, are shown in Table 9–F.

Two or more extinguishers of lower rating, except for foam extinguishers, should not be used to fulfill the protection requirements of Table 9–F. Up to three foam extin-

TABLE 9–F. Fire Extinguisher Size and Placement for Class B Hazards

Type of Hazard	Basic Minimum Extinguisher Rating	Maximum Travel Distance to Extinguishers ft	m
Light (low)	5-B	30	9.15
	10-B	50	15.25
Ordinary (moderate)	10-B	30	9.15
	20-B	50	15.25
Extra (high)	40-B	30	9.15
	80-B	50	15.25

Note: The specified ratings do not imply that fires of the magnitudes indicated by these ratings will occur, but rather they are provided to give the operators more time and agent to handle difficult spill fires that could occur.

Reprinted with permission from the National Fire Protection Association.

guishers holding 2.5 gal (9.5 L) each may be used to fulfill light-hazard requirements, and up to three AFFF (aqueous film-forming foam) solution extinguishers holding about 2.5 gal (9.5 L) each may be used to fulfill ordinary- or extra-hazard requirements. The protection requirements may be fulfilled with extinguishers of higher ratings provided the travel distance to such larger extinguishers does not exceed 50 ft (15 m).

For flammable liquids of appreciable depth, greater than 0.25 in. (6 mm), such as those in dip or quench tanks, the following recommendations apply:

- Provide Class B fire extinguishers on the basis of two numerical units of Class B extinguishing potential per square foot of flammable-liquid surface of the largest tank to be protected within the area.
- Two or more extinguishers of lower ratings, except for foam extinguishers, should not be used instead of the extinguisher required for the largest tank. However, up to three AFFF extinguishers may be used to fulfill these requirements.
- Portable fire extinguishers should not be installed as the sole protection for flammable light hazards of appreciable depth (0.25 in. or 6 mm) whose surface exceeds 10 ft² (0.9 m²). Where personnel are trained in the extinguishment of such fires, the maximum surface area should not exceed 20 ft² (1.9 m²). Use fixed fire protection systems for protecting tanks in excess of 20 ft² (1.9 m²). Portable extinguishers would then be used for putting out burning liquid spills outside the range of fire equipment or for putting out fires that originate outside the tank.
- Where approved automatic fire protection devices or systems have been installed for a flammable liquid, additional portable Class B fire extinguishers may be waived. When so waived, provide Class B extinguishers to protect areas near such protected hazards.
- Consider the travel distance between special hazards and extinguishers. Scattered or widely separated hazards should be individually protected if the specified travel distances in Table 9–F are exceeded. Likewise, extinguishers near a hazard should be easily reached during a fire without endangering the operator.

Class C Extinguishers. Locate extinguishers with Class C ratings close to energized electrical equipment, which would require a nonconducting extinguishing medium if fire either directly involves or surrounds the equipment. Because such a fire is Class A or Class B (once the electricity is deenergized), the extinguishers are sized and located on the basis of the anticipated Class A or Class B hazard. Whenever possible, deenergize electrical equipment before attacking a Class C fire.

Class D Extinguishers. Extinguishers used for Class D protection and other special fires are installed according to the size and type of the special hazard. Consider the type, quantity, and physical form of the combustible material when selecting the proper type of extinguishing agent and the method of applying it.

Class K Extinguishers. Class K extinguishers must be located within 30 ft (9 m) of cooking appliances where cooking oil, animal fats, and other such combustible materials are used in cooking.

Types of Portable Extinguishers
The following types of portable extinguishers are recommended for various types of fires: water solution, dry chemical, carbon dioxide, dry powder, and wet chemical extinguishers (Figure 9–13).

Water Solution Extinguishers. Fire extinguishers that use water or water solutions include pump tank, stored pressure, and AFFF. These extinguishers are effective against Class A fires because of the quenching and cooling effect of water. These units cannot be used on fires in or near electrical equipment because they can present a shock hazard to the operator. Frequently inspect the nozzle on water extinguishers for foreign particles (dirt, matchsticks, paper) that may prevent it from discharging. Also keep the pressure-relief hole in the cap of gas cartridge and foam units free from obstruction. This hole is designed to release any residual pressure that may injure a worker when removing the cap before recharging the extinguisher.

In locations exposed to freezing temperatures, install water solution extinguishers in heated cabinets, or charge them with a nonfreezing solution. However, do not use antifreeze or salt solutions unless the equipment has been designed for such use because these chemicals can cause rapid corrosion. Calcium chloride and other salt solutions are good conductors of electricity and, therefore, are especially dangerous if applied to live electrical apparatus. Check the manufacturer's recommendations in these instances. When charging any extinguisher with an antifreeze solution, thoroughly dissolve the chemical in warm water in a separate container. Then pour it into the extinguisher through a fine strainer to remove any foreign particles that may clog the unit. More maintenance is required for extinguishers containing antifreeze solutions because they may corrode more easily than those filled with plain water. When recharging an extinguisher with an antifreeze solution, thoroughly flush all parts, including the hose and nozzle, with plain water. To prevent freezing and clogging, be sure that all the plain water is drained off.

9 Fire Suppression and Fire Extinguishment   301

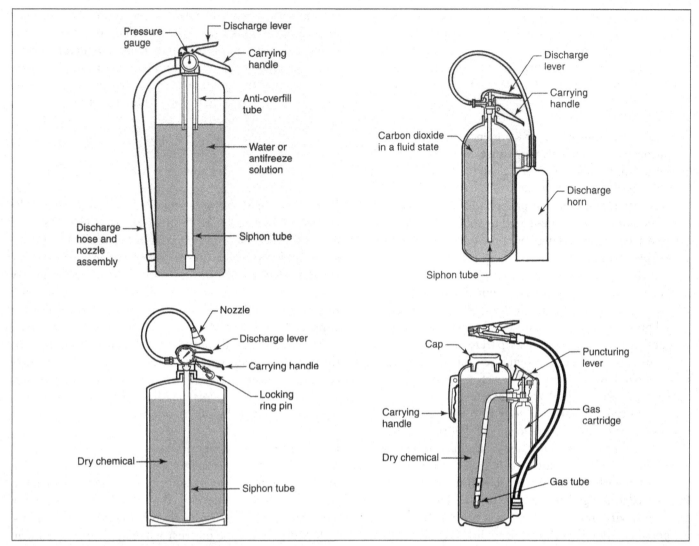

Figure 9–13. Among the major types of portable fire extinguishers are the stored-pressure water extinguisher, the large carbon dioxide extinguisher, the stored-pressure dry chemical extinguisher, and the cartridge-operated dry chemical extinguisher. The parts of these types of fire extinguishers are shown and labeled. *(NFPA 10, Standard for Portable Fire Extinguishers; reprinted with permission from the National Fire Protection Association.)*

Dry Chemical Extinguishers. The dry chemical extinguisher is one of the most versatile units available. It extinguishes by interrupting the chemical flame's chain reaction. Do not confuse it with a dry powder extinguisher.

There are four common types of base extinguishing agents used in dry chemical extinguishers: sodium bicarbonate, potassium bicarbonate (also marketed as urea potassium bicarbonate), potassium chloride, and ammonium phosphate. When recharging, use only the dry chemical agent recommended by the extinguisher's manufacturer.
- Sodium bicarbonate–based dry chemical, the most common agent, is available in ordinary and foam form. The ordinary form can be used simultaneously with foam without causing the foam blanket to break down.
- Potassium bicarbonate–based dry chemical is similar in extinguishing properties to sodium bicarbonate–based dry chemical. However, it has twice the effective fire-fighting capacity on a pound-to-pound basis. This agent has good moisture repellency and is compatible with the simultaneous use of water or foam. It does not allow fuel to reflash as easily or as rapidly as sodium bicarbonate. The absence of momentary flare-up also permits the user to approach fires more closely. In addition, the potassium compound extinguishes the leading edges of contained flammable-liquid fires more easily than does the sodium compound.
- On Class B and Class C fires, ammonium phosphate–based dry chemical (multipurpose) has operating characteristics similar to the other dry chemicals. When discharged into a Class A fire, its chemical reaction destroys the flames. Also, a coating formed when the

extinguishing agent softens adheres to the burning surface, thereby retarding further combustion. To obtain complete extinguishment on Class A materials, thoroughly expose all burning areas to the extinguishing agent. Because any small burning ember may be a source of reignition, properly applying multipurpose dry chemicals on Class A fires is more critical than with water solution extinguishers. In the presence of moisture, multipurpose dry chemicals may cause corrosion when discharged on metals. Therefore, clean up the multipurpose agent immediately after the fire is extinguished. Also, never mix an ammonium phosphate–based agent with a potassium bicarbonate–based or sodium bicarbonate–based agent because dangerous pressure can be developed by even a trace of moisture.

The two basic types of dry chemical extinguishers are defined by their propellant technique: gas cartridge or stored pressure. In the gas cartridge extinguisher, pressure is supplied by a gas stored in a separate cylinder.

In the stored-pressure extinguisher, the entire container is pressurized. Many models of dry chemical extinguishers have a high-velocity discharge. Therefore, take care not to aim the initial discharge directly into the burning area, because it may cause the fire to spread. For the best results when attacking a Class B fire, use a fanning action: rapidly move the nozzle from side to side to intermix the agent thoroughly with the flames. Start well in advance of the burning edge and go beyond the burning edge on each side to avoid leaving any burning pockets behind. To minimize the possibility of reflash, continue to discharge the chemical after the fire has gone out.

Carbon Dioxide Extinguishers. Carbon dioxide extinguishers put out fires by displacing the available oxygen. They do not leave a residue. The bell or nozzle from a carbon dioxide fire extinguisher can become very cold during discharge. Also, it is necessary to be quite close to the fire for a carbon dioxide fire extinguisher to be effective.

Dry Powder Extinguishers. Because the use of combustible metals, such as sodium, titanium, uranium, zirconium, lithium, magnesium, sodium-potassium alloys, and other less-common metals, has increased, dry powder extinguishers should be available to fight such fires. Although dry powders are very effective on combustible metal fires, they all have certain limitations. Consider the type, quantity, and form of the metal and the existing physical conditions when selecting the proper type of dry powder and the method of application.

G-1 Powdered Agent. The oldest powdered agent is the G-1 type, a graphite organic-phosphate compound. When it is applied with a scoop or shovel to a metal fire, the phosphate material generates vapors that blanket and smother the flames, and the graphite, being a good conductor of heat, cools the metal below its ignition temperature. Take care to ensure that the depth of the powder's cover is adequate to provide a smothering blanket. If hot spots should occur, cover them with additional powder. Allow the burning metal to cool before attempting to dispose of the material.

Met-L-X. Another dry powder is Met-L-X. It is composed of a sodium chloride base with additives to make it free flowing, to increase water repellency, and to create the property of heat caking. This material is dispensed from a 30-lb (13.6-kg) dry powder extinguisher similar in appearance and physical features to the cartridge-operated, dry chemical extinguisher or from larger wheeled or stationary units. The technique used to extinguish a metal fire with Met-L-X is to fully open the nozzle of the extinguisher and apply a thin layer of Met-L-X over the burning mass from a safe distance, until control is established. Then, throttle the nozzle to produce a soft, heavy stream, and completely cover the burning mass with a heavy layer from close range. The heat of the fire causes the Met-L-X to cake, thus forming a crust that excludes air.

Lith-X. Lith-X is another dry powder extinguishing agent. It is composed of a special graphite base with additives that make it free flowing so it can be discharged from an extinguisher. Lith-X was developed mainly for use on lithium fires, but it is also effective on other combustible metals. Lith-X does not cake or crust when applied over a burning metal. It excludes air and conducts heat away from the burning mass, thus extinguishing the fire.

Met-L-Kyl. A problem recently developed in fire fighting involves pyrophoric liquids, such as triethylaluminum. These liquids ignite spontaneously, and the resulting fires cannot be easily extinguished by dry powder or other commonly used agents. A special material, Met-L-Kyl, has been developed, consisting of a bicarbonate-base dry chemical and an activated absorbent. The principle of extinguishment involves the combination effect of the dry chemical, which extinguishes the flames, and the absorbent, which absorbs the remaining fuel and prevents reignition. Met-L-Kyl has been designed so that it can be discharged from an extinguisher similar to the standard cartridge-operated, dry chemical model.

Wet Chemical Extinguishers. These fire extinguishers can be used on Class A fires and on Class K fires. Typically, portable wet chemical fire extinguishers are available in 1.5-gal (6-L) and 2.5-gal (9.5-L) models. The chemical is a water-based solution containing one or more of the following chemicals: potassium acetate, potassium carbonate, or potassium citrate.

These fire extinguishers assist in suppressing Class K fires by creating a thick foam blanket over hot cooking surfaces and equipment while simultaneously cooling the cooking appliance and the fuel source.

Miscellaneous Portable Extinguishing Equipment

Besides the extinguishers themselves and the extinguishing agents, portable fire extinguishers sometimes require additional equipment. This is especially true if the extinguishers are large or mounted on vehicles.

Wheeled Equipment. Large portable units on wheels are commercially available. They include 50-, 75-, and 100-lb (23-, 34-, and 45-kg) carbon dioxide extinguishers; and 75- to 350-lb (34- to 160-kg) dry chemical extinguishers.

Wheeled, Twinned Extinguishers. With the development of water-soluble, fluorocarbon surface-active agents, foaming agents are available that give water the property of floating in thin layers on liquid fuel surfaces (light water). This characteristic provides excellent protection against reflash on liquid hydrocarbon fires with only one-fourth the volume as compared with protein air foam.

A wheeled extinguisher with both Purple K dry chemical and "light-water" fluorocarbon foam provides a synergistic extinguishing system. It rapidly knocks down flames and completely protects against reflash. The two extinguishing agents are simultaneously applied through dual pistol-grip nozzles.

Vehicle-Mounted Equipment. Water, foam, carbon dioxide, and dry chemical extinguishing agents are available in units that are mounted on vehicles. They range in size from in-facility fire vehicles able to turn in warehouse aisles to large trucks.

Fire Blankets

In some cases, fire blankets can be used to smother a small fire. Their major purpose is to extinguish burning clothing. However, they are also useful for smothering flammable-liquid fires in small, open containers. Flame-retardant blankets are also available. The most common size of blanket is 66 × 80 in. (1.7 × 2 m). Store blankets in containers mounted on a wall or column so they can be readily pulled out.

Miscellaneous Hand Equipment

Water or antifreeze backpack tanks are available in 2.5-gal (9.5-L) or 5-gal (19-L) sizes. Hand pumps are built into the hose nozzles' handles. Such units are carried on the back, and the slide-action pump is operated with both hands. The backpack unit is frequently used for combating brush fires.

Maintenance and Inspection

Consider giving one person the responsibility for maintaining and inspecting fire-extinguishing equipment. Whether weekly or monthly, as needed, the testing and repair work should be completed in accordance with NFPA 10.

If in-facility maintenance of fire-extinguishing equipment is chosen, then a stock of supplies of spare parts and refills will be necessary. In some facilities, specially trained personnel can test and refill extinguishers on a full-time basis. In others, this maintenance service is under contract with a service organization or the manufacturer's representative. Establish a record system and an organized plan for checking and repairing various types of extinguishers. The inspection and maintenance records should consist of at least durable tags fastened to the extinguishers. The tags must show dates of inspection (monthly) and dates of examination for recharge and other maintenance work (annually). The tags must indicate if the unit was recharged during maintenance.

Other records should be kept that list extinguishers by type, location, and recharge periods. The master record system should contain the history of each extinguisher, the type and quantity of refills on hand, and other pertinent information.

To discourage tampering and to make inspection easier, seal hand extinguishers or install them in cabinets. However, do not lock extinguishers in cabinets or store them in any way that would prevent immediate use in an emergency.

A mandatory requirement for keeping fire extinguishers in top condition is periodic hydrostatic pressure testing, set forth in NFPA 10. Injuries and deaths because of extinguishers rupturing during operation or recharging are the reasons for this requirement. Be sure to check NFPA 10 for details of testing and marking extinguishers.

Never allow a mass removal of extinguishers from a building without providing replacements. If the facility does not have an adequate supply of spares, then establish the testing program so only a few, scattered units are removed at a time. They should then be tested and returned to service promptly. Arrange for temporary replacement units or additional protection when extinguishers are being removed for testing.

Sprinkler and Water-Spray Systems

There are many types of sprinklers and water-spray systems for extinguishing fires. The type of building, operations performed in it, and materials used will help determine the type of sprinkler or water-spray system that should be used.

Water Supply and Storage

Sprinkler systems need a reliable water supply of ample

capacity and pressure for efficient fire extinguishment. Have the water supply engineered with the sprinkler protection to provide a hydraulically balanced system at the least cost. For example, to supply all sprinklers likely to open, in addition to hose streams, a flow rate of 500 to 3,000 gal/min (31.5 to 190 L/s), and sometimes more, may be needed. The precise need depends on the maximum flow of water through all openings. The pressure requirement will vary. However, it should be high enough to maintain a residual pressure of 15 psig (103 kPa) while required water volume is flowing in top-story sprinklers.

Water may be supplied from the following:

- underground supply mains from public waterworks
- automatically or manually controlled pumps drawing water from lakes, ponds, rivers, surface storage tanks, underground reservoirs, or similar adequate sources
- pressure tanks containing (1) water in a quantity determined by the formula in NFPA 13, Standard for Installation of Sprinkler Systems, and (2) compressed air for expelling the water into the piping supply system (the smallest tank for light-hazard occupancy is 2,000 gal [7,570 L])
- elevated tanks or reservoirs that depend on gravity to force water through the system.

Consider providing at least two of these four independent water sources. Then, in case of fire, the main source furnishes water to the system immediately; it is reinforced by the second source, which also can supply emergency protection if the primary source is out of service. The preferred source is a connection from a reliable public water system. Connection should be made to two different mains to provide greater volume and to give flexibility in case of failure of one water main.

A fire pump that can deliver water at high pressure over an extended time can be a second source. Locate the pump where it will not be put out of service by a fire. Be sure that this source has a reliable power supply, independent of the facility system and routed so that a fire anywhere in the protected areas will not expose the pump's power supply.

Do not use water stored for private fire protection for other purposes. Everyday use of tank-stored water necessitates constant refilling, hence the danger of accumulated sediment circulating into the hydrant and sprinkler system. Also, the varying water level may shorten the life of a wood tank or require frequent painting of a steel tank. In addition, take care that the water connections do not pollute drinking water, especially where emergency supplies are taken from a river or other nonpotable source.

Although construction and installation of water tanks are beyond the scope of this chapter, one basic consideration is worth mentioning: in areas with freezing temperatures, be sure that tanks are heated. Inspect the heating system daily during freezing weather; the control valves, weekly; and the entire system (tank, supporting tower, piping, valves, heating system, and all components and accessories), annually (see NFPA 22, Standard for Water Tanks for Private Fire Protection).

To make inspection easier, wood or steel gravity tanks supported on steel towers should be ringed by a platform that is protected by a substantial steel rail. The rail should meet the requirements of providing standard guardrails and toeboards. These tanks should also have substantial steel ladders equipped with approved steel cages or basket guards.

Protect ladders of water tanks that are accessible to the public with locked gates or fences. This will discourage unauthorized persons from climbing them.

Automatic Sprinklers

Automatic sprinklers are the most extensively used and most effective installations of fixed fire-extinguishing systems. These systems are so basic and have proved so effective that most fire protection engineers consider them the most important fire-fighting tool. Nationwide figures from NFPA indicate that sprinklers have a very high efficiency rating for satisfactory extinguishment, usually more than 95%.

The cost of automatic sprinkler protection is relatively small, averaging about 2% of the total facility's investment. Experience shows that a sprinkler system can often pay for itself in 10 years, and sometimes fewer. Insurance costs are usually 40% to 90% lower than for buildings without a sprinkler system. The cost of a sprinkler system is much less if it is installed when the building is built rather than later.

In addition to the economic factors, automatic sprinklers have an impressive lifesaving record. Loss of life by fire is rare where properly designed and maintained sprinkler systems have been installed.

However, when sprinklers do fail to operate, in perhaps 3% to 4% of the cases, the failure is caused by some readily preventable condition. More than one-third of all failures can be attributed to closed water-supply valves.

Although the primary function of the sprinkler system is to deliver water to a fire automatically, the system can also serve as a fire alarm. It can serve as an alarm when an electrical water-flow alarm switch is installed in each main riser pipe. When a fire occurs and the first sprinkler opens, the water rushing through the pipe sets off an alarm that alerts the control system or fire fighters. Dependable sprinkler protection requires a systematic maintenance and inspection program. Such a program includes periodic inspection of water-supply valves, water-supply tests, physical inspection of the system's piping for obstructions to distribution, and similar items (see NFPA 25, Standard for Inspection, Testing, and Maintenance of Water-Based Fire Protection

Systems, for additional maintenance requirements).

There are six basic types of automatic sprinkler systems: wet-pipe, dry-pipe, pre-action, deluge, combined dry-pipe and pre-action, and limited water-supply systems (see NFPA 13). The combination dry-pipe and pre-action systems are used on installations that are larger than can be accommodated by one dry-pipe valve. The limited water-supply system is used for installations that do not have access to a continual or large supply of water. The other four types of sprinkler systems are discussed in the following paragraphs.

Wet-Pipe Systems. The wet-pipe system represents the greatest percentage of sprinkler installations. All parts of the system's piping are charged to the sprinkler heads with water under pressure (Figure 9–14). Then, when heat fuses the fusible link on a sprinkler head, water is immediately sprayed over the area below.

If a portion of the wet-pipe system is subjected to freezing temperatures, as on a loading dock, fill the exposed sections with an antifreeze solution, or connect these sprinklers to a dry-pipe system. Use a water-soluble, liquid antifreeze that is proportioned to give low-temperature protection without producing a combustible mixture, as specified in NFPA 13 and NFPA 25. When the system is supplied from public water connections, use chemically pure glycerine or propylene glycol antifreeze, and then add it only as permitted by local drinking water safety and health regulations.

Dry-Pipe Systems. The dry-pipe system generally substitutes for a wet-pipe system in areas where piping is exposed to freezing temperatures. It is essential to locate the dry-pipe sprinkler control valve and water-supply line in a heated enclosure.

In the dry-pipe system, the piping contains compressed air that holds back the water by means of a dry-pipe control valve (Figure 9–15). When a sprinkler opens, the air pressure is released, the pressure drops, and the dry-pipe valve opens to admit water into the risers and branch lines. These sequential actions delay the actual suppression process when compared with a wet-pipe system. Because of this delay, extra-hazard buildings are difficult to protect with a dry-pipe system. In general, more water damage may result with the dry-pipe system because more sprinklers open than with the wet-pipe system. (More sprinklers open because the fire progresses further, hence more sprinklers are tripped before the extinguishing action of the water takes effect.) To reduce this delay, add quick-opening devices, such as exhausters or accelerators, to dry-pipe systems to expel the air more readily.

It is essential that all parts of a dry-pipe system be installed so that they can be thoroughly drained after the system is activated or tested.

Pre-Action Systems. Pre-action systems are similar to dry-pipe systems. However, they react more quickly. The pre-action valve, which controls the water supply to the system's piping, is actuated by a separate fire detection system. These fire detectors are located in the same area as the sprinkler but operate independently of the sprinkler. Because the detection system is more heat sensitive than the sprinklers, the water-supply valve opens sooner than in a dry-pipe system. The water-supply valve can also be operated manually.

Usually, an alarm is sounded when the valve opens and starts filling the system with water. There may then be time to put out the fire with portable equipment before one or more sprinkler heads open. A pre-action system is especially effective where valuable merchandise or equipment is handled or stored.

Deluge Systems. The deluge system wets down an entire area by admitting water to all sprinklers that are open at all times. Deluge valves that control the water supply to the system are actuated by a fire detection system located in the same area as the sprinklers. The water-supply valves can also be operated manually.

This type of system is primarily designed for extra-hazard facilities where great quantities of water may have to be applied immediately over large areas. Deluge systems are

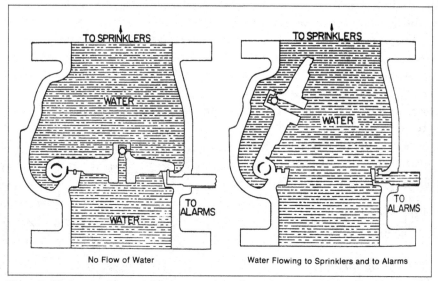

Figure 9–14. A wet-pipe sprinkler system is under water pressure at all times so that water will be discharged immediately when an automatic sprinkler operates. An alarm is triggered when water flows through the sprinkler piping. *(Reprinted with permission from the National Fire Protection Association.)*

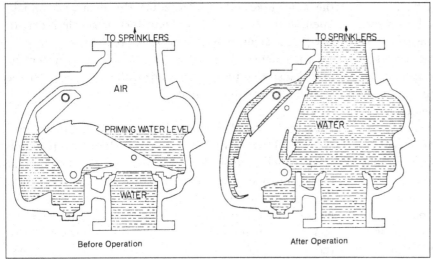

Figure 9–15. The principle of a dry-pipe system is illustrated. Compressed air in the sprinkler system holds the dry valve closed, preventing water from entering the sprinkler piping until the air pressure has dropped below a predetermined point. Like a wet-pipe system, an alarm is triggered when water flows through the sprinkler piping. *(Reprinted with permission from the National Fire Protection Association.)*

ordinarily used to best advantage where rapidly spreading fires, or flash fires, may be anticipated, such as in explosives facilities, those handling or processing nitrocellulose materials, lacquer facilities, and buildings that contain large quantities of flammable materials.

Another application of a deluge system is an open system of outside sprinklers for distributing water over the roof, the exterior of a building, and at windows and cornices. Such a system protects the building against fire from adjoining property. This system is usually manually operated and is used where construction is inadequately protected by design or by distance from adjacent fire hazards.

In special applications, where deluge protection is not needed over the entire area, open sprinklers and closed sprinklers may be combined in a single system. However, remember that (1) separate fire detectors are also required in the area covered by the closed sprinklers, (2) operation of a closed sprinkler will not activate the entire system, and (3) a fire in the area of the closed sprinkler will also cause water to discharge from all the open sprinklers.

Types of Sprinklers

There are many classifications of sprinklers, each of which is designed for specific applications. Upright sprinklers direct water upward against the deflector. Pendant sprinklers direct water downward against the deflector. Sidewall sprinklers discharge the major portion of water away from the nearby wall in a pattern resembling a quarter of a sphere. Sprinklers designated for early suppression and fast response are also available.

Water-Flow Alarms in Sprinklers

Automatic alarms are operated by the flow of water through the sprinkler system. Such alarms are treated exactly like any other fire detection and alarm system (see previous section in this chapter). Their purpose is to give prompt notice that the sprinkler system is operating. They also signal water leakage or discharge from causes other than fire. The water-flow alarm must be tested, inspected, and maintained on a frequent basis by qualified personnel.

Temperature Rating of Sprinklers

Sprinklers are selected on the basis of temperature rating and occupancy. Sprinklers are built either with heat-actuated elements that melt (fuse) or with special devices in which chemicals fuse or expand to open the flow of water through the sprinkler. Table 9–G shows the ratings and distinguishing colors of sprinklers.

Causes of Failure of Sprinkler Systems

Sprinklers seldom fail to control fires, but when they do, failure is usually caused by (1) not keeping all supply valves open and (2) shutting off the supply valves prematurely during a fire. Other causes of sprinkler failure, with corresponding remedies, are as follows:

- freezing of wet-pipe system sprinkler pipes

TABLE 9–G. Temperature Ratings, Classifications, and Color Codings

Maximum Ceiling Temperature		Temperature Rating				
°F	°C	°F	°C	Temperature Classification	Color Code	Glass Bulb Colors
100	38	135–170	57–77	Ordinary	Uncolored or black	Orange or red
150	66	175–225	79–107	Intermediate	White	Yellow or green
225	107	250–300	121–149	High	Blue	Blue
300	149	325–375	163–191	Extra high	Red	Purple
375	191	400–475	204–246	Very extra high	Green	Black
475	246	500–575	260–302	Ultra high	Orange	Black
625	329	650	343	Ultra high	Orange	Black

- defective dry-pipe system control valve, or slow operation of the dry-pipe system because of its excessive size
- foreign material obstructing the system
- improper drainage through faulty installation
- corrosion of sprinklers in such locations as bleacheries, dye houses, or chemical operations
- inadequate supply of water because of faulty design or poor maintenance.

Water-Spray Systems

Water spray may be effective on certain types of fires where there is no hazardous chemical reaction between the water and the materials that are burning. Although these systems are independent of and supplemental to other forms of protection, they are not a replacement for automatic sprinklers.

Fixed water-spray systems are similar to the standard deluge system except that the open sprinklers are replaced with spray nozzles. The water supply to the system may be controlled automatically or manually.

Water-spray systems are generally used to protect flammable-liquid and gas tankage, piping and equipment, cooling towers, and electrical equipment such as transformers, oil switches, and motors. Because of its low electrical conductivity, water spray applied through fixed-piping systems on electrical equipment with voltages as high as 345,000 volts has proved practical. When applied on some types of electrical equipment, however, water spray may cause short circuits by forming a continuous path of water between energized parts. In such cases, provide means for cutting off the electrical current before the water spray is applied (see NFPA 15, Standard for Water Spray Fixed Systems for Fire Protection).

The type of water spray required depends on the hazard and the purpose for which the protection is provided. The basic principle of water spraying is to wet the surface completely with a preselected water density, taking into consideration nozzle types, sizes, spacing, and water supply.

Water-spray systems can be designed effectively for any one, or any combination, of the following purposes:
- extinguishing fire
- controlling fire where extinguishment is not desirable, such as gas leaks
- exposure protection (absorbing heat transferred from equipment by the spray)
- preventing fire by having water spray dissolve, dilute, disperse, or cool flammable materials.

Because the passages in a water-spray nozzle are small in comparison with those in an ordinary sprinkler, they can easily be clogged by foreign matter in water. Therefore, strainers are ordinarily required in the supply lines of fixed-piping spray systems. The strainer baskets should have holes small enough to protect the smallest opening in the nozzles.

In cases where the nozzles have extremely small water passages, they should have their own internal strainer in addition to the supply line strainer.

Fire Pumps

Some fire sprinkler systems have sufficient hydraulic demand that the water supply cannot provide adequate pressure. This often occurs in the following systems:
- sprinkler systems in tall buildings, where the water must overcome gravity to get to upper floors
- deluge systems, where all heads are open and system demand for water is high
- sprinkler systems served by water districts with low or variable pressure.

Fire pumps require periodic operation to ensure consistent operation and routine annual service. This service can include pressure tests through test headers located on the exterior of the building, whether at street level or on the roof. The pressure testing might occur in a test loop inside the pump room. Protecting water pumps from incoming debris is essential long pump life.

Fire Hydrants

In large facilities where parts of the facility are a considerable distance from public fire hydrants or where no public hydrants are available, install hydrants at convenient locations in the facility's yard. The number needed depends on the fire exposure and the hose-laying distance to the built-up facility areas (see NFPA 24, Standard for Installation of Private Fire Service Mains and Their Appurtenances).

Keep exterior fire department connections that serve sprinkler or hose systems easy to reach. The discharge ports should be at least 18 in. (46 cm) above the ground or floor level. Keep vegetation, snow, and stored materials away from hydrants or hydrant houses. In areas subjected to heavy snow, attach to each hydrant a stiff metal wire with a flag on top. In this way, the hydrant can be found even after a heavy snow. Also, protect hydrants from mechanical injury; however, this protection cannot interfere with efficient use.

Before cold weather sets in, drain or pump out hydrants if they are not the type that normally drain. Check drainage whenever hydrants are used during freezing weather. To determine whether or not hydrants are frozen, (1) partially turn the hydrant's stem (if the hydrant is frozen, the stem will not turn) or (2) lower a weight on a string into the hydrant's barrel. Approximately 4 ft (1.2 m) of string should be attached to the weight. The weight with string should be dropped into the outlet opening. If less than 2 ft (0.6 m) of string can be advanced into the outlet opening, the hydrant is frozen.

Frozen hydrants can best be thawed with steam introduced through the outlet by means of a steam hose that is pushed slowly down the barrel, thawing as it goes. Do not use corrosive chemicals, such as calcium chloride, caustics, or salts, to thaw frozen hydrants.

Frequently test and maintain control valves. A number of persons, including members of the facility's fire brigade, should know the location of valves and the sections of the pipe controlled by them.

Have the local fire department check connections to be sure that they are of a size and thread that will fit its equipment. If special adapters are required, supply them to fire fighters and also have them available on the premises.

Fire Hose

Like other fire-extinguishing equipment, have hose lines available for immediate use. Be sure that they are easy to reach. Keep space around hose lines and control valves clear of obstruction. Be sure that fire brigade workers know where hose line equipment is located and how to operate it. Keep aisles and doorways clear and wide enough for rapid use of hose reel carts or other mobile equipment. If fire hoses are available, occupants must be trained how to use them (as with fire extinguishers).

Woven-jacket, lined hose with an outer rubber or plastic cover is chiefly used in industries where the hose jacket must be protected against chemicals and abrasions. Lined hose with a rubber or plastic cover is available in 0.75- and 1-in. (1.9- and 2.5-cm) sizes. It is generally used as a booster hose or as a hose on chemical engines, wheeled extinguishers, and wall-mounted or vehicle-mounted pressured hose reels.

For inside use by trained building occupants, as opposed to standpipe systems designed for the fire department, 1.5-in. (3.8-cm) unlined linen hose was often used. Only lined 1.5-in. (3.8-cm) hose has been specified by NFPA.

Keep the hose on an approved swinging rack or reel, approximately 5 ft (1.4 m) above the floor or high enough so that it will not be a hazard to passersby or damaged by trucking operations. Locate hose stations intended for use by employees at the exits (see NFPA 14, Standard for Installation of Standpipe and Hose Systems).

Arrange the hose so it will not kink or tangle when pulled out. Keep one end connected to the standpipe, and equip the other end with a 0.375- or 0.5-in. (9.5- or 13-mm) nozzle tip or a combination spray-straight stream nozzle. To prevent kinking in use, do not place more than 150 ft (46 m) of hose at a standpipe outlet.

Except for unlined linen hose, hydrostatically test all fire hoses annually. Thoroughly inspect them, dry them, and return them to service.

Maintenance of Fire Hose for Outdoor Use

Periodically inspect and test woven-jacket, lined hose to make sure that it will be in good condition when an emer-

gency arises. Water should be run through the hose at least twice a year. Store yard hose in standard hose houses for protection against weather. Use fire hose only for fighting fires. If hose is needed for other uses, provide separate hoses.

Mildew may attack untreated hose fabric containing cotton if the hose is stored in a damp location or if it is not thoroughly dried after being wet. Fire hose is available with chemically treated fabric. Treated fabric protects against mildew and rot. Treated jackets also absorb less water and, therefore, dry more quickly. The resistance to dampness and mildew is not 100% effective, however, even when the treatment is new, and the treatment deteriorates with age.

Jackets made entirely of synthetic warp and filler are impervious to mildew and rot. It is not necessary to dry such hose. However, wash it after using it and before storing it. Hose with all-cotton jackets and with jackets made from a combination of cotton and synthetic yarns must be carefully dried.

For facility yards containing rough services that will cause heavy wear or where working pressures are above 150 psig (1,060 kPa), use double-jacket, lined hose. If hose may be subjected to acids, acid gases, and other corrosive materials, such as those found in chemical facilities, use rubber-covered, woven-jacket, lined hose. For such conditions, also use hose with a neoprene-impregnated, all-synthetic jacket.

Maintenance of Fire Hose for Indoor Use

Maintain and test unlined hose and woven-jacket, lined hose as follows:

- Reserve the hose for fire fighting.
- Keep hose valves tight because leakage will rot linen hose.
- Examine hose visually each year for mildew, rot, and damage by chemicals, vermin, and abrasions. If the hose is in doubtful condition, give it a hydrostatic pressure test. Replace damaged hose.
- Give hose a pressure test after the fifth and eighth years of service. Then repeat the test every second year after the eighth year. The local fire department will often pressure-test hose. (Unlined hose cannot be pressure tested.)
- Keep hose clean. Wash woven-jacket, lined hose with laundry soap if necessary.
- Dry hose jackets thoroughly after use and keep them dry.

Hose Nozzles

Effective streams of water for fire fighting are controlled by the size and type of nozzle. The nozzle, in turn, must be supplied with the correct quantity of water at the discharge pressure for which it is designed. Nozzles are designed for solid streams, spray streams (frequently referred to as fog), or combination streams. Nozzles for special extinguishing agents, such as foam and dry chemical, are also available.

Spray Nozzles. Spray nozzles are widely used for both public and private fire protection. They make the application of water more effective under many conditions. A description of three types of nozzles follows.

1. Open nozzles of a fixed (nonadjustable)-spray pattern are usually attached to shutoff valves. Some nonadjustable nozzles can be equipped with an applicator (a long pipe extension, curved at the end and fitted with a fixed-spray nozzle) for fighting fires where an extended reach is necessary.
2. Adjustable nozzles provide variable discharges and patterns, from shutoff to straight-stream spray and from narrow- to wide-angle spray.
3. Combination nozzles in which a straight-stream spray, a fixed or adjustable spray, and a shutoff are selected usually have a two- or three-way control valve.

Monitor Nozzles. Permanently mounted monitor nozzles are frequently used to protect pulpwood storage piles at paper mills, in lumberyards, in stockyards, in railway-car storage yards, and near oil storage tanks. Nozzles are often elevated to clear obstructions so that the operator can stand on a shielded platform and direct a high-pressure stream of water over a wide area. Monitor nozzles are especially useful in large, congested areas where it is impractical to lay hose lines in an emergency.

Special Systems and Agents

Special hazards may require systems of extinguishment or control other than water. Each of the several systems available offers certain advantages and disadvantages to consider when making a selection. These systems are usually installed to supplement rather than replace the automatic sprinkler system. They should be engineered to fit the circumstances of the particular hazard. Install them so that their operation will shut down other processes, such as pumps and conveyors, which might intensify a fire.

The following special agents and systems are currently in use. The specific NFPA consensus standard related to each of these agents or systems is also shown (see also the latest edition of the NFPA *Fire Protection Handbook*).

- foam and foam systems: NFPA 11 and NFPA 16
- carbon dioxide extinguishing systems: NFPA 12
- dry chemical extinguishing systems: NFPA 17
- wet chemical extinguishing systems: NFPA 17A
- water spray and automatic sprinkler systems: NFPA 15 and NFPA 13
- explosion prevention systems: NFPA 69
- clean agent systems: NFPA 2001.

When considering the use of special extinguishing systems and agents, consulting a qualified fire protection engineer is essential.

FACILITY FIRE PROTECTION PROGRAM

Objectives

Developing and refining a comprehensive fire protection program protects employees and property and helps facilitate business continuity. A comprehensive fire protection program has several objectives, which have a natural sequence. A robust program will have initiatives and emphasis in each of the following.

1. preventing fires
2. detecting and responding to fires
   a. detecting fires early
   b. initiating appropriate alarms
   c. responding quickly to alarms
3. controlling, suppressing, and extinguishing fires
4. recovering from fires.

When developing these overall objectives within a comprehensive fire protection program, identify several statements of intent for each overall objective. For example, one statement of intent for objective 1 (prevent fires from occurring) could be, "All new facility construction and renovation projects should be designed and built to meet or exceed all related local, state, and national fire codes." This statement of intent then requires the program to describe the relevant fire codes and to identify particular areas in which the design and construction will exceed, or go beyond the requirements of, applicable fire code. Ideally, the fire protection program should identify several relevant statements of intent for each of these overall objectives. The detail of the program will follow from this structure.

Consider the following general rules of thumb while developing statements of intent for the comprehensive fire protection program.

- Nearly everything can burn, given ignition, adequate fuel, and sufficient oxygen. As a result, every facility has some degree of fire risk, and no facility is completely resistant to fire.
- Convection, conduction, radiation, and direct flame contact are all means of transfer of heat energy.
- Fire and flame will spread in a building both vertically and horizontally. Fire spreads vertically until the roof or ceiling confines the flames, followed by horizontal spread.
- Smoke and the toxic products of combustion present the greatest single danger to life, and they spread in much the same manner as fire. Inhalation of smoke and the toxic products of combustion cause 70% to 80% of deaths from structural fires.
- Early detection of a fire is essential to life safety, health, property protection, and business continuity.

- The use and occupancy type of a building greatly influence the degree of latent fire risk.
- The contents of a building are a more important factor in the start of a fire than the physical structure of the building.
- Only a few minutes may elapse between the beginning of combustion and the development of a destructive fire.
- What happens, or does not happen, in the first few minutes of a fire determines the degree of fire control.
- Every fire protection device involves some level of compromise. That is, a fire protection system usually represents some trade-off involving cost, reliability, or safety. The optimum level of fire protection is that which minimizes the cost from expected fire losses.
- The cost of fire protection should have a corresponding effect in reducing the amount of risk involved.
- Experience shows that an automatic sprinkler system is one of the best tools to reduce the risks of loss from fire. Such systems, when properly installed and maintained, are the only 24-hours-a-day, on-duty, and onsite fire detection and suppression systems.
- People and their actions are key to fire control. Approximately half of all fire losses result from human mistakes due to inadequate training, insufficient motivation, or improper action.
- Construction alone is not adequate to ensure adequate fire protection.

Preventing Fires

Fire prevention should be the first overall objective in a comprehensive fire protection program. This represents an ongoing process, from building design, through commissioning, to maintenance and use, ending with decommissioning, demolition, or sale.

Building design should incorporate program objectives for fire prevention. Incorporating appropriate (e.g., noncombustible) construction materials and configuring appropriate fire area separations are essential to fire prevention and control. To the extent possible, architects, engineers, and other designers should understand all expected uses of the building. This will help them consider and incorporate other facility-specific fire prevention features. Many design decisions that are intended to increase fire protection can also affect other overall objectives of a comprehensive fire protection program.

Design approaches that improve fire protection and prevention include emphasis on noncombustible building materials (e.g., concrete or steel instead of wood structure), and noncombustible materials for wall, floor, and ceiling finishes. If design objectives do not allow noncombustible finishes, finishes with limited combustibility improve fire prevention and protection outcomes while allowing a wider

selection of materials. Creating fire compartments using fire-resistive separations ("compartmentation") increases the time available for response before a fire begins to affect an entire building. Effective design for fire protection includes compartmentation for hazardous materials, utility equipment, plenums, and shafts. Often, new construction creates separations between occupancy types and/or uses. Building design often includes considerable attention to the separation of incompatible hazardous materials during storage.

Building commissioning helps to ensure that the fire protection features included in the design function per the design. Comprehensive building commissioning includes witnessed tests and inspections of all mechanical fire protection systems, from fire sprinkler and alarm systems to operation of fire/smoke dampers and fire doors. Comprehensive facility commissioning helps support the comprehensive fire protection program statements of intent relating to applicable codes—if the organization expects a building and systems to be designed and built to a standard, then commissioning helps ensure that the construction actually meets the standards. Commissioning also helps transfer knowledge from the construction team to the facility's operations and maintenance staff.

Maintenance and use of a facility dramatically affect fire prevention and protection outcomes. Establishing and enforcing rules or expectations for smoking, housekeeping, cooking and food preparation, and use of extension cords, small electrical appliances (including space heaters), and holiday decorations improve fire protection program outcomes. On the maintenance side, establishing procedures for evaluating, maintaining, and testing electrical and mechanical equipment will reduce the degree to which that equipment increases fire risk. A hot-work permit program—in which no employee or contractor can light an open flame or create sparks without an inspection of equipment and the work area—is a vital part of a comprehensive fire protection program. Finally, routine fire risk inspections of work areas will help discover fire risks and prevent fires.

Detecting and Responding to Fires

The second overall objective in a comprehensive fire protection program is detecting fires and responding appropriately to the alarms the fire detection process initiates.

Several fire detection methods exist. Of course, human observers sense some fires through sight, smell, feel, or hearing. Electrical and mechanical detection devices and systems also detect fires. Regardless of technological advances in detection systems, the purpose of a detection system remains straightforward: to quickly and accurately detect a threat and quickly sound an alarm.

Human response should follow the sounding of an alarm. For some, the response may be as simple as heading for the nearest exit and checking in. For others, the alarm may initiate emergency response and fire-fighting activities by trained employees. Alarm systems should also summon emergency forces, such as local police and fire departments, in a planned and prescribed manner.

Fire detection and alarm processes in a comprehensive fire protection program require planning at all stages. Building designers and building operations managers must consider the methods for detecting fires and how the facility and its occupants will respond to the ensuing alarms. Designers should plan a comprehensive, code-compliant fire/smoke detection and alarm system, with appropriate detectors and alarms throughout the facility. System design should include current needs and possible future uses of the facility. Finally, system design should include provision for summoning emergency forces and coordinating fire response activities. Often, this includes a manned monitoring station.

By itself, an alarm system does not create a successful outcome for a fire protection program. Building occupants must understand procedures for initiating and responding to an alarm, including evacuation procedures. Policies for fire alarm response, whether by local municipal forces or in-house fire brigades, should clearly state expectations for incident command, management, and recovery.

Fire alarm systems need periodic inspection, testing, and maintenance. These inspections should result in documentation of the serviceability of the systems and their components.

Facility procedures should include requirements for fire safety during alarm system impairments. When a fire protection system undergoes repair, expansions, or is out of service for any reason, the fire prevention program should include a process for manually detecting fires, such as dedicating personnel to act as fire watchmen. Fire watchmen tour the area affected by the impairment and actively look for fire or smoke, report it per procedure to summon emergency forces, and alert occupants in the space. An alternative to fire watch during periods of impairment to required alarm systems is relocation of affected occupants.

Controlling, Suppressing, and Extinguishing Fires

When fire prevention activities have failed and a fire starts, the fire protection program must consider control, suppression, and, ultimately, extinguishment of the fire. This third overall objective to a facility's comprehensive fire protection program assumes that the occupants or monitoring personnel have detected a fire and that extinguishing the fire is desirable. A defensive fire response procedure (letting the fire burn itself out, while protecting adjacent structures) might be a prudent response, in some cases.

The methods of controlling, suppressing, and extinguishing fires cover a broad base of knowledge. Many of the fire prevention design considerations described earlier in this chapter are also fire control methods. Likewise, many fire detection systems, by design, automatically engage fire control and suppression systems.

Fire control includes physical barriers to the spread of fire and the products of combustion. These barriers include fire-rated walls, doors, windows, and air-handling dampers. Fire suppression and extinguishment includes the actions of automatic water-based sprinkler systems and specialty chemical suppression systems. Fire suppression and control also include human intervention with portable fire extinguishers and conventional water-based fire attack, as with hoses.

Human intervention in fire suppression may involve employees trained in fire extinguisher use, employees trained in more formal fire-fighting techniques (e.g., fire brigades), and the use of local career and volunteer fire department forces.

Building Design and Construction Considerations

Based on the planned and possible uses and occupancies of the building, designers should include fire sprinkler systems, which might be wet-pipe, dry-pipe, pre-action, or deluge. To facilitate interior fire fighting, multistory or large area buildings should have standpipes for fire department and/or occupant use. If the facility contains certain specific hazards, it may need monitor nozzle systems, which can put a large volume of water over a large area very quickly. Depending on system demand and local water supply pressure, the automatic water-based fire suppression system may require a fire pump to boost system pressures. On the exterior of the building, fire departments will need hydrants to connect into for water supply and fire department connections to boost the performance of the building suppression system.

Some specific uses inside a building may require careful attention in the design phase. Commercial kitchens, which create grease-laden vapors in sufficient quantities to require automatic ventilation, require wet chemical suppression systems specifically designed to control grease fires. Expensive electronic equipment, like computer file servers, may require a clean agent suppression system to minimize downtime and equipment damage following a fire. Where fire suppression water is not available, such as in a remote outbuilding, dry chemical systems may be appropriate for fire suppression.

Compartmentation is an essential element of fire control. Robust construction keeps fires contained to a single portion of a building, allowing occupants valuable time to escape the fire and reducing the property damage in other parts of the building. Compartmentation often consists of fire-resistive barriers—which may be floors, doors, walls, or glazing. Most model building codes specify what kinds of compartmentation designers must include in new construction. Some examples include:

- exit access corridors, horizontal exits, exit stairs, and areas of refuge
- HVAC shafts, ducts, and plenums
- mechanical rooms, kitchens, laboratories, and other hazardous locations
- hazardous materials storage and use areas.

Building Maintenance and Use Considerations

- Install and maintain fire extinguishers.
- Establish and practice protocols for fire extinguisher use.
- Establish detailed testing and maintenance protocols for all automatic and manual fire suppression systems.
- Establish detailed testing and maintenance protocols for all fire brigade equipment.
- Plan and practice tactical fire suppression and extinguishment procedures with all designated employees and external fire-fighting response agencies.

Recovering from Fires

The facility's comprehensive fire protection program cannot omit the last overall objective. After a fire, a dedicated and planned effort is necessary to secure the scene, begin the investigation activities, and resume normal operations. Recall that the key reason for having a fire protection program is to protect employees and property and to facilitate business continuity.

Specific considerations associated with business continuity and emergency planning are covered in detail in the *Administration & Programs* volume.

Sample Fire Risk Analysis Questionnaire

1. property identification
   a. name and address
   b. building identification
   c. date of construction
   d. date of fire hazard analysis report
   e. names of surveyors/inspectors/analysts
2. property use
   a. general property use (e.g., educational, mercantile, industrial, warehouse)
   b. specific uses; state each principal use and its location; for storage areas, identify the product(s) stored
   c. names of tenants in a building or site of multiple occupancy, including their location and amount of space occupied
3. site information
   a. fire department access
   b. streets, roadways, parking, and traffic

   c. natural barriers
   d. bodies of water
   e. fences
   f. exposures: type and separation distances
      i. other buildings or structures
      ii. outside storage or processes
      iii. natural exposures (e.g., forest, grass, brush)
      iv. railways, highways, and runways
4. construction
   a. dimensions
      i. area
      ii. height and number of stories
      iii. ceiling heights
   b. classification of building construction—Types I, II, III, IV, and V (see NFPA 220, Standard on Types of Building Construction); for mixed construction, a table may be shown to report the location of each type
   c. compartmentation
      i. locations and ratings of fire areas, fire walls, and fire partitions
      ii. locations and ratings of vertical openings (e.g., atria, stairways, elevators, shafts, chutes)
      iii. protection of openings (e.g., fire doors, fire windows, fire dampers, fire stopping)
      iv. concealed spaces and voids
   d. building exterior
      i. doors
      ii. windows
      iii. loading docks
5. life safety
   a. exit facilities
      i. number and arrangement
      ii. occupant load capacity
      iii. door hardware, special locking arrangements, and security
   b. exit marking, illumination, and emergency lighting
   c. interior finish
      i. furnishings
      ii. decorations
   d. evacuation plans and drills
6. water supply and distribution
   a. general description including adequacy, deficiencies, and reliability
      i. fire flow requirements versus existing conditions, as determined by tests
      ii. storage requirements versus existing conditions
   b. storage tanks—gravity, suction, pressure
      i. capacity
      ii. construction
      iii. location and dimensions
      iv. pipe arrangement to yard mains or fire pump(s)

v. percentage of capacity for fire protection
c. public and private water distribution systems
  i. type of system—gravity, direct pumping, or a combination of both
  ii. size of water mains and arrangement in relation to the property
  iii. type of pipe (e.g., cement lined, cast iron, etc.)
  iv. connection arrangement to all supply sources available
d. other water storage methods (e.g., rivers, ponds, streams, harbors, wells)
e. hydrants
  i. locations of public and private hydrants
  ii. types—dry barrel or wet barrel
f. location and types of check valves and water meters
g. fire pumps
  i. location of pump and control valves
  ii. rated capacity
  iii. types—split case, horizontal end suction, in-line, or vertical shaft turbine
  iv. name of manufacturer
  v. type of motor drive—diesel, electric, gas, gasoline, or steam
  vi. suction pressure, head, or lift
  vii. starting mechanism—automatic, manual, or both
h. inspections, maintenance, and tests

7. extinguishing systems and devices
a. automatic sprinkler systems
  i. types—wet-pipe, wet-pipe with antifreeze, dry-pipe, deluge, pre-action
  ii. area covered by type
  iii. coverage, occupancy classification, and spacing of sprinklers
  iv. riser location and size
  v. temperature rating of sprinklers
  vi. control valves for water-supplying sprinklers—location, size, type, and status if normally shut
  vii. fire department connections—location, size, and area covered
  viii. inspection, maintenance, and tests
b. standpipe and fire hose systems
  i. classification—Class I, II, or III
  ii. types—wet or dry
  iii. outside system
  iv. inspections, maintenance, and tests
c. special hazard systems
  i. types (e.g., carbon dioxide, foam, dry chemical, wet chemical, clean agent)
  ii. type of hazard protected
  iii. location and capacity of storage containers
  iv. inspection, maintenance, and tests

d. portable extinguishers
  i. types of coverage
  ii. inspections and maintenance

8. fire alarm and detection systems
a. location and manufacturer of main control panel
b. monitoring and fire department notification
c. audible and visual device types (e.g., bells, horns, speakers, strobes, etc.)
d. products of combustion detection
  i. heat detectors (e.g., rate of rise, fixed temperature rate compensation, pneumatic line type)
  ii. smoke detectors (e.g., photoelectric, ionization, air sampling)
  iii. toxic/harmful gas detectors
  iv. flame detectors (e.g., ultraviolet, infrared)
e. water-flow detection on sprinkler risers
f. trouble and supervisory indications
g. power supplies
h. inspection, maintenance, and tests

9. electrical systems
a. condition of conduit, raceways, cables, conductors, and cords
b. clear space around switchboards and panel boards
c. grounding
d. overcurrent protection
e. ground-fault circuit interrupters
f. motors
g. hazardous (classified) locations
h. transformers—capacity, protection, liquids, and cutoffs
i. lightning protection

10. heating, ventilation, and air conditioning (HVAC) systems
a. location of equipment
b. smoke-control systems
c. inspection, maintenance, and tests

11. fire prevention
a. common hazards
  i. heat, light, power, and HVAC
  ii. housekeeping, brush, and grass
  iii. ordinary combustibles
  iv. flammable and combustible liquids (consumer quantities)
  v. electrical appliances
  vi. smoking
b. Hot-work permit program
c. internal employees
d. contractors
e. emergency permits
f. building inspection program
  i. frequency
  ii. scope

iii. records
g. employee fire safety training
    i. adequacy and frequency of training
    ii. reference materials
    iii. records
h. fire brigade program
    i. type of brigade
    ii. training
    iii. special equipment provided
    iv. records
12. special hazards and equipment
    a. flammable liquids (process quantities and storage)
    b. flammable gases (process quantities and storage)
    c. chemicals and hazardous materials (process quantities and storage)
    d. tanks and cylinders—location, size, contents, and construction
    e. finishing processes
    f. industrial processes
    g. welding and cutting
    h. cooking
    i. shops
    j. materials handling
    k. waste removal—incinerators, chutes, Dumpsters, compactors, recycling
    l. tunnels
    m. electronic equipment
    n. dust collectors
    o. boilers and furnaces
    p. water-cooling towers
    q. chimneys—location, construction, and height
    r. silos
    s. cranes
    t. conveyers
13. construction, demolition, and modifications
    a. fire protection—water supply, fire department access
    b. fire prevention—control of debris and combustibles, welding and cutting, roofing operations, fire watches, and storage and handling of hazardous materials
    c. life safety; temporary or alternative means of egress.

SUMMARY

- The chemistry of fire is highly complex. Fires can be controlled by cooling burning materials, removing oxygen from the fire, inhibiting the chemical chain reaction, and removing fuel. The objective of these methods is to interrupt the chain reaction in a fire. Fire-extinguishing agents help control fires in one or more of these ways.

- The National Fire Protection Association (NFPA) has developed five classifications of fires. Class A fires occur in ordinary materials and can be extinguished by water or dry chemical agents. Class B fires occur in the vapor-air mixture over the surface of flammable and combustible liquids and are extinguished by limiting air or applying dry chemicals. Class C fires occur in or near energized electrical equipment and can be put out by deenergizing the electricity feeding the fire and then extinguishing the fire using dry chemical agents or carbon dioxide. Class D fires occur in combustible metals and require special techniques and extinguishing agents and equipment to be put out. Class K fires occur in commercial kitchens and involve cooking oils or animal fats in the presence of hot cooking surfaces. Class K fires are extinguished by wet chemical extinguishing agents.

- Fire protection engineering is a highly developed engineering specialization. Achieving the most efficient fire protection system requires the involvement of qualified fire protection engineers as well as building users, other engineers, architects, interior designers, building contractors, fire detection and suppression system manufacturers, the local governmental authorities having jurisdiction, and even urban planners.

- Fire protection measures are effective only if they are based on a proper analysis and evaluation of the risk of fire. The best fire protection minimizes loss of life and property.

- A facility's fire risk can be evaluated, and actions to reduce the fire risk in the facility can be taken. This involves performing a comprehensive fire risk assessment, beginning with a fire hazard analysis.

- In the diamond-shaped symbol of the NFPA identification system, the three categories of hazards are identified for each material: health, flammability, and instability, with a space for special instructions. The order of severity in each category is indicated by numbers: 4 is a severe hazard and 0 indicates relatively no hazard.

- Facility construction methods can help confine fires and control smoke through proper design of stairways, fire walls, fire doors, separate units, ventilation ducts, physical barriers, and fire exits.

- Standpipes, traffic and transportation routes, and fire department and water access to the site should be designed to be as fire resistant as possible and should help minimize fire hazards.

- Fire protection systems for computer rooms should be designed to activate immediately after a fire is detected, to shut down the computer system automatically, to protect electrical connections and circuit boards, and to shut off automatically once the fire is extinguished.

- Companies must act to eliminate the causes of industrial

fires by using only approved equipment, establishing safe work practices, and enforcing good housekeeping procedures. Workers should be trained to spot unsafe conditions and report them immediately.

- When a fire occurs, good communication is vital to alert workers to the emergency and to mobilize fire protection forces.

- Fire detection must be part of every fire protection system. Its two main tasks involve (1) giving an early warning to enable building occupants to escape and (2) starting extinguishing procedures. Means of fire detection can be through a human observer; automatic sprinklers; smoke, flame, or heat detectors; or a combination of these.

- Facilities should be equipped with fire alarm signal systems that clearly communicate to all personnel where the fire is located and that summon appropriate fire-fighting units. Employees must be trained to respond properly to alarm signals. Spacing, location, and maintenance of fire detectors depend on the type of building, its operations, and its materials.

- Portable fire extinguishers are listed as Class A, B, C, D, K, or a combination of A, B, and C, depending on the type of fire they are designed to extinguish. This equipment must be approved by a recognized testing laboratory, located in accessible areas, clearly marked as to class and type of fire, and easily operated by workers.

- Sprinklers and water-spray systems come in many varieties, depending on the type of building, the operations performed in it, and the materials used. Sprinkler systems include wet-pipe, dry-pipe, pre-action, deluge, combined dry-pipe and pre-action, and limited water-supply systems.

- All systems require a reliable water supply of ample capacity and pressure. Automatic sprinklers are the most common and effective of all fixed fire-extinguishing systems and can serve as fire alarms as well as fire protection. Sprinkler systems and their water supplies must be inspected and tested regularly to ensure they function properly.

- Fire hydrants, fire hoses, and hose nozzles should be available for immediate use. Hydrants are particularly effective in large facilities where parts of the facility are far away from public fire hydrants or when no public hydrants are available. Fire hoses and nozzles should be inspected and maintained in good repair, and workers should be trained in their proper use during emergencies.

- Special fire hazards may require special fire-extinguishing or control agents other than water. These systems are usually installed to supplement, not replace, automatic sprinklers and other fixed or portable fire protection equipment.

- Fire protection includes everything related to preventing fires, detecting fires, responding to fires, suppressing and extinguishing fires, and recovering from a fire to resume normal business operations. To accomplish these goals,

companies must develop comprehensive fire protection programs.

- The primary purpose of a fire protection program is to prevent fires and to develop a system to help ensure that when fires occur, appropriate detection, response, and suppression activities occur. This involves a combination of facility design and engineering objectives along with occupant objectives such as training and drills.

- As a first step in fire prevention, every establishment should set up a system of periodic fire inspections for every operation and ensure that proper fire-extinguishing equipment is on hand and in good working order. The person in charge of fire prevention and protection should establish the schedule, determine routing reports, maintain a complete list of inspected items, and set up regular fire drills for all personnel.

- Employees should know their roles in detecting a fire and in transmitting an alarm, evacuating a building, confining the fire, and extinguishing the fire. Fire protection programs should enable companies to reduce fire risk significantly.

REFERENCES

National Fire Protection Association, 1 Batterymarch Park, Quincy, MA 02269.

Fire Protection Handbook, edited by A. Cote, C. Grant, J. Hall, R. Solomon, and P. Powell. 20th ed. 2008.

Principles of Fire Protection, edited by A. Cote and P. Bugbee. 1988.

NFPA 1, Fire Code.

NFPA 10, Standard for Portable Fire Extinguishers.

NFPA 11, Standard for Low-, Medium-, and High-Expansion Foam.

NFPA 12, Standard on Carbon Dioxide Extinguishing Systems.

NFPA 13, Standard for Installation of Sprinkler Systems.

NFPA 14, Standard for Installation of Standpipe and Hose Systems.

NFPA 15, Standard for Water Spray Fixed Systems for Fire Protection.

NFPA 16, Standard for Deluge Foam-Water Sprinkler and Foam-Water Spray Systems.

NFPA 17, Standard for Dry Chemical Extinguishing Systems.

NFPA 17A, Standard for Wet Chemical Extinguishing Systems.

NFPA 22, Standard for Water Tanks for Private Fire Protection.

NFPA 24, Standard for Installation of Private Fire Service Mains and Their Appurtenances.

NFPA 25, Standard for Inspection, Testing, and Maintenance of Water-Based Fire Protection Systems.

NFPA 30, Flammable and Combustible Liquids Code.

NFPA 45, Standard on Fire Protection for Laboratories Using Chemicals.

NFPA 51B, Standard for Fire Prevention during Welding, Cutting, and Other Hot Work.

NFPA 68, Standard on Venting of Deflagrations.

NFPA 69, Standard on Explosion Prevention Systems.

NFPA 70, National Electrical Code.

NFPA 72, National Fire Alarm and Signalling Code.

NFPA 80, Standard for Fire Doors and Other Opening Protectives.

NFPA 80A, Recommended Practice for Protection of Buildings from Exterior Fire Exposures.

NFPA 90A, Standard for Installation of Air-Conditioning and Ventilating Systems.

NFPA 101, Life Safety Code.

NFPA 211, Standard for Chimneys, Fireplaces, Vents, and Solid Fuel-Burning Appliances.

NFPA 220, Standard on Types of Building Construction.

NFPA 253, Standard Test Method for Critical Radiant Flux of Floor Covering Systems Using a Radiant Heat Energy Source.

NFPA 326, Standard for Safeguarding of Tanks and Containers for Entry, Cleaning, or Repair.

NFPA 551, Guide for Evaluation of Fire Risk Assessments.

NFPA 600, Standard on Industrial Fire Brigade.

NFPA 701, Standard Methods for Flame Propagation of Textiles and Films.

NFPA 704, Standard System for Identification of Hazards of Materials for Emergency Response.

NFPA 1081, Standard for Industrial Fire Brigade Member Professional Qualifications.

NFPA 1961, Standard on Fire Hose.

NFPA 2001, Standard on Clean Agent Fire Extinguishing Systems.

NFPA 5000, Building Construction and Safety Code®.

Society of Fire Protection Engineers, published by the National Fire Protection Association.

SFPE Fire Protection Engineering Handbook, edited by P. DiNenno. 3rd ed.

International Code Council, 500 New Jersey Avenue, NW, 6th Floor, Washington DC.

International Building Code®.

Occupational Safety and Health Standards (*Code of Federal Regulations*)

Electrical, 29 CFR 1910, Subpart S.

Exit Routes, Emergency Action Plans, and Fire Prevention Plans, 29 CFR 1910, Subpart E.

Fire Protection, 29 CFR 1910, Subpart L.

Fire Protection, Portable Fire Extinguishers, 29 CFR 1910.157, Subpart L.

Hazard Communication, 29 CFR 1910.1200.

Hazardous Waste Operations and Emergency Response, 29 CFR 1910.120.

Means of Egress, Emergency Action Plans, 29 CFR 1910.38, Subpart E.

Process Safety Management of Highly Hazardous Chemicals, 29 CFR 1910.119.

Welding, Cutting, and Brazing, 29 CFR 1910, Subpart Q.

REVIEW QUESTIONS

1. What are the four components of a comprehensive fire protection program?
   a.
   b.
   c.
   d.

2. Name at least three nongovernmental organizations from which information can be obtained on all aspects of fire protection and building codes.
   a.
   b.
   c.

3. What are the seven steps necessary in performing a comprehensive facility fire risk assessment?
   a.
   b.
   c.
   d.
   e.
   f.
   g.

4. When should a facility's fire prevention activities first occur?

5. List at least five things related to fire prevention that ought to be considered during the design of a facility.
   a.
   b.
   c.
   d.
   e.
6. What is the major cause of occupant injury and death in structural fires?
7. What are four essential fire prevention practices that must occur continually in existing buildings?
   a.
   b.
   c.
   d.
8. What are the four ways fires can be controlled?
   a.
   b.
   c.
   d.

9. When should an employee abandon his or her attempts at fighting a fire with a portable fire extinguisher?
   a.
   b.
   c.
   d.
10. Class A fires are associated with _____.
11. Class B fires are associated with _____.
12. What makes a Class C fire unique when compared with a Class A or Class B fire? What makes it the same?
13. What are Class D fires, and what should not be done to extinguish them?
14. What are Class K fires, and how are they different from Class B fires?
15. List four comprehensive national consensus standards and codes related to general facility fire protection.
   a.
   b.
   c.
   d.

Flammable and Combustible Liquids

10

John DeLaHunt, MBA, ARM

Philip E. Hagan, JD, MBA, MPH, ARM, CIH, CHMM, CET, CHCM, CEM

Definitions
Flash Point ▶ Flammable and Combustible Liquid ▶ Auto-Ignition Temperature ▶ Upper/Lower Flammable Limit (Flammable Range) ▶ Oxygen Level ▶ Propagation of Flame ▶ Rate of Diffusion ▶ Vapor Pressure ▶ Volatility

General Safety Measures
Preventing Dangerous Mixtures ▶ Smoking ▶ Static Electricity ▶ Bonding and Grounding ▶ Electrical Equipment ▶ Spark-Resistant Tools ▶ Health Hazards ▶ Combustible-Gas Indicators

Loading and Unloading
Fire Safety ▶ Personnel Safety ▶ Environmental Protection ▶ Electrical Considerations ▶ Inspection ▶ Relieving Pressure ▶ Removing Covers ▶ Connections ▶ Placards and Shipping Papers ▶ Spill Prevention, Control, and Countermeasures

Container (Nonbulk) Storage
On-Property Transportation ▶ Inside Storage and Mixing Rooms ▶ Inside Storage Cabinets ▶ Outside Storage Lockers

Cleaning Small Tanks and Containers
Steaming ▶ Inert Gas Displacement

Disposal of Flammable Liquids

Regulatory Issues

Bulk Storage: Installation and Maintenance
Tank Construction ▶ Tank Ventilation ▶ Pumps ▶ Gauging ▶ Tank Anchoring ▶ Underground Tanks ▶ Aboveground Tanks ▶ Spill Control ▶ Tank Fires and Their Control

Cleaning Tanks
Abandonment of Tanks

Common Uses of Flammable and Combustible Liquids
Dip Tanks ▶ Japanning and Drying Ovens ▶ Oil Burners ▶ Cleaning Metal Parts ▶ Internal-Combustion Engines ▶ Spray Booths ▶ Liquefied Petroleum Gases

Summary

References

Review Questions

319

DEFINITIONS

Any successful treatment of flammable-liquids safety must define the term. Many regulatory agencies define flammable liquids for their own purposes, so differences among these definitions should not surprise the reader.

Flash Point

Most definitions of the term *flammable liquid* refer to a test method for *flash point,* or the lowest temperature at which a liquid gives off enough vapor to create a flammable mixture near the surface of the liquid. As a temperature measurement, the relative hazard of fire and flame increases as the flash point decreases.

For instance, kerosene and diesel no. 1 fuel oil have flash points of 100°F (37.8°C), and, at temperatures below this level, do not give off vapor levels that will support combustion. On the other hand, gasoline has a flash point below –40°F (–40°C). In fact, refineries blend gasoline to enhance this characteristic so that car engines will start in very cold weather.

Testing for flash point uses "open-cup" and "closed-cup" methods, and there are several different sets of test equipment. While the technical differences between the two methods may not hold the general interest, the fact that regulations may specify one or the other method should.

Although other properties influence the relative hazards of flammable liquids, the flash point has the most significance in dealing with the flammable component.

Flammable and Combustible Liquid

The National Fire Protection Association (NFPA) defines the term *flammable liquid* in NFPA 30, Flammable and Combustible Liquids Code, and NFPA 321, Classification of Flammable and Combustible Liquids:

Any liquid having a closed-cup flash point below 100°F (37.8°C) and having a vapor pressure not exceeding 40 psia (276 kPa) at 100°F.

In those same standards, NFPA defines the term *combustible liquid* as:

Any liquid having a closed-cup flash point at or above 100°F (37.8°C).

Further, NFPA classifies flammable liquids as shown in the next column:

Hazard	NFPA Classification	Flash Point		Boiling Point
		From	To	
Flammable liquid	Class IA	Any	73°F (22.8°C)	Below 100°F (37.8°C)
	Class IB			Above 100°F (37.8°C)
	Class IC	73°F (22.8°C)	100°F (37.8°C)	Any
Combustible liquid	Class II	100°F (37.8°C)	140°F (60°C)	Any
	Class IIIA	140°F (60°C)	200°F (93.4°C)	
	Class IIIB	200°F (93.4°C)	Any	

Occupational Safety and Health Administration (OSHA) regulations use the same definitions of flammable liquid and combustible liquid as NFPA. In the Hazardous Materials Regulations, 49 CFR 170–179, the Pipeline and Hazardous Materials Safety Administration of the U.S. Department of Transportation defines the term *flammable liquid* as:

A liquid having a flash point of not more than 60°C (140°F), or any material in a liquid phase with a flash point at or above 37.8°C (100°F) that is intentionally heated and offered for transportation or transported at or above its flash point in a bulk packaging.

Some common flammable and combustible liquids include gasoline, crude oils, alcohols, various hydrocarbons, and their by-products. All of these materials are chemical combinations of hydrogen and carbon (thus, hydrocarbon). They may also contain oxygen, nitrogen, sulfur, and other atoms. However, almost every flammable and combustible liquid has a hydrocarbon component; an exception is hydrazine, a chemical used as a rocket propellant.

Any combustible liquid heated to a temperature at or above its flash point will produce ignitable vapors. Heavy fuel oil heated to several hundred degrees Fahrenheit, for example, may produce flammable vapors just as readily as gasoline. However, these vapors are heavier, are less volatile, and condense rapidly. Combustible liquids become more hazardous when atomized. When heating such liquids above their flash points or when atomized into airborne aerosols, treat them as flammable liquids.

Auto-Ignition Temperature

Generally, flammable liquids require a spark source to ignite. In some cases, however, the flammable gas-air or vapor-air mixtures will ignite in ambient heat or upon contact with a heated surface, even without the presence of an open spark or flame. This temperature is known as the *auto-ignition temperature*. Vapors and gases will spontaneously ignite at a lower temperature in an oxygen-rich environment (i.e., containing more than 20.9% oxygen, as found in common air). Chemical catalysts can lower the auto-ignition temperature of most flammable vapors and gases.

Upper/Lower Flammable Limit (Flammable Range)

Gases and vapors of flammable liquids have minimum concentrations in air below which propagation of flame does not occur, even upon contact with a source of ignition, known as the *lower flammable limit* (LFL). Similarly, gases and vapors of flammable liquids have a maximum proportion of vapor or gas in air above which propagation of flame does not occur, known as the *upper flammable limit* (UFL). The LFL and UFL represent the *flammable range* of the liquid or gas.

LFLs and UFLs vary by substance. For example, a gasoline vapor-air mixture with less than approximately 1% of gasoline vapor is too lean (not enough vapor) to ignite. Similarly, if there is more than approximately 8% of gasoline vapor, the mixture will be too rich (too much vapor) to ignite. Materials that are gases at standard temperature (room temperature) and pressure (sea level), such as hydrogen, acetylene, and ethylene, have a much wider range of flammable limits.

Convention determines flammable limits at pressures of 1 atmosphere (atm), or the typical pressure at sea level. Thus, the flammable limit range will increase as temperature or pressure increases. Generally, the UFL changes more than the LFL. Other factors, including oxygen content and chemical catalysts, can affect upper and lower flammable limits.

Oxygen Level

Usually, combustion requires sufficient oxygen and a vapor concentration in the flammable range. When ambient air contains a 12% to 14% *oxygen level*, for instance, flammable vapors may not burn. Actual limits depend on the inert gas used to decrease the oxygen level and other variables. For instance, increasing pressures and temperatures above normal will reduce the percentage of oxygen required for combustion.[1]

1 The Apollo 1 disaster occurred, in part, because of the elevated quantity of oxygen in the capsule. The capsule's design called for a 100% oxygen environment at a pressure of 5 pounds per square inch (psi). On the launch gantry, during a "plugs-out" test, NASA increased the interior pressure to almost 17 psi to maintain the design pressure differential of 5 psi above ambient. This pressure more than tripled the *amount* of oxygen in the capsule, which changed the flammable nature of the fixtures and equipment inside. When a spark ignited some wire strapping material, the enriched nature of the atmosphere caused an inferno, killing all three astronauts (Grissom, White, and Chaffee). At 5 psi, the combustion would not have proceeded at such a rapid pace.

Propagation of Flame

Propagation of flame is the spread of flame through the entire volume of the flammable vapor-air mixture from a single source of ignition. A vapor-air mixture above or below the upper or lower flammable limit may burn at the point of ignition without propagating (spreading away) from the ignition source, or vent.

Rate of Diffusion

Rate of diffusion indicates the tendency of a gas or vapor to disperse into or mix with another gas or vapor, including air. This rate depends on the density of the vapor or gas as compared with that of air, which is given a value of 1. Whether a vapor or gas is lighter or heavier than air determines, to a large extent, the design parameters of a ventilation system. When designing ventilation systems, place exhaust air duct openings where they will most effectively remove the vapors, based on rate of diffusion. If the vapor or gas is heavier than air, the exhaust air duct should be slightly above floor level. Conversely, if the vapor or gas is lighter than air, the exhaust air duct should be just below ceiling level.

Vapor Pressure

Vapor pressure is the pressure exerted by a volatile liquid under any of the equilibrium conditions that may exist between the liquid and its vapors. One testing method follows ASTM D323–2008, Standard Test Method for Vapor Pressure of Petroleum Products (Reid Method). Generally speaking, higher temperatures, larger volumes of liquids, and smaller volumes of air space above the liquid pool will create higher vapor pressures.

Volatility

Volatility is the tendency or ability of a liquid to evaporate or, in other words, turn into a gaseous state. Alcohol and gasoline evaporate rapidly and are widely known as volatile liquids. Generally speaking, liquids have higher volatility at higher temperatures and lower molecular weights.

GENERAL SAFETY MEASURES

Flammable and combustible liquids require careful handling. The vapor-air mixture—which forms when liquids evaporate—burns or explodes, not the liquids themselves. When the vapor-air mixture is below the LFL, it is too lean to ignite; the vapor concentration is too low. Conversely, a vapor-air mixture above the UFL is too rich to ignite because not enough oxygen is present to support ignition.

Be aware that vapor-air mixtures will pass in and out of the flammable range as conditions change, as when fuel vapor levels, oxygen levels, temperatures, or pressures vary.

When handling and using flammable liquids, minimizing exposure of large liquid surfaces to air will decrease the chances of producing a mixture that exceeds either the LFL or UFL and the resulting hazardous mixture. Therefore, store and handle these liquids in closed containers or systems,[2] and avoid exposing low-flash-point liquids during handling and use. Mixing and handling these materials in the presence of sparks, open flames, and/or high heat increases the potential for a fire or explosion. Flammable and combustible liquids vaporize and form flammable mixtures with air when left in open containers, when allowed to leak or spill, or during atomization or heating.

The degree of potential hazard depends largely on four elements:

- the flash point of the liquid
- the concentration of vapors in the air (i.e., whether the vapor-air mixture is within the flammable range)
- the availability of a source of ignition at sufficient temperature to enable ignition
- the degree to which ventilation prevents accumulation of vapors.

Manufactured liquids and fluid commodities that contain flammable liquids should be managed using the same precautions used for pure or concentrated product. Refer to Safety Data Sheets (SDSs) or other information sources to determine flash points and boiling points so that the materials can be classified and handled safely. Treat heated combustible liquids according to the requirements for flammable liquids.

Precautions for handling and using these liquids will vary by flash points, boiling points, and concentration of flammable liquid in the mixture. In addition, toxicity should be part of the evaluation. In many cases, toxicity will be the determining factor when identifying safety measures. Also, ensure that all containers have correct and appropriate labels, marking, and/or placards. (See Chapter 7 and Appendix B, Chemical Hazards, of the National Safety Council's *Fundamentals of Industrial Hygiene*, 6th edition, for details about many liquids.)

Some flammable liquids can accumulate dangerous levels of unstable peroxides or polymers upon prolonged storage. Review the Safety Data Sheets of all materials, and perform the necessary tests for peroxides or polymers at the recommended frequencies. Treat these materials with inhibitors regularly to reduce the accumulation of peroxides or polymers. Consider implementing a mandatory disposal process when these materials are not in active use.

Preventing Dangerous Mixtures

Avoid accidental mixing of flammable and combustible liquids. For example, a small amount of acetone put into a kerosene tank may lower the flash point of the tank's contents; acetone's high volatility will render the kerosene too dangerous to use. Similarly, gasoline mixed with fuel oil may change the flash point enough to make the fuel oil hazardous for home heating or similar uses. The lower-flash-point liquid can ignite, causing the higher-flash-point liquid to act as though it were a flammable liquid.

Keep lines from tanks of different types and classes of products separated, and, preferably, provide separate pumps for different types and classes of products. Use a portable container approved by Factory Mutual (FM) or listed by a nationally recognized testing agency (e.g., Underwriters Laboratories [UL] or Intertek [ETL]) for handling flammable liquids in quantities up to 5 gal (19 L).

Use colors or labels or both to identify fill openings, discharge openings, and control valves on equipment containing flammable and combustible liquids. Consider painting or banding pipelines with distinctive colors, and show the direction of a liquid's flow. The latest edition of ANSI/ASME A13.1, Scheme for Identifying Piping Systems, is an industry standard for such operations. Either mark each tank with the name of the product or identify the product with a label that communicates the contents clearly to users and first responders. It is also important to identify the container with the appropriate hazard communication labeling information.

Smoking

In a building or area in which personnel store, handle, or use flammable liquids, prohibit smoking and possession of strike-anywhere matches, lighters, or other spark-producing devices. The size of the restricted area will depend on the type of products handled, the design of the building, local codes, and local conditions. Conspicuously post approved NO SMOKING signs in buildings and other appropriate areas, and create a system of enforcement and honest accountability to help ensure employees and guests actually refrain from smoking.[3]

Prohibit smoking by truck drivers or their helpers while they are driving, loading or unloading, or attending to their unit. Drivers should keep smokers and all sources of ignition away from loading or unloading operations. Provide each tank vehicle with one or more portable fire extinguishers, each having at least a 10 BC rating.

---

2 NFPA provides a working definition of "closed system use" in its Hazardous Materials Code, NFPA 400 (2013 edition): "Use of a solid or liquid hazardous material in a closed vessel or system that remains closed during normal operations where vapors emitted by the product are not liberated outside of the vessel or system and the product is not exposed to the atmosphere during normal operations and all uses of compressed gases."

3 So-called e-cigarettes contain batteries, which may cause sparks during charging or use. Consider including controls on such devices in a no-smoking policy.

Static Electricity

The contact and separation of dissimilar materials generates static electricity, as when a fluid flows through a pipe or into a tank. Figure 10–1 shows several methods of generating static electricity. When a difference in electrical potential exists, a spark between two bodies can occur because no good electrical-conductive path exists between them.

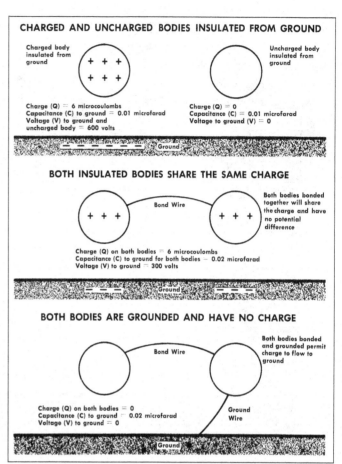

Figure 10–2. Bonding eliminates the difference in static charge potential between objects. Grounding eliminates the difference in static potential between objects and the ground. Both bonding and grounding apply only to conductive bodies and, when properly applied, can be depended on to remove the charge.

Nonconductive liquids, including many flammable liquids, can build up a static electrical charge when they flow or move, as through piping or filters or during stirring or splashing. The primary hazards of static electricity are fire and explosion caused by spark discharges that contain enough energy to ignite flammable or explosive vapors, gases, or dust particles. A static spark poses great danger where a flammable vapor may be present in air, such as at the outlet of a flammable-liquid fill pipe, at a delivery hose nozzle, near an open flammable-liquid container, around a tank truck's fill opening, or near a barrel bunghole.

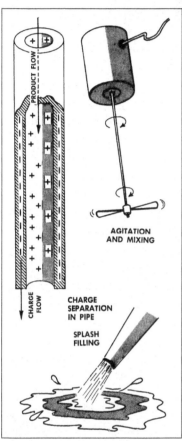

Figure 10–1. These are typical static-producing situations, including charge separation occurring in pipes.

Note that workers can react involuntarily to a static electric spark, causing injury and/or release of hazardous materials.

Bonding and Grounding

Static electricity is a common source of ignition for combustible and flammable materials. Grounding and bonding are two methods of controlling static electricity during fluid transfer. Bonding allows electrical charge to flow freely between objects and eliminates a difference in the static electric charge potential between them. This minimizes the chance of a static electric spark between the objects. Although bonding will eliminate a difference in charge potential between objects, it will not eliminate a difference in charge potential between these objects and the earth unless one of the objects has an adequate conductive path to the earth.

For this, containers need grounding. Grounding acts as a bridge and eliminates a potential difference between an object and the ground (earth) (Figure 10–2). An adequate ground will continuously discharge static electricity from a charged, conductive body. As a safety measure, use a grounding system when any doubt exists about a flammable-liquid handling situation. Note that some regulatory authorities require bonding *and* grounding in some circumstances.

Before transferring flammable liquids from one container to another, a means of bonding should be provided between the two containers, as shown in Figures 10–3, 10–4, and 10–5. Bond and ground a flammable-liquids tank truck or tank car to the loading rack (Figure 10–5). (See also NFPA 77, Recommended Practice on Static Electricity.)

Proper bonding and grounding of the transfer system usually drains off a static charge immediately. However, rapid flow rates in transfer lines can cause very high electrical potentials on the surface of liquids, regardless of grounding. Also, some petroleum liquids—especially pure, refined products—are poor electrical conductors.

10 Flammable and Combustible Liquids

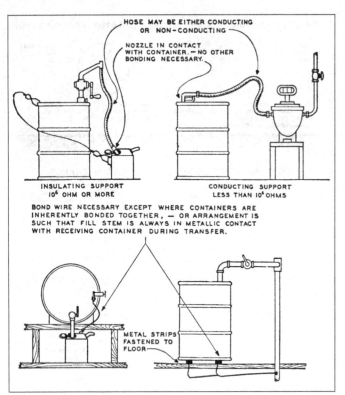

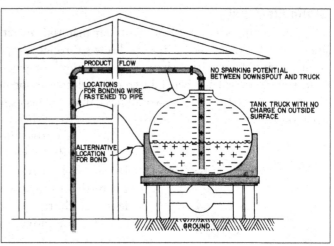

Figure 10-3. When a container is bonded during the filling process, any static electricity generated will be safely discharged. *(Reprinted courtesy of National Fire Protection Association.)*

Figure 10-4. Petroleum liquids can build up electrical charges as they flow through piping and become agitated when the tank or container is filled or emptied.

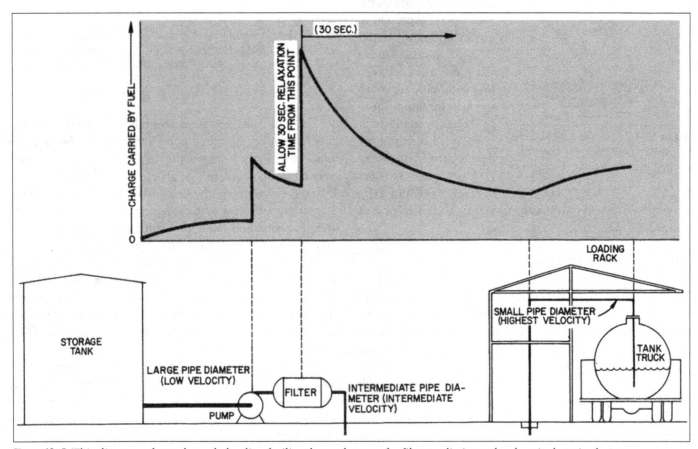

Figure 10-5. This diagram of a tank truck–loading facility shows the use of a filter to dissipate the electrical static charge.

Even though the transfer system has proper grounding, a static charge can accumulate on the surface of the liquid in the receiving container because static electric potential cannot flow through the liquid to the grounded metal. Accumulated charge can create a static spark with sufficient energy to ignite a flammable air-vapor mixture. This often occurs when the liquid approaches a body with different electrical potentials, like a metal probe during sampling or gauging. This situation is especially dangerous with liquids of intermediate volatility, like toluene or ethyl alcohol.

This high static charge can be controlled by reducing the flow rate, using side-flow spill lines to prevent violent splashing, and using relaxation time. In such cases, it is prudent to wait for a period of time—depending on the product, filling rates, and tank size—before gauging. The relaxation time (lag time between transfer operations and gauging) can last up to 4 hours after filling, especially in the case of large tanks.

When loading a flammable liquid through an open dome in a tank car or tank truck, use a downspout of sufficient length to reach the bottom of the tank. This prevents a static charge generated by the liquid's flow or splash during filling. Fill pipes should, therefore, maintain constant contact with the rim of the tank's upper opening. The loading rack should be grounded, and any tank trucks or cars should be bonded to the rack with a bonding wire (Figure 10–6). Ground wires should be uninsulated to permit easy inspection for damage (Figure 10–7).

To reduce the risk of stray currents, ensure that operators bond loading lines together. Install permanent bonding on flammable-liquid handling equipment where possible. Standards do not require grounding for metal aboveground tanks containing flammable liquids unless they rest on concrete or on nonconductive supports—metal frames and/or

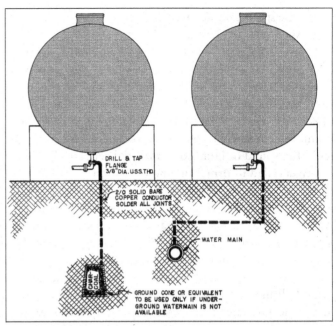

Figure 10–7. If an aboveground flammable-liquid storage tank is not inherently grounded, it should be grounded to a water main, a ground cone, or its equivalent.

tank bodies themselves act as the ground.

Regularly check bonding and grounding systems, depending on their use, for electrical continuity and damage. Bare-braided flexible wire facilitates inspection but increases risk of chemical and physical damage.

When cleaning with steam lines in the presence of flammable liquids, bond the nozzles on the lines to the surface of the object being cleaned. Also, be sure that steam will not impinge on any insulated conductive objects; this can create static charge accumulation.

Flat moving belts also are sources of static electricity. If they are made of a conductive material or have a conductive belt-dressing compound coating, however, static charges will not build up unless there is a defect in the system.

Nonconductive materials, such as fabric or rubber, can act as sources of charges of static electricity. Static from these materials, as well as from belts, can be discharged with plastic sheeting passing through or over rolls with grounded metal combs or tinsel collectors. Purpose-built radioactive devices and static neutralizers using electrical discharges also work for this purpose.

Bonding and grounding are effective only when the bonded objects are conductive materials. Some materials, such as some plastics and most glass, can accumulate significant static electrical charges. However, such a material does not inherently allow a charge to disperse enough throughout the material to permit effective grounding or bonding. This effect increases risk of fire and warrants caution when using plastic construction materials to transfer flammable liquids. Use only contain-

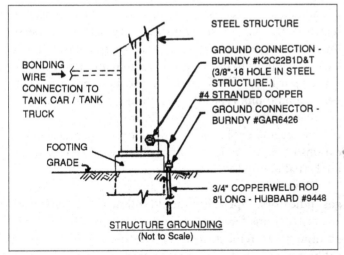

Figure 10–6. This is a detail of grounding/bonding loading rack structure. It is important that the loading rack–ground-cable connection to the ground rod be above grade to permit continuity and resistance testing to ensure that the rack structure is safely grounded.

ers that are specifically designed and approved for this type of usage; specify UL/ETL listing and/or FM approval.

To avoid a spark from the discharge of static electricity during flammable-liquid transfer operations, provide a wire bond between the storage container and the receiving container. If a metallic path between the two containers is otherwise present, however, this procedure is unnecessary. Again, remember that it is impossible, or at best very difficult, to ground or bond some materials effectively.

For detailed information and exceptions to these general rules, see current editions of NFPA 77, Recommended Practice on Static Electricity; American Petroleum Institute (API) RP 2003, Protection against Ignitions Arising Out of Static, Lightning, and Stray Currents; and NFPA 30, Flammable and Combustible Liquids Code.

To protect against stray voltage and ground faults, ground all motor frames, starting or control boxes, conduits, and switches according to the requirements for installing electrical and power equipment. See NFPA 70, National Electrical Code (NEC).

Electrical Equipment

Where flammable vapors exist, electricity can be an ignition source. This risk is higher when electrical equipment is not listed for these atmospheres, or is in poor repair. When using flammable liquids, consult NFPA 30, NFPA 70, and local design and construction codes.

Spark-Resistant Tools

Iron and steel tools can cause sparks when they strike flint-bearing stone, including most aggregates. Aluminum tools can spark when they strike rusted iron and steel, and rusty tools can spark when they strike aluminum. Steel or iron tools can spark when they strike flint, common chert, or other spark-yielding rocks. Some materials, such as carbon disulfide, acetylene, and ethyl ether, have low ignition-energy points. This increases the risk of ignition around such occasional and transient sparks.

A conservative safety measure when using these and similar materials is to use special tools designed to minimize the danger of sparks in hazardous locations. For example, leather-faced, plastic, and wood tools are free from the friction-spark hazard. Note, however, that use can embed metallic particles in these tools. Spark-resistant alloys, usually containing copper and/or beryllium, have lower tensile strength, but reduce the risk of these chemical sparks.

Health Hazards

Flammable and combustible liquids create health hazards when inhaled or when they contact the skin. Irritation results from the solvent action that these liquids have on the skin's natural oils and tissue. Intoxication as well as acute and chronic effects may result from breathing vapors of flammable liquids.

Atmospheres that are below flammable limits are not necessarily safe to breathe. For instance, a chemical with an LFL of 10% would be below the flammable limit if it were present at a concentration of 1% (10,000 ppm) and perhaps well above acceptable exposure levels.

Generally, vapors from flammable and combustible liquids are heavier than air. They will flow into pits, tank openings, confined areas, and low places where they may displace oxygen, contaminate the normal air, and perhaps create an explosive atmosphere.

Oxygen deficiency may also occur in closed containers, when rusting over a long period consumes the oxygen. Air out and test all confined spaces for toxic and flammable atmosphere, as well as for the level of oxygen, to determine the protective measures required before allowing personnel to enter. (See Threshold Limit Values, published annually by the American Conference of Governmental Industrial Hygienists; 29 CFR 1910.1000; and the NSC's Fundamentals of Industrial Hygiene.)

Combustible-Gas Indicators

Safe work around flammable and combustible liquids requires knowledge of whether atmospheres contain hazardous, flammable, or explosive levels of vapor. There are a multitude of instruments that can detect this and provide simple warning information and/or more complete analysis.

Assume that flammable vapors and toxic mixtures are present in all tanks that have contained or been exposed to flammable and combustible liquids. Test for flammable vapor-air mixtures in tanks and other vessels using a properly calibrated, approved combustible-gas indicator.

Most of these instruments read out measurements as a percentage of lower explosive level (LEL). These devices may have the capacity for adjusting alarm settings, depending on the make and model. Combustible-gas indicators use various methods to detect levels of flammable and combustible vapor levels. These indicators seldom report in parts per million (ppm, typically used for determining potential health impacts). As such, these indicators primarily describe risk of fire and explosion and do not provide precise indications of exposure levels. Always use approved chemical analytical methods to test for toxic substances.

Allow only experienced operators to use combustible-gas indicators. Operators should follow the manufacturer's instructions for calibrating the unit. To ensure a more accurate measurement, correct the reading often by using a calibration graph that adjusts the reading according to the atmosphere of concern. Manufacturers supply correlation graphs for many combustible gases.

Note: A combustible-gas indicator calibrated for the

LEL will not give any reading while sampling a very highly concentrated vapor, and exposure to high concentrations of flammable vapor can "poison" a detector, impairing it in the short or long term. Some instruments will give a negative reading while testing vessels containing high concentrations of inert gas.

The instrument operator will get useful information only after checking the conditions and calibration of the instrument according to the manufacturer's recommended procedures. The manufacturer will generally supply the appropriate calibration gas, which can also support daily operational checks if the instrument does not have internal circuitry to perform this function. At least once a year, thoroughly examine the instrument and test its integrity. The manufacturer, a recognized standards laboratory, an outside consultant, or a qualified employee on site can perform this maintenance.

When using the instrument, do not place the sampling hose where pumps could draw flammable liquid, steam, or water into the instrument. This kind of contamination could overload or damage the instrument's detector, putting it out of service. Generally, avoid using combustible-gas indicators to test vapors from heated combustible liquids because these vapors condense in the sampling lines or combustion chamber and could potentially result in false results. Use special filters from the manufacturer to prevent dust, particulates, moisture, or leaded gasoline from fouling the instrument and/or providing inaccurate readings.

Note that a tank can be vapor-free before cleaning, but disturbing sludge or scale can release flammable or combustible liquid and vapor. Therefore, consider tanks that contain sludge or scale to also contain flammable or combustible liquids or vapors.

If workers leave a tank for a period of time, such as for lunch or a shift change or overnight, test for oxygen and flammable and toxic gas before allowing workers to reenter the tank. Adhere to OSHA's permit-required confined-space standard, 29 CFR 1910.146, when workers must enter confined spaces—especially spaces that have contained flammable or combustible liquids or that require hot-work repairs, such as welding and cutting. (See Chapter 5, Legal and Regulatory Issues for the Safety Manager, in the *Administration & Programs* volume.)

Testing devices are available for specific measurement of carbon monoxide (CO), benzene, hydrogen sulfide, tetraethyl lead, and other specific toxic and flammable vapors.

LOADING AND UNLOADING

This section discusses general safety procedures for nonpressurized tank cars and tank trucks used to transport flammable and combustible liquids. Pressurized tank cars and trucks used to transport nonflammable and flammable gases, liquefied petroleum gas (LPG), and other similar materials will have additional and different safety constraints.

In rail situations, the serving railroad usually supervises spur track. For applicable specifications, refer to the Association of American Railroads (AAR) publication *Manual of Standards and Recommended Practices* (see References at the end of this chapter).

Responsibility for maintenance and serviceability of tanks, trailers, and other transport vessels falls to the owner or operator of the transport vessel. Facilities for loading or unloading are the responsibility of the facility owner. These parties must reconcile and agree to any loading or unloading procedures.

Ensure that manufacture and use of tank trucks, tank trailers, and tank semi-trailers used for the transportation of flammable liquids conform to appropriate regulatory standards. Carriers handling hazardous materials are subject to U.S. regulations under the Hazardous Materials Transportation Uniform Safety Act of 1990. In the United States, consult 49 CFR 171–178, especially 177.325 and 177.386–399. Revision and promulgation of different regulations occur often. Parties exempt from the U.S. Department of Transportation (DOT) Hazardous Materials Regulations should consult the latest version of NFPA 385, Standard for Tank Vehicles for Flammable and Combustible Liquids.

Fire Safety

Do not load or unload cars before ensuring that all sources of ignition are removed from the area, including exposed heating elements, fires, pilot lights, and inappropriate electrical equipment. As an added precaution, defer unloading and pumping operations during electrical storms.

Tank car and truck fires can be serious, especially if spilled product burns under the container. Shut off all loading lines to the rack and apply a spray of water to cool the tank car, structural steel, and piping. If conditions permit, flush spilled fuel from under the tank's body to an open area or into a closed drainage system. Avoid introducing water into the tank car and causing an overflow.

If a fire occurs at the dome of a container, and if the dome is unobstructed by a fill pipe, close the dome cover to extinguish the fire. If a fill pipe is in the dome opening, do not remove the fill pipe—removing it could splash the burning liquid and spread the fire. Use a dry chemical or carbon dioxide extinguisher to put out the fire. In some cases, the flames can be "blown out" with a straight stream of water directed across the opening. Take care, however, not to overflow the tank or to splash the contents and spread the fire.

To protect nearby facilities, provide drainage to limit the spread of a spill and to direct it to a safe area. The

Association of American Railroads can provide publications and training on this subject (see References at the end of this chapter).

Where burning vapors are escaping from leaks or vents, it may be better to let them burn until you can control their source. In all cases of fire, quickly inform affected personnel and report any emergency in accordance with established procedures. Establish and rehearse fire-fighting procedures so that in an actual emergency, employees and emergency crews can follow the procedures exactly.

In all cases of an unintended open flame at a facility, follow local fire authority rules for fire reporting. This might include calling the local fire department to confirm that the fire is, in fact, controlled. Expectations for this may vary based on the complexity of the facility, construction type of structures, whether the fire occurs inside or outside a structure, injuries, exposures, environmental impact, sophistication of responding agencies, and degree of damage to the facility.

Personnel Safety

Only trained employees should load or unload tank cars containing flammable or combustible liquids. These workers should understand the possible dangers of fire, explosion, asphyxiation, and toxic effects from exposure to flammable vapors. Provide fall protection in loading areas. A rack with a gangway or bridge to the car helps workers move safely when loading or unloading a railcar. The gangway or bridge should have guardrails (Figure 10–8).

When loading or unloading a tank truck, see that the

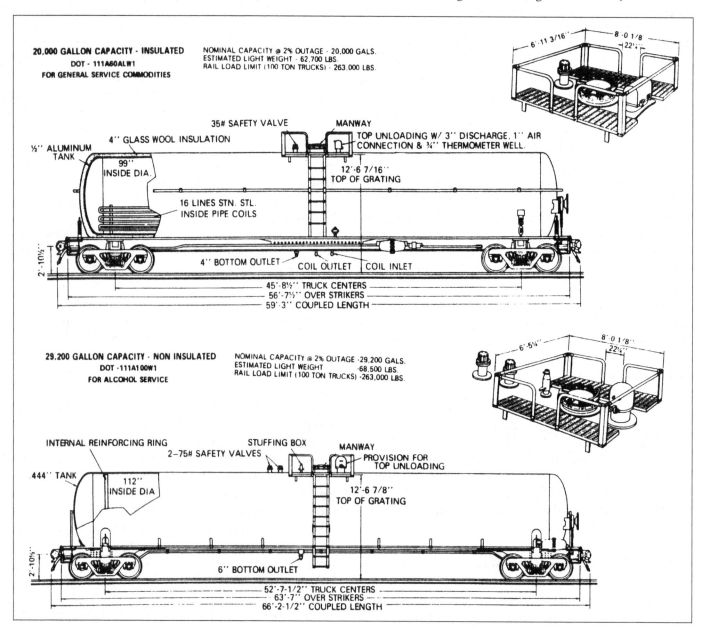

Figure 10–8. Tank cars are constructed to accommodate the various types of commodities transported.

brakes are set, the engine stopped (unless power takeoff [PTO] is required for unloading), the lights turned off, and the bonding connection made before opening the dome cover for inspection or gauging. Before moving trucks, be sure to close and latch all domes and remove all bonding connections before starting the engine.

Environmental Protection

Place approved containers under any leaks to prevent contamination and product loss. Frequently transfer accumulated product to storage to limit quantities of flammable liquid exposed to the atmosphere and to prevent release to ground. Prepare for spills and emergency responses, including complying with applicable regulations.

Avoid spills or overflows. If they do occur, immediately stop loading, shut off the valves, and clean up the overflow before resuming loading. Eliminate or control all sources of ignition in the area. Trap the liquid in containers, in an earthen or sand-diked area, or in a depression or pit, if possible. In case of a large spill, especially in urban areas, the best way to avoid endangering lives is to use portable hand pumps to discharge the product into approved drums or into another tank truck. Never flush spills into public sewers, drainage systems, or natural waterways.

Drivers should make every effort to park a damaged and leaking truck so that it will not endanger traffic or property. Park the truck off the highway—if possible, in a vacant lot or otherwise away from buildings and away from areas with high concentrations of people. Drivers should warn the public to keep away and should immediately notify the police and fire departments. Drivers should not leave the truck unattended. Allow adequate time for all flammable vapors to dissipate before starting the truck engine.

Drivers for carriers subject to the safety regulations of the U.S. DOT must report all broken, leaking, and damaged containers on Form F 5800.1. Immediate notification by calling 1-800-424-8802 is required if (1) a person is killed; (2) a person is injured and requires hospitalization; (3) the estimated carrier or property damage exceeds $50,000; or (4) fire, breakage, spillage, or contamination involving a radioactive material occurs.

Local environmental and emergency response agencies may have additional requirements in case of releases that expand past preengineered sumps or containment devices. Meet with local agencies and develop preincident plans where necessary.

Electrical Considerations

Provide approved electrical equipment on tank cars and tank trucks. Install electrical protection equipment according to the NEC and inspect regularly. Bond, ground, and insulate tank cars and tank trucks for protection from stray electrical currents (Figure 10–9).

The location of electrical power lines relative to a tank car's unloading position is essential to fire safety during loading and unloading. The Association of American Railroads' Circular 17.D, Recommended Practice for the Prevention of Electric Sparks That May Cause Fires in Tanks or Tank Cars Containing Flammable Liquids or Flammable Compressed Gases Due to Proximity of Wire Lines, supplies the following provisions:

1. Where any electric power line is within 20 ft (6 m) of the tank's opening, the use of a metallic gauging rod is prohibited.
2. When the contents are being gauged or transferred, tank cars, wherever possible, shall not be located under or near any electric power lines.
3. Where tanks or tank cars (the contents of which are being gauged or transferred) are necessarily located under or near power lines having a span length of 150

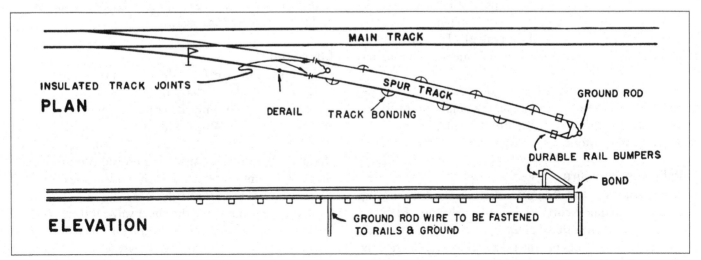

Figure 10–9. This siding is designed for unloading tank cars. It has rail joint bonding, insulated track joints, derailing, and track grounding.

ft (45 m) or less and operating at a voltage not exceeding 550 V between conductors, the following rules shall be observed:

   a. Where power lines pass overhead, there shall be a minimum vertical clearance of 8 ft (2.4 m) at 60°F (16°C) between the wires and the tank.

   b. Where power lines pass nearby and do not have the minimum vertical clearance specified in paragraph 3–a, there shall be a minimum horizontal clearance of 8 ft (2.4 m) between the wire lines and the tank.

   c. Openings in tanks shall be at least 6 ft (1.8 m) distant, measured horizontally from any overhead power lines.

4. Where tanks or tank cars (the contents of which are being gauged or transferred) are located under or near power lines having a span length in excess of 150 ft (45 m) or operating at a voltage in excess of 550 V between conductors, it is recommended that a special study be made by qualified persons. Based on the study, provide clearance as necessary to give adequate protection.

Inspection

The company receiving or shipping a tank car should make a general visual inspection of the unit and report any obvious defects to the carrier. Closely check invoices and shipping papers on incoming shipments to make certain they match the actual contents on containers. Do not rely exclusively on records checking; when in doubt, test the contents to prevent mixing of incompatible substances.

Shipping tank cars requires securing all valves and related parts and preparing shipping papers identifying the contents. In addition, place four placards, one on each side of the tank. Keep the placards clean and visible at all times.

Check inspection dates on containers and tanks to make sure testing matches the required intervals. Do not use tank cars that have not undergone proper testing. Inspect tanks and safety valves for defects in welds, piping, valves, gaskets, and interlocks. Stencil the tank of a U.S. carrier with the last current inspection date to indicate that the tank has been tested according to 49 CFR 177.

Ensure that trucks are kept in good repair and inspected daily. Place special emphasis on the condition of lights, brakes, horns, rearview mirrors, bonding straps, tires, steering, and motors.

Relieving Pressure

Relieve the tank car of interior pressure before removing the domed manhole cover or the outlet valve cap. To relieve pressure, raise the safety valves, open the air valve a small amount at a time, or cool the tank with water. Depress the vacuum-relief valve, if vacuum is a problem. Defer venting and unloading if a dangerous amount of vapor collects outside the car. If interior pressure is excessive, spray the car with water to cool it or allow it to stand overnight—then unload it early the next morning.

Removing Covers

Loosen a screw-type dome cover by placing a bar between the lug and the knob on the top of the cover. Make two complete turns to expose the ½-in. (1.3-cm) vent holes in the threaded portion of the dome cover. If vapors escape audibly, tighten the cover and release the pressure by raising the safety valve. Reduce exposure and risk of fire injury by standing upwind, away from vapor accumulation. The chemical and the physical layout of the unloading area may require appropriate respiratory protection, such as a self-contained breathing apparatus (SCBA) or an air-supplied respirator equipped with auxiliary SCBA.

Face the dome and remove the cover. With feet well braced, use short, vigorous pushes on the bar. Secure the cover and tools to stop them from falling, with either chains rated for overhead hanging or a walk platform.

On the bolted type of dome cover, unscrew all nuts one turn and lift the cover to break any adhesion between the cover and the dome's ring. If there is a sound of escaping vapors, tighten the dome and repeat the venting operation. On the interior types of dome covers, remove all dirt, cinders, and debris before unscrewing the yoke.

When the car is unloaded through the bottom outlet valve, adjust the dome cover to allow for venting, where vapor recovery is not required. Tighten a screw-type dome cover just enough to expose the vent holes in the threaded portion of the cover.

Place a small, thin wood block under the bolted type of cover, and tighten up the interior type of dome cover in the yoke to within ½ in. (1.3 cm) of the closed position. Place a metal cover or wet burlap or canvas over the tank manhole to prevent sparks or other sources of ignition from entering and to limit the vapor escaping.

Connections

Unloading a tank car from the bottom reduces effort and risk, as long as the car has an approved bottom outlet. A tank car can also be unloaded safely from the top, if it has appropriate outlets. When unloading a car through the bottom, close the tank's outlet valve before removing the outlet chamber cap or plug. Place an approved container under the outlet chamber and unscrew the outlet cap (see 49 CFR 174.67, Tank Car Unloading).

Check the condition of the outlet valve. If there is no serious leak, proceed with the unloading. If the plug does not loosen easily with a 48-in. (1.2-m) wrench, tap the bottom outlet cap, or plug, to loosen it. If the valve leaks so badly that any connection will spill the product, unload the car from the top.

Cold weather can block the chamber or valve with ice or frozen product. In this case, carefully examine the outlet chamber for cracks before attempting the unloading connection. If the outlet chamber is intact, make the connection for unloading, but wrap the outlet chamber with burlap or other similar material. Applying hot water or steam to the cloth wrapping will usually free the chamber of ice and/or frozen product.

Apply steam slowly when using heater pipes during unloading. Cease steam application when it begins to exhaust at the outlet pipe. Supervise steam pressure carefully. Pressure should not exceed that needed to bring the contents to the desired temperature and should never be high enough to vaporize or boil the contents.

Overheating combustible liquids causes them to emit flammable vapors. Also, overheating products containing certain additives can release dangerous amounts of flammable vapors.

Before opening the car valve, carefully check the condition of the connection from the car to the storage tank. Gauge the storage tank, and watch it to prevent an overflow. Frequently check the hose and unloading line. Before disconnecting loading lines, examine the tank car to make certain it is completely empty.

A worker should be in attendance during the entire loading or unloading operation, as well as any time a car is connected. If it is necessary to discontinue operations before they are complete, close the outlet valve and replace both the dome cover and outlet chamber cap. To prevent overflowing during transport, load cars only to the specified inches of outage. Remove loading or unloading connections as soon as possible after operations have finished.

For details on unloading or loading bulk containers, check with the car manufacturer, supplier, or railroad. See Chapter 12, Materials Handling and Storage, in this volume, for information on unloading chemical tank cars.

Shut off trucks with motor-driven pumps before connecting or disconnecting loading lines. Drivers must remain at the tank controls (within 25 ft [7.5 m]). If they must leave the site, they must remove the hose and stop the release of the flammable product without endangering themselves. When an unloading line must cross a sidewalk, provide suitable warning signs.

Make sure that the tank truck or railcar has proper venting and vapor-recovery units. Ensure that the tank can accommodate the amount of liquid to be loaded or unloaded. Operators should cross-check to be sure that they are handling the correct product. Ensure that the tank has previously contained the same liquid. If possible, contact personnel at the delivery site before unloading. Completely drain and flush all tanks to ensure removal of trapped liquids. This is especially important when changing tank contents.

Do not dispose of placards, rags, waste, or blocks in the tank or in the car's body.

Figure 10–10. These placards are used on vehicles and railcars transporting flammable or combustible liquids. The EMPTY placard can be either a separate placard, printed on the reverse side of a placard, or a composite made by covering the top triangle with a black triangle having the word EMPTY printed in white letters. FLAMMABLE and COMBUSTIBLE placards are red with white letters. *(Reprinted from 49 CFR 172, 525, 542, and 544.)*

Placards and Shipping Papers

The U.S. DOT and its counterparts in most other countries provide specific guidelines for using placards and shipping papers. Consult the appropriate requirements for details (Figure 10–10).

Spill Prevention, Control, and Countermeasures

The U.S. Environmental Protection Agency (EPA) has regulated bulk storage of oils since 1978, as part of the Clean Water Act. Operations within the jurisdiction of the EPA should check their storage and use patterns against the spill prevention, control, and countermeasure (SPCC) regulations for application and specific additional requirements. Individual facilities may have more comprehensive requirements for spill control. All parties should reconcile spill response roles, responsibilities, capabilities, and capacities with requirements prior to engaging in loading or unloading operations.

CONTAINER (NONBULK) STORAGE

Container storage relates to storage of hazardous materials, including flammable and combustible liquids, in containers below a certain size. Depending on the context, container storage might involve quantities smaller than 119 gal (450 L) or 60 gal (230 L).

NFPA 30 specifies the quantity of each class of flammable liquids that can be stored in various locations in a facility. Also, it describes the required conditions and procedures relating to such storage.

Keep or store Class I and Class II flammable liquids (see Definitions, earlier in this chapter) in a building used for public assembly (such as a school, church, or theater) only if they are in approved containers (Figure 10–11). Keep the containers in either an approved flammable-liquids storage cabinet or a storage room that does not open to the public portion of the building. Limit the quantities stored in such locations according to NFPA 30.

Do not store the containers where they will obstruct exits, stairways, or other areas used to safely leave the building. Also, never store them near stoves or heated pipes or expose them to sunlight or other sources of heat.

Do not store flammable or combustible liquids in open containers. Use approved containers for flammable liquids, and close them after each use and when empty. Ensure that containers are stored with openings or caps up, away from the liquid phase of the material being stored. Permanently mark high levels on containers, and place shutoffs or other control devices inside tanks to prevent overfilling.

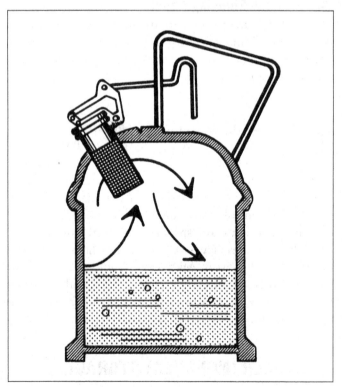

Figure 10–11. This spring-loaded cover is designed to open at 5 psi and relieve the internal vapor pressure. Only negligible losses are caused through evaporation of liquids stored in safety cans at ordinary temperature ranges.

On-Property Transportation

Transport flammable liquids in sealed containers in order to minimize damage to the containers. When employees are filling tanks and other containers, they should allow enough vapor space, or outage, above the liquid level so the liquid can safely expand when temperatures change. For example, gasoline expands at the rate of about 1% for each 14°F (8°C) rise in temperature. The recommended outage space for gasoline is 2% of a tank's or compartment's capacity. However, many jurisdictions recommend fill limits at 90% of the tank's capacity.

Inside Storage and Mixing Rooms

Flammable or combustible liquids in approved, sealed containers present a potential, rather than an active, hazard—the possibility of fire from outside. To reduce this risk, isolate inside storage rooms as much as possible. Locate them at or above grade, not immediately above a cellar or basement, and preferably along an exterior wall.

NFPA 30 specifies exactly the kinds of architectural and mechanical requirements of a flammable-liquids storage room. These requirements include:

- noncombustible construction with sufficient fire separation
- normally closed doors and other openings with fire ratings compatible with construction, not less than 90 min
- raised sills or sunken floors, at least 4 in., or an open-grate trench that drains to a safe location
- safe storage inside the room
- electrical fixtures compatible with room contents (Class I, Division 2 as described in NFPA 70)
- ventilation, either gravity or mechanical, for Class II and III liquids; mechanical for Class I liquids (generally, 100% fresh air supply—no recirculation)
- ducts conforming to NFPA 91, Standard for Exhaust Systems for Air Conveying of Vapors, Gases, Mists, and Noncombustible Particulate Solids
- air exchange rates as a function of floor area
- aisle and stacking provisions.

Where dispensing occurs inside flammable-liquid storage rooms, operations should comply with the provisions of NFPA 30 and locally applicable fire codes.

Inside Storage Cabinets

Several codes and regulations limit the amount of flammable liquids a facility can store indoors. The requirements are most stringent for Class IA liquids, with some limiting operations to 10 gal stored outside cabinets per control area.[4]

---

[4] A "control area" is a portion of a floor of a building separated from other areas by fire-rated construction. Modern building and fire codes limit the number of control areas per floor and reduce both the number of control areas and the quantities of materials allowed in control areas without additional precautions. These reductions become more severe the higher above, or lower below, grade plane the storage occurs.

NFPA 30 allows larger quantities of flammable-liquid storage in flammable-liquid storage cabinets. Manufacturers of these cabinets use metal or wood to build them and conform to ANSI standards and NFPA 30 in their design and fabrication. Use cabinets with a nationally recognized test laboratory listing and/or major insurance carrier approval.

In any event, codes limit the amount of flammable and combustible liquids, even in cabinets, to no more than 120 gal (454 L) of Class I, Class II, and Class IIIA liquids combined. Of this total, not more than 60 gal (227 L) may be of Class I and Class II liquids. Not more than three such cabinets may be in a single control area, except that additional cabinets can exist in industrial occupancies with sufficient separation (100 ft [30 m]).

Make sure that cabinets have labels with conspicuous lettering, FLAMMABLE—KEEP FIRE AWAY. There is no requirement in the fire code to provide ventilation to cabinets. Doing so can reduce occupational exposures to hazardous materials but may compromise the fire protection features of the cabinet. If ventilation is provided for a cabinet, ensure that the materials of construction for the ventilation system offer fire protection features at least as effective as the cabinet provides.

Consult local fire prevention authorities about the type of storage and handling that is used or planned for flammable liquids.

Outside Storage Lockers

If space permits, consider constructing or installing storage areas for flammable liquids as separate buildings set aside from the main plant. Construction may be similar to that described for inside storage rooms. Approved prefabricated lockers may also be used. The type of product stored and the proximity to other buildings and structures will determine the best design for outdoor storage lockers. Consult local regulations for location and use of outdoor storage lockers.

CLEANING SMALL TANKS AND CONTAINERS

Work on a container that has held flammable or combustible liquids should be supervised by a trained supervisor who can maintain a high degree of safety during operations. If the container has held reactive compounds such as nitrocellulose, pyroxyline solutions, nitrates, chlorates, perchlorates, or peroxides, take special precautions. Note that the container may contain enough oxygen to support combustion of residual materials. Contact the manufacturer or supplier, or refer to the manufacturer's Safety Data Sheets (SDSs), for specific information regarding cleaning procedures and other precautions.

Remove the covers, plugs, and valves, and permit the tank or drum to drain into an approved container. Environmental regulations may impose specific disposal requirements for this drained liquid.

Examine the inside of the tank or drum for rags, waste, or other debris that might interfere with draining or that could retain flammable vapors. Use only lights approved for use in hazardous locations by the NEC (e.g., Class I, Division I, Group D) for this inspection. See Chapter 8, Electrical Safety, in this volume. Mirrors can reflect light from the sun or flashlights into the tank without creating spark risk.

Clean and steam small tanks and drums that contain flammable or combustible products only in an approved area with ample ventilation. Preferably, clean them in an outdoor area away from ignition sources. Ensure that appropriate grounding and bonding connections have been installed between cleaning apparatus pipes and nozzles and the containers.

Steaming

Steaming, hot chemical washes, water filling, and use of inert gas are among the common methods for cleaning and vapor-freeing small tanks and drums. If the inside of a container is clean and steam is available, the easiest cleaning method is steaming. See the American Welding Society's document AWS A6.0, Safe Practices for Welding and Cutting Containers That Have Held Combustibles. See also NFPA 326, Standard for the Safeguarding of Tanks and Containers for Entry, Cleaning, or Repair.

After draining, place the tank or drum on a steam rack or over a steam connection with the outlet holes at the lowest point. The tank or drum must rest against a steam pipe, and an electrical bond must exist between the steam pipe and the tank or drum. Then apply an ample supply of live steam for an appropriate period of time, depending on the tank's size and contents.

Hot-water washes may be appropriate in advance of hot-work repairs. After cooling, test the atmosphere with a combustible-gas indicator. If the drum atmosphere is safe, proceed with hot-work repairs. If not, clean the drum again.

If steaming alone will not clean the tank or drum, use a cleansing compound of sodium silicate or trisodium phosphate (washing powder), dissolved in hot water and kept at a temperature of 170 to 190°F (77 to 88°C). Slowly add hot water to overflow the container until no appreciable amount of volatile liquid, scum, or sludge appears. If steam or hot water is unavailable, use a cold-water solution with an increased amount of cleansing compound. Agitate the solution to help ensure more thorough cleaning.

Exceedingly dirty containers can require preliminary treatment with a caustic-soda solution. Agitate the solution enough to clean the interior surfaces, and then drain, wash, and steam the container.

Do not use caustic-soda solutions to clean aluminum- or zinc-coated drums. The metal finishes on these materials can generate hydrogen gas when in contact with a caustic-soda solution. This practice creates an extreme risk of fire or explosion and impairs the metal finish.

Materials with extremely low auto-ignition temperatures and wide flammable ranges, such as carbon disulfide, create a special risk during cleaning operations. Steaming can create temperatures sufficient to cause ignition. Make these containers liquid-free with a suitable cleansing compound, and then test them for vapor.

Guard against burns, especially when using steam, hot water, and caustic soda. Ensure that workers have suitable clothing and personal protective equipment, such as boots, gloves, head coverings, face shields, and rubber aprons.

Inert Gas Displacement

Inert gases can make the atmospheres in containers, including small tanks, safe temporarily. Inert gas displaces oxygen and vapors but does not remove the residual source. Because the flammable vapor source remains inside the container, vapors and oxygen can reappear in the container. Moreover, purging with carbon dioxide can create static electricity. As such, this method is not as safe as steaming.

Inert gases can come from compressed gas cylinders. Often, portable inert gas generators that condense inert gases, most often nitrogen, from the atmosphere are available for special jobs. However, only well-trained and -equipped workers should use them to produce the proper atmosphere and to safeguard against the potential for fire and explosion.

When using an inert gas such as carbon dioxide or nitrogen, wash the tank or vessel so it is as free as possible of flammable liquids, and thoroughly flush it until the vessel overflows. When repairing, leave as much water in the tank as the work will permit.

Blankets of carbon dioxide concentrations for inert gas displacement should be at least 50%, by volume. If the tank has contained light, flammable gases such as hydrogen or carbon monoxide (CO), use a higher concentration (80%) to further reduce available oxygen. Nitrogen concentrations should be 60% or higher, depending on the previous contents of the container.

DISPOSAL OF FLAMMABLE LIQUIDS

Sealed and properly stored, most flammable liquids are stable and will last for years. Return unused, uncontaminated flammable liquids to the vendor; salvage them for resale; or use them for another, approved purpose.

Some mixtures of clean flammable liquids may require separation before they are useful. It is best to have a recov-

ery contractor do these separations. In most cases, environmental laws and regulations require the use of an approved contractor for offsite recycling and/or disposal of used or dirty flammable liquids.

If recycling or recovery of flammable liquids is not feasible, turn them over to a licensed hazardous-waste disposal contractor. Observe all regulatory requirements, such as those of the Resource Conservation and Recovery Act, including saving manifests and shipping papers indefinitely for future review.

REGULATORY ISSUES

Precautions must be observed when receiving, storing, handling, and using flammable and combustible liquids. Because the specific characteristics of flammable and combustible liquids vary, as do the handling precautions required for them, it is impossible to cover every detail of the safe handling and use of every liquid. Therefore, consult regulations, local codes, fire underwriters, the National Fire Protection Association (NFPA), trade associations, and specific handbooks for detailed information. Observe U.S. Department of Transportation (DOT) regulations and Occupational Safety and Health Administration (OSHA) standards. See the References at the end of this chapter.

BULK STORAGE: INSTALLATION AND MAINTENANCE

Tanks are large containers designed for storage of products prior to use or distribution. Generally, they exceed 55 gal in capacity and are installed above or below grade. Standards and regulations differ on tank location and construction. An earlier section discusses storage of flammable and combustible liquids in smaller containers (60 gal [230 L] or less).

Tank Construction

All aspects of tanks (construction, installation, testing, and maintenance) should conform to the latest edition of NFPA 30. See also NFPA 30A, Automotive and Marine Station Code, and API RP 652, Lining of Aboveground Petroleum Storage Tank Bottoms.

Store bulk quantities (generally more than 55 gal) of Class I flammable liquids in an underground tank or outside a building. Where permitted by local authorities, use properly protected, outside, aboveground storage tanks. No outlet from the tank should be inside a building, unless the outlet terminates in a flammable-liquids storage room, as defined by applicable building and fire codes (Figure 10–12).

10 Bulk Storage: Installation and Maintenance

Figure 10–12. A well-designed flammable-liquid storage room with both high- and low-level ventilation and automatic fire extinguishment. The ventilators are designed for 12 air changes per hour.

Tank Ventilation

Provide storage tanks with vents of a type and size recommended in NFPA 30. Vent pipes of underground tanks and vapor-recovery systems for Class I flammable liquids should terminate outside buildings. They should also be higher than the fill pipe opening and not less than 12 ft (3.7 m) above the adjacent ground level. They should discharge vertically and be located so flammable vapors cannot enter the openings of buildings or be trapped under eaves or other obstructions. Vent pipes from underground tanks storing Class II or Class III liquids should also terminate outside buildings and be higher than the fill pipe opening. Vent outlets should be above the normal level of snow. Consult U.S. Environmental Protection Agency (EPA) and state air pollution regulations concerning release of regulated chemicals into the atmosphere.

By design, flame arresters reduce the propagation of flame down a pipe or through an aperture. As such, flame arresters are common safety features on containers and vents. However, some authorities have questioned the effectiveness of brass mesh or copper screens as flame arresters in vent terminals.

Ventilation is an alternative to the flame arrester screen. Ideally, ventilation reduces the accumulation of vapors to a point below the LFL, thereby preventing a fire. A well-located vent can provide better protection than screens.

Place vents so that escaping vapors will disperse and be clear of ordinary sources of ignition or reentrainment into air intakes for buildings or processors. If the vents might become obstructed by mud wasps or other insects and their nests, provide a loosely attached screen of relatively coarse mesh (see NFPA 30 for recommended vent sizes). Note that containers of highly volatile materials may lose more product to evaporation when well ventilated.

Pumps

Locate flammable-liquid transfer pumps outside buildings and diked areas whenever possible. Use fire-resistant construction in buildings that house equipment for transferring flammable liquids. Provide ample ventilation, especially along the floor, where vapors might be present (Figure 10–13). Preferably, pump houses should not contain sunken pits, except small drain openings, because the danger of vapor concentrations in such areas is too great. A pump house should have a minimum of two exits, which should remain clear of obstructions at all times and have doors that swing in the direction of egress.

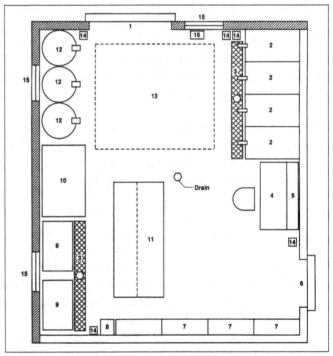

Figure 10–13. Some of the features of this oil house layout are as follows: 1 and 6: self-closing fire doors wide enough for passage of lift trucks or other materials handling equipment; 2: drum racks; 3: grating and drain; 4: desk; 5: filing and record racks; 7: individual lockers or storage cabinets; 8: waste disposal container; 9: solvent cleaning tanks; 10: purification equipment, soluble oil mixing equipment, or other special equipment; 11: cabinets and racks for equipment, supplies, and small containers; 12: grease drums with pumps; 13: parking area for oil wagons, etc.; 14: fire extinguishers; 15: ventilators; 16: container of sawdust or other absorbent.

In areas where flammable-liquid handling occurs, partition and seal off electrical equipment from the rest of the pump house, unless the electrical equipment is approved for use in flammable atmospheres. Keep the pump room well ventilated. Also, have a well-marked master cutoff switch outside the building. All electrical equipment inside the building must conform to the NEC, as well as to local codes for indoor use around flammable liquids and vapors.

Maintain valves and packing glands on pumps in good operating condition to prevent leaks. Excess-flow valves can also prevent large-scale spills and are sometimes required by local codes. Draining escaped liquids into a closed-pipe return system is preferable to simply catching leaking materials in drip pans because the former reduces the quantity of liquid exposed to the atmosphere. Repair leaks as soon as they are noticed.

If a centrifugal pump with priming bleeder is used, either keep the outlet of the bleeder plugged when not in use or keep it piped to a closed system for collection. Keep pump bearings well lubricated to prevent overheating caused by friction.

Because of the importance of good housekeeping in a pump house, provide approved containers for safe disposal of debris, rags, and other waste. See that these containers are emptied daily. Do not store tools, other than those required to operate a pump house, in the pump house. The pump house should not contain lockers or be used by employees to change clothes. Do not allow employees to loiter in the pump house. Keep all sources of ignition away from the pump house.

Locate fire extinguishers at convenient, easily identifiable points in pump houses. Have self-closing fire doors and the recommended type of automatic sprinkler or automatic fire suppression system available in pump houses.

Gauging

Tank gauging allows operators to keep track of liquid levels in tanks. Frequent gauging provides early indications of product loss. Automatic tank gauging takes place using sensitive electronic equipment. Manual tank gauging requires an operator to check liquid levels using a sight gauge glass or dipstick.

Operators should not walk on tank roofs to complete manual gauging; provide a walkway or platform that protects the worker and the tank. Provide safe means of entry and exit for persons gauging tanks. Personnel should stay off the tank roof while any pumping operation is in progress.

NFPA 30 requires vapor-tight caps on manual-gauging openings that are independent of fill pipes when they are located in a building or buried under a basement. Protect each such opening against liquid overflow and possible vapor release by means of a spring-loaded check valve or other approved device. Have only trained and qualified employees conduct manual gauging of tanks containing Class I liquids.

Aboveground storage tanks often use float- or pressure-operated gauges. Newer gauging systems use weight measurement to provide accurate indications of even small leaks. Some tanks require a sounding weight and tape or, if the hatch on top of the tank is small, a wooden sounding rod through a gauge. Some pressurized flammable-liquid tanks use sight gauge glasses.

Consider installing automatic tank gauging to protect employees and to provide earlier detection of releases. However, regulations may require that tank operators perform manual gauging to verify the accuracy of the automatic gauging systems.

Tank Anchoring

Where danger from floodwaters exists, install and anchor tanks according to NFPA 30.

Underground Tanks

Storing liquids underground reduces the danger of surface leakage and container damage but removes the possibility of frequent visual inspection for corrosion. Underground storage tanks have specific design and installation requirements, often found in local fire and environmental codes and practices.

Install underground tanks with firm foundations and a surrounding "jacket" of at least 6 in. (15 cm) of noncorrosive, inert materials. These can include well-tamped, clean sand, earth, or gravel. Do not use cinders or other acid-forming fills around the tank. Anchor the tank if there is any chance of it "floating" on rising groundwater.

NFPA 30 requires a separation between tanks storing Class I liquids and building foundations of at least 1 ft (0.3 m) and a separation between tanks and property lines of at least 3 ft (0.9 m). Separation space between tanks storing Class II and III liquids and buildings, supports, and/or property lines must be at least 1 ft (0.3 m).

NFPA 30 also specifies that corrosion protection for an underground tank and its piping should be provided by one or more of the following methods: (1) use of protective coatings or wrapping, (2) cathodic protection, or (3) corrosion-resistant construction materials. Select the type of protection according to the area's corrosion history and a qualified engineer's judgment. Some jurisdictions may require secondary protection and overfill safeguards for underground tanks. Under U.S. environmental regulations, these are mandatory measures for most underground tanks containing flammable and combustible liquids, excluding heating oil tanks. Note that local requirements are often more stringent than those required by the federal government. Some tank installations may require double-walled construction with sensing devices to detect leakage between walls. Before installing the tank, give written notification to the proper authorities (e.g., state and local) and get their approval. Keep records of tank installation approvals at least until the tank is removed or abandoned in place.

Regardless of the traffic pattern above an underground tank, protect it with covering specifically designed and installed to tolerate the traffic load. For light traffic loads, these coverings can include earth (2 ft [0.6 m] minimum) or earth (1 ft [0.3 m] minimum) plus reinforced concrete (4 in. [10 cm] minimum). For heavy traffic loads, these coverings can include earth (3 ft [0.9 m] minimum), tamped earth (18 in. [0.46 m] minimum) plus reinforced concrete (6 in. [15 cm] minimum), or tamped earth (18 in. [0.46 m] minimum) plus asphalt concrete (8 in. [20 cm] minimum). The concrete cover should extend at least 1 ft (0.3 m) beyond the outline of the tank.

Withdraw flammable liquids by pump. Arrange the pump and piping system so the liquid will flow back to the tank when the system is not in operation.

Aboveground Tanks

NFPA 30 sets minimum distances from property lines, public ways, and nearby buildings for aboveground tanks containing flammable or combustible liquids (Table 10–A). See NFPA 30 for distances for aboveground tanks containing unstable liquids. Some other recommended minimum distances are given in Table 10–B. Separate tanks storing Class I, II, or IIIA flammable or combustible liquids according to the distances shown in Table 10–C.

Although rarely the case, two tank properties owned by different parties might share a common boundary. If so, and if the owners and local authorities are in agreement, normal minimum shell-to-shell spacing can be used, such as that used within a tank farm.

Space tanks used only for storing Class IIIB liquids at least 3 ft (0.9 m) apart, unless they stand within a diked area or drainage path for a tank storing a Class I or Class II liquid. In such cases, the provisions of Table 10–C apply. Separate tanks of unstable liquids by at least one-half the sum of their diameters. Where end failure of a horizontal tank or vessel can expose property, place the tank with its longitudinal axis parallel to the nearest important structure's exposure.

Tank placement in a diked area can compromise spill control. NFPA 30 requires that tanks in a diked area containing Class I or Class II liquids have greater spacing. Similarly, tanks that stand in compacted rows or irregular patterns in a drainage path for tanks containing Class I or Class II liquids may require additional spacing. In addition, the local authority having jurisdiction might require extra tank spacing to facilitate fire response to tanks in the interior of the pattern.

NFPA codes have many specific requirements and exemptions, depending on the situation. Some examples include petroleum tanks in refineries and producing areas, unstable flammable or combustible liquids, liquefied petroleum gas, and tanks arranged in a compact or irregular pattern. Consult the NFPA and local fire codes most applicable to the situation.

Locate tanks to avoid danger from high water levels. Tanks beside a stream without tide should, where possible, stand downstream from combustible material.

Place truck-loading racks that dispense Class I liquids at least 25 ft (7.5 m) from tanks, warehouses, other plant buildings, and the nearest property line. Place truck-loading racks that dispense Class II or Class III liquids at least 15 ft (4.6 m) from tanks, warehouses, other plant buildings, and the nearest property line.

Piping materials for aboveground tanks should be made of steel, as recommended in NFPA 30. Piping can be made of materials other than steel (1) when underground or (2) when protected against fire exposure and located so leakage would not produce a hazard. Chemical properties of the materials in use may preclude the safe use of steel piping and require alternative materials. Avoid cast-iron pipe because it is brittle and can fracture under stress. When in doubt, consult the supplier, producer of the flammable or combustible liquid, or another competent authority about the suitability of the material to the application.

Piping from a storage tank should have an easy-to-reach shutoff valve at the tank. Make provisions to drain or pump the contents of the tank into another tank or to collect the liquid within dikes or retaining walls should the tank leak or be overfilled. Locate fill and vent pipes outside for any storage tanks inside buildings.

Small tanks often have an emergency self-closing valve, located inside the tank or at the entry point. Such valves automatically stop the flow of liquid in case of fire (see Factory Mutual System, Data Sheet 7–32, Ignitable Liquids Operations). Keep this valve closed except during loading and unloading operations. Be sure that the rope or wire and the fusible link attached to an emergency valve, normally open, are in good condition and tested regularly for easy operation.

Besides the normal vents that take care of vacuum and pressure during pumping operations, an aboveground storage tank must have some form of emergency relief venting. Such venting prevents buildup of excessive internal pressure in case a fire should surround the tank. In addition to a relief device, provide further protection either with a weak seam in the top or at the joint between the top and the shell of the tank or by some other recommended form. See NFPA 30 for guidance on sizing normal and emergency vents.

Mark tanks storing Class I liquids with the words FLAMMABLE—KEEP FIRE AWAY, in letters at least 2 in. (5 cm) high. Also, post NO SMOKING signs. Regulations

TABLE 10–A. Location of Outside Aboveground Storage Tanks from Adjoining Property or Public Way for Operating Pressures No Greater than 2.5 psig (17 kPa)

Type of tank	Protection	Minimum distance in feet from property line that may be built on, including the opposite side of a public way and not less than 5 ft (1.5 m)	Minimum distance in feet from nearest side of any public way or from nearest important building and shall be not less than 5 ft (1.5 m)
Floating roof	Protection for exposures	½ times diameter of tank	⅙ times diameter of tank
	None	Diameter of tank but need not exceed 175 ft (54 m)	⅙ times diameter of tank
Vertical with weak roof to shell seam	Approved foam or inerting system on the tank	½ times diameter of tank	⅙ times diameter of tank
	Protection for exposures	Diameter of tank	⅓ times diameter of tank
	None	2 times diameter of tank but need not exceed 350 ft (110 m)	⅓ times diameter of tank
Horizontal and vertical, with emergency relief venting to limit pressures to 2.5 psig	Approved inerting system on the tank or approved foam system on vertical tanks	½ times Table 10–B	½ times Table 10–B
	Protection for exposures	Table 10–B	Table 10–B
	None	2 times Table 10–B	Table 10–B
For Operating Pressures Greater than 2.5 psig (17 kPa)			
Any type	Protection for exposures	1 ½ times Table 10–B but shall not be less than 50 ft (15 m)	1 ½ times Table 10–B but shall not be less than 25 ft (7.5 m)
	None	3 times Table 10–B but shall not be less than 50 ft (15 m)	1 ½ times Table 10–B but shall not be less than 25 ft (7.5 m)
Floating roof	Protection for exposures	½ times diameter of tank	⅙ times diameter of tank
	None	Diameter of tank	¼ times diameter of tank
Fixed roof	Approved foam or inerting system	Diameter of tank	⅓ times diameter of tank
	Protection for exposures	2 times diameter of tank	⅔ times diameter of tank
	None	4 times diameter of tank but need not exceed 350 ft	⅔ times diameter of tank

From National Fire Protection Association No. 30, Flammable and Combustible Liquids Code.

TABLE 10–B. Reference Minimum Distances for Aboveground Outside Storage Tanks

	Floating Roof Tanks	Fixed Roof Tanks	
		Class I or II Liquids	Class IIIA Liquids
All tanks not over 150 ft (45 m) diameter	⅙ sum of adjacent tank diameters but not less than 3 ft (0.9 m)	⅙ sum of adjacent tank diameters but not less than 3 ft (0.9 m)	⅙ sum of adjacent tank diameters but not less than 3 ft (0.9 m)
Tanks larger than 150 ft (45 m) diameter			
Remote impounding area	⅙ sum of adjacent tank diameters	¼ sum of adjacent tank diameters	⅙ sum of adjacent tank diameters
Impounding (diking) around tanks	¼ sum of adjacent tank diameters	⅓ sum of adjacent tank diameters	¼ sum of adjacent tank diameters

From National Fire Protection Association No. 30.

TABLE 10–C. Minimum Tank Spacing (Shell-to-Shell)

Capacity tank (gallons) (1 gal = 3.78 liters)	Minimum distance in feet from property line which is or can be built upon, including the opposite side of a public way	Minimum distance in feet from nearest side of any public way or from nearest important building on the same property
275 or less	5	5
276 to 750	10	5
751 to 12,000	15	5
12,001 to 30,000	20	5
30,001 to 50,000	30	10
50,001 to 100,000	50	15
100,001 to 500,000	80	25
500,001 to 1,000,000	100	35
1,000,001 to 2,000,000	135	45
2,000,001 to 3,000,000	165	55
3,000,001 or more	175	60

From National Fire Protection Association No. 30.

or prudent practice may require installation of similar signs on tanks containing Class II and Class III liquids. Regulations usually require additional warning labels, such as for health hazards or special precautions. Some locations, like oil refineries, use different methods for providing fire safety warnings to employees and emergency responders.

Maintain good housekeeping around storage tanks. Do not permit debris to accumulate. Also, do not store combustible materials near tanks. Eliminate the grass around tanks (or under them, if they are off the ground), or at least keep it cut and remove the cuttings.

Make tanks used for storing flammable or combustible liquids vapor-tight to minimize evaporation losses and prevent fires. Most venting devices are normally closed and breathe automatically when liquid level in the tanks changes and as ambient temperature fluctuates. See API standard RP 2000, Venting Atmospheric and Low-Pressure Storage Tanks. Floating roofs with secondary seals are most effective in reducing evaporation losses and allowing expansion.

Equip large aboveground storage tanks with stairways and platforms, preferably made of steel. Having tanks with gauge tubes near the platform will reduce the need for personnel to walk on the roof of the tanks. Tanks that stand more than 1 ft (0.3 m) above ground should have foundations of noncombustible materials. However, wood cushions may be an acceptable alternative. For aboveground tanks, use supports made of concrete, masonry, or steel with fire protection from concrete or other approved fireproofing material. Such materials will help prevent supports from collapsing in case of a fire.

Overflowing of tanks presents a severe fire hazard because vapors might drift to a source of ignition, away from automatic vapor monitoring controls. Operators responsible for filling tanks should maintain a constant watch on the filling rate and the level of liquid. That way, they can stop operations or divert liquid to another tank upon reaching the required fill level. To prevent release of product, operators must remain at the open valve while water is draining from a tank.

It is common practice to apply aluminum, pastel, or white paint to flammable-liquid tanks with sun exposure. These colors reflect the heat and help reduce the internal vapor pressure. Some regulations require the use of such paint. In some highly volatile liquid storage installations, water sprays cool the system externally. Tanks that store viscous products, such as tar and heavy oils, may have insulation and heating coils to keep the product in a fluid state.

Spill Control

Release of flammable or combustible liquid from a tank car or storage container might have serious consequences because of topography or neighboring property. In these cases, install a curb, dike, or wall around a tank or group of tanks. Design such a structure to contain the capacity of the single largest container, including leakage from tanks, pipelines, and valves connected to these tanks. NFPA 30 gives construction details for capacity, drainage, dikes, and walls of aboveground storage tanks. Local environmental laws and regulations may have more stringent requirements.

Local conditions will determine the design of dikes. For example, if a tank is close to a building and the ground's slope could cause liquid to flow toward the building, it would be prudent to construct diversion walls to direct the escaping liquid to a safe location.

Consider installing a storm drain with a water-sealed catch basin for each dike enclosure. In larger plants, the drain often leads to an oil-water separator located far away from the main buildings. These devices allow plant operators to separate flammable and combustible liquids from process or rainwater. Equip drains with control valves and keep them closed, unless using them to drain water from the protected area. Be sure that the design of these systems prevents flammable liquids from entering natural water supplies, public sewers, or storm drains.

Tank Fires and Their Control

Prevention is the best way to reduce the risk of fires in storage tanks. One effective method is to provide tanks containing flammable liquids with a roof that floats upon the surface of the liquid—an internal or external open-top floating roof. Such a roof greatly reduces the headspace of the tank and thereby reduces the opportunity for vapors to accumulate within the flammable range.

It only takes a few seconds for a fire inside a fixed-roof tank either to extinguish itself or blow off the tank roof. Tank fires most commonly occur at one of the roof openings. Fire will occur if a source of ignition, such as lightning, comes in contact with expelled vapor that results from filling or heating. Operators can usually extinguish such fires without difficulty by removing the cause of the vapor's expulsion, as by cooling or by shutting off the filling operation. Seal fires frequently occur on floating-roof tanks. Operators can usually extinguish these fires using portable equipment or hand foam lines from the tank platform, wind girder, or roof.

If a fire occurs near a tank, apply water immediately to the exposed tank to cool it and to reduce vaporization. This action not only saves stock, but also makes ignition at the vents less likely. Because so many variables are involved in tank fires, ensure that only personnel with training in fighting these fires do so.

CLEANING TANKS

Cleaning tanks that contained flammable or combustible liquids is extremely hazardous work. Whenever personnel must access these tanks, ensure that they follow confined-space entry procedures, including appropriate training; take necessary safety precautions; wear protective equipment as needed; and provide for rescue.

Clean tanks and vessels that have contained flammable and combustible liquids before inspections, repairs, entry by personnel, or changes of product. Ensure that personnel who supervise the cleaning are competent to do so. These personnel should be familiar with fire and

incident prevention, as well as with the requirements for tank cleaning and the hazards of the products involved. API Standard 2015, Safe Entry and Cleaning of Petroleum Storage Tanks, Planning and Managing Tank Entry from Decommissioning through Recommissioning, and NFPA 326, Standard for the Safeguarding of Tanks and Containers for Entry, Cleaning, or Repair, provide recommendations for preventing fire, explosion, asphyxiation, and exposure to toxic materials.

Before beginning repair or cleaning operations, purge the tank of all flammable vapors using ventilation, displacement, or other effective means. Check the atmosphere inside the tank often to determine oxygen and flammable-liquid vapor levels. Recall that a chemical that tests well below the LFL can still be present in extremely hazardous concentrations.

Before ventilating, remove all sources of ignition from the surrounding area. Obtain appropriate confined-space entry and hot-work permits. Double-block and bleed, disconnect, or blank all piping; lock out all electrical equipment; and use only lighting approved for the specific atmospheric conditions in the tank.

Consider wind and weather conditions. Do not start work if wind might carry vapors into an area in which they could create a hazard or if an electrical storm is threatening or in progress. Prohibit employees from smoking and carrying matches and lighters.

Bond blast-cleaning equipment to the tank to prevent static sparks. Bond the nozzles on water or steam lines used to free tanks of vapor to the tank shells to prevent static accumulation. Avoid steaming large storage tanks to free gas. Do not use power-chipping tools and rivet busters when flammable vapors from tanks may be present in the area because these tools can create sparks. Keep motor trucks, gas engines, open flames, and portable electric equipment upwind at a safe distance from a tank undergoing cleaning.

Welding, cutting, burning, grinding, blast-cleaning, and other operations require a hot-work permit. Do not allow hot-work operations in a tank until the work area is clean, the tank's atmosphere is free from vapor and toxic hazards, and the area is clear of any flammable or combustible liquids or solids that could be ignited by cutting or welding operations. Where any vapor is present, continue to ventilate, displace, or reduce the flammable vapor levels through other means. If heavy scale is present, personnel should scrape it and probe it for flammable vapors while protected by confined-space entry procedures and equipment.

Unblanked lines or connections, breaks in the bottom of the tank, sludge, sediment, sidewall scale, and wood structures soaked with the liquid can allow flammable vapor levels to rise. Interior pontoons or center columns in floating-roof tanks can trap vapor-releasing liquid. Also, the seals of some floating-roof tanks and some internal floating-roof tanks may contain flammable or combustible materials. For this reason, hot work in these circumstances requires special safeguards; ensure that personnel periodically retest the atmosphere with a combustible-gas indicator during the work.

Abandonment of Tanks

Most authorities having jurisdiction require removal of tanks at the end of their useful life and generally will not accept abandoning tanks in place. Thoroughly clean obsolete tanks to remove flammable vapors. Dismantle and remove them from the premises according to national, state, and local underground storage tank regulations.

When tanks are temporarily out of service, cap and secure all fill lines, gauge openings, and pump suction lines. This will reduce tampering and accidental filling. Leave the vent lines open.

COMMON USES OF FLAMMABLE AND COMBUSTIBLE LIQUIDS

Flammable and combustible liquids have many uses in industry. When workers use these liquids, be sure they know and observe the necessary precautions.

Dip Tanks

By their nature, dip tanks have large exposed surfaces of flammable or combustible liquids. Because this increases the amount of flammable combustible vapor in a space, it increases risk of fire. NFPA 34, Standard for Dipping, Coating and Printing Processes Using Flammable or Combustible Liquids, and 29 CFR 1910.123–126 provide very specific fire prevention and safety expectations for these operations.

Where possible, conduct dipping operations above grade in a detached one-story building of noncombustible construction or in a separate one-story section. The room should be as large as possible to allow for vapor dilution and should be ventilated to reduce vapor accumulation. Moreover, the room should be free of sources of ignition and conspicuously marked as a flammable-liquid area (Figure 10–14).

Handling flammable or combustible liquids in open containers increases the rate of vapor accumulation. Keep the openings of such containers as small as possible, and use a cover where possible. Covers should be hinged-and-gravity closing or should slide on tracks and be held open by a fusible link or another heat-activated device (Figure 10–14).

NFPA 34 and NFPA 91, Standard for Exhaust Systems for Air Conveying of Vapors, Gases, Mists, and Noncombustible Particulate Solids, provide specific requirements for ventilation of dip tanks. Generally, these fire safety codes specify mechanical ventilation that directs fresh air over the vapor area, past the point of operation, and then to a safe outside location. Moreover, fire codes often require an interlock that shuts down dipping operations in the event of a ventilation system failure.

Tanks with capacities of more than 500 gal (1,892 L) should have bottom drains unless the viscosity of the liquids they contain makes this requirement impractical. Install overflow pipes on tanks with capacities of more than 150 gal (568 L) to carry off any overflow liquids to an approved holding tank. OSHA regulations and fire codes require some method of controlling fire, whether through automatic fire protection, self-closing covers with fusible links, or other means.

Japanning and Drying Ovens

Japanning is an enamel process that requires evaporating flammable and combustible liquids from varnishes. Ovens used for evaporating varnish, Japan enamel, and other flammable and combustible liquids can present serious fire and explosion hazards. Provide ample ventilation and explosion venting for these ovens.

Drying ovens come in two types: the box oven and the continuous-conveyor oven. The box oven remains closed dur-

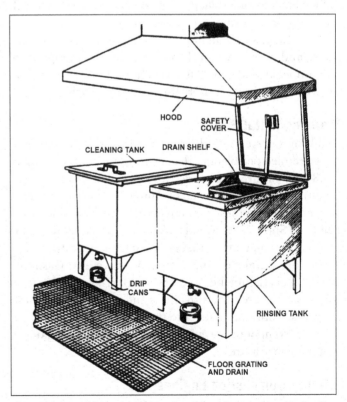

Figure 10–14. Although the overhead hood with forced ventilation is not mandatory in all cleaning tanks, it is generally good practice.

ing operation and is common for small-scale or batch applications. The continuous-conveyor oven, often open at both ends, supports process or large-scale production processes.

Provide ovens with the proper type of fire-extinguishing equipment and interlocks to shut down the process in case of fire. Consult NFPA 86, Standard for Ovens and Furnaces, for details about constructing and operating drying ovens. Also consult the local fire authority having jurisdiction.

Oil Burners

Oil and diesel can be fuel in heating applications. These burners may require preheating and/or atomizing of fuel to increase their ability to ignite easily. The primary hazard of oil burners is the possibility of a discharge of unburned oil into a hot firebox, where it can vaporize and form an explosive mixture. Provide approved automatic safeguards to control this hazard.

Use oil burners with listing or approval from a recognized testing laboratory. To prevent faulty ignition or accumulation of soot, with its attendant fire hazard, use the correct type of fuel oil, as recommended by the testing laboratory or manufacturer. Do not use a gravity feed to burners without safeguards that prevent abnormal discharges of oil at the burner. Fuel oil should not have a flash point lower than 100°F (37.8°C). It should be hydrocarbon oil free from acid, grit, and foreign matter likely to clog or damage the burners or valves. Some plants use acid sludge or waste oil for fuel. Such fuel requires special burning equipment and procedures.

Preferably, locate an industrial fuel oil supply tank outdoors and aboveground. Protect the tank and piping to contain leaks and spills, as required by local authorities. Provide all tanks with overfill-protection systems.

Cleaning Metal Parts

Cleaning grease and oil from metal parts is a common use of refined petroleum solvents (flash point above 100°F [37.8°C]). Ensure that work areas have sufficient ventilation to prevent accumulation of flammable and combustible vapors, and keep these areas free of sources of ignition (see FM Data Sheet 7–79, Metal Cleaning). Do not allow oil or grease to accumulate in these cleaning compounds.

Choose metal-cleaning compounds carefully. Gasoline is volatile, is flammable, and has health risks. Chlorinated solvents have chronic health effects and pose environmental concerns. Many alternatives to these cleaners are available. Alkaline compounds, available under several trade names, will not cause a fire.

Internal-Combustion Engines

Practice good housekeeping to prevent the accumulation of rubbish, oil or fuel, and rags around industrial internal-combustion engines. Provide proper receptacles for refuse.

See NFPA 37, Standard for the Installation and Use of Stationary Combustion Engines and Gas Turbines, for more specific practices.

Before filling a gasoline tank, shut down the engine and permit hot exhaust pipes to cool. Contact with hot metal will volatilize flammable liquids and may result in sufficient vapor concentration to ignite at or above the auto-ignition temperature. Use approved safety cans or a hand pump with a bonded filling hose, and keep the main fuel supply in approved containers outdoors.

When engines operate continuously, they cannot shut down for filling. In these cases, locate fuel storage tanks outside the engine room, away from sources of heat and spark. This minimizes exposing fuel vapors to hot engines and/or exhaust. Refuel lift trucks and other mobile equipment outside buildings or in areas with sufficient ventilation, overfill protection, and spill control.

Spray Booths

Conduct paint-spraying operations in detached buildings or away from other operations, when possible. Use approved spray rooms or booths with adequate ventilation for production areas. Provide an enclosed area of sufficient volume to prevent explosive mixtures of vapor and air. Eliminate heating units, air filters, and piping that might become coated with flammable materials, or protect them against such accumulations. See NFPA 33, Standard for Spray Application Using Flammable or Combustible Materials, for more specific requirements and practices.

Fires in spray booths and spray-booth operations most frequently result from spontaneous ignition of spray deposits. Prevent these fires by establishing a regular cleaning schedule. The frequency of cleanings depends on how quickly deposits accumulate. Water-wash booths, which trap excess spray before it enters exhaust ducts, are more fire safe than the dry type of booth.

Install only electrical devices rated for hazardous locations (UL Class I, Division 2 for flammable vapor locations).

Provide large spraying operations with automatic fire controls. Automatic sprinklers or inert-agent (including carbon dioxide) systems are most effective. Provide protection for the exhaust ducts, as well as for the spray booths. Protect the discharge heads of such equipment from overspray in a manner consistent with the manufacturer's instructions and the fire protection equipment's listing. Covering the heads with ordinary thin paper or plastic bags (0.003 in. [0.076 mm]) is a common solution. Ensure that any head protection method still allows the head to function as designed; check with the manufacturer and/or installer before beginning spray operations. Check head protection often; overspray accumulation may increase the time required to clear the fire protection heads.

Electrostatic spraying, usually automatic, introduces a possible source of ignition in the arcing of parts to the electrodes. Hold parts in tight-fitting fixtures to prevent them from coming close enough to the sprayer to induce a spark.

Liquefied Petroleum Gases

Liquefied petroleum gases (LPGs) include any material that is composed predominantly of any of the following hydrocarbons or mixtures of them: propane, propylene, butane (n-butane or iso-butane), and butylenes. These gases liquefy under moderate pressure but convert to a gaseous state upon relief of the pressure.

LPG vapor presents a similar hazard to that of any flammable gas. Because LPG vapor is heavier than air, it collects in low spots and requires specific ventilation designs.

Liquefied petroleum gases are used as fuel gases, as raw materials in chemical processes (such as the making of hydrogen), and to form special atmospheres in heat-treating furnaces. Train employees to understand the properties of these gases and in the safe practices for the handling, distributing, and operating processes that use them. Develop detailed safety programs to handle any emergencies, including release of gas inside and outside the production area.

Use only experienced and reliable manufacturers and installers who are thoroughly familiar with the hazards of LPG storage systems. Manufacturers should observe state or provincial and local codes and the recommendations of fire prevention organizations and insurance companies. See NFPA, the Factory Mutual System Research Organization, and the National Propane Gas Association listed in the References at the end of this chapter.

SUMMARY

- A flammable liquid is any liquid having a closed-cup flash point below 100°F (37.8°C) and a vapor pressure not exceeding 40 psia (276 kPa) at 100°F. These liquids are categorized by NFPA as Class IA through IC, depending on their flash points.
- Combustible liquids are those with flash points at or above 100°F (37.8°C) but below 200°F (93.3°C); are closed cup; and are divided into Classes II, III, IIIA, and IIIB. They do not ignite as easily as flammable liquids but must be handled with the same precautions.
- The degree of danger in these liquids is determined largely by (1) flash point, (2) concentration of vapors in the air, (3) risk of ignition at or above the flash point, and (4) amount of vapors present.
- Workers should always avoid exposing large surface areas of flammable or combustible liquids to the air because doing so creates a serious fire or explosion risk.

Other general safety measures include avoiding mixing flammable and combustible liquids, prohibiting smoking around or near these liquids, shielding the liquids from static electricity by bonding and grounding, and using only spark-resistant tools when working with or around these liquids.

- Health hazards associated with flammable and combustible liquids include skin irritation, intoxication or illness from inhaling their vapors and fumes, and oxygen deficiency in closed containers used to store these liquids. Unless tests prove otherwise, workers should assume that flammable vapors and toxic mixtures are present in all containers of these liquids.
- Loading and unloading of vessels used to transport flammable or combustible liquids should be done only by trained employees. When two or more organizations are involved in loading and unloading operations, they should plan these operations in advance, to ensure that roles and responsibilities are clearly communicated.
- Only Class I and Class II liquids can be safely stored in buildings used for public assembly. Where needed, a dike can be constructed around a tank or group of tanks for added protection. Transfer pumps should be located outside buildings and diked areas where possible to minimize hazards.
- Underground tanks should be protected against overhead traffic, built on a firm foundation, and shielded against corrosion. Aboveground tanks must be constructed an approved distance away from property lines, public ways, and nearby buildings. All tanks should be equipped with proper fire-extinguishing systems.
- Inside storage and mixing rooms should be isolated and protected as much as possible to guard against the hazard of fire from without. Outside storage lockers should be built away from the main plant whenever possible and must conform to regulations.
- Cleaning tanks that contained flammable and combustible liquids is extremely hazardous work and requires highly skilled and trained employees. Workers must wear protective equipment and understand the proper work and medical procedures to follow.
- Small tanks and containers can be cleaned by steam and made temporarily safe by means of an inert gas. Tanks to be abandoned must be thoroughly cleaned and then dismantled and removed from the premises.
- Proper disposal of flammable and combustible liquids is an important part of handling these materials safely. If recycling or recovery of these liquids is impossible, they should be burned in an approved incinerator or given to a disposal contractor.
- Because flammable and combustible liquids have many uses in industry, workers must know how to guard

against fire hazards, prevent unintentional incidents, and protect their health when using flammable and combustible liquids.

REFERENCES

American Conference of Governmental Industrial Hygienists, Bldg. D–7, Glenway Avenue, Cincinnati, OH 45211. *Threshold Limit Values*. Published annually.

American National Standards Institute, 11 West 42nd Street, New York, NY 10036.
ANSI/AIHA Z9.2–2006, Fundamentals Governing the Design and Operation of Local Exhaust Ventilation Systems.
ANSI Z49.1–2012, Safety in Welding, Cutting, and Allied Processes.
ANSI Z117.1–2009, Safety Requirements for Confined Spaces.

American Petroleum Institute, 1220 L Street NW, Washington DC 20005.
RP 652, Lining of Aboveground Petroleum Storage Tank Bottoms. 3rd ed. 2005.
RP 2000, Venting Atmospheric and Low-Pressure Storage Tanks. 7th ed. 2014.
RP 2003, Protection Against Ignitions Arising Out of Static, Lightning, and Stray Currents. 2008.
RP 2214, Spark Ignition Properties of Hand Tools. 4th ed. 2004.
Standard 640, Welded Steel Tanks for Oil Storage. 6th ed. 2013.
Standard 2015, Safe Entry and Cleaning of Petroleum Storage Tanks, Planning and Managing Tank Entry from Decommissioning through Recommissioning. 2006.

American Society for Testing and Materials, 1916 Race Street, Philadelphia, PA 19103.
Annual Book of ASTM Standards:
Section 5, "Petroleum Products, Lubricants, and Fossil Fuels."
Section 6, "Paints, Related Coatings, and Aromatics."
Section 15, "General Products, Chemical Specialties, and End Use Products."
ASTM D56-05–2010, Standard Test Method for Flash Point by Tag Closed Cup Tester.
ASTM D93-13e1, Standard Test Methods for Flash Point by Pensky-Martens Closed Cup Tester.
ASTM D323–2008, Standard Test Method for Vapor Pressure of Petroleum Products (Reid Method).

American Welding Society, PO Box 351040, 550 LeJeune Road NW, Miami, FL 33135.

AWS A6.0, Safe Practices for Welding and Cutting Containers That Have Held Combustibles. 1965.
Fire and Explosion Prevention, Fact Sheet 6. 2012.

Association of American Railroads (AAR), 425 3rd Street SW, Washington DC 20024. *Manual of Standards and Recommended Practices*. 2009.

Factory Mutual System Research Organization, 1151 Boston-Providence Turnpike, Norwood, MA 02062.
"Factory Mutual Loss Control Data Books." (Listing available).
Factory Mutual System Approval Guide.
Ignitable Liquids Operations, Data Sheet 7–32.
Metal Cleaning, Data Sheet 7–79.

Fawcett, H. H., and W. S. Wood, eds. *Safety and Accident Prevention in Chemical Operations*. 2nd ed. New York: Wiley-Interscience, 1982.

National Fire Protection Association, 1 Batterymarch Park, Quincy, MA 02269.
Fire Protection Handbook.
NFPA 1, Fire Code.
NFPA 10, Standard for Portable Fire Extinguishers.
NFPA 13, Standard for the Installation of Sprinkler Systems.
NFPA 30, Flammable and Combustible Liquids Code.
NFPA 30A, Automotive and Marine Station Code.
NFPA 30B, Code for the Manufacture and Storage of Aerosol Products.
NFPA 31, Standard for the Installation of Oil-Burning Equipment.
NFPA 33, Standard for Spray Application Using Flammable or Combustible Materials.
NFPA 34, Standard for Dipping, Coating and Printing Processes Using Flammable or Combustible Liquids.
NFPA 35, Standard for the Manufacture of Organic Coatings.
NFPA 36, Standard for Solvent Extraction Plants.
NFPA 37, Standard for the Installation and Use of Stationary Combustion Engines and Gas Turbines.
NFPA 45, Standard on Fire Protection for Laboratories Using Chemicals.
NFPA 51B, Standard for Fire Prevention During Welding, Cutting, and Other Hot Work.
NFPA 54/ANSI Z223.1–2012, National Fuel Gas Code.
NFPA 56, Standard for Fire and Explosion Prevention During Cleaning and Purging of Flammable Gas Piping Systems.
NFPA 58, Liquefied Petroleum Gas Code.
NFPA 59, Utility LP-Gas Plant Code.
NFPA 70, National Electrical Code.
NFPA 77, Recommended Practice on Static Electricity.

NFPA 80, Standard for Fire Doors and Other Opening Protectives.

NFPA 85, Boiler and Combustion Systems Hazards Code.

NFPA 86, Standard for Ovens and Furnaces.

NFPA 91, Standard for Exhaust Systems for Air Conveying of Vapors, Gases, Mists, and Noncombustible Particulate Solids.

NFPA 170, Standard for Fire Safety and Emergency Symbols.

NFPA 251, Standard Methods of Tests of Fire Resistance of Building Construction and Materials.

NFPA 301, Code for Safety to Life from Fire on Merchant Vessels.

NFPA 307, Standard for the Construction and Fire Protection of Marine Terminals, Piers, and Wharves.

NFPA 321, Classification of Flammable and Combustible Liquids.

NFPA 326, Standard for the Safeguarding of Tanks and Containers for Entry, Cleaning, or Repair.

NFPA 329, Recommended Practice for Handling Releases of Flammable and Combustible Liquids and Gases.

NFPA 385, Standard for Tank Vehicles for Flammable and Combustible Liquids.

NFPA 400, Hazardous Materials Code.

NFPA 407, Standard for Aircraft Fuel Servicing.

NFPA 472, Standard for Competence of Responders to Hazardous Materials/Weapons of Mass Destruction Incidents.

NFPA 484, Standard for Combustible Metals.

NFPA 497, Recommended Practice for the Classification of Flammable Liquids, Gases, or Vapors and of Hazardous (Classified) Locations for Electrical Installations in Chemical Process Areas.

NFPA 704, Standard System for the Identification of the Hazards of Materials for Emergency Response.

NFPA 1620, Standard for Pre-Incident Planning.

NFPA 2112, Standard on Flame-Resistant Garments for Protection of Industrial Personnel against Flash Fire.

NFPA 2113: Standard on Selection, Care, Use, and Maintenance of Flame-Resistant Garments for Protection of Industrial Personnel against Flash Fire.

National Institute of Occupational Safety and Health, c/o Government Printing Office, Washington DC 20402-9325.

NIOSH Pocket Guide to Chemical Hazards. 2005 (GPO stock number 017-033-00500-1).

NIOSH Pocket Guide to Chemical Hazards. 2010 ed. Downloadable version: cdc.gov/niosh/npg/.

National Propane Gas Association of America, 1600 Eisenhower Lane, Lisle, IL 60532.

National Safety Council, 1121 Spring Lake Drive, Itasca, IL 60143.

Chemical Hazards Fact Finder. 1992 (Out of print).

Fundamentals of Industrial Hygiene. 6th ed. 2012.

Underwriters Laboratories Inc., 333 Pfingsten Road, Northbrook, IL 60062.

U.S. Department of Labor, Occupational Safety and Health Administration, 200 Constitution Avenue NW, Washington DC 20210.

Code of Federal Regulations, Title 29.

"General Industry Standards," Part 1910.

Subpart H, "Hazardous Materials," Sections 106, 107, 120, 123–126.

Subpart J, "General Environmental Controls," Sections 146, 147.

Subpart L, "Fire Protection," Sections 155–165.

Subpart Z, "Toxic and Hazardous Substances," Section 1200.

U.S. Department of Transportation, 400 7th Street SW, Washington DC 20590.

Code of Federal Regulations, Title 49.

Hazardous Materials Regulations, Parts 170–179.

Motor Carrier Safety Regulations, Parts 393 and 397.

REVIEW QUESTIONS

1. What is a flammable liquid, as defined by NFPA 30, Flammable and Combustible Liquids Code?

2. When flammable and combustible liquids vaporize, forming flammable mixtures with air, the degree of danger depends on what four factors?

   a.

   b.

   c.

   d.

3. What is auto-ignition temperature?

4. Define *flash point*.

5. How is static electricity generated?

6. What is the difference between bonding and grounding?

7. Petroleum liquids can build up a static charge when they

   a. flow through piping or filters.

   b. are agitated in a tank or a container.

   c. are subjected to vigorous mechanical movement such as spraying or splashing.

   d. all of the above

10 Flammable and Combustible Liquids

8. Which of the following would put a combustible-gas indicator out of service if drawn into the instrument?
   a. steam
   b. oxygen
   c. water
   d. flammable liquid
   e. all of the above
   f. only a, c, and d

9. When a tank truck containing flammable liquids is being loaded or unloaded, what are the steps to take to ensure safety?
   a.
   b.
   c.
   d.

10. Which government agencies should operators of flammable and combustible liquid facilities contact to confirm requirements pertaining to their operations?
    a.
    b.
    c.
    d.

11. What type of extinguishing agents should be used on a flammable-liquid fire?

12. What are the three ways to protect an underground tank when it is buried under a heavily traveled roadway?
    a.
    b.
    c.

13. Why is it common practice to paint flammable-liquid tanks with aluminum, pastel, or white paint?

14. Even after a tank has been freed of vapor through proper cleaning, combustible mixtures may be formed again through admission of flammable vapors or liquids from what sources?
    a.
    b.
    c.

PART

3

Materials Handling

Part 3 addresses incident prevention in the various forms of materials handling. Unintentional releases of products or chemicals or falls in the workplace can have serious health consequences, and materials-handling incidents are some of the most common in the workplace. The control and management of forces generated by mechanically powered equipment; hoisting and conveying equipment; and ropes, chains, and slings can have significant impacts on overall workplace safety records. In addition, Chapter 11, Nanomaterials in the Workplace, addresses the growing debate concerning the adequacy of existing laws and regulations to deal with the emerging field of nanomaterials, which are being brought into the production facility. This chapter suggests that safety professionals pay close attention to regulatory agencies such as NIOSH, OSHA, and the EPA to ensure that requirements for worker exposure to these materials are addressed.

Nanomaterials in the Workplace

11

James T. O'Reilly, JD

Understanding Nanomaterials

Clearing Nanomaterials for Use at the Worksite

Preparing for Workplace Exposure

Monitoring Results of Nanomaterials Use

Engineering Controls and Personal Protective
Equipment

Explosion and Fire Risks

Regulatory Issues

Summary

References

Review Questions

UNDERSTANDING NANOMATERIALS

Nanomaterials are an emerging workplace risk. The creation, handling, and use of nano-scale materials represent rapidly growing workplace health concerns. *Nanoscience* is the creation, manipulation, and study of materials at the nano (one-billionth of a meter) scale, which is sometimes called the "near-atomic" scale. Scientists have known for decades that nano-scale materials such as carbon and titanium dioxide have properties of strength, electrical conductivity, and chemical reactivity that are distinct from their properties at their normal or "bulk" scale of traditional use.

Nanotechnology is the collection of technologies that deal with materials science on the nanometer scale; the creation of nanomaterials is one segment of nanotechnology. Defining the category can be attempted by referring to the National Nanotechnology Initiative's guidance documents. Nanotechnology includes materials in the length scale of 1 to 100 nanometers in any one direction. Materials of this scale produce unique phenomena, have novel properties, and pose difficulties of controlling and manipulating these particles. The subcategories of nanomaterials include those that are engineered for particular purposes, those that arise naturally, and those that are incidentally present in or with a conventional fiber or other material. The special concerns about inhalation into the human lung of very small "nano-aerosol" particles merit special attention from government, industry, and trade association technical centers that support workplace safety initiatives.

For the layperson, the primary difficulty in understanding nanomaterials arises from the inability to see these particles even with a microscope. A bacterium may be 1,000 nanometers, and a virus may be 100 nanometers; but some nanomaterials are even smaller than these more familiar particles. A representative example of a nanomaterial is carbon, which has physical forms such as coal or diamonds as well as nanoparticles that can be formed into "carbon nanotubes."

The areas of occupational health and safety concern largely focus on free, unbound nanoparticles and nanofiber materials. Although health hazard research is at a very early stage, there are concerns about passage of nanomaterials through the skin, especially abraded or compromised areas of skin; passage through the lungs into the blood; and passage from the blood into the brain and other organs of the body. In the absence of comprehensive hazard data, this chapter addresses emerging nanotechnology issues and makes general recommendations for protective measures in the workplace.

This chapter acknowledges at the outset that although there is much we do not know about nanomaterials, the number of industrial applications of nanomaterials is increasing so rapidly that the commercial need to adopt nano-technology may lead to safety concerns within a particular organization. Government has had a role in nanotechnology and workplace safety for many decades; for example, government agencies have worked with nanoparticles such as "ultrafine" automotive air emission particles, and these particles' effects on asthma and lung disease have been the basis of regulatory controls. But the role of government agencies is still evolving as well.

Generally, in a workplace where employees may be exposed to materials whose potential for health consequences is not well understood, well-established and prudent principles of occupational safety and health suggest minimizing exposure through the use of controls. These controls, in traditional order of preference, include engineering controls, administrative controls, and personal protective equipment. Because the amount of environmental health literature on handling nanomaterials continues to increase, the reader is encouraged to seek out the latest peer-reviewed journal research on using nanomaterials.

CLEARING NANOMATERIALS FOR USE AT THE WORKSITE

Nano-scale materials pose occupational safety issues that merit close attention prior to introducing the materials to a facility's production floor. Before allowing the preliminary use of these materials in the workplace, employers and worker safety committees should examine the toxicity and dose data available from the vendor and from independent sources. Several potential steps in this examination include:

- investigating and determining the physical and chemical properties (e.g., size, shape, solubility) that influence the potential toxicity of the nanoparticles
- evaluating the short- and long-term effects that nanomaterials may have on organ systems and tissues (e.g., lungs)
- determining the biological mechanisms of potentially toxic effects
- creating and integrating models to assist in assessing possible hazards
- determining whether a measure other than mass is more appropriate for ascertaining toxicity. (Different methods of making this determination have been debated within the toxicology field, and no clear answers have arisen. A metric other than mass may need to be factored into the reasoning to make safe decisions.)

Methods of measuring nanomaterials require special consideration so that the effects of the materials can be assessed accurately. After evaluating the measured results, modifications of the protective measures may be needed, including:

- evaluating methods of measuring the mass of respirable particles in the air and determining whether this measurement can be used to measure nanomaterials

- developing and field-testing practical methods to accurately measure airborne nanomaterials in the workplace
- developing testing and evaluating systems to compare and validate sampling instruments. (Conducting this comparison is beyond what the plant-level occupational safety professional can do, but advancing industry standards and technology should facilitate the evolution of testing systems.)

PREPARING FOR WORKPLACE EXPOSURE

Exposure assessment is a challenging first step in preparing the worksite and the workers for exposure to nanomaterials. Amounts of exposure at the machine and plant-floor level, among cleaning crews, in finished product testing, and in packaging and spray applications should be considered. Exposure information correlates with information about the probable size of the particles. Size is a key consideration; the ability to absorb potential toxins depending on surface area is another consideration. Several preparation steps include:

- determining key factors that influence the production, dispersion, accumulation, and reentry of nanomaterials into the workplace
- assessing possible exposure when nanomaterials are inhaled or settle on the skin
- determining how possible exposures differ by work process
- determining what happens to nanomaterials once they enter the body.

Five criteria for dealing with nanomaterials are recommended by the U.S. National Institute for Occupational Safety and Health (NIOSH). NIOSH advises companies to "anticipate, identify and track potentially hazardous nanomaterials in their workplace; assess the exposures of workers to the nanomaterials; assess risks and communicate hazards to their workers; manage the risks of occupational exposures," and "foster the safe development of nanotechnology and realization of its societal and commercial benefits" (Schulte 2014). To generate accurate risk assessment and regulatory evaluations, the emerging science of nanotechnology needs to "develop a means to improve the correlation of in vitro data to in vivo predictions, via enhanced cell models, relevant dosages (low vs high) and realistic exposure scenarios" (Gordon 2014).

Because the science of detection and measurement is making such great strides, the reader is encouraged to study published reports on current science trends in this field and to use reputable laboratory assistance to make accurate measurements at the worksite. Because ample sources of misinformation about nanomaterial "threats" can be found on various Internet sites, the prudent safety manager refers only to peer-reviewed scientific information.

Risk assessment is key to a reasonable and defensible justification for exposing workers to nanomaterials. Several risk assessment steps that should be taken include:

- determining the likelihood that current exposure-response data (human or animal) can be used in identifying and assessing potential occupational hazards
- developing a framework to evaluate potential hazards and predict potential occupational risks from exposure to nanomaterials.

MONITORING RESULTS OF NANOMATERIALS USE

After nanomaterials are put into use in the workplace, epidemiology and surveillance can be used to measure the consequences of this nanomaterials use. These monitoring steps include:

- evaluating existing epidemiological studies of workplaces where nanomaterials are used
- identifying knowledge gaps in which epidemiological studies could promote the understanding of nanomaterials and evaluating the likelihood of conducting such studies
- integrating nanotechnology health and safety issues into existing hazard surveillance methods and determining whether additional screening methods are needed
- using existing studies to share data and information about nanotechnology.

Although nanomaterials such as carbon nanotubes have wide-ranging uses, their potential toxicities include "pulmonary and systemic inflammation, fibrosis, immunosuppression, and cardiovascular dysfunction, and evidence is growing that [carbon nanotubes] may have properties that influence carcinogenicity" (Erdely 2014). Because little is known about nanomaterial exposure levels in the workplace, in contrast to the volumes of information available about other exposure situations, in vivo toxicology investigations are recommended in some studies (Erdely 2014). Inhalable breathing zone exposure figures have been calculated from some research projects that involved multiple facility exposure measurements (Dahm 2014). Some studies have recommended the use of direct online particle counters to map and identify critical exposure points inside a facility (Khaliullin 2014).

ENGINEERING CONTROLS AND PERSONAL PROTECTIVE EQUIPMENT

- Before, during, and after the introduction of nanomaterials, employers have both a legal duty and a pragmatic obligation to provide protection that will prevent avoidable harm to their workers. Just as a worksite should be regulated to reduce the possible occurrences of physical hazards like a falling scaffold, so should the employer provide reasonable means—such as a clean room, a glove box, respiratory protection, face shields, clothing, and so forth—to minimize air and skin exposures to the small set of nanomaterials that are deemed likely to cause harmful effects. Most plant-level professionals need higher-level corporate or trade association technical support that includes:

- evaluating the effectiveness of engineering controls in reducing occupational exposures to nanoparticles and nanoaerosols and developing new controls where needed
- evaluating and improving existing personal protective equipment
- developing recommendations to prevent or limit occupational exposures (e.g., respirator fit testing)
- evaluating the suitability of control banding techniques, the need for additional information, and the effectiveness of alternative materials.

In the United States, NIOSH Current Intelligence Bulletins can be used to gauge workplace exposure to carbon nanotubes and nanofibers (NIOSH 2014–12), and a representative sample of safety guidelines, "Safe Technology in the Workplace," is available online (NIOSH 2008–112, 2008).

EXPLOSION AND FIRE RISKS

Conventional-sized "bulk" chemicals have certain physical properties that are known or expected, and precautions are regularly taken to avoid fire or explosion hazards. But at the nanoscale, there may be different and perhaps greater likelihoods of reactions that lead to fire or explosion hazards. Although no overarching conclusion can be made that a nanomaterial is inherently be more hazardous than other types of chemicals, the reasonable steps that should be taken when dealing with nanomaterials include:

- identifying the physical and chemical properties that contribute to the dustiness, combustibility, flammability, and conductivity of nanomaterials
- recommending alternative work practices to eliminate or reduce workplace exposures to nanoparticles
- developing a disposal protocol that reduces the potential for harm to workers in the waste-handling system.

REGULATORY ISSUES

Substantial debate continues about the adequacies of existing laws and regulations to deal with nanomaterials issues. The reader should thus pay very close attention to the websites of government agencies such as the National Institute for Occupational Safety and Health (NIOSH), the Occupational Safety and Health Administration (OSHA), and the U.S. Environmental Protection Agency (EPA) in order to evaluate whether current regulatory requirements for worker exposure to nanomaterials can be satisfied before these materials are brought into the production facility. In December 2013, NIOSH announced an important study of the effects of nanotechnology ("Protecting the Nanotechnology Workforce: NIOSH Nanotechnology Research and Guidance Strategic Plan 2013–2016" [NIOSH 2014–106]). Nanotechnology is a rapidly evolving area of regulatory agency initiatives.

SUMMARY

- The small particle sizes of certain workplace materials will shrink even further with the evolving science of nanomaterials.
- The workplace safety professional needs to investigate, before nano-scale materials enter the workplace, whether the workplace is equipped to safely handle these small particles and whether the protection mechanisms, from entry to process to disposal, will adequately ensure the safety of workers.

REFERENCES

U.S. Department of Health and Human Services, National Institute for Occupational Safety and Health (NIOSH), 4676 Columbia Parkway, Cincinnati, OH 45226. cdc.gov/niosh

U.S. Department of Labor, Occupational Safety and Health Administration (OSHA), 200 Constitution Avenue NW, Washington DC 20210. osha.gov.

U.S. Environmental Protection Agency, 1200 Pennsylvania Avenue, NW, Washington DC 20460. epa.gov.

Cited articles are found in *Toxicologist Journal* 138, no. 4 (2014).

Cited NIOSH materials are available online at cdc.gov/niosh.

REVIEW QUESTIONS

1. What definitions are usually applied to nanoparticles?
2. Do nanoparticles have strength characteristics different from those of conventional-sized materials?
3. How does explosivity relate to nanoaerosol particle size?

Materials Handling and Storage

12

Philip E. Hagan, JD, MBA, MPH, ARM, CIH, CHMM, CET, CHCM, CEM
Salvatore Caccavale, CHMM, CPEA

Preventing Common Injuries
Personal Protection ▶ Materials-Handling Injuries
▶ Manual Lifting ▶ Back Belts

Fall Protection

Guidelines for Lifting
Guideline Limits ▶ Rules for Lifting ▶ Personnel
Selection for Materials Handling ▶ Team Lifting and
Carrying ▶ Handling Specific Shapes ▶ Machines and
Other Heavy Objects

Accessories for Manual Handling
Hand Tools ▶ Jacks ▶ Hand Trucks

Storage of Specific Materials
Planning Materials Storage ▶ Rigid Containers
▶ Uncrated Stock

Hazardous Materials
Containers for Liquids ▶ Containers for Solids
▶ Containers for Gases

Storage and Handling of Cryogenic Liquids
Characteristics of Cryogenic Liquids ▶ General
Safety Practices ▶ Special Precautions ▶ Inert
Gas Precautions ▶ Flammable Gas Precautions
▶ Asphyxiation ▶ Training ▶ Good Housekeeping
Practices ▶ Safe Storage and Handling
Recommendations ▶ Transfer Lines

Shipping and Receiving
Floors, Ramps, and Aisles ▶ Lighting ▶ Stock Picking
▶ Dock Boards ▶ Machines and Tools ▶ Steel and
Plastic Strapping ▶ Burlap and Sacking ▶ Glass and
Nails ▶ Pitch and Glue ▶ Barrels, Kegs, and Drums
▶ Boxes and Cartons ▶ Loading Railcars

Summary

References

Review Questions

353

Materials-handling operations are conducted in every department, warehouse, office, or facility of a company. Materials handling is a job that almost every worker in industry performs—either as a sole duty or as part of the regular work, either by hand or with mechanical help. On an average, industry moves about 45 tons (50 tonne) of material for each ton of product produced.

The work environment, the need for specific training, and proper materials-handling engineering should be examined when reviewing a facility's materials-handling needs. The specific amount and extent of manual lifting involved in a particular job should also be identified before selecting and placing employees.

Mechanized materials-handling equipment is commonly used (Figure 12–1a, b) in many industries, due to the fact that material could not be processed at low cost without efficient mechanical handling. Although mechanical handling creates a new set of hazards, the net result (entirely aside from increased efficiency) is fewer injuries, lower workers' compensation expenses, and a more productive workplace. (See also Chapter 13, Hoisting and Conveying Equipment; Chapter 14, Ropes, Chains, and Slings; Chapter 15, Powered Industrial Trucks; and Chapter 22, Automated Lines, Systems, or Processes, all in this volume, for hazards and safety techniques involved in the manual and mechanical handling of materials.) This chapter covers the following topics:

- safety measures for preventing common workplace injuries
- safety guidelines for lifting and materials handling
- types of accessories used for safe handling practices
- storage of hazardous and nonhazardous materials
- storage and handling of liquids
- safety practices for inert and flammable gas
- safety precautions in the shipping and receiving department.

PREVENTING COMMON INJURIES

Handling of material accounts for 20% to 45% of all occupational injuries. These injuries could occur in every part of an operation, not just the stockroom or warehouse. Strains, sprains, fractures, and contusions are the most common injuries.

Personal Protection

Overhead hoisting equipment operators and maintenance personnel can be exposed to falls. Fall protection and safe egress should be provided where workers will foreseeably be exposed to falls. For further information, see Fall Arrest Systems in Chapter 7, Personal Protective Equipment, in this volume.

Certain items of protective equipment are desirable for the prevention of various types of materials-handling injuries. Because toe and finger injuries are among the most common types, handlers should wear safety shoes and

Figure 12–1a. This power lift table helps prevent strains in a parts storage area.

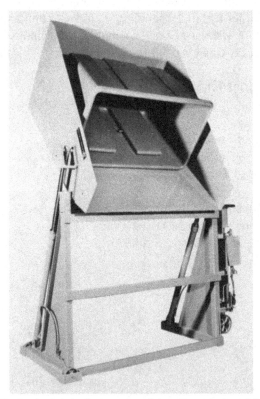

Figure 12–1b. A mobile dumper can lift materials from the floor level to table height and slide them onto the table.

stout gloves, preferably with leather palms. Other special protective clothing, such as goggles, aprons, and leggings, for the handling of certain types of materials should also be required. A personal protective equipment (PPE) hazard assessment should be conducted for each job task to effectively determine the required and recommended PPE.

Gloves should be dry and free of grease and oil. To prevent injury from splinters, handlers should wear gloves when handling wooden crates. Clean, leather-palmed gloves give better holding power on smooth metal objects than do cotton or other types of gloves. However, it may be unsafe for workers to wear gloves near conveyors or whenever there is a risk of catching in machinery. Take care not to bruise or squeeze the hands at doorways or other points where clearance is close. During inclement weather, special weather apparel may be necessary.

Where toxic or irritating solids are handled, workers should take daily showers to remove the material from their person before they leave the facility. Even though the exposure may not necessitate showers, encourage workers to wash thoroughly at the end of their shifts. Provide cleansing materials; clean, comfortable shower stalls; and washbasins. Also provide washable suits of tightly woven fabric, preferably full-length coveralls, and washable caps for workers to wear. Suits, caps, socks, and underwear should be laundered daily at the facility or through a laundry service. Clothing should be laundered less frequently only at the express direction of the company's safety and/or medical department.

Some industries, such as chemical and petrochemical, may require the use of flame-retardant coveralls in operational areas. Be aware that these suits, although they provide adequate protection for the workers, tend to be hot and, in some cases, uncomfortable.

Provide adequate breaks in extremely hot weather, and provide plenty of fluids for employees to replenish.

Materials-Handling Injuries

The primary type of injury experienced in materials handling is musculoskeletal disorders (MSDs). Bureau of Labor Statistics (BLS) reporting shows that MSDs accounted for 33% of all injury and illness cases in 2013.

From 2011 and forward, BLS defines musculoskeletal disorders to include cases where the nature of the injury or illness is a pinched nerve; a herniated disc; a meniscus tear; sprains, strains, tears; a hernia (traumatic or nontraumatic); pain, swelling, and numbness; carpal or tarsal tunnel syndrome; Raynaud's syndrome or phenomenon; musculoskeletal system and connective tissue diseases and disorders, when the event or exposure leading to the injury or illness is overexertion and bodily reaction, unspecified; overexertion involving outside sources; repetitive motion involving microtasks; other multiple exertions or bodily

reactions; and skin rubbed, abraded, or jarred by vibration.

Up through 2010, BLS defined musculoskeletal disorders to include cases where the nature of the injury or illness was sprains, strains, tears; back pain, hurt back; soreness, pain, hurt, except of the back; carpal tunnel syndrome; hernia; or musculoskeletal system and connective tissue diseases and disorders, when the event or exposure leading to the injury or illness is bodily reaction/bending, climbing, crawling, reaching, twisting, overexertion, or repetition. Cases of Raynaud's phenomenon, tarsal tunnel syndrome, and herniated spinal discs were not included. Although they may be considered MSDs, the survey classified these injuries and illnesses in categories that also included non-MSD cases.

A report by the Institute of Medicine (IoM) indicates that in 2010, the annual value of lost productivity from injuries ranged between $297.4 billion to $335.5 billion. The value of lost productivity is based on the following: days of work missed (ranging from $11.6 to $12.7 billion); hours of work lost (from $95.2 to $96.5 billion); and lower wages (from $190.6 billion to $226.3 billion).

According to the 2014 Liberty Mutual Workplace Safety Index, overexertion involving outside sources was ranked first among the leading causes of disabling injury in the workplace. This event category, which includes injuries related to lifting, pushing, pulling, holding, carrying, or throwing, cost businesses $15.1 billion in direct costs.

To gain insight into the injuries caused by materials handling, the safety professional should consider the following:
- Can the job be engineered to eliminate or reduce manual handling?
- Can the material be conveyed or moved mechanically?
- In what ways do the materials being handled (such as chemicals, dusts, rough and sharp objects) cause injury?
- Can employees be given handling aids—such as properly sized boxes, adequate trucks, or hooks—that will make their jobs safer?
- Would protective clothing or other personal equipment help prevent injuries?
- Would training and more effective management help reduce injuries?

These questions serve as a start for an overall appraisal of injuries caused by materials handling. Break each job into its separate tasks, and examine each task for ways to prevent injuries through job safety analysis (JSA) techniques.

Because most injuries occur to fingers and hands, give the following general pointers to employees who handle materials:
- Inspect materials for slivers, jagged or sharp edges, burrs, and rough or slippery surfaces.
- Grasp objects with a firm grip.
- Keep fingers away from pinch and shear points, especially when setting down materials.

- When handling lumber, pipe, or other long objects, keep hands away from the ends to prevent them from being pinched.
- Wipe off greasy, wet, slippery, or dirty objects before trying to handle them.
- Keep hands free of oil and grease.

In most cases, have employees use gloves, hand leathers, or other hand protectors to prevent hand injuries. Urge extra caution of employees who work near moving or revolving machinery. (See Chapter 7, Personal Protective Equipment, in this volume.) In other cases, have employees use handles or holders to move objects. Provide handles for moving auto batteries, tongs for feeding material to metal-forming machinery, or baskets for carrying control-laboratory samples.

Feet and legs also sustain a large portion of materials-handling injuries—the greater percentage occurring to the feet. Require that workers wear protective footwear. Safety shoes may need to be equipped with metatarsal guards for work activities that include heavy objects to be lifted. Workers' eyes, heads, and trunks can also be injured.

When opening a wire-bound or metal-bound bale or box, workers should take special care to prevent the ends of the binding from flying loose and striking the face or body. Workers should wear heavy gloves and protective eyewear equipped with side shields and heavy gloves. The same precaution applies to handling coils, wire strapping, or cable. In many cases, special tools are available to safely cut bands, strapping, and the like. Workers should always read the labels on packages for special handling instructions. If material has the consistency of dust or is toxic, the person handling it may need to wear a respirator or other suitable PPE to prevent injury to the lungs or use other appropriate controls provided, such as ventilation. Consult the Safety Data Sheet (SDS) or a technical data sheet.

Manual handling of materials increases the possibility of injury and adds to the product's cost. To reduce the number of injuries caused by materials handling and to increase efficiency, minimize manual handling of materials. For example, combine or eliminate operations and introduce ergonomic principles to job design. Mechanically move materials as much as possible.

Manual Lifting

Physical differences make it impractical to establish safe lifting limits applicable to all workers. A person's height and weight, although important, do not necessarily indicate lifting capability. Some small, thin individuals can handle heavier loads than some tall, heavy persons (Kroemer 1983). Conduct a job safety analysis and follow medical recommendations when establishing lifting standards. (See also the National Institute for Occupational Safety and Health [NIOSH] Work Practices Guide to Manual Lifting.)

Before workers lift a heavy or bulky object and carry it to another location, they should inspect the routes over which they will move the object. In that way, workers can make sure that there are no obstructions or spills that could cause them to slip or trip. If clearance is not adequate for handling the load, workers should choose an alternate route. Next, workers should inspect the object to decide how to grasp it and, thus, avoid sharp edges, slivers, or other things that might cause injury. Workers may have to turn an object over before attempting to lift it. Also, if the object is wet or greasy, workers should wipe it dry to prevent it from slipping. If this is not practical, workers should use a rope sling or other device that will firmly grip the object.

Most lower-back injuries come from tasks requiring lifting. Other activities—such as lowering, pushing, pulling, or carrying material or twisting the body—can also cause back injuries. Back injuries are second in number after injuries to the fingers and hands.

There are several techniques for manually lifting objects with reasonable safety. However, each of these lifting techniques has limitations. Consider all three main factors in manual lifting—load location, task repetition, and load weight—when determining what is safe or unsafe to lift.

The NIOSH has published an Applications Manual for the Revised NIOSH Lifting Guidelines (DHHS Publication 94-110) to assist in calculating safe lifts for most workers.

Back Belts

Back injuries account for nearly 20% of all injuries and illnesses in the workplace and cost the nation an estimated $20 billion to $50 billion per year (NIOSH). According to the Bureau of Labor Statistics, more than 1 million workers suffer back injuries each year.

Moreover, though lifting, placing, carrying, holding, and lowering are involved in manual materials handling (the principal cause of compensable work injuries), the annual BLS survey shows that four out of five of these injuries are to the lower back, and that three out of four occur while an employee is lifting.

Back belts have been a much debated response to alleviating back injuries. The discussion over back belts focuses on two issues: (1) employees are rarely trained in proper lifting techniques or even how to use the belts correctly, and (2) wearing a back belt can give a false sense of security—people think they can lift more than they can.

Most lifting injuries occur because of excessive load or pressure on the lower back. Excessive load can occur when a load is simply too heavy to lift or places weight too far from the spine, whether in front or to the side. Lifts that involve excessive reaching or twisting can produce injury.

Workers may also sustain injury when they fail to use proper lifting techniques or attempt to perform a job that is beyond their capabilities.

Back-belt manufacturers are the first to recommend that companies train employees in safe lifting techniques. They say companies should use ergonomic principles to determine whether they can reduce the lifting task by use of gravity or other lifting aids and whether changes in workstation layout can reduce lifting tasks. Job rotation and other administrative controls also can decrease the chance of injury.

Scientific studies have failed to demonstrate that back belts can prevent lifting injuries, although some studies have shown that the belts may be useful postinjury.

NIOSH summarized its review of the back-belt issue in a report entitled "Back Belts—Do They Prevent Injury?" (NIOSH Publication 94–127, issued in July 1994). According to the report, "After a review of the scientific literature, NIOSH has concluded that, because of the limitations of the studies that have analyzed workplace use of back belts, the results cannot be used to either support or refute the effectiveness of back belts in injury reduction." The report goes on to examine specific questions and issues related to back-belt use and concludes that the best way to protect workers is through ergonomic approaches designed to reduce the hazards of lifting. In NIOSH's view, belt use remains a personal decision.

A National Safety Council (NSC) technical advisory report on the use of back belts, issued in September 1995 by the Ergonomic/Human Factors standing committees of the Business and Industry and Labor divisions of the Council, comes to a similar conclusion. The NSC stresses that several conditions and controls should be in place before companies consider using back belts in the workplace. Companies should reduce or mitigate risk factors through ergonomic intervention. Also, they should use engineering controls, work-method analysis, and administrative controls to reduce worker exposure to lifting injuries.

Some scientific reports on back-belt use note increased blood pressure and heart rate in some belt users due to increased intra-abdominal pressure. Therefore, the NSC recommends that individuals who consider belt use be medically screened for cardiovascular and general health. To maintain strength and flexibility, companies should consider a simple exercise program for workers who use back belts. The program should be under the guidance of a health care professional.

FALL PROTECTION

Materials-handling equipment frequently gives rise to exposure to falls by operators and maintainers of overhead hoisting systems. Provision for fall protection and egress is needed where workers will foreseeably be exposed to fall hazards.

The approach to fall hazards is as follows:
1. recognition of each fall hazard including access to the workstation
2. control of the fall hazard through choice from a variety of fall solutions
3. selection and installation of fall protection including engineering if required
4. training sufficient to fulfill fall protection program
5. observation of workers and enforcement of fall protection program.

Recognition of fall hazards could be as follows: crane bridges, crane runways, ladder access to cranes, truss access, warehouse racks, platforms, operator cabs, access at heights, roofs, and mezzanines.

Fall protection measures include the following controls:
1. elimination of the hazard by reorganizing the work
2. prevention of falling by the use of guardrails, including aerial lifts
3. fall arrest systems for horizontal and vertical travel
4. warning lines 6 ft (1.8 m) from an edge.

Fall arrest systems are often valuable when workers overreach from aerial lifts, walk catwalks, climb trusses, walk on roof edges, climb fixed ladders, or use stepladders.

In case of fire and smoke, listed controlled descent devices may be beneficial in addition to fixed ladders typically supplied according to NIOSH 76–128. These systems may also be used for the rescue of individuals from height. Crane cabs and other workstations are examples of where installations may be made.

Training and inspection are vital for proper application of a fall protection program. Workers should be given a pretest and posttest to provide feedback to worker and manager alike. Observation is vital to monitor the effects of site-specific training. The training should cover regulations, standards, the site conditions, the work to be done, the sequence of the work, the materials, the work method, and fall arrest equipment and its proper use.

The fall protection for materials-handling-related applications must be applied at the very first thought about the work and when other initial planning is undertaken.

GUIDELINES FOR LIFTING

Regardless of the approach taken to evaluate the physical stresses of lifting, a large individual variability in risk of injury and lifting performance capability exists in the population today. This means that the resulting controls must be of both an engineering and an administrative nature. In other words, there are some lifting situations that are so hazardous that only a few people could be expected to be capable of safely performing them. These conditions need to be modified to reduce stresses through job redesign. On the other hand, some lifting conditions may be safely tolerated by some people, but others must be protected by an aggressive selection process, training program, and modifications of the workplace. To specifically define these conditions, two limits are provided based on epidemiological, biomechanical, physiological, and psychophysical criteria.

1. *Maximum permissible limit* (*MPL*) is defined to best meet four criteria:
   a. Musculoskeletal injury rates and severity rates have been shown to increase significantly in populations when work is performed above the MPL.
   b. Biomechanical compression forces on the L5/S1 disc are not tolerable over 1,430 lb (650 kg) in most workers. This would result from conditions above the MPL.
   c. Metabolic rates would exceed 5.0 Kcal/min for most individuals working above the MPL.
   d. Only about 25% of men and less than 1% of women workers have the muscle strengths to be capable of performing work above the MPL.
2. *Action limit* (*AL*), where the large variability in capacities among individuals in the population indicates the need for administrative controls when conditions exceed this limit based on the following:
   a. Musculoskeletal injury incidence and severity rates increase moderately in populations exposed to lifting conditions described by the AL.
   b. A 770-lb (350-kg) compression force on the L5/S1 disc can be tolerated by most young, healthy workers. Such forces would be created by conditions described by the AL.
   c. Metabolic rates would exceed 3.5 for most individuals working above the AL.
   d. More than 75% of women and more than 99% of men could lift loads described by the AL.

Thus, properly analyzed lifting tasks may be of three types:
1. Those above the MPL should be viewed as unacceptable and require engineering controls.
2. Those between the AL and MPL are unacceptable without administrative or engineering controls.
3. Those below the AL are believed to represent nominal risk to most industrial work forces.

Guideline Limits

With the large number of task variables (five in this case) that modify risk during lifting, it is virtually impossible to provide a simple, yet accurate procedure for evaluating all possible jobs. This problem is further complicated by the need to satisfy four separate criteria (epidemiological, biomechanical, physiological, and psychophysical). The following guideline is the simplest form known that best satisfies the four criteria.

In algebraic form:

$$AL(kg) = 40(15/H)(1 - 0.004|V - 75|)(0.7 + 7.5/D)$$
$$(1 - F/F_{max}) - \text{metric units}$$

or

$$AL(lb) = 90(6/H)(1 - 0.01|V - 30|)(0.7 + 3/D)$$
$$(1 - F/F_{max}) - \text{U.S. customary units}$$
$$MPL = 3(AL)$$

where H = horizontal location (centimeters or inches) forward of midpoint between ankles at origin of lift
V = vertical location (centimeters or inches) at origin of lift
D = vertical travel distance (centimeters or inches) between origin and destination of lift
F = average frequency of lift (lifts/minute)
F_{max} = maximum frequency that can be sustained

For purposes of this guide, these variables are assumed to have the following limits:
1. H is between 6 in. (15 cm) and 32 in. (80 cm). Objects cannot, in general, be closer than 6 in. (15 cm) without interference with the body. Objects farther than 32 in. (80 cm) cannot be reached by many people.
2. V is assumed between 0 and 70 in. (175 cm), representing the range of vertical reach for most people.
3. D is assumed between 10 in. (25 cm) and (80 – V) in. [(200 – V) cm]. For travel less than 10 in., set D = 10.
4. F is assumed between 0.2 (one lift every 5 minutes) and F_{max}. For lifting less frequently than once per 5 minutes, set F = 0.

Numerous attempts have been made to train material handlers to do their work, particularly lifting, in a safe manner. Unfortunately, hopes for significant and lasting reductions of overexertion injuries through the use of training have been generally disappointing. There are several reasons for the disappointing results:
- If the job requirements are stressful, "doctoring the system" through behavioral modification will not eliminate the inherent risk. Designing a safe job is basically better than training people to behave safely in an unsafe job.
- People tend to revert to previous habits and customs if practices to replace previous ones are not reinforced and refreshed periodically.
- Emergency situations, the unusual case, the sudden quick movement, increased body weight, or impaired physical well-being may overly strain the body because training does not include these conditions.

Proper training for safe materials handling (which is not limited to lifting) should be expected to generate results if the training is tailored to the work environment. The ultimate goal is to redesign or automate the job function, but if properly applied and periodically reinforced, training should help to alleviate some aspects of the basic problem.

The idea of training workers in safe and proper manual materials-handling techniques has been propagated for many years. Originally, it was advocated that a person lift with a straight back and to unbend knees while lifting. However, the frequency and intensity of back injuries was not reduced during the last 40 years while this lifting method was taught. Biomechanical and physiological research has shown that leg muscles used in this lifting technique do not always have the needed strength. Awkward and stressful postures may be assumed if this technique is applied when the object is bulky. It also implies employees are attempting lifts too heavy for themselves. Hence, the straight back/bent knees action evolved into the "kinetic" lift, in which the back is kept mostly straight and the knees are bent, but the positions of the feet, chin, arms, hands, and torso are prescribed. Another variant is the "free-style" lift, which, however, may be better for male (but not female) workers than the straight back/bent knee technique (Garg and Saxena 1985). It appears that no single lifting method is best for all situations (Anderson and Chaffin 1986), which is why employers must scope out the job in the design phase to minimize manual materials handling (MMH).

Training of proper lifting techniques is an unsettled issue. It is unclear what exactly should be taught, who should be taught, and how and how often a technique should be taught. This uncertainty concerns both the objectives and methods, as well as the expected results. Claims about the effectiveness of one technique or another are frequent but are usually unsupported by convincing evidence.

A thorough review of the existing literature indicates that the issue of training for the prevention of back injuries in manual materials handling is confused at best. In fact, training may not be effective in injury prevention, or its effect may be so uncertain and inconsistent that money and effort paid for training programs might be better spent on research and implementation of techniques for worker selection and ergonomic job design. Nevertheless, according to the NIOSH:

- The importance of training in manual materials handling in reducing hazards is generally accepted. The lacking ingredient is largely a definition of what the training should be and how this early experience can be given to a new worker without harm. The value of any training program is open to question as there appear to have been no controlled studies showing a consequent drop in the MMH accident rate or the back injury rate. Yet, so long as it is a legal duty for employers to provide such training or for as long as the employer is liable to a claim of negligence for failing to train workers in safe methods of MMH, the practice is likely to continue despite the lack of evidence to support it. Meanwhile, it may be worth considering what improvements can be made to existing training techniques.

- Currently, it appears that two major training approaches are most likely to be successful. One involves training in awareness and attitude through information on the physics involved in MMH and on the related biomechanical and physiological events going on in one's body. The other approach is the improvement of individual physical fitness through exercise and warm-ups (which can also influence awareness and attitude).

Rules for Lifting

There are no comprehensive and sure-fire rules for "safe" lifting. Manual materials handling is a very complex combination of moving body segments, changing joint angles, tightening muscles, and loading the spinal column. The following *dos* and *don'ts* apply, however:

- *Do* engineer manual lifting and lowering out of the task and workplace. If it must be done by a worker, perform it between knuckle and shoulder height.
- *Do* be in good physical shape. If a worker is not used to lifting and vigorous exercise, he or she should not attempt to do difficult lifting or lowering tasks.
- *Do* think before acting. Place material conveniently within reach. Have handling aids available. Make sure sufficient space is cleared.
- *Do* get a good grip on the load. Test the weight before trying to move it. If it is too bulky or heavy, get a mechanical lifting aid or somebody else to help, or both.
- *Do* get the load close to the body. Place the feet close to the load. Stand in a stable position with the feet pointing in the direction of movement. Lift mostly by straightening the legs.
- *Don't* twist the back or bend sideways.
- *Don't* lift or lower awkwardly.
- *Don't* hesitate to get mechanical help or help from another person.
- *Don't* lift with the arms extended.
- *Don't* continue lifting when the load is too heavy.

Personnel Selection for Materials Handling

Selecting persons who are unlikely to suffer an overexertion injury is one of the three methods to reduce the risk of musculoskeletal disorders in manual materials handling. The purpose of this assessment is to place on jobs only those individuals who can do them safely. Many employers now require physical functional capacity testing prior to employment. This enables the employer to determine if a potential employee can meet the physical demands of the job. The basic premise is that the risk of overexertion injury for manual materials handling decreases as the handler's capability to perform such activity increases. This means that the evaluation should be designed so that it allows the administrator to match a person's capabilities for manual materials handling to the actual demands of the job. This matching process and job safety analysis require that the administrator knows quantitatively both the job requirements and the related capabilities to be tested.

Scientists usually rely on the development and use of models. A model is an abstract (mathematical-physical) system that obeys specific rules and conditions and whose behavior is used to understand the real system (in this case, the worker-task) to which it is analogous in certain aspects (e.g., in physiological, biomechanical, psychophysical, or other traits). A model usually represents a theory. Without proper models, reliable and suitable methods cannot be developed. A method is a systematic, orderly way of arranging thoughts and executing actions. A technique is the specific, practical manner in which actions are done; it implements the methods that are derived from models. Evaluation techniques involve specific procedures and instruments used to obtain measurements with respect to the subject's capability to perform MMH activities.

Many models have been developed to describe the central and local limitations just discussed. In the following sections, these models are simplified and categorized by major disciplines for convenience.

Load Held Close to the Body

While lifting and carrying objects, the worker should keep the load close to the body (Figure 12–2). The closer the load is to the body, the less it affects the lower back. To do this, the arms should be close to the body and remain straight whenever possible. Flexing the elbows and raising the shoulders imposes unnecessary strain on the upper arm and chest muscles.

When carrying an object, the arms should stay in the same position. In the case of long-distance carrying, any assistance given by the body to support the weight of an object will lessen the tension in the muscles. Carrying an object with the arms lowered ensures that the weight of the object will rest against the body.

Correct Grip

An insecure grip may be caused by holding a load with the fingertips. Such a grip creates undue pressure at the ends of the digits and causes strain to certain muscles and tendons of the arm. A full-palm grip reduces local muscle stress in the arms and decreases the possibility that a load will slip (Figure 12–3). Handles or handholds are preferable. Because greasy surfaces often prevent a secure hold, wipe surfaces clean before grasping them. Use suitable, properly fitted gloves, if necessary. Here are some lifting techniques for specific situations:

1. If the object is too bulky or too heavy to be handled by one person, get help.
2. Workers should not "jerk-lift" loads, as this multiplies the stress to the lower back.

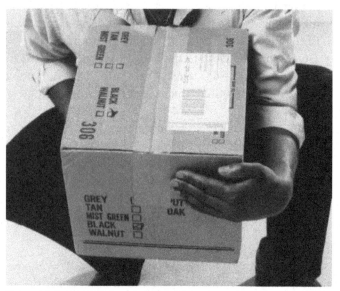

Figure 12–2. To lift, the load should be drawn close to the body, and the arms and elbows should be tucked into the side of the body. When the arms are held away from the body, they lose much of their strength and power. Keeping the arms tucked in also helps keep weight centered.

Figure 12–3. The palmar grip is one of the most important elements of correct lifting. The fingers and the hand are extended around the object to be lifted. Use the full palm; fingers alone have very little power. Pull the load in between the knees and as close to the body as possible. (The glove has been removed to show the finger positions better.)

3. Before lifting the load to be carried, workers should consider the distance to be traveled, any obstacles that need to be repositioned, and the length of time that the grip will have to be maintained. Workers should select a place to set the load down and rest to allow for the loss of gripping power. Pausing to rest is especially important when negotiating stairs and ramps.

4. To place an object on a bench or table, the workers should first set it on the edge and then push it far enough onto the support so that it will not fall. The object should be released gradually as it is set down. It should be moved in place by pushing with the hands and body from in front of the object. This method prevents pinched fingers.

5. Workers should securely position an object, placed on a bench or other support, so that it will not fall, tip over, or roll off. Supports should be correctly placed and strong enough to hold the load. Heavy objects should be stored at approximately waist height.

6. To raise an object above shoulder height, workers should:
   a. lift it to waist height
   b. rest the edge of the object on a ledge, stand, or hip
   c. shift hand positions, so the object can be boosted after the knees are bent. Workers should straighten their knees as the object is lifted or shifted to the shoulders.

7. To change direction, the worker should lift the object to the carrying position and turn the entire body, including the feet. The worker should avoid twisting the body. In repetitive work, the worker and the object should both be positioned so that twisting the body is unnecessary when moving the object.

8. To deposit an object manually into a tight space, the worker should slide it into place with the hands in the clear—thus avoiding pinched fingers.

Team Lifting and Carrying

When two or more people carry one object, they should adjust the load so that it rides level and so that each person carries an equal part of the load. Workers should do test lifts before proceeding. When two people carry long sections of pipe or lumber, they should walk one behind the other, carrying the material on the same shoulder and walking in step. Shoulder pads will prevent cutting into their shoulders and will also reduce fatigue.

When a team of workers carries a heavy object, such as a rail, the supervisor should make sure that proper tools are used and should provide direction for the work. Frequently, a whistle or direct command can signal "lift," "walk," and "set down." The key to safe carrying by gangs is to make every movement in unison.

Handling Specific Shapes

There are specific techniques for lifting and handling materials of various shapes.

Boxes, Cartons, and Sacks

The best way to handle boxes and cartons is to grasp the alternate top and bottom corners and to draw a corner between the legs. Banding materials should never be used to make a lift except in special cases.

Sacked materials are also grasped at opposite corners. Upon reaching an erect position, the worker should let the sack rest against the hip and belly and then swing the sack to one shoulder. As the sack reaches the shoulder, the worker should stoop slightly and put a hand on the hip so that the sack rests partly on the shoulder and partly on the arm and back. The other hand should be holding the sack at the front corner. When the sack is to be put down, the worker should swing it slowly from the shoulder until it rests against the hip and belly. While the sack is being lowered, the legs should be flexed and the back kept straight.

Barrels and Drums

Those who handle heavy barrels and drums require special training (Figures 12–4 and 12–5). A barrel is generally less hazardous to handle than a drum because the shape of the barrel aids in upending it and reducing hazards should it tip over. Workers should pay special attention to the weight and contents of barrels and drums because these factors vary greatly. Workers should wear safety shoes with metatarsal guards.

Frequently, only one person is available to handle a drum, in which case it is better to wait for help or use mechanical assistance. A commercially available drum tilter, equipped with wheels, is commonly used. An extension handle provides control and leverage during the tilting operation. The wheels allow for easy transport of the tilted drum over short distances (Figure 12–4). Another commercial device that makes tilting and transporting easier is a two-wheeled dolly equipped with large rubber tires.

Sheet Metal

Because sheet metal usually has sharp edges and corners, workers should handle it with leather gloves, hand leathers, or gloves with metal inserts. Gauntlet-type gloves or wristlets give added protection to wrists and forearms. Workers should use power equipment, such as a "grab" or a vacuum lifter, for bundles of sheet metal.

Plate Glass

Workers should wear gloves or hand laps when handling flat glass. Their wrists and arms should be protected with leather cuffs and safety sleeves. The worker should wear a

Figure 12–4. This spill containment caddy minimizes strains during drum transport. *(Courtesy Justrite Manufacturing Company LLC)*

Figure 12–5. Pneumatic devices may also reduce the need to manually handle drums. This device secures the drum by its sides and lifts, moves, and tilts the drum so it can be placed horizontally on a rack or vertically in storage.

leather apron, leggings, safety glasses, and also safety shoes with metatarsal guards. Unless the piece of plate glass is small, the worker should carry only one piece at a time and should walk with care. The plate should be lifted carefully and carried with its bottom edge resting in the palm (turned outward) and with the other hand holding the top edge to steady it. Plate glass should never be carried under the arm because a fall might break the glass and sever an artery. Plate glass must not be carried in such a manner that it bends.

To transport larger pieces of plate glass over any distance, workers should use handling equipment specifically designed for that purpose. Equipment such as cranes equipped with vacuum frames, C-frames or spreader bars, and special wagons or dollies is normally used to transport heavy glass. If large plates must be transported by hand, two workers wearing safety hats, safety glasses, safety sleeves, cuffs, gloves, and safety shoes should do the job.

Long Objects

A team of workers should carry long pieces of pipe, bar stock, or lumber on their shoulders. One of the workers should guide the object when going around corners. Workers should wear shoulder pads for this operation.

Irregularly Shaped Objects

Objects with irregular shapes present special handling problems. Often, the object must be turned over or up on end to secure the best possible grip. If the worker feels unable to handle the object, because of either its weight or shape, he or she should request assistance.

Miscellaneous Objects

Using slings, buckets, and bags is a safe method of securing miscellaneous material on overhead platforms. Workers should follow special safety measures when raising or lowering these materials to different elevations in order to protect workers below.

Scrap Metals

In a scrap storage area, require that the best possible housekeeping practices be observed. Irregularly shaped, jagged pieces can get tangled in such a way that strips or pieces may fly when a piece is removed from a pile. Therefore, provide workers with goggles, leather gloves or mittens, safety shoes, safety hats, and protection for their legs and bodies. Caution workers against stepping on objects that may roll or slide.

Heavy, Round, Flat Objects

If rolled by hand, heavy, round, flat objects, such as railcar wheels or tank covers, present considerable danger even to skilled personnel. The operation requires careful training and exacting precautions. Using a hand truck or power equipment designed for the purpose is preferred. Heavy rolls can be safely secured and handled by specially designed devices. (See Chapter 15, Powered Industrial Trucks, in this volume.)

Machines and Other Heavy Objects

Manual movement of heavy machinery and equipment requires special skill and knowledge. Sometimes machines or castings weighing 90.7 tons (100 tonnes) or more must be moved from freight cars to ground level and into permanent position without the use of heavy-duty cranes or similar equipment.

Only general safety principles for such jobs can be suggested. Each task presents its own problems and requires careful study and thorough planning. Some companies build scale models of the machines and the blocking, jacks, rollers, and other equipment to be used. They then work out the procedure in miniature. In all cases, determine the safe floor load limits for areas over which the machine or part will move, as well as for the place in which it is to be installed or stored.

Select blocking and timbers with great care. Hardwoods make the best blocking and timber—preferably oak and of the proper sizes to allow the machine to be safely blocked or cribbed as it is raised or lowered. Do not use wood that has round corners or that shows signs of dry rot.

For sufficient strength, cribbing should have a safety factor of at least 4. Be cautious about the natural tendency to underestimate the load. Cribbing must be placed on a foundation in such a manner that it can be removed readily as the machine is lowered.

ACCESSORIES FOR MANUAL HANDLING

In handling materials, a variety of hand-operated accessories are available. Each tool, jig, or other device should be kept in good repair and used only for the job for which it is designed.

Hand Tools
Of all the hand tools available for the manual handling of materials, hooks, crowbars, and rollers are the most common. Workers should know how to use these tools properly.

Hooks
Train the worker to use hand or packing hooks in such a way that they will not glance off hard objects and possibly injure the worker. If a hook must be carried in the worker's belt, be sure the point is covered.

Use hook handles made of hardwood, and keep them in good condition. Supervisors should check that hooks for handling logs, lumber, crates, boxes, and barrels are kept sharp and inspected daily and before each use.

Crowbars
The principal hazard in the use of a crowbar is slippage. Workers should keep their hands and gloves dry and free of grease and oil when using a crowbar. A dull, broken crowbar is more likely to cause injury than a sharp one.

The point or edge should have a good "bite." Workers should position themselves to avoid falling or pinching their hands if the bar slips or the object moves suddenly. Workers should never work astride an object. Crowbars not in use should be stored so that they will not fall or cause a tripping hazard.

Rollers
Heavy, bulky objects must often be moved by means of rollers. The principal hazard of rollers is that the fingers or toes may be pinched or crushed between a roller and the floor. Rollers should extend beyond the load to be moved and be sufficiently strong. Rollers under a load should be moved with a sledge or bar, not with a hand or foot.

Jacks
When a jack is used, workers should check the capacity plate or other marking on the jack to make sure the jack can support the load. If the identifying plate is missing, workers should determine the maximum capacity of the jack and paint it on the side. If a properly rated jack is used, it should not collapse under the load.

Workers should inspect jacks before and after each use. Any sign of hydraulic fluid leakage is sufficient reason to remove the jack from use. When a jack begins to leak, malfunction, or show any signs of wear or defects, it should be removed from service, tagged, repaired, and tested under load before being returned to service.

Workers using jacks should wear protective footwear (safety shoes) with metatarsal guards because jack handles may slip and fall or parts of machinery or equipment may become loose and drop while the load is being lifted or shifted. Furnish toweling to jack operators for removing oil from their hands and from the jack handles. In that way, they will always have a firm grip.

A heavy jack is best moved from one location to another on a dolly or special hand truck. If it has to be manually transported, it should have carrying handles. At least two workers should form a team to move it. The operating handle should never be left in the socket while a jack is being carried because it might strike other workers.

Workers should make certain that jacks are well lubricated, but only at points where lubrication is specified. They should also inspect them for broken teeth or faulty holding fixtures. Workers should never throw or drop a jack upon the floor. Such treatment may crack or distort the metal, causing the jack to break when a load is lifted.

The floor or ground surface upon which a jack is placed must be level and clean, and the safe floor load limit must not be exceeded. If the surface is earth, workers should set the jack base on substantial hardwood blocking (at least twice the size of the jack) so that it will not turn over, shift, or sink. If the surface is not perfectly level, workers should set the jack on blocking leveled by substantial shims or wedges. These devices should be placed so securely that they cannot be crushed or forced out of place.

To prevent the load from slipping, workers should avoid metal-to-metal contact between the jack head and the load. A hardwood shim, longer and wider than the face of the jack head, should be placed between the jack head and the contact surface of the load; 2-in. (5.1-cm) wood stock is suitable for this purpose. Workers should also remove oil that has collected in the bases of equipment or machines to be jacked before the operation is begun. This prevents spillage when the equipment or machines are tilted. Workers should immediately wipe up spillage of any residual oil.

Workers should never use wood or metal extenders. Instead, they should either obtain a larger jack or place higher blocking that is correspondingly wider and longer under the jack.

All lifts should be vertical—the jack correctly centered for the lift, the base on a perfectly level surface, and the head with its shim bearing against a perfectly level meeting surface. When an emergency requires that the lifting force be applied at an angle, workers must take the following extra precautions:

1. a base of blocking, securely fastened together and to the ground, to make an immovable surface at right angles to the lift for the jack base to sit on
2. cleats on the blocking to prevent shifting of the jack base
3. a meeting surface at right angles to the direction of the lift for the jack head with its shim to bear against
4. props or guys to the load to prevent the jack from swinging sideways when lifting begins.

When a jack handle is placed in the socket and before applying pressure, the worker should make sure that the area is clear and that there is ample room for an unobstructed swing of the handle. A faulty movement in the load may cause the handle to pop up and strike another worker. The person operating the handle should stand to one side so that, if the handle kicks, it will not strike any body or facial areas. When releasing a jack, the worker should keep all parts of the body clear of the movement of the handle.

After the load is raised, the worker should place metal or heavy wooden horses or blocking under it for support in case the jack should let go. The worker should never allow a raised load to remain supported only by jacks. The jack operator should immediately remove the handles of the jacks and should place them out of the way to prevent others from tripping over them.

The following special jacks require additional precautions:

- Hydraulic jacks may settle after raising a load. It is, therefore, especially important for workers to place blocking under a load that has been raised by such jacks.
- Screw jacks have a tendency to twist when a heavy load causes the floating head of the jack to bind. Jack operators, therefore, should anchor the base of a screw jack as securely as possible. In that way, the jack base will not twist and slip out from under the load when force is applied on the bars to raise the screw.

To raise a large piece of equipment with screw jacks, two or more jacks should be used. The load should be equally distributed on each jack. Operators should raise each jack a little at a time to keep the load level and the strain equal on each screw jack head. Operators should work out special signals to verify that all jacks are rising uniformly.

Hand Trucks

Many special types of hand trucks and dollies are available, such as two-wheeled trucks; flat trucks; lift trucks; and platform, refrigerator, and appliance dollies. Hand trucks can be purchased or designed for objects of various kinds and sizes. The type of hand truck most suitable for the work at hand should be used. No single hand truck is suitable for handling all types of material.

Two-Wheeled Trucks

Provide two-wheeled trucks and wheelbarrows equipped with knuckle guards to protect workers' hands from being jammed against door frames or other obstructions. Operators should also wear heavy gloves. Some two-wheeled trucks have brakes so that the worker need not hold the truck with a foot on the wheel or axle. Most commonly used trucks, however, do not have brakes.

To reduce the hazard to toes and feet, purchase trucks with wheels as far under the truck as possible. Wheel guards can be installed on many types of trucks, and operators should wear protective footwear.

Tongues of flat trucks should have counterweights, springs, or hooks to hold them vertical when not in use. If this is not possible, train workers to leave handles in such a position that they will not cause workers to trip over them.

Two-wheeled trucks look deceptively easy to handle, but workers should adhere to the following safe procedures:
- Tip the load to be lifted slightly forward so that the tongue of the truck goes under the load.
- Push the truck all the way under the load to be moved.
- Keep the center of gravity of the load as low as possible. Place heavy objects below lighter objects. When loading trucks, workers should keep their feet clear of the wheels.
- Place the load well forward so the weight will be carried by the axle, not by the handles.

- Place the load so it will not slip, shift, or fall. Load only to a height that will allow a clear view ahead.
- When a two-wheeled truck or wheelbarrow is loaded in a horizontal position, raise it to traveling position by lifting with the leg muscles and keeping the back straight. Observe the same principle in setting a loaded truck or wheelbarrow down—the leg muscles should do the work.
- Let the truck carry the load. The operator should only balance and push.
- Never walk backward with a hand truck.
- For extremely bulky items or pressurized items, such as gas cylinders, strap or chain the item to the truck.
- When going down an incline, keep the truck ahead so that it can be observed at all times.
- Move trucks at a safe speed. Do not run. Keep the truck constantly under control.
- Store hand trucks with the tongue under a pallet, shelf, or table.

Four-Wheeled Trucks or Carts

Operating rules for four-wheeled trucks or carts are similar to those for two-wheeled trucks. Place extra emphasis, however, on proper loading of four-wheeled trucks. They should be evenly loaded to prevent tipping. A truck's contents should be arranged so that the contents will not fall or be damaged in case the truck or the load is bumped. Four-wheeled trucks should be pushed rather than pulled. Pushing an object rather than pulling on it causes less stress to the lower back and protects the worker's heel from being caught under the truck back.

Four-wheeled trucks should not be loaded so high that operators cannot see where they are going. If there are high racks on the truck, two workers should move the vehicle— one to guide the front, the other to guide the back. Handles should be placed at protected places on the racks or body of the truck. In that way, passing traffic, walls, or other objects will not crush or scrape the operator's hands.

General Precautions

Be sure that handlers of two- and four-wheeled trucks are aware of three main hazards: (1) running wheels off bridge plates or platforms, (2) colliding with other trucks or obstructions, and (3) jamming their hands between the truck and other objects.

Workers should operate trucks at a safe speed and keep them constantly under control. Special care is required at blind corners and doorways. Properly placed mirrors can aid visibility at these places.

When not in use, trucks should be stored in a designated area—not parked in aisles or other places where they could obstruct traffic or cause someone to trip. Trucks with drawbar handles should be parked with handles up and out of

the way. Two-wheeled trucks should be stored on the chisel with their handles leaning against a wall or the next truck.

STORAGE OF SPECIFIC MATERIALS

Temporary and permanent storage of materials should be secure, neat, and orderly to eliminate hazards and conserve space. Materials piled haphazardly or strewn about increase the possibility of accidents to employees and of damage to materials.

The warehouse supervisor must direct the storage of raw material (and sometimes processed stock) that is kept in quantity lots for extended periods of time. The production supervisor is usually responsible for storage of limited amounts of material and stock that are kept near the processing operations for short periods of time. A good plan for storing materials reduces the amount of handling needed both to bring materials into production and to remove finished products from production to shipping.

Planning Materials Storage

When planning the storage of materials, allow adequate ceiling clearance under the sprinklers. The amount of clearance between storage and automatic sprinkler heads will vary with the material being stored and the height of storage. (See the National Fire Protection Association [NFPA] 13, Installation of Sprinkler Systems.) Be sure that automatic sprinkler system controls and electrical panel boxes are free and clear. Also, maintain unobstructed access to fire hoses and fire extinguishers.

Keep all the exits and aisles clear at all times. If materials are handled from the aisles, allow for the turning radius of power trucks. Employees should keep materials out of the aisles and out of the loading and unloading areas. These areas should be marked with painted or taped lines.

Use bins and racks to facilitate storage and reduce hazards (Figure 12–6). Consider adjustable storage containers to avoid placing a worker in an awkward position when he or she is returning a part. Material stored on racks, pallets, or skids is moved easily and quickly from one workstation to another with less damage to materials and fewer injuries to employees. When possible, have material that is piled on skids or pallets cross-tied.

Storage racks should be secured to the floor, the wall, and to each other. If flue spaces are provided, stock should never block the flue. If automatic sprinklers or fire prevention devices are provided in the racks, workers should exercise special care to avoid damage to them. Racks, when damaged, should be repaired. Never allow employees to climb on racks.

In an area where the same type of material is stored continuously, it is a good idea to paint a horizontal line on

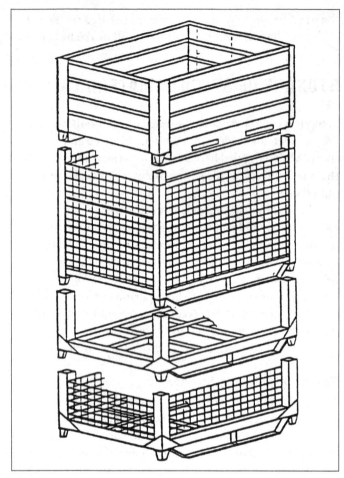

Figure 12-6. Interstacking containers like these are found in many industrial plants. To stack safely, be sure the surface is level and stable.

the wall to indicate the maximum height to which material may be piled. This will help keep the floor load within the proper limits and the sprinkler heads in the clear. Pickup and drop stations should also be clearly marked.

High-bay storage is the trend for storing containers or stock of uniform size. Some European and U.S. storage facilities have approximately 100-ft- (30.5-m-) high automated storage units.

High-bay facilities require unique, specially designed high-lift materials-handling equipment. Some, up to 30 ft (9 m) high, are operated manually. Others are operated through computer control. Standards that apply to high-bay storage include the American National Standards Institute (ANSI) and the American Society of Mechanical Engineers (ASME) ANSI/ASME B56.1-7 series, Safety Standards for Powered Industrial Trucks, and the Crawler Cranes section of ANSI/ASME B30.5, Mobile and Locomotive Cranes. High-bay storage presents not only unique materials-handling problems but also special fire protection problems. (See Chapter 15, Powered Industrial Trucks, in this volume, and NFPA 231C, Standard for Rack Storage of Materials.)

The protection of personnel who operate and maintain such facilities must include disaster and emergency planning, physical protection at the point of operation, lockout/tagout, and visible and audible warnings on moving equipment. Special procedures and equipment are also required for maintaining and taking physical inventories of high-bay facilities.

Rigid Containers

Each type of rigid container has its own storage requirements. Methods for storing several types of these containers are given here.

Large Metal Containers and Box Pallets

There are three general types of large containers commonly used in industry: wire mesh or expanded metal containers, solid-sided metal tubs, and skids and box pallets. There are also many variations on these types.

Safe stacking of large containers requires a level and stable stacking surface. These containers must be in good condition and be able to nest or interlock with the container below (Figure 12-6). Do not intermix different types of containers (e.g., metal tubs and wood box pallets) in a stack unless they are designed for each other and full safety is ensured.

A rule of thumb for stacking heights is three times the minimum base diameter of the container. For example, a container with a 2- to 3-ft (61- to 91-cm) base may be stacked 6 ft (1.8 m) high if the other stacking conditions have been met. Do not exceed weight capacities and weight-bearing capacities of the stacked containers. For visibility, do not stack containers near the corners of working (non-storage) aisles.

Fiberboard/Cardboard Cartons

Store loaded cartons on platforms to protect them against moisture. Even low piles, when wet, will collapse. Preferably store cartons on pallets or racks. If lower cartons show any signs of being crushed, restack them.

Because the height of piles of cartons is regulated by the kind of materials in the cartons, it is not possible to establish a standard height for such piles. An important factor to consider is that the sides of cartons do not support much load. Place sheets of heavy wrapping paper between layers of cartons, therefore, to help prevent the pile from shifting. Interlocking the cartons also increases the stability of the pile.

Do not stack certain bulky materials, such as skids of paper, to maximum allowable heights in rows next to aisles, especially aisles that carry hand-truck or power-truck traffic. A good rule is to make the first row one item high unless the material can be tightly interlocked. Stretch wrapping with plastic film or banding pallets will also assist in more stable storage.

Barrels and Kegs
Piles of barrels should be symmetrical and stable, preferably in the shape of a pyramid. Block the first or bottom row to prevent the barrels from rolling. If barrels or kegs are piled on end, place planks between rows. When barrels or kegs are piled on their sides other than in pyramid form, lay them on specially constructed racks. Otherwise, lay planks between rows and block the ends of the rows. (See the discussion of portable containers in the Containers for Liquids section later in this chapter for more information on drums.)

Rolled Paper and Reels
Clamp-type trucks can stack rolled paper and reels three or four high. Extreme care is required to ensure even stacking. If there is any physical damage to lower rolls or reels, the material must be restacked. Band paper rolls a few inches from each end.

Compressed Gas Cylinders
For safe handling and storage of compressed gas cylinders, refer to Chapter 19, Welding and Cutting, in this volume.

Uncrated Stock
Stock that does not come in its own container presents special storage problems. Examples of uncrated stock are lumber, bagged material, pipes, and sheet metal.

Lumber
Except for the amount that is needed at a given time, lumber is best stored outdoors or in a building separate from the general warehouse. Sort lumber by size and length, and store it in separate piles. When piling lumber outdoors, be sure that the ground is firm. The storage area should also be well drained to remove surface water and prevent softening of the ground. Periodically check stored lumber to make sure that it has not shifted position. Cover lumber stored outdoors to prevent checking or twisting. Use a well-ventilated building for storing lumber indoors.

For long-term storage, use substantial bearings or dunnage. Concrete with spread footing extending below the frost line is a good method. For temporary storage, heavy timbers may be used to support the crosspieces. Periodically inspect the timbers for signs of deterioration, which may cause the pile to list dangerously.

If the lumber must be removed by hand, store it in low piles or in racks, and provide galleries that permit workers to reach the top of the piles. If lumber must be moved by hand to or from a higher pile, the pile should be not more than 6 ft (1.8 m) high. Provide a safe means of access to the top of the pile.

When lumber is piled and removed mechanically by forklift trucks, 20 ft (6 m) is generally considered the maximum safe height for lumber piles.

Use tie pieces not only to stabilize the lumber pile but also to provide circulation of air. Do not let tie pieces extend into walkways; cut them flush with the pile. Use tie pieces on every layer of green lumber whether stored indoors or outdoors.

Bagged Material
Cross-tie bagged material with the mouths of the bags toward the inside of the pile. Stack bags neatly. Avoid overhangs that could be ripped, thus spilling the bags' contents. This precaution also applies to stacking on pallets and stacking several tiers high by lift truck. Stretch wrapping will help contain loose commodities.

Pipe and Bar Stock
Pipe and bar stock place a heavy load on the floor. Therefore, select the floor area with load-bearing strength in mind. Because removing pipe and bar stock from racks presents a hazard to passersby, do not locate fronts of pipe and bar stock racks on main aisles. Pile pipe and other round materials in layers with strips of wood or iron between the layers. Either the strip should have blocks at one end, or the end should be turned up.

Store larger sizes of bar stock in racks that are designed to rest the bars on rollers. The center distance between rollers will be governed by the sizes of the bars and should permit their easy withdrawal. Rollers with multiple sections make withdrawal easier. Racks should be set up so that bars cannot roll out (higher in the front). Light bar stock may be stored vertically in special racks. Special A-frame racks made of metal can hold a variety and quantity of pipes and bars safely, if they are loaded evenly and supported properly.

Materials such as lumber and pipe are particularly dangerous because of their tendency to roll or slide. Dropping, instead of placing, such objects on a pile frequently causes them to slide or bounce. Employees are likely to be injured if they attempt to stop rolling or sliding objects with their hands or feet.

Sheet Metal
Racks, similar to those used for bar stock, may be provided for plate and sheet stock, except that rollers are not always applicable. Oiled sheets require additional caution in handling.

Because sheet metal usually has sharp edges, workers should use hand leathers, leather gloves, or gloves with metal inserts when handling it. Large quantities of sheet metal should be handled in bundles by power equipment. These bundles should be separated by strips of wood to make handling easier when the material is needed for production. Using strips of wood also lessens the chance of the material shifting or sliding in the piles.

Tin-plate strip stock is heavy and razor sharp. Should a load or partial load fall, it could badly injure anyone in the way. Observe these two measures to prevent spillage and injuries: (1) band the stock after shearing, and (2) use wooden or metal stakes around the stock tables and pallets that hold the loads. The supervisor and all who handle the bundles are responsible for making sure the load is banded properly and that the stakes are in place when the load is on the table. (Special cutting tools are required to remove bands.)

Burlap Sacking and Other Materials

When burlap sacking is stored in high stacks, heat is generated by the weight. This sets the stage for spontaneous combustion. One way to reduce this hazard is to cut the size of the stack by breaking it up into smaller stacks. This can be done either by making smaller stacks (which would increase the number of stacks and take more space) or by placing blocks at intervals in the stack, so that, in effect, there would be a number of small piles, one atop the other. Provide additional protection by constructing the storage room of fire-resistant materials and by having sprinklers and dust-tight lights.

Straw, excelsior, and other packing materials are usually received baled and should preferably be stored that way. Store these materials either in a separate building or in a fire-resistant room provided with sprinklers and dustproof electric equipment. Because these materials are a fire hazard, only the amount necessary for immediate use should be taken into the packing room. Store this amount in bins made entirely of metal or made of wood lined with metal. Provide these bins with covers, and keep them closed. Large bins may have several compartments with counterweighted covers. The counterweight ropes should have fusible links to ensure automatic closing of covers in case of fire. Counterweights should be boxed in to prevent injury if the ropes break. Post NO SMOKING signs.

HAZARDOUS MATERIALS

Storage and handling of specific hazardous materials are discussed in other chapters in this volume: gases in Chapter 19, Welding and Cutting; flammable liquids (including refrigerants) and tank car and tank truck loading and unloading in Chapter 10, Flammable and Combustible Liquids; and NFPA hazard symbols in Chapter 9, Fire Protection. Advise local fire departments and emergency planning committees when storing hazardous materials. Comply with EPA requirements for hazardous material and waste storage.

Containers for Liquids

This chapter deals with handling containers in which hazardous materials are stored. Drums, tanks, piping, portable containers, and tank cars are sometimes used to store hazardous liquid materials. The following sections describe safety measures for handling these materials. All storage containers need identification of the contents. The label or other marking needs to be clear and clean.

Drums

Store filled drums containing volatile liquids in a protected area out of the sun. Heat from any source causes liquids to expand. The resulting buildup of pressure could cause leaks, with subsequent fire or explosion. It is recommended that an approved drum vent be installed in the bung opening as soon as a drum is received (Figure 12–7).

Reuse of drums causes problems if drums are not thoroughly cleaned beforehand, especially if incompatible materials are combined. Carefully purge drums and clean them out with water, steam, and appropriate solution. Consult a chemical expert to set up safe cleaning procedures. Often, the top of a drum is removed so the drum can be used as a receptacle. Never burn the top of a drum out with a torch

Figure 12–7. This simple device automatically relieves drum pressure and can be lifted by hand to relieve vacuum. *(Courtesy Justrite Manufacturing Company LLC)*

because some liquid or vapor left behind in the drum could cause an explosion. It is much safer to follow the drum-purging methods described in Chapter 19, Welding and Cutting, in this volume. Then remove the top of the drum with a mechanical opener. More details about handling and storing drums are given later in the section on portable containers. Following proper cleaning, some metal and plastic drums can be recycled by drum-recycling vendors.

Tanks

Design the structure of a new building to support the weight of storage tanks. However, have a structural engineer conduct an inspection before a tank is installed in an old building. Storage tanks for hazardous liquids are preferably installed outdoors. Be aware of both EPA aboveground and underground storage tank (UST) requirements.

There are many advantages in installing outdoor storage tanks underground. However, the UST regulations are very stringent, and the danger of leaks going undetected in underground tanks containing corrosive or toxic materials definitely outweighs the advantage of freedom from drips and sprays. When an aboveground outdoor tank is located in an enclosed dike, the dike must be large enough to permit easy access to all parts of the tank. Provide a permanent ladder and an access door that can be fastened shut. A confined-space entry permit may be required to enter pits and tanks.

Confined spaces would include pits and tanks (Permit-Required Confined Spaces, 29 CFR 1910.146). The rules for entering a confined space involve training of the employee, permits, and rescue procedures. (See also Confined Spaces in Chapter 2, Buildings and Facility Layout, and Precautions for Entering Boilers and Furnaces in Chapter 5, Fired Pressure Vessels (Boilers) and Unfired Pressure Vessels, both in this volume.)

Install process tanks that will contain volatile or corrosive liquids only at or above grade and in areas having adequate drainage. Separate process tanks from the processing area with construction materials having a fire-resistant rating of at least 2 hours.

Install tanks where traffic cannot pass under them. If people must walk beneath them, install drip pans and provide drainage of the pans to a safe disposal or recovery location.

Provide tanks with permanent stairs, or ladders, and walkways that have standard guardrails and toeboards. Empty, clean, and inspect tanks and piping for structural weaknesses (mechanical integrity) at regular intervals. All storage tanks should be adequately labeled or placarded with contents and hazard warnings (NFPA 704, Identification of Fire Hazards of Materials). Keep records of each inspection. A confined-space entry permit may be required.

Tanks for holding volatile materials should be bonded, grounded, and provided with emergency venting devices.

Venting should follow the provisions of NFPA 30, Flammable and Combustible Liquids Code. If tanks are inside buildings, vents should discharge outside the building at a location both free from any ignition source and not in contact with personnel (NFPA 704). Be sure to consider the effects of corrosion on venting devices for tanks that will contain corrosive liquids.

To minimize liquid loss and the possibility of injury from a broken fitting, connections for filling and emptying tanks are preferably made through the top. Plainly review labeled fill lines and equip them with a drain. According to NFPA 30, paragraph 5–2.4.5, use of compressed air is not permitted for the transfer of flammable and combustible liquids. To prevent priming of pumps that contain dangerous liquids and that have filling connections at the top of the tank, use self-priming pumps or pumps that generate enough suction to lift the liquid from the bottom of the tank.

Cleaning tanks can be an exceedingly dangerous operation and requires a confined-space entry permit. Establish, in written form, an exact and specific cleaning procedure to be strictly followed. Specifications for tank-cleaning procedure are set forth in NFPA 327, Cleaning and Safeguarding Small Tanks and Containers, and the American Petroleum Institute's (API) Requirements for, Safe Entry and Cleaning of Petroleum Storage Tanks.

The procedure should be modified for toxic compounds only to the extent that more complete cleaning and personal protective equipment may be required. Many chemicals can easily permeate the material of protective clothing and be absorbed through the skin. Liquid aromatic nitro compounds and amines are solvents for rubber and are absorbed through rubber gradually, as well as through the skin. Phenolic compounds, such as carbolic acid, cresylic acid, and the cresols, are also absorbed through the skin. For such exposures, use personal protective equipment made of one of the inert synthetic rubbers or plastics. (This equipment includes gloves, aprons, boots, and respiratory and eye protective equipment.)

Pipelines

Install pipelines in trenches or tunnels that carry chemicals. If they must be installed overhead, isolate them so they will not drip on anyone working underneath. Pipelines for carrying flammables, however, should not be installed in tunnels. Identify all pipelines as to their content.

The following are three major sources of chemical injury in pipeline work:

- *Failure of packing in valve stems or of gaskets in bolted flanges.* To minimize injuries from valve packing failure, surround the valve stem with a sheetmetal box or hood that will deflect spray away from the person operating the valve. So far as possible, renew packing without pressure on the valve.

- *Failure to check that valves are closed and locked and the lines drained before tension is released on flange bolts.* The opening between the flange faces may be temporarily covered with a piece of sheet lead while the flange bolts are being loosened and the faces separated. Loosen the bolts farthest away first so that drainage will tend to go away from the worker. Insert blinds in the flanges as soon as they are opened. For lines that are opened often, use blinds permanently pivoted on a flange bolt, with one end acting as a gasket and the other as a blind. Develop and implement a line-breaking procedure and standard.
- *Opening the wrong valve.* To prevent injuries and accidents from this source, identify pipelines and valves with tags, lettered markings, or distinctive colors.

Critical process valves should be automated and interlocked to prevent the chance of human error. Distinctive colors for identifying piping have been standardized in ANSI A13.1, Scheme for the Identification of Piping Systems, and are described in Chapter 2, Buildings and Facility Layout, in this volume. Provide specific identification of piping with a lettered legend that (1) names the material being piped, (2) summarizes the hazards involved, and (3) gives directions for safe use. Legends should be moisture resistant and contain pigments that are colorfast. Stencils or decals may be used to apply legends. Also have valves well separated and the immediate area well lit to ensure quick and easy identification.

At the conclusion of a job on a pipeline containing corrosive chemicals, wash tools and PPE thoroughly with a reagent to neutralize or remove the corrosive material. Then, rinse the tools and equipment in clean water.

Portable Containers

Where raw material liquid chemicals are used in quantity, it is generally better to install pipelines and outside storage tanks than to use portable containers. Spillage is reduced and localized, thereby making it easier to handle.

Properly store portable containers, such as drums, barrels, totebins, and carboys. Keep only a minimum amount of liquid at the point of operation—only enough for use on one shift is a common rule. Store the main stock in a safe, isolated place. If the liquid is corrosive or highly toxic, isolate the storage area from the rest of the facility by impervious walls and flooring. In the storage area, post a provisional plan for safe cleanup of spillage and safe disposal of contaminated materials. Otherwise, use a separate building.

Where corrosive liquids are stored, floors should be made of cinders, concrete treated to decrease its solubility, or other corrosion-resistant material. Concrete flooring is also satisfactory for storing flammable liquids. Be sure grounding is adequate if flammable liquids are involved. (See Chapter 10, Flammable and Combustible Liquids, in this volume.) Allow for good floor drainage in case a container in the storage area leaks or breaks. Governmental pollution-control regulations (under the Clean Water Act) prohibit draining these kinds of spills directly into sanitary or storm sewer systems.

The storage area must be well ventilated. Use natural ventilation whenever possible. Mechanical ventilation has no moving parts and fewer problems, but it may not always be adequate for the removal of vapors and fumes. Measurements should be made at a minimum annually to ensure adequate air changes per hour to prevent exceeding maximum contaminant levels in the room.

Stack full drums in racks, preferably with a separate rack for each material. Arrange these racks to permit easy access for moving the drums in and out, as well as for ready inspection of stock (Figure 12–8).

Barrels may be stacked vertically with dunnage between the tiers. However, for more ease in handling barrels, keep them in racks similar to those used for drums. The safest way to handle drums is to use mechanically powered lift equipment with drum-lifting clamps. Transporting drums on pallets is a common practice in many companies. Workers handling these drums should use nonsparking bung wrenches to tighten the bung and to prevent leaks during storage and transportation.

Different materials should be stored in separate designated areas, divided by wide aisles. Stack boxed carboys no higher than three. Do not use more than two tiers for carboys containing strong oxidizing agents, such as concentrated nitric acid or concentrated hydrogen peroxide. Before handling acid-filled carboys, inspect their nails for corrosion and the wood for weakening caused by acid. Before piling empty carboys, drain them thoroughly and replace the stoppers.

Special equipment for handling carboys should include a long-handled truck that picks up the boxed carboys under the handling cleats or between the bottom cleats provided on all

Figure 12–8. Special racks facilitate safe multiple tiering of drums. The rounded corners minimize the hazard of punctures. Bungs must always be tightened to prevent leaks.

standard 12- and 13-gal (45- and 49-L) boxed carboys. These trucks have handles long enough to keep persons who handle them away from splashes, in case carboys are dropped.

There is less danger, however, of dropping a carboy with a long-handled truck than with a standard two-wheeled truck. A long-handled truck becomes a much safer device for handling drums and barrels if it has a bed curved to fit the drum and a hook to catch the chime.

The safest way to empty a carboy is to suction the liquid with a vacuum pump or aspirator or to start a siphon with a rubber bulb or ejector. A satisfactory method of emptying a carboy is to use a carboy incliner, which holds the carboy firmly by the top, as well as by the sides, and automatically returns to the neutral position on being released.

Follow prescribed safety measures when handling carboys. Do not use compressed air, even from a hand pump, on a carboy unless the carboy is enclosed by another container so that the pressures inside and outside remain the same (Figure 12–9). Never permit pouring by hand or starting pipettes or siphons by mouth suction. Use mechanical pumps only if they are self-priming or have sufficient suction force to start themselves.

A corrosive or poisonous liquid requires a specially identified container. The simplest is a glass or plastic jug or jar, with a good closure, placed in a metal can.

Improvise, if necessary, a container resistant to shock by placing a jug in a metal pail and filling the space between the pail and the jug with pitch or foamed plastic. A container may be made to fit the jug with only a thin layer of padding, such as a layer of gasket rubber. Containers are also available commercially.

Figure 12–10. Tank cars should have wheels chocked and be marked by an approved visible sign, flags, or blue lights. A sign stating that the tank car is connected is another good safety precaution. *(Courtesy T & S Equipment Company.)*

Keep highly toxic substances, such as cyanides and soluble oxalates, in containers of a distinctive shape if they must be handled manually. All containers must be tagged or labeled, identifying the contents and specifying the appropriate hazard warning as required by OSHA CFR 1910.1200. Keep toxic materials locked up at all times, and allow them to be dispensed only by authorized personnel.

Where caustics or acids are stored, handled, or used, operable emergency showers or eyewash fountains must be available. Provide workers with chemical goggles, rubber aprons, boots, gloves, and other protective equipment necessary to handle a particular liquid. (See Chapter 7, Personal Protective Equipment, in this volume.)

Tank Cars

Isolate tank cars on sidings by derails and by blue stop flags or blue lights, chock the cars' wheels (Figure 12–10), and set hand brakes before the cars are loaded or unloaded. Before the car is opened, bond it to the loading line. Ground the track and the loading or unloading rack. Check all connections regularly. (See Chapter 10, Flammable and Combustible Liquids, in this volume, for a detailed discussion of loading and unloading tank cars.)

Unload tank cars holding chemicals through the dome connection rather than through the bottom connection. If the contents are nonflammable, air pressure (not to exceed 25 psi [172 kPa]) may be used for unloading. Equip the connections with a safety valve and gauge so that the pressure on the tank can be determined at any time (Figure 12–11).

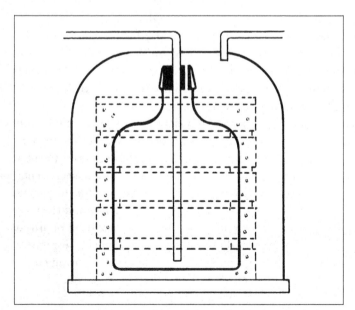

Figure 12–9. Air pressure admitted through the short pipe forces liquid from the carboy through the long pipe. Pressure is exerted on the bell cover, not the carboy.

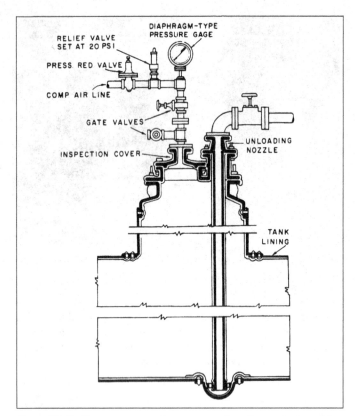

Figure 12–11. Cross-section of one type of tank car designed for unloading under air pressure. Note that air pressure should not be used for discharging flammable liquids.

Before the car is opened, the cap of the unloading pipe should be gently backed off without being completely removed. Allow pressure in the tank car to escape gradually before the cap is entirely removed.

Equip the unloading dock with a walkway at the height of the tank cars' domes and with drawbridges that can be lowered to make a firm walkway directly to the cars' domes. Install standard handrails and slip-resistant surfaces. If corrosive materials are handled, provide emergency showers and eyewash stations along the walkway.

Some materials normally shipped in tank cars solidify at temperatures reached during shipment. Tank cars used for such materials are ordinarily equipped with steam lines for melting the contents. Thoroughly thaw and clear the education lines and valves before unloading is begun.

The facility's line to the unloading dock must be completely drained after unloading. To facilitate this procedure, install the line with a slope toward the storage tank so that the line drains by gravity. Should vessel entry be necessary for maintenance or cleaning, check with the supervisor about special confined-space entry procedures at the facility.

Containers for Solids

Silos, portable containers, and magazines are sometimes used to store hazardous solids. The following sections describe safe ways of using these containers.

Silos

When new silos are to be installed in an old structure or when new materials are to be stored in old silos, check the mechanical strength of the structure or of the silos. Solids vary in unit weight, and the more dense materials produce higher unit loads.

Sufficient slope in the cone bottom of a tank or silo is a fundamental factor to consider when designing equipment for handling bulk solids. Sufficient slope permits the solid to run freely and prevents it from arching over. Where arching takes place, devise a method to restart the flow without having a worker enter the silo from either above or below the solid material. A vibrator to shake the bottom of a small metal silo or an agitator on the bottom of the silo can start the flow. Both are standard equipment, and one or the other can be applied to silos of almost any size or shape.

Sometimes, it is possible to work either from the bottom or from the top of a silo. If a person can break up the arch from the top with long tools without entering the silo, the job is reasonably safe. It is dangerous, however, when attempted from the bottom. Workers face an ever-present temptation to step inside the silo and work with a little more convenience until the material starts to flow. However, falls into open storage silos often result in injuries.

Surround silo openings at floor level or within 2 ft (61 cm) of it with standard guardrails and toeboards. If guardrails are impracticable or if the opening is not easily accessible (like the fill opening of a high silo), cover the opening with a grating that will not materially obstruct the opening but will prevent a worker from falling in. Many bin openings can be covered with a 2-in. (5.1-cm) mesh. Most silos can be covered by a 6-in. (15.2-cm) grating or by parallel bars on 6-in. (15.2-cm) centers.

Before entering a silo to perform any work, obtain a confined-space entry permit. This permit will outline the hazards associated with a silo, the tests required prior to entry, the names of qualified entrants and attendants, specific entry procedures, and an expiration time and date for the permit.

In most cases, tests for oxygen deficiency and the presence of flammable or toxic substances will be required. The use of approved fall protection equipment, respiratory or ventilation equipment, atmospheric monitoring equipment, and protective garments may be required. Lockout and tagout of all moving equipment and other associated energy sources is mandatory, is as continuous communication with a qualified attendant. Additional permits such as a hot-work permit may also be required.

Empty the silo before working inside. However, if this is not possible, be alert to the possibility of bridging of the material in the silo. Work activities can cause the bridging to fail, and if this occurs, an entrant positioned on the material may be engulfed.

Where silos are filled and emptied by continuous conveyors, the control of dust is likely to be a serious problem. In filling a silo, material is generally dropped from a belt conveyor. If the material is dropped through a chute from an elevator conveyor, even more dust may be produced. Prevent the escape of dust into the rest of the facility by enclosing the silo, except for the fill opening, with a skirt of either metal or fabric and by providing an exhaust ventilation up through the filling chute. If the material is scraped from a belt conveyor, cover the conveyor at the point of discharge with an exhaust hood. Also, provide a closed chute from the discharge point to the silo.

Because the dust in these cases is released at a low speed, an inward air velocity of about 50 fpm (0.25 m/s) through all the openings is sufficient. However, there can be no seriously disturbing air currents to blow dust out of the openings, and there must be enough velocity through the rest of the system.

The same general principles apply to the discharge of silos onto conveyors. There is seldom a serious dust problem, except at the loading and discharge points of the conveyors. Ventilate these points and cover them with hoods because it is seldom feasible to provide them with dust-tight enclosures or to reduce the dust by wetting it. Evaluate the air to determine if respiration protection is needed.

Combustible Solids

Where combustible materials are handled, keep the dust content of the air below the lower explosive limit. In addition to tight enclosures and dust-collection systems, good facility housekeeping will go far toward preventing disasters.

Dust explosions commonly occur as a series, not as a single shock. The first explosion uses up the dust in the air, but the shock stirs up more dust from the building. This dust is, in turn, set off. If the building is kept clean, this sequence cannot occur and damage will be minimal.

Exclude all sources of ignition from the area of a potentially explosive dust. Wiring, lights, and switches should comply with NFPA 61A–D on fire and dust explosions and with NFPA 70, National Electrical Code, for hazardous locations. Use electric motors that are totally enclosed and explosion-proof, or keep electric motors in a tight enclosure that is independently ventilated from a nonhazardous area and kept under positive pressure.

Use large and well-protected bearings. A hot bearing may ignite many types of dust. Use indirect heating systems only. Construct radiators for easy cleaning. Rigidly enforce the NO SMOKING rule.

Static electricity is the source of ignition in many fires. Prevent it from accumulating on most surfaces by maintaining relative humidity of 60% to 70%. If this cannot be maintained, use a ground to minimize static buildup. Also have workers wear electrostatic dispensating (ESD) protective footwear in these environments. (See ANSI Z41, Protective Footwear.) Remove static electricity from moving parts, such as conveyor belts and shafts, by using static collectors, grounding brushes, and conductive V-belts and leather belt dressings.

To avoid electrical shocks, ground metal silos to the conveyor frame. Check the electrical interconnection at loading and discharge points of conveyors. Use proper instrumentation to measure the effectiveness (resistance) of the grounding system. Install automatic sprinkler protection inside silos and processing equipment that contain combustible materials. Where water is undesirable—because of either reaction with or damage to the material—provide protection with inert gas extinguishers.

If a particularly hazardous material like metal powder is being handled, completely enclose the apparatus and flood it with an inert gas like carbon dioxide, nitrogen, or helium. This process removes or reduces the oxygen content and thus prevents ignition. Explosion-vent the area, preferably with windows and skylights that swing out on friction catches.

Portable Containers

The same general rules that apply to the storage and handling of portable containers of liquids apply also to the storage and handling of portable containers of solid materials. The most popular containers for solids are 50- and 100-lb (23- and 45-kg) paper bags. These bags are free from sifting or leaking. Handle them carefully, however, to prevent damage to machines. Have a few slipover bags available to cover the occasional broken or leaking container. This precaution both saves material and prevents skin contact with the dust.

Stack full bags on pallets or staging to prevent water damage. Interlocked stacking on pallets generally leads to better piling and fewer mechanical hazards when moving material. Protect bags from the weather—although some of the laminated bags are remarkably weather resistant.

Handle and ship large quantities of solid chemicals in bulk, cloth bags, barrels, and barrels with paper liners.

Filling bags or barrels with solids is always a potentially dusty operation. If the material being handled is finely divided or dangerous, the health and fire hazards may be serious. The simplest solution to this problem is to moisten the material so that it does not produce fine dust. You cannot use this solution, however, with some materials. Provide other methods of handling, or use hoods and exhaust ventilation.

The common way to open both cloth and paper bags is to slit one side crosswise and fold back the top and bottom. Reduce the hazard of knife cuts by keeping knives in scabbards when not in use and by using knives with hilts, guards, or returning blades.

Emptying bags and barrels involves health hazards similar to those of filling them. However, preventing skin contact with dust is somewhat harder when emptying such containers than when filling them. There is a tendency to dump them suddenly, which results in a rapid dispersion of dust that may not be trapped by the collecting systems.

Use exhaust hoods that are larger than the bags and barrels to solve this problem. In some instances, toxic materials can be handled by completely enclosing the barrel. For example, the head of the barrel is broken in after the enclosure is sealed. All dust is removed before the next barrel is put in.

Magazines

Store explosives in magazines of approved fireproof and bulletproof construction. Locate magazines at a safe distance from railroads and other buildings. Consult and follow federal, state, provincial, and local codes regarding storage of explosives. NFPA 495, Explosive Materials Code, gives detailed specifications for handling and storing explosives. (See also the Institute of Makers of Explosives in References at the end of this chapter.) Store explosives under lock and key, and maintain records of all explosives issued. Arrange storage so that the oldest explosives are used first. Advise local fire departments and emergency planning agencies when storing hazardous materials. (See also Chapter 18, Emergency Preparedness, in the *Administration & Programs* volume.)

Do not allow matches, flammable materials, metal, or metal tools in an explosives magazine. Keep floors clean and free from loose explosives. Have the floors, which are usually of wood, blind-nailed. Be sure that no nail or bolt head is exposed.

Keep magazines clean, dry, and well ventilated. Ventilation openings should not exceed 110 in.^{2} (710 cm^2) in area and should be screened to prevent the entrance of sparks and rodents. Permit only portable lights, approved for such use, in a magazine. Do not allow fire or sparks near a magazine. Keep the surrounding ground clear of brush, leaves, grass, debris, and other flammable material. Do not expose explosives to the direct rays of the sun.

Ammonium nitrate requires special precautions, including stacking limitations, air space, and ventilation. Do not permit any oils or hydrocarbons near ammonium nitrate.

Always open packages of explosives at least 50 ft (15 m) away from the magazine. Use only wood wedges and wood, fiber, rawhide, zinc, babbitt metal, or rubber mallets to open cases of explosives. Never keep blasting caps or detonators of any kind in the same magazine with other explosives.

Always keep magazines of explosives and blasting supplies in a place where access to them by animals, unauthorized persons, or children is impossible. Many children have been killed or crippled because they have obtained detonators from unwatched or unguarded sources.

Containers for Gases

There are many regulations regarding the storage and handling of gas cylinders. Several government agencies and private organizations and their standards should be consulted—OSHA, the National Fire Protection Association, the Compressed Gas Association, and others. Care should be taken to comply with all local government codes as well.

Compressed gas cylinders should be stored in the upright position on a smooth floor, and valve covers should be in place. All cylinders should be chained or otherwise fastened firmly against a wall, post, or other solid object. Different kinds of gases should either be separated by aisles or stored in separate sections of the building or storage yard. Empty cylinders must be stored apart from full cylinders.

Set up storage areas away from heavy traffic. Containers should be stored in a place where there is minimal exposure to excessive temperature, physical damage, or tampering. Never store cylinders of flammable gases near flammable or combustible substances. Before cylinders are moved, check to make certain all valves are closed. Always close the valves on empty cylinders. Never let anyone use a hammer or wrench to open valves. As stated earlier, cylinders should never be used as rollers to move heavy equipment. Handle all cylinders with care—a cylinder marked "empty" may not be.

To transport cylinders, use a carrier that does not allow excessive movement, sudden or violent contacts, and upsets. When a two-wheeled truck with rounded back is used, chain the cylinder upright. Never use a magnet to lift a cylinder. For short-distance moving, a cylinder may be rolled on its bottom edge, but never dragged. Cylinders should never be dropped or permitted to strike one another. Protective caps must be kept on cylinder valves when cylinders are not being used. When in doubt about how to handle a compressed gas cylinder, or how to control a particular type of gas once it is released, ask the safety professional or the safety department for advice.

STORAGE AND HANDLING OF CRYOGENIC LIQUIDS

Copyright © by Technical Publishing, a company of the Dun & Bradstreet Corporation. Reprinted with permission from the June 14, 1979, issue of Plant Engineering Magazine.

Most gases used in facilities are also available as cryogenic liquids. Among the most common are oxygen, nitrogen, argon, helium, and hydrogen.

Liquid oxygen is frequently delivered to a facility—and even to a construction site—and then vaporized for use in flame cutting, welding, metalizing, or heating. Other uses include oxygen injection into a foundry cupola and oxygen-based processes such as paper-pulp bleaching and steelmaking.

Liquid nitrogen is also very common. A variety of processes have been developed that use the liquid primarily because of its high refrigeration values. Examples include freezing food, stripping scrap rubber from tires and cables, and removing parting lines and risers from plastic injection-molded parts. It is even used as a super-cold quencher for high-alloy steels to transform retained austenite.

The availability of large volumes of liquid helium has made possible the rapid development of superconductivity. And these examples are only a few from only some of the major industrial gases. [Most recently, cryogenic liquids are being used in conjunction with specialized ceramics in superconductor technology.]

The key to expanding use of cryogenic liquids is economics. The cost of delivering and storing the liquid is often lower than buying the gas in compressed-gas cylinders. At room temperature (70°F or 20°C) and atmospheric pressure, nitrogen occupies 700 times as much space as the same amount of nitrogen in liquid form. The reduction in cost for containers, demurrage, shipping, and storage is enormous. However, handling liquefied gases that are stored and used at very low temperatures requires some special knowledge and special precautions. To use these gases safely, the facility engineer and employees must know the specific properties of each liquefied gas and its compatibility with other materials and must follow some commonsense procedures.

Characteristics of Cryogenic Liquids

A cryogenic liquid has a normal boiling point below –238°F (–150°C). [*Cryogenic* is defined in depth in the National Institute of Standards and Technology's (NIST) Handbook 44. See References at the end of this chapter.] The most commonly used industrial gases that are transported, handled, and stored at cryogenic temperatures are oxygen, nitrogen, argon, hydrogen, and helium. Three rare atmospheric gases—neon, krypton, and xenon—are used in the liquid state. Natural gas, liquefied natural gas (LNG) or liquid methane, and carbon monoxide also are handled as cryogenic liquids, although they are not usually classified as industrial gases. Liquefied ethylene, carbon dioxide, and nitrous oxide are transported and stored as liquids, but are not classified as cryogenic.

Handling cryogenic liquids in large volumes is not new. Liquid oxygen was first shipped by tank truck in 1932, and it is common to see portable liquid containers, cryogenic trailers and trucks, and railroad tank cars hauling large quantities of liquefied gases across the country. Cryogenic tanker ships transport LNG overseas, and aircraft move other liquefied gases, especially liquid helium, from one place to another.

Many safety precautions that must be taken with compressed gases [see Chapter 19, Welding and Cutting, in this volume] also apply to liquefied gases. However, some additional precautions are necessary because of the special properties exhibited by fluids at cryogenic temperatures.

Both the liquid and its boil-off vapor can rapidly freeze human tissue and can cause many common materials such as carbon steel, plastic, and rubber to become brittle or fracture under stress. Liquids in containers and piping at temperatures at or below the boiling point of liquefied air (–318°F or –194°C) can cause the surrounding air to condense to a liquid.

Extremely cold liquefied gases (helium, hydrogen, and neon) can even solidify air or other gases to which they are directly exposed. In some cases, even plugs of ice or foreign material will develop in cryogenic container vents and openings and cause the vessel to rupture. Following the supplier's operating procedures can help prevent plugging. If a plug should form, contact the supplier immediately. Do not attempt to remove the plug; move the vessel to a remote location.

All cryogenic liquids produce large volumes of gas when they vaporize. For example, 1 volume of saturated liquid nitrogen at 1 atmosphere vaporizes to 696.5 volumes of nitrogen gas at room temperature at 1 atmosphere. The volume expansion ratio of oxygen is 860.6 to 1. Liquid neon has the highest expansion ratio—1,445 to 1—of any industrial gas.

Vaporized in a sealed container, these liquids produce enormous pressures. For example, when 1 volume of liquid helium at 1 atmosphere is vaporized and warmed to room temperature in a totally enclosed container, it has the potential to generate pressure of more than 14,500 psig (100,000 kPa). Because of this high pressure, cryogenic containers usually are protected with two pressure-relief devices: a pressure-relief valve and a frangible disk.

Relief devices should function only during abnormal operation and emergencies. If they are triggered, the system should be checked for loss of insulating vacuum or for leaks. Do not tamper with the safety-valve settings. Report leaking or improperly set relief valves to the gas supplier and have them replaced or reset by qualified personnel. Similarly, all safety valves with broken seals or with any frost, ice formation, or excessive corrosion should be reported.

12 Materials Handling and Storage

Most cryogenic liquids are odorless, colorless, and tasteless when vaporized to a gas. As liquids, most have no color; liquid oxygen is light blue. However, whenever the cold liquid and vapor are exposed to the atmosphere, a warning appears. As the cold boil-off gases condense moisture in the air, a fog that extends over an area larger than the vaporizing gas forms.

General Safety Practices

The properties of cryogenic liquids affect their safe handling and use [Table 12–A]. The table presents data on flammability limits in air and oxygen, spontaneous ignition temperature in air at 1 atmosphere, and other information to help determine safe handling procedures. None of the gases listed is corrosive at ambient temperatures, and only carbon monoxide is toxic.

The liquids are listed by decreasing boiling point. Although xenon boils above –238°F (–150°C), it also has been included. Natural gas is not listed because it is a mixture of methane and other hydrocarbons; its boiling point depends on its composition. However, natural gas is primarily methane, and methane data are included.

General safety practices include:

- Always handle cryogenic liquids carefully. They can cause frostbite on skin and exposed eye tissue. When spilled, they tend to spread, covering a surface completely and cooling a large area. The vapors emitted by these liquids are also extremely cold and can damage delicate tissues.
- Stand clear of boiling or splashing liquid and its vapors. Boiling and splashing always occur when a warm container is charged or when warm objects are inserted into a liquid. These operations should always be performed slowly to minimize boiling and splashing. If cold liquid or vapor comes in contact with the skin or eyes, first aid should be given immediately [Figure 12–12].
- Never allow any unprotected part of the body to touch uninsulated pipes or vessels that contain cryogenic fluids. The extremely cold metal will cause the flesh to stick fast to the surface and tear when withdrawn. Touching even nonmetallic materials at low temperatures is dangerous.

Tongs should be used to withdraw objects immersed in a cryogenic liquid. Objects that are soft and pliable at room temperature become hard and brittle at extremely low temperatures and will break easily.

Workers handling cryogenic liquids should use eye and hand protection to protect against splashing and cold-contact burns. Safety glasses are also recommended. If severe spraying or splashing is likely, a face shield and chemical goggles should be worn. Protective gloves should always be worn when anything that comes in contact with cold liquids and their vapors is being handled. Gloves should be loose fitting so that they can be removed quickly if liquids are spilled into them. Trousers should remain outside of boots or work shoes.

Special Precautions

Some liquefied gases require special precautions. For example, when oxygen is handled, all combustible materials, especially oil or gases, should be kept away. Smoking or

TABLE 12–A. Physical Properties of Cryogenic Fluids

	Xenon (Xe)	Krypton (Kr)	Methane (CH$_4$)	Oxygen (O$_2$)	Argon (Ar)	Carbon Monoxide (CO)	Nitrogen (N$_2$)	Neon (Ne)	Hydrogen (H$_2$)	Helium (He)
Boiling point, 1 atm °F	–163	–244	–259	–297	–303	–313	–321	–411	–423	–452
°C	–108	–153	–161	–183	–186	–192	–196	–246	–253	–268
Melting point, 1 atm °F	–169	–251	–296	–362	–309	–341	–346	–416	–435	—*
°C	–112	–157	–182	–219	–189	–207	–210	–249	–259	—
Density, boiling point 1 atm lb/cu ft	191	151	26	71	87	49	50	75	4.4	7.8
Heat of vaporization, boiling point Btu/lb	41	46	219	92	70	98	85	37	193	10
Volume expansion ratio, liquid at 1 atm boiling point to gas at 60°F, 1 atm	559	693	625	861	841	—	697	1445	850	754
Flammable	No	No	Yes	No†	No	Yes	No	No	Yes	No

* Helium does not solidify at 1 atmosphere pressure.
† Oxygen does not burn, but will support combustion. However, high oxygen atmospheres substantially increase combustion rates of other materials and may form explosive mixtures with other combustibles. Flame temperatures in oxygen are higher than in air.

Treating Cold-Contact Burns

Workers will rarely come in contact with a cryogenic liquid if proper handling procedures are used. However, in the event of contact with a liquid or cold gas, a cold-contact "burn" may occur. Actually, the skin or tissue freezes.

Medical assistance should be obtained as soon as possible. In the interim, the following emergency measures are recommended:

- Remove any clothing that may restrict circulation to the frozen area. Do not rub frozen parts, as tissue damage may result.
- As soon as is practical, immerse the affected part in warm water (not less than 105°F or more than 115°F, or 40°C to 46°C). Never use dry heat. The victim should be in a warm room, if possible.
- If the exposure has been massive and the general body temperature is depressed, the patient should be totally immersed in a warm-water bath. Supportive treatment for shock should be provided.
- Frozen tissues are painless and appear waxy and yellow. They will become swollen and painful and prone to infection when thawed. Do not rewarm rapidly. Thawing may require 15 to 60 minutes. For fair-skinned people, thawing should continue until the pale blue tint of the skin turns pink or red. For dark-skinned people, assess frostbite by the swelling and blistering of the skin. Reduction of swelling indicates alleviation of frostbite. Morphine or tranquilizers may be required to control the pain during thawing and should be administered under professional medical supervision.
- If the frozen part of the body thaws before the doctor arrives, cover the area with dry, sterile dressings and a large, bulky protective covering.
- Alcoholic beverages and smoking decrease blood flow to the frozen tissues and should be prohibited. Warm drinks and food may be administered.
- As with any injury or illness, monitor vital signs.

Figure 12–12. Emergency treatment for contact with a cryogenic liquid or gas.

open flames should never be permitted where liquid oxygen is stored or handled. NO SMOKING signs should be posted conspicuously in such areas.

Oxygen will vigorously accelerate and support combustion. Because the upper flammable limit for a flammable gas in air is higher in an oxygen-enriched air atmosphere, fire or explosion is possible over a wider range of gas mixtures.

Liquid oxygen or oxygen-rich air atmospheres should not come in contact with organic materials or flammable substances. Some organic materials—oil, grease, asphalt, kerosene, cloth, tar, or dirt containing oil or grease—react violently with oxygen and may be ignited by a hot spark.

If liquid oxygen spills on asphalt or on another surface contaminated with combustibles (e.g., oil-soaked concrete or gravel), no one should walk on, and no equipment should pass over, the area for at least 30 minutes after all frost or fog has disappeared.

Other special precautions against electrostatic buildup can be taken. Clothing saturated with oxygen is readily ignitable and will burn vigorously. Any clothing that has been splashed or soaked with liquid oxygen, or exposed to a high gaseous-oxygen atmosphere, should be changed immediately. The contaminated systems should be aired for at least an hour until they are completely free of excess oxygen. Workers exposed to high-oxygen atmospheres should leave the area and avoid all sources of ignition until the clothing and the exposed area have been completely ventilated. Static dissipative (SD) protective footwear should be worn.

Finally, oxygen valves should be operated slowly. Abruptly starting and stopping oxygen flow may ignite contaminants in the system.

Inert Gas Precautions

The primary hazards of inert gas systems are rupture of containers, pipelines, or systems, and asphyxiation. A cryogen cannot be indefinitely maintained as a liquid even in a well-insulated container. Any liquid, or even cold vapor trapped between valves, has the potential for causing enough pressure buildup to cause violent rupture of the container or piping. The use of reliable pressure-relief devices is mandatory.

Loss of vacuum in vacuum-jacketed tanks will increase evaporation in the system, causing the relief devices to function and vent the product. Route the vented gases to a safe outdoor location. If the gases are not vented outdoors, maintain adequate ventilation; use instruments to monitor the area.

Flammable Gas Precautions

Do not permit smoking or open flames where flammable fluids are stored or handled. Clothes that minimize ignition sources should be worn in atmospheres that may contain concentrations of flammable gases.

Properly ground all major stationary equipment. Provide ground connections between stationary and mobile equipment before any flammable gas is loaded or unloaded. All electrical equipment used in or near flammable-gas loading and unloading areas, or in atmospheres that might contain explosive mixtures, should conform to NFPA 50B, Liquefied Hydrogen Systems at Consumer Sites, and NFPA 59A, Production, Storage, and Handling of Liquefied Natural Gas (LNG), or to Article 500 of the National Electrical Code [NEC] NFPA 70. When flammable cryogenic liquids and gases are handled inside, adequate positive mechanical ventilation is necessary. Electrical equipment and wiring must conform to Article 501 of the NEC. [See Chapter 8, Electrical Safety, in this volume.]

Pipe flash-off gas from closed liquid-hydrogen containers, used or stored inside, through a laboratory hood to the outside, or vent it by other means to a safe location. If hydrogen is vented into ductwork, the ventilation system should be independent of other systems, and sources of ignition must be eliminated at the exit.

Asphyxiation

All gases except oxygen have the potential to cause asphyxiation by displacing breathable air in an enclosed workplace. The presence of these gases can only be detected by using air-monitoring instruments. Asphyxiation can be sudden or may occur slowly with workers not being aware that they are in trouble. Use and store these gases in well-ventilated areas. [See SB-2, Oxygen-Deficient Atmospheres, the Compressed Gas Association, in the References at the end of this chapter.]

Unless large quantities of inert gas are present, the problem is easily prevented by using proper ventilation at all times. Vent nitrogen outside to safe areas. Install analyzers with alarms to alert workers to oxygen-deficient atmospheres. Constant monitoring, sniffers, and other precautions should be used to survey the atmosphere when personnel enter enclosed areas or vessels. When it is necessary to enter an area where the oxygen content may be below 19.5%, have workers wear self-contained breathing apparatus (SCBA) or combination supplied-air respirator with emergency escape SCBA. A conventional gas mask or other air-purifying respirator will not prevent asphyxiation.

Most personnel working in or around oxygen-deficient atmospheres rely on the buddy system for protection. But, unless equipped with a portable air supply, a co-worker may also be asphyxiated when entering the area to rescue an unconscious partner. The best protection is to provide both workers with a portable supply of breathable air. Lifelines are acceptable only if the area is free of obstructions and one worker is able to lift the other rapidly and easily.

Training

The best single investment in safety is trained personnel. Some workers will need detailed training in a particular type of equipment, cryogen, or repair operation. Others will require broader training in safe handling practices for a variety of cryogenic liquids. The following subjects should be familiar to everyone involved in using, handling, storing, or transferring cryogens:

- nature and properties of the cryogen in both its liquid and gaseous states
- operation of the equipment

- approved, compatible materials
- use and care of protective equipment and clothing
- first-aid and self-aid techniques to employ when medical treatment is not immediately available
- handling emergency situations such as fires, leaks, spills.

Good Housekeeping Practices

Good housekeeping is essential to safety when handling cryogens. Few cryogens are spontaneously hazardous, but each liquefied gas poses its own hazard.

Liquid oxygen may form mixtures that are shock sensitive with fuels, oils, or grease. Porous solids, such as asphalt or wood, can become saturated with oxygen and also become shock-sensitive. Ignition is more likely with weaker sparks and lower temperatures than would be required in air.

Flammable gases such as hydrogen and methane are lighter than air. At normal temperatures, they will rise. But at the first temperatures that exist just after these gases evaporate from the liquid state, the saturated vapor is heavier than air and tends to fall. Wind or forced ventilation will affect the direction of the released gases and must be considered during disposal of any leaking fluid.

The location and maintenance of safety and firefighting equipment are also important. Inform outside personnel also about all necessary safeguards before they enter a potentially hazardous area. In general, following good housekeeping rules and demanding that workers observe safety rules will minimize accidents.

Safe Storage and Handling Recommendations

Cryogenic liquids are stored and transported in a wide range of containers from small Dewar flasks to railroad tank cars and tank trucks [Figures 12–13 and 12–14]. Only use equipment and containers designed for the intended product, service pressure, and temperature. If any questions arise about correct handling or transporting procedures, or about the compatibility of materials with a given cryogen, consult the gas supplier.

Cryogenic liquids ordinarily should not be handled in open containers unless they are specifically designed for that purpose and for the product. Cryogenic containers should be clean and made from materials suitable for cryogenic temperatures—such as austenitic stainless steels, copper, and certain aluminum alloys. Cryogens should be transferred slowly into warm lines or containers to prevent thermal shock to the piping and container and to eliminate possible excessive pressure buildup in the system. When liquids are transferred from one container to another, the receiving container should be cooled gradually to prevent shock and reduce flashing. High concentrations of escaping gases should be vented so that they do not collect in an

12 Storage and Handling of Cryogenic Liquids   379

Figure 12–14. Large volumes of cryogenic liquids can be handled easily and safely. Here, more than 50,000 gal (189, 215 L) of liquid helium at −452°F are lifted by container carrier for loading aboard ship. Loss of helium from vaporization during a 2-week voyage is so small that it is nearly undetectable. The liquid helium is surrounded by a liquid nitrogen chamber and contained in a specially insulated outer jacket.

Figure 12–13. Cryogenic liquids are stored and transported in a wide range of containers. A typical example is this flask of liquid nitrogen, an open-mouthed, unpressured, vacuum-jacketed vessel. This flask is freestanding for photographing only. Note appropriately secured flasks behind displayed flask.

enclosed area. Be sure that workers observe the following rules when handling cryogens:

- Do not drop warm solids or liquids into cryogenic liquids. Violent boiling will result and liquid can splash onto personnel and equipment.
- Avoid breathing vapor from any cryogenic liquid source except for liquid-oxygen equipment designed to supply warm breathable oxygen. When cryogenic liquids are being discharged from drain valves or blowdown lines, open the valves slowly to prevent splashing. Smoking should never be permitted.

Two types of portable liquid-storage vessels are generally used to hold and dispense cryogenic liquids—nonpressurized Dewar containers and pressurized liquid cylinders [Figure 12–13]. Dewar containers are open-mouthed, nonpressurized, vacuum-jacketed vessels usually used to hold liquid argon, nitrogen, oxygen, or helium. Some of these containers are designed for lightweight liquids, such as helium, and for maximum holding times. Their internal support system cannot hold some of the heavier cryogens, such as argon. When using Dewar containers, be sure that no ice accumulates in the neck or on the cover. This could cause a blockage and subsequent pressure buildup. Laboratory Dewar flasks with wide-mouthed openings have no cover to protect the liquid. Most are made of metal, but some smaller units are of glass.

Liquid cylinders are pressurized containers, usually vertical vessels, designed and fabricated according to U.S. Department of Transportation (DOT) specifications. There are three major types of liquid cylinders: for dispensing liquid or gas, for withdrawing gas only, and for withdrawing liquid only. Each type of liquid cylinder has appropriate valves for filling and dispensing and is adequately protected with a pressure-control valve and a frangible disk.

An unusually cold outside jacket on a cryogenic vessel

indicates some loss of insulating vacuum. Frost spots may appear. A vessel in this condition should be drained, removed from service, and set aside for repair. Such repairs should be handled by the manufacturer or other qualified company.

Some liquid cylinders can be handled manually, but moving them by portable handcarts is strongly recommended. Secure the cylinder to the handcart with a strap to prevent the cylinder from slipping off the cart. Vessels containing cryogens must be handled very carefully. They should not be dropped or tipped on their sides.

Transfer Lines

Many types of filling or transfer lines are used to handle the flow of cryogenic fluids from one point to another. Some of these transfer lines are small, uninsulated copper or stainless steel lines; large-diameter rigid lines; flexible hose systems; vacuum-jacketed lines; and other insulated systems.

Cryogenic liquids can be transferred by two methods. The simpler of these is gravity. In this case, the height of the stored liquid serves as the transfer medium. The other method—pressurized transfer—uses the vapor pressure of the product, or pressure from an external source, to move the liquid to the lower-pressure receiving container.

Various types of cryogenic pumps are also available. Flow rates may vary from less than one to several hundred gallons (liters) per minute. The product should be in liquid form in the transfer lines. Any vaporization of liquid within the system may cause (1) excessive pressure drop, (2) two-phase (liquid and gas) flow, and (3) cavitation that can harm the operation of cryogenic pumps.

Short transfer lines used for intermittent service are normally not insulated. Lines used for continuous transfer of cryogens, however, are usually insulated. Be sure that all liquid transfer hoses have dust caps.

Because liquid hydrogen and liquid helium have extremely cold temperatures and low heats of vaporization, vacuum-jacketed lines are required to transfer them. To reduce costly line and flash-off losses, vacuum-jacketed lines are also occasionally used for in-facility transfer of atmospheric cryogenic fluids.

SHIPPING AND RECEIVING

In the United States, the supervisor of the shipping and receiving department must be aware of DOT regulations and labels. Bills of lading or shipping forms must identify the item. Items not properly labeled should not be accepted.

Floors, Ramps, and Aisles

Floors in warehouses, storerooms, and shipping rooms must be level. Unevenness of floors may lead to the toppling of piles of stored materials. Conspicuously post safe floor load capacities and maximum heights to which specific materials may be piled. Where bulk material, boxes, or cartons of the same weight are regularly stored, it is a good practice to paint a horizontal line on the wall indicating the maximum height to which the material may be piled.

Check the strength of floors prior to using powered industrial trucks. A structural expert should determine floor load capacity by studying architectural data of the facility, the age and condition of the floor members, the type of floor, and other pertinent factors.

Wherever materials are stored or transported, keep the surface of floors, platforms, and ramps in good condition. Repair damaged structures immediately; particularly watch the area around doorways and elevator entrances.

Ramps should have nonskid surfaces. When ramps are used for hand trucking, lay a slip-resistant foot strip in the center of the ramp, or in the center of each lane for two-way traffic. Ramps should have handrails and, where there is heavy trucking, substantial curbs. A separate pedestrian lane, divided from the truck lane by a handrail, is a good idea for ramps used by both pedestrians and trucks.

Aisles should be wide enough (1) to enable employees to move about freely while handling material or removing it from bins, racks, or piles and (2) to allow safe passage of loaded equipment. Use traffic control devices, such as stop signs, to help control in-facility traffic. Mirrors, placed at blind intersections, help to prevent collisions. Warning signs and signals at such locations also serve as useful reminders, particularly to operators of power equipment. Similarly protect doorways and entrances to tunnels and elevators. Equip mobile equipment used in storage areas with backup warning devices. Clearly mark aisles and unloading areas with white paint or black and white stripes. To prevent falling and tripping, trucks not in use, material, and other objects should not stand in or extend into aisles.

Keep aisles leading to sprinkler valves and fire-extinguishing equipment clear. Do not pile materials closer than 18 in. (45.7 cm) to sprinkler heads. Closer spacing may reduce the effectiveness of the heads in the event of fire. For overly large, closely packed piles of combustible cases, bales, cartons, and similar stock, provide up to 36 in. (91.4 cm) of clearance. (See NFPA 13, Installation of Sprinkler Systems.) Request administrative help to prevent workers from piling materials against fire doors.

Lighting

General illumination of warehouses and storage rooms should follow ANSI/IES RP7, Practice for Industrial Lighting, published by the Illuminating Engineering Society (IES). (See the IESNA Handbook.)

Provide special lighting for operations requiring greater

illumination. All lighting fixtures and wiring should meet the requirements of NFPA 70, National Electrical Code.

Stock Picking

To easily move full pallet loads, use powered industrial trucks. Shipments are typically made, however, by truck and/or railcar loads of mixed lots because many facilities produce a variety of small finished products.

In both operations, such stock is usually found in racks or bins. With the increase in high-bay storage, special order-picker equipment is required. Such operations lend new efficiency to the movement of material, as well as new hazards in the areas of traffic, personal injury, and fire protection. The worker operating from a mobile order-picker truck is exposed to falls from a height, as well as to falling objects and materials-handling accidents.

Often, workers are required to climb a ladder to get small parts or stock. Workers should use only heavy-duty, OSHA-approved, materials-handling ladders. These may be on rollers equipped with a braking mechanism with rubber feet that contact the surface as the worker's weight is imposed on the ladder. These ladders also have working platforms and standard guardrails that protect workers from falls as they reach for stock or parts.

Under no circumstances should employees be allowed to use ordinary stepladders (particularly the short two- or three-step stools). Heavy-duty, materials-handling equipment is required. Employees must never climb racks or shelves. Ladders that are damaged or not in good working condition must be tagged DO NOT USE and placed out of service until repaired or discarded.

Dock Boards

Dock boards are also known as bridge plates, dock plates, gangplanks, and bridge ramps. Design dock boards used in trailer and railcar loading and unloading to carry four times the heaviest expected load and to be wide enough to permit easy maneuvering of hand or power trucks. (Information given here applies to both hand- and power-truck operations.)

Many modern facilities use automatic dock levelers and fixed-position hydraulic dock boards on both truck and rail docks. Dock shelters, usually found in inclement weather zones, effectively keep moisture from the dock board and give shipping and receiving department personnel a safer working environment.

Dock boards require regularly scheduled maintenance. Most operate hydraulically and provide a solid working and walking surface for heavy industrial trucks, as well as for hand-truck operations. All shear points must be guarded. Paint the edges of movable sections yellow to denote a possible tripping hazard. When not in use, store dock boards in a safe, secure place provided for that purpose.

Trailers being loaded or unloaded by trucks must have wheel chocks at each wheel. The nose must have a jack stand if the trailer has been disengaged from the tractor and the equipment will operate in the forward portion of the trailer. Vertical restraints (locking mechanisms) can also be used. The truck backs into a loading dock bay, and the loader engages the system, which locks the ICC bar of the trailer. This enables the loader to move on and off the trailer with a powered industrial truck without the fear of the trailer being pulled away from the dock (Figure 12–15).

Design and maintain dock boards so they do not rock or slide when they are being used. Secure dock boards in position. Either anchor them down or equip them with devices to prevent slippage from the platform or the car threshold. The sides of dock boards should be turned up at right angles or otherwise designed to prevent trucks from running over the edge. Dock boards should also have a slip-resistant surface to prevent employees and trucks from slipping. They should be kept clean and free of oil, grease, water, ice, and snow.

Provide handholds, or similar devices, on dock boards to enable safe handling. Where practical, fit fork loops or lugs to the plates; in that way, they can be handled by forklift trucks. Another method for lifting steel dock boards uses a low-voltage magnet that is hung from the forks of a forklift truck and powered by the truck's battery.

When dock boards are handled manually, workers should lower or slip them into place and not drop them. Assign enough workers to the job to permit safe and easy handling. Take extra care to prevent foot injuries.

Protective devices (wheel checks) should be used to prevent engines or car pullers from moving railroad cars while dock boards are in position and workers are on them. To warn train crews, use standard blue flags for daytime and blue lights for night work. Consult local railroad authorities about the specific warning devices.

Machines and Tools

Machines used in receiving and shipping, such as shears, saws, and nailing machines, should have protective guards. Workers should wear protective clothing—goggles, for instance—when operating a nailing or banding machine.

Use high-quality tools and keep them in good condition. Keep tools sharp, and provide holsters for workers to carry them in. Provide files that have a handle on the tang.

When workers use a drawknife instead of a scraper to remove markings from cases, boxes, and barrels, they should never brace the work with their knees. Should the drawknife slip, injury would almost certainly occur.

Steel and Plastic Strapping

Train workers in both the application and removal of steel and plastic strapping—which may be either flat or round.

Figure 12–15. Two types of vehicle restraints to secure a truck. *(Courtesy Rite Hite Corp.)*

In all cases, the workers should wear safety goggles and leather gloves. Steel-studded gloves may be required for heavy strapping.

Use equipment designed for applying and removing strapping. When operating a strapping tool, workers should face in the direction of the pull, one foot ahead of the other. Then, if the strap breaks or the tool slips, workers are in a position to protect themselves. When final tension is being attained, the worker should get out of the direct line of the strapping so that, in case of breakage, the ends of the strapping will not strike the face or body. Never use an extension "cheater bar" or a hand tool to gain more tension. Keep others away from strapping application and removal.

Workers should break or cut off excess strap beyond the tension-holding seal before a bound shipment is considered safe for further handling or shipment. Before attempting to move bound merchandise or material, the worker should examine it for broken bands or loose ends. Broken bands should be removed and safely disposed of and, if possible, replaced to keep the shipment from coming apart.

Workers should never handle a box, carton, or package by the steel bands, either manually or with a lift truck or other mechanical-handling device. Workers should check stored packages, boxes, cartons, or other material for loose or protruding banding ends that might cut or otherwise injure passersby.

To remove strapping from bound containers, workers should use a cutter designed for the work. Workers should never break the steel strap by applying leverage with a claw hammer, chisel, axe, crowbar, or other tool. Strapping should be cut square, never at an angle. Strapping cut at an angle has much sharper ends. Have containers available to dispose of plastic and steel bands separately.

Before cutting strapping, workers should make sure that no one is standing close enough to be hit by loose ends of strapping. To cut bands safely, the workers should place one gloved hand on the nearest portion of the strapping. Then, if the strapping springs, it will be held to one side and fly away from the worker's face. In addition, the face should be held out of the direct line of the strap, and workers should wear goggles.

Burlap and Sacking

Burlap and sacking are often received baled. Opening these bales is a job requiring some skill and experience. The supervisor of the department should thoroughly instruct employees in the exact procedure to be used. Burred ends of wire used to tie sacks may cause many cuts. The employee should hold down one end of the wire when making the cut and should stand clear of the free end.

Glass and Nails

Broken glass, often found in containers being unpacked, is a serious hazard in the shipping room. When unpacking glass or crockery, the worker should assume that broken material may be present. If possible, workers should wear gloves.

Companies operating large shipping rooms report many injuries from flying nails. When driving a nail, workers should start it with a few light taps so that it will take a good hold and not fly. Workers should wear eye and face protection. Instruct workers to pull out nails that have been started at the wrong angle and drive them in properly. Nails should be driven flush so that no part of the nail projects above the surface. Not only fellow workers, but also employees of the customer or carrier, can be injured as they handle cases with poorly driven nails. Poorly driven nails may also cause loss of merchandise if the packages come apart while being shipped.

Do not let workers leave loose nails on the floors. Loaded trucks passing over the nails may drive them into the floor with the points up. In this position, they are a serious hazard because they can easily be driven through workers' shoes and can also puncture pneumatic tires. The best practice is to have workers pick up nails from the floor throughout the day.

Instruct employees who open packing cases to bend nails over. If the packing case is to be used again or to make smaller boxes, remove the nails.

If nails are used directly from kegs, supervisors should make certain that the nails holding the keg head in place are pulled out. Have kegs used for storing nails at the workstation placed in an inclined rack so that the nails will feed out of the kegs.

Pitch and Glue

Pitch is a sealing material. To protect export shipments, parts are wrapped first with paper, plastic, or cheesecloth and then often covered with burlap. Pitch or other material is used to seal the package. Hot pitch, however, can severely burn the skin—not only because it is hot, but also because it is difficult to remove.

Only skilled personnel who have been instructed in the hazards of using pitch and know how to avoid them should work with pitch. These workers should wear goggles, face masks, gloves with sleeves rolled over the gauntlets, and aprons. Wherever possible, it is better to use cold mastic asphalt instead of hot pitch. Plastics and other sealants and adhesives are often used instead of pitch and are safer to the employee.

Labeling glue, which often contains silicate of soda, causes discomfort when it splashes into the eyes. When working with labeling glue, workers should wear goggles and use the glue brush carefully.

Barrels, Kegs, and Drums

Projecting nails, jagged hoops and metal bands, ends of wire, and splinters and slivers cause many barrel-handling injuries, some of which lead to infection. Before handling barrels and kegs, employees should inspect them and take precautions against these hazards.

One method of opening a barrel is to use a lather's hatchet or a crate-opening tool to remove the nails. Then, when the top hoop is removed and the second hoop loosened, the head can easily be removed intact. To loosen nails on a barrel with wood hoops, simply strike the hoop sharply with a hammer or hatchet near the point where the nail is located. The nail can then be easily pulled. This method of opening barrels not only preserves the barrel for future use, but also prevents contaminating the barrel's contents.

Opening single-trip drums with a hammer and chisel is a frequent source of cuts and scratches. A commercial drum opener, however, will open these drums without hazard and leave a smooth, rolled edge. Drum handling, storage, and opening were discussed earlier in the Hazardous Materials section.

Boxes and Cartons

Employees who open boxes or cartons may incur wire punctures, nail punctures, or cuts from the device used for opening them. When wire-bound boxes or nailed boxes are opened, employees should bend back wires and turn over or remove nails. Employees should wear goggles. The boxes and covers should be piled neatly out of passageways in designated storage areas.

When opening cartons, employees should use safety openers made of a protected sharpened blade. The employee slides the blade along the edge of the carton. These tools are useful only, of course, when the carton will not be reused. Cartons that are to be reused should be pried open with a flat steel pry bar, so that the flaps will not be damaged.

Construction of Boxes and Crates

Corners of boxes and crates receive more blows than other parts and, therefore, should be strongly constructed. The interlocking-corner crate is stronger than any other type and requires less lumber. Use it whenever possible.

Diagonal braces help make crate sections rigid. One diagonal brace in the section will give more stiffening than several parallel slats. The diagonal brace should extend from corner to corner and should be placed so that it does not project beyond the sides of the crate.

Skids constructed as an integral part of a box should be made of sound lumber, free from knots, and of sufficient size to support the box without breaking. The skids should be firmly attached to withstand being dragged across the floor.

The principal reason for boxes and crates falling apart is poor nailing. Use cement-coated nails, which hold better than uncoated nails, to avoid this problem. Correct nailing is essential to safe shipment. Particular care is required when using power nailers to build skids and crates. Safety glasses with side shields are a must, and, in some cases, hearing protection is required.

Broken or Damaged Containers of Consumables

Handlers should not attempt to sample or distribute food or other commodities from damaged or broken containers. These commodities could have become contaminated or tainted, or made otherwise unsafe, while en route. In one documented case that occurred in Colombia, South America, contents of insecticide and food bags became mixed during rough shipment. The consequent use of the contaminated flour caused many deaths. Careless handling of samples could also endanger health.

Loading Railcars

Heavy machinery shipped on skids should be braced inside the railcar to prevent shifting. Because workers may drive in lag screws with a hammer rather than use a wrench, do not use lag screws to fasten a skid to the car floor. Using a hammer damages the wood, thus reducing the holding power of the lag screws. Skids with large knots are hazardous when used on shipments of heavy machinery. When rollers are used to move the object, the skid is likely to break when the roller comes under a knot. (See Association of American Railroads, Pamphlet No. 21, in the References at the end of this chapter.)

Before railcars are opened, carefully inspect the doors. If damaged, repair runners and take special precautions. Use door openers and have employees stand clear, in case an improperly loaded car is received. Do not use a forklift truck to open doors. The angle of the fork can lift the door off its track and risk injury to employees and damage to equipment.

Workers who are opening and closing railcar doors may catch their hands between the doors and the car's doorposts. Instruct workers never to grasp the leading end of the door. This might cause their fingers to catch between the railcar's door and the side of the car. Likewise, they should keep their hands and fingers away from the doorpost when they are closing the door.

To avoid leaving hazards for railroad employees or other workers, instruct employees to clean railcars after they have been unloaded. This also avoids contamination of or damage to future lading.

SUMMARY

- Handling of material accounts for 20% to 45% of all occupational injuries.
- To reduce the number of injuries caused by materials handling, companies should minimize the manual handling of material as much as possible.
- Fall hazards need to be recognized and solutions provided. Look for proper railings, platforms, stairs, aerial lifts, ladders, runways, and attached or detached slings.
- Although physical differences make it impractical to establish safe lifting limits for all workers, some general principles can be applied.
- NIOSH has established an equation to calculate action limits (AL) and maximum permissible limits (MPLs) for manual lifting based on vertical and horizontal lifts and the distance traveled.
- Guidelines for lifting must satisfy four criteria: epidemiological, biomechanical, physiological, and psychophysical.
- Accessories for manual lifting include hand tools (hooks, crowbars, rollers), jacks, and hand trucks.
- Temporary and permanent storage of materials should be neat and orderly to eliminate hazards and to conserve space.
- Rigid containers such as metal and box pallets, fiberboard/cardboard boxes, barrels and kegs, rolled paper and reels, and compressed gas cylinders must be stored to conserve space and to provide easy access when the material is needed.
- Hazardous liquids and combustible materials stored in containers require special handling and storage methods to ensure worker safety and health.
- The storage and handling of cryogenic liquids (oxygen, nitrogen, argon, helium, and hydrogen) require careful planning and worker training. These liquids can cause frostbite on contact with skin and (except for oxygen) displace breathable air in an enclosed workspace.
- Cryogenic liquids should be stored only in the containers designed for the particular gas. Containers should be transferred slowly into a warmer environment to prevent thermal shock to the containers and equipment.
- Supervisors of shipping and receiving areas must be aware of DOT regulations and labels.
- Workers in shipping and receiving must be trained in the proper use and handling of such common items as dock boards, machines and tools, steel and plastic strapping, burlap and sacking, glass and nails, pitch and glue, barrels, kegs, drums, and boxes and cartons.
- Common personal protective equipment used in materials handling includes safety shoes, gloves, goggles, aprons, and leggings to protect against the most common injuries to hands, feet, extremities, and eyes.

REFERENCES

American Chemical Council, 2501 M Street NW, Washington DC 20037.

American National Standards Institute, 1819 L Street NW, 6th Floor, Washington DC 20036.
Hazardous Industrial Chemicals Precautionary Labeling ANSI Z129.1–2006.

Mobile and Locomotive Cranes, ANSI/ASME B30.5–2007.

Personal Protection—Protective Footwear, ANSI Z41.

Practice for Industrial Lighting, ANSI/NECA/IESNA 502–1999 (R2006).

Safety Standards for Powered Industrial Trucks, ANSI/ASME B56.1–7.

Scheme for the Identification of Piping Systems, ANSI A13.1–2007.

American Petroleum Institute, 1220 L Street NW, Washington DC 20005.

Requirements for Safe Entry and Cleaning of Petroleum Storage Tanks, Standard 2014.

Anderson, C. K., and D. B. Chaffin. "A Biomechanical Evaluation of Five Lifting Techniques." Appl Ergon. 17, no. 1 (March 1986): 2–8.

Association of American Railroads, 50 F Street NW, Washington DC 20001.

Minimum Load Standard for Machinery in Closed Cars, Pamphlet No. 21. June 1995.

Bureau of Labor Statistics, OCWC/OSH, PSB Suite 3180, 2 Massachusetts Avenue NE, Washington DC 20212-0001.

www.bls.gov/iif/oshdef.htm.

Compressed Gas Association Inc., 1235 Jefferson Davis Highway, Arlington, VA 22202.

Oxygen-Deficient Atmospheres, Bulletin SB–2.

Garg, A., and U. Saxena. "Physiological Stresses in Warehouse Operations with Special Reference to Lifting Technique and Gender: A Case Study." *American Industrial Hygiene Association Journal* 46, no. 2 (1985): 53–59

Illuminating Engineering Society of North America, 345 East 47th Street, New York, NY 10017.

IESNA Handbook HB-9, 2000.

Institute of Makers of Explosives, 1120 19th Street NW, Suite 310, Washington DC 20036.

Safety in the Transportation, Storage, Handling and Use of Commercial Explosives Materials, SLP17, 2007.

Institute of Medicine of the National Academies Report. Relieving Pain in America: A Blueprint for Transforming Prevention, Care, Education, and Research. Washington DC: The National Academies Press, 2011.

International Society of Explosive Engineering, 30325 Bainbridge Road, Cleveland, OH 44139. 2003.

Kroemer, K. H. E. *Material Handling: Loss Control through Ergonomics*. 2nd ed. Schaumburg, IL: Alliance of American Insurers, 1983.

Lovested G. E. "Materials Handling Safety in Industry."

Materials Handling Handbook. 2nd ed. New York: John Wiley & Sons, 1985.

National Fire Protection Association, 1 Batterymarch Park, Quincy, MA 02269.

Explosive Materials Code, NFPA 495, 2013.

Flammable and Combustible Liquids Code, NFPA 30, 2008.

Identification of Fire Hazards of Materials for Emergency Response, NFPA 704, 2007.

Installation of Sprinkler Systems, NFPA 13, 2007.

National Electrical Code. NFPA 70, 2008.

National Fire Codes. Vol. 2, "Flammable and Combustible Liquids," 2006.

Powered Industrial Trucks, Including Type Designations, Areas of Use, Conversions, Maintenance, and Operations, NFPA 505, 2006.

Prevention of Fires and Dust Explosions in Agricultural and Food Products Facilities, NFPA 61, 2008.

Production, Storage, and Handling of Liquefied Natural Gas (LNG), NFPA 59A, 2006.

Standard for Liquefied Hydrogen Systems at Consumer Sites, NFPA 50B, 1999.

Standard for the Production, Storage, and Handling of Liquefied Natural Gas (LNG), NFPA 59A, 2013.

Standard for Rack Storage of Materials, NFPA 231C, 1998.

Standard for the Safeguarding of Tanks and Containers for Entry, Cleaning, or Repair, NFPA 326, 2015.

Venting of Deflagrations, NFPA 68, 2007.

National Institute for Occupational Safety and Health (NIOSH), 1600 Clifton Rd., Atlanta, GA 30329-4027.

Applications Manual for the Revised NIOSH Lifting Equation. DHHS Publication 94–110, 1994.

Back Belts—Do They Prevent Injury? NIOSH Publication 94–127, July 1994.

Criteria for a Recommended Standard: Emergency Egress from Elevated Workstations. NIOSH Publication 76–128, 1975.

Work Practices Guide for Manual Lifting. NIOSH Technical Report 81–122, 1981.

National Institute of Standards and Technology, 100 Bureau Drive, Gaithersburg, MD 20899.

Specifications, Tolerances, and Technical Requirements for Weighing and Measuring Devices. Handbook 44, 2008.

National Safety Council, 1121 Spring Lake Drive, Itasca, IL 60143.

Occupational Safety and Health Data Sheets: Belt Conveyors (Equipment), 12304–0569, 1990.

Construction Material Hoists, 12304–0511, 1987.

Dock Plates and Gangplanks, 12304–0318, 1990.

Electromagnetics Used with Crane Hoists, 12304–0359, 1985.

Front-End Loaders, 12304–0589, 1990.

Fuses and Torpedoes Used in Railroad Operations, Handling and Storage of, 12304–0639, 1990.

Handling and Storage of Sheet Metal, 12304–0434, 1991.

Handling and Storage of Solid Sulfur, 12304–0612, 1991.

Handling Bottles and Glassware in Food Processing and Food Service, 12304–0355, 1990.

Handling Large-Diameter Oil Field Pipe, 12304–0463, 1985.

Handling Liquid Sulfur, 12304–0592, 1993.

Handling Materials in the Forging Industry, 12304–0551, 1992.

Load-Haul-Dump Machines in Underground Mines, 12304–0576, 1990.

Motor Trucks for Mines, Quarries, and Construction, 12304–0330, 1990.

Pendant-Operated and Radio-Controlled Cranes, 12304–0558, 1991.

Powered Hand Trucks, 12304–0317, 1991.

Recommended Loads for Wire Rope Slings, 12304–0380, 1991.

Roller Conveyors, 12304–0528, 1991.

Scrap Ballers, 12304–0611, 2004.

Steel Plates, Handling for Fabrication, 12304–0565, 1990.

Truck Mounted Power Winches, 12304–0441, 1990.

Oresick, A. "Safety Techniques in Glass Handling." *ASSE Journal* (May 1973): 22–29.

Szymanski, E. "Safe Storage and Handling of Cryogenic Liquids." *Plant Engineering Magazine*, June 14, 1979.

U.S. Department of Health and Human Services, National Institute for Occupational Safety and Health (NIOSH), 4676 Columbia Parkway, Cincinnati, OH 45226.

Preemployment Strength Testing, DHEW (NIOSH) Publication 77–163.

U.S. Department of Labor, Occupational Safety and Health Administration (OSHA), 200 Constitution Avenue NW, Washington DC 20210.

Code of Federal Regulations, Title 29. Section 1910.1200, Hazard Communication Standard.

U.S. Department of Transportation, Office of Hazardous Materials, Washington DC 20590.

U.S. Department of the Treasury, Internal Revenue Service, 1111 Constitution Avenue, Washington DC 20224.

Published Ordinances: Explosives—State Laws and Local Ordinances Relevant to Title 18 USC, Chapter 40, Publication 740.

REVIEW QUESTIONS

1. Materials handling accounts for what percentage of all occupational injuries?
   a. 10% to 25%
   b. 20% to 45%
   c. 40% to 50%
   d. 50% to 55%

2. What are the three main factors to consider when determining the safety of manual lifting for a particular load?
   a.
   b.
   c.

3. What advantage is gained by holding a load close to the body?

4. What is the key to safe carrying by teams?
   a. Make every movement in unison.
   b. Grasp the load by opposite corners.
   c. Use a straight-back position.
   d. Twist the body toward the load.

5. What are the three most common hand tools for materials handling and the principal hazard for each?
   a.
   b.
   c.

6. What should be avoided to prevent slippage when using a jack, and how can this be achieved?

7. What is the proper method for using wood or metal jack extenders?

8. What special precaution should be taken when using a two-wheeled hand truck for moving pressurized items (e.g., gas cylinders)?

9. Why should four-wheeled hand trucks be pushed instead of pulled?

10. What are the three main hazards to be aware of when using two- or four-wheeled hand trucks?
    a.
    b.
    c.

11. What two purposes do tie pieces serve in the stacking of lumber?
    a.
    b.
12. To prevent injuries that are caused when the wrong valve is opened, piping should be clearly labeled with which three pieces of information?
    a.
    b.
    c.
13. What color flags and lights should be used to isolate railroad tank cars during unloading?
    a. red
    b. yellow
    c. blue
    d. orange
14. To prevent the accumulation of static electricity on most surfaces in areas with airborne dust, the relative humidity should be maintained at what level?
    a. 10% to 20%
    b. 30% to 50%
    c. 60% to 70%
    d. 90% to 100%

15. What is the simplest solution to the potentially dangerous production of fine dust during filling operations?
16. A cryogenic liquid has a normal boiling point below:
    a. –298°F.
    b. –238°F.
    c. –150°F.
    d. 0°F.
17. What is the main precaution that should be taken in areas where flammable fluids are stored or handled?
18. Workers should use a self-contained breathing apparatus in areas where the oxygen content may be below

    _____.
19. What is a Dewar container?
20. How much weight should dock boards used in railcar loading be designed to carry?

Hoisting and Conveying Equipment

13

John F. Montgomery, PhD, CSP, CHMM

Von M. Griggs-Laws

Hoisting Apparatus
Electric Hoists ▶ Air Hoists ▶ Hand-Operated Chain Hoists

Cranes
Design and Construction ▶ Guards and Limit Devices ▶ Ropes and Sheaves ▶ Crane and Hoist Signals ▶ Selection and Training of Operators ▶ Inspection ▶ Operating Rules ▶ General Maintenance and Safety Rules ▶ Overhead Cranes ▶ Electromagnets and Hook-On Devices ▶ Storage Bridge and Gantry Cranes ▶ Monorails ▶ Jib Cranes ▶ Derricks ▶ Tower Cranes ▶ Mobile Cranes ▶ Aerial Baskets ▶ Crabs and Winches ▶ Block and Tackle ▶ Portable Floor Cranes or Hoists ▶ Tiering Hoists and Stackers

Conveyors
General Precautions ▶ Belt Conveyors ▶ Slat and Apron Conveyors ▶ Chain Conveyors ▶ Shackle Conveyors ▶ Screw Conveyors ▶ Bucket Conveyors ▶ Pneumatic Conveyors ▶ Aerial Conveyors ▶ Portable Conveyors ▶ Gravity Conveyors ▶ Live Roll Conveyors ▶ Vertical Conveyors

Power Elevators
Types of Drives ▶ New Elevators ▶ Hoistways, Pits, and Machine Rooms ▶ Hoistway Doors and Landings ▶ Cars ▶ Hoisting Ropes ▶ Operating Controls ▶ Inspection and Maintenance ▶ Inspection Program ▶ Operation ▶ Emergency Procedures ▶ Requirements for the Disabled

Sidewalk Elevators
Operation ▶ Hatch Covers

Hand Elevators
Hoistway Doors ▶ Safety Devices and Brakes

Dumbwaiters
Hoistways and Openings ▶ Safety Devices and Brakes

Escalators
Safety Devices and Brakes ▶ Machinery ▶ Protection of Riders ▶ Fire Protection

Moving Walks

Man-Lifts
Construction ▶ Brakes, Safety Devices, and Ladders ▶ Inspections ▶ General Precautions

Summary

References

Review Questions

For centuries, hoisting apparatus have been used to raise, lower, and transport heavy loads for short distances. Many thousands of hoists (electric, air, and hand powered) are used in industry. Typically, they range from ¼ to 10 tons (226.8 kg to 9,072 kg) in capacity, but greater capacities are not unusual. Today, there are traveling cranes in steel mills, power plants, and naval shipyards able to lift hundreds of tons (9,072 kg). (See also the discussions of material and passenger hoists in Chapter 3, Construction of Facilities, and Chapter 4, Maintenance of Facilities, both in this volume.) This chapter covers the following topics:

- types of hoisting apparatus and general safety rules for operating hoisting equipment
- types of cranes and general safety rules for the safe operation of each crane
- importance of proper safety training for crane operators
- types of conveyors and general safety precautions in the operation of each conveyor
- general safety codes for operating and maintaining elevators, escalators, dumbwaiters, and moving walks
- general safety precautions in the construction and operation of man-lifts.

HOISTING APPARATUS

The safe load capacity of each hoist should be shown in conspicuous figures on the body of the machine. These figures should be clearly legible from the ground floor. In addition, all hoists must have a label, or labels, affixed to the hoist, hook block, or controls. These labels must explain safe operating procedures and be in a readable position.

It is very important to consider the load capacity of the support structure for the hoist. The rated lifting capacity of the hoist is directly proportional to the load capacity of the supporting structure. A certified structural engineer should certify the load capacity of the supporting structure. If the load capacity of the support structure of the lift is disproportional to the lift capacity, the lift capacity should be lowered to match the lifting capacity of the support.

All hoists should be securely attached to their supports (fixed member or trolley) with shackles.

Hoists can be either rigid suspended or hook suspended. If the hoist is hook suspended, the support hooks should be moused or have hook latches. Latches are also recommended for load hooks. Hoist supports should be designed to bear maximum loads. Overhead hoists, operating on rails, tracks, or trolleys, should have positive stops or limiting devices on the equipment, rails, tracks, or trolleys to prevent the overrunning of safe limits. Also, the maximum load capacity of the overhead rails should be posted on the rails in such a manner that the figures are clearly legible

Figure 13–1. To check the rope of this 1,000-lb (4,450-N) electric hoist, inspect 2- to 3-ft (61- to 91-cm) increments after the rope has come to a complete stop. To check for frays, wipe a rag up and down the rope. *(Courtesy Acco Hoist and Crane Division)*

from the ground floor. Keep flanges on hoist drums with single-layer spiral grooves free of projections that could damage a rope.

Hoist operators should pick up a load only when it is directly under the hoist. Otherwise, stresses for which the hoist was not designed could be imposed upon it. If the load is not properly centered, it can swing (upon being hoisted), and injury could result. Everyone must stay out from under raised loads. When operating an overhead hoist, operators should take care to avoid injuring themselves or any other nearby workers.

Caution: Do not use hoists or cranes to lift, support, or otherwise transport people. The standard commercial hoist or crane does not provide a secondary means of supporting the load should the wire rope or other suspension element fail. (See the section Aerial Baskets, later in this chapter.)

Examine hoists for evidence of wear, malfunction, damage, and proper operation of devices such as load hooks, ropes, brakes, clutches, and limit switches. Carefully examine deficiencies and, if determined hazardous, correct them immediately (Figure 13–1).

Electric Hoists

Rope-operated electric hoists should have nonconducting control cords unless they are grounded. Control cords should have handles of distinctly different contours. In that way, even without looking, the operator will know which is the hoisting handle and which is the lowering handle.

Clearly mark each control handle "hoist" or "lower." Some companies attach an arrow to each control cord, pointing in the direction in which the load will move when the rope is pulled. Also, pass the control cords through the spreader to keep them from becoming tangled. The spreader can be a 1-in. × 3-in. (2.5-cm × 7.5-cm) board or other nonconductive material with equally spaced holes, resting on the pull handles or lowest position. Periodically inspect control cords, usually made of fiber or light wire rope, for wear and other defects.

On the control, provide means for effecting automatic return to the OFF position. In that way, an operator must maintain a constant pull on the control rope or a push on the control button to raise or lower the load. Support the pendant station to protect the electrical conductors against strain. Also ground the station in case a ground fault occurs. Limit push-button control circuits to 150 V AC and 300 V DC.

Install a limit switch on the hoist motion. The minimum of two turns of rope should remain on the drum when the load block is on the floor, except when a geared lower limit switch is used. One turn is then permitted. If a load block can enter a pit or hatchway in the floor, the installation of a lower travel limit switch is recommended. If that switch trips, the turns of the rope remaining on the drum can be reduced to one when the block has been stopped.

Air Hoists
After a piston air hoist has been in operation for a time, the locknut that holds the piston on its rod may become loose. Should the nut come off, the rod could pull out of the piston, allowing the load to drop. Be sure to secure the locknut to the piston rod with a castellated nut and cotter pin. Whenever an air hoist is overhauled, check to see that the piston is well secured to the rod.

On a cylinder load balancer or hoist, do not use an ordinary hook to hang the hoist from its support. The cylinder may come unhooked if the piston rod comes in contact with an obstruction when the load is lowered. Use a clevis or other device to prevent the hook from being detached from the hoist support.

To prevent the hoist from rising or lowering too rapidly, place a choke (available from the hoist supplier) in the air-line coupling. For a rope-drum air hoist, provide a closing load-line guide.

Hand-Operated Chain Hoists
Some chain hoists are portable. Other chain hoists are either permanently hooked onto a monorail trolley or built into the trolley as an integral part. They are a good alternative for many operations that usually use a block and tackle fitted with fiber rope. Chain hoists are stronger, more dependable, and more durable than fiber-rope tackle.

There are three general types of chain hoists: spur geared, differential, and screw geared (or worm drive). The spur-geared type is the most efficient because it can pick up a load with the least effort on the part of the operator. Because the spur-geared type is free running, it tends to allow the load to run itself down. Therefore, an automatic mechanical load brake, similar to that on a crane, is provided to control the rate of descent of the load. The differential type is the least efficient. Screw-geared and differential hoists are self-locking and will automatically hold a load in position. (See Chapter 14, Ropes, Chains, and Slings, in this volume, for a description of chain hoist inspection.)

CRANES

Cranes raise, lower, and shift heavy objects through the use of a long movable arm. The hoisting apparatus of some cranes is supported on an overhead track.

Design and Construction
In the United States, the Occupational Safety and Health Administration (OSHA) requires that all overhead and gantry cranes constructed and installed on or after August 31, 1971, meet the design specifications of the American National Standards Institute's (ANSI) and American Society of Mechanical Engineers' (ASME) ANSI/ASME B30.2, Overhead and Gantry Cranes (Top Running Bridge, Single or Multiple Girder, Top Running Trolley Hoist). OSHA requirements do not cover single-girder cranes (29 CFR 1910.179[b][2]). Single-girder underhung cranes are covered in ANSI/ASME B30.11; single-girder top-running cranes in ANSI/ASME B30.17; and overhead hoists used on single-girder cranes in ANSI/ASME B30.16. (See References at the end of this chapter for specific titles.) Warning: Check for reverse phasing when installing a hoist or crane.

Designate a competent person to inspect all machinery and equipment prior to each use, and during use, to make sure it is in safe operating condition. Any deficiencies should be repaired, or defective parts replaced, before continued use.

All parts of every crane (Figure 13–2a and Figure 13-2b), especially those subject to impact, wear, and rough usage, should be of adequate strength for their rated service. Journals and shafts should be of sufficient strength and size to bring the bearing pressure to within safe limits.

Open hooks should not be used where there is danger of relieving the tension on the hook due to the load or to the

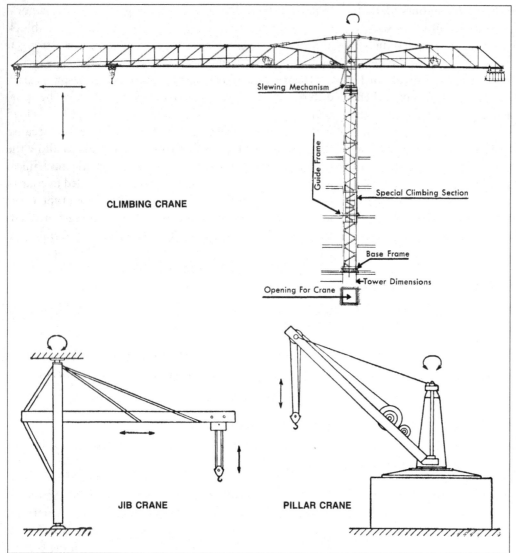

Figure 13–2a. Diagrammatic sketches of various types of cranes. *(Printed with permission from ANSI Standard Series B 30, "Safety Standards for Cableways, Cranes, Derricks, Hoists, Hooks, Jacks, and Slings," except for climbing crane.)*

hook catching or fouling. Each independent hoisting unit of a crane should have brakes complying with the requirements of the ANSI/ASME B30 Series, Safety Requirements for Cranes, Derricks, Hoists, Hooks, Jacks, and Slings.

Gantry cranes used for outdoor storage should be provided with remotely operated rail clamps or other equivalent devices. Parking brakes are considered as minimum compliance with this rule. Apply rail clamps only when the crane is not in motion. When rails are used as anchors, they should be secured to withstand the resultant forces applied by the rail clamps. If the clamps act on the rail, any projection or obstruction in the clamping area must be avoided. A wind-indicating device should be provided that will give a visible and audible alarm to the crane operator at a predetermined wind velocity.

The rated load of the crane should be plainly marked on each side of the crane. If the crane has more than one hoisting unit, each hoist should have its rated load marked on it or its load block. If the crane has more than one trolley unit, each trolley should have an identification marking on it or its load block. The marking should be clearly legible from the ground or floor. The crane should not be loaded beyond its rated capacity, except for testing.

Arrange the cab and locate control and protective equipment so that all operating handles are within convenient reach of the operator. When the operator faces either the area to be served by the load hook or the direction in which the cab is traveling, he or she should be able to reach the handles. The arrangement should allow the operator a full view of the load hook in all positions.

Mark each controller and operating lever with the action and the direction that it controls. These levers should have spring returns. In that way, if the operator releases a lever, it will automatically move into the OFF position. ANSI/ASME B30.2 requires that cranes not equipped with spring-return controllers, spring-return master switches, or momentary contact push buttons should be provided with a device that will disconnect all motors from the line, in the event of a power failure. Such devices should not permit any motor to be restarted until the controller handle is brought to the OFF position or a reset switch or button is operated.

Also observe the following regulations:

- All machinery, equipment, and material hoists operating on rails, tracks, or trolleys should have positive stops or limiting devices on the equipment, rails, tracks, or trolleys to prevent overrunning the safe limits.
- All points requiring lubrication during operation should have fittings so located or guarded as to be accessible

without hazardous exposure.
- Platforms, footwalks, steps, handholds, guardrails with intermediate rails, and toeboards should be provided on machinery and equipment to provide safe footing and accessways. Platforms and steps should be of slip-resistant material.
- Access to the crane cab and/or bridge walkway should be provided by conveniently placed fixed ladders, stairs, or platforms whose steps leave no gap exceeding 12 in. (30.5 cm). Fixed ladders should be in conformance with ANSI A14.3, Safety Requirements for Fixed Ladders.
- A dry chemical or equivalent fire extinguisher should be kept in the crane cab.
- Accessible areas within the swing radius of the rear of the rotating superstructure of the crane, either permanently or temporarily mounted, should be barricaded in such a manner as to prevent an employee from being struck or crushed by the crane.

Guards and Limit Devices

If contact can be made with gears and other moving parts during normal operating conditions, they should be totally enclosed, covered by screen guards, or placed out of reach. No overhung gears should be used, unless means are provided to prevent their falling—should they break or work loose. The bolts in shaft couplings should be recessed so that the tops of the nuts do not project.

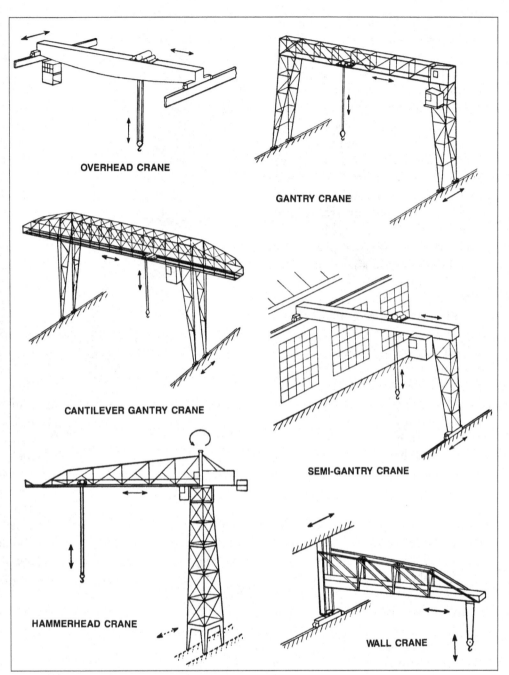

Figure 13–2b.

To prevent crushed fingers, large load hooks on cranes should have handles. In that way, a person can hold or guide the hooks when slings are being placed on them. Also, on small cranes, the pinch points where cables pass over sheaves in the load block should be guarded (Figure 13–3).

The hoisting motion of every crane, with the exception of boom-type cranes and derricks, must have an overtravel limit switch in the hoisting direction to stop the hoisting motion. Use lower-travel limit switches for all hoists if the load block enters pits or hatchways in the floor.

Limit devices should always operate on a normally closed circuit and should be tested at the beginning of each shift. To make this test, the unloaded block should be carefully run up to actuate the device. Here is a suggested testing procedure:
1. Inch block into limit switch.
2. Lower load approximately 10 ft (3 m).
3. Stop and then operate at full speed into limit switch.
4. Always conduct the test away from equipment and employees.

Hoist limit switches are operational safety devices that prevent unintended overtravel of the load block. They are not intended for constant duty. Refer questions about requirements for hoists with constant-duty limit switches to a crane or hoist manufacturer.

The hook block should be designed to lift vertically without the wire ropes of the cable twisting around each other. The hook should be of solid forged steel or built-up steel plates. Bronze and stainless steel hooks are frequently used for fire protection. Large hooks should swivel on rollers or ball bearings.

Wiring for electric equipment should be installed according to Article 610, Cranes and Hoists, of NFPA 70, National Electrical Code. On electric-power-operated cranes, the power supply to the runway conductors should be controlled by a switch or circuit breaker. Locate the switch or circuit breaker on a fixed structure that is accessible from the floor and that can be locked in the OFF position.

Figure 13-3. This 1-ton hoist is designed for light operations. Note that the push-button pendant is shaped for easy holding. *(Courtesy Delmac Inc.)*

Other precautions include the following:

- No guard, safety, appliance, or device should be removed or otherwise be made ineffective in machinery or equipment, except for the purpose of making immediate repairs, lubrications, or adjustments, and then only after the power has been turned off, except when power is necessary for making adjustments.
- All guards and devices should be replaced immediately after completion of repairs and adjustments.
- Traveling cranes should be equipped with a warning device that can be sounded continuously while the crane is traveling. Traveling cranes should also have a rotating or strobe light that will alert other workers in the area to the crane's movement.

Before starting maintenance or repair work on a crane, workers should apply their personal tags and padlocks to the main power switch while it is in the OFF position. (See Glossary: lockout/tagout, energy isolation, and zero mechanical state [ZMS].) When repairing cranes on a multiple-crane runway, workers should make provisions to prevent other cranes from running into the crane being repaired.

Ropes and Sheaves

Use hoisting ropes of recommended construction for crane or hoist service. The crane or rope manufacturer should be consulted whenever a change is contemplated. The rated load divided by the number of parts of rope should not exceed 20% of the nominal breaking strength of the rope.

A written systematic procedure of inspection testing and maintenance of shafts, hoisting systems, conveyances, hoist ropes, and rigging should be developed and followed. Such inspection, testing, and maintenance should be done by a qualified person. If it is found or suspected that any component is not functioning properly, hoisting should not be conducted until the malfunction has been located and corrected.

A thorough inspection of all ropes should be made at least once a month, and a certification record, which includes the date of inspection, the signature of the person who performed the inspection, and an identifier for the ropes that were inspected, should be kept on file where readily available to appointed personnel. Any deterioration resulting in appreciable loss of original strength should be carefully observed and determination made as to whether further use of the rope would constitute a safety hazard.

Inspect sheaves and drums for wear. If the grooves become enlarged from wear or corrugated from excessive rope pressure, replace them. These conditions will cause rapid wear and loss of rope strength. In cases where considerable material must be removed to regroove the drum or sheave, the strength of these parts may be impaired. In such cases, consult the hoist or crane manufacturer prior to regrooving. (See ANSI/ASME B30.2.)

To reduce the strain on the hoist rope where it enters the socket or anchorage on the drum, the minimum of at least two wraps should remain on the drum when the load block is at the lowest elevation. However, one turn is permitted when a geared lower limit switch is used. Anchor the drum end of the rope to the drum by a socket arrangement approved by the crane or rope manufacturer or both. As with electric hoists, if the load block can enter a pit or hatchway in the floor, use a lower-travel limit switch. If that switch trips, reduce the turns of the rope remaining on the drum to one after the block has been stopped.

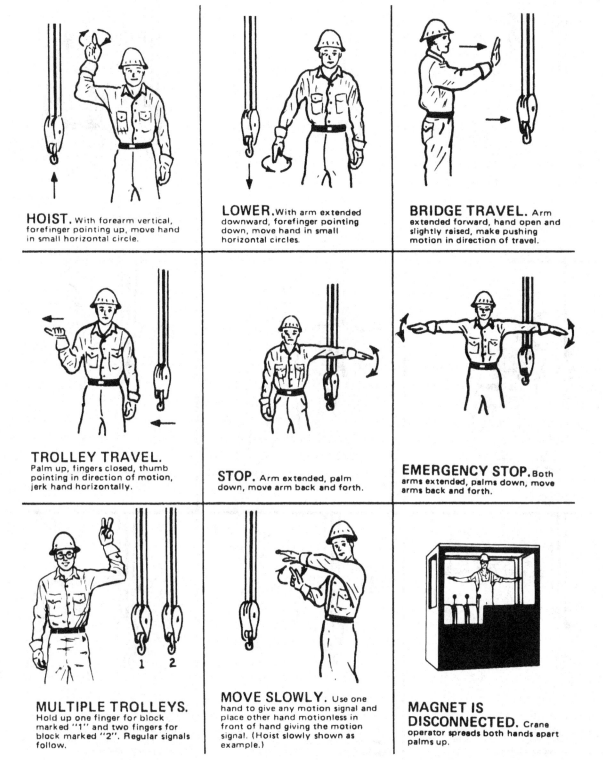

Figure 13-4. Standard hand signals for controlling operation of overhead gantry cranes. *(Printed with permission from ANSI Standard Series B 30, "Safety Standards for Cableways, Cranes, Derricks, Hoists, Hooks, Jacks, and Slings," The American Society of Mechanical Engineers, New York.)*

Crane and Hoist Signals

Crane movements should always be governed by a standard of code signals that are transmitted to the crane operator by the crane director (signaler). Signals may be given by any mutually understood and officially adopted method.

Hand signals, however, are preferred. A simple code of

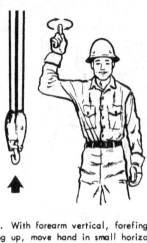 **HOIST.** With forearm vertical, forefinger pointing up, move hand in small horizontal circle.	 **LOWER.** With arm extended downward, forefinger pointing down, move hand in small horizontal circles.	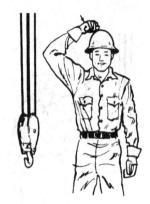 **USE MAIN HOIST.** Tap fist on head; then use regular signals.
 **USE WHIP LINE.** (Auxiliary Hoist) Tap elbow with one hand; then use regular signals.	 **RAISE BOOM.** Arm extended, fingers closed, thumb pointing upward.	 **LOWER BOOM.** Arm extended fingers closed, thumb pointing downward.
 **MOVE SLOWLY.** Use one hand to give any motion signal and place other hand motionless in front of hand giving the motion signal. (Hoist Slowly shown as example)	 **RAISE THE BOOM AND LOWER THE LOAD.** With arm extended thumb pointing up, flex fingers in and out as long as load movement is desired.	 **LOWER THE BOOM AND RAISE THE LOAD.** With arm extended, thumb pointing down, flex fingers in and out as long as load movement is desired.

Figure 13-5. Standard hand signals suitable for crawler, locomotive, and truck boom cranes. One-hand signals for extending or retracting boom (not shown in figure): *extend boom*—one fist in front of chest with thumb tapping chest; *retract boom*—one fist in front of chest, thumb pointing outward and heel of fist tapping chest. (Printed with permission from ANSI Standard Series B30, "Safety Standards for Cableways, Cranes, Derricks, Hoists, Hooks, Jacks, and Slings," The American Society of Mechanical Engineers, New York.)

one-hand signals is appropriate for an overhead crane or bridge crane. The ANSI set of signals, adopted by many companies, is shown in Figure 13-4. (See the ANSI/ASME B30 series.) A set of one- and two-hand signals for a locomotive or crawler crane, or any other boom rig, is shown in Figure 13-5.

13 Cranes

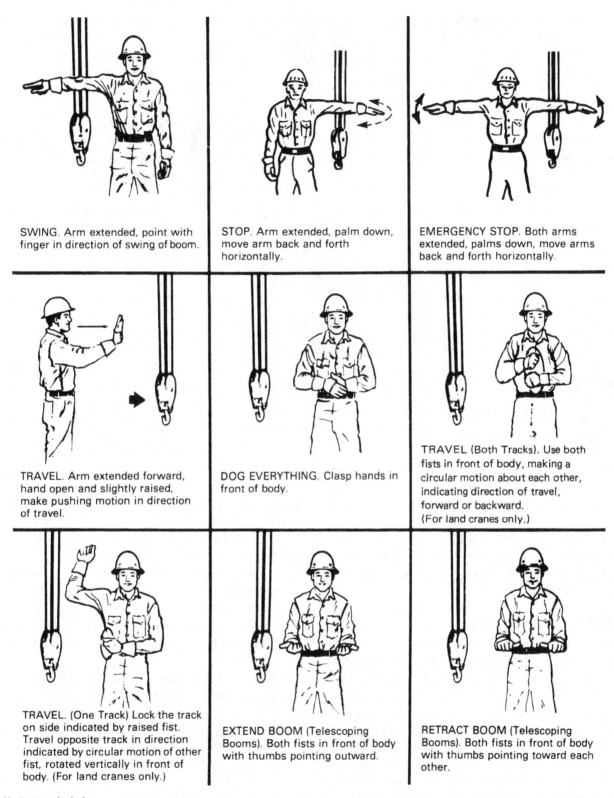

Figure 13-5. Concluded.

Signals must be discernible or audible at all times. Where visual or audible signals are inadequate, use a telephone or portable radio to communicate. A remote radio-control system for overhead cranes, which eliminates the need for hand signals, is available (Figure 13-6). The operator controls all movements of the bridge, trolley, and hoist from the plant floor. Circuits are so designed that failure of any one system component causes all crane motions to stop.

There should be only one designated person who is qualified to give crane signals to the operator. However, if signalers are changed frequently, they should be provided with one (and only one) conspicuous armband, hat, glove, or other badge of authority. This badge of authority must be worn by the signaler currently in charge.

Operators should not move equipment unless signals are clearly understood. The operator should move the hoisting apparatus only on signals from the proper person. A STOP signal, however, should be obeyed regardless of who gives it. Unless obedience would result in an injury, the operator should be governed absolutely by the signal. However, if an injury seems unavoidable by obeying the signal, the operator should notify the signaler at once so that corrective measures can be taken immediately.

Employees who work near cranes or assist in hooking on or arranging loads should be instructed to keep out from under loads. Supervisors should see that this rule is strictly enforced. One manufacturing company publishes the following warning:

> From a safety standpoint, one factor is paramount: Conduct all lifting operations in such a manner that, if there were an equipment failure, no personnel would be injured. This means keep out from under raised loads!

Selection and Training of Operators

Cranes, like many other pieces of equipment, present certain hazards that cannot be removed through engineering. In such instances, it is only through the exercise of intelligence, care, and good judgment that the associated risks can be reduced to an acceptable level. It is, therefore, imperative that only employees who are physically and mentally fit operate cranes. Qualified operators of hoisting equipment should meet the following minimum requirements:

- *Age:* The legal age for crane operations as determined by the local governing agency—generally 18 years of age.
- *Language:* Understand spoken and written English, as well as any other language generally used at the location.
- *Physical:* Pass physical examinations, including vision test for depth perception.
- *Knowledge:* Have basic knowledge and understanding of equipment-operating characteristics, capabilities, and limitations including equipment-rate capacity; the effect of variables on that capacity; safety features; required operating procedures; and requirements established by local, state, and federal agencies (U.S. Department of Energy 1988).
- *Skill:* Demonstrate skill in manipulating and controlling equipment through all phases of the operation.

Figure 13–6. Operator controls overhead crane by means of hand-held remote radio (lower right).

The crane operator should have a preemployment physical examination that emphasizes acuity of vision, depth and color perception, hearing, muscular coordination, and reaction time. Screening for drugs should also be done at this time. The physical examination should be required annually thereafter.

Employees selected to operate a crane must be able to learn and understand basic safety information for equipment and personnel. They must also learn the special requirements for the safe handling and use of the equipment that they will operate. Training for crane operators requires two parts: (1) an information exchange in which rules, regulations, requirements, limits, and dos and don'ts are discussed and explained, and (2) onsite, practical training in which safe operation is explained, demonstrated by the instructor, and tried by the trainee.

The initial operator-qualification process should include the following:

1. Actual hands-on training on the equipment for which the employee plans to qualify. The training should be conducted under the direction of a qualified crane operator.
2. Review of the trainee's knowledge through both written and oral examinations. The trainee should also demonstrate his or her skills for the instructor.
3. Placing a record of the training course, examination scores, and authorization permit to operate the specific hoisting equipment in the individual's training file. The record should identify the specific equipment that the individual is qualified to operate.
4. Some companies require both operators and local rigging personnel to have authorization permits. These are renewable at intervals of 1 or 2 years upon reexamination. Finally, establish a system for documenting training and proficiency levels. Maintain this system to ensure that operators' skills are up-to-date.

Inspection

Overhead and gantry cranes must be inspected according to the procedure given in ANSI/ASME B30.2, Overhead and Gantry Cranes. Hooks must be inspected in accordance with the procedure in ANSI/ASME B30.10.

OSHA and various states are developing revised inspection and certification rules. Washington State law HB 2171 was signed on April 10, 2007. Effective January 1, 2010, no employer or contractor may permit a crane operator engaged in construction work to operate a crane unless the crane operator is qualified. The new law also requires cranes to be certified at least annually by a certified crane inspector. Many U.S. states have similar requirements.

Hooks having any of the following deficiencies should be removed from service unless a qualified person approves their continued use and initiates corrective action. Hooks approved for continued use should be subjected to periodic inspection.

1. crack(s)
2. wear exceeding 10% (or as recommended by the manufacturer) of the original sectional dimension
3. a bend or twist exceeding 10 degrees from the plane of the unbent hook
4. for hooks without latches, an increase in throat opening exceeding 15% (or as recommended by the manufacturer); for hooks with latches, an increase of the dimension between a fully opened latch and the tip section of the hook exceeding 8% (or as recommended by the manufacturer)
5. for a provided latch that becomes inoperative because of wear or deformation, and that is required for the service involved, replace or repair before the hook is put back into service; if the latch fails to fully close the throat opening, the hook should be removed from service or moused until repairs are made.

A crane operator should not attempt to make repairs. Any condition that might make the crane unsafe to operate should be reported to the supervisor. Certain faults may be so dangerous that the crane should be shut down at once and not operated until the faults are corrected.

A list of unsafe conditions to be checked by operators of overhead traveling cranes prior to each shift follows. Many companies have developed checklists to be completed and signed daily before work.

- bearing: loose, worn
- brakes: shoe wear
- bridge: alignment out of true (indicated by screeching or squealing wheels)
- bumpers on bridge: loose, missing, improper placement of
- collector shoes or bars: worn, pitted, loose, broken
- controllers: faulty operation because of electrical or mechanical defects
- couplings: loose, worn
- drum: rough edges on cable grooves
- end stops on trolley: loose, missing, improper placement of
- footwalk: condition
- gears: lack of lubrication or foreign material in gear teeth (indicated by grinding or squealing)
- guards: bent, broken, lost
- hoisting cable: broken wires
- hook block: chipped sheave wheels
- hooks: straightening
- lights (warning or signal): burned out, broken
- limit switch: functioning improperly
- lubrication: overflowing on rails, dirty cups
- mechanical parts (rivets, covers, etc.): loose
- overload relay: frequent tripping of power
- rails (trolley or runway): broken, chipped, cracked
- wheels: worn (indicated by bumpy riding).

Many companies believe in performance (operating) tests for all hoisting equipment. They make sure that all hoisting equipment satisfactorily completes a performance test before it is placed in service. The test should be repeated (1) at least every year; (2) prior to unusual or critical lifts; and (3) after alteration, modification, or reassembly. Record the test results and keep them available for review. In addition, regular preshift and on-shift inspections should be made and periodic load tests performed to provide added assurance that the equipment is safe to operate.

Operating Rules

The following operating rules for crane operators are from the Crane Manufacturers Association of America Inc.:

One measure of a good crane operator is the smoothness of operation of the crane. Jumpy and jerky operation, flying starts, quick reversals, and sudden stops are the "trademarks" of the careless operator. The good operator knows and follows these tried and tested rules for safe, efficient crane handling.

1. Crane controls should be moved smoothly and gradually to avoid abrupt, jerky movements of the load. Slack must be removed from the sling and hoisting ropes before the load is lifted.
2. Center the crane over the load before starting the hoist to avoid swinging the load as the lift is started. Do not swing loads suspended from the crane to reach areas not under the crane.
3. Crane hoisting ropes should be kept vertical. Cranes shall not be used for side pulls.
4. Never lower the block below the point where less than two full wraps of rope remain on the hoisting drum. Should all the rope be unwound from the drum, be sure it is rewound in the

correct direction and seated properly in the drum grooves, otherwise the rope will be damaged and the hoist limit switch will not operate to stop the hoist in the high position.

5. Be sure everyone in the immediate area is clear of the load and aware that a load is being moved. Activate the warning device (if provided) when raising, lowering, or moving loads wherever people are working to make them aware that a load is being moved.

6. Do not make lifts beyond the rated load capacity of the crane, sling chains, rope slings, etc.

7. Do not operate the crane if limit switches are out of order or if ropes show defects or wear.

8. Make certain that before moving the load, load slings, load chains, or other load-lifting devices are fully seated in the saddle of the hook.

9. When a duplex hook (double saddle hook) is used, a double sling or choker should be used to assure that the load is equally divided over both saddles of the hook.

10. On all capacity or near capacity loads, the hoist brakes should be tested by returning the master switch or push button to the OFF position after raising the load a few inches off the floor. If the hoist brakes do not hold, set the load on the floor and do not operate the crane. Report the defect immediately to the supervisor.

11. Check to be sure that the load is lifted high enough to clear all obstructions and personnel when moving bridge or trolley.

12. At no time should a load be left suspended from the crane unless the operator is at the master switches or push button with the power on, and under this condition keep the load as close as possible to the floor to minimize the possibility of an injury if the load should drop. When the crane is holding a load, the crane operator should remain at the master switch or push button.

13. When the hitcher is used, it is the joint responsibility of the crane operator and the hitcher to see that the hitches are secure and that all loose material has been removed from the load before starting a lift.

14. Do not lift loads with any sling hooks hanging loose. (If all sling hooks are not needed, they should be properly stored or a different sling should be used.)

15. All slings or cables should be removed from the crane hooks when not in use. (Dangling cables or hooks hung in sling rings can inadvertently snag other objects when the crane is moving.)

16. Crane operators should not use limit switches to stop the hoist under normal conditions. (These are emergency devices and not to be used as operating controls.)

17. Do not block, adjust, or disconnect limit switches in order to go higher than the switch will allow.

18. Upper limit switches (and lower limit switches, when provided) should be tested when stopping the hoist at the beginning of each shift, or as frequently as otherwise directed.

19. No loads should be moved or suspended over people regardless of the attachment, mechanical, magnetic, friction, or vacuum.

20. Molten metal shall never be carried overhead where it could splash onto personnel.

21. If the electric power goes off, place controllers in the OFF position and keep them there until power is again available.

22. Before closing main or emergency switches, be sure that all controllers are in the OFF position so that the crane will not start unexpectedly.

23. If plugging protection is not provided, always stop the controllers momentarily in the OFF position before reversing—except to avoid injuries. (The slight pause is necessary to give the braking mechanism time to operate.)

24. Whenever the operator leaves the crane, this procedure should be followed:
    Raise all hooks to an intermediate position.
    Spot the crane at an approved designated location.
    Place all controls in the OFF position.
    Open the main switch to the OFF position.
    Make a visual check before leaving the crane.

Note: On yard cranes (cranes on outside runways), operators should set the brake and anchor it securely so the crane will not be moved by the wind.

25. When two or more cranes are used in making one lift, it is very important that the crane operators take signals from only one designated person.

26. Never attempt to close a switch that has an OUT OF ORDER or DO NOT OPERATE card on it, regardless of whether it locked out or not. Even when a crane operator has placed the card, it is necessary to make a careful check to determine that no one else is working on the crane, before removing the card.

27. In case of emergency or during inspection, repairing, cleaning, or lubricating, a warning sign or signal should be displayed and the main switch should be locked in the OFF position. This should be done whether the work is being done by the crane operator or by others. On cab-operated cranes when others are doing the work, the crane operator should remain in the cab unless otherwise instructed by the supervisor.

28. Never move or bump another crane that has a warning sign or signal displayed. Contacts with runway stops or other cranes shall be made with extreme caution. The operator shall do so with particular care for the safety of persons on or below the crane, and only after making certain that any persons on the other cranes are aware of what is being done.

29. Do not change fuse sizes. Do not attempt to repair electrical apparatus or to make other major repairs on the crane unless qualified and specific authorization has been received.

30. Never bypass any electrical limit switches or warning devices.

31. Load limit or overload devices shall not be used to measure loads being lifted. This is an emergency device and is not to be used as a production operating control.

General Maintenance and Safety Rules

The following maintenance safety rules are taken from ANSI/ASME B30.2:

- To be repaired, a crane must be moved to a location where there will be minimum interference with other cranes and operations in the area.
- All controllers should be in the OFF position.
- The main power source should be disconnected, de-energized, and locked, tagged, or flagged in the de-energized position. All other sources of energy should be neutralized so that they are in a state of energy isolation.
- WARNING or OUT OF ORDER signs should be placed on the crane, on the floor beneath, or on the hook where they are visible from the floor.
- If other cranes are in operation on the same runway, rail stops or other suitable devices shall be provided to protect the idle crane.
- Where rail stops or other devices are not available or practical, a person should be located where he [or she] can warn the operator of reaching the limit of safe distance from the idle crane.
- Where there are adjacent crane ways and the repair area is not protected by wire mesh or other suitable protection, or if any hazard from adjacent operations exists, the adjacent runway must also be restricted. A signaler shall be provided when cranes on the adjacent runway pass the work area. Cranes shall come to a full stop and may proceed through the area on being given a signal from the designated person.
- Trained, qualified, and authorized personnel shall be provided to work on energized equipment when adjustments and tests are required.
- After all repairs have been completed, guards shall be reinstalled, safety devices reactivated, and maintenance equipment removed before restoring crane to service.

Overhead Cranes

An overhead crane (Figure 13–7) may be operated either from a cab or from the floor. In the latter case, control devices may be either pendant push buttons or pull ropes. (In some cases, they are radio controlled.) The control handles should be clearly identified by signs and by shape or position so that the operator, while maintaining visual contact with the signaler, can identify each control by touch. Identify controls on all floor-operated overhead traveling cranes. Likewise, identify the controls in cab-operated cranes. If there are several cranes on the same runway or in the same building, all should have controls in identical positions so that a substitute operator will not be confused.

Safe means should be provided for the operator to pass from the cab to the footwalk. Stairs are preferred to ladders. If ladders are used, they should meet the appropriate safety standards. Furthermore, the space that the operator must step across in going from the landing or the runway girder to the crane should not exceed 12 in. (30 cm). Safe access also should be provided to the bridge motor and brake and to the equipment on the crane trolley. OSHA requires that a clearance of not less than 3 in. (7.6 cm)

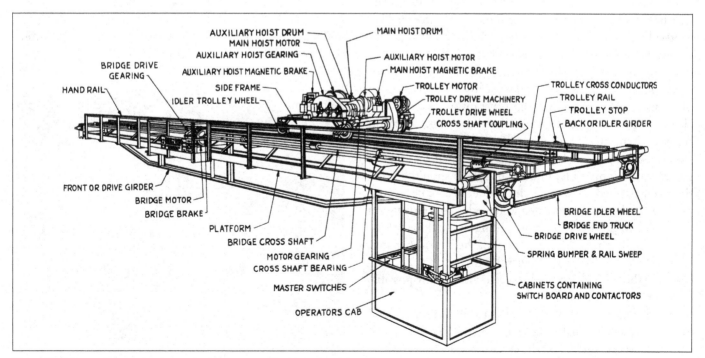

Figure 13–7. Essential parts of a typical cab-controlled overhead traveling crane. *(Courtesy Shaw Box Crane and Hoist Division of Manning, Maxwell & Moore Inc.)*

overhead and 2 in. (5.08 cm) laterally be provided and maintained between the crane and obstructions. Finally, stanchions or grab irons should be installed to enable a person to climb safely onto the trolley.

In case of an emergency, the operator must be able to escape from the crane regardless of its location on the runway. If a fire were to occur while the crane is attached to a load, the operator would be in particular danger if he or she could not travel at once to the access landing. Install an emergency means of escape in the cab, unless a means of escape via the bridge and runway is provided. A CO_2 dry chemical or equivalent fire extinguisher should also be installed in the cab.

Strict precautions should be taken to restrict personnel from servicing or riding the crane while it is in motion. This will prevent a person from being brushed off the crane by low beams or trusses. Service personnel must stay on the footwalk.

Footwalks and platforms should be substantial, rigidly braced, and protected on open sides with standard railings and toeboards. The footwalk should be reached by one or more fixed ladders not less than 16 in. (40 cm) wide. The outside edge of the walk should not be less than 30 in. (76 cm) from the nearest part of the trolley. The bridge walkway should have a 42-in.- (1-m-) high handrail, an intermediate rail, and a 4-in. (10-cm) toeboard. The space at the squaring shaft should be guarded so that a person cannot fall between the walkway and the crane girder.

A footwalk should also be provided, if headroom permits, along the entire length of the bridge of any crane that has a trolley running on the tops of the girders. This footwalk should be on the drive side of the bridge and should have toeboards and metal handrails. Safe access to the opposite side of the trolley also should be provided.

Flooring of walkways should be neatly fitted, leaving no openings, and should have a slip-resistant surface. Vertical clearance between the floor of the walkway and overhead trusses, structural parts, or other permanent fixtures should be at least 6.5 ft (2.0 m). Where such clearance is structurally impossible, built-in members should be distinctively painted, or striped and padded where necessary. Toeboards should be provided at the edges of flooring on the trolley to prevent tools from falling to the floor below.

To guard against electric shock, a heavy rubber mat should be provided at the control panel in the cab. The operator should have an unobstructed view of any possible position of the load hook.

The bridge truck wheels and the trolley wheels should all have sweeps to push away a person's foot or hand. To prevent a serious impact on the crane should a bridge wheel fail, the end frames or trucks should have safety lugs not more than 1 in. (2.5 cm) above the top of the rails.

After years of service, runway rails may become distorted. Or, the span between them may be altered from settling of the column footings. Wear on the flanges of the bridge wheels thus results. Rail alignment, therefore, should be checked every few years. If the tread of bridge truck wheels is tapered, the crane will run constantly square with the runway. Rail stops or bumpers must be so located that, when contacted, the crane bridge is square with the runway.

Electromagnets and Hook-On Devices

Electromagnets are often used with electric overhead cranes and gantry cranes, as well as with several other types of cranes. Electromagnets handle scrap iron and hot or cold ingots; they also move iron and steel products.

Do not use magnets either close to steel machines or parts or near iron materials being processed. Also keep watches and other delicate instruments out of the electromagnetic field. In that way, these objects will not become magnetized.

Label switches or switchboxes controlling power to the magnet DANGER—DO NOT OPEN SWITCH—POWER TO ELECTROMAGNET. The magnet switch must have a means of discharging the inductive load of the magnet.

The metal body of an electromagnet should be grounded. The magnet's power supply circuit should have a battery backup system. Even with a backup system, however, never move a load, suspended from a magnet, over personnel. The switchboard, wiring, and all other electrical equipment should comply with NFPA 70, National Electrical Code. (See also National Safety Council Data Sheet 359, Electromagnets Used with Crane Hoists.)

Special hook-on or clamping devices can be designed and made for handling special shapes. A grab for positioning steel coils is shown in Figure 13–8.

Figure 13–8. Mounted on a 20-ton- (18,144-kg-) capacity crane, this grab has full access to the coil storage area. *(Courtesy Harnischfeger Industries Inc.)*

Storage Bridge and Gantry Cranes

Storage bridge and gantry cranes, while similar to traveling cranes, travel on rails at ground level instead of on elevated runway girders. Gantry cranes have relatively short spans. Storage bridge cranes, however, may have a span of 300 ft (92 m) or more—sometimes with a cantilever on one or both ends. Storage bridge cranes are usually used for handling coal or ore. (See the section Conveyors, later in this chapter.) Ordinarily, a caged ladder on one of the legs of the crane provides access to the cab. Provide cranes that travel on rails with substantial rail scrapers or track clearers at each end of the tracks. Rail scrapers should be effective in both directions of travel.

There may be a serious shearing or crushing hazard in the area between cab-operated cranes and adjacent structures, or stored material. Provide a gong or other warning device that will sound intermittently from the time the travel controller handle is first moved from the OFF position until it is returned to the OFF position.

The wheel truck of gantry cranes should have adequate side clearance. On storage bridge and gantry cranes, provide bumpers made of cast steel that are at least one-half the diameter of the truck wheels in height. Fasten both wheel truck and trolley bumpers to the girder and not to the rail.

Spring bumpers are usually provided where bridge axles have antifriction bearings. If compression springs are used, they should be at least 5 in. (12.7 cm) in diameter at the point of contact. Arrange compression springs so no part can fall on a crane if a spring or guide pin breaks.

To prevent the crane from being moved down the track by a strong wind, the operator should apply rail clamps before leaving the cab, even for a short time. The holding power of the clamps should be sufficient to withstand wind pressure of 30 lb/ft^2 (1.4 k/Pa) of projected area of the crane. The electric contact rails or wires should be so located, or so guarded, that persons normally could not come into contact with them.

So that a bridge crane can be squared to the track, the squaring shaft should have a clutch. One end of the bridge can then be moved to bring the crane into proper position, while the other end remains stationary.

All cranes used outside should have the following features:
- floors of the footwalk constructed to provide drainage
- an operator's cab that (1) is constructed of fire-resistant material, (2) is weatherproof, (3) has provision for heating and ventilation, (4) has ample space for control equipment, and (5) allows the operator to see signals clearly
- the floor of the cab extended to an entrance landing and equipped with a handrail and toeboard of standard construction
- a rope ladder or other means of emergency escape from the cab

- locking ratchets on wheel locks, rail clamps, and brakes so that the crane will not move in a high wind
- skew switches to prevent excessive distortion of the bridge
- a screen or other barrier (preferably nonconductive) between the contact bars and the bridge walkways to prevent unintentional contact with the current conductor.

The crane's main line switch should be so constructed that it can be locked in the OPEN position. Mount the main line switch above the cab so it can be reached conveniently from the footwalk.

Monorails

A monorail system consists of one or more independent trolleys, supported from or within an overhead track, from which hoists are suspended. All applicable safety features should conform to ANSI/ASME B30.11.

Many industries extensively use monorail hoists to raise, lower, and transport materials. There are three major groups of monorails: hand-operated monorails, semi-hand-operated monorails, and power-operated monorails. On the hand-operated monorail, material is raised with a hand-powered hoist, and the trolley is propelled by hand. The semi-hand-operated monorail has a power hoist and is moved horizontally by hand. The power-operated monorail is electrically actuated for both vertical and horizontal movements.

No attempt should be made with a monorail hoist to lift or otherwise move an object by a side pull, unless the hoist has been designed for such use. Monorail hoists, operated in swivels, should have one or more safety catches or lugs that will support the load should a suspension pin fail. Provide stops at open ends of tracks such as interlocked cranes, track openers, and track switches.

Design all trolleys for monorail systems and underhung cranes to accommodate the maximum load. Design monorail track supports and track to carry the intended loads safely. Post the maximum load capacity on the monorail system. Frequently inspect both the track and its support for signs of weakening and wear. Make necessary repairs as soon as possible.

If an electric monorail carrier or crane is operated from a cab, a fixed platform or ladder should be provided to give the operator access to the cab. A means must also be provided for the emergency escape of the operator. The electric contact wires and the current collectors should be so located that an operator cannot inadvertently make contact with them upon entering or leaving the cab or while in the normal operating position inside the cab.

Jib Cranes

A jib crane is a crane capable of lifting, lowering, and rotating a load within a circular arc covered by a rotating arm or a jib. The jib, and the trolley running on it, are usually supported from a building wall, column, or pillar. A hoist (chain, air, or electric), with which the loads are lifted, is usually suspended from the trolley that travels on the jib boom.

Before a jib crane is mounted to a building wall or column of a building, the strength of the structure should be checked by a qualified engineer to determine whether or not the column or wall to be used is strong enough to support the jib, hoist, and load. Freestanding jibs must also have good foundation support. The jib should be braced or guyed, if necessary, to withstand the loads it is expected to carry.

A stop plate or angle iron should be installed at the outboard end of the jib to prevent the load trolley from running off the beam. Frequently inspect this end stop to see that it is not becoming loose or rusting off.

Derricks

The principal types of derricks are the A-frame derrick, the stiff-leg derrick, and the guy derrick. Although there are other types of derricks, these are the most common. All derricks must have every part firmly anchored.

A-Frame Derricks

As the name implies, the A-frame derrick (Figure 13–9) has a frame of steel or timber shaped like the letter "A" and is erected in a vertical plane. It has a brace, or leg, that extends from the top of the A at a 45-degree angle to the ground. The sills, or lowest part of the framework, tie this brace to the bottom of the A-frame. The boom is hinged at the horizontal member of the A. The base of the A-frame and the rear brace must be firmly weighted down.

Stiff-Leg Derricks

The stiff-leg derrick (Figure 13–10) has a mast with two braces at a 90-degree angle to each other and at a 45-degree angle to the ground. Usually, steel or timber sills tie the

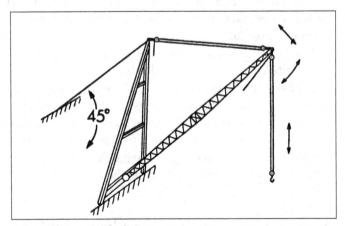

Figure 13–9. A-frame derrick. Brace is set at 45 degrees to the ground.

mast and the braces together at the ground. To withstand the uplift caused by a heavy load on the boom, use sandbags, cast-iron weights, or concrete blocks to hold down the stiff legs.

The hoist engines for both stiff-leg and A-frame derricks are usually bolted to the sills. On smaller derricks, the suspended loads may be slewed by being pushed manually. The boom of a large derrick may be swung by a "bullwheel" to which cables from another drum on the hoist engine are attached. A loaded cable may whip considerably and cause severe injury. Therefore, the horizontal cables between the hoist engine and the boom hinge should be barricaded, and workers should be prohibited from crossing over or under them.

Guy Derricks

The guy derrick (Figure 13–11) is used largely for erecting structural steel in tall buildings, especially those more than 10 stories high that cannot be reached by the boom of a crawler crane operating on the ground. Such derricks usually are made of latticed steel and have an odd number of equally spaced wire rope guys, each equipped with a turnbuckle and attached to the steel beams or columns on the erection floor. If the guy derrick is erected on the ground, however, the guys should be secured to heavy steel anchors buried deep in the ground, with additional weights placed at the anchor points. Steel beams or heavy timbers, 12 × 12 or 12 × 16 in. (30 × 30 or 30 × 40 cm), should be placed on the floor beams to support the foot of the mast. These foot blocks must be braced against the stubs of the building columns. This prevents their being "kicked" out of position when a heavy load is picked up with the boom at a low angle. Wire rope and turnbuckles may be used in place of 8 × 8 in. (20 × 20 cm) timbers to secure the base of the mast.

The hoist engine, whether on the same level with the derrick base or on the ground many floors below, should be securely anchored by steel cables or shoring timbers. This prevents it from being pulled toward the base of the mast by the tension on the cables.

A unique feature of the latticed-steel guy derrick is its ability to lift itself from one erection floor to the next. First, the hinge pin is removed to disconnect the boom from the mast. The boom hoist cable (or topping lift) then lifts the boom in a vertical position and sets it on the foot blocks close to the mast. The boom is then rotated 180 degrees, the normally unused guys are secured, and the boom then stands as a guyed gin pole. (See the section on gin pole derricks.) The load hoist of the boom then is used to pick up the mast, the mast guys being slackened off a few at a time and reattached at the upper level. When the mast is secured at the new erection floor, the boom is raised and again connected at the hinge.

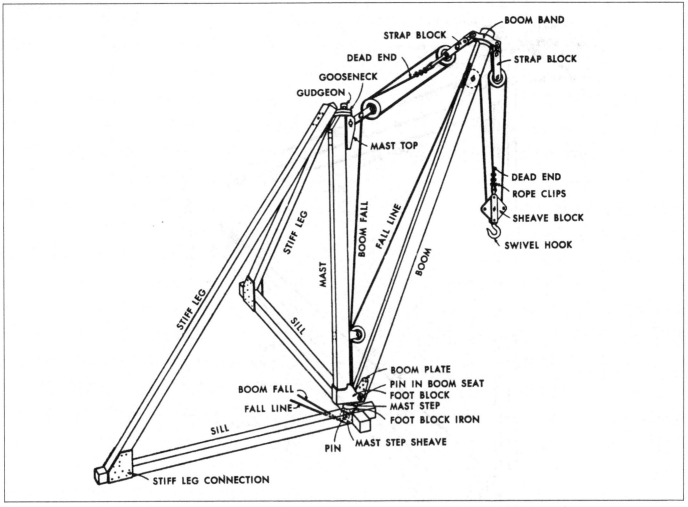

Figure 13-10. Diagram of stiff-leg derrick shows names of various parts. *(Courtesy Travelers Insurance Co.)*

To swing guy derricks, the workers push either on the suspended load or on a pipe "bull stick" attached near the base of the mast. A bullwheel can also be used as a mechanical means to swing the derrick.

Gin Pole and Breast Derricks

Other types of derricks are the gin pole and the breast derrick. The gin pole is merely a mast slightly out of plumb, with a hoisting tackle suspended from its upper end. The gin pole is supported by a number of guys, most of which are on the side away from the load. This rig is used for raising and lowering a load that needs to be moved horizontally only a few feet.

The breast derrick is a small, portable A-frame with a winch attached to it. Like the gin pole, it is erected in a nearly vertical position, with one or two guys to support it. Care must be exercised to prevent the base from slipping and causing the load to fall. Warn workers to watch their fingers when they operate the winch.

Tower Cranes

There are many design variations for the tower crane, depending on the manufacturer and the intended use. Tower cranes can be erected on a minimum of ground area or within a building, such as within an elevator shaft or other floor opening. To increase their range and versatility, some tower cranes are mounted on undercarriages that run on rails rather than on a fixed base. There is also a truck-mounted tower crane.

The following procedures are causes of tower crane incidents:
- improper erection, climbing, or dismantling of the crane
- lifting loads above the rated capacity of the crane, or lifting eccentric loads
- improper bracing of the crane
- bracing against, or attaching to, material or structural members that are insecure or unable to provide the needed support
- erection within a building not designed to support the crane's weight

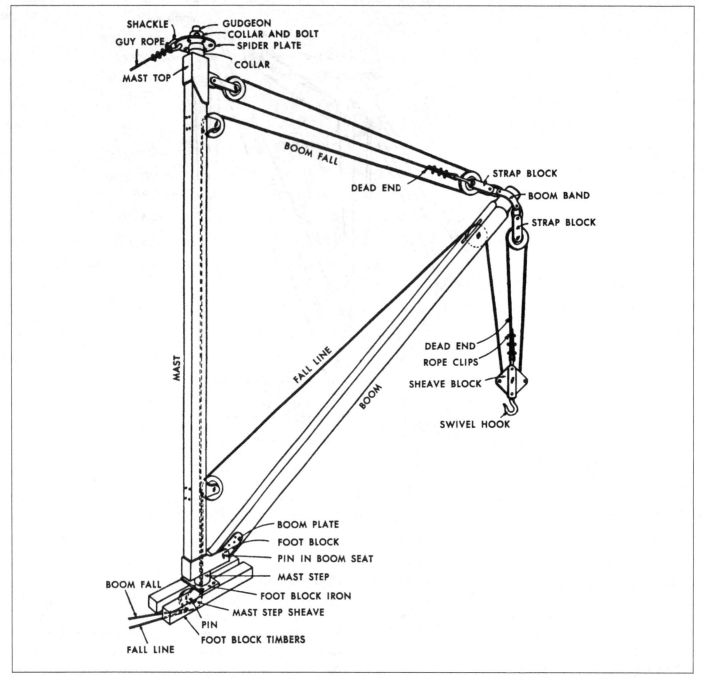

Figure 13–11. Diagram of a guy derrick. (For simplicity, only one guy is shown.) *(Courtesy Travelers Insurance Co.)*

- operators not knowing the limitations or operating characteristics of tower cranes because of inadequate training or follow-up
- tampering with limit switches or other safety devices
- operators not receiving instructions in their native language, or in clear English
- using tower cranes during high winds.

The following are general guidelines for preventing incidents with tower cranes:

- stresses for steel used for fabrication and construction conforming to American Institute of Steel Construction (AISC) specifications (if special materials such as high-tensile steel or aluminum alloys have been used in the crane structure, the crane should bear a notice to this effect)
- a secure attachment of counterweights plus safety ropes, rods, or chains to hold the counterweights, in addition to the basic attachment
- the strength of the system used to anchor the rope on a

winding drum with an ample safety factor exceeding the normal working load of the rope
- flanges of winding drums projected well above the height of the highest layer of rope wound on the drum in normal operation
- nonrotating hoist rope (except on receiving systems that do not require it)
- guarding of all moving parts including pulley block and sheave guards
- operators who are fully trained and certified to operate the crane and familiar with the work environment.

Operation

Only personnel of recognized ability should operate a tower crane. They should be mature in attitude, have quick responses, and be in good health. Their backgrounds should include both training and experience in the operation of this type of equipment.

Operators should possess a general knowledge of the crane's construction and the necessary knowledge of electricity, hydraulics, trade terms, parts identification, and the maintenance needs for this work. They also should have a knowledge of safety codes and standards applicable to crane operation and of any special safety recommendations of the crane manufacturer. Operators also should know the procedures involving visibility and signaling, lifting and lowering, swinging, and shutting down.

The crane operator should never stand on, or climb upon, the framework outside the cab while the crane is in operation. Prohibit climbing to the end of the jib, except when absolutely necessary. At such times, use prescribed special precautions and equipment. Operators should use fall protective equipment and other necessary personal protective equipment when necessary.

A final point needs to be made about tower crane operation and management. A tower crane is often rented, but the rental agency will not operate the crane on the job. Therefore, crane inspections prior to and during loading, and when climbing and dismounting, are extremely important.

Management needs to plan how the crane will be used—particularly when unusual loads, such as air conditioning units, are lifted. Load weight should always be known before lifting. Management plans should also include moving the crane. Numerous unintended incidents have occurred when the crane is being lifted or climbed to another level. Finally, the plan should provide that, after the job is done and the crane is being dismantled and removed, proper procedures are followed.

Communication between the crane operator and those at the loading and unloading areas must be clear, concise, and direct. While arm and hand signals are commonly used, direct voice communication can be even more effective.

Prior to initial use and each alteration or modification, cranes should be tested by a qualified person, with 125% of rated load, unless the manufacturer recommends otherwise. Frequent and periodic inspections are to be made dependent on the service requirements of the crane. Frequent inspections consist of the following items:
- all control mechanisms for maladjustment interfering with proper operation—daily, when in use
- all control mechanisms for excessive wear of components and contamination by lubricants or other foreign matter
- all crane function operating mechanisms for maladjustment interfering with proper operation and excessive wear of components
- motion-limiting devices for proper operation with the crane unloaded; each motion should be included into its limiting device or run at slow speed with care exercised
- load-limiting devices for proper operation and accuracy of settings
- all hydraulic and pneumatic hoses, particularly those that flex in normal operation
- electrical apparatus for malfunctioning, signs of excessive deterioration, dirt, and moisture accumulation
- hooks and latches for deformation, chemical damage, cracks, and wear (refer to ANSI/ASME B30.10)
- braces supporting crane masts (towers) and anchor bolt base connections for looseness or loss of preload
- hydraulic system for proper fluid level—daily, when in use.

Periodic inspections should include the requirements of a frequent inspection plus an examination by a designated person to check for hazards such as the following:
- deformed, cracked, or corroded members in the crane structure and boom
- loose bolts or rivets
- cracked or worn sheaves and drums
- worn, cracked, or distorted parts such as pins, bearings, shafts, gears, rollers, locking and clamping devices, sprockets, and drive chains or belts
- excessive wear on brake and clutch system parts, linings, pawls, and ratchets
- load, wind, and other indicators for inaccuracies outside the manufacturers' recommended tolerance
- power plants for performance and compliance with safety requirements
- electrical apparatus for signs of deterioration in controllers, master switches, contacts, limiting devices, and controls
- crane hooks inspected per ANSI/ASME B30.10
- travel mechanisms for malfunction, excessive wear, or damage
- hydraulic and pneumatic pumps, motors, valves, hoses, fittings, and tubing for excessive wear or damage.

Light service consists of the following: irregular operation with loads generally about one-half or less of the rated load, frequent inspections monthly, and periodic inspections annually. Normal service consists of the following: operations at less than 85% rated load and not more than 10 lift cycles per hour, except for isolated instances; frequent inspections weekly to monthly; and periodic inspections semiannually to annually. Heavy service consists of the following: operations at 85% to 100% of rated load, or in excess of 10 lift cycles per hour, as a regular specified procedure; frequent inspections daily or weekly; and periodic inspections quarterly.

Frequent inspections are most often done by the operator with no records required. Periodic inspections require records and are done by an appointed person.

Mobile Cranes

Mobile cranes are an engineering marvel. The construction industry, in particular, is able to perform tasks once nearly impossible to do. A mobile crane may have up to a 1,000-ton capacity and a length of 600 ft plus boom and jib.

Canadian (Ontario) data suggest that about one in five construction fatalities is crane-related. In general, 90% of mobile crane injuries are attributed to operator error. Other factors are:

- support failure—30%
- failure to use outrigger—20%
- crane failure—10% to 20%
- rigging—4% to 15%.

See Tables 13–A and 13–B for statistics on crane-related fatalities.

In addition, of all injuries to employees, at least 25% occurred to those involved with the load, and 10% to 15% involved maintenance, refueling, and so forth, around the crane.

Mobile cranes include locomotive cranes, crawler cranes, wheel-mounted cranes, and industrial truck cranes. The first three types have standard designs, but industrial truck cranes have various designs, depending on intended use.

All mobile cranes have booms with load hoists and boom hoists. In most instances, the crane swings or rotates on a turntable, which rests on a railroad car, crawler, or wheel chassis. Power is provided by electric motor, steam, gasoline, or diesel engine.

Electric-powered equipment should be grounded as specified in ANSI/IEEE C2, National Electrical Safety Code. Repairs or adjustments should be made only by qualified personnel. Power should be disconnected before repairs are made. Trailing cables should be kept off the ground whenever possible and should be handled only with insulated hooks.

TABLE 13–A. Causes of Crane-Related Deaths in Construction, 1992–2006

Cause of Death	Number of Deaths	% of Total
Overhead power line electrocutions	102	32%
Crane collapses	68	21%
Struck by crane booms/jibs*	59	18%
Struck by crane loads	24	7%
Caught in/between	21	7%
Struck by cranes**	18	6%
Other causes***	31	10%
Total	323	****

*52 of 59 struck by crane booms/jibs were due to falling booms/jibs.
**Includes 10 run over by mobile cranes.
***Other causes includes 14 struck by other crane parts and 9 highway incidents.
****Does not add to 100 due to rounding.

Source: U.S. Bureau of Labor Statistics Census of Fatal Occupational Injuries Research File

Gasoline-operated cranes require protection against the hazards of fire and explosion. Engines should not be refueled while running. If refueling is done by hose connected to a tank truck or connected to drums by means of pumps, use a metallic bonding connection between the hose nozzle and the fill pipe.

If fuel is carried to the crane by hand, use safety cans. Open lights, flames, and sparks should be eliminated, and lights on the equipment should be of an approved explosion-proof type. A fire extinguisher of 5BC rating should be kept in the cab of the rig.

To prevent the boom from being dropped unintentionally, operate the boom hoist mechanism by gearing or chain and by lowering the engine's speed. Provide a self-setting brake and locking pawl or other positive locking device. Furthermore, with the exception of some hydraulic cranes, install boom stops—preferably of the shock-absorbing type—on all cranes.

If possible, install a hoist-line, two-blocking limiting device. This device could be controlled in conjunction with the load-line hoist clutches and brakes. The operator must have full control of all crane functions at all times. No attempt should be made to lift the boom by means of the hoist's load cable. Do not lift a load with the boom's hoist line unless the crane is designed for this purpose.

Load Chart

Every crane should have a capacity plate or a sign that states its safe load capacity at various radii from the center pin of the turntable. Install this sign so it is plainly visible to the crane operator, signaler, and rigger. Mount a boom angle

TABLE 13–B. Fatal Occupational Injuries Involving Cranes,* 1997–2006

	1997	1998	1999	2000	2001	2002	2003	2004	2005	2006
Crane-related fatalities	97	93	80	90	72	80	62	87	85	72

*Includes fatalities where the source of the injury was a crane, where the secondary source of the injury was a crane, or where the worker activity was operating a crane.

Note: Totals for 2006 are preliminary. Totals for previous years are revised and final.

Source: U.S. Department of Labor, Bureau of Labor Statistics (in cooperation with state, New York City, District of Columbia, and federal agencies) Census of Fatal Occupational Injuries.

indicator with a freely suspended pointer actuated by gravity in front of the safe load capacity sign on the side of the boom near the hinge. The pick must be within the limits prescribed by the load chart.

A capacity chart for the operators should indicate boom length, boom angle, and capacity. A load indicator device enables the operator to handle the load better. For locomotives using outriggers fully extended, the maximum load rating is 80% of tipping load. Where structural competence governs lifting performance, load ratings are reduced and the rating chart should indicate this. When handling loads that are limited by structural competence rather than by stability, the person responsible for the job should make sure that the weight of the load has been determined within ±10% before it is lifted.

The operator must have safe access to and exit from the crane's cab or seat. To operate the crane safely, the operator must have an unobstructed view of the load hook and the point of operation at all times or must rely on someone giving signals. The operator must also be able to see ahead of the crane when it is traveling on the ground, whether the chassis is moving forward or backward. On some cranes, visibility to the rear is obstructed. In such cases, the operator must use extreme caution and rely on a signaler. Other workers should always stay beyond the range of the cab's swing and out from under the boom and the load.

Other precautions include good lighting and warning devices. A crane operated after dark should have clearance lights. Floodlights should illuminate the area beneath the boom, and lights mounted on the underside of the boom are recommended. A warning bell or horn and an automatic backup alarm are necessary equipment for a wheel-mounted crane. Also, provide a warning bell or horn on crawler cranes.

When other ways are not available, it may be necessary to hoist employees to another work level or area. To do so with a crane, a number of practices need to be followed. The following information is taken from ANSI A10.28, Safety Requirements for Work Platforms Suspended from Cranes or Derricks for Construction and Demolition Operations:

Preparation is the key to avoiding an unintentional incident. The work platform must be designed and rigged by someone qualified to do so. A safety factor of at least five is to be built in. Identification shall include empty weight and capacity. There shall be an access gate, perimeter protection, and a grab rail. The suspension systems shall minimize tipping during use.

At each new job site, prior to hoisting employees, the platform and rigging shall be proof-tested to 125% of the platform-rated capacity for five minutes. Any deficiencies must be corrected and another proof test conducted before hoisting employees.

Prior to hoisting personnel, a trial run shall be made with the platform and rigging to the proposed work elevation to ensure (1) that boom configuration and load lines are adequate, and (2) that no interference of any kind exists and the anti-two-block protection is working, if the crane is so equipped.

The operator shall demonstrate ability to operate the crane and derrick before hoisting personnel in a suspended work platform. The operator must be comfortable in the procedure and not feel upset or unduly anxious.

Some of the crane's requirements are that live booms are not allowed and crane travel must be only on tracks unless using a portal crane. An occupied work platform must not be allowed to fall free. In addition, cranes used for hoisting personnel must have the swing brake or lock engaged and be inspected prior to the hoist, including wire rope, hook brakes, boom, and all other equipment. And finally, the total weight of the work load must not exceed 50% of the rated crane capacity.

Crawler and Wheel-Mounted Cranes

These types of cranes are frequently used in erecting structural steel for tall buildings. Extension sections may be inserted to lengthen the booms of these cranes. When the boom is assembled, check all parts of the structure for damaged or missing parts. In one documented case, a lattice bar was unintentionally omitted. Later, when a heavy load was picked up, the boom buckled at this point and several men were killed.

Locomotive Cranes

In the cab of a locomotive crane, a clear passageway should be provided from the operator's platform to an exit door near the operator's side of the cab. Doors should be hinged at the rear edge and should open outward. Sliding doors should slide to the rear to open.

The motor and all power-transmission apparatus should be guarded. Enough light should be provided in the cab to permit the operator to work safely and to see the gages and indicators plainly.

Install steps and handholds for safe access to the cab. Some state laws require that footboards and handholds be provided at each end of the truck bed or that the truck have a pilot or fender. Permit no one except necessary operating personnel on a rig while it is operating.

An on-track crane should have standard automatic couplers and uncoupling levers. It should also have air brakes as well as hand brakes. It should have rail clamps at each corner to hold it in a stored position. Provide a guard at the end of the boom to prevent the thimble on the cable from coming into contact with the sheave.

An on-track crane should be moved only on signal from an authorized signaler or switchman. This person should walk ahead of the crane to warn others and to see that switches are properly set and that the track is free of obstructions. When there is no signaler or switchman, the crane operator should move the crane only on order from the supervisor of the department in which the crane is working. The crane should not be swung across another track until the crane operator and signaler have made sure that cars are not on that track and will not be moved to it.

When moving the crane about the yard or worksite, the operator should keep the crane and boom parallel to the track, to avoid striking buildings or other structures, and should carry the boom low enough to clear overhead wires. Operators should not carry buckets and magnets on the boom when the crane is going from one location to another.

Operation of Cranes

Extended outriggers are considered a part of the counterweight on the load charts of new cranes. Separate charts state crane capacity for a traveling load and for lifting a load without using outriggers. Whether traveling or stationary, the crane's turntable should remain level.

A boom must never be swung too rapidly. If it is, the suspended load will be swung outward by centrifugal effect. This action could cause the crane to rock or even tip. If this occurs, the load may swing and strike a person or object.

Operating a crane on soft or sloping ground or close to the sides of trenches or excavations is dangerous. Be sure the crane is level before it is put into operation.

Outriggers can be relied upon to give stability only when used on solid ground. Use heavy timber mats whenever there is doubt as to the stability of the soil on which a crane is to be operated. Do not permit the use of makeshift

methods to increase the capacity of a crane, such as using timbers with blocking or adding counterweight. If the crane tips when hoisting or lowering a load, the operator should lower the load as quickly as possible by snubbing it lightly with the brakes. Therefore, never allow workers to ride a load that is being hoisted, swung, or transported.

When operating a crane with the boom at a high angle, the operator should take care that the suspended load does not strike the boom and bend the steel lattice bars on its underside. Bending these bars will weaken the boom so that when it picks up the next heavy load, it may collapse. Likewise, if the main parts of the boom are bent even slightly, the strength of the boom may be materially reduced.

When an extended boom is used on a crane, as for erecting structural steel, the operator must use extreme care in lowering it to the ground at the end of the job. An extended boom should never be lowered to one side of the chassis or crawler. The stability of the crane is greatly reduced in that position, and the crane may even tip.

When using a boom tip extension or jib, the allowable load on the jib is limited. The operator must know its capacity. The operator must refer to the crane's capacity chart in order not to exceed load limits. The crane operator must use care when swinging with a load, especially when the jib is lowered at an angle to the main boom.

The operator must center the hook over the load to keep it from swinging while it is being lifted. When holding the hook or slings in place while the slack is taken up, employees should keep their hands out of pinch points. A hook, or even a small piece of board, may be used for the purpose. If a person must use his or her hand, it should be placed flat against the sling to hold it. The hook-on person, the rigger, and everyone else must be in the clear before the load is lifted. Use a tag line for guiding loads. While the crane is operating, the area inside the counterweight and body frame should be roped off to prevent entry by ground personnel into the pinch point formed by these two crane parts.

Operators should never remove a heavy load from a truck by hooking a crane to the load and then having the truck pull out from under the load. If the load should prove too heavy for the crane, the crane will tip before the operator can lower the load to the ground. The load should be lifted clear of the truck body. The operator should make sure that the crane can handle it safely before the truck is moved out from under it.

A crane should never be used for jerk piling. If piling cannot be pulled by a straight, steady pull—limited to rated capacity—a pile extractor should be used. When a pile extractor is used, keep the boom angle at or less than 60 degrees above horizontal.

Consult the crane's manufacturer before modifying the equipment. Such changes should maintain at least the same factor of safety as the originally designed equipment. Maintenance and repair work should be performed only by trained and qualified personnel. The operator, however, is responsible for keeping the unit clean. Before leaving the crane at the end of the workday, the operator should lower the load block or bucket to the ground in such a manner that it cannot be upset.

Travel of Cranes

Except for very short distances, a crane should not travel with a load suspended from the boom. When a crawler crane must travel on a public thoroughfare, the boom should point forward, and someone with a flag should walk ahead of it.

A wheel-mounted crane and a crawler crane on a semi-trailer should be transported with the boom pointing toward the rear and high enough to clear an automobile. Wheel-mounted cranes with short booms may travel with the boom forward in the boom rest. One of the work crew should follow in a car or truck to keep other vehicles from traveling beneath the boom. Otherwise, if the semi-trailer's wheels should roll into a low spot in the pavement, the boom might suddenly crash through a car's roof. If not disassembled, a crane being transported should have its engine running and the operator should be in the cab. In that way, the operator can swing the boom, when necessary, thus avoiding damage to trees, poles, or buildings when the semi-trailer turns corners.

Before moving heavy, slow-moving equipment or heavy equipment on a low-slung trailer over any public or private railroad grade crossing, notify a representative of the railroad company. The railroad company can then provide flag signaling to guard against a train's striking the equipment while it is moving over the crossing or if the equipment becomes stalled or "hung up" on the crossing. This is an important precaution for the benefit of both the equipment owner and the railroad company.

Electric Wires

Consider any overhead wire as an energized line until either the person who owns the line or the electric company indicates that it is not energized. Compliance with recommended practices, and not reliance on other devices, should be followed in determining how close the crane and its extensions, including load, are to electric power lines. A qualified signaler should be assigned to observe the clearances and give warning before the crane approaches the stated limits. The boom's load line and the cables of the crane should be kept away from all electric wires, regardless of their voltage. In the United States, OSHA requires

that, except where the electrical lines have been de-energized and visibly grounded at the point of work, or where insulating barriers, not a part of or an attachment to the crane, have been erected to prevent physical contact with the lines, cranes should be operated near power lines only in accordance with the following:

- For lines rated 50 kV or below, minimum clearance between the lines and any part of the crane or load must be 10 ft (3 m).
- For lines rated above 50 kV, minimum clearance between the lines and any part of the crane or load must be either 10 ft + 0.4 in. (3 m + 10 mm) for each 1 kV over 50 kV, or twice the length of the line insulator but never less than 10 ft (3 m) (29 CFR 1910.180[j][1]).
- In transit and with no load and boom lowered, the clearance should be a minimum of 4 ft (1.2 m).
- At construction sites, the 4-ft (1.2-m) minimum applies only to voltages less than 50 kV. Clearance must be 10 ft (3 m) for voltages above 50 kV and up to 345 kV and 16 ft (4.9 m) for voltages up to 750 kV (29 CFR 1926.550[a][15]).

If cage-type boom guards, insulating links, or proximity warning devices are used on cranes, such devices must not be a substitute for the requirements of a specifically assigned signal person, even if such devices are required by law or regulation. In view of the complex, invisible, and lethal nature of the electrical hazard involved, and to lessen the potential of false security, devices should be used and tested in the manner and at the intervals prescribed by the devices' manufacturers.

If the boom's load line or cables unintentionally come into contact with a wire, the operator should swing the crane to get clear. If the wire has been broken and the boom cannot be cleared from it, the operator should stay on the crane and remain calm.

If the ground is wet or damp, a crawler crane will be electrically grounded so that when the boom touches a power line, the wire will, in turn, be grounded and the power company's circuit breaker will open. Some arcing may occur. After a few seconds, however, many power line circuit breakers feature a recloser that will automatically close and reenergize the wire to determine if the fault was created by a nuisance. Again, the circuit breaker may open, and again, it will close. Thus, the wire may be "dead" at one instant but live a few seconds later.

If the boom of a wheel-mounted crane on rubber tires should become tangled with a "hot" electric wire, the entire crane may be energized. In such a case, the rubber tires may or may not insulate the crane and chassis from the ground. Depending on the voltage and the soil conditions, the tires on the crane may burn and melt—and thus lose any insulating qualities. Hence, the circuit breaker may not open, and the wire and the crane may remain energized.

Stepping from the crane to the ground is often fatal because one hand and one foot may be in contact with the crane when the other foot touches the ground. Therefore, the operator should remain on the crane until the emergency crew from the electric company arrives and frees the crane from the live wire. However, if the gasoline tank should ignite, or if for any other reason it is impossible to remain on the crane, the operator should jump, making sure that all body parts are clear of the crane before touching the ground. Should the operator find it necessary to jump from the equipment, the best practice is to land with both feet together and hop or shuffle away from the crane to prevent electric shock due to step potential because there may be a difference in voltage potential at each of the operator's feet as a normal step is taken, allowing current to pass through the operator's body from leg to leg. (See Figure 13–12 and the NSC's Occupational Safety and Health Data Sheet 12304-0743, Mobile Cranes and Power Lines.)

A crane that has been idle for a period of 1 month or more, but less than 6 months, should be given an inspection by a qualified person conforming to frequent inspection criteria both of the crane and rope before being placed in service.

A crane that has been idle for a period of more than 6 months should be given a complete inspection by a qualified person of both the crane and rope conforming to the requirements for both frequent and periodic inspections.

Frequent inspections are done by a designated person, usually the operator, on a daily or monthly interval depending on the crane's usage. A good practice is for the operator to perform the frequent inspection routine each day before he or she starts to operate the crane. The inspection consists of the following:

- all control mechanisms for maladjustment interfering with proper operation—daily, when used
- all control mechanisms for excessive wear of components and contamination by lubricants or other foreign matter
- all safety devices for malfunctions
- visual inspection of all hydraulic hoses, and particularly those that flex in normal operation of crane functions, once every working day, when used
- hooks and latches for deformation, chemical damage, cracks, and wear (refer to ANSI/ASME B30.10)
- rope reeving for compliance with crane manufacturer's specifications
- electrical apparatus for malfunctioning, signs of excessive deterioration, dirt, and moisture accumulation
- hydraulic system for proper oil level—daily, when used
- tires for recommended pressure
- visual inspection of all running rope in service once each working day to discover gross damage that may be an immediate hazard, such as:
  ○ distortion of the rope such as kinking, crushing, unstranding, bird-caging, main strand displacement, or core protrusion; loss of rope diameter in a short rope length or unevenness of outer strands should provide evidence that the rope or ropes must be replaced
  ○ general corrosion
  ○ broken or cut strands
  ○ number, distribution, and type of visible broken wires (see Chapter 14, Ropes, Chains, and Slings, in this volume).

Care should be taken when inspecting sections of rapid rope deterioration such as flange points, crossover points, and repetitive pickup points on drums. Care should be taken when inspecting certain ropes such as:

- rotation-resistant ropes because the internal deterioration of rotation-resistant ropes may not be readily observable
- boom hoist ropes because of the difficulties of inspection and the important nature of these ropes.

Periodic inspections are performed at intervals of 1 to 12 months or as specifically recommended by the manufacturer or by a qualified person. Dated records for periodic

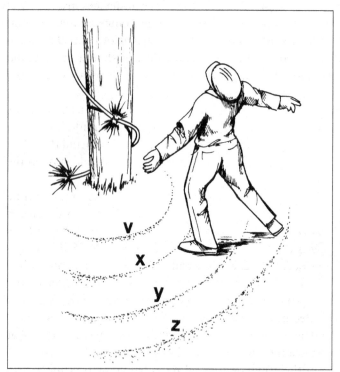

Figure 13–12. An operator on an electrified crane should jump with feet side by side and should not run (but hop with feet together) away from it, not stopping until well away. Hopping avoids the ground gradient effect: if one foot is at y voltage, and the other at z voltage, the difference in voltage will cause a flow of electricity through the body.

inspections should be made on all critical items such as brakes, crane hooks, ropes, hydraulic and pneumatic cylinders, and hydraulic and pneumatic pressure valves. Records should be kept where available to appointed personnel. This inspection includes all the elements of a frequent inspection plus the following. Any deficiencies, such as those listed, should be examined and determination made as to whether they constitute a hazard:

- deformed, cracked, or corroded members in the crane structure and entire boom
- loose bolts or rivets
- cracked or worn sheaves and drums
- worn, cracked, or distorted parts such as pins, bearings, shafts, gears, rollers, and locking devices
- excessive wear on brake and clutch system parts, linings, pawls, and ratchets
- load, boom angle, and other indicators over their full range, for any significant inaccuracies
- gasoline, diesel, electric, or other power plants for performance and compliance with safety requirements
- excessive wear of chain-drive sprockets and excessive chain stretch
- crane hooks inspected for cracks
- travel steering, braking, and locking devices, for malfunction
- excessively worn or damaged tires
- hydraulic and pneumatic hose, fittings, and tubing inspected for the following:
  - evidence of leakage at the surface of the flexible hose or its junction with the metal couplings
  - blistering or abnormal deformation of the outer covering of the hydraulic or pneumatic hose
  - leakage at the threaded or clamped joints that cannot be eliminated by normal tightening or recommended procedures
  - evidence of excessive abrasion or scrubbing on the outer surface of a hose, rigid tube, or fitting.

Means should be taken to eliminate the interference of elements in contact or otherwise protect the components.

Hydraulic and pneumatic pumps and motors should be inspected for the following:
- loose bolts or fasteners
- leaks at joints between sections
- shaft seal leaks
- unusual noise or vibration
- loss of operating speed
- excessive heating of the fluid
- loss of pressure.

Hydraulic and pneumatic valves should be inspected for the following:

- cracks in valve housing
- improper return of spool to neutral position
- leaks at spool or joints
- ticking spools
- failure of relief valves to attain correct pressure setting
- relief valve pressure as specified by manufacturer.

Hydraulic and pneumatic cylinders should be inspected for the following:
- drifting caused by fluid leaking across the piston
- rod seals leakage
- scored, nicked, or dented cylinder rods
- dented case (barrel) loose or deformed
- rod eyes or connecting joints.

Hydraulic filters should be inspected for evidence of rubber particles on the filter element, which may indicate hose, "O" ring, or other rubber component deterioration. Metal chips or pieces on the filter may denote failure in pumps, motors, or cylinders. Further checking will be necessary to determine the origin of the problem before corrective action can be taken.

Aerial Baskets
Aerial lift equipment is now commonly used for working above ground. These boom-mounted buckets, baskets, or platforms are used extensively in constructing and maintaining electric and telephone lines. Because of their capabilities, however, their use is increasing in harbor and port work; in the aircraft industry; in highway sign and lighting construction; and in maintenance, painting, sandblasting, and fire-fighting work.

Hazards of Aerial Baskets
The most frequent causes of unintentional incidents while using mobile aerial baskets include the following:
- not observing proper precautions against electrical hazards to personnel both in the basket and on the ground
- improper positioning of vehicle or outriggers, lack of sufficient blocking under outriggers, or overloading the boom, causing the apparatus to overturn or fail
- overreaching from basket or other improper work procedures
- not using proper personal protective equipment, including safety belts
- moving the truck while the boom is raised or moving it where there is inadequate clearance for the boom
- structural or mechanical failure or control jamming
- swinging the boom or basket against overhead obstructions or energized equipment
- moving the boom into positions that interfere with traffic
- inadequately trained personnel.

Operation of Aerial Baskets

The operating and maintenance instruction manuals issued by the manufacturer should be followed. Lift controls should be tested each day prior to use to determine that such controls are in safe working condition. Aerial baskets should be inspected daily for defects. Mechanical equipment should be inspected each day before using it to make sure it is safe to operate. Additional safe operating procedures include the following:

- Load limits of the boom and basket should be posted and should not be exceeded.
- A warm-up period and test of the hydraulic system is required.
- Basket equipment approved for use on energized equipment should be dielectrically tested periodically.
- When working near energized lines in aerial-basket trucks and aerial-ladder trucks, the trucks should be grounded. Where grounds are not permitted by the company, barricading should be required.
- The insulated portion of an aerial-lift boom and basket should not be altered in any manner that might reduce its insulating value.
- Drivers of trucks with mounted aerial equipment should be constantly aware that the vehicle has exposed equipment above the truck cab, and they should provide necessary traveling clearance.
- The truck should not be moved unless the boom is lowered and the basket or ladder is cradled.
- Riding in the basket while a truck is traveling should not be permitted. (Employees may ride in the basket at the work location for short moves if the basket is returned to the cradled position for each move.)
- Available footing for the truck's wheels and outriggers should be examined carefully to ensure a stable setup. Hand brakes, chocks, and/or cribbing, when needed, should also be used to ensure stability. The truck should sit approximately level when viewed from the rear.
- Before lowering stabilizers, outriggers, or hydraulic jacks, the operator should be certain there is no one in an unsafe position.
- When the boom must be maneuvered over a street or highway, necessary precautions should be taken to avoid unintentional incidents with traffic and pedestrians.
- The operator should always face the direction in which the basket is moving and be sure that the path of the boom or basket is clear when it is being moved.
- An employee should not stand or sit on the top or the edge of the basket. Ladders should not be used in the basket. While in the basket, the employee's feet should always be on the floor of the basket.
- Employees should not belt themselves to an adjacent pole or structure.

- Employees should always belt themselves to the boom or ladder. Belting to the basket equipment should be done upon entering the basket.
- When working with rubber protective equipment on energized circuits or apparatus above 300 V, the following minimum conditions should be met in addition to all other rules governing the use of protective equipment:
  - Properly rated, electrically insulated and tested rubber gloves with leather protectors and rubber sleeves should be worn.
  - An employee should make physical contact with protective devices installed on energized primary conductors only with rubber gloves and rubber sleeves.
  - Employees should be isolated from all grounds by using approved protective equipment or other approved devices.
  - When employees are working on the same pole or substation structure, or from a bucket truck, in no case should they work simultaneously on energized wires or on equipment of different polarities.
  - An employee should not enter or leave the basket by walking the boom.
  - Employees should not transfer between the basket and a pole. On dual-basket trucks, employees should not transfer between the baskets.
  - Employees in baskets should not wear climbers.
  - When two workers are in the basket or baskets, one of them should be designated to operate the controls. One employee should give all signals and make sure these signals are thoroughly understood by all people concerned.
  - Baskets should be located under or to the side of conductors or of working equipment. Raising the basket directly above energized primary conductors or equipment should be kept to a minimum.
  - Only nonconductive attachments should be allowed on baskets.
  - The operator should be sure that hoses or lines attached to tools cannot become entangled with the levers that operate the boom.
  - Air- or hydraulic-operated tools should be disconnected from the source when not in use.
  - When employees are working from the basket, take extreme care to avoid contacting poles, crossarms, or other grounded or live equipment.

Inspection of Aerial Baskets

An effective daily inspection should cover the following points. If defects are found, the inspector should report them and see that they are corrected before they develop into dangerous conditions.

- visual inspection of all attachment welds between actuating cylinders and booms or pedestals
- visual inspection of all pivot pins for security of their locking devices
- visual inspection of all exposed cables, sheaves, and leveling devices both for wear and for security of attachment
- visual inspection of hydraulic system for leaks and wear
- inspection of lubrication and of fluid levels
- visual inspection of boom and basket for cracks or abrasions
- operation of boom from ground controls through one complete cycle, listening for unusual noises and looking for deviations from normal operation.

Basket Safeguards

Equip aerial baskets with (1) full-body safety harnesses, safety belts, and lanyards to be worn by all persons working from the baskets and (2) a means for attaching the lanyard to the equipment. In general, it is more satisfactory to anchor the lanyard to the boom. However, if this will interfere with the controls or if other considerations are involved, then anchor the lanyard to the basket. Lanyards should be only long enough to allow movement within the basket and should prevent climbing onto the rim. Such lanyards limit free fall but do not restrict work or entangle workers' feet. When aerial baskets are used near energized conductors, the harness and lanyard should be arc rated, and the person(s) in the basket should wear the proper level of arc-rated protective equipment, as well as use insulated tools and the proper category of test equipment, based on the anticipated electrical hazards.

Each basket operator and driver should be thoroughly trained in the use of the equipment before operating it on a job. He or she should know not only the particular equipment involved, but also the type of work it will do in the field.

If the public is exposed to possible contact with the vehicle, set up barricades. A boom coming in contact with energized equipment might electrically charge the truck and could injure people who are walking or standing nearby.

Many incidents with aerial baskets have been caused by inadequate footing, so solid footing for the wheels and outriggers should be provided. Snow, ice, mud, and soft ground call for extra caution because additional firm footing under the outriggers may be necessary.

Crabs and Winches

Crabs and winches may be either hand operated or electrically driven. Install some form of brake or safety lowering device, and anchor portable units securely against the pull of the hoisting rope or chain.

Install barricade guards to protect the operator against flying strands of wire and the recoil of broken ropes. Design such guards to protect the operator from the whipping effect of a broken rope if the point of operation cannot be totally guarded (Figure 13–13). The best defense against injury is to be well away from the direct line of pull. Also, be sure that gears are fully guarded. Power-driven crabs and winches should have their moving parts encased and should be electrically grounded.

The locking pawl on the ratchet of a winch frequently presents a serious hazard to fingers, particularly when the operator attempts to disengage the pawl. To reduce this hazard, weld a small lever to the pawl so that it can be safely grasped.

Hand-operated equipment that has a crank handle instead of a handwheel poses a major danger. Operators may be struck by the revolving crank handle if they lose control while lowering a load. Provide a dog to lock the gears. A pin through the end of a crank will keep it in the socket during hoisting operations.

To lower loads rapidly, use a strap brake. Before using the brake, remove the crank or take other steps, such as replacing a spur gear and dog with a worm gear, to prevent the crank handle from flying around.

Block and Tackle

A safety factor of 10 is recommended for determining the safe working load of Manila rope (falls) in a block-and-tackle assembly. This large safety factor allows for (1) error in estimating the weight of the load, (2) vibration or shock in handling the load on the tackle, (3) loss of strength at knots and bends, and (4) deterioration of the rope due to wear or other causes.

The governing factor is usually the safe working load of the blocks rather than of the falls (rope). By multiplying the number of sheaves and rope parts, the weight of the

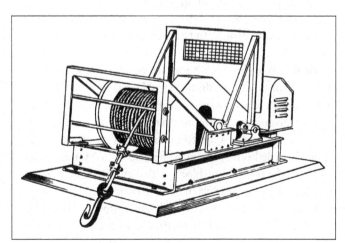

Figure 13–13. The operator of this car puller winch stands behind the shield, which protects the employee if the rope breaks.

load that can be handled by the rope multiplies but does not correspondingly increase the strength of the blocks. Calculations show that, in most instances, using a safety factor of 10 for the rope automatically makes the load on the blocks correspond to the rope size within safe work load limits. (Mark blocks with their safe working load, as specified by their manufacturers.) The total weight on the tackle should never exceed this safe load limit. (See Chapter 14, Ropes, Chains, and Slings, in this volume.) Safe workloads for rope used in block-and-tackle assemblies are conversely 1/10 of the block's breaking strength, based on a safety factor of 10.

To find the required breaking strength for new rope, proceed as follows:
1. For each sheave 3 in. (7.6 cm) in diameter or larger, add 10% to the weight of the load to compensate for friction loss.
2. Divide this figure by the number of ropes or parts running from the movable block.
3. Multiply the resultant figure by a safety factor of 10.

An example for working out the procedure just given is as follows:
1. A load to be lifted weighs 2,000 lb (900 kg), and the tackle consists of two double blocks—four sheaves, four rope parts at the movable block.
2. Friction loss (10% for each sheave) = 40% or 800 lb (363 kg).
3. 2,000 + 800 = 2,800 lb (1,270 kg), which divided by 4 (the number of parts at the movable block) is equal to 700 lb (318 kg).
4. Applying the safety factor of 10 (10 × 700) gives 7,000 lb (3,200 kg), the required breaking strength of the rope.
5. New Manila rope of 7/8 in. (2.2 cm) has a breaking strength of 7,700 lb (3,500 kg) and, therefore, is the proper size for the load. Synthetic fibers would have greater tensile strength. (See Chapter 14, Ropes, Chains, and Slings, in this volume.)

The safe workload limit for two double blocks made for rope of 7/8-in. (2.2-cm) diameter is 2,000 lb (900 kg)—the equivalent of the total load in the example. This information is from a prominent manufacturer for one series of its standard blocks (regular mortise, inside iron-strapped blocks, with loose side hooks, intended for use with Manila rope).

Attach the rope to the block with a thimble and a proper eye splice. A mousing of yarn or small rope should be placed on the upper hook of a set of falls as a precaution against its unintentional detachment. Inspect blocks thoroughly and frequently, paying particular attention to parts that are subject to wear.

Figure 13–14 shows how tackle blocks should be reeved. If the sheave holes in blocks are too small to permit sufficient clearance, excessive surface wear of the rope will occur. Likewise, excessive internal friction on the fibers will occur if the diameter of the sheave is too small for the rope.

When using block and tackle in confined spaces, provide guards on the pulley block so that a person's hands cannot be caught between the pulley and the rope. When blocks and falls are used to lift heavy materials or to keep heavy loads in suspension, as on heavy-duty scaffolds, wire rope is more serviceable than fiber rope.

Portable Floor Cranes or Hoists

Portable floor cranes or hoists are hoists mounted on wheels. They can be moved from place to place, either by hand or under their own power. They can raise and lower loads in a vertical line but cannot rotate around a fixed point.

Portable floor hoists are useful where overhead construction, belting, or shafting prevents the use of overhead hoists or cranes and where more expensive equipment is not justified because it would be used infrequently. These hoists are handy for placing work on machines, loading heavy material on trucks, and moving material from one location in the shop to another.

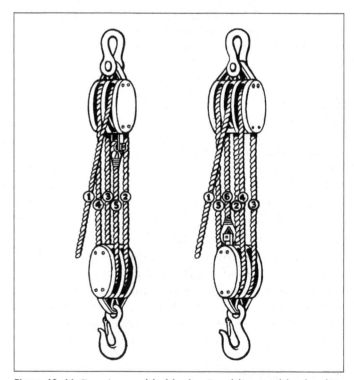

Figure 13–14. Reeving tackle blocks. Lead line and becket line should come off a middle sheave when blocks contain more than two sheaves. The upper and lower blocks will then be at right angles to each other. This eliminates tipping and the accompanying loss in efficiency.

Portable floor hoists are usually operated by hand or by electric power. The lifting mechanism is ordinarily either a chain hoist or a winch with wire rope and block. Hoists operated by electric power should be effectively grounded to prevent shock in case of a short circuit. If the power cord does not have a separate ground conductor, use a special grounding wire with one end of the wire fixed permanently to the frame of the hoist. The other end of the wire is equipped with a device that can be attached to a grounded building column, water line, or other direct-to-ground connection.

To prevent foot injuries when the hoist is moved, have employees wear protective footwear and install sweep guards on the truck's wheels, where conditions permit. Truck handles on hoists should be designed to stand upright when not in use. If they project horizontally from the hoist or lie on the floor, these handles can cause workers to trip.

Tiering Hoists and Stackers

The tiering hoist—sometimes called a stacking elevator, portable elevator, tiering machine, or platform hoist—is designed to raise material in a vertical line on a moving platform. This hoist is portable and is used extensively in warehouses for piling and storing materials. It is operated either manually or electrically. Large-capacity hoists usually are power driven.

Tiering hoists should have a braking device that permits safe lowering of the platform. They should also have a rachet lock (or dog) to lock the platform in position for loading and unloading operations. Do not permit workers to ride the platform because they could be crushed if the platform meets an obstruction.

Tiering machines, especially the revolving type, should be operated so that they will not tip over. Basic precautions include having the machine solidly on the floor, making sure that its safe load capacity is not exceeded, and placing the load properly on the platform.

Before the machine is used, lift the casters off the floor. One type of hand-operated hoist is arranged so that the platform cannot be moved unless the machine stands solidly on the floor—on the frame and not on the casters. Materials such as rolls and other round objects should be blocked to prevent them from rolling off the platform.

Ground power-operated tiering hoists and safeguard the wiring, preferably with armored cable. On two-section machines, provide locks to prevent the release of the upper section. To prevent shearing of fingers, install guards at gears and channel-iron guides. Some tiering machines have almost all the safeguards that a freight elevator has, including limit stops for top and bottom travel on the hoisting cable drum and for the shipper rope, if one is provided.

CONVEYORS

There has been much confusion as to what a conveyor is and what it is not, as well as confusion about the correct names for different types of conveyors. Therefore, the general definition of conveyor, given in ANSI/ASME B20.1A, Safety Standards for Conveyors and Related Equipment, is quoted here in full:

> Conveyor. A horizontal, inclined, or vertical device for moving or transporting bulk material, packages, or objects, in a path predetermined by the design of the device, and having points of loading and discharge, fixed or selective. Included are skip hoists, and vertical reciprocating and inclined reciprocating conveyors. Typical exceptions are those devices known as industrial trucks, tractors, and trailers; tiering machines; cranes, hoists, power shovels; power scoops, bucket drag lines; platform elevators designated to carry passengers or the operator; man-lifts, moving walks; moving stairways; highway or rail vehicles; cableways; or tramways, pneumatic conveyors, or integral transfer devices.

By industry agreement, nomenclature and definitions have been standardized and published in ANSI/Conveyor Equipment Manufacturers Association (CEMA) 102, Terms and Conveyor Definitions. Terms and definitions in this chapter follow this standard.

General Precautions

The most common conveyors are of the belt, slat, apron, chain, screw, bucket, pneumatic, aerial, portable, gravity, live roll, en masse, flight, mobile, and vertical types. Design and construct these conveyors to conform with applicable codes and regulations.

Place a highly visible sign at each loading point of manually loaded conveyors traveling partially or entirely in a vertical path. The sign should show the safe load limit that can be raised or lowered. Gears, sprockets, sheaves, and other moving parts must be protected either by standard guards or positioned in such a way to ensure against personal injuries.

Periodically inspect the entire conveyor mechanism. Immediately replace any part showing signs of excessive wear. Pay particular attention to brakes, backstops, antirunaway devices, overload releases, and other safety devices to ensure that they are working and in good repair.

Lubricate all machine parts according to the manufacturer's instructions. Install grease nipples on long tubes or pipes. This not only permits oilers to keep a safe distance from moving parts, but also prevents shutting down the conveyor for greasing.

All conveyors within 6 ft 8 in. (2 m) of a floor or walkway surface that is a means of exit must have alternate passageways that comply with NFPA 101, Life Safety Code, requirements. Frequently, a work platform on a movable conveyor tripper can be used as a crossover, if properly railed.

Underpasses should have sheetmetal ceilings. Where overhead conveyors dip down at workstations, install guards or handrails. Also install guards below all conveyors passing over roads, walkways, and work areas.

For conveyors that run in tunnels, pits, and similar enclosures, provide adequate drainage, lighting, ventilation, guards, and escapeways. These features are required wherever it is necessary for persons to work in or enter such areas. Provide sufficient side clearance to allow safe access and operating space for essential inspection, lubrication, repair, and maintenance operations.

Where conveyors pass through building floors, the openings should be guarded by standard handrails and toeboards. As a fire precaution, protect each opening against the spread of fire or superheated gases from one floor to the next. To prevent the spread of fire, install doors that close automatically or install fog-type automatic sprinklers. Place the sprinklers so they provide a curtain of water fog across the opening. Where a conveyor passes through a fire wall, provide similar protection. Conveyor tunnels under stockpiles of materials should be open at both ends.

If the top of a loading hopper is at or near the level of a floor or platform, protect the hopper with standard railings and toeboards or with a bar guard with openings not greater than 2 in. (5 cm) in one dimension, such as 2 in. × 12 in., 2 in. × 14 in., 2 in. × 16 in. (5.1 cm × 30.5 cm, 5.1 cm × 35.5 cm, 5.1 cm × 40.6 cm, respectively).

Elevated conveyors should have access platforms or walkways on one or both sides (Figure 13–15). Handrails should be 42 in. (1 m) high with an intermediate rail. Platforms should have 4-in. (10-cm) toeboards. Use checkered plate flooring or other slip-resistant surface, particularly on sloping walkways.

Sideboards along edges and at corners and turns of overhead conveyors, along with screen guards underneath high runs, will protect workers from falling material. Provide crossovers or underpasses with proper safeguards for passage over or under all conveyors. Prohibit workers from crossing over or under conveyors, except where safe passageways are provided. Forbid workers from standing or riding on conveyors.

Operating Precautions

Locate the start button or switch for a conveyor so that the operator can see as much of the conveyor as possible. If the conveyor passes through a floor or wall, equip each side of the floor or wall with starting and stopping devices. Require that all starting buttons or switches be operated

Figure 13–15. This elevated conveyor system in a quarry has a clear walkway with handrails. *(Courtesy Vulcan Materials Co.)*

simultaneously to start the conveyor. Clearly mark these start–stop devices. Keep the area around them free of obstructions so they can be seen and reached easily. Instruct all personnel working on or near the conveyor about the location and operation of all stopping devices.

Provide electrical or mechanical interlocking devices—or both types—that will automatically stop a conveyor when the unit it feeds (another conveyor, bin, hopper, or chute) has been stopped or is blocked. Thus, the conveyor cannot receive additional loads. If two or more conveyors operate in a series, design the controls so that if one conveyor is stopped, all conveyors feeding it are also stopped.

Locate emergency-stopping devices not more than 75 ft (23 m) apart along walkways by the conveyor. For some installations, a good solution is to have a lever-operated emergency-stopping device at the end of the conveyor, with a strong wire cord strung on each side of the conveyor for its entire length. A pull on the wire cord will stop the conveyor.

On conveyors where there is a possibility of reversing or running away, provide anti-runaway and backstop devices. In addition, design the conveyor track so that the load (or conveyor parts) cannot slide or fall in case of mechanical or electrical failure. If such a design is not practical, install guards able to withstand the impact of shock and to hold the falling load. Design electric machines with brakes that are applied and released by the movement of operating devices so that, if the power is interrupted with the brakes in the OFF position, the load will descend at a controlled speed.

In addition to the overload protection usually provided for electric motors, include an overload device to protect the conveyor and mechanical drive parts. Shear pins and slip or fluid couplings are examples of overload protective devices. In the event of an overload, the device must shut off the electric power quickly, disconnect the conveyor or drive parts from the power, or limit the applied torque. When a conveyor has stopped because of an overload, lock out all starting devices and remove the cause of the overload. Inspect the entire conveyor before restarting it.

With exhaust hoods, cover the loading and discharge points of a conveyor carrying material in fine or powdered form. To prevent the formation of dust clouds, provide good ventilation. Dust removal must be in accordance with government regulations. If the material is combustible, the concentration of dust must be kept below the lower explosive limit. Use only approved explosion-proof electrical fixtures. Exclude all sources of ignition from the dusty area. Ground the conveyor, and bond its parts to prevent differences in polarity. Also, ground the container into, or from, which the material is conveyed, and bond its parts.

Persons working near or on conveyors should wear close-fitting clothing that cannot become caught in moving parts. Safety shoes are recommended. If the conveyor galleries are dusty, have workers wear goggles and, if necessary, respirators.

Maintenance Precautions

Before maintenance personnel start working on a conveyor, they should lock out the main power control in the OFF position. The workers should carry the only key to their lock. If two gangs work on one conveyor, the supervisor of each gang should lock out the master switch per OSHA standard 29 CFR 1910.147 for group lockout/tagout.

Maintenance personnel should be able to change the position of pulleys, sprockets, or sheaves to compensate for normal working conveyor stretch and wear. Provide guards for the on-running belt at least 18 in. (55 cm) from where the belt and the head and tail pulleys touch and from where the belt and the tripper and hump pulleys touch. If the hazard is out of reach (more than 8 ft [2.4 m] above the floor or platform) or close to a wall or other obstruction, workers in the normal course of their duty should not be exposed to it.

If the skirtboards of the loading boot are close to the upper surface of the belt, a person's arm could get caught. In such a case, the belt could not be raised sufficiently to allow the arm to ride over an idler pulley under the belt. Thus, the arm undoubtedly would be badly mangled. Therefore, install guards at the sides of the conveyor at the loading boot.

Also guard the points of contact where the wheels of movable trippers roll on the rails. Where the operator travels on the tripper, provide a platform. Locate and construct the platform to protect the operator from slipping and falling and to prevent contact between the operator or his or her clothing and any moving parts. To help prevent the operator from falling into the hopper, install handholds

and railings on the platform.

At the conveyor floor above the coal bunkers in power plants, the slot—through which coal from the tripper chute is discharged into the bunker—is protected by bars placed across it about 12 in. (30 cm) apart at floor level. A piece of discarded belting of the required width and length can be placed to cover the slot, with its ends securely anchored. At the tripper platform are four pulleys that raise this belt vertically so that it passes over the access to the tripper platform. This device not only provides safety, but also seals the slot and thus keeps the dust in the bunker.

If a person should fall onto a conveyor, suspend a gate or paddle as low as possible above the belt near the head pulley. In that way, the person riding on the belt could automatically pull a stop rope and quickly stop the conveyor. On belt conveyors that are at floor level or on balconies or galleries, a shield guard or housing should completely enclose each end. Guardrails and toeboards should extend the length of the conveyor.

To help remove static from belt conveyors, place tinsel or needlepoint static collectors close to the out-running sides of the drive pulleys and idlers. Ground the pulleys and idlers, along with the shafting, through carbon or bronze brushes running on the shaft. A belt that does not move too fast can be grounded to the drive gear by a continuous strip of copper foil on the pulley side. Other belt conveyors can be grounded by being treated with conductive belt dressings.

One of the dangers of belt conveyors is that workers are tempted to clean off material that sticks to the tail drums or pulleys while they are in motion. Fixed scrapers and revolving brushes eliminate the need for cleaning by hand. To protect workers should they attempt to clean or dress the belts while they are moving, place the barrier guards directly in front of the pinch points of belts and drums. Guard the belt and drum on the side, also at a sufficient distance from the drum to prevent contact. When catwalks are provided to access raised or elevated portions of conveyors, they should be equipped with standard railing, including a midrail and toeboard. When located outdoors, if such areas are subject to buildup of ice, snow, or debris, a standard railing and toeboard may not provide sufficient protection to keep workers from sliding beneath the midrail. If a slip and fall occurs, consider installing a horizontal cable, screen, or panel between the midrail and toeboard for added protection.

Belt Conveyors

A belt conveyor is an arrangement of mechanical parts that supports and propels a conveyor belt, which in turn carries bulk material. The five principal parts of a typical belt conveyor are (1) the belt, which forms the moving and supporting surface on which the conveyed material rides; (2)

the idlers, which form the supports for carrying the belt; (3) the pulleys, which support and move the belt and control its tension; (4) the drive, which imparts power to one or more pulleys to move the belt and its load; and (5) the structure, which supports and maintains the alignment of the idlers and pulleys and supports the driving machinery.

Like all moving machinery, belt conveyors present hazards to workers, who must be safeguarded. Operators must be trained in safe work procedures around belt conveyors. Stress that workers stand clear of moving conveyors. Train them to avoid pinch points and other areas where their fingers or hands may get caught when the conveyor is moving. Place guards at all pinch points.

Other frequent causes of conveyor incidents are improper cleanup of conveyors and shoveling of spillage back on the belt. Severe injuries also arise from (1) attempting repairs or maintenance on moving conveyors, (2) attempting to cross moving belts where no crossover exists, and (3) attempting to ride a moving belt. Moreover, workers standing or working on conveyors, particularly maintenance personnel, can be injured by falls or crushed against stationary objects, particularly if a conveyor is unintentionally started.

The following injuries and hazards could be reduced by using mechanical and environmental controls around belt conveyors:

- fires from friction, overheating, static, or other electrical sources
- explosions of dust raised by combustible materials at transfer points, where belts are loaded or discharged
- respiratory and eye irritations from toxic dusts
- electrical shock from ungrounded or improperly installed controls or conductors.

Provide guards for transmission equipment and other power-driven parts in accordance with ANSI/ASME B15.1, Safety Standard for Mechanical Power Transmission Apparatus, and with all state or provincial and federal regulations governing the safety, health, and welfare of employees. ANSI/ASME B15.1 stipulates how pulleys, chains, sprockets, belts, couplings, and other parts of conveyor drives should be enclosed.

Suitable sweeps should be provided for shuttle conveyors, movable trippers, traveling plows, and hoppers and stackers. The sweeps push objects ahead of the moving pinch points between the wheels and the rails, thus guarding against nips. Again, it is imperative to comply with all state or provincial and federal regulations. Provide guards at pinch points at the head, tail, and take-up pulleys. An idler pulley becomes a hazard when skirt plates and chute skirts are so positioned as to force the belt against the idler, thus creating a pinch point.

Mechanical belt cleaners, such as fixed or tension scrapers, revolving brushes, or rubber disks, sometimes eliminate manual cleaning. They also eliminate a major reason for working around moving pulleys.

To lubricate a conveyor that is in continuous operation, install extension grease lines. In that way, workers cannot get caught by rollers and bearings when working around them. All grease fittings inside a guard enclosure (except those that move with the part they serve) should be fitted with extension pipes to make them accessible from outside the guard.

Slat and Apron Conveyors

The slat conveyor has one or two endless chains operating on sprockets and usually runs horizontally or at a slight incline (Figure 13–16). Attached to the chain or chains are nonoverlapping, noninterlocking slats that are closely spaced.

Apron conveyors have overlapping or interlocking plates that form a continuous moving bed. They vary greatly in size: one may be part of a bottling machine, and another may handle billets and castings in a steel mill.

Place guards at pinch points between slats or plates, and between them and the chain, sprockets, and guides. Where slats are spaced farther apart than 1 in. (2.5 cm), a serious shearing hazard exists between the slats and the conveyor substructure. When a slat conveyor is located at floor level or in working areas, the space under the top run of the slats should be filled in solid.

Figure 13–16. This high-speed, slat-type conveyor system uses pushers to gently sweep items onto sort lanes. Note the access stile with handrails at the far right. *(Courtesy Western Atlas Inc.)*

When designed for handling heavy material, slat and apron conveyors are usually installed flush with the floor to make loading and unloading easier. When so located, conveyors should be guarded by handrails (except at loading stations) so that workers will not step onto the moving conveyor. Openings in the floor at the loading platforms should be guarded or covered.

Chain Conveyors

Chain conveyors take many forms. However, they all carry, pull, push, haul, or tow the load either (1) directly by the chain or (2) by means of attachments, pushers, cars, or similar devices. Conveyors in which the chain itself directly moves the load are drag, rolling, and sliding chain conveyors.

Tow Conveyors

Conveyors consisting of an endless chain that is supported by trolleys from an overhead track or that runs in a track above, flush with, or under the floor, with attachments for towing trucks, dollies, or cars, are known as tow conveyors. Tow conveyors use four-wheeled carts that are held in place by a pin through the chain. These pins should be equipped with an automatic disengaging mechanism. Many pins have a horizontal bumper or kick mechanism that will disengage with 3 to 5 lb (1.4 to 2.3 kg) of pressure. Workers are consequently protected from being caught between carts.

Trolley Conveyors

An endless chain that propels a series of trolleys (Figure 13–17) supported from or within an overhead track, with the loads suspended from the trolleys, is a trolley conveyor. On trolley conveyors, guard well the return portion of the chain. Whenever possible, install guardrails along both sides of the trough to prevent a person from stepping or falling into it. Clearly designate the path of travel of such overhead chain conveyors. Provide emergency stop switches every 75 ft (23 m).

Shackle Conveyors

A shackle conveyor consists normally of a chain-type conveyor, with suspended shackles spaced evenly along the line to convey poultry or other meat products through a processing plant. In a poultry processing plant, the shackle conveyor or conveyors carry poultry (1) from the beginning of the line, where live poultry is hung from each shackle, head down; (2) through all the processing phases; and (3) to the end of the line, where the finished product is removed for packaging and shipping. In some plants, one continuous conveyor travels through all departments and processes, while in other plants two or more similar conveyors carrying suspended shackles are used.

The speed of such conveyors varies from plant to plant, depending on the number of workers on any given conveyor

Figure 13–17. A trolley can be used to support hoists in trolley conveyors, in overhead tow conveyors, in overhead chain conveyors, and on jibs and monorails. This overhead monorail delivery system can carry weights up to 6,000 lb (2,722 kg). *(Courtesy Litton)*

line. The speed generally increases with more workers on the line. Poultry is normally processed by being hung head down and, after being cut and bled, repositioned feet down before going through scalders and beaters and then on to viscerating and cleaning.

The hazard posed by such shackle conveyors involves workers unintentionally placing a thumb or finger in one of the shackles and being dragged through a beater or scalder or to the end of a platform where, suspended by a finger, a worker could lose a thumb, a finger, or his or her life. Limit this hazard by placing EMERGENCY STOP switches at critical points, such as at the end of a platform or before entry into hazardous areas. The emergency switches are actuated by pull wires positioned below and perpendicular to the line of travel.

Screw Conveyors

Screw conveyors move bulk materials horizontally, on

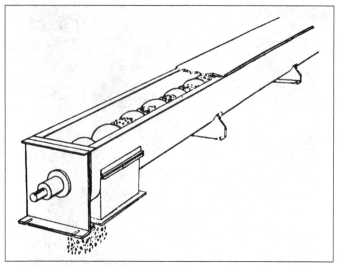

Figure 13-18. A screw conveyor with the cover cut away to show the interior. *(Courtesy Conveyor Equipment Manufacturer Association)*

inclines, and vertically. They are employed in many industries and handle a variety of materials. A screw conveyor essentially consists of a continuous spiral mounted on a pipe or shaft. The screw rotates in a stationary trough, generally U-shaped. Material introduced into the trough is conveyed by the screw's rotation (Figure 13-18).

A screw conveyor that is not enclosed presents the hazard of entrapment. Feet, hands, or other portions of the body may get caught between the rotating screw and the stationary trough (Figure 13-19). When such unintended incidents occur, they can result in serious injury or death. These kinds of incidents can be avoided by setting up and enforcing an in-plant safety program and by training employees to observe the following basic safety rules:

- Covers, gratings, and guards should be securely fastened before operating the conveyor. Conveyors should be provided with solid covers that are securely fastened by bolts, spring clamps, or screw clamps. When an exposed feed opening is required, cover it with a securely fastened grating. If an open housing is functionally necessary, guard the entire conveyor with a fence or railing. Moving drive components should be guarded.
- Never step or walk on covers, gratings, or guards.
- Lock out power before removing covers, gratings, or guards by padlocking the main disconnect in the OFF position.

Figure 13-19. This warning label is used by screw conveyor manufacturers to illustrate the nature of the hazard, the severity of an injury, and precautions to avoid it. Serious injury or death may result due to failure of personnel to follow basic safety rules.

More detailed information on screw conveyor operation and safety is available in Book No. 350, Screw Conveyors, by the Conveyor Equipment Manufacturers Association; in ANSI/ASME B20.1A, Safety Standards for Conveyors and Related Equipment; and in ANSI Z244.1, Safety Requirements for Conveyors and Related Equipment.

Bucket Conveyors

The three general types of bucket conveyors all have an endless belt, chain, or chains that carry either fixed or pivoted elevator buckets (Figure 13-20). Descriptions of these three types of bucket conveyors follow:

- A bucket elevator carries fixed buckets in a vertical or inclined path. The buckets discharge by the force of gravity as they pass over the head sheave or drum.
- A gravity-discharge conveyor elevator has fixed buckets and operates in vertical, inclined, and horizontal paths. The buckets act as flights while carrying material along a trough in the horizontal plane to a point of gravity discharge.
- A pivoted-bucket conveyor also operates in horizontal, inclined, and vertical paths. The buckets remain in the carrying position until they are tipped or inverted to discharge.

To protect operating personnel, totally enclose bucket conveyors in a housing. Do not attempt to take samples when the conveyor is in motion. The pivoted-bucket conveyor has tripping devices for emptying the buckets. Frequently, these tripping devices are movable so materials can be distributed evenly in the storage bins. Arrange for the remote control of levers for shifting and locking the trippers. In that way, workers will not have to go on the conveyor to change the position of the trippers.

For the safety and convenience of repair personnel and operators, build a permanent footwalk alongside a conveyor that hoists material and carries it over stokers or bins. Equip the footwalk with standard handrails and toeboards, and be sure it is well lit.

Pneumatic Conveyors

A pneumatic conveyor is an arrangement of tubes or ducts through which solid objects such as cash, mail, and other small items are moved by using compressed air or a vacuum. Solid objects are placed inside a cylindrical cartridge. Compressed air, injected into the tubing system, pushes the cylinder forward at a relatively high velocity.

13 Conveyors

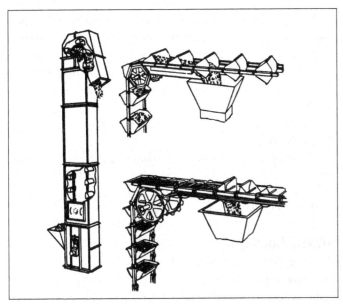

Figure 13-20. Three types of bucket conveyors. Left: bucket elevator. Upper right: gravity-discharge conveyor elevator. Lower right: pivoted-bucket conveyor. *(Printed with permission from Conveyor Equipment Manufacturers Association.)*

Bulk materials such as grain and dust are also moved by pneumatic conveyors. To convey bulk material, compressed air is injected into the piping through a nozzle below the loading chute. The air mixes with the material, making it flow like a fluid through the piping.

When transporting material that could cause a dust explosion, the air velocity should exceed the critical velocity for starting a fire. In this way, an explosion from one point to another within the system will be prevented. Bond equipment electrically, and ground it to prevent static electricity from being a source of fire. Additional safety precautions for working with pneumatic conveyors follow:

- A constant-volume, variable-pressure (positive-type) blower should have a relief valve on or adjacent to it. Keep doors of blower conveyors interlocked so they cannot be opened if there is positive internal pressure. To keep conveyed material from being thrown against workers or into the working area if a gasket leaks, shield gaskets that hold the line pressures.
- Where suction lines are large enough to draw a person in, place bar guards or screening over the intake. Instruct employees to stay at a safe distance from such lines.
- Receivers and storage bins should have full-bin indicators or controls to prevent overfilling.
- A pneumatic conveyor serving an area containing contaminated air must be arranged so that no contaminated air enters the conveyor tube and is thus carried to other areas.

Aerial Conveyors

A description of the two types of aerial conveyor systems follows:

- A cableway aerial conveyor is a wire rope–supported system in which the materials-handling carriers are not detached from the operating span and the travel is wholly within the span.
- A tramway aerial conveyor is a wire rope–supported system in which the travel of the materials-handling carriers is continuous over the supports of one or more spans.

Aerial conveyors are used frequently in industrial plants, and particularly on large construction sites, to carry material from point to point. They also move coal and ore. Workers are generally injured by falling material when inspecting and oiling the cables and carriages, or if signals are misunderstood.

No one should work directly under the conveyor except, of course, at the loading and unloading stations. Take exceptional precautions at these points. Wherever workers must pass under the conveyor, provide a covered passageway. Provide heavy wire screens, suspended under the conveyor from pole to pole, that are wide enough to catch material that would otherwise fall on roadways or work areas.

Inspect the equipment regularly. Pay special attention to the sheave wheels and bearings, the rope fastening, the bucket latch and trunnions, and all load-sustaining parts.

For operation efficiency and protection against the weather, keep all ropes well oiled. Hauling rope should be continuously lubricated by means of a controlled drop feed from an oil reservoir at the point (one or both ends of the line) where the rope leaves the drive sheave and passes over a support sheave. In one facility, a worker had to ride a trolley carriage to oil a tramway rope used for conveying coal. An automatic lubrication system was devised to eliminate this dangerous practice.

For night work and for repairs, provide suitable lighting. Also provide a suitable signaling system, such as a telephone or electric push-button system, for every cableway aerial conveyor. On large construction jobs, a portable telephone system allows a signaler to direct the raising and lowering of loads that are out of the operator's sight.

For a tramway aerial conveyor, provide the following three control systems:

1. push-button stations and a bell signal code that indicates stop, start, slow speed, high speed, and reverse
2. an all-metallic, aerial wire-circuit telephone with instruments at certain points along the line in addition to terminal sets
3. a second telephone circuit, which may be grounded if desired.

The U.S. Army Corps of Engineers suggests taking the following precautions when using cableway aerial conveyors:

- Keep the control console compartment locked when the cableway is in use. Permit no one in the compartment with the operator while the cableway is in operation.
- Continuously maintain at least two communication and control systems between the signaler and the cableway operator. This dual system should include voice communication by telephone and radio. Lights or bells may be included with, or substituted for, one of the voice systems.
- Permit only authorized inspection and maintenance personnel to ride cableway carriages.
- Do not permit the riding of cableway load blocks.

Portable Conveyors

Belt, flight, apron, extendable, and fixed-bucket conveyors are made as inclined portable units set on a pair of large wheels. They are used to load railroad cars and trucks with bulk materials and to raise construction material from one level to another.

Electrical equipment on portable conveyors should be weatherproof. With three-phase power, the flexible cord that is connected to the power outlet should be a four-conductor cord, the fourth wire being grounded in all plugs and receptacles. Arrange the cable so it cannot be run over by trucks or other machines. If it must cross a driveway, hang it on poles at a minimum height of 14 ft (4.3 m). If two or more sections of cable are required, the connectors should be kept above the ground.

Provide portable conveyors with guards as specified for the corresponding types of fixed conveyors. Skirtboards or sideboards (not less than 10 in. [25 cm] high) keep heavy material from falling over the sides and light or loose material from blowing off. This safety precaution applies to belt conveyors, as well as to other types of conveyors, because troughing of the belt gives insufficient protection against such spillage.

Fit the conveyor with a locking device to hold the conveying unit at various fixed levels, thus keeping the conveyor stable. Closely check the mechanism that raises and lowers the boom on all types of portable conveyors. This is the major hazard area of such machinery. While any positive type may be used, the self-walking worm or jack-screw type is preferred. Gear drives should be completely housed and run in oil. Thoroughly guard all chains within easy reach of workers. Devise a system for conveniently oiling chains.

To work in small spaces in warehouses, portable conveyors usually have booms consisting of two sections. Design these conveyors so material will not roll back from one section to the other at the transfer point.

When loading or unloading railroad cars with a portable conveyor, use a suitable safety device to prevent the cars from shifting during the operation. Do not remove the device until the crew is sure no one is in the car. A red banner or a standard blue warning sign at each end of the car is recommended as a warning to the switching crew. (See Chapter 2, Buildings and Facility Layout, in this volume.)

When using portable conveyors to raise or lower construction materials, provide ample stairs or ladders in the immediate vicinity of the conveyor. Prohibit workers from walking on idle or moving conveyors.

Gravity Conveyors

Because gravity conveyors (Figure 13-21) depend wholly upon the natural pull of gravity, workers often disregard some necessary safe practices when working around them. If employees climb on a conveyor to release a blockage, they may slip on the rollers or be knocked down should the jam suddenly be released. Therefore, prohibit workers from climbing onto conveyors. In addition, the installation of steel or wood plates between rolls helps ensure that neither a worker's body nor limb can fit between the rolls.

Chute Conveyors

Chute conveyors made of polished metal sheets or bars lower packing cases, cartons, and crates from one floor to another. They also move material from the sidewalk to the basement of a building through an elevator shaft in the sidewalk, or other opening.

Inclined chute conveyors may be straight or have a vertical curve with radius large enough to deliver packages onto the lower floor without impact or damage. Some gravity-chute conveyors are built like a spiral around a vertical pipe, with a slope at the outer edge between 18 and 30 degrees.

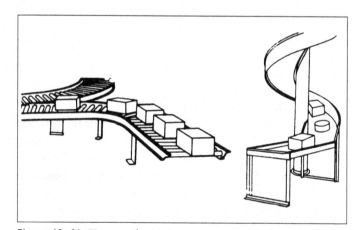

Figure 13-21. Types of gravity conveyors. Left: roller conveyor. Right: spiral chute. *(Courtesy Conveyor Equipment Manufacturers Association)*

In removing packages from the delivery end of spiral chutes, workers frequently injure their hands when they are caught by descending packages or are crushed against other packages on the delivery table. Where the chute is enclosed, place a warning sign over the delivery end. Install a simple mechanical or electrical device that signals when a package is about to be delivered from the chute. This is especially needed where the descending packages cannot be plainly seen.

Spiral chutes present a serious fire hazard because they form flues from lower floors to upper floors through which fire can quickly spread. Two methods for eliminating this hazard are to (1) enclose the chute in a tower made of fire-resistant material, such as steel, concrete, or masonry, or (2) provide automatic fire doors (draft checks) where the chute passes through floors. The enclosed tower has doors at each charging station and a door at the delivery end. Keep the charging station doors closed, except when charging is being done. The door at the delivery end should close automatically in case of fire. Automatic fire doors are of two types—vertical sliding and shutter. Both types should have fusible links so that they will close automatically in case of fire.

Where an open chute is used, provide a guardrail and toeboard at each floor. Use either a movable railing or a hinged door or gate at the charging stations.

Roller or Wheel Conveyors

Roller or wheel conveyors are similar to chute conveyors, except that the angle of slope is much less (2% to 4%). These conveyors, therefore, can be used to convey packages for considerable distances on one floor. If the rollers or wheels are placed radially instead of parallel, the course of travel can be changed from a straight line to a curve.

The principal hazards in the use of roller conveyors are that (1) material may run off the edge of the rollway and fall to the floor and (2) loads may run away. Provide a guard railing on each side of the roller conveyor-way to guide the material and prevent it from running off. Such guardrails are especially advisable at corners and turns and on elevated conveyors under which workers must work or pass. When heavy loads are conveyed, retarders, brakes, or similar devices help prevent the loads from running away. A power conveyor on which speed can be controlled could also be substituted.

Hinge a vertically swinging, hinged section of a roller or wheel conveyor to the end of a stationary section of the conveyor from which the material is flowing. This helps block the oncoming material. The open end of the conveyor line should have a stop that (1) automatically projects above the level of the rollers or wheels when the hinged section is opened and (2) automatically retracts when the hinged section is closed.

Where a horizontally swinging, hinged section occurs in a conveyor, equip the open ends of stationary sections (the two ends adjacent to the hinged section) with retractable stops. The stops prevent loads from dropping off when the hinged section is open.

Live Roll Conveyors

A live roll (or roller) conveyor consists of a series of rolls over which objects are moved by power applied through belts or chains to all or some of the rolls. Where installed at floor level or when used in work areas, design live roll conveyors to eliminate hazards from pinch points and moving parts. Such a precaution will prevent personnel from coming in contact with or crossing the conveyor.

Vertical Conveyors

Vertical conveyors handle packages or other objects in a vertical, or substantially vertical, direction. In some cases, a hinged section, interlocked to the power in the main system, will be provided for access to the workstation. The interlock should shut down the entire system until the section is restored to its position. Descriptions of three basic types of vertical conveyors follow:

- Vertical reciprocating conveyors are power- or gravity-actuated units that receive objects on a carrier or car bed, usually constructed of a power or roller conveyor, and that elevate or lower objects to other locations.
- Suspended tray conveyors are vertical conveyors having one or more endless chains with pendant trays, cars, or carriers. These carriers receive objects at one or more levels and deliver them to another or several levels. To prevent materials from falling on people, install guards on the underside of suspended tray conveyors.
- Vertical chain conveyors (as opposed to shelf types) are two or more vertical-elevating conveying units opposed to each other. Each unit consists of one or more endless chains whose adjacent-facing runs operate in parallel paths. Thus, each pair of opposing shelves or brackets receives objects (usually dish trays) and delivers them to any number of levels.

Where vertical conveyors are automatically loaded and unloaded, provide guards to protect personnel from contact with moving parts. Where they are manually loaded and unloaded, install guards and safety devices, such as lintel and sill switches and deflectors.

Carriages of vertical reciprocating conveyors designed to register at a floor, balcony, gallery, or mezzanine level never should have a solid bed. This type of conveyor is not intended to carry passengers or operators or to have its car or carriage called to a station by a manually operated push button.

POWER ELEVATORS

Before drawing up specifications for new elevators or planning major alterations for existing ones, refer to the latest edition of ANSI/ASME A17.1, Safety Code for Elevators and Escalators. For the rest of this chapter, this standard will be referred to as the Elevator Code.

Specifications for new equipment conform to Elevator Code requirements, unless federal, state, provincial, or local codes are more stringent. In such cases, observe the governmental regulations as the minimum requirements. Usually, however, the Elevator Code will be stricter.

Two commonly used wordings in the Elevator Code are, "The elevator and associated equipment shall meet the requirements of A17.1, Safety Code for Elevators and Escalators, latest edition, except as hereinafter specifically exempted or modified"; and "Except as changed or modified herein, the elevator and associated equipment shall meet the requirements of A17.1, Elevators, Escalators, and Moving Walks, latest edition, and also shall comply with all applicable local laws and/or ordinances." Such paragraphs will generally take care of items not specifically spelled out in the specifications and drawings.

Types of Drives

There are two major types of power elevators: (1) electric-drive elevators and (2) hydraulic-drive elevators. However, some belt- and chain-drive machines are still in use, and repairs to existing installations are permitted by the Elevator Code. Their new installation is now prohibited by the Elevator Code.

Electric Elevators

The two general types of drives for electric elevators are traction drive and winding-drum drive. The winding-drum type, however, is now obsolete and is presently being used only in dumbwaiter elevators and in freight elevators, with restrictions as specified by the Elevator Code.

The Elevator Code requires that all drives be of the traction type. However, it permits the use of winding-drum drives on freight elevators that travel not more than 40 ft (12.2 m) at speeds not exceeding 50 fpm (0.25 m/s) and that are not provided with counterweights.

In the winding-drum drive, the hoisting rope is anchored in and winds on a spirally grooved drum. This is a positive drive. If the machine is not stopped at the limits of travel, the car may be pulled into the overhead structure, and, should the motor be powerful enough, the ropes may be pulled from their anchorage.

In the traction-drive elevator (Figure 13–22), the hoisting rope is not attached to the drive. The elevator is moved by the traction (friction) of the ropes in grooves on the

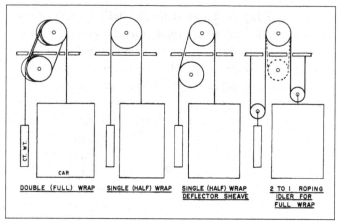

Figure 13–22. Common elevator traction drives showing double (full) and single (half) wraps.

drive sheave. The grooves may be semicircular (U-groove) or undercut U-groove. Some V-groove sheaves are still in use. Generally, the V-groove or undercut U-groove is used on geared machines. The U-groove is used on gearless machines, but there are exceptions.

A simple rule to remember is that when the drive sheave has twice as many grooves as there are hoisting ropes, the drive is a double (full) wrap. The total angle of contact of each rope with the drive sheave generally is between 300 and 360 degrees. The traction relation changes very little with wear.

Most companies use some form of undercut U-groove for their single-wrap drives. The friction is higher than an ordinary U-groove because of the pinching action. The width of the undercut is varied to suit the needs of traction. The traction remains substantially constant until the groove has worn to near the bottom of the undercut. Periodically check the groove to make sure that it is not worn so much that the rope bottoms.

The V-groove likewise has higher friction when new. As the V-groove wears, however, the rope seats deeper and deeper, increasing the area of contact of rope with groove and decreasing the unit pressure. When the rope reaches the bottom of the groove, much of the driving traction probably will be lost, and a loaded (or empty) car may "break traction" and slide with the brake locked. For this reason, frequently check such grooves for wear, and machine them before the rope wears the groove down to the bottom. Observe a minimum grooving bottom diameter.

Traction machines have an inherent safety feature: when the descending member (counterweight or car) bottoms, driving traction is lost, and, in most cases, the ascending member cannot be pulled into the overhead. With extremely high rises, the weight of rope hanging on the down-run side may be sufficient to maintain traction after the car or counterweight has landed. Because the compensative sheave is tied down, however, there is no danger.

For safe and successful operation, the rope's tension on the car's side must bear a definite relation to the rope's tension on the counterweight's side. Normally, this ratio ranges from 2:1 to 1:2. (See the discussion of overloads, under the heading Operation, later in this section.)

Hydraulic Elevators

Hydraulic elevators are power elevators in which the energy is applied by means of a liquid under pressure in a cylinder that is equipped with a plunger or piston. Hydraulic elevators are being installed in many new buildings averaging up to six stories in height. New elevators are usually of the electrohydraulic type. In these elevators, the problems associated with older hydraulic elevators (generally lower efficiency, the difficulty of keeping the valves and stuffing boxes tight, etc.) have been eliminated by using modern technology and materials and by following proper maintenance programs.

The Elevator Code requires that new elevators be equipped with (1) anti-creep-leveling devices, (2) hoistway-door locking devices, (3) electric car-door or car-gate contacts, (4) hoistway access, and (5) parking devices. This equipment is the same as that required for electric-drive elevators.

Because most electrohydraulic elevators do not have a counterweight, the motor must supply pressure to lift the entire weight of the car and the load. In addition, it must be more powerful than the motor of the traction drive of an electric-drive elevator, on which the weight of the car and part of the load is compensated for by the counterweight, to maintain the same speed. As built, the electrohydraulic elevator has all the electrical protective devices (interlocks, car-gate contacts, limit switches, and similar devices) found on an electric elevator.

Car safeties are not required on electrohydraulic elevators because they can come down no faster than the fluid can be forced out of the cylinder by the descending plunger. If counterweights and/or car safeties are on the elevator, they must comply with the Elevator Code.

Belt-Drive and Chain-Drive Elevators

The Elevator Code prohibits belt-driven and chain-driven machines from being installed. Existing elevators should be provided with electrically released brakes, terminal stopping, and safety devices as required for electric elevators. Some jurisdictions have outlawed these types of elevators, while others allow their continued use.

New Elevators

The Elevator Code specifies the general requirements and load classifications for new elevators.

Requirements

When new elevators are being planned, be sure that all requirements of the latest edition of the Elevator Code are met. For example, check that there are (1) safe and convenient access to the machine room and the pit, (2) adequate lighting in the machine room and overhead spaces, and (3) convenient electric outlets on the crosshead and in the pit.

An inspection station, with slow-speed UP and DOWN operating buttons and an EMERGENCY STOP switch, is required on the top of the car for maintenance personnel and inspectors to use.

The elevator must have normal- and final-limit stops, interlocks on all hoistway doors, a contact on the car door, and emergency exit or exits for the car. If a single-elevator hoistway exists, emergency access doors must be provided to the blind portions of such a hoistway. Unless the elevator is the hydraulic-plunger type, it should have a governor-actuated safety that meets the latest code requirements—city, state, provincial, federal, and/or ANSI standards.

Buffers are required to absorb the energy of descending cars and counterweights at the limits of travel. The main type of buffer in use is the spring buffer. Spring buffers are permitted only for cars whose rated speed is not in excess of 200 fpm (1.0 m/s). A few states and cities, however, permit them for higher speeds. At best, a spring buffer is a poor absorber of energy. Because a good spring will return approximately 95% of the stored energy, a spring-buffer stop affords only a series of decreasing surges.

Capacity Ratings

The size, capacity, and speed of new freight elevators will depend on the purposes for which they are to be used. When preparing specifications for new freight elevators, anticipate future load requirements as well as present ones. Therefore, find out from production personnel what size units (both freight package and freight carrier) will probably be in use 20 years hence.

The Elevator Code classifies freight elevators as follows:
- Class A—elevators with a distributed load that is loaded by hand or by hand trucks. Here, the weight of any single piece of freight or of any single hand truck and its load is limited to a maximum of one-fourth the rated load capacity.
- Class B—elevators used solely to carry passenger trucks and passenger automobiles up to rated capacity of the elevator.
- Class C1—industrial truck loading in which the truck is carried by the elevator.
- Class C2—industrial truck loading in which the truck is not carried by the elevator but used only for loading and unloading.
- Class C3—other loading of heavy objects in which a truck is not used.

For Class C1, C2, and C3 loadings, the Elevator Code requires that the rated load of the elevator be not less than the load (including any truck) to be carried. It also requires that the elevator be provided with a two-way automatic-leveling device.

Palletized loads increase in size year by year, and lift trucks are made larger to handle them. Because of these factors, old elevators that are overloaded can start downward when the last loaded lift truck is run onto the car. In some cases, the brakes do not hold. In others, the traction relation is broken, and the hoisting ropes slip through the drive sheave. Where old elevators are used to handle heavy palletized loads, however, determine and strictly enforce safe load limits and safe operating procedures to prevent serious unintended incidents. Observe the Elevator Code to prevent such cases in new elevators.

Hoistways, Pits, and Machine Rooms

Hoistways, pits, and machine rooms are dangerous work areas. However, by following code specifications, these places can be safe for workers.

Hoistways

Most codes for building elevators require that new elevators be installed in 2-h fire-resistant hoistways. These hoistings should have 1.5-h fire doors that fill the entire opening to prevent the rapid spread of fire from floor to floor.

At frequent intervals, clean hoistways, guide rails, and all other parts of grease and dirt to eliminate fire hazards. Projections in the hoistways should be beveled at an angle of not less than 75 degrees with the horizontal. Windows in the walls of the hoistway enclosures are prohibited.

Do not install in or under any elevator or counterweight hoistway any pipe conveying gas or liquids that—if discharged into the hoistway—would endanger lives. However, low-pressure steam (5 psig or less) or hot-water pipes used only for heating the hoistway and the machine room (or penthouse) are permitted, if certain conditions are met. All electrical equipment and wiring should conform to NFPA 70, National Electrical Code. Only such electrical wiring and equipment used directly in connection with the elevator may be installed in the hoistway.

Pits

To protect persons working in elevator pits, ensure that the elevator is deenergized and all sources of energy are controlled prior to entering the pit. Release or restrain any stored energy, and consider treating the pit as a confined space. Keep a minimum clearance of 2 ft (60 cm) between the lowest projection on the underside of the car's platform (not including guide shoes and aprons attached to the sill) and any obstruction in the pit (exclusive of compensating

devices, buffers, buffer supports, and similar devices). Take these measurements when the car is resting on fully compressed buffers.

Enclose counterweight runways from a point not more than 1 ft (30 cm) above the pit floor to a point at least 7 ft (2 m) above the pit floor and adjacent pit floors, except where compensating chains or cables are used. Screen partitions, at least 7 ft (2 m) high between adjacent pits, will protect persons in one pit from cars and counterweights in adjacent pits. They will also protect employees from hazards when adjacent pits are at different levels.

Never use an elevator pit as a thoroughfare or storage space. It should be fully enclosed and the entrances kept locked. To remove water, provide a sump pump. Drains are not to be connected directly to sewers.

Keep pits clean and free of debris. Never sweep rubbish into pits. Provide vertical ladders for easy access for cleaning pits.

Provide lighting of at least 5 fc (54 lux) at the pit's floor level. A light switch must be reachable from the pit's access door. An EMERGENCY STOP switch should be installed in every pit and should be reachable from the pit's access door.

Machine Rooms

Provide safe and convenient access to machine rooms. So that persons repairing or inspecting elevator hoisting machinery have sufficient room and are safe, allow at least 7 ft (2 m) of headroom between the machinery platform and the machine room's roof.

As with pits, never use elevator machine rooms as thoroughfares. The one exception would be if the elevator's equipment were in a separate locked enclosure. Rooms should be well ventilated and lighted with not less than 10 fc (108 lux) at floor level. Keep doors locked and affix a warning sign to prevent entry by unauthorized persons.

Keep machine rooms clean and do not use them for storage. Place small quantities of ordinary maintenance supplies in a wall cabinet. Keep a portable, Class C fire extinguisher within reach of someone standing at the door. (See Chapter 9, Fire Protection, in this volume.)

Overhead Protection

If the elevator's drive is located in the penthouse, provide a substantial grating or floor of fire-resistant construction under the drive. On all installations of overhead drives, include a cradle below the secondary sheaves if they extend below the floor or grating. For detailed information about flooring, consult the Elevator Code.

Do not hang elevator machinery, except the idler or deflecting sheaves, underneath the supporting beams at the top of the hoistway. When the governor or other device (other than terminal-stopping switches) must be installed below the machine's floor, set it on a substantial secondary floor.

For winding-drum drives, a substantial beam or bar should be placed at the top of the counterweights' guide rails and beneath the counterweights' sheaves. This will prevent the counterweights from being drawn into the sheaves.

Hoistway Doors and Landings

The Elevator Code specifies that openings at hoistway landings of all elevators must be provided with hoistway doors that guard the full height and width of the openings.

Hoistway Doors for Passenger Elevators

Records show that most unintended elevator incidents occur at the hoistway door. Most are "tripping" incidents. Others occur when people are caught and crushed between doors or when they fall down the hoistway. These types of incidents are likely to happen on old elevators. The interlocks required on new elevators by the Elevator Code, if properly installed and maintained, practically eliminate these types of incidents.

Many serious injuries and fatalities have occurred on older elevators, with doors that needed a special key to open the doors on the corridor side, regardless of the elevator's position in the hoist. Injuries occurred when the victim assumed the elevator to be at the landing, opened the door, and stepped into the shaft. The Elevator Code permits using unlocking devices on elevators equipped with doors that are unlocked for entering the car in case of an emergency. Assign all emergency keys only to qualified personnel who will guard them and use them carefully.

To help prevent unintended incidents in hoistways of older elevators, install conduit in each hoistway. In that way, electric light can be installed opposite each opening. If the door is opened with an emergency key and the car is not at the landing, the light, which burns continuously, will help a person see that the shaft is empty. Thus, the person will not step into the shaft (Figure 13–23).

Requirements for entering hoistways can be met in either of the following ways:
1. Provide only two means of entry to the hoistway. One entrance should be at an upper landing to permit access to the top of the car. The other entrance should be at the lowest landing—if this landing is the normal point of entry to the shaft.
2. Where elevators operate in a single hoistway, provide hoistway doors that can be (1) unlocked when closed with the car at the floor or (2) locked but possible to open from the landing only when the car is in the landing zone. Check on this requirement with the local authorities.

Figure 13–23. View from the pit looking up hoistways; note illumination (arrows). *(Printed with permission from Architect U.S. Capitol.)*

In general, three types of doors are used at landings: vertical- or horizontal-sliding doors, combination sliding and swinging doors, and swinging doors. On all three types of doors, use nothing less than direct-acting mechanical interlocks. Power-operated vertical-sliding doors and gates must operate in sequence if the elevator is used for passengers and, in all cases, if the doors close automatically.

For new elevators, according to the Elevator Code, the distance between the hoistway side of the hoistway door, opposite the car opening, and the hoistway edge of the landing threshold should be not more than ½ in. (13 mm) for swinging doors and 2½ in. (63 mm) for sliding doors. The face of the hoistway door should not project into the hoistway beyond the edge of the landing sill. On existing elevators, if this distance exceeds 1½ in. (38 mm) for swinging doors and 2½ in. (63 mm) for sliding doors, fill in the excess space.

Do not design automatic fire doors, whose functioning depends on the action of heat, to lock (1) any landing opening in the hoistway of any elevator or (2) any exit leading to the outside of the building. Taking these precautions could save lives.

Construct and maintain the loading platform for at least 2 ft (60 cm) back from the door so persons will not readily slip. Some operators use firmly secured rubber mats, adhesive abrasive strips, or abrasive-surfaced concrete directly in front of all hoistway doors. (See NSC Data Sheet 12304–0595, Floor Mats and Runners.) To eliminate tripping hazards, make such surfaces flush with the surrounding floor.

Hoistway Doors and Gates for Freight Elevators

Like passenger elevators, most freight elevators are installed in 2-h fire-resistant hoistways. These hoistways have 1.5-h fire doors that fill the entire opening, as required by practically all codes. However, there are some older elevators—mostly freight—that have hoistways enclosed only to a 6- or 7-ft (1.8- or 2.1-m) height. These older elevators also have hoistway gates of hardwood slats that are 5 or 6 ft (1.5 or 1.8 m) high with a clearance under them of as much as 8 to 10 in. (20 to 25 cm). Many of these elevators are shipper-rope operated, so it is necessary to reach into the hoistway to bring the car to a landing. This procedure is prohibited in new elevators.

Hoistway doors (Figure 13–24) should comply with applicable state and municipal requirements for fire resistance. Arrange doors that are closed by hand so that it is not necessary to reach in back of any panel, jamb, or sash to operate them. To ease the movement of trucks, install a doorsill to fill the gap between the landing and the car.

Replace hoistway gates with doors in older elevators as soon as possible. Where gates are used, maintain them at the highest safety standard. For example, the openings in gates made of grille, lattice, or other openwork should reject a ball 2 in. (5 cm) in diameter.

The bottom of the gate should come down within 1 in. (2.5 cm) of the threshold to prevent objects from sliding under the gate into the hoistway. Where lack of headroom precludes a standard gate at the lowest landing, make the gate in two sections.

Gates should have convenient handles or straps for manual operation. However, a power attachment for closing the gate, controlled from the landing buttons, saves time and labor and decreases the possibility of leaving or propping the gate open. Power-operated doors and gates are recommended. For freight elevators, these are usually of the continuous-pressure-operation type.

The growing weight and speed of lift trucks requires that hoistway doors and gates of freight elevators be protected. A 5-ton (4,536-kg) load moving only 2 mph (2.93 fps) has an impact of 2,690 ft-lb ($F = \frac{1}{2} MV^2$). If it is assumed that the door or gate can deflect 1 in., a force of 32,300 lb (144 kN) results. This is certainly more than any hoistway door can stand. Solutions to this problem include the following:

- Install a heavy wire rope across the opening. The truck operator must lower the rope before the truck can enter the car's platform.
- Place one person in charge of operating the elevator.

Interlocks and Electric Contacts for Both Passenger and Freight Elevators

To prevent a car from moving away from the landing, unless the hoistway door is locked in the closed position, equip hoistway doors with interlocks (Figure 13–25) that comply

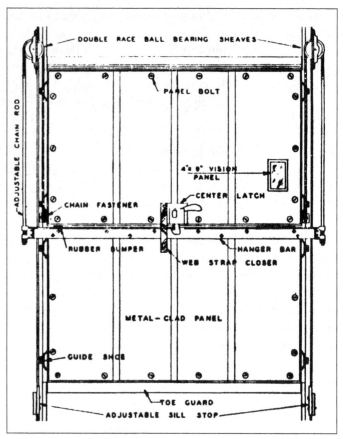

Figure 13–24. Design of a typical vertical-sliding biparting steel hoistway door for freight elevators, as seen from inside the car. *(Courtesy Peele Co.)*

with the Elevator Code. Interlocks should be direct-acting mechanical/activated devices. All interlocking devices should be incapable of being plugged or made inoperative in any way. In addition, the interlock also prevents opening of the hoistway door from the landing side, except by emergency key, unless the car is within the landing zone and is either stopped or being stopped.

Locks and contacts (not interlocks) are permitted in a few cases. Contacts are required for the car door or car gate, which is considered closed if within 2 in. (5 cm) of the nearest face of the jamb. The Elevator Code defines the closed position of hoistway doors as being 3/8 in. (9.5 mm) from the jamb or between panels of center-opening doors. For vertical-slide biparting doors, however, the dimension is 3/4 in. (19 mm). Under certain conditions, the elevator may be started when the doors are 4 in. (10 cm) from full closure. Design all interlocks so that the door must be locked in the closed position, as defined earlier, before the car can be operated.

Like any safety device, an elevator interlock or electric contact will be useless if it goes out of order easily. Therefore, use only those devices that comply with the Elevator Code and that have been either tested and listed by

Figure 13–25. An electromechanical interlock designed for use on biparting, manually operated steel doors on push-button-controlled freight elevators.

competent, designated testing laboratories or approved by the proper city or state authorities. Also, frequently inspect interlocking devices and see that they are maintained in proper working order.

Cars

Observe the Elevator Code safety requirements for the enclosures, top covers, doors and gates, and floors of elevator cars.

Car-Leveling Devices

Car-leveling devices automatically bring the car to a stop when the platform is level with the desired landing. A tripping hazard from uneven surfaces is thus eliminated. Moreover, in the case of freight elevators, the level surface that is automatically provided for the passage of trucks saves wear and damage to sills. It also reduces the possibility of material being jarred off trucks.

Enclosures

All elevator cars are required to be enclosed on the sides and top, except for the side or sides used to exit and enter. Openings for ventilation, an emergency exit, signals, and operating or communication equipment are allowed.

The sides and top of every passenger elevator car should be made of metal or of an approved, fire-retardant material. Although wood and openwork have been used for freight elevator cars, the Elevator Code now specifies a solid metal enclosure to a height of at least 6 ft (1.8 m) for enclosures. However, they may have perforations above the 6-ft (1.8-m) level, except in the area where the counterweight passes the car. That portion of the enclosure, for 6 in. (15 cm) on each side of the counterweight and up to the crosshead or car top, should be solid.

Fasten the enclosure securely to the floor and to the suspension sling. If the enclosure is cut away at the front to provide access to the hand rope, keep the opening low enough to prevent injury to the operator's hand.

Keep enclosures in good condition to prevent injuries to people loading and unloading the car. Broken wooden wainscoting or torn sheet metal can cause severe lacerations. One way to prevent such damage is to install bumper strips at the level of the truck's platforms and push bars. Bumper strips should be made of oak, ash, hickory, or similar tough, resilient wood. They should be at least 1 in. (25 mm) thick—2 in. (50 mm) thick where heavy trucks are used. Steel channels are also effective. Bumper strips should also be wide enough to match all heights of trucks' platforms and trucks' push bars.

Painting the inside of the car a flat black, from the floor to the top of the upper guard strip, will make bumps and scrapes less conspicuous. If the sides above the upper guard strip are painted aluminum, the high reflectance value of the paint will help overall illumination.

Top Covers

Freight elevator cars may have either solid or openwork top covers. The cover should be strong enough to sustain a load of 300 lb (135 kg) on any square area, 2 ft (61 cm) on a side, or 100 lb (45 kg) applied at any one point. If openwork is used, the openings should be small enough to reject a ball 1½ in. (3.8 cm) in diameter. Some companies, however, recommend ¾-in. (19-mm) openings.

Wire mesh is recommended for openwork covers because it combines maximum strength with minimum weight. It also offers little interference with light and air. Do not use less than No. 10 gauge (2.5-mm-diameter) steel wire.

An emergency exit of at least 400 in.^{2} (0.258 m^2) in area and not less than 16 in. (40.6 cm) on any one side should be in the top of every elevator car. Such an exit provides a clear passageway for easy escape should the car become stalled between floors or should some other emergency arise.

The exit cover should open outward and be so hinged or attached that it can be opened easily from the top of the car only. A heavy cover should be counterweighted or divided into several hinged sections, which should be kept down except when being used.

The Elevator Code now requires a clear refuge space of not less than 650 in.^{2} (0.4 m^2) on top of the car's enclosure. It should not measure less than 16 in. (40.6 cm) on any side. The top emergency exit may open into this space provided that there is an area adjacent to the opening available for standing when the emergency exit cover is open. The minimum vertical distance in the refuge area between the top of the car's enclosure and the overhead structure or other obstruction should not be less than 3.5 ft (1 m) when the car has reached its maximum upward movement.

Doors or Gates

To conform to requirements for new elevators, a door or gate must be provided at each entrance to the car. If the elevator is electrically operated, the door or gate should have electric contacts. The electric contacts will then prevent movement of the car unless the door or gate is within 2 in. (5 cm) of the fully closed position.

It is recommended that existing elevator cars having more than one opening be equipped with a gate at each entrance. Provide the gate with a contact to prevent its being opened when the car is in motion. When closed, the gate should guard the full width and height of the opening, except for vertical-sliding gates. They should extend from a point not more than 1 in. (2.5 cm) above the car's floor to a height of at least 6 ft (1.8 m). Collapsible gates, when fully expanded, should reject a ball 3 in. (7.6 cm) in diameter. They should have convenient handles with guards that protect the operator's fingers.

Floors

Car floors should be kept in good condition. Do not use ordinary metal sheets for surfacing or repairing floors because they soon become smooth and slick. In time, even sheet metal with raised surface markings may wear smooth and then have to be replaced.

Do not let the edge of the car's platform or the edge of the landing become slippery or badly worn. At such points, cast-iron and steel sills are slipping hazards unless provided with slip-resistant surfaces. Apply abrasive strips or welded beading at those points to avoid slipperiness.

Loads

Even though the rated capacity (safe working load) of an elevator includes a safety factor, do not exceed the rated capacity. Indicate the rated capacity with a conspicuous sign inside the car. Metal signs with stamped, etched, or raised letters and figures, not less than ¼ in. (6 mm) high, are satisfactory for passenger elevators. Use a minimum of 1-in.- (25-mm-) high letters and figures, however, for freight elevators.

Lighting

All cars and landings should be well lit at all times when in use. The following are minimums and can be exceeded:

- At the landing edge of the car platform, when the car and loading doors are open, a car should have at least two lights that provide a minimum illumination of 5 fc (54 lux) for passenger elevators; 2 fc (27 lux) for freight elevators.
- Car lights may be omitted on freight elevators with open-

work car tops on cars that travel no more than 15 ft (4.57 m). However, install at least two electric lights at the top of the hoistway to furnish the minimum specified illumination.
- Illumination on landing thresholds should be at least 5 fc (54 lux). New elevators, in addition to the regular lighting, should have emergency lighting that automatically goes on 10 seconds after the regular lighting fails.

Hoisting Ropes

Records show that it is relatively unusual for an elevator incident to be caused by the parting of the hoisting ropes. When it does happen, though, it is more likely to occur with winding-drum drives than with traction drives. Periodic resocketing intervals for winding-drum elevators are specified in the Elevator Code. Although ropes usually are installed to give the highest safety factor specified in the Elevator Code, closely inspect them to avoid the possibility of an incident. (See ANSI/ASME A17.2, Guide for Inspection of Elevators, Escalators, and Moving Walks, for wire rope inspection.)

Rope life is shortened by improper brake action. Unduly sudden stops may result from brake defects, such as heavy spring pressure or brakeshoe wear. Unnecessary starting and stopping also shorten rope life. Avoid inching for landings. In some cases, inching can be eliminated by properly adjusting the stopping devices. In most cases, however, it is caused by faulty operation; instruct the operator how to properly operate the elevator.

Ropes and rope fastenings used to hoist elevator cars must follow certain requirements. Safe operating conditions result when these requirements are observed.

Car and Counterweight Ropes

Car and counterweight ropes must be of iron (low-carbon steel) or steel having the commercial classification "Elevator Wire Rope." They can also be of rope specifically constructed for elevator use. Do not use ropes less than ⅜ in. (9.5 mm) in diameter on passenger elevators. Traction-drive elevators should have not fewer than three hoisting ropes. Winding-drum drive elevators should have not fewer than two hoisting ropes and should have two ropes for each counterweight used.

The diameter of sheaves or drums for hoisting, or counterweight ropes, should be not less than 40 times the diameter of the ropes. In practical application, however, this ratio is usually higher than 40.

On winding-drum elevators, rope should be long enough so there will be not less than one full turn of rope on the drum when the car is at the extreme limit of its overtravel. Drum ends of ropes are usually secured by babbitt-filled, tapered sockets or by clamps on the inside of the drum.

Rope Fastenings

Ends of car and counterweight-suspension ropes are usually fastened by individual babbitt-filled, tapered sockets. (See ANSI/ASME 17.1, Figure 212.9d and Table 212.9d, for sketch and dimensions.) Rope sockets must develop at least 80% of the breaking strength of the rope used. Shackle rods, eyebolts, and other means used to connect sockets to the car or counterweight ropes must have a breaking strength at least equal to that of the rope.

Governor Ropes

Governor ropes should be made of iron, steel, monel metal, phosphor bronze, or stainless steel. They should be of regular-lay construction not less than $3/8$ in. (9.5 mm) in diameter. Rope used as a connection from the safety to the governor rope, including rope wound on the safety drum, must be corrosion resistant.

Operating Controls

Stopping Devices

Every electric elevator must have an EMERGENCY STOP switch in the car, adjacent to the operating device, that will cut off the power. The button should be clearly identified and be red in color. Contacts of such switches should be directly opened mechanically and should not depend solely on springs for opening contacts. Some local codes have outlawed the EMERGENCY STOP switch; consult the local authority.

Winding-drum elevators are required to have a direct-driven, adjustable, automatic, machine-limit stop mechanism. Such a mechanism will stop the car if it overruns the highest and lowest landings. In addition, limit switches are required either in the hoistways or on the car. Winding-drum elevators are also required to have a slack-rope device to shut off power to the drive and to brake in case the rope becomes slack. This device must be such that it will not reset automatically when slack in the rope is removed.

Electric elevators with traction drives are required to have stopping switches in the hoistway, on the car, or in the machine room if the switches are operated by the motion of the car.

Grounding

The motor frame, and the operating cable, if insulated from the motor frame, should be grounded. All switches and wiring should conform to NFPA 70, National Electrical Code.

Safety Devices

Every elevator car suspended by wire ropes is required to have one or more safety devices that will catch and stop the car in the event of overspeed or failure of the hoisting ropes. The safeties must be attached to the car's frame. One safety must be located within or below the lower members of the car's frame (safety plank).

Counterweight safeties should conform to the car safety requirements and are classified as follows:

- Type A, instantaneous, safeties rapidly increase their pressure on the guide rails to give a very short stopping distance. They are permitted on elevators having a rated speed of not more than 150 fpm (0.76 m/s).
- Type B safeties apply limited pressure to the guide rails, and the retarding forces are reasonably uniform after full application. They are permitted on elevators of any speed and may be used in multiples.
- Type C safeties—Type A with oil buffers between the safety plant and elevator car—are permitted on elevators with a rated speed of not more than 500 fpm (2.5 m/s).

Car Switches

Design the handle of the car switch (operating control) to return to the STOP position and lock when the operator's hand is removed.

Signal System

Every elevator, except automatic-operation and continuous-pressure-operation elevators, should have a signal system that can be operated from any landing. In this way, the car can be signaled when it is wanted at that landing.

Inspection and Maintenance

Inspections and tests must conform to requirements and regulations set forth by the particular regulatory agency. Much of the discussion in this section is based on ANSI/ASME A17.2, Guide for Inspection of Elevators, Escalators, and Moving Walks.

Requirements for inspecting elevators, as given in ANSI/ASME A17.2, follow:

1. Acceptance inspections and tests of all new installations and alterations should be made by an inspector employed by the regulatory agency.
2. Routine inspections and tests of all installations should be made by a person qualified to perform such services, in the presence of an inspector employed or authorized by the enforcing authorities. It is recommended that periodic inspections and tests be made at intervals not longer than 6 months for power passenger elevators and escalators and 12 months for power freight elevators, hand elevators, and power and hand dumbwaiters.
3. Periodic inspections and tests should be made by a person qualified to perform such service in the presence of an inspector employed or authorized by the regulatory agency. Car and counterweight safeties, governors, and oil buffers should be given periodic inspections and

tests at least every year with no load and every 5 years with a full load.

Inspection Program

Careful maintenance is essential to the safe operation of elevators and all their parts. It also reduces the need for repairs. Frequent and thorough inspections by qualified outside personnel are the first requisite of an efficient maintenance program. However, minor day-to-day inspection and maintenance can be performed by qualified plant personnel.

Maintenance on a regular basis includes the following: (1) inspection of hoisting and counterweight wire ropes; (2) lubrication of oil buffers; (3) cleaning and lubrication of guide rails, controller contactors and relays, and car safety mechanisms; and (4) cleaning of hoistways, pits, machine rooms, and tops of cars.

Hoistways and Landings

Incidents occurring or originating at landings usually are due to (1) tripping or slipping at the car entrance or landing, (2) being caught inside the car, (3) falling into the hoistway, and (4) being caught by the doors.

To help prevent tripping and slipping incidents at hoistway landings, pay special attention to the following points:

- Check the leveling of the car. Improper leveling may be due to careless operation or, where cars are leveled automatically, to improper adjustment. Observe the operation of elevators and, if necessary, give instructions to the operators. With an automatic car-leveling device, the operator has no control over the final stop. Have a competent mechanic, preferably one trained or employed by the elevator manufacturer, adjust the automatic car leveler.
- Watch the condition of landing sills and floors. Landing sills should be of slip-resistant material; if they are not, they should be replaced or roughened if worn smooth. Repair broken sills, holes in flooring, worn floor coverings, and other conditions that create tripping hazards. The finished surface should be flush with the surrounding floor.
- Check the illumination of landings. Give special attention to those near building entrances where the difference between the intensity of outdoor and indoor light is noticeable. Clean globes and reflectors, and provide lamps of adequate size.
- Maintain and correctly adjust interlocks and contacts to provide the required protection at landings.

Track grooves for hoistway doors should be clean so that the doors will move freely. There should be no excessive play. Vision panels should be clean and unbroken.

Counterweights should operate freely, and sheaves should be properly aligned.

Because the entire load is transferred to the guide rails when the safety operates, the guide rails must be properly aligned at the joints and securely attached to the brackets. In addition, the bolts or welds with which the brackets are attached to the building's structure must remain tight. Alignment of rails can be checked easily by sighting along the faces; bracket bolts must be tested individually.

Check elevator cars for structural defects, such as loose bolts and other fastenings, excessive play in guide shoes, and worn or damaged flooring. Keep guide rails properly lubricated to help reduce wear between the rails and the guide shoes, except where roller guide shoes, which run on dry rails, are provided. Use the lubricant specified by the elevator manufacturer. Keep roller guide shoes clean and, if necessary, adjust them for pressure against the rails.

Test each emergency exit by opening it. Also check each exit panel to ensure that it is securely fastened in the closed position. If panels are held in place by locks, keep the key in a location not reached by the public, but available to personnel for emergency use. If the panels are held by thumbscrews, they should be removable without having to use pliers.

Subject car doors or car gates to the same inspection as hoistway doors, and follow the same maintenance standards. Check contacts for adjustment in the same manner as hoistway door contacts.

Test the car-operating switch, when provided, to see that it returns to, and locks in, the neutral position, when released by the operator. On a cable-operated car, check the cable lock to make sure that, when it is locked, the cable cannot be operated. Examine all switch contacts in the car-operating device, and place the glass cover over the emergency-release switch. Emergency-release switches, however, are prohibited on automatic elevators by the Elevator Code.

Check car lights for proper operation, loose or missing screws, and broken or cracked lamps. Keep lamps and bulbs clean. Periodically check the emergency lighting.

Safety Devices

Maintenance of safety devices is often neglected because they do not affect normal operation of the car. However, in case of an emergency, the safety of passengers depends entirely on the proper performance of these devices. It is of the utmost importance, therefore, that they be maintained in proper working condition.

At frequent intervals, clean and lubricate all the safety devices and equipment; inspect them for worn, cracked, broken, or loose parts; and test them to determine their ability to stop and hold the car. Safety devices should be tested and adjusted only by a person qualified to perform such services. Tests should be done in the presence of an inspector employed, or authorized, by the enforcing authorities.

Limit Switches

The inspector should check limit switches for proper alignment. The mounting of switches and cams should be checked for rigidity.

Buffers

Because oil buffers lose some oil during normal usage, check the oil levels at least once every month and each time the buffer is known to have been compressed. For refilling oil buffers, use an oil of the type specified by the manufacturer. Empty, clean, and refill with fresh oil buffers that have been submerged by floods or pit leakage. Check the alignment and the tightness of bolts in the anchorage.

Check spring buffers for alignment and for proper seating in the cups or mountings. Examine springs for deformation and permanent set.

Hoisting Machines

Examine hoisting machines carefully at each inspection. Check the machine's base for misalignment and cracks; immediately repair defects. The inspection should also include the following:

- the oil level in the motor, if used
- brake operation
- the oil level in the gear housing, if provided
- sheaves and drums for cracks and wear in the grooves.

Belted Machines

At frequent intervals, examine belts for proper tension, wear, burns, condition of splices, and cuts and breaks in the surface. Check chains for excessive wear. Regularly check machine-fastening bolts, belt guards, and the fastenings of platforms under any ceiling machinery.

Inspection Routine

Persons making elevator inspections should take all necessary precautions for their own safety. ANSI/ASME A17.2 contains not only detailed instructions for the conduct of all tests but also comprehensive information on personal safety for inspectors working in machine rooms, in and on top of cars, and in pits.

Inspectors should wear close-fitting clothing, preferably one-piece overalls with all buttons fastened and without cuffs. Inspectors should not wear gloves, except when checking wire rope.

Inspectors must pay close attention to moving objects such as counterweights, to hoistway projections, and to limited overhead and pit clearance. Before inspecting electrical parts, be sure that the main line disconnect switches are locked in the open position. In that way, current-carrying parts cannot be energized. When an inspector is on top of a moving car, he or she should keep one hand free to hold onto the crosshead or another part of the top of the car's frame. Inspectors should not hold hoisting ropes for support. They should attach safety belts to a fixed structure of the car or frame, not to the ropes.

If controls are not provided on top of the car for the inspector's use, the person who operates the elevator car should receive specific instructions on what to do while the inspector is inspecting the top of the car.

The order in which the various parts of the car are inspected depends on the type of elevator and the preference of the inspector. However, inspectors can reduce the inspection time and the amount of downtime for the elevator by planning. Determine which parts of the job can be done from each part of the elevator.

A written report of each inspection should be prepared and kept on file. Such a record is especially valuable for checking the progress of defects in ropes. Each report, therefore, should give definite details concerning the ropes' condition—diameter, number of broken wires per unit length, and estimated percentage of wear.

The mechanic who corrects the trouble should make a report of each break in service. The mechanic should enter the report on the log sheet for the particular elevator. In that way, the maintenance engineer can spot defective equipment that causes repeated breaks in service and thus correct the basic fault.

Operation

Safety and health professionals can eliminate many common causes of unintended elevator incidents by insisting on safe operating practices. Best results are obtained by assigning a properly instructed operator to full-time duty. In any case, permit only employees who have received proper instruction to act as elevator operators.

Selecting Elevator Operators

Select elevator operators with extreme care. Select operators who are mentally alert, not easily excited, and capable of carrying out instructions and of insisting on compliance with rules.

In many cases, the lives of others depend on the operator's efficiency. Moreover, an incompetent or poorly trained operator may damage valuable and indispensable equipment.

Faulty practices, such as starting a car before the doors and gates are closed, blocking gates open, and permitting crowding on cars not provided with gates, have caused many serious injuries. Actions such as improper loading, unnecessary starting and stopping, and reversing of the controller have caused damage to equipment.

Some companies have found that elevator operators work more safely if they are given cards that state they have completed the training course and are authorized to operate the equipment. In some cases, company rules require that operators have such cards.

Operating Rules

Adopt definite operating rules. In the case of industrial freight elevators, post the rules in the car, and require operators to know and observe them. Many companies that employ a number of elevator operators prepare pocket-sized rulebooks for their guidance. Plants with a number of elevators that are used under varying conditions prepare specific rules for specific elevators or groups of elevators.

Incident Investigations

Carefully investigate minor elevator incidents. Such investigations can disclose conditions or practices that, if left uncorrected, could later cause a serious injury. Therefore, determine the conditions or practices that caused the incident, and take prompt corrective action.

Overloads

Overloading elevators may result in injury to personnel, mechanical failure of the drive or of the car, or both. Many elevators still in service were built with much lower safety factors than are now required. With such elevators, the hazard of overloads is particularly great. However, even though elevators meet current safety factor requirements, the danger of overloads cannot be overemphasized. Consider the following engineering factors:

- In a traction-drive elevator, the ratio between rope tension on the driving sheave (car side) and rope tension on the counterweight side must be kept within certain limits. The counterweight normally is equal to the weight of the car plus 40% of the rated capacity. The motor torque and the brake are designed to handle this difference in weight.
- In the case of a traction-drive freight elevator, with a car weighing 8,000 lb (3,600 kg) and a rated capacity of 10,000 lb (4,500 kg), the counterweight would equal 12,000 lb (5,450 kg)—that is, the weight of the car, or 8,000 lb (3,600 kg), plus 40% of the rated capacity, or 4,000 lb (1,800 kg). The motor would be designed to lift, and the brake to hold, a 6,000-lb (2,700-kg) load.
- If the elevator is overloaded by 50% of its rated capacity (for this example, 5,000 lb or 2,250 kg), the total platform load will be 15,000 lb (6,800 kg). If 4,000 lb (1,800 kg), or overbalance of the counterweight, is subtracted from the 15,000 lb (6,800 kg) on the platform, a net weight of 11,000 lb (5,000 kg) must be handled by a motor and brake designed to handle only 6,000 lb (2,725 kg). This is an 83% overload.

In such a case, the traction relation may be broken, and the motor may not even pick up the load. As the brake is lifted, the car may start to move downward, a fuse might blow, and other mechanical failures, as well as injury to personnel, may result.

In no case should an elevator be overloaded unless the manufacturer has checked the entire installation for its ability to handle the load. Failure to do so may result in serious injury or death to employees and in serious damage to valuable equipment.

For an overload of a one-piece load (e.g., a transformer) as heavy as, or heavier than, the rated capacity of the car, consult the company that installed the elevator. The company's representative should check that:

- the machine is strong enough to handle the load
- the elevator structure, including the car frame (sling), platform, and undercar safeties, is adequate
- the traction relation will not be exceeded.

If the machine is otherwise strong enough for the overload, the elevator company's crew may increase the counterweight. This will maintain the traction relation while the overload is being lifted.

Emergency Procedures

Implement an emergency procedure similar to the one shown in Figure 13–26 for safely removing persons from elevators stalled between floors. As is illustrated, emergency instructions for persons involved in such incidents are spelled out on self-adhesive stickers. Mount these stickers behind protective transparent material to preserve their legibility. In addition, when deemed appropriate, post stickers in languages other than English. In gaseous or toxic environments, take tests to determine incident and injury potential. If necessary, implement other emergency procedures. (See ANSI/ASME A17.4, Guide for Emergency Personnel.)

ELEVATOR EMERGENCY INSTRUCTIONS

1. When the alarm sounds, the elevator has stopped between floors.

2. Notify the Fire Department or Rescue Team, at once, by calling

   _____
   (Telephone Number)

3. Notify the elevator maintenance company at once by calling

   _____
   (Telephone Number)

4. Do not attempt to rescue until trained, authorized maintenance personnel arrive. Assure the stuck passengers that help is on the way and that they are safe.

Figure 13–26. This sign should be posted in each elevator car.

A telephone or other means of communication is recommended for all existing cars and on new elevators. Having this equipment may prevent occupants from panicking and may help rescuers coordinate emergency procedures. Operating elevators in an emergency and emergency signal devices are discussed in the Elevator Code.

Requirements for the Disabled

To help further increase the mobility of disabled individuals, incorporate the requirements found in the National Elevator Industry's Minimum Passenger Elevator Requirements for the Handicapped for all new elevators. Where practical, incorporate the requirements on existing elevators.

SIDEWALK ELEVATORS

A sidewalk elevator is a freight elevator for carrying material, exclusive of automobiles, between a landing in a sidewalk, or other area outside a building, and the floors below the sidewalk or grade level. Sidewalk elevators present hazards that are not easy to eliminate. Therefore, it is best to locate them inside the building line or in an area not open to the public. Sidewalk elevators should conform to the requirements of the Elevator Code.

Except by permission of the local authorities, the maximum dimensions of openings for sidewalk elevators should be 5 ft (1.5 m) at right angles to, and 7 ft (2.1 m) parallel with, the building line. The side of the opening nearest the building should be not more than 4 in. (10 cm) from the building wall.

Where hinged doors or covers that can be lifted vertically are provided at the sidewalk or at other areas outside the building, bow-irons or stanchions should be provided on the car to operate such doors or covers. A loud audible signal should sound when the car is ascending.

A sidewalk elevator with winding-drum drive should have a normal terminal-stopping device on the drive. A similar device should be either in the hoistway or on the operating device.

Operation

The Elevator Code requires that a sidewalk elevator be raised and lowered only from the sidewalk or other outside area. Use either a key-operated, continuous-pressure up-and-down switch or continuous-pressure up-and-down buttons on the free end of a detachable, flexible cord, 5 ft (1.5 m) or less in length.

Hatch Covers

Automatic hatch covers, when closed, should sustain not less than 300 lb (1,136 kg) applied on any area 2 ft (610 mm) on a side and not less than 150 lb (68 kg) applied at any point. Hatch covers must be self-closing. They are not to be fastened or held open when the car is away from the top landing.

The covers should be made of metal and should lift vertically. For hinged covers, the line of the hinges should be at right angles to the building's wall. When the covers are fully open, there should be minimum clearance of 18 in. (46 cm) between them and any obstruction.

Hatch covers should be secured to the walkway's surface. To avoid tripping hazards to passersby, no hinges, locks, or flanges should project above the closed covers.

HAND ELEVATORS

Hand-powered elevators were once used widely in storage rooms, warehouses, and private residences. Few companies manufacture hand elevators. Do not install a hand elevator where it is to be used constantly during the working day. Instead, some form of power equipment should be installed wherever an elevator is a basic part of the manufacturing process or service function.

Never apply mechanical power to hand elevators by means of rope grips or similar attachments. If power is to be used, change the entire installation, and use all the protective devices as required for other elevators.

Hand-elevator cars in which persons can ride should not have more than one compartment nor be arranged to counterbalance another car. Hoistways, hoistway openings, pits, machinery spaces, supports, and foundations for hand elevators should conform to the requirements for power elevators.

Hoistway Doors

Conspicuously display on the landing side of hoistway doors: DANGER—ELEVATOR—KEEP CLOSED. The letters should be not less than 2 in. (5.1 cm) high. Hoistway openings may have (1) self-closing doors that extend to the floor; (2) doors made in two parts, one above the other, and so arranged that the lower part can be opened only after the upper part has been opened; or (3) doors equipped with two spring locks or latches, one located 6 ft (1.8 m) above the floor.

Safety Devices and Brakes

Hand elevators should have hand or automatic brakes that operate in either direction of motion. When the brakes are applied, they should remain locked in the ON position until released.

A hand elevator that travels more than 15 ft (4.5 m) must have a safety attached to the underside of the car's frame. The safety must be able to stop the car and sustain it and its load. If the car is to travel more than 40 ft (12 m), it must have a hand-operated brake that automatically slows the car down.

DUMBWAITERS

A dumbwaiter is defined as a hoisting and lowering mechanism equipped with a car that (1) moves in guides and has a floor area not exceeding 9 ft² (0.8 m²), (2) has a compartment height not exceeding 4 ft (1.2 m), (3) has a rated capacity not greater than 500 lb (225 kg), and (4) is used exclusively for carrying materials. A dumbwaiter may be hand or power operated.

Hoistways and Openings

The requirements for dumbwaiter hoistways are almost the same as those for elevator hoistways. Design landing openings and doors for dumbwaiters to protect persons from falling into the hoistways. Provide the landing openings of power-driven dumbwaiters with hoistway doors that guard the full height and width of the openings.

With certain specified exceptions, power-operated dumbwaiter doors must be equipped with hoistway-unit-system, hoistway-door interlocks. They will prevent the dumbwaiter from moving if any hoistway door or gate is open.

Conspicuously display on the landing side of hoistway doors of hand-operated dumbwaiters: DANGER—DUMBWAITER—KEEP CLOSED. The letters should be not less than 2 in. (5.1 cm) high.

Safety Devices and Brakes

Power-operated dumbwaiters, except hydraulic dumbwaiters, should have brakes that are automatically applied when the power is cut off or fails. The brakes should also be able to stop the car automatically within the limit of overtravel at each terminal. Hand-operated dumbwaiters should have hand-operated or automatic brakes that can sustain the weight of the car and its load.

A power-operated dumbwaiter having winding-drum drive, a travel greater than 30 ft (9 m), and a capacity in excess of 100 lb (45 kg) requires a slack-rope device. This device will cut off the power from the motor and stop the car if it is obstructed in its descent.

ESCALATORS

An escalator is a power-driven, inclined, continuous stairway for raising or lowering passengers.

Safety Devices and Brakes

EMERGENCY STOP buttons, or other types of hand-operated switches having red buttons or handles, should be accessibly located in the right-hand newel base on new escalators, at or near the top and bottom landings. Emergency-stopping devices should be protected from being unintentionally operated. An escalator STOP button with an unlocked cover over it that can readily be lifted or pushed aside should be considered accessible. The operation of STOP buttons or switches should interrupt the power to the drive. It should not be possible to start the drive by pressing these buttons or switches.

Use key-operated buttons or switches to start the units. Locate them within sight of the escalator's steps. Also, provide a way to cut the escalator's power in case an ascending escalator unintentionally reverses its travel.

Each escalator should have a speed governor that will interrupt the power if the predetermined speed is exceeded—not more than 40% greater than the rated speed. The speed governor is not needed if a low-slip, alternating-current, squirrel-cage induction motor is used and if the motor is directly connected to the drive.

If a tread chain breaks, a broken chain–sensing device should cut the power. If an escalator has tightening devices that are operated by tension weights, make provisions to retain these weights in the escalator truss should they fail.

Each escalator should have an electrically released and mechanically applied brake able to stop the fully loaded escalator when it is traveling either up or down. The brake should automatically stop the escalator as soon as any safety devices begin to function.

Machinery

Every escalator machine room should have a light that can be lit without workers having to pass over or reach over any part of the machinery. The lighting control switch should be located within easy reach of the access to such rooms. Where practicable, the light control switch should be located on the lock, jambside of the access door. Provide reasonable access to the interior of the escalator for inspection and maintenance purposes. For the protection of maintenance personnel, install guards around all chains in escalator machinery compartments. Full lockout/tagout procedures must be followed in servicing an escalator. (See Chapter 6, Safeguarding, in this volume, for lockout/tagout procedures.)

While bearings on escalators are sealed and require no oiling, the chains do require lubrication. Take care not to overlubricate. Provide an oil pan in the bottom of the truss to catch oil or grease that may drip from moving parts and to catch dust and dirt that fall between the treads.

Clean the oil pan periodically to eliminate a fire hazard

from the accumulated oil-soaked dust and dirt. Attach a brush, made for this purpose, to a step axle. Draw the brush down over the drip pan to brush all the dirt to the lower end of the truss. The sweepings then can readily be removed. Reverse the unit to return the brush to the top, and then disconnect the brush.

The moving handrail will show some stretch over a period of time. Check it at intervals, and use the handrail's drive adjustment to take up the slack.

The balustrades must have handrails that move in the same direction and at about the same speed as the steps. Each moving handrail is to extend at normal handrail height, not less than 12 in. (305 mm) beyond the points of the comb plate's teeth at the upper and lower landings. Provide hand or finger guards at the point where the handrail enters the balustrade.

Inspect all parts of escalators and their drives' machinery at regular intervals to keep them well maintained. Test all safety devices for proper functioning.

Protection of Riders

Most escalators are installed in public places, and their principal hazards arise from misuse by the public.

Most escalator incidents occur when heels of shoes, fingers, or toes are caught between the surface grooves or slots on the treads and the comb plate. The width of each slot is to be not more than ¼ in. (6.3 mm) and the depth not less than ⅜ in. (9.5 mm), with a center-to-center spacing of not more than ⅜ in. (9.5 mm) between adjoining slots. In some incidents, edges of shoe soles have been caught between the step and the vertical side member (skirt guard), which should have a maximum of ³⁄₁₆ in. (4.8 mm) on each side.

Because barefoot passengers have been injured on escalators, post signs warning barefoot persons not to ride the escalator. Additionally, a caution sign should be located at the top and bottom landing of each escalator. This sign should include the words: CAUTION, PASSENGERS ONLY, HOLD HANDRAIL, ATTEND CHILDREN, and AVOID SIDES (Figure 13–27). The sign used should be the standard depicted in Rule 805.2 of the Elevator and Escalator Code. Umbrella tips are frequently caught between the grooves and the comb plate. This type of incident sometimes results in minor injuries and damage to equipment. Other incidents have resulted from riders mishandling baggage on escalators. Riders should not place suitcases and handbags on the steps. They should always carry such items parallel to the run of the escalator.

Escalator treads and landings should be made of noncombustible material that provides a secure foothold. Some riders, through inexperience or infirmity, have trouble seeing the parting point where treads rise or descend and the point where they level off. To aid these riders, illumination should be provided for all tread surfaces. Mount additional warnings in green demarcation lights inside the truss at top and bottom to shine through the treads where they break away and come together at the trouble points.

Some manufacturers mark the edges of the steps to emphasize the lines between adjacent steps. One manufacturer, for example, adds a distinctive color strip to the edge of the step adjacent to the riser of the next step.

Post signs reading PLEASE HOLD HANDRAIL at the top and bottom of the escalator. Do not place distracting signs, such as advertising, near these critical points. At each level, use directional arrows and mark the level's number on the floor to improve the flow of traffic from the escalators.

It is extremely important that no object or construction of any kind obstruct the free flow of passengers from the area at the exit of an escalator. (This area is not a part of the escalator.) Serious injuries have occurred where the flow of traffic from the exit was restricted by a fence or barrier placed at some distance from the escalator.

Fire Protection

Protection of escalator floor openings against the spreading of fire and smoke is required by local building codes. One of the best safeguards against the spread of fire is to divide buildings made of fire-resistive construction into limited areas in which fire can be readily controlled. Protect

Figure 13–27. Signs should be prominently displayed at the top and bottom of an escalator.

vertical openings from the passage of fire from story to story. This principle often is disregarded when escalators are installed.

Escalators approved as a required means of entry must be fully enclosed in accordance with local laws and ordinances. Escalators not approved as a required means of entry must have the floor openings protected in accordance with national and local standards and regulations. (See NFPA 101, Life Safety Code.)

MOVING WALKS

A moving passenger-carrying device, on which persons stand or walk and in which the passenger-carrying surface is uninterrupted and remains parallel to its direction of motion, is a moving walk. Criteria for the design, construction, installation, operation, inspection, and testing of moving walks are given in the Elevator Code, Part IX.

Moving walks may operate in a horizontal plane or in a slope up to a maximum of 12 degrees. However, the slope should not exceed 3 degrees within 3 ft (91 cm) of the moving walk's entrance or exit. Operating speed and treadway width are governed by the slope.

Protection of passengers on moving walkways is the same as that on escalators. (See the section Escalators, in this chapter.)

MAN-LIFTS

The following are the principal hazards in the use of man-lifts:

- The rider may be carried over the top.
- The rider may be unable to make an emergency stop.
- The rider may jump on or off after the step has passed the floor.
- His or her head or shoulders may strike the edges of floor openings if there is not a conical hood.
- The rider may be unable to reach the landing because of power failure and belt stoppage.
- Parts of the man-lift may fail or operate unsafely.

Because using cranes or derricks to hoist personnel poses a serious risk to the employees being lifted, any cranes and derricks that hoist personnel should conform to the following:

- be placed on a firm foundation
- be uniformly level within 1% of level grade
- have a minimum safety factor of 7 for the load line (wire rope) of the crane or derrick (this means it must be capable of supporting seven times the maximum intended load)

- move the personnel platform slowly and cautiously without any sudden jerking of the crane, derrick, or platform
- have rotation-resistant rope with a minimum safety factor of 10
- have all brakes and locking devices on the crane or derrick set when the occupied personnel platform is in a stationary working position.

In addition, the combined weight of the loaded personnel platform and its rigging must not exceed 50% of the rated capacity of the crane or derrick for the radius and configuration of the crane or derrick.

Construction

Construct, maintain, and operate man-lifts in strict compliance with the recommendations of ANSI/ASME A90.1, Safety Standard for Belt Manlifts. Use a safety factor of 6, based on a 200-lb (90-kg) load, on each step, on both the up and down runs. Brace all equipment securely at top, bottom, and intermediate landings. Secure the man-lift's rails to prevent their spreading apart, vibrating, and becoming misaligned. Suspend the entire man-lift from the top to prevent bending or buckling of the rails.

Handholds

Paint handholds, of either the open or closed type, in a bright color, such as orange or yellow. Place handholds not less than 48 in. (1.2 m), nor more than 56 in. (1.4 m), above each step. Locate steps not less than 16 ft (5 m) apart. Use slip-resistant material on the steps.

Landings

Emergency Landings. Where there is a travel of 50 ft (15 m) or more between floor landings, one or more emergency landings should be provided so that there will be a landing (either floor or emergency) for every 25 ft (7.5 m) or less of man-lift travel.

Between the surface of any landing and the lower edge of the conical guard suspended from the ceiling, allow at least 7 ft (2.3 m) of clearance. Allow a minimum clearance of 5 ft (1.5 m) between the center of the head pulley and the roof or other obstruction. The bottom landing on the side that is up should have steps to a platform level with the man-lift's step as it rises to a horizontal position.

Control of Illumination. Lighting of man-lift runways should be by means of circuits permanently tied in to the building circuits (no switches) or should be controlled by switches at each landing. Where separate switches are provided at each landing, any switch should turn on all lights necessary to illuminate the entire runway.

Floor Openings and Conical Guard Openings

At floor landings, provide standard 42-in. (1.06-m) guardrails and 4-in. (10-cm) toeboards around floor openings. Install them in such a way as to permit a landing space at least 2 ft (0.6 m) wide. Guardrails should have maze or staggered openings or self-closing gates that open away from the man-lift (Figure 13–28). At each floor opening on the side that is up, install funnel-shaped (conical) guards (Figures 13–29 and 13–30).

Brakes, Safety Devices, and Ladders

Install on the motor shaft of direct-connected units, and on the input shaft of belt-driven units, a brake that automatically works when the power is shut off. The brake should get its power or force from an external source. The brake must be capable of quickly stopping the man-lift and holding it when the side that is down is loaded with 250 lb (110 kg) on each step. The brake should release electrically.

Provide a control rope, not less than $3/8$ in. (9.5 mm) and made of Manila and/or cotton with a bronze wire center, within easy reach of both the up and down runs. When pulled in the direction of the belt's travel, the rope should cut off the power and apply the brake. Install an

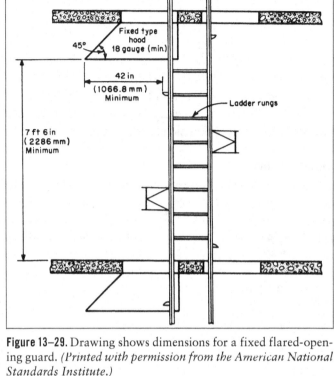

Figure 13–29. Drawing shows dimensions for a fixed flared-opening guard. *(Printed with permission from the American National Standards Institute.)*

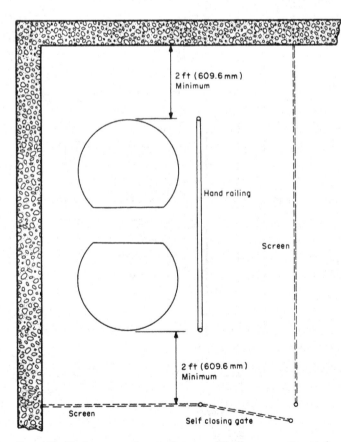

Figure 13–28. Screen enclosure for man-lift floor openings (plan view). *(Printed with permission from the American National Standards Institute.)*

Figure 13–30. In this installation, the guard is counterbalanced and will yield slightly if hit. *(Courtesy Humphrey Elevator Co.)*

electromechanical device that will automatically shut off the motor's power supply and apply an electric brake should the rider fail to alight at the top landing. RESET buttons, located at the top and bottom terminals, will permit restarting the man-lift after it has been shut off by the electrical safety devices.

A secondary safety control located on the top operating floor is also required. It should be set to operate when the belt has traveled 6 in. (15 cm) beyond the point of operation of the primary safety switch, in case the latter fails. The device should stop the man-lift before the loaded step reaches a point 24 in. (0.6 m) above the top landing.

Provide a fixed metal ladder, accessible from both the up and down runs, for emergency exiting if the vertical distance between landings exceeds 20 ft (6 m). This ladder should meet the requirements of the local codes. However, do not provide an enclosing cage, because the ladder should be accessible from either side throughout the entire run.

Inspections

All man-lifts should be inspected by a competent designated person at least every 30 days. Use an inspection report similar to the one in Figure 13–31. Indications of a defect are such things as (1) unusual or excessive vibrations; (2) continual misalignments; and (3) "skips" when mounting steps, which indicate worn gears. Upon discovery of a defect, immediately take the man-lift out of operation and do not use it until it is repaired. Each periodic inspection should cover, but not necessarily be limited to, the following items:

- steps
- gears
- lubrication
- belt and belt tension
- step fastenings
- drive pulley
- illumination
- handholds and fastenings
- rails
- warning signs and lights
- bottom (boot) pulley and clearance
- rail supports and fastenings
- floor landings
- pulley supports
- signal equipment
- guardrails
- motor
- belt splice joint
- limit switches
- brake
- driving mechanism
- step rollers
- electric switches
- key coupling keyway.

Inspect the safety mechanism of a man-lift daily. Check that it is operating freely and that it is free of dirt and grease. Dismantle a man-lift once every year, and replace defective or excessively worn parts.

A certification record should be kept of each inspection that includes the date of the inspection; the signature of the person who performed the inspection; and the serial number, or other identifier, of the man-lift that was inspected.

General Precautions

The maximum speed of a man-lift belt should not exceed 80 fpm (0.4 m/s). This speed should be uniform on all man-lifts throughout the plant. If a man-lift carries a great deal of traffic, a maximum speed of 60 fpm (0.3 m/s) is recommended.

Number floors with large figures in full view of both ascending and descending riders. Place a constantly illuminated sign, TOP FLOOR—GET OFF, placed not more than 2 ft (0.6 m) above the top landing, in full view of ascending passengers. Use block letters at least 2 in. (5 cm) high for the sign. In addition, locate at least a 40-W red warning

BELT MANLIFT INSPECTION REPORT (Weekly & Monthly)

Location _____   Date _____

Manlift Make & Serial No. _____   Code _____

	ITEM	Weekly	Monthly		ITEM	Weekly	Monthly
1	OBSERVE MANLIFT IN OPERATION FOR POSSIBLE DEFECTS	( )	( )	19	TOP LANDING SAFETY SWITCHES	( )	( )
2	STEPS AND ROLLERS		( )	20	TOP BAR SAFETY	( )	( )
3	HAND HOLDS		( )	21	PHOTO EYE SAFETY (IF APPLICABLE)	( )	( )
4	BELT JOINT		( )	22	ON-OFF SWITCH & CONTROL ROPE (TEST UP & DOWN RUN)	( )	( )
5	BELT (LOOK FOR CUTS OR DAMAGE)	( )	( )	23	STEP CLEARANCE AT DRIVE BELT TRACKING AT DRIVE	( )	( )
6	BELT TENSION AND BELT TAKE-UP AT BOTTOM PULLEY	( )	( )	24	MOTOR		( )
7	STEP CLEARANCE AT BOOT	( )	( )	25	BRAKE		( )
8	BELT TRACK ON BOTTOM PULLEY	( )	( )	26	GEAR REDUCER & CHECK OIL LEVEL & CHANGE PER MANUAL		( )
9	BOTTOM BEARING LUBRICATION & SUPPORT		( )	27	COUPLINGS – COLLARS – KEYS		( )
10	GUIDE RAILS, PROPER ALIGNMENT FASTENINGS & SUPPORT		( )	28	HEAD SHAFT BEARINGS AND LUBRICATION		( )
11	FLOOR BRACES FOR GUIDE RAILS		( )	29	TOP PULLEY AND LAGGING	( )	( )
12	FLOOR HOODS AND OPENINGS		( )	30	OVERALL DRIVE ASS'Y AND SUPPORTS		( )
13	SAFETY SWITCHES ON MOVEABLE HOODS	( )	( )	31	SKIPPING WHEN MOUNTING STEP (CHECK DRIVE TRAIN)		( )
14	GUARD RAILS AND GATE OPERATION	( )	( )	32	VIBRATION OR MISALIGNMENT IN DRIVE		( )
15	FLOOR LANDINGS (CLEAR OF OBSTRUCTION)	( )	( )	33	GREASE BEARINGS PER MAINTENANCE MANUAL		( )
16	ILLUMINATION OF MANLIFTS AND FLOOR LANDINGS	( )	( )	34	GUIDE TRACK FREE OF FOREIGN MATERIAL AND LUBRICANTS	( )	
17	TOP & BOTTOM FLOOR WARNING SIGNS & LIGHTS	( )	( )	35			
18	BOTTOM SAFETY SWITCHES TREADLE &/ OR DROPOUTS	( )	( )	36			

IS MANLIFT BEING PROPERLY USED? _____ BY AUTHORIZED PERSONNEL? _____

ITEM NUMBERS ABOVE THAT WERE CORRECTED OR ARE IN NEED OF SUCH (GIVE COMPLETE DESCRIPTION ON BACK SIDE OF THIS FORM)

INSPECTED BY _____   DATE _____

Figure 13–31. A typical checklist for inspecting a man-lift. (Printed with permission from the American National Standards Institute.)

light immediately below the top landing so it will shine in an ascending rider's face. The entire man-lift should be illuminated at all times while it is in operation with at least 1 fc (11 lux) at all points and at least 5 fc (54 lux) at landings.

Prominently, at each landing, display signs carrying instructions for use of the man-lift. Permit only authorized employees to ride man-lifts, and display signs stating this at each landing. Riders should not carry on a man-lift anything that cannot be placed entirely inside a pocket, a sling, or a pouch. Carefully instruct employees, particularly new ones, in the safe use of the man-lift. Tell them to report immediately any defects or any irregularity in the operation of the man-lift or its safety devices. Supervisors should take corrective action immediately.

SUMMARY

- Hoisting equipment must never be overloaded or used to transport people. Operators should examine hoists regularly and repair or replace any damaged or malfunctioning parts.
- Cranes should have adequate safeguards to provide safe footing and accessways for the operator, to prevent injuries, and to limit the action of the crane arm and hoisting devices. All hoisting ropes, sheaves and drums, and other equipment should be appropriate for the crane being used. The operator's cab must protect the operator against fire and weather, be well ventilated, contain ample control equipment, and allow a clear view of signals.
- Other types of cranes—including monorails, jib cranes, derricks, tower and mobile cranes, and portable floor cranes—all have specific guidelines for safe operation and transport.
- Aerial-basket lifts are commonly used for working aboveground. The manufacturer's operating and maintenance instructions should be followed for safe handling of these machine parts.
- Conveyors are generally defined as a horizontal, inclined, or vertical device for moving or transporting objects. Loading points must be clearly marked, showing the safe load limit that can be raised or lowered, with safeguards used along the entire length. Maintenance personnel must lock out all power before working on a conveyor. Operators must stand clear of moving conveyors, avoid pinch points and other areas where hands or fingers can get caught, and never attempt to repair a moving belt.
- Power elevators should conform to the Elevator Code for electric-drive and hydraulic-drive elevators. The Elevator Code prohibits belt-driven and chain-driven machines from being installed. Interlocks and electric contacts for both passenger and freight elevators should be direct-acting mechanical/activated devices that cannot be inactivated.
- Companies should establish a regular program of inspection, testing, and maintenance of all elevator parts according to city, state, and federal regulations. Companies also should select elevator operators carefully and establish proper working procedures, safety rules, and emergency procedures for operators to follow.
- Dumbwaiters may be hand or power operated and must be equipped with hoistway-unit-system, hoistway-door interlocks. Landing openings and doors should be designed to protect people from falling into the hoistways.
- Escalators must have EMERGENCY STOP buttons or other types of hand-operated switches in easily accessible locations. Each escalator should have sensing devices that will interrupt power if the preset speed is exceeded or a tread chain breaks. The principal hazards on escalators and moving walks arise from their misuse by the public.
- Injuries on man-lifts frequently occur when riders are carried over the top, jump off or on after the step has passed a floor, or are unable to reach the landing because of power failure. All man-lifts must have emergency brakes, safety devices, and ladders to prevent injuries and incidents. Inspectors should test and examine man-lifts every 30 days, with daily inspection of safety devices.

REFERENCES

American National Standards Institute, 1819 L Street NW, 6th Floor, Washington DC 20036.

ANSI A10.4–2007, Safety Requirements for Personnel Hoists and Employee Elevators.

ANSI A10.28–1998 (R2004), Safety Requirements for Work Platforms Suspended from Cranes or Derricks for Construction and Demolition Operations.

ANSI A14.3–2002, Safety Requirements for Fixed Ladders.

ANSI A1264.1–2007, Safety Requirements for Workplace Walking/Working Surfaces and Their Access; Workplace Floor, Wall and Roof Openings; Stairs and Guardrails Systems.

ANSI Z244.1–2003, Safety Requirements for Conveyors and Related Equipment.

ANSI/ASME A17.1–2007, Safety Code for Elevators and Escalators.

ANSI/ASME A17.2–2007, Guide for Inspection of Elevators, Escalators, and Moving Walks.

ANSI/ASME A17.4–1999, Guide for Emergency Personnel.

ANSI/ASME A90.1–2003, Safety Standard for Belt Manlifts.

ANSI/ASME B15.1–2000, Safety Standard for Mechanical Power Transmission Apparatus.

ANSI/ASME B20.1A–2006, Safety Standards for Conveyors and Related Equipment.

ANSI/ASME B30 Series, Safety Requirements for Cranes, Derricks, Hoists, Jacks, and Slings.

ANSI/ASME B30.2–2005, Overhead and Gantry Cranes (Top Running Bridge, Single or Multiple Girder, Top Running Trolley Hoist).

ANSI/ASME B30.10–2014, Hooks, Safety Standards for Cableways, Cranes, Derricks, Hoists, Hooks, Jacks and Slings.

ANSI/ASME B30.11–2004, Monorails and Underhung Cranes.

ANSI/ASME B30.16–2007, Overhead Hoists (Underhung).

ANSI/ASME B30.17–2006, Overhead and Gantry Cranes (Top Running Bridge, Single Girder, Underhung Hoist).

ANSI/IEEE C2–2002, National Electrical Safety Code.

ANSI/CEMA 102–2012, Terms and Conveyor Definitions.

Conveyor Equipment Manufacturers Association, 6724 Lone Oak Boulevard, Naples, FL 34109.
Screw Conveyors, Book No. 350.

Crane Manufacturers Association of America, 8720 Red Oak Boulevard, Suite 201, Charlotte, NC 28217.

International Union of Operating Engineers, 1125 17th Street NW, Washington DC 20036.

National Elevator Industry Inc., 1677 County Route 64, PO Box 838, Salem, NY 12865-0838.

Minimum Passenger Elevator Requirements for the Handicapped.

National Fire Protection Association, 1 Batterymarch Park, Quincy, MA 02269.
Life Safety Code, NFPA 101, 2006.
National Electrical Code, NFPA 70, 2008.

National Safety Council, 1121 Spring Lake Drive, Itasca, IL 60143-3201.
Occupational Safety and Health Data Sheets (available in the Council Library):
Electromagnets Used with Crane Hoists, Data Sheet 359.
Floor Mats and Runners, 12304–0595.
Mobile Cranes and Power Lines, 12304–0743, 2005.

U.S. Department of Commerce, National Institute of Standards and Technology, Gaithersburg, MD 20899.
National Electrical Safety Code, NBS Handbook H30.

U.S. Department of Energy. *DOE Hoisting and Rigging Manual*. Idaho Falls, ID: EG&G Idaho Inc., October 1988.

U.S. Department of Labor, Bureau of Labor Statistics, Postal Square Building, 2 Massachusetts Avenue NE, Washington DC 20046. Census of Fatal Occupational Injuries, 2006.

U.S. Department of Labor, Occupational Safety and Health Administration, 200 Constitution Avenue NW, Washington DC 20210. *Code of Federal Regulations*, Title 29. Sections 1910.179(b)(2), 1910.180(j)(1), and 1926.550(a)(15).
29 CFR 1910.147, The Control of Hazardous Equipment (Lockout/Tagout).

REVIEW QUESTIONS

1. Why should a load be lifted only when it is directly under the hoist?
2. What are the three general types of chain hoists, and which is most efficient?
   a.
   b.
   c.
3. What is the purpose of a spring return on an operating lever?
4. Which crane movement control hand signal should be obeyed even if it is being given by someone other than the signaler in charge?

5. What is the maximum bend or twist from the plane of the unbent hook in a crane hook that is allowable before corrective action must be taken?
   a. any bend or twist is not permissible
   b. 5 degrees
   c. 10 degrees
   d. 20 degrees
6. What are the first three steps that should be taken before maintenance work can be performed on a crane?
   a.
   b.
   c.

7. Name and describe the three groups of monorail hoists.
   a.
   b.
   c.

8. Why should hoists or cranes not be used to lift, support, or otherwise transport people?

9. Before the first use and after modification, cranes must be tested to _____ of the rated load unless the manufacturer recommends otherwise.
   a. 100%
   b. 125%
   c. 150%
   d. 200%

10. What is the difference between light service and heavy service?

11. What is the difference between frequent inspections and periodic inspections?

12. What are five factors that are implicated in unintended mobile crane incidents?
   a.
   b.
   c.
   d.
   e.

13. What rating should the fire extinguisher that is stored in the cab of a crane have?
   a. BC
   b. ABC
   c. 5BC
   d. UL

14. What three pieces of information should be on a crane operator's capacity chart?
   a.
   b.
   c.

15. For loads limited by structural competence, the weight should be determined to what level of precision before the load is lifted?
   a. 80%
   b. 25% of rated load
   c. 20 lb
   d. 10% of the load weight

16. Name two purposes of a trial run before lifting personnel.
   a.
   b.

17. What three things should the authorized signaler for an on-track crane do while walking ahead of the crane?
   a.
   b.
   c.

18. For a crane operating near power lines that are rated between 50 kV and 345 kV, what is the minimum necessary clearance between the lines and any part of the crane?
   a. 4 ft
   b. 4.9 ft
   c. 10 ft
   d. 16 ft

19. What are the four benefits/purposes of using a safety factor of 10 when determining the safe working load of Manila rope?
   a.
   b.
   c.
   d.

20. What is the maximum distance between emergency-stopping devices for a conveyor that operates near a walkway?
   a. 10 ft
   b. 16 ft
   c. 23 ft
   d. 75 ft

21. What are two types of chain conveyors?
   a.
   b.

22. What is the difference between chute conveyors and roller conveyors?

23. What code governs the use and design of elevators?

24. Why are car safeties not required on electrohydraulic elevators?

Ropes, Chains, and Slings

14

Bradley A. McPherson, MS, CSP

Wayne Clifton, CSP, CPCU, ARM, PE, CIE

Philip E. Hagan, JD, MBA, MPH, ARM, CIH, CHMM, CET, CHCM, CEM

Fiber Rope
Types of Fiber Ropes ▶ Working Load ▶ Inspections ▶ Care of Fiber Rope in Use ▶ Care of Fiber Rope in Storage

Wire Rope
Types of Wire Ropes ▶ Design Factors for Rope Used in Hoisting ▶ Inspections and Replacement ▶ Care of Wire Rope in Use ▶ Sheaves and Drums ▶ Wire Rope Fittings

Rigging
Choker Hitch ▶ Basket Hitch

Fiber and Wire Rope Slings
Types of Fiber and Wire Rope Slings ▶ Methods of Attachment ▶ Working Load ▶ Inspections ▶ Safe Operating Practices for Slings

Chains and Chain Slings
Types of Chain Slings ▶ Properties and Working Load of Alloy Steel Chain ▶ Hooks and Attachments ▶ Inspections ▶ Safe Practices for Chain Slings

Synthetic Web Slings and Metal Mesh Slings
Synthetic Web Slings ▶ Inspections of Synthetic Web Slings ▶ Metal Mesh Slings ▶ Safe Practices for Metal Mesh Slings ▶ Inspections of Metal Mesh Slings

Summary

References

Review Questions

447

Special safety precautions apply to using and storing fiber ropes, wire ropes, rope slings, chains, and chain slings. The safety and health professional should know the properties of the various types of ropes, chains, and slings used and the precautions for both use and maintenance. Figure 14–1 gives a checklist of important factors to consider when obtaining rope for a specific application. This chapter covers the following topics:

- types of fiber rope and its performance and care
- characteristics of wire rope and safety issues in its use
- safety practices in using rigging
- safety issues in using fiber and wire rope slings
- hazards and safety practices for chains and chain slings
- safety issues in using synthetic web and metal mesh slings.

FIBER ROPE

Fiber rope is used extensively in handling and moving materials. Natural fiber ropes are generally made from Manila (abaca) or sisal (sisalena or henequen). Hemp and, sometimes, coir, cotton, and jute are other types of fiber rope used; however, these are relatively unimportant in the heavy cordage field. Synthetic fiber ropes on the market include those made from nylon, polyester, and polyolefin. Manila or nylon ropes give the best uniform strength and service.

Types of Fiber Ropes

Manila Fiber

The properties of Manila fiber make it the best-suited natural fiber for cordage. Manila rope is often recommended for capstan work because of its ability to pay out evenly

---

ROPE CHECKLIST

Strength	Friction melting
Stretch with load	Combustibility
Impact load	Sunlight resistance
Permanent stretch	Latitude and altitude
Recovery from stretch	Color and type
Length	Diameter and construction
Size	of rope
Yardage	Frequency of use
Floatability	Storage methods
Flexibility	Marine growth resistance
Twist direction and torque	Rot resistance
Flex life in bending	Chemical resistance
Slipperiness	Color
Texture	Aging
Water repellency	Contamination
Hygroscopicity	Uniformity
Ruggedness in shape	Service cost
Temperature resistance	Toughness against wear

Figure 14–1. Factors that may be of significance when obtaining rope for a specified use.

---

when so used. High-grade Manila rope, when new, is firm but pliant, varies in color from ivory to light yellow, and has considerable luster. The manufacturer treats the rope with chemicals to make it more mildew resistant, which increases rope quality. Its good reputation in fresh water and saltwater is well established.

Sisal Fibers

The properties of sisal fibers do not give sisal ropes the high general acceptance of Manila. The sisals lend themselves to use mostly in smaller ropes. Their breaking strengths are generally about 20% lower than those of Manila. Sisal rope varies in color from white to yellowish white and lacks the gloss of high-grade Manila. The fibers are stiff and harsh and tend to splinter. This makes the ropes uncomfortable to handle.

Sisal and Manila fibers deteriorate when in direct contact with acids and caustics as well as in their mists or vapors. This deterioration is accelerated by hot, humid conditions. Both fibers lose 50% of their strength at 180°F (80°C) and burn at temperatures greater than 300°F (150°C).

The strength of sisal and henequen varies from grade to grade, where Manila has less variability. Sisal rope is about 80% as strong as Manila; henequen is about 50% as strong, but it resists deterioration from exposure to the air better than sisal.

Other Natural Fibers

Other natural fibers are also used in ropes but to a lesser or negligible degree and only for special reasons. These fibers include cotton, flax, coir, straw, asbestos, istle, jute, kenaf, silk, rawhide, and sansevieria.

Synthetic Fibers

Nylon, polyester, and polyolefin ropes are the major types of synthetic fiber ropes. Synthetic fiber ropes are used more often than natural fiber ropes for the following reasons:

- More is known about the properties of various synthetics. Successful use of synthetic fiber rope depends largely on selecting the synthetic with the physical properties and characteristics that most closely match the requirements of the job.
- Splices can be made readily in synthetic fiber rope and can develop nearly the full strength of the rope. Tapered splices are highly recommended for rope sizes with a diameter 1 in. (2.5 cm) or larger.

Nylon Rope

Nylon rope has more than two and a half times the breaking strength of Manila rope and about four times its working elasticity. It is, therefore, well suited to shock loading, such as is required for restraint lines. Its resistance to abrasion is remarkably high compared with that of other ropes. When nylon rope is wet or frozen, its breaking strength is reduced by 10% to 15%. The advantage of using nylon rope is that it is waterproof and has the ability to stretch, absorb shocks, and resume normal strength.

Nylon rope also is highly resistant to organisms that cause mildew and rotting and to attack by marine borers in seawater. Prolonged exposure to air results in little loss of strength. Because there is no swelling, wet nylon rope runs through blocks as easily as dry nylon rope. Although resistant to petroleum oils and most common solvents and chemicals, nylon's strength is affected by drying oils, such as linseed oil or the phenols. Nylon rope is also vulnerable to strong mineral acids, phenolic compounds, and heat.

Nylon loses some of its strength at 300°F (150°C) and all of it at 482°F (250°C), its normal melting point. Short of melting, most of nylon's strength is regained upon cooling to normal temperature. Nylon of a higher melting point is available.

Nylon, more than any other rope material, will absorb and store energy in the same manner as a spring. When nylon rope breaks, this energy makes the rope's moving ends as dangerous as a projectile. Exercise caution, therefore, when working nylon lines around corners, capstans, timber heads, and the like.

Polyester Rope

Probably the best general-purpose rope available, especially for critical uses, is made of polyester. Polyester stretches about half as much as nylon, so energy absorption is also about half as much. It is not weakened by rot, mildew, or prolonged exposure to seawater. In addition, polyester retains its full strength when wet because it does not absorb moisture. It shows little deterioration from long exposure to sunlight and resists abrasive wear well. Polyester is somewhat vulnerable to alkalis, but its resistance to ultraviolet light is good to excellent. It burns at about 480°F (250°C) and loses strength at temperatures greater than 390°F (200°C). OSHA specifies that a safe operating temperature range for synthetic fiber rope slings is between –20°F and 180°F (–29°C and 80°C) [29 CFR 1910.184(h)(2)].

Polyolefin Rope

In general, polyolefin rope is strong and inexpensive. It floats and is unaffected by water. Polyolefin, like polyester, does not absorb moisture; therefore, it does not shrink or swell with water. It is unaffected by rot, mildew, and fungus. Polyolefin rope is also highly resistant to a wide variety of acids (except nitric acid) and alkalis, as well as to alcohol-type solvents and bleaching solutions. However, it swells and softens when exposed to hydrocarbons, particularly at temperatures above 150°F (66°C). The movement of crossed ropes, as well as other types of abrasion, must be avoided because even modest loads will cause a rapid friction sawing motion that leads to deterioration. Descriptions of two types of polyolefin ropes follow:

- Polypropylene rope, with a specific gravity of 0.91 and a softening point of 300°F (150°C), is made in several sizes of filaments and from film with or without longitudinal fracturing. Polypropylene rope is about 50% stronger than Manila rope, size for size. Pure polypropylene rope has relatively poor rendering properties. It burns at 330°F (166°C) and loses some strength at 150°F (66°C).
- Polyethylene rope, with a specific gravity of 0.95 and a softening point of 250°F (120°C), is characteristically slippery and has very little springiness. It is strong and has little stretch. Polyethylene rope also has a comparatively low softening point and low coefficient of friction.

Composite Rope

Rope made by combining several types of synthetic fibers or by combining synthetic and natural fibers is also available. Composite rope results from attempts to give the surface of the rope or strand more wear resistance, greater internal tensile strength, or more structural strength to retain its shape. Composite rope can be made to match the requirements of specific jobs.

Other Types of Rope

Rope made of paper, glass, acrylic, rayon, polyvinyl chloride, fluorocarbon, rubber, cellulose acetate, and polyurethane is also available. These types of ropes enjoy only a small percentage of the market for reasons of cost, limited use, or short supply.

Working Load

Table 14–A lists linear density, new rope tensile strength, safety factor, and working load for Manila, sisal, and synthetic ropes. Table 14–B gives the same information for double-braided nylon rope. Because the safety factor is not the same for all ropes and is based on static loading, exercise caution when using this number.

Also be cautious when using the working load figures. Rope use, rope condition, exposure to several factors affecting rope behavior, and the degree of risk to life and property vary widely. Therefore, it is impossible to make blanket recommendations as to working loads. However, in order to provide general guidelines, working loads are tabulated for rope (1) in good condition, (2) with appropriate splices in noncritical applications, (3) under normal service conditions, and (4) under very modest dynamic loads.

Select a higher working load only with expert knowledge of conditions and a professional estimate of the risks involved. Factors to consider include (1) whether the rope

TABLE 14–A. Specifications for Synthetic and Natural Fiber Rope

Three-Strand Laid and Eight-Strand Plaited—Standard Construction

Nominal Size		New Manila Rope				New Polypropylene Rope				New Polyester Rope			
Diameter	Circum.	Linear Density[1] (lbs/100 ft)	Tensile Strength[2] (lbs)	Safety Factor	Working Load[3] (lbs)	Linear Density[1] (lbs/100 ft)	Tensile Strength[2] (lbs)	Safety Factor	Working Load[3] (lbs)	Linear Density[1] (lbs/100 ft)	Tensile Strength[2] (lbs)	Safety Factor	Working Load[3] (lbs)
3/16	5/8	1.50	406	10	41	.70	720	10	72	1.20	900	10	90
1/4	3/4	2.00	540	10	54	1.20	1,130	10	113	2.00	1,490	10	149
5/16	1	2.90	900	10	90	1.80	1,710	10	171	3.10	2,300	10	230
3/8	1 1/8	4.10	1,220	10	122	2.80	2,440	10	244	4.50	3,340	10	334
7/16	1 1/4	5.25	1,580	9	176	3.80	3,160	9	352	6.20	4,500	9	500
1/2	1 1/2	7.50	2,380	9	264	4.70	3,780	9	420	8.00	5,750	9	640
9/16	1 3/4	10.4	3,100	8	388	6.10	4,600	8	575	10.2	7,200	8	900
5/8	2	13.3	3,960	8	496	7.50	5,600	8	700	13.0	9,000	8	1,130
3/4	2 1/4	16.7	4,860	7	695	10.7	7,650	7	1,090	17.5	11,300	7	1,610
13/16	2 1/2	19.5	5,850	7	835	12.7	8,900	7	1,270	21.0	14,000	7	2,000
7/8	2 3/4	22.4	6,950	7	995	15.0	10,400	7	1,490	25.0	16,200	7	2,320
1	3	27.0	8,100	7	1,160	18.0	12,600	7	1,800	30.4	19,800	7	2,820
1 1/16	3 1/4	31.2	9,450	7	1,350	20.4	14,400	7	2,060	34.4	23,000	7	3,280
1 1/8	3 1/2	36.0	10,800	7	1,540	23.8	16,500	7	2,360	40.0	26,600	7	3,800
1 1/4	3 3/4	41.6	12,200	7	1,740	27.0	18,900	7	2,700	46.2	29,800	7	4,260
1 5/16	4	47.8	13,500	7	1,930	30.4	21,200	7	3,020	52.5	33,800	7	4,820
1 1/2	4 1/2	60.0	16,700	7	2,380	38.4	26,800	7	3,820	67.0	42,200	7	6,050
1 5/8	5	74.5	20,200	7	2,880	47.6	32,400	7	4,620	82.0	51,500	7	7,350
1 3/4	5 1/2	89.5	23,800	7	3,400	59.0	38,800	7	5,500	98.0	61,000	7	8,700
2	6	108	28,000	7	4,000	69.0	46,800	7	6,700	118	72,000	7	10,300
2 1/8	6 1/2	125	32,400	7	4,620	80.0	55,000	7	7,850	135	83,000	7	11,900
2 1/4	7	146	37,000	7	5,300	92.0	62,000	7	8,850	157	96,500	7	13,800
2 1/2	7 1/2	167	41,800	7	5,950	107	72,000	7	10,300	181	110,000	7	15,700
2 5/8	8	191	46,800	7	6,700	120	81,000	7	11,600	204	123,000	7	17,600
2 7/8	8 1/2	215	52,000	7	7,450	137	91,000	7	13,000	230	139,000	7	19,900
3	9	242	57,500	7	8,200	153	103,000	7	14,700	258	157,000	7	22,400
3 1/4	10	298	69,500	7	9,950	190	123,000	7	17,600	318	189,000	7	27,000
3 1/2	11	366	82,000	7	11,700	232	146,000	7	20,800	384	228,000	7	32,600
4	12	434	94,500	7	13,500	276	171,000	7	24,400	454	270,000	7	38,600

[1]Linear density (lbs/100 ft) shown is "average." Maximum is 5% higher.

[2]New rope tensile strengths are based on tests of new and unused rope of standard construction in accordance with Cordage Institute Standard Test Methods.

[3]Working loads are for rope in good condition with appropriate splices, in non-critical applications, and under normal service conditions. Working loads should be exceeded only with expert knowledge of conditions and professional estimates of risk. Working loads should be reduced where life, limb, or valuable property are involved, or for exceptional service conditions such as shock, loads, sustained loads, etc.

[4]Composite rope. Materials and construction of this polyester/polypropylene composite rope conform to MIL-R-43942 and MIL-R-43952. For other composite ropes, consult the manufacturer.

Source: The Cordage Institute.

has been subject to dynamic loading or other excessive use; (2) whether it has been inspected and found to be in good condition; (3) whether it is to be used in a recommended manner; and (4) whether the application involves high temperatures, extended periods under load, or obvious dynamic loading, such as sudden drops, snubs, or pickups. For all such applications and for applications involving more severe conditions of exposure, or for recommendations on special applications, consult the manufacturer.

Many uses of rope involve serious risk of injury to personnel or of damage to valuable property. This risk is often obvious—for example, a heavy load supported above one or more workers. An equally dangerous situation occurs if personnel are in line with a rope that is under excessive tension. Should the rope fail, it may recoil with considerable force—especially if the rope is made of nylon. Workers should be warned against standing in line with the rope. In all cases where such risks are present, or if there is any question about loads or other conditions of use, greatly reduce the working load and properly inspect the rope. Consult

TABLE 14–A. (Continued)

Nominal Diameter (in.)	New Composite[4] Rope				New Nylon Rope				New Sisal Rope			
	Linear Density[1] (lbs/100 ft)	Tensile Strength[2] (lbs)	Safety Factor	Working Load[3] (lbs)	Linear Density[1] (lbs/100 ft)	Tensile Strength[2] (lbs)	Safety Factor	Working Load[3] (lbs)	Linear Density[1] (lbs/100 ft)	Tensile Strength[2] (lbs)	Safety Factor	Working Load[3] (lbs)
3/16	.94	720	10	72	1.00	900	12	75	1.50	360	10	36
1/4	1.61	1,130	10	113	1.50	1,490	12	124	2.00	480	10	48
5/16	2.48	1,710	10	171	2.50	2,300	12	192	2.90	800	10	80
3/8	3.60	2,440	10	244	3.50	3,340	12	278	4.10	1,080	10	108
7/16	5.00	3,160	9	352	5.00	4,500	11	410	5.26	1,400	9	156
1/2	6.50	3,960	9	440	6.50	5,750	11	525	7.52	2,120	9	236
9/16	8.0	4,860	8	610	8.15	7,200	10	720	10.4	2,760	8	345
5/8	9.50	5,760	8	720	10.5	9,350	10	935	13.3	3,520	8	440
3/4	12.5	7,560	7	1,080	14.5	12,800	9	1,420	16.7	4,320	7	617
13/16	15.2	9,180	7	1,310	17.0	15,300	9	1,700	19.5	5,200	7	743
7/8	18.0	10,800	7	1,540	20.0	18,000	9	2,000	22.5	6,160	7	880
1	21.8	13,100	7	1,870	26.4	22,600	9	2,250	27.0	7,200	7	1,030
1 1/16	25.6	15,200	7	2,170	29.0	26,000	9	2,880	31.3	8,400	7	1,200
1 1/8	29.0	17,400	7	2,490	34.0	29,800	9	3,320	36.0	9,600	7	1,370
1 1/4	33.4	19,800	7	2,830	40.0	33,800	9	3,760	41.7	10,800	7	1,540
1 5/16	35.6	21,200	7	3,020	45.0	38,800	9	4,320	47.8	12,000	7	1,710
1 1/2	45.0	26,800	7	3,820	55.0	47,800	9	5,320	59.9	14,800	7	2,110
1 5/8	55.5	32,400	7	4,620	66.5	58,500	9	6,500	74.6	18,000	7	2,570
1 3/4	66.5	38,800	7	5,500	83.0	70,000	9	7,800	89.3	21,200	7	3,030
2	78.0	46,800	7	6,700	95.0	83,000	9	9,200	108	24,800	7	3,540
2 1/8	92.0	55,000	7	7,850	109	95,500	9	10,600	—	—	7	—
2 1/4	105	62,000	7	8,850	120	113,000	9	12,600	146	32,800	7	4,690
2 1/2	122	72,000	7	10,300	149	126,000	9	14,000	—	—	7	—
2 5/8	138	81,000	7	11,600	168	146,000	9	16,200	191	41,600	7	5,940
2 7/8	155	91,000	7	13,000	189	162,000	9	18,000	—	—	7	—
3	174	103,000	7	14,700	210	180,000	9	20,000	242	51,200	7	7,300
3 1/4	210	123,000	7	17,600	264	226,000	9	25,200	299	61,600	7	8,800
3 1/2	256	146,000	7	20,800	312	270,000	9	30,000	—	—	7	—
4	300	171,000	7	24,400	380	324,000	9	36,000	435	84,000	7	12,000

the manufacturer for recommendations on working loads.

Dynamic loading voids the working load. Working load figures do not apply when rope is subject to significant dynamic loading. Whenever a load is picked up, stopped, moved, or swung, there is an increased force due to dynamic loading. The more rapidly or suddenly such actions occur, the greater this increase will be. In extreme cases, the force put on the rope may be two, three, or even more times the normal load involved, such as when picking up a tow on a slack line or using a rope to stop a falling object. Therefore, in applications such as towing lines, lifelines, safety lines, climbing ropes, and the like, working load as given in Tables 14–A and 14–B does not apply.

Dynamic effects are greater on a rope with little stretch, such as Manila, than on a rope with higher stretch, such as nylon. Dynamic effects are also greater on a shorter rope than on a longer one. The working load listed provides for very modest dynamic loads. This means that when a working load has been used to select a rope, the load must be handled slowly and smoothly to minimize dynamic effects and to avoid exceeding the provision for them.

Inspections

Before placing new rope in service, it should be thoroughly inspected along its entire length to determine that no part of it is damaged or defective. Any irregularity in its appearance is evidence of the possibility of degradation or weakness. Experts disagree on what determines when a rope should be removed from service. Synthetic rope damage is not always visible.

TABLE 14–B. Specifications for Double-Braided Nylon Rope

Diameter (in.)	Circum. (in.)	Linear Density[1] (lbs/100 ft)	New Rope Min. Tensile Strength[2] (lbs.)	Safety Factor	Working Load[3] (lbs)
¼	¾	1.56	1,650	11	150
5/16	1	2.44	2,570	11	234
⅜	1 ⅛	3.52	3,700	11	336
7/16	1 ¼	4.79	5,020	10	502
½	1 ½	6.25	6,550	10	655
9/16	1 ¾	7.91	8,270	9	919
⅝	2	9.77	10,200	9	1,130
¾	2 ¼	14.1	14,700	8	1,840
13/16	2 ½	16.5	17,200	8	2,150
⅞	2 ¾	19.1	19,900	8	2,490
1	3	25.0	26,000	8	3,250
1 1/16	3 ¼	28.2	29,300	8	3,660
1 ⅛	3 ½	31.6	32,800	8	4,100
1 ¼	3 ¾	39.1	40,600	8	5,080
1 5/16	4	43.1	44,700	8	5,590
1 ½	4 ½	47.3	49,000	8	6,130
1 ⅝	5	56.3	58,300	8	7,290
1 ¾	5 ½	66.0	68,300	8	8,540
2	6	76.6	79,200	8	9,900
2 ⅛	6 ½	100	103,000	8	12,900
2 ¼	7	113	117,000	8	14,600
2 ½	7 ½	127	131,000	7	18,700
2 ⅝	8	156	161,000	7	23,000
2 ⅞	8 ½	172	177,000	7	25,300
3	9	225	231,000	7	33,000
3 ¼	10	264	271,000	7	38,700
3 ½	11	329	338,000	7	48,300
4	12	400	410,000	7	58,600

[1]Linear density (lbs/100 ft) shown is "average." Maximum is 5% higher.
[2]New rope tensile strengths are based on tests of new and unused rope of standard construction in accordance with Cordage Institute Standard Test Methods.
[3]Working loads are for rope in good condition with appropriate splices, in non-critical applications, and under normal service conditions. Working loads should be exceeded only with expert knowledge of conditions and professional estimates of risk. Working loads should be reduced where life, limb, or valuable property are involved, or for exceptional service conditions such as shock, loads, sustained loads, etc.

If rope is being used under ordinary conditions, inspect it every 30 days. If it is used in critical applications, such as to support scaffolding on which employees work, inspect it more often. If used to connect a load to a material-handling device, OSHA requires that it be inspected each day before being used. Inspection involves examining the entire length of the rope, inch by inch, for wear, abrasions, powdered fiber between strands, broken or cut fibers, displacement of yarns or strands, variation in size or roundness of strands, discoloration, and rotting.

To inspect the inner fibers, untwist the rope in several places to see whether the inner yarns are bright, clear, and unspotted. Mildewed rope has a musty odor, and the inner fibers of the strands have a dark, stained appearance. Broken strands or broken yarns ordinarily are easy to identify. Dirt and sawdust-like material inside a rope, caused by chafing, indicate damage. In rope having a central core, the core should not break away in small pieces when examined. If it does, this is an indication that the rope has been overstrained. If exposed to acids, natural fiber ropes, such as Manila, and synthetic ropes should be scrapped or retired from critical operations. Visual inspections do not always reveal acid damage. A rope, like a chain, is only as strong as its weakest part—in the case of rope, its cross-section. If there is a visible core or core damage, replace or splice out the rope. When unsatisfactory conditions are found, destroy the rope or cut it up in short pieces to prevent its being used in hoisting. The short pieces may be used for other purposes.

Natural fiber rope loaded to more than 50% of its breaking strength will be permanently damaged; synthetics loaded to more than 65% may be damaged. Damage from overloading may be detected by examining the inside fibers. These will be broken into short lengths in proportion to the degree of overload. To make a good estimate of the strength of fibers, scratch the fibers with a fingernail—fibers of poor strength will readily part. This "fingernail test" is a quick test for chemical damage.

If the diameter of a rope is worn away more than 5%, replace the rope. In a small rope (up to ¾ in. [19 mm] in diameter), surface wear that has progressed to the center of the twisted element (yarn) may account for a loss of more than 80% of the rope's strength. In a rope with a diameter of ¾ in. (19 mm) or more, diameter surface wear may destroy the strength of the cover yarns, yet not affect the original strength of the core yarns. The remaining strength of the rope will depend on the proportion of the core yarns to the original total of yarns. If fiber samples can be secured from the rope, an estimate of the rope's strength can be made. Manually break the fiber samples, and estimate the distribution of fibers in a cross-section, quartered to allow for twist configuration.

Due to slippage on a supporting surface when under high tension, synthetic ropes sometimes melt on the surface and form a skin. This skin is evidence of wear and degradation. Rope having multifilament synthetic fiber on the surface will often "fuzz." Fuzzing results from minute fiber breakage. If a rope is very fuzzy, replace it and look for the source of abrasion.

Care of Fiber Rope in Use

Safe use of rope involves recognizing the effects of such factors as chafing, cutting, elasticity, diameter-strength ratio, and general anticipated mishandling. To keep rope in good condition, observe the following precautions:

- Do not drag rope. Dragging rope wears away the outer fibers. Sand or grit between the fibers cuts them and reduces the rope's strength.
- Handle twisted rope so it retains the amount of twist (called *balance*) that the rope seeks when free and relaxed. If rotating loads and improper coiling and uncoiling change the balance, restore it by properly twisting the rope at either end. Severe imbalance can cause permanent damage; localized overtwisting causes kinking or hocking.
- Kinking strains the rope and may overstress the fibers. It may be difficult to detect a weak spot made by a kink. To prevent a new rope from kinking while it is being uncoiled, lay the rope coil on the floor with the bottom end down. Pull the bottom end up through the coil, and unwind the rope counterclockwise. If it uncoils in the other direction, turn the coil of rope over, and pull the end out on the other side.
- Avoid sharp bends over an unyielding surface because this causes extreme tension on the fibers. To make a rope fast, select an object with a smooth, round surface of sufficient diameter. If the object does have sharp corners, pad the corners. To avoid excessive bending, use sheaves or surface curvatures of suitable size for the rope's diameter (Table 14–C).
- Splice lengths of rope that must be joined. Do not knot them. A well-made splice will retain up to 100% of the strength of the rope, but a knot retains only 50% (Table 14–D).
- Thoroughly dry rope that has become wet; otherwise, it will quickly deteriorate. Hang up a wet rope, or lay it in a loose coil in a dry place until thoroughly dry. Rope deteriorates more rapidly if it is alternately wet and dry than if it remains wet.
- Do not allow wet rope to freeze. If a wet rope has frozen, completely thaw it out before using it; otherwise, the frozen fibers will break as they resist bending.
- Do not use wet rope or rope reinforced with metallic strands near power lines or other electrical equipment. Use of such rope could inflict electric shock on workers.

Care of Fiber Rope in Storage

To maintain the strength of any rope, store it away from fumes, heat, chemicals, moisture, sunlight, and rodents. Store rope in a dry place where air can circulate freely about it. Air should not be extremely dry, however. Hang up small ropes. Lay larger ropes on gratings so air can get underneath and around them.

Do not store or use rope in an area containing acid or acidic fumes because the rope will quickly deteriorate. Signs of deterioration from acid are dark brown or black spots on the rope.

TABLE 14–C. Sheave Sizes for Fiber Ropes of Varying Thickness

Diameter of Rope (in.)	Diameter of Sheave (in.)
¾	6
⅞	7
1	8
1¼	10
1⅜	11

Conversion factor: 1 in. = 2.54 cm.

TABLE 14–D. Efficiency of Manila Fiber Rope with Splices, Hitches, and Knots

Jointure	Percent Efficiency
Full strength of dry rope	100
Eye splice over metal thimble	90
Short splice in rope	80
Timber hitch, round turn, half hitch	65
Bowline, slip knot, clove hitch	60
Square knot, weaver's knot, sheet bend	50
Flemish eye, overhead knot	45

Do not store rope unless it has been cleaned. Hang dirty rope in loops over a bar or beam, and then spray it with water to remove the dirt. The spray should not be so powerful that it forces the dirt into the fibers. After washing, allow the rope to dry, and then shake it to remove the rest of the dirt.

WIRE ROPE

Wire rope is more widely used than fiber rope. This is because wire rope has greater strength and durability under severe working conditions. The physical characteristics of wire rope do not change when used in varying environments. Wire rope has controlled and predictable stretch characteristics.

Types of Wire Ropes

Wire rope consists of steel wires, strands, and a core. The individual wires are cold drawn to predetermined size and breaking loads, according to required grades. Grades include iron, tractor, mild plow steel, plow steel, improved plow steel, and extra-improved plow steel. The wires are then laid together in various geometrical arrangements, according to construction requirements for strands and classifications of wire rope (6 × 19, 6 × 37, and so on).

To make a strand, carefully selected lengths of pitch or lay are used. These lengths have a definite ratio to the length of lay or pitch used in forming the finished wire rope. After the individual strands are made, the required number is coiled around the core, which supports the load-carrying strands. The core can be made of sisal or synthetic fiber, or it can be a metallic strand core or independent wire rope core (IWRC). The intended use of a wire rope determines the size, number, and arrangement of wires; the number of strands; the lay; and the type of core.

Classifications

The most widely used constructions of wire rope are six-strand ropes of these two classifications: 6 × 19 and 6 × 37. The 6 × 19 classification includes a variety of constructions, ranging from 15 to 26 wires per strand. Typical constructions are 6 × 19 Seale (Figure 14–2), 6 × 25 filler wire, and 6 × 19 Warrington. The 6 × 37 classification also covers a large number of designs and constructions, ranging from 27 to 49 wires per strand. Typical constructions of this classification are 6 × 41 filler wire (Figure 14–2), 6 × 37, 6 × 36 Warrington Seale, and 6 × 49 filler wire.

Generally speaking, the more wires there are per strand, the more flexible the wire rope is. However, the fewer wires there are per strand, the more abrasion resistant and crush resistant the rope is. In ropes with large diameters, 2 in. (6.4 cm) and larger, practically all wire rope is produced in the 6 × 37 or 6 × 61 class. Therefore, because of the large number of possible rope constructions available, exercise care to select the right one (Table 14–E). (See American Iron and Steel Institute [AISI], *Wire Rope Users Manual*, in References at the end of this chapter.)

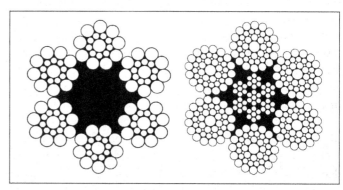

Figure 14–2. Wire rope is made from a number of individual wires grouped in strands, and then laid together over a core member (fiber core, IWRC, or strand core). The number of wires per strand and the number of strands per core depend on the expected working conditions and the amount of flexibility required. The cross-section at the left shows one construction of the 6 × 19 fiber core (FC) classification containing 114 wires. The cross-section at the right shows one construction of the 6 × 37 classification containing 343 wires, including those of the independent wire core (IWRC).

TABLE 14–E. Nominal Breaking Strength in Tons of Improved Plow Steel (IPS) and Extra-Improved Plow Steel (EIPS) Ropes

Diameter (in.)	6 × 9 & 6 × 37			8 × 19			19 × 7	
	IPS[1]		EIPS[2]	IPS		EIPS	IPS	EIPS
	FC[3]	IWRC[4]	IWRC	FC	IWRC	IWRC	FC	IWRC
⅜	6.10	6.56	7.55	5.24	5.76	6.63	5.59	6.15
⁷⁄₁₆	8.27	8.89	10.2	7.09	7.80	8.97	7.58	8.33
½	10.7	11.5	13.3	9.23	10.1	11.6	9.85	10.8
⁹⁄₁₆	13.5	14.5	16.8	11.6	12.8	14.7	12.4	13.6
⅝	16.7	17.9	20.6	14.3	15.7	18.1	15.3	16.8
¾	23.8	25.6	29.4	20.5	22.5	25.9	21.8	24.0
⅞	32.2	34.6	39.8	27.7	30.5	35.0	29.5	32.5
1	41.8	44.9	51.7	36.0	39.6	45.5	38.3	42.2
1 ⅛	52.6	56.5	65.0	45.3	49.8	57.3	48.2	53.1
1 ¼	64.6	69.4	79.9	55.7	61.3	70.5	59.2	65.1
1 ⅜	77.7	83.5	96.0	67.1	73.8	84.9	71.3	78.4
1 ½	92.0	98.9	114	79.4	87.3	100	84.4	92.8
1 ⅝	107	115	132					
1 ¾	124	133	153					
1 ⅞	141	152	174					
2	160	172	198					
2 ⅛	179	192	221					
2 ¼	200	215	247					

[1]IPS = Improved plow steel.
[2]EIPS = Extra-improved plow steel.
[3]FC = Fiber core.
[4]IWRC = Independent wire rope core.

Service Requirement

Depending on service requirements and conditions, six-strand ropes may have a fiber core (FC), a wire strand core (WSC), or an IWRC. Wires in the strands may be laid in the opposite direction (regular lay) from that of the strands in the rope, or they may be laid in the same direction (lang lay) as those of the strands in the rope.

Where maximum flexibility is required, eight-strand hoisting ropes are used. They are usually of the 8 × 19 classification, with regular lay and an FC or an IWRC. Where such flexibility is not required, as in guy wires and highway guards, wire rope of 6 × 7 construction (six strands with seven wires per strand) is suitable. When selecting wire rope for a particular job, consult engineers from reliable wire rope manufacturers.

Some conditions require rope with special qualities. Fiber cores are affected by temperatures above 250°F (120°C). Under such conditions, a metallic core provides greater efficiency and safety. In wire rope with IWRC, the wire rope core can add 7.5% to the strength of the rope compared to a fiber core (Dickie 1975, 8). A zinc-coated or stainless-steel wire rope effectively resists some types of corrosion. Refer to rope manufacturers for information on specific corrosion problems.

Because preformed wire rope does not unravel, it has advantages for certain services, such as for slings to hoist heavy construction equipment. Preformed wire rope is less likely to set or kink than other types of wire rope; thus, broken wires are less likely to protrude and create a hazard to workers.

Design Factors for Rope Used in Hoisting

The operating or design factors for rope used in hoisting are calculated by dividing the nominal catalog strength of the rope by the sum of the maximum loads to be hoisted. It is normal practice to base this sum on static loads. For rope used in hoisting in mines, the maximum loads to be hoisted include the weight of the skip or car or cage plus the weight of the material plus the weight of the suspended rope when the skip or cage is at the lowest point in a shaft. In some cases, acceleration stresses are also considered. It is recommended that hoisting rope have at least the strength of improved plow steel. For some applications, use extra-improved plow steel, which is the strongest steel, to provide an adequate design factor and better service.

The minimum design factors for rope used in hoisting depend on the type of service required and the federal, state or provincial, or local codes covering the particular hoisting operation. Many of these codes describe exactly how the design and operating factors should be figured. Therefore, check what codes are in force before making a final selection of wire ropes to be used in hoisting. Also, obtain the advice of a reliable wire rope manufacturer.

Inspections and Replacement

The frequency of inspections and replacement of wire rope depend on service conditions. In the United States, OSHA minimum inspection requirements for wire rope or cable include installation and yearly inspections with documentation when used as a lifting device. At regular intervals, a specially trained inspector should examine ropes on which human life depends and document such inspections. Some facilities and mines, for instance, make a daily inspection for readily observable defects, such as kinking and loose wires, and a thorough inspection weekly. For the latter

WIRE ROPE WEAR AND DAMAGE

The evidence in these illustrations will aid the inspector in determining the actual cause of wear or damage in any wire rope.

A wire rope which has been kinked. A kink is caused by pulling down a loop in a slack line during improper handling, installation, or operation. Note the distortion of the strands and individual wires. Early rope failure will undoubtedly occur at this point.

Localized wear over an equalizing sheave. The danger of this type wear is that it is not visible during operation of the rope. This emphasizes the need of regular inspection of this portion of an operating rope.

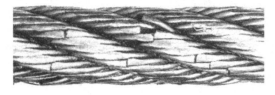

A typical failure of a rotary drill line with a poor cut-off practice. These wires have been subjected to excessive peening causing fatigue type failures. A predetermined, regularly scheduled, cut-off practice will go far toward eliminating this type of break.

A single strand removed from a wire rope subjected to "strand nicking." This condition is the result of adjacent strands rubbing against one another and is usually caused by core failure due to continued operation of a rope under high tensile load. The ultimate result will be individual wire breaks in the valleys of the strands.

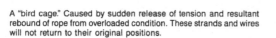

A "bird cage." Caused by sudden release of tension and resultant rebound of rope from overloaded condition. These strands and wires will not return to their original positions.

An example of a wire rope with a high strand—a condition in which one or two strands are worn before adjoining strands. This is caused by improper socketing or seizing, kinks or dog legs. Picture A is a close-up of the concentration of wear and B shows how it recurs in every sixth strand (in a six-strand rope).

Figure 14–3. Typical characteristics and causes of broken wires in wire ropes. *(Courtesy Wire Rope Corp. of America Inc.)*

inspection, the rope speed is generally less than 60 feet per minute (fpm).

The inspector checks specifically for wear of the crown wires, kinking, high strands, corrosion, loose wires, nicking, and lubrication (Figure 14–3). Rope calipers (Figure 14–4) and micrometers are used to determine changes in the cross-section of rope at various locations. In most cases, a sudden change in rope length and/or diameter is a warning that the wire rope is nearing the end of its useful life and that it should be removed from service. The reason for this change is general deterioration of the structure of the interior rope, such as corrosion of wires that cannot be inspected and general deterioration of the core. A decrease in the rope's diameter is often difficult to determine.

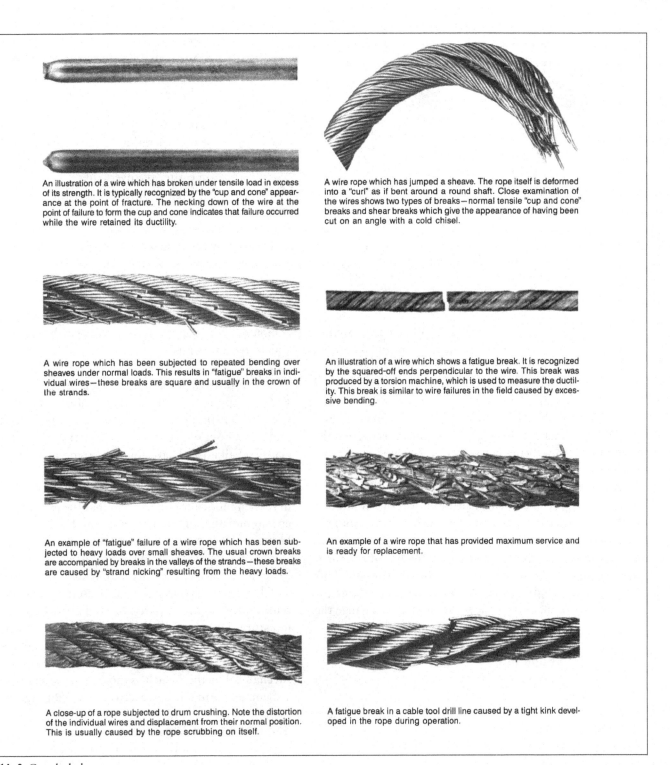

Figure 14–3. Concluded.

The number of broken wires per lay is one of the principal bases for judging the condition of a rope. If most of the broken wires in a lay are concentrated in several strands, that section of the rope is weaker than it would be if the broken wires were uniformly distributed throughout all strands and along the length of the rope. If, however, the number of broken wires along the length of a rope increases rapidly between inspections, the rope is becoming fatigued and is nearing the end of its useful life.

Inspection codes may vary from state to state or from province to province with regard to rope inspection and allowable degrees of deterioration. Usually, replacement is based on the number of broken wires per strand in one rope lay or on the number of broken wires per rope lay in

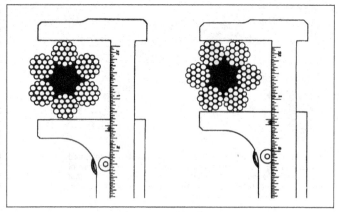

Figure 14–4. Always read the widest diameter when measuring wire rope. The correct way is shown on the left; the wrong way is shown on the right. *(Courtesy Armco Steel Corporation)*

all strands. (For running ropes, OSHA currently specifies 3 broken wires in one strand in one rope lay and 6 random broken wires in one rope lay as unacceptable. For wire rope slings, they currently specify 5 broken wires in one strand in one rope lay and 10 random broken wires in one rope lay as unacceptable.) Consult state or provincial codes and the OSHA regulations for specific information covering the type of operation being performed. Electronic inspection devices are available for determining loss of strength due to corrosion, loss of metallic area, and broken wires.

Applying the specified inspection criteria, combined with the length of time the rope has been in service and the previous experience of the rope, determine when it should be replaced. At intervals throughout the life of the rope, a short section should be cut off at the socket end. This practice has two purposes: (1) to remove wires damaged by vibration dampened at the socket and (2) to change the positions of critical wear points throughout the system.

Care of Wire Rope in Use

Deterioration of wire ropes has a number of causes, which vary considerably in importance depending on the conditions of service (Figure 14–3). For example, corrosion often is the principal cause of deterioration of wire rope used for hoisting in wet mine shafts. Moisture and the presence of acid in the water lead to corrosion. Corrosion, particularly of the interior wires, is indicated by pitting. Corrosion accelerates wear. This highly dangerous condition is difficult to detect. Other causes of deterioration include the following:

- Wear, particularly on the crown or outside wires, can result from contact with sheaves and drums.
- Kinks are acquired from improper installation of a new rope or from hoisting with slack in the rope. A kink cannot be removed without creating a weak place.
- Fatigue, indicated by a square fracture at the end of a wire, can be caused by bending stresses from sheaves and drums with small radii; by stresses from whipping, vibration, and pounding; or by torsion stresses.
- Drying out of lubrication is often hastened by heat and operating pressure.
- Overloading, including dynamic overloading, can damage wire rope if acceleration and deceleration are factors of importance. Such damage may not become known until some time after the overload.
- Overwinding, when rope length is greater than the drum can accommodate in a single layer, can cause heavy abrasion and excessive wear at crossover points. However, successful overwinding can be achieved by using specially engineered drum grooving.
- Mechanical abuse, such as running over wire rope with equipment and permitting obstructions to remain in a rope's path of travel, can ruin the rope. It is more common for wire rope to be thrown away because of abuse than because of use.

When possible, clean a wire rope before lubricating it. Regular application of a suitable lubricant to wire rope used for hoisting prevents corrosion, wear from friction, and drying out of the core. Good lubricants are free from acids and alkali and have adhesive strength. They also have the ability to penetrate the strands. The lubricant should be insoluble under the prevailing conditions. Ropes should be dry when lubricant is applied so that the lubricant will not entrap moisture. Thin lubricants can be applied by hand. However, it is better to apply them by providing some means of dripping them on the rope or using a spray device to apply the proper quantity automatically.

Clean wire rope monthly, as is done in mine shafts, to remove dirt, abrasive particles, and corrosion-producing moisture. Do not use cleaning fluids on wire rope; they harm the core's lubricant. Light oils are sometimes used to loosen the coating of lubricant and harmful materials. Mechanical methods such as those using compressed air or a steam jet clean a rope effectively and thoroughly.

Sheaves and Drums

Fatigue of wire rope resulting from bending stresses depends on the diameter of drums and sheaves: the larger the diameter, the more favorable a rope's service life will be. However, sometimes operators sacrifice diameter to accommodate designs and considerations for other equipment. Consider the recommendations in Table 14–F as a design base only. Many types of equipment successfully operate with smaller drum-and-sheave rope ratios, while others use much larger ratios. A case in point is the drum-and-sheave requirements in most mining codes; elevator codes; and shovel, hoist, and crane codes.

TABLE 14–F. Recommended Tread Diameters of Sheaves and Drums for Wire Rope

Rope Classification	Average Recommended (times rope diameter)	Minimum
6 × 7	72	42
6 × 19	45	30
6 × 37	27	18
8 × 19	31	21

Printed from ANSI/ASME A17.1, Elevators, Escalators, and Moving Walks.

TABLE 14–G. Groove Diameter in Relation to Wire Rope Diameter

	Amount that Groove Diameter Should Be Larger than Nominal Rope Diameter (inches)	
Rope Size (inches)	Used	New
¼ and ⁵⁄₁₆	¹⁄₁₂₈	¹⁄₆₄
⅜ to ¾ incl.	¹⁄₆₄	¹⁄₃₂
¹³⁄₁₆ to 1 ⅛ incl.	³⁄₁₂₈	³⁄₆₄
1 ³⁄₁₆ to 1 ½	¹⁄₃₂	¹⁄₁₆
1 ⁹⁄₁₆ to 2 ¼ incl.	³⁄₆₄	³⁄₃₂
2 ⁵⁄₁₆ and larger	¹⁄₁₆	⅛

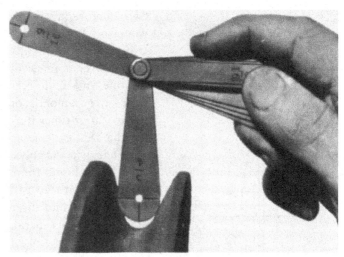

Figure 14–5. Check sheave grooves with a gauge designed with one-half the allowable oversize. If light is seen between the gauge and sheave, replace the sheave. *(Courtesy Armco Steel Corporation)*

The safety and the service life of installations for hoisting rope can be greatly increased by using sheaves and drums of suitable size and design and by properly lubricating them. The rope and the hoisting equipment also require good maintenance to assist in the safety and service life of the rope.

Heads, idlers, knuckles, curved sheaves, and grooved drums must have grooves that support a rope properly. Before installing a new rope, inspect the grooves, and, where necessary, machine them to proper contour and groove diameter. The diameter should exceed the nominal rope diameter by the amount shown in Table 14–G. For recommended grooving for drums and sheaves, consult a wire rope manufacturer's handbook.

Sheaves

The condition and contour of sheave grooves is important for the service life of wire rope. Periodically check sheave grooves (Figure 14–5), and do not let them wear to a diameter smaller than those shown for used grooves in Table 14–G. If the grooves become more worn than this, expect a reduction in the rope's service life. Reconditioned sheave grooves should conform to the tolerance shown in Table 14–G for new or remachined grooves.

On all new sheaves, ensure that the grooves are made for the size of rope specified. The bottom of the groove should have a 150-degree arc of support, and the sides of the groove should be tangent to the ends of the bottom arc. The depth of the groove should be 1 times the nominal diameter of the rope. The radius of the arc should be one-half the nominal rope diameter plus one-half the value shown in Table 14–G for new or remachined grooves.

Check sheaves for proper alignment when they are installed. During rope changes, check the sheaves for worn bearings, broken flanges, proper groove size, smoothness, and contour. Recondition or replace heavily worn or damaged sheaves.

Sheave groove-bearing pressures can become very high, depending on operating conditions and rope loadings. High pressures can cause excessive sheave wear and shorten the life of wire rope. It is necessary, therefore, to consider this factor and to select proper sheave materials and liners at the time of installation. For information on this subject, consult wire rope manufacturers' handbooks. (See also AISI, *Wire Rope Users Manual*, in References at the end of this chapter.)

Drums

Avoid multiple-layer winding of rope on drums, if possible. Multiple layering causes the rope to wear, thus shortening the rope's life, particularly at the point where the rope rises to the next layer. Where practical, use drums with enough diameter and length that they can take all the rope in a single layer.

Minimize crushing and excessive wear of wire rope by using spirally grooved drums that can accommodate one layer of rope. In any case, limit the number of layers to three. Rope lifters at the flanges are recommended when

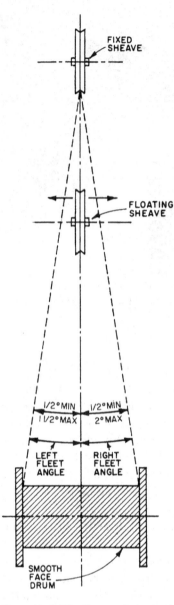

Figure 14–6. The fleet angle is graphically defined in this illustration of wire rope running from a fixed sheave, over a floating sheave, and then on to a smooth drum.

two or more layers are wound on drums. To distribute wear uniformly at crossover points, cut off one-and-a-quarter wraps every 6 months or three or four times during the life of the rope. According to OSHA, in no case should there be fewer than two full wraps on a drum; three are preferred. In general, avoid reverse bending of wire rope (bending first in one direction and then in the opposite) over sheaves or drums. This wears out the rope faster.

Correct fleet angle is important for even, efficient winding of wire rope. The fleet angle is the included angle between the rope winding on the drum and a line perpendicular to the drum shaft and running through the head or lead sheave (Figure 14–6).

To reduce any tendency for the rope to open-wind, do not let the fleet angle exceed 1°30'. To ensure that the rope starts back on the next layer, use a minimum angle of 0°30' for smooth drums and 2° for grooved drums. Adhering to these specifications helps achieve uniform winding on smooth-faced drums and also increases the winding efficiency of grooved drums. For smooth-faced drums, proper direction of lay of rope for specified winding conditions helps achieve uniform winding.

Installing a wire rope on a plain-faced or smooth-faced drum requires great care. The starting position should be at the drum end so that each turn of the rope winds tightly against the preceding turn (Figure 14–7). Maintain close supervision during the entire installation process to make sure of the following:
1. The rope is properly attached to the drum.

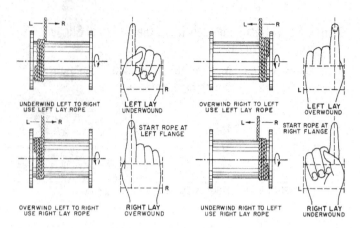

Figure 14–7. By holding the left or right hand with the index finger extended, palm up or palm down, the proper procedure for installing left- or right-lay rope on a smooth drum can be easily determined. *(Courtesy American Iron and Steel Institute)*

2. Appropriate tension on the rope is maintained as it is wound on the drum.
3. Each turn is guided as close to the preceding turn as possible so that there are no gaps between turns.
4. There are at least two dead turns on the drum when the rope is fully unwound during normal operating cycles.

Loose and uneven winding on a plain-faced or smooth-faced drum usually causes excessive wear, crushing, and distortion of the rope. Such abuse results in lower operating performance and a loss in the rope's effective strength. Also, on jobs that require moving and spotting a load, the operator will encounter control difficulties because the rope will pile up, pull into the pile, and fall from the pile to the drum surface. The ensuing shock can break or otherwise damage the rope.

Wire Rope Fittings

There are several ways to attach wire rope to fittings: by using pressed fittings, mechanical sleeve splices, hand-tucked splices, clips and clamps, sockets, or knots. Fittings are important for safety because they develop from 75% to 100% of the breaking strength of the rope. Manufacturers specify fittings of suitable size and design for ropes of different sizes. The strength of an attachment is attained only when the connection is made exactly according to the manufacturer's instructions (Figure 14–7). Some types of attachments, such as pressed fittings and mechanical sleeve splices that are used in making slings, must be made at either a wire rope manufacturer's facility or at a properly equipped commercial sling shop.

Efficiencies of properly made hand-tucked splices vary according to the splicer's ability and the rope's diameter, but can be as high as 90% (Figure 14–8). The efficiency of mechanical sleeve splices varies from 90% to 95% when IWRC wire rope is used.

Rope often is connected to the fittings of conveyances by means of clips and clamps. Clips and clamps are rated to develop 75% to 80% of the rope's breaking strength. Figure 14–8 shows how the clips should be attached, and Table 14–H gives the number of clips and the spacing required for ropes of different sizes. It is important to retighten the nuts on all clips after the rope's first load-carrying use as well as at all subsequent regular inspection periods.

Socketing with zinc and a thermostatic plastic resin will develop 100% of the rope's breaking strength. Figure 14–9 shows zinc-poured and swaged sockets. Because there is no

TABLE 14–H. Number of Spacing of Clips for Ropes of Various Sizes

Diameter of Rope (inches)	Minimum Number of Clips	Length of Rope Turned Back Exclusive of Eye (inches)	Torque (ft-lb)
1/8	2	3 1/4	—
1/4	2	3 1/4	—
1/2	3	11 1/2	65
5/8	3	12	95
3/4	4	18	130
7/8	4	19	225
1	5	26	225
1 1/8	6	34	225
1 1/4	7	44	360
1 3/8	7	44	360
1 1/2	8	54	360
1 5/8	8	58	430
1 3/4	8	61	590
2	8	71	750
2 1/4	8	73	750

1 in. = 2.54 cm.

1 ft-lb = 1.36 newton-meter.

The number of clips is based upon using right regular of Lang lay wire rope, 6 × 19 class or 6 × 37 class, fiber core of IWRC, IPS, or EIPS. If Seale construction or similar large outer wire type construction in the 6 × 19 class is to be used for sizes 1 in. (2.5 cm) and larger, add one additional clip.

The number of clips shown also applies to right regular lay wire rope, 8 × 19 class, fiber core, IPS, nominal sizes 1 1/2 in. and smaller; and right regular lay wire rope 18 × 7 class, fiber core, IPS or EIPS, nominal sizes 1 3/4 in. and smaller.

For other classes of wire rope not mentioned above, it may be necessary to add additional clips to the number shown.

Turn back the specified amount of rope from the thimble. Apply the first clip one base width from the dead end of the wire rope (U-bolt over dead end—live end rests in clip saddle). Tighten nuts evenly to recommended torque. Apply the next clip as near the loop as possible. Turn on nuts—take up rope slack—tighten all nuts evenly on all clips to recommended torque.

Note: Apply the initial load and retighten nuts to the recommended torque. The rope will stretch and shrink in diameter when loads are applied. Inspect periodically and retighten.

The efficiency rating of a properly prepared termination for clip sizes 1/8 to 7/8 in. is approximately 80 percent and for sizes 1 to 3 in. is approximately 90 percent. This rating is based on the catalog breaking strength of wire rope. If a pulley is used in place of a thimble for turning back the rope, add one additional clip.

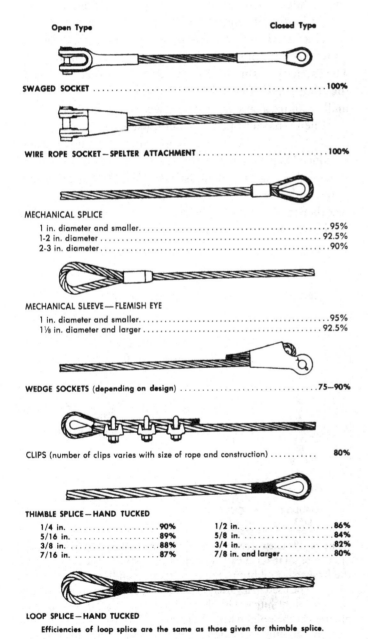

Figure 14–8. Typical efficiencies of attaching wire rope to fittings in percentages of strength of rope.

ready way to detect flaws in the finished job, follow the recommended procedure exactly. In high-speed hoisting, fatigue is especially likely to develop with this type of attachment. Therefore, the section adjacent to the conveyance should be cut off and discarded at frequent intervals. Some state mining laws include a required interval of every 6 months.

Square knots and other types of knots have low and unpredictable efficiencies of 40% or less. Using them will

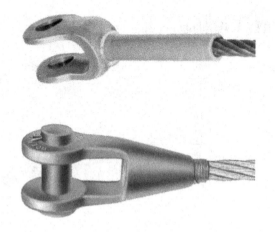

Figure 14–9. The end fittings should be of the best possible type for the specific use. Zinc-poured sockets (bottom) are efficient in straight tension but are not as fatigue resistant as swaged sockets (top).

likely result in the failure of a rope assembly and, under certain conditions, can result in a serious accident. In the United States, OSHA regulations and other industrial and construction codes prohibit the use of knots in wire rope.

RIGGING

In lifting various materials and supplies, a number of standard chokers, slings, bridle hitches, and basket hitches can be used. Because loads vary in physical dimension, shape, and weight, a rigger needs to know what method of attachment can be used safely. It is estimated that 15% to 35% of crane accidents may involve improper rigging.

Employers must see that the personnel responsible for rigging loads receive thorough training. Riggers must (1) know the load, (2) judge distances, (3) properly select tackle and lifting gear, and (4) direct the operation.

The most important job of any lifting operation is rigging the load. Poor rigging may result in injury to personnel, property damage, and other hazards. Rigging is the most time-consuming of any lifting operation and represents the greatest hazard potential of a lifting operation. The single most important rigging precaution is to determine the weight of the load before attempting to lift it. The weight of the load will, in turn, determine the lifting device (such as a crane) and the rigging gear to be used. It is also important to rig a load so that it will be stable—that is, motionless while being lifted. Properly maintaining, storing, and protecting the rigging gear will increase its life and safety.

Choker Hitch

Figure 14–10 shows this simplest of sling hitches. The sling passes entirely around the load, and the other end attaches to the hook. Due to stress created at the choke point, the choker hitch, used singly or with others, has a capacity of about 75% when the choke angle is 120 degrees. A choker hitch should always be pulled tight before the lift, never during the lift.

Basket Hitch

This hitch can be made with the same sling by passing the choker sling under the load, both loop eyes going to the crane hook (Figure 14–10). Slings are used in pairs when the entire load is suspended and singly when one end of the load must be raised. Rated capacities vary. A safe figure when two slings are used is to double the load rating of a choker hitch where square corners are encountered. Reduce the load rating if sling legs deflect from a vertical position.

A basket hitch with a cradle configuration of 90 degrees allows each leg of the sling to function as a separate sling. The capacity of this sling is twice that of a vertical sling if the sling angle of each leg is 90 degrees. This would normally require a spreader bar or two lifting devices.

When a basket hitch has a cradle configuration of less than 90 degrees, the sling capacity is reduced. How much it is reduced depends on the sling angle. The rated capacity of a 30-degree basket is about half that of a 90-degree basket. Sling angles less than 30 degrees are greatly discouraged. Another method of calculating the load rating of a basket hitch is to double the load rating of a single directly con-

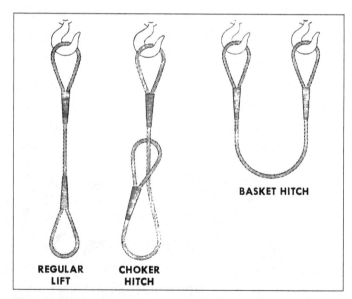

Figure 14–10. The three most common hitches for all types of slings are the regular (also called straight or vertical), the choker, and the basket. In addition, the grommet (or endless loop) type of sling can be used in a variety of configurations. The most efficient is the straight hitch. The choker hitch places a certain amount of stress at the point where the end loop encircles (or chokes) the body of the sling. Although the basket hitch results in two legs, the result is not a doubling of load capacity of the sling. *(Courtesy Bethlehem Steel Co.)*

nected sling. Reduce this figure for sharp bends or other conditions that do not occur in a vertical sling. On most slings when a basket hitch is used, the rated capacities at the various angles to the vertical apply (Tables 14–I and 14–J), provided that the radius of curvature where the sling makes contact with the load is at least 20 times the diameter of the individual component rope. Shorter radii or contact with any sharp corner will reduce the sling capacity accordingly.

FIBER AND WIRE ROPE SLINGS

The safety of a rope sling assembly depends on the following factors: the material used (fiber rope or wire rope), fittings of suitable strength for the load, the method of fastening the rope to the fittings, the type of sling (such as single-legged or three-legged), the type of hitch, and regular inspection and maintenance. Keep these factors in mind when using rope slings.

Types of Fiber and Wire Rope Slings

Because the strength of fiber rope is affected by chemicals, freezing, high temperatures, and sharp bends, consider these factors when selecting rope for slings. OSHA 29 CFR 1910.184(h) stipulates, "Only fiber rope slings made from

new rope shall be used. Use of repaired or reconditioned fiber rope slings is prohibited." Fiber rope is particularly suitable for handling loads that might be damaged by contact with metal slings.

Wire rope slings are usually made of extra-improved plow steel rope. If this grade of wire rope is unavailable, improved plow steel rope is used. The difference between the two grades is 15%. Normally, wire rope of IWRC construction is used in extra-improved plow steel and improved plow steel slings where mechanical loop endings or swaged or pressed-on terminations are used. In the smaller wire rope diameters up to and including 1 in. (2.54 cm), use the 6 × 19 classification wire rope. For rope diameters larger than 1 in. (2.54 cm), use the 6 × 37 classification rope. The most popular type of sling in use is the strand-laid sling made from the 6 × 19 or the 6 × 37 wire rope constructions.

Another popular type of wire rope sling is the cable-laid sling. Made from multiple wire ropes laid into one rope structure, cable-laid slings offer greater flexibility than strand-laid slings. Braided slings (Figure 14–11) consist of a number of ropes braided into a single unit. They are used where flexibility, high strength, and resistance to rotation are essential. Because braided slings are braided in an open manner, they are fairly easy to inspect.

TABLE 14–I. Rated Capacity Limits (in tons) of Wire Rope Slings, Using Preformed Improved Plow Steel Wire Rope (depending on method of attaching the rope to the fittings)

Rope Diameter (inches)	Single Leg						Two-Leg Bridle or Basket Hitch											
	Vertical			Choker			Vertical*			30 Degrees Vertical			45 Degrees			60 Degrees Vertical		
	S	MS	HT	S	MS	HT	S	MS	HT	S	MS	HT	S	MS	HT	S	MS	HT
6 × 19 Classification Construction																		
⅜	1.3	1.2	1.2	.92	.92	.92	2.6	2.4	2.4	2.3	2.1	2.0	1.8	1.7	1.7	1.3	1.2	1.2
½	2.3	2.2	2.0	1.6	1.6	1.6	4.6	4.4	4.0	4.0	3.8	3.5	3.3	3.1	2.8	2.3	2.2	2.0
⅝	3.6	3.4	3.0	2.5	2.5	2.5	7.2	6.8	6.0	6.2	5.9	5.2	5.1	4.8	4.2	3.6	3.4	3.0
¾	5.1	4.9	4.2	3.6	3.6	3.6	10.0	9.8	8.4	8.7	8.5	7.3	7.1	6.9	5.9	5.0	4.9	4.2
⅞	6.9	6.6	5.5	4.8	4.8	4.8	14.0	13.0	11.0	12.0	11.0	9.5	9.9	9.3	7.8	7.0	6.6	5.5
1	9.0	8.5	7.2	6.3	6.3	6.3	18.0	17.0	14.0	16.0	15.0	12.0	13.0	12.0	10.0	9.0	8.5	7.2
1⅛	11.0	10.0	9.0	7.9	7.9	7.9	22.0	20.0	18.0	19.0	17.0	16.0	16.0	14.0	13.0	11.0	10.0	9.0
6 × 37 Classification Construction																		
1¼	14	13	11	9.7	9.7	9.7	28	26	22	24	23	19	20	18	16	14	13	11
1⅜	17	15	13	12	12	12	34	30	26	29	26	23	24	21	18	17	15	13
1½	20	18	16	14	14	14	40	36	32	35	31	28	28	25	23	20	18	16
1¾	27	25	21	19	19	19	54	50	42	47	43	36	38	35	30	27	25	21
2	34	32	28	24	24	24	68	64	56	59	55	48	48	45	40	34	32	28
2¼	43	40	34	30	30	30	86	80	68	74	69	59	61	57	48	43	40	34

*If slings are used to handle loads with sharp corners, pads or saddles should be used to protect the rope. The radius of bend should not be smaller than five times the diameter of the rope. If the radius of the bend is smaller, a choker hitch rating should be used.

S = Socket or swaged terminal attachment

MS = Mechanical sleeve attachment

HT = Hand-tucked splice attachment

Note: Table is based on a design factor of 5, sling angles formed by one leg and a vertical line through the crane hook, and uniform loading. For 3-leg bridle slings, multiply safe load limits for 2-leg bridle slings by 1.5, and for 4-leg bridle slings, multiply by 2.0.

TABLE 14–J. Rated Capacity Limits (in tons) of Wire Rope Slings, Using Preformed Extra-Improved Plow Steel (depending on method of attaching the rope to the fittings)

Rope Diameter (inches)	Single Leg							Two-Leg Bridle or Basket Hitch											
	Vertical			Choker			Vertical*			30 Degrees Vertical			45 Degrees			60 Degrees Vertical		
	S	MS	HT	S	MS	HT	S	MS	HT	S	MS	HT	S	MS	HT	S	MS	HT
6 × 19 Classification Construction																		
3/8	1.5	1.4	1.3	1.1	1.1	1.1	3.0	2.8	2.6	2.6	2.4	2.3	2.1	1.0	1.8	1.5	1.4	1.3
1/2	2.7	2.5	2.3	1.9	1.9	1.9	5.4	5.0	4.6	4.7	4.3	4.0	3.8	3.5	3.3	2.7	2.5	2.3
5/8	4.1	3.9	3.5	2.9	2.9	2.9	8.2	7.8	7.0	7.1	6.8	6.1	5.8	5.5	4.9	4.1	3.9	3.5
3/4	5.9	5.6	4.8	4.1	4.1	4.1	12.0	11.0	9.6	10.0	9.7	8.3	8.3	7.9	6.8	5.9	5.6	4.8
7/8	8.0	7.6	6.4	5.6	5.6	5.6	16.0	15.0	13.0	14.0	13.0	11.0	11.0	11.0	9.0	8.0	7.6	6.4
1	10.0	9.8	8.3	7.2	7.2	7.2	20.0	20.0	17.0	17.0	17.0	14.0	14.0	14.0	12.0	10.0	9.8	8.3
1 1/8	13.0	12.0	10.0	9.1	9.1	9.1	26.0	24.0	20.0	23.0	21.0	17.0	18.0	17.0	14.0	13.0	12.0	10.0
6 × 37 Classification Construction																		
1 1/4	16	15	13	11	11	11	32	30	26	38	26	23	23	21	18	16	15	13
1 3/8	19	18	15	13	13	13	38	36	30	33	31	26	27	25	21	19	18	15
1 1/2	23	21	18	16	16	16	46	42	36	40	36	31	33	30	25	23	21	18
1 3/4	31	28	24	21	21	21	62	56	48	54	49	42	44	40	35	31	28	24
2	40	37	32	28	28	28	80	74	64	69	64	55	57	52	45	40	37	32
2 1/4	49	46	40	35	35	35	98	92	80	85	80	69	69	65	57	49	46	40

*If slings are used to handle loads with sharp corners, pads or saddles should be used to protect the rope. The radius of bend should not be smaller than five times the diameter of the rope. If the radius of the bend is smaller, a choker hitch rating should be used.

S = Socket or swaged terminal attachment
MS = Mechanical sleeve attachment
HT = Hand-tucked splice attachment

Note: Table is based on a design factor of 5, sling angles formed by one leg and a vertical line through the crane hook, and uniform loading. For 3-leg bridle slings, multiply safe load limits for 2-leg bridle slings by 1.5, and for 4-leg bridle slings, multiply by 2.0.

Figure 14–11. Braided slings are resistant to kinking. Be sure loads are hoisted uniformly and that all slings have a minimum safety factor of 5.

Methods of Attachment

All hooks and rings used as sling connections should develop the full rated capacity of the wire rope sling. Sockets and compression fittings, when properly attached, should develop 100% of the rated strength for wire rope. Swaged-sleeve sling endings should develop 92% to 95% of the wire rope's strength. Compression fittings and swaged-sleeve fittings are available from wire rope manufacturers or from any properly equipped sling shop.

Hand-tucked splices develop about 90% of the rope's strength in rope diameters less than 1/2 in. (1.27 cm) and 80% for larger diameters. There are tables in the OSHA standards that specify the capacities with respect to hand-tucked splices. Fittings used with hand-tucked slings should develop the same strength efficiencies as those used with mechanical slings. The recommended load rating for a sling assembly is usually based on one-fifth the calculated strength of the assembly. However, there may be cases where engineered lifts are made that do not meet this value.

Working Load

The rated load capacities as given in various sling catalogs and tables are based on newly manufactured slings. As the sling is used, factors such as abrasion, nicking, distortion, corrosion, and bending around small radii will affect the load rating. Consider these factors before lifts are made.

Here are two tips to increase the wear of wire rope slings:
- If loads that have sharp edges or sharp corners must be lifted, use pads or saddles to protect the ropes or chains.
- Thimbles spliced in the ends of slings will materially reduce wear.

Because slings can be used at various angles and because the rope stress increases rapidly with the angle of lift, it is essential to keep this in mind when ordering slings. Fortunately, most catalogs for wire rope slings have tables that give the load ratings for the most used and most critical angles of lift. Consult these tables for safe rigging practices.

When the rope is made into a sling and placed in position on a load, determine and carefully consider the angle formed by the ropes and the horizontal. The rated load capacity of the sling decreases sharply as the angle formed by the sling's leg and the horizontal becomes smaller. When this angle is 45 degrees, the rated load capacity has decreased to 71% of the load that can be lifted when the legs are vertical. As this angle decreases, the rated load capacity continues to decrease (Tables 14–I and 14–J).

Figure 14–12 shows how tension on a leg of a sling increases as the angle decreases from the vertical. When the angle formed by a leg and the vertical is 30 degrees, the rated load capacity is only 87% of that if both legs were vertical. For an angle of 60 degrees, the rated load capacity is only 50% of that if both legs were vertical. These losses are proportional to the cosine of the sling's angle with the vertical. The actual stress is equal to the amount of the load that a leg must support, divided by the cosine of the angle that the leg is from the vertical. To avoid excessive angles, use longer slings, if head room permits.

Rated load capacities tables for slings should be posted throughout the shop. Also, each sling should be tagged to indicate its rated load capacity (Figure 14–13).

Also consider the rated load capacity for different types of hitches. For example, a decrease of at least 25% in rated load capacity for a single-legged vertical sling occurs when a choker hitch is used. The suitable load for a basket hitch is based on the angle of the legs. For strand-laid and braided wire rope slings used in a basket hitch, the minimum diameter of curvature of the sling in contact with the

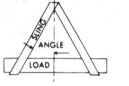

ANGLE STRENGTH LOSS FROM RATED CAPACITY

ANGLE	FACTOR			FACTOR	ANGLE
70°	.3420			.2588	75°
60°	.5000			.4226	65°
50°	.6428			.5736	55°
40°	.7660			.7071	45°
30°	.8660			.8192	35°
20°	.9397			.9063	25°
10°	.9848			.9659	15°
0°	1.0000			.9962	5°

The increased angle of the sling leg reduces its capacity. See chart for loss factor. Determine the angle between the sling leg and the vertical plane. Then multiply the sling rating by the appropriate loss factor from the chart. This will determine the sling's reduced rating.

EXAMPLE:
Assume sling capacity 2,000#
If angle = 50° then loss factor = .6428
Multiply: 2,000# × .6428
1,286# = rated capacity of sling at 50°

Figure 14–12. Increasing the angle between the sling leg and vertical increases the stress on each leg of the sling and reduces its capacity. *(Courtesy Web Sling Association)*

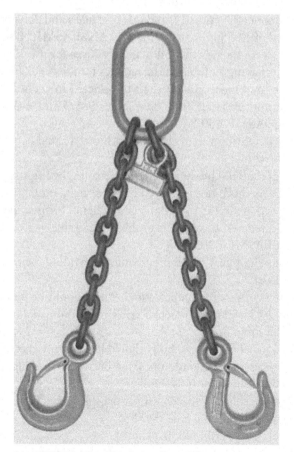

Figure 14–13. Typical double-chain sling. All components, such as the oblong master link, the body chain, and the hook, are carefully matched for compatibility. Note permanent identification tag. *(Courtesy Columbus McKinnon Corporation)*

load should be at least 20 times the rope's diameter; for cable-laid slings, at least 10 times.

Special clamps are used to handle steel plate, flanged castings, and similarly shaped products. Slings using horizontal and vertical clamps have the same rated load capacity as other bridle slings, provided the strength of the clamp is equal to the other components of the sling.

Inspections

Train employees to check slings daily, during a lift, and whenever they suspect damage after a lift. Employees should promptly report any questionable conditions in the equipment or in the assembly. In the United States, use OSHA inspection requirements for industrial slings as a guide in evaluating the sling's condition. A trained and competent person must make a thorough inspection at least every 12 months according to OSHA and more frequently according to some manufacturers. Documentation of such inspections should be maintained.

Immediately withdraw from service slings that fail inspection requirements. Make them unusable by further destroying them before they are discarded.

Safe Operating Practices for Slings

The American National Standards Institute's and American Society of Mechanical Engineers' (ANSI/ASME) B30.9, Slings, recommends the following practices for all slings:

- Slings having suitable characteristics for the type of load, hitch, and environment shall be selected in accordance with appropriate tables. (See Tables 9–5.3 and 9–5.5 in ANSI/ASME B30.9.)
- The weight of a load shall be within the rated capacity of the sling.
- Slings shall not be shortened or lengthened by knotting or other methods not approved by the sling manufacturer.
- Slings that appear to be damaged shall not be used unless inspected and accepted as usable under Table 9–5.6. (See ANSI/ASME B30.9.)
- Slings shall be hitched in a manner providing control of the load.
- Sharp corners in contact with a sling should be padded with material of sufficient strength to minimize damage to the sling.
- Portions of the human body should be kept from between the sling and the load, and from between the sling and the crane hook or hoist hook.
- Personnel should stand clear of the suspended load.
- Personnel shall not ride the sling.
- Shock loading should be avoided.
- Slings should not be pulled from under a load when the load is resting on the sling.
- Slings should be stored in a cool, dry, and dark place to prevent environmental damage.

- Twisting and kinking the leg branches shall be avoided.
- A load applied to the hook should be centered in the bowl of the hook to prevent point loading on the hook.
- During lifting, with or without a load, personnel shall be alert for possible snagging.
- In a basket hitch, the load should be balanced to prevent slippage (Figure 14-10). When using a basket hitch, the sling's legs should contain or support the load from the sides above the center of gravity.
- Slings should be long enough so that the rated capacity is adequate when the angle of the legs is taken into consideration.
- Slings should not be dragged on the floor or over an abrasive surface.
- In a choker hitch, slings shall be long enough so the choker fitting chokes on the webbing and never on the other fitting (Figure 14–10).
- Nylon and polyester slings shall not be used at temperatures above 194°F (90°C).
- When extensive exposure to sunlight or ultraviolet light is experienced by nylon or polyester web slings, the sling manufacturer should be consulted for recommended inspection procedure because of loss in strength.

CHAINS AND CHAIN SLINGS

The safety of a chain sling assembly depends on the following factors: the kind of material used, the strength of the material for the load, the method of fastening the chain to its fittings, and proper inspection and maintenance. Consider these factors when using chain slings.

Types of Chain Slings

Alloy steel has become the standard material for chain slings. Chain made from alloy steel has high resistance to abrasion and is practically immune to failure because the metal is cold worked. Special-purpose alloy chains are made from stainless steel, Monel metal, bronze, and other materials. They are designed for use where resistance to corrosive substances is required, or where other special properties are desirable.

Proof coil chain, also known as common or hardware chain, is used for miscellaneous purposes where failure of the chain would not endanger human life or result in serious damage to property or equipment. Never use proof coil chain for slings.

Properties and Working Load of Alloy Steel Chain

Alloy steel chain is produced from heat-treatable alloy steel in accord with American Society for Testing and Materials (ASTM) Specifications for Alloy Steel Chains, A391–1975.

After heat treatment, this chain has the following mechanical properties:

Tensile strength = 115,000 psi minimum

Elongation = 15% minimum

The tensile strength of alloy steel chain increases in proportion to its hardness (produced by heat treating). Resistance to abrasion also increases proportionally with hardness.

Table 14–K shows the recommended working load limits, proof test loads, and minimum breaking strengths of alloy steel chain. The working load limit (safe load strength) is arrived at by dividing the breaking strength (ultimate strength) by a specified safety factor. Any number of chains can be used to lift equipment (Figure 14–14).

The values shown in Table 14–K represent the maximum loads that should be applied in direct tension to a length of alloy steel chain. Prior to final inspection and shipment, all alloy steel chain is tested in direct tension under the proof test loads shown in the table. (See the National Association of Chain Manufacturers in References at the end of this chapter for data on other types of chains.)

Impact conditions caused by faulty hitches, bumpy crane tracks, and slipping hookups can add materially to the stress in the chain. Despite efforts to avoid it, if severe impact loading may be encountered, use a chain with a higher working load limit or reduce the weight of the load where possible.

Alloy steel chains are suitable for high-temperature operations. However, continuous operation at a temperature of 800°F (425°C) (the highest temperature for which continuous operation is recommended) requires a reduction of 40% in the regular working load limit. These chains may serve at temperatures up to 1,000°F (540°C) at 50% of the regular working load limit, but only for intermittent service. However, the working load limit of the chains must be permanently reduced by 25% after they have served at this high temperature. On the other hand, the general strength and working load limits of alloy steel chains are not altered appreciably by low temperatures. See Table 14–L on the effect of elevated temperature on the working limit of alloy chain.

Hooks and Attachments

As a general rule, hooks, rings, oblong links, pear-shaped links, coupling links, and other attachments should be made of heat-treatable alloy steel identical or equivalent to that of the chain itself. In most cases, attachments will be installed on the chain by the chain manufacturer, who will then heat-treat and proof-test the assembly.

If emergency conditions make it necessary for users to replace an attachment, they should select the grade and size with extreme care. Use high-strength, heat-treatable, alloy-connecting links of the same type as those used by the chain manufacturer. Do not use unalloyed carbon-steel hooks, repair links, rings, pear-shaped links, or other attachments. OSHA regulation 29 CFR 1910.184 (e)(2)(ii) states, "Makeshift links or fasteners formed from bolts or rods, or other such attachments, shall not be used."

Standard items produced from alloy steel include sling hooks, grab hooks, foundry hooks, grab links, rings, oblong links, pear-shaped links, and repair links. All such attachments, used with the recommended chain size, provide a safety factor equal to or greater than that of alloy steel chain itself. Specifications for the dimensions of these attachments will vary somewhat with the individual manufacturer.

Other useful items are handles. Many injuries have resulted from employees catching their fingers between the hook attachment and the load. To prevent such injuries, attach handles to the assembly hook or end attachment. To increase operating efficiency, use handles on large hooks, master links, and other attachments.

Inspections

Following an inspection procedure can reveal most of the causes of chain failures before failure occurs. Chain slings require three types of inspection:

Figure 14–14. Double, triple, or quad chain slings, as in this application, can be rigged to handle loads of virtually any shape or size. *(Courtesy Campbell Chain Company)*

TABLE 14–K(a). Working Load Limits, Proof Test Loads, and Minimum Breaking Loads for Grade 80 Alloy Steel Chain

Nominal Chain Size		Material Diameter		Working Load Limit (Max.)		Proof Test* (Min.)		Minimum Breaking Force*		Inside Length (Max.)		Inside Width Range	
in.	mm	in.	mm	lbs	kg	lbs	kN	lbs	kN	in.	mm	in.	mm
7/32	5.5	0.217	5.5	2,100	970	4,200	19.0	8,400	38.0	0.69	17.6	0.281 – 0.325	7.14 – 8.25
9/32	7.0	0.276	7.0	3,500	1,570	7,000	30.8	14,000	61.6	0.90	22.9	0.375 – 0.430	9.53 – 10.92
5/16	8.0	0.315	8.0	4,500	2,000	9,000	40.3	18,000	80.6	1.04	26.4	0.430 – 0.500	10.92 – 12.70
3/8	10.0	0.394	10.0	7,100	3,200	14,200	63.0	28,400	126.0	1.26	32.0	0.512 – 0.600	13.00 – 15.20
1/2	13.0	0.512	13.0	12,000	5,400	24,000	107.0	48,000	214.0	1.64	41.6	0.688 – 0.768	17.48 – 19.50
5/8	16.0	0.630	16.0	18,100	8,200	36,200	161.0	72,400	322.0	2.02	51.2	0.812 – 0.945	20.63 – 24.00
3/4	20.0	0.787	20.0	28,300	12,800	56,600	252.0	113,200	504.0	2.52	64.0	0.984 – 1.180	25.00 – 30.00
7/8	22.0	0.866	22.0	34,200	15,500	68,400	305.0	136,800	610.0	2.77	70.4	1.080 – 1.300	27.50 – 33.00
1	26.0	1.020	26.0	47,700	21,600	95,400	425.0	190,800	850.0	3.28	83.2	1.280 – 1.540	32.50 – 39.00
1 1/4	32.0	1.260	32.0	72,300	32,800	144,600	644.0	289,200	1,288.0	4.03	102.4	1.580 – 1.890	40.00 – 48.00

*The proof test and minimum breaking force loads shall not be used as criteria for service and design purposes. Courtesy of National Association of Chain Manufacturers, adopted September 2005.

TABLE 14–K(b). Working Load Limits, Proof Test Loads, and Minimum Breaking Loads for Grade 100 Alloy Steel Chain

Nominal Chain Size		Material Diameter		Working Load Limit (Max.)		Proof Test* (Min.)		Minimum Breaking Force*		Inside Length (Max.)		Inside Width Range	
in.	mm	in.	mm	lbs	kg	lbs	kN	lbs	kN	in.	mm	in.	mm
7/32	5.5	0.217	5.5	2,700	1,200	5,400	23.8	10,800	47.6	0.69	17.6	0.281 – 0.325	7.14 – 8.25
9/32	7.0	0.276	7.0	4,300	1,950	8,600	38.5	17,200	77.0	0.90	22.9	0.375 – 0.430	9.53 – 10.92
5/16	8.0	0.315	8.0	5,700	2,600	11,400	51.0	22,800	102.0	1.04	26.4	0.430 – 0.500	10.92 – 12.70
3/8	10.0	0.394	10.0	8,800	4,000	17,600	79.0	35,200	158.0	1.26	32.0	0.512 – 0.600	13.00 – 15.20
1/2	13.0	0.512	13.0	15,000	6,800	30,000	134.0	60,000	268.0	1.64	41.6	0.688 – 0.768	17.48 – 19.50
5/8	16.0	0.630	16.0	22,600	10,300	45,200	201.0	90,400	402.0	2.02	51.2	0.812 – 0.945	20.63 – 24.00
3/4	20.0	0.787	20.0	35,300	16,000	70,600	315.0	141,200	630.0	2.52	64.0	0.984 – 1.180	25.00 – 30.00
7/8	22.0	0.866	22.0	42,700	19,400	85,400	381.0	170,800	762.0	2.77	70.4	1.080 – 1.300	27.50 – 33.00

*The proof test and minimum breaking force loads shall not be used as criteria for service and design purposes. Courtesy of National Association of Chain Manufacturers, adopted September 2005.

TABLE 14–L. The Effect of Elevated Temperature on the Working Load Limit of Alloy Chain

Temperature		Grade of Chain			
		Grade 80		Grade 100	
(°F)	(°C)	Reduction of Working Load Limit WHILE AT Temperature	Permanent Reduction of Working Load Limit AFTER EXPOSURE to Temperature	Reduction of Working Load Limit WHILE AT Temperature	Permanent Reduction of Working Load Limit AFTER EXPOSURE to Temperature
Below 400	Below 204	None	None	None	None
400	204	10%	None	15%	None
500	260	15%	None	25%	5%
600	316	20%	5%	30%	15%
700	371	30%	10%	40%	20%
800	427	40%	15%	50%	25%
900	482	50%	20%	60%	30%
1,000	538	60%	25%	70%	35%
Over 1,000	Over 538	OSHA 1910.184 requires all slings exposed to temperatures over 1,000°F to be removed from service			

1. *Initial inspections.* Both new and repaired slings should be inspected before use to determine (a) that each sling meets the requirements of the purchase order; (b) that it is the correct type and has the proper rated capacity for the application; and (c) that it has not been damaged in shipment, unpacking, or storage.
2. *Frequent inspections.* The sling should be inspected by the person handling it each time it is used.
3. *Periodic inspections.* A semiannual or more frequent inspection should be performed by a competent person who is experienced in the inspection of chain slings. The frequency of periodic inspections should be based on the following factors: frequency of use, severity of service conditions, and knowledge about the service life of slings used in present or similar conditions.

Documentation of such inspections should be maintained. The user of a chain should be able to detect links and hooks that have become visibly unsafe because of overloading, faulty rigging, or other unsafe practices. The competent person should have the authority to remove damaged assemblies from service so they can be reconditioned or replaced.

The best way to detect wear and stretching is by a visual, link-by-link inspection. Overall measurements of sling length—and even measurements of 1- to 3-ft (30- to 91-cm) lengths—are inadequate because not all links are affected the same. Likewise, caliper readings of only certain links can also miss wear and stretching (Table 14–M and Figure 14–15). A link-by-link inspection should be made to detect the following:
- bent links

- cracks in weld areas, in shoulders, or in any other section of links
- transverse nicks and gouges
- corrosion pits
- stretching caused by overloading.

When inspecting the hook, measure between the shank and the narrowest point of the hook opening. Whenever the throat opening exceeds 15% of the normal opening, replace the hook. Pay special attention to slings to which hooks have been added; make sure the hooks are secure.

Safe Practices for Chain Slings

Follow these recognized safe practices to prevent chain failures:
- Purchase chain slings complete from the manufacturer. Whenever repairs are required, send them back to the manufacturer.
- Never anneal or normalize alloy steel chains and hooks. These processes reduce their hardness and, therefore, greatly reduce their strength.
- Never splice a chain by inserting a bolt between two links.
- Never put a strain on a kinked chain. Train workers to take up the slack slowly so they can see that every link in the chain seats properly.
- Do not use a hammer to force a hook over a chain link.
- Never remove the permanent identification tags that have been attached to chain slings by the manufacturer.
- Remember that decreasing the angle between the legs of a chain sling and the horizontal increases the load of the legs.

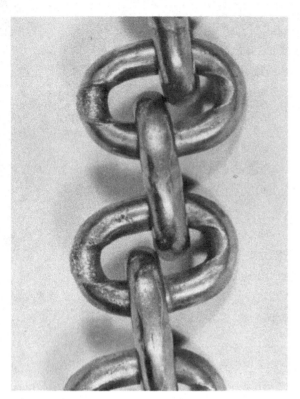

Figure 14–15. The links are turned to show the extreme wear at the bearing surfaces.

| TABLE 14–M. Maximum Allowable Wear at Any Point of Link ||
Size of Chain (in.)	Maximum Allowable Wear (in.)
1/4	3/64
3/8	5/64
1/2	7/64
5/8	9/64
3/4	5/32
7/8	11/64
1	3/16
1 1/8	7/32
1 1/4	1/4
1 3/8	9/32
1 1/2	5/16
1 3/4	11/32

- Use chain attachments (rings, shackles, couplings, and end links) designed for the chain to which they are fastened.
- See that the load is always properly set in the bowl of the hook. Loading on or toward the point (except in the case of grab hooks or others especially designed for the purpose) overloads the hook and leads to spreading and possible failure.

Figure 14–16. This "out-of-balance" load is secured by a double sling with chain leg adjusters (the two short chains with grab hooks attached to the master link). The chain leg passing through the bore of this casting has been protected from damage by adequate padding. *(Courtesy American Chain Division of Acco)*

- Store chains not in use in a suitable rack. Do not let them lie on the ground or floor where they can be damaged by lift trucks or other vehicles.
- Secure "out-of-balance" loads properly (Figure 14–16).

SYNTHETIC WEB SLINGS AND METAL MESH SLINGS

Two widely used slings are the synthetic web sling and metal mesh sling. They are strong and dependable slings if properly used.

Synthetic Web Slings

Nylon and polyester are the fibers most often used for synthetic web slings. Each has specific advantages and disadvantages as to stretch, strength, and chemical resistance. Synthetic web slings are useful for lifting loads that need their surfaces protected by the soft, supple web sling's

Type I. Triangle and choker end fittings usable in vertical, choker, and basket hitches.

Type III. Flat eye ends usable in vertical, choker, and basket hitches.

Type V. Endless (or grommet) usable in vertical, choker, and basket hitches.

Type II. Triangle fittings each end usable in vertical and basket hitches only.

Type IV. Twisted eye ends usable in vertical, choker, and basket hitches.

Type VI. Reversed (or return) eye. Essentially an endless sling, butted on the sides with wear pad(s) on body. Usable in vertical, choker, and basket hitches.

Figure 14–17. Basic synthetic web sling types.

surface. This usage is well suited for tubular, nonferrous, ceramic, painted, polished, and highly machined products with fine or delicate surfaces. To this end, a synthetic web sling's service life is secondary to load protection. Be warned, therefore, that synthetic web slings can be cut relatively easily and have little resistance to abrasion compared with chain or wire rope. Figure 14–17 shows several types of synthetic web slings.

Observe the following requirements when using synthetic web slings:

- According to ASTM B783-04, Specifications for Materials for Ferrous Powder Metallurgy Structural Parts, the minimum breaking strength for synthetic web slings should be five times the rated capacity (Tables 14–N and 14–O).
- Every synthetic web sling should bear, in a legible manner, the following identification information: the name or identification of the manufacturer, the sling's code number, the rated load capacities for usable types of hitches, and the type of material (such as polyester or nylon).
- When two slings, or one sling in a basket hitch (Figure 14–10), are used to lift a load from one crane hook, the sling's capacity is reduced. The load-carrying capacity of the sling is determined by applying the appropriate factor times the hitch's capacity.

Inspections of Synthetic Web Slings

As with chain slings, synthetic web slings should have three types of inspections:

1. *Initial inspections.* Both new and repaired slings should be inspected before use to determine (a) that each sling meets the requirements of the purchase order; (b) that it is the correct type and has the proper rated capacity for the application; and (c) that it has not been damaged in shipment, unpacking, or storage.

TABLE 14–N. Rated Capacity in Pounds for 1,600 lb/in. Web Slings

	Single-Ply Capacities for Various Type Slings in Vertical, Choker, and Vertical Basket Hitches								
	Types 1, 2, 3, & 4			Types 5 Endless			Type 6 Reversed Eye		
	Hitches			Hitches			Hitches		
Web Width in inches	Vertical	Choker	Basket	Vertical	Choker	Basket	Vertical	Choker	Basket
1	—	—	—	2,600	2,100	5,200	—	—	—
2	3,200	2,400	6,400	5,100	4,100	10,200	4,500	3,600	9,000
3	4,800	3,600	9,600	7,700	6,200	15,400	—	—	—
4	6,400	4,800	12,800	10,200	8,200	20,400	7,700	6,200	15,400
5	8,000	6,000	16,000	12,800	10,200	25,600	—	—	—
6	9,600	7,200	19,200	15,400	12,300	30,800	11,000	8,800	22,000

Rated capacities should never be exceeded.
See manufacturer's rated capacities for multiple-ply slings.

TABLE 14–O. Rated Capacity in Pounds for 1,200 lb/in. Web Slings

	Single-Ply Capacities for Various Type Slings in Vertical, Choker, and Vertical Basket Hitches								
	Types 1, 2, 3, & 4			Types 5 Endless			Type 6 Reversed Eye		
	Hitches			Hitches			Hitches		
Web Width in inches	Vertical	Choker	Basket	Vertical	Choker	Basket	Vertical	Choker	Basket
1	—	—	—	1,900	1,500	3,800	—	—	—
2	2,400	1,800	4,800	3,800	3,000	7,600	3,500	2,800	7,000
3	3,600	2,700	7,200	5,800	4,600	11,600	5,000	4,000	10,000
4	4,800	3,600	9,600	7,700	6,200	15,400	6,800	5,400	13,600
5	6,000	4,500	12,000	9,600	7,700	19,200	—	—	—
6	7,200	5,400	14,400	11,500	9,200	23,000	8,000	6,400	16,000

2. *Frequent inspections.* The sling should be inspected by the person handling it each time it is used.
3. *Periodic inspections.* A semiannual or more frequent inspection should be performed by a competent person who is experienced in the inspection of synthetic web slings. The frequency of periodic inspections should be based on the following factors: frequency of use, severity of service conditions, and knowledge about the service life of slings used in present or similar conditions.

Synthetic web slings are required by OSHA to be inspected each day before and during usage by a competent person [1910.184(d)]. It is recommended that a person trained to use web slings also be trained to competently inspect them.

For guidelines on inspecting synthetic slings for ultraviolet light damage, consult the manufacturer. Onsite inspections should identify the following kinds of wear or damage:

• excessive abrasive wear on webbing and any fittings
• cuts, tears, snags, punctures, holes, and crushed fabric
• worn or broken stitches, particularly that of the laps
• burns, charring, melting, or weld spatter damage
• acid, caustic, or other chemical damage
• broken, distorted, or excessively worn fittings
• knots that cause doubt about the sling's safety.

Keep records of all inspections. Such records should identify the sling, the dates of inspection, and the sling's condition at the time of each inspection. If the web sling was not manufactured with individual identification numbers, a system of identification should be devised to facilitate record keeping. Most manufacturers today will provide an ID system with new slings.

Although generally not repaired, web slings may be repaired by the manufacturer or qualified person, who should identify the work and certify the rated load capacity.

All repaired slings should be proof-tested to two times their newly rated load capacity. No temporary repairs should be made.

Metal Mesh Slings

Metal mesh slings can safely handle sharp-edged materials (Figure 14–18), concrete in its many prestressed forms, and high-temperature materials up to 500°F (260°C). Metal mesh slings are classified as heavy duty, medium duty, or light duty. Figure 14–19 shows the sling structure and terms to identify parts of a metal mesh sling.

All metal mesh slings should have their safe working load limits stamped on them for vertical, basket, and choker hitches (metal handle). They can be used efficiently with all weights that fall below that limit. The design factor of metal mesh slings is 5 to 1, or five times the amount stated on the sling. All metal mesh slings are proof-tested to a minimum of 200% of their rated load capacity. This removes all permanent stretching when used at the rated load capacity.

Safe Practices for Metal Mesh Slings

The safe use of metal mesh slings depends primarily on two factors: (1) use of the right sling for the right load and (2) the construction of the sling. Any danger in the use of metal mesh slings stems mainly from improper use.

• Use elastomer-coated slings only at temperatures up to 200°F (93°C).
• Damaging the slings at the edges by faulty loading or dragging a sling out from under a load may eventually cause wear to the spirals that compose the mesh. Such abuse will reduce the sling's wire diameter and thus require the sling to be taken out of service.
• Never shorten metal mesh slings by using knots, bolts, or other unapproved methods. If shortening becomes necessary, consult the sling manufacturer on how to do it.
• Tampering with the surface of any sling can weaken it and make it highly dangerous.

Figure 14-18. A four-legged basket-hitch sling of steel mesh can take the sharp edges of lumber without failure and without damaging the wood.

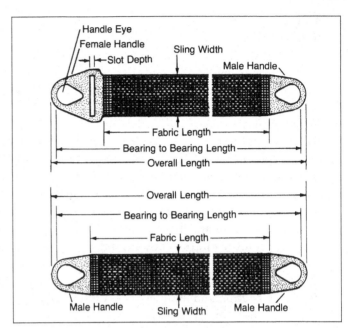

Figure 14-19. Structure and nomenclature of typical metal mesh slings.

- Never twist or kink the legs of a metal mesh sling or use one when the spirals are locked. A sudden jolt may break the spirals, cause the load to shift, and create havoc on the floor below. With metal mesh slings, as well as with all other slings, follow rules for certain hitches to ensure the safe handling of a load.

Inspections of Metal Mesh Slings

One of the most important precautions in the use of metal mesh slings is regular inspection by a qualified person. Base the frequency of inspections on the requirements of OSHA or other applicable governing body, as well as on the amount of use that a sling receives and the severity of conditions of use.

As with other types of slings, metal mesh slings should have three types of inspections:

1. *Initial inspections.* Both new and repaired slings should be inspected before use to determine (a) that each sling meets the requirements of the purchase order; (b) that it is the correct type and has the proper rated capacity for the application; and (c) that it has not been damaged in shipment, unpacking, or storage.
2. *Frequent inspections.* The sling should be inspected by the person handling it each time it is used.
3. *Periodic inspections.* A semiannual or more frequent inspection should be performed by a competent person who is experienced in the inspection of metal mesh slings. The frequency of periodic inspections should be based on the following factors: frequency of use, severity of service conditions, and knowledge about the service life of slings used in present or similar conditions.

Keep written inspection records that identify each sling and the items that were inspected.

Remove metal mesh slings from service if a broken weld or brazed joint is discovered along the sling edge. Also, watch for the following signs of wear. Any one of these conditions or a combination of them, if ignored, could eventually result in sling breakdown:

- broken wires in any part of the mesh
- a loss of 25% in wire diameter due to abrasion
- a lack of sling flexibility
- cracked end fitting
- visible distortion.

SUMMARY

- Workers should be trained in the special safety precautions required when using and storing various ropes, chains, and slings.
- Natural or synthetic fiber ropes are durable but tend to deteriorate when in contact with acids, caustics, and high temperatures. Nylon, polyester, and polyolefin ropes are more resistant to corrosive materials, wear and tear, and temperature extremes.
- Supervisors and workers must know the working load of the rope they are using to ensure safety. Dynamic effects

are greater on ropes with little stretch and can cause them to break, thus endangering workers.

- Ropes must be regularly inspected for wear or damage, and the results must be documented. Workers should be trained to handle ropes with care and store them properly, away from harmful substances.
- Wire rope has great strength and durability under severe working conditions and maintains its characteristics in widely different environments. Inspectors should check wire ropes for deterioration.
- Fiber rope slings are used for loads that might be damaged by contact with metal, while wire rope slings provide extra strength and durability.
- The safety of a chain sling assembly depends on the material used, its strength for the load handled, the method of attaching the chain to fittings, and proper inspection and maintenance. Chains should be inspected daily by workers and every 6 months by a trained professional.
- Synthetic web slings are useful for lifting loads that need their surfaces protected. Metal mesh slings can handle sharp-edged material, concrete, and high-temperature loads. Their safe use depends on the use of the right sling for the right load and on their construction.

REFERENCES

American Iron and Steel Institute, Committee of Wire Rope Producers, *Wire Rope Users Manual*. 1988.

American National Standards Institute, 11 West 42nd Street, New York, NY 10036.

Elevators, Escalators, and Moving Walks, ANSI/ASME A17.2–2007.

Safety Code for Cableways, Cranes, Derricks, Hoists, Hooks, Jacks, and Slings, ANSI/ASME B30 series, 1990–1993.

Slings, ANSI/ASME B30.9–2006.

American Petroleum Institute, 1220 L Street NW, Washington DC 20005.

Recommended Practice on Application, Care, and Use of Wire Rope for Oil-Field Service, Code No. API–RP–9B.

American Society for and Testing Materials, 1916 Race Street, Philadelphia, PA 19103.

Specification for Alloy Steel Chains, ASTM A391/A391M–07–2012.

Specifications for Materials for Ferrous Powder Metallurgy Structural Parts, ASTM B783-04–2013

Broderick & Bascom Rope Co., Rt. 3, Oak Grove Industrial Park, PO Box 844, Sedalia, MO 65301.

Rigger's Handbook. 1986.

Wire Rope Handbook. 1966.

Cordage Institute, 42 North Street, Hingham, MA 02043.

Dickie, D. E. *Rigging Manual*. Construction Safety Association of Ontario, 1975.

National Association of Chain Manufacturers, PO Box 3143, York, PA 17402.

Alloy Steel Chains Specifications, No. 3001.

National Fire Protection Association, 1 Batterymarch Park, Quincy, MA 02269.

Flammable and Combustible Liquids Code, NFPA 30, 2003.

National Electrical Code, NFPA 70, 2008.

National Safety Council, 1121 Spring Lake Drive, Itasca, IL 60143-3201.

Occupational Safety and Health Data Sheet: Recommended Loads for Wire Rope Slings, 12304–0380, 1991.

U.S. Department of the Interior, Bureau of Mines, 2401 E Street NW, Washington DC 20241.

Recommended Procedures for Mine Hoists and Shaft Installation, Inspection, and Maintenance, Information Circular 8031.

U.S. Department of Labor, 200 Constitution Avenue NW, Washington DC 20210.

29 CFR 1910.184 Subpart N, Materials Handling and Storage: Slings.

Wire Rope Technical Board, 801 North Fairfax Street, Suite 211, Alexandria, VA 22314-1757.

REVIEW QUESTIONS

1. Which of the following fiber ropes gives the best uniform strength and service?
   a. Manila or nylon
   b. polyester or rayon
   c. sisal or polyethylene
   d. polyolefin or henequen

2. The properties of _____ fiber make it the best-suited natural fiber for cordage, and it is often recommended for capstan work.

3. List the three major types of synthetic fiber ropes.
   a.
   b.
   c.

4. How often should fiber rope, being used under ordinary conditions, be inspected for damage?
   a. every 2 weeks
   b. once a month
   c. every 2 months
   d. none of the above
5. What is a quick way to make a good estimate of the strength of fibers in a rope and to test for chemical damage?
6. When lengths of fiber rope must be joined, a well-made splice will retain up to 100% of the strength of the rope, but a knot retains only
   a. 25%.
   b. 50%.
   c. 75%.
   d. 85%.
7. Wire rope is more widely used than fiber rope because
   a. wire rope has greater strength and durability under severe working conditions.
   b. the physical characteristics of wire rope do not change when used in varying environments.
   c. wire rope has controlled and predictable stretch characteristics.
   d. all of the above
   e. only a and c
8. Generally speaking, when there are more wires per strand, the wire rope is more
   a. flexible.
   b. crush resistant.
   c. abrasion resistant.
   d. all of the above
   e. only a and b
9. The minimum design factors for wire rope used in hoisting depend upon what two conditions?
   a.
   b.

10. List the causes of deterioration of wire ropes.
    a.
    b.
    c.
    d.
    e.
    f.
    g.
    h.
11. In the United States, OSHA regulations and other industrial and construction codes prohibit the use of _____ in wire rope.
12. In the United States, OSHA requires wire rope or cable to be inspected how often?
13. The safety of a rope sling assembly depends on what six factors?
    a.
    b.
    c.
    d.
    e.
    f.
14. Why has alloy steel become the standard material for chain slings?
15. What is the best way to detect wear and stretching of chains and chain slings?
    a. overall measurements of sling length
    b. even measurements of 1- to 3-ft (30- to 91-cm) lengths
    c. caliper readings of only certain links
    d. visual, link-by-link inspection
16. What is the difference between the usage of synthetic web slings and metal mesh slings?

Powered Industrial Trucks

15

David E. Marquette, CSP

Bonnie Steward

John F. Montgomery, PhD, CSP, CHMM

Introduction

Types of Trucks
Rider-Controlled Trucks ▸ Other Types of Trucks

General Safeguards
Safety Requirements ▸ Industrial Trucks in Hazardous Locations ▸ Masts, Backrests, and Overhead Protection ▸ Straddle Trucks ▸ Crane Trucks ▸ Tractors and Trailers ▸ Motorized Hand Trucks ▸ Automated Guided Vehicles

General Operating Guidelines
Speed ▸ Elevators, Bridge Plates, and Railroad Tracks ▸ Loading and Unloading ▸ Proper Care of Trucks ▸ Operator and Pedestrian Safety ▸ Maneuvering ▸ Driving on Grades ▸ Load Capacity ▸ Pallets

Inspection and Maintenance
Electric Trucks ▸ Gasoline-Operated Trucks ▸ Liquefied Petroleum Gas Trucks ▸ Storage and Handling of Liquefied Petroleum Gases

Operators
Selection ▸ Training

Summary

References

Review Questions

Appendices
Appendix A: OSHA FAQ ▸ Appendix B: Example Operator's Daily Checklist ▸ Appendix C: Example Operator Training Program Outline

477

INTRODUCTION

Factories, warehouses, docks, and transportation terminals use powered industrial trucks (PITs) to carry, push, pull, lift, stack, and tier material. Establishing safe practices for the operation, maintenance, and inspection of powered industrial trucks is essential. The topics covered in this chapter include:

- discussion of types of powered industrial trucks
- methods to reduce hazards associated with powered industrial truck operation
- general safety principles in operating powered industrial trucks
- safe practices in operating powered industrial trucks
- important issues in inspection and maintenance
- selection and training of powered industrial truck operators.

This chapter does not apply to nonflammable, compressed-gas-powered industrial trucks, farm vehicles, or vehicles intended for earth moving or over-the-road or highway use or hauling. Although this chapter also does not discuss powered industrial trucks employed in airports and air terminal areas, many of the same procedures apply. For information about trucks and operations in these areas, refer to the *Aviation Ground Operation Safety Handbook*, published by the National Safety Council. (See References at the end of this chapter.)

TYPES OF TRUCKS

Powered industrial trucks may be classified by power source, operator position, or means of engaging the load. Power sources include electric motors powered by storage batteries; engines using gasoline, liquefied petroleum gas (LPG), or diesel fuel; and trucks using a combination of gas, diesel, and/or electricity. Provisions for safe operation, maintenance, and design of powered industrial trucks should meet the design and construction requirements for powered industrial trucks established by the American National Standards Institute/Industrial Trucks Standards.

The OSHA standard for powered industrial truck training, found at 29 CFR 1910.178, requires employers to provide truck operators with training on a variety of topics. Among these topics are vehicle inspection and required maintenance. The OSHA standards for powered industrial trucks must be reviewed to ensure compliance, and all vehicles should bear a label or identifying mark indicating their classifications.

Rider-Controlled Trucks

One class of powered industrial truck is designed to be controlled by an operator who rides on the truck. The widely used lift truck, with its cantilevered load engager, vertical masts, and elevating mechanism, is considered a rider-controlled truck. Some rider-controlled trucks use a telescoping platform to engage the load. Styles of this truck are either high-lift trucks—with an elevating mechanism that permits the tiering of one load on another load (Figure 15–1)—or low-lift trucks—with a mechanism that raises the load only enough to permit horizontal movement.

Figure 15–1. A high-lift fork truck for pallet storage on racks proceeds down this aisle. *(Courtesy Allstate Insurance Company, Training Division)*

Figure 15–2. Straddle trucks are designed to carry loads of pipes, lumber, and other long materials. *(Printed with permission from Towmotor Corp.)*

Figure 15–3. The special clamp on this lift truck permits handling paper rolls without damaging them. *(Courtesy Clark Equipment Company, Industrial Truck Division)*

These are not the only types of rider-controlled trucks. Straddle carriers (Figure 15–2) carry long material, such as pipes or lumber, under the truck's body, which rides on four legs above the wheels. Powered industrial tractors draw trailers, nonpowered trucks, and other mobile loads. In some warehouse operations, order-picker trucks are used to raise the operator to the desired height.

The use of attachments has increased the versatility of today's powered industrial trucks. Clamps, rotators, shifters, stabilizers, pushers, pullers, up-enders, bottom dumpers, top lifters, rams, cranes, scoops, and other modifications have been engineered to meet specific application needs. Two or more motions have been built into some attachments. For example, an attachment could clamp and rotate, or it could side-shift and push and pull (Figure 15–3). It is possible to change attachments so that one truck can be used to handle various types of loads. Note that the truck's manufacturer must approve, in writing, any modification to the powered industrial truck.

Other Types of Trucks

Another category of powered industrial truck is the motorized hand truck, which is controlled by an operator who walks or rides behind the truck (Figure 15–4). It may have a platform or lifting forks to engage the load and may be either a high-lift truck, for tiering, or a low-lift truck, to raise the load only enough for horizontal movement.

A noteworthy powered industrial truck is the electronically controlled vehicle (ECV) or automated guided vehicle (AGV), which does not need an operator (Figure 15–5). These trucks travel over a prearranged route and are controlled by frequency sensors, a light beam, or induction tape that is placed on or embedded under the floor.

GENERAL SAFEGUARDS

Consider the worksite when purchasing or leasing powered industrial trucks. Working outdoors might necessitate lengthy travel in less-than-perfect conditions. Operators often have little control over their environments or situations and thus suffer fatigue from noisy, cramped compartments and strain from having to deal with blind spots or rough terrain. Aids proven to reduce fatigue and strain include the following: backup alarm lights or audible signals; headlights; turn signals; enhanced front and rear vision; noise-reducing insulation; fail-safe brakes; and comfortable, wraparound seats that provide protection similar to safety belts for the operator.

Some manufacturers offer operator-restraint systems on new vehicles, and other manufacturers can provide the system on request by retrofitting older lift trucks with operator-restraint systems.

Each company needs to develop its own policy of lift-truck safety-belt usage. In developing a policy, the operators' safety should be the most important concern. Consider such factors as forklift truck accident data, type and design of the facility, operators' duties, ease of entering and exiting the truck,

Figure 15–4. The walking operator controls this electric pallet-lift via handlebar-mounted controls. *(Courtesy Clark Equipment Company, Industrial Truck Division)*

Figure 15–5. Some automated guided vehicles should have clearly marked aisles that are free of any obstacles. *(Courtesy Litton)*

work area conditions, materials-handling requirements and exposure, and so forth. When the manufacturer has provided an operator safety-restraint system on the truck, that system should be used as specified in the owner's manual.

Safety Requirements

A powered industrial truck capable of lifting loads higher than the operator's head or operated in areas where there is a hazard from falling objects must be equipped with an overhead guard (see Figure 15–1). This guard should not interfere with good visibility. Be sure that the openings in the guard are small enough that the operator will not be struck by material falling from an overhead load or stack. Overhead guards should extend beyond the operator's position and should conform to ANSI/ITSDF B56.1–2012.

A load backrest extension should always be used when the load presents a hazard to the operator. The top of the load should not exceed the height of the backrest, in accordance with U.S. Occupational Safety and Health Administration (OSHA) standard 29 CFR 1910.178.

To prevent particles from being thrown at the operator, install guards over exposed tires. To protect the operator in the normal operating position, place guards over hazardous moving parts such as chain-and-sprocket drives and exposed gears.

Although lift trucks may come with steering wheel knobs, many companies prohibit the use of these knobs, and their use is not recommended. If the knobs are used, they should be of the mushroom type and engage the palm

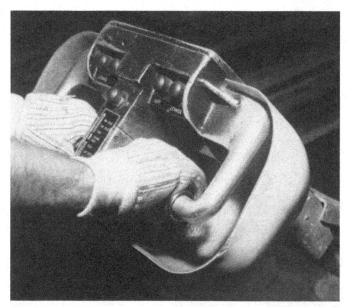

Figure 15–6. This guard protects the operator's hands from coming into contact with obstacles when the truck is maneuvered in tight spaces.

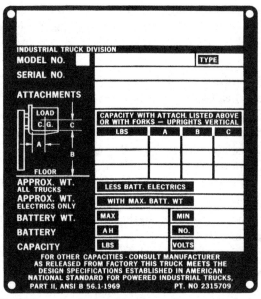

Figure 15–7. The manufacturer must attach a nameplate to every lift truck. This nameplate must include the weight of the truck, its rated capacity, and its model and serial numbers.

of the operator's hand in the horizontal position and should be mounted within the periphery of the steering wheel. The steering mechanism should also minimize the transmission of road shock to the steering wheel. Confine all steering controls within the perimeter outline of the truck. When this is not possible, provide guards to prevent injury to the operator when passing obstacles (Figure 15–6).

Powered industrial trucks should have horns or other warning devices that make a sound that is distinctive and loud enough to be distinguished from other noises. A backup alarm that signals whenever a truck backs up is a best practice. Where excessive noise could cause confusion, flashing lights mounted on overhead guards can effectively warn employees of approaching trucks.

Every powered industrial truck must display a nameplate indicating the weight of the truck and its rated capacity, as specified by ANSI/ITSDF B56.1–2012 (Figure 15–7).

Various parts of the vehicle may have limitations that also should be acknowledged by the operator. Specifications of steering, braking, and other controls should conform to ANSI/ITSDF B56.1–2012.

In addition, powered industrial trucks should be constructed and equipped to comply with the Underwriters Laboratories Standards for Safety, Nos. 558 and 583. These standards are specified in the National Fire Protection Association's NFPA 505, Fire Safety Standard for Powered Industrial Trucks Including Type Designations, Areas of Use, Conversions, Maintenance, and Operation. Users are not permitted to modify trucks without the written approval of the trucks' manufacturers.

Industrial Trucks in Hazardous Locations

The definitions of hazardous locations are given in NFPA 70, the National Electrical Code.
- Class I locations are those in which flammable gases or vapors are or may be present in quantities sufficient to produce explosive or ignitable mixtures.
- Class II locations are those in which combustible dusts are present.
- Class III locations are those in which easily ignitable fibers or filings are present but are not likely to be in suspension in quantities sufficient to produce ignitable mixtures.

ANSI/ITSDF B56.1–2012 stipulates that electric- or gasoline-powered trucks should not be used in hazardous locations unless the trucks either comply with NFPA requirements or are specifically approved by the inspection authority for use in the location involved. (See NFPA 505, Fire Safety Standard for Powered Industrial Trucks Including Type Designations, Areas of Use, Conversions, Maintenance, and Operation, for definitions of hazardous locations.)

If a lift truck is operating out of doors or away from stationary fire extinguishers in the facility, an appropriate fire extinguisher must be installed on the lift truck. Train all truck operators to use the fire extinguisher. An important advantage of having operators trained to use truck-mounted fire extinguishers is that the operators can quickly combat small fires anywhere on the premises.

Each type of powered industrial truck requires specific safeguards. To ensure the safety of operators and other workers, select only the lift trucks, straddle trucks, crane trucks, tractors and trailers, motorized hand trucks, and AGVs that are properly equipped.

Masts, Backrests, and Overhead Protection

If an overhead guard is attached to protect the operator from falling materials, the guard must be attached to the front of the truck's body and not to the mast. The overhead guard must completely cover the operator's compartment. An exception is made for those trucks in which a tilt cylinder is an integral part of the overhead guard construction. In such cases, the overhead guard should be designed so that it can prevent injury to the operator if the mast-tilting mechanism fails.

Forks should be locked to the carriage, and the fork extension, if used, should be designed to prevent unintentional lifting of the toe or displacement of the fork extension. Lift trucks also should be equipped with mechanical hoist and tilt mechanisms that prevent overtravel of hoist and tilt motions. If the lifting systems are hydraulically driven, a relief valve should be installed in the system and suitable stops should be provided to prevent overtravel.

Straddle Trucks

Straddle trucks (see Figure 15-2) should have horns or other warning devices, as well as headlights and tail lamps, to make working at night possible. Straddle trucks also should have safe-access ladders, wheel guards, and chain-drive guards. Some types of work may require a rigid overhead guard for the operator.

Operators can determine overhead clearances by looking for warning devices posted ahead of overhead obstructions across railroad tracks and other passageways where straddle trucks operate. Gauge rods may be mounted on the truck at the front and at the rear to guide the operator.

Straddle trucks present special hazards because they sit so high off the ground that the operator's view of objects immediately to the front or to the rear is reduced. Although precautions must always be taken to avoid striking pedestrians, precautions must especially be taken when carrying long loads. Such precautions include attaching red flags to the ends of such loads and stationing signal persons in congested areas. Special care must be taken when a truck is used after dark.

Crane Trucks

Although some crane trucks have three wheels, most have four wheels (Figure 15-8). One model is designed so that the operator sits behind a small, pillar-type jib crane mounted on a chassis. In another model, the operator stands on a platform and operates a fully or a partially rotating crane.

Figure 15-8. This truck-mounted hydraulic crane is capable of swinging through a full 360 degrees. Outriggers provide stability, and the enclosed cab offers the operator protection and clear visibility. *(Courtesy Grove Worldwide)*

Still another type has a fixed boom; in this case, the entire rig must be moved from one position to another in order to make side motions.

An operator should drive a crane carrying a load at the lowest possible speed and as low as possible. The operator should also have a helper to hook on the load and give signals. When a long load is being carried, the helper should walk alongside the load and, by means of a tag line, keep the load from swinging and striking against objects. When the crane travels without a load, the operator should fasten the hook to the lower end of the boom to prevent the hook block from swinging.

Tractors and Trailers

The coupling used to make up the tractor-trailer train should incorporate all necessary safeguards. The type of coupling used depends on the construction of the trailer, the loads carried, and whether the route traveled includes sharp curves, ramps, or inclines. The coupling must be one that will not come unhitched on grades or permit the trailer to whip or cut in on curves. Loads on trailers should be secured to the trailer to avoid the widespread scattering of material should the load shift or the trailer roll over.

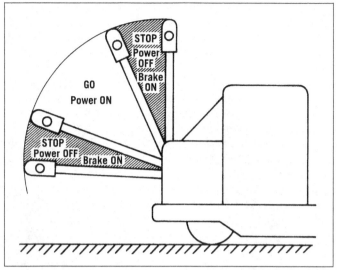

Figure 15–9. The hazards from touching the frame or wheels of a walkie lift truck can be minimized if the "power on" area is limited to approximately the area shown.

Motorized Hand Trucks

In operating a motorized hand truck, the principal hazards are that the operator may be pinned between the truck and a fixed object and that the truck may run up on the operator's feet. Operators should thus walk ahead of such trucks, leading them from either side of the handle and facing the direction of travel.

When a truck must be driven close to a wall or other obstruction, down an incline, or onto an elevator, the operator should put the truck in reverse and walk behind it while facing the direction of travel. Some motorized trucks are designed to be ridden over longer distances as well as guided while walking.

Guards are required for steering handles to prevent the operator's hand or the controls from coming into contact with obstacles when the truck must be maneuvered in tight spaces (see Figure 15–6). A motorized hand truck should be equipped so that its brakes are applied when the steering handle is in either the fully raised or the fully lowered position (Figure 15–9). The wheels of many motorized hand trucks can be considered guarded by location because they are under the frame or under the lift platform.

Hand trucks that have a platform on which the operator rides in a standing position must be designed so that the platform extends beyond the operator's position and is strong enough to meet the requirements of ANSI/ITSDF B56.1–2012. If an enclosure for operators, similar to the one in Figure 15–3, is provided, the enclosure will provide easy entry to and exit from the platform. To discourage operators from sitting on the truck during the truck's operation, install a prism-shaped cover over the battery box.

Figure 15–10. Two automatically guided, driverless vehicles electronically decide which one will have the right-of-way, thus eliminating potential collisions. *(Courtesy Jervis B. Webb Co.)*

Automated Guided Vehicles

Because trucks guided by remote control (AGVs) function without an operator, they must be equipped with some means to stop them completely should someone step in front of them (Figure 15–10). Such trucks should be furnished with a lightweight, flexible bumper or proximity sensor that, when contacted or triggered, shuts off the power and applies the brakes. Sufficient clearance between the bumper and the front of the truck is needed so that the truck can come to a full stop before contacting anything in its path.

Using AGVs requires the aisles where the trucks operate to be clearly marked and free of material. Forbid employees from jumping on or off or from riding these trucks. Do not allow the loading or unloading of AGVs that are in motion.

Design transfer conveyors so that they can be moved out of the way except when transferring loads to and from AGVs. This positioning prevents a pinch point from resulting between the vehicle and the conveyor. The same reasoning should apply when laying out a route next to machines and columns.

GENERAL OPERATING GUIDELINES

Operators of powered industrial trucks can prevent traffic accidents by using the same safe practices that they use in highway traffic. Specifically, operators should observe rules regarding speed, maneuvering, and the loading and unloading of other vehicles. Operators should also consider their own safety as well as that of other workers and equipment when using powered industrial trucks.

Operating a powered industrial truck has some basic differences from operating an automobile or truck on the highway, and the prospective operator should realize that a lift truck:

- is generally steered by the rear wheels
- steers more easily when loaded than when empty
- is driven in the reverse direction as often as in the forward direction
- is often steered with one hand, with the other hand used to operate the controls.

Speed

Excessive speed can lead to accidents both inside and outside the facility. A safe speed is the rate of travel that permits a truck to stop well before a certain area ahead or to make a turn without overturning. Wet or slippery floors require a slower-than-ordinary speed. Depending on their reasons for use and operating conditions, establish specific speed limits for trucks. For example, the speed limit for a yard will be too high for the speed limit of a congested warehouse. Many companies install governors to control vehicle speed.

Collisions between trucks and stationary objects often occur when trucks are backing up, usually while they are turning and maneuvering. In such moments, operators may be so focused on handling the load that they forget to watch where the rear of the truck is going. Because accidents caused while backing up usually result from this failure to look to the rear, operators should always look in the direction of travel and maintain a clear view of it (Figure 15–11). Some trucks permit the operator to sit sideways, which makes looking backward and forward easier to do. Operators must be especially careful when turning the truck because the truck's rear wheels may project beyond the truck's enclosure, thus creating a hazard.

Operators should stop and sound the horn at blind corners and before they pass through doorways. They should proceed forward only when they can see that the way is unobstructed. Many companies have installed large convex mirrors at blind corners so that operators and pedestrians can see others approaching. To be effective, these mirrors must be kept clean and properly adjusted.

Operators should avoid making quick starts, jerky stops, or quick turns. They should be extremely cautious when steering trucks on turns, railroad crossings, ramps, grades, or inclines. On descending grades, operators should keep the trucks under control so that they can be brought to an immediate stop in the clear space in front of them. Operators should never use the reverse control to brake.

Figure 15–11. An operator must be alert when backing up and must face the direction of travel. *(Courtesy Clark Equipment Company)*

Operators should keep trucks a safe distance apart during operation; some companies require operators to adhere to a minimum separation of three truck lengths. Operators should not pass other trucks traveling in the same direction at intersections, blind spots, or other dangerous locations. They must keep to the right, if aisle width permits, and not pass dangerously close to machine operators and others. Where aisles are not wide enough for continuous two-way traffic, trucks should travel in the middle of the aisle except in a passing situation, where the vehicle should move to the side and pass with caution.

Elevators, Bridge Plates, and Railroad Tracks

Operators should not drive trucks onto an elevator unless they have been authorized to do so. An operator should approach the elevator slowly and at a right angle to the elevator's door. The operator should enter the elevator only after the car is properly leveled and after checking to make sure that the total weight of the truck, the load, and the driver does not exceed the capacity of the elevator. Before exiting the elevator and after making sure that the load is lowered to the floor, the brakes are set, the power is shut off, and the controls are in neutral, the operator should get off the truck. Other personnel should stay off any elevator occupied by a lift truck.

Powered industrial trucks should be driven carefully and slowly over bridge plates that are properly secured (Figure 15–12). Trucks should cross railroad tracks diagonally whenever possible and park at least 8 ft (2.1 m) from the centerline of the tracks.

Loading and Unloading

Highway trucks, trailers, and railroad cars should have their brakes set and their wheels securely blocked while they are being loaded or unloaded by powered industrial

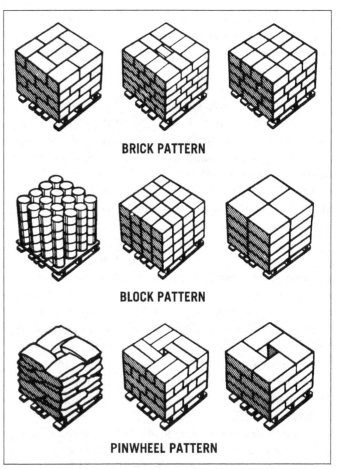

Figure 15–13. Typical pallet-loading patterns. *(Printed with permission from Industrial Truck Association.)*

Figure 15–12. This hydraulic dock leveler prevents forklifts and other vehicles from rolling or being driven off the dock when the truck or trailer is not present. Note that the forklift truck has a liquefied petroleum fuel storage container. *(Courtesy Rite Hite Corp.)*

trucks. Before entering a trailer, lift-truck drivers should make sure that wheel chocks are squarely placed in front of the rearmost tires on dual-axle trailers. On tri-axle or quad-axle trailers, two additional chocks should be squarely placed in front of the foremost tires that are still on the ground. Operators should inspect trailer floors for defects that might affect safe loading procedures.

When a trailer is next to a ramp or a wall, place at least two wheel chocks squarely in front of the outside rear- and foremost tires. When trailers are parked without the tractor, some companies also place jacks under the front of the trailer. Trailer-restraint systems also are available.

Loads, whether on trucks, trailers, skids, or pallets, should be stable. Neatly pile and cross-tie objects, if their shape permits (Figure 15–13). Load irregularly shaped objects so that they cannot roll or fall off the truck. Place heavy, odd-shaped objects with the weight as low as possible. Block round objects, like pipe or shafting, and, if necessary, tie them so that they cannot roll off the truck. Loading to an excessive height not only blocks the view ahead but also makes it likely that part of the load will fall.

When standard forks are used to pick up horizontal cylindrical objects such as rolls or drums, care must be taken to ensure that the fork tips do not damage the load or push it against workers. First tilt the mast so that the tips of the forks touch the floor. Then move forward so that the forks can slide under the objects. Tilting the mast backward will then cause the load to roll back against the vertical face of the backrest and/or carriage and the load backrest extension—a secure carrying position. Place a block or wedge against the drum or roll.

To unload a large case or a similar object without a pallet, the operator should first drive into position for stacking. The operator should then place a block near the edge of a base, lower the load onto the base, and withdraw the forks so that only their tips hold up the end of the load. Next, the operator should withdraw the block, tilt the mast forward, and back away.

To pick up a palletized load, keep the forks fully and squarely seated in the pallet, an equal distance from the center stringers and well out toward the sides. Forks inserted into a pallet should be level, not tilted forward or backward. If the forks are placed close together, the pallet will tend to drop at the sides and seesaw, resulting in strain and instability (Figure 15–14).

When raising or lowering loads with the truck standing still, the operator should not leave the truck in gear with the clutch depressed. Rather, the shift should be returned to neutral and the clutch disengaged.

Tell lift-truck operators to refuse to pick up improperly loaded skids or pallets, broken pallets, or loads too heavy for the truck. Operators should also insist on the proper identification of all chemicals before moving them and should observe the safety guidelines and regulations for handling chemicals.

When a lift truck is parked, the forks should be placed flat on the floor. No one should be permitted to stand or walk under elevated forks.

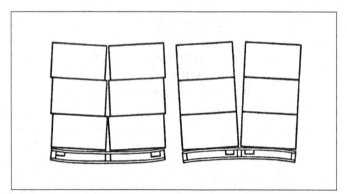

Figure 15–14. With forks spread wide (left), the load is well distributed and tends to bind itself together. The effect of placing the forks too close together is shown at the right.

Using a lift truck to elevate employees (e.g., to service light fixtures) should be done only if an approved safety work platform with guardrails and toeboards is secured to the forks and mast. The truck should also have an overhead guard to protect the operator. The lift-truck operator must not leave the controls while the truck is used to lift a person. Lift trucks without special attachments are generally not designed to lift personnel; instead, special trucks are built to lift personnel.

Proper Care of Trucks

Operators should not use a powered industrial truck for any purpose other than the one for which the truck was designed. Common dangerous misuses include moving skids, pushing piles of material out of the way, using makeshift connections to move heavy objects, using the forks as a hoist, and moving other trucks. Disabled trucks should not be pushed or carried by other lift trucks. They should, rather, be towed with a tow bar and safety chain. Never use powered industrial trucks to tow or push freight cars, and never use them to open or close freight-car doors unless the truck has a device specifically designed for this purpose.

An operator should leave a truck unattended only after the controls have been put in neutral, the power has been shut off, the brakes have been set, the key has been removed, the connector plug has been pulled, and the load-engaging mechanism has been placed in a lowered and inoperative position. (OSHA defines an "unattended truck" as one whose operator is more than 25 ft [7.6 m] from it or cannot see it.) It is unsafe to park a truck on an incline; when such an action is unavoidable, block the wheels as an added precaution.

It is the operator's responsibility to refrain from parking a truck in an aisle or doorway and obstructing material or equipment to which another worker may need access. Accidents often happen when a truck is blocking a passageway and an unauthorized employee tries to move it.

Operators should not permit gasoline engines to idle for long periods in enclosed or semi-enclosed areas because exhaust vapors and combustion gases will accumulate. Concentrations of carbon monoxide (CO) in areas where powered industrial trucks are operated should not exceed the safe levels established by regulatory authorities. A qualified industrial hygienist should sample for CO. Direct reading instrumentation may be needed in some locations.

Catalytic exhaust purifiers, which considerably reduce the levels of carbon monoxide and other noxious gases in engine exhaust, are obtainable. However, even when exhaust purifiers are installed on lift trucks, management must provide adequate ventilation for enclosed areas and proper maintenance of the trucks in order to maintain clean air. It is also necessary to properly maintain exhaust purifiers.

Operator and Pedestrian Safety

Operators should keep their feet and their legs inside the guard or inside the operating station of the truck. Driving with a foot or a leg outside the station is unsafe, as is placing a hand, an arm, or a leg on or between the uprights of the truck. In tight spaces, operators should keep their hands where they cannot be pinched between the steering or control levers and projecting stationary objects. Steering handles on motorized hand trucks should have guards to protect against injuries of this kind.

No one other than the operator should be permitted to ride on a truck, coupling, or trailer. It is the operator's responsibility to keep unauthorized personnel off the truck unless a safe, designated place to travel is provided and authorized by the employer.

Looking out for pedestrians is also the truck operator's responsibility. Operators should make eye contact and sound the horn when approaching pedestrians or intersections. Discourage excessive horn-blowing, however. Having sounded the warning, the operator should proceed with caution, passing only when pedestrians are aware of the truck's presence and are out of danger. The operator should not use the horn to "blast" a way through. He or she should never drive a truck directly toward anyone who is standing in front of a bench or other fixed object.

Pedestrians also have a responsibility to watch out for trucks and to get out of their way promptly. Ill feelings and accidents can result if pedestrians refuse to move out of the way when a truck approaches—consideration from both sides is needed.

Maneuvering

Because a lift truck is generally steered by its rear wheels, operators must always carefully observe the swing of the rear of the truck (Figure 15–15). Beginners usually try to turn too sharply. When traveling forward, some lift trucks exhibit a peculiarity known as "free turning." That is, once a turn is started, the truck tends to turn more and more sharply in smaller and smaller circles. To counteract this tendency and the sharpness of the turns, the operator must apply force on the steering wheel in the opposite direction. When a lift truck is traveling in reverse, the opposite holds true—the operator must apply force in the direction of the turn. Turns should be made smoothly and gradually and at a safe speed.

Operators should learn to judge the correct aisle width for the truck's size and load. They should also adhere to general operating safety rules as well as the specific lift-truck rules that are discussed next. All starts and stops should be easy and gradual to prevent the load from shifting.

When traveling or maneuvering, the operator should be particularly careful to avoid striking overhead structures

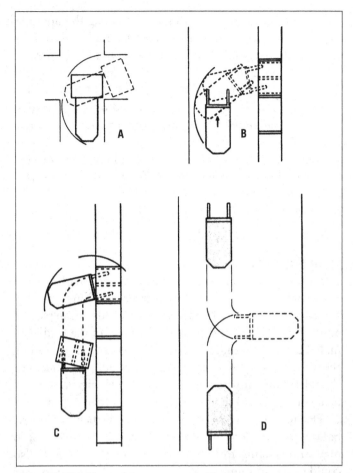

Figure 15–15. Lift-truck maneuvers. A: Turning a sharp corner. B: Turning across an aisle. C: Turning in an exceptionally narrow aisle. D: Turning around in a narrow passage. The driver should allow ample space for rear-end swing and make turns carefully.

and nearby objects such as sprinkler piping, electrical conduit, and fixed structures. To protect critical equipment and materials such as electrical panels, fire equipment, fire doors, and load-supporting structures, strong barriers, posts, and curbing should be installed.

Operators should raise or lower loads only when stopped. Loaded or empty, the forks should be carried as low as possible but still high enough that they will not strike any raised or uneven surface. Tilting back the mast will keep the load stable and secure.

When carrying a bulky load that cannot be lowered enough to prevent an obstructed view, the operator should drive the truck backward. By doing so, the operator can see where the truck is going. Some companies have the policy that trucks are to be driven in reverse any time they are carrying a load, regardless of the load's size. Use a spotter when necessary.

Driving on Grades

Trucks should ascend or descend grades slowly. When ascending or descending grades in excess of 10%, loaded trucks should be driven with the load facing the upgrade. Low gear or the slowest speed should be used when a truck is descending a grade. Unloaded trucks should be operated on grades with the load-engaging mechanism facing the downgrade. Keep in mind that high-lift and order-picker trucks are not designed to be operated on steep grades. Consult the manufacturer's operating instructions for recommended safety procedures.

On all grades, the load and the load-engaging mechanism should be tilted backward, if applicable, and raised only as far as necessary to clear the road's surface. The operator should also keep clear of the edge of loading docks and ramps and never make a sharp turn on a ramp.

Load Capacity

Lift trucks are rated by capacity in pounds and load center in inches. The capacity of a truck might be 5,000 lb (2,268 kg) at a 24-in. (60-cm) load center. In other words, it can pick up 5,000 lb (2,268 kg) if the center of gravity of the load is 24 in. (60 cm) from the face of the load arm backrest. Exceeding the rated capacity is considered an overload. Every operator should be familiar with the maximum load limits of the truck being operated and should be required to observe them.

Placing extra weight on the rear of a lift truck to counterbalance an overload is never permitted because doing so may strain chains, forks, tires, axles, and the motor. It also could cause an injury. ANSI/ITSDF B56.1–2012 provides more information on the stability of lift trucks. Most lift-truck manufacturers use the stability values in this standard as criteria for design factors and for determining the rated capacity of various truck models. Side stability is also a critical factor in making turns at high speed or on a slope or ramp. Tilting back the mast reduces side stability on high lifts and may cause tipping; allow for this factor.

Operators should never, under any circumstances, attempt to operate an overloaded truck. Such a load is dangerous because it removes weight from the wheels that steer and affects the steering. Standing on a truck or adding counterweights to compensate for an overload should never be permitted.

Particular care should be taken not to exceed floor load limits. The force exerted by a truck on a floor varies with the truck's speed, load, and total weight distribution. The number of wheels, the wheelbase, and other factors also affect the load the floor can carry. Refer all questions about floor capacities to a qualified architect or structural engineer.

Pallets

Most companies buy ready-made pallets (Figure 15–16). Establish procedures to inspect pallets, both before they are put into service and at regular intervals, to be sure that they are safe to use. The top boards of pallets should be

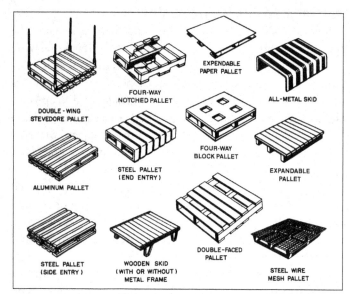

Figure 15–16. Lift-truck operators should be familiar with the different pallet and skid types. *(Printed with permission from the Industrial Truck Association.)*

sound and securely fastened to runners. Repair or replace splintered, broken, or loose parts. Loose nails or chunks of wood can cause injury to workers and damage to trucks.

Provide a safe place, out of the way of traffic and work areas, to store pallets. Neatly stack pallets and limit their height (no more than 6 ft) so that they are stable and not likely to slide or collapse. Also, do not leave pallets standing on end or in a position from which they might topple onto people or other objects. To prevent overworking the sprinklers in case of a fire, avoid storing large blocks of pallets within a building. Keep large stacks in outside storage, well away from buildings.

INSPECTION AND MAINTENANCE

Qualified and authorized maintenance personnel should thoroughly and regularly inspect powered industrial trucks and give them complete overhauls after specific periods of operation. Prohibit operators from making repairs on trucks. Before repairs are made to any part of a powered industrial truck, the operating mechanism should be locked off.

Operators should make daily inspections of their trucks' controls, brakes, tires, and other moving parts. They must do this at the start of each shift in multi-shift operations. They should also use checklists (Figure 15–17) to record any conditions that require corrections. Defective brakes, controls, tires, lights, power supplies, load-engaging mechanisms, lift systems, steering mechanisms, and signal equipment should be repaired before trucks are allowed to go back into service. Keep a detailed schedule of inspections and repair records for

Operator's Daily Report

Battery-Powered Lift Trucks

Truck No. _____ Make _____ Date _____ Shift_____

Hour Meter Reading: Start _____ End _____ Hours for Shift _____

CHECK EACH ITEM If OK write OK	SHIFT			Explain below if not OK or any other action taken
	Start	During	End	
1. Battery plug connection				
2. Battery charge				
3. Battery load test				
4. Brakes – service and seat brake				
5. Lights – head, tail, and warning				
6. Horn				
7. Hour meter				
8. Steering				
9. Tires				
10. Hydraulic controls				
11. Other conditions				

Remarks and additional explanation or suggestions _____

_____

_____

2C97567   Printed in USA   Stock No. 199.74

Figure 15–17. Operators should use a checklist to make a daily inspection of their powered industrial trucks.

each vehicle (Figure 15–18). The truck's operator's manual should be kept with the vehicle to facilitate daily inspections.

Forks should be magnafluxed or subjected to other nondestructive testing on a regularly scheduled basis as determined by use.

Electric Trucks

Only trained and authorized personnel may perform battery changing and charging operations on electric trucks. Depending on individual facility situations, truck operators may or may not be so authorized. The handling and charging of storage batteries for electric trucks incorporates several hazards.

Operators of charging equipment can be protected from experiencing acid burns when refilling or handling batteries by wearing chemical safety goggles, rubber gloves,

NATIONAL SAFETY COUNCIL
FORKLIFT TRUCK OPERATORS TRAINING COURSE

Inspection and Maintenance Log

Truck Number	Make	Work Done		Work Description or Remarks	Cost
		Date	Hour Meter		

2C97567   Printed in USA   Stock No. 199.74

Figure 15–18. A detailed inspection and repair record should be kept for each truck.

aprons, and rubber boots. Do not allow the sulfuric acid used for refilling to touch cast-iron, lead, steel, or brass drains. Wood-slat mats, rubber mats, or clean floorboards will help prevent slips and falls and will prevent electric shocks from the charging equipment. When racks are used to support batteries, the racks should be made of materials that do not generate sparks.

Take precautions to prevent open flames, sparks, or electric arcs from developing in battery-charging areas. Electrical installations should conform to local codes and NFPA 70, the National Electrical Code. Never lay tools or metal parts on a battery, and prohibit smoking in the charging area. (See NFPA 505, Fire Safety Standard for Powered Industrial Trucks Including Type Designations, Areas of

Use, Conversions, Maintenance, and Operation, and NSC Data Sheet 12304–0635, Lead-Acid Storage Batteries.)

Charge batteries only in the areas designated for that purpose. Also provide facilities for:
- flushing and neutralizing spilled electrolyte
- fire protection and extinguishers
- protecting charging apparatus from possible damage by trucks
- emergency eyewash and shower stations
- adequate ventilation to disperse flammable hydrogen gases, vapors, and fumes from batteries.

Properly position trucks and apply their brakes before attempting to change or charge their batteries. Reinstalled batteries should be properly positioned and secured in the truck. When charging batteries, wear eye protection and rubber gloves. Pour acid into water, never the reverse. Provide a carboy tilter or siphon for handling electrolyte. If acid or electrolyte is spilled on the worker's skin or clothing, it should be washed off immediately with plenty of water. When charging batteries, keep the vent caps in place to prevent electrolyte spray. Take care when determining whether vent caps are functioning. Open the battery (or compartment) cover(s) to dissipate heat. Refer to the manufacturer's recommendations before performing charging and maintenance procedures.

Provide a roller conveyor, an overhead hoist, or equivalent materials-handling equipment to prevent overexertion injuries resulting from manually handling heavy or awkward batteries. To prevent short-circuiting, use insulated or nonconductive chains, hooks, and yokes in the hoisting mechanism. To reduce wear on insulation, which may produce arcing, watch the points where cables contact reels and suspension attachments. Battery chargers must be in the OFF position while being connected and disconnected.

Gasoline-Operated Trucks

Handle and store gasoline for trucks according to the provisions of NFPA 30, Flammable and Combustible Liquids Code.

Safety cans used for fuel handling should be tested and approved by Factory Mutual (FM) or listed by Underwriters Laboratories (UL). Safety cans should have a flame arrester and a self-closing lid. (See Chapter 10, Flammable and Combustible Liquids, in this volume.)

Fill fuel tanks on gasoline-operated trucks and tractors only at designated locations, preferably out in the open, with the filling hose and equipment properly grounded and bonded. Select locations distant from outside main buildings to lessen the chances of a fire starting.

Engines must be stopped and operators must be off trucks before the trucks are refueled. Prohibit smoking during refueling. Workers should avoid spilling gasoline or letting the gas tank overflow during refueling. Before an attempt is made to start the engine, the gas tank's cap should be replaced and any spilled fuel allowed to vaporize. Gasoline tanks should be drained only into grounded, self-closing cans that are out of harm's way.

Liquefied Petroleum Gas Trucks

The use of liquefied petroleum gas (LPG) as a fuel for powered industrial trucks is increasing. A properly adjusted engine burning LPG generally produces a substantially lower concentration of carbon monoxide (CO) in the exhaust than does a similar engine that uses gasoline as fuel. Only air sampling, however, can determine whether the CO concentration in an area is below the maximum allowable level. The 2013 Threshold Limit Values (TLV) listing from the American Conference of Governmental Industrial Hygienists (ACGIH) indicate that 25 parts per million (ppm) over an 8-hour exposure is the maximum exposure allowable. The U.S. OSHA has set the permissible exposure level (PEL) at 50 ppm. The Canadian exposure limit is 25 ppm. Be sure to check and comply with maximum permissible exposure limits in the relevant jurisdiction. As a best practice, CO exposures should be controlled to the lowest feasible level practical.

Fittings incorrectly installed or not recognized by a nationally respected agency and connections not properly tightened before refueling begins may fail and release combustible gas into the air. Use only the conversion units and fittings authorized by an agency such as Underwriters Laboratories or Factory Mutual. The units and fittings should include all the safety features that are incorporated into the LPG-fueled trucks recognized by the testing agency. Install the units and fittings in strict conformity with the requirements specified in NFPA 58, Liquefied Petroleum Gas Code, and the Underwriters Laboratories Standard for Safety No. 558, Industrial Trucks, Internal Combustion Engine-Powered. Adhering to these requirements will provide maximum protection against damage to the truck's system from vibration, shock, or objects striking against it.

Storage and Handling of Liquefied Petroleum Gases

Only FM- or UL-listed fuel containers, designed in accordance with U.S. Department of Transportation (DOT) or ASME standards, should be used. Fill permanently mounted and removable fuel containers outdoors. However, this filling may be done indoors if one of the two methods specified in NFPA 58 is used. Refueling of LPG trucks with permanently mounted containers should be done outdoors and away from any building ventilation ducts or ignition sources.

A special building or outside storage area is recommended for the storage of fuel containers. When fuel cylinders must be stored inside a building, keep them in a special room or designated safe area in accordance with NFPA 58. NFPA standards permit no more than two containers of LPG on each powered industrial truck. They also require this storage to be inside a building that is not frequented by the public and to be limited to a total of 300 lb (135 kg) of gas. Enclose containers in a separate room that is well ventilated and of ample size. The walls, floor, and ceiling of the room must have fire-resistant construction. Use self-closing, rated fire doors to shield openings that lead to other parts of the building.

The proper filling of containers is of the utmost importance. The person filling the containers must be trained in the safe handling of LPG. Filling containers from bulk storage must be done at least 10 ft (3 m) from the nearest masonry-walled building and at least 25 ft (7.5 m) from nonmasonry buildings and openings in masonry and nonmasonry buildings (Figure 15–19). The filling facility must conform to NFPA 58 and to applicable state, provincial, or local insurance regulations.

Trucks themselves must comply with NFPA 505, Fire Safety Standard for Powered Industrial Trucks Including Type Designations, Areas of Use, Conversions, Maintenance, and Operation. Park or garage LPG-fueled trucks in a well-ventilated area. Because LPG is heavier than air, provide ventilation at floor level. Do not park or garage trucks in the same room that LPG cylinders are stored.

Follow a thorough and documented inspection and maintenance procedure for LPG-fueled trucks. LPG trucks can be stored or serviced inside garages provided that:

1. the fuel system is leak free and the container is not filled beyond the limit specified in NFPA 58
2. the container shutoff valve is closed except when the engine is operated
3. the truck is not parked near inadequately vented pits or sources of heat, open flames, or other sources of ignition.

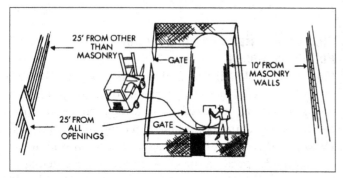

Figure 15–19. The schematic shows the minimum distances allowed for powered industrial truck refueling operations. *(Printed with permission from the LP-Gas Association.)*

Wear eye and hand protection when changing tanks. For trucks with permanently mounted fuel containers, make major repairs outdoors or in a well-ventilated, fire-resistive area provided for this purpose.

OPERATORS

Operators should have valid driver's licenses, good driving records, and few, if any, traffic violation tickets. They should have good attitudes toward the responsibility of operating expensive, heavy industrial equipment in new and difficult circumstances. For this reason, verify a prospective operator's previous experience, both off and on the job, whenever possible. Driving a car is different from driving a forklift.

The Fair Labor Standards Act (FLSA) prohibits minors under age 18 from working in any occupation that the FLSA deems to be hazardous. Among these occupations are excavating, manufacturing explosives, mining, and operating many types of power-driven equipment. Certain industries allow minors under age 18 to perform tasks whose primary activity is dangerous as long as these tasks are very specific and the state and federal governments closely monitor compliance with the FLSA.

Child labor laws vary from state to state. Regulations provide very specific information on these occupations and other safety standards for minor employees. Please consult your state or provincial Department of Labor for this information.

Selection

Operators should meet certain physical requirements and be examined by a qualified physician familiar with the job's requirements. Minimum requirements are 20/40 vision, corrected if necessary; good reaction time; depth perception of no less than 90% of normal; and good hearing.

Operators must acknowledge the importance of the training program. The proper attitude will lead to good performance, especially if supervision is not as attentive as it should be.

Operators must show good judgment as well as respect for the safety of both personnel and property. Operators should understand that they have considerable responsibilities and that reckless or careless work will not be tolerated. Evaluate operators on the job from time to time to see whether they observe this rule. To readily identify employees, many companies issue badges to authorized truck operators. Badges also remind operators of their responsibilities and encourage pride in their jobs.

To familiarize operators with their responsibilities, adopt a set of rules governing the operation of powered industrial trucks. Because facility conditions and equipment vary widely, set up rules that cover the specific conditions the company's operators face.

Training

Only persons with the necessary knowledge, education, and experience to train powered industrial truck operators and to evaluate their competence should conduct all training and evaluation. An example of a qualified trainer is a person who, by possessing a recognized degree, certificate, or professional standing or by having extensive knowledge, training, and experience, has demonstrated the ability to train and evaluate powered industrial truck operators. In a 2003 response to a letter requesting the interpretation of what the term qualified trainer meant, Richard E. Fairfax, Director of the OSHA Directorate of Enforcement Programs stated that U.S. OSHA defines trainer competency as follows:

> A trainer must have the "knowledge, training, and experience" to train others how to safely operate the powered industrial truck in the employer's workplace. In general, the trainer will only have sufficient "experience" if he has the practical skills and judgment to be able to himself operate the equipment safely under the conditions prevailing in the employer's workplace. For example, if the employer uses certain truck attachments and the trainer has never operated a truck with those attachments, the trainer would not have the experience necessary to train and evaluate others adequately on the safe use of those attachments. However, the standard does not require that the trainers operate a PIT regularly (i.e., outside of their operator training duties) as part of their job function or responsibility.

Many resources are available to employers that choose not to perform this training in-house. For example, truck manufacturers, local safety and health organizations such as local chapters of the National Safety Council, private consultants with expertise in powered industrial trucks, and local trade and vocational schools are some resources that can provide this training.

Several Internet sites are devoted to forklift safety. Private companies that provide forklift safety training services, including videos and printed programs, can be located via Internet searches. Most videos can be either leased or purchased. One important thing to remember is that simply showing employees a video or videos on some aspect of forklift safety does not meet the complete requirements of the U.S. OSHA standard. Facility-specific information must also be conveyed, and there must be a method, following the training, to evaluate an employee's acquired knowledge.

U.S. OSHA standards require employers to develop and implement an operator-training program that is based on the general principles of safe truck operation, the types of vehicle(s) being used in the workplace, the hazards of the workplace created by the use of these vehicle(s), and the general safety requirements of the U.S. OSHA standard. Topics to be covered by training include:

- operating instructions, warnings, and precautions for the types of truck the operator will be authorized to operate
- differences between the truck and an automobile
- truck controls and instrumentation: where they are located, what they do, and how they work
- engine or motor operation
- steering and maneuvering
- visibility (including restrictions due to loading)
- fork and attachment adaptation, operation, and use limitations
- vehicle capacity
- vehicle stability
- any vehicle inspection and maintenance that the operator will be required to perform
- refueling and/or charging and recharging batteries
- operating limitations
- any other operating instructions, warnings, or precautions listed in the operator's manuals of the types of vehicles that the employee is being trained to operate
- surface conditions where the vehicle will be operated
- composition of loads to be carried and load stability
- load manipulation, stacking, and unstacking
- pedestrian traffic in areas where the vehicle will be operated
- the characteristics of narrow aisles and other restricted places where the vehicle will be operated
- details about the hazardous (classified) locations where the vehicle will be operated
- the characteristics of ramps and other sloped surfaces that could affect the vehicle's stability
- details about closed environments and other areas where insufficient ventilation or poor vehicle maintenance could cause a buildup of carbon monoxide or diesel exhaust
- other unique or potentially hazardous environmental conditions in the workplace that could affect safe truck operation.

Trained operators must know how to do the job properly and do it safely, as demonstrated during workplace evaluation. Formal (lecture, video, etc.) and practical (demonstration and practical exercises) training must be provided. Employers must also certify that each operator has received the training and must evaluate each operator at least once every 3 years. Prior to allowing an operator to operate a truck in the workplace, the employer must evaluate that operator's performance and determine that the operator is competent to operate a powered industrial truck safely. New hires with previous truck-operating experience should be retrained and recertified to account for the unique characteristics of the worksite.

Refresher training is also needed whenever an operator demonstrates a deficiency in the operation of a truck. In addition, operators must receive some form of pedestrian safety training because a truck operated incompetently can cause severe injury or substantial property damage. To be most effective, a training program should focus on the company's policies, operating conditions, and types of trucks used.

Before establishing a training program, companies should determine the problems they have experienced with industrial trucks such as the numbers and types of accidents, the extent of economic losses from accidents, and the operating habits of operators. This information can then be incorporated into the training program.

As with any program, management's support is essential for truck operator training to be effective and lasting. Management must also understand and accept that training programs cost money because of the necessary materials and equipment, the payment for nonproduction time and lost production, and the possible damage to materials. Practical training can be conducted on obstacle courses (Figure 15-20).

Supervisors also should fully accept the need for the training program; they must be willing to adjust their employees' schedules to make people available for training either during or after work hours.

Maintenance personnel should be involved in the training program from the start. Mechanics are possible instructors because they know the trucks and how they operate. Mechanics also will understand why truck operators request additional repairs and adjustments.

Other factors to consider when setting up a training program include determining:
- who will be in charge
- the necessary qualifications of both instructor(s) and trainees
- the number and length of both classroom and hands-on sessions
- the location of both classroom and hands-on sessions
- the optimal number of trainees in a class
- whether operators who are taking a refresher course should be in the same class as first-time trainees
- how to notify all employees about the program and convey its importance
- how to establish and maintain a training record-keeping system.

An effective training program does not end with the presentation of a certificate of completion. Instead, management must continue to maintain safe operating condi-

Figure 15-20. Having to maneuver through a mockup of a crowded aisle tests the driving ability of this lift-truck operator trainee. All training programs should include a driving test as well as a written examination. *(Courtesy Clark Equipment Co., Industrial Truck Division)*

tions and to insist on safe performance by all employees. To ensure continued safe work habits, keep a record of each operator's performance (Figure 15-21).

SUMMARY

- Powered industrial trucks require safeguards for the operator's protection and for the safety of other workers. Management must establish guidelines for the operation, maintenance, and inspection of this equipment.
- Powered industrial trucks are classified by power source, operator position, or means of engaging the load.
- Factors to consider when purchasing trucks include worksite conditions, operator comfort, backup systems, and safety features such as safety belts and wraparound seats.
- NFPA standards specify certain hazardous locations, Classes I through III, in which various types of trucks should not be used unless they comply with NFPA requirements or are officially approved.
- Lift trucks should have overhead guards that are designed to prevent injury.

OSHA Sample Performance Test for Forklift Operators

EMPLOYEE _____ DATE _____ TIME _____a.m./p.m.

Item Evaluated	Pass/Fail	Comments
Shows familiarity with truck controls		
Gave proper signals when turning		
Slowed down at intersections		
Sounded horn at intersections		
Obeyed signs		
Kept a clear view of direction of travel		
Turned corners correctly – was aware of rear end swing		
Yielded to pedestrians		
Drove under control and within proper traffic aisles		
Approached load properly		
Lifted load properly		
Lifted load properly		
Maneuvered properly		
Traveled with load at proper height		
Lowered load smoothly/slowly		
Stops smoothly/completely		
Load balanced properly		
Forks under load all the way		
Carried parts/stock in approved containers		
Checked bridgeplates/ramps		
Did place loads within marked area		
Did stack loads evenly and neatly		
Did drive backward when required		
Did check load weights		
Did place forks on the floor when parked, controls neutralized, brake on set, power off		
Followed proper instructions for maintenance – checked both at beginning and end		

Total Rating: _____ Evaluator: _____

osha.gov/dte/library/pit/test.html

Figure 15–21. OSHA Website Form for PIT Testing

- Operators should realize that lift trucks are generally steered by the rear wheels, handle more easily when loaded, are driven in the reverse direction as often as in the forward direction, and are often steered with only one hand.
- Straddle trucks should have horns, flags, and other warning devices to alert pedestrians.
- Crane trucks should be driven at the lowest possible speed when carrying a load to maintain balance.
- Motorized hand trucks must be equipped for proper operation.
- Automated guided vehicles must have some means of stopping if someone steps in front of them. Such trucks should also be equipped with flexible bumpers or proximity sensors that shut off power when triggered or contacted.

- Operators of powered industrial trucks can prevent accidents by using the same safe-driving techniques they use when driving on highways.
- When loading and unloading trailers, operators should make sure that the brakes are on, the wheels are blocked, and the loads are neatly stacked and stable and are fastened to the trailer securely.
- Powered industrial trucks should not be used for any purpose other than the one for which they were designed.
- Operators are responsible for the care of trucks and should never leave a truck unattended, park in an aisle or doorway, idle engines for too long, or ignore mechanical problems.
- Powered industrial trucks should be inspected and maintained regularly. Repairs, replacements, and other work should be performed only by trained mechanics wearing proper protective equipment, particularly when working on electrically powered industrial trucks.
- Only authorized fuel and fuel tank equipment should be used in these trucks.
- Training programs should focus on company policies, operating conditions, and types of trucks used. Management should maintain a record of each employee's driving performance.

REFERENCES

American National Standards Institute/Industrial Trucks Standards Development Foundation, ANSI/ITSDF B56.1–2012, Standard for Low Lift and High Lift Trucks.

American Conference of Governmental Industrial Hygienists, 6500 Glenway Avenue, Bldg. D-7, Cincinnati, OH 45211.

Threshold Limit Values, latest ed.

American Insurance Association, 1130 Connecticut Avenue NW, Suite 1000, Washington DC 20036.

National Fire Protection Association, 1 Batterymarch Park, Quincy, MA 02269.

Fire Safety Standard for Powered Industrial Trucks Including Type Designations, Areas of Use, Conversions, Maintenance, and Operation, NFPA 505, latest ed.

Flammable and Combustible Liquids Code, NFPA 30, latest ed.

Liquefied Petroleum Gas Code, NFPA 58, latest ed.

National Electrical Code®, NFPA 70, latest ed.

National Propane Gas Association of America, 1600 Eisenhower Lane, Lisle, IL 60532.

National Safety Council, 1121 Spring Lake Drive, Itasca, IL 60143-3201.

Aviation Ground Operation Safety Handbook, latest ed.

National Safety Council Data Sheet 12304–0635, Lead-Acid Storage Batteries, nsc.org/Pages/Home.aspx.

Superintendent of Documents, Government Printing Office, Washington DC 20420.

OSHA. 29 CFR 1910.178, Powered Industrial Trucks.

Swartz, G. Forklift Safety: A Practical Guide to Preventing Powered Industrial Truck Incidents and Injuries. Rockville, MD: Government Institutes, 1997.

Underwriters Laboratories, Inc., 333 Pfingsten Road, Northbrook, IL 60062.

"Industrial Trucks, Electric-Battery-Powered," Standard for Safety, No. 583, latest ed.

"Industrial Trucks, Internal Combustion Engine-Powered," Standard for Safety, No. 558, latest ed.

U.S. Department of Health, Education, and Welfare, National Institute for Occupational Safety and Health, Division of Technical Services, 4676 Columbia Parkway, Cincinnati, OH 45226.

U.S. Occupational Safety and Health Administration, Office of the Directorate of Enforcement Programs, OSHA. 29 CFR 1910.178, July 23, 2003, Powered Industrial Trucks, osha.gov/pls/oshaweb/owadisp.show_document?p_table=INTERPRETATIONS&p_id=25021.

REVIEW QUESTIONS

1. Name five types of powered industrial trucks.
   a.
   b.
   c.
   d.
   e.

2. What aids should leased or purchased industrial trucks have to reduce driver fatigue and strain?
   a.
   b.
   c.
   d.
   e.

3. Which of the following should conform to ANSI/ITSDF B56.1?
   a. overhead guards on trucks
   b. nameplates indicating truck weight and rated capacity
   c. specifications for steering, braking, and other controls
   d. stability of lift trucks
   e. all of the above

4. Because operators in straddle trucks sit so high off the ground, their angle of sight is reduced immediately to the front and to the rear, posing a hazard to _____.

5. What are the two principal hazards of operating a motorized hand truck?
   a.
   b.

6. Operators of powered industrial trucks can prevent traffic accidents by using the same safe practices that they apply to _____.

7. Describe the common dangerous misuses of powered industrial trucks that operators should never perform.
   a.
   b.
   c.
   d.
   e.

8. Operators should be aware of what basic differences between lift trucks and automobiles or highway trucks?
   a.
   b.
   c.
   d.
   e.

9. To be effective, a training program should focus on what three aspects?
   a.
   b.
   c.

APPENDICES

Appendix A
OSHA FAQ

Occupational Safety and Health Administration, U.S. Department of Labor Information Card

- What vehicles are considered to be powered industrial trucks?

  The American Society of Mechanical Engineers (ASME) defines a powered industrial truck as a mobile, power-propelled truck used to carry, push, pull, lift, stack, or tier materials. Powered industrial trucks, often called forklifts or lift trucks, can be ridden or controlled by a walking operator. Excluded from the OSHA standard are vehicles used for earth moving or over-the-road haulage.

- What industries are covered by the standard?

  The standard covers general industry, maritime, and construction. The general industry standard is 29 CFR 1910.178(l).

- Where can an operator obtain the training required to become a certified forklift operator?

  The employer is responsible for implementing a training program and ensuring that only trained drivers who have successfully completed the training program are allowed to operate powered industrial trucks. An evaluation of each trained operator must be conducted during the initial training, at least once every 3 years, and after refresher training. The training and evaluation may be conducted by the employer, if qualified, or by an outside training organization.

- What type of training is required?

  The training must be a combination of formal (lecture, video, etc.) and practical (demonstration and practical exercises) instruction and include an evaluation of operator performance in the workplace. Truck-related and workplace-related topics must be included, along with the requirements of the OSHA standard. The specific training topics are listed in the standard.

- Who should conduct the training?

  All training and evaluation must be conducted by a person with the necessary knowledge, education, and experience to train operators and evaluate their competencies. This person may be the employer, another employee, or another qualified person. The training and evaluation do not have to be conducted by a single individual but can be conducted by several individuals, provided that each one is qualified.

- Is refresher training required?

  Refresher training is required when an operator has been observed driving unsafely, has been involved in an accident or near miss, has received an evaluation that indicates unsafe operation, or is assigned to drive a different type of truck, or when a workplace condition affecting safe operation changes. An operator evaluation is required after refresher training.

- What does *certified* mean?

  The employer must certify that each operator has been trained and evaluated as required by the standard. The certification must include the name of the operator, the date of the training, the date of the evaluation, and the identity of the person(s) who performed the training or evaluation.

- Does an operator who has already been trained as a powered industrial truck operator have to be retrained under the standard?

  If an operator has received training in a required topic and the training is appropriate to the truck and the working conditions encountered, additional training in that topic is not required as long as the operator has been evaluated and found competent.

- Where can I get additional information about OSHA standards?

  For more information, contact your local or regional OSHA office or go to osha.gov.

15 Appendices

Appendix B
Example Operator's Daily Checklist

OPERATOR'S DAILY CHECKLIST: INTERNAL-COMBUSTION ENGINE INDUSTRIAL TRUCK—GAS/LPG/DIESEL TRUCK

Have a qualified mechanic correct all problems.

Date		Operator		Fuel	
Truck number		Model number		Engine oil	
Department		Serial number		Radiator coolant	
Shift		Hour meter		Hydraulic oil	

Engine off Checks	OK	Maintenance
Leaks—fuel, hydraulic oil, engine oil, or radiator coolant		
Tires—condition and pressure		
Forks, top clip retaining pin and heel—check condition		
Load backrest—securely attached		
Hydraulic hoses, mast chains, cables, and stops—check visually		
Overhead guard—attached		
Finger guards—attached		
Propane tank (LPG truck)—rust corrosion, damage		
Safety warnings—attached (refer to parts manual for location)		
Battery—check water/electrolyte level and charge		
All engine belts—check visually		
Hydraulic fluid level—check level		
Engine oil level—dipstick		
Transmission fluid level—dipstick		
Engine air cleaner—squeeze rubber dirt trap or check the restriction alarm (if equipped)		
Fuel sedimentor (diesel)		
Radiator coolant—check level		
Operator's manual—in container		
Nameplate—attached and information matches model, serial number, and attachments		
Seat belt—functioning smoothly		
Hood latch—adjusted and securely fastened		
Brake fluid—check level		

Engine on Checks—Unusual Noises Must Be Investigated Immediately	OK	Maintenance
Accelerator or direction control pedal—functioning smoothly		
Service brake—functioning smoothly		
Parking brake—functioning smoothly		
Steering operation—functioning smoothly		
Drive control—forward/reverse—functioning smoothly		
Tilt control—forward and backward—functioning smoothly		
Hoist and lowering control—functioning smoothly		
Attachment control—operating		
Horn and lights—functioning		
Cab (if equipped)—heater, defroster, wipers—functioning		
Gauges: ammeter, engine oil pressure, hour meter, fuel level, temperature, instrument monitors—functioning		

Appendix C
Example Operator Training Program Outline

1. Introduction
   a. Overview of the program
   b. Goal of the program: to provide a training program based on the trainee's prior knowledge, the types of vehicles used in the workplace, and the hazards of the workplace
   c. Course will utilize video, group discussion, and hands-on practice. Each operator must obtain the knowledge and skills needed to do his or her job correctly and safely.
2. Types, Features, and Physics
   a. Familiarize each operator with the basic types and functions of powered industrial trucks.
   b. Develop an understanding of the information shown on a data plate.
   c. Understand the critical truck measurements that affect safety.
   d. Understand the forces that cause tip-overs and the truck design considerations and safety ratings that help prevent them, including the "stability triangle."
3. Inspecting the Vehicle
   a. Understand the purpose and importance of preoperational checkouts.
   b. Provide a basic understanding of areas investigated during a preoperational checkout.
   c. Familiarize each operator with a checklist for preoperational checkouts and what to do if a problem is discovered.
4. Driving the Truck
   a. Understand the elements of safe movement of a powered industrial truck.
   b. Understand the differences between an automobile and a powered industrial truck.
   c. Recognize the safety hazards associated with operating a powered industrial truck.
5. Load Handling
   a. Understand the elements of load-lifting safety.
   b. Understand the safe operating procedures for raising and lowering loads in aisles.
6. LPG for Lift Trucks
   a. Discuss LPG and its properties.
   b. Understand the elements and procedures of safely refueling internal-combustion vehicles.
   c. Describe tank components: service valve, surge valve, relief valve, and so forth.
   d. Discuss related safety issues.
7. Battery and Charging
   a. Understand the elements and procedures of safely changing and charging batteries.
   b. Discuss filling procedures and maintenance.
   c. Discuss related safety issues.
8. Safety Concerns
   a. Review/reinforce the potentials of serious injury.
   b. Review/reinforce the safety procedures in your facility.
9. Specific Truck and Workplace Training/Hands-On Training
   a. Review the features of the specific PITs to be operated.
   b. Review the operating procedures of the specific PITs to be operated.
   c. Review the safety concerns of the specific PITs to be operated.
   d. Review the workplace conditions and safety concerns of areas where PITs will be operated.
   e. Learn/practice the actual operations of the specific PITs to be operated and the specific workplace conditions where PITs will be operated.
   f. Demonstrate proficiency with performing the powered industrial truck operator duties specific to the trainee's position and workplace conditions.
10. Certification of Completion of the Course

Haulage and Off-Road Equipment

16

John F. Montgomery, PhD, CSP, CHMM

General Requirements
Operator-Restraint System (Seat Belt) Use
▶ Pre-Operational Equipment Safety Inspection (POI)
▶ Haul Roads ▶ Driver Qualifications and Training
▶ Operating Vehicles Near Workers ▶ Procedures for
Dumping ▶ Protective Frames for Heavy Equipment
▶ Transportation of Workers ▶ Towing

Power Shovels, Cranes, and Similar Equipment
Grounding Systems for Powered Equipment
▶ Maintenance Practices ▶ Operating Practices
▶ Mobile Cranes

Graders, Bulldozers, and Scrapers
Maintenance ▶ General Operating Practices

Summary

References

Review Questions

Heavy-duty trucks are used extensively for important off-the-road operations in industries such as quarrying, mining, and construction (Figure 16–1). When on the road, these trucks are governed by the same safe-driving rules and regulations that apply to other types of automotive equipment. This chapter discusses general safety issues and operating practices that can help companies achieve their safety goals. Topics covered include:

- general safety requirements for operating heavy-duty equipment
- operating power shovels, cranes, and similar equipment safely
- safety issues in the maintenance and operation of graders, bulldozers, and scrapers.

Figure 16–1. Various types of heavy-duty equipment. *(Courtesy Caterpillar, Inc.)*

GENERAL REQUIREMENTS

The use of heavy-duty trucks, mobile cranes, tractors, bulldozers, and other motorized equipment always presents the possibility of accidents. Workers near this equipment can be injured or killed; the equipment may slip over embankments, and so on (Figure 16–2). In addition, personnel who service and maintain this equipment must recognize the hazards involved.

There are four basic safety components related to haulage and the operation of off-road equipment:

1. the working environment around the equipment or machine
2. the machine itself
3. the worker
4. the work process.

Even accidents that do not cause injuries can result in serious damage to equipment, loss of production, and high replacement costs. In general, preventing accidents related to heavy equipment use requires:

1. maintenance of safety features on the equipment
2. systematic equipment maintenance and repair
3. trained operators
4. trained repair and maintenance personnel
5. trained employees
6. well-planned work processes.

Safe and proper equipment operation instructions can be found in manufacturers' manuals. Many driving practices are the same as those necessary for the safe operation of highway vehicles. Off-the-road driving, however, involves particular hazards and requires particular training and safety measures.

Figures 16–2. Here's what a scraper looked like after it had rolled down a canal berm (bottom). The operator was belted in and was only slightly hurt; if he had tried to jump off the scraper, he would have been thrown into the path of the rolling scraper. The close-up photo (top) shows the rollover protective structures (ROPSs) that protected the windshield and the operator's area from damage and probably saved the operator's life. *(Courtesy U.S. Department of the Interior, Bureau of Reclamation)*

Operator-Restraint System (Seat Belt) Use

The Occupational Safety and Health Administration, under 29 CFR 1926.602 (2014), requires operator-restraint systems (seat belts) to be provided on all equipment covered by the Society of Automotive Engineers' standard (SAE J386–1969), Seatbelts for Construction Equipment. Seat belts for agriculture and light industrial tractors should meet the requirements of SAE J333A–1970, Operator Protection for Agriculture and Light Industrial Tractors. In 29 CFR 1926.602(a)(2)(ii), OSHA states that seat belts need not be provided for equipment that is designed for stand-up operations, and in 29 CFR1926.602(a)(2)(iii), OSHA states that seat belts need not be provided for equipment that has a rollover protective structure (ROPS) or adequate canopy protection.

On June 20, 2003, the Mine Safety and Health Administration issued a direct final rule (MSHA 68 FR 19344, 2014) that updates the Administration's requirements for operator-restraint systems in off-road work machines and wheeled agricultural tractors at metal and nonmetal mines. The direct final rule states that seat belts for off-road work machines must meet the requirements of SAE consensus standard (SAE J386, 1997), Operator Restraint System for Off-Road Work Machines, as applicable. It also states that seat belts for wheeled agricultural tractors must meet the requirements of SAE J1194 (1999), Roll-Over Protective Structures (ROPS) for Wheeled Agricultural Tractors, as applicable. The direct final rule makes compliance easier by allowing mine operators to use the operator-restraint systems that manufacturers provide on new equipment, as long as the operator-restraint systems comply with the most recent revisions of the incorporated SAE standards. These revisions reflect advances in seat belt design and materials. Therefore, the direct final rule does not reduce protection for miners.

Pre-Operational Equipment Safety Inspection (POI)

Under 29 CFR 1926.601(b)(14), OSHA (2014) requires all vehicles in use to be checked at the beginning of each shift to ensure that the following parts, equipment, and accessories are in safe operating condition and are free of apparent damage that could cause failure while a vehicle is in use: service brakes, including trailer brake connections; parking system (hand brakes); emergency stopping system (brakes); tires; horn; steering mechanisms; coupling devices; seat belts; operating controls; and safety devices. All defects should be corrected before the vehicle is placed back into service. This requirement also applies to equipment such as lights, reflectors, windshield wipers, defrosters, fire extinguishers, etc., when such equipment is necessary.

This Pre-Operational Equipment Safety Inspection (POI) should be conducted before any equipment is used or operated. The driver/operator of the equipment or a designated person should use a POI checklist to conduct the inspection. Good safety record keeping suggests that POI records be retained for 90 days (3 months). If damage or a mechanical defect affecting the safe operation of a vehicle is found, the vehicle must be removed from service and tagged out of service until repaired.

Haul Roads

Both temporary and permanent roads are often too narrow to accommodate heavy equipment and oncoming traffic, especially at curves and fills. Make sure haul roads have enough space at curves so that large trucks do not need to cross the centerline of the road. Bank all curves toward the outer roadside.

1. Both temporary and permanent roadways require regular patrolling and maintenance. Serious accidents, breakdowns, delays, and unnecessary maintenance expenses can often be traced to neglected roadways. Provide members of road patrols with a means of summoning help, such as warning signs, flags, or flares, and with protective equipment, such as barricades.

2. Seasonal conditions create road hazards that require prompt attention. Some companies provide sprinkler trucks to mitigate the effects of dusts and other airborne hazards during dry and windy periods.

3. Skidding on snow and ice is a serious hazard. Make sure that snowplows and blade graders remove snow and ice as promptly and as completely as possible.

4. When roadways are built close to high banks, inspect the slopes of the banks for loose rocks, especially after rain and freezing or thawing weather. Remove all loose rocks, and verify the stability of the banks.

5. Where trucks enter public highways, install signs warning both the highway traffic and other off-road vehicles of the possible emergence of these trucks. The design, color, and placement of the signs should conform to the U.S. Department of Transportation's *Manual on Uniform Traffic Control Devices for Streets and Highways* (also published by the AASH, 2009). If operations are conducted at night, make these signs of a light-reflecting material or directly illuminate them. Where temporary roads cross railroad tracks, especially the tracks of high-speed trains, contact the railroad representatives, and place a warning device at the crossing.

Driver Qualifications and Training

The modern heavy-duty vehicle or other off-the-road equipment is a carefully engineered and expensive piece of machinery. Only drivers who are qualified physically, mentally, and by training and experience should operate such machinery.

No employees should be allowed to drive for work duties until management has ascertained their knowledge, experience, and abilities. The amount of time necessary for a prospective driver or operator to become thoroughly acquainted with the equipment, safety rules, driver reports, and emergency conditions varies. Even experienced operators should not be permitted to operate equipment until the instructor or supervisor is satisfied with their abilities.

Because accidents resulting from unsafe practices outnumber accidents resulting from unsafe equipment and roadway conditions, the time required to thoroughly train and check the performances of drivers and mechanics is well warranted. After employees have been trained, they should be closely supervised to make sure that they continue to work safely.

Operating Vehicles near Workers

Workers especially face the possibility of being struck or run over by power shovels, concrete mixers, and other equipment in garages, shops, dumps, and construction areas.

Backing—The "Crunch Zone"

The most dangerous movement is backing (Figure 16–3). The OSHA Construction Standards CFR 1926.601(b)(4) and 1926.602(a)(9) (2014) require that all vehicles with restricted vision to the rear have an automatic audible signaling device to warn workers when the vehicles are backing up, unless the vehicles are backed up only when an observer signals that it is safe to do so.

The OSHA General Industry Standard 29 CFR 1910.269(p)(1)(ii) (2014) states that no vehicular equipment

Figure 16–3. Beware of the "crunch zone." All equipment operators must signal when backing up. All other persons must give plenty of room to vehicles, especially those with limited operator visibility (e.g., this large front-end loader, whose driver cannot see anything that is within 45 ft [14 m] of the rear). This photograph also illustrates the safe practice of lowering the bucket or blade on all equipment not being used. *(Courtesy Inland Steel Corporation)*

having an obstructed view to the rear may be operated on off-highway job sites unless (1) the vehicle has a reverse signal alarm that is audible above the surrounding noise level or (2) the vehicle is backed up only when a designated employee signals that it is safe to do so.

For additional safety, company policies should require drivers of vehicles that are not equipped with backup alarms to signal with three horn blasts their intention to back up.

Where several employees are working, a driver should ask another employee to signal when the path is clear for backing or moving the vehicle. The person giving the signals should always position him- or herself within sight of the driver, and a standard set of signals should be used to ensure unambiguous communication.

Moving Forward

Serious accidents also may occur during forward movement. The hazards to workers increase as the heights and capacities of the trucks increase. Because drivers often fail to see workers crossing from the right and immediately in front of the truck, drivers should be required to blow two blasts on the horn before starting forward. Construction vehicles with attachments such as front-end loaders and dozers should have the attachments positioned for maximum visibility.

Procedures for Dumping

Vehicle operations on dumps and banks invite the danger of the vehicle going over the crest while dumping a load. A person trained in proper dumping procedures is probably the best insurance against injury to the driver and damage to the equipment. Drivers are required to follow this person's instructions and signals, especially when backing to dump. Signalers should always use a standard set of signals to ensure unambiguous communication.

The person responsible for dumping must know how close to the edge a vehicle can safely approach under various weather conditions. The signaler should stand on the driver's side of the vehicle, clear of the truck and falling material but visible to the driver. To protect the signaler further, the driver should turn and look over his or her left shoulder when backing to have the most expansive view of the area in the direction the truck is moving. To avoid hitting overhead lines or other low clearances, drivers should lower the dump box as soon as they dump their loads.

Left-hand driving also reduces the danger of going over the crest, especially in the operation of side dump trucks, because the driver is on the crest side. To help prevent the crest from caving in, stockpiles and dumps are frequently graded toward the crest so that vehicles back up the slope. Drivers may also dump loads a safe distance from the crest; the material is then pushed over the crest by a grader or dozer.

Solidly built cabs, cab protectors on canopies, and safety belts are effective in preventing injuries if vehicles overturn. An operator's inclination to jump away from a vehicle that is beginning to roll over or slide down an embankment is not advised; it is far safer to remain in the cab.

Holes, ruts, and similar uneven places on dumps and roadways may cause the front wheels of a truck to lock up, resulting in the steering wheel spinning and injuring fingers, arms, and ribs, particularly if the truck is not equipped with power steering. Gripping the steering wheel on its outside and not by the spokes, driving at reduced speed, and surveying the ground ahead for rough places will help the driver prevent such injuries.

Using floodlights during nighttime dumping operations also helps prevent accidents.

Protective Frames for Heavy Equipment

All bulldozers, tractors, and similar equipment used in clearing operations must be equipped with extensive guards, shields, canopies, and grilles to protect operators from falling and flying objects. Equip crawlers and rubber-tired vehicles; self-propelled, pneumatic-tired earth movers; water tank trucks; and similar equipment with steel canopies and safety belts to protect operators from the hazards of rollover (Figures 16–2 and 16–4). Managers should train drivers in the use of safety equipment and require them to wear safety belts.

A canopy and its support should be able to bear at least two times the weight of the prime mover. This calculation is based on the maximum strength of the metal and on the integrated loading of support members, with the resulting load concentrated at the point of impact. In addition, there should be a vertical clearance of 52 in. (132 cm) from the deck to the canopy where the operator enters or leaves the seat.

For more details, see National Safety Council, Occupational Safety and Health Data Sheet 622, Tractor Operation and Roll-Over Protective Structures (2005).

Transportation of Workers

Some jobs require workers to be transported to and from the worksite. Because transporting employees can involve particular risks, management should take appropriate precautions.

Hazards are more likely to occur when transportation of workers is infrequent. Often in this situation, a pickup, an open cargo truck, or another vehicle not designed for transporting personnel is used as the transporting vehicle. Although such transportation is not recommended, when it is used, advise workers of the following safety procedures:
- A seat belt should be available for each rider in the vehicle.
- A flat surface where employees get on and off the vehicle should be provided.

Figure 16–4. The scraper (top) and bulldozer (bottom) have protective bars and screens to protect operators. *(Courtesy Caterpillar, Inc.)*

- The vehicle brake should be set prior to workers boarding or exiting the vehicle.
- When getting on the vehicle, workers should look before and where they step and use every handhold available, even a helping hand from someone already on the vehicle. Workers should also step squarely, never at an angle. No one should ever attempt to board a moving vehicle, no matter how slowly it is traveling.
- If possible, benches should be provided for passengers. In no case, however, should passengers remain standing while the vehicle is in motion. If necessary, they should sit on the truck bed.
- Employees should avoid horseplay. Of all the undesirable actions that a group of people may engage in, horseplay is one of the most foolish and dangerous. Some companies fire anyone caught taking part in this type of activity. Already inherently dangerous, horseplay may be fatal in a moving vehicle.
- When getting off the vehicle, workers should look before they step and then get off slowly and carefully, using every available handhold. Under no circumstances should anyone attempt to jump off a moving or even a stopped

vehicle. Many injuries occur when jumping off a vehicle, whether it is moving or not.
- Employees should be prohibited from riding on top of any load that could shift, topple, or otherwise become unstable.

Towing

Towing is a hazardous operation, especially when coupling or uncoupling equipment. Workers can be crushed when a truck or other piece of equipment moves unexpectedly while they are between the two machines.

The following safe practices are essential for preventing accidents during the coupling or uncoupling of motorized equipment:

1. No one should move between the vehicles while either one is in motion.
2. Parked vehicles must be secured against movements by setting the brakes, blocking the wheels, or both.
3. A driver should not move a vehicle while someone is between the vehicle and another vehicle, a wall, or anything else that is reasonably solid and immovable. Before moving the vehicle, the driver should receive an all-clear signal.
4. Tow bars are more secure than tow ropes or chains or cables. If ropes, chains, or cables are used, they must be in good condition and of sufficient size and length for the towing job.
5. Equipment towed on trailers should be secured to the trailer.
6. When a trailer is uncoupled from a towing vehicle, use a jack on the front and block the rear to prevent tilting. Unexpected tilting of a trailer can cause worker injury and vehicle damage.

POWER SHOVELS, CRANES, AND SIMILAR EQUIPMENT

Safe operation of power shovels, draglines, and similar equipment begins with the machines' purchase. A good policy is to require in the equipment specifications that guards cover gears and that the manufacturer provide oiling devices, handholds, slip-resistant steps, and other safeguards. In any case, before equipment is put into operation, managers or supervisors should inspect it thoroughly and install necessary safety devices.

To keep workers from being caught between truck frames, crawler tracks, cabs, and counterweights of cranes and shovels, erect a barricade to warn nearby employees that they are close to a hazardous area. Signs, flashing lights, and other warning devices can also be used to alert people to the hazards. Barricades can easily be taken down when the equipment is moved.

Figure 16–5. Instruct both experienced and new operators and maintenance personnel in manufacturer-recommended procedures for inspection, maintenance, and repairs.

Instruct experienced and new operators and maintenance personnel in manufacturer-recommended procedures that pertain to lubrication, adjustments, repairs, and operating practices, and make sure workers observe these procedures (Figure 16–5). A preventive maintenance program for industrial shovels and other equipment is essential for safe and efficient operations. Frequent inspections and prompt repairs are the bases of effective preventive maintenance. Generally, the operator is responsible for inspecting the condition of such items as hold-down bolts, brakes, clutches, clamps, hooks, and similar essential parts.

Workers should keep wire ropes, hooks, clamps, and pulleys lubricated in accordance with the manufacturers' instructions and should inspect them daily. Equipment failures can cause serious accidents. Ropes are particularly likely to develop deficiencies at fastenings, at crossover points on drums, and in sections that frequently contact sheaves. Construction, installation, operation, inspection, and maintenance of crawler cranes, locomotive cranes, wheel-mounted cranes, and any variations that retain the same fundamental characteristics should comply with ANSI/ASME B30.5–2004, Mobile and Locomotive Cranes (2004).

Grounding Systems for Powered Equipment

Proper protection from electrical shocks is essential to the safe operation of powered equipment. To prevent electrical shocks, electrically powered equipment must have a grounding system that protects workers from electrical faults in trailing cables and at the equipment. Although the cable might have physical contact with the ground's surface, the resistance to the current's flow from the equipment frame and the cable to the earth is usually significant because of the cable's insulation.

A machine-to-ground fault resistance of 100 ohms and a current of 10 amps translates to an electric shock of about 1,000 volts. A leakage fault current of 7–15 mA is very painful and can lead to the loss of muscular control. As little current as 70 to 200 mA coursing through the body can result in death. (See Chapter 8, Electrical Safety). Because low-leakage currents can be sent through wet skin by commercial 120-V AC, no employee should be exposed even momentarily to this electrical hazard.

A good earth–ground system can be created by driving electrodes or copper-clad steel rods into suitable soil to a depth of at least 8 ft (2.4 m). Rods that are available commercially for this purpose lower the earth's resistance and provide better grounding. Because the appropriate number of rods and their appropriate distribution depend considerably on soil conditions, a company may have to increase soil conductivity by applying common salt, sodium nitrate, or a similar chemical that is then carried into the soil by rain. A grounding system having a total resistance of 1 ohm, including the cable, can be obtained in many areas through proper design and construction.

The pole-line ground wire should have at least the same wire gauge size as the power wires. Wherever current is tapped from the power line, a connection is made from the pole ground wire to a ground wire in the cable. A good cable has metal shielding outside the insulating material that surrounds each conductor, and the ground conductor is in complete electrical contact with the shielding. The shielding should have electrical continuity that, if broken, is bridged.

The ground wire in the cable is connected to the equipment frame, where there should be good metal-to-metal contact. If necessary, scrape off paint or another coating to permit good contact. A resistor between the pole-line ground wire and the transformer's neutral system limits the amount of current to no more than 50 amps, thus preventing dangerous voltages at the powered equipment and permitting sufficient current to open the circuit breakers.

The neutral ground system permits the operation of all equipment except the machine where the fault occurred. That machine is segregated from the rest of the system as soon as the fault activates the circuit breaker. Suitable switching equipment in the neutral grounded system eliminates the danger of several faults occurring at once in different phases and at different locations, except for the space required for the circuit breakers to open. Do not connect the equipment ground to the substation ground in any way; this lack of connection prevents a fault in the power supply from energizing the equipment.

Inspect circuit breakers and other devices in the ground-ing system monthly, and regularly check the resistance of ground rods. A megohmmeter test of the cable insulation is a recommended part of cable inspection. Workers should not tape defective cable insulation but instead replace the cable. Workers should wear rubber gloves and use insulated tongs or hooks for handling trailing power cables.

Minimal wear and damage to trailing cables is important to ensure safety. Shield the cables from blasting operations as much as possible while keeping them as close to the operations as practical so that only a minimum length of cable is required. Workers can use tripods and wooden construction horses to keep cables off the ground. Tripods are preferable to trenches when cables cross roads.

Assign an electrician to regularly inspect and maintain the electrical parts of shovels and similar equipment, including trailing cables.

Maintenance Practices

If a power shovel is being used in a deep excavation, workers should make needed repairs or adjustments where the shovel will not be endangered by falling or sliding rocks or earth. The operator is responsible for setting the brakes; securing the boom; lowering the dipper or bucket to the ground; taking the machine out of gear; and, before leaving the machine, performing additional precautions to prevent unexpected movement.

Before repairing any vehicle, maintenance personnel should notify the operator about the problem's nature and location. If the work is to be done on or near moving parts, the operator should turn off the equipment using regular shutdown procedures, and the maintenance personnel should lock out the controls. Only the maintenance person that installed the lock can remove the lock. This requirement is essential to prevent the operator from starting the equipment while it is still being worked on. Before inspecting equipment or checking lubricant levels, the worker should make sure that the machine is safe. A caution sign such as INSPECTOR UNDERNEATH can be used.

If machine parts must be in motion while employees are working on them, make sure the parts are turned slowly, by hand if possible, in response to guidance or a prearranged signal if two or more persons are involved. This precaution applies particularly to those who work around gears, sheaves, and drums. Workers who grab ropes just ahead of the sheaves risk having their hands yanked into the sheaves. To prevent hand injuries, use a bar to guide any rope being wound onto a drum.

If guards must be removed for ease of making repairs, the job should not be considered completed until workers replace the guards, plates, and other safety devices. Repair personnel should wear snug-fitting clothing, eye protection, and protective shoes.

Operating Practices

Rocks and other materials falling from steep faces and banks cause some of the worst accidents involving power shovels. In some instances, rockslides have buried the shovel and the operator, and in others, falling materials have struck and injured workers who were working around the equipment.

Some quarries that have high faces limit the height of banks to 25 ft (7.6 m) by benching and by blasting. These procedures force enough rock from the face to reduce the rock pile's height. As a result, the shovel operator is able to keep the bank at a safe slope. When loading under a high face, the operator should swing the shovel to the clear side and away from the face, thereby providing a better view and reducing injury potential.

Undercutting banks of earth, sand, gravel, and similar materials is dangerous, especially during winter and spring months. Freezing and thawing can result in a collapse of the overhanging material. To maintain a safe slope, the work crew should blast the overhanging material and remove it.

The shovel operator is responsible for the safety of employees who must work near the shovel. These workers may be struck by falling rocks, squeezed between the shovel and the bank or similar pinch points, or struck by the dipper. No worker should enter a dangerous location without first notifying the operator, who, in turn, should not move the equipment while the employee is passing by.

Operators should fill dippers to capacity but not to overflowing in order to prevent excess material from spilling out of the dipper and falling on workers. Whenever possible, loading should be done on the blind side. The operator should not swing a load over a vehicle nor load a truck until its driver has dismounted and is in the clear. The only exception is when the truck has a canopy designed to protect the driver and the truck. Railcars should be loaded evenly so that earth or rocks do not spill over the sides.

Supervisors should instruct workers to observe good housekeeping practices on and around the shovel. The operator should store tools in a designated place and keep the cab floor free of grease and oil. Ice and snow should be removed promptly, and a dozer should keep the area around the shovel free of rocks and ruts.

Workers should get on or off a shovel or dragline only after notifying the operator of their intentions. The operator, in turn, should swing the platform so that workers can grasp the sides and use steps or tread. It does not matter whether the operator is swinging a load or the equipment is stationary; no one should get on or off by jumping onto the tread. Do not permit any unauthorized person on a shovel or dragline.

Mobile Cranes

The noteworthy characteristic of accidents involving crawlers and similar cranes is the severity of the injuries.

Although these accidents occur relatively infrequently, the resulting injuries are about twice as serious as those resulting from accidents involving other types of heavy equipment. For this reason, supervisors and managers should select crane operators based on their intelligence, reliability, and willingness to follow instructions. Because accidents usually occur when an operator is performing more than one task and becomes confused, distracted, or agitated, operators' duties should be limited to those essential for safe crane operation.

Operators are primarily responsible for the safe conditions of their cranes. They must regularly inspect brakes, ropes and their fastenings, and other essential parts and promptly report worn, broken, or otherwise defective parts. Like other equipment, mobile cranes should be maintained on a regular schedule.

Operators are also responsible for the safety of the oiler and for preventing injuries to hookers, riggers, and others working around the equipment. However, anyone working in the vicinity of a crane should learn to stay clear of the boom and, in no case, work or cross under the boom.

Some of the most severe accidents result from overloading cranes. Operators should never exceed the manufacturer-specified load limits for the various positions of the boom. Post these load limits prominently in the crane cab. If an operator has any doubts about a load's weight, he or she should test the crane's capability by first lifting the load slightly off the ground. Because operating a crane on soft or sloping ground is dangerous, always make sure the crane is level before putting it into operation. Outriggers provide reliable stability only when used on solid ground. Using makeshift methods, such as timbers for blocking, to increase the lifting capability of a crane is extremely dangerous and should not be permitted.

Clutch linings may swell during wet weather, and the master clutch, boom clutch, or both may drag and cause the boom to be pulled over backward. Operators should thus test the clutches before starting work on rainy days and adjust the clearances if necessary.

The sudden release of a load when the boom angle is high (e.g., from the parting of a sling) is another cause of crane accidents. Boom stops limit the travel of the boom beyond the angle of 80 degrees above the horizontal plane and prevent the boom from being pulled backward and over the top of the machine by the boom-hoisting mechanism or by the sudden release of a heavy load suspended at a short radius. Either of these occurrences usually results in serious damage to the equipment and serious injuries to the operator or other workers.

Boom stops are best suited to medium-sized cranes (the 5- to 60-ton range). Boom stops should disengage the master clutch or kill the engine and stop the boom before it reaches the maximum permissible angle. One type of stop meeting these requirements has a piston and cylinder. It is spring or pneumatically actuated and is mounted on the A-frame to intercept the boom as high above the boom hinges as possible. By positive displacement of an actuator mounted on the A-frame, the boom action disengages the master clutch (or ignition breaker or compression release) by means of light rope reeved over a few small sheaves.

When a mobile crane operator must work near electric power lines, the supervisor should consult with the power company to determine whether the lines can be deenergized. Most mobile crane operator fatalities result from contact with power lines, and often the power company's service is seriously disrupted because of this contact. Various states and OSHA have enacted legislation specifying the distances operators must keep their booms and wire ropes from power lines. A minimum of 10 ft (3 m) is often specified; however, the necessary distance increases with increasing voltage. OSHA 29 CFR 1926.550(a)(15) (2014) states that—except where electrical distribution and transmission lines have been deenergized and visibly grounded at the point of work or where insulating barriers that are not part of or attached to the equipment or machinery have been erected to prevent physical contact with the lines—equipment or machines must be operated near power lines in accordance with the following:

1. For lines rated 50 kV or below, the minimum clearance between the lines and any part of a crane or load should be 10 ft.
2. For lines rated over 50 kV, the minimum clearance between the lines and any part of a crane or load should be 10 ft + 0.4 in. for each 1 kV over 50 kV, or twice the length of the line insulator but never less than 10 ft.
3. The recommendations of the power company and all legal requirements should be observed (Figure 16–6).

An experienced operator working with an untrained or relatively inexperienced hooker or rigger should determine the details of lifts, such as the type of sling and hitch to be used.

Although an operator can usually rely on the knowledge of an experienced rigger, the operator is ultimately responsible for the safety of a lift and should ask the supervisor for any needed assistance. The following safe practices are essential when handling loads:

1. Operators should center the hook over the load to keep the load from swinging when lifted.
2. Employees should keep their hands out of the pinch point when holding the slings in place while the slack is taken up. A hook or even a small piece of board may be used to hold the slings in place. If workers must use their hands, they should hold the slings in place with the flat of their hands.
3. The hooker, rigger, and all other personnel must be in the clear before a load is lifted.
4. Tag lines should be used for guiding loads.
5. Hookers, riggers, and others working around cranes also must keep clear of the swing of the boom and the cab.
6. No load should be lifted or moved before the operator receives a signal. When the operator cannot see the entire movement of a load, as when lowering a load into a pit, a trained person should be posted as a guide. To prevent confusion, use only standard hand signals.

Figure 16–6. When mobile cranes must be operated in the vicinity of electric power lines, first consult with the utility company to determine whether the lines should be deenergized. OSHA and many state regulations require that booms and wire ropes be kept a minimum of 10 ft (3 m) away from the power lines. *(Courtesy Washington State Department of Transportation)*

GRADERS, BULLDOZERS, AND SCRAPERS

Many of the fundamental safety measures recommended for off-road trucks also apply to graders and other types of earth-moving equipment. Operators should regularly inspect their machines and promptly report any defects and malfunctioning systems or parts. Scheduled maintenance should be carried out to increase the safety and efficiency of the equipment. Management should select as operators only physically and mentally qualified individuals and train them in the correct operating practices specified in the manufacturers' manuals and according to company requirements. Specific precautions will help prevent injury when workers are servicing and repairing these machines. In addition, employees should adhere to the safety procedures that apply to other types of motorized equipment.

Maintenance

Regularly inspect brakes, controls, engines, motors, chassis, blades, blade holders, tracks, drives, hydraulic mechanisms, transmissions, and other essential parts. Likewise, frequently check wheel- and engine-mounted bolts for tightness.

Making adjustments and repairs while the engine is running is dangerous, particularly when employees are working near the engine fan or are adjusting a tractor clutch. Refueling should be done only when the engine is stopped.

When inflating split rim tires, workers must guard against a locking ring blowing off. OSHA 29 CFR 1910.177 and 1926.600 require restraining devices, barriers, safety tire racks, cages, or equivalent protection to be provided and used when inflating, mounting, or dismounting tires installed on split rims or on rims equipped with locking rings or similar devices. All employees servicing multi-piece wheels must receive training in safe operating procedures.

When replacing cutting edges, operators should always block up the scraper bowl or dozer blade. After the scraper has been lifted to the desired height, place blocks underneath the scraper's bottom near the ground plates. Raise the apron arms to the greatest height and place a block under each arm so that the apron can drop enough to wedge each block firmly in place.

Before receiving wire rope on a drum or through sheaves, the operator should disengage the master clutch, idle the engine, and lock the brakes. The engine should be at a complete stop before working with the rope on a front-mounted drum.

If an operator assisting a maintenance worker is working behind the scraper and with the tailgate in the forward position, a block should be placed behind the tailgate so that it cannot fall. This precaution is necessary in case someone releases the power control brake, which would permit the tailgate to fall backward. When replacing ropes on scrapers, make sure the tailgate is upright and at the end of its movement.

General Operating Practices

Operators must look to the front, sides, and rear before moving their machines. They should also be constantly alert for employees on foot when operating near other equipment, offices, tool and supply buildings, and similar locations.

Speed should be governed largely by the conditions. Slow speeds are essential when driving (1) off the road and beyond the shoulder, on steep grades, and in rough terrains in order to prevent violent lurching that may throw the driver off the machine or against levers and cause serious injury; (2) in congested areas; and (3) in icy and other slippery conditions. Jumping off a standing machine can result in sprained ankles and other injuries. The best practice is to step down after checking to make sure that footing is secure and that no other vehicles are approaching. Ice, mud, stones, holes, and similar factors also present falling hazards. Likewise, deck plates and equipment steps should be free of grease and other slippery substances.

An operator should not drive equipment onto a haul road without first stopping and looking both ways, whether or not the entranceway includes a stop sign. Generally, loaded equipment is given the right-of-way on job or haul roads.

Before an operator leaves equipment even for a short time, the bucket or blade should be lowered to the ground and the engine stopped. A safe parking location is on level ground, off a roadway, and out of the way of other equipment. An operator should never leave equipment on an inclined surface or on loose material with the engine running; the engine's vibration can put the equipment in motion.

Procedures on Roadways

When graders, scrapers, and other earth-moving equipment are in operation along a section of road, the precautions discussed next will help prevent harm to the public, to employees, and to the equipment.

Warn oncoming traffic of the danger ahead by placing barrier signs at both ends of the section that is undergoing construction. Primary warning signs such as ROAD UNDER CONSTRUCTION or BARRICADE AHEAD should be placed 1,500 ft (460 m) from the starting point of the construction.

Orange flags or markers at the ends of blades, which may project beyond the tread of a machine, serve to warn traffic and other equipment operators. An orange flag on a staff that projects at least 6 ft (1.8 m) above the rear wheel of a blade grader or mowing vehicle is recommended for operations in hilly terrain (Figure 16–7). Affix slow-moving vehicle emblems at the rear of these vehicles if they must be driven even short distances on public roads (Figure 16–8).

Operators of motor graders should keep to the right side of a roadway. When blading against traffic is necessary, use flags and barricades to warn oncoming traffic. Warning signs must be placed at a significant distance from the work area. This distance should increase as highway speed increases. Suggestions are given in the American Association of State Highway and Transportation Officials' *Manual on Uniform Traffic Control Devices*. Most states use this standard as the minimum requirements, with additional traffic control devices when conditions dictate. When operations are extensive, personnel with warning devices should stand at each end of the

Figure 16–7. Mowers should have ROPSs, a slow-moving vehicle emblem, and an orange flag projecting at least 6 ft (1.8 m) high. *(Courtesy Navistar International Corp.)*

Figure 16–8. Close-up of a slow-moving vehicle (SMV) emblem.

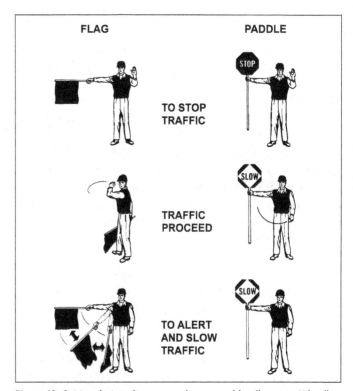

Figure 16–9. Hand-signaling procedures used by flaggers. The flag and the background of the STOP sign are bright red; the diamond of the SLOW sign is orange. *(From Manual on Uniform Traffic Control Devices)*

working area so that they are visible to oncoming traffic at least 500 ft (152.5 m) away.

When earth-moving equipment will be stopping, turning, or backing at curves, crests of hills, and similar dangerous locations, station signals or personnel with flags or other devices to warn motorists. For maximum safety, such movements generally require an unobstructed view of approaching traffic up to a distance of about 1,000 ft (305 m).

Use signal personnel when the working area is congested with other equipment, workers, buildings, excavations, and similar hazards. See Figure 16–9 for hand-signaling procedures.

Coupling and Towing Equipment

An operator should not back up to couple a tractor to a scraper, sheepsfoot roller, or other equipment without first checking to make sure that all nearby personnel are in the clear. An operator assisted by a person on the ground should not move the equipment until signaled to do so.

Before an employee is allowed to couple the trailing equipment, he or she should stop the tractor, put the shift lever in neutral, and set the brakes. He or she should also block the wheels of the equipment to be coupled.

All equipment being towed should be secured not only with a regular hitch or drawbar but also with a safety chain attached to the pulling unit. A drawbar failure can result in a serious accident. When towing a scraper from one job to another, the operator should use a scraper bowl safety latch or place a safety bolt in the beam to allow maximum clearance for road projections, such as at crossings. This precaution prevents the bowl from striking the ground or pavement and injuring persons or damaging equipment.

Clearing Work

Work taking place near low-hanging tree limbs or high brush involves serious hazards. These hazards can be readily eliminated by using suitable protective measures and following safe practices.

When using a bulldozer, equip it with a heavy, well-supported, arched steel-mesh canopy to protect the operator. (See the Protective Frames for Heavy Equipment section earlier in this chapter.) The operator also should wear goggles to shield his or her eyes from whipping branches.

Head protection guards against injuries from falling branches. When a bulldozer shoves hard against the butt of a large dead tree, the tree may crack in the middle, or limbs may fall onto the machine. Dead branches also can drop from live trees. A safe procedure for eliminating this danger is to cut tree roots on three sides and then apply power to the fourth side. Use a long rope to pull over large trees, but make sure in advance that the tractor and the operator will be in the clear when the tree falls. Before pushing over any trees, bulldozing rock, or rolling logs, operators should ensure that all workers in the area are out of harm's way.

Special Hazards

Fatalities can occur easily when equipment is used in dumps and fills, near excavations, and on steep slopes. The operator should keep the bulldozer blade close to the ground for balance when the machine is traveling up a steep slope.

When a worker is driving a tractor-dozer down a slope, he or she should bulldoze three or four loads of dirt to the bottom of the slope, keeping the loads in front of the blade. If any dirt is lost on the way down, the operator should not lower the blade to regain the load because of the possibility of overturning. Never use the blade as a brake on a steep slope except in cases of extreme emergency.

Ground conditions will determine how close to an excavation or the crest of a dump a driver can safely operate a machine. Wet weather means the operator must work at a greater distance from the edge or crest. When the ground is treacherous, assign someone to signal the driver as to where it is safe to operate the machine.

Sometimes employees, the public, livestock, and property are endangered when material is pushed over the edge during side hill work. In such cases, make sure there is sufficient clearance below before the work begins.

SUMMARY

- To prevent damage to heavy equipment, companies must maintain safety features and equipment, train operators, and train repair and maintenance personnel. Heavy-duty equipment requires haul roads that have been constructed especially for these vehicles.
- Only qualified drivers should operate heavy-duty equipment. All heavy-duty vehicles must be equipped with extensive guards, shields, canopies, and grilles to protect the operator.
- When workers are transported to and from worksites, supervisors must make sure that employees know safe procedures for riding in and getting on and off the vehicles.

- During towing operations, safe practices should be used to prevent accidents in the coupling and uncoupling of motorized equipment.
- The operation of power shovels, cranes, and similar equipment requires such safeguards as oiling devices, handholds, and slip-resistant steps. Operators are responsible for the conditions of their equipment, for inspecting all parts and reporting any problems, and for the safety of those working around the machines.
- Electric-powered equipment must have a good grounding system to protect workers from electrical hazards and shocks. Workers should also wear protective equipment when handling power cables or other potentially dangerous equipment.
- Companies must carefully train employees in safe practices for maintaining and repairing powered equipment and in safe operating procedures. All employees should observe good housekeeping practices.
- Mobile crane accidents are infrequent but result in severe injuries to workers. To help prevent accidents, operators must be trained to handle loads safely.
- Graders, bulldozers, and scrapers must be inspected regularly, and any problems must be corrected immediately. As with other heavy-duty equipment, only those employees who have been carefully selected and trained should drive these vehicles.
- Operators and other employees must be trained in the proper procedures for using towing and coupling equipment and for performing clearing work.
- Operators and other workers on a construction site or a worksite must be aware of the particular hazards associated with heavy-duty equipment. Supervisors must ensure that employees observe safety practices at all times.

REFERENCES

Mobile and Locomotive, Cranes, ANSI/ASME B30.5-2007, American Society of Mechanical Engineers, Three Park Avenue, New York, NY 10016.

Manual on Uniform Traffic Control Devices, American Association of State Highway and Transportation Officials, 444 N. Capitol Street NW, Suite 249, Washington DC 20001, 2009.

National Safety Council, 1121 Spring Lake Drive, Itasca, IL 60143-3201.

Occupational Safety and Health Data Sheets:
Berms in Pits and Quarries, 12304–680, 1990.
Falling or Sliding Rock in Quarries, 12304–332, 1990.
Motor Trucks for Mines, Quarries, and Construction, 12304–330, 1990.

Operation of Power Shovels, Draglines, and Similar Equipment, 12304–271, 1992.

Snow Removal and Ice Control on Highways, 12304–638, 1992.

Tractor Operation and Roll-Over Protective Structures, 12304-622 May, 2005.

U.S. Department of Labor, Mine Safety and Health Administration, 68 FR 19344, Seat Belts for Off Road Work Machines and Wheeled Agriculture Tractors at Metal and Nonmetal Mines.

U.S. Department of Labor, Occupational Safety and Health Administration, 200 Constitution Avenue NW, Washington DC 20210.

29 CFR 1910.269(p)(1)(ii), Electric Power Generation, Transmission, and Distribution.

29 CFR 1910.269(p)(1)(iii), Electric Power Generation, Transmission, and Distribution.

29 CFR 1910.177, Servicing Multi-Piece and Single Piece Rim Wheels.

29 CFR 1926.550(a)(15), Clearance between Electrical Power Lines and Cranes.

29 CFR 1926.600, Motor Vehicles, Mechanized Equipment, and Marine Operation 29 CFR

1926.601(b)(4), Construction Standard.

29 CFR 1926.601—Motor Vehicles.

29 CFR 1926.602(a)(9)—Material Handling Equipment.

29 CFR 1926.602(a)(2)(ii), Electronic Code of Federal Regulations.

29 CFR1926.602(a)(2)(iii), Electronic Code of Federal Regulations.

U.S. Department of Transportation, Bureau of Motor Carrier Safety Regulations, 400 Seventh Street SW, Washington DC 20590.

Manual on Uniform Traffic Control Devices for Streets and Highways. 3rd ed. 2003.

SAE International Washington Office, 1200 G St. NW, Suite 800, Washington DC 20005.

SAE J1194, Roll-Over Protective Structures (ROPS) for Wheeled Agricultural Tractors, 1999.

SAE Consensus Standard J386, Operator Restraint System for Off-Road Work Machines, 1997.

SAE J333A–1977, Operator Protection for Agriculture and Light Industrial Tractors.

SAE J386, Seat Belts for Construction Equipment, 2012.

REVIEW QUESTIONS

1. What signal should be made before a truck goes forward?
   a. The driver should flash the truck's lights.
   b. The driver should blow two blasts on the truck's horn.
   c. The driver should turn on the truck's hazard light.
   d. The driver should wait for the signaler's hand signals.

2. What are the three purposes of having the signaler stand on the driver's side of the vehicle during dumping procedures?
   a.
   b.
   c.

3. How much weight should the canopy and the canopy support of heavy machinery be able to bear?

4. What are five guidelines that should be followed when coupling or uncoupling motorized equipment?
   a.
   b.
   c.
   d.
   e.

5. Grounding rods or electrodes should be driven at least _____ into the ground.
   a. 3 ft
   b. 5 ft
   c. 8 ft
   d. 12 ft

6. What is the purpose of boom stops?

7. List three of the six essential safe practices for handling loads.
   a.
   b.
   c.

8. What three conditions make driving at slow speeds extremely important?
   a.
   b.
   c.

9. What are the precautions that should be taken before stepping down from standing machinery?

10. When traveling up a slope, the operator should keep the bulldozer blade _____ for balance.
    a. even with the top of the cab
    b. at a 75-degree angle
    c. hitched to a tow bar
    d. close to the ground

PART 4

Production Operations

The chapters in Part 4 explore the hazards and injury prevention strategies associated with selected production activities. Hand-held and portable power tools are significant sources of exposure to workers in a variety of production operations. Potential hazards in the manufacturing sector arise from tools and machines that weld and cut hot and cold metals. Other sources of potential exposure to hazards have developed in automated lines, systems, and processes. These chapters describe administrative controls, safe procedures, and mechanical safeguards that can reduce the hazards and incidence rates in these production operations.

The last chapter in Part 4 explores the increasing use of computers as a safety information tool. As computers become more powerful and easier to use, they have also become a valuable asset to any safety program. This chapter explores how the safety professional can use electronic information for reference, networking, marketing and education, and information management.

Hand and Portable Power Tools

Cristine Z. Fargo
Alan Barr, MS Eng.
John Kurtz
Philip E. Hagan, JD, MBA, MPH, ARM, CIH, CHMM, CET, CHCM, CEM

Preventing Incidents
Tool Selection ▶ Changing Tools ▶ Safety Practices ▶ Central Tool Control ▶ Toolboxes ▶ Carrying Tools

Maintenance and Repair
Inspection and Control ▶ Redressing Tools ▶ Handles

Use of Hand Tools
Screwdrivers ▶ Hammers ▶ Punches ▶ Tools for Cutting Metal ▶ Tools for Cutting Wood ▶ Miscellaneous Cutting Tools ▶ Tools for Materials Handling ▶ Wrenches ▶ Spark-Resistant Tools ▶ Soldering Irons

Portable Power Tools
Hazards and Safety Precautions ▶ Selecting Tools ▶ Inspection and Repair ▶ Electrical Tools ▶ Air-Powered Tools ▶ Percussion Tools ▶ Special Power Tools

Personal Protective Equipment

Summary

References

Review Questions

Hand and power tools enable employees to apply additional force and energy to accomplish a task. These tools improve efficiency and make better products. Because of the increased force of hand and power tools, the objective of safety with these tools is to protect users from inflicting harm to themselves and others, as well as to provide ergonomically designed tools (see References). Through proper selection, use, care, and supervision of hand and power tools, injuries related to the use of tools can be prevented.

This chapter covers the following topics:

- safety practices and work procedures for preventing unintentional incidents using hand and power tools
- safety issues in the repair and maintenance of these tools
- safe work methods for handling various hand tools
- hazards and safety precautions for handling portable power tools
- personal protective equipment (PPE) used with hand and power tools.

PREVENTING INCIDENTS

Disabilities resulting from either the misuse of tools or using damaged tools include loss of eyes and vision; puncture wounds from flying chips; severed fingers, tendons, and arteries; broken bones; contusions; infections from puncture wounds; ergonomic stress; and many other injuries.

Tool Selection

Consider all aspects of the work situation when selecting hand and portable power tools. The tools selected should (1) perform the job and (2) should not cause the employee any physical pain or discomfort. Ergonomics factors should be considered in tool selection. The checklist in Figure 17–1 will assist in evaluating hand tools from an ergonomics perspective. Seeking employee input in the selection of tools is also helpful.

Consider the following key points when selecting tools:

- The handle's shape and form and the material used to make it can minimize stress to employees' hands and upper body. An ergonomically well-designed tool will reduce the incidence of fatigue and injury and may improve productivity as well. Antivibration handles can be used to effectively reduce vibration-related stress.
- Quality of the tool, including sharpness of cutting edges, affects the amount of force needed to do a job. Jobs using the same tool can vary

greatly in the amount of force that employees must apply.

- Power tools designed to have minimal vibration will be more comfortable to use and less likely to result in hand-arm vibration syndrome (HAVS). Also known as Reynaud's syndrome and vibration white finger, HAVS is a disease entity characterized by circulatory disorders that include blanching and numbness of fingers; loss of dexterity; and, in more severe cases, permanent disability affecting a worker's muscle, bone, nerves, and joints of fingers and hands. Provide pads and gloves for relief from the vibration. Administrative controls, such as work rotation and more frequent rest breaks, can be used to reduce the time workers use vibrating tools.
- Vibration levels of particular tools are often published on the information/data sheet of the particular tool, and the information can be used to estimate appropriate exposure limits or the effectiveness of integrated vibration controls.
- If possible, set up some foot-operated controls to avoid repeated use of hands and fingers. Excessive repetitive motion can also cause damage to the foot and leg muscles, nerves, and/or tendons.

Hand Tool Analysis

No responses indicate potential problem areas which should receive further investigation.

	Yes	No
1. Are tools selected to limit or minimize		
• exposure to excessive vibration?	❏	❏
• use of excessive force?	❏	❏
• bending or twisting the wrist?	❏	❏
• finger pinch grip?	❏	❏
• problems associated with trigger finger?	❏	❏
2. Are tools powered where necessary and feasible?	❏	❏
3. Are tools evenly balanced?	❏	❏
4. Are heavy tools suspended or counterbalanced in ways to facilitate use?	❏	❏
5. Does the tool allow adequate visibility of the work?	❏	❏
6. Does the tool grip/handle prevent slipping during use?	❏	❏
7. Are tools equipped with handles of textured, nonconductive material?	❏	❏
8. Are different handle sizes available to fit a wide range of hand sizes?	❏	❏
9. Is the tool handle designed not to dig into the palm of the hand?	❏	❏
10. Can the tool be used safely with gloves?	❏	❏
11. Can the tool be used by either hand?	❏	❏
12. Is there a preventive maintenance program to keep tools operating as designed?	❏	❏
13. Have employees been trained		
• in the proper use of tools?	❏	❏
• when and how to report problems with tools?	❏	❏
• in proper tool maintenance?	❏	❏

Source: NIOSH. Elements of Ergonomics Programs. *March 1997.*

Figure 17–1. A checklist such as this sample can assist in the evaluation of ergonomics factors of hand tools.

Before placing tools with employees, also consider the workstation and the work methods. Workstations should be adjustable for each employee. In addition, workstations should allow for the full range of movements required for the job and tools. When possible, provide mechanical means of handling materials. To reduce employees' stress, allow employees to perform a variety of jobs that require use of different muscle groups.

Work methods should achieve maximum production while causing a minimum of stress to employees. Consider the following factors when deciding on appropriate work methods:

- force needed to hold and use a tool
- direction of the force
- weight of materials
- duration and number of repetitions of an activity
- employee posture.

Changing Tools

To determine if tools should be changed, consider the following factors:

- employees' concerns about tool problems
- the facility's injury and medical records that implicate tools
- work methods
- the setup of workstations
- trends for particular jobs.

Some changes that may be necessary include providing personal protective equipment, especially gloves or pads for hand tools. Other possible changes might include rest breaks, adding employees to difficult jobs, job rotation, and body conditioning to relieve stress and fatigue. The ergonomic aspects of hand-tool design and use are discussed in many of the resources listed in the chapter References. (Electrical concerns and the effects of tool vibration are addressed later in this chapter. See the sections titled Electrical Tools and Pneumatic-Impact Tools, respectively.)

Safety Practices

By observing the following six safety practices, most unintentional incidents with hand tools and portable power tools can be eliminated:

1. Provide proper PPE, and ensure employees wear it. Eye and face protection prevents injuries from flying objects or liquids. Hand and arm protection prevents injuries from flying or sharp objects. Respiratory protective equipment can provide protection from particulates and fumes, but should be used only in conjunction with a respiratory protection program as described in 29 CFR 1910.134. Power tools can also create noise levels above regulatory standards. When noise monitoring indicates that noise levels exceed OSHA standards, it will be necessary to include noise-exposed employees in a hearing conservation program with annual audiometric testing and use of hearing protection. Companies also should consider use of less noisy models of tools or other noise controls where noise levels exceed 90 dBA.

2. Select the right tool for the job. Examples of unsafe practices include (a) striking hardened faces of hand tools together, such as using a hammer to strike another hammer or hatchet; (b) using a claw hammer to strike a steel chisel; (c) using a file or a screwdriver to pry; (d) using a wrench instead of a hammer; and (e) using pliers instead of the proper wrench.

3. Know if a tool is in good condition and keep it in good condition. Tools should be replaced if the following conditions exist: wrenches with cracked or worn jaws; screwdrivers with broken tips or split or broken handles; hammers with chipped, mushroomed, or loose heads and broken or split handles; mushroomed heads on chisels; dull saws; and extension cords or electrical tools with broken plugs, improper or removed grounding systems, or split insulation.

4. Properly ground power tools. Use a ground-fault circuit interrupter (GFCI)–protected circuit. Per 29 CFR 1926.404(b)(1)(i), the employer should use either GFCI or an assured equipment ground program for branch circuits at a construction site.

5. Use tools correctly. Some common causes of injuries are (a) screwdrivers applied to objects held in the hand, (b) knives pulled toward the body, (c) failure to ground electrical equipment, (d) nail hammers striking hardened tools, and (e) using tools when work is not properly secured. (For proper uses of hand tools, see Use of Hand Tools, in this chapter.)

6. Keep tools in a safe place. Many injuries are caused by tools falling from overhead. Another source of injuries is leaving the cutting edge of knives, chisels, and other sharp tools exposed when carrying them in pockets or leaving them in toolboxes.

A safety program designed to control unintentional incidents involving hand and power tools should include the following activities:

- Train employees to select the right tools for each job, and see that the tools are available.
- Establish regular tool inspection procedures (including inspection of employee-owned tools), and provide good repair facilities to make sure that tools are kept in safe condition.
- Train and supervise employees to correctly use tools.
- Establish a procedure to control company tools. For example, set up a check-in and check-out system that

evaluates tools' condition and ensures that only properly conditioned tools leave the tool-storage area.

- Provide proper storage areas in the toolroom and at the workstation.
- Enforce the use of proper personal protective equipment.
- Plan each job well in advance.

Each supervisor should also check all operations to determine if special tools will do the work more safely than ordinary tools. Have special tools readily available for employees.

Central Tool Control

From the standpoint of incident prevention, central tool control ensures uniform inspection and maintenance of tools by a trained employee. This employee can also distribute the correct type of personal protective equipment, such as eye and face protection, when a tool is issued.

A central control area and effective record keeping on tool failure and other causes of injuries help locate hazardous conditions. A central area also ensures better control than does scattered storage. In that way, tools are exposed to less damage and deterioration and are not as likely to fail or create other hazards. Some companies issue each employee a set of numbered checks that are exchanged for tools from the supply room. With this system, the attendant knows the location of each tool and can recall it for inspection at regular intervals. Tools used and assigned to a workstation should be checked either between shifts by a qualified person or prior to each shift by the tool user.

The tool control attendant can help promote safety by:
- recommending or issuing the right type of tool
- encouraging employees to turn in damaged or worn tools
- tagging and removing from service defective tools
- encouraging the safe use of tools.

Set up a procedure so the tool control attendant can send tools in need of repair to a department or service firm thoroughly familiar with methods of repairing and reconditioning.

Companies that perform work at scattered locations may find that it is not always practical to maintain a central tool control area. In such cases, the supervisor should frequently inspect all tools and remove from service those found to be damaged. Many companies have each supervisor check all tools every week. Draw up a checklist for the most hazardous hand tools so inspection will be consistent (Figure 17–2).

Some workers prefer to use their own tools even though tools are furnished by the company. In this case, supervisors should examine these tools frequently to prevent the use of any that are unsafe. If worker-owned tools are found to be damaged or unsafe, supervisors should insist that they be

---

PORTABLE ELECTRIC TOOLS
Inspection Checklist

GENERAL
Low voltage or battery powered equipment used in tanks and wet areas?	❑
Tools well maintained?	❑
Motors in good condition?	❑
Approved tools used in explosive atmospheres?	❑
Tools left where they cannot fall?	❑

CORDS
Insulation and plugs unbroken?	❑
Cords protected against trucks and oil?	❑
Cords not in aisles?	❑

GROUNDING
Ground wire fastener in safe condition?	❑
3-wire plug extension cord (if a 3-wire tool)?	❑
Ground wire used?	❑
Defects or minor shocks reported?	❑
Ground fault circuit interrupter used?	❑

GUARDING
Guards used on grinders and saws?	❑
Movable guards operate freely?	❑
Eye or face protection worn?	❑

Figure 17–2. An inspection checklist card can be used for portable electric tools. Such a card encourages workers to inspect equipment before and after use. More specific cards or tags simplify the prompt recording of defects and result in better maintenance records.

---

replaced. Regulatory requirements usually state that the employer is responsible for seeing that safe tools are used, including tools and equipment that are furnished by the employee.

Toolboxes

Toolboxes are meant to hold tools, not to be stood on or used as an anvil, a sawhorse, or a storage place for lunches. Lightweight toolboxes are made of plastic or steel, but strong, heavy-duty toolboxes are usually made of steel.

Portable Toolboxes

Portable toolboxes may have up to five drawers, a lift-out tray, and possibly a cantilevered tray that automatically opens out when the cover is lifted. All seams should be welded and smooth with no protruding edges to catch clothing or hands. In addition to the handle on the top of the toolbox's cover, look for handles at each end for those boxes designed to hold an extra-heavy load of tools. A good toolbox should have a catch or a hasp at each end and should be able to be locked with either a padlock or its own built-in lock. For outdoor use look for weatherproof construction that will allow rain to drain away without entering the toolbox.

Tool Chests

Tool chests are usually heavier, stronger, and, of course, have a greater capacity than toolboxes. On some models, the drawers, sometimes more than 10, can be secured with their own built-in locks. Some have a tote tray that can be removed for carrying only those tools needed for a particular job. Most tool chests are designed to be placed on top of tool cabinets.

Mobile Tool Cabinets

Mobile tool cabinets—the kind on wheels—may have 10 or more drawers. If they are designed to hold a chest, they may sometimes have 20 or more drawers. Look for a locking arrangement in which all drawers lock automatically. Also look for construction that will allow drawers, no matter how heavily loaded, to roll out freely. To prevent rolling, two wheels should lock by means of a brake. Casters should be of ball-bearing construction.

Gang Boxes

Used primarily by the construction industry, gang boxes are generally less organized than toolboxes and cabinets. For this reason, workers and supervisors should perform daily checks on all tools to ensure that any needed repairs are made.

Carrying Tools

Workers should carry their tools to and from work or the workstation in a toolbox, cabinet, or other appropriate tool holder or pouch. In these ways, the worker is protected, as well as the tools themselves. Employees should never carry chisels, screwdrivers, and pointed tools with the edges or points up either in their pockets or by hand. They should carry such tools with the points and cutting edges away from their bodies.

Employees should never carry tools in any way that might interfere with the free use of both hands when climbing a ladder or other structure. Instead, have them put tools in a toolbelt, pouch, or holder or hoist tools from the ground to the job in a strong bag, bucket, or similar container. Tools should be returned in the same manner—not brought down by hand, carried in pockets, or dropped to the ground.

Mislaid and loose tools cause a large number of injuries. Tools are laid down on scaffolds, on overhead piping, on top of stepladders, and in other locations from which they can fall on persons below. Leaving tools overhead is especially hazardous where there is vibration or where people are moving about.

Employees should hand tools to one another, never throw them. Workers should pass edged or pointed tools, preferably in the tool's carrying case, with the handle toward the receiver.

Workers carrying tools on their shoulders should pay close attention to clearances when turning around. They should handle the tools so they will not strike others.

MAINTENANCE AND REPAIR

When metal tools break during normal use, the causes are usually related to the tools' quality. Therefore, purchase tools of the best quality obtainable.

Inspection and Control

The toolroom attendant or tool inspector should be qualified to determine the condition of tools for further use. Never return a dull or damaged tool to stock. Keep enough tools of each kind on hand so that when a damaged or worn tool is removed from service, it can be replaced immediately with a safe tool.

Efficient tool control requires periodic inspections of all tool operations. These inspections should cover housekeeping in the tool supply room, tool maintenance, service, number of tools in the inventory, handling routines, and condition of tools. Responsibility for such periodic inspections is usually placed with the department head and should not be delegated to others.

Hand tools receiving the heaviest wear, such as chisels, punches, wrenches, hammers, star drills, and blacksmith's tools, require frequent maintenance on a regular schedule.

Proper maintenance and repair of tools require adequate work space and equipment, such as workbenches, vises, eye and face protection, tools for repair and sharpening, and good lighting. Employees specially trained in the care of tools should be in charge of these work spaces; otherwise, send tools out to a qualified shop for repairs.

Redressing Tools

Always follow the tool manufacturers' recommendations to repair, shape, and maintain tools. A properly dressed tool is a safe and efficient tool. Do not redress or reshape tools having chipped, battered, or mushroomed striking or struck surfaces. When a tool has reached this stage through normal use or abuse, discard it.

When performed, redressing should be done by a person who knows the proper dressing information for that particular tool and has the skill to use that information. If a grinder is used, take care to keep from destroying the temper of the cutting edge.

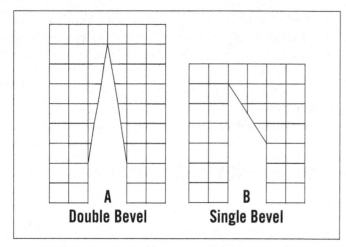

Figure 17-3. Redressing hatchets.

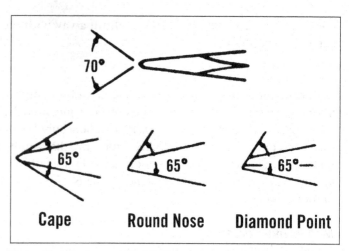

Figure 17-4. Redress cold chisels to restore the angle.

Hatchets
Redress hatchets with double bevels as illustrated in Figure 17–3a. Redress hatchets with single bevels as illustrated in Figure 17–3b.

Flat Cold Chisels
Cold chisels are hardened on the cutting edge. Redress them with care to restore them to their original shape or to an included angle of approximately 70 degrees (Figure 17–4).

Hot Chisels
Hot chisels are tools with handles used for cutting hot metal. Redressing instructions are the same as for cold chisels.

Other Machinist's Chisels
Round nose, diamond point, and cape are other chisels commonly used in metalworking. Redressing instructions for them are the same as for flat cold chisels except that the bevel angles are different (Figure 17–4).

Punches
The working end of pin-and-rivet punches and blacksmith's punches should be redressed flat and square with the axis of the tool. The point of center punches should be redressed flat and square with the axis of the tool and should be redressed to an included angle of approximately 60 degrees. Prick punches should be redressed to an included angle of approximately 30 degrees.

Bricklayer's Tools
Bricklayer's tools should be redressed as follows: hammer blade, 40 degrees; brick chisel, 90 degrees; and brick set, 45 degrees. Bevel slightly to remove feather edge (Figure 17–5).

Screwdrivers
Screwdrivers with a flat blade should be redressed to a flat end at a right angle to the blade. Both sides of the blade at screw contact should also be at right angles to the blade.

Files
To keep files' cutting surfaces from clogging, use a file card often during the filing process. Never use a file without a handle. When a file becomes dull, discard it or have it resharpened.

Star Drills
Hand file all cutting edges of star drills to an included angle of approximately 70 degrees (see Figure 17–4 for included angle of 70 degrees).

Handles
Wooden handles of hand tools used for striking, such as hammers and sledges, should be made from the best straight-grained material. Hickory, ash, or maple, neatly finished and free from slivers, is preferred. Alternate mate-

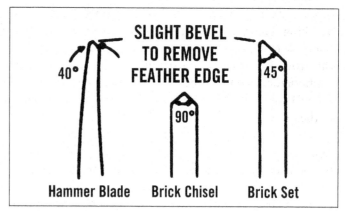

Figure 17-5. Bricklayer's tools.

rials such as fiberglass or steel with a rubber sleeve may be used.

To make sure that handles are properly attached to the tools, have them fitted to tools only by an experienced person. Poorly fitted handles make it difficult for workers to control their tools, and such handles can be dangerous as well.

Loose wooden handles in sledges, axes, hammers, cold cutters, and similar tools create hazards. No matter how tightly a handle may be wedged at the factory, both use and shrinkage can loosen it. In some cases, tapping the wedges will take up the shrinkage. In others, the head of the tool can be driven back on the handle, the wedges reset, and the protruding end of the handle cut off.

Eventually, any wooden-handled tool will need its handle replaced. Drive the old handle out. Select a new but equivalent handle. If the wood of the new handle does not bear against the head eye at all points, shave the handle until it fits snugly. Then, replace the handle in the tool, and sight along it to be sure that the head is properly centered. After the handle is firmly fitted into the head eye, wedge it according to the original pattern.

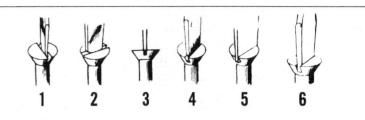

1. This tip is too narrow for the screw slot; it will bend or break under pressure.
2. A rounded or worn tip. Such a tip will ride out of the slot as pressure is applied.
3. This tip is too thick. It will only serve to chew up the slot of the screw.
4. A chisel ground tip will also ride out of the screw slot. Best to discard it.
5. This tip fits, but it is too wide and will tear the wood as the screw is driven home.
6. The right tip. This tip is a snug fit in the slot and does not project beyond the screw head.

Figure 17–6. Select the correct screwdriver to fit the screw.

USE OF HAND TOOLS

The misuse of common hand tools is a source of many injuries to industrial workers. In many instances, injuries result because supervisors assume that everyone knows how to use common hand tools. Observations and the records of injuries show that this is not the case.

A part of every training program should, therefore, be detailed instruction in the proper use of hand tools. The following sections describe safe practices for using hand tools.

Screwdrivers

The screwdriver is probably the most commonly used and abused tool. Discourage the unsafe practice of using screwdrivers as punches, wedges, pinch bars, or pries. If used in one of these ways, they can cause injury and become unfit for the work they are intended to do. Furthermore, a broken handle, bent blade, or dull or twisted tip may cause a screwdriver to slip out of the slot and cause a hand injury.

Select a screwdriver tip that fits the screw (Figure 17–6). A sharp, square-edged blade will not slip as easily as a dull, rounded one, and it requires less pressure. By redressing the tip to its original shape, it may be kept clean and sharp to permit a good grip on the head of the screw. Phillips screwdrivers and many other types of screwdrivers are safer than the flat blade screwdriver because they have less tendency to slip.

When putting in a screw, hold the work in a vise or lay it on a flat surface. This practice lessens the chance of injury to the hands should the screwdriver slip from the workpiece.

When it is necessary to work around electrical-current-bearing equipment, use an insulated screwdriver. However, the handle, insulated with dielectric material, is intended only as a secondary protection. Insulated blades are also intended only as a protective measure against shorting out components. Be sure that electrical current is shut off before beginning work.

Hammers

Hammers are made in different shapes and sizes, with different configurations and varying degrees of hardness. Each hammer has a specific purpose. Select hammers for their intended uses, and use them only for those purposes. Proper use of most hammers involves the following basic rules:

- Always wear eye protection.
- Always strike a hammer blow squarely, with the hammer's striking face parallel with the surface being struck. Always avoid glancing blows, overstrikes, and understrikes.
- When striking another tool (chisel, punch, wedge, etc.), the striking face of the hammer should have a diameter approximately 3/8 in. (0.9 cm) larger than the struck face of the tool.
- Always use a hammer of suitable size and weight for the job. Do not use a tack hammer to drive a spike nor a sledgehammer to drive a tack.
- Never use a hammer to strike another hammer.
- Never use a hard-surface hammer to strike another, harder surface.
- Never use a hammer with a loose or damaged handle.
- Discard any hammer if it shows dents, cracks, chips, mushrooming, or excessive wear. Redressing is not recommended.

Common Nail Hammers

Designed for driving unhardened common nails, finishing nails, and nail sets, common nail hammers use the center of the hammer's face. Their shape, depth of face, and balance make them unsuitable for striking against metal, especially heavier objects, such as cold chisels.

Nail Hammers

Nail hammers are made in two patterns: curved claw and straight or ripping claw. The face is slightly crowned with the edges beveled. However, certain heavy-duty patterns may have checkered faces designed to reduce glancing blows and flying nails. Handles may be made of wood, tubular or solid steel, or fiberglass. These materials are generally furnished with rubber grips that also occasionally use wooden handles. When drawing a nail from a piece of wood with a nail hammer, place a block of wood under the head to increase the leverage, to reduce the strain on the handle, and to prevent marring of the wood.

Ball Peen Hammers

Ball peen hammers of the proper size are designed for striking chisels and punches. They are also used for riveting, shaping, and straightening unhardened metal.

Sledgehammers

Many types and weights of sledgehammers are designed for general sledging operations such as striking wood, metal, concrete, or stone. Consult the manufacturer for specific recommendations.

Hand-Drilling Hammers

Hand-drilling hammers are designed for use with chisels, punches, star drills, and hardened nails.

Bricklayer's Hammers

The striking face of bricklayer's hammers is flat with beveled edges. The blade has a sharp, hardened cutting edge. Handles are made of wood, solid steel, or fiberglass. They also may have rubber grips.

Bricklayer's hammers are designed for setting and cutting (splitting) bricks, masonry tile, and concrete blocks. They are also used for chipping mortar from bricks. Never use a bricklayer's hammer to strike metal or to drive struck tools, including brick sets and chisels.

Riveting and Setting Hammers

Riveting hammers are designed for driving and spreading unhardened rivets on sheetmetal work. The setting hammer is designed for forming sharp corners, closing and preening seams and lock edges, and glazing points.

Other Hammers

Consult manufacturers for designs, uses, and safety programs for hammers not mentioned in this section. Other types of hammers include scaling, chipping, soft-face, nonferrous, magnetic, engineer's, blacksmith's, and spalling hammers and woodchopper's mauls.

Punches

Hand punches are made in various patterns from square, round, hexagonal, or octagonal steel stock. Punches are designed to (1) mark metal and other materials that are softer than the punch's point end, (2) drive and remove pins and rivets, and (3) align holes in different sections of materials.

Never use a punch with a mushroomed struck face or with a dull, chipped, or deformed point. Discard any punch if it is bent, cracked, or chipped. Redress the cutting edge's point to its original contour as required.

Tools for Cutting Metal

Cold chisels, hacksaws, files, and snips are some hand tools used for cutting metal. Because of the material being worked with, workers must be especially careful when using metal-cutting tools.

Cold Chisels

Cold chisels have a cutting edge at one end for cutting, shaping, and removing metal that is softer than the cutting edge itself, such as cast iron, wrought iron, steel, bronze, and copper. There is a struck face on the other end of the cold chisel.

A chisel can be used to cut metal in hard-to-reach places; it will cut any metal that is softer than its own cutting edge. The cutting edge is hardened and kept sharp, at a 70-degree angle; for softer materials, a 60-degree angle is satisfactory because less pressure is required.

Types of Chisels. The following four principal kinds of chisels are used for bench metal work:

1. A flat cold chisel is most commonly used for cutting, shearing, and chipping. The width of the cutting edge determines the size. For ordinary work, a ¾-in. (1.9-cm) chisel and a 1-lb (0.45-kg) hammer are used. The flat chisel should be ground with a slightly convex cutting edge. This reduces the tendency for its corners to dig into the surfaces being chiseled and concentrates the force directly to the material being cut.
2. A diamond-point chisel cuts V-grooves and sharp interior angles.
3. A cape chisel is used for cutting keyways, slots, or square corners.
4. A round-nose chisel cuts rounded or semicircular

grooves and corners that have fillets. It can draw back a drill that wandered from its intended center.

Selecting Chisels. Factors determining the selection of a cold chisel are the materials to be cut, the size and shape of the tool, and the depth of the cut to be made. The chisel should be heavy enough so that it will not buckle or spring when struck. Select a chisel just large enough for the job. In that way, the blade is used rather than only the point or corner.

As discussed earlier, use the proper hammer for the job. The striking face of the appropriate type and size hammer should have a diameter approximately ⅜ in. (0.95 cm) larger than the struck face of the chisel.

Holding Chisels. Some workers prefer to hold the chisel lightly in the hollow of the hand with the palm up, supporting the chisel with the thumb and first and second fingers. They claim that if the hammer glances from the chisel, it will strike the soft palm rather than the knuckles. Other workers think that a grip with a loose fist, keeping the fingers relaxed, holds the chisel steadier and minimizes the chances of being hit by glancing blows. Moreover, in some positions, this is the only grip that is natural or even possible. For regular use, a sponge rubber pad, forced down over the chisel, provides a protective cushion for the hand. Punch and chisel holders are commercially available. When shearing and chipping with a cold chisel, the worker should hold the tool at an angle that permits one bevel of the cutting edge to be flat against the shearing plane.

Bull chisels held by one employee and struck by another require the use of tongs or a chisel holder to guide the chisel. In that way, the person holding the chisel will not be exposed to injury. Protective holders are also commercially available.

Maintenance of Chisels. Discard any chisel that is bent, cracked, or chipped. Redress the cutting edge, or struck end, to its original contour as required.

When grinding a chisel, do not apply too much pressure to the head. The heat generated from grinding can draw the temper. Periodically immerse the chisel in cold water when grinding.

Stamping and Marking Tools

Stamping and marking tools of special alloy and design are available. If possible, workers should use holders for marking tools so they do not have to hold their fingers close to the face of the tool being struck.

Tap and Die Work

Tap and die work requires certain precautions. Firmly mount the workpiece in the vise. Use a tap wrench of the proper size. Workers should keep their hands away from broken tap ends. If a broken tap is removed by using a tap extractor or a punch and hammer, the worker should wear eye protection. When a long thread is being cut with a hand die, workers should keep their hands and arms clear of the sharp threads coming through the die.

Hacksaws

Adjust and tighten hacksaw blades in the frame to prevent buckling and breaking. However, they should not be so tight that the pins that support the blade break off. Install blades with teeth pointing forward.

Use blades with 14 teeth per inch (per 2.5 cm) to cut soft metal; 18 teeth for tool steel, iron pipe, hard metal, and general shop use; 24 teeth for drill rods, sheet metal, copper and brass, and tubing; and 32 teeth for thin sheet metal, less than 18 gauge (0.12 cm), and for tubing. When thin metals are cut, make sure that at least three teeth are in contact with the surface being cut.

Apply pressure on the forward stroke only. Lift the saw slightly and pull back lightly in the cut to protect the teeth. Cutting speeds of 40 to 60 strokes per minute are recommended. If the blade is twisted or too much pressure is applied, the blade may break and injure the hands or arms of the user. Do not continue an old cut after changing to a new blade. The new blade may bend and break because the set of the teeth on the new blade is thicker than that of a used blade.

Files

Select the right kind of file for the job to prevent injuries, lengthen the life of the file, and increase production. Use only files with secure handles. Because the extremely hard and brittle steel of the file chips easily, never clean the file by striking it against a vise or other metal object. Instead, use a file-cleaning card.

For the same reason, do not use a file as a hammer or as a pry. Such abuse frequently causes the file to chip or break, thus resulting in injury to the user. Do not convert a file into a center punch, chisel, or any other type of tool because the hardened steel may fracture.

Clamp the work to be filed in a vise at about waist height. Grasp the handle firmly in one hand and use the thumb and forefinger of the other to guide the point. This technique gives good control and ensures better and safer work. To file, push the file forward while bearing down on it. Release the pressure and bring the file back to its original position. If pressure is not released, the teeth will wear excessively.

Always use a file with a smooth, crack-free handle. Otherwise, if the file should slip or be struck by a revolving part of a machine, the tang may puncture the palm of the hand, the wrist, or other part of the body. Under some conditions, a clamp-on, raised-offset handle may be useful to give extra clearance for the hands.

When work to be filed is placed in a lathe, the job should be done left-handedly, with the file and hands clear of the chuck jaws or the dog. Use a fine mill file or long-angle lathe file. Take long, even strokes across the rotating work.

Hand Snips

Hand snips are divided into two groups—those for straight cuts and those for circular cuts. Snips for thicker sheets and harder materials have longer handles, alloy steel blades, and, sometimes, special arrangements of levers to make cutting easier. Do not hammer on the handles or jaws of the snips. Use hand snips that are heavy enough to cut the material easily. In that way, the worker needs only one hand on the snips and can use the other to keep the edges of the cut material pulled aside. Be sure that the material is well supported before the last cut is made. In that way, the cut edges do not press against the hands. When cutting long sheetmetal pieces, push the sharp ends down next to the hand holding the snips. File off any jagged edges or slivers after cutting. (Hand snips are not designed to cut wire.) Keep jaws of snips tight and well lubricated. Select a hand snip that cuts easily and is not tiresome to use.

Workers should wear eye protection because small particles often fly with considerable force. Wearing leather or heavy canvas work gloves prevents cuts or scratches to hands caused by handling sharp edges of the sheet metal.

Cutters

Cutters used on wire, reinforcing rods, or bolts should be heavy enough for the stock. Otherwise, the jaws may be sprung or spread. Also, a chip may fly from the cutting edge and injure the user.

Cutters are designed to cut at right angles only. Do not "rock" cutters to make the cut. They are not designed to take the resulting strain. This practice can also cause the knives to chip. Cutters require frequent lubrication. To keep cutting edges from becoming nicked or chipped, do not use cutters as nail pullers or pry bars.

Cutter jaws should have the hardness specified by the manufacturer for the particular kind of material to be cut. Adjust the bumper stop behind the jaws to set cutting edges for a clearance of 0.003 in. (0.076 cm) when closed.

Tools for Cutting Wood

Workers should wear eye protection when using wood-cutting tools. When cutting wood, workers should hold tools so that if a slip should occur, the direction of force will be away from the body. For efficient and safe work, keep wood-cutting tools sharp and ground to the proper

angle. A dull tool does a poor job and may stick or bind. A sudden release may throw the user off balance or cause a hand to strike an obstruction.

Wood Chisels

Instruct inexperienced employees in the proper method of holding and using chisels. Keep wooden handles free of splinters. If the wooden handle of a chisel is designed to be struck by a wood or plastic mallet, protect the handle with a metal or leather band to prevent it from splitting. Heavy-duty or framing chisels are made with solid or molded handles and can be struck with a steel hammer.

Clamp or otherwise secure the work so it cannot move while cutting is being done. Make finish or paring cuts with hand pressure alone. Be sure that the chisel's edge is sharp and that both hands are back of the cutting edge at all times. To avoid damage to the blade of the chisel, be sure that the workpiece is free of nails. Should metal be struck by the cutting edge, a chip from the chisel might hit nearby workers.

Do not use the wood chisel as a pry or a wedge. The steel in a chisel is hard so that the cutting edge will hold. However, the steel may break if the chisel is used as a pry. When not in use, keep chisels in a rack, on a workbench, or in a slotted section of the toolbox. In that way, the sharp edges are out of the way and cannot come in contact with metal surfaces.

Saws

Select saws for the work they must do. For cutting across the grain of the wood, use a crosscut saw; for cutting with the grain, use a ripping saw. The difference between them is the angle and shape of their teeth. For fast crosscut work on green wood, use a coarse saw (4 to 5 points per inch [per 2.5 cm]); for smooth, accurate cutting of dry wood, use a fine saw (8 to 10 points per inch [per 2.5 cm]). When ripping, use a coarse saw for thick stock and a fine saw for thin stock. The number of points per inch is stamped on the blade.

To prevent binding, keep blades sharp and the teeth well set. When not in use, wipe saws off with an oily rag, and keep them in racks or hung by their handles to prevent their teeth from becoming dull.

Avoid sawing nails or drywall with a saw. Use a keyhole saw with a metal-cutting blade to cut nails; use an old saw for cutting drywall. Be careful not to drop a saw because the handle can break, or become loose, or the teeth can get nicked.

Axes

The double-bit axe is usually used to fell, trim, and prune trees and to split and cut wood. It is also used for notching and shaping logs and timbers. The single-bit axe, in addition to the preceding uses, can be used to drive wooden stakes with its face.

The cutting edges of axes are designed for cutting wood and equally soft materials. Use a narrow-bladed axe for hard wood and a wide axe for soft wood. A sharp, well-honed axe yields better chopping speed and is much safer to use because it bites into the wood (Figure 17–7). A dull axe will often glance off the wood being cut and may strike the user in the foot or leg. Also observe the following precautions:
- Never strike an axe against metal, stone, or concrete.
- Never use an axe as a wedge or maul.
- Never strike with the side of an axe.
- Never use an axe with a loose or damaged handle.
- Use steel wedges for splitting wood, such as splitting logs for fireplace wood. Use a sledgehammer or maul for driving the wedges.

To use an axe safely, workers should lift it properly, swing it correctly, and place the stroke accurately. The proper grip for a right-handed person is to have the left hand about 3 in. (7.6 cm) from the end of the handle and the right hand about three-fourths of the way up. A left-handed person should reverse the position of the hands.

Before starting to chop, a worker should make sure that there is a clear circle in which to swing the axe. Also, all vines, brush, and shrubbery within the range should be removed, especially overhead vines that may catch or deflect the axe. When using an axe, workers should wear safety shoes, eye protection, and pants of durable material.

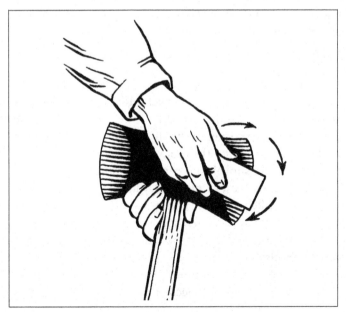

Figure 17–7. Good honing saves labor and makes an axe safer to use. The axe should be honed after each sharpening and each use. Correct honing motion is shown here. A double-bit axe can be ground to different cutting bevels for various types of work.

Protect axe blades with a sheath or metal guard when possible. When the blade cannot be guarded, workers should carry axes at their sides. They should carry single-edged axes with the blade pointed down.

Hatchets

Hatchets are used for many purposes and frequently cause injuries. For example, a worker attempting to split a small piece of wood while holding the wood in his or her hand may cut his or her hand or fingers with the hatchet.

To properly start a hatchet cut, strike the wood lightly with the hatchet. Then force the blade through by striking the wood against a solid block of wood.

Miscellaneous Cutting Tools

Permit only trained employees to use planes, scrapers, bits, and drawknives. Keep these tools sharp and in good condition. When not in use, place them in a rack on the bench, or in a toolbox, in such a way that will protect the user and prevent damage to the cutting edge. If knives are used on the job, consider using resistance gloves and sleeves.

Knives

Knives are more frequently the source of disabling injuries than any other hand tool, particularly in the meat-packing industry. To prevent injuries, (1) keep knives sharp, (2) replace knives that have worn handles, and (3) use knives with retractable blades when possible.

The principal hazard in the use of knives is that workers' hands may slip from the handle onto the blade or that the knife may strike the body or the free hand. A handle guard, or a finger ring and swivel on the handle, can reduce these hazards.

When cutting, stroke away from the body. Wear a cut-resistant glove on the hand opposite the knife if that hand is used to hold the material being cut. Wear a heavy leather apron or other protective clothing when it is not possible to cut away from the body. When possible, use a rack or holder for the material to be cut. To help maintain balance, avoid jerky motions.

Be sure employees are trained and supervised. Employees who must carry knives with them on the job should keep them in sheaths or holders. They should never carry a sheathed knife on the front part of a belt. They should always carry it over the right or left hip, toward the back. This prevents severing an artery or vein in a leg in case of a fall.

Safe placement and storage of knives and other sharp hand tools is important to knife safety. When not in use, keep knives in racks and guard their edges. To protect the employee, as well as the cutting edge of the knife, keep knives separate from other tools. Supervisors should make sure that employees do not leave knives hidden under the product, under scrap paper or wiping rags, or among tools in workboxes or drawers.

Supervisors should also make certain that employees who handle knives have ample room in which to work. In that way, they are not in danger of being bumped by trucks, by the product, by overhead equipment, or by other employees. For instance, a left-handed worker should not stand close to a right-handed person. Place the left-handed person at the end of the bench or give him or her more room.

Careful job and incident analysis may suggest ways to make operating procedures in which knives are used safer. For instance, on some jobs special rigs, racks, or holders may be provided so it is not necessary for the operator to stand so close to the piece being cut. Also discourage the practice of wiping a dirty or oily knife on the apron or clothing. Instead, have workers wipe the blade with a towel or cloth with the sharp edge turned away from the wiping hand. Sharp knives should be washed separately from other utensils and in such a way that they will not be hidden under soapy water.

Supervisors should make sure that nothing is cut that requires excessive pressure on the knife, such as frozen meat. Food should be thawed before it is cut, or else it should be sawed. Do not use knives as substitutes for can openers, screwdrivers, or ice picks.

To cut corrugated paper, use a hooked linoleum knife. This permits good control of pressure on the cutting edge and eliminates the danger of the blade suddenly collapsing, which might happen if a pocket knife is used. Be sure that hooked knives are carried in a pouch or heavy leather, or plastic, holder. The sharp tip must not stick out.

Prohibit horseplay around knife operations. Throwing, "fencing," trying to cut objects into smaller and smaller pieces, and similar practices are not only dangerous but reflect poor training and inadequate supervision.

Ring Knives

Ring knives—small, hooked knives attached to a finger— are used where string or twine must be cut frequently. Supervisors should make sure that the cutting edge is kept outside the hand, not pointed inside. A wall-mounted cutter or blunt-nose scissors would be safer.

Carton Cutters

Carton cutters are safer than hooked or pocket knives for opening cartons. They not only protect the user but also eliminate deep cuts that could damage the carton's contents. Frequently, damage to contents of soft plastic bottles may not be detected immediately. The subsequent leakage could cause chemical burns, damage to other products, or a fire.

Brad Awls

Brad awls should be started with the edges across the grain

to keep the wood from splitting. Hold them at right angles to the surface to prevent them from slipping.

Tools for Materials Handling

Crowbars, hooks, shovels, and rakes are used for materials handling. Observe safety precautions when using them.

Crowbars

Whenever a crowbar is needed, use the proper size and kind of bar for the job. The crowbar should have (1) a point or toe of such shape that it will grip the object to be moved and (2) a heel to act as a pivot or fulcrum. In some cases, a block of wood under the heel will prevent the crowbar from slipping and injuring the worker's hand.

If crowbars are stood on end when not in use, secure them so they will not fall. If they are laid on the ground when not in use, place them where they will not create a stumbling hazard.

Hooks

Keep hand hooks sharp so they will not slip when applied to a box or other object. The handle should be strong and securely attached and shaped to fit the hand. The handle and the point of long hooks should be bent on the same plane so that the bar will lie flat when not in use and, thus, not cause a tripping hazard. Shield the hook's point when it is not in use.

Shovels

Keep the edges of shovels trimmed, and check the handles for splinters. Workers should wear heavy shoes with sturdy soles, preferably safety shoes. To have good balance and spring in the knees, workers should keep their feet well separated. The leg muscles should take much of the load when shoveling.

To reduce the chance of injury, use the ball of the foot— not the arch—to press the shovel into clay or other stiff material. If the instep is used and the foot slips off the shovel, the sharp corner of the shovel may cut through the worker's shoe and into the foot.

Dipping the shovel into a pail of water occasionally keeps it free from sticky material, thus making it easier to use and less likely to cause strain. Greasing or waxing the shovel's blade also prevents some kinds of material from sticking. Chemically coated shovels prevent sticking of certain materials.

When shovels are not in use, hang them or stand them against a wall. Or, install special racks or boxes for shovels.

Rakes

Never leave a rake with the prongs turned upward where they may be stepped on, thus causing a foot injury or causing the handle to fly up and hit a worker's head. Place rakes in racks when they are not in use.

Wrenches

To ensure the safe use of all wrenches, workers must always be alert. They should be prepared for the possibility of (1) a wrench slipping off the fastener, (2) a fastener suddenly turning free, (3) a wrench breaking, or (4) a fastener breaking. Therefore, workers should brace themselves in such a way that should the wrench become free for any reason, they will not lose their balance and be injured by falling into moving machinery or falling off a platform. Workers should always try to pull the wrench toward themselves. This will provide better wrench control and less likelihood of slipping or falling.

Workers should always inspect wrenches for flaws because a previous overloading or misuse of a tool may have weakened it to the point that it will not carry a normal load. Do not grind wrenches to change their sizes or to reduce their dimensions to fit into close quarters. Instead, use a wrench of the correct size and fit. It is an unsafe practice to use shims to make the wrong wrench fit. The great variety of wrenches used for turning nuts, bolts, and fittings makes it important that workers know the purpose and limitations of each type and size.

Open-End Wrenches

Open-end wrenches have strong jaws and are satisfactory for medium-duty turning. They are susceptible to slipping if they do not fit properly or are used incorrectly. A wrench with an offset provides hand clearance and allows the worker to reach into recesses and over obstructions. A combination wrench has both jaws the same size.

Box and Socket Wrenches

Box wrenches and socket wrenches are used when a heavy pull is necessary and safety is a consideration. The greater gripping strength of the box helps to remove the nut quickly. Box wrenches (Figure 17–8) and socket wrenches (Figure 17–9) completely encircle the nut, bolt, or fitting. They grip the nut, bolt, or fitting at all corners, as opposed to the two corners that are gripped by an open-end wrench. Box and socket wrenches will not slip off sideways, and they eliminate the dangers of sprung jaws.

Socket wrenches and box wrenches normally come in two styles of openings—single hex or double hex (Figure 17–8). On square-headed bolts, nuts, and fittings, use either a single- or double-square design. Single- or double-hexagonal wrenches are designed for hexagonal-shaped fittings, bolts, and nuts. Ratchet handles and universal-joint fittings for socket wrenches allow them to be used where space is limited (Figure 17–9).

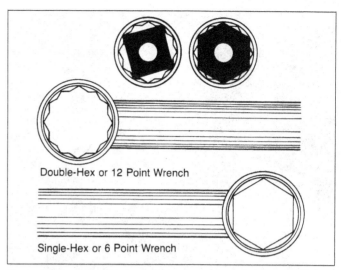

Figure 17–8. The ring of a box has 6 or 12 points. The 12-point wrench need be turned only 30 degrees before engaging a new set of flats and is, therefore, handy for use in confined spaces. Never use a 12-point box wrench on a square nut or bolt because it may slip.

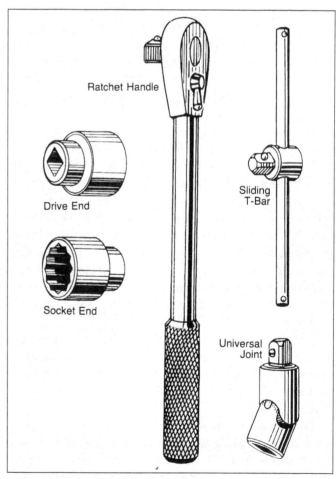

Figure 17–9. Socket wrenches (left) have a square hole in the drive end into which various handles can be fitted. The ratchet handle increases turning speed because the socket does not have to be removed from the nut between turns. The universal-joint fitting provides flexibility where space is limited.

Figure 17-10. Using a striking-face wrench designed to be hammered is a safe approach where large nuts must be set tight or frozen nuts loosened.

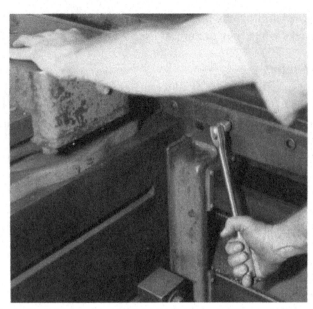

Figure 17-11. The correct tool at the right time makes an otherwise hard job safe and easy.

Never overload the capacity of the wrench by using a pipe extension on the handle or by striking the handle of a wrench with a hammer. Abuse by hammers weakens the metal of the wrench and can cause the tool to break. Special heavy-duty wrenches are available that can be used with handles as long as 3 ft (91 cm). For extra-stubborn bolts and nuts, heavy-duty, sledge-type box wrenches are available. These are of a heavy design, are properly tempered, and have a striking surface for the hammer or sledge (Figure 17–10). When possible, first use penetrating oil to loosen tight nuts.

There is a correct box or socket wrench for every nut and bolt. Oversize openings will not grip the corners securely, and shims should not be used to compensate for an oversize opening. The use of wrenches of the wrong size can round the corners of the bolt or cause slippage, as well as make it difficult to then apply the proper size. Be sure to use the proper tool, and do not try a makeshift approach (Figure 17–11).

Keep the insides of sockets clean of dirt and grime. Dirt prevents the socket from seating fully. Thus, the concentration of the pulling force at the end of the socket's opening, even with a moderate pull, can easily damage the socket or nut.

Cocking is a common cause of breakage for a socket or box wrench. Cocking is a situation in which the tool does not fit securely on the bolt or nut but instead fits at an angle. This concentrates the entire strain on a smaller area, making the tool or nut vulnerable to fracture.

Combination Wrenches

Combination wrenches have a box end and an open end. They are very handy for speeding the turning with the open end and for initial loosening or final tightening with the box end.

Torque Wrenches

Torque wrenches measure the amount of twisting force that is applied to a nut or bolt by means of a dial or calibrated arm. The torque, or twisting force, exerted on a nut or bolt is directly proportional to the length of the wrench handle and the pulling force exerted on it. Torque is usually measured in foot-pounds or newton-meters; that is, the force used times the length of the lever used to apply it. For example, a 15-lb (6.7-kg) force applied to a 2-ft (61-cm) lever gives 30 ft-lb torque. The metric unit is the newton-meter (1 kgf × m = 9.8 N × m; 1 lbf × ft = 1.356 N × m).

Torque wrenches are used when the torque has been specified for the job or when it is important that all fasteners be fully and uniformly tightened. If a torque has been specified for a particular application, it is unsafe not to measure that torque with a torque wrench. Carefully use torque wrenches and recalibrate them frequently in accord with the manufacturer's instructions.

Adjustable Wrenches

Adjustable wrenches are generally recommended for light-duty jobs or when the proper-size fixed-opening wrench is not available. Adjustable wrenches can slip, however, because it is difficult to set the correct opening size. In addition, the jaws have a tendency to work loose as the wrench is being used. These wrenches do possess one advantage—they are easily adjusted to fit metric system nuts and bolts.

Place an adjustable wrench on the nut with the open jaws facing the user, unless the space in which the job is being done makes this method impractical. With the open jaw facing the user, the pulling force applied to the handle tends to force the movable jaw onto the nut. For that reason, and for reasons of safety, pull, do not push, wrenches (Figure 17–12). The adjusting nut is used to adjust the movable jaw. When used to grip a pipe, the jaws should grip about midway. According to the manufacturers, the movable jaws on adjustable wrenches are weaker than the fixed jaws.

Pipe Wrenches

Workers, especially those on overhead jobs, have been seriously injured when pipe wrenches have slipped on pipes or fittings, thus causing workers to lose their balance and fall. Pipe wrenches, both straight and chain tong, should have sharp jaws. Keep them clean to prevent them from slipping.

Frequently inspect the adjusting nut of the wrench. If it is cracked, the wrench should be taken out of service. A cracked nut may break under the strain, thus causing complete failure of the wrench and possible injury to the user.

The handle of every wrench is designed to be long enough for the maximum allowable safe pressure. Do not use handle extensions, also known as cheater bars, to gain extra turning power, unless the wrench is so designed. A piece of pipe slipped over the handle to give added leverage can strain a pipe wrench or the workpiece to the breaking point. Using a makeshift extension to secure greater leverage may easily cause the wrench's head to break. Using a wrench of the wrong length is also a source of incidents. A wrench handle too small for the job does not give proper grip or leverage. An oversized wrench handle may strip the threads or suddenly break the work, thus causing a worker to slip or to fall.

Never use a pipe wrench on nuts or bolts. The corners of nuts and bolts will break the teeth of the wrench, thereby making it unsafe to use the latter on pipes and fittings. A pipe wrench also ruins the heads of nuts and bolts. Do not use a pipe wrench on valves or small brass, copper, or other soft fittings that may be crushed or bent out of shape. Do not strike a wrench with a hammer nor use it as a hammer, unless the pipe wrench is specifically designed for such use.

Tongs

Tongs are usually bought, but some companies make their own to perform specific jobs. Often, they are designed in such a way that the hands are pinched when the tongs are closed. To prevent pinching, the end of one handle should be up-ended toward the other handle, to act as a stop, or projections can be welded on the handle ends to allow clearance for the user's hands.

Pliers

Pliers are often considered a general-purpose tool and, therefore, are often used for purposes for which they were not designed. Pliers are meant for gripping and cutting operations. They are not recommended as a substitute for wrenches because their jaws are flexible and frequently slip when used for this work. Pliers also tend to round the corners of bolts' heads and nuts and to leave jaw marks on the surface. This makes it difficult to use a wrench at some future time.

Side-cutting pliers sometimes cause injuries when short ends of wire are cut. A guard over the cutting edge and the use of eye protection prevent short ends from causing injuries.

Be certain that pliers used for electrical work are insulated. In addition, employees should wear electrician's gloves, if company policy requires them to.

Warning: The cushion grips on handles are primarily for comfort. Unless specified as insulated handles, they are not intended to give any degree of protection against electric shock and should not be used on live electric circuits.

Special Cutters

Special cutters for heavy wire, reinforcing wire, and bolts are safer than makeshift tools. The cutting edges should

Figure 17–12. Correct use of a wrench. The wrench is in good condition and securely gripped. The hand of the worker is braced and clear in the event that the nut should suddenly turn. To protect the worker's hand, the wrench is pulled, not pushed.

apply force at right angles to the wire or other work being cut. Do not use the cutter near live electrical circuits, and use it only for the rated capacity specified by the manufacturer. Workers should wear eye protection.

Special cutters include those used for cutting banding wire and strap. Do not use claw hammers and pry bars to snap metal banding material. Only cutters designed for the work provide safe and effective results.

Pullers

Pullers or knockers are the only quick, safe, and easy way to pull a gear, wheel, pulley, or bearing from a shaft. Do not use pry bars and chisels because they concentrate the force at one point and tend to cock the part on the shaft. Select the correct-sized puller. The jaw capacity should be such that the jaws press tightly against the part being pulled. Use a puller with as large a pressure screw as possible.

Spark-Resistant Tools

So-called spark-resistant tools made of nonferrous materials, such as beryllium-copper alloy, are used where flammable gases, highly volatile liquids, and other explosive substances are stored or used. There is some question, however, about the ability of these materials to prevent the hazard of friction sparks igniting gasoline vapors and petroleum products. Nonferrous tools reduce the hazard from sparking but do not eliminate it.

The working edges of these tools are not as hard as steel tools. Therefore, inspect them more often before each use. Be sure that they have not picked up foreign particles that could produce friction sparks, thereby negating their value.

Soldering Irons

Soldering irons are the source of burns and of illnesses that result from inhaling the fumes. Insulated, noncombustible holders practically eliminate the fire hazard and the danger of burns from unintentional contact. Ordinary metal coverings on wooden tables are not sufficient because the metal conducts heat and may ignite the wood below.

Holders should be designed so that employees cannot unintentionally touch the hot irons if they reach for them without looking. The best holder completely encloses the heated surface and is inclined so that the weight of the iron prevents it from falling out. (See National Safety Council Data Sheet 12304–0445, Soldering and Brazing.)

Fumes from soldering can be toxic and/or irritating. Remove soldering fumes through local-exhaust ventilation, especially in a continuous-production operation. Lead oxides and chlorides are released when soldering with lead-tin solder and zinc-chloride flux. Lead oxides and aldehydes are released when soldering with rosin-core solder. There are different types of solder, and the hazards from each

should be known before beginning work. Conduct air sampling to determine if hazardous amounts of contaminants are present.

Do not let particles of lead solder accumulate on the floor and on worktables. If the operation is such that the solder or flux may spatter, employees should wear face shields or do the work under a transparent shield. Because a primary route of exposure to lead can be ingestion, do not allow workers to eat, drink, smoke, chew gum, apply cosmetics, and so forth, in the work area. They should also practice good personal hygiene (e.g., washing hands before eating or leaving the workplace).

PORTABLE POWER TOOLS

Portable power tools are divided into five primary groups according to their power source: electrical, pneumatic, gasoline, hydraulic, and powder actuated. Several types of tools, such as saws, drills, and grinders, are common to the first three groups. Hydraulic tools are used mainly for compression work, and powder-actuated tools are used exclusively for penetration work, cutting, and compression.

Hazards and Safety Precautions

Portable power tools present hazards similar to stationary machines performing the same functions. In addition, portable power tools also have inherent risks. Because of the extreme mobility of power-driven tools, they can easily come in contact with the operator's body. At the same time, it is difficult to guard such equipment completely. There is also the possibility of breakage because the tool may be dropped or roughly handled. Furthermore, the source of power—electrical, mechanical, air, hydraulic, or powder cartridge—is brought close to the operator, thus creating additional potential hazards.

Typical injuries caused by portable power tools are burns, cuts, and strains. Sources of injuries include electrical shock, particles in the eyes, fires, falls, explosions of vapors or gases, and falling tools and other objects.

Observe the following precautions when using portable power tools:

- Always disconnect the tool from the source of power before changing accessories. Replace or put guards in correct adjustment before using the tool again.
- Never leave a tool in an overhead place where there is a chance that the cord or hose, if pulled, will cause the tool to fall. The cord or hose and the tool may be suspended by a tool balancer that keeps them out of the operator's way. Cords and hoses on the floor create a stumbling or tripping hazard. Suspend them over aisles or work areas

in such a way that they will not be struck by other objects or by material being handled or moved. An unexpected pull might cause the tool to jam or be dislodged from its holder and cause injury or damage to the tool. Protect tools with wooden strips or special raceways if they are laid across the floor. Do not hang cords or hoses over nails, bolts, or sharp edges. Also keep them away from oil, hot surfaces, and chemicals.

- When using powder-loaded equipment for driving anchors into concrete, or when using air-driven hammers or jacks, use proper hearing protection.

All companies and manufacturers of portable power tools attach to each tool a set of operating rules or safe practices. Their use will supplement the thorough training that each power tool operator should receive. The operator manual is an excellent source of information and should be followed.

In addition, observe the following safe practices when using portable power tools:

- Store power-driven tools in secured places. Do not leave them in areas where they may be struck by passersby or be otherwise activated.
- Keep work areas heated, clean, and well lit.
- Secure, or clamp, workpieces. Normal tool use does not require a great deal of force.
- When working on a ladder, a scaffold, or in other high places, do not reach out too far. Keep the body in balance.
- Wear proper clothing for the job. Loose clothing, jewelry, and long hair may add risk to the job.
- Never use a power tool with a malfunctioning switch or part. Remove it from service, and repair or discard the tool.
- Use only accessories recommended by the manufacturer.

Selecting Tools

Replacing a hand tool with a power tool designed for the same purpose may mean substituting electrical or mechanical hazards for relatively less serious manual hazards. Therefore, anticipate the new problems and avoid as many as possible. Insist on purchasing tools designed to meet regulatory standards for safety, and be sure that workers are properly trained.

Provide the tool supplier with complete information about the job on which a tool is to be used. In that way, the supplier can recommend the most appropriate tool. Factors to consider when selecting power tools include clearance in the working quarters, the type of job to be done, and the nature and thickness of materials.

Portable power tools designed to be used intermittently or on light work are generally designated as "homeowner's grade." Those intended for continuous operation and production service, or for heavy work, are usually identified as industrial duty.

For safe operation of portable power tools, train workers to select tools according to the tool's limitations. Workers should never tackle a job with an undersized tool. A tool that is too light may not only fail but may also cause undue fatigue to the operator and create a risk of injury.

Inspection and Repair

To maintain power tools, periodically inspect them. In addition to uncovering operating defects and preventing potentially costly breakdowns, inspecting and repairing tools may prevent hazardous conditions from developing. A portable power-tool tester is an excellent means of keeping tools in good condition. Set up an inspection schedule and a system for keeping records for each tool. Tag defective tools and withdraw them from service until they are repaired.

Check electrical tools periodically (Figure 17–13). Provide a visual or external inspection at the toolroom each time a tool is returned. Also provide a thorough knockdown inspection at specified intervals. Inspect as specified by the OSHA electrical safety-related work standard.

Use colored tags to tell when the tool was last inspected. The important thing is to record the condition of the tool and to correct any unsafe conditions.

Instruct and train employees to inspect tools and to recognize and to report (and, if authorized, to correct) defects. Clearly outline the extent of this inspection and the responsibilities for correcting defects. In that way, there is neither unnecessary duplication of effort nor misunderstanding about who is responsible for maintenance. A convenient reminder of points to check is a card similar to that shown

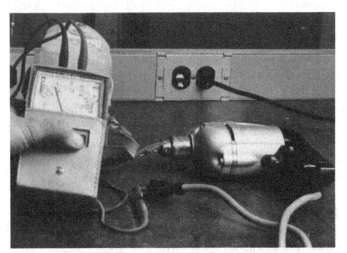

Figure 17–13. This insulation-resistance tester has test leads attached to a grounding pin of the attachment plug and the chuck of the portable tool. The meter indicates the condition of the insulation of the portable tool and permits identification of impending failures. *(Reprinted with permission from Daniel Woodhead Company.)*

in Figure 17-2. Warn employees not to do makeshift repairs and to do no repair work unless authorized.

Clean power tools with a recommended nonflammable and nontoxic solvent. Use air drying in place of blowing with compressed air.

Electrical Tools

There are a number of rechargeable, battery-powered tools available. For safety from electrical shock, they are the best possible tool. Not needing an extension cord adds to their mobility and is a convenient feature of battery-powered tools.

Electric shock is one hazard of electrically powered tools. Types of injuries include electrical flash burns, minor shock that may cause falls, and shock that results in death. Serious electrical shock is not entirely dependent on the voltage of the power input. The ratio of the voltage to the resistance determines the amount of current that will flow and the resultant degree of hazard.

The current is regulated by the resistance to the ground of the body of the operator and by the conditions under which he or she is working. It is possible for a tool to operate with a defect or short in the wiring. The use of a ground wire protects the operator and is mandatory for all but double-insulated electrical power tools. GFCI should be used with all electrical power tools, whether grounded or double insulated. (See Chapter 8, Electrical Safety, in this volume.)

Wet Locations

Low voltage of 6, 12, 24, or 42 V through portable transformers will reduce the shock hazard in wet locations. Issue standing orders to supervisors and employees to use any available low-voltage equipment or GFCIs when working in wet locations.

Electrical tools used in wet areas or in metal tanks expose the operator to conditions favorable to the flow of current through the body, particularly if a person is wet with perspiration. Most electrical shocks from tools are caused by the failure of insulation between the current-carrying parts and the metal frames of the tools. Insulating platforms, rubber mats, and rubber gloves provide an additional factor of safety when tools are used in wet locations, such as in tanks and boilers or on wet floors.

Double-Insulated Tools

Protection from electrical shock while using portable power tools depends upon third-wire grounding. Double-insulated tools, however, are available that generally provide equivalent shock protection without third-wire grounding. Paragraph 250-45 of National Fire Protection Association (NFPA) 70, National Electrical Code, permits double insulation for portable tools and appliances. Tools in this category are permanently marked by the words "double insulation" or "double insulated." Units designated to this category have been tested and listed by a nationally recognized testing lab such as Underwriters Laboratories and also use a listing mark. Many U.S. manufacturers are also using the symbol shown in Figure 17-14 to denote double insulation. This symbol is also widely used in most European countries.

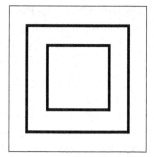

Figure 17-14. Symbol for double insulation for electrical appliances and tools.

Conventional grounded electrical tools have a single layer of functional insulation and are metal encased. For small-capacity tools, double insulation can be provided by encasing the entire tool in a nonconductive material that is also shatterproof. The switch and gripping surface are also nonconductive. Therefore, no metal part comes in contact with the operator.

Large-capacity electrical tools require a more rigid design to provide for greater stress requirements where more power and high-torque gearing are involved. Double-insulated tools with metal housings have an internal layer of protecting insulation that completely isolates the electrical components from the outer metal housing. This is in addition to the functional insulation found in conventional grounded tools, thus incorporating a reinforced or protecting insulation into the tool. This extra, or reinforced, insulation is physically separated from the functional insulation. It is arranged so that deteriorating influences—such as temperature, contaminants, and wear—will not affect both insulations at the same time. Unless subject to immersion or extensive moisture that might nullify the double insulation, a double-insulated or all-insulated tool does not require separate ground connections. In other words, the third wire or ground wire is not needed.

Failure of insulation is harder to detect than worn or broken external wiring. Thus, frequent inspection, testing with an insulation-resistance tester, and thorough maintenance are needed. Care in handling the tool and frequent cleaning will help prevent the wear and tear that cause defects.

Grounding

Grounding portable electrical tools and using a GFCI provide a convenient way of safeguarding the operator. Thus, if there is any defect or short circuit inside the tool, the current is connected from the metal frame through a ground wire and does not pass through the operator's body. If a GFCI is used, the current is shut off before a serious shock can occur. Effectively ground all electrical power tools, except the double-insulated and cordless types. Correctly grounded tools are as safe as double-insulated or low-voltage tools, especially when used with a GFCI. Check the continuity of the ground to avoid a false sense of security.

Electrical Cords

Periodically inspect electrical cords prior to use, and keep them in good condition. Use heavy-duty plugs that clamp to the cord to prevent strain on the current-carrying parts, in case the cord is unintentionally pulled. Cover terminal screws, or connections on plugs and connectors, with proper insulation. Instruct employees not to jerk cords or wind them tightly around the tool. Also instruct workers to protect cords from sharp objects, heat, and oil, or from solvents, that might damage or soften the insulation. To ensure the continuity of grounding, use extension cords of the three-wire, grounded-connection type with tools and equipment that require grounding. Check the wire size to ensure that the current's needs can be met. (See Cords in this chapter and Extension Cords in Chapter 8, Electrical Safety, in this volume.)

Cord Length (ft)	Average Wire Size for Amp Rating of Tool					
	0–2.0	2.1–3.4	3.5–5.0	5.1–7.0	7.1–12.0	12.1–16.0
25	16	16	16	16	14	14
50	16	16	16	16	14	12
100	16	16	14	12	10	—

Electric Drills

Electric drills cause injuries in several ways: (1) the torque and twisting of the tool can cause sprains and related injuries; (2) a part of the drill can be pushed into the hand, leg, or another part of the body; (3) the drill can be dropped when the operator is not actually drilling; and (4) material being drilled or parts of a broken drill can strike the eyes. Using proper eye protection will reduce the possibility of eye injuries. Although no guards are available for drill bits, some protection is afforded if drill bits are carefully chosen for the work to be done. Do not pick drill bits longer than needed to do the work.

When the operator must guide the drill with a hand, equip the drill with a sleeve that fits over the drill bit. The sleeve protects the operator's hand and also serves as a limit stop if the drill should plunge through the material.

Do not grind down oversized bits to fit small electric drills.

Observe the following precautions when using portable electrical drills:

- Be sure the trigger switch works properly. The trigger switch should turn the tool on and return it to the off position after the switch is released. If equipped with a lock-on, be sure it releases freely.

- Check carefully for loose power-cord connections and frays or damage to the cord. Replace damaged tools and extension cords immediately.
- Be sure the chuck is tightly secured to the spindle. This is especially important on reversible drills.
- Tighten the drill bit securely, as prescribed by the owner/operator's manual if there is a chuck key. Remove it from the chuck before starting the drill. A flying key can cause injuries.
- Check auxiliary handles, if they are part of the tool. Be sure they are securely installed. Always use the auxiliary drill handle when provided. It provides more control of the drill. Grasp the drill firmly by insulated gripping surfaces.
- Always wear safety goggles, or safety glasses with side shields, that comply with current national standards and a full-face shield when needed. Use a dust mask for dusty working conditions. Wear hearing protection during extended periods of operation. (See 29 CFR 1910.133, Eye and Face Protection.)
- Always hold or brace the tool securely. Brace against stationary objects for maximum control. If drilling in a clockwise (forward) direction, brace the drill to prevent a counterclockwise reaction.
- If the drill binds in the work, release the trigger immediately, unplug the drill from the power source, and then remove the bit from the workpiece. If the drill operation could potentially bind, then do not actuate any switch lock-on.
- Never attempt to free a jammed bit by starting and stopping the drill.
- As the hole is about to be broken through, grip or brace the drill firmly, reduce pressure, and allow the bit to pass easily through the hole.
- Unplug the drill before changing bits, accessories, or attachments.
- Do not raise or lower a drill by its power cord.

Electrical Circular Saws

Electrical circular saws must have guards above and below the faceplate. The lower guard must retract automatically when cutting material. Instruct employees to use the guard as intended. Frequently check the guard to be sure that it operates freely and encloses the teeth completely when it is not cutting (Figures 17–15 and 17–16).

In the United States, OSHA standards require that frames and exposed metal parts of portable saws and other portable electrical woodworking tools that operate at greater than 90 V be grounded, unless they are double insulated or battery powered.

Blades

Use sharp blades. Dull blades cause binding, stalling, and possible kickback. They also waste power and reduce the

Figure 17-15. Portable band saw has guard that encloses the blade except for the actual work area. Hands are kept above the blade and are occupied.

Figure 17-16. The workpiece must be securely clamped. To maintain control, use both hands to properly and safely guide the saw. *(Courtesy Power Tool Institute)*

life of the motor and the switch. Use the correct blade for the specific application. Check that the blade has the proper size and arbor hole. Also check that the speed marked on the blade is at least as high as the no-load rpm on the saw's nameplate.

Before each use, check blades carefully for proper alignment and possible defects. Be sure that the blade's washers (flanges) are correctly assembled on the shaft and that the blade is tight.

Before each cut, be sure that the blade's guard is working. Check often to ensure that guards return to their normal position quickly. If a guard seems slow to return or hangs up, repair or adjust it immediately. Never tie back or remove the guard to expose the blade.

Cords
Before starting a circular saw, be sure that the power cord and extension cord are out of the blade's path and are long enough to freely complete the cut. On the job, stay constantly aware of cord location. A sudden jerk or pulling on the cord can cause loss of control of the saw and perhaps a serious injury.

Using the Saw
For maximum control, hold the saw firmly with both hands after securing the workpiece. Clamp workpieces if they are small enough. Check frequently to be sure that the clamps remain secure. Never hold a workpiece in the hand or across the leg when sawing. Avoid cutting small pieces of material that cannot be properly secured, as well as materials on which the saw's shoe cannot properly rest.

When making a blind cut (the operator cannot see behind what is being cut), be sure that hidden electrical wiring, water pipes, or any mechanical hazards are not in the blade's path. If wires are present, have a qualified person disconnect them at the power source. Contact with live wires could cause lethal shock or fire. Drain and cap water pipes. If the tool is double insulated, hold it by the insulated grasping surfaces or the handles as provided.

Also observe the following precautions:
- Set the blade's depth to no more than ⅛ to ¼ in. (0.32 to 0.64 cm) greater than the thickness of the material being cut.
- When starting to saw, allow the blade to reach full speed before contacting the workpiece or material.
- Be alert to the chance that the blade may bind and that kickback may occur.
- If a fence or guide board is used, be certain that the blade is kept parallel with it.
- When making a partial cut, or if power is interrupted, immediately release the trigger and do not remove the saw until the blade has come to a complete stop.
- Never reach under the saw or workpiece.

Kickback

Kickback is a sudden reaction to a pinched blade that causes an uncontrolled portable tool to lift up and out of the workpiece toward the operator. Kickback is the result of misusing a tool and/or of incorrect operating procedures or conditions.

Take the following specific precautions to help prevent kickback when using any type of circular saw:

- Keep blades sharp. A sharp blade cuts more easily and tends to cut its way out of a pinching condition.
- Make sure the blade has an adequate set in the teeth. Tooth set provides clearance between the sides of the blade and the workpiece, thus minimizing the chance of binding. Some saw blades have hollow-ground sides instead of tooth set to provide clearance.
- Keep blades clean. A buildup of pitch or sap on the surface of the blades increases the blades' thickness and also increases the friction on their surface. These conditions cause an increase in the likelihood of a kickback.
- Do not cut wet wood. It produces higher friction against the blade. The blade also tends to load up with wet sawdust and has a greater chance of kickback.
- Be cautious of stock that is pitchy, knotty, or warped. Such stock is most likely to create pinching conditions and possible kickback.
- Release the switch immediately if the blade binds or the saw stalls.
- Never remove the saw from a cut while the blade is rotating.
- Never use a bent, broken, or warped blade. The chance of binding, and the resulting kickback, is greatly increased by these conditions.
- Overheating a blade can cause it to warp and, thus, result in a kickback. Buildup of sap on the blades, insufficient set, dullness, and unguided cuts can all cause an overheated blade.
- Never use more blade protrusion than is required to cut the workpiece—1/8 to 1/4 in. (0.32 to 0.64 cm) greater than the thickness of the stock is usually sufficient. On some blades, the blade gullet needs to be clear of the material surface to kick off sawdust. This minimizes the amount of the blade's surface that is exposed and reduces the chance of kickback, and severity, if any kickback does occur.
- Minimize blade pinching by placing the saw's shoe on the clamped, supported portion of the workpiece and by allowing the piece that is cut off to fall away freely.

Reciprocating Saws

The versatility of the reciprocating saw in cutting metal, pipe, wood, and other materials has made it a widely used tool. By design, it is a simple tool to handle. Its few demands for safe use, however, are very important. Observe the following precautions:

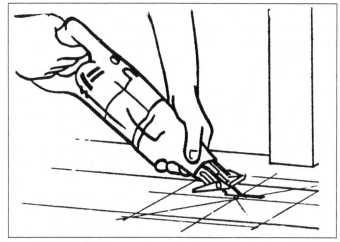

Figure 17-17. When plunge cutting, maintain firm contact between the shoe and the workpiece. Plunge cutting requires blades designed for the purpose. (*Courtesy Power Tool Institute*)

- On some blades, the blade gullet needs to be clear of the material surface to kick off sawdust.
- Without exception, use the blade specifically recommended for the job being done. Follow the owner/operator's manual recommendations.
- Operators should position themselves to maintain full control of the tools. They should avoid cutting above shoulder height.
- Use sharp blades. Dull blades can produce excessive heat, make sawing difficult, result in forcing the tool, and possibly cause an injury.
- To minimize blade flexing and to provide a smooth cut, use the shortest blades that will do the job.
- When plunge (pocket) cutting, use a blade designed for that purpose. Maintain firm contact between the saw's shoe and the material being cut (Figure 17–17).
- When making a blind cut, be sure that hidden electrical wiring or water pipes are not in the path of the cut. If wires are present, have a qualified person disconnect them at their power source to prevent the chance of lethal shock or fire. Drain and cap water pipes.
- Always hold the tool by the specific gripping surfaces.
- When making anything other than a through cut, allow the saw to come to a complete stop before removing the blade from the workpiece. This prevents both breakage of the blade and possibly losing control of the saw.
- Remember that the blade and blade clamp may be hot immediately after cutting. Avoid contact until they have cooled.

Miter Saws

These saws are used for crosscutting, mitering, and beveling wood, nonferrous metals, and plastics. They cut through the workpiece at a predetermined angle or miter.

When using miter saws, operators should always wear safety goggles or safety glasses with side shields that comply with standards, and a full-face shield when needed. Provide dust masks in dusty working conditions and hearing protection during extended periods of operation. Operators should not wear gloves, loose clothing, jewelry, or any dangling objects—including long hair—that may catch in rotating parts or accessories of the saw.

Be sure that all guards are in place and working. If a guard seems slow to return to its normal position or hangs up, adjust or repair it immediately. Because of the downward cutting motion, operators' safety requires that they keep their hands and fingers away from the path in which the blade travels—especially during repetitive, monotonous operations. Do not be lulled into carelessness because of a false sense of security. Blades are extremely unforgiving.

Observe the following safety precautions:
- Clean the lower guard frequently to help visibility and movement. Unplug before adjusting or cleaning.
- Use only recommended sizes of blades and rpm-rated blades.
- Do not use abrasive cutoff wheels on miter saws. Miter-saw guards are not appropriate for abrasive cutoff wheels.
- Remember that loose blades can fly off. Regularly check and tighten the blade and blade-attachment mechanism.
- When installing or changing a blade, be sure that the blade and its related washers and fasteners are correctly positioned and secured on the saw's arbor.
- To avoid losing control or placing hands in the blades' path, hold or clamp all material securely against the fence when cutting. Do not perform operations freehand.
- Never recut small pieces, and support long material at the same height as the saw table.
- Never place hands or fingers in the path of the blade or reach in back of the fence.
- Use the brake, if one is provided, because blades coast after being turned off. To avoid contact with a coasting blade, do not reach into cutting areas until the blade comes to a full stop.
- After completing a cut, release the trigger switch and allow the blade to come to a complete stop, and then raise the blade from the workpiece. If the blade stays in the cutting area after a cut is complete, injury can occur from unintentional contact.

Miter saws have spring-loaded saw heads that return the saw's head to its up position. Adjust, repair, or replace the spring mechanism if the saw's head does not automatically return to its up position when released (Figure 17–18).

Jig/Saber Saws

Observe the following precautions when using jig/saber saws:

Figure 17–18. Lock the miter head in the down position when not in use (back view shown). *(Courtesy Power Tool Institute)*

- Check carefully that the blades are adequately secured in position before plugging the saw in.
- Make sure the cord is out of the way and not in the line of the cut.
- Firmly position the saw's base plate/shoe on the workpiece before turning on the tool.
- Keep hands and fingers clear of moving parts.
- After making partial cuts, turn the tool off. Remove the blade from the workpiece only after the blade has fully stopped.
- Know what is behind a cut before making it. Be sure that hidden electrical wiring, water pipes, and hazardous objects of any kind are not in the path of the cut. If wires are present, have a qualified person disconnect them at their power source to prevent the chance of lethal shock. Drain and cap water pipes.
- Always hold the tool by the specific grasping surfaces.
- When plunge cutting, use a blade designed for the purpose and follow the manufacturer's recommended procedures.
- Throughout cutting procedures, maintain firm contact between the base and the material being cut.
- Remember that the blade and blade clamp may be hot immediately after cutting. Therefore, keep hands away from them until they have cooled down.
- Do not leave saws unattended. Unplug and secure the tool immediately after use.

Rotary Die Grinders

Rotary die grinders perform a wide variety of jobs. Be sure

that operators have a thorough understanding of the procedure for which they are using these tools.

Because grinders operate at high speeds, be alert and cautious to avoid injuries from contacting the working end or from thrown objects. Always wear safety goggles or safety glasses with side shields that comply with standards and a full-face shield, when needed. Use a dust respirator for dusty working conditions. Wear hearing protection during extended periods of operation. Air sampling may have to be performed to determine if exposure to air contaminants caused by grinding is within acceptable limits.

Be sure that the switch is in the OFF position before plugging the grinder in. Caution operators to hold the wheel or cutter away from themselves and co-workers when starting a grinder. Also warn operators not to use a rotary die grinder with the cutter pointing toward them. If the grinder should slip, the cutter could cause injury. Have operators remove from the area all materials and debris that might be ignited by sparks.

Use grinding wheels when working with hard materials, and use rotary files for soft materials, such as aluminum, brass, copper, and wood. Using grinding wheels on soft materials will excessively load the wheel and could cause the wheel to shatter or disintegrate. Parts could fly off and cause injuries.

Before each use, check the cutter or wheel for tightness. A loose cutter or wheel can be thrown from the rotary grinder and cause serious injuries. However, do not overtighten the collet. It can damage the collet cutter or wheel.

If the grinder is dropped, inspect it for damage, such as a cracked wheel, broken collet, or bent mandrel. Repair or replace damaged parts to prevent further breakage or other objects from being thrown.

Excessive pressure during use can bend or break the collet, mandrel, or wheel/cutter. If the grinder runs smoothly when not under load, but does not run smoothly under load, then excessive pressure is being used.

When placing a mounted grinding wheel, burr, or cutter in the collet, keep the distance between the back of the wheel and the front of the collet at a minimum. This prevents both bending the shank and wheel damage that could cause injuries. Make sure the shaft is engaged in the collet at least ½ in. (1.25 cm).

Grinding Wheels, Buffers, and Wire Brushes

These tools have special functions and thus require extra care when used. Grinding wheels present especially unusual hazards. Consequently, the storage, mounting, and use of grinding wheels require thorough training and extensive knowledge. Persons using or maintaining grinding wheels should observe safety recommendations conforming to ANSI B7.1, Safety Requirements for the Use, Care, and Protection of Abrasive Wheels, and ANSI B74.2, Specifications for Shapes and Sizes of Grinding Wheels, and Shapes, Sizes, and Identification of Mounted Wheels. Improper mounting, storage, or use of abrasive wheels could turn an important, useful tool into a lethal device. (See also Chapter 20, Metalworking Machinery, in this volume.)

Sanders

Belt and disk sanders cause serious skin burns when the rapidly moving abrasive touches the body. Because it is impossible to guard sanders completely, thoroughly train employees in their use. For example, the motion of the sander should be away from the body. Keep all clothing clear of moving parts. Wear dust-type safety goggles or plastic face shields (Figure 17–19). If harmful dusts are created, use a respirator approved for this type of exposure. Work in well-ventilated areas. Refer to 29 CFR 1910.1000, Air Contaminants, Table Z-1, Limits for Air Contaminants, to regulate dust exposure. OSHA regulates wood dust exposure under the requirements for Particulates Not Otherwise Regulated (PNOR). In addition, ensure that the recommendations in OSHA's Combustible Dust National Emphasis Program are implemented. Also, every affected employee should receive

Figure 17–19. Be sure you have proper ventilation, eye protection, and a dust mask, if necessary. Check the owner/operator's manual. *(Courtesy Power Tool Institute)*

applicable hazard communications training in accordance with OSHA's revised Hazard Communication Standard, 29 CFR 1910.1200.

Cleaning

Sanders require especially careful cleaning because of the dusty nature of the work. If a sander is steadily used, periodically dismantle it. Thoroughly clean it every day by blowing it out with low-pressure air.

Caution: Compressed air used for cleaning must be 30 psig (200 kPa) or less. The operator should wear safety goggles or work with a transparent shield between his or her body and the air blast.

Fire and Explosion Hazards

Because of the dust created in wood sanding, the fire and explosion hazard is considerable. Keep the dust to a minimum through adequate ventilation or, if the sander is so designed, with a dust collector or vacuum bag. Because of the extreme combustibility of wood dust and wood finishes, do not dispose of such dust in incinerators. There it would burn with almost explosive force.

To minimize the explosion hazard, if much wood sanding must be done, use electrical equipment designed for this exposure. Provide fire extinguishers approved for electrical or Class C fires, and tell employees what to do in case of fire. (See Chapter 18, Woodworking Machinery, in this volume.)

Operating

Observe the following precautions when using sanders:
- Before connecting a portable sander to the power supply, be sure that the switch and switch lock, if provided, are in the OFF position. If not, the sander will start immediately and loss of control could result in an injury.
- Stay constantly aware of the cord's location. Keep power supply cords and extension cords from getting entangled with the moving parts of the sander. Damaged cords can result in an electrical shock. A cord that is contacted by a moving belt can cause loss of tool control and, thus, possible injury.
- Use abrasive belts that are the width recommended by the manufacturer.
- Always keep face and hands clear of moving parts such as belts and pulleys.
- Never lock a portable sander in the ON position when the nature of a job may require stopping the sander quickly, such as using a disk sander on an automobile's fender. The rotating disk could become jammed and result in an injury.
- Never force a portable sander. The sander's weight applies adequate pressure. Forcing the sander and providing too much pressure can cause stalling, overheating of the tool, burning of the workpiece, and possible kickback of the tool or workpiece.
- Be careful not to expose the sander to liquids. Do not use it in damp or wet locations.
- When adjusting the tracking of the belt on a portable belt sander, be sure that the sander is supported and positioned to avoid unintentional contact with the operator or adjacent objects.
- Do not work with a faulty tracking sander. Discontinue work until the problem is corrected.
- The work area should be at least 3 to 4 ft (0.9 to 1.2 m) larger than the length of stock being sanded.
- Use jigs or fixtures to hold the workpiece, whenever possible.
- Always unplug sanders, and store them after use.
- Remove from the area material or debris that might be ignited by sparks from sanded metal. (See Chapter 20, Metalworking Machinery, in this volume.)

Routers

The widespread use of routers is based on their ability to perform an extensive range of smooth finishing and decorative cuts. Safety in operating a router starts by understanding that it runs at a very high speed—in the range of 20,000 rpm. This is 15 to 25 times faster than a drill.

Observe the following precautions when using a router:
- Install router bits securely and according to the owner/operator's manual.
- Always use the wrenches provided with the tool.
- Keep a firm grip with both hands on the router at all times (Figure 17–20). Failure to do so could result in loss of control and, thus, lead to possible serious injuries.
- Hold only those gripping surfaces designated by the manufacturer.

Figure 17–20. Grip routers firmly with two hands before turning on the switch. *(Courtesy Power Tool Institute)*

- Always face the cutter blade's opening away from the body.
- When starting a router equipped with carbide-tipped bits, start the router beneath a workbench. This protects the operator from a possible flying cutter, should the carbide be cracked.
- If the router is equipped with a chip shield, keep it properly installed.
- Keep hands away from bits and the cutter area when the router is plugged in.
- Do not reach underneath the work while bits are rotating. Never attempt to remove debris while the router is operating.
- Always disconnect the plug from the electrical outlet before changing bits or making any adjustments. When changing a bit immediately after use, be careful not to touch the bit or the collet with hands or fingers. They could be burned because of the heat built up from cutting.
- Make cutting-depth adjustments only according to the tool manufacturer's recommended procedures for these adjustments. Tighten adjustment locks. Make certain that the cutter's shaft is engaged in the collet at least ½ in. (1.25 cm). Check the owner/operator's manual.
- Be certain to secure clamping devices on the workpiece before operating the router.
- Be sure that the switch is in the OFF position before plugging the router into the power outlet.
- For greater control, always allow the motor to reach full speed before feeding the router into the work.
- Never force a router into the material.
- When removing a router from the workpiece, always be careful not to turn the base and bit toward the body.
- Unplug and store the router immediately after use.
- Always wear eye protection with side shields or full-face protection.

Air-Powered Tools

Air hoses, air-powered grinders, and pneumatic-impact tools each have specific hazards. By observing certain precautions, however, using these tools will be much safer.

Air Hoses

An air hose presents the same tripping or stumbling hazards that cords on electric tools do. A number of manufacturers offer self-storing, recoiling air hoses that work well when suspended above workstations. Persons or material unintentionally hitting the hose may cause the operator to lose his or her balance, or cause the tool to fall from an overhead place. Protect air hoses on the floor from trucks and pedestrians by laying two planks on either side of them or by building a runway over them. It is preferable, however, to suspend hoses over aisles and work areas.

Warn workers against disconnecting the air hose from the tool and using it for cleaning machines or removing dust from clothing. In the United States, regulations mandate that air pressure in excess of 30 psig (200 kPa) must not be used to clean machines and that low pressure must be used only with effective chip guarding or PPE, such as safety goggles. To remove dust from clothing, brush or use vacuum equipment.

Incidents sometimes occur when the air hose becomes disconnected and whips about. A short chain attached to the hose and to the tool's housing will keep the hose from whipping about should the coupling break. In some cases, couplings should also have such chains between the sections of hose.

Before attempting to disconnect the air hose from the air line, shut off the air. Also release any air pressure inside the line before disconnecting it.

A safety-check valve installed in the air line at the manifold will shut off the air supply automatically if a fracture occurs anywhere in the line. However, the safety-check valve must be compatible with the air-flow rate.

If kinking or excessive wear of the hose is a problem, protect the hose with a wrapping of strip metal or wire. One objection to armored hose is that it may become dented and thus restrict the flow of air. This applies only to heavy-duty hose used in construction work, however.

Air-Powered Grinders

Air-powered grinders require the same type of guarding as electrical grinders. Maintenance of the speed regulator or governor on these machines is of particular importance to avoid overspeeding the wheel. Have a qualified person inspect the grinder at each wheel change.

Pneumatic-Impact Tools

In pneumatic-impact tools, such as riveting guns and jackhammers, the tool proper is fitted into a gun and receives its impact from a rapidly moving reciprocating piston. The piston is driven by compressed air at about 90 psig (600 kPa) pressure.

Determine noise levels caused by pneumatic-impact tools to see if protective-noise devices should be provided for workers. Check the time limits and sound-level requirements of the regulatory standards. Consider isolating such operations or substituting quieter methods.

Pneumatic-impact tools require two safety devices. The first device is an automatic-closing valve that is actuated by a trigger located inside the handle, where it is reasonably safe from being unintentionally operated. The machine can operate only when the trigger is depressed. The second device is a retaining device that holds the tool in place so that it cannot be unintentionally fired from the barrel (Figure 17–21). Impress on all operators of small air

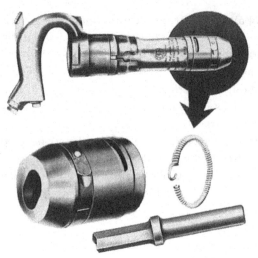

Figure 17–21. Chipping hammer safety retainer prevents the discharge of the tool. *(Reprinted with permission from Chicago Pneumatic Tool Co.)*

hammers not to squeeze the trigger until the tool is on the workpiece.

When using pneumatic-impact tools, there is the hazard of flying chips. Have operators wear the correct eye protection for this hazard. If other employees must be in the vicinity, they should be similarly protected. Where possible, set up screens to shield other workers when chippers, riveting guns, or air drills are being used.

When two chippers are working, they should stand back to back to prevent face cuts from flying chips. Workers should not point a pneumatic hammer at anyone, nor should they stand in front of operators who are handling pneumatic hammers.

Handling heavy jackhammers causes fatigue and may cause strains. Provide jackhammer handles with heavy rubber grips to reduce vibration and fatigue. Operators should wear protective footwear with metatarsal guards to reduce the possibility of injury should the jackhammer fall. Because hand-arm vibration syndrome (HAVS) is the potential result of working with heavy vibrating equipment, supervisors should become familiar with the methods used to reduce exposure to vibration and to symptoms of vibration-induced disease. The National Safety Council publication *Occupational Vibration: Preventing Injuries and Illness*, discusses vibration-related problems.

Many incidents are caused by the steel drill breaking because the operator loses balance and falls. Also, if the steel is too hard, a particle of metal may break off and strike the operator. Follow the manufacturer's instructions for sharpening and tempering steel.

Impact Wrenches

Impact wrenches are widely used in garages, repair shops, field work, and manufacturing industries during disassembly and assembly operations. Electrical and pneumatic power are commonly used. Pneumatic power is favored in heavy-duty operations. Pneumatic-powered tools frequently generate high-impact noise levels. If noise cannot be controlled to meet noise regulatory requirements, then include workers in a hearing conservation program with annual audiometric testing and use of hearing protection. Also, enforce use of eye protection near the workstation.

Electrical power is used most often with single-drive socket operations. On single-socket operations, extension cords and air hoses left lying on the floor can cause tripping hazards. To reduce operator fatigue on repetitive operations such as assembly lines, a common practice is to suspend the unit above the point of operation or with a spring-loaded overhead on trolleys or a similar balancing device. This practice of suspending devices on balancing units could cause injury by striking the upper body during a pendulum-motion swing.

Use sockets that are specifically designated as impact-wrench sockets (Figure 17–22). Sockets and accessories that are made only for hand use will not stand up to impact-wrench use. They are subject to premature failure and breaking and, thus, could cause injuries. Impact-wrench sockets usually are identified by a black finish on the outside and have heavier section thickness.

Observe the following precautions when using impact wrenches:

- Never use a wire, soft pin, or nail to hold the socket onto the square spindle of the impact wrench. If the proper retaining device on the tool is broken, repair the tool.
- Avoid excessive impacting, particularly on small bolt sizes. Small bolts could easily be broken or the threads stripped. Overtorquing can cause premature failure of fasteners or other damage and lead to incidents.

Figure 17–22. For impact wrenches, use only sockets designated as impact-wrench sockets. *(Courtesy Power Tool Institute)*

- On applications where a low or critical level of torque is required, lightly impact each fastener. Then perform the final tightening with a hand torque wrench.
- If the owner/operator's manual recommends using wood-boring bits with an impact wrench, be sure to unplug the tool before changing the bits.
- Do not use an impact wrench in wet or damp environments.
- Do not use nails, bolts, or other makeshift items as substitutes for safety pins.

Safety pins are usually designed to shear at definite preset pressure levels. Secure pins with retainers. Substituting poor-quality, inadequately designed shear pins or improperly using sockets could cause sudden failure and result in parts flying off.

Power Nailers and Staplers

Power nailers and staplers are fastening tools used primarily for securing wood products and other materials to wood by rapidly driving a nail or staple fastener into the materials being assembled. These tools are most commonly used in the building construction industry but are also found in mass production applications located in manufacturing and industrial facilities. Tool capabilities and features vary with the wide variety of fastening applications.

Although established safety practices apply to all power nailers and staplers, additional precautions may be required depending on the tool's source of power. Most of these tools are pneumatic; that is, they are powered by compressed air from an air compressor. Some are electric. Others use a gas fuel cartridge, sometimes with a battery. Tools with a gas fuel cartridge should be used only in well-ventilated areas. Manufacturers of tools using gas fuel have guidelines for tool and gas cartridge storage. Power-driven nailers and staplers are not intended for use in explosive atmospheres.

Pneumatically powered tools must be connected only to a source of clean, dry air regulated to the manufacturer's safe operating pressure. Bottled gases must never be used with pneumatic tools because they can result in the explosion of the tool. This is due to the fact that gas cylinders, bottles, and tanks are filled to very high pressures for efficiently transporting large quantities of the gas. The tank pressures far exceed the safe operating pressures of power nailers and staplers. Failure of pressure regulators used to reduce tank pressure could result in the delivery of high-pressure gas directly to the tool, possibly resulting in an explosion. Furthermore, the use of many bottled gases, such as oxygen, can result in combustion—also presenting the possibility of explosion if used with these pneumatic tools.

Power nailers and staplers vary in the types of operating controls and how the controls are activated before driving a fastener. The tool may have one or more trigger controls,

a "workpiece contact," or both. If so equipped, the tool's workpiece contact must be brought into contact with the materials being assembled before the fastener can be driven. To minimize the possibility of unintentionally activating a power nailer or stapler, tool operators must always keep their fingers away from the trigger or triggers until they are ready to drive a fastener.

When selecting a power-driven nailer or stapler for a given fastening application, consideration should be given to the type of operating control activation sequence desired. For example, tools may require a specific sequence of the operating controls, including activation of the trigger or triggers, and workpiece contact before the tool will drive a fastener. The operating control activation sequence can vary depending on the tool or application and may be changeable by the operator via the tool's built-in features or by a "modification kit" available from the manufacturer.

It is mandatory that all power-driven nailer and stapler operators, and other people in the immediate work area, wear eye protection. Airborne dust and debris are commonly generated during the operation of these tools. Therefore, it is imperative that both front and side protection be provided to the eye. Tool manufacturers and the applicable American National Standards Institute (ANSI) safety standard for these tools require use of eye protection meeting ANSI/ISEA Z87.1, American National Standard for Occupational and Educational Personal Eye and Face Protective Devices.

Because power-driven nailer and stapler designs vary among manufacturers—and sometimes within the manufacturers' product line—tool operators must have a thorough understanding of the operating and safety instructions for each tool they use. Besides the subjects discussed earlier, tool instructions typically address other topics such as installation, maintenance, inspection before use, loading, use, clearing of jams, handling between fastenings, and additional personal protective equipment. Tool users must read and understand all instructions—they will get the best results and performance by doing so.

Listed here are six basic rules found on power nailer and stapler warning labels and recommended by ANSI SNT 101–2002, which is the ANSI safety standard for these tools. Nailer and stapler users should always remember:

1. Read and understand tool labels and manuals. Failure to follow warnings could result in death or serious injury.
2. Operators and others in the work area must wear safety glasses with side shields.
3. Keep fingers away from triggers when not driving fasteners to avoid accidental firing.
4. Choice of triggering method is important. Check the manual for triggering options.

5. Never point a tool at yourself or at others in the work area.
6. Never use oxygen or other bottled gases. Explosions may occur.

Percussion Tools

Percussion tools, such as hammers, rotary hammers, and hammer drills, are primarily associated with masonry applications as varied as chipping, drilling, anchor setting, and breaking of pavement. They range from pistol-grip tools to large demolition hammers. Normal operating modes include hammering, hammering with rotary motion and rotation, or drilling only. Many models incorporate a combination of these modes.

The capacity of these tools is normally rated in maximum diameter, which is displayed on the nameplate. Do not attempt to use a bit larger than that specified unless otherwise recommended in the owner/operator's manual. Before operating one of these tools, compare the data on the nameplate with the voltage source. Be sure that the voltage and frequency are compatible.

Observe the following precautions when using percussion tools:

- For maximum control, use the auxiliary handles provided with the tool.
- Do not tamper with clutches on those models that provide them. Have the clutch settings checked at the manufacturer's service facility, at the intervals recommended in the owner/operator's manual.
- Check for subsurface hazards, such as electrical conductors or water lines, before drilling or breaking blindly into a surface. If wires are present, have a qualified person disconnect them at the power source, or be sure to avoid them. Otherwise, there is the risk of lethal shock or fire. Drain and cap water pipes.
- Always hold the tool by the insulated grasping surfaces.
- Do not force or overdrive the tool. Percussion tools are designed to hit with a predetermined force. Added pressure by the operator only causes operator fatigue, excessive bit wear, and reduced control.
- Keep the work area clear of debris.
- Always have firm footing.
- Remember to unplug the tool before changing bits or servicing.
- For percussion tools with rotating features, comply with all of the operating considerations referred to under Electric Drills in this chapter and Drills in Chapter 20, Metalworking Machinery, in this volume.

Special Power Tools

Tools with a flexible shaft require the same type of PPE as do direct-powered tools of the same type. Install and operate abrasive wheels to conform with ANSI B7.1, Safety Requirements for the Use, Care, and Protection of Abrasive Wheels, and federal regulations. Protect the flexible shaft against denting and kinking, which may damage the inner core and shaft. Whenever the tool is not in use, shut off the power. When the motor is being started, hold the tool end with a firm grip to prevent injury from sudden whipping. The abrasive wheel or buffer of the tool is difficult to guard. Because it is more exposed than the wheel or buffer on a stationary grinder, use extra care to avoid damage. Place grinder wheels on the machine or put them on a rack—do not leave them on the floor.

Hydraulic Power Tools

Hydraulic power tools are used in some industries, notably the electric utility industry, where employees work aloft from a hydraulically powered aerial-lift device. The power is obtained from the source used to operate devices such as hydraulic chain saws and compression devices. Some compression devices have a small hydraulic press that is pumped by the operator. Small leaks in the hydraulic hose or around fittings are hazards in the use of such equipment. There have been instances in which employees have put a hand over a pinhole leak and had oil forced into their finger by the high pressure. Also take care to always use a hose built for the pressure involved because a rupture can have serious consequences.

Gasoline-Powered Tools

Gasoline-powered tools are widely used in logging, construction, and other heavy industries. The best-known and most commonly used such tool is the chain saw.

Selecting a Chain Saw. Chain saws can be purchased in a variety of horsepowers and sizes. Some points to consider before purchasing include the size of the job, the balance of the saw, hand guards, kickback features, vibration reduction systems, and the convenience or ease of refueling.

Before beginning an operation, read the manufacturer's manual. Be sure that persons using chain saws are trained to operate the equipment according to the manufacturer's specific instructions.

Chain Saw Hazards. Operators of chain saws are exposed to hazards similar to those encountered by workers using handsaws. Kickback is the single biggest cause of chain saw injuries. A kickback is the sudden and potentially violent rearward and/or upward movement of the chain saw. It is often caused by the chain striking wood or other objects on the top quadrant on the tip of the chain guide bar. It can also be caused by binding or pinching in the cut.

Three types of anti-kickback devices are found on chain

saws: a safety nose or guard, a safety chain, or a chain brake. The safety nose prevents contact with the chain at the end of the chain. The safety chain is designed to reduce the tendency for the chain to catch or "hang up" in the wood. A chain brake stops the chain as the chain bar rises upward and the hand (at the handle) pivots against the brake switch.

In addition, the following hazards are specific to chain saws:
- falling while carrying a saw or when sawing
- sprains and strains from carrying, and working with, a heavy saw
- hand-arm vibration syndrome
- being cut by contact with the chain while it is in motion
- being cut by the chain when it is not in motion, either on or off the saw
- injuries from starting the gasoline engine
- inhaling exhaust fumes
- being struck by wood from overhead because a tree is vibrating
- sawdust in the eyes, especially when holding the tail stock or "stinger" end of a saw above the head
- burns from contact with a hot muffler or cylinder head
- injuries due to saws binding and kicking back at an operator
- injuries from falling trees and snags or rolling logs because the operator could not hear them above the noise of the saw's engine.

Carefully select saw operators. During their training period, inexperienced employees should work with an experienced faller under constant supervision.

Prevention of Fires
Take gasoline to the job in a sturdy, capped, UL-listed container that is painted red and labeled GASOLINE. The container should have a suitable spout for pouring gasoline into the tank, or provide a funnel for this purpose. Under no circumstances should the gasoline tank be replenished while the engine is running. Carefully wipe off any gasoline spilled on the tank or engine before starting the engine.

Additional precautions to prevent fires include the following:
- Turn off engine before refueling.
- Fill tanks only in an area with bare ground.
- Do not fill tanks where another tank was previously filled.
- Keep saws clean of gasoline, oil, and sawdust.
- Keep mufflers in good condition.
- Keep spark plugs and wire connections tight.
- Have fire extinguishers near power saws at all times.
- Keep flammable materials away from the point of the saw's cut.

Powder-Actuated Tools
Powder-actuated tools are used to make forced-entry fastenings in various construction materials (Figure 17-23). The systems are simple to use. However, there are precautions and safeguards that must be observed. Allow only trained and qualified operators to use this equipment, and then under close supervision. To become a qualified operator, a person must be thoroughly trained under supervision of a manufacturer's authorized instructor. Upon completion of the background training, the person is required to demonstrate competence through use of the system in varied applications and to pass a written examination. Upon successful completion of the examination, the instructor will issue a Qualified Operator's Card (Figure 17-24), records of which may be kept at or required by certain regulatory agencies.

Also, the possessor of a Qualified Operator's Card must be familiar with any regulations that apply to the use, maintenance, and storage of the system. Obtain more information about the use of these systems by writing to the Powder Actuated Tool Manufacturers' Institute (see References), and from the NSC Occupational Safety and Health Data Sheet 12304-0236, Powder-Activated Hand Tools. [See also 29 CFR 1910.243(d)(1)(i), Explosive Actuated Fastening Tools.]

Figure 17-23. This powder-actuated tool drives studs into concrete slab. The operator should wear eye, ear, and head protection. Note the holster (left) for carrying the tool. *(Reprinted with permission from Hilti Inc.)*

QUALIFIED OPERATOR OF POWDER-ACTUATED TOOLS

Make(s) _____ Model(s) _____

This certifies that _____
(Name of Operator)

Card No. _____ Soc. Sec. No. _____

Has received the prescribed training in the operation of powder actuated tools manufactured by

_____
(Name of Manufacturer)

Trained and issued by _____
(Signature of Authorized Instructor)

I have received instruction in the safe operation and maintenance of powder actuated fastening tools of the makes and models specified and agree to conform to all rules and regulations governing their use. Failure to comply shall be cause for immediate revocation of this card.

_____   _____
(Signature)   (Date)

Figure 17–24. A wallet card for qualified powder-actuated tool operators is available from tool manufacturers. A list of instructors should be maintained by each manufacturer. *(Based on ANSI A103–1984.)*

PERSONAL PROTECTIVE EQUIPMENT

Workers using revolving tools, such as drills, saws, and grinders, should not wear ties, loose clothing, and jewelry. Clothing should be free of oil, solvents, or frayed edges to minimize the fire hazard from sparks. The weight of most power tools makes it advisable for users to wear safety shoes to reduce the chances of injuries should the tools or workpiece fall or be dropped.

When power tools are used in overhead places, the operator should wear fall protection devices to minimize the danger of falling should the tool break suddenly or shock the operator or should the operator slip. Also, attach a safety line to the tool to keep it from falling on persons below should it be dropped. (See 29 CFR 1926, Subpart M, Fall Protection.)

Workers operating chain saws and carrying them through the woods must be surefooted. Because falls are among the most common incidents, operators should wear proper footwear to minimize this hazard. Operators should wear sharp, caulked boots. In some parts of the country, hobnailed shoes are preferred. In the winter, operators should wear rubber-soled shoes. Protective footwear is a good investment for members of a cutting crew.

Fallers and buckers should always wear protective hel-

mets because many people have been killed by falling trees. Many other lives have been saved, however, because workers wore safety hats. Make their use mandatory and select products meeting ANSI/ISEA Z89.1, American National Standard for Industrial Head Protection.

A ballistic nylon patch that covers part of the leg has been shown to reduce injuries to that part of the body when using a chain saw. The patches increase the time that an operator has to shut off the saw in the event a saw's chain should come against the leg.

On buffing, grinding, and sanding jobs that produce harmful dusts, provide workers with approved dust-type respirators. For operators of powder-actuated tools or jackhammers, provide hearing protection if more positive noise controls are not possible.

Employees should wear eye and face protection where flying particles present a hazard. Some companies require eye protection for all power-saw operators. In all operations where striking and struck tools are used, or where the cutting action of a tool causes particles to fly, provide eye protection that conforms to ANSI/ISEA Z87.1, American National Standard for Occupational and Educational Personal Eye and Face Protective Devices. Minimize the hazard of flying particles by using nonferrous, soft striking tools and by shielding the job site with metal, wood, or canvas. However, eye protection is still required.

Wear eye protection or face shields when using woodworking or cutting tools, such as chisels, brace and bits, planes, scrapers, and saws. There is always the chance of particles falling or flying into the eyes. Also wear eye protection or face shields when working with grinders, buffing wheels, and scratch brushes. The unusual positions in which the wheel operates may cause particles to be thrown off in all directions. For this reason, eye protection is even more important than it is when working with stationary grinders.

Do not overlook eye protection on the following jobs:

- cutting wire and cable
- striking wrenches
- using hand drills
- chipping concrete
- removing nails from lumber
- shoveling material
- working on the leeward side of a job
- using wrenches and hammers overhead
- working on other jobs where particles of materials or debris may fall.

Provide first-aid kits at the job site. At a minimum, the kit should include items specified in ANSI/ISEA Z308.1, American National Standard, Minimum Requirements for Workplace First Aid Kits. Employers should also evaluate the work environment to determine other first-aid supplies that may be needed. In areas where poisonous snakes are known to exist, provide a snake bite kit. Be sure that at least one individual in each work crew is trained in first aid.

Take noise-level readings of the chain saws and related equipment. If necessary, provide earmuffs or earplugs for workers. Protective clothes such as vibration damping gloves may need to be used to prevent HAVS from developing. Consider hand protection products complying with ANSI/ASA S2.73/ISO 10819, Mechanical Vibration and Shock—Hand-Arm Vibration Measurement and Evaluation of the Vibration Transmissibility of Gloves at the Palm of the Hand. Keeping warm and dry is also an important precaution, so adequate rainwear and cold-weather gear are necessary. Smoking also exacerbates vibration-induced illness and should be discouraged.

SUMMARY

- Proper selection, use, care, and supervision of hand and power tools can prevent abuse of these tools and eliminate or reduce employee injuries.
- Management should select proper hand and power tools and change tools only after careful consideration. These changes may include adding personal protective equipment, job rotation, or other adjustments.
- Six safety practices include (1) provide proper PPE, (2) select the right tool for the right job, (3) keep tools in good condition, (4) properly ground power tools, (5) train workers to use tools correctly, and (6) store tools in a safe place.
- Toolboxes should be used only for storing tools and not as stools, anvils, sawhorses, or lunchboxes. Make sure toolboxes and cabinets are locked after each workday and that all tools are accounted for.
- A good maintenance and repair program includes tool control through periodic inspection of all tool operations. Make sure employees have adequate work space and equipment for repairs.
- Misuse of common hand tools is a source of many injuries. Workers should be trained in safe work habits and proper use of tools.
- Soldering irons can be the source of burns and illnesses that result from inhaling fumes. Soldering irons must have adequate holders to prevent burns, and workers must have proper protective gear and ventilation to eliminate hazards.
- Hazards of portable power tools are associated with their mobility and energy sources. Workers should be trained in proper tool use, to select the right tool for the job, and to inspect and repair their equipment.
- The risk of electrical shock from electrically powered

tools can be reduced by using battery-operated tools, properly grounding equipment, and using only approved wiring and current. All parts of electrical equipment should be inspected regularly.
- To prevent injuries associated with air hoses, workers should make sure hoses do not present tripping hazards, avoid using hoses as cleaners, and prevent unintentional disconnection of hoses from the tools.
- Pneumatic-impact tools require an automatic-closing valve and a retaining device to hold the tool in place. Workers must protect their hearing and eyesight when using these tools.
- Special power tools include hydraulic, gasoline-powered, and powder-actuated equipment. Each type has its own hazards and safety precautions for safe operation.
- Workers operating power tools must wear the correct clothing, personal protective equipment, and fall protection equipment where appropriate.

REFERENCES

American National Standards Institute, 11 West 42nd Street, New York, NY 10036.

ANSI B7.1, Safety Requirements for the Use, Care, and Protection of Abrasive Wheels.

ANSI B74.2, Specifications for Shapes and Sizes of Grinding Wheels, and Shapes, Sizes, and Identification of Mounted Wheels.

ANSI/ASA S2.73–2014/ISO 10819:2013, Mechanical Vibration and Shock—Hand-Arm Vibration—Measurement and Evaluation of the Vibration Transmissibility of Gloves at the Palm of the Hand.

ANSI SNT 101–2002, Safety Requirements for Power Tools—Portable, Compressed-Air-Actuated, Fastener Driving Tools.

Budzik, R. S. *Precision Sheet Metal Shop Theory*. 2nd ed. Chicago, IL: Practical Publications, 1988.

Compressed Air and Gas Institute, 1230 Keith Building, Cleveland, OH 44115.

Grandjean, E. *Fitting the Task to the Man: An Ergonomic Approach*. 4th ed. New York: Taylor & Francis, 1988.

Grinding Wheel Institute, 30200 Detroit Road, Cleveland, OH 44145.

International Safety Equipment Association, 1901 N. Moore Street, Suite 808, Arlington, VA 22209

ANSI/ISEA Z87.1, American National Standard for Occupational and Educational Personal Eye and Face Protective Devices.

ANSI/ISEA Z89.1, American National Standard for Industrial Head Protection.

ANSI/ISEA Z308.1, American National Standard, Minimum Requirements for Workplace First Aid Kits.

Jackson, A. *Tools and How to Use Them—An Illustrated Encyclopedia.* Avenal, NJ: Outlet Book Company, 1992.

McDonnell, L. P., and A. Kaumeheiwa. *The Use of Hand Woodworking Tools.* 2nd ed. Albany, NY: Delmar Publications, 1978.

National Fire Protection Association, 1 Batterymarch Park, Quincy, MA 02269.
National Electrical Code, NFPA 70, 1993.

National Safety Council, 1121 Spring Lake Drive, Itasca, IL 60143-3201.
Fundamentals of Industrial Hygiene. 6th ed. 2012, Chapter 13, Ergonomics.
Occupational Vibration: Preventing Injury and Illness. 1991.
Occupational Safety and Health Data Sheets:
12304–0236, Powder-Activated Hand Tools.
12304–0445, Soldering and Brazing.

Powder Actuated Tool Manufacturers' Institute, 1000 Fairgrounds Boulevard, St. Charles, MO 63301.
Basic Training Manual.

Power Tool Institute, Inc., PO Box 818, Yachats, OR 97498.
Power Tool Safety Is Specific.

Putz-Anderson, V., ed. *Cumulative Trauma Disorders: A Manual for Musculoskeletal Diseases of the Upper Limbs.* New York: Taylor & Francis, 1988.

TPC Training Systems, 750 Lake Cook Road, Buffalo Grove, IL 60089.
Maintenance Fundamentals: Hand Tools and Portable Power Tools. 1980.

Underwriters Laboratories Inc., 333 Pfingsten Road, Northbrook, IL 60062.
Electric Tools, UL 45.

U.S. Bureau of Naval Personnel. *Tools and Their Uses.* Rate Training Manual, NAVPERS 10085-B. New York: Dover, 1973.

U.S. Consumer Product Safety Commission, Washington DC. Product Safety Fact Sheet No. 51, Chain Saws.

U.S. Department of Commerce, Office of Technical Service, Washington DC 20234.
Sparking Characteristics and Safety Hazards of Metallic Materials, Technical Report No. NGF-T–1–57, PB 131131.

U.S. Department of Health and Human Services, Public Health Service, Centers for Disease Control, NIOSH.
Criteria for a Recommended Standard, Occupational Exposure to Hand-Arm Vibration. September 1989.
Elements of Ergonomics Programs: A Primer Based on Workplace Evaluations of Musculoskeletal Disorders. March 1997.

U.S. Department of Labor. Occupational Safety and Health Administration, 200 Constitution Avenue NW, Washington DC 20210.
Code of Federal Regulations, Title 29. Section 1910.242. Chapter XVII.
29 CFR 1910.133, Eye and Face Protection.
29 CFR 1910.134, Respiratory Protection.
29 CFR 1910.243(d)(1)(i), Explosive Actuated Fastening Tools.
29 CFR 1910.1000, Air Contaminants, Table Z-1, Limits for Air Contaminants.
29 CFR 1910.1200, Hazard Communication Standard.
29 CFR 1926, Subpart M, Fall Protection.
29 CFR 1926.404, Subpart K, Electrical, Wiring Design and Protection.
Combustible Dust National Emphasis Program (Reissued). CPL 03-00-008 (March 11, 2008).

Walker, J. R. *Exploring Metalworking: Basic Fundamentals.* South Holland, IL. Goodheart-Wilcox Co., 1987.

Zinngrabe, C. J., and F. W. Schumacher. *Sheet Metal Hand Processes.* Albany, NY: Delmar Publications, 1974.

REVIEW QUESTIONS

1. Most incidents involving hand tools and portable power tools can be eliminated by observing what six safety practices?
   a.
   b.
   c.
   d.
   e.
   f.

2. What is the advantage of having centralized tool control in an industrial setting?

3. When metal tools break during normal use, the causes are usually related to the tools' _____.
   a. size
   b. quality
   c. handle

4. Which of the following is probably the most commonly used and abused tool?
   a. screwdriver
   b. hammer
   c. wrench
   d. pliers
5. When striking another tool, the striking face of the hammer should have a diameter approximately _____ larger than the struck face of the tool.
6. Identify the tool that is more frequently the source of disabling injuries than any other hand tool.
7. Based on their power source, portable power tools are divided into what five primary groups?
   a.
   b.
   c.
   d.
   e.
8. What are the inherent risks of portable power tools?
   a.
   b.
   c.
   d.

9. What are three precautions that should be observed when using portable power tools?
   a.
   b.
   c.
10. Name seven ways to properly maintain power tools.
    a.
    b.
    c.
    d.
    e.
    f.
    g.
11. What is the most convenient way of safeguarding the operator of portable electrical tools?
12. Name 7 of the 12 precautions that should be observed when using portable electric drills.
    a.
    b.
    c.
    d.
    e.
    f.
    g.

Woodworking Machinery

18

Patrick J. Conroy, OHST, CHST

Philip E. Hagan, JD, MBA, MPH, ARM, CIH, CET, CHMM, CHCM, CHSP, CEM

General Safety Principles
Electrical Equipment ▶ Guards ▶ Work Areas
▶ Materials Handling ▶ Inspection ▶ Hearing
Protection and Conservation ▶ Personal Protective
Equipment ▶ Standards and Codes

Saws
Circular Saws ▶ Selecting and Maintaining Circular-
Saw Blades ▶ Overhead Swing Saws and Straight-Line
Pull Cutoff Saws ▶ Underslung Cutoff Saws ▶ Radial
Saws ▶ Power-Feed Ripsaws ▶ Band Saws ▶ Jigsaws

Woodworking Equipment
Jointer-Planers ▶ Shapers ▶ Power-Feed (Thickness)
Planers ▶ Sanders ▶ Lathes and Shapers

Summary

References

Review Questions

Each piece of woodworking equipment poses its own hazard potential. To reduce the possibility of serious injury, management should provide workers with correctly guarded equipment, adequate jigs and fixtures, appropriate training, and proper enforcement of established safety rules. The topics covered in this chapter include:

- general principles to be included in a safety program for woodworking machinery
- hazards and safe work practices for various types of power saws
- safety precautions and worker protection for the use of woodworking equipment.

Because woodworking equipment is used in many industries, this chapter covers only the equipment and not specific woodworking industry operations.

The woodworker should know proper and safe procedures: how to choose the right machinery for the job and how to use the machinery correctly. The well-trained operator recognizes the potential for incidents and knows what to do when warning signs arise. For example, a change in noise, pitch, or any other operating characteristic of mechanical equipment should alert the trained worker to follow approved procedures for reporting or correcting a potentially hazardous situation. Supervisors and line managers should observe novice operators frequently to ensure that they are following established procedures.

GENERAL SAFETY PRINCIPLES

Companies should provide employees with equipment that meets the existing standards and regulations of the U.S. Occupational Safety and Health Administration (OSHA), the American National Standards Institute (ANSI), and the National Fire Protection Association (NFPA) National Electrical Code. Purchasers must specify on their purchase orders whatever optional or accessory parts are necessary to meet the requirements for mechanical and electrical safeguarding.

Management must observe the following general safety principles. A summary of safety rules for specific machines is given in Figure 18–1.

- Maintain all machines so that while they are running at full or idle speed and with the largest cutting tool attached, they are free of excessive noise and vibration.
- Level all machines, including portable or mobile ones, and, where necessary, dampen their vibration. Secure machines to the floor or other suitable foundations to eliminate all movement or "walking."
- Secure small units to benches or stands of adequate strength and design (Table 18–A).

- Make sure the machine is constructed so that tools that are too large for the machine's design cannot be mounted on it.
- Ensure that all arbors and mandrels have firm and secure bearings and are free from slip or play.
- Regularly check the adjustment of all safety devices. Those involving electrical circuits should be actuated to make sure they operate properly. Operators should always stop and securely lock out machines (i.e., via energy isolation) before cleaning, adjusting, or maintaining them. (See Control of All Energy Sources in Chapter 6, Safeguarding, in this volume.)
- Keep loose clothing, long hair, jewelry, and gloves away from rotating parts of machinery, especially from nip points and the point of operation.
- After the equipment has been completely stopped, clean work surfaces with a brush, not with the hand or a compressed-air nozzle.
- If possible, make adjustments only while the machine is not running. (See Chapter 6, Safeguarding.)

Electrical Equipment

All of the metal framework on electrically driven machines should be grounded, including the motor. The framework should comply with NFPA 70, National Electrical Code, and other applicable standards. There may be other local, state, provincial, and federal codes that apply to these machines. The NFPA code includes the following provisions:

- The machine should have a cutoff device (an easily identified EMERGENCY STOP switch, panic bar, or deadman's switch) within reach of the operator in the normal operating position.
- Electrically driven equipment should be controlled with a magnetic switch or other device that will prevent automatic restarting of the machine after a power failure. This is needed in cases where the machine, should it start automatically, would create a hazard (Figure 18–2).
- Clearly marked power controls and operating controls should be located within easy reach of the operator and away from a hazardous area. They should be positioned so the operator can remain at the regular work location while operating the machine.
- Each operating control should be protected against unexpected or accidental activation (Figure 18–2).
- Each machine operated by an electric motor should be provided with a positive means (lockout) for rendering the controls inoperative. If more than one person is involved in the maintenance or repair of the machine, each should install a separate padlock with a hasp. In addition to locking out the machine, the machine should be identified as inoperative. If the machine does not have a power discon-

SUMMARY OF RULES FOR SAFE OPERATION OF WOODWORKING TOOLS

Every operator should be trained in the safety rules covered in this chapter. As a summary, safety rules that demand close attention are listed below. Always read and understand the tool's operator's manual, tool markings, and the instructions packaged with the accessory before starting any work.

Table Saw

- Always wear safety goggles or safety glasses with side shields complying with current national standards, and a full face shield when needed
- Use appropriate mask or respirator in dusty work conditions.
- Use clean, sharp blades and check them carefully before each use. Damaged or dull blades could throw teeth, posing serious injury risk.
- Use the correct blade for your tool and for the job.
- While cutting, take specific precautions to avoid kickback (i.e., keep the fence parallel to the blade, always push the workpiece through the cut, set blade height to no more than ⅛ in. to ¼ in. greater than the thickness of the workpiece, do not cut "freehand."

Circular Saw

- Do not use a circular saw that is too heavy for you to easily control.
- Know your workpiece and what is behind the workpiece before you do the job. Clamp workpieces securely and check frequently to be sure clamps remain secure. Never hold a workpiece in your hand or across your leg when sawing.
- Use a straight edge or rip fence as a guide for gripping. If a fence or guide board is used, be certain the blade is kept parallel to it.
- Grip saw with both hands, keeping hands away from the blade.
- Keep the cord away from the blade and kerf.

Radial Arm Saw

- Keep your radial arm saw in correct adjustment and alignment. Use only sharp accessories that are designed for your saw.
- Position the workpiece so that the cut-off piece falls away from the table.
- Anti-kickback devices may not work when cutting smooth, hard surfaces. Always cut with the smooth, hard surface down on the table.
- A spreader should always be used when rip cutting.
- Always hold the workpiece firmly against the fence when crosscutting. Hold onto the saw handle until the blade comes to a complete stop.

Band Saw (Portable and Stationary)

- Always place workpiece securely in a vise or clamp when making cuts and never try to remove the workpiece while the blade is rotating.
- Never attempt to cut materials larger than the rated capacity listed in the manual and always check maximum operating speeds established for blades against band saw speed.
- With portable band saws, do not bear down on the blade while cutting. The weight of the band saw will supply adequate pressure for the fastest cutting.
- With stationary band saws do not make curved cuts with too small a radius for the width of the blade being used. This can cause unnecessary binding and possible blade breakage.
- Be sure all guards are in place and working properly before each use. Do not defeat guards.

Jointer/Planer

- Always keep cutter blades (knives) sharp and clean of rust and pitch to avoid excessive friction and use only blades recommended by the tool manufacturer.
- Never operate the tool without the cutter blade cover securely in place and make ensure that the blade flange fits in the arbor hole when installing the blade.
- Never joint or plane wood narrower than ¾ in., thinner than ¾ in., or shorter than 12 inches.
- Never feed the workpiece in the direction of the the the cutting blade rotation.
- Use push blocks to hold down the workpiece to protect hands and fingers.

Wood Shaper

- Examine the workpiece carefully before cutting. Do not shape chipboard, panel board, or any stock containing nails, paint, or varnish.
- Maintain proper adjustments for infeed and outfeed tables.
- Adjust the fence halves so that the cutter opening is more than is required to clear the cutter blade and be sure that the fence is locked into position after making all fence adjustments.
- Shaping narrow pieces can be hazardous. Always use fixtures, featherboards, push blocks and/or other jigs to hold down the workpiece.
- Always use a miter gauge and clamp for "end shaping" to maintain safe control of the workpiece.

Sander (Portable and Stationary)

- Choose the right tool and proper accessory for the job. Don't use a small sander for a big job or a large sander for a small job.
- Abrasive belts should be the width recommended by the manufacturer and follow the manufacturer's recommendations when selecting sanding paper.
- Always support you[r] workpiece on a stationary sander with a table of backstop and maintain a ¹⁄₁₆-inch maximum clearance between the table and the sanding disc or belt.
- Hold portable sanders firmly with both hands and never lock a portable sander in the ON position.
- Adequate ventilation is required when using any type of sander. If a sander is equipped with a dust bag, empty it frequently and when you are done sanding to avoid spontaneous combustion, which may result from a mixture of some wood finishing chemicals with dust particles.

Wood Lathe

- A lathe should not be altered in any way, or set to perform any operation not covered in the operator's manual.
- Clear the lathe bed of all objects before turning on the tool.
- Make sure the belt guard or cover is in place and the workpiece is free but firmly mounted between centers.
- Never adjust the tool rest with the lathe turned on.
- Always use the lowest speed when starting a new workpiece. Lathes should be operated at slow speeds until the workpiece is cylindrical.

Figure 18–1. Summary of rules for safe operation of woodworking tools. *(Used with permission from Power Tool Institute Inc.)*

TABLE 18–A. Typical Heights and Work Space

Machine	Table Heights Inches	Table Heights Centimeters	Work Area
Band saws	46	115	On three sides—a radius equal to twice the band-saw diameter (as measured from the point cut).
Circular saws	36 (hand feed) 32 (power feed)	90 80	Clearance on the working side should be 3 ft (90 cm) plus the length of stock.
Jointers	33	85	3 ft plus the length of stock.
Lathes	41	100	Clearance of at least 30 in. (75 cm) from stand, with smaller distances on ends and backside allowable.
Radial saws	39	85	Ripping—saw table equal to twice the length of the stock. Crosscutting—saw table equal to length of the stock plus 3 ft.
Sanders	36	90	3 ft plus the length of stock.
Shapers	36	90	3 ft plus the length of stock.

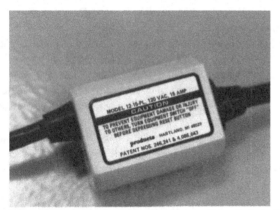 

Figure 18–2. Left: When installed in the electrical cord of a machine, this device will prevent automatic restarting. Right: This magnetic switch has a ring guard around the START button to protect against accidental reactivation.

nect to lock it in the OFF position, unplug the cord and insert a small padlock through the holes in the plug (Figure 18–3). (See Chapter 6, Safeguarding, in this volume.)
- Install an electronic motor brake on machines that have excessive coasting time (Figure 18–4). This device can greatly reduce the exposure at the point of operation.

Guards

Enclose or guard all belts, shafts, gears, and other moving parts so no hazard is present for the operator. (See ANSI/ASME B15.1, Safety Standard for Mechanical Power Transmission Apparatus.) Because most woodworking operations involve cutting, it is necessary, although often difficult, to provide guards at the point of operation. On most machines, the point-of-operation guard must be (1) movable to accommodate the wood, (2) balanced so as not to impede the operations, and (3) strong enough to provide protection to the operator. Whenever possible, completely cover blades and cutting edges at the point of operation. Not all such areas can be fully covered while the tool is in the workpiece; for example, radial saws cannot be guarded in this manner. Management should use another method for protecting the worker and bystanders in these cases. (See Chapter 6, Safeguarding.)

Work Areas

Provide ample work space around each machine. Suggested typical heights and minimum work spaces are given in Table 18–A. The working surfaces of the machine should be at a height that will minimize fatigue. Make adjustments if the worker is taller or shorter than average (see Chapter 16, Ergonomics Yesterday, Today, and Tomorrow, in the *Administration & Programs* volume). All accessory or feed tables should be the same height as the working surface of the machine.

Perform routine floor maintenance in the work area to prevent splintering and protruding nails. Keep floors level and free from holes and other irregularities. Install slip-resistant flooring in the work area near the machines. Mark aisleways with paint, railings, or other approved markings.

Maintain good housekeeping to prevent the accumulation of dust and chips. For instance, many exhaust vacuum

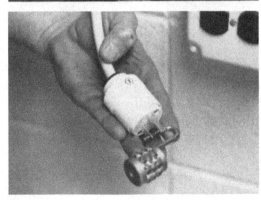

Figure 18–3. Top: Single-pole breaker lockout device. *(Courtesy W. H. Brady Co., Signmark Division.)* Center: Lockout with hasp, separate padlocks, and a tag. Bottom: Lockout for machines without a power disconnect.

systems are desirable and effective. A clean operation makes work easier and helps prevent fire and dust explosions. Because a number of fires originate and spread through ductwork, management should install automatic extinguishing systems in ducts, as well as in the collecting systems.

Adequately light the work area and the adjacent stock areas. Generally, 50 fc (538 lux) will be needed for work, but fine work may require 100 or more fc (1,076 lux). General illumination of 80 to 100 fc (861 to 1,076 lux) will pay dividends

Figure 18–4. On machines with excessive coasting time, this electronic motor brake greatly reduces exposure at the point of operation.

in both accident prevention and efficiency. There should be no shadows or reflected glare on the working surface.

Materials Handling

The facility's layout should encourage an even flow of materials and keep backtracking and crisscrossing to a minimum. Operators should not have to stand in or near aisles.

Arrange the machines so that the material handled by the operator and others requires a minimum of movement and changes of heights. This applies to both incoming supply and outgoing stock.

Inspection

Make safety checks by putting machines through trial runs before beginning a job and after each new setup. This usually is the responsibility of the setup or maintenance person.

The operator should inspect the machine at each new setup and at the start of each shift. The inspection process should follow the manufacturer's recommendations and the requirements and flow patterns of the workplace. This process would include inspecting the operating controls, safety controls, power drives, and sharpness of cutting edges and other parts. All cutting edges and tools must be kept sharp and be properly adjusted and properly secured.

Hearing Protection and Conservation

Most woodworking machinery creates high noise levels, requiring employers to establish and maintain effective hearing conservation programs. Because some woodworking machines, especially saws, are noisy, management should have a qualified person or industrial hygienist conduct sound-level measurements. If the reading of the sound level decibels (dBA) (slow response) exceeds 85 over an 8-hour period, that worker must be included in an effective hearing conservation program. If the sound level equals or exceeds 90 dBA over an 8-hour period, then the worker's exposure to the sound level must be reduced. If the measured levels are less than 85 dBA (the OSHA level triggering a hearing protection program in the United States), no action will be required. However, ear protection may be desirable at 85 dBA and less, and management should try to control the employee's exposure or somehow reduce the noise level. Some circular-saw blades are specifically designed to reduce noise levels. In other cases, large sound-dampening washers can be used to keep noise at a safe level.

When a woodworking process creates fine dust, the safety and health professional should have the amount sampled. The Threshold Limit Values (TLVs) and maximum permissible exposure (MPE) levels have been established for many materials and should be observed. Fine dust can be a health, fire, or explosion hazard. For workers' protection, respirators that reduce inhalation of various types of nuisance dust are available. (See National Safety Council, *Fundamentals of Industrial Hygiene,* 6th ed., in References at the end of this chapter.)

Personal Protective Equipment

All individuals in the work area should wear eye protection. Safety goggles complying with ANSI Z87.1, Practice for Occupational and Educational Eye and Face Protection, are excellent for operations that may generate flying objects. Face shields are not adequate if there are flying objects but do help if there is dust. On some operations, workers may need to use face shields in addition to goggles. Safety glasses with side shields may also be effective.

Workers should not wear loose clothing, gloves, and jewelry (especially rings, bracelets, and chains) that can become entangled in moving machinery. They should wear hair nets or caps to keep long hair away from moving parts and should keep their beards trimmed. Workers can protect their hands from splinters and rough lumber by wearing gloves. However, gloves should not be worn near rotating parts of the machine. Workers should wear approved protective footwear when handling heavy material or when there is a danger of injuring their feet. Where there is danger of a kickback—especially in ripping operations—workers should wear proper abdominal guards or anti-kickback aprons and always stand to one side of the saw.

Standards and Codes

There are a number of OSHA standards, such as 29 CFR 1910.213, Woodworking Machinery Requirements, that indicate required safety features for woodworking machines. Additionally, some states and other jurisdictions have codes that specify requirements. Management should consult all of these sources. The National Safety Council library has available an Occupational Safety and Health Data Sheet on woodworking machines. (See References at the end of this chapter.)

SAWS

All saws pose potential hazards for operators. Safety and health professionals can minimize these hazards by (1) providing training for operators; (2) ensuring that all machinery is properly guarded; and (3) making sure that all ANSI, NFPA, and government regulations are followed.

Circular Saws

Blade cuts or abrasions and kickbacks are among the most frequent incidents involving circular saws. These can be minimized by proper guarding and training and by enforcing safe working procedures.

Circular-saw operators are often injured when their hands slip off the stock while pushing it into the saw or when holding their hands too close to the blade during cutting operations. Other personnel can be injured by coming into contact with the blade when removing scrap or finished pieces from the table. Poor housekeeping practices and slippery floors are other sources of incidents involving circular saws.

Circular saws are designed to permit a wide range of cutting tasks. The problem with saws, as with most multiple-use equipment, is the difficulty in designing one guard that offers maximum protection for all types of tasks. The object is to prevent contact with the blade by using the proper type of hood guard, jigs, fixtures, combs, or other devices. Figures 18–5, 18–6, and 18–7 illustrate safety features and procedures.

Kickbacks and Ripping

A kickback occurs during a ripping operation when part or all of the workpiece is violently thrown back to the operator. Operators should keep their faces and bodies to one side of the blade, out of line with a possible kickback. To avoid kickbacks—and possible injury from them—operators should do the following:

- Maintain the rip fence parallel to the blade so the stock will not bind on the blade and be thrown.
- Keep the blade sharp. Replace or sharpen anti-kickback pawls when points become dull.
- Keep blades' guards, spreaders, and anti-kickback pawls in place and operating properly. The spreader must be in alignment with the blade, and the pawls must stop a

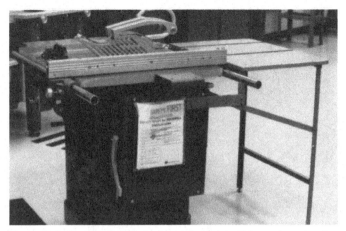

Figure 18–5. Features of this industrial model, tilting-arbor circular table saw include (1) posted safety rules, (2) push stick, (3) tail-off table, (4) rip fence, (5) crosscut guide, (6) self-adjusting point-of-operation guard, and (7) enclosed power transmission.

Figure 18–6. This operator is following safe operating procedures by (1) standing to the side while ripping; (2) keeping sleeves rolled up; and (3) using the rip fence, blade guard, splitter and anti-kickback device, and tail-off table. When using a tail-off table, the operator is less likely to reach over the blade to catch the stock before it falls to the floor.

kickback once it has started. Check their action before ripping.
- Cut only material that is seasoned, dry, and flat and that has a straight edge to guide it along the rip fence.
- Release work only when it has been pushed completely past the blade.
- Use a push stick for ripping widths of 2 to 6 in. (5 to 15 cm) and an auxiliary fence and push block for ripping widths narrower than 2 in. (5 cm).
- Allow the cutoff piece to be unconfined when ripping or crosscutting.
- Apply the feed force to the section of the workpiece between the blade and the rip fence.

A safe ripping procedure is described in Figure 18–7.

Supplying Proper Equipment
Guards (Figures 18–8 and 18–9) greatly reduce the likelihood of injury and are now considered standard equipment. If guards do not come with a saw, supply them when the saw is installed. The protection gained by using guards makes them essential. Management should be sure that the guards are practical and correct for the job, or operators may be tempted to remove them.

Provide a circular table saw used for ripping with a spreader to prevent wood with internal stresses from clamping down or binding at the out-feed edge of the blade. In this way, a spreader helps prevent kickbacks. It also keeps chips and slivers away from the back of the saw, where they might be caught by the saw's teeth and be thrown.

Supervisors and operators should check that the spreader is (1) rigidly mounted, not more than ½ in. (13 mm) in back of the blade when the blade is fully elevated and (2) at least 2 in. (5 cm) wide at table level. It should conform to the radius of the saw as nearly as possible and be high enough above the table to penetrate the full thickness of the stock. The spreader should be attached so it will remain in true alignment with the blade, even when the table or arbor is tilted.

Guard a circular table saw used for cutting with a hood that completely covers the blade projecting above the table. Operators should let the guard ride the stock being cut, adjusting to the thickness of the stock (Figure 18–10). The hood should be strong enough to resist any blows it might sustain during reasonable operation, adjusting, and handling. It should be made of shatter-resistant material and should be no more flammable than wood. To be effective, the hood must remain in true alignment with the blade, even if the table or arbor is tilted.

The hood may be suspended from a post attached to the side of the machine or supported on the spreader. However, operators should secure and support the mounting so it will not wobble and strike against the blade. In strength and design, the hood must protect the operator against flying slivers or broken saw teeth. The mounting should resist reasonable side thrust or force.

Guard the part of the blade underneath the table so the operator cannot accidentally contact the blade. The enclosure, which may be part of the exhaust hood, should be constructed with a hinged cover so the blades can be easily changed.

Rabbeting and Dadoing
When rabbeting and dadoing, it is impossible to use a spreader and often impractical to use the standard hood guard. These operations can be effectively guarded by a jig that slides in the grooves of the transverse guide. In this

18 Woodworking Machinery

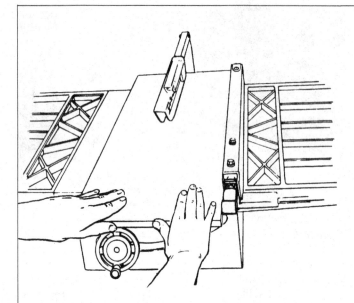

When the width of the rip is 6 in. or wider, use your right hand to feed the workpiece until it is clear of the table. Only use the left hand to guide the workpiece—do not feed the workpiece with the left hand.

When the width of rip is 2 in. to 6 in., use the push stick to feed the work

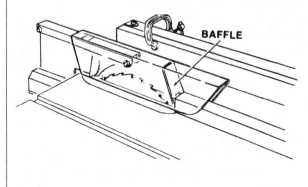

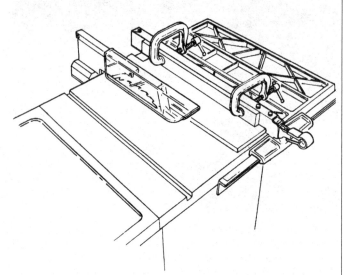

When the width of the rip is less than 2 in., the push stick cannot be used because the guard will interfere. Use the auxiliary fence-work support and push block. Use two C clamps to attach the auxiliary fence-work support to the rip fence.

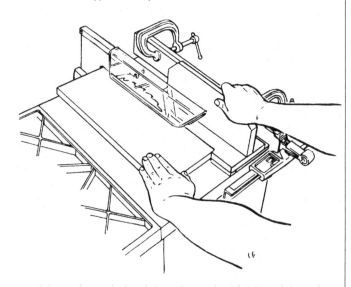

Feed the workpiece by hand along the auxiliary fence until the end is about 1 in. beyond the front edge of the table. Continue to feed using the push block. Hold the workpiece in position and install the push block by sliding it on top of the auxiliary fence-work support (this might raise the guard).

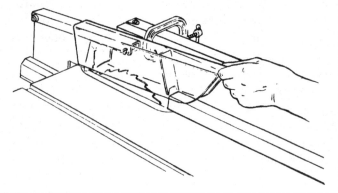

Figure 18-7. Safe ripping procedure.

Figure 18–8. Close-up of a properly functioning splitter and anti-kickback device during a ripping operation.

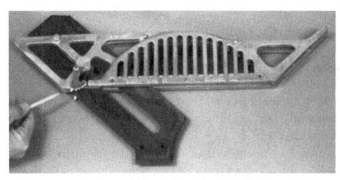

Figure 18–9. When sawing large pieces of stock, the support for the guard (see Figures 18–5 and 18–6) can be in the way and prevent sawing through the wood. The guard shown here permits sawing large pieces without interference. The splitter and anti-kickback dogs are built into the guard.

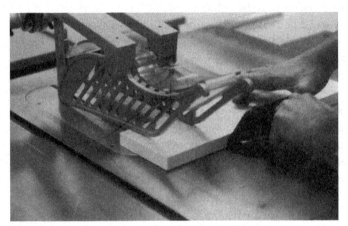

Figure 18–10. The overhead self-adjusting guard on this circular saw rides the stock as it is being cut and automatically adjusts to its thickness.

Figure 18–11. When crosscutting several pieces to the same length, it is important to use a small block of wood clamped to the rip fence to allow room for clearance when the piece is cut off. If this is not done, the piece being cut off will bind between the fence and the blade and be thrown back toward the operator.

way, the work is locked in the jig, and the operator's hands are kept well away from the saws or cutting head.

Because rabbeting and dadoing jobs vary, special jigs may be needed (Figure 18–11). The hazards of these jobs justify special guarding, especially when work is being done on small stock. If a shop does a lot of dadoing and rabbeting, supervisors should set aside one or more machines for this work. This will eliminate frequently removing standard guards from machines that are normally used for cutting and ripping.

Operators can use feather boards to hold the work to the table and against the fence as it is fed past the dado head. Feather boards are suitable for short runs because they can be quickly set up and are inexpensive. A feather board can be made from straight-grained stock, preferably hardwood. When using the board, the operator should make the parallel saw cuts (the comb) in the direction of the grain. The feather board should bear against the stock at an angle of 45 to 60 degrees.

Proper Operating Methods

Only authorized persons should operate circular saws. A saw in good condition and running at the correct speed should cut easily. Operators should not have to cut freehand or crowd the saw (push it hard) by forcing the stock faster than it can be cut. If the saw does not cut as fast as it should, or if it does not saw a clean, straight line, the saw blade or the running speed may be improperly set. Operators should check and correct these conditions—which are potential sources of incidents—before proceeding with the job.

Most hand- and power-feed saws run at about 3,450 rpm. This will give a 12-in. (0.3-m) blade a rim speed of 10,839 sfm (55 m/s); a 16-in. (0.4-m) blade, 14,451 sfm (74 m/s); and an 18-in. (0.46-m) blade, 16,258 sfm (83 m/s). Workers should always follow the manufacturer's instructions when operating these saws. Supervisors must find a way to prevent operators from placing a larger blade on the mandrel than is allowable for the mandrel's speed.

When the collar of a saw is tightly clamped in position, only the collar's outer edge should come in contact with the blade. If the inside of the collar has not been machined properly, it will force the rim of the saw out of line. When the blade comes in contact with the stock being cut, there will be a buckling effect on the saw. After the loose collar is securely fastened in place, the operator should test the saw with a straightedge. This test is important to conduct on circular saws as well as on edgers and trimmers.

When feeding a table saw, operators should keep their hands out of the line of the cut. Although the guard offers protection from the sides and from above, it does not protect workers from the front. When operators are ripping with the rip fence close to the saw, they should use a push stick between the blade and the fence to keep their fingers away from the blade. Operators should keep push sticks or blocks of various sizes and shapes near the machine. To make push sticks long enough to keep hands well away from the blade, add 6 in. (15.2 cm) to the blade's diameter. Because of kickbacks, operators should stand to the side of the stock they are ripping. A heavy leather or plastic apron or abdomen guard gives additional protection.

Operators should hold stock against a gauge; they should never saw freehand. Freehand sawing endangers the hands and may cause work to get out of line and bind on the saw. When ripping stock with narrow clearance on the fence side, the operator can gain more clearance by clamping a filler board flat to the table between the fence and the blade and by guiding the stock against it. Use of a filler makes the hazardous practice of removing the hood guard because of lack of clearance unnecessary.

The best height for the blade above the workpiece depends on the following considerations: (1) high-blade silhouette (the blade is as high as possible) and (2) low-blade silhouette (the saw blade extends just through the stock). See Table 18–B for a list of advantages to using a high-blade silhouette versus a low-blade silhouette.

Operators should use the right saw for the job. Using the wrong saw for the job makes the work harder and requires additional force when feeding the stock. Do not use a cross-cut blade for ripping or a ripsaw for crosscutting. Using a general-purpose table saw for work that should be done on a special machine is also a poor practice. Work that can be done on special or power-feed machines should not be done on hand-feed, general-purpose machines. For example, a table saw is often used for hand-feed ripping operations. Instead, operators should perform this work on a power-feed ripsaw, which virtually eliminates the dangers of kickback and injury to the hands.

Long stock is sometimes crosscut on a table saw. Unless the stock is adequately supported, this, too, is a

TABLE 18–B. Advantages of High-Blade Silhouette and Low-Blade Silhouette

Advantages	
High-Blade Silhouette	**Low-Blade Silhouette**
• Reduced kickback potential • Saw tooth cuts down nearly vertical to table • Saw blade is closest to spreader • Less power needed • Faster cutting • Less saw blade wear	• Less exposure of the blade to the operator • Smoother cut • More table support needed for the workpiece in front of the saw blade

dangerous practice. The long stock extends beyond one or both ends of the table, interferes with other operations, and may be a hazard to other workers or trucks. Also, it is difficult to guide long pieces. Operators must exert considerable pressure while their hands are close to the saw. Such stock is more easily cut on a swing saw, pull saw, or radial saw.

Under no circumstances should operators adjust the fence while the saw is running. Parallel setting of the fence is particularly important. To enable the operator to set the fence accurately, mark the top of the saw table with a permanent, distinct line or other suitable device. Make the mark directly in front of and in line with the blade.

Operators should stop a circular table saw before leaving it. It is not sufficient to cut the switch and walk away. Workers have suffered amputations caused by saws still coasting with the power off. An electric brake attached to the motor's arbor offers fast, positive action.

Selecting and Maintaining Circular-Saw Blades

The characteristics and conditions of circular-saw blades are important safety factors for the operators who use them. Manufacturers of circular-saw blades have published valuable information on the selection, use, and care of these blades (Figures 18–12 and 18–13).

During designing, building, and tensioning, the maker gives a saw enough rigidity and tensile strength to cut without harmful distortion. When operators or others alter its original design, operate it at other than the rated speed, or change the balance or tension, they seriously affect the saw's efficiency and safety.

In addition, the following conditions of blades may cause unsafe, difficult, or unsatisfactory operation:

- *Blade out of round.* If some teeth are longer than others, the long teeth do most of the work. An unequal strain is imposed on the blade, which may cause it to run out of line, to heat up, and to warp.
- *Blade not straight, out of plane.* Lumps or warps can be checked with a straightedge across the length of the

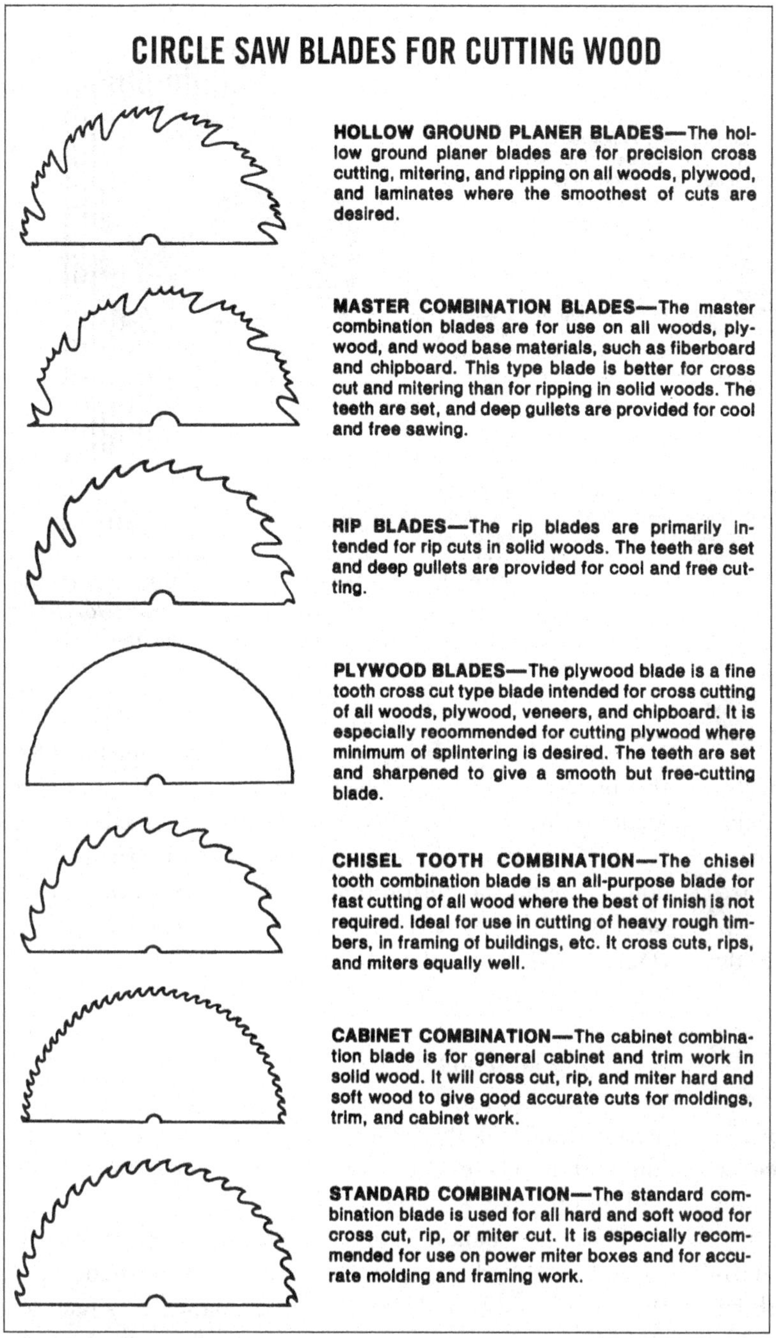

CIRCLE SAW BLADES FOR CUTTING WOOD

HOLLOW GROUND PLANER BLADES—The hollow ground planer blades are for precision cross cutting, mitering, and ripping on all woods, plywood, and laminates where the smoothest of cuts are desired.

MASTER COMBINATION BLADES—The master combination blades are for use on all woods, plywood, and wood base materials, such as fiberboard and chipboard. This type blade is better for cross cut and mitering than for ripping in solid woods. The teeth are set, and deep gullets are provided for cool and free sawing.

RIP BLADES—The rip blades are primarily intended for rip cuts in solid woods. The teeth are set and deep gullets are provided for cool and free cutting.

PLYWOOD BLADES—The plywood blade is a fine tooth cross cut type blade intended for cross cutting of all woods, plywood, veneers, and chipboard. It is especially recommended for cutting plywood where minimum of splintering is desired. The teeth are set and sharpened to give a smooth but free-cutting blade.

CHISEL TOOTH COMBINATION—The chisel tooth combination blade is an all-purpose blade for fast cutting of all wood where the best of finish is not required. Ideal for use in cutting of heavy rough timbers, in framing of buildings, etc. It cross cuts, rips, and miters equally well.

CABINET COMBINATION—The cabinet combination blade is for general cabinet and trim work in solid wood. It will cross cut, rip, and miter hard and soft wood to give good accurate cuts for moldings, trim, and cabinet work.

STANDARD COMBINATION—The standard combination blade is used for all hard and soft wood for cross cut, rip, or miter cut. It is especially recommended for use on power miter boxes and for accurate molding and framing work.

Figure 18–12. Choosing the correct saw blade for the job will increase operator safety. *(Printed with permission from Sears, Roebuck and Co.)*

diameter of the unmounted blade. However, because of blade tensioning, it may not be flat, except under power.
- *Blade out of balance.* Too much of the blade was removed from one side during improper sharpening.
- *Improper hook or pitch of teeth.* The teeth of ripsaws and cutoff saws have different designs for different kinds of wood and for different purposes. Combination saws may be used for both crosscutting and ripping. There are other blades for certain kinds of woods, wood material, metals, and plastics.

- *Improper or uneven set.* A blade has to cut a kerf thicker than the blade to give adequate clearance for the saw to pass through the wood. The teeth can be given set or swage by bending alternate teeth right and left or by spreading the point of every tooth so each is slightly wider than the blade.
- *Dull blades.* Keep blades sharp so the saw works at top efficiency and the operator exerts minimum force when feeding the saw.
- *Gummed blades.* Also keep blades clean and free of pitch buildup so they will run at top efficiency and safety. A gummed blade can cause a kickback.
- *Improper bushings.* Bushings are provided to match the blade to the arbor's diameter. A bushing that is too large will cause a blade to be unstable at high-rotation speeds.
- *Cracked blades.* As soon as a crack is detected, remove the blade from service. Inspect blades for cracks every time the teeth are filed or set. Some cracks are so small that they may be invisible to the naked eye. For such cracks, use a nondestructive testing method, such as Magnaflux. If cracked blades are left in service, the crack frequently grows larger and eventually will cause partial fragmentation. Most cracks start in the gullets. Excessive heat and vibration cause saw blades to crack.

To prevent cracking, operators should follow these precautions:
- Tighten the blade on the arbor for which it is designed.
- Operate the saw at the speed specified by the manufacturer. If the saw is not tightened and operated according to the manufacturer's instructions, it may wobble, vibrate, heat, expand, and crack.
- Allow sufficient clearance (set or hollow grinding) for the teeth to prevent burning, and, thus, heating and cracking.
- Keep the blade in perfect round and balance.
- Keep the blade sharp at all times. A dull blade will not cut; rather, it will pound or burn itself through the wood, so that vibration, heating, and then cracking result.

After repairs have been made, the blade must be retensioned. This is a job for a sawsmith. Unless the company has the services of such a person, the blade should be repaired by the manufacturer.

Overhead Swing Saws and Straight-Line Pull Cutoff Saws

Overhead swing saws and straight-line pull cutoff saws cause hand injuries in several ways. Hands can be cut (1) while the

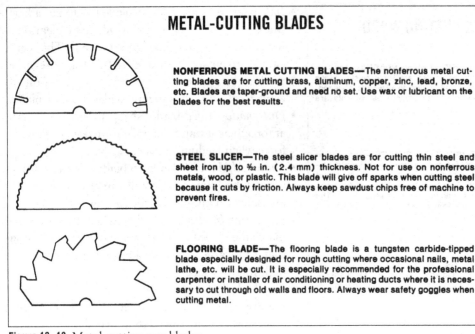

Figure 18-13. Metal-cutting saw blades.

Figure 18-14. START button for a cutoff saw should be protected so accidental contact will not start the saw. (*Printed with permission from Machine Design Magazine.*)

blade coasts or idles, (2) when operators attempt to remove a sawed section of board or a piece of scrap, and (3) when operators measure boards or place them in position for the cut.

Operators' hands can be struck by a saw if it bounces forward from a retracted position or if it moves forward should the return device fail. Operators may pull the saw against their hands in the cutting path or may suffer body cuts from a saw that swings beyond its safe limits.

Guards

Cutoff saws must be guarded with a hood guard. The hood should extend at least 2 in. (5 cm) in front of the saw blade when the saw is in the back position. Some guards cover the lower half of the saw when the saw is not cutting and ride on the top of the stock as the saw cuts.

Provide a counterweight or other device to automatically return the swing saw to the back of the table without rebounding when released. Secure the counterweight with a device designed to hold twice its weight. The counterweight should be guarded, if within 7 ft (2.1 m) of the floor.

Install a limit chain or other device to prevent the saw from swinging beyond the back or front edges of the table. Another device should likewise keep the saw from rebounding from its idling position. A latch with a ratchet release on the handle is best, but, in some instances, a nonrecoil spring or bumper is adequate. A magnetic latch provides another way to prevent rebounding. Provide the saw table with a wood bumper to prevent bodily contact with the blade when it is extended the full length of the support arm.

Place STOP and START buttons for quick and easy access. STOP buttons should be easily contacted in an emergency. Mushroom-type buttons are recommended for this purpose. The STOP button should also have an easily identifiable color. Use protected START buttons, however, so accidental contact will not cause the saw to start. A collar around the button that extends $1/8$ to $1/4$ in. (3 to 6 mm) above the top of the button is a recommended method (Figure 18-14). (See also Chapter 13, Ergonomics, in NSC's *Fundamentals of Industrial Hygiene*, 6th ed.)

Operating Methods

If the saw is pulled by a handle, the handle should be attached either to the right or left of the saw rather than in line with it. The operator should stand to the handle side and pull the saw with the hand nearer it.

Thus, if the handle is on the right side of the saw, boards should be pulled from the right with the right hand, and the saw should be pulled with the left hand. This method (1) makes it unnecessary for operators to bring their hands near the saw's path while the saw is cutting and (2) keeps the operators' bodies out of the line of the saw.

Saws may be ordered with either right or left handles. For a new saw, order it with the handle on the side from which the stock is to be pulled. If it is necessary to pull stock from the opposite side, place the handle on that side so the operator can stand in the correct position.

To measure boards, place their ends against a gauge stop. When it is necessary to measure the board with a scale while the board is on the table, move the board away from the blade.

At the completion of each cut, the operator should put the saw back to the idling position and make sure that all bounce has stopped before putting his or her hand on the table. Do not use automatic or constant-stroking saws unless the point of operation is guarded and there is no hazard to the operator.

Underslung Cutoff Saws

An underslung cutoff saw is usually operated by a treadle. Because its forward movement is fast, it should be completely enclosed in the noncutting position. For general work, it should also be covered by a movable hood guard that slides forward or drops to rest on the stock while the saw is cutting. A guard on the treadle ensures that the treadle will not be used accidentally.

Underslung cutoff saws are commonly used to cut knots out of narrow pieces such as flooring and molding. The stock is placed on the table by hand, and the hands are customarily held close to the line of the cut on either side. The movable guard gives little protection because the saw's action is so fast that the guard can ride over the top of the hands.

On either side of the line of travel, construct a barrier guard with enough clearance between the guard and the table top to admit the stock, but not the operator's hands or fingers. With practice, an operator can rapidly feed stock under this type of guard.

Radial Saws

When crosscutting, radial saws cut downward and pull the wood away from the operator and against a fence. These saws, like straight-line pull cutoff saws, require many adjustments to permit their full use. Adjustments should not be made when the saw is running. Lockout means should be provided.

The radial saw's head can be tilted to cut a bevel, or the supporting beam and track can be swung at an angle to make a miter cut. Both adjustments may be used to cut a compound bevel or miter. Likewise, the head may be turned parallel to the length of the table so the saw can be used for ripping. In this case, it is an overhead, stationary saw against which the stock is fed by hand.

Always guard the upper half of the saw, including the arbor end. The lower half of the saw should have an articulating guard for 90-degree crosscut operations. The lower guard should automatically adjust itself to the stock's thickness and should remain in contact with the stock being cut for the full working range. This prevents accidental contact with the sides of the blade (in an axial direction) when the cutting head is at rest (not in the cut) behind the fence and in the 90-degree crosscut mode. Under certain conditions, lower blade guards can cause additional hazards.

Provide some means so the cutting head will not roll or move out on the arm away from the column because of gravity or vibration (Figure 18–15). For repetitive crosscut operations, provide an adjustable stop to limit forward travel of the cut-

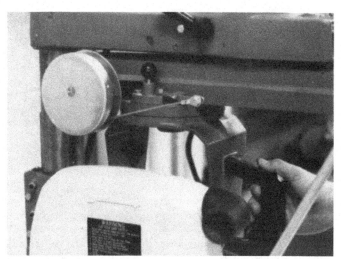

Figure 18–15. A spring return is installed on this saw so the cutting head will not roll or move out on the arm—away from the column—because of gravity or vibration.

ting head to that necessary to complete the cut.

The saw table should be large enough to cover the blade in any position (miter, bevel, or rip). Therefore, workers should never operate the saw with the blade in a position where it protrudes or extends beyond the table.

Obviously, only competent woodworkers should operate a saw with so many features. Operators should be well trained and aware of possible hazards. They need to know what to do when the machine is performing below standards.

The principal sources of injury connected with operating radial saws are those common to other power-driven saws. They include cuts to the arms and hands caused by the blade, by flying wood and chips, and by handling materials. As with other power saws, prevention of injuries requires proper use of the equipment.

Ripping

When ripping, rotate the radial saw's head 90 degrees so the blade is parallel to the fence and is clamped in position. Then, lower the blade until it will cut through the stock. Before ripping, position (1) the nose of the guard (or drop the guard down), (2) the spreader, and (3) the anti-kickback devices (Figures 18–16 and 18–17). Feed the stock against the direction of rotation of the revolving blade from the nose of the guard—the side at which the blade rotates upward toward the operator. For in-rip, feed material from right to left; for out-rip, feed from left to right (Figure 18–18).

Use a spreader in ripping to prevent the wood from immediately coming together after being cut. This action reduces the chances of the blade binding and causing a kickback. Mount the spreader in direct line with the blade. When properly adjusted, the spreader prevents wrong-way feed.

Figure 18-16. Before ripping, lower the guard on the in-feed side until it almost touches the workpiece. Adjust the splitter/anti-kickback device on the out-feed side so the splitter rides within the saw kerf. Feed the workpiece against the blade rotation.

The anti-kickback device must be used in the ripping operation. This device is positioned so the anti-kickback fingers ride on the stock. Adjust the angle or height of the fingers so that if the stock is pulled out by hand, it will jam under the fingers and the stock cannot be moved. The anti-kickback device is adjustable for different thicknesses of stock. Check the anti-kickback fingers regularly for sharpness, making sure none of them are bent or in contact with the stock when sawing.

Caution: Operators should always follow the proper direction of blade rotation. The saw blade should always rotate downward as viewed from the operator position. Feeding from the wrong side (wrong-way feed) tends to grab the material away from the operator and throw it toward the in-feed (nose) end of the guard. Wrong-way feed is prevented if a spreader is properly installed and correctly positioned.

Two possibilities of severe injury arise from feeding from the wrong side. The blade's direction of rotation makes it easy for the operator's hands to be drawn into the revolving saw. There is the additional danger to the helper and to other people on the opposite, or in-feed, side of the saw. Flying stock can be thrown with enough force to drive the stock through a 1-in. (2.5-cm) board. There have been serious and fatal injuries because of failure to observe this precaution. Power-feed rolls are available for ripping operations. These not only greatly reduce kickback, but also speed up production.

When feeding the stock, hold it firmly against the table and fence. The blade should be sharp and parallel to the fence. Apply feed pressure between the blade and the fence. Use a push stick, longer than the blade's diameter by 6 in. (15.2 cm), when ripping narrow or short stock. Operators should never release the feed pressure until the cut is completed and the workpiece has fully cleared the blade.

Operators should exercise special care when ripping material with thin, lightweight, hard, or slippery surfaces because of the reduced efficiency of anti-kickback devices. When ripping, the operator should wear an anti-kickback apron.

Crosscutting

Radial saws used for crosscutting are pulled across the cutting area by means of a handle located to one side of the blade, rather than in line with the blade. Whenever possible, the operator should stand on the handle's side, pull the cutting head with the hand nearer the handle, and maneuver the lumber with the other hand. In this way, the operator's body is not in line with the blade. At the same time, the operator's hands are not near the blade's cutting area. Operators should never pull the blade beyond the point necessary to complete the cut because the back of the blade could lift the workpiece and throw it over the fence. Operators should always place the workpiece against either the fence or a special jig. They should never crosscut freehand.

Operators should never remove short pieces from the table until the saw has been returned to its position at the rear of the table. They should always use a stick or brush—never their hands—to remove scrap from the table.

Under normal circumstances, operators should measure by placing the boards to be cut against a stop gauge. However, in instances in which it is necessary to measure with a rule, operators should turn the saw off until they have finished measuring.

At the conclusion of each cut, operators should always return the cutting head to the full-rearward position behind the fence. Operators should never remove their hands from the operating handle unless the cutting head is in this position.

Because the blade's direction of rotation and the feed's direction tend to cause the blade to feed itself through the work, operators should develop the habit of holding their right arms straight from the shoulder to the wrist. This will prevent the blade from grabbing and possibly stalling while in the workpiece.

Power-Feed Ripsaws

Because long stock is often ripped on power-feed ripsaws, the clearance at each working end of the saw table should be at least 3 ft (0.9 m) longer than the length of the longest material handled. Operators should adjust feed rolls to the thickness of the stock being ripped. Insufficient pressure on the stock can contribute to kickbacks (Figure 18-19).

Where multiple-cut power-feed ripsaws are used, install a dado head alongside the last blade. This head disposes of the edging. The offbearer then does not have to handle any scrap pieces and can pay more attention to material coming from the saw.

18 Saws 565

Figure 18–17. Warning sign on the out-feed side reads CAUTION—NEVER RIP FROM THIS END. Not to heed this warning would cause the workpiece to be pulled from the operator's grasp and sent flying across the room.

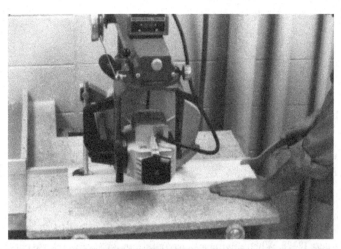
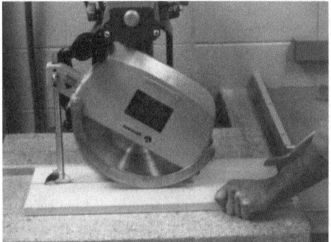

Figure 18–18. Top: In-ripping. Bottom: Out-ripping.

A common hazard occurs with a power-feed chain ripsaw that uses an overhead cutting saw with a solid chain having a rabbeted center trough. Unless care is taken to maintain the rabbet in the trough, very thin slivers may drop from a ripped edge into the center trough of the chain. These slivers can fly out like bullets toward the operator. Regular inspection is necessary.

Band Saws

Although injuries from band saws are less frequent and less severe than those from circular saws, they still occur. The usual cause of band-saw injuries is operators' hands coming into contact with the blade. For instance, when hand-feeding stock, the operators' hands must come dangerously close to the blade. Operators should, therefore, use a push stick to control the workpiece when it is close to the blade. For this reason, it is especially important that the saw's table be well lit, yet free from glare.

The band saw's point of operation cannot be completely

Figure 18–19. Anti-kickback fingers should spread at least the full width of the feed rolls between the operator and the saw blade on a power-feed saw. This saw is equipped with a double set of anti-kickback fingers. *(Printed with permission from Western Electric Company.)*

Figure 18-20. When sawing a sharp radius, make several release cuts up to the cutting line. This prevents saw blade binding and possibly breaking the blade or causing it to jump off the guide wheels.

Figure 18-21. This jointer is equipped with two guards: one on the working side of the fence and the second on the backside of the blade.

covered. However, install an adjustable guard—designed to prevent operator contact with the front and right side of the blade above the upper blade guides—as close as possible to the workpiece. Be sure that the wheels and all nonworking parts of the blade are encased. Professional band saws have outside enclosures made of solid metal. The front and back of the enclosure should be made of solid material or sturdy mesh material. Some smaller, lightweight band saws are made of metal and plastic materials.

A band saw should have a tension-control device to indicate proper blade tension. If it does not, the operator should test the blade for correct tension before beginning a job. An automatic tension-control device will help to prevent breakage of blades. Another device prevents the motor from starting if the tension on the blades is too tight or too loose.

A band saw, especially a large one, will run for a long time after the power is shut off. For this reason, it should have a brake that operates on one or both wheels to minimize the potential hazard of coasting when the machine is shut off and left unattended. Serious injuries occur when operators take hold of the running blade, not realizing it is in motion. Another reason for brakes is to stop the wheels in case the blade should break. The safety device shown earlier in Figure 18-15 stops the saw if the blade breaks.

On band saws, be sure that a guard, formed to the curvature of the feed rolls, covers the nip point. Install this guard so the edge is $3/8$ in. (9 mm) from the plane that is formed by the inside face of the feed roll, in contact with the stock.

When small pieces of stock are cut, use a special jig or fixture. A technique for safely sawing a sharp radius is shown in Figure 18-20.

Jigsaws

Jigsaws are not normally considered as hazardous as other woodworking machinery. Occasionally, however, they also cause injuries, especially to the fingers and hands. Safe operating procedures for jigsaws require

1. the blade to be properly attached and secured
2. the threshold rest, or slotted foot, to be on the stock
3. the guard to be in an effective position
4. the operators to keep their hands a safe distance from the blade.

Also, all drive belts, pulleys, and other moving parts should be guarded.

In addition, operators should make turns slowly, with no sharp- or small-radius turns if working with a wide blade. Narrow blades should be used when small-radius curves are needed. Operators should plan clearance cuts to eliminate the need to back out of curves. Operators should clean the table with a long-handled brush after the blade has stopped.

WOODWORKING EQUIPMENT

Woodworking equipment such as jointer-planers, shapers, power-feed planers, sanders, and lathes all poses safety hazards for those operating it or working around or near the equipment. Workers must be trained in safe work practices and in emergency first aid and other procedures to prevent or minimize incidents and injuries.

Jointer-Planers

Second only to circular saws, hand-feed jointers or surface planers are the most dangerous woodworking machines (Figure 18-21). Most of the injuries are caused when the

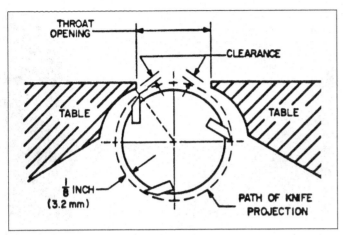

Figure 18–22. Table clearance for jointers. *(Printed with permission from American National Standards Institute; Underwriters Laboratories Inc.)*

Figure 18–23. Note the difference between a hold-down (left) and a push block (right). The push block has a piece of wood acting as a positive stop against the end of the workpiece; the hold-down is flat on the bottom. Both are used to keep the operator's thumbs and fingers away from the cutter head.

hands and fingers of operators come in contact with these machines' knives. Many of these incidents occur when short lengths of stock are being jointed.

A jointer should be equipped with a horizontal cutter head, the knife projection of which extends beyond the body of the head not more than 1/8 in. (3.2 mm). The clearance between the path of the knife projection and the rear table should be not more than 1/8 in. (3.2 mm). The clearance is measured radially from the path of the knife's projection to the closest point on the table and with the rear table level with the path of the knife's projection. The clearance between the path of the knife projection and the front table should not be more than 3/16 in. (4.8 mm). This clearance is measured the same as that for the rear of the table but with the tables in the same place (Figure 18–22).

The openings between the table and the head should be just large enough to clear the knife. In addition, the openings should be not more than 2 in. (5.1 cm) when the front and rear tables are set or aligned with each other for zero cut.

Cover the table opening on the working side of the fence with a spring-loaded, self-closing guard that adjusts itself to the moving stock. For good protection when edge jointing, install a swinging spring-loaded, self-closing guard or a guard that moves away from the fence along the axis of the other head. For surface planing, use only the swinging guard because it permits the use of hold-down push blocks, such as those pictured in Figure 18–23, to feed material smoothly over the other head. Use hold-down push blocks whenever the operator joints wood that is narrower than 3 in. (7.6 cm) (Figure 18–24).

Be sure that the unused end of the head, which is behind the fence, is enclosed at all times. A sheetmetal telescoping guard is acceptable for this purpose.

Jointer-planers are commonly used for planing off cupped or warped stock to make it flat. To do this job, power-feed attachments are available with resilient hold-down devices that simulate the pressure of the hands. If stock is properly conditioned, however, very little of it should have to be trued in this manner.

For doing surfacing work on the jointer, operators should have both hands on top of the stock, if it is thicker than 3 in. (7.6 cm). They should never place their hands over the front or back edges, where they can easily come in contact with the head.

Shapers

The principal danger in using wood shapers is that operators' hands and fingers might strike against the revolving knives. Severe incidents can also result when broken knives are thrown by the machine. When one knife breaks or is thrown from the collar, the other knife is usually thrown as well. Thus, four or five pieces of heavy, sharp steel are thrown about the shop with sufficient speed to kill a person.

The greatest number of injury incidents occurs when shaping narrow stock, which, if held in the hand, brings it close to the knives. Use hold-down push blocks or jigs in these instances. Each shop should have a well-understood rule that stock narrower than a specified width must be held in a jig. Some shops put the limit at 6 in. (15 cm), others as high as 12 in. (30 cm). A cardinal rule is always feed against the direction of rotation of the cutter. Never back up the workpiece, or a kickback can occur.

Eliminate the danger from broken or thrown knives by using solid cutters that fit over the spindle. The initial cost is greater for cutters than for knives. However, on moderately long runs, cutters are less expensive. Carbide-tip or solid-carbide cutters are available. In all cases, cutters are safer than knives.

When knives are used, operators should take the following precautions to keep the blades from breaking or flying:
- Knives must meet rigid specifications for shaper steel.
- Knives must be sharpened and installed only by a fully qualified person.
- Knives and the grooves in the collars must fit perfectly and be free of dust.
- The two knives must balance perfectly. They must be weighted against each other each time they are set.
- A knife must not be used after it has become so short that the butt end does not extend beyond the middle point of the collar.
- Deep cuts should be avoided. It is safer and more efficient to take two light cuts than one heavy cut.
- During start-up, operators should apply the power in a series of short starts and stops to slowly bring the spindle up to operating speed. They should listen carefully for chatter and should watch for other evidence that the knives are out of balance.

Various types of safety collars are in use that help prevent knives from flying. Although nothing should be done to discourage the use of such collars, do not consider them to be substitutes for perfectly balanced and fitted knives.

Another safety measure is to use some type of braking device to stop the spindle after the power is shut off. With double-spindle shapers, there should be starting and stopping devices for both spindles. The spindles should be started one at a time.

Shaper work must be held against guidepins for curved shaping or a fence for straight-line shaping. A feather board may be clamped to the fence (Figure 18–25). Keep the portion of the cutter, or knives, behind the fence covered. For curve shaping, adjust the overhead guard to just clear the stock. Use a starting pin when curve shaping. Operators should use a long-handled brush, or a vacuum system, to remove chips and scraps from the worktable.

A number of guards are available to protect operators' fingers. For straight-line shaping, the fence frame or housing should contain the guard. A fence should have as small an opening as possible for the knives and should extend as far as possible, at least 18 in. (46 cm), on either side of the spindle. This ensures good support at the start and finish of the cut. If the entire edge of the stock is to be shaped, adjust the portion of the fence beyond the cutter to receive the thinner stock and provide a stable bearing. This cannot be accomplished with a flat, continuous fence. It is best done with a split, adjustable fence.

Adjust the cutting-head guard for minimum head exposure. Provide a way to contain wood dust and chips. Jigs

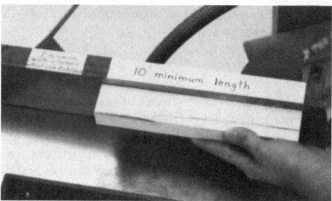

Figure 18–24. Painting a 3-in. (7.6-cm) strip (top) and a 10-in. (25-cm) strip (bottom) on top of the fence, and labeling them accordingly, serves as a quick reference for checking the width, thickness, or length of stock before jointing it.

and hold-down clamps should be maintained to hold the stock securely.

Power-Feed (Thickness) Planers

Operators can reduce planer vibration by anchoring the planer on a solid foundation and by insulating it from the foundation with cork, springs, or other vibration-absorbing material. The planer should have a three-point bearing and be bolted down without distortion. Distorting any woodworking machine will ultimately cause it to malfunction.

Because of the noise planers create, they should be isolated in a separate room or in a special soundproof enclosure built for this equipment. Helical cutter heads also will substantially reduce the noise levels. If neither soundproofing nor using helical cutter heads is practical, provide hearing protection for those working in the immediate area.

Completely enclose cutter heads in solid metal guards. These guards should be kept closed when the planer is running. Provide good local exhaust from the cutting heads.

Have operators stop, lock out, and tag out feed rolls, cutter heads, and cylinders before placing their hands in

18 Woodworking Equipment

Figure 18-25. Top: A pair of feather boards clamped to the tabletop hold the workpiece down on the table and in against the fence. The operator is using a push stick. Bottom: This ring guard is installed over the cutter at the point of operation.

the bed plate to remove wood fragments, to make adjustments, or for any other reason. If planer parts are driven by belts running on the backside of the planer, completely enclose the belts and sheaves with sheet metal or heavy-mesh guards, even though the planer is fenced at the back or is next to the wall.

Guard feed rolls with a wide metal strip or bar that will allow boards to pass but that will keep operators' fingers out of the rolls. Install anti-kickback fingers that operate on the in-feed side across the entire throat (width of the machine).

Danger of kickbacks cannot be entirely overcome by mechanical means. Wood splinters and knots are frequently thrown out. Therefore, operators should not look into the backside to "watch" the operation. Operators also should always stand out of the line of the board's travel. Other people should not work or walk directly behind the feeding end of the planer. Use a barrier or guardrail when the machine is running. Operators must avoid feeding boards of different thicknesses at the same time. Thinner boards are not held by the feed rolls and can be kicked back from

the heads. To keep workers from being struck by long, fast-moving boards, fence or mark off the space at the out-running end. Operators should always wear safety goggles.

Sanders

Supervisors and others should make sure that drum, disk, or belt-sanding machines are enclosed by dust exhaust hoods. The hood should enclose all portions of the machine except the portion designed to feed the stock. Personnel who operate sanders should wear goggles and dust respirators during sanding operations and cleanup.

On a belt-sanding machine, place a guard at each in-running nip point on both power-transmission and feed roll parts. Guard feed rolls with a wide metal strip or bar that will allow boards to pass but that will keep the operators' fingers out. Guard the unused run of the abrasive belt on the operator's side of the machine to prevent contact with the operator.

All hand-feed sanders should have a work rest and be properly adjusted (1) to provide minimum clearance between the belt and the rest and (2) to secure support for the work. Small workpieces should be held in a jig or holding device.

Abrasive belts on sanders should be the same width as the pulley drum. When material is brought into contact with the moving abrasive belt, operators should adjust the drums to keep the abrasive belt taut enough to turn at the same speed as the pulley drum. Operators should inspect abrasive belts before using them and replace those found to be torn, frayed, or excessively worn.

Lathes and Shapers

Supervisors or operators should ensure that the rotating heads of lathes, whether running or not, are covered as completely as possible by hoods or shields. Hinge these hoods so they can be thrown back when adjustments are needed. The clearance between the workpiece and the tool rest should be about 1/8 in. (3 mm). Hold turning chisels firmly on the tool rest.

Operators should have lathes used for turning long pieces of stock that are held only between the two centers with long, curved guards. The guards should extend over the tops of the lathes to prevent the workpiece from being thrown out should it come loose.

Management should select and train lathe operators with care. Operators should carefully inspect all parts of the lathe for any defects and correct them before starting a job. Operators must not use stock that has checks, splits, cracks, or knots. They must give constant attention to stock being turned in worder to discard any material likely to break. Sometimes, stock can be trimmed out before turning. Operators must carefully place stock in the machine and feed the cutting tool slowly into the stock. They must wear safety goggles; a face shield is also required if the operation is dusty or if rough stock that may produce splinters is used. Operators should tie up long hair and not wear loose-fitting clothing to prevent becoming entangled in the machine. Figure 18–1 provides a summary of the safety rules for lathe use, as well as for other woodworking tools.

With shapers, operators should use jigs with handles to keep their hands at least 6 in. (15 cm) away from the cutting bit. Mount a ring guard cupguard around the cutting bit to reduce contact with it. Always feed stock into the rotation of the cutter. Always use a fence or a fixture with the stock.

SUMMARY

- All electrically driven machines should be adequately grounded, have a cutoff switch, and have some means of rendering the controls inoperative to prevent injuries. All belts, shafts, gears, and other moving parts must be guarded.
- Employees using woodworking machines should read the operator's manual prior to working with a new piece of equipment.
- Employees should maintain ample work areas around each machine, keep floors and work surfaces clear and free from hazards, provide adequate lighting and ventilation, and arrange machines to ensure a steady flow of materials. All machine controls, cutting edges, and power drives should be inspected at each new setup and at the start of each shift.
- Management should guard against hearing loss, airborne dust and contamination, explosion and fire hazards, and related problems. Workers should use personal protective equipment.
- Management can minimize the hazards posed by saws by (1) training operators, (2) ensuring all machines are guarded, and (3) making sure operators follow all safety procedures.
- To operate a circular saw safely, workers should keep their hands out of the cut line and use a holder to guide the stock. Hood guards are used for cutoff saws, while a counterweight will return swing saws to the proper position. Underslung cutoff saws can be guarded by constructing a barrier on either side of the line of travel. The upper half of radial saws should always be guarded, while the lower half should have an articulating guard for 90-degree crosscut operations.
- Because power-feed ripsaws are often used to cut long stock, operators should adjust feed rolls to the thickness of the stock being ripped to prevent kickbacks. Band saws must be adequately lighted and the point of operation completely covered.

- When operating wood shapers, workers should use push blocks or jigs to keep their hands away from the knives and use guards to prevent injuries from thrown knife blades.
- Power-feed planers should be secured to reduce vibration and isolated to help control noise hazards. Solid metal guards will enclose cutter heads to prevent injuries, and proper guards will also prevent kickbacks.
- To operate sanders safely, workers should use dust exhaust hoods, wear goggles and dust respirators during operation, inspect all belts, and be sure that hand-feed sanders have the proper distance between the sander and work rest.
- The rotating heads of lathes should be covered as completely as possible by hoods or shields. Lathe operators must be selected and trained with care to use, inspect, and maintain their equipment and to wear protective gear.
- Operators should clean their machines and work surfaces with a long-handled brush after the equipment has stopped and not use their hands or air nozzles.

REFERENCES

American National Standards Institute, 11 West 42nd Street, New York, NY 10036.
  Practice for Occupational and Educational Eye and Face Protection, ANSI Z87.1.
  Safety Standard for Mechanical Power Transmission Apparatus, ANSI/ASME B15.1–1992.
  Safety Standard for Stationary and Fixed Electric Tools, ANSI/UL 987–1990.
International Safety Equipment Association, 1901 North Moore Street, Arlington, VA 22209,
National Fire Protection Association, 1 Batterymarch Park, Quincy, MA 02269.
  2015 NFPA 70E®: Standard for Electrical Safety in the Workplace.
  Fire Protection Handbook. 20th ed. Vols. 1 and 2.
  Prevention of Fires and Explosions in Wood Processing and Woodworking Facilities, NFPA 664, 1993.
National Safety Council, 1121 Spring Lake Drive, Itasca, IL 60143-3201.
  Fundamentals of Industrial Hygiene. 6th ed. 2012.
  Safeguarding Concepts Illustrated. 7th ed. 2002.
  Occupational Safety and Health Data Sheet: Tilting-Arbor and Tilting-Table Saws, 12304–0605, 1991.
Power Tool Institute, Inc. PO Box 818, Yachats, OR 97498.
U.S. Department of Health and Human Services, National Institute for Occupational Safety and Health, 4676 Columbia Parkway, Cincinnati, OH 45226.

Health and Safety Guide for Manufacturers of Woodworking Machinery, DHEW (NIOSH) Publication 79–131.

Health and Safety Guide for Millwork Shops, DHEW (NIOSH) Publication 76–111.

Health and Safety Guide for Plywood and Veneer Mills, DHEW (NIOSH) Publication 77–086.

Health and Safety Guide for Prefabricated Wooden Building Manufacturers, DHEW (NIOSH) Publication 76–159.

Health and Safety Guide for Sawmills and Planing Mills, DHEW (NIOSH) Publication 78–102.

Health and Safety Guide for Wooden Furniture Manufacturers, DHEW (NIOSH) Publication 75–167.

U.S. Department of Labor, Occupational Safety and Health Administration, 200 Constitution Avenue NW, Washington DC 20210.

29 CFR 1910.213, Woodworking Machinery Requirements.

29 CFR 1926.304, Woodworking Machinery Requirements.

REVIEW QUESTIONS

1. Companies should provide employees with equipment that meets the existing standards and regulations of what three organizations?
   a.
   b.
   c.

2. On most machines, the point-of-operation guard must be:
   a. movable to accommodate the wood.
   b. balanced so as not to impede the operations.
   c. strong enough to provide protection to the operator.
   d. all of the above
   e. only b and c

3. The working surfaces of the machine should be at a height that will minimize _____.

4. The machine inspection process should include inspecting what elements?
   a.
   b.
   c.
   d.

5. Name the two most frequent hazardous incidents involving circular saws.
   a.
   b.

6. When feeding a table saw, make push sticks long enough to keep hands well away from the blade by adding _____ to the blade's diameter.

7. What three actions can the operator of a saw take to seriously affect the saw's efficiency and safety?
   a.
   b.
   c.

8. Name three ways overhead swing saws and straight-line pull cutoff saws can cause hand injuries.
   a.
   b.
   c.

9. What two possibilities of severe injury arise from feeding from the wrong side of a saw?
   a.
   b.

10. Because long stock is often ripped on power-feed rip-saws, the clearance at each working end of the saw table should be at least _____ longer than the length of the longest material handled.

Welding and Cutting

Denis E. Clark, PE, CWI

Gary A. Higbee, EMBA, CSP

Health Hazards
Toxic Gases and Vapors ▶ Primary Pulmonary Gases ▶ Nonpulmonary Gases ▶ Particulate Matter ▶ Pulmonary Irritants and Toxic Inhalants ▶ Cleaning Compounds ▶ Chlorinated Hydrocarbons ▶ Asbestos

Safety Hazards
Fire Protection ▶ Drums, Tanks, and Closed Containers

Controlling Hazardous Exposures
Ventilation ▶ Fume Avoidance ▶ Nonionizing Radiation ▶ Noise ▶ Chipping

Personal Protective Equipment
Respiratory Protection ▶ Eye Protection ▶ Protective Clothing

Training in Safe Practices

Oxyfuel Welding and Cutting
Welding and Cutting Gases ▶ Compressed Gas Cylinders ▶ Manifolds ▶ Distribution Piping ▶ Portable Outlet Headers ▶ Regulators ▶ Hoses and Hose Connections ▶ Torches ▶ Powder Cutting

Resistance Welding
Power Supply ▶ Cables ▶ Machine Installation

Arc Welding and Cutting
Power Supply ▶ Voltages ▶ Cables ▶ Electrodes and Holders ▶ Protection against Electric Shock ▶ Gas-Tungsten Arc Welding, Plasma-Arc Welding, and Cutting ▶ Gas-Metal Arc Welding ▶ Flux-Cored Arc Welding ▶ Other Welding and Cutting Processes

Summary

References

Review Questions

The purpose of welding is to join metal parts. Most welding processes require heat, and sometimes the addition of other substances, to produce a weld. Because high heat is used to make the weld, a number of by-products result from the process—including fumes and gases—that can be a serious health hazard to workers. Other health and safety hazards are also associated with welding, such as the potential for fire or explosion and injuries from arc radiation, electrical shock, or materials handling.

Definitions used in this chapter are those of the American Welding Society, as specified in ANSI/AWS A3.0, Standard Welding Terms and Definitions. *Welder* and *welder operator* refer to the individual worker only. The machine used to perform the welding operation is referred to as the *welding machine*. Equipment supplying current for electric welding is generally called the *welding power supply*, or sometimes a *welding generator* or a *welding transformer*. (See the References at the end of this chapter for the standards for welding and cutting operations. Also consult other regulations, where applicable.) The following topics are discussed in this chapter:

- health hazards from airborne and liquid toxins
- combustible materials and hazardous locations
- exposure to fumes, noise, and radiation
- protective equipment for workers
- safety training
- handling gases
- resistance welding
- arc welding.

HEALTH HAZARDS

The most significant health hazard in the welding process is the generation of toxic metal fumes, vapors, and gases. The amount and type of fumes, vapors, and gases involved will depend on the welding process, the base material, the filler material, and the shielding gas used, if any. The toxicity of the contaminants depends primarily on their hazardous properties and concentrations and on the physiological responses of the human body. Sampling by an industrial hygienist may be necessary to fully identify the toxic materials actually being given off in a specific operation.

Toxic Gases and Vapors

Most welding processes deliberately introduce shielding gases into the area around the arc to protect the metal from oxidation. This can be accomplished by flowing gas (e.g., argon, helium, or CO_2) or by gases generated from welding fluxes specifically for shielding purposes. Gases may also be produced by chemical reactions in the arc area or by the effects of ultraviolet (UV) radiation distant from the arc. The combustion products of oxyfuel-welding and -cutting processes, which provide the heat required by these processes, can include CO and CO_2.

Gas generation rates are typically not high compared with the flows in normal ventilation schemes. For example, most gas-shielded processes use less than 50 ft^3/h (0.39 L/s). Nonetheless, if particularly toxic gases are being generated, or if the welding is being done in a confined space, precautions must be taken.

Exposure to large amounts of toxic gases, vapors, and particulates generated during welding may produce one or more of the following effects:

- inflammation of the lungs (chemical pneumonitis)
- pulmonary edema (swelling and accumulation of fluids)
- emphysema (loss of elasticity of the lungs; only a small percentage of emphysema cases are caused by occupational exposure)
- chronic bronchitis
- asphyxiation.

The major toxic gases associated with welding are classified as (1) primary pulmonary and (2) nonpulmonary.

Primary Pulmonary Gases

Primary pulmonary gases can impair or injure the lungs and pulmonary system of workers who inhale these substances in hazardous amounts. Approximate exposure limits mentioned in this section are ACGIH (American Conference of Governmental Industrial Hygienists) and/or OSHA (Occupational Safety and Health Administration) values—8-hour time-weighted averages—and are intended as an informative guide; refer to the latest standards for precise values and measurement techniques.

Ozone

Ozone is formed by electrical arcs and corona discharges in the air or by ultraviolet photochemical reactions. In welding, it is formed from atmospheric oxygen wherever UV radiation from the arc travels, which may be some distance from the arc. Good general ventilation is important, especially for inert-gas-shielded processes such as gas-tungsten arc welding (GTAW) and gas-metal arc welding (GMAW), which produce higher intensities of UV radiation. Exposure limits are approximately 0.1 parts per million (ppm). Welders who have had acute exposure at an estimated 9 ppm of ozone, plus exposure to other air pollutants, have developed pulmonary edema.

Oxides of Nitrogen

Oxides of nitrogen are very irritating to the eyes and mucous membranes. Exposure limits are approximately 5 ppm, but exposure to high concentrations may immediately produce coughing and chest pain. Death may occur within 24 hours if pulmonary edema develops.

Phosgene

Phosgene is produced when metals that have been cleaned with chlorinated hydrocarbons are heated to the temperatures used in welding and by the ultraviolet light emitted by welding arcs. Low concentrations of phosgene can have a sweet smell like newly mown grass, but the odor detection limit is typically above the TLV (Threshold Limit Value, a trademark of ACGIH) for toxicity. Exposure limits are approximately 0.1 ppm. Inhalation of high concentrations of phosgene gas will produce pulmonary edema, frequently preceded by a latent period of several hours' duration. Death may result from respiratory or cardiac arrest.

Phosphine

Phosphine or hydrogen phosphide is generated when steel that has been coated with a phosphate rustproofing is welded. Exposure limits are approximately 0.3 ppm. High concentrations of the gas are irritating to the eyes, nose, and skin. Phosphine is flammable and explosive in high concentrations.

Nonpulmonary Gases

Although these gases do not directly injure the lungs or pulmonary systems of workers, they can threaten workers' lives by displacing oxygen in the air or replacing it in the human bloodstream. Management should routinely monitor the workplace to ensure these gases are not present in harmful amounts.

Carbon Monoxide

In some welding processes, carbon dioxide is reduced to carbon monoxide. In carbon-dioxide-shielded-metal arc welding, carbon monoxide concentrations exceeding the recommended levels have been detected in the fumes near the arc; however, the concentration level rapidly decreases farther from the arc. With adequate ventilation, the carbon monoxide concentration in the welder's breathing zone can be maintained at acceptable levels. Approximate exposure limits are 20 to 25 ppm.

Carbon Dioxide

Carbon dioxide is not usually considered a toxic gas. It is present in the atmosphere in a concentration of about 0.040%, or 400 ppm, and can be at a higher concentration in occupied structures. OSHA's 8-hour time-weighted average (TWA) exposure limit is 5000 ppm.

Asphyxiants

Carbon dioxide causes *suffocation*: a buildup of CO_2 causes the breathing changes associated with a feeling of "running out of air." Any gas that displaces oxygen (CO_2, argon, helium, and nitrogen would commonly be found in welding situations) can cause *asphyxiation*, instead. Argon, helium, and nitrogen (and other, less common gases) are sometimes referred to as "simple asphyxiants" because their main effect is not physiological toxicity but the displacement of oxygen. Unconsciousness can occur almost instantly in such a situation, and the lack of a breathable atmosphere poses problems for rescue and first aid. For example, the instinct to rush to the aid of a fallen co-worker may prove fatal. This is a particular problem in welding in confined spaces. Oxygen sensors should be used where asphyxiants may accumulate, set to alarm at levels before breathing would be impaired.

Particulate Matter

The deposition of particulate matter into the lungs is called *benign pneumoconiosis*. Benign pneumoconioses associated with welding are aluminosis (aluminum), anthracosis (carbon), siderosis (iron oxide), and stannosis (tin oxide).

Aluminosis

The inhalation of aluminum, aluminum oxide, and aluminum hydrate does not appear to injure the pulmonary system. This conclusion is supported by both industrial experience and animal experimentation. Most of the complications associated with the inhalation of aluminum are probably produced by some co-inhalant such as silica. However, welding in inadequately vented confined spaces may produce ozone and oxides of nitrogen as well as aluminum oxides. The combined contaminants may produce effects on the lung.

Anthracosis

The term *anthracosis* refers to a blackish pigmentation of the lungs caused by the deposition of carbon particles. Carbon dust is uncommon in most welding and cutting operations but may be encountered in carbon arc cutting or gouging operations. Some investigators still feel that this pneumoconiosis is due to impurities in coal such as silica. In most instances, inhalation of pure carbon does not produce a significant lung tissue reaction.

Siderosis

Siderosis is a benign pneumoconiosis resulting from the deposition of inert iron oxide dust in the lung. In general, there is neither fibrosis nor emphysema associated with this condition unless, as often occurs, there is a concomitant exposure to silica dust. Siderosis does not result in disability nor does it show any predisposition to pulmonary tuberculosis. Most welding is the arc welding of steels, and iron oxides are the most common constituent of welding fumes.

Stannosis

The inhalation of tin oxide dust over a long period of time will produce a benign pseudonodulation in the lungs known as stannosis. It is considered to be nonprogressive and nondisabling. Tin is a component of many solders, but extensive exposure to tin dust would be rare in most welding and brazing situations.

Metal Fume Fever

Metal fume fever is a scientifically imprecise term that has been used in the welding and metalworking industries for many decades. It is a shorthand term for an acute exposure to metal fumes, has a variety of symptoms, and can be caused by a variety of metal exposures, some more toxic than others. Any acute response to metal fumes indicates a serious failure of worker protection, and work conditions should be mitigated immediately.

Pulmonary Irritants and Toxic Inhalants

The majority of the metal components contained in welding fumes do not produce radiographic changes in the lungs. Depending on the definition of pneumoconiosis, they can be classified in the separate categories of pulmonary irritants or toxic inhalants. These materials include cadmium, chromium, lead, magnesium, manganese, mercury, molybdenum, nickel, titanium, vanadium, zinc, and the fluorides.

One thing to bear in mind is that most welding exposures are to "welding fume," complex particles that may contain toxic constituents, the exact proportion of which will depend on the welding process and materials used. While the most toxic of these constituents can definitely cause acute effects, welder exposures are typically much lower than those found, for example, in primary metal manufacturing or industrial situations where high concentrations of pure materials are handled.

Another thing to consider is that when fumes are controlled for a toxin with very low limits (e.g., hexavalent chromium and, more recently, manganese, which may require a respirator), it is often the case that other fume constituents are pushed far below their allowable limits. The entire spectrum of fume constituents must be considered for user protection.

Approximate exposure limits mentioned in this section are ACGIH and/or OSHA values—8-hour time-weighted averages—and are intended as an informative guide; refer to the latest standards for precise values and measurement techniques.

Particulate matter is classified as *inhalable* or *respirable*. Inhalable particles are defined to be less than 10 μm in diameter and can be inhaled into the upper parts of the respiratory tract (and can be physically removed from the body by the motion of cilia in these passages), while respirable particles are defined to be less than 2.5 μm in diameter and are assumed to be capable of reaching the alveoli, the farthest reaches of the lungs. Where a distinction is made, exposure limits are lower for respirable particles than for inhalable particles. Most welding fume is less than 1 μm in diameter and is, thus, in the respirable category.

Fumes are typically measured by industrial hygienists by collecting them on filter papers in cassettes with vacuum pumps whose inlets are placed in the breathing zones of welders—for example, under the welding helmet. Using standard analytical methods, they are analyzed for hazardous constituents, and the total fume load to which the welder is exposed is calculated, usually as an 8-hour time-weighted average.

Beryllium

The only nonfibrotic type of harmful pneumoconiosis is associated with the inhalation of beryllium. Beryllium is sometimes welded and can be a constituent of welded or brazed alloys, but general exposure is rare in welding. The OSHA TWA for beryllium is 0.002 mg/m^3. Inhalation of beryllium dust or fumes may result in an acute or chronic systemic disease. Chronic beryllium poisoning has been reported as resulting from exposure in facilities producing beryllium phosphors or beryllium compounds from the ore, in beryllium-copper founding, in ceramics laboratories, and in metallurgical shops. This disease has been reported as occurring among individuals exposed to atmospheric pollution in the vicinity of facilities processing beryllium and in persons dwelling in the same household as beryllium workers. To date, there have been no known cases of acute or chronic beryllium poisoning caused by inhalation of beryl dust (the dust of beryllium ore).

There is a specialized medical screening test for beryllium exposure. This test is known as the lymphocyte transformation test (LTT) and is useful for screening beryllium-exposed workers. Beryllium is a suspected human carcinogen (as defined by ACGIH).

Cadmium

Operations that involve the heating of cadmium-plated or cadmium-containing parts—such as welding, brazing, and soldering—may, due to cadmium's high vapor pressure, produce a high concentration of cadmium and cadmium oxide fumes, which are very toxic. The OSHA PEL is 5 μg/m^3 with a 2.5 μg/m^3 action level. Inhalation of these fumes may cause respiratory irritation with a tender, sore, dry throat and a metallic taste, followed by cough, chest pain, and difficulty in breathing. A worker's liver or kidneys and bone marrow may be injured by the presence of this metal.

A single exposure to cadmium fumes may cause a severe lung irritation that can be fatal. Most acute intoxications have been caused by the inhalation of cadmium fumes at concentrations that did not produce warning symptoms of irritation. Continued exposure to lower levels of cadmium has resulted in chronic poisoning characterized by irreversible lung injury and urinary excretion of multiple low-molecular-weight proteins that may be indicative of potentially progressive kidney impairment. Cadmium exposure via inhalation is associated with the development of cancer of the respiratory tract; however, direct oral ingestion of cadmium is not associated with carcinogenic toxins.

The use of cadmium, in both pigment and plating applications, has decreased due to its toxicity. It may be encountered in repair welding or on legacy materials, in which case it can be chemically stripped from areas to be welded. Even if this is done, workers should realize that cadmium can vaporize at relatively low temperatures, and at a substantial distance from the edge of the weld fusion zone.

Chromium

Chromium is a major constituent of stainless steels and other corrosion-resistant alloys and a minor constituent of other metals. Some welding processes produce fumes containing substantial amounts, up to a few percent of the fume by weight, of hexavalent chromium. The oxidation of chromium alloys can produce chromium trioxide fumes that are often referred to as chromic acid. These fumes react with water vapor to form chromic and dichromic acid. Contact with these fumes in high concentrations can produce small, painless cutaneous ulcers as well as dermatitis from primary irritation or allergic hypersensitivity. Inhalation of these fumes can produce bronchospasm, edema and hypersecretion, bronchitis, and hyperreaction of the trachea. Perforation of the nasal septa can also occur. Exposure to hexavalent chromium compounds seems to be related to an increased risk of lung cancer.

Trivalent chromium compounds [Cr(III)] are suspected by some sources to be potentially carcinogenic, though they are found in many more natural settings, including food. OSHA Cr(VI) exposure limits are 5 $\mu g/m^3$, while Cr(III) limits are higher, 500 $\mu g/m^3$ or 0.5 mg/m^3.

Copper

In most welding, copper is a minor constituent—for example, copper plating on bare electrodes—but copper and copper-based alloys are sometimes welded or brazed. Exposure limits for copper fumes are about 0.2 mg/m^3 and 1 mg/m^3 for copper dusts. Inhalation of copper fumes at high concentrations has been reported to produce symptoms of metal fume fever in welders. In chronic exposure, the liver, kidneys, and spleen may be injured and anemia may develop, although chronic poisoning like that from lead poisoning is unknown. Excessive exposure can cause nasal congestion and ulceration with perforation of the nasal septum.

Fluoride

Fluorides are a minor constituent of some welding fluxes and slags. Exposure limits for fluorides are about 2.5 mg/m^3. The use of electrodes with fluoride-containing coatings creates a definite hazard. Fluoride compounds—fumes and gases—can burn eyes and skin on contact. When excessive amounts are inhaled, excretion lags behind the daily intake, resulting in a buildup of fluorides in the bones. If storage of fluorides continues over a sufficiently long period, the bones may show an increased radiographic density.

Lead

Lead poisoning in industry is almost always a result of inhalation of lead fumes from lead-containing materials or materials protected by lead-based paints. Lead was previously a component of solders; lead-based solders have largely been phased out because of toxicity concerns but may be encountered in the repair or recycling of older items. The OSHA PEL for lead is 50 $\mu g/m^3$ with an action level of 30 $\mu g/m^3$. Inadvertent oral ingestion can also occur if poor workplace hygiene is present. The most common symptom of lead toxicity is abdominal pain. Lead levels can be easily measured in blood, and chelation therapy may be necessary for acute toxicity.

Magnesium

Magnesium is not commonly encountered in welding, but it can be welded with inert gas processes. Oxide fumes from magnesium can produce metal fume fever, which can result in an irritation of mucous membranes. Exposure limits for magnesium oxide are about 10 mg/m^3. Experimental work with animals has failed to show any detrimental response in the lungs.

Manganese

At high concentrations, manganese fumes are highly toxic and can produce total disability after exposure as short as a few months—a condition known as maganism. Such high-level exposures are usually caused by inhalation of manganese dioxide dust and are more common in battery workers and manganese miners than in welders, whose exposures are typically lower. Chronic low-level exposure may have long-term neurological effects similar to, but distinct from, those of parkinsonism. In 2012, the ACGIH exposure limits for manganese were reduced from 0.2 to 0.02 mg/m^3 (20 $\mu g/m^3$) for respirable particles.

Mercury

The welding of metals coated with protective materials containing mercury compounds produces toxic mercury fumes. Systemic mercury poisoning could result. Exposure levels for mercury and its compounds are around 25 $\mu g/m^3$.

Molybdenum

Molybdenum is a minor constituent in some commonly welded alloys. Little is known concerning the exposure of humans to molybdenum or its compounds. Animal studies indicate molybdenum may produce nasal irritation, diarrhea, and weight loss at high concentrations. Exposure limits are currently about 15 mg/m^3.

Nickel

Nickel is a minor constituent of some steel and stainless steel alloys and a major constituent of nickel-based alloys. Nickel and its compounds are carcinogenic and toxic, with exposure levels from 0.1 to 1.5 mg/m^3. A significant increase in cancer of the lungs and sinuses has occurred among employees in nickel smelting and refining facilities, although exposures under common welding conditions are typically much lower. Nickel fumes have been known to cause severe pneumonitis. Nickel carbonyl is highly toxic and can produce cyanosis, delirium, and death between 4 and 11 days after exposure. "Nickel itch" is a dermatitis that results from sensitivity to nickel.

Titanium

It is uncommon to cut or weld titanium except in well-controlled conditions. Titanium itself is not particularly toxic (it is used for surgical implants), but high concentrations of titanium dioxide dust may produce irritations in the respiratory tract. OSHA exposure limits for TiO$_2$ are about 5 mg/m^3. Slight fibrosis, without disabling injury, has been observed in the lungs of industrial workers exposed to titanium dioxide dust.

Vanadium

Vanadium is present in some welding filler wires, but it is typically a very minor constituent of commonly welded materials, and exposures are more likely to occur in chemical and petroleum processing. The exposure limits for vanadium (as the pentoxide V$_2$O$_5$) are 50 µg/m^3. It irritates the eyes and respiratory tract and may be responsible for asthmatic reactions.

Zinc

In welding, zinc is most commonly encountered during the welding, brazing, or cutting of galvanized metals. This exposure can be mitigated by removing the galvanized coating where feasible, and by proper ventilation. Zinc exposure limits are quite variable, depending on the particular compound involved, but are 5 mg/m^3 for zinc oxide. The inhalation of freshly formed fumes may produce a brief, self-limiting illness known variously as zinc chills, metal fume fever, brass chills, and brass founders' fever. This condition is characterized by chills, fever, nausea, vomiting, muscular pain, dryness of mouth and throat, headache, fatigue, and weakness. These signs and symptoms usually abate in 12 to 24 hours, with complete recovery following. Immunity from this condition is rapidly acquired if exposure occurs daily but is quickly lost during holidays or over weekends. Because of this behavior, metal fume fever is sometimes known as "Monday morning sickness."

Any such symptoms are not normal, however, but indicate an exposure situation that needs to be corrected.

Cleaning Compounds

Because of their chemical properties, cleaning compounds can create health hazards if improperly mixed. They often require special ventilation precautions. Follow the manufacturer's instructions.

Chlorinated Hydrocarbons

Degreasing operations often employ chlorinated solvents that can decompose to toxic phosgene gas in the presence of the ultraviolet radiation emitted by the welding arc. Because the UV radiation can travel considerable distances, degreasing operations should be shielded from arc radiation.

Asbestos

If welding or cutting involves asbestos, the regulations of the agency having authority must be consulted before beginning the job. Asbestos is associated with fibrotic lung disease, lung cancer, and a tumor known as a mesothelioma.

SAFETY HAZARDS

Management and workers should know the safety hazards involved in the workplace. They should also be trained to avoid, reduce, or eliminate them through the use of safe work practices, personal protective equipment, and safety equipment. The most common workplace safety hazards are discussed in the following sections.

Fire Protection

Because portable welding and cutting equipment creates special fire hazards, welding is preferably done in a permanent welding and cutting location that can be designed to provide maximum safety and fire protection, as well as good ventilation. Otherwise, the welding and cutting site should be inspected to determine what fire protection equipment is necessary (see American National Standards Institute [ANSI] Z49.1 and National Fire Protection Association [NFPA] 51B).

It is advisable, particularly in hazardous locations, to require "hot-work" permits issued by the welding supervisor, a member of the facility fire department, or some other qualified person before welding or cutting operations are started. Specifications for hot-work permits are outlined in Chapter 9, Fire Protection, in this volume.

Floors and Combustible Materials

Where welding or cutting must be done near combustible materials, special precautions are necessary to prevent sparks or hot slag from reaching such material and starting fires. If the work itself cannot be moved, the exposed combustible material should, if possible, be moved a safe distance away. Otherwise, it should be covered with sheet metal. Spray booths and ducts should be cleaned to remove combustible deposits. Before welding or cutting is started, wood floors should be swept clean and, preferably, covered with metal or other noncombustible material where sparks or hot metal may fall.

If gas welding or oxygen cutting is done inside a booth provided for arc welding, the gas cylinders should be placed in an upright and secured position away from sparks to prevent contact with the flame or heat.

Hot metal or slag should not be allowed to fall through cracks in the floor or other openings, nor into machine tool pits. Cracks or holes in walls, open doorways, and open or broken windows should be covered with sheetmetal guards. Because hot slag may roll along the floor, it is important that no openings exist between the curtain and the floor. Similar protection should be installed for wall openings through which hot metal or slag may enter when welding or cutting operations are conducted on the outside of the building.

If it is necessary to weld or cut close to wood construction or near combustible material that cannot be removed or protected, a fire hose, water pump tank extinguisher, or fire pails should be conveniently located. Portable extinguishers for specific protection against Class B and Class C fires should also be provided (see Chapter 9, Fire Protection). Pails of limestone dust or sand may be useful. It is good practice to provide a fire extinguisher—dry chemical, multipurpose chemical, or carbon dioxide—for each welder.

A fire watcher equipped with a suitable fire extinguisher should be stationed at or near welding or cutting operations conducted in hazardous locations to see that sparks do not lodge in floor cracks or pass through floor or wall openings. This person should have no other responsibilities during the hot work but should be devoted full time to fire-watch duties. The fire watch should be continued for at least 30 minutes after the job is completed to make sure that smoldering fires have not been started. The fire watch must also inspect adjoining areas (ventilating ducts, adjacent rooms, floors above and below, etc.) into which hot items may have escaped unnoticed during welding.

Hazardous Locations

Welding and cutting operations should not be permitted in or near rooms containing flammable or combustible vapors, liquids, or dusts. Nor should they be permitted on or inside closed tanks or other containers that have held such materials until all fire and explosion hazards have been eliminated. All of the surrounding premises should be thoroughly ventilated and frequent gas testing provided. Sufficient draft should be maintained to prevent accumulation of explosive concentrations. Local exhaust equipment should be provided for removal of hazardous gases, vapors, and fumes (present in the surroundings or generated by the welding or cutting operations) that ventilation fails to dispel.

Drums, Tanks, and Closed Containers

Workers should thoroughly clean closed containers that have held hazardous substances before welding or cutting on them. Guidance may be found in NFPA 326 and 327, Safeguarding of Tanks and Containers for Entry, Cleaning, or Repair, and ANSI/AWS F4.1–2007, Safe Practices for the Preparation of Containers and Piping for Welding and Cutting. Also see Chapter 10, Flammable and Combustible Liquids, in this volume.

According to ANSI/AWS F4.1, hazardous substances include, but are not limited to, those that are explosive, combustible, toxic, or corrosive. They may be present in a container having previously held one of the following:

1. volatile liquid that can release potentially hazardous vapors
2. an acid or alkaline material that reacts with metals to produce hydrogen
3. a nonvolatile liquid or solid that will release hazardous materials if the container is heated (Combustible vapors or hazardous decomposition products may be generated by the heat of welding or cutting.)
4. dust clouds or finely divided airborne particles that may reach an explosive concentration
5. flammable or toxic gas
6. corrosion by-products due to reaction of the container with its contents.

Even if the containers are not known to have contained combustible materials as listed here, a number of hazards need to be considered.

ANSI/AWS F4.1 applies to containers and piping *except*:
1. containers and confined spaces that can be entered by workers (see ANSI Z117.1, Safety Requirements for Confined Spaces)
2. containers that have held radioactive substances
3. compressed gas containers
4. containers that have held explosive substances

5. tanks, bunkers, or compartments on ships (see NFPA 306, Control of Gas Hazards on Vessels, and consult the local fire marshal or Coast Guard marine inspection officer)
6. gasometers or gas holders for natural and manufactured gases (consult the American Gas Association)
7. outside, aboveground, vertical petroleum storage tanks
8. containers holding flammables that are to be repaired while in service.

For containers of these types, specialized requirements apply, and the relevant standards and regulations should be consulted. It is important to have a qualified employee who can anticipate, recognize, and evaluate employee exposure to hazardous substances or other unsafe conditions.

ANSI/AWS F4.1 identifies three stages to consider in the process of welding and cutting on containers: (1) preparation of the container for cleaning, (2) methods of cleaning and guidelines for selection, and (3) preparation for welding and cutting.

Preparation of the Container for Cleaning
1. Determine the hazardous characteristics of contents.
2. Do not attempt to clean containers holding unknown substances; instead, the container should be properly disposed of.
3. Designate the cleaning procedure to be performed by a qualified person.
4. Identify a qualified person to clean the container.
5. Ensure that proper PPE precautions for cleaning personnel are available.
6. Evaluate container location (e.g., ventilation).
7. Determine removal methods for contents, including liquids and sludges and all internal piping.
8. The same practices are to be applied to connected vessels or piping even if they are not to be welded.
9. Consider use of coatings such as plastic and refractory materials.

Methods of Cleaning and Guidelines for Selection
1. Use water cleaning for materials known to be readily soluble in water.
2. Use hot chemical-solution cleaning (e.g., trisodium phosphate).
3. If steam cleaning, use appropriate protective gear.
4. Mechanical cleaning has the disadvantage of requiring access for cleaning and inspection operations, but it may be needed when dry, scaly residues are to be removed.
5. Chemical cleaning may be needed for materials insoluble in water and when mechanical cleaning cannot be used; the hazards of residual chemicals need to be considered.

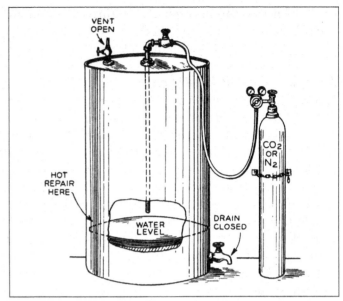

Figure 19–1. As an added precaution after cleaning, a container to be welded or cut may be filled with either carbon dioxide or nitrogen to dilute and render nonhazardous any remaining combustible gas or vapor. *(Reprinted with permission from the American Welding Society.)*

6. A combination of methods may be required, with due precautions taken for personnel protection and hazardous reactions.

Preparation for Welding and Cutting
1. Immediate area should be cleared of obstacles and hazardous materials.
2. Appropriate PPE should be used.
3. Ensure there is adequate ventilation, including testing for toxic or flammable vapors as appropriate, both before and periodically during welding.
4. Perform a prewelding inspection to ensure that cleaning was adequate.
5. Provide a pressure relief.
6. Inert the area using water filled to within a few inches of the welding point, or with sand (see Figure 19–1).
7. Use inert gas, with due caution for maintaining the inert environment and awareness of asphyxiation hazards.
8. Isolate the vessel from further entry of hazardous substances, such as by blanking off open pipe ends and closing valves.

CONTROLLING HAZARDOUS EXPOSURES

Certain materials, sometimes contained in the consumables, base metals, coatings, or atmospheres of welding or cutting operations, have low OSHA permissible exposure

limits (PELs) and/or low ACGIH Threshold Limit Values (TLVs). Among these materials are:
- antimony chromium mercury
- arsenic cobalt nickel
- barium copper selenium
- beryllium lead silver
- cadmium manganese vanadium.

Refer to Safety Data Sheets (SDSs; formerly Material Safety Data Sheets [MSDSs]) provided by the manufacturer to identify any of the materials just listed that may be contained in the consumable. Because the Safety Data Sheets for welding consumables such as electrodes typically contain a list of potential exposures for that specific item, they are a good guide for estimating hazards from welding operations. Whenever these materials are encountered as designated constituents in welding, brazing, or cutting operations, ventilation precautions must be taken to ensure that the level of contaminants in the atmosphere is below the limits allowed for human exposure. Unless atmospheric tests under the most adverse conditions have established that the exposure is within acceptable concentrations, management and workers should observe the precautions discussed in the following sections.

Ventilation

To keep fumes, vapors, and other toxic by-products of welding at safe levels, management must ensure that the workplace is properly ventilated. Natural and mechanical ventilation can be used to reduce the levels of airborne contaminants to acceptable levels.

Natural Ventilation

Natural ventilation is acceptable for welding, cutting, and related processes where the necessary precautions are taken to keep the welder's breathing zone away from the plume and where sampling of the atmosphere shows that concentrations of contaminants are below mandated or recommended levels (discussed earlier). Taking air samples in workers' breathing zones is the only way to be sure that airborne contaminant levels are within allowable limits.

Natural ventilation often meets standards if all of the following specifications are met:
- Space of more than 10,000 ft³ (284 m³) per welder is provided.
- Ceiling height is more than 16 ft (5 m).
- Welding is not done in a confined space.
- "Welding space" refers to a building or an enclosed room in a building, not a welding booth or screened area that is used to provide protection from welding radiation; the welding space does not contain partitions, balconies, or other structural barriers that obstruct cross-ventilation.
- Potentially hazardous materials, covered earlier, are not present as known constituents.

Mechanical Ventilation

Mechanical ventilation includes local exhaust, local forced, and general-area mechanical air movement. Local exhaust ventilation is preferred. It means fixed or movable exhaust hoods placed as near as practical to the work and able to maintain a capture velocity sufficient to keep airborne contaminants below regulatory limits (Figure 19–2).

Local forced ventilation means a local air-moving system (such as a fan) placed so that it moves the air at right angles (90 degrees) to the welder (across the welder's face). It should produce an approximate velocity of 100 fpm (30 m/min), and be maintained for a distance of approximately 2 ft (0.6 m) directly above the work area. Precautions must be taken to ensure that contaminants are not dispersed to other work areas.

General-area mechanical ventilation includes roof exhaust fans, wall exhaust fans, and similar large area air movers. General-area mechanical ventilation is not usually sufficient for spaces where welding is occurring. It is often helpful, however, when used in addition to local ventilation. General mechanical ventilation may be necessary to maintain the background level of airborne contaminants below regulatory levels (particularly as regulatory levels continue to become more stringent) or to prevent the accumulation of explosive gas mixtures.

Figure 19–2. This welder wears personal protective equipment and is using a local exhaust system and fan. The protective curtain is pulled back for this photograph only. *(Courtesy International Stamping Co./Midas International Corp.)*

Air Cleaners

Where permissible, air cleaners that have high efficiencies in the collection of submicron particles may be used to recirculate a portion of air that would otherwise be exhausted. Most filters do not remove gases, however. Therefore, adequate monitoring must be done to ensure that concentrations of harmful gases remain below allowable limits.

Other Factors

In addition to the factors listed earlier, there are a number of other items that must also be considered and evaluated to determine ventilation requirements:

- dimensions and layout of working areas
- number of welding stations or welders or both
- fume emission rates of welding processes being used
- tendency of air currents to dissipate or concentrate fumes in certain areas of the working space.

The size and arrangement of the working facilities are major factors in ventilation requirements, especially when the working area is confined or divided into sections that limit air circulation. The number of welding stations and welders is also important to consider in designing ventilation systems. Welders who are paid by piece rate or incentive basis may weld a significantly higher number of pieces than those who are paid a fixed daily rate. As a rule of thumb, general ventilation should be provided at a minimum rate of 2,000 ft³/min per welder, except where local exhaust hoods are available.

The local exhaust hoods can be fixed or movable and should be provided with an air flow sufficient to maintain a velocity of 100 fpm near the breathing zone of the welder.

Special Equipment—Fume Extraction Welding Guns

In addition to conventional ventilating devices, special equipment has been developed to remove welding fumes at their source. For example, fume extraction nozzles using high-velocity, low-volume exhaust units are designed to fit over the end of the hand-held welding torch. This sleevelike fixture is usually connected to a small exhaust fan by a flexible hose.

Fume Avoidance

Welders and cutters must take precautions to avoid breathing the fume plume directly. This can be done by positioning the work and the head or by ventilation that directs the plume away from the face. Tests have shown that fume removal is more effective when the air flow is directed across the face of the welder, rather than from behind the person.

Nonionizing Radiation

Electric arcs and gas flames produce ultraviolet and infrared radiation, which has a harmful effect on the eyes and skin upon continued or repeated exposure. The usual effect of ultraviolet is to "sunburn" the surface of the eye, which is painful and disabling but temporary in most instances. The effects of visible and near-infrared radiation, however, may cause permanent eye injury if the worker looks directly into a very powerful arc or furnace without eye protection. Ultraviolet may also produce the same effects on the skin as a severe sunburn.

Production of ultraviolet radiation is much higher in gas-shielded arc welding than in flux-based processes such as stick electrode welding (shielded-metal arc welding [SMAW]). There are few fume particles to block its passage, and UV intensities may be several times as great. This also means that ozone production is higher in these processes.

Infrared radiation's only effect is that it heats the tissue with which it comes in contact. If the heat is not enough to cause an ordinary thermal burn, there is no harm. Exposure to certain intensities of infrared radiation is associated with the development of cataracts.

Whenever possible, arc-welding operations should be isolated so that other workers will not be exposed to either direct or reflected radiation. Arc-welding stations for regular production work can be enclosed in booths if the size of the work permits. The inside of the booth should be coated with a paint that is nonreflective to ultraviolet radiation. Portable flameproof screens similarly painted or flameproof curtains should also be provided. Booths should be designed to permit circulation of air at the floor level and adequate exhaust ventilation.

Chapter 27, Laboratory Safety, in the *Administration & Programs* volume, provides further coverage on ionizing and nonionizing radiation.

Noise

In welding, cutting, and the associated operations, noise levels can exceed the permissible limits. Hearing protective devices may be needed. (See Chapter 7, Personal Protective Equipment, in this volume.)

Chipping

Welders should never use an ordinary carpenter's hammer as a chipping hammer because the head of a carpenter's hammer can splinter and split. Special slag-chipping hammers are the appropriate tool.

Because slag can be sharp and shatter, welders should always wear safety glasses whenever they chip; it is good practice always to wear safety glasses under the welding helmet, which may be raised for chipping operations. Chipping can also be very noisy, especially when air-driven hammers are used. For this reason, welders should wear hearing protection as well as eye protection when chipping. (See ANSI/AWS F6.1 [R1989], Method for Sound Level Measurement of Manual Arc Welding and Cutting Processes.)

PERSONAL PROTECTIVE EQUIPMENT

A baseline physical, including a chest x-ray and pulmonary function testing, is recommended for all persons engaged in welding. Periodic reexaminations should be made as recommended by the company or facility physician. In addition, workers should be trained in the type of personal protective equipment each job requires and in proper use and care of the equipment.

Respiratory Protection

If gases, dusts, and fumes cannot be kept below the applicable PEL or TLV, welders should wear respiratory protective equipment certified for the exposure by the National Institute for Occupational Safety and Health (NIOSH). Where oxygen is also deficient, workers should wear a self-contained breathing apparatus.

Precautions for proper respiratory protection must be provided for workers, even those doing inert-gas-shielded arc welding. Depending on a number of factors, including the particular variety of welding to be done, the nature of the materials to be welded, and whether or not the work must be done in a confined space, workers will need positive ventilation, local exhaust removal, approved respirator equipment, or a combination of these precautions.

With increasingly low TLVs for many welding fume constituents (such as hexavalent chromium and manganese), respirators are becoming increasingly common and, properly used, allow the welder to completely avoid fume exposure, to be more comfortable, and to do a better job.

Eye Protection

Goggles, helmets, and shields that give maximum eye protection for each welding and cutting process should be worn by operators, welders, and their helpers. These items should conform to ANSI Z87.1–1989, Practice for Occupational and Educational Eye and Face Protection, and ANSI Z89.1–1986, Protective Headwear for Industrial Workers. Table 19–A is a guide for selecting the correct filter lens for various welding and cutting operations. Goggles or spectacles should have side shields (see guidance for lens care given in this volume, Chapter 7, Personal Protective Equipment).

Protective Clothing

Welding processes vary greatly in their protective clothing requirements. Light-duty welding processes such as gas-tungsten arc welding may require only a lab coat, along with appropriate gloves and eye protection, although the hazards of arc radiation and fumes always need to be considered. Some of the items of protective clothing needed by welders for heavier-duty welding processes (Figure 19–3) might include:

- flame-resistant gauntlet gloves—leather or other suitable material (may be insulated for heat)
- aprons made of leather or other flame-resistant material to withstand radiated heat and sparks
- for heavy work, fire-resistant leggings, high boots, or similar protection
- safety shoes, wherever heavy objects are handled (because of spark hazard, avoid using low-cut shoes with unprotected tops)
- for overhead work, capes or shoulder covers of leather or other suitable material (skullcaps of leather or flame-resistant fabric may be worn under helmets to prevent head burns; also, for overhead welding, ear protection is sometimes desirable)
- safety hats or other head protection against sharp or heavy falling objects.

Operators and other persons working with or near welding processes should keep all parts of the body that could be exposed to the ultraviolet and infrared radiation covered to protect against skin burns and other types of injuries. Dark clothing, particularly a dark shirt, is preferable to light-colored clothing in order to reduce reflection to the operator's face underneath the helmet. Woolen clothing is preferable to cotton because it is more resistant to deterioration and is not readily ignited.

Caution: Welders should never wear synthetic clothing and/or synthetic blends, including synthetic insulated underwear.

For gas-shielded arc welding, woolen clothing is also preferable to cotton. It is not readily ignited and protects the welder from changes in temperature. Cotton clothing, if used, should be chemically treated to reduce flammability. In either case, clothing should be thick enough to keep radiation from penetrating it.

Outer clothing should be reasonably free from oil and grease. Sleeves and collars should be kept buttoned. Aprons and overalls should have no front pockets, where sparks could be caught. For the same reason, trousers or overalls should not have turned-up cuffs.

Butane cigarette lighters should never be carried while welding. They may inadvertently open and saturate the welder's clothing with flammable gas—with disastrous results.

Thermal-insulated underwear is designed to be worn only under other clothing and should not be exposed to open flames, sparks, or other sources of ignition.

TABLE 19–A. Guide for Selection of Lens Shade

Operation	Electrode Size 1/32 in. (mm)	Arc Current (A)	Minimum Protective Shade	Suggested* Shade No. (comfort)
Shielded-metal arc welding	Less than 3 (2.5)	Less than 60	7	—
	3-5 (2.5-4)	60-160	8	10
	5-8 (4-6.4)	160-250	10	12
	More than 8 (6.4)	250-550	11	14
Gas-metal arc welding and flux-cored arc welding		Less than 60	7	—
		60-160	10	11
		160-250	10	12
Gas-tungsten arc welding		250-500	10	14
		Less than 50	8	10
		50-150	8	12
		150-500	10	14
Air carbon	(Light)	Less than 500	10	12
Arc cutting	(Heavy)	500-1,000	11	14
Plasma-arc welding		Less than 20	6	6 to 8
		20-100	8	10
		100-400	10	12
		400-800	11	14
Plasma-arc cutting	(Light)**	Less than 300	8	9
	(Medium)**	300-400	9	12
	(Heavy)**	400-800	10	14
Torch brazing		—	—	3 or 4
Torch soldering		—	—	2
Carbon arc welding	in.	Plate thickness	—	14
		mm		
Gas welding				
Light	Under 1/8	Under 3.2		4 or 5
Medium	1/8 to 1/2	3.2 to 12.7		5 or 6
Heavy	Over 1/2	Over 12.7		6 or 8
Oxygen cutting				
Light	Under 1	Under 25		3 or 4
Medium	1 to 6	25 to 150		4 or 5
Heavy	Over 6	Over 150		5 or 6

* As a rule of thumb, start with a shade that is too dark to see the weld zone. Then go to a lighter shade that gives sufficient view of the weld zone without going below the minimum. In oxyfuel gas welding or cutting where the torch produces a high yellow light, it is desirable to use a filter lens that absorbs the yellow or sodium line in the visible light of the (spectrum) operation.

** These values apply where the actual arc is clearly seen. Experience has shown that lighter filters may be used when the arc is hidden by the workpiece. ANSI Z49.1–1983.

Figure 19–3. Proper personal protective equipment for a welder includes flash goggles worn under a helmet, a chrome leather jacket, an apron, gauntlet gloves, and leggings. *(Courtesy Westinghouse Electric Corp.)*

TRAINING IN SAFE PRACTICES

Management should make sure that welders and cutters are well trained in the safe practices that apply to their work. The standards for training and qualification of welders set up by the American Welding Society (AWS) are recommended. A training program should particularly emphasize that a welder or cutter can best provide for the safety of co-workers, as well as the operator, by observing safe practices that include the following:

- When possible, place work at an optimal height to avoid back strain or shoulder fatigue (Figure 19–4).
- For work at more than 5 ft (1.5 m) above the floor or ground, use a platform with railings or with fall protection equipment.
- Wear respiratory protection as needed and a safety harness with attached lifeline for work in confined spaces,

Figure 19–4. This welder uses a scissor lift table to place the work at an optimal height for improved posture, which will reduce strain and fatigue. *(Courtesy Presto Corp.)*

such as tanks and pressure vessels. The lifeline should be tended by a similarly equipped helper whose duty is to observe the welder or cutter and effect rescue in an emergency.
- Take special precautions if welding or cutting in a confined space is stopped for some time. Disconnect the power of arc welding or cutting units, and remove the electrode from the holder. Turn off the torch valves of gas welding or cutting units. Shut off the gas supply at a point outside the confined area. If possible, remove the torch and hose from the area.
- After welding or cutting is completed, mark hot metal or post a warning sign to keep workers away from heated surfaces.
- Follow safe housekeeping principles. Do not throw electrode or rod stubs on the floor—discard them in the proper waste containers. Keep tools and other tripping hazards off the floor—put them in a safe storage area.
- Use equipment as directed by the manufacturer's instructions and practices.

Operators and management should recognize their joint responsibilities for safety. Management should ensure that welders and supervisors are trained and should establish and enforce procedures to be used at off-plant locations and/or in hazardous locations. Only approved welding equipment should be used, and the manufacturer's instructions must be followed. Supervisors should be responsible for the safe operation of welding equipment and operators.

OXYFUEL WELDING AND CUTTING

An oxyfuel-welding process unites metals by heating; the heat source is a flame produced by the combustion of a fuel gas or gases. This process sometimes includes the use of pressure and a filler metal. The use of these processes in the welding industry has been declining in recent decades as more efficient, arc-based processes have been developed. They are still used, however, because they do not require electricity and because the oxyfuel torch is a versatile tool that can provide heat for cutting and brazing as well as welding.

An oxygen-cutting process severs or removes metal by the chemical reaction of the base metal with oxygen at an elevated temperature. The temperature is maintained by heat from the combustion of fuel gases or from an arc. In the metal-powder-cutting process, a finely divided material, such as iron powder, is added to the cutting-oxygen stream. The powder bursts into flame in the oxygen stream and starts cutting without preheating the material to be cut. Metal-powder cutting is used on stainless and other steels, on many nonferrous metals, and on concrete in construction and demolition jobs. Plasma-arc cutting is now replacing metal-powder cutting in most applications.

Welding and Cutting Gases

Fuel gases are usually hydrocarbons that can be oxidized (burned) in air or with added oxygen to provide heat. The addition of pure oxygen increases the flame temperature and energy output of combustion processes. In addition to its function in creating a useful flame for welding or cutting, oxygen can form explosive mixtures in certain proportions with acetylene, hydrogen, and other combustible gases.

The acetylene molecule (C_2H_2) contains a triple bond, which gives it a very high energy content when oxidized. Acetylene burned with oxygen produces a higher flame temperature (approximately 6,000°F or 3,300°C) than any other gas used commercially. Acetylene, like other combustible gases, ignites readily and, in certain proportions, forms a flammable mixture with air or oxygen. Acetylene is also unstable by itself at high pressures, so storage and delivery require special equipment and precautions (discussed later).

Other fuel gases are also used with oxygen in torches, primarily for oxygen cutting. For example, propane, propylene, and gas mixtures are supplied in cylinders in liquid or compressed form, generally under various trade names. These gases and liquefied petroleum gas are discussed in Chapter 10, Flammable and Combustible Liquids, in this volume. Welding equipment needs to be compatible with the particular gas combinations being used.

Compressed Gas Cylinders

Most of the oxygen and fuel gases used for welding and cutting are supplied in pressurized cylinders. These cylinders should be constructed and maintained in accordance with regulations of the U.S. Department of Transportation (DOT). The purchaser should make sure that all cylinders bear DOT, ICC (Interstate Commerce Commission), or CTC (Canadian Transport Commission) specification markings. The contents should be legibly marked on each cylinder in large letters.

Oxygen is supplied in steel cylinders; the usual size for welding contains 244 ft³ (6.9 m³) of oxygen under pressure of 2,200 psi (15.2 MPa) at 70°F (21°C). A cap should be provided to protect the outlet valve when the cylinder is not connected for use. Similar cylinders at similar pressures are used for argon, CO_2, and other gases commonly used in welding. Hydrogen is also furnished in similar cylinders, though it is seldom used in oxyfuel systems for welding.

Any of the fuel gases have flammability ranges when mixed with air, as seen in Table 19–B. The ranges will be wider (more dangerous) with pure oxygen. Within the flammability ranges are narrower explosive ranges, in which a spark could cause a detonation. In addition, if the flammable or explosive gas is denser or less dense than air, it may accumulate in hazardous concentrations if ventilation is insufficient.

TABLE 19–B. Flammability Limits for Selected Materials

Gas	Lower Flammability Limit (vol %)	Upper Flammability Limit (vol %)
Acetylene	2.5	100
Carbon monoxide	12	75
Hydrogen	4	75
Methane (natural gas)	4.4	15–17
Propane	2.1	9.5–10.1

This information should be considered, for example, in confined spaces or vessels, where these flammable or explosive concentrations might commonly be reached. It is not intended for precise calculation but to indicate how easily flammable mixtures may inadvertently be achieved.

Oxygen has other dangers. Although it is not flammable or explosive by itself, because it is such a strong oxidizer, it causes fuel materials to burn more intensely or to explode. For example, regulators typically carry the notation "Use no oil" because when oxygen is suddenly introduced under pressure, an explosion may result. For the same reason, oxygen should never be used as a substitute for compressed air because (1) compressed air systems usually contain lubricating oil and (2) the oxygen emerging from compressed air appliances can so severely enhance flammability.

Acetylene for welding and cutting is usually supplied in cylinders having a capacity up to about 300 ft³ (8.5 m³) of dissolved acetylene under pressure of 250 psi (1.7 MPa) at 70°F (21°C). Acetylene cylinders are completely filled with a porous material impregnated with acetone, the solvent for acetylene. The porous material should have no voids of appreciable size so that acetylene can be safely stored at the prescribed full-cylinder pressure. Because acetylene is highly soluble in acetone at cylinder filling pressure, large quantities of acetylene can be stored in comparatively small cylinders at relatively low pressures.

Acetylene is either supplied in cylinders or generated as needed. It is a product of the reaction between water and calcium carbide, a gray crystalline substance made commercially by fusing lime and coke in an electric furnace. Calcium carbide is neither flammable nor explosive. It is sold and stored in air- and watertight cans or drums. If the drums are damaged in handling and if water comes in contact with carbide, acetylene will be generated, leading to a danger of ignition and explosion.

Handling Cylinders

Serious accidents may result from the misuse, abuse, or mishandling of compressed gas cylinders. Workers assigned to the handling of cylinders under pressure should be properly trained and should work only under competent supervision. Observance of the following rules will help control hazards in the handling of compressed gas cylinders.

- Accept only cylinders approved for use in interstate commerce for transportation of compressed gases.
- Do not remove or change the marks and numbers stamped on the cylinders.
- Cylinders that are difficult to carry by hand should not be rolled on their bottom edges; they are heavy, and it is easy for them to get out of control and fall, possibly damaging a valve or injuring personnel. Instead, use a cylinder cart or hand truck for cylinders weighing more than 40 lb (18.2 kg).
- Keep the cylinders clean, and protect them from impact or abrasions.
- Do not lift compressed gas cylinders with an electromagnet. Where cylinders must be handled by a crane or derrick, as on construction jobs, carry them in a cradle or suitable platform, and take extreme care that they are not dropped or bumped. Do not use slings.
- Do not drop cylinders or allow them to strike each other violently.
- Do not use cylinders for rollers, supports, or any purpose other than to contain gas.
- Do not tamper with safety devices in valves or on cylinders.
- Consult the supplier of the gas when in doubt about the proper handling of a compressed gas cylinder or its contents.

- Clearly identify empty cylinders with appropriate tags or labels for return to the vendor. Do not completely empty cylinders because this allows air and moisture to contaminate the remaining contents, but, instead, return them with a low but positive pressure.
- Close cylinder valves and replace valve protection caps, if the cylinder is designed to accept a cap.
- Load cylinders to be transported to allow as little movement as possible. Secure them to prevent violent contact or upsetting.
- Always consider cylinders as being full, and handle them with corresponding care. Accidents have resulted when containers under partial pressure were thought to be empty.

The fusible safety plugs on acetylene cylinders melt at about the boiling point of water. If an outlet valve becomes clogged with ice or frozen, it should be thawed with warm (not boiling) water, applied only to the valve. A flame should never be used.

Storing Cylinders

Cylinders need to be secured in the upright position in a safe, dry, well-ventilated place prepared and reserved for the purpose. Flammable substances, such as oil and volatile liquids, should not be stored in the same area. Cylinders should not be stored near elevators, gangways, stairwells, or other places where they can be knocked down or damaged.

Oxygen cylinders should not be stored within 20 ft (6 m) of highly combustible materials or cylinders containing flammable gases. If closer than 20 ft (6 m), cylinders should be separated by a fire-resistive partition at least 5 ft (1.5 m) high with a fire-resistance rating of at least 30 minutes.

Acetylene and liquefied fuel gas cylinders should be stored and used with the valve end up. If storage areas are within 100 ft (30.5 m) of each other and not protected by automatic sprinklers, the total capacity of acetylene cylinders stored and used inside the building should be limited to 2,500 ft³ (70 m³) of gas, exclusive of cylinders in use or connected for use. Quantities exceeding this total should be stored in a special room built in accordance with the specifications of NFPA 51, Design and Installation of Oxygen-Fuel Gas Systems for Welding, Cutting, and Allied Processes, in a separate building or outdoors. Acetylene storage rooms and buildings must be well ventilated, and open flames must be prohibited. Storage rooms should have no other occupancy.

Cylinders should be stored on a level, fireproof floor. One common type of storage house consists of a shed room with side walls extending approximately halfway down from the roof and a dividing wall between one kind of gas and another.

To prevent rusting, cylinders stored in the open should be protected from contact with the ground and against extremes of weather—accumulations of ice and snow in winter and continuous direct rays of the sun in summer.

Cylinders are not designed for temperatures in excess of 130°F (54°C). Accordingly, they should not be stored near sources of heat, such as radiators or furnaces, or near highly flammable substances like gasoline.

Cylinder storage should be planned so that cylinders will be used in the order that they are received from the supplier. Empty and full cylinders should be stored separately, with empty cylinders being plainly identified as such to avoid confusion. Group together empty cylinders that have held the same contents.

Storage rooms for cylinders containing flammable gases should be well ventilated to prevent the accumulation of explosive concentrations of gas. No source of ignition should be permitted. Smoking should be prohibited. Wiring should be in conduit. Electric lights should be in fixed position and enclosed in glass or other transparent material to prevent gas from contacting lighted sockets or lamps. Electric lights should also be equipped with guards to prevent breakage. Electric switches should be located outside the room.

Using Cylinders

Safe procedures for the use of compressed gas cylinders include the following:
- Use cylinders, particularly those containing liquefied gases and acetylene, in an upright position, and secure them against being accidentally knocked over.
- Keep the metal cap in place to protect the valve when the cylinder is not connected for use unless the cylinder valve is protected by a recess in the head. A blow on an unprotected valve might cause gas under high pressure to escape.
- Make sure the threads on a regulator or union correspond to those on the cylinder valve outlet. Do not force connections that do not fit.
- Open cylinder valves slowly. A cylinder not provided with a handwheel valve should be opened with a spindle key or a special wrench or other tool provided or approved by the gas supplier.
- Do not use a cylinder of compressed gas without a pressure-reducing regulator attached to the cylinder valve, except where cylinders are attached to a manifold, in which case the regulator will be attached to the manifold header.
- Before making connection to a cylinder valve outlet, "crack" the valve for an instant to clear the opening of particles of dust or dirt. Always point the valve and opening away from the body and not toward anyone else. Never crack a fuel gas cylinder valve near other welding work or near sparks, open flames, or other possible sources of ignition.

- Small fires at the cylinder should be extinguished, if possible, by closing the cylinder valve. In case of a larger fire or if extinguishment is not possible, evacuate the area and immediately contact trained fire-fighting personnel.
- Use regulators and pressure gauges only with gases for which they are designed and intended. Do not attempt to repair or alter cylinders, valves, or attachments. This work should be done only by the manufacturer.
- Unless the cylinder valve has first been closed tightly, do not attempt to stop a leak between the cylinder and the regulator by tightening the union nut.
- Leaking fuel gas cylinders should be taken out of use immediately and handled as follows:
  1. Close the valve, and take the cylinder outdoors well away from any source of ignition.
  2. Properly tag the cylinder, and notify the supplier. (A regulator attached to the valve may be used temporarily to stop a leak through the valve seat.)
  3. If the leak occurs at a fuse plug or other safety device, take the cylinder outdoors, well away from any source of ignition; open the cylinder valve slightly; and permit the fuel gas to escape slowly. Tag the cylinder plainly.
  4. Post warnings against approaching with lighted cigarettes or other sources of ignition.
  5. Promptly notify the supplier, and follow instructions for returning the cylinder.
- Do not permit sparks, molten metal, electric currents, excessive heat, or flames to come in contact with the cylinder or attachments.
- Never use oil or grease as a lubricant on valves or attachments of oxygen cylinders. Keep oxygen cylinders and fittings away from oil and grease, and do not handle such cylinders or apparatus with oily hands, gloves, or clothing.
- Never use oxygen as a substitute for compressed air in pneumatic tools, in oil preheating burners, to start internal-combustion engines, or to dust clothing. Use it only for the purpose that it is intended.
- Never bring cylinders into tanks or unventilated rooms or other closed quarters.
- Do not refill cylinders except with the consent of the owner and then only in accordance with DOT (or other applicable) regulations. Do not attempt to mix gases in a compressed gas cylinder or to use it for purposes other than those intended by the supplier.
- Before a regulator is removed from a cylinder valve, close the cylinder valve and release the gas from the regulator.
- Cylinder valves must be closed when work is finished.

Manifolds

Cylinders are manifolded to centralize the gas supply and to provide gas continuously and at a rate in excess of that which may be obtained from a single cylinder. Manifolds

Figure 19–5. This is a well-designed manifold system for acetylene cylinders. *(Courtesy Linde Co., Division of Union Carbide Corp.)*

must be of substantial construction and of a design and material suitable for the particular gas and service for which they are to be used. Manifolds should be obtained from and installed under the supervision of a reliable manufacturer familiar with safe practices in construction and use of manifolds.

Portable manifolds connect a small number of cylinders (usually not more than six) for direct supply to a consuming device. The cylinders may be connected by individual leads to a single, common coupler block, or individual cylinders may be connected to a common line with coupler tees attached to the cylinder valves. A properly supported regulator serves the group of connected cylinders.

Stationary manifolds connect a larger number of cylinders for supply through piped distribution systems. This type of manifold consists of a substantially supported stationary pipe header that connects the cylinders by individual leads (Figure 19–5). One or more permanently mounted regulators serve to reduce and regulate the pressure of the gas flowing from the manifold.

Oxygen manifolds should be located away from highly flammable material and oil, grease, and the like. They should not be located in acetylene generator rooms or in close proximity to cylinders of combustible gases. There should be a 5-ft- (1.5-m-) high, 30-min fire-resistant partition between an oxygen manifold and combustible gas cylinders, unless the manifold and such cylinders are separated at least 50 ft (15.2 m). Regulations of NFPA 51 should be followed.

Distribution Piping

All piping should be color-coded or clearly identified as to type of gas. Distribution piping carrying oxygen from a manifold or other centralized supply should be of steel, wrought iron, brass, or copper, as outlined in NFPA 51.

All pipe and fittings for oxygen service lines should be examined before use and, if necessary, tapped with a hammer to free them from dirt and scale. They should, in every case, be washed out with a suitable nonflammable cleaner—hot-water solutions of caustic soda and trisodium phosphate are effective.

Only steel or wrought-iron piping should be used for acetylene distribution systems. Under no circumstances should acetylene gas be brought into contact with unalloyed copper, except in a torch treated to prevent chemical reaction, because explosive copper acetylides may be formed. Joints in steel or wrought-iron pipe should be welded or made up with threaded or flanged fittings. Flanged connections in acetylene lines should be electrically bonded. Gray or white cast-iron fittings should not be used.

Joints in brass or copper pipe may be welded, threaded, or flanged. A socket joint may be brazed with silver solder or similar high-melting-point material. Threaded connections in oxygen piping should be tinned or made up with litharge and glycerine or other joint compound approved for oxygen service.

In fuel gas distribution systems, a backflow check valve or hydraulic seal should be used to prevent backflow at every point where gas is withdrawn from the piping system to supply a torch or machine. Such devices should be listed (or approved) by an agency such as Factory Mutual or Underwriters Laboratories.

Portable Outlet Headers

Portable outlet headers are assemblies of valves and connections used for service outlet purposes and are connected to a permanent service piping system by means of hose or other nonrigid conductors. Devices of this nature are commonly used at piers and dry docks in shipyards, where the service piping cannot be located close enough to the work to provide a direct supply. Their use should be restricted to outdoor locations and to temporary service where conditions preclude a direct supply, and they should be used in accordance with regulations in NFPA 51.

Regulators

Pressure regulators must be used on both oxygen and fuel gas cylinders to maintain a uniform gas supply to the torches at the correct pressure. The oxygen regulator should be equipped with a safety relief valve or be so designed that should the diaphragm rupture, broken parts will not fly. Workers should stand to one side and away from regulator gauge faces when opening cylinder valves.

Only regulators listed by agencies such as Underwriters Laboratories or Factory Mutual should be used on cylinders of compressed gas. If unlisted regulators are used, they should be fully checked by a competent welding engineer. Each regulator (oxygen or fuel gas) should be equipped with both a high-pressure (contents) gauge and a low-pressure (working) gauge.

High-pressure oxygen dial gauges should have safety vent covers to protect the operator from flying parts in case of an internal explosion. Each oxygen dial gauge should be marked OXYGEN—USE NO OIL OR GREASE.

Serious, even fatal, accidents have resulted when oxygen regulators have been attached to cylinders containing fuel gas, or vice versa. To guard against this hazard, it has been customary to make connections for oxygen regulators with right-hand threads and those for acetylene with left-hand threads, to mark the gas service on the regulator case, and to paint the two types of regulators different colors. Cylinder valve outlet threads have been standardized for most industrial and medical gases (see Compressed Gas Cylinder Valve Outlet and Inlet Connections, ANSI/CGA V-1–1987). Different combinations of right-hand and left-hand threads, internal and external threads, and different diameters to guard against wrong connections are now standard.

The regulator is a delicate apparatus and should be handled carefully. It should not be dropped or pounded on. Regulators should be repaired only by qualified persons or sent to the manufacturer for repairs.

Leaky or "creeping" regulators are a source of danger and should be withdrawn from service at once for repairs. If a regulator shows a continuous creep, indicated on the low-pressure (delivery) gauge by a steady buildup of pressure when the torch valves are closed, the cylinder valve should be closed and the regulator removed for repairs.

If the regulator pressure gauges have been strained so that the hands do not register properly, the regulator must be replaced or repaired before it is used again.

When regulators are connected but are not in use, the pressure-adjusting device should be released. Cylinder valves should never be opened until the regulator is drained of gas and the pressure-adjusting device on the regulator is fully released.

These procedures should be followed in detail when regulators or reducing valves are being attached to a gas cylinder:

1. To blow out dust or dirt that otherwise might enter the regulator, "crack" the discharge valve on the cylinder by opening it slightly for an instant and then closing it. On a fuel gas cylinder, first see that no open flame or other source of ignition is near; otherwise, the gas might ignite at the valve.

2. Connect the regulator to the outlet valve on the cylinder. Be sure the regulator inlet threads match the cylinder valve outlet threads. Never connect an oxygen regulator to a cylinder containing fuel gas, or vice versa. Do not force connections that do not fit easily. Be sure that the connections between the regulators and cylinder valves are gastight.

3. Release the pressure-adjusting screw on the regulator

Figure 19–6. How to test for leaks: With the pressure on and the torch valves closed, hold the hose (top left) and the torch tip under water. Bubbles indicate leaks. Use soapsuds to test for leaks in the torch valves and hose-to-torch connections, as shown by the arrows. Separately test the oxygen cylinder (top right) and the acetylene cylinder and regulator connections (bottom) for leaks at points marked by arrows. *(Bottom photo courtesy of J. I. Case.)*

to its limit—turn it counterclockwise until it is loose. Engage the adjusting screw and open the downstream line to the air to drain the regulator of gas.

4. Open the cylinder valve slightly to let the hand on the high-pressure gauge move up slowly. On an oxygen cylinder, and on similar high-pressure cylinders for other gases, gradually open the cylinder valve to its full limit, so the double-seated valve can serve its function in preventing leaks. Acetylene cylinders are the exception; make no more than 1.5 turns of the valve spindle.
5. Leaks may be checked for using the methods shown in Figure 19–6.

Beyond this point, the procedure will depend on the type of equipment and its intended use. For example, there are different kinds of torches (injector versus positive-pressure type, cutting versus welding or heating heads, etc.). A detailed, illustrated guide can be found in the *AWS Welding Handbook* (9th ed., vol. 2, pp. 467–99).

Hoses and Hose Connections

Oxygen and acetylene hoses should be different colors or otherwise identified and distinguished from each other. Red is the generally recognized color for the fuel gas hose and green for the oxygen hose (Figure 19-7). Black is used for inert gas and air hoses. The hose connections are usually marked STD-OXY for oxygen and STD-ACET for acetylene. The acetylene union nut has a groove cut around the center to indicate left-hand threads.

Connections for joining the hose to the hose nipple on the torches and regulators may be either the ferrule or clamp type. Gaskets should not be used on these connections. Special torch connectors with built-in shutoff valves are available.

Following are suggestions for the safe use of hose in welding and cutting operations:

- Do not use an unnecessarily long hose.
- Repair leaks at once. Besides being a waste, escaping fuel gas may become ignited and start a serious fire; it may also set fire to the welder's clothing. Escaping oxygen is equally hazardous. Repair hose leaks by cutting the hose and inserting a splice. Do not try to repair leaky hose by taping.
- Examine hoses periodically and frequently for leaks and worn places, and check hose connections. Test for leaks by immersing the hose in water under normal working pressure. (Refer to Figure 19-6.)
- Protect hoses from flying sparks, hot slag, other hot objects, and grease and oil. Store hose in a cool place.
- A single hose having more than one gas passage should not be used. When oxygen and acetylene hoses are taped together for convenience and to prevent tangling, not more than 4 in. (10 cm) of each 12 in. (30.5 cm) of hose should be taped.
- The use of hoses with an external metallic covering is not recommended. In some machine processes and in certain types of operations, hoses with an inner metallic reinforcement that is exposed neither to the gas passage nor to the outside atmosphere are acceptable.
- Flashback devices (Figure 19-7) between the torch and hose can prevent burn-back into hoses and regulators. If a flashback occurs and burns the hose, discard the burned section.
- A hose that has been subject to flashback, or that shows evidence of severe wear or damage, should be tested to twice the normal pressure, but in no case less than 300 psi (21.097 kg/cm). A defective hose, or a hose in doubtful condition, should not be used.

Torches

Torches are constructed of metal castings, forgings, and tubing. Usually, they are made of brass or bronze, but stainless steel may also be used. They should be of substantial design to withstand the rough handling they sometimes receive. It is best to use only those torches listed by an agency such as Underwriters Laboratories or Factory Mutual.

The gases enter the torch by separate inlets, go through valves to the mixing chamber, and then go to the outlet orifice, located in the torch tip. Several interchangeable tips are provided with each torch and have orifices of various sizes according to the work to be done.

The cutting torch, unlike the welding torch, uses a separate jet of oxygen in addition to the jet or jets of mixed oxygen and fuel gas. The jets of mixed gases are for preheating the metal, and the pure oxygen jet is for cutting. The flow of oxygen to the cutting jet is controlled by a separate valve. The *AWS Welding Handbook* (9th ed., vol. 2, pp. 598-636) contains further illustrated information about oxygen cutting.

In the operation of torches, several precautions should be observed:

- Select the proper welding head or mixer and tip or cutting nozzle (according to charts supplied by the manufacturer), and screw it firmly into the torch.
- Before changing torches, shut off the gas at the pressure-reducing regulators and not by crimping the hose.
- To discontinue welding or cutting for a few minutes, closing only the torch valves is permissible. If the welding or cutting is to be stopped for a longer period (during lunch or overnight), proceed as follows:
  1. Close oxygen and acetylene cylinder valves.
  2. Open torch valves to relieve all gas pressure from hose and regulator.
  3. Close torch valves and release regulator pressure-adjusting screws.
- Do not use matches to light torches. Use a friction lighter, stationary pilot flame, or other suitable source of ignition. When lighting, point the torch tip so no one will be burned when the gas ignites.

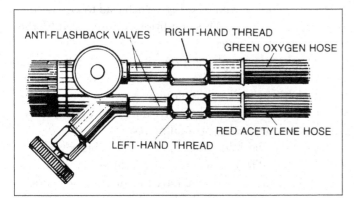

Figure 19-7. When attaching hoses to a welding or cutting torch, use the red hose for acetylene and the green hose for oxygen; then test connections for leaks.

- Never put down a torch until the gases have been completely shut off. Do not hang torches from a regulator or other equipment so that they come in contact with the sides of gas cylinders. If the flame has not been completely extinguished or if a leaking torch ignites, it may heat the cylinder or even burn a hole through it.
- When extinguishing the flame, close the acetylene and oxygen valves in the order recommended by the torch manufacturer. If the oxygen valve is closed first, carbon soot will be deposited in the air. However, this ensures that the acetylene valve is closed tight when the flame is extinguished. If the acetylene valve is turned off first, no soot is formed, but there is no assurance that the fuel gas valve is closed and that it is not leaking.

Powder Cutting

Powder-cutting processes for metal and concrete use similar equipment and gas supplies as oxygen-cutting operations use. The precautions previously discussed for safe handling and use of compressed gas equipment and cutting torches therefore apply. Manufacturers' recommendations for the operation and maintenance of the powder-dispensing apparatus—both pneumatic and vibratory—should be followed.

RESISTANCE WELDING

Because resistance welding equipment is normally permanently installed, the hazards are usually minimized if the equipment has been properly designed and safe operating practices have been established.

Certain hazards in the operation of this equipment—lack of point-of-operation guards, flying hot metallic particles, improper handling of materials, unauthorized adjustments and repairs—may cause eye injuries, burns, and electrical shock. Most of these hazards can be eliminated by safeguarding the equipment, wearing protective clothing, and strictly controlling operating practices.

Resistance welding is a metal-joining process whereby welding heat is generated at the joint by the resistance to the flow of electric current. The three fundamental parameters of resistance welding are current magnitude, current time, and tip pressure. Each of these must be accurately controlled (Figure 19–8).

Power Supply

Resistance welding usually employs a 60-Hz alternating current that is fed to the primary of the water-cooled welding transformer. The primary can vary from 150 to 10,000 amp at 240, 440, or 550 V. The output at the secondary of the transformer is a low-voltage (max 30 V) and high-amp current (up to 200,000 amp) used for welding.

The welding current is sometimes furnished by "stored-energy" equipment. Energy is built up and stored either in capacitors or in a combination transformer-reactor during the nonwelding period and then is discharged to form the weld. This process involves low primary currents and high voltages that must be guarded against.

To facilitate servicing the equipment, a safety-type disconnecting switch or circuit breaker of the correct rating for opening supply circuits should be installed near the welding machine. (See Control of All Energy Sources in Chapter 6, Safeguarding, in this volume, for lockout/tagout procedures.) Permanent injuries and several fatalities have resulted from neglecting to use the line-disconnecting switch before making adjustments inside enclosures. This precaution is imperative because the use of single-pole primary circuit breakers and electronic contactors leaves one line to the welder "hot."

Figure 19–8. This is an automatic-resistance welding machine with a dial feed. The operator removes and places the work when the proper dial fixture comes to the front. *(Courtesy General Electric Co.)*

Cables

Abuse of the cables for resistance welding is severe. The production requirements demand the utmost of the cable materials used, and even the best cables need frequent replacement. In use, the cables are subjected to electrical pulsation, bending, and twisting, which lead to fatigue and eventual breakdown. This condition is minimized by the use of concentric cables.

The secondary voltage presents little shock hazard because the maximum voltage is about 30 V, but the operator can be hurt by a cable blowout such as is caused by steam pressure due to overheating from faulty water-cooling circulation or from electrical failure.

A periodic check for weak spots in the cable covering is good practice. The use of concentric welding cables is now common because they do not have the undesirable features of the pulsating cables. Portable welding machines, including the cables, should have proper weight balance to permit operation without undue strain to the operator.

Machine Installation

Installation of resistance welding equipment should conform to the National Electrical Code, NFPA 70. Some items worthy of special attention are listed here:

- Control circuits should operate on low voltage, not exceeding 24 V maximum for portable spot welders.
- Stored-energy equipment (capacitor discharge or resistance welding) having control panels involving high voltage (more than 550 V) should be completely enclosed. Doors should have locks and contacts wired in the control circuit to short-circuit the capacitors when the door or panel is opened. A manually operated switch will serve as an additional safety measure, ensuring complete discharge of the capacitors.
- Back doors of machines and panels should be kept locked or interlocked to prevent tampering.
- A fused safety switch or circuit breaker should be located conveniently near the welding machine so that power supply circuits may be opened before servicing the machine and its controls.
- The point-of-operation hazard should be eliminated by suitable guards. Enclosure guards, gate guards, two-hand controls, and similar standard guards as designed for punch press operations are applicable.
- A flash-welding machine should have a shield or hood to control flash and fumes and a ventilating system to carry off the metallic dust and oil fumes.
- Where flying sparks are not confined, the operators and nearby persons should be protected by shields of safety glass or other fire-resistant material or by the use of personal eye protection.
- Foot switches, air or electrical, should be guarded to prevent accidental operation.

ARC WELDING AND CUTTING

Arc welding and cutting use an electric arc—a high-current electrical discharge typically operating at 10 to 40 V and 10 to 2,000 amp. The arc is very hot, 5,000°C (9,032°F) or more, and it is this heat that is used to melt, weld, and cut materials. The electrical circuit that maintains the arc generally attaches one lead to the workpiece (the "ground" cable) and the other lead to the "torch" or electrode holder. These leads are of heavy cable, usually copper, to carry the high currents without overheating. The ground current may be carried for some distance through a large structure such as a steel frame building.

Power Supply

Either alternating current (AC) or direct current (DC) may be used for arc welding or cutting of any kind. With small-diameter electrodes used on thin sheets for manual arc welding, current values can be as low as 10 amp. For most manual welding, because the welder must withstand the heat, current values should not exceed 500 or 600 amp. Automatic machine arc welding may use current values up to 2,000 amp or even higher on special applications.

A gasoline- or diesel-powered generator may be used to energize the welding circuit. If used inside a building or in a confined area, the engine exhaust should lead to the outside atmosphere to prevent the accumulation of carbon monoxide and other toxic gases.

Voltages

The voltage across the welding arc varies from 10 to 40 V, depending on the process and the type and size of electrode used. The welding circuit must supply somewhat higher voltage to strike the arc. This voltage is called the open-circuit or "no-load" voltage. After the arc is established, the open-circuit voltage drops to a value about equal to the arc voltage plus the lead voltage drop. The open-circuit voltages on DC welding machines should be less than 100 V—more typically, 80 V or less. Constant-voltage power supplies (welder or converter) are now also being widely used, as continuous wire processes become more common, and these have even lower open-circuit voltages—typically less than 50 V.

Welding power supplies typically have maximum open-circuit voltages of 80 V or less. In some electrically hazardous conditions, it is desirable, according to ANSI Z49.1, to reduce this as much as possible, for example, through the use of automatic controls. More detail can be found in ANSI/NEMA EW1, Electric Arc Welding Power Sources.

Cables

Several lengths of welding cable may be used in one circuit. To splice or connect cables, workers should use substantially insulated connectors of a capacity at least equivalent to that of the cable. Cable lugs used for ground and machine connections should be securely fastened to give good electric contact.

Welding cable is subjected to severe abuse if it is dragged over work under construction, dragged across sharp corners, or run over by shop trucks. Workers should use special cable with high-quality insulation. The fact that welding circuit voltages are low may make workers lax about keeping the welding cable in good repair. Operators and maintenance personnel should make sure that defective cable is immediately replaced or repaired.

On large jobs, there is likely to be much loose cable lying around. Welders should keep this cable orderly and out of the way, placed overhead as necessary to permit the passage of persons and vehicles. Welding cables should not lie in

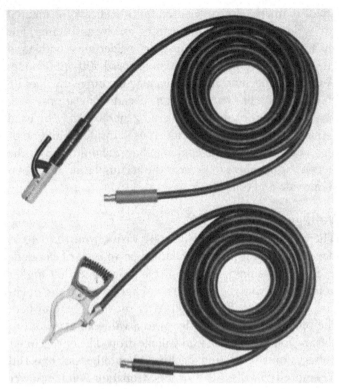

Figure 19–9. The fully insulated electrode holder on the electrode lead and the ground clamp on the work lead have insulated locking plugs for connection to receptacles on the welding machine. *(Courtesy Westinghouse Electric Corp.)*

water or oil, in ditches, or in bottoms of tanks. Management should require that rooms where arc welding is to be done regularly be permanently wired with enough outlets to prevent the excessive use of extension cables.

Electrodes and Holders

Electrode holders for shielded-metal arc welding (SMAW) are used to connect the electrode to the welding cable supplying secondary current. Fully insulated holders (Figure 19–9) are preferred because there is less likelihood of shocking the welder or of accidentally striking an arc with such holders, particularly in close quarters.

Electrode holders will become hot during welding operations if holders designed for light work are used on heavy welding or if connections between the cable and the holder are loose. If workers cannot use a holder of the correct size for the electrode, they should have an extra holder so that one can cool while the other is in use. These are, however, undesirable conditions, and the holder should be sized to the job for safety. Under no circumstances should hot electrode holders be dipped in water.

On light- or medium-duty work for which workers use light, extremely flexible cables, holders may be attached directly to the work lead running to the machine. On heavier work, welders generally prefer a short length of lighter cable attached to the holder, which is more flexible than the main work lead. Properly insulated cables of weight and flexibility that will not inconvenience the welder are available. Fully insulated connectors should be used to attach the short flexible cable to the main work lead.

Protection against Electric Shock

The voltage between the electrode holder and the ground, during the off-arc or no-load period, is the open-circuit voltage. Although open-circuit voltages on standard arc-welding units are not high compared with those of other processes, they should be considered a potential hazard. Normally, the work setup is such that the work is grounded; unless care is exercised, the welder or operator can easily become grounded.

The welder or welding operator should be insulated from both the work and the metal electrode and holder. The bare metal part of an electrode or electrode holder should never be permitted to touch the operator's bare skin or wet clothing. Consistent use of well-insulated electrode holders and cables, dry clothing on the hands and body, and insulation from ground will be helpful in preventing contact.

Pacemaker wearers should check with the manufacturer or medical personnel regarding welding operations.

Some specific precautions for prevention of electric shock are the following:

1. In confined places, cover or arrange cables to prevent contact with falling sparks.
2. Never change electrodes with bare hands or wet gloves or when standing on wet floors or grounded surfaces.
3. Ground the frames of welding units, portable or stationary, in accordance with NFPA 70. A primary cable containing an extra conductor, one end of which is attached to the frame of the welding unit, can be used with a small welding unit. This ground connection can be carried back to the permanently grounded connection in the receptacle of the power supply by means of the proper polarized plug.
4. Arrange receptacles of power cables for portable welding units so that it is impossible to remove the plug without opening the power supply switch, or use plugs and receptacles that have been approved to break full-load circuits of the unit.
5. If a cable (either work lead or electrode lead) becomes worn, exposing bare conductors, it may be repaired if the insulation repair on work-lead cables is equivalent in insulation to the original cable covering. Such repair operations may fall under regulatory controls, and the appropriate regulations should be consulted.
6. Keep welding cables dry and free of grease and oil to prevent premature breakdown of the insulation.
7. Suspend cables on substantial overhead supports if the

cables must be run some distance from the welding unit. Protect cables that must be laid on the floor or ground so that they will not interfere with safe passage or become damaged or entangled.
8. Take special care to keep welding cables away from power supply cables or high-tension wires.
9. Never coil or loop welding cable around the body.
10. In repair welding, the welder may need to lie on the floor to get the necessary access to the weld joint. The large body-contact area makes this electrically hazardous, increasing the danger of electric shock, and this situation should be avoided, or special electrically insulating techniques should be used.

Gas-Tungsten Arc Welding, Plasma-Arc Welding, and Cutting

In gas-tungsten arc welding (GTAW), the electrode does not melt and is not used for filler metal. The electrode is tungsten, which is highly resistant to heat and is nonconsumable in the welding process. Filler metal may be added by using a cold (nonelectrical) welding rod that is introduced into the arc or molten weld puddle. The process can be used to weld nearly any material and is most commonly used on aluminium, stainless steel, and titanium structures.

Either AC or DC welding units can be used for GTAW. AC welding is commonly used for welding aluminium or magnesium because the half-cycle when the electrode is positive provides a "cleaning action" for the refractory oxides found on such materials. It is limited in current capacity, however, and is used more on thin materials. DC with the electrode negative (DCEN) is the more common situation (used mainly for other materials) and provides more power to the workpiece and better penetration. A high voltage at high frequency may be imposed on the circuit to start the arc and stabilize it during AC welding. Installation must be carefully done to avoid interference with radio transmission. For gas-metal arc, DC is used with the electrode positive (DCEP).

Argon, helium, and gas mixtures are supplied by manufacturers in cylinders similar to oxygen cylinders. Because cylinder pressures can range as high as 2,640 psig (18.2 MPa), argon and helium cylinders should be stored and handled like other high-pressure gas cylinders.

Carbon dioxide, although not an inert gas at arc temperatures, is sometimes used as a shield gas when steel is welded by the gas-metal arc-welding process. It is usually supplied in partially liquid and partially gaseous form in cylinders at approximately 835 psig (5.8 MPa). These cylinders should therefore be handled like other high-pressure gas cylinders.

To supply gas to the welding torch, a regulator must be used to lower the pressure to 25 psig (172 kPa) or less, and a flowmeter should measure the volume of gas being used. If more than one torch is used from the same gas line, a flowmeter should be installed at each torch connection.

Air is used to cool the torch and electric current cables. Water is also used for cooling, generally where the welding current is more than 250 amp. The water-supply line, even if city water is used, should be equipped with a strainer to keep out impurities that might plug the water-cooling passages. Self-contained recirculating water cooler systems can provide more consistent cooling and avoid the necessity of a water supply and drain system.

In the gas-shielded tungsten arc torch (Figure 19–10), gas is conducted to the welding point through orifices in the torch around the electrode holder. Cooling water goes through passages through the torch handle and about the holder. In the smaller torches, ceramic cups are used. A torch for heavier work generally has a water-cooled gas cup.

Gas-Metal Arc Welding

Gas-metal arc welding (GMAW) is defined by the AWS as "an arc-welding process that uses an arc between a continuous filler metal electrode and the weld pool. The process is used with shielding from an externally supplied gas and without the application of pressure." (See ANSI/AWS A3.0, Standard Welding Terms and Definitions.)

This welding process is also known as the metal inert gas (MIG) process. When GMAW was first introduced, it used inert gas to shield the arc. The gas was usually argon, or argon with a small amount of oxygen added. As the GMAW process was developed, other gases came into use, some of them not inert but chemically active. In Europe, the process using active gases came to be known as the metal active gas (MAG) process.

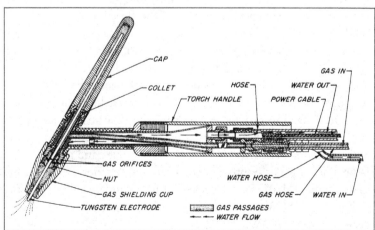

Figure 19–10. The electrode is nonconsumable in this gas-shielded tungsten arc-welding torch. (*Reprinted with permission from* Welding Handbook, *American Welding Society.*)

One version of the MAG process uses CO_2 as the shielding gas. CO_2 is an active gas, but this variation of the process became known separately as the CO_2 welding process. All variations of the process use a consumable wire electrode that is fed through a welding gun. Most of the processes use an externally applied shielding gas of some type or a mixture to protect the weld zone.

GMAW and its variants have become increasingly common in industry. The process is relatively easy to learn, and the continuous wire feed leads to increases in productivity over SMAW, where the electrode must be changed often.

Flux-Cored Arc Welding

Flux-cored arc welding (FCAW) is defined by the AWS as "an arc-welding process that uses an arc between a continuous-filler metal electrode and the weld pool. The process is used with shielding gas from a flux contained within the tubular electrode, with or without additional shielding from an externally supplied gas, and without the application of pressure." (See ANSI/AWS A3.0.)

FCAW is sometimes called cored-wire welding by mistake; actually, the term *cored wire* describes any wire with a core. For example, it could be a metal-core, low-carbon steel wire containing metal powder to produce a stainless steel weld. The term *flux-cored wire*, however, is a wire with a core of flux. The wire is like an inside-out shielded-metal arc electrode. Instead of the flux covering the outside of the electrode as it does in SMAW, the flux is on the inside. Because of its similarity to GMAW wire, FCAW wire is fed into the arc with similar equipment. Only slight variations are necessary because of the tubular and collapsible nature of the wire.

FCAW competes with SMAW as well as with GMAW because it is used mostly to weld steels. The flux acts to reduce porosity, control penetration, and stabilize and shield the arc. When used properly, the flux gives high-quality welds that have a clean, smooth appearance and that can be made in all positions. Just as in SMAW, the flux produces a slag covering that must be removed.

Some flux-cored electrodes do not provide enough gases to protect the weld metal from air. Then, gas shielding, such as CO_2, is used to provide extra protection. Because of its similarity to GMAW, it is easy to provide gas shielding to assist the flux. Usually, a large amount of fumes and gases is produced, and the process is considered "dirtier" than GTAW and GMAW. Both shielded (with gas) and unshielded (without gas) FCAW are currently used in commercial processes. (See ANSI/AWS A3.0.)

As the continuous wire feed processes have been further developed, they have come to occupy a larger place in the welding industry.

Figure 19–11. A self-standing safety shield provides close-quarter protection and rolls up for storage.

Other Welding and Cutting Processes

There are several relatively new heat sources for welding and cutting, such as friction, ultrasonics, and lasers. Each of these special heat sources requires guarding and safe practices.

For example, the laser (light amplification by stimulated emission of radiation) presents the hazard of eye damage from the optically amplified light beam. Because of its intensity, the beam can do damage even at great distances. All employees working with or near laser devices should be given preemployment and periodic follow-up eye examinations. Most companies require that employees work in pairs when using laser equipment.

To confine the laser beam, the company should develop and install suitable shields (Figure 19–11). Shields and curtains designed for arc welding typically do not meet laser shielding requirements. Because a reflected laser beam is also hazardous, the work area should contain no glossy surfaces. It is becoming more common for lasers to be located in enclosures that are protected from access by interlocked doors, so personnel are not exposed to the beam at all.

Power supplies for lasers and the electron beam are high-voltage equipment that should be operated with the precautions developed for this type of equipment. (See the discussion in Chapter 7, Personal Protective Equipment, in this volume; in Chapter 27, Laboratory Safety, under Ionizing and Nonionizing Radiation, in the *Administration & Programs* volume; and in NSC's *Fundamentals of Industrial Hygiene*.)

SUMMARY

- In welding operations, the leading health risks are from toxic gases, particulate matter, pulmonary irritants and toxic inhalants, cleaning compounds, chlorinated hydrocarbons, and asbestos.
- Safety hazards common to all welding and cutting processes require measures such as fire protection, shielding, safe working areas, and safe handling and storage of all combustible or flammable materials in drums, tanks, and closed containers. Safety Data Sheets list the minimum exposure levels and operating standards for many hazardous welding materials.
- Personal protection for welding and cutting operations usually includes respiratory protection against dusts, gases, and fumes. Operators should be trained in the kind of protective gear to wear.
- The American Welding Society has established standards for training and qualification of welders. Supervisors are responsible for the safety of operators and the safe operation of welding equipment during all phases of work.
- In oxyfuel-welding and oxygen-cutting operations, workers must know how to handle and store cylinders safely to avoid explosion and fire. They should inspect their cylinders regularly to ensure that all parts are in good working order.
- Workers must strictly observe precautions for selecting and using welding equipment and for preventing fires and explosions.
- Risks associated with resistance welding can be eliminated by safeguarding equipment, using protective clothing, and strictly controlling operating practices.
- For arc welding or cutting operations, the power supply and welding structures should be grounded and insulated to prevent shock hazards, and all cables, connectors, and electrode holders should be well insulated.
- New heat sources for welding and cutting require guarding and safe practices to protect workers from risks posed by high-frequency sound waves and amplified light waves.

REFERENCES

American Conference of Governmental Industrial Hygienists, Committee on Industrial Ventilation, 1330 Kemper Meadow Dr., Cincinnati, OH 45240-1634.
Industrial Ventilation—A Manual of Recommended Practice.

American Gas Association, 1725 Jefferson Davis Highway, Suite 1004, Arlington, VA 22202-4102.

American Insurance Association, 85 John Street, New York, NY 10038.
Z-125, Lasers and Masers, Special Hazards Bulletin.

American National Standards Institute, 11 West 42nd Street, New York, NY 10036.
ANSI Z49.1–2012, Safety in Welding and Cutting; available without charge at aws.org.
ANSI Z87.1–1989, Practice for Occupational and Educational Eye and Face Protection.
ANSI Z88.2–1992, Practices for Respiratory Protection.
ANSI Z89.1–1986, Protective Headwear for Industrial Workers.
ANSI Z117.1, Safety Requirements for Confined Spaces.
ANSI Z136.1–1993, Safe Use of Lasers.
ANSI/ASME B15.1–1992, Safety Standard for Mechanical Power Transmission Apparatus.
ANSI/AWS A3.0, Standard Welding Terms and Definitions.
ANSI/AWS F4.1–2007, Safe Practices for the Preparation of Containers and Piping for Welding and Cutting.
ANSI/AWS F6.1 (R1989), Method for Sound Level Measurement of Manual Arc Welding and Cutting Processes.
ANSI/CGA V-1–1987, Compressed Gas Cylinder Valve Outlet and Inlet Connections.
ANSI/NEMA EW1, Electric Arc Welding Power Sources.
ANSI/UL 551–1993, Safety Standard for Transformer-Type Arc-Welding Machines.

American Petroleum Institute, 1220 L Street NW, Washington DC 20005.
Cleaning Mobile Tanks Used for Transportation of Flammable Liquids, Accident Prevention Manual, No. 13. 1958.
RP 2009, Gas and Electric Cutting and Welding.
RP 2015, Cleaning Petroleum Storage Tanks.

American Society of Mechanical Engineers, 3 Park Ave., New York, NY 10016-5902.
ASME Boiler and Pressure Vessel Code, Section IX, "Qualification Standard for Welding and Brazing Procedures, Welders, Brazers, and Welding and Brazing Operators." 1965.

American Welding Society, 550 NW 42nd Ave., Miami, FL 33126.
AWS 249.2, Fire Prevention in Arc Welding and Cutting Processes.
AWS C.5.2 (R1989), Recommended Practices for Plasma Arc Cutting.
AWS C.5.5 (R1989), Recommended Practices for Gas Tungsten Arc Welding.
AWS Welding Handbook. 9th ed. Vol. 2.
Characterization of Arc Welding Fume. 1983.

The Welding Environment. 1973.

Compressed Gas Association Inc., 1725 Jefferson Davis Highway, Arlington, VA 22202-4100.

SB–2: Oxygen-Deficient Atmospheres.

Regulator Connection Standard.

Safe Handling of Compressed Gases in Cylinders.

Specification for Rubber Welding Hose.

Linde Division, Union Carbide Corp., 39 Old Ridgebury Road, Danbury, CT 06817.

Plasma-Arc Process Bulletins.

National Fire Protection Association, Batterymarch Park, Quincy, MA 02269.

NFPA 50, Bulk Oxygen Systems at Consumer Sites, 1996.

NFPA 50A, Gaseous Hydrogen Systems at Consumer Sites, 1999.

NFPA 50B, Liquefied Hydrogen Systems at Consumer Sites, 1999.

NFPA 51, Design and Installation of Oxygen-Fuel Gas Systems for Welding, Cutting, and Allied Processes, 1997.

NFPA 51B, Fire Prevention during Welding, Cutting, and Other Hotwork, 1999.

NFPA 70, National Electrical Code, 1996.

NFPA 306, Control of Gas Hazards on Vessels, 1997.

NFPA 326 and 327, Safeguarding of Tanks and Containers for Entry, Cleaning, or Repair, 1999.

Solon, L. R. "Occupational Safety with Laser (Optical Maser) Beams." *Archives of Environmental Health* 6 (March 1963): 414–17.

U.S. Department of Health, Education, and Welfare, National Institute for Occupational Safety and Health, Division of Technical Services, Cincinnati, OH 45226.

DHEW 73–11009, Criteria for a Recommended Standard: Occupational Exposure to Ultraviolet Radiation.

DHEW 75–115, Engineering Control of Welding Fumes.

DHEW 78–138, Safety and Health in Arc Welding and Gas Welding and Cutting.

U.S. Navy, The Pentagon, Washington DC 20350.

NAVSHIPS 250–00–92, *Bureau of Ships Manual,* Chapter 92, "Welding and Allied Processes."

NAVSHIPS 250–692–9, *Underwater Cutting and Welding Manual.*

REVIEW QUESTIONS

1. What is the most significant health hazard in the welding process?

2. List five health conditions a welder can experience as a result of exposure to various toxic gases generated during welding.
   a.
   b.
   c.
   d.
   e.

3. Ozone, one of the primary pulmonary gases that can injure the lungs, is formed by
   a. electrical arcs and corona discharges in the air.
   b. welding metals that have been cleaned with chlorinated hydrocarbons.
   c. ultraviolet photochemical reactions.
   d. all of the above
   e. only a and c

4. Which of the following inhalants is highly toxic in high concentrations and can produce total disability after exposure as short as a few months?
   a. manganese
   b. titanium
   c. magnesium
   d. chromium

5. If closed containers that have held flammable liquids cannot be removed for standard cleaning procedures prior to welding or cutting them, what two other practices can be followed?
   a.
   b.

6. Name the three general categories of personal protective equipment that should be used by welders.
   a.
   b.
   c.

7. Why is dark, woolen clothing preferred when welders work with inert-gas-shielded arc-welding machines?

8. The standards for training and qualification of welders were established by which organization?

9. What are the five steps in handling a leaking fuel gas cylinder?
   a.
   b.
   c.
   d.
   e.

10. Which of the following piping should be used for acetylene distribution systems?
    a. unalloyed copper
    b. steel
    c. wrought iron
    d. all of the above
    e. only b and c
11. The color that is generally recognized for a fuel gas hose is
    a. green.
    b. red.
    c. black.
    d. blue.
12. What device must be used on both oxygen and fuel gas cylinders to maintain a uniform gas supply to the torches at the correct pressure?
13. Define *resistance welding*.
14. Name three relatively new heat sources for welding and cutting.
    a.
    b.
    c.
15. Installation of resistance welding equipment should conform to what standard?

Metalworking Machinery

David A. Felinski
Philip E. Hagan, JD, MBA, MPH, ARM, CIH, CHMM, CET, CHCM, CEM

General Safety Rules
Electrical Controls on Machine Tools ▶ Rules for Safely Operating Machine Tools ▶ Safely Removing Chips, Shavings, and Cuttings ▶ Personal Protection

Turning Machines
Engine Lathes ▶ Turret Lathes and Screw Machines ▶ Spinning Lathes

Boring Machines
Drills ▶ Boring Mills

Milling Machines
Basic Milling Machines ▶ Metal Saws ▶ Gear Cutters ▶ Electrical Discharge Machining

Planing Machines
Planers and Shapers ▶ Slotters ▶ Broaches

Grinding Machines
Grinding Machine Hazards ▶ Abrasive Disks and Wheels ▶ Surface Grinders and Internal Grinders ▶ Grindstones ▶ Polishing Wheels and Buffing Wheels ▶ Wire Brush Wheels

ANSI Standards for Metalworking Machinery

Summary

References

Review Questions

20 Metalworking Machinery

Metalworking machinery includes all power-driven machines not movable by hand. They are used to shape or form metal by cutting, impact, pressure, electrical techniques, chemical techniques, or a combination of these processes. This and the next chapter can help managers develop safety programs for metalworking machinery. The topics covered include:

- general safety rules for operating and maintaining machine tools
- specific hazards and safety practices for turning, boring, milling, planing, and grinding machines
- importance of proper safety training for operators of machine tools.

The Association for Manufacturing Technology (AMT—formerly the National Machine Tool Builders' Association) has classified some 200 types of machine tools into two major categories: metal removal (cutting) machines and metal forming machines (covered in Chapter 21, Working with Hot and Cold Metals, in this volume. Of the metal removal machines, there are five basic groups: turning, boring, milling, planing, and grinding. Another classification contains electrodischarge, electrochemical, laser, and machining tools. Some machines combine the functions of two or more groups.

Power presses, press brakes, and power squaring shears are also classified as metalworking tools. However, these tools are covered in Chapter 21, Working with Hot and Cold Metals. Portable power tools, normally hand-held during operation, are not considered to be machine tools. They are covered in Chapter 17, Hand and Portable Power Tools, in this volume.

Injuries with machine tools are most often caused by unsafe work practices or incorrect procedures. Insufficient training and inadequate supervision usually give rise to these problems. More rarely, injuries result when a machine fails mechanically or is operated after an unsafe condition develops. Only qualified and competent personnel should operate hazardous equipment with exposed moving parts and cutting edges. In addition, properly maintaining and operating equipment would reduce injuries from machine tools. Installing effective guarding devices that do not hamper operation or lower production would further reduce the number of injuries. Of course, certain guards are required by federal, state or provincial, and local regulations.

Good housekeeping in the work area can help establish good work habits when operating machine tools. These two factors—good housekeeping and good work habits—result in fewer accidents.

GENERAL SAFETY RULES

Emphasize safely operating metalworking machinery. Establish a written policy to eliminate unsafe practices by operators. Include the following provisions in the safety policy. Maintenance and repair personnel should also comply with this policy.

- Restrict operation, adjustment, and repair of any machine tool to authorized, experienced, and trained personnel.
- Ensure proper lockout/tagout procedures are followed whenever work is performed on equipment.
- Closely supervise all personnel during training.
- Establish and maintain safe work procedures.
- Prohibit shortcuts and chance taking.
- When purchasing new equipment, make sure that specifications conform to all applicable standards, codes, and regulations concerning guarding, electrical safety, and other safeguards.
- Inspect new and modified equipment and make safety innovations before allowing operators to use the equipment.
- Devote full-time attention to the work in progress. If the operator must leave the machine, it should be shut down and locked out unless the machine tool has been designed to operate in this mode. For example, have interlocked guarding encompassing the machine, and equip the machine with automatic shutdowns that work as soon as there is any deviation from normal operation.
- Make supervisors responsible for the strict enforcement of all safety policies.

Provide a tool rack for the convenience of operators and repair and maintenance personnel. Include all wrenches and tools needed for operation or adjustment as standard equipment. Provide necessary material-handling equipment to avoid strains (Figure 20–1).

ANSI/NFPA 70, National Electrical Code, by the American National Standards Institute and National Fire

Figure 20–1. This machinist avoids back strain by sliding heavy materials onto a mobile lifter adjusted to the worktable height.

Protection Association, and ANSI C2, National Electrical Safety Code, should govern installation of electrical circuits and switches. The actual electrical and electronic component safety of the machine is covered in NFPA 79, Electrical Standard for Industrial Machinery.

Electrical Controls on Machine Tools

In addition to the manufacturer-installed electrical controls on machine tools, each machine must have a disconnect switch that can be locked in the OFF position to isolate the machine from the power source. Do not permit maintenance or repair on any machine until the disconnect switch serving the equipment has been shut off, padlocked in the OFF position, and tagged. (See the discussion of lockout/tagout in Chapter 6, Safeguarding, in this volume.)

Rules for Safely Operating Machine Tools

The following rules apply to safely operating any machine tool. Be sure that operators know and follow these rules:

- Never leave machine tools running unattended, unless the machine has been designed to do so.
- Never wear jewelry or loose-fitting clothing, especially loose sleeves, loose shirt or jacket cuffs, and neckties.
- Cover or tie long hair that could be caught by moving parts.
- Wear appropriate eye protection. This rule extends to others in the area, such as inspectors, stock handlers, supervisors, and especially visitors.
- Do not contaminate the metal removal fluid (e.g., discard refuse or spit into the machine tools' coolant sump or reservoir). Such actions create a chemical imbalance in the metal removal fluid that can contribute to unhealthy working conditions.
- Do not manually adjust and gauge (caliper) work while the machine is running.
- Use brushes, vacuum equipment, or special tools for removing chips. Do not use hands.
- Understand the differences in machining ferrous and non-ferrous metals, and know the health and fire hazards related to working with these metals.
- Use the proper hand tools for each job.

Safely Removing Chips, Shavings, and Cuttings

One of the major causes of accidents from machine tools, especially drilling equipment, is the careless use of high-pressure compressed air to blow chips, cuttings, or shavings from machines or workers' clothing. Brushes provide a less dangerous method, as do hand-held or air-operated vacuum units.

In cases where neither a brush nor a vacuum system is practical, it may be necessary to use compressed air. Keep all unnecessary employees out of the area. Ensure employees wear applicable personal protective equipment (PPE): safety goggles, face shield, and full-body protective clothing as determined by a PPE job evaluation. Many companies have found a nozzle pressure of 10 to 15 psig (70 to 100 kPa) is sufficient for most operations. The U.S. Occupational Safety and Health Administration (OSHA) specifies that the nozzle pressure be less than 30 psig (207 kPa). Nozzles meeting the OSHA requirements are available.

Isolate machine tool operations so that nearby employees are not endangered. Place baffles or chip guards around the machine to shield the operator. Permit only reliable and properly trained employees to use compressed air. These employees should wear cup-type goggles and other PPE. Prohibit employees from using high-pressure compressed air to blow dust or dirt from their clothing or out of their hair. Such actions can cause damage to, or inject foreign materials into, their ears and eyes.

One company has almost eliminated injuries caused by removing chips, shavings, or cuttings by using proper PPE and tools. Employees wear leather-palmed gloves and use a 3-ft- (910-mm-) long rod to pull shavings and cuttings from machines. The ¼-in.- (6-mm-) diameter round rod has a handle on one end for the operator to hold and a crook at the other to pull the shavings.

Personal Protection

The machine tool operator's safety largely depends on following established safe working procedures and wearing proper protective clothing and equipment. Obviously, all machine tool operators should wear eye protection with side shields. Wearing close-fitting clothing is also vitally important to the operator's safety. Many serious injuries and fatalities have resulted when neckties, loose shirtsleeves, gloves, or other clothing has gotten caught in a belt and sheave, between gears, in a revolving shaft, or in the revolving workpiece held in the chuck (Figure 20–2). Operators also should not wear rings, necklaces, or other jewelry because of the potential for getting caught in moving machinery. Because most machine operations involve handling heavy stock or heavy machine parts, such as faceplates and chucks, every operator should wear protective footwear.

Operators with long hair should wear caps, snoods, hairnets, or other protection that completely covers their hair. There have been many instances of people being partially, or entirely, scalped when their hair became entangled in the moving parts of a machine.

Use splash guards, shields, PPE, and other means to minimize exposure of workers to the irritating metalworking fluids and mineral spirits used to clean metal parts. Provide barrier creams, and encourage personal hygiene measures, such as thorough washing, to minimize skin irritations.

Figure 20–2. This lathe chuck shield has a semicircular construction and is made from high-impact-resistant, transparent acrylic. The shield is mounted to a chromium-plated extension tube fastened to the lathe headstock by mounting brackets. The shield can be lifted up and out of the way for quick and easy access to the piece-part.

TURNING MACHINES

Shaping a rotating piece with a cutting tool, usually to give a circular cross-section, is known as *turning*. This procedure is done on machines such as engine lathes, turret lathes, chuckers, semiautomatic lathes, and automatic screw machines.

Engine Lathes

To prevent accidents with engine lathes, allow only qualified personnel to operate them. Injuries are likely to result from incorrectly operating engine lathes in the following ways:
- having contact with projections on work or stock, faceplates, chucks, or lathe dogs, especially those with projecting setscrews
- being hit by flying metal chips
- hand braking the machine
- filing right-handed, using a file with an unprotected tang, or using the hand instead of a stick to hold an emery cloth against the work
- failing to keep both the center holes of taper work clean and true and the lathe's center true and sharp
- leaving the machines running unattended
- handling chips by hand instead of using a hooked rod
- calipering or gauging the job while the machine is running
- attempting to remove chips when the machine is running
- having contact with rotating stock projecting from turret lathes or screw machines
- leaving the chuck wrench in the chuck
- catching rings, loose clothing, gloves, or rags for wiping on revolving parts.

Several preventive measures will help operators run engine lathes safely. Use faceplates and chucks without projections whenever possible. Otherwise, install a simple shield formed to the contour of the chuck or plate and hinged at the back. This will prevent contact with the revolving plate or chuck.

Because safety-type lathe dogs are relatively inexpensive to install, substitute them for those with projecting setscrews. Chip shields, particularly on high-speed operations, help control flying chips. Use plastic or fine-mesh screen chip guards because they allow operators to see through them while confining the flying chips (Figure 20–2). These shields do not eliminate the need for protective eye equipment, however. Shields may need to be frequently replaced because of the abrasive action of the chips.

Provide mechanical means—such as an overhead hoist or a swinging, welded pipe fixture—to lift heavy faceplates, chucks, and stock on both lathes and screw machines.

Turret Lathes and Screw Machines

Hazards associated with turret lathes and screw machines (Figure 20–3) are similar to those listed for other lathes. Additional hazards are caused by operators not moving the turret back as far as possible when changing or gauging work or using machine power to start the faceplate or chuck onto the spindle. Other hazards result when operators fail to keep their hands clear of the turret's slide or when they

Figure 20–3. This automatic screw machine has a chip shield to protect against metal removal fluid splashes and flying chips. (*Printed with permission from the Delco-Remy Division of General Motors.*)

permit a hand, arm, or elbow to strike the cutter while adjusting or setting up.

Install splash shields, especially on automatic machines, and keep them in good condition. Enclosure shields over the chuck confine hot metal chips and oil splashes. They also act as exhaust hoods to remove fumes.

When steel and some other materials are turned on lathes, the chip produced is in a continuous spiral that frequently causes injuries to hands and arms. Install chip breakers to protect against this.

Screw machines can create noise that exceeds noise limits. Try to reduce the noise level through engineering methods, such as commercially available sleeves for rotating bar stock. Use hearing protection whenever engineering and administrative controls are not sufficient to maintain noise exposure below regulated levels. (See the National Safety Council's publication *Hearing Conservation in the Workplace: A Practical Guide*.)

Spinning Lathes

A spinning lathe is a forming tool rather than a cutting tool. It usually requires a specially skilled and qualified operator.

Unsafe practices that should be prohibited when operating a spinning lathe include:

- inserting blanks and removing the processed part without first stopping the machine
- failing to fully tighten the tailstock's handle and risking that the blank will work loose or ruin the stock or tool
- allowing the swarf (cuttings, turnings, particles) to build up into a long coil when trimming copper and certain grades of steel.

This last practice has resulted in severed hands and severely cut arms. One fatality occurred when a coil became snarled around the operator's neck. Operators should remove the tool, when necessary, to allow the swarf to break off.

The spinning lathe's chuck is usually a form built up of hardwood, shaped exactly like the finished part. If the piece being worked must be "necked down" so the chuck cannot be taken out after the piece is formed, the chuck is made in sections held together by locking rings or locking grooves. A substantial hazard is possible if a chuck flies apart because the grooves have become worn or the rings break. Because spinning lathes reach speeds of 500 to 2,000 rpm, the danger of flying chuck sections is obvious. Prevention of injuries from this cause lies in frequent inspection and maintenance of chucks. Similarly, inspect lathe tools frequently for cracks in the tools or handles.

The tailstock of spinning lathes can loosen during operation because of vibration or worn parts. The piece then will work loose or fly off. The solution to this problem is also frequent inspection and maintenance.

BORING MACHINES

Boring consists of cutting a round hole using a drill, boring cutter, or reamer. Drilling machines are equipped with rotating spindles, handles, and chucks that carry pointed or fluted cutting tools. Operations performed with drilling machines include countersinking, reaming, tapping, facing, spot facing, and routing. Boring mills use a cutter, either single- or multi-edged, that is mounted on a supporting spindle or shaft. The cutter trues up or enlarges a hole that has already been rough formed by drilling, casting, or forging.

Drills

Drill press accidents are more likely to occur during unusual jobs because special jibs or vises for holding the work are not usually provided. Radial drill accidents are frequently caused by incorrect manipulation. Properly clamp the drill's head and arm, as well as the workpiece, prior to cutting metal. The most common hazards in drilling operations are:

- contacting the rotating spindle or the tool
- being struck by a broken drill
- using dull drills
- being struck by insecurely clamped work
- catching hair, clothing, or gloves in the revolving parts
- sweeping chips or trying to remove long, spiral chips by hand
- leaving the key or the drift in the chuck
- being struck by flying metal chips
- failing to replace the guard over the speed-change pulley or gears.

Observing safety precautions and using good operating habits can protect operators from these hazards. To guard an operator from contact with a spindle, use a plastic shield, a simple wire mesh guard, or other barrier (Figure 20–4). When necessary, guard the tool with a telescoping guard that covers the end of the tool. This leaves only enough of the tool exposed to allow easy placement into the piece being worked. The telescoping drill shield shown in Figure 20–4 has a stationary, ribbed cage. The outer sleeve rides up and down on the inner cage. This shield provides high visibility.

A frequent cause of breakage is using a dull drill. A thin drill, smaller than 1/8 in. (3 mm) in diameter, will often break and cause injury. A larger drill may "fire up," freeze in the hole, and then break. Furthermore, a frozen tool may cause unclamped or insecurely clamped work to spin and injure the operator.

To avoid having a drill catch in thin material and spin it, clamp the work between two pieces of metal or wood before drilling. When drilling thin ferrous stock, grind the drill point to an included angle of about 160 degrees, and thin the point of the drill by grinding the flutes. With non-

Figure 20-4. This transparent drill shield protects against flying chips and pieces of broken drills.

ferrous metals, a negative rake will further reduce chances of the drill grabbing or digging in.

When deep holes are being drilled, frequently remove the drill and clean out the chips. If chips are allowed to pile up, the tool may jam, with results similar to those of freezing. Maintain counterweight chains in good condition and install a shield around the counterweight.

Boring Mills

Some common causes of injury in boring mill operations are listed here:
- being struck by insecurely clamped work or by tools left on or near a revolving table
- catching clothing or rags for wiping in revolving parts
- falling against revolving work
- calipering or checking work while the machine is in motion
- allowing turnings to build up on the table
- removing turnings by hand.

Horizontal Boring Mills

The same accident prevention measures are effective on both table and floor types of horizontal boring mills. While the machine is in motion, the operator should never attempt to make measurements near the tool, reach across the table, or adjust the machine or the work.

Frequently inspect clamps and blocking to make certain the clamping is positive. Always avoid makeshift setups.

Before attempting to raise or lower a boring mill's head, the operator should make sure that the clamps on the column have been loosened. Otherwise, the boring bar can be bent or the clamps or bolts broken. This can cause damage to the machine and injury to the operator.

Before the boring bar is inserted into the spindle, the operator should make certain that the spindle's hole and the bar are clean and free from nicks. The operator should not attempt to drive the bar through the tailstock's bearing with a hammer or other heavy tool. Instead, the operator should use a soft metal hammer to drive the bar into the spindle. If a steel hammer or piece of steel must be used, the operator should hold a piece of soft copper or brass against the bar while driving it into the spindle.

Vertical Boring Mills

The same procedures apply to the safe operation of vertical boring mills. Each mill's table, particularly those tables 100 in. (2.5 m) or less in diameter, should have the rim enclosed in a metal band guard to protect the operator from being struck by the revolving table or by projecting work. Such guards should be hinged so they can be easily opened during setting up and adjustment (Figure 20-5).

If the table is flush with the floor, install a portable fence, usually of iron pipe sections. Such fencing should conform to the state code or the specifications of ANSI A1264.1, Safety Requirements for Workplace Walking/Working Surfaces and Their Access; Workplace, Floor, Wall and Roof Openings; Stairs and Guardrails Systems.

While the machine is in operation, the operator should never attempt to tighten the work, the tool, or the caliper nor measure the work, feel the edges of the cutting tool, or oil the mill. The operator should never ride the table while it is in motion. There is one exception. On some large mills, such as those used for boring turbine castings, the operator may have to ride the table in order to observe the work's progress. In such cases, he or she should always make sure that no portion of the body will come in contact with a stationary part of the mill.

In addition, steps or stairs that provide access to the machine or to the work should have a pitch of not more than 50 degrees and should have slip-resistant treads. Stairs with four or more risers must have a handrail.

MILLING MACHINES

Machining a piece of metal by bringing it into contact with a rotating multi-edged cutter is *milling*. This procedure is done by horizontal and vertical milling machines; by gear hobbers, profiling machines, circular and band saws; and by a number of other types of related machines (Figure 20-6).

Many accidents with milling machines occur when operators unload or make adjustments. Other causes of injuries include:
- failure to draw the job back to a safe distance when loading or unloading
- using a jig or vise that prevents close adjustment of the guard

Figure 20–5. This guard for a vertical boring mill is made in two sections of sheet metal. The sections are hinged to the machine. Left: The guards are closed. Right: The guards are opened to allow setup or adjustment.

- placing the jig- or vise-locking arrangement in such a position that force must be exerted toward the cutter
- leaving the cutter exposed after the job has been withdrawn
- leaving hand tools on the worktable
- failing to securely clamp the work
- reaching around the cutter or hob to remove chips while the machine is in motion
- removing swarf cuttings by hand instead of with a brush
- adjusting the coolant's flow while the cutter is turning
- calipering or measuring the work while the machine is operating
- using a rag to clean excess oil off the table while the cutter is turning
- wearing gloves, rings, ties, or loose clothing; catching fingers, gloves, or clothing in power clamps
- using incorrectly dressed cutters
- incorrectly storing cutters
- attempting to remove a nut from the machine's arbor by applying power to the machine
- striking the cutter with a hand or an arm while setting up or adjusting the stopped machine
- misjudging clearances between the arbor or other parts
- cleaning the machine while it is in motion.

Basic Milling Machines

Regardless of the classification, direction of movement, or special attachments that make varied operations possible on a milling machine, the safeguarding requirements are basically the same. To guard the cutter, employ one of several methods (Figure 20–7).

Mount hand-adjusting wheels on the shaft by either clutches or ratchet devices for quick or automatic traverse on some models. In this way, the wheels do not revolve when the automatic feed is used. As an alternative, provide compression-spring wheels with removable handles. These handles cannot remain in the wheels unless held in place by the operator.

The horizontal milling machine should have a splash guard and pans for catching ejected metalworking fluids or lubricants running from the tools. Direct the lubricant

Figure 20–6. The adjustable shield on this vertical milling machine protects against both flying metal chips and splashing coolant.

 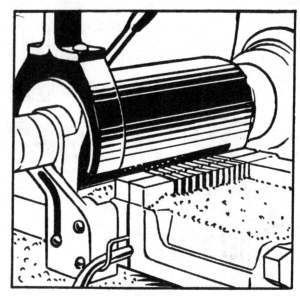

Figure 20–7. Self-closing guard for milling machine cutter. Left: The cutter is completely enclosed when the table is withdrawn. Right: As the table moves forward, the guard automatically opens.

on the work so the distribution setup will not be drawn into the cut by the cutter's rotation. When possible, make all cuts into the direction of travel of the table, rather than away from the direction of travel.

Metal Saws

Circular, swing, and band saws are metalworking, as well as woodworking, machinery. Provide suitable guarding for each type of saw. Chapter 18, Woodworking Machinery, in this volume, contains additional information about guarding circular, swing, and band saws.

Circular Saws

A circular saw for cutting cold metal stock should have a hood guard at least as deep as the roots of the teeth. The guard should automatically adjust itself to the thickness of the stock being cut.

Use a sliding stock guard when tube or bar stock is cut. Guard the portion of the saw under the table with a complete enclosure that provides for disposal of scrap metal. A plastic or metal guard placed in front of and over the saw provides protection against flying pieces of metal. Do not, however, consider a guard as a substitute for eye protection.

Swing Saws

For swing saws, adjust the length of the stroke so the blade will not pass the table at its most forward point. Locate the control so the saw can be operated with the left hand when fed from the left or with the right hand if fed from the right. In this way, the operator is positioned to the side away from the moving blade.

Band Saws

Completely enclose the upper and lower wheels of metal-cutting band saws with sheet metal or a heavy, small-mesh screen mounted on angle-iron frames. To make the changing of blades convenient and safe, provide access doors equipped with latches. Except for the point at which the cut is made, completely enclose the portion of the blade between the upper wheel and the saw table with a sliding fixture attached to the slide.

The length of blade exposed should not be more than the thickness of the stock plus 3/16 in. (9 mm). Confine flying particles of metal with a metal or transparent plastic guard installed in front of the saw. On a hand-fed operation, take care at the end of a cut. Use a push block to advance the workpiece through the blade, not hands.

Gear Cutters

During operation of gear cutters and hobbers, both the tool and the workpiece move. Therefore, keep the point-of-operation guards simple and easily adjustable.

On operations where the workpiece (gear blank or rough-cut gear) is moved to the tool, a simple barrier guard, formed to cover the point of operation and sized to fit the workpiece, is satisfactory. Mount the guard on a spindle that carries the workpiece. This causes the guard to fit over the point of operation when the workpiece is brought into position.

When the tool is brought to the workpiece, and when both the tool and workpiece are adjustable, attach an encircling type of guard to the tool's head. Such a guard can be an automatic drop-gate device. This device can be equipped with both a release latch to open the guard's enclosure and a spring release to return the guard to a position clearing the work. Each guard should have an automatic interlock so the machine will not operate except when the guard is in place.

On some makes of machines, the lever that controls the spindle's direction of operation is located so operators can catch their hands on the back gears driving the spindle. In such cases, install an auxiliary lever that can be operated at a point outside the danger zone created by these gears.

On large machines where the operator is not close to the regular control switch, install a pendant switch mounted on an arm or sweep. This switch acts as a magnetic brake to stop the machine instantly.

When operators insert an arbor into the spindle, they should make sure that both the arbor and spindle holes are clean and free from nicks. Operators should draw the arbor firmly into place by a sleeve nut and securely tighten the nut. Before removing the arbor from the spindle, operators should make certain the machine is at a standstill.

Electrical Discharge Machining

The electrical discharge machine (EDM) process is designed to perform a variety of machining operations. This process makes simple or complex machining possible through hole boring or cavity sinking in any electrically conductive work material, including carbide, high-alloy steels, and many types of hardened metals.

The EDM Process

During the machining process, the workpiece is normally clamped to the table, and an electrode is fastened to the vertical ram platen above the workpiece. The electrode is then brought near the workpiece so that an accurately controlled electrical discharge takes place between the electrode and the workpiece. This discharge removes metal from the workpiece at the point where the gap is smallest.

As metal is removed, the electrode tool is fed into the workpiece and held in the correct cutting relationship by electrohydraulic servo control of the ram's workhead. The servo control automatically maintains a gap between the electrode and the workpiece.

During the process, a dielectric fluid should completely cover the workpiece in the work tank. During EDM, a flow of dielectric fluid through the machining gap should be employed whenever possible to increase machining efficiency (Figure 20-8).

Hooking Up EDM Machines

Only a qualified electrician designated to work on machine tool circuits should maintain or connect the electrical system of EDM machines. Before attempting any work, the electrician should read and completely understand the electrical schematics for the machine. In addition, appropriate lockout/tagout procedures should be followed in accordance with 29 CFR 1910.147.

After the machine has been connected, operators should test all aspects of the electrical system to make sure it is functioning properly. Before considering the electrical system connection complete, operators should do the following:
- Be sure the machine is properly grounded, and check that all exposed electrical systems are properly covered.
- Place all selector switches in the OFF or neutral (disengaged) position.
- Be sure that the machine's push buttons, manual limit switches, or controls are set for a safe setup.
- Check that the doors of the main electrical cabinet are closed and that the main disconnect switch is in the OFF position.

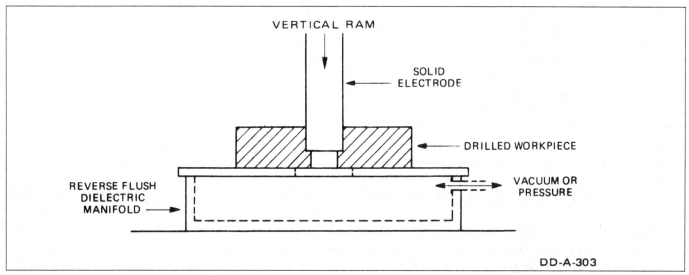

Figure 20-8. Electrical discharge machining into manifolds (including a reverse-flush dielectric manifold), tanks, domes, or any other structures capable of trapping discharge gases. *(Printed with permission from Cincinnati Milacron Company.)*

Safety Precautions for EDM

To keep the operator from accidentally brushing against the live electrode or platen when the machine is operating at high voltages, install a clear plastic safety shield on the work tank. The shield must be in place before the power supply is turned on. Do not attempt to block out the wire around any electrical safety interlock. Before removing any cover, or before working on the machine or power supply, turn off and lock out the main electrical disconnect device. Tag the device and all START buttons with an OUT OF ORDER or DO NOT START tag. Always turn the electrical disconnect switches to the OFF position at the end of the working day.

Maintain the oil level above the highest portion of the electrode workpiece's working gap. Once the safe oil level above the part has been determined, adjust the safety float's switch to make certain that the appropriate oil level is maintained. Maintain the dielectric level at a minimum of 1 in. (25 mm) per 100 amp of average current for flat geometry work. Because EDM is a heat-producing process, install an EDM machine only where there is adequate ventilation. The dielectric oil removes the concentrated heat from the machining gap and distributes it in the available oil. Air conditioners and electronic precipitators do not constitute adequate ventilation.

Discharge Gases Hazard

Operators and maintenance personnel should read and completely understand all the precautions before operating, setting up, running, or performing maintenance on EDM machines. Failure to comply with instructions can result in serious or fatal injury. The EDM machine operator must be aware of the possibility of discharge gases igniting. Turn off the electrical power to stop additional gas or hot metal particles from forming to extinguish the flame. Operators should have received training in how to use a carbon dioxide (CO_2) foam fire extinguisher. It is necessary that a CO_2 fire extinguisher be kept in the vicinity of the EDM machine.

Because all discharge gases are flammable, keep them away from sparks or flame. Allow discharge gases to escape without being trapped in a closed area. Avoid any setup that can result in trapped gases. Therefore, be sure that the EDM machine is adequately ventilated. When the dielectric level in a storage tank is suddenly raised (e.g., by dumping a work tank), a large volume of discharge gases will escape into the outside air. Ignition of such displaced gases can cause a fire that could backflash into the enclosed area and cause an explosion.

Currently, there is no specific ANSI safety standard for the EDM process; however, there is a European standard. The current European (CEN) standard is designated BS EN ISO 28881:2013, Machine Tools, Safety, Electro-Discharge Machines.

PLANING MACHINES

Planing machines come in several forms. Planers machine metal surfaces. The cutting tool is held stationary while the workpiece is moved back and forth underneath it. With shapers, generally classified as planing machines, the process is reversed. The workpiece is held stationary while the cutting tool is moved back and forth. Other machine tools classified as planing machines are slotters and broaches.

Planers and Shapers

Accidents with planers frequently result from unsafe practices caused by inadequate training and supervision, such as the following:

- placing hand or fingers between the tool and the workpiece
- running the bare hand over sharp metal edges
- measuring the job while the machine is running
- failing to clamp the workpiece or tool securely before starting the cut
- riding the job
- having insufficient clearance for the workpiece
- coming in contact with reversing feed dogs
- failing, when magnetic chucks are used, to make sure the current is turned on before starting the machine
- unsafely adjusting the tool holder on the cross head.

To avoid these accidents, install guards on planers. Cover the reversing feed dogs on planers. If the planar bed travels within 18 in. (46 cm) of a wall or fixed object(s) when fully extended or when any stock on the bed is being processed, close the space between the end of the travel and the obstruction with a guard on either side of the planer (Figure 20–9). Construct the guard so it will not cause an accident when the bed is extended.

Accidents with shapers have essentially the same causes as those with planers. In addition, injuries frequently result from contact with projections on the workpiece or with projecting bolts or brackets, especially when the table is being adjusted vertically. Leave the shaper's ram projecting over the table to alert the operator that the table is high enough.

Failure to properly locate the stops or dogs can also injure shaper operators. Rigidly bolt the stops to the table, especially on heavy jobs.

The shaper operator should make sure the tool is set. That way, if it shifts away from the cut, it will rise away from the cut and not dig into the work. Remove the handle of the stroke-change screw before starting the shaper. To prevent injury to the operator and workers nearby from flying chips, install guards. Also, cover the reversing feed dogs on shapers.

20 Grinding Machines   611

work, the ram's head puller, and the follow rest, if used, should all line up. Check fixtures to make sure the workpiece is securely held. Do trial runs at slow speeds to make certain the chips do not pack tightly between the teeth.

During normal operations, safeguard broaches (like heavy production machine tools) through supplementary controls. The most widely used safeguard for electrically controlled broaches is a standard, two-hand, constant pressure control. Install an EMERGENCY STOP button, preferably of the mushroom type, adjacent to one of the two-hand controls. Another type of safeguard for broaches is a foot-operated EMERGENCY STOP bar with a wide surface plate.

All pneumatically or hydraulically powered clamping equipment should be actuated by two-hand controls. Locate these controls so the operator's hand cannot reach the pinch area before the clamps close. Shield controls if they could be tripped by other parts of the body. If the hands are exposed in the clamping area, use tongs for loading and unloading.

GRINDING MACHINES

Grinding machines shape material by bringing it into contact with a rotating abrasive wheel or disk. *Grinding* includes surface, internal, external, cylindrical, and centerless operations, as well as polishing, buffing, honing, and wire brushing. Portable machines that use small, high-speed grinding wheels are discussed in Chapter 17, Hand and Portable Power Tools, in this volume.

The text and illustrations in this section have been adapted with permission from ANSI B7.1, The Use, Care, and Protection of Abrasive Wheels. Specifications for operation of grinding machines and construction of guards and safety devices were revised in 2010 and primarily address the safety aspects of abrasive products and flanges.

Grinding Machine Hazards

Hazards associated with grinding machines include the following:
- failure to use eye protection in addition to the eye shield mounted on the grinder
- incorrectly holding the work
- incorrectly adjusting or not using the work rest
- using the wrong type, a poorly maintained, or imbalanced wheel or disk
- grinding on the side of a wheel not designed for side grinding
- taking too heavy a cut
- applying work too quickly to a cold wheel or disk
- grinding too high above the wheel's center

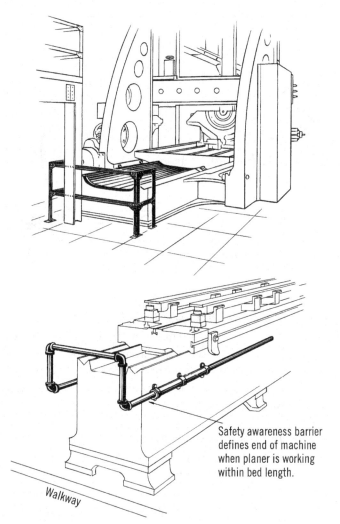

Figure 20–9. Top: A guardrail or similar barrier should close off any space 18 in. (460 mm) or less between a fixed object and parts of a fully extended planer or its stock. Any openings in the planer should be filled to eliminate shear hazards. Bottom: This self-adjusting planer table guard moves out with the table and is retained in position by friction sleeves.

Slotters

In slotter operations, the most serious accident is catching the fingers between the tool and the workpiece. Fingers can also be caught between the ram and the table when the ram is at the end of the downstroke. Because the ram works at a slow speed and the platen or machine table is small, operators may instinctively reach across the table and under the ram to pick up a tool or other object—thus catching their fingers. To avoid this kind of accident, enclose the ram's eccentric with a hinged guard made of sheet metal or cast iron.

Broaches

The broach's rated capacity should be equal to or greater than the force required for the job. The centerline of the

- failure to use wheel washers (blotters)
- vibration and excessive speed that lead to bursting a wheel or disk
- using bearing boxes with insufficient bearing surface
- using a spindle with incorrect diameter or with the threads cut so the nut loosens as the spindle revolves
- installing flanges of the wrong size, with unequal diameters, or with unrelieved centers
- incorrect wheel dressing
- contacting unguarded moving parts
- using controls that are out of the operator's normal reach
- using an abrasive blade instead of a grinder disk
- failure to run a wet wheel dry, without coolant, for a period of time before turning off the machine
- using an untested, broken, or cracked grinding wheel
- reaching across or near the rotating grinding wheel to load, unload, or adjust the machine during setup.

Abrasive Disks and Wheels

An abrasive disk is made of bonded abrasive, with inserted nuts or washers, projecting studs, or tapped plate holes on one side of the disk. This side is mounted on the faceplate of a grinding machine. Only the exposed flat side of an abrasive disk is designed for grinding.

An abrasive wheel is made of bonded abrasive and is designed to be mounted, either directly or with adapters, on the spindle or arbor of a grinding machine. Only the periphery or circumference of many abrasive wheels is designed for grinding.

Inspecting Abrasive Disks and Wheels

When unpacking abrasive disks and wheels, inspect them for damage from shipment and have a qualified person give them the "ring" test. This test can be used for both light and heavy disks or wheels that are dry and free of foreign material. To conduct the ring test, suspend a light disk or wheel from its hole on a small pin or the fingertip, and place a heavy one vertically on a hard floor. Then gently tap the wheel or disk with a light tool, such as a wooden screwdriver's handle. A mallet may be used for heavy wheels or disks. Make the tap at a point 45 degrees from the vertical centerline and about 1 or 2 in. (25 or 50 mm) from the periphery (Figure 20-10). A wheel or disk in good condition will give a clear, metallic ring when tapped. Wheels and disks of various grades and sizes give different pitches.

Daily inspection of grinding machines should include those points shown in Figure 20-11. Thoroughly investigate grinding wheel and disk failures, preferably with the manufacturer's representative. This type of investigation, along with immediate corrective action, greatly reduces the possibility of recurrent failures.

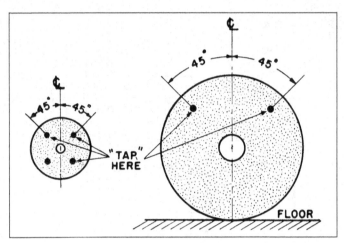

Figure 20-10. Tap points for ring test.

Handling Abrasive Disks and Wheels

Abrasive disks and wheels require careful handling. Do not drop or bump them. Do not roll large disks and wheels on the floor. Transport disks and wheels too large or heavy to be manually carried by hand truck or other means that provide the correct support.

Storing Abrasive Disks and Wheels

Store abrasive disks and wheels in a dry area not subject to extreme temperature changes, especially below-freezing temperatures. Wet wheels might break or crack if stored below 32°F (0°C). Breakage can occur if a wheel or disk is taken from a cold room and work is applied to it before it has warmed up.

Store abrasive disks and wheels in racks in a central storage area under the control of a specially trained person. The storage area should be as close as possible to the grinding operations to minimize handling and transportation.

The length of time abrasive disks and wheels may be stored and still be safe to use should be in accord with manufacturers' recommendations. Conduct the ring test with disks and wheels taken out of long storage. Follow this by a check for recommended speed and a speed test on the machine that will be mounted. Check the speed of all grinding wheels against the spindle speed of the machine—some are designed only for low-speed use. The grinding wheel must be rated at the same revolutions per minute (rpm) value as the machine. If it is not, the wheel may explode and throw fragments at high velocity into the work area.

Mounting Wheels

Mount all abrasive wheels between flanges. Exceptions to this rule include mounted wheels; threaded wheels (plugs and cones); plate-mounted wheels; and cylinder, cup, or segmental wheels mounted in chucks.

```
                    GRINDER CHECKLIST
TYPE _____ RPM _____
SIZE _____ PERIPHERAL SPEED _____

                        Item                          OK
    WHEEL GUARD: securely fastened                    ☐
                 properly aligned                     ☐
    GLASS SHIELD: clean                               ☐
                  unscored                            ☐
                  in place                            ☐
    WORK REST: within ⅛ in. (3.2 mm) of wheel         ☐
               securely clamped                       ☐
    FRAME: securely mounted                           ☐
           no vibration                               ☐
    WHEEL FACE: well lighted                          ☐
                dressed evenly                        ☐
    FLANGES: equal size                               ☐
             correct diameter (½ wheel diam.)         ☐
    SPEED: correct for wheel mounted                  ☐
    GUARD FOR POWER BELT OR DRIVE:
             in place                                 ☐

DATE _____ DEPARTMENT _____
INSPECTED BY _____
```

Figure 20-11. A summary of checkpoints for safe grinder operation.

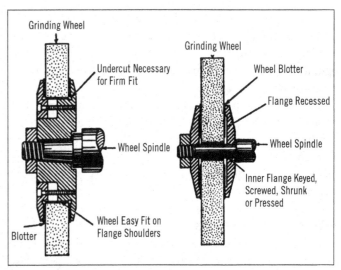

Figure 20-12. Correct methods of mounting abrasive wheels with large holes (left) and wheels with small holes (right).

Flanges should have a diameter not less than one-third of the wheel's diameter and preferably should be made in accordance with ANSI B7.1, clause 5. Flanges for the same wheel should be of the same diameter and thickness, accurately turned to correct dimensions, and in balance. The requirement for balance does not apply to flanges made out of balance to counteract an unbalanced wheel.

Key, screw, shrink, or press the inner or driving flange onto the spindle. The bearing surface of the flange should run true with the spindle. The outer flange's bore should easily slide onto the spindle.

Schedule flange inspections frequently. Remove from the spindle a flange found to be sprung, not bearing evenly on the wheel, or defective in any other way. Replace it with a flange that is in good condition.

An incorrectly mounted abrasive wheel is the cause of much wheel breakage. Because rotational forces and grinding heat cause high stresses around the wheel's central hole, follow safety regulations concerning size and design of mounting flanges and mounting techniques.

Before a wheel is mounted, conduct the same inspection and ring test as was conducted when the wheel was originally received and stored. In addition, check the bushings, particularly on wheels that have been rebushed by the user, for shifting or looseness.

Use compression washers to compensate for unevenness of the wheel or flanges. Blotting paper, not more than 0.025 in. (0.6 mm) thick, or rubber or leather compression washers, not more than 0.125 in. (3.2 mm) thick, may be used for this purpose.

Make allowance for the wheel mounting's fit in the wheel hole rather than in the arbor or wheel mount (Figure 20-12). Do not force the wheel onto the spindle. Forcing can loosen, or otherwise damage, the wheel's bushing, or it can crack the wheel. A wheel that is too loose on the spindle will run off-center, causing stress and vibration. Spindle end nuts should hold the wheel firmly but not too tightly. Too much pressure can spring or distort the flange or even break the wheel. If rebushing is necessary to make the wheel fit the spindle, have the manufacturer do it, or have an experienced employee with suitable equipment do it.

Immediately after mounting the wheel and before turning on the power, the operator should turn the wheel by hand for a few revolutions. At this time, check to make sure the wheel clears the hood guard and machine elements, such as work rests on work-holding equipment.

Operating a Grinding Machine

When starting a grinding machine, stand to one side away from the grinding wheel. Allow at least 1 minute of warm-up time before truing or grinding with the wheel. Always use coolant when truing the wheel or during normal grinding. Never allow coolant to flow on a stationary grinding wheel; coolant might collect on one portion of the wheel, causing an unbalanced condition. This unbalanced condition can cause the wheel to disintegrate upon restarting.

While the machine is running, never remove a guard fastener or guard. The guards are on the machine for the operator's safety; if they are removed, serious injuries to the operator or others can result.

Do not touch any moving part of the machine or the rotating grinding wheel to determine its smoothness or condition. Do not attempt to physically operate a machine that is in its automatic mode. Never alter or try to alter the machine, its wheel speed, or any of its safety equipment at any time.

Adjusting Safety Guards

The guard should enclose the wheel as completely as the nature of the work will permit. Adjust the peripheral guard to the constantly decreasing diameter of the wheel with an adjustable tongue or similar device. Doing this prevents the maximum distance between the wheel's periphery and the tongue or end of the peripheral band at the top of the opening from exceeding ¼ in. (6 mm) (Figure 20–13). Also, it maintains the angle of exposure specified for bench and floor-stand grinders throughout the life of the wheel. The maximum exposure angle varies with the type of grinding (Figure 20–14). Safety guards should also cover the exposed arbor ends.

On machines used for cutting, grooving, slotting, or coping stone or other materials, the safety guard or hood seldom offers adequate protection. On machines that permit a horizontal traverse between the wheel and the workpiece greater than 10 in. (250 mm) and on those that use solid cutting wheels 10 in. (250 mm) or more in diameter, provide an auxiliary enclosure in addition to the guard. This auxiliary enclosure can be a set of heavy screen panels, suspended approximately 8 ft (2.5 m) above the floor to or below the worktable. The panel screens should be ½ in. (13 mm) mesh or smaller, and the wire should be ⅛ in. (3.2 mm) or more in diameter. The framework of the panels should be made of 1- to 1¼-in. (25- to 31-mm), or heavier, structural steel angles or channels.

Safe Speeds

Do not operate abrasive wheels and disks at speeds exceeding those recommended by the manufacturer. In particular, unmarked wheels of unusual shape, such as deep cups with

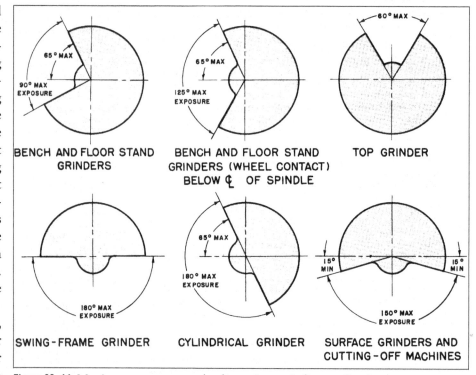

Figure 20–13. The correct wheel exposure can be maintained with an adjustable tongue (left) or a movable guard (right).

Figure 20–14. Maximum exposure angles for various grinding applications.

thin walls or backs with long drums, should be operated according to the manufacturer's recommendations.

As the wheel wears down, the spindle's speed (in revolutions per minute [rpm]) is sometimes increased to maintain the surface speed (in surface feet per minute [sfpm]). When the wheel is nearly worn down, the spindle is running at the highest rpm. When the worn wheel is replaced, adjust the spindle's speed. If the spindle's speed is not adjusted, the new wheel might break because the surface speed exceeds manufacturer's recommendations.

Grinding equipment for high-speed operation should be specially designed. Give special attention to spindle strength, guards, and flanges to eliminate mounting stresses. Such things as side-grinding pressure and the wheel's shape must also be considered. Proper maintenance

20 Grinding Machines   615

have broken, thus causing serious injury. The work rest should be substantially constructed and securely clamped not more than 1/8 in. (3.2 mm) from the wheel (Figure 20-15). Check the work rest's position frequently. The work rest's height must be on the horizontal centerline of the machine's spindle.

Never adjust the work rest while the wheel is in motion. The work rest might slip and strike and break the wheel, or the operator might catch a finger between the wheel and the work rest. To prevent work from adding twisting and bending stress to the wheel, operators should use guides to hold the work in position when slot grinding or performing similar operations.

Dressing Abrasive Wheels

Abrasive wheels that are not true or not in balance (Figure 20-16) will produce poor work. They can damage the machine and injure the operator. Keeping the wheels in good condition eliminates these possibilities, decreases wheel wastage, and lengthens the wheel's life.

To recondition a rutted or excessively rough wheel, it is often necessary to dress it by removing a large area of the face. Equip wheel-dressing tools with hood guards over the tops of the cutters. They will protect the operator from particles flying from the wheel or pieces of broken cutters. The operator of a wheel dresser should use a rigid work rest set close to the wheel. The operator should move the wheel dresser back and forth across the wheel's face while firmly

Figure 20-15. Top: With a properly adjusted work rest, the operator can keep hands away from the wheel and still firmly hold the work in place. Bottom: There should be a safe space of no more than 1/8 in. (3 mm) between the tool rest and the wheel.

and protective devices are also important for safe high-speed operations. Obtain the manufacturer's approval for all high-speed wheel and disk operations.

Work Rests

Because work has become wedged between the work rest and the wheel, many bench and floor-stand grinder wheels

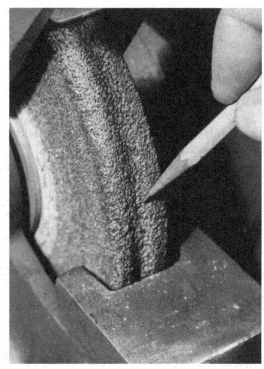

Figure 20-16. A badly rutted or out-of-balance grinding wheel should be taken out of service and dressed.

20 Metalworking Machinery

> **BENCH AND STAND GRINDERS**
>
> **DRESSING**
>
> 1. Wear a face shield over your safety glasses for protection against heavy particles.
> 2. Use a dressing tool approved for the job. Never use a lathe cutting tool.
> 3. Inspect star dressers for loose shaft and worn disks.
> 4. Round off the wheel edges with a hand stone before and after dressing to prevent the edges from chipping.
> 5. Use the work rest to support and guide the tool. Use a tool holder if one is available.
> 6. Apply moderate pressure slowly and evenly.
> 7. Always apply diamond dressers at the center or slightly below the center, never above.

Figure 20-17. Wheel dressing operations should adhere to these safety measures.

holding the heel or lug—on the underside of the dresser's head—against the edge and not on top of the work rest (Figure 20-17).

Occasionally test wheels for balance, and rebalance them if necessary. Wheels that are too worn, or too out of balance, and that cannot be balanced by truing or dressing should be taken out of service.

Surface Grinders and Internal Grinders

Operating requirements for surface grinders (Figure 20-18) and internal grinders differ from those for other types of wheels. Insecurely clamped workpieces and unenergized magnetic chucks are common sources of injury to operators of surface grinders. Under these conditions, workpieces can be thrown with considerable force.

If the operator takes too deep a cut or traverses the table or wheel too quickly, the wheel can overheat at the rim and crack. Therefore, train operators and supervise them to clamp work tightly. They must always properly adjust and turn on a magnetic chuck before applying the wheel. They must also control the work's speed and depth.

Baffle plates on each end of a surface grinder are usually standard equipment. They should also include some provision for exhausting the grinding dust.

Internal grinders can often be guarded with an automatic positioning hood. This kind of hood covers the grinding wheel when it is in the retracted or idling position.

Grindstones

When using grindstones, follow the manufacturer's suggested running speeds and operating procedures. Never run stones of unknown composition or manufacture at more than 2,500 sfpm (12.5 m/s)—and ordinarily not more than 2,000 sfpm (10 m/s). The size and weight of grindstones require a stand that is rigidly constructed, heavy enough to hold the stone securely, and mounted on a solid foundation to withstand vibration.

Because grindstones are run wet, take all possible precautions to prevent slipping accidents near the stones. Use rough concrete or other slip-resistant floor material in grindstone-operating areas.

Carefully inspect grindstones for cracks and other defects as soon as they arrive from the manufacturer. Store those not to be used immediately in a dry, uniformly heated room where they will not be damaged.

Many grindstone failures result from faulty handling and incorrect mounting. Do not leave grindstones partially submerged in water. This practice causes an unbalanced stone that can break when rotated. Do not use wooden wedges on power-driven stones. Often, these wedges are too tightly driven or can become wet and swell. In either case, cracks start in the corners of a square center hole, radiate outward, and weaken the stone, causing ruptures to occur when operated at normal speeds.

After the stone has been centered, fill the central space about the arbor with lead or cement. Use double thicknesses of leather or rubber gaskets, rather than wood washers, wherever possible. If wood washers are used between the flanges and the stone, the washers should be ½ to 1 in. (13 to 25 mm) thick, and the flanges should be clamped in place by heavy nuts.

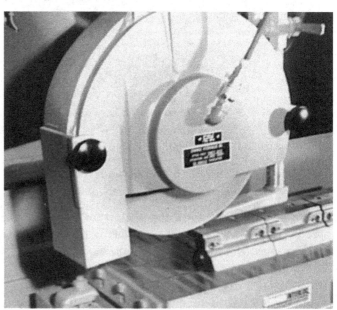

Figure 20-18. Guarded horizontal surface grinder.

To remove dust and wet spray or mist when dressing or operating power-driven grindstones (either wet or dry), provide an adequate exhaust system. Work rests should comply with the same requirements as those for grinding wheels.

Polishing Wheels and Buffing Wheels

Polishing wheels are either wood faced with leather or are made of stitched-together disks of canvas or similar material. A coat of emery or other abrasive is glued to the periphery of the wheels.

Buffing wheels are made of disks of felt, linen, or canvas. The periphery is given a coat of rouge, tripoli, or other mild abrasive.

The softness of the wheel—built up of linen, canvas, felt, or leather—is determined by the size of the flanges used: the larger the flange, the harder the surface. When large flanges are used, it is often necessary to soften the working surface of the built-up wheels to conform to the contour of the object being polished. A safe procedure for softening the working surface is to place the wheel on the floor or other flat surface and pound the edges of the wheel with a hammer or mallet. Do not place the wheel on the spindle with a file or other object held against it. The file could catch in the wheel and be thrown with such force that the operator or nearby workers are injured.

Mounting

Mount polishing wheels and buffing wheels on rigid and substantially constructed stands that are heavy enough for the wheels used. Mounting procedures for polishing wheels and buffing wheels are the same as those for grinding wheels.

Speed

The polishing and buffing peripheral speed range is from 3,000 to 7,000 sfpm (15 to 35 m/s), with 4,000 sfpm (20 m/s) in general use for most purposes. If the motors that drive the polishing and buffing wheels are equipped with adjustable speed controls, install the controls in a locked case and have only an authorized person change the speed.

Safeguards

Hood guards should be designed to prevent the operator's hands or clothing from catching on protruding nuts or the ends of spindles. If working conditions require a hood that does not give the needed protection, then use a spanner wrench to install smooth nuts over the spindle's ends. Never substitute a prick punch and hammer for a spanner wrench. Exhaust hoods should be designed to capture particles thrown off by the wheels.

Operators of polishing wheels and buffing wheels should not wear gloves. A glove can catch and drag the operator's hand against the wheel. Operators should not attempt to hold a small piece against the wheel with the bare hands. Small pieces being polished or buffed can frequently be held in a simple jig or fixture. Some operators use a piece of an old linen or canvas wheel for holding small pieces.

When applying rouge or tripoli to a revolving wheel, hold the side of the cake lightly against the wheel's periphery. If a stick is used, apply the side of the stick to the off side so, if thrown, it will fly away from the wheel.

Wire Brush Wheels

Wire brush wheels or, more commonly, scratch wheels are used to remove burrs, scale, sand, and other materials. These wheels are made of various kinds of protruding wires, with different thicknesses.

The same machine setup and conditions that apply to polishing and buffing wheels apply to wire brush wheels. Use flanges or nuts to hold scratch wheels rigidly in place. Do not exceed the speed recommended by the manufacturer. The hood on scratch wheels should enclose the wheel as completely as the nature of the work allows. The hood should also be adjustable so protection will not lessen as the diameter of the wheel decreases. The hood should also cover the exposed arbor ends. If not, install a smooth nut on them. Adjust the work rest to about $1/8$ in. (3 mm) from the wheel.

Personal protective equipment is especially important when operating scratch wheels because the wires tend to break off. It should be mandatory for operators to wear aprons made of leather, heavy canvas, or other heavy material; leather gloves; face shields; and goggles.

ANSI STANDARDS FOR METALWORKING MACHINERY

The American National Standards Institute has developed ANSI B11, Machine Tools Safety Package, which gathers together specific standards that delineate guidelines for the safe function and operation of a wide assortment of types of machine tools. These standards specify appropriate methods of design and utilization of a variety of machine tools in order to manage hazards and reduce risk.

The ANSI B11 Machine Tools Safety Package contains more than 30 standards, with each addressing a particular machine tool. These standards address safety principles for the use of turning machines, work-holding chucks, electrical discharge machines, drilling machines, pneumatic presses, milling machines, hydraulic power presses, mechanical power presses, power press brakes, shears, manual milling/drilling machines, grinding machines, metal sawing machines, and others. This package is supported with a guide to estimate, evaluate, and reduce risks associated with machine tools and the performance criteria for safeguarding them.

SUMMARY

- Injuries on machine tools are most often caused by unsafe work practices or incorrect procedures. Proper safeguarding on machines, good housekeeping in the work area, and good work habits help to reduce injuries and accidents.
- General safety rules should be founded on a policy to eliminate unsafe practices by workers and to train workers to operate machines safely.
- Turning machines should be properly shielded to prevent flying or spiral chips from striking or cutting operators. Screw machines should be shielded for excessive noise levels.
- Boring machine operators should be protected by shielding and guarding the drills and ensuring that drill bits are sharp and firmly attached to the drill head. Operators should never attempt to adjust work while the machine is running.
- Most accidents with milling machines occur when operators unload or make adjustments to the work while the machines are running. Machines in this class should be guarded to prevent injuries from contact or thrown chips and lubricant and have emergency shutoff switches.
- Only qualified electricians should maintain or hook up the electrical system of electrical discharge machines. Workers must maintain proper oil levels, make sure the area is adequately ventilated, prevent buildup of discharge gases, and be well trained in fire-extinguishing procedures.
- Injuries from planers and shapers generally result from contact with projections on the workpiece or projecting bolts or brackets. To avoid accidents, the cutting edge should be guarded, and operators should secure the tool or workpiece.
- Most accidents on slotters result from workers catching their fingers between the tool and workpiece. Broaches should be safeguarded through supplementary controls, which should be shielded if they can be tripped by any part of the worker's body.
- Surface and internal grinders require different operating methods than do other grinding wheels. Injuries often occur when workpieces are thrown from the grinder. To prevent injuries, workpieces must be securely clamped and magnetic chucks energized.
- Grindstones, polishing and buffing wheels, and wire brush wheels must be operated at the manufacturer's suggested running speeds, be frequently inspected, and be mounted and handled correctly.

REFERENCES

American National Standards Institute, 11 West 42nd Street, New York, NY 10036.

Accuracy of Engine and Tool Room Lathes, B5.16 (R2002).

Machine Tools Safety Package, B11.

Markings for Identifying Grinding Wheels and Other Bonded Adhesives, B74.13–1990 (R2007).

Milling Machine Arbor Assemblies, B5.47–1972 (R2013).

Milling Machines, B5.45 (R1998).

National Electrical Safety Code, ANSI C2–2012.

Rotary Table Surface Grinding Machines, B5.44 (R1998).

Safety Requirements for Gear and Spline Cutting Machines, B11.11–2001 (R2012).

Safety Requirements for Iron Workers, B11.5–1988 (R2008).

Safety Requirements for Machine Tools Using Lasers for Processing Materials, B11.21–2012.

Safety Requirements for Machining Centers and Automatic Numerically Controlled Milling, Drilling and Boring Machines, B11.23–2002 (R2012).

Safety Requirements for Manual Drilling, Milling, and Boring Machines, with or without Automatic Control, B11.8–2001 (R2012).

Safety Requirements for Manual Turning Machines with or without Automatic Control, B11.6–2001 (R2012).

Safety Requirements for Metal Sawing Machines, B11.10–2003 (R2009).

Safety Requirements for the Construction, Care, and Use of Grinding Machines, B11.9–2010.

Safety Requirements for the Construction, Care, and Use of Single- and Multiple-Spindle Automatic Screw/Bar and Chucking Machines, B11.13–2007.

Safety Requirements for Transfer Machines, B11.24–2002 (R2012).

Safety Requirements for Turning Centers and Automatic Numerically Controlled Turning Machines, B11.22–2002 (R2012).

Safety Requirements for Workplace Walking/Working Surfaces and Their Access; Workplace, Floor, Wall and Roof Openings; Stairs and Guardrails Systems, ANSI A1264.1–2007.

Specifications for Shapes and Sizes of Grinding Wheels, and Shapes, Sizes, and Identification of Mounted Wheels, B74.2–1992 (R2003).

Spindle Noses and Tool Shanks for Horizontal Boring Machines, B5.40–1970 (R2013).

The Use, Care, and Protection of Abrasive Wheels, B7.1 (R2010).

BS EN ISO 28881:2013, Machine Tools, Safety, Electro-Discharge Machines.

National Fire Protection Association, 1 Batterymarch Park, Quincy, MA 02269.

Electrical Safety Requirements for Employee Workplaces, NFPA 70E, 2015.

Electrical Standard for Industrial Machinery, NFPA 79, 2015.

National Electrical Code, NFPA 70, 2014.

National Machine Tool Builders Association (NMTBA), now known as the Association for Manufacturing Technology, 7901 Westpark Drive, McLean, VA 22102.

National Safety Council, 1121 Spring Lake Dr., Itasca, IL 60143-3201.

Hearing Conservation in the Workplace: A Practical Guide. 1992.

Safeguarding Concepts Illustrated. 7th ed. 2002.

Occupational Safety and Health Data Sheets:

Engine Lathes, 12304–0264, 1992.

Fire-Resistant, Water-in-Oil Emulsion Hydraulic Fluids, 12304–0543, 1990.

Gear-Hobbing Machines, 12304–0362, 1991.

Horizontal Metal Boring Mills, 12304–0269, 1991.

Manganese, 12304–0306, 1990.

Metal Planers, 12304–0383, 1991.

Metal Shapers, 12304–0216, 1991.

Metal-Working Drill Presses, 12304–0335, 1991.

Metal-Working Milling Machines, 12304–0364, 1991.

Portable Grinders, 12304–0583, 1990.

Vertical Boring Mills, 12304–0347, 1991.

Zinc and Zinc Oxide, 12304–0267, 1991.

Zirconium Powder, 12304–0729, 1991.

United Abrasive Manufacturers Association, 30200 Detroit Road, Cleveland, OH 44145.

Safety Recommendations for Grinding Wheel Operations, 1991.

U.S. Department of Health and Human Services, National Institute for Occupational Safety and Health, Division of Technical Services, 4676 Columbia Parkway, Cincinnati, OH 45226.

Control of Exposure to Metalworking Fluids, DHEW (NIOSH) Publication 78–165.

Health and Safety Guide for the Screw Machine Products Industry, DHEW (NIOSH) Publication 76–165.

U.S. Department of Labor, Francis Perkins Building, 200 Constitution Ave. NW, Washington DC 20210.

The Control of Hazardous Energy (Lockout/Tagout), 29 CFR 1910.147.

REVIEW QUESTIONS

1. What two factors result in fewer accidents when operating machine tools?
   a.
   b.
2. Identify one of the major causes of accidents from machine tools, especially drilling equipment.
3. List three preventive measures that will help operators run engine lathes safely.
   a.
   b.
   c.
4. What are six common causes of injury in boring mill operations?
   a.
   b.
   c.
   d.
   e.
   f.
5. Operators should do what four things after the electrical discharge machine (EDM) has been hooked up?
   a.
   b.
   c.
   d.

6. Injuries from shapers and planers frequently result from contact with projections on the workpiece or with projecting bolts or brackets, especially when the table is being adjusted _____.
7. What type of machine shapes material by bringing it into contact with a rotating abrasive wheel or disk?
   a. planing
   b. boring
   c. grinding
   d. milling
8. Specifications for operation of grinding machines and construction of guards and safety devices are in what code?
9. What effect do cold and wetness have on grinding wheels?
10. Many grindstone failures result from faulty handling and incorrect _____.

Working with Hot and Cold Metals

Gary A. Higbee, EMBA, CSP

Philip E. Hagan, JD, MBA, MPH, ARM, CIH, CET, CHMM, CHCM, CHSP, CEM

PART 1—COLD FORMING OF METALS

Power Presses
Primary Operations (Blanking) ▶ Secondary Operations

Point-of-Operation Safeguarding
Guards ▶ Point-of-Operation Safeguarding Devices
▶ Safeguarding Full-Revolution Clutch Power Presses
▶ Safeguarding Partial-Revolution Clutch Power Presses

Auxiliary Mechanisms
Feeding and Extracting Tools ▶ Foot Control and
Shielding for Protection ▶ Single-Stroke Attachments

Feeding and Ejecting Mechanisms
Primary Operations ▶ Secondary Operations
▶ Semiautomatic Feeds ▶ Ejector Mechanisms

Kick Presses
Types of Injuries ▶ General Precautions
▶ Guards ▶ Eye Protection ▶ Fatigue and Strain
▶ Maintenance

Electrical Controls on Power Presses
Installation ▶ Safeguarding Exclusions for Two-Hand
Systems ▶ Construction Features for All Two-Hand
Tripping Systems ▶ Construction Features for Two-
Hand Control Systems ▶ Stroking Control Systems'
Component Failure Protection ▶ Interlocks for Two-
Hand Control Systems ▶ Control-Circuit Voltage and
Ground Protection for Two-Hand Control Systems
▶ Foot Operation of Two-Hand Control Systems

Power Press Setup and Die Removal
Transferring Dies Safely ▶ Procedure for Setting Dies
▶ Removing Dies

Inspection and Maintenance
Troubleshooting with Power On ▶ Maintenance

Metal Shears
Power Squaring Shears ▶ Alligator Shears

Production System
Planning the Production System ▶ Power Press Brakes
▶ Mechanical Press Brakes ▶ Hydraulic Press Brakes
▶ General-Purpose Press Brakes ▶ Special-Purpose
Press Brakes ▶ Tooling ▶ Responsibility for Guarding and
Safeguarding

PART 2—HOT WORKING OF METALS

Health Hazards in Foundries
Hazardous Materials ▶ Medical Program ▶ Personnel
Facilities

Work Environment in Foundries
Housekeeping ▶ Floor Loading ▶ Ventilation ▶ Noise
Control ▶ Lighting ▶ Inspection and Maintenance ▶ Fire
Protection ▶ Facility Structures ▶ Compressed Air Hoses

Materials Handling in Foundries
Handling Sand, Coal, and Coke ▶ Ladles ▶ Hoists
and Cranes ▶ Conveyors ▶ Scrap Breakers
▶ Storage ▶ Slag Disposal

621

Cupolas

Charging ▶ Charging Floor ▶ Carbon Monoxide ▶ Blast Gates ▶ Tapping Out ▶ Dropping the Cupola's Bottom Doors ▶ Repairing Linings

Crucibles

Storing ▶ Annealing Process ▶ Charging ▶ Handling ▶ Crucible Furnaces

Ovens

Gas-Fired Ovens ▶ Ventilation ▶ Inspection

Foundry Production Equipment

Sand Mills ▶ Dough Mixers ▶ Sand Cutters ▶ Sifters ▶ Molds and Cores ▶ Molding Machines ▶ Core-Blowing Machines ▶ Flasks ▶ Sandblast Rooms ▶ Tumbling Barrels ▶ Shakeout Machines

Cleaning and Finishing Foundry Products

Magnesium Grinding ▶ Chipping ▶ Welding ▶ Power Presses

Forging Hammers

Open-Frame Hammers ▶ Gravity-Drop Hammers ▶ Steam Hammers and Air Hammers ▶ Hazards of

Forging Hammers ▶ Guarding ▶ Safety Props ▶ Hand Tools ▶ Die Keys ▶ Design of Dies ▶ Setup and Removal of Dies ▶ Safe Operating Practices ▶ Personal Protection ▶ Maintenance and Inspection ▶ Lead Casts

Forging Upsetters

Design of Dies ▶ Setup and Removal of Dies ▶ Inspection and Maintenance ▶ Auxiliary Equipment

Forging Presses

Basic Precautions ▶ Die Setting ▶ Maintenance

Other Forging Equipment

Nondestructive Testing

Magnetic Particle Inspection ▶ Penetrant Inspection ▶ Ultrasonic Methods ▶ Triboelectric Method ▶ Electromagnetic Tests ▶ Radiography

Summary

References

Review Questions

This chapter covers both cold forming and hot working of materials. Although these two areas of manufacturing have some processes in common, they often require different safety systems. These safety systems are not only necessary for safe performance of each operation, but they also help with improved quality and productivity. Cold forming of materials will be addressed first, followed by hot working of materials. To properly understand this subject, it is important to know the meanings of various terms used in this chapter. Definitions can be found in Appendix 3, Glossary.

PART 1
COLD FORMING OF METALS

Fundamental to many products is the operation known as *cold forming*: the forming of blank metal into a different shape. A variety of machines are used for this operation, depending on the type of metal and the end product. The most common machines, power presses, metal shears, and press brakes are discussed in this part of the chapter, along with the following topics:

- primary/secondary operation of power presses

- safeguarding devices
- auxiliary tools and attachments
- feeding and ejecting mechanisms
- kick-press hazards, safety precautions, and maintenance
- electrical control operations, construction, and safeguarding
- safe procedures for die press setup and removal
- maintenance and troubleshooting
- metal shears
- guarding and safeguarding press brakes.

POWER PRESSES

Power presses, with different dies attached, perform many types of metalworking operations that produce a variety of products. Because power press machines are so versatile, safeguarding the point of operation depends on (1) the die or tool component; (2) the type of press selected to power the die; and (3) the method selected to insert materials, to remove parts, and to dispose of scrap. (See Chapter 6, Safeguarding, in this volume.)

When designing the die or tool component, consider the following items:

- material used

- configuration of the finished part
- method of feeding the material
- method of removing parts
- method of scrap disposal
- ways to reduce noise.

The method of feeding the material may be manual, semi-automatic, or automatic. Scrap disposal may be manual or automatic, and recycling should be considered.

The die component's design is very important because it determines the physical characteristics of the press, such as the tonnage required, dimensions of the die's space, and the speed and stroke of the press component.

Safety of power presses depends on (1) adequately safeguarding the point of operation, (2) properly training press operators, and (3) enforcing safe working practices. Setup and maintenance personnel must be trained to ensure their safety while working in or around a press.

Primary Operations (Blanking)

Power press operations fall into two basic categories: primary operations and secondary operations. A primary operation is one in which stock material is processed to produce suitably sized and shaped flat blanks. Generally, these blanks require more forming and shaping, which takes place during a secondary operation.

Primary operations are easier to guard than secondary operations at the point of operation. The use of flat material permits construction or adjustment of the guards so the opening is only large enough for material to pass through the guard into the die. The trailing edge of strips can often be processed by pulling the stock through the die with the scrap skeleton or by pushing the stock through the die with the leading edge of the next strip. If this is not possible, bend the guard inward to meet the die at the point of the stock's entry. In this way, the next strip is fully used to advance the processed strip manually. If the guard cannot be customized in this way, scrap the balance of the material unless other appropriate point-of-operation protection is used.

The remaining hazard of primary operations occurs during die setup or repair when guards, necessary for the operator's safety, have been removed. Use the following suggestions to avoid this hazard:

- On partial-revolution clutch presses, operate the machine in the inch or single-stroke mode, using two-hand buttons to stroke the press when the dies are being tested. Do not repair or modify work in the die when the machine is capable of being stroked. Provide protection by using interlocked safety blocks and energy isolation procedures (see Chapter 6, Safeguarding, in this volume).
- On full-revolution and partial-revolution clutch machines, stop the flywheel. In addition, for partial-revolution clutch machines, shut off the control and the motor when work is to be done in the dies.
- If foot controls are used for production, disconnect and remove them from the area while work in the die is being performed. Before turning on the main motor to start a full-revolution clutch press after the work in the die has been performed, and at any other time, be sure that the slide is adjusted correctly and that everything has been removed from the die, including the operator's hands. This precaution is required in case the foot pedal is accidentally tripped while the machine is shut down. In such a case, the press would stroke immediately when the motor is turned on.

Stock Material for Primary Operations

The following materials are used in primary operations:

- *Coiled material.* Stock material initially in coiled form is frequently fed directly into a press that, with one or more dies, produces a finished part. Coiled material is sometimes fed into a cut-to-length line that produces strips. It can also be fed into a blanking line where shaped and sized blanks are produced for use in secondary operations.
- *Strip material.* Cut-to-length line is processed as stock material for subsequent primary operation or a combination primary and secondary operation.
- *Scrap and drop-off material.* Select scrap and drop-off material is a valuable source of primary material for producing some products. The material extending from the die varies in length and position with each stroke; this forces the operator to constantly reposition his or her hands. All die hazards, pinch points, and guide posts, as well as the point of operation at which blanking is performed, are potential hazards. To protect the operator, use a die with an attached guard that covers every hazard and allows only enough of an opening for material to pass through. This guard should also allow the operator to position the material close to the die for maximum yield. Such an operation requires constant ejection of parts and freedom of movement of the material within the die. Such a setup makes it absolutely unnecessary to remove guards except for sharpening dies.

Material-Handling Hazards

Hazards exist when handling the sharp edge or burr on strips, coiled stock, or scrap stock. A material-handling hazard is frequently encountered when unstrapping stock. The sudden release of a strap on a bundle of strips or on a restrained coil can injure any part of the body, including the eyes. When handling stock material, use gloves, eye protection, and arm gauntlets.

Feeding of Stock Material into Machines

Observe the following suggestions when feeding various stock material into machines:

Coil Feeds. When using coiled stock, place it into an uncoiler that will pay out, on demand, the required length of stock. Some uncoilers are powered to maintain a free loop of material from which the powered rolls on roll feeds or gripper jaws on gripper feeds pull a measured amount of stock. Other uncoilers are simple reels from which material is pulled directly. Heavy coils, fast feeding, and payout of long strips require powered uncoilers to keep the pulling effort to a minimum.

Strip Feeds. Although strip material is usually fed directly into the die by hand, it can be fed by machine. Although a roll feed can be used, gripper feeds are generally preferred. Gripper feeds can be automatically supplied with strips from an unstacker, or each strip can be manually loaded by an operator.

Blank Feeds. Blanks are usually inserted by hand into the die. However, blanks can be and frequently are loaded by robots.

Hand-Fed Material. Material is sometimes hand-fed into the die from a coil or from a stack of strips. In each case, material can be fed through a guard opening that is small enough so that the operator's hands cannot enter the point of operation. This type of operation is easy to safeguard by proper die-guard design with part ejection.

Feed-Machine Hazards. Pinch points, crush points on any type of feeder, and nip points at the in-running side of rolls or gripper jaws are feed-machine hazards to be guarded. Keep openings for material large enough only for free movement of stock and too small to permit fingers to enter (Figure 6–7 in Chapter 6, Safeguarding, in this volume). To guard in-running points of rolls, it is sometimes necessary to use bell-mouthed guides up to the rolls for threading in the end of the stock.

Part and Scrap Ejection

Design the removal of parts and scrap into the operation so that there is never any accumulation within the die. Accumulation of parts and scrap makes guards difficult to use and discourages their use. Efficient, reliable ejection of parts and scrap is an important consideration in a fully guarded die because it eliminates access to the point of operation. At the first evidence of ejection failure, shut down and lock out the operation to resolve the problem. It is frequently necessary to provide inclined chutes and/or to incline the press, if possible, to ensure ejection of parts and scraps.

The die's design should permit material to be removed or retracted from the side in which it enters, and the side at which it exits. Sometimes this requires beveled edges at both sides of component parts of the die so that edges in the material do not hang up on edges within the die. This would prevent the withdrawal or backup of the strip or scrap.

In the event that a die is used to recover usable drop-off material and scrap, the material should be fully mobile in the die without any hang up. Consequently, all edges in the die should be beveled. On some dies, it may be necessary to use supporting steel runners within the die.

Secondary Operations

Many secondary power press operations are adaptable for various feeding methods, including manual. With adequate safeguarding, the feeding operator will not allow the primary operator to reach into the danger area.

Gravity Feed

One such method uses gravity feed in which the part is placed on a chute and slides into the lower die (Figure 21–1). Provide pins, gauges, and stock guides to ensure that parts nest in the proper position. Open back, inclinable presses, because they can be inclined, can use gravity feed. The part can be placed into the die by gravity and ejected from the back of the press by gravity or air. Even if the opening in the chute is small enough to prevent a hand from entering the die area, install a full-barrier guard to

Figure 21–1. There are advantages to inclining a press. The fixed-barrier guard is simple and economical to make. This operation allows either hand feed or automatic gravity feed direct from the blanking operation (via the chute).

prevent operators from reaching into the danger area. If the opening is large enough to allow the entry of a hand, use safeguarding devices such as type-A or type-B movable barriers, as defined in American National Standards Institute (ANSI) B11.1–2001, Safety Requirements for the Construction, Care, and Use of Mechanical Power Presses; two-hand controls; presence-sensing devices; or pull-back devices. These devices are explained in Point-of-Operation Safeguarding Devices, later in this chapter.

Use various adaptations of gravity feeds to ensure that only one part is placed in the die at a time, such as through the use of a single-piece feeder. When oil or other lubricant on the parts causes the parts to stick in the chute or slide, install wire or metal rods in the chute. They reduce friction and allow the parts to slide into the die without sticking.

Follow or Push Feed

A common type of semiautomatic feeding is the follow or push feed. This feed allows the operator to push parts that are on a tray into the die by pushing one part into the preceding part or using a stick to push the part into the die. This keeps the operator's hands well out of the danger area. This die is easily guarded with a die-enclosure guard. Even irregularly shaped parts can frequently be fed with a push feed to keep the operator's hands out of the danger area.

Magazine Feeds

Adding a magazine on the push feed can increase the rate of production as well as lessen the manual handling of blanks at the press operation. Blanks with various shapes and sizes are adaptable to magazine feeds. Magazine slide feeds can be made nearly automatic in operation by powering the slide with mechanical devices attached to the press ram or crankshaft. The operator then only needs to keep the magazine filled with blanks to feed the press. Other ways of providing power movement of slide feeds include air or hydraulic cylinder power that is controlled by solenoid valves timed with the press's cycle.

Hand Tools

Where parts must be placed and/or removed from the die manually, hand tools, such as soft metal pliers, tongs, tweezers, and suction cups, provide effective ways to keep the operator's hands out of the danger area (Figure 21–2). It is sometimes necessary to provide clearance holes or slots for hand tools to assist the operator when inserting or removing the workpieces. Providing clearance makes the job of grasping the parts safer and more efficient. Hand tools are not, by themselves, a means of safeguarding. For the operator's safety, install a barrier guard or some type of safeguarding device.

Cleaning Dies and Clearing Jammed Parts

Consider that these activities are going to happen during any press run, and plan how to do them safely. Use cleaning and prying tools that keep hands out of the point of operation. If safeguards will need to be removed, have a lockout procedure or a well-planned alternative if lockout is not feasible.

Tool Identification

Figure 21–2 shows 34 simple, safe hand tools—all can be made in a shop, and all can save the hands and fingers of power press operators.

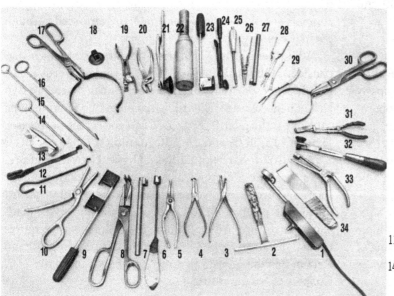

1. 110-V electric magnet for picking up sheet metal
2. steel or brass pusher
3. pliers with extra-long handles
4. pliers with adapters for grasping vertical edges
5. pliers with long handles and long grips
6. Alnico magnet on a stick
7. fiber stick with Alnico magnet
8. pliers with adapters for grasping vertical edges
9. fiber stick with Alnico magnet
10. pistol-grip pliers
11–12. push stick
13. sheet-edge gripper
14–16. hooks with 90-degree bend
17. pliers with adapter for grasping vertical edges
18. vacuum cup
19–20. pliers with high-pressure grip
21. releasable vacuum gripper
22. cylindrical holding tool
23. fiber stick with magnet
24. releasable vacuum gripper
25. hook with 90-degree bend
26. normally closed pliers
27. fiber stick with Alnico magnet
28. pliers with adapters for grasping vertical edges
29. long-nosed pliers
30. pliers with adapters for grasping vertical edges
31. high-pressure pliers
32. vacuum cup with handle
33. adjustable-handled pliers with Alnico magnet
34. push stick

Figure 21–2. A sample of 34 "mechanical hands" that permit a NO HANDS IN DIE operation.

Figure 21-3. An adjustable barrier fits any size die used. This barrier is especially practical for short-run jobs.

The tools shown include pliers up to 12 in. (30 cm) long, which permit a hand to be kept out of the danger zone. Other pliers are designed to grasp a vertical flange with bent jaws, to pick up thin pieces, or to hold material in work. Tools also include vacuum cups for handling sheet metal at slitting and shearing machines, permanent magnets and electromagnets, pliers with magnets, steel hooks, and steel or brass pusher sticks.

Not only do tools save workers' fingers and hands, but they also contribute to speed of operation. Studies show that it takes 1.4 seconds to load a press with 12-in. (30-cm) pliers compared with 1.8 seconds by hand.

Safeguarding

Base a power press's safeguarding program on an evaluation of the specific problems. Formulate a definite company policy to cover (1) use of guards or devices for all operations, (2) consideration of the safety factor in new operations, and (3) enforcement of safe operating standards. (See Chapter 6, Safeguarding, in this volume.) Such a program should include adhering to ANSI B11.1 and to applicable regulatory standards.

Many press rooms are small, with only a foreman, a die setter, and a few press operators. In such a shop, a simple program consisting of personal supervision, the guarding of some dies, and the provision of proper safeguarding devices and hand tools will do.

Large power press shops that are divided into departments under various supervisors with several die setters usually have split responsibilities. These shops should develop companywide safeguarding standards. These standards should apply to and be followed by all groups concerned with power presses: supervisors, operators, maintenance personnel, and electricians.

The number and productivity of the dies are major factors in deciding the number of point-of-operation guards to be installed. First, guard high-production, long-run dies individually. Consider adjustable barriers for safeguarding only for dies used infrequently and for short-run jobs (Figure 21-3). If a shop has short runs, and thus many die changes each day, or uses dies owned by other manufacturers, install safeguarding devices for secondary operations. Install adjustable-barrier guards for primary operations.

Other factors that require attention for the safe operation of power presses include the following:
- layout of machines
- machine and aisle space
- light and visibility
- containers for handling scrap and processed parts
- effective preventive maintenance program.

POINT-OF-OPERATION SAFEGUARDING

Safeguarding the point of operation means protecting operating personnel, including helpers, after the dies have been installed, tested, and operated and are ready for production. Protection of the die setter, who must have access to the point of operation while setting dies, is discussed later in this chapter (see Power Press Setup and Die Removal). Companies must also protect passersby from power press hazards. Although engineering controls should be used to protect passersby, if not feasible, then administrative controls should be clearly outlined and enforced through company policy.

When determining point-of-operation safeguarding, consider all hazards in the die's space that may crush, cut, punch, sever, or otherwise injure personnel. There are two basic categories of safeguarding the point of operation. One category is the guard; the second category is the device (see Figure 6-13, in Chapter 6, Safeguarding).

A guard, or barrier guard, is a physical barrier that prevents access to a die's hazard during a production run of successive cycles. If a barrier allows access to a die's hazard during the production run, it is not a guard; it is an inadequate enclosure. An inadequate enclosure always requires using a device to form an acceptable safeguarding system. Recommended guard openings are shown in Figure 6-7, Chapter 6, Safeguarding.

A safeguarding device controls access to the point of operation. Devices can be divided into three types: (1)

press-controlling devices, (2) operator-controlling devices, and (3) devices that control both the operator and the power press. Some devices require the use of enclosures to form an acceptable guarding system.

Guards

There are four main types of guards for safeguarding power presses at the point of operation: fixed die-enclosure guards, fixed-barrier guards, interlocked press-barrier guards, and adjustable-barrier guards.

Fixed Die-Enclosure Guards

Fixed die-enclosure guards provide the most complete protection for the operator because the die is completely enclosed and the guard is a permanent part of it. Die enclosures can be used on many types of press operations to prevent operators from putting their hands into the point of operation. Die enclosures are attached to the die's shoe, or stripper, in a fixed position. They are designed so that the operator's hands cannot reach over, under, through, or around the guard into the point of operation (Figures 21–4 and 21–5).

Fixed-Barrier Guards

Fixed-barrier guards, when used, should be attached to the frame of the press or to the bolster plate.

Interlocked Press-Barrier Guards

Barrier guards can be designed with a pivoting, sliding, or removable section to allow ready access to the die. The pivoting or sliding section interlocks with the press's clutch control to prevent the machine from operating when the section is open.

This type of interlocked fixed-barrier guard is used successfully on automatic presses where the point of operation must occasionally be exposed to relieve jams. As a further safety measure, if the interlock fails to function, use hand tools or picks to relieve jams, or support the slide with safety blocks. Do not use the pivoting, sliding, or removable section for feeding on single-stroke applications because the simple interlocking is too easily bypassed.

Adjustable-Barrier Guards

When a die-enclosure guard or fixed-barrier guard would take too much time to complete or would be impractical, or both, provide an adjustable-barrier guard on each press. This type of guard prevents the operator's hands from entering the point of operation (Figure 21–3).

Adjustable-barrier guards are available commercially or may be made in the facility. They are attached to the frame of the press and have adjustable front and side sections for dies of almost any size. They are especially practical for short-run jobs.

This type of guard is usually constructed of rod stock or perforated metal. Interlock any pivoting or sliding sections, used for occasional access, with the press's control for maximum safety.

Unless feeding or ejection is automatic, it may be necessary to leave an opening in the barrier to insert a tool to remove a part from the die. The opening should not be wide enough to allow a hand or finger to extend into the point of operation.

In the case where the whole guard is removed to change dies or to adjust a job, the guard is usually not interlocked with the press's control. When adjusting sections of an adjustable-barrier guard, be sure that die setters are instructed to follow the dimensions for permissible openings given in Figure 6–7, Chapter 6, Safeguarding. Never allow the operator to make changes in the adjustments without the supervisor's approval. Also, be sure that appli-

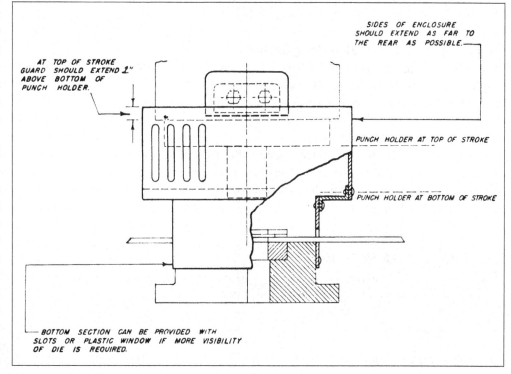

Figure 21–4. This die-enclosure guard for strip or coil stock shows the vertical clearances required. *(Courtesy Liberty Mutual Insurance Co.)*

21 Working with Hot and Cold Metals

Figure 21-5. The slide feed allows loading of the die outside the danger zone. The permanent plastic barrier guard permits full visibility of the operation. Note the separation of the guard at the top to permit die maintenance. Overlap of the guard at the separation eliminates shear hazard during travel of the slide. *(Courtesy Allis-Chalmers Manufacturing Co.)*

Figure 21-6. A movable-barrier device protects the operator by enclosing the point of operation before a press stroke can be initiated.

cable regulatory guidelines are also followed: 29 CFR 1910, Subpart O—Machinery and Machine Guarding.

Point-of-Operation Safeguarding Devices

There are several types of point-of-operation safeguarding devices for power presses. All of the following devices control access to the point of operation.

Type-A Movable-Barrier Devices

The type-A movable-barrier device protects the operator by enclosing the point of operation before a press stroke begins. This barrier remains in the enclosed position until motion of the slide has ceased at the top of stroke (Figure 21-6).

Type-B Movable-Barrier Devices

The type-B movable-barrier device also protects the operator by enclosing the point of operation before a press stroke begins. This device makes it impossible to reach into the point of operation prior to the die's closing during the downstroke.

Two-Hand Tripping Devices

Two-hand tripping devices are used on a full-revolution clutch machine.

Two-Hand Control Devices

Two-hand control devices must meet the requirements for protected buttons for two-hand tripping devices (Figure 21-7). They must also be arranged so concurrent pressure from both hands is required during a substantial part of the die-closing portion of the stroke. Where more than one operator is required on a press, two-hand controls must be provided for each operator.

Pull-Back Devices

Pull-backs, or pull-outs, when properly used, adjusted, supervised, and maintained, always remove the operator's hands from the point of operation as the slide descends. This type of device is usually limited to secondary operations and jobs where the operator can remain at the feeding position. When more than one operator is required on a press, provide pull-backs for each operator (Figure 21-8).

Restraints

Restraints, or holdouts, prevent the operator from reaching into the point of operation at any time. This is achieved by providing wrist straps and firmly anchored restraint cords or cables. When more than one operator is required on a press, provide restraints for each operator (Figure 21-8).

21 Auxiliary Mechanisms   629

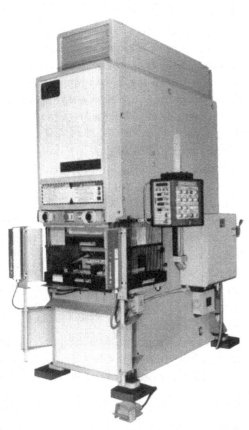

Figure 21-7. This hydraulic open back stationary (OBS) press has protected buttons for two-hand control, hairpin side guards, and a photoelectric presence-sensing device at the front. Notice the interlocked slide blocks mounted at the rear of the side frame. *(Courtesy Cincinnati Inc.)*

Presence-Sensing Devices

Presence-sensing devices are designed, constructed, and arranged to create a sensing field or area. The clutch's control of the press deactivates when an operator's hand or any other body part is within such field or area (Figure 21-7).

Use a presence-sensing device only on a partial-revolution clutch machine.

Always supplement these devices with hand tools or feeding and ejection devices (Figure 21-2). In that way, operators need not place their hands in the point of operation. These devices, however, cannot protect against repeats.

Safeguarding Full-Revolution Clutch Power Presses

Power presses where stroking cannot be interrupted—that is, controlled—during the closing or opening of the stroke cannot use a press-controlling device. However, a properly installed two-hand tripping device can serve as a means of both press activation and safeguarding.

Figure 21-8. Properly set pull-back devices will remove the operator(s) hands from the area near the dies to a safe distance by the time the ram has traveled through the first half of its downstroke. *(Courtesy Cincinnati Inc.)*

Safeguarding Partial-Revolution Clutch Power Presses

On a press whose stroke can be interrupted during the closing or opening of the stroke, use a press-controlling device, as well as an operator-controlling device.

AUXILIARY MECHANISMS

Many auxiliary mechanisms to protect operators of power presses are available. They can be purchased or designed and made in a plant's shop.

Feeding and Extracting Tools

A variety of special tools has been developed for feeding and extracting parts. These tools are made of soft metal, aluminum, or magnesium—some are magnetized.

Special tools for feeding and extracting provide protection only if operators always use them. Operators should never substitute them for proper safeguarding, however.

One way to make sure tools are used properly is to provide convenient storage for them in the workplace. This encourages operators to keep tools in their correct place.

Foot Control and Shielding for Protection

Foot operation, by itself, is only an operating means and provides no protection for the operator unless hands and other body parts are kept a safe distance from the point of operation. When the operator is within reach of the point of operation, safeguard the point of operation. When barrier guards are used on foot-controlled presses, always ensure that they are replaced after die changes, cleaning, or jam clearing. Shield and position a foot-operated mechanism (1) to control its accessibility and (2) to afford protection under circumstances, such as die setting, when safeguarding is necessarily removed from the point of operation.

Use shields that are large enough to allow room for operators to place a foot in the operating position without undue fatigue and without striking or scraping their leg against sharp edges of the cover.

Foot Controls for Full-Revolution Clutch Presses

A mechanical foot pedal trips the clutch whenever it is depressed. It remains tripped until a stroke is made. Control access to the pedal to prevent it from being turned on unintentionally.

An electric foot switch can be turned off to reduce the possibility of accidentally starting the press during off periods. However, during operating periods, the switch is more easily turned on than a pedal. Therefore, shield the switch so an operator's foot cannot accidentally turn the machine on.

Foot Controls for Partial-Revolution Clutch Presses

When using foot switches with clutches able to be disengaged anywhere in the stroke, control the foot switches through an antirepeat circuit. This prevents a second stroke if the foot switch is depressed for the full cycle so that release of the foot switch on the downstroke will cause the ram to stop.

Single-Stroke Attachments

A press with a full-revolution clutch should have a single-stroke attachment that disconnects the pedal or operating lever after each stroke. A single-stroke spring device should depend on spring action only if it is a compression spring encased in a close-fitting tube or closely wound on a rod. Of necessity, a single-stroke device is made inoperative when the press is used for continuous operation.

FEEDING AND EJECTING MECHANISMS

A significant percentage of power press injuries, such as puncture wounds, lacerations, strains, and amputation or crushing of the hands or feet, occur when work is handled. Thus, any mechanism that eliminates handling of work should reduce exposure to those hazards, especially in operations that place the operator's hands in a danger zone.

Primary Operations

Primary operations are normally conducted with random lengths of strip stock or coiled stock. Strip stock is usually fed by hand. Coiled stock can be fed either by hand or by means of a roll or hitch feed.

Automatic roll feeds are often used on continuous operations of blanking from strip stock. Enclose gears on the feed rolls, and also guard the in-running nip point of the roll feeds. These precautions are especially important when operators wear gloves or have long hair, even with hairnets or close-fitting caps.

Automatic push or pull feeds are similar to roll feeds and are used mostly in blanking larger pieces. When coiled stock is used with a reel and roll feed or hitch feed, a feed table is generally unnecessary. When coiled stock is used with a stock reel but not with a roll feed, a feed table helps the operator backgauge the stock and feed it to the die with minimum effort.

When feeding strip stock by hand, a feed table eliminates unnecessary motion and reduces operator fatigue. Adjust the feed table to the height at which the operator can work with minimum effort.

Use oiling rolls or pressure guns instead of a paintbrush system to lubricate strip or roll stock. Provide automatic or manual-controlled pressure guns to lubricate the punch and the die.

Secondary Operations

In secondary operations, selecting feeding and ejection methods that keep the operator's hands out of the point of operation is more difficult. When automatic feeding methods are used, safeguard the point of operation by limiting the slide stroke to ¼ in. (0.6 cm) or less. Also, provide a guard or device, according to ANSI B11.1.

Semiautomatic feeding mechanisms place the workpiece under the slide by a mechanical device that requires the attention of the operator at each stroke of the press. Such feeds have a distinct advantage because the operator is not required to reach into the point-of-operation area to feed the press. This feeding method also permits complete enclosure of the die. Semiautomatic feeding may not be adaptable for certain blanking operations or for nesting of odd-shaped pieces.

Semiautomatic Feeds

The six principal types of semiautomatic feeds are chute (both gravity and follow), slide or push, plunger, sliding die, dial, and revolving die.

Chute Feed

Of the six semiautomatic methods, the chute feed is probably the most widely used. It is a horizontal or inclined chute into which each workpiece is placed by hand. The workpieces then slide or are pushed, one at a time, into position in the lower die. The entire die may be enclosed because it is unnecessary for operators to place their hands in the point-of-operation area if automatic ejection is also provided (Figure 21–9).

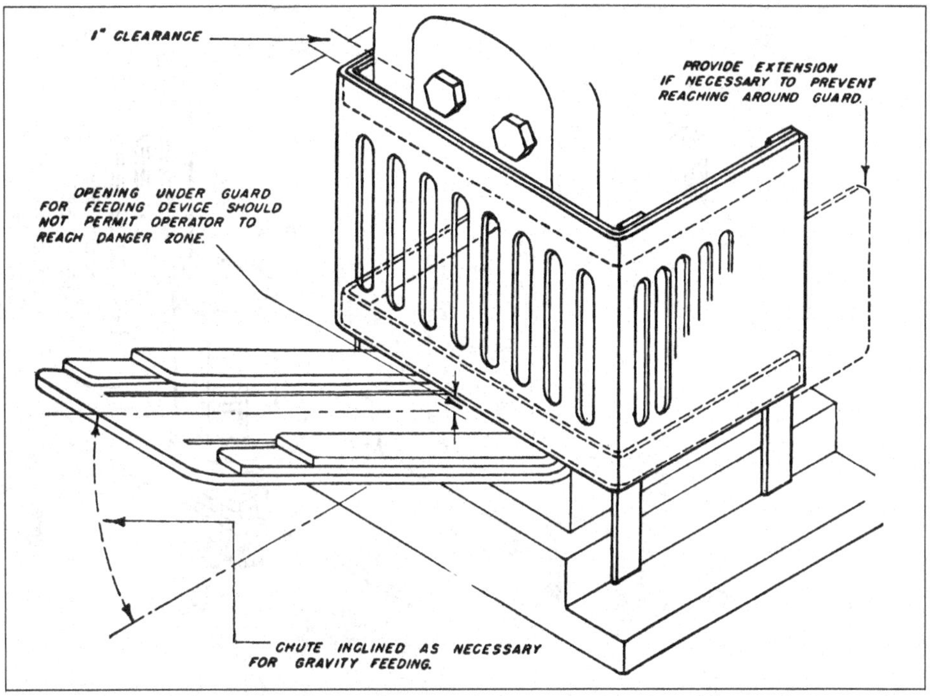

Slide or Push Feed

A variation of the chute feed, the slide or push feed is combined with magazines and plungers. The workpieces are stacked in the magazine. As each piece reaches the bottom, it is pushed into the die by means of a hand-operated plunger (Figure 21–10).

Plunger Feed

A variation of the push feed, the plunger feed may be semiautomatic or manual in operation. The semiautomatic plunger feed is a magazine or chute in which blanks or partly formed workpieces are placed. The blanks or pieces are fed, one at a time, by a mechanical plunger or other device that pushes them under the slide (Figure 21–11).

Sliding-Die Feed

With a sliding-die feed, the die is pulled toward the operator for safe feeding and then pushed into position under the slide, or ram, for the downstroke. The die may be moved in and out by hand or by means of a foot lever. Regardless of how the die is moved, interlock it with the press to prevent tripping when the die is out of alignment with the slide (Figure 21–12). Provide stops to prevent the die from being inadvertently pulled out of the slides.

Dial Feed

With a dial feed, two or more nests arranged in the form of a dial revolve with each stroke of the press so that the operator can safely feed the machine. The part to be processed is placed in a nest on the dial that is positioned in front of the die. The dial is indexed with each upstroke of the press to deliver the next nested part into the die.

When using a dial feed, enclose the point-of-operation area. Motion of the dial when operators' hands are free may create a hazard. An idle station may be required beyond the load station to prevent injury at the station that enters the guard.

Revolving Die Feed

Operating on the same principle as the dial feed, the revolving die feed may consist of two dies or multiple dies.

Figure 21–9. This die-enclosure guard with an inclined chute for gravity feeding can also be used on an incline press with a straight chute. *(Courtesy Liberty Mutual Insurance Co.)*

Ejector Mechanisms

Properly designed and installed ejector mechanisms will eliminate many common hazards. Ejector mechanisms can automatically clear the press faster than humans can and with greater safety. In every facility, however, there will be times when parts will be moved manually. Not only is manual removal less efficient, it causes greater operator fatigue and exposes operators to more injuries than does mechanical removal. Follow the basic principle of keeping the operator out of danger zones even if parts are removed by hand. Properly guard the danger zones and have operators use suitable hand tools.

When designing ejection mechanisms, two problems must be solved: how to strip the piece from the punch or the die and how to eject the piece from the die and the press to a container or conveyor. These problems can be solved in many ways. In some cases, a single mechanism performs both operations. In other cases, a separate mechanism is used for each operation.

Air Jets

Single or multiple air jets can be used for effective removal of small pieces. Air ejection can be combined with other means of mechanical release or with gravity removal. Anchor all jets securely to direct the air stream effectively and to prevent jet tubes from shifting into the die's working area.

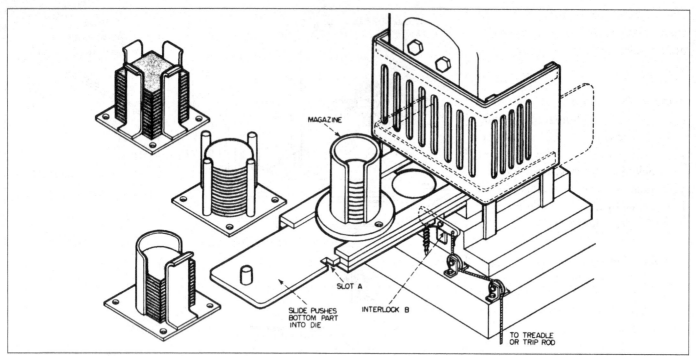

Figure 21-10. This magazine on a push feed enables the operator to catch every press stroke. Slot A in the pusher must be in alignment with interlock B before the press can be tripped. This feature ensures proper positioning of the part in the die. *(Courtesy Liberty Mutual Insurance Co.)*

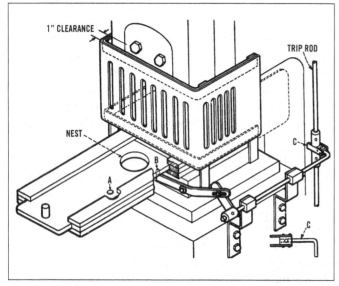

Figure 21-11. In this manually operated plunger feed, note that the press cannot be tripped until the part is pushed to the next location. At this point, hole A is directly over tapered pin B; it can rise and release yoke C so that the press can be tripped.

Figure 21-12. In this adjustable barrier with a sliding die, a locating mechanism is provided on the slide bolster to locate and lock the die slide in alignment with the punch holder. An interlock should be provided to prevent tripping of the press until the die slide is in the proper location.

There are always hazards from flying particles in press operations, and the addition of air ejection increases these hazards. Using compressed air also increases the overall noise level. Be sure all operators wear eye and ear protection at all times.

Pneumatically Powered Cylinders

Pneumatically powered cylinders—operating sweeps or kickout pins timed with the upstroke of the press—are more effective than air jets for removing large and heavy pieces. For safety, when clearing jams or doing tryout operations, equip the pneumatic equipment with a valve to shut off the flow of air and to bleed any residual air pressure between the valve and the cylinders.

Clamp and Pan Shuttle Extractors

Clamp and pan shuttle extractors are used on presses of all sizes. Pivoting or straight-line clamps grip the part and remove it from the die's area to a pallet, bin, or conveyor for further processing. A pan shuttle (1) catches the piece as it is stripped from the upper die by knockout pins or other means and (2) removes it from the die's area.

Some extractors are mechanically powered by connection with the press's slide or shafting. However, independently powered extractors are used more frequently because they are more versatile and can be used with several presses. Interlock every independently powered extractor with the press's control circuit so the press cannot operate unless the extractor is in the home or out position.

Steel Chutes

Simple sheet steel chutes are generally provided along with air ejection to guide the parts into a container. Use chutes or slides for a controlled movement of pieces to subsequent operations. Eliminate sharp bends and inside projections (which may cause parts to pile up) from enclosed chutes.

Elevators and Conveyors

Use elevators and power and gravity conveyors to transport pieces of work to containers or to subsequent press operations.

Support all conveyors for stability on the sides and ends. Enclose the drive mechanisms of power conveyors and elevators as completely as possible. Eliminate pinch points formed by moving belts at pulleys or at conveyors' structures.

KICK PRESSES

All power presses have counterparts in presses that are hand or foot powered. Many shops still use kick presses, foot shears, hand folders, and hand rollers. Because hazards do exist even though the operator is the source of power, guard such presses.

Safeguard foot-operated presses with an interlocking tripping mechanism. This mechanism requires the operator to use both hands simultaneously to release the slide's head before the pedal can be used.

Types of Injuries

The principal types of injuries resulting from unsafe operation of kick presses include the following:

- finger amputations and finger punctures from unguarded points of operation
- fatigue and abdominal strain resulting either from pressure required to perform the work or from improper posture
- strains from lifting materials
- eye injuries caused by small flying particles.

General Precautions

When a new kick press is installed or an old one is moved to a new location, see that it is securely fastened to the floor or the bench.

Provide good general lighting. In some cases, local lighting may be required. Good visibility not only will increase production and reduce fatigue but also will lessen the possibility of injuries. To improve visibility, have the press painted in contrasting colors.

Guards

When a kick press is used for piercing and similar operations, guard the point of operation. Make a guard that fits around the punch from a small piece of perforated sheet metal, transparent plastic, or other material. The guard should allow enough room to insert the work into the press, but not enough to permit the fingers to come within the danger zone.

On some assembling jobs, when operators must place their hands under the punch, the point of operation cannot be guarded. In such cases, use a two-handed safety device, such as the ratchet mechanism shown in Figures 21–13 and 21–14.

To operate the kick press, the operator must use both hands. Should the operator release either hand, the pawls will engage the ratchet and stop the punch instantly. During the loading and unloading periods, the ram is locked in the top position.

Eye Protection

Operators on kick-press operations should wear eye protection. When hard materials, such as brass and spring steel, are being shaved or notched, small particles thrown off with great force may cause serious injuries. Require goggles or spectacles with full side shields or face shields on such operations.

Fatigue and Strain

Because most kick presses are operated from a sitting position, provide a seat or chair that enables the operators to work with minimum fatigue. Operators should not have to make a long reach with their foot to perform the operation. In addition, arrange the seat so operators do not sit in a cramped position. The seat should have a comfortable backrest, and the base of the chair's legs should be wide

Figure 21-13. In this ratchet mechanism for kick presses, the operator must use both hands to disengage the pawls from the ratchet. The ratchet teeth are cut into the plate attached to the slide head. Below the ratchet teeth, the plate tapers to the width of the slide and is secured to the slide by two hex-head bolts and washers. *(Courtesy General Electric Co., Pittsfield, MA.)*

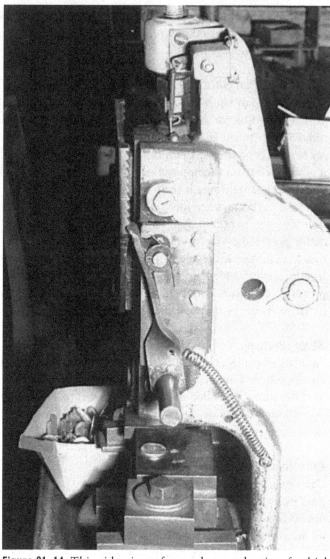

Figure 21-14. This side view of a ratchet mechanism for kick presses shows the simple ratchet principle of the guard. *(Courtesy General Electric Co., Pittsfield, MA.)*

enough so the chair will not tip over easily. (See Chapter 16, Ergonomics Yesterday, Today, and Tomorrow, in the *Administration & Programs* volume.)

Proper adjustment of the counterweight and stop also lessens operator fatigue. Proper adjustment also reduces the travel distance of the pedal and results in a balanced operation.

If the pedal is worn smooth, the operator's foot can readily slip, thus causing a strained or sprained ankle. Placing flanges on both ends of the pedal and rubber or abrasive surfacing on the pedal will minimize this hazard.

There is also a possibility that the operator may suffer abdominal strain from operating a kick press. Occasionally, heavy operations are performed that cause unusual fatigue.

In such cases, permit the operator to alternate kick-press work with light work. Allow only physically able employees to work a kick press. Do not permit any employee who is known to have weak abdominal walls, such as from a recent operation, to operate a kick press.

Lifting heavy containers filled with parts may result in severe muscle strain, especially in continuous operations where fatigue may make lifting increasingly difficult. To eliminate this hazard, install conveyor belts to carry material along the bench to the press. If the volume of work does not warrant the use of a conveyor, fasten brackets to the press at truck height. This makes it possible to slide containers from trucks to the press and back again with little effort.

Maintenance

Check the kick press and its safeguard frequently. Lubricate and adjust all working parts. Replace immediately parts of the press or the guard that show wear or defects.

ELECTRICAL CONTROLS ON POWER PRESSES

Properly designed, applied, and installed electrical controls are an important element in press safety. This is particularly true when a two-hand control device or a two-hand tripping device is used for point-of-operation safeguarding.

Electrical controls may range from a simple power disconnect switch and a motor starter on a small full-revolution clutch press to an extensive system on a large partial-revolution clutch machine.

Installation

The following information applies to all installations of two-hand controls and two-hand tripping devices, regardless of the means of point-of-operation safeguarding. Build controls for all machines in accord with NFPA 79, Electrical Standard for Industrial Machinery, as well as recognized industry standards. Installation of machines should conform to NFPA 70, National Electrical Code, and any applicable local codes. The purpose of these codes and standards is to safeguard personnel and equipment. Consult ANSI B11.3–2002, Safety Requirements for the Construction, Care, and Use of Power Press Brakes, for specific application to press brakes.

Control Panel

The machine's control panel should contain the main disconnect switch, motor starters, clutch-control relays, and control transformer. Locate this panel on or immediately next to the machine it serves, but out of the operator's way during normal operation. Properly identify the control panel and have it easy to reach for maintenance (Figure 21–15).

Control Components

Ground cases or frames of all control components to the press's frame. Mounting bolts are satisfactory for grounding, provided that all paint and dirt are removed from the joint's surfaces before assembly. Ground movable control components not secured to the press's frame by means of a grounding conductor in the connecting cable or conduit.

Rotary Limit Switch

The rotary limit switch is a vital element in the control system (Figure 21–16). Mechanical or electrical failure of

Figure 21–15. Presses must also be made safe for electricians, millwrights, pipe fitters, and others who must maintain them in a safe condition. Identification of components of this machine control panel is given in the chart on the inside of the door. Components are readily accessible for maintenance.

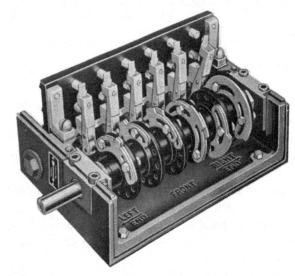

Figure 21–16. All rotary cam limit switches should be the type in which the contacts are forced open by the actuating cam.

this switch may cause the press to malfunction. The switching-driving mechanism—including all couplings, gears, sprockets, and associated parts—should be securely assembled and be of a rugged design.

Flywheel Brake

When using a flywheel brake, electrically interlock it with the main drive's starter. This prevents simultaneous operation.

Safety Blocks

Provide safety blocks with interlock plugs. They disconnect the clutch- and motor-control circuits when blocks are removed from their storage pockets.

Magnetic Air Valves

For actuating the clutch, use three-way, normally closed, poppet-designed magnetic air valves. Never use spool valves for this purpose. Their pistons slide in close-fitting sleeves and can easily stick because of dirt, corrosion, or improper lubrication.

Safeguarding Exclusions for Two-Hand Systems

In the continuous mode of stroking, two-hand control and two-hand tripping do not qualify as point-of-operation safeguarding devices. Use a die-enclosure guard or a fixed-barrier guard to keep operators' hands out of the point of operation. In that way, the operators can use the hand controls to initiate strokes.

If a foot control is connected in place of any hand-operated button or buttons, do not use the operator's foot control for point-of-operation protection—one or more hands are free to enter the die's area at any time. Instead, use a die-enclosure guard; a fixed-barrier guard; or a type-A movable-barrier device for point-of-operation safeguarding on power presses.

Construction Features for All Two-Hand Tripping Systems

The individual operator's hand controls—RUN buttons or other two-hand tripping mechanisms—for tripping the clutch must meet all of the following requirements:

- Protect each control against unintentional operation. Use protective rings around RUN buttons or suitable barriers.
- Arrange each pair of controls so both hands are required. This means that buttons must be far enough apart and located so that a hand and elbow of the same arm, a knee, or any other part of the body cannot be used instead of both hands.
- When two-hand tripping is used on a multiple-operator press, provide each operator with a separate set of two buttons or other hand mechanism. Bypass buttons, not needed for particular operations, should be provided in complete sets of two—not by individual buttons. Supervisors should monitor bypassing.
- When two-hand tripping is used as a device for safeguarding the point of operation, fix buttons in position with a safe distance between the point of operation and each button. This distance should be great enough to prevent the operator from moving either hand from a button into a point of operation prior to the die's closure. Minimum safe distances are shown in Table 21–A and are based on the following formula:

$$D_m = 63 \text{ in. (160 cm) per second} \times T_m$$

where:

D_m = minimum safety distance in inches; 63 in. (160 cm) per second
= common hand movement speed (higher speeds have been measured)

T_m = maximum time in seconds for die closure after the press has been tripped.

TABLE 21–A. Minimum Safety Distance for Two-Hand Trip between Hand Controls and Nearest Point-of-Operation Hazard

Press Speed in Strokes per Minute	Time in Seconds (T_m) for One Revolution of Crankshaft	Minimum Safety Distance (D_m) in inches					
		Full-Revolution Clutch Engaging Points per Revolution					Part-Revolution Clutch
		1	2	3	4	14	Infinite
30	2.0	189	126	105	95	72	63
45	1.33	126	85	70	63	48	42
60	1.0	95	63	53	48	36	32
75	0.8	76	51	42	38	29	26
90	0.67	63	42	35	32	24	21
105	0.57	54	36	30	27	21	18
120	0.5	48	32	27	24	18	16
135	0.44	42	28	24	21	16	14
150	0.4	38	26	21	19	15	13
165	0.36	35	23	19	18	13	12
180	0.33	32	21	18	16	12	11
210	0.29	27	18	15	14	11	9
240	0.25	24	16	14	12	9	8

With adjustable-speed drive, use slowest speed. (Based on OSHA regulations 1910.217(c)(3)(viii) of December 3, 1974, for presses with full- or partial-revolution clutches.)

For full-revolution clutch presses with only one engaging point, T_m is equal to the time necessary for 1½ revolutions of the crankshaft. For full-revolution clutch presses with more than one engaging point, T_m is calculated as follows:

$$T_m = \frac{1}{2} + \frac{1}{\left(\begin{array}{c} \text{number of} \\ \text{engaging points} \\ \text{per revolution} \end{array}\right)} \times \left(\begin{array}{c} \text{time in seconds} \\ \text{necessary to complete} \\ \text{one revolution of} \\ \text{the crankshaft} \end{array}\right)$$

The two-hand tripping system must include the following features:

- concurrent operation of all RUN buttons or other hand mechanisms
- an antirepeat feature for single-stroke operations

- electrical control circuit and valve-coil voltages not exceeding a nominal 120 V AC or 240 V DC isolated from higher voltages
- ground-fault detection to locate an accidental ground, and circuitry to prevent false operation due to an accidental ground
- features to minimize failures that would cause an unintended stroke
- a control system to prevent actuation of the clutch, on a multiple-operator press, if all operating stations are bypassed.

Construction Features for Two-Hand Control Systems

The individual operator's hand controls—RUN buttons or other hand-control mechanisms—for engaging the clutch/brake must meet all of the following requirements:

- Protect each control against unintentional operation. Use protective rings around RUN buttons or suitable barriers.
- Arrange each pair of controls so both hands are required. This means that buttons must be far enough apart and located so that a hand and elbow of the same arm, a knee, or any other part of the body cannot be used instead of both hands.
- When a two-hand control is used on a multiple-operator press, provide each operator with a separate set of two buttons or other hand mechanisms. Bypass buttons, not needed for particular operations, should be provided in complete sets of two—not by individual buttons. Supervisors should monitor bypassing.
- When a two-hand control is used with a partial-revolution clutch press as a device for safeguarding the point of operation, fix buttons in position with a safe distance between the point of operation and each button. This distance should be great enough to prevent moving either hand from a button into a point of operation while the slide (ram) is in motion during the die-closing portion of a stroke. Minimum safe distances shown in Table 21–B are based on the following formula:

$D_s = 63$ in. (160 cm) per second $\times T_s$

where:

D_s = minimum safety distance in inches; 63 in. (160 cm) per second
= possible hand movement speed

T_s = longest stopping time of the slide (ram) in seconds, usually measured at approximately the 90-degree position of the crankshaft's rotation.

The stopping time includes the operating time of the press's control system, air exhaust from the clutch/brake, and braking time of the slide. Stopping-time measuring units are available, or a stopwatch can sometimes be used. If the minimum safe distance is not practical for production, use a guard or another safeguarding device.

TABLE 21–B. Minimum Safety Distances for Two-Hand Control between Hand Controls and Nearest Point-of-Operation Hazard

Stopping Time of Slide in Seconds (T_s) at 90° Point of Stroke	Minimum Safety Distance (D_s) in Inches
0.100	7
0.125	8
0.150	10
0.175	11
0.200	13
0.225	15
0.250	16
0.275	18
0.300	19
0.325	21
0.350	22
0.375	24
0.400	26
0.450	29
0.500	32
0.600	38
0.700	45
0.800	51
0.900	57
1.000	63

With adjustable-speed drive, use the stopping time for highest speed. (Based on OSHA regulations 1910.217(c)(3)(viii) of December 3, 1974, for presses with partial-revolution clutches only.)

Stop Control

Provide a red STOP button with the clutch/brake control system. Momentary operation of this button must immediately deactivate the clutch and apply the brake. This button must override all other controls. To put the clutch in motion again must require use of the RUN buttons or other operating mechanisms. Make a STOP button available to each operator. At least one STOP button must be connected and operative, regardless of whether a hand or foot control is being used.

Stroking Selector

Supply a means of selecting OFF, INCH, SINGLE STROKE, or CONTINUOUS (when continuous stroking is provided) with the clutch/brake control. Also include other positions when required for special features. Keep hand–foot selection separate, however, from the stroking selector.

Fixing the selection must be monitored by a supervisor. Key-operated selectors are commonly used to lock selectors in position. Additional precautions are needed for continuous stroking.

Inch Control

Never use the INCH mode of operation for production. There is generally no antirepeat circuitry, automatic top stop, or drive-motor interlock as there is in the single-stroke mode.

To prevent exposure of personnel within the point of operation, require two-hand controls to move the clutch. Use a single-hand button only if it is protected from being started accidentally and is located so that the worker cannot reach into the point of operation while pressing the button. Never use a foot control for inching.

Single-Stroke Control

In addition to the RUN button features listed in this section, provide RUN-button holding time and interrupted stroke protection for the control system. Always try to use dies and feeding methods that do not require placing hands in the point of operation at any time. Control reliability and brake monitoring are also required with hands-in-dies operations.

Holding Time. If all RUN buttons are held down until the dies have closed sufficiently for safe release of the buttons, the slide must stop before a hand can enter the point of operation. The control should provide for an adjustment of the rotary limit switch's contacts to bypass the RUN buttons at a safe point in the stroke.

Interrupted Stroke Protection. Before an interrupted stroke can be resumed, again by pressing all RUN buttons, all such buttons should be released. The purpose of interrupted stroke protection is to minimize the possibility of accidentally restarting the slide after a button has been released during the holding time.

Control Reliability. Construct the control system so that any control failure does not interfere with the normal stopping action when required. However, the initiation of successive strokes should not occur. Design the control system so a failure is easily detectable.

Brake System Monitoring. Incorporate a brake monitor into the control. It should prevent activation of a successive stroke if the stopping time or distance deteriorates to a point where the safe distance no longer complies with distances shown in Table 21–B. The monitoring action must take place during each cycle in single-stroke operations. The monitoring system should indicate any unsafe deterioration.

Automatic Single Stroke. The press control is given single-stroke actuating signals by an automatic feeding mechanism or other auxiliary equipment without action by an operator after initial start. Proper safeguarding is required for this type of automatic single-stroke operation because using only a two-hand control is not acceptable.

Continuous Control

The following three different types of continuous operations are used:

1. continuous—continuous stroking of the slide after initiation without the operator's controls being activated
2. maintained continuous—continuous stroking of the slide only as long as the operator's controls are activated
3. continuous on demand—continuous stroking of the slide after initiation without the operator's controls being activated, and with periodic stopping and automatic restarting controlled by auxiliary equipment.

All of the continuous modes of operation have special control requirements.

Multiple-Operator Machine

On a multiple-operator machine, the control system must prevent activation of the clutch/brake if all operator stations are bypassed. This is frequently called dummy-plug protection because supervised dummy plugs or key-operated selectors are generally used to bypass unneeded stations.

Stroking Control Systems' Component Failure Protection

In stroking control systems (clutch/brake-control circuits), incorporate design features to lessen the chance of failures that could cause an unintended stroke.

Design the controls for air-clutch machines to prevent significant increases in the normal stopping time if an operating valve's mechanism fails. They should also be designed to inhibit further operation if such failure does occur. A self-checking assembly of two-valve elements in a common housing is usually the best way to meet this requirement.

If a machine has separate clutch and brake systems that require individual valves connected to a common manifold, both valves should be of the self-checking, two-valve type. The protective valve arrangement is not needed on machines intended only for continuous operation with automatic feeding.

Interlocks for Two-Hand Control Systems

The clutch/brake control must automatically deactivate in case electrical power or proper air pressure is lost, or if the clutch is air operated. If the machine has air counterbalance cylinders, loss of proper counterbalance air pressure must also deactivate the control. Reactivation of the clutch must require (1) restoration of normal electrical and air supply and (2) use of the RUN buttons or other tripping means.

The control must include an automatic means to prevent initiation or continued activation of the single-stroke and continuous functions unless the drive's motor is energized and set forward. To meet this requirement, connect an auxiliary contact on the main drive's forward contactor in the single-stroke and continuous control circuits.

Control-Circuit Voltage and Ground Protection for Two-Hand Control Systems

The AC electrical control circuits and valve coils on all machines must be powered by not more than a nominal 120-V supply obtained from a transformer with an isolated secondary winding. Isolate voltages above 120, which may be needed for particular mechanisms, from any control component handled by an operator. All DC control circuits must be powered by not more than a nominal 240-V DC supply isolated from any higher voltage.

Protect all clutch/brake control circuits against the possibility of an accidental ground in the control circuit. This could cause false operation of the machine (Figures 21–17a and b).

Foot Operation of Two-Hand Control Systems

If foot operation is an alternative to two-hand operation, provide point-of-operation protection and the required controls. A die-enclosure guard; fixed-barrier guard; or a type-A movable-barrier device is recommended for safeguarding presses.

POWER PRESS SETUP AND DIE REMOVAL

Power press dies must remain rigidly accurate in spite of the pressure and stress they transmit during metal-stamping operations. As a result, they are usually heavy and difficult to handle. They vary in weight from a few pounds for small dies to several thousand pounds for large dies.

Handling, setting up, and removing these dies are hazardous unless operators use proper equipment and methods. Entrust these operations only to experienced setup personnel whom the supervisor has instructed in detail about safe procedures.

Transferring Dies Safely

Very light dies can be handled and carried manually. If proper die trucks are provided, setup personnel can generally handle dies weighing up to about 100 lb (45 kg) without using lifting apparatus (Figure 21–18). Die trucks should have elevating tables adjustable to the height of storage shelves and press bolster plates. When transporting dies,

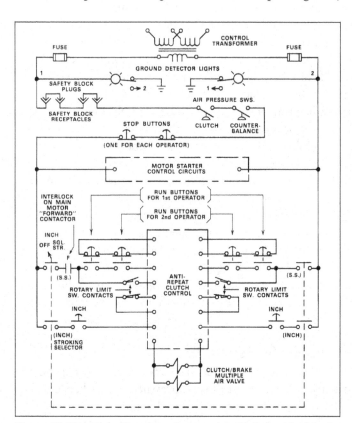

Figure 21–17a. This diagram shows a typical single-break, grounded press-control connection.

Figure 21–17b. This diagram shows a typical double-break, ungrounded press-control connection.

Figure 21-18. A die truck with adjustable height and an adjustable angle die table can match the angle of inclined presses. The truck is secured to the press before moving the die. The holes in the upper die shoe indicate the drilled-hole-and-pin method of lifting.

personnel should carry them at the truck's lowest elevation. Use rollers, balls, or windows mounted in the top surface of the table to aid in sliding the die on or off the truck.

Heavy dies require more equipment for safe handling. Because they are often lifted and moved by hoists, these dies should have tapped holes and eyebolts (or lifting hooks), drilled holes and pins, chain slots, cast lugs in lower shoes, or clamping lugs to ease hookup and transfer. For moving dies that are about 1,000 lb (450 kg) or heavier, use special die-handling power trucks. These trucks have special equipment for pushing or pulling the dies, including power winches, roller tables, and hydraulic or pneumatic clamps. To minimize the effect of uneven floor surfaces, use trucks with large-diameter wheels.

When transferring dies, place the truck close to the storage shelf or press. Adjust the table to the same height as the storage shelf or bolster plate. Then, either chock the wheels, lock the brake, and chain the truck to the press, or use some other method to prevent movement (Figure 21-18). Next, push the die off the truck onto the storage shelf or bolster plate. To bring a die onto a truck, engage a hook in the die so it cannot slip, and exert a steady pull—do not jerk or tug on the hook. Where lifting is needed to transfer the die, use a hoist and never lift higher than is necessary for minimum clearance. At no time should an employee have hands, feet, or another body part underneath a suspended load. Only the person in charge should give signals for movement.

Procedure for Setting Dies

Safe procedures for setting dies vary slightly and depend on the press's size. Most of the difference occurs because the slide on light presses can be moved manually by turning the flywheel or crankshaft, while on heavy presses it must be power operated. Use the following safe method for setting and removing dies for all presses. Special procedures are given for light presses and heavy presses.

1. Dismantle or disconnect a point-of-operation safety device only if it is absolutely necessary. (See Chapter 6, Safeguarding, in this volume, for lockout/tagout procedures.)
2. Clean off the bolster plate, preferably with a vacuum system or brush. Keep all bolt holes clear of all obstructions.
3. Check the die to make sure that it contains no chips, tools, or parts and that it is in good operating order. Then transfer the die from the truck to the press. (See Transferring Dies Safely, earlier in this section.)
4. Line up the die in the correct operating position, and remove the posts or blocks from under the slide. Then, lower the slide until it fits firmly against the top die. It is extremely important not to put too much pressure on dies. Tighten all bolts and clamps to secure the top half of the die. Bolt the die to the ram with bolts through holes in the upper die's shoe (Figure 21-19). On heavy presses, if air-cushion pads are used, keep the ram close to the top of the die until the die is properly seated on the cushion's pins.
5. Shim or block up the lower half of the die to the proper level; bolt and clamp it to the bolster plate. Bolting the die's shoe to the bolster produces the most secure die setup. If clamps must be used, block up their outer ends slightly higher than the die's surface on which their inner ends will rest. Clamp-fastening bolts should be closer to the die than to the block end of the clamp.
6. Check all bolts and clamps to see that they are tight and that dies are securely fastened in the press. Remove all tools and equipment from the dies, bolster plates, or other areas on the press.

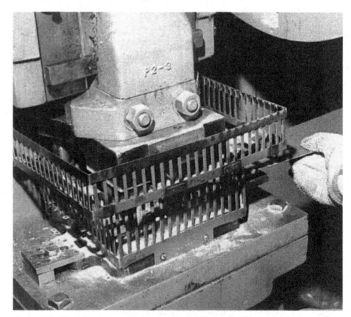

Figure 21-19. Mounting bolts pass directly through the ram to fasten the upper die securely. The lower die is clamped to the bolster.

7. Raise the ram to its highest point and block it in this position. On heavy presses, disconnect the power before proceeding further. Wipe out the die and remove the safety blocks. Replace the safety device and check it for adjustment and operation. Properly adjust pull-out devices, when used. This will minimize the chance of an accident occurring to an operator who has longer hands or arms than the previous operator.

8. Reconnect the power and try out several actual operations, using the proper stock. Make any necessary adjustments only after shutting off the power and blocking up the ram. After completing the adjustments, turn on the power and again try several actual operations on the press to ensure safe operation.

Light Presses

Disconnect or shut off the electric power (and/or that of other energy sources) and lock out the switch. Lower the slide to its lowest position by turning the flywheel or crankshaft. If a safety bar is used as a lever to turn the crankshaft, it should have a spring-and-collar arrangement that will prevent the safety bar from being accidentally left in the crankshaft. When physical access is necessary, block the slide with timber, metal blocks, or posts provided for this purpose. Equip blocks with an electrical receptacle plug to hold the circuit open when blocks are in the die. Thus, they will not be left in the press when power is applied.

Heavy Presses

Jog or inch the press to bring the slide down to its lowest position. Measure the clear height between the slide and the bolster plate. If this distance is not slightly greater than the height of the die in the closed position, adjust the slide until such clearance is ensured. Raise the slide, block it in position, shut off the power, and lock out the switch.

Removing Dies

Follow safe die-removal methods at all times. Although modifications may be necessary in special cases, the safe procedure is as follows:

1. Make sure that the working space is cleared of all stock, containers, tools, and other items.

2. Disconnect or shut off the power and lock out the switch. Turn the flywheel by hand or by the safety bar until the ram is at the bottom of the stroke. If the press cannot be turned over by hand, jog it under power, shut off the power, and lock out the switch.

3. Dismantle or disconnect the point-of-operation safety devices as required. Store the parts of the dismantled safety device so it can be reinstalled in good condition when the new die is in place.

4. Clean off the bolster plate, preferably with a vacuum system or brush.

5. If the die is to be operated with an air pad, shut off the air supply and open the release valve to permit the pins to go down. Also, shut off the air supply to the automatic blowout system used in the die.

6. Remove bolts and clamps holding the die in the press.

7. Make certain that the die is loose and that bolts, nuts, clamps, and other obstructions have been removed.

8. Raise the ram slowly—by hand on light presses and by jogging or inching under power on heavy presses—and make sure that the die does not hang in the slide.

9. Block the ram in its highest position. If power was used, shut off the power and lock out the switch.

10. Place the die truck close to the press, adjust the table to the same height as the lower bolster plate, and chock the wheels or set the truck's brake to prevent movement. To pull the die onto the truck, use a device engaged so the die cannot slip.

11. Inspect, repair, and protect dies before storing them for the next run. Also, inspect the pins and bushings. Store hardened dies and punches in the closed position with a piece of soft wood between the edges to protect them. Injuries have occurred when hardened dies were handled when open or partially open. A sudden jolt can cause the dies to close and pinch a hand or finger.

INSPECTION AND MAINTENANCE

The best safety program for power presses cannot succeed, nor can maximum production be met, without good inspection and maintenance of presses and their safeguards. Proper inspection, adjustment, and repair of power presses and related equipment can be done only by competent, thoroughly trained employees. Be sure that these employees are completely familiar with the construction and operation of the equipment for which they are responsible. Also, provide them with proper tools and equipment. Clear the work area of all personnel not directly involved in maintenance. Erect flashing warning lights or other barriers or barricades used to mark the temporary maintenance area.

Troubleshooting with Power On

When it is necessary to locate and define problems with the power on, the employee can work on power presses with guards removed or work within areas protected by barriers. However, such action should not place any body part in the path of any movable part of the press. A press may have to be stopped or locked out before removing a guard or barrier so that the press may subsequently be observed with power on. Ensure that a written policy is in place to address these situations, and ensure compliance with OSHA's lockout/tagout standard, 29 CFR 1910.147.

Power presses, like all machinery, are subject to wear, breakage, and malfunction. Therefore, to prevent costly accidents and repairs and to promote maximum production, inspect the entire machine and its related equipment periodically. Make the required adjustments and repairs. The type of press, its related equipment, and its usage should determine the frequency of inspections.

Regulatory standards may require that each press be inspected weekly to determine the condition of the clutch/brake mechanism, antirepeat feature, and single-stroke mechanism. Necessary maintenance or repair or both should be performed and completed before the press is operated. The weekly inspection is not required on those presses meeting the requirements for "hands-in-the-dies." Employers should maintain records of these inspections and the maintenance performed.

Set up a checklist that details the frequency of inspection and maintenance for each press. Such a checklist gives immediate knowledge of the press's condition and makes it easier to schedule production and avoid downtime due to equipment failures. These checklists need not be complicated but should provide inspection frequencies for the following items: frames, bearings, drives, electrical controls, rams or slides, clutches and brakes, cushions and springs, die carriages, lubrication, and guards and safeguarding devices.

Frames

Visually inspect the press's frame for cracks and broken parts. Check the fastenings of all brackets, guides, cylinders, covers, and other auxiliary parts. Check the tie-rods and nuts for fractures or stretching. Because the tie-rod's nuts on top of the press may fall if the tie-rod fractures, chain these nuts to the press's frame. A metal strap under each bottom tie-rod's nut will prevent the rod from dropping, should it fail. If a machine is bolted to a foundation, check hold-down fasteners for looseness and fractures.

Bearings

Check crankshaft, pinion-shaft, eccentric-gear, and toggle-link bearings for snug, nonrotating fit and for any loose caps or fastenings. Replace badly scored bearings. Check for proper lubrication.

Examine the slide or ram guide and gib surfaces for dirty or clogged lubricant grooves. Check for proper running clearances. These vary with the type of work. These materials require more clearance than bronze and steel or brass and steel because they have a greater tendency to "pick up." Tighten all screws wond locknuts to hold the setting.

Check antifriction bearings for proper lubrication. Overlubrication can cause swelling of oil seals and over-heating of the shaft. This can result in failure of bindings and shafts. Both this failure and the failure of flywheel bearings can cause unexpected descent of the ram. If lubrication of noisy flywheel bearings fails to silence bearings, replace bearings immediately.

Motor or Power Source and Drive

Properly adjust drive belts to prevent excess slippage or excess loads on the bearings and shaft, which could cause premature failure. Properly adjusted drive belts should slip slightly when the motor first starts but not when the press is operating. Check that all pins, slides, turnbuckles, jack screws, or other means of adjusting the motor are secure. Tighten all hold-down screws. Attach the motor to the press by a chain or wire rope for maximum safety. Inspect and lubricate the motor's shaft bearings.

Check gears for worn, pitted, and broken teeth and for proper lubrication. Check bores and shafting for worn keys and keyways.

Whenever a crankshaft, or shaft carrying the flywheel or clutch and brake, is removed from a press, inspect it for fatigue cracks. Some companies inspect all drive shafts once a year for cracks, bending, or deformation. Fatigue cracks can be detected by numerous methods such as ultrasonic, radiographic, magnetic, or dye-penetrant techniques.

Have a supervisor monitor the selection of turnover bar operations. Use a separate push button to activate the clutch. Activate the clutch only if the drive motor is deenergized. Also use turnover bars that are spring loaded to prevent the possibility of leaving the bar in the bar's hole.

Electrical Controls

Check all operating buttons for proper operation. All buttons must be depressed to start the press cycle and released at the end of the stroke. Holding time should be adequate for the operation and tooling involved.

Check for defective lamps in ground detector circuits where provided. Check ground connections on grounded controls or ground detector connections on ungrounded controls.

Check the physical condition of wiring, relays, rotary limit switch drives, pressure switches, valves, and other electrical and pneumatic devices. Follow manufacturers' recommendations for preventive maintenance.

Rams or Slides

Check the slide's structure visually for cracks, and check the die's mounting surface for evidence of overload or improper die-mounting damage. Inspect the slide's adjustment lock to be sure the die's setting can be maintained. On a machine with a motorized ram adjustment, check the motor for loose mounting bolts, loose drive chain or gears, excessive grease in the motor, and worn or frayed flexible electrical lead-in wires for motor control. Check that the slide's adjustment limit switches are operating properly.

Pin or chain knockout bars to prevent them from falling. Install finger guards along the top of the bar if needed.

If the ram is counterbalanced by springs, check them for breaks. If the ram is counterbalanced by air, check it for air leaks, air-line restrictions, correct operating pressure, loose piston rods, lubrication, and proper operation of pressure switches. Tighten all brackets. Note any rise in air pressure on the downstroke of the press. Periodically drain all surge tanks.

Visually check for fatigue cracks in the ram-adjusting screw and the connection. Check that the slide is securely fastened to the adjustment mechanism and that the connection cap is fastened to the connection. Look for evidence of rams being adjusted too high, with subsequent interference with the frame. All mechanical power presses are capable of producing an overload force several times the press's tonnage at the top of the stroke as well as at the bottom. Sudden failure of any of the parts that attach the ram to the crank may cause an equally sudden and dangerous dropping of the slide.

Maintenance

Clutches and Brakes

There are several types of clutch and brake units. Each requires special inspection and maintenance.

Dry Friction. All dry-friction clutch units are air engaged and spring released. All brake units, except constant-drag types, are spring set and air released. When the clutch and brake are combined into a single unit operated by a single air cylinder, only one unit can be engaged at a time due to the mechanical interlock. When the clutch is separated from the brake and each has its own air cylinder, both can be engaged at the same time. Prevent this from happening by either restricting the air flow into the clutch and out of the brake or by limit-switch and/or pressure-switch timing of two air valves. Check to see that both the clutch and brake are not engaged at the same time.

Check the unit for loose fasteners, broken parts, lubrication leaks, air leaks, faulty or loose wiring, excessive accumulation of particles on the friction lining, and broken springs. Replace springs that have changed in free length more than 5%.

During inspection, check the action of the clutch and brake, both at rest and in motion. Travel on the friction disks will indicate the amount of wear. Adjust the disks according to the manufacturer's instructions. If not adjusted properly, they can cause a malfunction. It is also important

to properly adjust the brake. There should be little or no coasting of the press's slide (ram) when the brake engages. The clutch and brake should operate smoothly and engage and disengage quickly. If the press is equipped with a brake-monitoring device, check it for proper operation according to the manufacturer's recommendations.

Check sliding surfaces, which keep parts in alignment, for excessive wear that might allow the parts to cock or wedge.

Inspection of the clutch and brake units will readily disclose leaks in the air-cylinder packings and in air glands. Check traps and strainers and clean them frequently with proper cleaning materials. Refill lubricators regularly with the type of oil recommended by the press's manufacturer.

Magnetically operated air valves should operate smoothly without sticking or leaking. Valves may stick because of dirt or scale in the air line. Inspect, clean, and repair valves according to the manufacturer's recommendations.

Electrical controls, although usually not part of the clutch and brake unit, affect the operation of the unit. Inspect push buttons, limit switches, relays, and contactors for excessive wear, broken springs, loose parts, loose or broken wires, bent magnetic-field surfaces, badly burned contacts, and dirt. Check circuit-grounding connections for continuity. Replace badly worn contacts. Specifically, inspect the rotary cam's limit-switch drive.

Oil-Wet Clutches. Oil-wet clutches are a newer type of air-actuated friction clutch. They have many of the same maintenance characteristics as the dry-friction clutch, with the following exceptions: (1) friction surfaces last much longer, (2) units are usually physically smaller because of higher rpm and may contain a set of planetary or other type of gears, and (3) the unit may require an oil cooler and pump. To prevent damage to the unit, lubricating oils should meet the manufacturer's requirements. These units may be serviced by qualified maintenance personnel.

Electric Clutch. The eddy-current electrical clutch has no friction surfaces to maintain. However, it does have slip rings and a special electrical control to maintain the torque and slip characteristics. Proper maintenance consists of lubricating bearings and taking care of electrical apparatus according to the manufacturer's recommendations.

Full-Revolution Clutch Presses. A full-revolution clutch is one that, when tripped, cannot be disengaged until it has completed its cycle. It is known by many names—such as pin, jaw, dog, positive, key, or spline—depending on the type and the manufacturer. Usually associated with this type of clutch is a drag brake on the crankshaft.

Typically, this clutch couples the flywheel to the press's crank by releasing the spring-loaded means of coupling, such as a pin, rolling key, or jaw arrangement. These are normally disengaged by a cam's action that extracts the engaging part, thus disconnecting the flywheel from driving the crank. The crank is held in its disengaged position by a braking system.

Examine the clutch for loose parts, worn pins, worn dogs, broken or weak springs, damaged lubrication seals, and excessive wear in the bearings. Replace worn or broken parts, and adjust the clutch to throw out at the top or just before the top center stroke. This adjustment will affect the brake's setting.

On some presses, the clutch is tripped by foot or hand levers or by levers with a spring return. Other methods of tripping the clutch include electric, air, or a combination of air and electric. Examine all elements of the clutch's trip. Sources of trouble include broken or weak springs; worn pins and bushings; loose fasteners; leaking air packings, connections, or valves; loose or broken wires; poor electrical contacts; and defective relays and limit switches. After replacing defective parts, readjust the tripping mechanism and check it for smooth operation.

Drag or Band Brake. The continuous band brake is used on a large majority of full-revolution clutch presses. If the brake is not set or operating properly, the press may repeat and cause a serious accident. Therefore, replace worn, glazed, or oil-soaked brake linings. Rivets should not project above the linings. Remove, inspect, and replace anchor and drag force–applying bolts, studs, springs, and other parts found defective.

Pneumatic Die Cushions and Springs
Examine pneumatic die cushions and springs for foreign or scrap material between the pressure pad and the bolster. Also check for faulty air packings, air leaks, lubrication, and fasteners on the supporting rods or plates.

Rolling-Bolster Die Carriages
Die carriages vary from small, two-position, manually moved die holders to large-capacity, eight-wheeled, four-directional holders. The latter are self-powered to change tracks.

The most common source of power for rolling-bolster die carriages is pneumatic; a less common power source is electricity. Inspect all hydraulic and air valves, hydraulic and air cylinders, and air or electric motors for normal operation. Inspect air-line filters and oilers. Inspect and lubricate all gears and bearings. Inspect all springs and latches for wear and breakage. Also inspect locating keys in the bed, retractable locating pins in the carriage, and locating pin holes in the floor, bed, and tracks.

Lubrication
Proper press and air-line lubrication is essential. Failures producing accidents and downtime are frequently directly traceable to either lack of lubrication or overlubrication. It is well known that lack of lubrication cannot be tolerated in machine parts. The results of overlubrication are less well known. Excess lubrication to flywheel and shaft bearings in the vicinity of dry-friction clutch and brake linings is undesirable. It causes a loss of work capacity, and prevents stopping in emergencies.

Overlubrication of air cylinders can lead to sluggish clutch and brake action. Improperly mounted clutch and brake surge tanks will accumulate oil and water. This results in increased clutch slippage and wear. For specific lubrication information, refer to the press manufacturer's service manual.

Guards and Safeguarding Devices
Cover all gearing, belting, or other drive parts that can be accidentally contacted. Many safeguarding devices are synchronized with the action of the press. Because most of these devices will go out of adjustment through wear and vibration, they require periodic checking.

Wire ropes, leather straps, or steel springs used as parts of safeguarding devices will in time need replacing because of wear. In such cases, use nothing but the proper replacement parts. Follow the manufacturer's recommendations for maintenance and adjustment.

Keep all guards and covers on a power press in place. Properly adjust them after completion of each inspection and any necessary repairs. Check each press for all modes of operation before releasing for production.

METAL SHEARS

The following sections cover safeguarding methods and safe operating procedures for power squaring shears and alligator shears.

Power Squaring Shears
Equip the power squaring shear with safeguarding that will (1) prevent operators from placing their hands into the point of operation, (2) prevent or stop the operation of the shear if any part of the operator's body approaches the point of operation, or (3) provide awareness to operating personnel upon entry into a hazardous area. Follow the safeguarding requirements in ANSI B11.4–2003, Safety Requirements for the Construction, Care, and Use of Shears. According to ANSI B11.4, the point of operation includes the area between the upper and lower blades and the area between the hold-down, clamping mechanism, and the shear table.

Safeguarding should allow operators to see clearly into the point of operation to position material for shearing to a scribed line. Removal of the safeguarding is normally

Figure 21–20. Material transfer conveyors are used at the rear of shears to eliminate the need for personnel to enter this area to remove the product. *(Courtesy Cincinnati Inc.)*

Figure 21–21. This large-capacity plate shear is safeguarded with an awareness barrier at the point of operation. *(Courtesy Cincinnati Inc.)*

necessary for changing or adjusting the blades. Reinstall the safeguarding when this is completed. If a guard is used, provide sufficient clearance to allow the material to be fed. Normally, a recognized guideline for clearance is double the metal thickness of the material being sheared.

The design of the fixed guard should meet the requirements of ANSI B11.4. (See Figure 1 and Table 1 from ANSI B11.4. This table provides dimensional guidelines for a guard that will prevent operators from placing their hands into the point of operation.) When it becomes impractical to adhere to the guarding dimensions given in Table 1, ANSI B11.4 suggests using an awareness barrier. The dimensions for this barrier are in Figure 2 and Table 2 of ANSI B11.4. The design should ensure that the barrier's movable sections are heavy enough so that operators would be aware of their hands entering the safeguarding. Operators should know, however, that this safeguarding may not prevent them from forcing their hands into the point of operation.

On shears with a throat, or end guard, provide a guard. It may be removed to provide for slitting material longer than the shear. However, it must be replaced when the slitting work is completed.

Provide a work chute or conveyor to discourage or eliminate the need for employees to be at the rear of the shear while it is being operated (Figure 21–20). Position no one at the rear of the shear (i.e., within the area of moving machine parts).

New shears should be manufactured to comply with the construction and safeguarding requirements of ANSI B11.4. Users should update machines already in the field to meet these requirements (Figure 21–21).

At the start of each shift, check the shear for the following items:

- safeguarding at the point of operation properly adjusted
- pinch-point guarding properly installed
- operator station working properly
- operating modes functioning properly
- ram starting and stopping properly
- warning plates clean and easy to read
- electrical wiring in good condition
- caution color coding in good condition and clearly visible
- auxiliary equipment checked and working properly
- hand tools and personal protective equipment in good order and readily available
- safety manuals or operator manuals available
- normal maintenance work completed.

Alligator Shears

Alligator shears perform a variety of cutoff operations. Their principal use is for cutting rods and bar stock to length.

Alligator shears can operate continuously. Therefore, under continuous operation, the operator must be trained to time movements with the opening and closing of the cutter. Because the machine is relatively simple and comparatively slow in its movement, this machine's hazards are often disregarded. Consequently, alligator shears are responsible for far more injuries than their inherent hazards or frequency of use warrants.

If possible, build a long bench to the right or left of the shear, depending on the type of machine. The material should slide along the bench and through the cutter. Because the ragged edges are hazardous to handle, use care in piling the material on the bench.

The wide variety of sizes and shapes of material to be cut makes it difficult at times to closely guard the point of

21 Working with Hot and Cold Metals

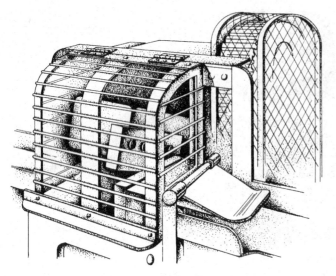

Figure 21-22. A well-designed guard for an alligator shear should permit easy maintenance and adjustment. Hinged section of the guard should be interlocked electrically to prevent shear operation if the hinged section is not in place. *(Courtesy Jones & Laughlin Steel Corp.)*

operation. However, installation of an adjustable guard can often be used. When it is, set it far enough from the knife area to prevent the fingers from entering this danger zone (Figure 21-22).

When stock size is such that the end held by operators may fly up and strike them, use hold-down guards or bars. They can be adjusted to fit any type of shear.

Keep material to be cut within the capacity of the machine. Do not attempt to cut hardened steel. Such action can result in damage to the machine and injury to the operator.

PRODUCTION SYSTEM

Planning the Production System

All parts of the system—power press brake, tools, feeding and safeguarding components, and operating personnel—must be brought together to perform any metalworking operation on a piece-part component. A human factors engineering approach should be used to provide for the most efficient and safest method of performing a piece-part bending operation.

The power press brake is the power component of such a system. Depending on the tooling component selected by the user, press brakes can bend, form, notch, punch, pierce, or perform other operations on the piece-part component. The piece-part component and the product being produced determine the feeding component of the production system.

Feeding can be either mechanical or manual. Included in this element of the system is removal of parts and scrap. Follow ANSI B11.3 and consult appropriate regulatory agencies along with this standard when working with press brakes.

The component that completes a functioning production system is the safeguarding component. Before selecting a suitable safeguarding component, the user should complete a thorough risk assessment, based on all the elements of the production system. Each new combination of production system elements requires that the user perform a new risk assessment to select a suitable safeguarding component.

A safe combination of components for one production system may not be a safe combination of components for another piece-part production system. Also, it may be necessary to change more than one component to provide a safe piece-part bending production system, once it is determined that a change must be made.

POWER PRESS BRAKES

Power press brakes have been classified into two basic categories: general-purpose press brakes and special-purpose press brakes. General-purpose press brakes, both mechanical and hydraulic, are operated by one individual with a single operating control station (Figure 21-23). Special-purpose press brakes include all other types hav-

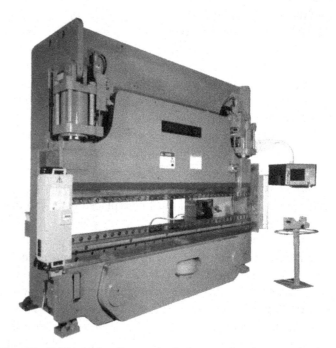

Figure 21-23. This hydraulic press brake has a pedestal-mounted palm button, operator control station, and photoelectric presence-sensing device for point-of-operation safeguarding. Note the computer control for programming the ram motion, backgauge, safeguarding device, and operator controls. *(Courtesy Cincinnati Inc.)*

ing mechanical, hydraulic, and other drive arrangements. A power press brake is also sometimes called a bending brake or a brake press. Over the years, its design has evolved from the hand or folder brake because of the need for a power machine with enough capacity to bend thick sheet and heavy plate products. The primary function of the press brake is to cold-form angles, channels, and curved shapes in plate, strip, or sheetmetal stock. Press brakes can also be used for punching, trimming, embossing, corrugating, and notching, when manufactured and arranged to do so, even though these operations are considered power press operations.

Power press brake beds are typically long and narrow and are located in front of, and often extend beyond, the machine side frames. The frames are gapped (cut out) to permit full-length use of the bed and ram. The piece-part component typically extends in front of the press brake and moves during the bending operation. Both the bed (or lower die holder) and the ram are equipped with a die-clamping arrangement along their full length to accept a standardized die tongue. Press brake beds are often equipped with an adjustable die holder that provides for aligning and adjusting the upper and lower dies. Backgauges or material-position gauges and stops in the front or rear are used with power press brakes to gauge the distance from the edge of the piece-part component's blank to the forming or bend line.

Mechanical Press Brakes

A limited range of strokes is available in mechanical press brakes. Ram position is adjustable to accommodate the closed height of dies. This is accomplished by changing the length of the connections from the drive to the ram through the use of die-height adjustment screws.

Hydraulic Press Brakes

On a hydraulic press brake (Figure 21–23), stroke length is variable. Speed changes, from high-speed advance to low-speed press, and upper and lower limits of ram travel are generally established by limit switches. Operating strokes per minute can approach that of a mechanical press brake of equal capacity due to their variable stroke length. Rated tonnage can be exerted through the full downstroke on a hydraulic press brake.

General-Purpose Press Brakes

General-purpose press brakes are designed and built to be operated by one person, who controls the speed and movement of the ram by use of the operator's control, usually a mechanical foot pedal. The ability and skill of the operator to control the speed of the ram permits slow bending of wide sheets, using general-purpose dies, without

fast "whip-up" of the extended edge of the sheet. Precise ram speed control is also required to permit the operator to control the ram to a partially closed position for line gauging—that is, bending to a previously scribed line. On single-speed mechanical press brakes, this is accomplished by slipping the mechanically actuated partial-revolution friction clutch to bring the ram to a partially closed position. Variable-speed and two-speed mechanical drive units and general-purpose hydraulic press brake units permit the same type of control. Operating a press brake at reduced speed (1) makes handling of the piece-part component by the operator easier and (2) can minimize the exposure of the operator to sheet or piece-part whip-up.

Stroking control on a general-purpose mechanical press brake is managed by a foot pedal. A foot pedal–operated machine should be operated only from a safe distance. Determine the safe distance by the size and shape of the piece-part, unless point-of-operation safeguarding is provided. Position a foot control so the operator is not able to reach into the point of operation, unless safeguarding is provided. Locate the foot pedal above a "step-high" position to minimize the chance of operators accidentally stepping on it. Adjust the foot pedal (1) to require enough force to avoid accidentally running the machine and (2) to return the linkage to its normal OFF position. Stroking control on a hydraulic press brake may be either by foot control or two-hand operator station.

Special-Purpose Press Brakes

Special-purpose mechanical or hydraulic press brakes can be constructed with many operational features or stroking options. The user must select a press brake component with the features that are suitable for its safe intended use in each and every piece-part operation.

Like the general-purpose press brake component, the special-purpose press brake can be used to bend, form, notch, punch, and pierce, if it is machined and constructed to do so. Special-purpose press brake components, however, can be operated by one or more operators. Each operator should have an operator control station appropriate to the piece-part production system in use. In this way, each operator and helper is able to exercise concurrent control of the press brake's ram cycle by activating an operator control station.

Stroking Controls

A variety of stroking controls and drive options are available for special-purpose press brakes, such as hydraulic-electric, air-electric, hydraulic-mechanical, single-speed, and two-speed brakes. Hydraulic-electric controls generally are two-speed brakes having a high-speed ram advance, a slow-speed work-forming portion of the stroke, and a high-speed ram return. Limit switches are used to control the speed changeover points in the stroke.

Hydraulic-electric-controlled press brakes have an infinite number of stroke lengths within their range.

Air-electric clutch/brake controls are generally used on mechanical press brakes, both single speed and two speed. The mechanical two-speed drive is similar to the hydraulic stroke's control. However, the stroke's length is constant. The changeover point from high speed to slow speed is also adjustable.

Operating Modes

Special-purpose press brakes have various employer/supervisor-controlled modes of operation. They are designated in ANSI B11.3 as OFF, INCH, SINGLE STROKE, and CONTINUOUS. OFF shuts off the operator's control station and stops the press brake. The person who sets up the die uses the INCH control; it is not used in production. In this mode, the ram may be inched down and up but only with a two-palm button, or a single control, firmly secured and located a safe distance from the point of operation.

The single-stroke mode is the standard operating mode during production. It can be initiated by the foot control or by a two-palm button control. The setup person determines which of the controls to use after considering the various components of the production system in use. In this mode, the press brake is under the operator's control in the descent portion of the stroke. It automatically returns to the top position, where it must stop. The operator's control station must be deactivated and then reactivated in order to initiate the next stroke.

In continuous operation control, used only with automatic feeds, the press brake does not stop after each stroke. It operates continuously until the STOP button is activated. There are several methods of initiating operation in this mode. Each method requires a positive, separate action on the part of the operator to minimize inadvertently placing the special-purpose press brake in the continuous-operating mode.

Tooling

Press brake tools, or dies, are generally divided into two categories: general-purpose tools and special-purpose tools. General-purpose tools are widely available, universal dies used to perform bending and forming operations on a wide variety of piece-parts and products. Special-purpose tools are designed and built to perform specialized work on a specific piece-part of a product. Many special-purpose tools are designed to eliminate the need for an operator to hold the piece-part component's blank while one or more forming operations are performed on it during the ram's stroke of the press brake component.

When operating a general-purpose press brake, protect the operator's hands by locating them along the extended edge of the piece-part component's blank at a safe distance from the point of operation. If a safe distance and point-of-operation safeguarding are used, the production system should involve the following three elements: (1) the operator must support the workpiece with both hands; (2) the operator must use material-position backgauges, and their stops must be large enough to keep the workpiece from slipping past them; and (3) operators must be instructed to remove their foot from the foot pedal after each stroke of the ram.

The operator's controls should be in place or operable only when the operator intends to start the machine. At other times, the foot pedal should be removed and/or the operating linkage locked to prevent the machine from starting.

If a general-purpose press brake is used because a special-purpose press brake with two-hand controls is unavailable, provide safeguarding such as pull-backs or restraints for feeding by hand. Tool setups for these types of parts require arrangements that support the part before and after it is formed. This prevents the part from falling behind the die and the operator from reaching between the dies for the fallen part. Material-position gauges that locate front and back edges are frequently required so operators do not place their hands near the point of operation during forming. If forming is done to a scribed line, use supports and a backstop to help operators maintain control over the part and to prevent their hands from entering the die. Their hands could enter the die if they follow a part that could otherwise slip past the scribed line.

When the piece-part component is large and extends some distance in front of the die, the operator must hold the sheet so that hands or fingers are not exposed (1) to injury from impact with the moving piece-part or (2) to pinch points or the point of operation while the bend is being made.

When operating a special-purpose press brake, protect the operator with either a two-hand operation control station or a safeguarded foot control (Figure 21–24). Depending on the piece-part bending system, use a presence-sensing device, movable-barrier device, or other means of safeguarding. A presence-sensing device may be useful for large or small parts for point-of-operation safeguarding. However, the piece-part's movement or requirements for holding during forming should not interfere with its function.

Special-purpose mechanical press brakes have air-electric, clutch-control mechanisms that provide a base for adapting many means of safeguarding. It is easier to provide and use many safeguarding means not available for a general-purpose press brake. Although some modifications are costly, they will (1) extend the use of the machine by

21 Hot Working of Metals   649

ards are described in B11.3 Section 6.1.4, Safeguarding the Point of Operation, which states:

> It shall be the responsibility of the employer, after selecting the tooling and the specific type of power press brake for producing a piece-part, to evaluate that operation before the piece-part is worked (bent, etc.) and to provide point-of-operation safeguarding according to provisions of section 6.1.4 (1).

Methods used to provide point-of-operation safeguarding for press brake–forming work include the following:

1. a point-of-operation guard, such as fixed barriers, die guards, and other means that do not allow access to the point of operation
2. a point-of-operation device, such as presence-sensing devices, gates or movable barriers, pull-backs, restraints, and two-hand operator controls
3. safe-distance methods when guards and/or devices cannot be used.

PART 2—HOT WORKING OF METALS

This part of the chapter discusses how to control materials-handling hazards and environmental stresses (dust, fumes, gases, heat, and noise) that are present in foundries and permanent mold and die-casting facilities. Also covered are safeguarding methods and safe operating practices for forging and hot metal stamping operations. A survey of the use of nondestructive testing methods supplements these discussions.

An understanding of and knowledge about the environmental stresses are important if the safety and health of employees are to meet regulatory standards. If the operation is safely implemented, very specific precautions and procedures must be followed, and personal protective equipment must be worn. This part of the chapter will cover the following topics:

- hazardous materials in foundries
- work environment, including housekeeping, ventilation, inspection, and maintenance
- substance handling and storage
- maintenance and repair of cupolas
- crucible storage and handling
- oven safety and inspection
- foundry production equipment
- cleaning and finishing foundry products
- hammer safety, inspection, and maintenance
- safe handling of dies and inspection and maintenance of forging upsetters
- basic precautions for forging presses
- nondestructive testing.

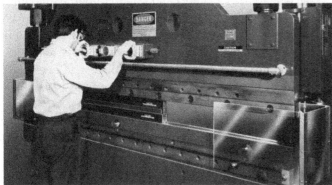

Figure 21–24. Top: Hand tools are used for inserting and removing small piece parts. Bottom: The operator uses a two-hand control station as a point-of-operation safeguarding device. Note the use of partial barriers that guard the unused portion of the brake point of operation. *(Courtesy Cincinnati Inc.)*

permitting point-of-operation safeguarding devices, with piece-part bending systems; (2) provide protection for operators; (3) reduce costs from injuries; and (4) increase production.

Responsibility for Guarding and Safeguarding

The following hazards associated with power brakes require installing protective covers or other means of protecting operators and others in the vicinity:

- rotating components, such as flywheels, gears, sheaves, and shafts in close proximity to operating personnel
- in-running pinch points associated with meshing gears, belts, and chains
- pinch points between the moving and stationary components of the power press brakes or auxiliary equipment

The manufacturer should warn against the hazard, if the hazard cannot be eliminated or otherwise safeguarded.

ANSI B11.3 describes two major areas of hazards associated with power press brakes: (1) those related to the design and manufacture of the brake and (2) those associated with the point of operation. The manufacturing haz-

HEALTH HAZARDS IN FOUNDRIES

In foundries and in permanent mold and die-casting facilities, metals are formed into finished castings. The overall foundry operation usually includes a pattern shop and sometimes a machine shop.

Safety and health professionals should refer to the National Safety Council's *Fundamentals of Industrial Hygiene* (6th ed., 2012) for principles and practices used to recognize, evaluate, and control health hazards in foundries. This book gives details on toxic and flammable hazards; general and local ventilation; and specific problems such as silicosis, dermatitis, and radioactivity. Before designing and installing equipment to control these hazards, consult with persons who are technically familiar with these hazards. Also consult these technical experts to test procedures and analyze new processes for the foundry's safety and health program. Besides NSC, other sources of help include the American Foundrymen's Society, the National Institute for Occupational Safety and Health, the American Industrial Hygiene Association, the American Conference of Governmental Industrial Hygienists (ACGIH), insurance carriers, safety and health consultants, and state or provincial and local governmental departments of industrial safety and health.

Hazardous Materials

Dust, solvents, and other materials present a health hazard in foundries. Their hazards and mode of production in foundry operations are discussed here.

Dust

Dust is generated in many foundry processes and presents a twofold problem: (1) cleaning to remove deposits and (2) control at the point of origin to prevent further dispersion and accumulation. Vacuum cleaning is the best way to remove dust in foundries. The special equipment needed is well worth the investment. Once dust has been removed, prevent further accumulation by using local exhaust systems that remove it at the point of origin.

Solvents

Evaluate each solvent on the basis of its chemical ingredients. Proper labeling, substituting less hazardous for more hazardous chemicals, limiting the quantities in use, and using other methods of control can help minimize the toxic and flammable hazards involved in using solvents.

Other Materials

Many metals, resins, and other substances present safety and health hazards in foundries. Several of them are listed here:

- Acrolein occurs in foundry operations as a result of thermal decomposition of core oil and is highly irritating.
- Aluminum, while usually not a toxic hazard in casting processes, presents a fire and explosion hazard in dust-collecting systems.
- Beryllium may produce pulmonary and skin disease, especially where plants cast beryllium-copper alloy.
- Carbon, as sea coal, is a common ingredient of molding sand used for facing. Carbon dust may cause anthracosis, a relatively harmless condition, but one that produces characteristic lung shadows in an x-ray.
- Carbon monoxide gas is generated (1) during some cycles in the operation of a cupola and (2) after pouring into green sand molds. Carbon monoxide is a toxic gas that preferentially binds to the hemoglobin molecule and significantly decreases the amount of oxygen to vital organs and tissue.
- Chromium is encountered in stainless-steel castings as the element or the oxide. Exposures occur during melting, gate and head burning, and grinding. Of the two common forms of chromium, trivalent and hexavalent, hexavalent is far more toxic and has a variety of health effects ranging from irritation to carcinogenicity.
- Fluorides, sometimes in cryolite form (sodium aluminum fluoride), are used in manufacturing ductile iron and magnesium castings. They are respiratory, skin, and eye irritants that can also cause diarrhea and abdominal pain. Further, fluorides are associated with calcification—hardening of ligaments and bones.
- Iron oxide fumes and dust are created during melting, burning, pouring, grinding, welding, and machining of ferrous castings. Exposure may be particularly high where manganese-steel castings or oxygen-lancing of the furnace is involved. Use local exhaust to vent these fumes (Figure 21–25).
- Lead is a health hazard in nonferrous foundries. It forms the oxide in melting, pouring, and welding operations. Elemental lead dust is produced in cleaning and machining operations. Lead is a toxic material capable of adversely affecting a variety of organs and systems, including kidneys and nervous system.
- Magnesium dust or chips create hazards of fire and explosion. Physiological effects are confined to a form of metal fume fever from inhaling finely divided magnesium. Magnesium-oxide fumes are generated when burned.
- Manganese is usually associated with steel castings and bronze alloys in foundry work. Constant exposure to high levels of manganese is associated with neurological disease.
- Phosphorus is used in the production of phosphor-copper. Acute cases of poisoning have not been reported, and chronic cases are rare. The drying of phosphor-copper shot may produce phosphine gas, which is highly toxic.
- Resins—phenolformaldehyde and ureaformalde-

Figure 21–25. In these fume-diverting baffle furnaces, the hoods roll on a trolley so that correct positioning is made easy. *(Courtesy American Brake Shoe Co.)*

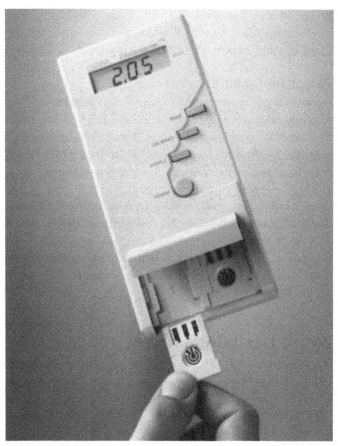

Figure 21–26. This hand-held meter detects trace metals in solution. It can be used for monitoring metal plating baths, industrial effluent, and water quality. It can measure a wide concentration range, meeting various international regulatory standards.

hyde—are used in shell molding. They create several hazards. Phenolformaldehyde resins contain hexamethylenetetramine (hex), a skin irritant that is highly explosive. When heated, this type of resin decomposes to produce a mixture of phenol and formaldehyde vapors. Ureaformaldehyde decomposes to give off ammonia and carbon dioxide. These materials are known to have good warning properties, but industrial hygiene evaluations should be conducted to evaluate potential exposure levels.

- Some types of resin dust, however, are highly explosive when suspended in air and require wet dust collectors. Alcohol, sometimes used for cold coating sand with resin, must be controlled to keep its concentration well below the lower flammable limit.
- Silica is usually encountered in the use of silica flour in molding sand or in core washes and sprays. Zircon, which is more dense and therefore settles more rapidly, is an effective substitute for silica flour in some applications. Sand-handling and conditioning systems, shakeout operations, and sand slinging constitute other sources of exposure to silica dust. Repeated exposure to high levels of silica dust is associated with the lung disease silicosis, and exposures should be controlled to ensure compliance with regulatory exposure limits.
- Silicones are used as mold-release agents in shell molding. Hydrolyzing silicones are highly corrosive and hazardous if touched or inhaled. Care in handling them can eliminate the dangers of skin or eye contact. However, nonhydrolyzing silicones (methyl, mixed methyl, and phenylpolysiloxane) are better choices because they can be just as effective as mold-release agents, yet less toxic.
- Sulfur dioxide is the result of the oxidation of sulfur used in magnesium castings. In concentrations normally present in foundries, it is an irritant that can produce transient changes in pulmonary function.

Medical Program

Base the safety and health program for foundry workers on the recommendations and guidance of a safety and health professional, an industrial hygienist, and/or a physician (Figure 21–26). Make sure employees are aware of specific hazards to which they may be exposed and the proper control or emergency responses to those hazards. Make Safety Data Sheets (SDSs) available to all employees. (See, in the *Administration & Programs* volume, Chapter 6, Loss Control Programs; Chapter 9, Identifying Hazards; Chapter 12, Occupational Health Programs; Chapter 14, Environmental Management; and Chapter 18, Emergency

Preparedness. See also, in this volume, Chapter 9, Fire Protection; Chapter 10, Flammable and Combustible Liquids; and Chapter 12, Materials Handling and Storage.)

Personnel Facilities

Coreroom workers whose hands and arms may be exposed to sand and core oil mixtures are candidates for dermatitis. Prolonged contact with oil, grease, acids, alkalis, and dirt also can produce dermatitis. Encourage frequent washing with soap and water, and install adequate facilities.

Recommendations for toilets, washrooms, showers, locker rooms, and food service are given in Chapter 21, Industrial Sanitation and Personnel Facilities, in the *Administration & Programs* volume. Sanitary food preparation and service are especially important in nonferrous foundries. Prohibit eating in work areas where toxic materials are handled.

WORK ENVIRONMENT IN FOUNDRIES

Good housekeeping, ventilation, and lighting help maintain a safe and healthy work environment. Proper inspections, maintenance, and fire protection increase workers' safety in foundries.

Housekeeping

To achieve good housekeeping, hold each individual responsible for maintaining order in the work area. Set aside a specific time for housekeeping. Provide the necessary housekeeping equipment, and see that trash cans and special disposal bins are kept handy and emptied regularly.

Each worker should do the following chores:
* Clean machines and equipment after each shift, and keep them reasonably clean during the shift.
* Place all trash in the proper trash bins.
* Keep the floors and aisles in the work area unobstructed.
* Properly stack and store materials.

Floor Loading

Many buildings are used for purposes for which they were not designed. Mechanized movement of material introduces floor load problems caused by deadweight of platforms and lift trucks. Suspension of overhead cranes and hoists from wood ceiling joists severely taxes roof and floor members. Insurance engineers or local building inspectors can help determine safe floor load limits.

Ventilation

Control of air contaminants is the primary purpose of ventilation in foundries. The need for controlling ventilation may be determined by one or more of the following:

* applicable federal, state, provincial, and local regulations or standards, codes, and recommendations
* comparison with similar operations in a like environment
* collection and analysis of representative air samples taken by qualified personnel in the breathing zone of workers.

Noise Control

Controlling excessive levels of noise, more than 85 dBA, may sometimes be difficult. Controlling noise through engineering is not always possible because of a lack of technology or is impractical because of short-term, infrequent exposures or economic considerations. In such cases, develop a hearing conservation program that provides approved hearing protection for each worker, and minimize exposure to identified high-noise-level hazards. Chapter 7, Personal Protective Equipment, in this volume, discusses elements of a successful hearing protection program as does NSC's *Fundamentals of Industrial Hygiene.*

Lighting

Good lighting is difficult to achieve in foundries because of the nature of the operations. Where craneways are used, light fixtures must be placed high and at considerable distances from the work areas. Nevertheless, provide good lighting for each work area. (See Appendix 1, Safety and Health Tables, for recommended levels of illumination.) Foundries having difficulty maintaining recommended levels of light can call on their local power companies or illumination consultants for expert information.

Inspection and Maintenance

Follow standard inspection and maintenance procedures in foundries. Carefully select maintenance personnel. Train them in safe practices, particularly in procedures for locking out controls and isolating other energy sources. (See Chapter 6, Safeguarding. See also ANSI Z241.1, Safety Requirements for Sand Preparation, Molding, and Coremaking in the Sand Foundry Industry.)

Fire Protection

Foundries make periodic fire inspections and perform emergency fire-fighting drills. (Emergency Services, if available, should participate in drills.) A fire brigade, if present, will also aid the safety program by keeping its members, as well as other employees in the foundry, safety conscious.

Facility Structures

Entrances and exits, stairways, floors and pits, galleries, gangways, and aisles of foundries must meet special requirements. Observe the following suggestions and requirements for facility structures in foundries.

Entrances and Exits

To prevent drafts from reaching employees, provide entrances and exits to heated buildings with vestibules or enclosures. Make these enclosures large enough to permit the passage of trucks regularly used inside the facility. This provision does not apply to entrances used for railroad or industrial cars handled by locomotives or for traveling cranes, trucks, and automobiles.

All doors, particularly double-acting, swinging doors, should have a window opening approximately 8 × 8 in. (20 × 20 cm). Locate the window at normal eye level to permit a view beyond the door.

Stairways

Provide substantial handrails, standard guardrails, and toeboards for all permanent and portable stairways having four or more risers. (See ANSI A1264.1–1995, Safety Requirements for Workplace Floor and Wall Openings, Stairs, and Railing Systems.)

Floors and Pits

Have the floor beneath and immediately surrounding foundry melting units pitched away from the melting units to provide drainage and to prevent incidents, especially those caused by spills and "run-outs" of molten metal. Clean floors frequently and keep them in good condition—that is, firm and level. Workers should report worn spots, holes, or other defects, and maintenance workers should repair them immediately. Install special types of flooring where fires or explosions may occur or where other serious hazards may exist. To prevent an explosion hazard, keep the floor free of pools of water. Where water is needed to reduce dusty operations, use only enough to hold down the dust. Where tram or standard-gauge railroad tracks run into or through a foundry, keep the top of the rails flush with the foundry's floor, which should be maintained at this level.

Because of the danger of explosion, keep pits and other containers in which molten metal is handled or poured free from dampness. Protect pits connected with ovens, furnaces, and floor openings with either a cover or a standard guardrail when not in use.

Locate pig molds and receiving stations for excess molten metal from ladles clear of passageways and at least 1 ft (0.3 m) above floor level. Never allow pig holes in the floor near pouring areas; it is inherently unsafe.

Galleries

Where molten metal is poured into molds, provide the galleries with solid, leak-proof floors—concrete or sheet steel covered with sand—and with partitions of sheet steel. The partitions should be approximately 42 in. (1 m) high. Install them on the open sides of such galleries.

Where floor space is cramped, construct galleries to store ladles, flasks, flask boards, and other equipment. Equip these galleries with standard handrails and toeboards, and provide them with sturdy stairways—not ladders.

Keep concrete pavements around pouring floors coated with sand during pouring operations. This will reduce spalling of cement in case of a molten-metal spill.

Gangways and Aisles

Keep every gangway and aisle in good condition. They should be firm enough to withstand the daily traffic for which they are intended. Make sure they are uniformly smooth, without obstructions, and free from pools of water.

Compressed Air Hoses

The compressed air hose presents another foundry hazard. Do not use air hoses to clean clothes. Improper use of the air hose and "horseplay" have caused severe injuries to internal organs and eardrums. To prevent injuries, reduce compressed air to less than 30 psig (210 kPa). Install whip checks at all joints, especially at quick-release couplings.

Prohibit unsafe practices such as blowing and brushing sand from new castings without regard for the cloud of dust produced, blowing dust off patterns, and removing parting compounds and other light materials. Substitute vacuum methods for cleaning molds with compressed air. Carefully instruct workers in the safe use of air hoses if the latter method is employed. Use of nonsilica partings eliminates the possibility of silicosis from this source.

MATERIALS HANDLING IN FOUNDRIES

Improper materials handling in foundries results in a wide variety of injuries. Injuries involving fingers, hands, toes, feet, legs, arms, and the back can occur during the manual handling of scrap metals, pig iron, and similar materials.

Many foundries have replaced manual handling of materials with mechanical means, reducing exposure to manual-handling hazards. However, mechanical material handling usually brings with it hazards of its own. For instance, do not use magnets for lifting over areas where people are working. A break in the magnetic circuit would cause the load to be released without warning. In addition, pieces dangling from the magnet can be jarred loose.

Some of the precautions that can be taken to prevent material-handling injuries include the following:

- Instruct workers in the safe methods of manual and mechanical materials handling.
- Provide personal protective equipment—eye protection; safety hats; face shields; leather mitts or gloves, preferably studded with steel, unless hot metal is to be handled;

hand pads, aprons; protective footwear, including metatarsal shoes; and other items such as flame-retardant clothing. Respiratory protection may also be needed.
- Plan the sequence and method of handling materials to eliminate unnecessary handling.
- Safeguard mechanical devices, and set up inspection procedures to ensure their proper maintenance.
- Keep good order at storage piles and bins, and pile materials properly.
- Keep ground and floor surfaces level so that workers handling materials will have good footing.
- Install side stakes or sideboards on tramway or railroad cars to prevent materials from falling off.
- Chock railroad cars and flag tracks as required. Use dock plates with boxcars when loading or unloading.

Handling Sand, Coal, and Coke

Avoid certain hazards in handling materials such as sand, coal, and coke, as follows:
- Prevent falls through hoppers while unloading bottom-dump railroad cars by requiring the use of fall protection equipment. Be sure that observers are on the scene; they should be prepared to perform rescues and/or summon help in emergencies.
- Use safety ratchet wrenches for hopper doors to keep the doors from swinging and striking workers.
- Prevent hand and foot injuries by using safety car movers instead of ordinary pinch bars to spot cars by hand. However, use a locomotive when available.
- To keep dump cars under repair from being moved, use locking switches and car chocks; at night use warning targets, derails, and red lanterns.
- To reduce the danger of cave-ins of loose material, prohibit the undermining of piles and avoid material overhangs.
- Prevent electric shock by grounding portable belt-conveyor loaders.

Some foundries have eliminated double handling of materials by having raw materials taken directly from the cars, storage piles, or bins and placed into unit charging trays or boxes. The trays or boxes are then taken to the point of use and dumped mechanically. Properly trim trays or boxes to be carried overhead.

Ladles

Handshank ladles and mixing ladles are used for distributing molten metal or reservoir. Ladles are mounted on stationary supports or trucks or are handled by overhead cranes or monorails. Such ladles have a capacity of not more than 2,000 lb (900 kg). Provide ladles with a manually operated safety lock (Figure 21–27). Construct the

Figure 21–27. This tilting ladle is equipped with a manually operated antitilt level. *(Courtesy American Foundrymen's Society Inc.)*

shanks of the ladles from solid material, and install shields. Provide suitable covers for portable ladles.

When dealing with more than a 2,000-lb (900-kg) capacity, use gear-operated ladles. Equip such ladles and those that are mechanically or electrically operated with an automatic safety lock or brake to prevent overturning or uncontrolled sway.

Thoroughly dry out and heat ladles before use. Provide local exhaust to control vapors or fumes produced during ladle drying. Some foundries perform all ladle-drying operations in a shed located outside the foundry. This segregates the exposures and typically makes control measures easier to implement.

Equip monorail ladles and trucks used to transport molten metal ladles with warning devices, such as bells or sirens. Sound the bell or siren whenever molten metal is being transported.

Construct trunnions and devices used to attach them to flasks, buckets, ladles, and other equipment with a safety factor of at least 10. The diameter of the head on the outside end of the trunnion shaft should be not less than 1.5 times the diameter of the trunnion shaft. Fillet the inside corners where the trunnion shaft joins the base and the head to prevent the sling or hook from riding the trunnion's base or head.

Inoculation, or treatment, of molten metal to desulfurize it or to change its composition or type, as in the making of an alloy or ductile iron, is done in the reservoir or in a pouring ladle. Install a hood to cover this operation so that the workers are effectively shielded from possible spatters of metal caused by the violence of the reaction. The hood should draw off the fumes that result and should exhaust them through a baghouse filtering system and then through a stack.

Hoists and Cranes

Hoists and cranes that handle molten metal require a preventive maintenance program, conducted by personnel trained on and thoroughly familiar with the equipment. This program is in addition to ongoing observations and inspections made by supervisors and operators. The degree to which the program is carried out depends on both the equipment being used and the tonnage moved. Gear the program to ensure that the operation is safe—much safer than simply to comply minimally with existing regulations. For example, an effective program for a 300-employee gray-iron foundry could require weekly visual inspections of crane and hoist structures, as well as an inspection of wire ropes and hooks before every shift.

Because of the severe stresses and demanding service in some high-tonnage operations, these operations may require more elaborate inspection programs. Conduct inspections on a weekly basis by trained specialists. Some programs regularly schedule nondestructive testing: ultrasonic testing of the crane's hoist shafts and parts and dye-penetrant inspection for surface cracks on bales, dumping chains, clevises, and pins. (See Nondestructive Testing later in this chapter.)

Conveyors

Conveyor systems are typically used in foundries. Sand mixed in the mixing room is carried by a belt conveyor to hoppers at molding stations, where each hopper is filled by a movable plow. Surplus sand is carried to the end of the belt and returned by bucket to the storage bin.

An endless conveyor is used to handle molds. Empty flasks come from the shakeout machine to the mold operators, who remove them and make their molds on molding machines, taking sand from the overhead hoppers as required. Spilled sand goes through a grating onto a belt conveyor that returns the sand to the mixing room. Thus, all shoveling operations are eliminated.

The molds are placed on a conveyor and passed into the pouring area. The pourers get their metal from the cupola or other furnaces and then step onto a moving platform geared to an endless, single-rail conveyor that moves at the same rate of speed as the platform. Hand ladles can be supported on the conveyor. Pouring is done as the workers move along. Install an electric switch near the end of the conveyor so that if workers ride that far, their feet will come into contact with the switch and the conveyor will stop.

The mold conveyors then pass into a cooling zone. Weights can be removed from the molds by a mechanical device and returned by another conveyor to the place where they were originally used. The molds then move to the shakeout machine, where sand is dumped from flasks onto a vibrating grating. The sand falls to a belt conveyor that returns it to the mixing room. Using a hook, a worker pulls the castings onto another conveyor, which takes them to the tumbling barrels. Empty flasks are brought back to the molding section by another conveyor.

This is a complete system for mass production in which each worker performs one function rather than several. When installing a system, guard shear points, crush points, and moving parts. Where conveyor systems run over passageways and working areas, the employees beneath them must be protected with screens, grating, or guards. These protective devices should be strong enough to resist the impact of the heaviest piece handled by the conveyor.

Where chain conveyors operate at various levels other than in a fixed-horizontal plane, install a mechanism of safety dogs on both the upgrade and the downgrade, in accordance with applicable standards. In case the chain fails, the safety dogs will hold the chain and prevent the load from piling up at the bottom of the incline.

Scrap Breakers

Guard shears to protect operators and passersby from flying particles. Keep the working floor clear and level.

Prohibit the use of a drop to break castings or scrap inside foundry buildings during working hours, unless such operations are performed within a permanent enclosure made of planking or equivalent materials. The enclosure should be strong enough to withstand the most severe impacts from the drop or from flying scrap. Construct the enclosure high enough to protect workers in the vicinity from flying fragments of metal.

If a rope is used, extend it over pulleys to a point clear of the breaking area. This ensures that operators will be at a safe distance, preventing their entanglement in the rope.

Storage

Store foundry materials and equipment that are not in regular use in a safe and orderly manner on a level and firm foundation. When workers remove materials from bins located at floor level or from storage piles, they should not undermine the piles and thereby cause cave-ins.

Cover hopper bins containing material that is fed out at the bottom, either by hand or by mechanical means, with a grating that prevents workers from entering the bin. Allow no one to get on the rails of a bin or to enter a hopper to break down bridged material. A worker who must enter a bin should wear fall protection equipment with an attached lifeline. A second worker may be equipped for the rescue or to summon aid.

Buildings that store patterns should have racks and shelves strong enough to hold the loads placed upon them. Provide pattern keepers with a sound ladder so they can safely reach the patterns. The pattern keeper, who is likely to be alone in the pattern storage area, should report at regular intervals to the supervisor. Design the floors and stairways of pattern storage areas well, and keep them in good condition. The storage area should also be well lit.

Store flammable liquids in accord with National Fire Protection Association standard NFPA 30, Flammable and Combustible Liquids Code, and mark them according to 29 CFR 1910.1200, the Hazard Communication Standard. (See Chapter 10, Flammable and Combustible Liquids, in this volume.)

Slag Disposal

Design furnaces and pits with removable receptacles into which slag and kish (separated graphite) may flow or be dumped. Unless slag is disposed of in the molten state, provide enough of these receptacles so slag can solidify before it is dumped.

To decrease the amount of slag that goes into the slag pits, use slag or cinder pots. The pots can be set aside and allowed to cool, eliminating the danger of explosion when they are emptied.

Dump slag where there is absolutely no water or dampness. Slag coming in contact with water might cause an explosion if some of the slag is still molten. Before breaking up slag, allow it to stand for several hours to prevent encountering still-molten slag in the center.

CUPOLAS

Cupolas are vertical cylindrical furnaces used to melt iron in a foundry. The charging and blasting that take place in cupolas and generate carbon monoxide (CO) present several hazards.

Charging

The dangers in the charging of cupolas are principally confined to handling material. Never unevenly load or overload barrows or buggies. "Tip-up" barrows, used for charging coke, are sometimes so poorly balanced that they will not stay in the tipped-up position after being emptied. Instead, they could fall back on the chargers' feet at the slightest touch. Lowering the center of gravity minimizes this hazard.

To prevent an explosion in the cupola, break open scrap cylinders, tanks, and drums before charging. Be sure these containers are empty.

The use of mechanical devices for charging cupolas not only saves labor, but also reduces the number of material-handling injuries. Most foundry cupolas are now charged either by fully automatic charging machines equipped with crane- and cone-bottom buckets or by lift trucks equipped with tilting boxes.

The charging opening on some cupolas is covered by a door or chain curtain, which should be kept closed except during charging. To prevent material from dropping onto workers during charging operations, install railings or other safeguards for the space underneath the cupola's charging elevators, machines, lift hoists, skip hoists, and cranes.

Occasionally, during idle periods, workers might rest under the charging platforms close to the warm chambers and flues. Prohibit this practice because of the danger of objects falling from the platform and the possibility of carbon monoxide escaping from the flues. Also, do not place employees' lockers under these platforms.

Charging Floor

For charging floors, use steel floor plates that are heavy enough not to turn up; securely fasten these plates in place. Steel floor plates in the immediate vicinity of the furnace, however, become extremely hot. Therefore, install brick flooring laid on a solid steel framework in those areas.

Keep charging floors free from loose materials, and provide storage racks for equipment not in use.

Provide standard railings, 42 in. (105 cm) high, and 4-in. (10cm) toeboards around all floor openings. [See 29 CFR 1910.179(d)(3),(4)(ii).] Because railings on the charging floor receive much abuse, construct them of angle iron, which is more easily repaired than pipe railings. At the tapping platforms, provide hinged gates or chains that may be hooked in place.

Where cupolas are manually charged, place a guardrail across the charging opening. Where cupolas are charged with wheelbarrows or cars, provide a curb of a height equal to the radius of the wheel of the barrow or car. This prevents the barrow or car from pitching over and falling into the cupola.

Carbon Monoxide

Carbon monoxide (CO) is generated during some cycles in the operation of a cupola. CO is an explosion hazard if it gets into the wind boxes and blast pipes when the blowers are shut down. To eliminate this hazard, supply adequate natural or mechanical ventilation in back of the cupola, and open two or more tuyeres after the blowers are shut down.

The large amount of blast air in the cupola generally carries the CO out the stack. In some cases, it is burned in the stack before it can be discharged. Sometimes, however, CO may escape. Because CO gives no warning, locate CO indicators around the cupola that light and give a loud sound. Also, post signs that show the proper procedures to follow should the CO indicators' alarm sound.

See that approved breathing equipment is close by. Train workers in its use, and be sure that this equipment is in good condition. If the concentration of CO is more than 200 parts per million (ppm) (0.02%), an engineering assessment should be considered. In addition, positive pressure, self-contained breathing equipment, or an air-line respirator with an emergency escape bottle should be provided. The OSHA permissible exposure limit (PEL) for an 8-hour time-weighted average (TWA) is 50 ppm, and all efforts should be directed toward managing exposure through the use of engineering controls.

Blast Gates

Blast gates and explosion doors are successfully used to prevent damage from gas explosions. They are sometimes placed in front of the tuyeres, so they can be opened to admit fresh air when the blowers are shut down. Never close blast gates or explosion doors until the blast has entered the wind box and driven out all gas.

Provide blast gates in the blast pipe that supplies air to the melting equipment. Close the blast gates when the air supply fails or when the melting equipment is shut down. This prevents the accumulation of combustible gases in the air-supply system. In the cupola, omit the blast gate if alternate tuyeres are opened to permit circulation of air.

Locate blast gates in relation to the cupola's wind box to keep the duct's volume at a minimum. Install motorized dampers at centrifugal blowers so they will close automatically when the air supply fails.

Equip positive-pressure blowers with safety valves having liberal discharge areas. If these are not provided, clogging of the cupola with slag, or quick closing of the gate or damper, may produce sufficient pressure in the blast pipe to cause it to burst.

Every cupola should have at least one safety tuyere, with a small channel 1 or 2 in. (2.5 or 5 cm) below the normal level of a tuyere. This channel has a fusible plate that will melt through should the slag and iron rise to an unsafe level.

Tapping Out

Tapping out with safety requires skill; have only experienced and dependable operators perform tapping out. In "botting-up" the hole, operators should not thrust the bott directly into the stream of molten metal because that would cause spattering. To eliminate this hazard, have operators place the bott immediately over the stream of metal, close to the hole, and aim it down toward the hole at a sharp angle. Keep a supply of botts ready for use within convenient reach of the operator who does the tapping.

When the cupola is tapped, hold the back end of the tapping bar below the level of the hole. This prevents puncturing the sand bed and causing molten metal to run out through the bottom.

A tilting spout placed with one end directly beneath the stationary cupola's spout and mounted on trunnions on a stand increases safety and efficiency. Operators can tilt it back and forth with a foot lever. The rear end of the tilting spout is closed so that when that end is tilted down, it forms a reservoir to receive the molten metal from the cupola. When the supplementary spout is tilted forward, the metal runs from it into the waiting ladle. At the same time, more metal continues to run into the spout from the cupola. Thus, the stream of metal runs from the cupola continuously, and the tilting spout acts as a reservoir between loads from the ladle.

Equip the slag spout of the cupola with a shield or guard to protect workers from sprays of molten slag and to form a hood to collect slag wool. The slag wool is sometimes collected through a wet-slagging system in which slag is thrown off into a water-filled container, or trough, and flushed away.

Dropping the Cupola's Bottom Doors

When the cupola is in operation, support its bottom doors with one solid prop and two adjustable screw props (of the required structural strength) on a metal prop base. The base should be set on a concrete footing or other fabricated footing of equivalent strength (Figure 21–28).

Dropping the bottom doors of a cupola requires extraordinary care. One of the best methods for doing this is to use a block and tackle with a wire rope and a chain leader attached to the props that support the doors. The props can then be pulled out by means of the block and tackle from a safe distance or from behind a suitable barrier. Special locking devices for bottom doors may be used if the cupola's drop is to be caught in a container, car, or skid.

Before the bottom doors are dropped, carefully inspect the area underneath the cupola to see that no water has seeped under the sand. One worker should make sure that no one is in the danger zone and that workers stay away during the operation. Warn employees by means of a whistle or other signal before the bottom doors are dropped.

If the cupola's bottom doors fail to drop, or if the remaining charge inside the cupola bridges over, do not permit employees to enter the danger zone to force the

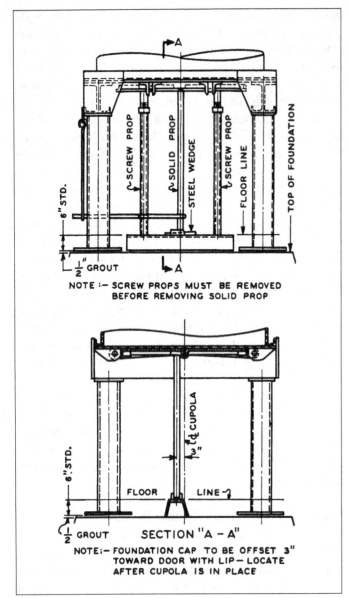

Figure 21-28. This is the proper method of supporting cupola bottom doors.

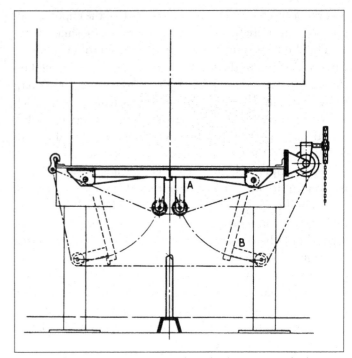

Figure 21-29. This is the suggested method of raising the bottom doors of the cupola by mechanical means.

Figure 21-30. A screen placed over the charging door prevents falling objects from dropping on a worker who is repairing the cupola lining. *(Courtesy Hamilton Foundry & Machinery Co.)*

doors or relieve the bridging. Relieve the bridging, instead, by turning on the blast fan. The vibration produced usually corrects the condition. A mechanical vibrator attached to the bottom doors is also effective. Another method of relieving the bridging is to drop a demolition ball from the charging door. If these methods fail, flame cut the doors with a lance, but only after the cupola has cooled to a safe temperature.

Repairing Linings

Allow only careful and experienced workers to repair a cupola's linings using appropriate procedures and permits (Figure 21-30).

CRUCIBLES

The principal danger in handling refractory clay crucibles is that one may break when full of molten metal. Therefore, have a trained inspector check all new crucibles for cracks, thin spots, and other flaws. Return to the manufacturer those showing signs of dampness. Examine the packages and the car in which they were shipped to find out whether or not they were exposed to moisture in transit.

Storing

Store crucibles in a warm (about 210°F or 120°C), dry place, and protect them from moist air as much as possible. It is generally best to place them in an oven built on top of a core oven or at some other point where waste heat can be used. If all crucibles in stock cannot be kept in ovens, accurately date those stored elsewhere and use the oldest and best-seasoned crucibles first.

Annealing Process

In the annealing process, crucibles are brought up to red heat very slowly and uniformly, usually over a period of 8 to 10 hours. Do not allow crucibles to cool before they are charged because, as they cool, they may again absorb moisture. Moisture in the walls of crucibles that are heated quickly is converted into steam. The steam expands and causes cracks or ruptures and may also cause pinholes or "skelping." Do not use damp or high-sulfur coke or coal, or fuel oil containing excessive moisture, to heat crucibles.

Too high a percentage of sulfur in the fuel used in the drying or annealing process is also likely to cause fine cracks, sometimes called "alligator cracks." Too little oil, or too much air or steam, used at the burners of oil furnaces tends to oxidize a portion of the graphite in the crucible's wall. This leaves the binding material somewhat porous.

Charging

Proper care of crucibles is good economy as well as good safety. Because crucibles are costly, they should be made to last through as many "heats" as possible. To protect the crucible's lining or structure from damage, establish a process for cleaning crucibles. Improper cleaning may result in early failure of the crucible and, thus, injury or loss of the product.

Charge crucibles carefully. Do not throw in ingots with such force that they bend the bottom or walls of the crucible out of shape. Also do not force the ingots into the crucibles so they become wedged or jammed. Heat the new crucibles very slowly for the first few runs, especially the first run. Because crucibles are soft at white heat and easily forced out of shape, handle them with great care.

Handling

To prevent damage to a crucible, select tongs of the proper size and shape for the particular crucible. Tongs should fit well around the bilge or belly of the crucible and should extend to within a few inches of the bottom. Provide at least two pair of tongs for each size of crucible so that if one pair becomes bent, the other will be available.

Before applying the tongs, check the sides of the crucible to see that no clinkers are adhering to them. Never drive tong rings down tight with a skimmer or other tool. This practice is almost certain to squeeze the crucible out of shape and produce cracks and fissures.

The blacksmith should have a complete set of cast-iron forms in the exact shapes of the crucibles used. Then the smith will have only to heat the tongs to red heat, clamp them onto the forms, and bring them into the exact shape with a heavy hammer.

Avoid ramming the fuel bed around a crucible. Should this become necessary, have it done cautiously and only by experienced workers. Support crucibles on foundations or pedestals of firebrick, graphite, or other infusible material.

The removal of heavy crucibles from furnaces calls not only for special skill, but also for physical strength. If the workers are overstrained, serious injuries are likely to result. Where possible, therefore, use a mechanical device to remove heavy crucibles—those exceeding 100 lb (45 kg) in combined weight of crucible, tongs, and metal. When using air or electric hoists to move large crucibles, have one person at each sling and one operator controlling the hoist.

Crucible Furnaces

To make the operation of crucible furnaces relatively free from hazards, install suitable exhaust hoods on all furnaces used to melt metals that give off harmful fumes. Equip upright furnaces, having crown plates more than 12 in. (30 cm) above the surrounding floor, with metal platforms having standard rails.

Many crucible furnaces are oil fired. Unless the air supply for these furnaces and the motors driving the oil pump are connected to the same source of power, a considerable quantity of oil may flow onto the floor if the air line loses its power. One remedy for this is to put a gate valve in the oil supply line so that in case the air supply fails, oil can be shut off from the entire battery of crucible furnaces in one operation. Another preventive measure is to install a gate or lever-operated valve in the oil line.

OVENS

The principal hazards in the construction and operation of core ovens and mold-drying ovens are excess smoke, gases, and fumes. Other unsafe conditions are unprotected firing pits; unguarded, vertical sliding doors or their counterweights, which may drop on workers; and flashbacks from fireboxes.

Gas-Fired Ovens

Separate gas-fired ovens, whenever possible, from the molding floors and from the core-making room by a partition. This measure helps prevent equipment failures caused by sand in the controls.

Equip blast-tip pipe burners with baffles to keep sand out of the tip and also to spread the flame. Place tips horizontally to protect them from sand.

Install safety pilot valves on every gas-burning furnace or oven. They prevent the flow of unburned gas into the oven's combustion chamber, should the burner's pilot light go out, or should a cock or burner be opened unintentionally.

Install a bleeder valve in the line between two control valves close to the burner as an additional safety device. The operator can then allow gas to escape safely into the atmosphere instead of to the firebox, should there be leakage past the main control valve when the burner is not in use.

Ventilation

Where fumes, gases, and smoke are emitted from drying ovens, install hoods and ducts, exhaust fans, or other means of removing these hazards near the oven's doors. Such devices should be designed to keep the concentrations of fumes, gases, and smoke below toxic and irritant levels. Be sure that the composition of any emissions discharged outside the building complies with air pollution regulations.

To prevent flashbacks, install the proper-size flues, and keep them free of soot. Then, using oil burners, the type of equipment and the arrangement and control of drafts should ensure perfect combustion as much as possible. In some installations, forced-draft equipment may be needed.

Equip core ovens with explosion vents. Lightweight panels may be installed on the top of the oven, or the oven may have hinged doors with explosion latches.

Natural-draft ventilation, however, is usually considered adequate for ovens under 500 ft³ (14,000 L) in volume. Larger ovens, especially those with vertical sliding doors and other heavy construction, should have forced-draft ventilation. Interlock the ventilation system with the gas supply through a time relay. Arrange the relay to allow for at least three complete changes of air in the oven before the burners are lit.

Inspection

Before a foundry's core ovens are lit, thoroughly inspect the ovens and burners. Only trained and qualified personnel should do this work. Establish an inspection and preventive maintenance program for core ovens.

FOUNDRY PRODUCTION EQUIPMENT

On production-line equipment, fully guard moving parts (such as belts, pulleys, gears, chains, and sprockets) and other common machine hazards (such as projecting setscrews) in accord with standard practices (see Chapter 6, Safeguarding). Ground electrical equipment to eliminate shock hazards. Allow repairs only on equipment that is locked in the OFF position and after all other sources of energy have been eliminated.

Some operations require mills, mixers, and cutters of such size that an employee can enter the machine to clean or repair it. In these cases, set up and enforce a lockout procedure. (See Lockouts in Chapter 6, Safeguarding; and in Chapter 8, Electrical Safety, both in this volume.)

Sand Mills

The principal danger of sand mills, or mullers, exists when operators reach in for samples of sand or attempt to shovel out sand while the mill is running. In doing so, they may be caught and pulled into the mill. To protect against this hazard, one or more of the following measures may be used:

- Provide screen enclosures for charging and discharging the openings of mills.
- Install self-discharging mills, or equip mills with discharge gates or scoops.
- Provide sampling cones for taking samples of the sand during the mixing operation.
- Prohibit the shoveling of sand out of mills while they are running.
- Install an interlocking device so the mill cannot be operated until the doors are closed.

Dough Mixers

To prevent operators from reaching into a dough mixer while the blades are in motion, cover the top of the mixer with a sturdy grating made of ⅜-in. (1-cm) round bars, or an equivalent. Another method is to attach an interlocking device arranged so that the cover cannot be opened nor the bowl tilted until the blade's drive mechanism has been shut off. In such a setup, the blades cannot be set in motion again until the cover is in place.

If the dough mixer is driven by an individual motor, attach a small steel cable to the cover and extend it over a pulley to a counterweight. This cable is attached to a ring on the motor control switch's handle. In that way, when the cover of the mixer is lifted a predetermined distance, the switch is pulled open and cannot be closed again until the cover is back in place.

Sand Cutters

Sand cutters throw sand and pieces of tramp metal with bulletlike force, sometimes causing serious puncture wounds. If a guard that would not seriously impair the efficiency of the operation cannot be devised, then have operators wear suitable personal protective equipment.

It is often difficult to operate a power-driven cutter on a sand floor. Therefore, install parallel concrete strips to act as runways for the cutter's wheels.

Sifters

Guard rotary sand sifters with enclosures or with angle iron or pipe railings. Place belt shifters and motor control switches within convenient reach of the operators. The control switches should be designed so that they cannot be unintentionally started.

Portable sand sifters equipped with pneumatic vibrators usually move slower than those equipped with electric vibrators. Oscillation of their heavy parts causes the entire machine to move around the floor in jerky fashion. Often, the machine's travel is limited only by the air hose. If the hose coupling breaks, the hose flails around and blows sand in every direction, presenting a hazard to workers' eyes. To prevent such incidents, anchor the sifter with a rope a little shorter than the hose.

Molds and Cores

The principal hazards in hand molding and core making include letting flasks down on feet, pinching fingers between flasks, dropping heavy core boxes on feet, cutting hands on nails and other sharp pieces of metal in the sand, and stepping on nails. Minimize hand and foot injuries by training workers to handle flasks and core boxes properly and to wear foot protection with stout soles. Screening or magnetic separation to remove nails and other sharp metal from the sand is also essential to safety.

In general molding and core making, gagger rods— pieces of iron used in a mold to keep the sand or core in place—and core wires are cut, straightened, and bent using hammers and cutting sets. This operation presents danger from flying pieces of metal and dirt. Machines are available for performing this work, but many of their hazards are similar to those found in the use of hand tools.

As the work progresses, carefully brace heavy cores in large molds to keep the cores from toppling over. Prohibit work underneath molds suspended from cranes. Sturdy tripod supports or wooden or steel horses will provide greater safety and efficiency.

Venting molds properly is essential to avoid explosions during pouring. However, when the sand in an undried mold is too wet, metal can boil and explosions may occur even though the molds are well vented.

In ramming a mold, do not place the peen of the ram too close to the pattern. Otherwise, a hard spot in the sand will be made, and molten metal coming in contact with it will boil and tear the sand away to the depth of the hard spot. This will also occur if a gagger iron is rammed against a pattern and the sand between is pressed into a hard spot. When the molten metal reaches the wet gagger iron, an explosion usually results.

Molding Machines

Three types of molding machines are used in foundries: straight, semiautomatic, and automatic. Equip all molding machines with two-hand controls for each operator assigned to the machine. On automatic molding machines, install shields or apron-type metal guards to protect pinch points.

Core-Blowing Machines

Straight, semiautomatic, and automatic core-blowing machines are used in foundries. On semiautomatic and automatic machines, guard core-box push cylinders, counterweight cable pulleys, wheel guides, and table-adjusting footpads. Install an automatic barrier guard between the operator and the machine. If the drier is lowered automatically from the rollover and then pushed and raised toward the operator, there can be a pinch point between the lowering table and the raised table. This hazard is eliminated by an automatic barrier guard.

Equip automatic and semiautomatic core machines with double-solenoid valves. Maintain the slide valve well, and lubricate it to prevent recycling or other malfunctions.

General Suggestions

To prevent sand blows, maintain parting lines in good condition. Also, guard the parting line of the core box with a dike seal (Figure 21–31).

Safeguarding

Where practical, provide two-hand operating controls to prevent the operator from placing a hand or fingers between the top of the core box and the ram. Where two operators are employed on a core-blowing machine, provide two sets of two-hand control buttons. Equip all core boxes with handles so employees can move their boxes without placing their hands on top of them. If driers are located above the roll-over area for each core, place them high enough so they will not become entangled during the roll-over process.

Cleaning

Some core dips contain substances capable of producing dermatitis on sensitive persons. Rubber gloves and plastic sleeves usually provide adequate protection. However, check employees engaged in core dip operations at frequent intervals for sensitivity to the core dip solution. Materials used in cleaning core boxes may also be toxic. Therefore,

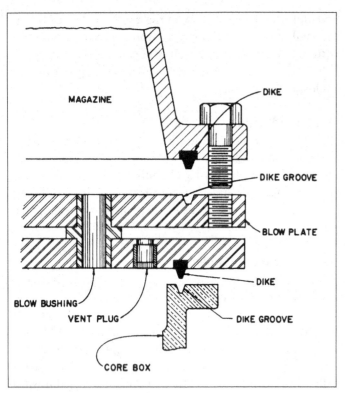

Figure 21-31. This section of a core box shows a rubber dike seal, which prevents sand blows and abrasion of the box. *(Courtesy Dike-O-Seal Corp.)*

remove their emissions with a properly designed ventilation system.

Flasks

Iron or steel flasks are preferable to wood flasks. Wood flasks become worn, burned, or broken so that they do not fit together well and may let molten metal run out during pouring. Do not leave defective flasks in the foundry building or in outdoor storage piles because they may be put back into service without first having been repaired. Have competent inspectors carefully inspect flasks at frequent intervals. Inspectors should have the authority to have the defective ones destroyed or sent to the repair shop.

Flask trunnions should have end flanges at least twice the diameter of the trunnions to minimize the danger of hooks slipping or jumping. Trunnions should preferably be turned, or otherwise be smooth castings. It is sometimes best to cast the trunnions separately and bolt or weld them in place. This procedure speeds up machining operations and permits reuse of trunnions recovered from broken flasks. Trunnions cast separately should be of steel.

When trunnions are bolted or welded in place, the nuts should be inside the flask. If they project on the outside, slings are likely to catch on them and slip off with a jerk, which subjects both the sling and the trunnion to severe strain.

Large flasks should have loop handles made of wrought iron. On steel flasks, cast handles at frequent intervals to make chaining possible.

Design trunnions and handles for the loads they are to carry, and construct them with a safety factor of at least 10. Make sure that the bolts that fasten trunnions and handles to the flasks are of sturdy enough construction.

Sandblast Rooms

Each foundry should have dust-tight sandblast rooms. Keep the doors to these sandblast rooms closed, and dust castings before they are removed from the rooms. Even small cracks in the walls or under doors will allow fine dust to escape and to contaminate air in the foundry. Equipment for workers in sandblast rooms should include air-supplied hoods and full-body protection.

Tumbling Barrels

Tumbling barrels need frequent care to keep them dust-tight. Enclose barrels that cannot be maintained dust-tight in booths connected to an exhaust system. Barrels may be equipped with exhaust ducts through the trunnions. Safety precautions to observe with tumbling barrels include (1) placing a removable guardrail around the machine and (2) locking barrels in a stationary position during loading and unloading.

Shakeout Machines

Shaking out castings presents the danger of hands and feet being crushed or arms and legs being broken. For this reason, workers must wear steel-toed or metatarsal-guarded foot protection. If steel hooks or rakes are used to pull castings from the screen, instruct workers to stand with one foot behind the other. That way, they can keep their balance in case the hook slips from the casting while they are pulling.

Because this operation is also often a source of dust, install hoods on shakeout machines and provide local exhaust to draw the dust to a collector. In fact, many foundries perform shakeout operations at night so that as few people as possible are exposed to dust.

Design shakeout machines so that the flasks cannot fall off the plunger. Do not allow foundry workers to retrieve gagger irons while these machines are in operation.

The hazards of sand conveyors are also found at shakeout machines because the sand is collected under the machines on a conveyor belt that moves the sand to storage for reclaiming and reuse. Keep the area around the shakeout machine free of sand and scrap. Guard the conveyor belt's opening at the sides.

CLEANING AND FINISHING FOUNDRY PRODUCTS

Install and operate grinding, polishing, and buffing equipment for foundry use as recommended in Chapter 20, Metalworking Machinery, in this volume. Have qualified personnel mount and change abrasive grinding wheels. Closely supervise the use of correct washers and wheel-mounting procedures. Keep required wheel guarding intact. Speed-test new wheels before allowing them to be used on the job.

Require operators to wear full personal protective equipment for eyes, face, hands, and feet. Excessive dust generated by dry abrasive wheels is a potential health hazard. Remove this dust with an exhaust system at the point of origin. Precleaning castings in a barrel, mill, or abrasive chamber also minimizes dust from grinding. Keep the space around the machines dry, clean, and as free as possible of castings and other obstructions. Note that silica dust is considered a suspect carcinogen.

Magnesium Grinding

The fundamental hazards of grinding magnesium are the possibilities of fire and explosion. To eliminate these hazards, use a proper dust-collection system.

Dust-Collecting System

In a dust-collecting system for magnesium, the dust should be wetted by a heavy spray of water and immediately washed into a sludge pit in which the dust is collected under water. Keep sludge pits or pans well ventilated because hydrogen evolves from the reaction of the collected dust with water. Frequently clean sludge pits or pans. Do not let wet magnesium dust stand and become partially dried because fire or explosion could result.

The dust-collecting system must not have filters or obstructions that allow dust to accumulate. Install pipes and ducts, and use the shortest possible route to eliminate bends or turns in which magnesium dust or fines could collect. As often as necessary, clean pipes and ducts connecting the grinder and the collecting device. Disconnect pipes and ducts while wheels are being dressed.

Also provide the following safeguards when grinding magnesium:

- a means for immediate quenching of sparks from grinding wheels, disks, or belts
- dustproof motors to prevent the accumulation of static charges
- explosion doors on the collection system
- an automatic interlocking control on the collection system to ensure its operation whenever grinding is started.

General Housekeeping

Good housekeeping is essential for safe handling of magnesium. Prevent accumulations of magnesium dust on benches, floors, window ledges, overhead beams and pipes, and other equipment. Do not use vacuum cleaners to collect the dust. Have it swept up and placed in covered, plainly labeled, iron containers, and if it is not recycled into operations, dispose of it in accordance with applicable federal, state, and local regulations. Do not allow magnesium dust to be mixed with regular floor sweepings.

Because sparks can be produced, it is dangerous to use equipment that grinds magnesium to grind other metals. Mark equipment for magnesium grinding FOR MAGNESIUM ONLY. Use benches made of wood grating for rough finishing operations.

Prominently display warning signs inside and outside the grinding rooms or areas. Post signs that warn against smoking and against the use of water on magnesium fires. Signs should instruct how to use powdered graphite, limestone, or dolomite as an extinguishing agent.

Keep close to each grinding unit an ample supply of powdered graphite in plainly labeled and covered metal containers. Place a scoop inside each container. Keep the container's lid loose for easy access.

Chipping

Where castings are cleaned or chipped, provide tables, benches, and jigs or fixtures specially designed and shaped to hold the particular casting. Install screens or partitions to protect other employees from flying chips. Install hoods and exhaust systems in these areas to remove dust. Require workers to wear eye and face protection when cleaning or chipping castings.

Welding

Consider where welding is done when cleaning or reclaiming castings. To help prevent fires in areas where welding operations are conducted, spread sand on the floor to a depth of 2 in. (5 cm). Sand, one of the best noncombustible materials, is plentiful in all foundries, but is also a health hazard.

Powder washing is a method of cleaning castings in which a stream of powdered iron oxide is introduced into a gas flame to intensify the heat produced. Perform powder washing according to the same safe practices as other carbon-steel or cast-iron welding and cutting. However, when this method is used to clean or cut sprues, gates, and risers from alloyed castings, use exhaust ventilation. (See Chapter 19, Welding and Cutting, in this volume, for more safe welding practices.)

Power Presses

Power presses are used widely in finishing departments of foundries. For safety in power press operations, provide sufficient aisle space, good housekeeping, and effective lighting. Properly guard and maintain machines in good working order. Carefully select operators, and train them in the efficient and safe operation of power presses. Use mechanical feed and ejection equipment whenever practical. These topics are fully discussed in Part 1 of this chapter, Cold Forming of Metals.

FORGING HAMMERS

There are several types of forging hammers: open-frame, gravity-drop, and steam and air hammers. They have similar hazards in common and require special safeguarding and safe work practices.

Open-Frame Hammers

Open-frame or Smith forging hammers are constructed so the anvil's assembly is separate from the foundation of the frame and operating mechanism of the hammer. They may be single or double frames. Flat dies are generally used in Smith hammers, and the work done allows for more machining of material.

Gravity-Drop Hammers

Drop forgings in closed-impression dies are produced on gravity-drop hammers—both board-drop hammers and steam- or air-lift drop hammers. Both types of gravity-drop hammers shape the hot metal in closed-impression dies. The impact of the hammer's blows shapes the forging through one or more states to the finished shape. On gravity-drop hammers, the ram and the upper die are raised to the top of the hammer's stroke. The impact blow comes from the free fall of the ram and the die.

Steam Hammers and Air Hammers

Steam hammers are also classified as drop hammers. Most steam hammers are double acting. They use steam pressure, or air pressure, that goes through a piston and cylinder to raise the ram and the die and to assist in striking the impact blow. Because steam or air power is used in addition to the weight of the falling ram and die, the steam hammer strikes a heavier blow than a gravity-drop hammer using an equivalent falling weight.

Hazards of Forging Hammers

For the most part, all types of forging hammers have identical hazards. The most frequent causes of injury include the following:

- being struck by flying drift and key fragments or by flash or slugs
- using feeler gauges to check the guides, wear, or the matching of dies

- using material-handling equipment improperly, such as tong lifts
- having fingers, hands, or arms crushed between the dies
- having fingers crushed between tong reins
- receiving kickbacks from tongs
- using swabs or scale-blowing pipes with short handles
- being burned by hot scale
- dropping stock on the feet
- getting foreign objects, such as iron dust or scale, in the eye
- noise-induced hearing loss.

A hearing conservation program that includes proper hearing protection and annual audiometric examinations, as well as engineering controls, will greatly reduce or limit noise-induced hearing loss. (See NSC's *Fundamentals of Industrial Hygiene*, 6th ed., 2012.

Injuries may also occur from a steam drop hammer when the ram pulls off a new piston rod. Sometimes the rod must be set in the ram several times before it holds. If the piston rod breaks, the ram will fall. This hazard emphasizes the importance of operators using a safety prop to support the ram before reaching under it.

Operating a hammer with a worn cylinder sleeve is also hazardous. When the sleeve is so worn that the swing of the ram cannot be controlled at the throttle control, shut the hammer down and repair it.

Operating a hammer with broken piston rings is also dangerous. A piece of broken piston ring passing through the steam ports and lodging in the throttle's valve can cause the ram to drop out of control. When this happens, the operator's tongs or the transfer tool is often caught, thus causing serious injury.

Guarding

Maintenance personnel, in particular, are exposed to the potential danger of crushing injuries when they remove and install parts on the top of the hammer and when they remove sow blocks, anvils, and columns. To avoid these injuries, provide and use means for locking out the power. To provide safe footing for personnel, install catwalks and guardrails on all hammers (Figure 21–32).

Gravity-Drop Hammers

On gravity-drop, steam-lift, or air-lift hammers, use a hand lever rather than a treadle for cold restrike operations. Provide two-hand tripping controls (1) if the material being forged is not held by the hands or by hand tools or (2) if a safety stop or tripping lever cannot be installed.

On board-drop hammers, provide a substantial guard around the boards above the rolls. This prevents the boards from falling should they break or come loose from the ram (Figure 21–33). Other standard protective features for a

Figure 21-32. Permanent catwalks installed along the row of board-drop hammers make repair and servicing of hammers easy and safe.

board-drop hammer include the ram stop and safety chain for the tie bolt and nut.

Steam and Air Hammers
Steam and air hammers should have a stop valve or quick-opening and -closing valve. Also, provide a safety head in the form of a steam or air cushion (if not already standard on the hammer) to prevent the piston from striking the top of its cylinder. Connect the cylinder's head and safety bolt head to an anchored wire rope.

Key-Driving Rams
A pneumatic key-driving ram is superior to a manually operated one and offers a far greater margin of safety (Figure 21–34). Make the key-driving ram of properly hardened steel so that it will not chip on impact. Keep it in shape or replace it—do not burn it off with a cutting torch.

Scale Guards
Install a scale guard (to confine pieces of flying scale) as standard equipment on the back of every hammer. The guard should allow ample clearance for the ram and easy access to the dies. It may be installed in one of the following ways:

- hinged on one side to an upright post so that the guard can be swung closed or open, out of position, when access to the die area is required; considered the most efficient and widely used throughout the industry
- supported on a floor standard
- suspended from the ceiling or anchored to a rail.

Treadles and Pedals
Provide treadles and pedals with ample clearance. Guard them to prevent them from being unintentionally activated by a falling object. Also guard any portion of a treadle or pedal at the rear of the hammer so that scrap or other material cannot interfere with the treadle's action.

Flywheels and Pulleys
Enclose flywheels or drive pulleys with a guard that is strong enough to prevent the pulley from falling to the floor should the shaft break. In this installation, the strength and location of the guard's bracket or the frame are more important safety factors than is the gauge of the sheet metal used for the enclosure. Bolt the brackets to the column of the hammer. In some instances, the guard enclosure is sup-

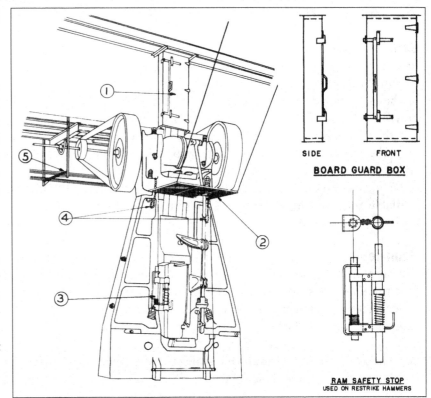

Figure 21-33. A well-guarded board-drop hammer features (1) sheet steel board guard box, (2) screen platform made from No. 9 expanded metal, (3) steel ram safety stop that swivels on the left column, (4) safety chain to restrain tie bolt and nut, and (5) catwalk and belt catcher. Details of board guard box and ram safety stop are shown in drawings at the right. (*Courtesy American Brake Shoe Co.*)

 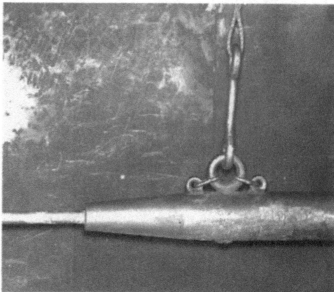

Figure 21-34. Mechanical key-driving rams, like the pneumatic model on the left, are preferred to manual ones. A manually operated key driver is shown on the right. *(Right photo courtesy Tractor Works, International Harvester Co.)*

ported from the floor by an I-beam. Restrain all cylinder bolts, gland bolts, and guide bolts and liners, as well as the head assembly over the operator's working position, with wire ropes or chains.

Safety Props

Provide safety props equipped with handles at the middle. Require workers to use them when repairing, adjusting, or changing dies. The props should be held in place while power is released. This permits the weight of the upper die and the ram to rest on the props. Operators should never place their hands on top of a prop. The props can either be chained to the hammer, so they cannot slip out of position, or be hinged to the side of the hammer, so they are readily available and easily moved into and out of their blocking position (Figure 21–33).

Hand Tools

Use pliers, tongs, and other devices specially designed to feed the material so the operators need not place their hands under the hammer at any time. Tongs should be long enough that they can be held at the side of the body rather than in front. Tongs should fit the shape of the materials being held for forging. Oil swabs and scale brushes or pipes should also have handles long enough so operators do not have to place their hands or arms underneath the die.

Die Keys

Use die keys made of a suitable grade of medium, carbon-alloy steel that has been properly heat treated so it will not crack or splinter. Both ends of the key should be tapered for clearance in driving and removing the key. Never use mushroomed keys.

Die keys must be the correct length. Use shims if necessary. If keys project farther, they become a hazard to the operator working in front. They may also break off while the hammer is operating and fall between the dies in back.

Stock an adequate supply of die keys so drifts will be needed only when the end of a key becomes distorted and must be cut off before the key can be driven out. Block or securely hold the drift with a drift holder.

Design of Dies

Hammer dies are usually made of chrome, nickel, or molybdenum stellite—materials that have high resistance to heat, shock, and abrasion. Die blocks are commercially supplied in four different tempers. Selection of the proper die steel in the correct range of hardness is important in controlling checking and breakage of dies.

Size, amount of striking surface, and height are other pertinent factors in the safe design of dies. Allow the correct amount of striking surface in relation to the size of the die. Too little striking surface may cause breakage or an undersized forging when the dies pound down. Too much striking surface, especially if it is unbalanced, may cause a pull or misalignment.

Specify correct die height, especially for resinking forge dies. The dies must be made so they meet in precise alignment. Lay out the dies so the major portion of the heavy forge work is done in the center of the die under the center of the ram, where the maximum force of the hammer is transmitted. Each impression in the die must be backed up with enough die material to reduce the possibility of breakage, especially where multiple impressions or nesting methods are employed.

Systematically arrange preliminary or breakdown operations so they do not create a hazard for the operator as the forging cycle is completed. Avoid radical bends or severe reductions in volume that might tend to jerk the tongs from the hands of the operator. Modify such operations or have them completed in additional operations.

Make the thickness and width of flash, gutter, and sprue ample enough so the flash or tong holds are not sheared off. The size of the gates is important—design in relation to the size of the stock and the tongs used. Gates should have enough width, depth, and clearance to allow safe handling.

Some dies, especially for smaller hammers, are designed with cutoffs that shear the completed forging from the end of the bar. If possible, place such cutoffs on one of the rear corners of the die for the operator's safety. If cutoffs are placed on the front of the die and are used by placing the stock across the knife's portion at an angle, provide enough clearance between the die and the hammer's frame or gib.

Because of the nature of forging work, and the abnormal abuse to which the dies are subject, maintenance of hammer dies is important.

Provisions for storing dies, such as racks and rails, are essential to safety, good housekeeping, and efficiency. Store dies in an area separate from the forge shop and away from vibration.

Setup and Removal of Dies

When forge dies are set up or removed, the hammer operator should act as leader of the group. The operator should see that all efforts are coordinated and that all safety rules are observed. That way, the work will be done efficiently and safely.

Pre-Setup Activities

Before setting up dies, clean the immediate area around the hammer and clear it of obstructions. Do not perform maintenance work on the equipment when setting up a die.

The hammer crew should check the equipment between setups.

Good lighting is essential for accurate setting of dies. It gives the operating crew a better view of potential hazards. Portable lights may be used. They should have heavy-duty cords, with bulbs protected by heavy-screen guards.

If lift trucks are used, be sure that the floor is level, in good condition, and free of obstructions. If cranes are used, check that the lift chains are in good condition and that the die pins have a snug, but free, fit.

Setting Up Dies

Dies are usually heavy and hazardous to handle without proper equipment.

Do not use transfer boards to move dies between the workbench and the machine. Transfer trucks, preferably of the elevating type, are safer and more efficient. Use power lift trucks or die trucks for moving and installing dies. Block or secure lift trucks to the base of the hammer when dies are to be set or removed. Otherwise, the truck may slip from the hammer, causing the die to slip and fall. In addition, the operator should check the truck's safety controls prior to starting the die set. See that operators are trained to safely operate this equipment.

Many methods are used in setting dies in hammers. The type and size of dies and the type of hammer determine the method to be selected.

Using shims on the dowel in the top die creates an extra hazard. Normalized spring steel is used to shim dowels that must be set so they will fall into place when the ram engages the die. The hammer operator should record the number and location of shims (whether front or back) so succeeding shifts or different hammer crews can refer to the record of the setup for that specific set of dies.

If a die must be moved to match, use a prop after the ram is raised and before the operator reaches under the hammer to reset the shims. This prop must be strong enough to support the ram and long enough to extend from the top of the die to the ram.

If allowance is made for moving the dies, make the allowance on a steam hammer in the top die only—the bottom die should have a tight fit. On gravity-drop hammers, however, the general safe practice is to have the top die tight, allowing for movement in the bottom die.

Take extra precautions and use special equipment for abnormally large or long dies. In setting such dies, the regular safety procedures for propping and handling may have to be changed. Get the approval of proper facility authorities for any changes.

After driving the die keys and before adjusting the gibs or column wedges, apply heaters to the dies if they have not been preheated. On deep impression jobs, it is a good practice to preheat dies in special low-temperature furnaces, in hot-water baths, or with hot scrap steel before setting them up. After heating the dies to proper temperatures, drive the die keys tight again by means of either a pneumatic ram or a light, suspended ram (Figure 21–34). If any further adjustment to the hammer is required, it can be done after a tryout forging has been made.

The hammer crew should use any waiting time to make a final check before getting ready for production.

Removing Dies

Before dies are removed, clear the immediate area around the hammer of overhead trolleys, suspended tongs, portable conveyors, tool and billet stands, and other equipment. Tie down overhead trolleys so they will not creep back into the work area. Move the scale guard back, and remove accumulated scale that would interfere with safe footing. Immediately move forgings away from the unit, and place them in the next workstation.

If another set of forging equipment has been delivered, place it nearby but not directly in the area where the hammer crew will work. To eliminate unnecessary handling, make sure that service personnel (truckers, crane operators, and hookers) are familiar with the proper procedure.

Shut off and lock out the hammer's energy sources (electrical, air, steam, or hydraulic) before loosening the die keys. The top key is generally loosened first, usually with a mounted pneumatic ram (Figure 21–34). A light, well-balanced ram suspended from a cross beam or from an overhead crane or chain fall can also be used successfully.

Using a manually held drift pin or a knockout on a die key after it has been loosened and driven to a position even with the face of the ram or the sow block is a hazardous operation. Instead, use a special type of adjustable knockout that is held in position mechanically rather than manually.

After the die keys have been driven out, raise the ram and prop it at once. The prop must be in good condition and must be placed on a clean surface. On a gravity-drop hammer, use a jack to raise the ram. A special prop may be required.

Do not attempt to raise the hammer to propping level if the top die has a tendency to "hang." Instead, first free the die within the shortest possible distance from the face of the bottom die. Prop the ram on an air-lift drop hammer with special care. After securely positioning the prop under the ram, shut off and lock out the power. Use special platform trucks with winches for this operation. They are practical and safe because the dies are horizontally winched or pulled out directly onto the table of the truck. Do not permit the dies to be dumped out of the hammer onto the floor.

After removing the dies from the hammer, extract the dowels. Two workers should drive out the dowels with the proper tools, usually a drift and a sledge. These tools should be in good condition and have sound handles. Because there is metal-to-metal contact, see that workers are careful and wear personal protective equipment.

Load the removed dies onto low, steel pallets and take them from the area as soon as possible. If dies need repair or modification, the hammer operator should notify the supervisor. The supervisor should then have the die-servicing department take care of any repairs before the next run.

Safe Operating Practices

Make the supervisor who directs the activities of workers in the hammer crew responsible for safe work practices.

Personal Protection

Operators of forging hammers and other employees in the vicinity of equipment should wear suitable personal protective equipment. This includes full eye protection, safety hats, protective footwear, leather leggings and aprons, and hearing protection in accordance with the plant's policy.

Operators should also wear cotton gloves. When the gloves get wet, they should be removed and allowed to dry. Operators should not wear leather gloves because perspiration may cause steam burns.

Maintenance and Inspection

A well-planned preventive maintenance program for forging hammers helps reduce the number and severity of incidents by minimizing the breakage and wear of parts. Regular inspections disclose production units that are not properly operating so repairs or adjustments can be made.

The results of a good maintenance program can be measured in reduced operating costs that include:

• cost of machine downtime, breakage, and lost production
• cost of replacement parts and labor
• cost of incidents due to faulty equipment.

Maintenance checklists for hammers (Figures 21–35 and 21–36) can be the basis for formulating a definite, planned inspection program. A written checklist avoids the errors resulting from verbal reports that are often forgotten or misunderstood.

Set up a work schedule for repairs based on data recorded on the checklists. Transfer data to the permanent records of the equipment, and use it for future planning in the maintenance program and to compare costs.

Because steam hammers constitute a considerable portion of the forging equipment, establish a definite maintenance program for them. Many steam hammers are not kept in as efficient condition as possible. Usually, the cost of operation is not known, but upkeep costs are higher for units in poor condition. Waste of steam usually results from worn piston rings or sleeves, loose heads, blown head gaskets, and leaky glands. Replacing worn rings reduces costs. Worn piston sleeves and sloppy linkage make the hammer

BOARD HAMMER MAINTENANCE CHECK			
Date_____ Hammer No._____ Location_____			
ITEM CHECKED	CONDITION	TYPE OF REPAIR	EST. HRS. TO REPAIR
TIE BARS & SPRINGS			
FRAME STUDS & SPRINGS			
DIE KEYS & SHIMS			
SOW BLOCK KEY & SHIMS			
RAM CLEARANCE			
GUIDE BOLTS & ADJ.			
MOTOR MOUNTS			
MOTOR COUPLING			
ROLLSHAFT BEARINGS			
WIRING & CONTROLS			
FLYWHEEL & BEARINGS			
DRIVE GEARS			
ROLL ADJUSTMENT			
LUBRICATION			
STEAM LINES			
BOARDS & WEDGES			
AIR LINES			
AIR FOOT SWITCH			
AIR CYL. & LINKAGE			
TREADLE & LINKAGE			
BOARD CLAMPS & LINKAGE			
DOGS & STOPS			
KNOCKOUT ARM			
FRICTION RODS			
SAFETY RODS			
REMARKS:			

Figure 21–35. A maintenance checklist for board-drop hammers.

STEAM HAMMER MAINTENANCE CHECK			
Date_____ Hammer No._____ Location_____			
ITEM CHECKED	CONDITION	TYPE OF REPAIR	EST. HRS. TO REPAIR
CYL. HEAD BOLTS			
MOTION VALVE STEM			
MOTION VALVE CRANK			
MOTION VALVE CONNECT.			
WIPER BAR & CRANK			
THROTTLE CRANK			
THROTTLE LINKAGE			
RAM & SOW BLOCK			
DIE KEYS & SHIMS			
SOW BLOCK KEYS & SHIMS			
GUIDE BOLTS			
GUIDE WEAR			
GUIDE ADJUSTING BOLTS			
GUIDE WEDGE POSITION			
HOUSING BOLTS & SPRINGS			
TREADLE			
TIE PLATE LINER			
PISTON ROD GLAND PLATE			
GLAND BOLTS			
TREADLE PLATFORM			
STEAM CUSHION LINE			
SCALE HOSE			
SAFETY LINER			
SAFETY PROP			
STEAM LINES			
STEAM SHUTOFF VALVES			
REACH ROD			
COLUMNS FLAT ON BASE			
COLUMN WEDGE & BOLTS			
SPOOL BOLT & PIN			
SAFETY CABLES			
CRACKS IN BASE			
REMARKS:			

Figure 21–36. A maintenance checklist for steam hammers.

hard to control and create a hazard. Loose cylinder heads also are dangerous.

Periodic inspection of every forging hammer helps ensure the proper condition of bolts, screws, keys, valves, and other parts that may be loosened by vibration. Similarly, make thorough periodic inspections and adjustments of all parts of the treadle or pedal, clutch, and other operating mechanisms. Worn or loose treadle linkage, motion arm, crank arm, and treadle can cause the hammer to go out of control. Keep these parts in especially good repair.

The clutch is also a vital part of the forging press. Keep it in good condition if the press is to operate efficiently. Replace a broken spring or part that shows wear at once.

Lead Casts

Lead casts are taken in practically every conceivable manner in the forging industry. If possible, take casts only in an isolated area, where there is no likelihood of interference from, or injury to, other workers. Make sure die impressions are dry because hot metal that contacts water produces flying particles of molten metal. Make sure that lead pots are properly ventilated. (See NSC's *Fundamentals of Industrial Hygiene*, 6th ed., 2012.

FORGING UPSETTERS

The upsetter is a horizontal forging machine that forges hot bar stock, usually round, into a great many forms. The forms are made by squeezing action instead of impact blows, as in the case of forging hammers. Although numerous hazards are involved in the operation of an upsetter, the most serious problems are encountered in changing the dies.

Enclose the entire machine as much as possible, except for the feed area. Use heavy wire mesh, or expanded or sheet metal reinforced with structural steel. Cut doors into the enclosure to service the flywheel, brake, and other moving parts. Install a guard over the operating pedal.

For safe operating conditions, keep the area around the machine clean and clear of obstructions and litter. Especially keep the top of the machine clear of any objects—such as loose bolts, bars, nuts, or shims—that might fall into it or from it.

Note: Before attempting to adjust dies, heading tools, stock gauges, or backstops, the operator should shut off the power, lock the main power switch, and, after the flywheel has stopped completely, immobilize the flywheel.

Design of Dies

Dies and heading tools used in an upsetter, or horizontal forging machine, do not usually receive the severe abuse that hammer dies receive. For gripper dies, use a good grade of chrome, nickel, or molybdenum steel of the correct hardness.

For abnormally heavy jobs (or jobs that would create an unbalanced condition when running), design and use balancing equipment that eases handling and reduces operator fatigue. Make sure that the grip sometimes provided on the upsetter's die impressions is strong enough to hold the stock securely. Check the grip after every run. This precaution is important for the operator's safety, especially on jobs where the heading tool could push the bar stock out of the impression toward the operator.

Setup and Removal of Dies

At the end of a run and before further work is done, move all skids of stock or forgings out of the area to allow as much room as possible for changing the dies. Die setters should inch the header slide forward to make a complete setup by measuring headers, strokes, and dies. See that this practice is especially followed on a worn machine that requires special shimming for proper alignment.

Inspection and Maintenance

Because worn or defective upsetters can be dangerous to operate, keep these machines in top working order. Establish a definite program of inspection and maintenance.

At least once a week, the maintenance crew should check all working parts for wear and proper adjustment. Daily, the crew should inspect the air clutch and brake. Upsetters also require daily lubrication. If possible, install a means for automatic lubrication.

Operators should daily inspect air gauges, air lines, water lines, water valves, belts, pulleys, and tools. They should also check daily, and immediately report, any abnormal function. At each use, inspect all equipment for handling dies, such as chains, cables, and eyes or swivels.

Auxiliary Equipment

Design all equipment needed to safely operate upsetters, such as stock gauges, tongs, oil swabs, and scale removers, for the particular job.

Keep tools in good condition. Provide a complete set of wrenches to fit all sizes of nuts or bolts that are on the machine.

There are three basic types of stock gauges. The front gauge locates and swings away; the backstop gauge locates and helps hold the stock in place; and the special tong gauge, or finger gauge, locates and helps control the stock.

Use tongs made of tough, low-carbon steel so they will not harden from repeated quenching in water. The jaws of tongs should conform to the shape of the stock being handled.

Use oil swabs and scale removers with long handles. Thus, operators can reach the full length of the dies without having to put an arm or hand between the dies.

Air, electrical, water, and oil lines should have distinctly marked shutoffs. Locate safety valves or switches in a spot convenient for the operator.

FORGING PRESSES

Forging presses, because of their basic design, are similar to power presses, discussed earlier in Part 1, Cold Forming of Metals. Forging presses range from 500- to 6,000-ton (453,600- to 5,443,200-kg) capacity. However, because their work is quite different from the conventional cold-stamping operation, forging presses have their own operating technique, die setup, and maintenance problems.

Compared with the cold-stamping press, the forging press is designed with a faster-acting slide. The speed of the downward motion and of the pickup of the slide is one of the factors that determine the life of forging dies. Another factor is temperature control of the dies. Therefore, the action of the press should be fast enough to minimize the length of time that the dies are exposed to the billet's forging temperature.

Basic Precautions

The rapid action of the slide creates certain hazards that the operator must recognize and control. The size and shape of the forgings and the method of moving them into and out of the dies limit the use of point-of-operation guarding. The single most important factor for preventing injuries is the operator's control of the tools and methods used. Provide safety hats, hearing protection, eye protection, and protective footwear for operators at all forging press operations.

Tongs

Tongs, die swabs, and special handling tools are ordinarily the only tools used by an operator during the forging operation. Keep these tools in good condition at all times.

Maintain proper clearance at the front of the press so operators have enough room for their hands in case of upward or downward motion of the tongs. This motion can be caused by improper spotting of the billets or tongs at the striking surfaces of the forging dies.

Scale

Properly locate steam lines, air lines, water headers, and splash aprons for scale and oil. If too near the working area of the operator's hands, they create pinch and shear points when tong handles are forced against them.

Unless properly confined, hot scale that is produced during the forging operation can cause serious burns and eye injuries. To prevent scale from coming out of the front of the press, locate air or steam curtains at the front of the die and direct them onto the die's facing. Install combination scale-and-smoke exhaust hoods at the back of the press to confine the scale and exhaust the smoke created by die lubricant.

Guarding

Equip all forging presses with pedal guards and antirepeat devices. Never operate a forging press in the continuous-stroke mode. The operating controls should require depression of the pedal for every cycle of the slide. The press's controls should also permit inching of the slide for die setting or other press adjustments.

Die Setting

Because the setting of dies in forging presses is very different from the setting of dies in cold-stamping presses, certain extra precautions are required. Dies in forging presses are set in die holders designed in sets, each set consisting of an upper and a lower section. The upper section of the holder is secured to the press's slide by threaded bolts; the lower section, to the press's bed by threaded bolts. The forging dies are set or nested in die pockets that are recessed in both the upper and lower sections of the holder.

To solve the problem of setting the dies in the recessed pockets and of removing them, use an eyebolt attachment at the face of the die and a die truck with a boom attachment. Carefully use pry bars and blocking because a pry bar could easily slip from the die and injure the operator's hands or fingers.

Secure the dies in the pockets of the holder with sectional flat clamps and a series of cap screws placed through the holder at all four sides of the die. These screws are primarily used to shift the die for matching and to prevent the die from floating after it has been properly aligned.

Before removing the clamps that secure the top die in the holder, install suitable blocking to hold the top die in the pocket. Do not rely on the adjustment cap screw to hold the die after removing the clamps. To prevent the motor from starting when the props are in use, equip the press with safety props that are interlocked with the press's motor circuit.

Keep the table wedges that are used for making vertical adjustment of the dies free of scale. In this way, the wedge can be raised and lowered with minimum pressure. Do not use trucks or driving rams for this operation. If wedges are kept free of scale, they can be raised or lowered by hand or by an air-powered motor wrench.

Maintenance

Maintenance of forging presses requires the same precautions used in maintenance of power presses used in cold stamping. First, lock out the energy source so the equipment cannot run while the maintenance is being done. Place safety props under the ram and surge tanks, and bleed the pressure lines so that the press's parts cannot be unintentionally activated. However, if adjustments must be made with the power on, see that the work is carried out under the direct supervision of the maintenance supervisor.

Install permanent work platforms for making brake adjustments and doing repair work at surge tanks and booster cylinders. To prevent falls, maintenance crews should not use portable straight ladders nor stand on parts of the press, such as the press's crown or the backshaft.

Major repair work on forging presses usually requires removal of bulky, heavy parts. Tearing down the rolling clutch, flywheel, slide, pitman, and crankshaft requires special heavy-duty rigging. Have only skilled workers, trained in tearing down, do this type of work.

OTHER FORGING EQUIPMENT

The following are examples of additional forging equipment:

- *Hot-trim dies and punches.* These dies and punches can be made from hardened chrome, nickel, carbon, and vanadium steels. They can also be made from medium-carbon steels with cutting edges that have been hard faced with a rod similar to a Haynes Stellite No. 6 rod. Use stellite on the trimmer only and not on both the trimmer and the punch. A stellite punch should never work against a stellite punch die. One or the other, preferably the punch, should be softer.

- *Cold-trim dies and punches.* These dies and punches are usually made from a high-carbon, high-chrome, molybdenum-steel hardened and drawn to a 60 to 62 Rockwell C hardness. In both hot- and cold-trim dies, the designer should equalize the trim so that stresses are equally distributed. Provide proper working clearance for unloading cold-trim dies and punches. If possible, design cold-trims to work on guide pins for proper alignment.

- *Padding, bending, or straightening equipment.* This equipment is made from a good grade of wear-resistant, carbon-nickel steel. Hot-pad dies are usually made with an opening between the die faces so that the hot forging acts as a cushion. Provide additional clearance in a hot-pad or restrike die because the forging is constantly cooling and shrinking.

- *Cold-coin dies.* These dies are made from high-carbon, high-chrome steels hardened to a 62 to 64 Rockwell C hardness. The forging is coined cold. For best results, use either interlocking dies or dies with guide pins. Additional safety measures, such as magazine-type loaders, are sometimes made an integral part of the die for the operator's convenience.

  Grind the faces of cold-coin dies as smooth as possible. Do not use any lubricant—it could cause the forging to stick to one die face. The opening between the dies is sometimes limited mechanically so that no more than one forging can be loaded at a time.

- *Bulldozers.* The greatest danger in operating bulldozers is the possibility of a worker getting caught between the dies. To decrease this hazard, (1) attach a guard to the side of the moving head that travels with the moving head past the stationary head, (2) use telescoping rods or rails, (3) notch out the base plate to leave room for the operator's leg, (4) keep the clutch in good order so the machine will not repeat, and (5) guard the power-transmitting mechanisms.

- *Cold-heading machines.* Provide screen shields on cold-heading machines to protect workers from flying pieces. Guard relief springs to prevent the bolts and nuts from being thrown out should they break.

- *Bolt headers and riveting machines.* Install treadle guards on these machines to prevent them from being unintentionally operated. Stop and block these machines before changing dies or making adjustments.

- *Hot saws.* Place tanks of water below the saws. Install 8-in. (20-cm) sheetmetal guards to stop flying sparks.

NONDESTRUCTIVE TESTING

Visual observation, even with magnification, cannot locate all small, below-the-surface defects in cast and forged metals, or in weldments, such as found in pressure vessels, boilers, and nuclear components. Proper nondestructive testing, however, reveals all such defects without damaging the parts being tested. Nondestructive testing methods locate the following defects:

- defects that are inherent in the metal, such as nonmetallic inclusions, shrinkage, and porosity
- defects that result from processing, such as high-residual stresses, cracks, and checks caused by handling, spruing, or grinding, or casting and forgings
- in-service defects, such as corrosion, erosion, and sharp changes in section.

The types of testing most commonly used for forged and cast metals are the following:

1. magnetic particle inspection
2. penetrant inspection
3. ultrasonic methods
4. triboelectric method
5. electromagnetic tests
6. radiography.

These methods, as well as others that apply to nonmetallic substances, are fully discussed in NSC Occupational Safety and Health Data Sheet 12304–0662, Ultrasonic Nondestructive Testing for Metals. Recommendations for installation, inspection, and maintenance of the electrical equipment used in many of these testing procedures are given in Chapter 8, Electrical Safety, in this volume.

Magnetic Particle Inspection

Magnetic particle inspection is the most widely used testing method for forgings. It uses magnetism to attract and hold very fine magnetic particles right on the part itself. If a defect is present, it interrupts the magnetic field and is clearly shown by the pattern made by the particles. The part is magnetized in suitable directions by DC-line voltages transformed to low-voltage (4 to 18 V), high-amperage AC, half-wave current, or three-phase full-wave current. Install and ground all electrical circuits according to NFPA 70, National Electrical Code.

Local exhaust is required to control the dust particles used for testing. If local exhaust is not feasible, operators should wear respiratory protection. They should also wear eye protection to guard against the irritating effects of the dust particles and of arcing. In addition, operators should wear personal protective equipment to prevent skin irritations from the dry powder and wet material used with this testing method.

Penetrant Inspection

Penetrant inspection is useful for revealing cracks, pores, leaks, and similar defects that are open to the surface in a metal or other solid material. The penetrant inspection process is as follows:

1. Clean the part to be inspected.
2. Apply penetrant to the surface. Within a few minutes the penetrant is drawn into defects by capillary action.
3. Remove the penetrant from the surface. Depending on the sensitivity of the material, the penetrant is removed by a water wash, a solvent cleaner, or an emulsifier followed by a water wash. The penetrant remains in the surface opening, however, until it is removed by the developer.

Use fluorescent penetrants to reveal defects under ultraviolet black light. Effectively shield ultraviolet equipment, or wear filter lenses of the correct shade. Defects may also be detected by a dye penetrant that contrasts with the surface color.

Because most penetrants are organic compounds that may cause dermatitis, avoid skin contact with penetrants. Wash exposed skin before smoking, eating, or drinking. Do not use or store smoking materials, food, or drink in the test area.

Ultrasonic Methods

Ultrasonic waves (above the audible range of 20,000 Hz) are created by an electronic generator that supplies high-frequency voltage to a piezoelectric crystal transducer. Three basic ultrasonic methods have been developed for nondestructive testing of forged and cast metals: reflection method, through-transmission method, and resonant-frequency method.

Reflection Method

In the reflection method, that portion of the ultrasonic beam that strikes a flaw or discontinuity in the material is reflected; the rest of the beam goes on. The piezoelectric crystal transducer radiates these waves through a coupling medium into the material. It also acts as a receiver to detect reflections, which are then picked up by an electronic amplifier and applied to a cathode-ray oscilloscope. The time intervals between the outgoing and the incoming waves are measured on the oscilloscope.

Through-Transmission Method

In the through-transmission method, a beam or wave is directed through a piece of material. If a flaw or discontinuity is found, the energy is absorbed, and the beam or wave does not get through. Fluids such as water, oil, and glycerine give better coupling than air. For that reason, they are generally used as the coupling medium among the transmitter, the material, and the receiver. In some applications, however, air or other gases can be used.

Resonant-Frequency Method

The resonant-frequency method is used primarily to measure the thickness of material. The equipment for resonant frequency consists of an electronic oscillator that supplies voltage of ultrasonic frequencies to a piezoelectric transducer. The transducer is pressed into contact with the part to be tested and includes lengthwise vibrations in the test piece under the area of contact. Another type of resonance instrument displays the thickness reading on a cathode-ray tube as a pip on a calibrated scale. Caution: Disconnect equipment from the power supply, and discharge the condensers whenever a cathode-ray tube must be adjusted or removed.

Triboelectric Method

The triboelectric method is used to detect minute quantities of current generated when two metallurgically or chemically unlike conductors are moved into frictional contact. If the conductors are alike, no current is generated. This method is designed to sort and identify metal parts of not more than four alloy types.

The equipment for the triboelectric method consists of a control unit and a portable sorting head, which are connected by means of a cable. The sorting head contains the main controls for conducting the test and is designed as a one-hand operation.

Electromagnetic Tests

Two types of electromagnetic tests are currently being used in industry: magneto-inductive and eddy current. A third type, employing radar frequency, is also being used, but only to a limited extent.

Magneto-Inductive Tests

This method of electromagnetic testing uses variations in the porosity of magnetic materials to create variations in a pickup coil or probe.

Eddy-Current Tests

The second and most common type of electromagnetic testing uses alternating current in a coil or probe to induce eddy current into the part being tested. Defects and variations in properties or shape cause changes in the strength and distribution of the eddy current. The readout of eddy-current tests is presented on a cathode-ray tube, on a meter, by audible or visible alarm, or by a combination of these methods. Follow the manufacturer's recommendation to determine any leakage of electromagnetic radiation that may be present.

Radar-Frequency Tests

This third method uses high-frequency radar waves to measure the electromagnetic properties of thin coatings and surface layers of material. To make such tests, (1) a wave guide or cavity oscillator is coupled to the test object, and (2) high-frequency waves are then reflected from the object, thus indicating the surface's electrical resistance and the thickness of the nonconducting coatings.

In some radar-frequency testing installations, operators have been burned internally when they passed between the object being tested and the testing device. Formulate and enforce special regulations, and set up barriers to prevent operators and other workers from entering such areas. Explicitly follow the recommendations of the manufacturer of the equipment.

Radiography

Radiography uses x-rays and gamma rays. X-rays are unidirectional and their wavelengths can be varied, within certain limits, to suit the condition. Gamma radiography differs from x-ray radiography in that the gamma rays are multidirectional and their wavelengths, being characteristic of the source, cannot be regulated. Gamma rays for radiography usually are obtained from isotopes of cobalt-60 or iridium-192.

In some instances, gamma-ray exposures are inferior to x-ray exposures in sensitivity and contrast. Gamma radiography, however, has several advantages. Because of the nature of isotopes, a number of tests can be made at the same time, provided that specimens can be suitably located. Moreover, isotopes are independent of electrical power, their sources are portable, and the small size of the sources makes it possible to obtain radiographs in tight quarters.

Devices used to transform differences in intensity of the penetrating radiation into visible images are x-ray films, fluorescent screens, proportional-scintillation Geiger counters, and ionization gauges. All sources of ionizing radiation are potentially dangerous. X-ray and gamma-ray sources may also produce hazardous secondary radiation. In addition, x-ray units involve both low and high potential electrical hazards. Appropriate radiation controls must be implemented.

SUMMARY

- Whether you are cold forming or hot working, the goal is to keep employees out of harm's way. If we can prevent entry into the work zone (point of operation), we can eliminate most injuries.
- In cold forming, the guarding options are many—with new technology being developed regularly. The key is to match the proper guarding design and controls to the application, and then provide appropriate training for operators, setup people, and maintenance workers. In addition, each power press should also have a stop control, stroking selector, inch control, and stroking control system. The clutch/brake control must automatically deactivate in case of loss of electrical power or proper air pressure.
- In hot working, there are fewer options when it comes to guarding, given the nature of the work and material, but we still have options, as explained in the material provided.
- A significant number of injuries can occur when work is handled hot or cold. Feeding and ejecting mechanisms can help to eliminate handling work and reduce exposure to those hazards. The proper use of personal protective equipment, while it seldom prevents an incident, can compensate by preventing an injury.
- In addition, the process of setting up or tearing out dies, fixtures, and the like can easily cause injury. Only experienced personnel should perform these operations.
- During maintenance work, equipment should be locked out to prevent accidental operation or electrical shock.

When employees must work on a machine with the power on and guards removed, they should not place any part of their body in the path of the equipment.

- Health hazards in foundries involve toxic and flammable materials; general and local ventilation; and specific problems such as silicosis, dermatitis, and radioactivity. Dust, solvents, and other materials present serious health hazards in foundries. Companies must design and implement safety and health programs, safeguards, and safety devices to recognize, evaluate, and control or eliminate these hazards.
- Because of the charging and blasting that take place in cupolas and the carbon monoxide generated by them, these furnaces present several hazards. Only experienced workers should do tapping out operations or attempt to repair cupola linings.
- Complete safety of power presses depends on adequately safeguarding the point of operation, training press operators and setup/maintenance personnel, and enforcing safe working practices.
- When determining point-of-operation safeguards, consider all hazards in the die's space that may crush, cut, punch, sever, or otherwise injure workers. A safeguarding device controls access to the point of operation and can be press controlling, operator controlling, or a combination of the two.
- Many auxiliary mechanisms to protect power press operators are available either from manufacturers or from a company's own shop.
- Kick presses present serious hazards to workers and must be carefully guarded and well lighted to prevent injuries. Kick-press operators should also wear eye protection and should guard against fatigue and strain, which can affect their performance.
- Electrical controls on power presses should be properly designed, applied, and installed as an important part of press safety. This is particularly true when a two-hand control device or two-hand tripping device is used for point-of-operation guarding.
- Good inspection and maintenance of power presses are essential in achieving a company's safety goals and productivity levels. Workers must be completely familiar with the construction and operation of the equipment for which they are responsible.
- Management should set up a checklist that outlines the frequency of inspection and maintenance for each press.
- Point-of-operation safeguards and backgauges for power press brakes should be specially designed for each press braking task. Operators must be able to control the speed and stroke of press brakes and have the correct hand tools to work the press brake safely.
- Improper materials handling in foundries results in a

wide variety of injuries. Supervisors and other managerial staff should instruct workers carefully in the use of proper safety practices and protective equipment.

- Refractory clay crucibles and ovens must be carefully inspected, operated, and maintained to prevent worker injuries. Good ventilation and guarding can protect workers from harmful fumes and hot materials.
- For safe operation of foundry production-line equipment; all grinding, polishing, and buffing equipment; and forging equipment, all exposed moving parts must be properly guarded and all electrical equipment grounded to eliminate shock hazards. Workers should wear personal protective equipment when using these machines.
- Nondestructive testing reveals all below-surface defects without damaging the parts being tested. This type of testing can locate defects inherent in metals and other solid materials or those that result from processing or in-service use.

REFERENCES

American Conference of Governmental Industrial Hygienists, 6500 Glenway Avenue, Bldg. D7, Cincinnati, OH 45211.

American Foundrymen's Society, Golf and Wolf Roads, Des Plaines, IL 60016.
    Engineering Manual for Control of In-Plant Environment in Foundries.
    Health Protection in Foundry Practice.
    Recommended Practices for Grinding, Polishing, and Buffing Equipment.
    Safety in Metal Casting.

American Industrial Hygiene Association, 475 Wolf Ledges Parkway, Akron, OH 44311.

American National Standards Institute, 11 West 42nd Street, New York, NY 10036.
    Safety Requirements for the Cleaning and Finishing of Castings, Z241.3–1999.
    Safety Requirements for the Construction, Care, and Use of Hydraulic Power Presses, ANSI B11.2–1995 (R2005).
    Safety Requirements for the Construction, Care, and Use of Mechanical Power Presses, ANSI B11.1–1988 (R2001).
    Safety Requirements for the Construction, Care, and Use of Power Press Brakes, ANSI B11.3–1982 (R2002).
    Safety Requirements for the Construction, Care, and Use of Shears, ANSI B11.4–1993 (R2003).
    Safety Requirements for Melting and Pouring of Metals in the Metalcasting Industry, Z241.2–1999.

    Safety Requirements for Sand Preparation, Molding, and Coremaking in the Sand Foundry Industry, Z241.1–1999.
    Safety Requirements for Workplace Floor and Wall Openings, Stairs, and Railing Systems, A1264.1–1995.

McMaster, R. Nondestructive Testing Handbook. 2nd ed. Columbus, OH: American Society for Nondestructive Testing, 1982.

National Fire Protection Association, 1 Batterymarch Park, Quincy, MA 02269.
    Electrical Standard for Industrial Machinery, NFPA 79, 1997.
    Flammable and Combustible Liquids Code, NFPA 30, 1996.
    National Electrical Code, NFPA 70, 1996.

National Safety Council, 1121 Spring Lake Dr., Itasca, IL 60143-3201.
    Fundamentals of Industrial Hygiene. 6th ed. 2012.
    Power Press Safety Manual. 4th ed. 1989.
    Safeguarding Concepts Illustrated. 6th ed. 1993.
    Occupational Safety and Health Data Sheets (available in the Council Library):
        Alligator Shears, 12304–0213, 1990.
        Coated Abrasives, 12304–0452, 1991.
        Concepts of Mechanical Power Press Point-of-Operation Safeguarding, 12304–0710, 1993.
        Electrical Controls for Mechanical Power Presses, 12304–0624, 1983.
        Handling Materials in the Forging Industry, 12304–0551, 1992.
        Handling Steel Plates for Fabrication, 12304–0565, 1990.
        Inspection and Maintenance of Mechanical Power Presses, 12304–0603, 1986.
        Kick (Foot) Presses, 12304–0363, 1990.
        Mechanical Forging Presses, 12304–0728, 1992.
        Mechanical Power Press Safeguarding: Movable Barrier Devices, 12304–0712, 1993.
        Metal Cutting Shears, 12304–0328, 1993.
        Power Press Safeguarding: Presence-Sensing Devices, 12304–0711, 1993.
        Power Press Safeguarding: Pullbacks and Restraint Devices, 12304–0713, 1986.
        Power Press Safeguarding: Two-Hand Tripping Devices, 12304–0714, 1990.
        Press Brakes, 12304–0419, 1993.
        Scrap Ballers, 12304–0611, 1989.
        Setting Up and Removing Power Press Dies, 12304–0211, 1991.
        Setup and Removal of Forging Hammer Dies, 12304–0716, 1993.

Steam Drop Hammers, 12304–0720, 1993.
Ultrasonic Nondestructive Testing for Metals, 12304–0662, 1990.
Upsetters, 12304–0721, 1993.

United Auto Workers, 8000 East Jefferson Avenue, Detroit, MI 48214.
General.
Nondestructive Testing.

U.S. Department of Health and Human Services, National Institute for Occupational Safety and Health, 4676 Columbia Parkway, Cincinnati, OH 45226.

U.S. Department of Labor, Occupational Safety and Health Administration, 200 Constitution Avenue NW, Washington DC 20210.
Code of Federal Regulations, Title 29.
29 CFR 1910.147, The Control of Hazardous Energy (Lockout/Tagout).
29 CFR 1910.179, Overhead and Gantry Cranes.
29 CFR 1910.211–219, Subpart O—Machinery and Machine Guarding.
29 CFR 1910.1200, Hazard Communication Standard.

Wilson, F. W., ed. *Handbook of Fixture Design*. New York: McGraw-Hill, 1962.

REVIEW QUESTIONS

PART 1

1. Describe the primary means of safeguarding power presses.
2. Briefly describe the antirepeat function of the clutch/brake control system.
3. Identify and describe the two types of clutches.
4. What is a presence-sensing device?
5. What is a pinch point?
6. What is the point of operation?
7. List the two basic categories of safeguarding the point of operation.
   a.
   b.
8. Name three of the four types of guards.
   a.
   b.
   c.
9. Where is the two-hand tripping device used?
10. If more than one operator is used on a machine, how many pull-back devices are necessary?
11. What is a kick press? Give two uses.
    a.
    b.
12. What is the recommended safeguard for a kick press?
13. Setup, removal, and handling of power press dies can be very hazardous. Identify three of the six typical injuries that can be suffered by maintenance personnel.
    a.
    b.
    c.
14. How should a power squaring shear be guarded?
15. What type of guard is commonly used on alligator shears?
16. What is the function of a power press brake?
17. When is the safe-distance method used on power press brakes?
18. What is the combined stroking control system?

Part 2

19. Identify two ways of eliminating dust as a health hazard.
    a.
    b.
20. List three ways of minimizing the toxic and flammable hazards involved in using solvents.
    a.
    b.
    c.
21. Which of the following presents a fire and explosion hazard in dust-collecting systems?
    a. aluminum
    b. antimony
    c. beryllium
    d. phosphorus
    e. all of the above
22. What aspects should be included in the safety and health program for foundry workers?
    a.
    b.
    c.
    d.
    e.
23. Prolonged contact with oil, grease, acids, alkalis, and dirt can produce _____.
24. The need for controlling ventilation in foundries is determined by what three factors?
    a.
    b.
    c.

25. What hazard is sometimes generated when charging and blasting takes place in cupolas?
26. To eliminate a carbon monoxide explosion hazard, supply adequate natural or mechanical _____ in back of the cupola, and open two or more tuyeres after the blowers are shut down.
27. What is the principal danger in handling refractory clay crucibles?
28. List four hazards in the construction and operation of core ovens and mold-drying ovens.
    a.
    b.
    c.
    d.
29. Which of the following is usually considered adequate for ovens under 500 ft³ (14,000 L) in volume?
    a. forced-draft ventilation
    b. no ventilation
    c. natural-draft ventilation
30. Describe four safeguards that should be provided when grinding magnesium.
    a.
    b.
    c.
    d.

31. To help prevent fires in areas where welding operations are conducted, spread sand on the floor to a depth of
    a. 1 in. (2.5 cm).
    b. 1.5 in. (3.75 cm).
    c. 2 in. (5 cm).
    d. 3 in. (7.5 cm).
32. What are the differences between the conventional cold-stamping press and the forging press?
33. Identify the most widely used nondestructive testing method for forgings.
34. The resonant-frequency method falls under what general type of nondestructive testing for forged and cast metals?
    a. ultrasonic
    b. triboelectric
    c. electromagnetic
    d. radiography

Automated Lines, Systems, or Processes

Anne Germain, PE, BCEE

Philip E. Hagan, JD, MBA, MPH, ARM, CIH, CHMM, CET, CHCM, CEM

Manufacturing Philosophies
Process Safety Management ▶ Up-Front Planning for
Safety ▶ Design-in Safety ▶ Just-in-Time Method
▶ Computerized Maintenance Management Systems

Hazard Identification and Controls
Types of Hazards ▶ Boundaries between Restricted and
Nonrestricted Areas ▶ Visual/Mechanical Warnings

Barriers and Interlocked Barriers
Hazard Controls ▶ Maintenance and Safety

Automated Production
Automated Materials-Handling and Transport Systems
▶ Robotic Equipment

Hazards of Robotics
Safeguarding Robots ▶ Protecting the Robot Teacher
▶ Protecting the Robot Operator ▶ Protecting the
Robot Maintenance/Repair Personnel ▶ Computer-
Integrated Controls

Chemical Processes
Process Safety Information ▶ Hazard/Risk Analysis
▶ Pre-Start-up Reviews ▶ Operating Procedures
Manuals ▶ Management of Change ▶ Auditing

Summary

References

Review Questions

In a highly competitive manufacturing environment, organizations are increasingly installing automated equipment and processes or expanding the use of this equipment and these processes. A company can choose to set up islands of automation or to operate a dedicated, hard-automated assembly line that has virtually no human intervention. Either way, the company's safety and health professional needs to address the specific safety concerns associated with the machines involved, the nature of automation, the machine–person interface, and the resulting change in manufacturing philosophies. Within the manufacturing community, any cultural shift or philosophical change has the potential to impact worker safety.

The following topics will be discussed in this chapter:
- safety planning and management
- identification of hazards
- hazard controls, maintenance, and safety
- automated tasks and robotic equipment
- chemical processes.

MANUFACTURING PHILOSOPHIES

The way that a company addresses its manufacturing processes directly affects safety in the workplace. If safety is regarded as a secondary concern and is thus minimally factored into the costs to comply with regulations, any decision to increase production by means of redesigning a process or speeding up an assembly or production line could actually result in less safe working conditions.

Process Safety Management

One way to address the apparent conflict between maximizing production and keeping workers safe is to institute process safety management (PSM). This approach to safety includes everyone involved in the production process—from operators and mechanics to the facility manager to members of the housekeeping staff. As in the chemical-processing industry, PSM goes beyond traditional precautions: it is a means of managing process safety by recognizing and understanding production risks and by having employees operate in a safe manner so that potential vulnerabilities do not result in injuries and loss of life or property. PSM incorporates process hazard analysis (PHA), a careful review of what could go wrong and what safeguards must be implemented to prevent releases of hazardous chemicals.

Elements of PSM include:
- analyzing hazards and managing risks
- managing change in facility design or operation
- maintaining the integrity of equipment
- training and performance

- investigating incidents
- responding to and controlling emergencies
- auditing
- taking corrective actions.

PSM involves employers and contractors and clarifies the responsibilities for work that affects or takes place near covered processes. Additional requirements include written operating procedures, employee training, pre-start-up safety reviews, evaluation of the mechanical integrity of critical equipment, and written procedures for managing change. PSM also specifies a permit system for hot work, investigations of incidents involving releases or near misses of covered chemicals, emergency action plans, compliance audits at least every 3 years, and trade secret protection.

As part of PSM, hazards are analyzed by the following methodologies:
- what-if
- checklist
- what-if/checklist
- hazard and operability study (HAZOP)
- failure mode and effects analysis (FMEA)
- fault tree analysis
- an appropriate equivalent methodology.

These methodologies would be appropriate for analyzing many automated processes. A discussion of these methods of analysis is contained in the publication, OSHA 3133, Process Safety Management Guidelines for Compliance. See also Chapter 24, Process Safety Management, in this volume, for a more detailed analysis of implementing process safety management.

Up-front Planning for Safety

Up-front planning for safety is another manufacturing philosophy that has safety implications. One way of addressing the safety concerns associated with automation is to include safety costs as part of the original installation, factor these costs into management's buy/no-buy decision, and capitalize them as part of the original investment. At one company, worker teams (which are gradually replacing supervisors at some of the plants) are involved in determining how increased production goals can be met safely. They ask and answer questions such as:
"Where is the line going to be placed?"
"What equipment is needed?"
"How is the line going to be staffed?"

For example, if a new line has additional ventilation requirements, the associated up-front costs would be factored in at the beginning of the project. This approach assumes that it is easier to get costs approved up front rather than after installation. Additionally, the costs of retrofitting are usually significantly higher than the installation costs included as part of up-front costs. Training costs, as well as the time required for training, should be part of the capital investment go/no-go decision. When a project is approved, it should be approved in its entirety—including appropriate safety needs (e.g., costs of future programming for safety support functions).

In addition, a company's safety and health professional should be a proactive safety advocate and argue that concerns for worker safety during automated production should be included in the budget prioritization process. Such arguments can be more effective when they are backed up with lost-time data, workers' compensation costs, and legal expense figures from the facility's own history or from similar-industry estimates. Management decisions must also factor the costs of compliance with current or forthcoming OSHA regulations and other industry standards into financial decisions on installation and upkeep of automated equipment.

Design-in Safety

Another up-front approach to automation-related safety concerns ensures that safety factors are primary considerations during the design process and prior to the purchase or modification of automated equipment. Some companies encourage engineering and operations personnel to incorporate design-in safety into machine and tool procurement. It is also important to design to support maintenance activities, mechanical interlocks, programmable controllers, and safety controls.

Although the comparisons in Table 22–A are not all-inclusive and are not appropriate for all automated operations, they clarify the reasons that companies believe engineering can take a leading role in safety issues in the concept and design phases of both product and processes. Design-in safety can contribute to safer working conditions because safety planning is targeted toward specific work procedures instead of based on a one-size-fits-all philosophy.

Just-in-Time Method

Just-in-time (JIT) production, a manufacturing philosophy that reduces inventories and relies heavily on computerized scheduling, permits manufacturers to be more flexible and to reduce costs. The objective of JIT is to produce the needed part in the needed place at the needed time. Yet this increasingly popular manufacturing philosophy has safety implications that safety and health professionals should consider.

In traditional manufacturing practices, the following assumptions about production were widespread:

- Large lots are efficient.
- Faster production is more efficient.

- Queues are necessary.
- Inventory "smooths" production and represents a "comfortable" position.

However, these assumptions have come under scrutiny, and many companies have adopted JIT in varying degrees. Among the claims made for JIT:

- Inventories are reduced.
- Quality control problems are discovered as they occur.
- Manufacturers can react more quickly to demand changes.

Using a JIT system, a company can eliminate waste by adopting practices such as total quality management (TQM) and effective use of technology. Another important aspect of JIT is respecting people, which includes employee training, employee participation, and employee teamwork. JIT techniques also reduce lead times and lot sizes because machines and worker idle time are used to conduct preventive maintenance. In addition, JIT creates a flexible work force in which workers are trained to operate different machines and perform different maintenance tasks and quality inspections. It is important to ensure that adequate training is provided to workers to support the flexible work concept and that enough time and resources are allocated to maintenance tasks. Note that some of the aspects that make JIT successful—training, TQM, technology, and teamwork—also support a safe environment.

However, a downside of automation that is designed, implemented, or modified around JIT principles is that machines may be interdependent, with little or no allowance for inventory to back up between any two stations. Consequently, if one automated element—a robot, for example—malfunctions and a line is shut down, there may be substantial pressure to fix the robot and get the line moving again. However, if safety and health professionals do not insist that the robot be carefully synchronized with the rest of the line before production resumes, worker safety may be compromised.

Computerized Maintenance Management Systems

Because of increasing automation, computerized maintenance management systems (CMMSs) are yet another manufacturing philosophy that can affect safety decisions. Under CMMS, maintenance management is automated; companies use microcomputers and sophisticated software to plan and control maintenance. Personnel, work orders, materials, purchasing, and scheduling are kept track of in computer databases. Because machine history can be traced through a database search of work orders, equipment failures can be analyzed. Should equipment be repaired or replaced? CMMS reporting can provide production

TABLE 22–A. Design-in Safety versus Traditional Philosophies

Traditional Approaches	Design-in Safety Approach
• common safeguards regardless of task	• safeguards designed to suit intended work
• zero risk	• recognize that zero risk does not exist—deal with task requirements, exposure, hazards, and appropriate safeguards
• depend on employee to take corrective action	• where feasible, use automatic, passive controls such as a motion control timer, presence sensing, interlocked access, etc.
• local Health and Safety Committee attempts to oversee safeguards for each new machine	• determine feasible controls before awarding contract; assure that controls and cost are well defined
• heavy effort at runoff	• develop specifications—attend machine runoff as required
• periodic safety meetings for supervisors and employees	• periodic safety meetings for engineers, also
• compare injury statistics	• analyze injury statistics to determine specific problems
	Machine Safeguarding
• disconnects	• use control circuits for isolating energy, coupled with disconnects; provide for quick recovery to productive operation
• two-hand controls to initiate cycle	• use light screws and mats on many machines. Note: Mechanical power presses are special consideration.
• die blocks	• automatic side locks
• guard pinch points	• eliminate pinch points in design
• use of perimeter guards or large guarded areas	• locate guards as close to hazard as possible; provide observation of operation where required
• power is either ON or OFF	• design controls to isolate drive power and leave power on to limit switches for diagnostic work
• totally enclose robot and argue about power being locked out	• design multiple levels of safeguarding; i.e., walking through first interlocked door moves robot to home and allows ready access to change weld tip; going through second door into working envelope drops out drive
• guards located and interlocked such that necessary maintenance tasks cannot be performed	• guarding designed to facilitate maintenance in a safe manner
• no plan for access to elevators	• design in access and safeguarded platforms required for periodic maintenance
• service points for lubrication and adjustment inside safeguarded areas	• design in location of service points so safeguarding need not be interrupted

Source: General Motors

downtime figures and cost data to support decisions and to ensure that safety measures are factored into the process.

As the use of automation increases, CMMS, which can be customized for industries and facilities, is becoming more popular because of its money-saving potential. Because a single factory cell may handle two to six or more processes, cell downtime is expensive. CMMS can shorten that downtime by providing easy-to-retrieve details on replacement parts, vendor contact information, and repair instructions.

As part of CMMS, companies often develop modules on preventive maintenance (based on performing routine tasks at fixed intervals) and predictive maintenance (based on monitoring process conditions). Workers can use techniques such as vibration analysis, infrared scanning, thermography, and ultrasonic detection to identify potential trouble spots. They can then remove equipment from service, examine it, and, if necessary, repair it before it becomes a serious safety hazard. In older, traditional methods of maintenance management, work orders for machine repairs were often filed chronologically, with multiple copies scattered throughout various offices. Thus, the work history of a specific machine was often difficult to trace, and potential safety problems were not always spotted in time to prevent incidents.

HAZARD IDENTIFICATION AND CONTROLS

Safety in automated manufacturing processes can be greatly increased by following these guidelines:

- Carefully identify hazards during design, installation, and operation.
- Use interlocking principles and devices where possible.
- Design so that maintenance issues can be addressed safely.
- Develop strategies to control the environment where processes take place.

Although the primary issues are divided into different categories, their details often overlap. Safety devices must be designed to account for any future maintenance needs, to use interlocking principles and elements in order to prevent the operator from damaging the machine or personnel (and to prevent the machine from damaging itself), and with new and improved control methods that can considerably lower hardware and cabling costs and lead to reduced troubleshooting time as well as increased diagnostic capabilities.

For maximum effectiveness, such strategies also should include a training program that addresses the specific safety precautions required to operate and maintain various automated equipment and any precautions regarding the entire system that workers need to know.

Ensure that strategies and training programs include the requirements of applicable regulatory standards (e.g., the provisions of the Hazard Communication Standard, 29 CFR 1910.1200, for exposure to hazardous materials).

Types of Hazards

Safety precautions related to automation within a factory vary for several reasons, including the particular industry, the types and complexities of the machines on the line, the way workers interact with the machines, and the success of the automation used to integrate the various components of the production line. All of these factors influence the kinds of safety precautions a company must incorporate into its safety planning. Also, safety planning must address a wide range of hazards. Good ergonomic design becomes an important part of addressing these hazards because proper ergonomic design provides the flexibility to effectively deal with the many hazards found in the workplace. See Chapter 16, Ergonomics Yesterday, Today, and Tomorrow, in the *Administration & Programs* volume, for information on ergonomic design and principles.

Following are examples of hazards that the food industry has to address. Each industry has its own set of safety issues that stem from the use of automation.

Chemical Hazards

Consider, for example, food processing and the complex safety precautions necessary to protect workers when production lines are automated. Spray dryers used to produce coffee, dairy substitutes, and sucrose can generate dust and powder, and workers who inhale the dust may experience irritation or have a higher risk of developing an occupational illness. In addition, if the concentration of dust achieves a critical level and has a high electrostatic charge, all employees in the area will be affected by an explosion.

Workers on a line can also be exposed to potentially hazardous chemicals. Sodium hydroxide, phosphoric acid, and hydrochloric acid are used in food processing to adjust pH levels. Although facilities have in-line pH meters and automatic-injection mechanisms for chemical agents, workers who handle the chemicals for process preparation and for feeding the line must wear appropriate personal pro-

tective equipment (PPE). Automatic processes that distill vegetable oil use extraction methods that incorporate hexane or benzene, which give off hazardous vapors even at low concentrations. Although the gases are part of a closed system, there may be leakage. Management should thus ensure that vapor concentrations are monitored carefully and provide ventilation that will eliminate exposures.

In aseptic packaging, product and package are sterilized separately and are brought together in an aseptic environment. Some sterilizing is done by heat, approximately 300 to 400°F (150 to 205°C), but plastic packaging materials for food products are frequently sterilized by hydrogen peroxide. Typically, an automated line has a roll sheet of the package material that enters an aseptic environment for forming. After forming, the container is sprayed with hydrogen peroxide, which is later evaporated by hot air. These lines are off-limits to employees so that the aseptic environment is not compromised. However, the hydrogen peroxide vapors expended from the aseptic unit could lead to hazardous exposures if leaks or breakdowns of the equipment develop. PPE and appropriate ventilation safeguards are thus necessary for those involved in feeding the line or handling the chemical in any way. Many industries have potential chemical exposure issues that are unique to their particular situations and need to be addressed on a case-by-case basis.

Burn Hazards

Few industries lack the potential for burn hazards of some form. Products such as pet foods and some cereals are often produced by automated extrusion lines, which present challenging technical safety problems. If pressure builds in the extruder to a level that is too high, if the extruder die is blocked, or if a plug builds up inside the barrel of the extruder, the material being extruded can escape at a high temperature and pressure. However, extrusion units usually have a safety switch that shuts off the main power when a sensor detects that torque exceeds predetermined limits.

Evaporating, drying, cooking, sterilizing, and retorting—common automated operations in many food-processing facilities—often require steam as a heating agent. A leaking or broken steam line is a serious hazard, and the supervisor should have the line repaired immediately. In order to prevent steam leaks, maintenance should include regular inspections to identify and replace seals and gaskets that are degrading.

Cleaning in place (CIP) of food-processing equipment is frequently automated, and storage tanks with cleaning agents and pumps are connected directly to the automatic food-processing lines. Often, pneumatic valves close the lines and switch them from the processing mode to the cleaning mode. Because sulfuric acid and various alkalis are common cleaning solvents, the chemicals and the automatic switching process must be safeguarded against incidents, a safety concern that the engineer must take into account during design, installation, and operation.

Utility regulation and safety also are concerns in safety planning. Electricity, steam, air, hot or cold water, compressed air, liquid carbon dioxide, and liquid ammonia—all of which are used in various aspects of food-processing automation—have their own potential dangers, and each can become a hazard. For instance, using steam to heat a kettle of tomatoes may require injecting the steam into the jacket of the kettle. As a result of this heat exchange, the steam becomes condensate, which is then emitted by a steam trap. If a flowmeter, a device that measures the amount of condensate going through a pipe per unit of time, malfunctions or fails to indicate a problem, the hazard might go unnoticed until an emergency situation develops. Maintenance schedules should thus incorporate processes to ensure that devices used to regulate the flow of utilities are checked and maintained regularly.

Radiation Hazards

Equipment that uses both ionizing and nonionizing radiation poses several hazards in food-processing automation if the equipment is not properly designed, installed, or operated or is functioning improperly. Microwaves are often where the final stage of drying noodles and macaroni takes place in automated lines. Workers should check the equipment periodically to be sure that safety interlocks are functioning and that there is no microwave leakage. If a company uses gamma radiation to sterilize spices, employees involved with the process should wear radiation badges and be checked regularly for radiation exposure. Management should take this safety measure even if the company has a well-designed automation process and complies with the regulations to keep employees a safe distance from the radiation chamber. (See also Chapter 27, Laboratory Safety, in the *Administration & Programs* volume, for more information on ionizing and nonionizing radiation.)

Pinch Hazards

In the traditional primary packaging of food products, most production lines use heavily mechanized, automated equipment, especially to convey, form, and package products. Because the combination of a machine's moving and non-moving parts can create hazardous conditions for workers, moving parts should be enclosed wherever possible. Sensors that automatically stop the line whenever something gets caught or jams in the moving parts of the machinery provide an additional level of safety.

Many of the American National Standards Institute (ANSI) standards for machine tools furnish information on hazard identification and recommendations for a control strategy. Safety and health professionals should be sure that all equipment complies with the latest standards. Be sure to check whether revisions have been published (see the References). Chapter 6, Safeguarding, has more detailed explanations of and requirements for addressing the hazards of moving parts in automated machinery.

Boundaries between Restricted and Nonrestricted Areas

It is important to identify the boundaries that separate restricted areas from the areas in which workers can move without threat of injury from the automated process. These boundaries can be established by determining the specific clearances required around each machine or component in the automated system. For instance, in laying out a system for automated guided vehicles (AGVs) to travel, all areas that include pedestrian travel or interaction with workers require a minimum clearance of 18 in. (46 cm) between a vehicle with its load and any fixed object. That clearance allows a person to stand safely while a vehicle passes by.

However, AGVs often transfer loads automatically via powered roller belt or chain to and from fixed stations through a lift/lower fork truck design or through an automatic push or pull mechanism. At these load transfer stations, it is not always possible to maintain the 18-in. (46-cm) clearance. Consequently, load transfer stations need to be designated as restricted areas in which no one can walk or stand. Areas used for AGV battery charging and battery changing also require special markings, designated by the following standards: Standard for Industrial Trucks, Internal-Combustion Engine-Powered, UL 558–2012; Standard for Electric-Battery-Powered Industrial Trucks, UL 583–2012; and Fire Safety Standard for Powered Industrial Trucks Including Type Designations, Areas of Use, Conversions, Maintenance, and Operation, NFPA 505–2013. Typically, restricted areas are designated by posted signs and striped floors.

Some ANSI standards specify the areas necessary for machine motion. For instance, Safety Standard for Industrial Robots and Industrial Robot Systems, ANSI/RIA R15.06–2012, replaces the engineering term *envelope*, which refers to a robot's work space, with the more understandable term *space*:

- *Envelope (space), maximum.* The volume of space encompassing the maximum designed movements of all robot parts, including the end-effector, workpiece, and attachments.
- *Restricted envelope (space).* That portion of the maximum

envelope to which a robot is restricted by limiting devices. The maximum distance that the robot can travel after the limiting device is actuated defines the boundaries of the restricted envelope (space) of the robot.

- *Operating envelope (space)*. That portion of the restricted envelope (space) that is actually used by the robot while performing its programmed motions.

The National Safety Council's Occupational Safety and Health Data Sheet 12304-D717–1991, Robots, recommends that companies place warning signs around each robot at points of access to the operating space. These signs warn those who enter the operating space of the usual hazards of the robot and/or any unusual hazards that may exist. Examples of unusual hazards include overlapping operating spaces of two or more robots and other automated devices or machinery that can move into the work area.

Visual/Mechanical Warnings

Several systems of visual and mechanical warnings are commonly used to alert workers to present, or approaching, hazards. These warnings include signs, flashing lights, barriers, line markings on the floor, railings, or the equivalent. See Chapter 2, Buildings and Facility Layout, in this volume. Be sure to consult the latest regulations for your situation.

Signs and in-Process Warnings

Make sure that employees whose native language is not English understand and follow the warnings. Wherever possible, use international signs. If color blindness is a factor for some workers, management should consider installing flashing lights. Workers, supervisors, and maintenance personnel should be screened, trained, and tested to make sure they understand the meanings of signs, tags, and color codes. Remember that helpers and passersby may be exposed to hazards, so plan to use additional safeguards.

Audible warning devices such as horns, bells, and electronic beepers must have a sound that is louder than the machines' noise and louder than the noise level within the facility. These devices should not be used in conjunction with paging systems or as signals to indicate the start and stop of work.

Visible or audible warnings can also alert employees that work is in progress. Automatic turn signals on AGVs can show workers which way the vehicles will turn at an intersection. An amber light installed on a robot that is visible from any angle should be lighted any time the robot is energized. This light signifies that the robot is "live" and is thus a potential hazard, even if the robot is not moving.

Inspect warning signals and devices often to make sure they are operating properly. Workers may not always realize that such devices have failed.

BARRIERS AND INTERLOCKED BARRIERS

Awareness barriers keep operators from reaching into hazardous areas unless they make a conscious effort to do so. Such awareness barriers should not only indicate the hazardous area but also provide visible boundaries for the operator's movements. Inspect and test these parts regularly.

Perimeter guarding signs and barriers restrict entrance to certain enclosed areas and protect the personnel who work with and around robots. Management should post signs prominently at all entrances to prohibit unauthorized personnel from entering restricted areas. In addition, identify the entrance so that authorized personnel can enter the area without inadvertently entering a restricted work space.

Interlocks are arrangements in which the operation of one control or mechanism automatically leads to or prevents the operation of another control or mechanism. Some of the most common forms of interlocks include contact switches that are activated by electrical, pneumatic, hydraulic, or magnetic mechanisms. Interlocked barrier guards ensure that a machine tool will not cycle or continue to cycle unless the guard itself, or its hinged or movable sections, encloses the hazardous area.

For maximum safety, if an interlocking device is disconnected, the machine and any automated process within the workstation should stop promptly. Further, connecting the device again should not start the automation cycle. A movable barrier device can function as an interlocked barrier guard when a machine tool has been set to operate continuously.

Hazard Controls

Once hazards have been identified in automated production, appropriate controls should be installed to facilitate worker safety. The ANSI B11 Machine Tool Safety Standards are excellent sources of machine tool guarding information for use in the United States, specifically:

B11.20: Safety Requirements for Integrated Manufacturing Systems.

B11.21: Safety Requirements for Machine Tools Using Lasers for Processing Material.

B11.22: Safety Requirements for Turning Centers and Automatic, Numerically Controlled Turning Machines.

B11.23: Safety Requirements for Machining Centers and Automatic, Numerically Controlled Milling, Drilling, and Boring Machines.

B11.24: Safety Requirements for Transfer Machines.

Also of use is B11.19–2010, Performance Requirements for Safeguarding, which provides recommendations for the placement and effectiveness of control devices—such as an

EMERGENCY STOP button, a START button, or a selector switch—in various situations. Workers should follow these recommendations and manufacturers' instructions, and all personnel should be properly trained in the use of these controls.

If a machine or any other part of an automated production line has been modified, check the control devices to ensure that they still meet ANSI and other industry standards as well as the recommendations from the manufacturer. In particular, make sure that electromagnetic fields (EMF) or radio frequency interference (RFI) does not impede operation of the controls and that the power supply is constant and uninterrupted.

Controls outside Boundaries

Whenever possible, controls should be located outside boundaries. For instance, if a worker will need to reach a control for a robot while a machine is in automatic operation, the control should be placed outside the restricted space. This space is the maximum area to which a robot is restricted by its limiting devices. In that way, a person who needs to activate the control is also outside the restricted space. Locating controls outside the restricted space can ensure that if a robot is out of the automatic mode for maintenance, repair, or teaching purposes, personnel cannot reactivate the robot until they have left the restricted space, restored all safeguards required for automatic operation, and begun deliberate start-up procedures.

Controls inside Boundaries

Other controls can be located inside boundaries. For instance, a hostage control device (such as a two-hand control, a single-hand control, a foot switch, or a safety mat) can keep the operator at a control station during the hazardous portion of a machine cycle. This type of switch is also known as a deadman's switch or a kill switch. When a person is inside the robot's work space to teach, maintain, or repair the robot and the robot's power must be on, the robot system—as well as the movements of other equipment in the restricted work space—should be under the sole control of the person in the space.

Controls that stop the system should override all other controls. The controls should interact so that if one component fails, the STOP control will shut down the system immediately. For more information on guarding and controls, see Chapter 6, Safeguarding, in this volume.

Maintenance and Safety

Whether an automated manufacturing facility chooses to have maintenance and repair performed by its own personnel, by vendor personnel, or by an outside contract service, safety should be a major goal of all concerned. Be sure to follow the manufacturers' recommendations for inspec-

tion and maintenance of all components in the automated system—each part individually and as a component of an integrated assembly process. Each machine and automated materials-handling device must not only perform safely as a stand-alone unit but also interface properly at the appropriate speed.

Safety training for maintenance and repair personnel should include, but not be limited to, a review of all applicable industry safety procedures and standards; manufacturers' recommendations; equipment-specific training, tasks, and responsibilities for each person; and identification of the hazards associated with each task. Keep accurate records and documentation of all changes to components in the system as well as a history of maintenance and repairs for each piece of equipment. Make sure that maintenance personnel can easily access this information.

Also, make sure that the maintenance and repair personnel working on the machines or system components have copies on site of the operations and maintenance manuals from vendors, as well as any updates. Keep a readily accessible, second copy of the manuals and updates at a supervisor's station or in the safety and health professional's office. Repair and maintenance workers should also know how to reach the appropriate person if problems develop that exceed the workers' responsibilities or training.

Companies that move from crisis or reactive maintenance to planned maintenance greatly enhance safety in the workplace. Preventive maintenance can help supervisors anticipate and schedule production downtime. Consequently, there will be less pressure to get a line or cell up and running again by cutting corners on safety.

Predictive maintenance, which gathers data automatically through monitoring conditions and integrates the data with manufacturers' guidelines on tolerances and ranges, can also help boost safety procedures and practices by reducing catastrophic failures. In this approach, computerized systems predict when a piece of equipment is likely to fail, thus allowing personnel to repair or replace the equipment before the failure occurs.

Companies should train and retrain workers, as necessary, to ensure that they are competent in and knowledgeable about safety practices. If outside contract personnel or services are used for maintenance or repair, provide them with the information on company safety procedures at the time the service is contracted. Make it clear that *all* personnel, not just the company's own employees, must follow the facility's safety practices at all times. If the firm uses an outside service for maintenance, consider writing such contracts so that they require the personnel to receive both classroom and on-the-job training. Their training should be documented and meet industry standards and legal requirements.

Maintenance personnel also should be aware of the

regulations affecting the overall factory environment as well as the regulations dealing with particular equipment. For instance, the use and characteristics of protective coatings in manufacturing facilities now are subject to volatile organic compound (VOC) restrictions, which stipulate the solvent content percentage per liter of coating, and to regulatory right-to-know legislation. Although the federal government has set VOC limits for coatings, some state governments have enacted their own legislation. Consequently, companies with facilities in several states need to follow coating specifications that meet the restrictions established by the governments in all of the states where the company has facilities.

Energy Isolation—Lockout/Tagout

When workers are doing maintenance or repair work, make sure that they lock out and tag out all equipment in accordance with OSHA lockout/tagout standards (29 CFR 1910.147 and ANSI Z244.1–2003 [R2008]). For a detailed explanation and discussion of required procedures, see Chapter 6, Safeguarding, and Chapter 8, Electrical Safety, both in this volume.

Do not allow workers to perform servicing, maintenance, or repair work on any equipment unless that equipment is isolated from all hazardous energy sources. Only designated, authorized employees should complete the lockout/tagout procedures, carefully following the steps established by the company.

Machine-specific lockout steps are difficult to impose because the appropriate steps vary from machine to machine. In general, such steps incorporate the various types of hazardous energy and their magnitudes; the locations and identification numbers of switches and valves; whether locks, tags, or both are needed in each machine's lockout procedure; additional safety measures; a list of "designated, authorized employees"; and a phone number to call if there are questions. Post the comprehensive lockout specifications at each machine. See the Energy Isolation section in Chapter 6, Safeguarding.

Diagnostic Aids and Procedures

Various diagnostic aids and procedures can help solve maintenance problems. For example, vibration measurement and analysis monitor the condition of machinery and help diagnose machine problems. Measurement data—collected by portable, programmable, microprocessor-controlled vibration meters or permanently installed, embedded sensors—are transmitted to computers, whose ongoing analyses warn the operator when problems arise or maintenance is necessary. If needed, workers can later analyze the information to determine specific component faults. Laser monitors that continuously check dust levels are available in the process-

ing industry, and software packages have been developed to track steam traps and steam trap maintenance. These computer programs support ultrasonic, infrared, or standard pyrometry inspections of steam traps.

Companies also can use various techniques to help determine why equipment fails. Bearings, which are difficult to evaluate from a visual perspective, can be analyzed through a variety of techniques. These techniques include measuring and analyzing the wear track, identifying and classifying surface damage, analyzing lubricant, and analyzing the fretting corrosion patterns that form on bearing inner ring bores and outer ring surfaces.

Staff can compile equipment performance trends by evaluating equipment output. A standard statistical process control run chart enables a facility manager to collect data. If a machine's demonstrated capability shifts significantly, adjustments may be necessary.

Employees who repair machines or other components of the automated production line should document their actions. If problems recur, supervisors can evaluate those actions to determine where a solution can be implemented to prevent similar problems in the future. Supervisors should also maintain a history of problems with and repairs for each machine to help identify performance trends and problem components. Some microcomputer-based maintenance software that runs on certain personal computers as well as on local area networks (LANs) or wide area networks (WANs) allows users to "build" a problem list by entering appropriate keywords into a database. When a piece of equipment goes down, the user selects a particular problem from the list. Pop-up displays on screen, automatically generated reports, or similar trigger mechanisms then show possible causes and suggested methods of correction.

AUTOMATED PRODUCTION

Automated manufacturing requires safety precautions beyond those needed with single machines. Whenever workers and equipment interact—during operation, programming, and maintenance or through unintentional contact—the potential for injury exists. The best way to reduce injuries is to incorporate the following into job and workstation design: diligently complying with existing industry standards; monitoring the development and implementation of new standards; and establishing safety training, testing, and retraining (not only for specific equipment but also for the automated system as a whole).

Effective safety training is especially important in automated production. Because automated manufacturing processes often require workers to interface with more than one system component, such as a program, terminal, machine, materials-handling system, or robot, each worker needs to understand more than just his or her own part of the system. Programmable automation, which lets a machine or group of machines make a wide range of parts or products, and automated manufacturing systems require operators, maintenance personnel, and supervisors to develop many skills in order to work safely and productively. Nevertheless, safety and health professionals cannot assume that all workers have the skills necessary to read and understand technical manuals and specifications. A company may thus need to establish procedures to test and train workers in reading and comprehension or to develop safety training programs that do not rely solely on written or English-language communications.

Some automated manufacturing involves continuous processes, such as those used in the food- and beverage-processing, textiles, pulp and paper, metal refining, printing, pharmaceutical, and petrochemical industries. Another type of automated production, usually referred to as automated manufacturing systems/cells, generally involves machine tools and related machines and equipment being brought together to form a new or modified manufacturing system. Such a system often is linked to a materials-handling procedure that is interconnected with and operated by an electronic system. Management can program and reprogram the electronic system to control the manufacture of single parts or assemblies.

Safety and health professionals involved with manufacturing systems or cells should use applicable standards, including the Standard for Electrical Safety in the Workplace, NFPA 70E–2012; the Electrical Standard for Industrial Machinery, NFPA 79–2012; and the Safety Standard for Industrial Robots and Robot Systems, ANSI/RIA R15.06–2012.

Automated Materials-Handling and Transport Systems

Company safety and health professionals can help make materials-handling and transport systems safer by being familiar with current regulations and instituting safe work practices. Two systems that involve special risks are conveyors and automated guided vehicles.

Conveyors

See Chapter 13, Hoisting and Conveying Equipment, in this volume, for safety recommendations regarding conveyors. Facility managers and supervisors should be familiar with Safety Standards for Conveyors and Related Equipment, ASME B20.1–2012, discussed in detail in Chapter 13. Management should safeguard all conveyors to protect employees and should train employees in safe work procedures, including the location and operation of all START/STOP devices. Because the unit that a conveyor feeds can be stopped or become blocked, make sure that electrical or mechanical interlocking devices are provided so that workers can stop the conveyor automatically. Guard pinch points adequately, and ensure that workers use proper lockout/tagout procedures when performing maintenance and repairs.

Automated Guided Vehicles

Automated guided vehicles (AGVs), which provide a transport system for materials handling, are one of the key elements of automation in various industries (Figure 22–1). Since 1985, AGV usage has increased in each of the following industries: automobile, pharmaceutical, chemical, electronics, paper, aircraft, textile, food, metal fabrication, farm machinery, printing, primary metals, rubber/plastics, and lumber/wood.

AGVs can tow pallets and integrate conveyors with other systems. They can also collect and dispose of refuse (Figure 22–2). In assembly-line applications, AGVs often carry major subassemblies between assembly workstations. Because AGVs can be used in parallel operations, their use is often cost effective in flexible manufacturing. In certain industries such as the manufacture of printed circuit boards, AGVs offer advantages over conveyor systems in product routing because AGVs can easily move material between "islands" of automatic insertion equipment.

Traditional AGVs follow electromagnetic wires buried in the floor. However, "free-roaming" AGVs use technologies such as optical guidance (which uses a chemical guidepath [see Figure 22–1] or painted line), infrared guidance, inertial guidance (which uses a gyroscope), position-referencing beacons, or computer programming (Figure 22–3).

Safety and health professionals must be attentive to a variety of system conditions. Loads that the AGVs carry

Figure 22–1. Automated guided vehicles use a guidance system to follow an invisible chemical path that is bonded to the floor surface. This AGV is receiving material from a conveyor system. (*Courtesy Litton Corp.*)

Figure 22-2. This self-guided lift truck can collect and dispose of chips from a machining cell. *(Courtesy Caterpillar Inc.)*

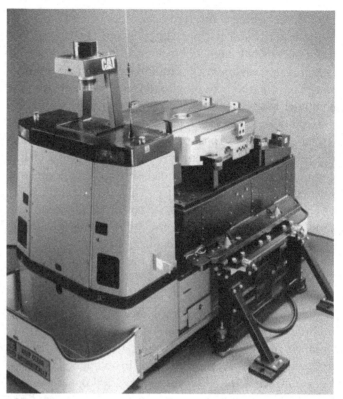

Figure 22-3. This free-roaming, self-guided AGV requires no wires or special flooring. Instead, it is guided by an on-board computer, which can be reprogrammed for flexible service. *(Courtesy Caterpillar Inc.)*

should be appropriate for each vehicle's capabilities. Check to be sure that employees are following the manufacturers' specifications for loading AGVs; workers should not stack loads any higher than the specifications indicate. In addition, if a load is stacked too loosely, it might tip over when on an incline. As a result, the vehicle will not be able to carry its rated load capacity safely.

Every time the products, packaging, or loads that an AGV carries change or the operating procedures are altered, workers should check the center of gravity of the load against the AGV system design parameters and make appropriate adjustments.

For the transport system to operate safely, the company needs to establish a program that monitors the floor conditions of the routes the AGVs travel. Floors should not sag or move. If the AGVs need to travel over or around expansion joints or rails in the floor, make sure, before the AGV system is installed, that the vehicles can do so.

Floors should be as dry as possible because AGVs can lose traction and behave unpredictably on wet floors, especially when the vehicles have to stop suddenly. If floors are wet or slippery, apply slip-resistant floor coatings. Keep floors as clean as possible, and do not allow residues to build up; residues may prevent the AGVs from steering properly or stopping completely.

Floor conditions are especially important for AGVs that travel via chemical guidepaths or painted lines. Because the lines can become obscured if dirt, oil, or grease accumulate, keep floors clean, and repaint lines as often as needed, following manufacturers' recommendations. Keep all guidepath areas free of obstructions such as boxes, stacked parts, or other vehicles.

Companies can design safety factors into the AGV system before installation. Establish provisions during the design phase that not only safeguard the transport of materials but also protect personnel in the immediate area of the vehicle system. Design-in safety has two objectives: to keep AGVs from colliding with people and other vehicles and, if an anticollision device fails to work, to prevent an AGV from operating until it is repaired.

Safety devices for an AGV can include monitors for guidance and velocity, guards, deadman's switches, turn signals, and sensors (infrared optics, lasers, and ultrasonics) that anticipate and prevent collisions. Traffic control techniques include bumper blocking, in-floor zone blocking (which uses sensors in the floor at zone boundaries and energizes a path to allow an AGV to enter a zone), and computer zone blocking.

Traditional AGVs are regulated by programmable controllers that "talk" to the vehicles through signals transmitted via a wire guidepath buried in the floor. Other communication technologies include radio frequency and infrared guidance systems; however, the route must be clearly marked if guidewires are not used. The company should also make provisions for communicating with an AGV when it is off the guidepath.

Refer to ANSI B56.5, Safety Standard for Guided Industrial Vehicles and Automated Functions of Manned Industrial Vehicles, for safety requirements for powered, unmanned, automatic guided industrial vehicles, which include standards on bumper design and activation. Bumpers should not need software, hardware logic, or signal conditioning in order to operate. In addition, if an AGV loses guidance in the automatic mode, ANSI B56.5 requires the vehicle to stop moving immediately. Also covered are vehicles originally designed to operate in a manned mode but that are later modified to operate in an unmanned, automatic mode or in a semiautomatic, manual, or maintenance mode.

Robotic Equipment

Organizations considering the installation or expansion of automated manufacturing soon learn of the increasing number of safety problems associated with robotics. In 2012, the North American robot population was just shy of 200,000 units, second only to Japan's robot population. The International Federation of Robotics (IFR) estimates that North American companies order more than 25,000 industrial robots annually and that the world population of industrial robots was close to 1.25 million at the end of 2012.

In terms of the number of robots per 10,000 production workers, however, the United States trails Korea, Japan, Germany, Italy, Sweden, and Denmark, according to the IFR. Other information about robots can be obtained from the Robotic Industries Association (RIA). RIA is a trade association of U.S.-based manufacturers, distributors, systems integrators, accessory equipment suppliers, users, research organizations, and consulting firms.

RIA defines an industrial robot as "a reprogrammable multifunctional manipulator designed to move material, parts, tools, or specialized devices, through variable programmed motions for the performance of a variety of 'tasks.'" In addition, RIA classifies robots in the following manner:

- handling devices with manual control
- automated handling devices with predetermined cycles
- programmable, servo-controlled robots with continuous point-to-point trajectories
- robots capable of type C specifications, which also acquire information from the environment for intelligent motion.

Other definitions of robots are used throughout the international arena. The International Organization for Standardization gives a definition of a robot in ISO 8373: "an automatically controlled, reprogrammable, multipurpose, manipulator programmable in three or more axes, which may be either fixed in place or mobile for use in industrial automation applications."

Three basic parts make up a robot: a manipulator, a power supply, and a system for controlling the robot. The robot arm, from the base of the robot through the wrist, is called the "manipulator." Actuators, drives, bearings, and feedback (all within the arm) make it possible for the arm to move in different directions. The extent to which the robot hand, or working tool, can reach in all directions is called the robot's "work space." The dimensions of the work space depend on how the manufacturer arranged the robot's axes of motion, or degrees of freedom.

Usually a robot has three major axes of motion: a vertical stroke, which determines how high the robot can reach; a horizontal reach, which lets the robot move in and out; and a swing, or rotation, around the robot's base. Some robots have additional axes of motion: pitch, yaw, and roll. A "wrist" at the end of a robot's arm contains components that allow for these additional movements. Manufacturers who want to increase a robot's reach and ability to manipulate objects can provide additional axes of motion in end-of-arm tooling.

Industrial robots are powered pneumatically, hydraulically, or electrically. Three types of robot control systems exist: non-servo-controlled point-to-point robots (often used in pick-and-place applications), servo-controlled point-to-point robots (often used for loading and unloading), and servo-controlled continuous-path robots (often used for spray painting and other finishing operations).

An industrial robot is almost never a stand-alone device. Rather, it is part of a system that includes industrial robots, end-effectors, and any equipment, devices, and sensors—including communication interfaces that sequence or monitor the robot—that are required for the robot to perform its tasks.

Companies use most industrial robots to perform common repetitive tasks. Often, robots are the automation of choice in factory environments in which working conditions are potentially hazardous or overly strenuous for human workers. Spray painting on an automotive assembly line or loading molten steel in extremely hot temperatures are ideal jobs for robot units to perform. Robotics is also used to assist in handling, moving, and assembling filter systems when employees inspect products. This hazard control greatly decreases the risks of developing musculoskeletal disorders and ergonomic-related injuries.

Robots have many other functions in industry. In the automotive industry, for example, robots are used for arc and resistance welding, assembling, dispensing and applying sealants and adhesives, painting, inspecting vehicles, loading and unloading machines, transferring parts from one conveyor to another, packaging, and palletizing.

Robots also load and unload other machines in such fields as die casting, loading presses, forging and heat treating, and plastic molding. Robots that can handle tools or that have grippers to hold special tools are especially useful for grinding, drilling, and riveting in machining. The strong sales growth of assembly robots that are used primarily in nonautomotive markets indicates that new uses for robots are emerging in a diverse range of industries. From aerospace facilities to appliance manufacturing to printed circuit board assembly, robots have important roles in automated manufacturing. Currently, the IFR estimates that robot use is increasing in the metal, food and beverage, glass/ceramics, pharmaceutical, medical devices, and photovoltaic industries.

HAZARDS OF ROBOTICS

Those concerned with protecting workers in an environment in which industrial robots are used have special factors to consider because of the nature, installation, and operation of robots. Because robots have limited reach and mobility, they are generally placed close to each other. Yet the immense area around a robot, which includes any area within reach of its arm, is a potential danger zone (Figure 22–4). The danger zone is even larger if a robot loses control of the object it is holding.

In 1982, the Labor Ministry of Japan conducted a survey of 190 factories, which had a total of 4,341 robots installed. The factories reported 11 incidents, including 2 deaths; 37 noninjury incidents (operators nearly contacting robot arms); and 300 robot-related problems, including those unrelated to injuries. In 8 of the 11 incidents, operators reportedly entered the robot's arm range while the robot was stationary and were hit by the arm when it moved unexpectedly. When 300 robot-related problems were analyzed to determine the relationship between these problems and the actions of the robot manipulator, the results showed that in 30.6% of the cases, the robot operated unnaturally; in 28.3% of the cases, the robot's

Figure 22–4. This robot is guarded by a chain-link fence with gates. Safe areas for the maintenance and repair of robots must be designed into the worksite.

procedure started unexpectedly; and in 8.9% of the cases, the piece the robot was working on was released or fell.

A Japanese survey of 18 near misses found the following causes: erroneous action of the robot during normal operation (5.6%); erroneous action of peripheral equipment during normal operation (5.6%); worker' carelessly approaching the robot (11.2%); erroneous action of the robot during teaching and testing operations (16.6%); erroneous action of peripheral equipment during teaching and testing operations (16.6%); erroneous action during manual operation (16.6%); erroneous action during checking, regulation, and repair (16.6%); and other (11.2%).

In 1987, a cause-and-effect analysis of 32 industrial robot incidents reported in the Swedish, West German, Japanese, and American literature indicated that in 24 of the cases, robot–human interaction was a definite cause: 17 incidents were caused by line workers, 5 by maintenance workers, and 2 by programmers.

The researchers then grouped the incident causes into four categories: human error, workplace design, robot design, and "other." Of the 32 incidents, the researchers attributed 13 to human error, 20 to workplace design (18 related to guarding and 2 related to interfacing), 7 to robot design, and 0 to "other." Note that the total number of causes is 40, not 32, because some of these incidents had more than one cause. Of the 32 reported incidents, 72% involved the robot operator or a nearby worker. None of the 32 incidents involved an unauthorized individual.

OSHA groups robotic incidents into these four categories: a robotic arm or controlled tool causes the incident, a robot places an individual in a risky situation, an accessory of a robot's mechanical parts fails, and the power supply to the robot is not controlled. These causes lead to the following types of accidents.

Impact or Collision Accidents

Unpredicted movements, component malfunctions, or unpredicted program changes related to a robot's arm or peripheral equipment can result in contact accidents.

Crushing and Trapping Accidents

A worker's limb or other body part can be trapped between a robot's arm and other peripheral equipment, or the individual may be physically driven into and crushed by peripheral equipment.

Mechanical Part Accidents

The breakdown of a robot's drive components, tooling or end-effector, peripheral equipment, or power source is a mechanical accident. The release of parts, the failure of gripper mechanisms, or the failure of end-effector power tools (e.g., grinding wheels, buffing wheels, deburring tools, power screwdrivers, and nut runners) are also mechanical failures.

Other Accidents

Other accidents can result from working with robots. Equipment that supplies robot power and control can present potential electrical and pressurized fluid hazards. Ruptured hydraulic lines could become high-pressure cutting streams or whipping hose hazards. Environmental accidents from arc flash, metal spatter, dust, electromagnetic, or radio frequency interference can also occur. In addition, equipment and power cables on the floor are tripping hazards.

Safeguarding Robots

Robot safety can be divided into three major areas:
- safety in the processes of manufacturing, remanufacturing, and rebuilding robots
- safety when installing robots
- safeguarding workers exposed to the hazards associated with the use of robots.

Safety and health professionals need to be involved in each of these areas. Although one might assume that safety and health professionals should focus primarily on installation and safeguarding, this is not always true. Because robot users also rebuild robots, safety and health professionals must be sure that rebuilt robots also perform according to safety guidelines.

At the heart of any robot safety program is risk assessment. In conducting a risk assessment, the end user must determine the types of hazards that a specific robot presents and develop procedures to minimize those hazards. Risk

assessment is appropriate not only in production but also in the installation and testing of the robot before production begins (Table 22–B). Although a risk assessment is probably the best approach, a safe installation can also result from following a series of safeguarding steps.

TABLE 22–B. Robot Safety Program Risk Assessment

A risk assessment for a robot safety program should consider:
1. size, capability, and speed of the robot
2. application/process
3. anticipated tasks that will be required for continued operation
4. hazards associated with each task
5. anticipated failure modes
6. probability of occurrence and probable severity of injury
7. level of expertise or exposed personnel and the frequency of exposure.

Source: Robotic Industries Association.

In addition, the appropriate precautions to safeguard personnel vary depending on the stage of development of the robot and the robot system. As RIA points out, elaborate safeguarding while a robot is being integrated into its manufacturing site may impede the development of the overall system. As a result, it is generally impractical to debug the robot at this stage. However, when the robot is operating on the production line, full safeguarding and interlocks may be required for the entire system.

The industry recognizes four stages of robot and robot system development:
- integration at the manufacturer or system developer
- verification and buy-off testing
- installation and testing at the site of operation
- operation in production.

Used robots have become a prominent part of the automation environment. To address the potential safety issues of used robots (which include robots that are remanufactured and robots that are rebuilt), the U.S. robotics safety standard includes detailed requirements that any robot with a change in ownership must conform to. It is important to evaluate all used robots to be sure that they conform to current robot safety standards.

Protecting the Robot Teacher

Often, a robot is programmed by being physically guided by an operator through a desired sequence of tasks. Typically, a servo-controlled point-to-point robot is programmed through a TEACH pendant, which is a control box resembling the remote control device for a television. The operator uses the pendant to slowly "walk" the robot through the program steps and makes sure to record each step. When the operator is finished, he or she can switch the robot from TEACH to REPEAT.

A different method of teaching by guiding is often used with servo-controlled continuous-path robots. To teach such a robot, the operator typically takes the end of the robot's arm and leads it through a pattern of motions. Meanwhile, the robot's control system records feedback data from the position sensors on the axes and stores the data on a mass memory storage system.

In either case, the teacher must be within the operating space of the robot to program the robot's movements. Hazardous situations are most likely to develop when a robot is in the TEACH mode. A Japanese survey in 1977 indicated that the greatest risk of injuries involving robots occurs when robots are being programmed, taught, and maintained. These are the times when a person is within a robot's operating space.

ANSI/RIA R15.06–2012 contains recommendations for the safe installation and operation of industrial robots in the TEACH mode. Consult this publication for complete details.

Before starting any operation, anyone who is teaching the robot should visually inspect the machine and its work space to be sure that no hazardous conditions exist. Operators should check the TEACH pendant's EMERGENCY STOP and motion controls to be sure that they are operating properly and should repair them if they are damaged or malfunctioning. The EMERGENCY STOP control should be hardwired into the drive-power stop circuit and should not be interfaced through a computer's input/output register.

During teaching, preferably only the teacher is in the restricted work space of the robot. If more than one person is present, only the teacher should control the robot's motion. However, anyone else within the restricted space should use an enabling device that, when released, stops the robot's motion. All personnel should leave the restricted space and restore all safeguards for automatic operation before starting the robot's AUTOMATIC mode.

Protecting the Robot Operator

Special precautions may be needed when the operator interacts with the robot during each operating cycle. Safeguards should keep the worker out of the restricted space while the robot is moving automatically or stop the robot's motion while any part of the operator's body is inside the restricted space. These safeguards include devices such as photoelectric cells, pressure-sensitive mats, laser and visual monitor sensors, and light or sound curtains, which can sense a person's presence. Additional safeguards are barriers, awareness barriers, awareness signals, safety training, and detectors that indicate when robots or machines malfunction.

All persons involved with robots must remember two important points:

- If the robot is motionless, do not assume that it will remain motionless; many programs have delays or wait periods during which the robot "sits" until told to do something.
- If the robot is repeating a pattern, do not assume that it will continue to repeat the pattern; computers can instantly modify the path a robot has been programmed to follow, which might trap personnel within the work space because the robot moves in an unexpected direction.

Analysis of robot-related fatalities has led to recommendations for ergonomic and system design, training, procedures, and supervision. The National Institute for Occupational Safety and Health (NIOSH) conducted a comprehensive cause-and-effect analysis of a 1984 fatal pinch-point injury and found that both human error and workplace design problems were to blame. Similar analyses were conducted for 1999 and 2001 fatalities that were associated with robotic systems.

In the 1984 incident, the worker had received 1 week of training 3 weeks before the incident; had been warned by others to follow safety rules but had disregarded their advice and applicable safety procedures; and had entered the robot's area while the robot was in normal operating mode. The worker died after being pinned between the back end of the industrial robot and a steel safety pole.

A fatal accident in 1999 occurred when a worker tripped a light sensor, which caused a computer-controlled robotic platform to descend. The worker was killed by the platform. The company had instructed personnel not to enter any area around the cycling robot when it was in automatic mode.

NIOSH found in its 2001 investigation that the injury resulted when a cycling single-side gantry robot hit an employee who was inside the robot's work space. The employer had a written lockout/tagout procedure for each piece of equipment, and the employees had received documented training. However, the robot was not safeguarded in accordance with ANSI/RIA R15.06–2012. Also, the employer had not conducted a risk assessment of the operation.

Although workplace design was a factor in all three accidents, human error was a primary factor—training and enforcement of safety rules would have prevented these accidents from occurring.

Training specific to a particular robot should be provided to workers. This training should emphasize safety and the technological developments that address programming, operating, and/or maintaining robots or associated systems.

Recommendations

Robotic and associated systems designs should:

- include physical barriers that use gates with electrical interlocks to stop the robot when the gate is open

- include motion sensors, light curtains, or floor sensors that stop the robot whenever a worker crosses the barrier
- provide barriers, as appropriate, between robotic equipment and any freestanding objects that limit robot arm movement so that workers cannot get caught between any part of the robot and the "pinch points"
- provide adequate clearance around all moving components of the robotic system
- use remote "diagnostic" instrumentation so that troubleshooting can be done from areas outside the operating space of the robot
- provide adequate illumination in the control and operational areas of the robotic system so that written instructions, as well as buttons, levers, etc., are clearly visible
- proactively design good ergonomic engineering elements into each system
- allow for the installation of a protective gate around access areas to the robotic platform
- include on floors or working surfaces clearly visible marks that indicate the zones of potential movement of the robot.

Training should emphasize the following:
- Before operating or performing maintenance in robotic workstations, workers must be familiar with all working aspects of the robot, including full range of motion, known hazards, how the robot is programmed, emergency stop buttons, and safety barriers.
- To prevent hazards during programming, programmers, operators, and maintenance workers should operate robots at speeds that are consistent with adequate worker response and should be aware of all conceivable pinch points, such as poles, walls, and other equipment, in the robot's operational area.

Procedures should be documented and each operator should be trained on the systems he or she will be working with:
- All equipment should be properly locked out/tagged out prior to performing maintenance on it.
- A spot inspection program ensures that all employees are complying with safety requirements.
- Procedures should be developed to ensure that individuals not involved in maintenance activities are not in the immediate area of the maintenance being performed.
- The robot and the point of operation should be safeguarded to prevent entry during automatic operation.
- Whenever it is necessary for a worker to be within the operating space of a robot, additional safety provisions should be taken, including, at minimum, the presence of another worker (buddy system), who can turn off the robot should an emergency situation develop.

- Users should conduct a risk assessment of the robot/robot system to identify equipment, installation, standards, and process hazards so that adequate employee safeguards can be provided.
- Users should ensure that personnel who interact with the robot or robot system, such as programmers, teachers, operators, and maintenance personnel, are trained on the safety issues associated with the task, robot, and robot system.

Close supervision is imperative to ensure the safety of such operations:
- Supervisors should ensure that no one enters the operational space of a robot without first putting the robot on "hold," in a "power down" condition, or at a reduced operating speed.
- Experienced workers performing automated tasks may, as time passes, become complacent, overconfident, or inattentive to the hazards inherent in complex automated equipment. Supervision and ongoing training should be used to overcome this problem.
- Operators should not be in the robot's working space while the robot is operating.
- Supervisors should ensure that workers follow safety rules and operating procedures in all cases and should enforce applicable company policies for anyone found not doing so.

Protecting the Robot Maintenance/Repair Personnel

Because robot design differs among manufacturers, the specific instructions provided by the manufacturer for maintaining and repairing a particular robot should be followed. If the robot has been modified in any way or if its procedures or programming has changed, maintenance personnel must be informed of the modifications and their implications for safety. Rebuilt or previously owned robots should have all changes from their original design documented and available to maintenance personnel. Before robots are maintained or repaired, provide the corresponding manuals, procedures, and instructions to the employees who perform such work. Updated drawings and schematics should be available to train workers properly in safety procedures and applicable standards.

Workers should use proper lockout/tagout procedures to shut off and lock out power sources and should follow appropriate testing procedures before performing maintenance work on and while repairing a robot. Refer to 29 CFR 1910.147 and the more detailed explanation in Chapter 6, Safeguarding, in this volume. Maintenance personnel should make sure that any potentially hazardous stored energy is released or blocked before they service the robot. This energy may be in the form of air and hydraulic accumulators, springs, counterweights, flywheels, and loads held by the robot.

When lockout/tagout is not possible, provide appropriate alternate safeguarding. These safeguards could include:

- EMERGENCY STOP and SLOW SPEED controls
- a second person monitoring the robot control panel, who can immediately react to potential hazards
- a system that gives the maintenance employee within the work space control over robot movement or blocking devices such as blocks and pins.

If any safeguards will need to be bypassed while workers are maintaining or repairing a robot, provide alternative safeguarding in accordance with lockout/tagout regulations. When the maintenance task is complete, check any bypassed safeguards and return them to their active state before operations begin or before the robot is energized.

Users of robots and robot systems should establish and document an effective inspection and maintenance program. This program should include any preventive maintenance recommended by the robot's or the system's manufacturer.

Computer-Integrated Controls

In automated production, machines and equipment are coordinated by means of computer-integrated controls. Such controls facilitate information flow, coordinate factory operations, and increase efficiency and flexibility. The flexibility of programmable automation is becoming more desirable to firms seeking to compete effectively because the equipment for design, production, and management is linked together and can be reprogrammed.

However, the ramifications of computer-integrated manufacturing (CIM) on safety cannot be ignored. A 1984 report by the Office of Technology Assessment indicates that the absence of standard programming languages, data formats, communications protocols, teaching methods, controls, and well-developed offline programming capabilities are barriers to buying or using CIM. Even though manufacturers are addressing these issues, this lack of standardization, which includes the fact that different machines have different human–system interfaces, makes effective safety training in CIM even more necessary than it is in other manufacturing processes.

CHEMICAL PROCESSES

Automated chemical processes include many worker hazards that must be handled through written safety policies, engineering and administrative controls, worker training, and PPE. Safety responsibilities are particularly wide ranging in the chemical industry because of the many significant operations that workers must handle and because the nature and the number of chemicals—and their by-products—constantly change. Companies can fulfill their safety responsibilities by informing workers of process safety information, conducting ongoing hazard/risk analyses and pre-start-up reviews, developing operating procedures manuals, managing changes, and auditing processes. (See also Chapter 24, Process Safety Management, in this volume.)

Process Safety Information

Communicating the design basis of equipment and processes to all affected personnel is not always easy. This is especially true in large, sophisticated chemical-processing facilities, where pressures, temperatures, and other control parameters must be maintained within safe tolerance levels at thousands of critical junctures.

Process Document

One method that managers use to explain design basis to their employees is to produce a "cookbook" of chemical processing, which is commonly known as the facility's chemical process document. This basic manual, which appropriate employees must understand as a fundamental part of their job training, includes the parameters for the safe operation of facility processes. Management must be aware of the consequences of operating outside those parameters in order to develop process controls that ensure safe operation.

Suggested Contents

A process document should contain at least the following process safety information.

General Information

- an assessment of the hazards posed by the materials used in chemical processing
- toxicity information
- permissible exposure limits
- physical data
- thermal and chemical stability data
- reactivity data
- corrosivity data
- data on the hazardous effects of inadvertently mixing materials.

Process Design Information

- block flow diagrams or simplified process flow diagrams
- process chemistry

- maximum needed inventory
- acceptable upper and lower limits, where applicable, for items such as temperatures, pressures, flows, and compositions.

Mechanical Design Information
- piping and instrument diagrams, electrical area classification, and design basis of relief systems
- design of the ventilation system
- equipment and piping specifications and a description of the shutdown and interlock systems
- design codes employed
- required or mandatory inspections and maintenance activities.

When mechanical design deviates from applicable consensus codes and standards, the deviation and its design basis should be documented.

Information concerning additional data that could be included in a facility's process document is provided in Chapter 24, Process Safety Management, in this volume, and in Chapter 9, Identifying Hazards, in the *Administration & Programs* volume.

Hazard/Risk Analysis

Safe operation of any chemical process requires the person in charge to ask the three "whats":
- What can go wrong?
- What is the probability that something will go wrong?
- What are the consequences if something does go wrong?

In other words, identify the risks and the resulting dangers in the system. By following the procedure recommended by Daniel A. Crowl and Joseph F. Louvar in their book, *Chemical Process Safety: Fundamentals with Applications*, managers can identify the hazards and risks and judge whether the dangers and risks are acceptable (Figure 22–5).

If the dangers are deemed acceptable, management can create and operate the process. When management deems the risks or hazards unacceptable, design engineers must modify the design and repeat the hazard identification and risk assessment processes.

Popular Methods

Hazard identification can be achieved using any of several established methods. Four of the most widely used methods are:
1. hazard surveys
2. process checklists
3. hazard and operability studies (HAZOP)
4. safety reviews.

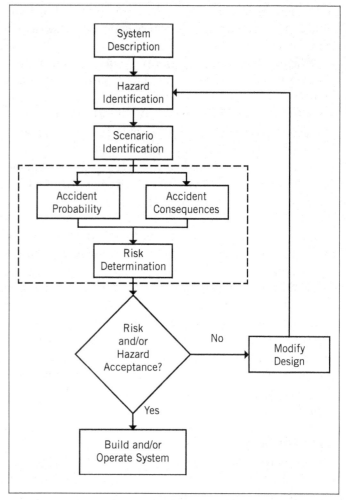

Figure 22–5. This procedural flow chart illustrates the process of hazard and/or risk assessment.

Hazard Surveys. In small firms, a chemical hazard survey might consist of an inventory of hazardous materials. Managers can refer to this inventory to verify, for example, that dangerous substances are stored safely and that plans are in place to prevent an explosion or escape of toxic substances in the event of a fire or other unplanned incident.

Many large companies, however, use Dow's stringent Fire and Explosion Index (Figure 22–6). The Dow index provides a relative ranking of a hazard and a mechanism for estimating dollar losses resulting from that hazard. Crowl and Louvar's *Chemical Process Safety* (see the References at the end of this chapter) provides a concise explanation (with examples) of how the index works.

Although surveys can help identify hazards associated with material storage and design, surveys cannot alert management to hazards associated with human error. Still, most such surveys can be conducted quickly and easily and, often, do not need to be conducted by individuals who are highly trained in conducting surveys.

FIRE AND EXPLOSION INDEX ◆DOW◆

			LOCATION		DATE

PLANT	PROCESS UNIT	EVALUATED BY	REVIEWED BY	

MATERIALS AND PROCESS

MATERIALS IN PROCESS UNIT

STATE OF OPERATION START-UP SHUT-DOWN NORMAL OPERATION	BASIC MATERIAL(S) FOR MATERIAL FACTOR

MATERIAL FACTOR (SEE TABLE I OR APPENDICES A OR B) Note requirements when unit temperature over 140 F)

	PENALTY	PENALTY USED	
1. GENERAL PROCESS HAZARDS			
BASE FACTOR ——————————➤	1.00	1.00	
A EXOTHERMIC CHEMICAL REACTIONS (FACTOR .30 to 1.25)			
B ENDOTHERMIC PROCESSES (FACTOR .20 to .40)			
C MATERIAL HANDLING & TRANSFER (FACTOR .25 to 1.05)			
D ENCLOSED OR INDOOR PROCESS UNITS (FACTOR .25 to .90)			
E ACCESS	.35		
F DRAINAGE AND SPILL CONTROL (FACTOR .25 to 50) ___ Gals			
GENERAL PROCESS HAZARDS FACTOR (F₁) ——————————➤			
2. SPECIAL PROCESS HAZARDS			
BASE FACTOR ——————————➤	1.00	1.00	
A TOXIC MATERIAL(S) (FACTOR 0.20 to 0.80)			
B SUB-ATMOSPHERIC PRESSURE (500 mm Hg)	.50		
C OPERATION IN OR NEAR FLAMMABLE RANGE (.) INERTED (.) NOT INERTED			
1 TANK FARMS STORAGE FLAMMABLE LIQUIDS	.50		
2 PROCESS UPSET OR PURGE FAILURE	.30		
3 ALWAYS IN FLAMMABLE RANGE	.80		
D DUST EXPLOSION (FACTOR .25 to 2.00) (SEE TABLE II)			
E PRESSURE (SEE FIGURE 2) OPERATING PRESSURE ___ psig RELIEF SETTING ___ psig			
F LOW TEMPERATURE (FACTOR .20 to .30)			
G QUANTITY OF FLAMMABLE/UNSTABLE MATERIAL QUANTITY___ lbs. H_c ___ BTU/lb			
1 LIQUIDS, GASES AND REACTIVE MATERIALS IN PROCESS (SEE FIG 3)			
2 LIQUIDS OR GASES IN STORAGE (SEE FIG 4)			
3 COMBUSTIBLE SOLIDS IN STORAGE, DUST IN PROCESS (SEE FIG 5)			
H CORROSION AND EROSION (FACTOR 10 to 75)			
I. LEAKAGE – JOINTS AND PACKING (FACTOR .10 to 1.50)			
J. USE OF FIRED HEATERS (SEE FIG 6)			
K. HOT OIL HEAT EXCHANGE SYSTEM (FACTOR .15 to 1.15) (SEE TABLE III)			
L ROTATING EQUIPMENT	.50		
SPECIAL PROCESS HAZARDS FACTOR (F₂) ——————————➤			
UNIT HAZARD FACTOR (F₁ x F₂ F₃) ——————————➤			
FIRE AND EXPLOSION INDEX (F₃ x MF F & EI) ——————————➤			

Figure 22–6. This sample index can be used to assess hazards and estimate losses.

Process Checklists. Good checklists tend to become longer and more specific with use and might even develop into extensive documents in a large chemical facility. One way to keep the hazard identification method from getting out of hand is to limit its scope by creating separate checklists as incident prevention requires. For example, one checklist might be geared toward starting up a new process and another geared toward modifying a process already under way. However, checklists are useful only for preliminary hazard identification and should be followed by a more complete hazard identification method.

Hazard and Operability Studies. Commonly known as a HAZOP study, this is a formal procedure used to identify hazards in a chemical-processing facility and often requires the dedication of a committee made up of experienced chemical and facility process personnel. A complete, thorough HAZOP review can require a significant investment in time and resources.

Briefly stated, this method encourages the committee members to imagine all the ways that process failures might occur and all the ways that these failures could be mitigated or minimized. Although a tedious process, it is nonetheless highly effective and is gaining wider use. For a complete explanation of how a HAZOP review works, see the American Institute of Chemical Engineers' *Guidelines for Hazard Evaluation Procedures* or Crowl and Louvar's book.

Safety Reviews. There are at least two types of safety reviews: (1) an informal review, which involves a few people who are concerned about a minor matter such as a change in procedure, and (2) a formal review, which usually results in a more detailed report from a formal committee. Of course, a report might also be written following an informal review. However, the primary goal of an informal review is to provide a forum in which ideas can be voiced and safety improvements can be developed.

Conversely, before undertaking substantial updating or changes in processes, a company should establish a formal review committee and request an official written report. The report should contain relevant technical data, equipment descriptions, explanations of the process and attendant hazards, a safety checklist, a material safety data sheet for each hazardous material used in the process under review, and a full accounting of all procedures. Crowl and Louvar, in their book, *Chemical Process Safety*, provide a good example of a full report following a formal safety review.

In many cases, a formal safety review committee can be convened with little difficulty. Even though the formal review process may take substantially more time than an informal review, the formal review often produces beneficial results. If experienced safety reviewers are not available to form a committee, a company could use less experienced personnel and conduct a more structured, HAZOP study and a follow-up analysis.

Other Methods
Additional methods for pinpointing hazards include:
- *"What if?" analysis.* In this analysis, a hazard review team poses hypothetical questions about potential problems in the process, such as, "What if the pipe breaks?" The team then ponders the consequences and proposes solutions.
- *Failure modes, effects, and criticality analysis (FMECA).* In this analysis, a review team catalogs all the equipment used in the process under review and notes how each piece of equipment could fail. The team then considers the consequences of each type of failure and recommends appropriate safety measures. FMECA is often a good approach for analyzing electromechanical systems and single-fault events.
- *Failure modes and effects analysis (FMEA)* is a step-by-step approach for identifying all possible failures in a design, a manufacturing or assembly process, or a product or service. Failures are ranked according to how serious their consequences are, how frequently they occur, and how easily they can be detected. The purpose of the FMEA is to lead to the elimination or the reduction of failures, starting with the highest-priority failures.
- *Fault tree analysis (FTA)* is conducted by looking at a main system failure and analyzing its cause. This analysis is usually effective when analyzing combinations of failure events and human-use errors. FTA and failure modes analysis are often used together to create a comprehensive, top-down and bottom-up risk analysis of complex systems.
- *Event tree analysis (ETA)* begins by looking at an initiating event and following that event through a series of potential paths. These paths then are assigned probabilities of occurring. This process also allows the probabilities of the possible outcomes to be calculated. Specific combinations of events that lead to an adverse event can then be analyzed by other methods.
- *Human error analysis.* When conducting this analysis, the hazard review team identifies the part of a process that is most likely to fall prey to human error. A review committee then makes safety recommendations or suggests preventive measures. This process is particularly useful in the layout and design of a control room's panels and instruments because the large amount of interactions among workers conducting processes in this location greatly increases the risk of injury.

Timing
Hazard identification and risk assessment should be ongoing processes that begin during the earliest conceptual phase of a new chemical process: when the firm is gathering basic data about the anticipated process. This is the time to ask fundamental questions such as, "Why are we considering the use of hazardous materials?" and "Are less hazardous alternatives available?"

Another good time to ask questions is during preauthorization, when the design is less than half completed and before the project has been officially authorized. Ask questions about hazards that could affect cost, such as, "How are we going to handle steam venting?"

Inserting Risk Assessment

In one prominent chemical company's process safety management plan, a four-part risk assessment is undertaken during the design stage. Reviewers ask:
- What potentially catastrophic incidents could occur?
- What is the downwind dispersion likely to be in the event of a toxic gas release?
- What would the impact of a catastrophic event be on the workplace and the community?
- Can we quantify or conduct a probability analysis for incidence occurrence?

Hazard identification also should occur during pre-start-up. Before chemical processing begins, look over the hazard reviews, and use a pre-start-up checklist to make sure that the elements of process safety management have all been properly addressed.

Pre-Start-Up Reviews

All the elements of a process safety management program should be in place and fully functioning before the warm-up phase of facility operation begins. A company should call on persons with no vested interest in facility start-up to assist in this effort.

A checklist of the various elements that comprise a facility's process safety management plan should be created and should indicate who is responsible for reviewing each item, who is responsible for executing each item, and who is responsible for following through on each item (Table 22–C).

The American Petroleum Institute (API) recommends that pre-start-up safety reviews ensure that:
- construction conforms to specifications
- safety, operating, maintenance, and emergency procedures are in place and are adequate

- process hazard analysis recommendations have been considered and implemented
- training of operating personnel has been completed.

Analysis of Existing Facilities

The API recommends that process hazard analyses be reviewed in order of priority and be updated every 3 to 10 years. In addition, the API says that the following factors should be considered when establishing priorities:
- a high substance hazard index value or large quantities of toxic, flammable, or explosive substances at the facility
- proximity to a populous area or a facility location that has a large work force
- harsh operating conditions, such as high temperatures or pressures, or conditions that cause severe corrosion or erosion.

The hazard analysis team should include persons knowledgeable in engineering, operations, design, and chemical processing.

Operating Procedures Manuals

Every chemical facility needs an operating manual that details the rules and guidelines of safe operation. The manual should also contain the relevant safety, health, and environmental information needed to operate the facility without unintentional injuries.

Much has been written about what an operating manual should contain. The API, for example, recommends that operating manuals include the following points:
- the position(s) of the person(s) responsible for each of the facility's operating areas
- clear instructions for safe operation
- operating conditions and steps for initial start-up
- procedures for normal, temporary, and emergency operations
- normal shutdown procedures
- procedures for start-up following a turnaround
- operating limits (where safety considerations are important)

TABLE 22–C. Pre-Start-Up Checklist

Is construction in accordance with design specifications?	Are general subjects adequately addressed?
Are key elements of process hazard management in place? • Safety information package • Completed process hazard reviews • Rules and procedures for operating, maintenance, safety, spill control, and emergencies • Training • Equipment tests and inspections	• Fire protection • Means of exit • Availability and location of safety equipment • Equipment guards • Ventilations • Tripping hazards • Proper drainage

Does the checklist call for the inspection team to decide whether the facility is ready for start-up? Where recommendations are made, are responsibilities assigned to appropriate personnel for execution and follow-up?

TABLE 22–D. Ten Rules for Safe Chemical Processing Operations

1. Never start adding a reactant at a lower or higher temperature than is called for in the process.	6. Be extremely cautious with spill or low-flash-point materials. Make sure the proper personnel understand and practice those plans.
2. Do not add at a rate faster than specified.	7. In flammable and toxic liquid service, eliminate as much as practical the use of glass and other breakable materials.
3. During scale-up, make sure enough cooling capacity has been supplied.	8. Use nitrogen purges or other equivalent, recognized means on vapor spaces containing explosive materials.
4. Have specific, written plans on what to do if an emergency occurs (see Chapter 18, Emergency Preparedness, in the *Administration & Programs* volume).	9. Do not undercharge or overcharge reactants if a mistake has been made without consulting with technical personnel.
5. If material is being added and the agitator has stopped, do not restart without first consulting with technical personnel.	10. Watch for unusual incidents and for temperature, pressure, or other process parameter extremes that could make the reaction uncontrolled or dangerous.

- descriptions of the consequences of deviation
- steps to correct or avoid deviation
- safety systems and their functions
- occupational safety and health considerations.

The operating procedures manual should explain to personnel the hazards of chemical operations and what the consequences could be if operations exceed safe limits. It is important that operating manuals be updated on a regular basis, as changes occur in the workplace, and/or as new knowledge is gained. The company should not authorize any change in technology, the facility, or procedural steps until it updates the operating procedures manual and trains the relevant personnel appropriately. A complete set of operating procedures, such as those given in Table 22–D, should be the fundamental set of documents for all operator training.

Management of Change

Whenever someone proposes a change to established technology, management should thoroughly analyze the proposal in light of its potential impact on the workplace and on the environment. Seemingly insignificant changes can have far-reaching effects.

For example, a worker repairs a piece of equipment using a bolt that is different from the original bolt. Later, different materials are used or process conditions are changed to improve efficiency. Perhaps raw materials begin to come from a different supplier. If there are no controls in place, change can build upon change until the equipment being operated is so different from the original equipment's design that the design review is obsolete.

Change must be consistent with established process technology, which means that management must establish requirements for all field modifications before the modifications are implemented.

If management follows the suggestions of the API in this regard, management must:

- consider the process and the mechanical design basis for change
- analyze the safety, health, and environmental issues related to the proposed change, as well as the potential effects on facilities upstream and downstream
- consider how modifications will affect operating procedures
- inform appropriate personnel about the pending change and its potential consequences
- consider the duration of the change
- acquire the necessary authorizations.

Whenever a chemical process changes, the corresponding equipment and personnel often change as well, thus requiring supplemental employee training. Make sure that all affected personnel understand the change, the reason for the change, how operations have been affected, and how employees are to perform their new responsibilities. Each of these aspects is important to ensure safe facility operations.

Any modifications to facilities or operating procedures should be considered carefully, and the appropriate manager should document and authorize all changes.

Auditing

Auditing is a way for management to make sure that a process safety management program is operating as designed. One authoritative body, the API, recommends that audits be conducted at 3- to 5-year intervals. Recurring audits should be able to pinpoint deficiencies caused by such factors as changes in personnel or confusion over priorities. When a deficiency is discovered at an early stage, only minor corrective action will be needed.

Auditing process safety management is primarily a line responsibility. It is also easier to implement corrective changes when workers regard the changes as coming from an in-house source. Of course, outside audits are extremely useful and are often used in conjunction with internal audits to ensure an unbiased result. But internal auditing reinforces employees' understanding of priorities and the corporate mission or philosophy and also lays the groundwork for a successful outside audit. (For more information, see Process Safety Audits in Chapter 24, Process Safety Management, in this volume.)

SUMMARY

- Safety in automated manufacturing processes can be enhanced by (1) careful identification of hazards and (2) development of appropriate strategies and worker training to regulate the workplace environment. Risks associated with automated processes include inhalation, burn, radiation, and pinch hazards.
- Companies should provide visual and mechanical warning signs, awareness barriers, interlocks, and controls to protect workers from the hazards of automated equipment or systems.
- Preventive and predictive maintenance programs can enhance worker safety, particularly if maintenance procedures follow manufacturer and industry/government standards for the equipment and processes.
- Automated production requires special precautions because of the nature of the machinery, the variety of human–machine interactions on the job, and the rapid development of automated technologies.
- Many automated manufacturing systems or cells use materials-handling and transport systems (automated guided vehicles) and robots. Each type of machine has its own hazards, and workers must be trained in the proper procedures and housekeeping practices in order to operate this equipment safely.
- When safeguarding workers around robotic machinery, companies should focus on (1) the manufacturing, remanufacturing, and rebuilding of robots; (2) robot installation; and (3) safeguards for workers exposed to the hazards associated with the use of robots.
- Automated chemical processes include considerable safety hazards for workers. Companies should thus conduct ongoing risk analyses to identify the hazards in current, new, and redesigned processes and should also design effective safety procedures and training programs.
- Proactive steps that companies can take in hazard analysis and safety planning include conducting pre-start-up reviews, compiling pre-start-up checklists, carrying out preliminary analyses of existing facilities, updating operating procedures manuals, and creating strategies to manage change.
- Once a process safety management program is in place, a firm should conduct recurring audits to keep the process operating as designed.

REFERENCES

American Institute of Chemical Engineers 120 Wall Street, Foor 23, New York, NY 10005-4020.

Dow's Fire and Explosion Index Hazard Classification Guide. 7th ed. New York: 1994.

Guidelines for Hazard Evaluation Procedures. 3rd ed. New York: Wiley 2008.

American National Standards Institute, 11 West 42nd Street, New York, NY 10036.

American National Standard for Industrial Robots and Robot Systems—Safety Requirements, ANSI/RIA R15.06-2012, (revision of ANSI/RIA R15.06-1999).

American National Standard for Plastics Machinery—Horizontal Injection Molding Machines—Safety Requirements for Manufacture, Care, and Use, ANSI/SPI B151.1–2007.

American National Standard for Plastics Machinery—Safety Requirements for the

Integration of Robots with Injection Molding Machines, ANSI/SPI B151.27–2013.

Cold Headers Safety Requirements for Construction, Care and Use, ANSI B11.07–1995 (R2010).

Machine Tools—Iron Workers—Safety Requirements for Construction, Care, and Use, ANSI B11.5–1988 (R2008).

Machine Tools—Single- and Multiple-Spindle Automatic Bar and Chucking Machines—Safety Requirements for Construction, Care, and Use, ANSI B11.13–1992 (R2007).

Performance Requirements for Safeguarding, ANSI B11.19–2010.

Control of Hazardous Energy—Lockout/Tagout and Alternative Methods, ANSI Z244.1–2003 [R2008]).

ANSI B11 Machine Tool Safety Standards:

Safety Requirements for Integrated Manufacturing Systems, B11.20.

Safety Requirements for Machine Tools Using Lasers for Processing Material, B11.21.

Safety Requirements for Turning Centers and Automatic, Numerically Controlled Turning Machines, B11.22.

Safety Requirements for Machining Centers and Automatic, Numerically Controlled Milling, Drilling, and Boring Machines, B11.23.

Safety Requirements for Transfer Machines, B11.24.

Safety Requirements for Gear and Spline Cutting Machines, ANSI B11.11–2001 (R2012).

Safety Requirements for Grinding Machines, ANSI B11.09–2010.

Safety Requirements for Hydraulic and Pneumatic Presses, ANSI B11.2–2013.

Safety Requirements for Machines Processing or Slitting Coiled or Non-Coiled Material, ANSI B11.18–2006.

Safety Requirements for Manual Milling, Drilling, and Boring Machines with or without Automatic Control, ANSI B11.8–2001 (R2012).

Safety Requirements for Manual Turning Machines with or without Auto Control, ANSI B11.6–2001 (R2012).

Safety Requirements for Mechanical Power Presses, ANSI B11.1–2009.

Safety Requirements for Pipe, Tube, and Shape Bending Machines, ANSI B11.15–2001 (R2012).

Safety Requirements for Power Press Brakes, ANSI B11.3–2012.

Safety Requirements for Roll-Forming and Roll-Bending Machines, ANSI B11.12–2005 (R2010).

Safety Requirements for Shears, ANSI B11.4–2003 (R2013).

Safety Standard for Industrial Robots and Industrial Robot Systems, ANSI/RIA R15.06–2012.

Safety Standard for Guided Industrial Vehicles and Automated Functions of Manned Industrial Vehicles, ANSI B56.5.

American Petroleum Institute (API), 1220 L Street, NW, Washington, DC 20005-4070.

Bachelor, B. G., and F. M. Waltz, eds. *Proceedings on Machine Vision Systems Integration in Industry.* SPIE—Society of Photo-Optical Instrumentation Engineers, Boston, 1990.

Chiantella, N. A., ed. *Management Guide for CIM (Computer-Integrated Manufacturing).* Dearborn, MI: The Computer and Automated Systems Association Society of Manufacturing Engineers, 1986.

Crowl, D. A., and J. F. Louvar. *Chemical Process Safety: Fundamentals with Applications.* New York: Prentice-Hall, 2011.

Gainer, C. A., Jr., and B. C. Jiang. A Cause-and-Effect Analysis of Industrial Robot Accidents from Four Countries. SME Technical Paper MS87–204. Society of Manufacturing Engineers, Dearborn, MI, 1987.

Gargeya, V. B., and J. P. Thompson. "Just-in-Time Production in Small Job Shops." *Industrial Management* (July/August 1994): 23–26.

HAZOP and HAZAN: Identifying and Assessing Process Industry Hazards. 4th ed. and English ed. Warwickshire, England: The Institution of Chemical Engineers, 1999.

Holland, J. R., ed. *Flexible Manufacturing Systems.* Dearborn, MI: Society of Manufacturing Engineers, 1984.

International Labour Office. *Major Hazard Control: A Practical Manual.* Geneva, Switzerland: 1988.

ISO 8373, Robots and robotic devices—Vocabulary, International Organization for Standardization, ISO Central Secretariat, 1, ch. de la Voie-Creuse. CP 56, CH-1211, Geneva, Switzerland, 2012.

Kayes, P. J., ed. *Manual of Industrial Hazard Assessment Techniques.* London: Technical Ltd., The World Bank, 1985.

Kletz, T. A. *Learning from Accidents in Industry.* London: Butterworth, 1988.

Lane, J. D., ed. *Automated Assembly.* 2nd ed. Dearborn, MI: Society of Manufacturing Engineers, 1986.

Lees, F. P. *Loss Prevention in the Process Industries.* London: Butterworth, 1980.

Miller, R. K. *Automated Guided Vehicles and Automated Manufacturing.* Dearborn, MI: Society of Manufacturing Engineers, 1987.

National Fire Protection Association, 1 Batterymarch Park, Quincy, MA 02269.

Electrical Standard for Industrial Machinery, NFPA 79–2012.

Fire Safety Standard for Powered Industrial Trucks Including Type Designations, Areas of Use, Conversions, Maintenance, and Operation, NFPA 505–2013.

Standard for Electrical Safety in the Workplace, NFPA 70E–2012.

National Institute for Occupational Safety and Health, U.S. Department of Health and Human Services, 4676 Columbia Parkway, Cincinnati, OH 45226.

Fatal Accident Summary Report: Die Cast Operator Pinned by Robot. In-House Fatality Assessment and Control Evaluation (FACE) Report 8420, 1984.

Machine Operator Crushed by Robotic Platform. Nebraska Fatality Assessment and Control Evaluation (FACE) Report 99NE017, 1999.

Mold Setter's Head Struck by a Cycling Single-Side Gantry Robot. Michigan Fatality Assessment and Control Evaluation (FACE) Report 01MI002, 2001.

National Safety Council, 1121 Spring Lake Drive, Itasca, IL 60143.

Occupational Safety and Health Data Sheet 12304–D717, 1991, Robots.

Noro, K., ed. *Occupational Health & Safety in Automation & Robotics.* London: Taylor & Francis, 1987.

Occupational Safety and Health Administration. 29 CFR 1910.147, The Control of Hazardous Energy. 29 CFR 1910.1200, Hazard Communication Standard.

Plog, B., ed. *Fundamentals of Industrial Hygiene.* 6th ed. Itasca, IL: National Safety Council, 2012.

Roffel, B., and J. E. Rijnsdorp. *Process Dynamics, Control and Protection.* Ann Arbor, MI: Ann Arbor Science, 1982.

Sohal, A. S., L. Ramsay, and D. Samson. "JIT Manufacturing: Industry Analysis and a Methodology for Implementation." *International Journal of Operations and Production Management* 13, no. 7 (1993): 22–56.

Strubhar, P. M., ed. *Working Safely with Industrial Robots.* Dearborn, MI: Robotics International Society of Manufacturing Engineers, 1986.

Suzaki, K. *The New Manufacturing Challenge: Techniques for Continuous Improvement.* New York: Free Press, 1987.

UL 558–2012; Standard for Industrial Trucks, Internal-Combustion Engine-Powered,

UL 583–2012; Standard for Electric-Battery-Powered Industrial Trucks.

Wantuck, K. A. *The Japanese Approach to Productivity.* Southfield, MI: Bendix Corporation, 1983.

What Went Wrong? Case Histories of Process Facility Disasters. 2nd ed. Houston: Gulf Publishing Co., 1988.

Zuech, N., ed. *Machine Vision Capabilities for Industry.* Dearborn, MI: Society of Manufacturing Engineers, 1986.

REVIEW QUESTIONS

1. What is the disadvantage of automation being designed, implemented, or modified using the just-in-time (JIT) philosophy?

2. How can a computerized maintenance management system (CMMS) reduce company costs?

3. Safety in automated manufacturing processes can be greatly improved in what two ways?
   a.
   b.

4. List four factors that influence the kinds of automation-related safety precautions that companies must incorporate into their safety planning.
   a.
   b.
   c.
   d.

5. A minimum clearance of _____ is required between an automated guided vehicle (AGV) with its load and any fixed object in all areas that include pedestrian travel or interaction with workers.

6. Name the standard that specifies the areas, or spaces, necessary for machine motion.

7. What two areas of automated materials-handling and transport systems have particular risks?
   a.
   b.

8. Refer to _____ for standards on bumper design, activation, and requirements for AGVs.

9. List the three basic parts that make up a robot.
   a.
   b.
   c.

10. List the three major areas of robot safety.
    a.
    b.
    c.

11. Why are hazards most likely to occur when a robot is in the TEACH mode?

12. List the three "whats" that the person in charge is required to ask to ensure the safe operation of any chemical process.
    a.
    b.
    c.

13. Name the formal procedure used to identify hazards in a chemical-processing facility.

The Computer as a Safety Information Tool

Ralph B. Stuart, CIH

Introduction

The Internet as a Reference Tool
Web Tools ▶ An Internet Search Strategy

The Internet as a Networking Tool
General Considerations ▶ Styles of Participation
▶ Finding Discussion Groups

The Internet as a Safety Culture Tool
Online Training ▶ Developing a Reference Library
▶ Maintaining Awareness

The Internet as a Safety Program Management Tool
Collecting Data ▶ Management System Development
▶ Connecting to Other Management Information
Systems

Summary

Resources

Review Questions

INTRODUCTION

Computers and computer networks are becoming increasingly important tools for safety professionals. First used to facilitate data storage, retrieval, and calculations, they have gradually emerged as core communication and educational tools throughout society. The rapid development of the Internet, particularly its powerful search engines, has increased the importance of electronic information in the daily life of professionals and organizations alike. It is difficult to imagine managing the variety and amount of information that flows through a 21st-century safety program without strong computer skills.

Another aspect of computer use by safety professionals is the collection and analysis of safety program data. This is a skill that can require significant practice to master. Many safety and health professionals have developed valuable tools for managing their work, only to see these tools become obsolete as program needs, computer platforms, and organizational support change. As options for collecting, managing, and distributing electronic data proliferate, this concern is becoming an increasingly significant issue. For this reason, computer solutions proposed to serve a safety program must be carefully evaluated not only for their current functionality, but also for their scalability and long-term sustainability. The best approach to addressing this concern is developing a coherent information architecture for the data being collected; development of such architectures will benefit from partnerships with information professionals who have experience in addressing these concerns. However, the good news is that computer systems are continuing to become more powerful and easy to use; when their use is carefully planned and maintained, they can be vital assets to safety programs.

This chapter focuses on four aspects of the use of electronic information by safety professionals:

1. *As a reference tool.* Computer networks provide convenient access to a wide variety of information resources of value to the safety professional. This section describes the use of these resources at the conceptual level, and specific resources are listed in the Resources section at the end of this chapter.
2. *As a networking tool.* It is increasingly difficult for a safety professional to effectively be a "jack of all trades." The hazards and protection strategies of the 21st-century workplace are evolving rapidly as new technologies (e.g., biological tools and nanoparticles) result in new safety-related challenges and as traditional hazards are better recognized and understood. The ability to consult peers on both technical and management issues is a valuable professional tool. While professional organizations have tradition-

ally facilitated this aspect of professional work, the development of the Internet, particularly e-mail, has become another important way to connect with colleagues for discussions of mutual benefit. This second section discusses ways to maximize the value of this use of the Internet.

3. *As a safety culture tool.* A key aspect of a successful safety program is the ability to distribute information effectively and efficiently to stakeholders. This involves not only identifying the information that specific individuals and groups need, but also providing it in a form that is useful for that audience. This may involve providing text documents to some people while developing video forms of the same information for other people. The use of the Internet provides important opportunities and challenges for this function.
4. *As a safety program management tool.* In addition to collecting and sharing information about their workplaces, safety professionals need to fashion this information to the needs of a variety of audiences, including upper management, workers, and regulators. Often, this means restating the same information in a different way according to the question being asked. In addition, information about safety conditions in the workplace needs to be connected to other information that the organization manages, such as financial or human resource records, in order to be used effectively. This last section of the chapter considers this aspect of a safety program's use of computers and computer networks.

THE INTERNET AS A REFERENCE TOOL

One of the biggest challenges facing safety and health professionals is finding relevant information when they need it. Changing regulations, new data, and emerging technologies can make paper-based resources an unreliable medium for researching occupational safety and health issues. Fortunately, the Internet now provides a legitimate alternative to an extensive library of books, manuals, and regulations. This section describes some of the tools that can be used to become familiar with the types of information available on the Internet. A strategy is then described for answering specific questions that arise in daily safety and health work.

Just as being familiar with how a library is organized makes searching a paper library for information much easier, being familiar with the Internet also makes searching for particular information easier. It is a good idea to spend some time building familiarity with a variety of websites before attempting to answer a particular question.

Web Tools

While it is tempting to rely on an Internet search engine to search for whatever information is available on a particular topic as the need arises, this approach will often result in finding information that falls short of professional quality. It is a worthwhile investment of time to explore the Internet using web directories and careful search engine searches to identify reliable sources of information. There are websites that act as subject guides that list a number of Internet sites related to a specific subject. A good example of one such website for occupational safety and health audiences is that of the Canadian Centre for Occupational Health and Safety (CCOHS). This site includes a page that organizes a large number of occupational safety and health links at ccohs.ca/oshlinks/. By using this site as a starting point, a sense of the types of resources on the Web can be obtained. There are many other sites that can serve as starting points for this process; the Resources section of this chapter includes many different subject-specific web-based sources that can be useful to the safety professional.

Search Engines

When interested in finding specific information on a safety subject, using a subject-based directory may not be the most efficient way of approaching the Internet. Rather, it is desirable to conduct a keyword search to find information of interest. There are several search engines available for this purpose; some search the entire Internet; others restrict their searches to specific databases on the Web. It is important to understand the scope and syntax of the search engines available for use.

In general, a large number of hits in a search indicate that the search needs to be further refined if looking for specific information. It can become quite time-consuming to check out a long list of links unless in an exploratory mode. It is also possible that the search engine's weighting of the search results is at odds with identifying the subject of the researcher's concerns. For example, the order of results from some search engines is based on sponsorship deals with specific vendors, rather than the value of the information the site contains.

It is important to remember that general-purpose web search engines are changing rapidly and competing with each other for users, so trying more than one for a particular search is likely to find a variety of different resources. Practical experience with the various search engines will help determine which one works best for a specific search.

Refining the Search

Simply putting in the first words that one thinks to search can be rather inefficient (e.g., a simple search using the words "confined space hazards" generated more than

2,000,000 websites with those words). Fortunately, Internet search engines provide ways of refining the search so that it can be more selective and the results more useful. While the precise format of these refinements varies from search engine to search engine, the concepts they use are similar.

The first step in refining a search involves phrasing the question in a way that clarifies the needed answer. This is often rather easy (Does OSHA have any regulations that cover the use of this chemical?), but other times this can be more difficult (Is this workplace situation a confined space?).

If having trouble coming up with a question that describes the need, it may be helpful to think up the name of a magazine article that would be just what is needed. A four- or five-word phrase is a good place to start a search. It is important to think about possible other meanings for the words selected. For example, "safety" may refer to chemical concerns in one searcher's mind, while it refers to law enforcement issues in many other people's minds. An ambiguous word such as this is usually a poor choice to include as a search term unless used with connected descriptors or logical connectors: "safety in confined spaces" or "safety and confined spaces."

Subject-Specific Indexes

It is best to test out an initial set of keywords by using it in one search engine, with the idea of seeing how many useful responses are produced, before using other search engines. In this phase of research, it is better to start with a more subject-focused index, such as the OSHA or NIOSH web pages.

By searching through these locations, it can be determined whether other safety professionals use the chosen phrase to describe the situation being researched. If the results are not related to the researcher's specific concerns, change the words used until the results are more appropriate. Once the keywords are refined in this search, they are likely to be more effective when using larger search engines.

An Internet Search Strategy

Although the preceding tips can make searching the Internet for specific information easier, it is still easy to get distracted from the original question as the Internet search proceeds. To conduct research on the Internet efficiently, it is important to have a search strategy in mind while looking. An outline of such a strategy is provided here.

Refine the Question

The first step in using the Internet successfully is to clearly state the question to be answered. While it is likely that useful information will be found that does not directly answer that question, it is important to know the end result before starting the search.

Decide What Kind of Information to Look For. Looking for a specific piece of data (e.g., the flash point of acetone) is different from looking for a technical interpretation of that data (e.g., whether the use of acetone requires adequate ventilation due to its flammability), which is different from looking for informal knowledge (i.e., use acetone in a fume hood if using more than 500 mL). These different types of information will be found in different places on the Internet.

- *Formal databases.* For specific pieces of data, formal databases are often the best places to look. Formal databases are usually maintained by public or private entities. It is important to verify that these databases contain accurate information before basing decisions on their content. There are a variety of such sources, such as Safety Data Sheet (SDS) collections and databases containing government regulations. The OSHA website is a good example of one of these formal databases. These databases are usually indexed to allow for keyword searches. As with more general search engines, selecting keywords carefully will make any search at these sites more efficient.
- *Professional interpretations.* Professional interpretations are likely to be less specific than raw data and organized in less structured ways. For technical interpretations of raw data, the best places to look are likely to be in collections of policies and procedures that are available online. Such collections are usually associated with websites that companies and institutions put online for the convenience of their employees or customers.
- *Informal knowledge.* Because informal knowledge requires technical expertise to apply that knowledge appropriately in specific situations, it is unlikely to be found in the formal information sources on the Web. However, the Internet has many informal information collections available in the archives of electronic mailing lists. These are the first places to check for informal information. Even if the desired information is not found there, a reference to another Internet resource that has the needed information might be located.

Select Keywords to Use for the Search. The result of refining a question should be a set of keywords to use for the search. These keywords will be used in performing searches at various websites that are likely to contain appropriate information. For example, if simply wanting to determine the flash point of acetone, "flash point" and "acetone" are appropriate keywords. On the other hand, if concerned about ventilation requirements for using acetone, "flash point" is not likely to be helpful and "flammable liquid" may be a reasonable substitute for "acetone."

Keywords need to be as specific as possible while allowing for variations in terminology that are likely to arise.

Using keywords such as "safety" or "health" are likely to produce too many sites for most purposes. Most website indexes allow the use of logical connectors such as "and," "or," and "not" when conducting a search. This can help refine the keyword search until there are about 20 to 40 hits. Lists of hits longer than that are probably too long to effectively search and are an indication that the keyword strategy should be refined.

Select a Website to Start the Search

Once the researcher has a good idea of what kind of information is needed to answer the posed question and what keywords are likely to be associated with that information, it is time to start searching the file libraries on the Web. Start with familiar websites. If not familiar with any websites that would have the type of desired information, consult the website listings in the Resources section in this chapter.

Ask a Discussion Group

If a search of the file libraries fails to produce the desired information, or if looking for more informal information than is available at websites, it is time to post a request for information to an appropriate e-mail list or list-serv. To increase the chances of success when asking a question of a list, be sure to follow the Internet etiquette guidelines appropriate to that group.

If possible, review the archives of the group's discussions to see if it is the right group of which to ask the question and to be sure that it is not a question that has been asked and answered repeatedly. When framing the question, be as specific as possible, so those who read it can determine what type of answers are appropriate (i.e., general pointers to the professional literature versus specific interpretations of your information).

It is most helpful, if possible, to monitor the traffic on the list for about a week before asking a question to see what sorts of questions are appropriate for the list.

Check the Information

Always be sure to assess the information obtained from the Internet before acting on it. Remember that the information available on the Internet was written based on someone else's assumptions, in ignorance of the details of a particular situation. There may be specific, critical differences between the situations faced in this specific instance and those of the person writing the information. The effort involved in confirming Internet information may range from asking, "Does this make sense?" to checking a paper reference source to consulting with a professional with more expertise. It is also important to check sources as to whether they are relevant and credible.

THE INTERNET AS A NETWORKING TOOL

In addition to the ability to access file libraries for research purposes, the Internet provides valuable opportunities for professional networking. The Internet can serve the same purposes as any other professional network. Review of technical issues of a specific field, tips about how to approach specific problems, forewarning of new issues developing in the field, identifying prospective partners or consultants, and celebrating (or commiserating) with others in similar circumstances are just a few of the benefits of developing a professional network through the Internet.

A major advantage of using the Internet for networking activities is that it provides a convenient way to have ongoing discussions with geographically dispersed colleagues. These discussions can take place either in a group or individually. This section describes some of the considerations involved in using the Internet for this purpose.

General Considerations

As the Internet has grown in popularity, the time required to remain current with its content has increased as well. While the technical details of using e-mail and websites have simplified significantly since the 1990s, the task of wading through all the possibly relevant information sources has become more complex. It is important for the researcher to have a clear idea of what his or her goals are for using the Internet; otherwise, there is a high probability of devoting a lot of time to using it without much payback.

Using the Internet to network with other people with similar interests can minimize this learning curve. A benefit of networking on the Internet is that it is a low-budget way to be involved in a professional community. Productive professional relationships can be developed with a wide range of people without face-to-face meetings. These relationships usually start in discussion areas such as e-mail lists or newsgroups. They often develop into private correspondence that is able to be more speculative than public discussions.

The primary costs of developing a network of professional contacts over the Internet are time, patience, and a network connection. Fortunately, a powerful Internet connection is not required, as most networking happens via e-mail with little graphical content and small files.

It is important to note that such electronic networking will not replace personal contacts made at professional conferences. While e-mail and other collaborative Internet tools can facilitate follow-up on ideas developed face-to-face, professional communication and brainstorming within the safety community require a level of personal understanding and trust that cannot be generated electronically.

Styles of Participation

The way in which a researcher participates in an Internet discussion group will vary depending on many factors. There are three primary styles of participating:
- daily participation
- selective participation
- lurking.

With experience, different styles become appropriate for different lists.

Daily Participation

Some people enjoy e-mail discussions and make a daily habit of responding to many of the discussion topics that arise. These people are usually less than 10% of a list's subscribers and often account for 30% to 50% of the postings to a list. While this can be annoying if someone's e-mail style is objectionable, these people are important in keeping the list active. Not only do they provide topics of discussion, they also provide some sense of what questions are likely to be answered by the list's subscribers. It is usually easy to identify these people by following the list for a week or so.

Selective Participation

Another way to participate in a list is to read the postings regularly, but avoid responding publicly unless an issue of special interest arises. This is a more common practice than daily participation, and about 25% of a list's members fall in this category. These people provide an important "error-checking" function for the list, in that they will usually point out occasions when incorrect information is presented as fact. They also broaden the range of questions that can be answered successfully by the list.

Lurking

The most common use of e-mail lists is "lurking," or simply reading the postings of a list without responding, unless an answer is needed to a specific question. For most lists, the large majority of subscribers do this. Most people find active participation in more than one or two lists to be too time-consuming to manage. The presence of lurkers is important to a list because they provide questions and comments that normally would not come out of the general flow of discussion among more regular posters.

Finding a comfortable level of participation for a particular list will take some experience with the list and the population of participants. However, the more active a researcher is on the list, the more likely it is that questions to that list will generate useful responses. In addition, by actively participating in discussions, a thread (a sequence of messages on the same subject) can be moved into a direction that is useful to the researcher's particular interests.

Finding Discussion Groups

There are thousands of discussion groups operating on the Internet. Some are formally organized; others are simply collections of e-mail addresses being held together by someone's personal e-mail software. For this reason, finding valuable discussions can be a bit of a challenge. However, there are several good places to check.

The first place to check is professional organizations, trade magazines, and professional journals. Many sponsor e-mail lists about their technical specialties, although some are limited to members. Other sources for relevant discussions are the various "lists of lists" that exist on the Internet. One such list is available at CCOHS (ccohs.ca/resources/listserv.html). This site includes descriptions of a wide variety of mailing lists, organized by subject area.

Remember that formal descriptions of discussion groups can often be significantly different from the actual subjects talked about within the group. It is often possible to search the actual text of many discussions within the group's archives.

THE INTERNET AS A SAFETY CULTURE TOOL

Since 2000, the Internet has become an important form of general communication throughout society. This has led to significant cultural changes and has demonstrated the impact of information technologies on culture. System safety has become increasingly important as industrial disasters have grown in size, so the importance of organizational safety culture has been highlighted. This connection between the organization's electronic presence and safety culture must be carefully considered as web materials are developed. Developing a consistent, user-friendly tone in the interface used for safety functions will support the development of a sustainable safety culture. For this reason, it is becoming increasingly important for the safety professional to establish a useful presence within his or her organization's website so that it is clear what information applies to situations within the organization. This section briefly describes some strategies for achieving this objective.

Online Training

The first step in any safety program is to establish a strong educational and training presence within the organization. The style and content of this presence will vary depending on the needs of the organization, but online educational tools are increasingly being used for this purpose. Online training systems are available commercially and can also be built with relatively simple website-building tools. The challenge comes in choosing the most effective approach for the population being served. For example, experience has shown that a highly educated population such as a laboratory work force will respond quite well to online training, while use with a custodial work force tends to be less effective.

A complete discussion of the considerations to be included in selecting an online training strategy is beyond the scope of this chapter. Issues to be considered include the type of presentations to be made (text documents versus animated sequences versus interactive games), the comfort level with computer use of the population in question, how specific the information to be provided must be, and how often refresher training must be provided. Implementation of a selected strategy should include identification of a pilot group of workers to test and evaluate the system as it is developed and a follow-up plan for determining how successful the training provided is.

Developing a Reference Library

In addition to introducing new employees to the safety program through training, an organization's website can serve as a reference source for both operational (procedures and specific facility information) and strategic (policies and plans) safety information. In developing such a library of information, the usability of the information posted is the key concern. This means that the information should be adapted to the applicable electronic medium.

Web pages are not well suited to delivering multiple screens of detailed information on the same page. Research has consistently shown that people using web pages browse text on the page quickly, scanning for specific details they need rather than reading carefully for comprehension. Therefore, web pages must be carefully planned and linked together in ways that allows the logic of the system to help rather than challenge the reader's use of the system.

An important source of information about usability research can be found at useit.com. This site describes how the design of a web page can make it more or less successful at providing information to a target audience.

Maintaining Awareness

Safety professionals face a special challenge in maintaining a successful program because there is a tendency for both individuals and the organization as a whole to take safety issues for granted over time. Maintaining an appropriate level of safety awareness is an ongoing challenge, and the organization's website, in conjunction with careful use of e-mail, can be used to meet this challenge. Accomplishing this means that the safety program's web pages must maintain a modern appearance and the content should be updated regularly. Major changes in the safety program are probably best communicated by other means (face-to-face meetings or paper manuals), but the web page should support those changes by providing a reference source for follow-up questions. Small, regular changes to the web page are useful in reminding people of the presence and value of the safety program within the organization. Since 2005, the emergence of social media platforms such as Facebook and Twitter has provided another opportunity to serve this function.

THE INTERNET AS A SAFETY PROGRAM MANAGEMENT TOOL

In addition to being an important medium for professional communication, computers and the Internet can be used to collect and manage data about the activities involved in a safety program. This use of the Internet is still emerging as organizations develop a better understanding of the strengths and weaknesses of their internal information architecture, but three aspects of this use of the Internet can be briefly mentioned here.

Collecting Data
The advance of electronic technologies has enabled data from specific environmental sampling instruments to be collected and organized more effectively and at lower costs than a few short years ago. Air-quality monitors, digital cameras and recorders, geographic positioning systems, bar coding, and other related techniques enable safety professionals to provide a more accurate and meaningful assessment of workplace hazards than could be previously considered. As with other forms of electronic information, using these capabilities must be carefully planned and frequently reviewed for the full potential of such capabilities to be realized. However, safety issues—including chemical exposure determinations, documentation of accidents, oversight of workplace inspections, and emergency response protocols—have all benefited from this trend.

Management System Development
As electronic information proliferates, it is easy to be swamped by data collected and stored. A useful tool in selecting which data are appropriate to be collected and how the data should be stored is a safety management system. Such a system, which can use a variety of models (such as the ANSI Z10 Standard for Occupational Health and Safety Management Systems, OSHA's Voluntary Protection Program, or the International Standard Organization's ISO 18000) can provide an overall architecture for the documentation necessary to manage and evaluate the safety program.

There are several Internet-based software packages commercially available for maintaining such systems for quality or environmental management aspects of an organization. It is possible that such systems can also be used for a safety management system. However, the safety professional should always be careful when choosing and implementing a proprietary software package because of potential issues with support or changes in technology. For this reason, open-source architectures for the safety data being managed should be considered as an alternative to proprietary packages. While such open-source systems may not be cheaper to implement in the short term, they tend to be more stable over time because their support is not dependent on the business viability of a single company.

Connecting to Other Management Information Systems
The development of a safety management system will greatly benefit from being connected to other management systems within an organization, such as those containing facility information, financial information, and human resources data. Implementing these connections can be more challenging than first anticipated because the software involved is designed for other purposes.

For this reason, the organization's information technology (IT) professionals should be consulted on how best to succeed in this endeavor—or if it is even possible. Oftentimes, much of this other information has significant security considerations involved, so using the data fully is not always possible. However, as the safety information management system is piloted and as it grows, attention should be given to opportunities for synergy between electronic safety information and other organizational data.

SUMMARY

- Computers have become a core professional tool for the safety community; while a comprehensive discussion of the ways in which a safety professional can use computers and computer networks may be beyond the scope of a single chapter, forethought and planning can turn the Internet into a valuable professional asset.
- The challenge is that the power of these systems requires that a clear understanding of the safety program's role be established and careful planning and maintenance of the computer hardware and software systems being used be in place.
- The good news is that as the power of the systems increase, they enable significant increases in the effectiveness of safety professionals to serve their audiences.
- Becoming familiar with the web resources specific to the safety field, rather than simply relying on search engines, can elevate the Internet from a casual source of confusing information to a professional tool.

RESOURCES

Agency for Toxic Substances and Disease Registry (ATSDR), atsdr.cdc.gov

The ATSDR website has a variety of information about hazardous chemicals in the environment. This information includes ToxFAQs, fact sheets on the hazards associated with a variety of chemicals; a list of the Top 20 Hazardous Substances, based on Superfund experience; and a Science Corner, which includes ATSDR Special Report(s), Health and Environment Resources, Science Corner History, and other information.

American Biological Safety Association (ABSA), absa.org

The ABSA was founded in 1984 to promote biosafety as a scientific discipline and serve the growing needs of biosafety professionals throughout the world. This web page provides links to information on biosafety issues.

American Conference of Governmental Industrial Hygienists (ACGIH), acgih.org

The ACGIH promotes excellence in environmental and occupational health and provides access to high-quality technical information through this website. This web page provides access to information about the ACGIH Threshold Limit Values.

CAMEO, cameochemicals.noaa.gov/

CAMEO Chemicals is a tool designed for people who are involved in hazardous material incident response and planning. This tool is part of the CAMEO software suite, and it is available as a website and as a downloadable desktop application that you can run on your own computer. CAMEO Chemicals contains:

- a library with thousands of data sheets containing response-related information and recommendations for hazardous materials that are commonly transported, used, or stored in the United States
- a reactivity prediction tool that you can use to predict potential reactive hazards between chemicals.

Canadian Centre for Occupational Health and Safety (CCOHS), ccohs.ca

The CCOHS is a Canadian federal government agency based in Hamilton, Ontario, that serves to support the vision of eliminating all Canadian work-related illnesses and injuries. Its website includes a variety of information—some free, some available on a subscription basis.

Centers for Disease Control and Prevention (CDC), cdc.gov

This U.S. government site contains a large variety of health and disease information. There are numerous documents and data sets, most of which are related to public health, biosafety, occupational health and safety, and infectious diseases.

Consumer Product Safety Commission (CPSC), cpsc.gov

CPSC is an independent U.S. federal regulatory agency. CPSC works in a number of different ways to reduce the risk of injuries and deaths by consumer products: it develops voluntary standards with industry, issues and enforces mandatory standards or banning of consumer products, obtains the recall of products, conducts research on potential product hazards, and informs and educates consumers.

Cornell Ergonomics website (CUErgo), ergo.human.cornell.edu

CUErgo presents information from research studies and classwork by students and faculty in the Cornell Human Factors and Ergonomics Research Group. CHFERG focuses on ways to enhance usability by improving the ergonomic design of hardware, software, and workplaces to enhance people's comfort, performance, and health in an approach called Ergotecture.

Environmental Protection Agency (EPA), epa.gov

The U.S. EPA's website contains a wide variety of information related to the environment and public health. Some of the many categories of available information include technical documents, research funding, and more; assistance for small businesses and entire industries; projects and programs; news and events; laws and regulations; databases and software; and publications.

Ergoweb, ergoweb.com

Ergoweb® provides ergonomics consulting and ergonomics training services and carefully selected products, and it publishes applicable news and information. Its information is a mix of free and subscriber-only documents.

European Agency for Safety and Health at Work, osha.europa.eu

This site links to more than 30 national websites maintained by the agency's focal points (usually the lead occupational safety and health [OSH] organization in the EU member states, candidate countries, and other international partners). This is a single-entry point to an overview of information that the network has to offer, from current campaigns to popular links. It is a database-driven, multilingual portal providing access to OSH information in your preferred language. You can personalize the site and access the European and international network.

Haz-Map: Information on Hazardous Chemicals and Occupational Diseases from the National Institutes of Health, hazmap.nlm.nih.gov

Haz-Map is an occupational health database designed for health and safety professionals and for consumers seeking information about the health effects of exposure to chemicals and biologicals at work. Haz-Map links jobs and hazardous tasks with occupational diseases and their symptoms.

Health Environment and Work, agius.com/hew/index.htm

This website is self-funded by the author, who is a professor of occupational and environmental medicine at the University of Manchester in the United Kingdom. This site consists of hundreds of files about environmental and occupational health.

International Agency for Research on Cancer (IARC), iarc.fr

The IARC is part of the World Health Organization. IARC's mission is to coordinate and conduct research on the causes of human cancer and to develop scientific strategies for cancer control. It is involved in both epidemiological and laboratory research and disseminates scientific information through meetings, publications, courses, and fellowships.

International Centre for Genetic Engineering and Biotechnology (ICGEB), Biosafety Web Pages, icgeb.org/~bsafesrv/

This site, based in Trieste, Italy, provides access to a variety of biosafety-related resources, including the ICGEB bibliographic database on biosafety studies, an index of selected scientific articles published on biosafety and risk assessment since 1990. Official documents on biosafety produced by international agencies as well as scientific findings, articles, proceedings, and workshops on international biosafety regulations from Europe, the United States, and other countries are available.

International Labour Organization (ILO), ilo.org

The ILO is the tripartite UN agency that brings together governments, employers, and workers of its member states in common action to promote decent work throughout the world. Its website includes:

- Encyclopedia of Occupational Health and Safety (subscription fee required). The encyclopedia covers the technical fields encompassing occupational health and safety. With contributions by internationally renowned experts, this reference answers questions involving health and safety in the workplace.
- CISILO Database (subscription fee required). CISILO is a bilingual, bibliographic database that provides references to international occupational health and safety literature. The database is created by the International Occupational Safety and Health Information Centre/Centre international d'informations de sécurité et d'hygiène du travail (CIS) in Geneva.

National AgSafety Database (NASD), nasdonline.org/browse/1/topic.html

The information contained in NASD was contributed by safety professionals and organizations from across the nation. Specifically, the objectives of the NASD project are (1) to provide a national resource for the dissemination of information; (2) to educate workers and managers about occupational hazards associated with agriculture-related injuries, deaths, and illnesses; (3) to provide prevention information; (4) to promote the consideration of safety and health issues in agricultural operations; and (5) to provide a convenient way for members of the agricultural safety and health community to share educational and research materials with their colleagues.

National Institute of Environmental Health Sciences (NIEHS), niehs.nih.gov

The NIEHS conducts basic research on environment-related diseases. Its web pages outline the Institute's history and research highlights; it also provides complete contact and visiting information.

National Institute for Occupational Safety and Health (NIOSH), cdc.gov/niosh

This website provides access to information resources, programs, and news from NIOSH, the federal agency responsible for conducting research and making recommendations for the prevention of work-related injury and illness.

North American Emergency Response Guidebook (ERG), hazmat.dot.gov/pubs/erg/gydebook.htm

The Emergency Response Guidebook was developed jointly by the U.S. Department of Transportation, Transport Canada, and the Secretariat of Communications and Transportation of Mexico for use by fire fighters, police, and other emergency services personnel who may be the first to arrive at the scene of a transportation incident involving a hazardous material. It is primarily a guide to aid first responders in (1) quickly identifying the specific or generic classification of the material(s) involved in the incident and (2) protecting themselves and the general public during this initial response phase of the incident. The ERG is updated every 3 to 4 years to accommodate new products and technology.

Occupational Safety and Health Administration (OSHA), osha.gov

The website for the primary workplace regulatory body in the United States includes access to its regulations, information about safe work practices for a variety of industries, and access to statistics about its enforcement programs and actions.

Public Health Agency of Canada, Pathogen Safety Data Sheets and Risk Assessment, phac-aspc.gc.ca/msds-ftss

These SDSs are produced for personnel working in the life sciences as quick safety reference materials relating to infectious micro-organisms.

SafetyLine, safetyline.wa.gov.au

SafetyLine is an information service providing online access to the major publications issued by the WorkSafe Western Australia Commission and WorkSafe Western Australia. The objective of SafetyLine is to provide people in the workplace with access to safety and health information that can be used to help improve their working environment.

Society for Chemical Hazard Communication (SCHC), schc.org

SCHC is a nonprofit organization with a mission to promote the improvement of the business of hazard communication for chemicals.

TOXNET, toxnet.nlm.nih.gov

TOXNET, provided by the U.S. National Library of Medicine, is a collection of databases in the areas of toxicology, hazardous chemicals, environmental health, and toxic releases.

Typing Injury FAQ, tifaq.com

The Typing Injury FAQ (frequently asked questions) and Typing Injury Archives are sources of information for people with typing injuries, repetitive stress injuries, carpal tunnel syndrome, etc. It is targeted at computer users suffering at the hands of their equipment. You will find pointers to resources all across the Internet, general information on injuries, and detailed information on numerous adaptive products.

United Kingdom Health and Safety Executive, hse.gov.uk

The Health and Safety Executive ensures that risks to people's health and safety from work activities are properly controlled. As the website says, "The law says employers have to look after the health and safety of their employees; employees and the self-employed have to look after their own health and safety; and all have to take care of the health and safety of others, for example, members of the public who may be affected by their work activity. Our job is to see that everyone does this." The site includes a variety of useful tools for the safety professional.

University of Michigan Health Physics Website, umich.edu/~radinfo

The site, maintained at the University of Michigan, contains information and links related to radiation safety issues.

University of Minnesota Environmental Health and Safety, dehs.umn.edu

Many colleges and universities have health and safety department websites with valuable information available on them. Larger universities have nearly every general safety or health hazard associated with some part of their operation. Thus, policies and procedures for many different situations can be found on their sites. The University of Minnesota site is typical of a well-maintained university website.

Where to Find Material Safety Data Sheets on the Internet, ilpi.com/msds/index.html

This commercial site provides a free list of sites useful in finding SDSs and related software.

World Health Organization (WHO), who.int

This website provides descriptions of WHO's major international programs, a list of the organization's publications, the full text of the WHO *Weekly Epidemiological Record*, access to the WHO Statistical Information System, newsletters, and international travel and health information.

Young Worker Awareness, Workplace Health and Safety Agency, Ontario, Canada, yworker.com

This site contains health and safety information for young workers, their parents, teachers, principals, employers, and others. Although the information is specific to the province of Ontario, Canada (the Young Worker Awareness school program is available only to Ontario high schools), many safety professionals are likely to find the information here useful.

REVIEW QUESTIONS

1. Discuss the pros and cons of using Internet-based tools for gathering safety information.
2. Describe the process for conducting an effective Internet-based search for safety-related information.
3. Compare and contrast the different roles list-serv participants engage in as part of an Internet-based safety discussion group.
4. Discuss issues related to online training.
5. Discuss areas where the Internet can be used to support safety program management elements.

PART

5

Industry-Specific Safety Issues

Rapidly evolving engineering and technology issues can impact safety on many levels. As industries face new challenges, it is important to understand the scope of those challenges as related to safety. The aviation, oil and gas, and waste and recycling industries are examples of those facing unique obstacles in maintaining safe working environments. Some of these industry-specific challenges have been around since the birth of the industry (e.g., human error, hazardous material exposures, the physics of flight, proximity of workers and heavy mobile equipment), while others have emerged with the evolution of the industry and the passage of time (e.g., traffic dangers for waste-collection workers, terrorist acts, technology failures). It is important to determine whether solutions used by some industries can be implemented successfully by others, or whether new approaches need to be developed and implemented due to the unique nature of the industry-specific hazards.

Process Safety Management

24

Anne Germain, PE, BCEE

Philip E. Hagan, JD, MBA, MPH, ARM, CIH, CHMM, CEM, CET, CHCM

Historical Perspective

Industry and Labor Process Safety Involvement
Center for Chemical Process Safety ▶ American
Petroleum Institute ▶ American Chemistry Council
▶ Other Initiatives

Government Process Safety Involvement
Occupational Safety and Health Administration
▶ Environmental Protection Agency

Process Safety Management Program
Basic Process Safety Management Requirements
▶ Process Safety Accountability ▶ Process Risk
Management

**Elements of the Process Safety
Management Program**
Process Safety Information ▶ Employee Involvement
▶ Conducting Process Hazard Analyses ▶ Management
of Change Programs ▶ Operating Procedures ▶ Safe
Work Practices and Permits ▶ Employee Information
and Training ▶ Contractor Personnel ▶ Pre-Start-Up
Safety Reviews ▶ Design and Quality Assurances
▶ Maintenance and Mechanical Integrity ▶ Emergency
Planning, Preparedness, and Response ▶ Compliance
Audits ▶ Process Incident Investigation ▶ Standards
and Regulations ▶ Confidentiality and Trade Secrets

**Enforcement of OSHA Process Safety
Management Standard**

Process Safety Audits
Elements in Process Safety Audits ▶ Self-Evaluation
Process Safety Audit Checklist ▶ Confidentiality of
Self-Evaluation Audit Reports

Summary

References

Review Questions

Releases of flammable liquids, vapors, and gases or chemicals have probably occurred for as long as there have been processes that use temperature and pressure to produce hazardous and toxic materials. Beginning in the early 1980s, a number of serious major incidents occurred involving highly hazardous chemicals. These incidents resulted in considerable numbers of fatalities and injuries and significant property losses (Tables 24–A and 24–B). They spurred government agencies, labor organizations, and industry associations throughout the world to develop and implement codes, regulations, procedures, and safe work practices aimed at eliminating or reducing these undesirable events. This chapter covers the following topics:

- regulatory, industry, and labor efforts to control the processing of toxic and hazardous substances
- government involvement in regulating the manufacture of toxic and hazardous substances
- components of an effective process safety management (PSM) program
- OSHA standards for process safety management
- conducting efficient, confidential process safety audits.

HISTORICAL PERSPECTIVE

The term *process safety* is most commonly used to describe various regulations and activities designed to protect employees, the public, and the environment from the consequences of major chemical accidents involving highly hazardous materials. In practice, the term has many definitions. According to OSHA, process safety management is:

the proactive identification, evaluation, and mitigation or prevention of chemical releases that could occur as a result of failures in processes, procedures, or equipment.

The American Petroleum Institute (API) states that

the goal of process safety management is to develop plant systems and procedures to prevent unwanted releases that may ignite and cause toxic impacts, local fires or explosions.

The American Chemistry Council (ACC), formerly the Chemical Manufacturers Association (CMA), has developed a Process Safety Code of Management Practices.

Companies must consider all of the following in the systematic identification and evaluation of hazards: the process design, technology, changes in the process, materials and changes in materials, operations and maintenance practices and procedures, training, emergency preparedness, and other elements affecting the process. Such an analysis will help determine whether these items have the potential to cause a catastrophic incident in the workplace and surrounding community.

INDUSTRY AND LABOR PROCESS SAFETY INVOLVEMENT

During the 1980s, the petroleum, petrochemical, and chemical industries recognized that process safety technology without PSM would not prevent catastrophic incidents. A number of industry associations, such as the American Institute of Chemical Engineers (AIChE), the American Petroleum Institute (API), and the ACC, initiated programs to develop and provide PSM guidelines for use by their members who process hazardous materials. (See also the References at the end of this chapter.)

Center for Chemical Process Safety

The Center for Chemical Process Safety (CCPS) was formed by the AIChE in 1985 to promote the improvement of PSM techniques among those who store, handle, process, and use hazardous materials. The CCPS has developed books on process safety. The following documents are examples of some materials developed and issued by the CCPS to promote PSM systems:

TABLE 24–A. Examples of Serious Hazard Release Incidents (1984-1991)*

Year	Location	Incident	Fatalities	Injuries
1984	Bhopal, India	Toxic chemical release	2,000+	20,000+
1984	Mexico City	LPG release & explosion	542	many
1985	West Virginia	Toxic chemical release	0	135
1988	Louisiana	Vapor cloud explosion	7	unknown
1989	Texas	Vapor cloud explosion & fire	23	132
1990	Texas	Wastewater tank explosion	17	unknown
1991	Louisiana	Explosion and fire	8	128

*Such incidents led to the development and passage of the OSHA PSM standard.

TABLE 24–B. Major Worldwide Process Industry Explosions and Fires, 1965-1995 (Losses in Millions of Dollars Adjusted to 1995 Values)

Local Facility		Number of Incidents and Property Loss ($ MM)				
		Refineries	Petrochem.	Gas Plants	Misc.	Totals
North & South America	No.	29	26	2	1	58
	Amt.	$1,930	$2,010	$55	$25	$4,020
Europe	No.	7	6		1	14
	Amt.	$650	$460		$45	$1,155
Asia & Australia	No.	3	2	1		6
	Amt.	$440	$100	$65		$605
Middle East & Africa	No.	2		2	1	5
	Amt.	$180		$245	$15	$440
Total No.		41	34	5	3	83
$ Amount		$3,200	$2570	$365	$85	$6,220

Source: David Mahoney, ed. *Large Property Damage Losses in the Hydrocarbon-Chemical Industries, A Thirty-Year Review.* 16th ed. Chicago: M&M Protection Services, 1995.

- *Guidelines for Implementing Process Safety Management Systems.* This manual describes the principles to apply in multifaceted technical management systems to determine and eliminate the complex and interrelated causes of catastrophic accidents in the process industries.
- *Guidelines for Engineering Design for Process Safety.* This book focuses on process safety issues in the design of chemical, petrochemical, and hydrocarbon-processing facilities.
- *Guidelines for Risk Based Process Safety.* This document provides guidelines for industries that manufacture, consume, or handle chemicals by focusing on new ways to design, correct, or improve PSM practices.

American Petroleum Institute

Beginning in 1990, the API initiated an industrywide program titled Strategies for Today's Environmental Partnership (STEP), aimed at addressing public concerns by improving industry's environmental, health, and safety performance. Since then, they have developed a number of documents and programs related to process safety, including a variety of recommended practices.

- *Management of Hazards Associated with Location of Process Plant Buildings* (RP 752). This recommended practice, co-developed by API and ACC, provides a methodology for assessing and evaluating the hazards associated with the location of process buildings.
- *Process Safety Performance Indicators for the Refining and Petrochemical Industries* (RP 754). This recommended practice was developed after new ANSI standards were issued in the wake of a U.S. Chemical Safety and Hazard Investigation Board review of the 2005 BP Texas City Incident.

American Chemistry Council

In 1988, the ACC, formerly the CMA, initiated its Responsible Care Program in answer to public concerns about the manufacture and use of chemicals. Each Responsible Care company commits to systematic, continuous improvement in process safety through applying the Process Safety Code's seven management practices, which describe virtually every aspect of chemical manufacturing, transporting, and handling.

Other Initiatives

Additional examples of industry associations that have developed materials and programs providing guidance to their members on PSM include, but are not limited to, the following:

- An independent federal agency, the Chemical Safety and Hazard Investigation Board (the Board) was created by an amendment to the U.S. Clean Air Act to serve as a new resource in the effort to enhance industrial safety. Congress modeled the Board after the highly respected National Transportation Safety Board. The Board's mission is to provide industries that manufacture, use, or otherwise handle chemicals with information to enable identification and mitigation of operational conditions that compromise safety.
- American Fuel and Petrochemical Manufacturers' (AFPM) document *Advancing Process Safety*
- Synthetic Organic Chemical Manufacturers Association's (SOCMA) program *Chemical Process Operator Certification Training*
- UN International Labour Organization's (ILO) *Code of Practice on the Prevention of Major Accident Hazards*
- International Chamber of Commerce's (ICC) *Charter for Sustainable Development*

GOVERNMENT PROCESS SAFETY INVOLVEMENT

In 1985, the Environmental Protection Agency (EPA) responded to public concern over the hazards of chemicals by launching its Chemical Emergency Preparedness Program (CEPP), now EPA Emergency Management. This evolved into the Emergency Planning and Community Right to Know Act of 1986 (EPCRA), known as SARA Title III, which required companies to develop emergency preparedness, recognition, knowledge, and inventory of hazardous chemicals and reporting of toxic releases.

About the same time, OSHA promulgated its Hazard Communication Standard, which required manufacturers and users to identify hazardous materials in the workplace and inform employees and consumers of the hazards presented by their manufacture, use, storage, and handling. The Clean Air Act Amendments (CAAA), passed in 1990, mandated that both the EPA and OSHA develop regulations for chemical safety management to prevent catastrophic accidents. In response to this mandate, OSHA proposed standards to control process safety within the workplace, and EPA issued regulations covering releases primarily affecting areas outside the workplace.

Occupational Safety and Health Administration

OSHA's PSM standard evolved from a number of previous industry and government guidance documents and initiatives, including OSHA's Hazard Communication Standard and two OSHA directives, *Systems Safety Evaluation* and *PETROSEP*.

- Instruction CPL 2-2.45, *Systems Safety Evaluation of Operations with Catastrophic Potential*. This 1988 directive provided guidance for systems safety programs and inspections in workplaces focusing on chemical-related operations with potential for fires, explosions, and hazardous releases. The directive was superseded by OSHA's PSM standard.
- OSHA *Petrochemical Industries (PETROSEP) Compliance Directive*. This program, initiated in 1990 and canceled in 1992, targeted large petroleum refining and petrochemical processing facilities to determine whether management systems governing safety and health procedures for operations, maintenance, and contractor activities were in place and were able to control risks and prevent disasters. In 1992 PETROSEP was replaced by OSHA's PSM standard.
- 29 CFR 1910.119, Process Safety Management of Highly Hazardous Chemicals. The major objective of this 1992 PSM standard is to prevent releases of hazardous chemicals that could expose employees and others to danger.

The PSM standard covers a wide range of manufacturing industries involved in handling, storing, processing, or transporting flammable liquids and gases in quantities of 10,000 pounds (4,500 kg) or more and specific toxic and reactive chemicals in regulated quantities. Exempted from the PSM standard are retail facilities; oil and gas well drilling or servicing operations; normally unoccupied remote facilities; hydrocarbons used solely as fuel in the workplace; and flammable liquids stored in atmospheric tanks or transferred at temperatures below their normal boiling point without cooling, which are not connected to a process. All facilities in the United States that store, use, manufacture, transport, or handle flammable liquids and highly hazardous chemicals should be familiar with the requirements of the PSM standard and whether it applies to their workplaces.

A key provision of the PSM standard requires covered facilities to gather written information concerning process technology, process equipment, and chemical hazards. This information is used to identify the processes that pose the greatest risks in the workplace. Employers then establish priorities for conducting process hazard analyses based on risk and improvement and conduct these analyses within prescribed time limits. The PSM standard required that at least 25% of the covered processes were to be evaluated by May 26, 1994, with an additional 25% evaluated each following year. All process hazard analyses were to be completed no later than May 1997.

The PSM standard recommends that companies consider one of six commonly used methodologies (discussed later in this chapter) to conduct process hazard analyses. The PSM standard also has specific requirements regarding the experience and qualifications of the persons who are to conduct the process hazard analysis.

Other requirements of the PSM standard cover operating procedures, employee participation, training, contractors and contractor employees, pre-start-up safety reviews, mechanical integrity, hot-work permit system, management of change, accident and incident investigation, emergency preparedness, compliance audits, and confidentiality agreements. Information concerning OSHA requirements for PSM in the United States can be found in the following two booklets: OSHA 3132, *Process Safety Management*; and OSHA 3133, *Process Safety Management Guidelines for Compliance*. Copies of each booklet may be obtained on the OSHA website, www.osha.gov.

Environmental Protection Agency

The EPA's PSM regulations and risk management program evolved from the requirements of the Clean Air Act Amendments (CAAA) of 1990, which led to the following initiatives:

- 40 CFR 68 Sections 301(r) and 304, Clean Air Act (CAA) Provisions for Process Safety Management. The CAA contains provisions requiring the EPA to establish a list of substances and thresholds and promulgate accident prevention rules. The list promulgated in 1994 included 77 acutely toxic chemicals and 63 flammable gases and volatile flammable liquids. The program requirements established by EPA to prevent releases affecting areas outside the facility are similar to those contained in the OSHA PSM standard covering the workplace and include the following:

- reviewing and documenting the facility's chemicals, processes, and procedures
- conducting detailed process hazard analyses to identify hazards, assess the likelihood of releases, and evaluate the consequences of such releases
- developing standard operating procedures and training employees on procedures
- implementing preventive maintenance and management of change in operations programs
- conducting reviews prior to initial start-up and prior to start-up following a modification
- conducting periodic safety audits to ensure compliance.

- 40 CFR 68 Section 112(r), Risk Management Programs for Chemical Accidental Release Prevention. Section 112(r) of the CAA requires facilities that store, handle, manufacture, distribute, or use more than the threshold quantity of a listed substance in a process to develop risk management plans (RMPs) to identify hazards, maintain safe facilities, and minimize the consequences of accidental releases of listed hazardous substances, regardless of quantity. These RMPs are to be registered with the EPA and made available to state and local emergency response agencies and the public. In 1999, Public Law 106-40 amended the Clean Air Act to remove flammable fuels from the list of substances with respect to which reporting and other activities are required under the risk management plan program.

The EPA's risk management plans' regulation requires facilities to conduct hazard assessments that define the offsite impacts of potential releases, including worst-case scenarios. Facilities must also document a 5-year history of releases and develop and implement protection programs that build on OSHA's PSM standard. Information on EPA process safety requirements can be found in Bulletin EPA 510, *Managing Chemicals Safely*, which is available online at the EPA's website, www.epa.gov.

PROCESS SAFETY MANAGEMENT PROGRAM

A PSM program should be among the top priorities in a company's organization. All employees must be encouraged to adopt such a program as their own and to ensure that new employees understand its importance. Initially, man-

agement must identify the basic requirements for a safety management program, assign accountability for implementing and monitoring it, and continue to look for ways to improve the company's risk management efforts.

Basic Process Safety Management Requirements

PSM is an integral part of a company's overall facility safety program. An effective PSM program requires the leadership, support, and involvement of top management, facility management, supervisors, engineers, employees, contractors, and contractor employees. Much of the PSM program relies on access to good records and documentation. Therefore, management of information is one of the key factors to a good program.

When developing a PSM program, management should consider several important items, including the following:

- incident-prevention objectives
- existing employer and contractor PSM programs
- use of internal resources versus outside consultants.

Process Safety Accountability

Because PSM is the basis for all other safety efforts within the process facility, the company needs to establish accountability for all of its management systems. This step requires that the company clearly define management and supervisory responsibility and accountability and establish specific goals and objectives in order for the program to work. Components of PSM accountability identified by the CCPS include the following:

- interdependent continuity of operations, systems, and organization
- control of process quality, deviations, exceptions, and alternate methods
- management and supervisory accessibility and communications
- company expectations, compliance audits, and measuring performance.

Process Risk Management

Prior to program implementation, top management should establish both long- and short-term goals and objectives for each of the elements of its PSM program. Long-term goals should be based on the regulatory requirements for program development and implementation. Short-term objectives should be based on the hazard analyses conducted within each facility for each specific unit, process, or material; the determination of its hazard potential; the degree of risk that management is willing to accept; and the timeliness and economic feasibility of instituting necessary changes.

Companies should consider the following items when developing process risk management parameters:

- hazard identification and potential impact
- risk analysis to determine degree of risk reduction
- residual risk management capabilities
- available emergency process management and shutdown procedures.

Managing highly hazardous chemicals by reducing inventory, controlling inventory more efficiently, and dispersing inventory to several locations on site can help reduce PSM risks by reducing the risk or potential for a catastrophic incident. These and similar options should always be considered as potential courses of action in the development of a successful PSM program.

ELEMENTS OF THE PROCESS SAFETY MANAGEMENT PROGRAM

All PSM programs should cover the same basic program requirements, although the number of elements may vary depending on the criteria used. Whether a company uses OSHA's PSM standard, EPA's risk management program, or another company, association, or government source document, certain basic requirements should be included in every PSM program (Table 24–C).

Process Safety Information

Process safety information is used to determine which processes, materials, and equipment are critical to safety efforts. Process safety information includes compiling all available written information concerning process technology, process equipment, materials, and chemical hazards before conducting a process hazard analysis. This information may exist in various locations throughout the facility, may be in a central location, or may remain at the design or supply source. Other critical information includes documentation of capital project reviews and design criteria. The following items are particularly important.

Chemical Information

This includes the chemical and physical properties, reactivity and corrosive data, and thermal and chemical stability of the highly hazardous materials in the process, as well as the hazardous effects of inadvertently mixing incompatible materials. Safety Data Sheets (SDSs; formerly Material Safety Data Sheets [MSDSs]) or other sources of information should be available covering each chemical and material, including its toxicity and permissible exposure limits, so that safety professionals can evaluate the potential hazard.

Management must have information on the maximum intended inventory of highly hazardous chemicals within the process or facility at any one time and their typical usage to determine whether company or regulatory quantity limits for listed chemicals are exceeded. Chemical information also includes data that may be needed to conduct environmental hazard assessments of toxic and flammable releases.

TABLE 24–C. Comparison of Selected Government and Industry Process Safety Management Program Elements

Program Elements	OSHA PSM Std	RMP	API RP 750	CCPS Elements
Process safety information	✓	✓	✓	✓
Employee involvement	✓			✓
Process hazard analysis	✓	✓	✓	
Management of change	✓	✓	✓	✓
Operating procedures	✓	✓	✓	
Safe work practices and permits	✓		✓	
Employee information and training	✓	✓	✓	✓
Contractor personnel	✓			
Pre-start-up safety reviews	✓	✓	✓	✓
Design and quality assurance	✓		✓	✓
Maintenance and mechanical integrity	✓	✓	✓	✓
Emergency planning, preparedness, and response	✓	*		
Compliance audits	✓	✓	✓	✓
Process incident investigation	✓	✓	✓	✓
Standards and regulations				✓
Trade secrets	✓			

*EPA covers emergency response separate from RMP.

Process Technology Information

This includes block-flow diagrams and/or simple process-flow diagrams. The safety professional should provide a description of the chemistry of each specific process, including safe upper and lower limits for temperatures, pressures, flows, compositions, and, where available, process design material and energy balances. In addition, the safety professional should evaluate consequences of deviations, including comprehensive effects on employee safety and health. Whenever processes or materials are changed, the information must be updated and reevaluated.

Process Equipment and Mechanical Design Information

This includes all documentation covering the design codes employed and an evaluation as to whether company equipment complies with recognized engineering practices. There should be written documentation verifying that existing equipment designed and constructed in accordance with codes, standards, and practices that are no longer in general use has been maintained, operated, inspected, and tested to ensure its continued safe operation. In addition, up-to-date piping and instrumentation diagrams should be available, construction materials for all equipment and piping systems specified, the design basis and sizing calculations for pressure relief and safety-sensitive ventilation systems documented, the boundaries of hazardous (classified) electrical locations defined, and safety systems (shutdown, interlocks, detectors, alarms, etc.) described in the plan. Whenever equipment or systems are changed, the company should update or revalidate this information. If original design information is not available, the necessary information may be reconstructed from equipment and inspection records or determined in conjunction with a process hazard analysis action item.

Employee Involvement

Management should consider developing and implementing a program that enables employees to participate in process safety analyses and other elements of the PSM program. This program may give employees and contractor employees (or their representatives) access to all process safety information, incident investigations, and process hazard analyses. The following are elements that should be included in employee involvement programs:

- requiring development of a written plan of action regarding employee participation
- consulting with employees and their representatives on the development and implementation of process hazard analyses and other elements of PSM required under 29 CFR 1910.119
- providing to employees and their representatives access to process hazard analyses and all other information required to be developed under the rule

- training and educating employees
- informing affected employees of conclusions from incident investigations conducted in keeping with requirements of the PSM program.

Conducting Process Hazard Analyses

After process safety information is compiled, the company should conduct a thorough and systematic multidisciplinary process hazard analysis appropriate to the complexity of the process to identify, evaluate, and control the hazards of all processes. Persons performing the process analyses must have experience in engineering and general process operations. The analysis team should include at least one person who is thoroughly familiar with the process being analyzed and one person who is competent in the hazard analysis methodology being used.

Each facility or company should determine and document the priority order it will use to begin conducting process hazard analyses, based on the following criteria:

- extent of the process hazards
- number of potentially affected employees
- operating history of the process
- age of the process.

Some of the more frequently used methods of conducting process safety analyses include the following.

Fault Tree Analysis and Event Tree Analysis

Both methods are formal deductive techniques used to estimate the quantitative likelihood of events occurring. Fault tree analysis works backward from a defined incident to identify and display the combination of operational errors and/or equipment failures involved in the incident. Event tree analysis (the reverse of fault tree analysis) works forward from specific events, or sequences of events, to pinpoint those that could result in hazards and to calculate the likelihood of an event occurring.

Hazard and Operability Study

The hazard and operability (HAZOP) study method is commonly used in the chemical and petroleum industries. HAZOP requires a multidisciplinary team, guided by an experienced leader. The team uses specific guide words (such as *no*, *increase*, *decrease*, and *reverse*) that are systematically applied to parameters (e.g., temperature, pressure, flow) to identify the consequences of deviations (e.g., reduced flow) from design intent for various processes and operations.

Failure Mode and Effect Analysis

The failure mode and effect analysis (FMEA) is a method used to correlate each system or unit of equipment with (1) its potential failure modes, (2) the effect of each potential failure on the system or unit, and (3) how critical each failure could be to the integrity of the system. The failure modes and the effects are then ranked according to criticality to determine which ones are most likely to occur and possibly cause a serious incident.

"What If ... ?"

This method works by asking a series of questions to review potential hazard scenarios and possible consequences. "What if ... ?" is a good method to use when analyzing proposed changes to materials, processes, equipment, or facilities.

Checklist

This method is similar to the "What if ... ?" method, except that a predeveloped checklist is used that is specific to an operation, process, or unit. Companies often use the checklist method when conducting pre-start-up reviews at the time of initial construction or after completing a major turnaround or addition to the process or facility.

Checklist methodology works as long as the process is stable. Unless the checklist is updated regularly, processes that have undergone recent change may have new process elements that may not be evaluated effectively. The "yes–no" aspect of checklist evaluation tools can preclude detailed input from the evaluation team, usually an important aspect of safety process management. It is important to ensure selected evaluation tools are capable of producing the desired result.

"What If ... ?"/Checklist

This method is a combination of the "What if ... ?" and checklist methods and is typically used when a company analyzes units that are identical in construction, materials, and process.

No matter which method of process hazard analysis is used, all analyses should include the following information:
- process location and hazards of the process
- identification of any prior incident with potential catastrophic consequences
- engineering and administrative controls applicable to the hazards
- interrelationships of controls and appropriate application of detection methodology to provide early warnings
- consequences of human factors, facility siting, and failure of the controls
- consequences of safety and health effects on workers within areas of potential failure.

Management should develop a schedule for acting on process hazard analysis findings and recommendations. This schedule may include timetables for planned actions,

documentation of actions to be taken (or not taken), and notification of employees involved. Process hazard analyses, updates, and revalidation, including documentation of resolution of recommendations, should be maintained for the life of each process. A good PSM practice is to review, update, and reevaluate process hazard analyses at least every 5 years, even if no changes to the processes, equipment, or materials have occurred.

Management of Change Programs

Facilities should develop and implement a program to revise process safety information, procedures, and practices as necessary. The program should include a system of management authorization and written documentation for all changes to materials, chemicals, technology, equipment, procedures, personnel, or facilities that affect each process.

The management of change program should consider and document the following information, as a minimum:
- change of process technology
- changes in facility, equipment, or materials
- management of change personnel and organizational and personnel changes
- temporary changes, variances, and permanent changes
- enhancement of process safety knowledge, including the following:
  - the technical basis for the proposed change
  - the impact of the change on safety and health
  - modifications to operating procedures and safe work practices
  - modifications required to other processes
  - time required for the change
  - authorization requirements for the proposed change
  - updating documentation relating to process information, operating procedures, and safety practices
  - training or education required because of the change.
- management of subtle change (anything that is not a replacement in kind)
- nonroutine changes.

Before start-up of the process or affected part of the process, managers and supervisors should make sure that employees and maintenance and contractor personnel involved in the process are informed of the changes. They should also provide employees with updated operating procedures, process safety information, safe work practices, and training as needed.

Operating Procedures

Process facilities should develop and implement written operating instructions and appropriate work practices and procedures that enable employees to safely conduct activities involved in every process. Consistent with the management

of change program, management should review and update or amend operating instructions and procedures as changes occur and certify them annually for completeness and accuracy. Operating instructions should specify detailed procedures and cover the process unit's operating limits, including the following three areas:
- consequences of deviations
- steps to avoid or to correct deviation
- functions of safety systems related to operating limits.

Operating instructions should be clear, precise, and understandable and should incorporate existing safe work practices such as hot-work permits, lockout/tagout, personal protection, fire prevention, and confined-space entry. Operating instructions should be accessible to all employees and contractor employees involved in the process and cover the following areas, as a minimum:
- initial start-up and start-up after turnaround
- normal start-up, normal and temporary operations, and normal shutdown
- emergency operations and emergency shutdown
- start-up after emergency and after temporary operations
- conditions under which emergency shutdown is required and assignment of shutdown responsibilities to qualified operators
- nonroutine work
- operator/process and operator/equipment interface
- administrative controls versus automated controls.

Safe Work Practices and Permits

Facilities should develop and implement a program that uses hot-work and safe work permits to control all work conducted in or near process areas. These permits, as a minimum, should include fire prevention and protection, lockout/tagout, safe work practices, personal protective equipment, emergency response, and confined-space entry requirements applicable to the work, nature of work, materials involved, and object or equipment on which the work is to be performed. Supervisors, employees, and contractor personnel should be trained in the requirements of the permit program, including permit issuance and expiration and appropriate safety, materials-handling, and fire protection and prevention measures. Facilities may consider retaining copies of permits on file until the work is completed to provide information in the event that an incident occurs during the course of the work.

Examples of work to be considered for inclusion in a permit program include, but are not limited to, the following:
- hot work
- lockout/tagout of electrical, mechanical, pneumatic energy and pressure
- confined-space entry

Figure 24–1. Following safe procedures and wearing protective equipment, this worker checks a process vessel.

- opening process vessels, equipment, lines, and the like (Figure 24–1)
- control of entry into process areas by nonassigned personnel.

Facilities should develop and implement safe work practices that apply to both employees and contractor personnel. These practices should cover appropriate safety and health considerations to control potential hazards during process operations including, but not limited to, the following:
- properties and hazards of chemicals used in the process
- engineering, administrative, and personal protective equipment controls to prevent exposures
- measures to be taken in event of physical contact or exposure
- quality control of raw materials and inventory control of hazardous chemicals
- safety (interlock, suppression, detection, etc.) system functions
- special or unique hazards in the workplace.

Employee Information and Training

For the protection of themselves, their fellow employees, and the citizens of nearby communities, all employees and contractor employees involved with highly hazardous chemicals need to fully understand the safety and health hazards of the chemicals and processes encountered in their workplace. One legally mandated requirement—training conducted in accordance with the provisions of 29 CFR 1910.1200 (Hazard Communication Standard)—can help employees be more knowledgeable about the chemicals. However, additional training with clearly defined goals and objectives should cover subjects such as operating procedures and safe work practices, emergency evacuation and response, safety procedures, routine and nonroutine work authorization activities, and other areas pertinent to process safety and health.

Management should develop and implement a formal safety training program covering all incumbent, reassigned, and new supervisors and employees. Employers may initially certify that experienced employees are qualified by knowledge, skill, and experience to conduct operating procedures safely. However, requirements for refresher, remedial, and skills-improvement training for each employee involved in a particular process should be reviewed regularly (e.g., every 3 years or sooner, if necessary). The review schedule will be determined by management in consultation with the employees involved. The company should keep detailed training documentation on file, including identity of the employee, date of training, and means of verifying the employee's knowledge, understanding, and skills. (See also Chapter 31, Safety and Health Training, in the *Administration & Programs* volume.)

Management should consider the following areas, as a minimum, when implementing the training program:

- required skills, knowledge, and qualifications of process employees
- design of process operating and maintenance procedures
- selection and development of process-related training programs
- measuring and documenting employee performance and effectiveness
- hands-on training
- selection and training of instructors knowledgeable in the process.

Training should be provided for all operating and maintenance supervisors and personnel, both employee and contractor. Initial and refresher process training programs should include the following, as a minimum:

- overview of process operations and process hazards
- availability and suitability of materials and spare parts for the processes in which they are to be used
- process operating procedures
- process emergency and shutdown procedures
- safety and health hazards related to the process
- facility and process area safe work practices and procedures.

Employers need to periodically evaluate their training programs. The means or methods for evaluating the effectiveness of the training program should be developed along with the training program goals and objectives. If the evaluation indicates that employees are not retaining the necessary level of knowledge and skills, the employer will need to revise the training program until the goals and objectives can be met.

Contractor Personnel

Management should regard contractor personnel performing maintenance, repair, turnaround, major renovation, or specialty work on a process unit or within a process area as temporary employees of the company. Employers should develop and implement procedures to control the entry, presence, and exit of contractor personnel in covered process areas. Contractor personnel working in process areas must be fully aware of the hazards, processes, and operating and safety procedures and equipment in the area. Employers and contractors should periodically evaluate contractor personnel performance to ensure that contractor employees are trained and qualified and follow all safety rules and procedures. As a minimum requirement, all contractors and contractor personnel working in the process area should be informed and aware of the following:

- potential fire, explosion, and toxic release hazards related to their work
- facility safety procedures and contractor safe work practices
- emergency plan and contractor personnel actions
- controls for contractor personnel entry, exit, and presence in process areas.

When selecting contractors, employers should obtain and evaluate their safety records, performance, and programs. To provide safety information for future evaluation, employers may consider maintaining separate contractor injury and illness logs for all work conducted in process areas. Contractor employers should assure management of the following for all contractor personnel:

- Contractor employees are trained to perform work safely and know and follow the safety rules of the facility.
- Contractors are instructed in the hazards related to the job and process and in applicable provisions of the emergency plan.
- Contractors are instructed to advise facility management of any hazards found or created as a result of contractor work.
- Contractors document names of their employees who have been trained, dates of training, and means used to verify employees' understanding and knowledge of the training.

(See also Chapter 28, Contractor and Customer Safety, in the *Administration & Programs* volume.)

Pre-Start-Up Safety Reviews

Process facility management should institute a program to ensure that pre-start-up process safety reviews are conducted (1) before start-up of new process facilities, (2) before introduction of hazardous chemicals into new facilities, (3) following a major turnaround, or (4) where facilities have had process modifications that affect the process safety information.

The pre-start-up safety reviews, as a minimum, should make sure the following items have been accomplished:

- Construction, materials, and equipment are in accordance with design criteria.
- Process systems and hardware, including testing of computer control logic, have been inspected, tested, and certified.
- Alarms and instruments, relief and safety devices, signal systems, and fire protection and prevention systems have been inspected, tested, and certified.
- Safety, fire prevention, and emergency response procedures have been developed and reviewed, have been put in place, and are appropriate and adequate for the job-related hazards.
- Necessary start-up procedures are in place and proper actions have been taken to ensure safe start-up and operation.
- A process hazard analysis has been performed; all recommendations have been addressed, implemented, or resolved; and actions have been documented.
- All required initial and/or refresher operator and maintenance personnel training is completed, including emergency response, process hazards, and health hazards.
- All written procedures, including operating procedures (normal and upset), operating manuals, equipment procedures, and maintenance procedures are completed and in place, have been reviewed, and are appropriate and accurate for the job-related hazards.
- Management of change requirements have been met for new processes and modifications to existing processes.

Design and Quality Assurances

When management contemplates new processes or major changes to existing processes, it needs to establish a formal design and quality assurance program. Review teams, similar to the teams that perform process hazard analyses, should be established and directed to provide process safety design reviews before and during construction (before the pre-start-up review). Depending on the construction, modifications, and complexity of the project, the appropriate times to perform these design and quality assurance reviews may vary from facility to facility and from process to process. There are, however, some obvious interim review periods common to all projects. For example, the first design control review for a capital project would be conducted just before plans and specifications are issued as "final design drawings," and typically would cover the following areas:

- plot plan, siting, spacing, electrical classification, drainage, and so on
- hazards analysis and process chemistry design review
- project management
- process equipment and mechanical equipment design and integrity
- piping and instrumentation drawings (P&IDs)
- reliability engineering, alarms, interlocks, reliefs, and safety devices
- materials of construction and compatibility.

Another review is normally conducted just before the start of construction. It would cover such areas as the following:

- demolition and excavation procedures
- control of raw materials
- control of construction personnel and equipment entry into the facility and site
- fabrication, construction, and installation procedures and inspection.

One or more reviews may be conducted during the course of construction or modification to verify that the following areas are in accordance with design specifications and facility requirements:

- materials of construction provided and used as specified
- proper assembly and welding techniques, inspections, verifications, and certifications provided
- chemical and occupational health hazards considered during construction
- physical, mechanical, and operational safety hazards considered during construction and facility safety practices followed
- interim protective and emergency response systems provided and working
- facility permit program in place and used during construction.

Maintenance and Mechanical Integrity

Process facilities should develop and implement a program with written procedures to maintain the ongoing integrity of process equipment. This program would ensure and document the periodic inspection, testing, performance, maintenance, corrective action, and quality assurance of process-related equipment. Management should make sure that the mechanical integrity of equipment and materials is reviewed and certified and deficiencies corrected before start-up or that provisions have been made for appropriate safety measures. Mechanical integrity requirements apply to the following equipment and systems:

- pressure vessels and storage tanks
- emergency shutdown systems
- relief and vent systems and devices
- process safeguards such as controls, interlocks, sensors, and alarms
- pumps and piping systems (including components such as valves)
- quality assurance, materials of construction, and reliability engineering
- maintenance and preventive maintenance programs.

The company should develop and implement an inspection and testing program that complies with regulatory requirements or recognized good engineering practices. New and replacement equipment should be inspected during fabrication and installation to verify that it meets design requirements and is suitable for the process application intended. The program also should cover inspection and testing of maintenance materials, spare parts, and equipment to ensure proper installation and adequacy for the process application involved. The acceptance criteria and frequency of inspections and tests should conform with manufacturers' recommendations, good engineering practices, regulatory requirements, industry practices, facility policies, or prior experience.

Inspections and test results should be documented and include the following information, as a minimum:

- date of inspection or test
- name of person performing test
- identification of equipment
- description of work performed
- acceptance limits or criteria and results of test or inspection
- steps required and taken to correct or mitigate deficiencies outside acceptable limits.

Emergency Planning, Preparedness, and Response

Management should develop and implement an emergency preparedness and response plan covering the entire process facility. The company should also make sure that all employees and contractor employees are trained or educated in emergency notification, response, and evacuation procedures. An emergency control center should be established that includes the appropriate process safety information and adequate means of communication should there be an emergency in any process unit within the facility or a hazardous substance release affecting areas outside the facility.

The emergency preparedness plan should comply with applicable company and regulatory requirements and include the following minimum requirements:

- distinctive employee and/or community alarm system
- preferred method of internal reporting of fires, spills, releases, and emergencies
- requirements and methods of reporting process-related incidents to appropriate government agencies
- evacuation, emergency escape procedures, and route assignments
- procedures to account for all personnel
- emergency response capabilities including public safety, contractors, and mutual aid
- employee and contractor employee rescue and emergency duties
- procedures for handling small releases of hazardous chemicals
- Hazardous Waste Operations and Emergency Response (HAZWOPER) procedures.

Compliance Audits

Process facilities should develop and implement a PSM audit program. The program would be used to measure facility performance and ensure compliance with internal and external (regulatory and industry) PSM requirements. Management should establish audit objectives and set a schedule to audit process units periodically. The frequency of audits should be determined by management; however, for compliance with the OSHA PSM standard, employers must certify that all process operations have been audited at least every 3 years.

When establishing a compliance audit program, facility management should be prepared to accomplish the following:

- Establish goals, schedules, and methods of verification of findings prior to the audit.
- Determine the methodology (or format) to be used in conducting the audit, and develop appropriate checklists or audit report forms.
- Prepare to certify compliance with government, industry, and company requirements.
- Assign a knowledgeable audit team (internal and/or external expertise).
- Promptly address all findings and recommendations, and document actions taken.
- Keep a copy of at least two of the most recent compliance audit reports on file (OSHA requires the last two reports to be maintained on file).

Because the objectives and scope of audits can vary, the compliance audit team should include at least one person knowledgeable in the process being audited, one person with applicable regulatory and standards expertise, and other persons with the skills and qualifications necessary for conducting the audit. Management may decide to include one or more outside experts on the audit team because of lack of facility personnel or expertise or because of regulatory requirements. (See also Process Safety Audits later in this chapter.)

Process Incident Investigation

Management should establish a system to (1) thoroughly investigate and analyze incidents and near-misses associated with process systems, (2) promptly address and resolve findings, and (3) review recommendations with employees and contractors whose jobs are relevant to the incident findings. Any incident (or near-miss) resulting in a catastrophic hazardous chemical release (or potential release) should be thoroughly investigated as soon as possible by a team trained in investigation techniques. This team should include at least one person knowledgeable in the process operation involved, a contractor employee if the work involved a contractor, and others with appropriate knowledge and experience.

A detailed report of each investigation should be prepared and maintained on file (OSHA requires records to be kept for 5 years) together with documentation of corrective actions (or reasons for no action). The investigation report should include at least the following information:
- date of incident or near-miss
- date investigation began
- description of incident
- contributing factors
- root cause assessments
- recommendations, follow-up, and actions resulting from the investigation.

(See also Chapter 6, Loss Control Programs, and Chapter 10, Incident Investigation, Analysis, and Costs, both in the *Administration & Programs* volume.)

Standards and Regulations

Process facilities are subject to two distinct and separate forms of standards and regulations: external and internal.

External

External codes, standards, and regulations applicable to the design, operation, and protection of process facilities and employees typically include government regulations and association and industry standards and practices.

Internal

Internal policies, guidelines, and procedures are those developed or adopted by the company or facility to complement the external requirements and to cover distinct or unique processes. Internal standards should be reviewed periodically and changed when necessary, in accordance with the facility's management of change system.

Confidentiality and Trade Secrets

Process facility management needs to provide process information, without regard to possible trade secrets, to persons with the following responsibilities:
- gathering and compiling process safety information
- conducting process hazard analyses
- developing maintenance, operating, and safe work procedures
- assisting in incident (near-miss) investigations
- developing emergency plans and response
- conducting compliance audits.

The OSHA PSM standard provides for confidentiality agreements to protect trade secrets. The standard states that nothing should prevent employers from requiring that persons who receive information about the processes sign agreements not to disclose the information.

ENFORCEMENT OF OSHA PROCESS SAFETY MANAGEMENT STANDARD

Following the promulgation of OSHA's standard 29 CFR 1910.119, Process Safety Management of Highly Hazardous Chemicals, OSHA issued Instruction CPL 02-02-045. This regulation establishes uniform policies and procedures for enforcement of the PSM standard within the United States. Four types of process safety compliance inspections may be conducted to determine whether a facility is covered by the standard and/or to assess the facility's compliance with the standard:

- *Inspections Resulting from Response to Accidents and Catastrophes.* Unprogrammed inspections, especially those triggered by a serious incident or by a complaint, can occur any time. OSHA will follow normal procedures in responding to serious events of potential catastrophic significance. A program quality verification (PQV) inspection may be recommended as a result of OSHA's response. (See *Program Quality Verification Inspections* later in this list.)
- *Unprogrammed Process Safety Management-Related Inspections.* Any investigation conducted as a result of a complaint or referral will include a determination as to whether the facility is covered by the PSM standard. A PQV inspection may be recommended if major deficiencies exist. OSHA, however, will not wait for a company to conduct a PQV to follow through on obvious process safety violations.
- *Programmed General Industry Inspections.* Whenever any other type of programmed general industry safety and health inspection is conducted, OSHA will determine whether the facility should be added to the PQV inspection list.

- *Program Quality Verification Inspections.* Candidates for PQV inspections will be determined by criteria that include the facility's accident/incident history; facility age; number of affected employees; known catastrophic potential of chemicals used in the facility's processes; and past EPA, OSHA, and local fire department experience. A facility may be deleted from the list if its PSM systems have been audited during the past 5 years, if it is included in a corporate settlement agreement covering process safety, if it is a Voluntary Protection Program (VPP) participant, or if it was previously screened and inspectors determined that a PQV inspection would not be conducted. Because of limited resources, OSHA may be able to conduct only a few PQV inspections each year in each region.

Program quality verification inspections are used to evaluate procedures used by facilities and their contractors to manage hazards associated with highly dangerous chemicals. These inspections have the following three goals:
- evaluate the employer's and contractors' PSM programs
- compare the quality of the programs to acceptable industry practices
- verify effective implementation of the programs.

PROCESS SAFETY AUDITS

The two basic principles of conducting self-evaluation audits are (1) gather all relevant documentation covering PSM requirements at a specific facility and (2) determine the program's implementation and effectiveness by following up on its application to one or more selected processes. Facility management should develop a report of the audit findings and recommendations and should maintain documentation noting how deficiencies were corrected or mitigated and, if not, reasons why no corrective action was taken.

Elements in Process Safety Audits
The following items should be considered, as a minimum, when conducting process safety audits.

Orientation
A detailed overview of the facility—including block diagrams indicating chemicals, materials, equipment, and processes—should be given to the audit team. During the orientation, noncompany or nonfacility team members should become familiar with the facility's requirements for personal protective equipment, respiratory protection, alarm systems, and emergency response procedures.

Process Safety Management Program Overview
A person capable of explaining the company's PSM program should address the audit team. He or she should cover how each of the elements of the program is implemented, who is responsible for the implementation, and what records are used to verify implementation.

Preliminary Walkthrough
A scheduled walkthrough should be considered to give the audit team a basic overview of the facility; to look for potential hazards; and to solicit preliminary input from supervisors, employees, and contractor employee representatives.

Documentation Review
The audit team should have unlimited access to or be provided with copies of general safety, fire protection, health, and process safety–related documents, including, but not limited to, the following:
- accident, injury, and illness records (OSHA logs) for the past 3 years for the facility and contractors working in processes
- copies of other facility permits along with any violations received
- a written plan of action for implementing employee participation
- written current and accurate operating procedures, maintenance procedures, process integrity schedules, and process safety information for the units selected to be reviewed for verification of the program
- documented prioritization, rationale, and method used to conduct process hazard analyses
- safety process hazard analysis results, action plan, and resolution documentation
- records for initial, updated, and refresher training and skills evaluations for supervisors, employees, and contractor employees working in processes to be reviewed
- pre-start-up safety review criteria and documentation for any new and modified processes in the units to be reviewed
- hot-work, safe work, and entry permit programs and permits for units to be reviewed
- written management of change procedures, including work orders and contractor work notices for the units to be reviewed
- incident investigation reports, actions recommended, and results for the units to be reviewed
- a written emergency action plan, including emergency shutdown and nonroutine start-up, documentation of known hazards, and employee and contractor emergency response actions
- the two most recent compliance audit reports for the units to be reviewed, including findings, recommendations, and actions taken

- information related to contractor employee safety performance and programs, including contractor employee safe work practices in process areas and methods of informing contractor employees of potential hazards
- written evaluation of contractor employee performance toward PSM requirements in the units to be reviewed.

Review of Selected Process Units

The company may conduct a supplemental detailed audit to verify the effectiveness of the process safety program within one or more process units. Selection of the units to be reviewed may be based on the following criteria:

- factors observed and employee and contractor employee representative input during the preliminary walkaround
- incident reports, age of process units, and other historical information
- units whose process hazard analyses are completed or prioritized for completion
- nature and quantity of materials and chemicals involved in the process
- units with current hot-work, replacement, turnaround, and maintenance activities
- number of employees and contractor employees working in the process or activity.

Self-Evaluation Process Safety Audit Checklist

A company may develop a process safety checklist tailored specifically to the facility and/or process units. The checklist would cover such areas as the following:

- prior incidents and near-misses (in the facility or unit)
- process chemistry hazards
- process unit construction
- process unit documentation
- process unit controls
- process unit maintenance, repairs, testing, and inspection
- process unit operations
- process-related training and employee involvement
- process management of changes
- process emergency procedures.

Confidentiality of Self-Evaluation Audit Reports

Maintaining confidentiality of the process safety audit report is always a concern to management. Many companies that conduct normal, routine internal process safety audits have questioned whether they should continue to do so for fear that agencies would use the results against them in a civil or criminal proceeding. Reports, recommendations, and resulting actions (or no actions) may be subject to review by OSHA or others in the event of an inspection or incident investigation.

In an effort to mitigate this situation, a number of states have passed (or are considering) audit privilege legislation that protects internal audits from regulators and third parties. Internal audit privilege legislation typically contains these requirements:

- Companies that decide to assert the confidentiality of their process safety audits bear the burden of proof and must actively (legally) assert the privilege, which may be waived by act or omission or otherwise forfeited. Waiver can be expressed or implied, such as when a supposedly privileged internal process safety report is widely distributed internally or among third parties.
- The internal process safety audit report must have attributes common to all internal safety audits conducted by the company. It should be marked or titled as an internal safety audit document. In some states, it must also be labeled *privileged*.
- Violations discovered during the internal process safety audit must be voluntarily disclosed and promptly addressed. Where this has been done, some states will provide a company with immunity from civil penalties or criminal prosecution and may allow that the report cannot be used by regulators for enforcement.

In some instances, common law may also provide protection from involuntary disclosure of internal process safety audit reports. A degree of confidentiality may be maintained by conducting the self-evaluation audits at the request of the company attorney. In addition, team members and those with access to the report and its findings must follow the rigorous requirements of maintaining attorney–client confidentiality and work-product privilege. Common-law concepts that may be applicable include the following:

- *attorney–client privilege*—covers confidential legal advice between companies and their attorneys
- *work-product privilege*—protects work products, such as process safety reports, that have been prepared at the direction of an attorney in anticipation of litigation
- *self-evaluation privilege*—a new concept recognizing that a company should be able to examine its own process safety compliance confidentially without threat of disclosure.

SUMMARY

- Process safety describes various regulations and activities designed to protect employees, the public, and the environment from the consequences of major chemical accidents involving highly hazardous materials.
- A number of petroleum, petrochemical, and chemical associations recognize the need for safety management of chemical processing operations. These organizations

have expended considerable time and effort to promote safety guidelines, practices, and regulations in this area.

- The EPA created EPCRA in response to public pressure over the hazardous release and transport of chemicals in communities. OSHA promulgated its Hazard Communication Standard at the same time, along with the Clean Air Act Amendments and the PSM standard.
- PSM should be an integral part of a company's overall safety program. To establish a PSM program, management should establish specific goals and conduct process risk assessments.
- Elements of an effective PSM program include process safety information; conducting process hazard analyses; establishing management of change programs; developing operating procedures; establishing safe work practices and permits; conducting employee training; running pre-start-up safety reviews; and other administrative, regulatory, and operational elements.
- Four types of process safety compliance inspections can be conducted to determine whether a facility is covered by the OSHA PSM standard: inspections in response to accidents and catastrophes, unprogrammed PSM inspections, programmed general industry inspections, and program quality verification inspections.
- Two basic principles of conducting self-evaluation audits of PSM are (1) gather all relevant documentation and (2) evaluate the program's application to one or more selected processes.
- Elements in a process safety audit include orientation, overview, preliminary walkaround, documentation review, and review of selected process units.

REFERENCES

American Institute of Chemical Engineers, 345 East 47th Street, New York, NY 10017. www.aiche.org.
> *Guidelines for Engineering Design for Process Safety.* 2012.
> *Guidelines for Implementing Process Safety Management Systems.* 1994.
> *Guidelines for Risk Based Process Safety.* 2007.

American Petroleum Institute, 1220 L Street NW, Washington DC 20005. www.api.org.
> *Management of Hazards Associated with Location of Process Plant Buildings* (RP 752), 2009.
> *Process Safety Performance Indicators for the Refining and Petrochemical Industries* (RP 754), 2010.

Occupational Safety and Health Administration. 29 CFR 1910.119, Process Safety Management of Highly Hazardous Chemicals, Explosives and Blasting Agents, Final Rule, May 1992.

REVIEW QUESTIONS

1. According to OSHA, what is process safety management?
2. What two things must manufacturers do, according to the Hazard Communication Standard?
   a.
   b.
3. What is the main difference between the OSHA and EPA chemical safety management regulations?
4. The PSM standard applies to the handling, storage, processing, and transport of flammable gases and liquids in quantities of
   a. 500 gallons or more.
   b. 1,000 gallons or more.
   c. 10,000 gallons or more.
   d. none of the above
5. What are the three purposes of a risk management program?
   a.
   b.
   c.
6. Which two people (i.e., what proficiencies) must be present for a hazard analysis team to be successful?
   a.
   b.

7. When deciding the priority order for hazard process analysis, the company should take what four criteria into account?
   a.
   b.
   c.
   d.
8. What is the difference between fault tree analysis and event tree analysis?
9. What does HAZOP stand for?
10. Briefly define *failure mode and effect analysis*.
11. When is the "What if ... ?" method of process hazard analysis appropriate?
12. Operating procedures need to be certified for completeness and accuracy every
    a. 6 months.
    b. 1 year.
    c. 2 years.
    d. 5 years.

13. List five of the eight areas that operating instructions must address.

    a.

    b.

    c.

    d.

    e.

14. List four times that a pre-start-up process safety review should be conducted.

    a.

    b.

    c.

    d.

15. List three of six items of information that inspection and test result documentation should contain.

    a.

    b.

    c.

16. What is the purpose of a process safety management audit program?

17. How long does OSHA require accident investigation records to be kept?

    a. 1 year

    b. 2 years

    c. 5 years

    d. 7 years

18. What are the four types of OSHA process safety compliance inspections?

    a.

    b.

    c.

    d.

19. What are the three goals of PQV inspections?

    a.

    b.

    c.

20. What are two basic principles of conducting self-evaluation audits?

    a.

    b.

21. Mr. Smith is the head of the audit team in charge of reviewing documentation for your area. As a manager, you should provide him with access to which compliance audit reports for the area?

    a. all reports

    b. the most recent

    c. the two most recent

    d. none

Aviation Safety

Curt L. Lewis, PhD, CSP
John F. Montgomery, PhD, CSP, CHMM

History

Governmental Organizations
The Federal Aviation Administration ▸ The National Transportation Safety Board

International Aviation
International Civil Aviation Organization ▸ European Aviation Safety Agency

Stakeholders
The Airlines—Flight Safety Department ▸ Data Collection and Analysis ▸ Information Dissemination ▸ Safety Education

Accident/Incident Preparedness
Aircraft and Systems Engineering ▸ Cabin Safety ▸ Safety Audits ▸ Aircraft Accident/Incident Investigation ▸ Interaction with the NTSB ▸ The Flight Safety Foundation ▸ Trade Organizations ▸ IOSA and ISAGO

Concepts in Practice
Training ▸ Safety Management System ▸ Flight Operational Quality Assurance

The Future
Next Generation ▸ Aircraft Telemetry

Summary

References

Review Questions

This chapter provides a general overview of air transportation safety, with emphasis on the following areas:
- a brief historical overview of the aviation field in the United States
- the governmental organizations involved in air transportation safety in the United States
- international aviation organizations
- air transportation safety stakeholders such as commercial airlines and trade organizations
- aircraft accident investigations
- safety concepts in practice
- the future of air transportation safety

HISTORY

The modern air transportation system is one of the safest methods of transport available, and safety is a crucial part of this system. Yet barely more than a century ago, most people would not have dreamed of being able to fly, let alone dreamed of the system of aviation transportation that is possible today. How did aviation evolve so quickly? The work of many individuals and organizations—as well as trial and error—led to the advantages that air travelers enjoy today.

Although hot air balloons and airships have been around since the 1700s, this chapter focuses on the form of air transportation that originated at the beginning of the 20th century: engine-powered flight. The Wright brothers' first successful powered flight took place in 1903, only a little more than a century ago. A few years later, in 1907, Paul Cornu flew the first helicopter. Since that time, humans have flown faster than the speed of sound and have created a worldwide network of air travel that involves thousands of flights per day.

GOVERNMENTAL ORGANIZATIONS

In the United States, one of the first aviation organizations to incorporate extensive safety practices was the U.S. Air Mail Service. The Air Mail Service completed its first official delivery in 1915. (See Figure 25–1.) Some of its safety practices included abiding by strict criteria when selecting pilots and requiring regular pilot medical examinations; performing stringent, by-the-book aircraft inspections; using a 180-item checklist at the end of virtually every flight; and overhauling engines and aircraft regularly. These practices resulted in a fatality rate of 1 per 789,000 miles flown between 1922 and 1925, with a comparable figure for itinerant commercial fliers (for 1924 only) of 1 per 13,500. However, flying was still very dangerous. Pilots could determine their locations only by studying the terrain beneath them. Airfields and farmers would light fires at night to create visible navigation points for pilots.

An important milestone during the early years of flight was the Air Commerce Act of 1926. This act called for the causes of civil aviation accidents to be made public. It also called for aircraft airworthiness requirements, air traffic rules, periodic examinations of aircraft, and ratings of pilots. These roles were tasked to the Department of Commerce, which established its own aeronautics branch. This legislation was an important step that allowed the industry to continue growing. As safety improved and as

Figure 25–1. A Curtiss JN-4H *Jenny* is prepared for takeoff during the inauguration of U.S. airmail service in Washington DC on May 15, 1918. *(Library of Congress, Prints & Photographs Division, photograph by Harris & Ewing [LC-DIG-hec-10830])*

business increased, the number of commercial air traffic passengers increased from 6,000 in 1926 to 173,000 passengers in 1929.

The aeronautics branch, renamed the Bureau of Air Commerce in 1934, advanced air safety by encouraging airlines to create air traffic control centers. In 1938, the Civil Aeronautics Act was passed. This act created the Civil Aeronautics Authority (CAA), which included a three-member Air Safety Board that investigated aircraft accidents and recommended actions to prevent future accidents.

In 1940, the CAA was divided into two agencies, the CAA and the Civil Aeronautics Board (CAB). Although the CAB became part of the Department of Commerce, it remained an independent agency. Safety responsibilities such as air traffic control, pilot and aircraft certifications, safety enforcement, and airway development fell to the CAA, whereas safety rulemaking, accident investigation, and economic regulation of the airlines were the responsibilities of the CAB.

Despite these advances in air safety, accidents still occurred with some frequency, including high-profile accidents such as the midair collision between a DC-7 and a Super Constellation Aircraft over the Grand Canyon in 1956, which killed all 128 occupants. (See Figure 25–2.) Such accidents put a spotlight on the need for a more comprehensive approach to air safety. The federal government responded by creating what is now called the Federal Aviation Administration (FAA) (Hansen, McAndrews, and Berkeley 1–37).

Figure 25-2. The severed tail section of the Lockheed L-1049 Super Constellation operating as TWA Flight 2 on June 30, 1956. TWA Flight 2 collided with United Flight 718 in what became known as the 1956 Grand Canyon mid-air collision. The photo was taken by National Park Service employees in the course of the Civil Aeronautics Board's investigation of the crash. *(Photo courtesy of the National Park Service)*

The Federal Aviation Administration

In 1958, the Federal Aviation Act created the Federal Aviation Agency (FAA). The FAA was a new, independent federal agency that was responsible for civil aviation safety. In 1966, the FAA was included under the direction of the newly created Department of Transportation and renamed the Federal Aviation Administration. The role of the Civil Aeronautics Board in accident investigation was transferred to the National Transportation Safety Board (NTSB).

The FAA's current responsibilities include a broad range of air transportation safety measures. For example, the FAA is responsible for registering aircraft, issuing airworthiness certifications, approving aircraft designs and productions, etc.

The FAA also has authority over airports in the form of regulating safety, environmental programs, engineering design and construction, and airport compliance. Through its Airport Safety Program, the FAA is involved in such areas as runway safety, fire fighting, safety management systems, and wildlife strike prevention.

The FAA administers air traffic control services through its operational arm, the Air Traffic Organization. This service covers more than 17% of the world's airspace, which includes all of the United States, large portions of the Atlantic and Pacific Oceans, and the Gulf of Mexico. This extensive network is made up of 22 air route traffic control centers in 19 U.S. states.

The FAA's air traffic control operations consist of three main parts: Terminal Radar Approach Control Facilities (TRACON), air traffic control towers, and en-route centers. TRACON facilities are locations where air traffic control professionals utilize radar displays and radios to guide aircraft that are approaching or leaving the airspace within a 30- to 50-mile radius of an airport. Air traffic control towers, also located at airports, coordinate takeoff and landing procedures (see Figure 25–3). En-route centers manage the air traffic between airports. Because of these operations, very little land in the continental United States is without aviation radar coverage.

An additional role of the FAA is to provide data and research to the public and the aviation industry. These data often take the form of accident and incident reports, aviation statistics, forecasts, safety data, and other information. The Office of Accident Investigation and Prevention is one of the main organizations within the FAA that provides such data. This organization provides preliminary accident and incident data for the previous 10 working days.

Another important way in which the FAA provides data to the public is through its Aviation Safety Information Analysis and Sharing (ASIAS) system. This Internet-based system provides lessons learned from aircraft accidents.

The FAA also ensures that proper certifications have

Figure 25-3. Air traffic control towers coordinate takeoff and landing procedures. *(Comstock/Stockbyte/Thinkstock)*

been obtained and oversees the certification types for airmen, aircraft, airlines, airports, and even commercial outer space transportation. For example, airmen certifications establish the eligibility, training, experience, and testing requirements for pilots and mechanics. The certifications for pilots are known as pilot's licenses, and they do not apply only to those who fly airplanes. In fact, pilot's licenses are also required for those who fly helicopters, gyroplanes, gliders, balloons, and airships. Maintaining these standards is one of the first measures that the FAA takes to ensure a safe air transportation system.

Certifications for aircraft are known as Airworthiness Certifications, and they require the determination that the aircraft are in safe operating condition. Several different types of Airworthiness Certificates are available, depending on the type of aircraft and its intended use (Figure 25-4).

In addition, all airports serving commercial airlines in the United States are required to have an FAA-issued Airport Operating Certificate. These airports must undergo certification inspections; a few components of such an inspection are an administrative inspection, runway area inspections, aircraft rescue and fire-fighting inspection, and a nighttime inspection.

Airlines are also required to be certified. The purpose of this certification is to determine whether an airline is able to conduct business in a manner that complies with all applicable regulations and safety standards and can deal with hazard-related risks in the operating systems environment. Airlines based outside of the United States must also be audited by the FAA before they are allowed to provide commercial routes to U.S. destinations.

As part of its safety mission, the FAA offers training and testing programs. Training is available for many different occupations within the aviation field, and many specific subjects are covered. To appreciate the vital roles that the FAA plays in aviation safety, one needs only imagine how dangerous a nationwide air transportation system would be without a regulating body having the FAA's authority.

The National Transportation Safety Board

The National Transportation Safety Board (NTSB) is an independent agency within the U.S. federal government and is tasked with investigating all U.S. aviation accidents; serious accidents in other transportation areas such as rail, highway, and marine; hazardous materials; and the nation's pipelines. The NTSB is responsible for making safety recommendations based on these investigations. In addition, the NTSB administers the federal assistance program for families of aviation accident victims.

With the passage of the Air Commerce Act, the

Figure 25-4. Twin engine turboprop aircraft. *(RyanFletcher/iStock/Thinkstock)*

Department of Commerce was given the responsibility to investigate the causes of aircraft accidents; however, the NTSB was not formally established until 1967, when Congress made it part of the Department of Transportation (DOT). In 1974, the NTSB was removed from the DOT and was made into its own agency. Congress did so because there was an inherent conflict of interest with the NTSB being part of the DOT. If the NTSB's investigative findings showed that the transportation system itself had been at fault, then the NTSB essentially would have been criticizing its supervising department. Thus, the agency was given its independence so that it could create unbiased safety recommendations. This independence gives the NTSB a unique position among other aviation stakeholders.

The NTSB has established four strategic goals for itself. These goals help to explain the functions of the NTSB. The first goal is to conduct effective accident investigations. The general population hears about the NTSB only when a major transportation accident has occurred. However, the NTSB investigates *all* aviation accidents in the United States, and in order to be effective, the NTSB must identify the accidents that represent the greatest potential for investigative opportunity. The NTSB must also select the appropriate response to each accident. In addition, the NTSB investigates accidents that occur abroad if they involve U.S. carriers or U.S.-manufactured or -designed equipment.

The second goal of the NTSB is to increase advocacy for its safety recommendations and to maintain and advocate for items on its "Most Wanted List." Somewhat similar to the well-known most wanted list of the Federal Bureau of Investigation (FBI), the NTSB's list consists of the most essential changes needed to reduce transportation accidents and save lives. This list is available to the public on the NTSB's website. The NTSB advocates for its safety recommendations through publications, dialogue with relevant government agencies and other stakeholders, testimony, and other public communications.

The third strategic goal is to conduct fair and expeditious adjudication of airmen and mariner appeals of the FAA and the U.S. Coast Guard's enforcement actions and certificate denials. The NTSB manages this appeals process and must balance the interests of the airmen and the mariners with aviation and marine safety.

The fourth strategic goal of the NTSB is to provide outstanding mission support. This goal has an internal focus and concerns making effective use of agency resources and supporting employees.

The structure of the NTSB consists of five board members, each of whom is nominated by the president of the United States and serves 5-year terms. These board members report to the chairman of the NTSB. Several offices are within the organization, and these offices report to the managing director, who is one of the five board members.

An NTSB investigation begins as soon as the NTSB learns of a serious accident and calls a "Go Team" to action. This team is a group of three to more than a dozen specialists from the NTSB's headquarters in Washington DC, who are assigned on a rotation basis to respond as quickly as possible to the scene of an accident. Go Teams travel by commercial airliner or government aircraft, depending on circumstances and availability. Go Team members have 24-hour shifts of being on call, during which they await notification to head to an accident scene. The purpose of the Go Team is to begin an investigation as quickly as possible. This team brings all the specialized equipment that it needs as well as cameras, tape recorders, film, etc.

Organizations such as airlines and aircraft manufacturers may have their own Go Teams. These teams are allowed to assist the NTSB in an investigation if the NTSB designates them as parties to the investigation. The NTSB does not have to accept any person or group as a party; rather, it selects only those organizations that have the needed expertise and/or technical knowledge. All such parties to an investigation report to the NTSB official known as the Investigator-in-Charge (IIC). The IIC, a senior investigator with years of NTSB and industry experience, manages the individual investigators at an accident scene. During an aviation investigation, for example, the individual investigators are divided into groups that have the following specialties: operations, structures, power plants, systems, air traffic control, weather, human performance, and survival factors. These groups are often assisted by non-NTSB personnel. For example, a representative from an aircraft engine manufacturer may assist the power plants group. If it is determined that the accident is the result of criminal activity, the NTSB surrenders control over the investigation to another government agency such as the FBI. However, the NTSB may still assist with the investigation if requested to do so.

Another NTSB member at the scene is one of the five board members, who represent the team as a media spokesperson. The NTSB does not speculate over the cause of an accident. A transportation disaster assistance specialist administers the NTSB's duties under the Aviation Disaster Family Assistance Act of 1996.

One of the actions that the Go Team takes at a crash scene is locating the cockpit voice recorder and the flight data recorder, which is commonly referred to as the "black box." This term is actually a misnomer because the "box" is painted orange so that it can be easily located. The team also documents the scene and takes the required measurements. The NTSB has its own manuals regarding major investigations, the flight data recorder, and the cockpit voice recorder.

After the initial onsite investigation has taken place, hearings may take place. These hearings are open to the public, and their main objective is to gather sworn testimony from subpoenaed witnesses regarding the accident. The NTSB then analyzes the data and other collected infor-

mation and issues a final report, which is public information. These reports contain recommendations on how to prevent similar accidents in the future. However, the NTSB does not have the authority to force any party to act upon its recommendations. Thus, if legal cases arise from an accident, the findings of the NTSB cannot be entered as evidence at trial.

The NTSB has a lengthy history of investigating aviation accidents and making safety recommendations. To date, the NTSB has published more than 13,700 safety recommendations.

INTERNATIONAL AVIATION

International Civil Aviation Organization

One of the early dilemmas that arose with the beginning of flight was how to safely fly from one country to another when each country could have its own aviation standards. This is the dilemma that the attendees of the 1944 Chicago Convention sought to resolve. The Chicago Convention is also known as the Convention on International Civil Aviation. The International Civil Aviation Organization (ICAO) came into being as a specialized agency of the United Nations in 1947.

At the convention, ICAO stated that "... the undersigned governments hav[e] agreed on certain principles and arrangements in order that international civil aviation may be developed in a safe and orderly manner and that international air transport services may be established on the basis of equality of opportunity and operated soundly and economically." In other words, the organization was established to secure international cooperation and the highest possible degree of uniformity in regulations, standards, procedures, and organizations regarding civil aviation matters. The original convention was signed by 52 countries. At the time of this writing, there are 191 signatory states.

According to the ICAO's website, the current mission of the ICAO is "to serve as the global forum of States for international civil aviation. ICAO develops policies and standards, undertakes compliance audits, performs studies and analyses, provides assistance and builds aviation capacity through many other activities and the cooperation of its Member States and stakeholders." It is important to note that the ICAO is not a regulatory body. Although the ICAO develops aviation standards, the legal regulations in each country are set by that country itself.

The ICAO has five main objectives: safety, security, environmental protection, air navigation capacity and efficiency, and economic development of air transport. This section focuses on the safety objective.

To reach this objective, the ICAO develops global strategies, which are contained in the Global Aviation Safety Plan (GASP) and the Global Air Navigation Plan (GANP). The Global Aviation Safety Plan lays out the strategies for member states and regions to follow so that they remain focused on establishing, updating, and addressing their safety priorities as they encourage expansion of their air transport services. The GASP also sets target dates and objectives so that member states can remain up-to-date with the large amount of predicted growth in the aviation industry. The GANP complements the GASP and deals specifically with current and future air navigation priorities, aviation system performance, strategies for investing in system upgrades, and technology road maps.

The ICAO also develops and maintains standards, recommended practices, and recommended procedures. These take the form of 16 annexes and four procedures for air navigation services and numerous manuals and circulars that provide guidance on implementation. In addition, the ICAO collects and analyzes data in order to identify emerging risks.

The ICAO has a program called the Universal Safety Oversight Audit Program (USOAP). This audit of an individual member state checks for primary aviation legislation and civil aviation regulations, civil aviation organizations, personnel licensing and training, aircraft operations, airworthiness of aircraft, aircraft accident and incident investigations, air navigation services, and aerodromes and ground aids. The USOAP also responds to interruptions in the aviation system that are created by natural disasters and implements targeted safety programs.

The ICAO is overseen by the one-person position of Secretary General. The Secretary General heads a body called the Secretariat, which consists of five main divisions. Another body is the Council, which is a permanent part of the organization and is responsible to the Assembly. Thirty-six member states are elected to the Council for 3-year terms. The largest body is the Assembly, which consists of all the member states. The Assembly meets once at least every 3 years.

European Aviation Safety Agency

The FAA is not the only governmental body involved in aviation safety. In Europe, one of the largest governmental bodies is the European Aviation Safety Agency (EASA).

The European Aviation Safety Agency is a multinational, collaborative agency that promotes the highest standards of safety and environmental protection in civil aviation in Europe and worldwide. The organization was created in 2003 as an independent European Community agency, and it is currently headquartered in Cologne, Germany. Although the EASA was created under European public law, it is independent of European Community institutions such as the Council, the Parliament, and the Commission.

Figure 25-5. Jet turbine *(pablographix/iStock/Thinkstock)*

The need for such an agency emerges from the need for standardization among the member states. The organizers found that the best way to create uniformly interpreted standards and freedom for numerous political influences was to create an independent agency.

The EASA has been given specific regulatory and executive tasks in the fields of civil aviation safety and environmental protection. These tasks include:
- drafting aviation safety legislation and providing technical advice to the European Commission and to member states
- safety and environmental-type certification of aircraft, engines, and parts (Figure 25-5)
- inspections, training, and standardization programs to ensure the uniform implementation of European aviation safety legislation in all member states (Figures 25-6 and 25-7)
- approval of aircraft design organizations worldwide and of production and maintenance organizations outside the EU
- authorization of third-country (non-EU) operators
- coordination of the European Community program SAFA (Safety Assessment of Foreign Aircraft) regarding the safety of foreign aircraft using European Community airports
- data collection, analysis, and research to improve aviation safety.

The EASA's organizational structure starts with an Executive Director, who leads the agency and makes the agency's safety decisions. An independent Board of Appeal functions as a control to ensure that the Executive Director correctly applies European legislation. Also part of the EASA's structure is a Management Board, which consists of the member states, and the Commission. Currently,

Figure 25-6. Airplane mechanics work on a jet engine. *(Iagereek/iStock/Thinkstock)*

Figure 25-7. A mechanic repairs a 747 aircraft's engine at JFK Airport. *(Kim Steele/Photodisc/Thinkstock)*

more than 30 countries are on the Management Board. This Board appoints the Executive Director, sets the agency's priorities, and takes care of other internal tasks. An EASA Advisory Board assists the Management Board. The Advisory Board is made up of organizations such as aircraft manufacturers, pilots, and maintenance personnel.

Much like the FAA in the United States, the EASA has certain requirements for non–member states who wish to fly to and from EASA countries. For example, foreign aircraft operators may be required to comply with ICAO standards or receive an authorization issued by the EASA. The agency also releases numerous technical and general publications.

STAKEHOLDERS

The Airlines—Flight Safety Department

What role do commercial airlines have in the realm of safety? Many major airlines and aviation corporations realize the importance of safety and have developed competent, professional departments devoted to aviation safety. These flight safety departments are responsible for creating policies, procedures, and programs that ensure a safe environment for aircrews and provide the world with the safest form of transportation.

Flight safety departments include expert safety personnel who interact with national and international flight operation regulatory agencies such as the FAA and the NTSB. These departments are responsible for investigating incidents and accidents throughout the corporate system and identifying the causative factors and measures to prevent a recurrence.

It is imperative that the aviation industry maintain and expand safety programs that shift investigative activities from cause-analysis investigation to prevention-analysis investigation. Flight safety departments maximize profit for the corporation and present a positive message to customers, employees, and governmental regulators when they take potent actions to prevent incidents and accidents. A safe aviation operation reduces aircraft accidents, avoids fines, and saves money by controlling losses and managing risk.

The major activities and responsibilities described earlier must be assumed by all aviation corporations' flight safety departments. In most cases, regulatory mandates also include compliance, surveillance, and reporting.

An airline safety department should act as an independent department that reports to the highest level of management. This reporting structure, promoted by the FAA, enables the department to take an unbiased approach when investigating incidents/accidents. This structure also enables the flight safety department to use multi-departmental resources to conduct an accident prevention analysis that is not biased by any particular departmental affiliation. These reasons are similar to the ones behind the NTSB being separated from the Department of Transportation. Because of this lack of bias, the safety reports on and the recommendations of a safety department's investigation are more likely to be accepted as true reflections of the corporation's safety concerns and policies.

Data Collection and Analysis

Collecting data from various sources helps flight safety departments research, document, and analyze pertinent information relating to safety-of-flight operations. These data are then analyzed, statistically and graphically, to identify developing trends that may either contribute to incidents or help prevent them.

Data management provides managers with information on operational irregularities from various points of view. Such management requires a corporate reporting system that allows the following:
- integration of all available reporting sources
- creation or streamlining of the initial data entry
- a comprehensive database that is capable of accurate, efficient, and timely reporting of all safety events.

Corporate-wide reporting systems reduce inefficiencies in various departments throughout the organization. Cost savings can be achieved when management at all levels

takes responsibility for collecting and handling the safety-related data to be entered into the system. The department manager is then responsible for ensuring adequate control or elimination of identified hazards.

Trend analysis is used to provide documentation for industry issues currently under scrutiny, to support operational procedures, and to develop new policies arising from internal concerns and from communication with other aviation corporations. For example, using trend information, various aviation safety departments were able to determine that portable electronic devices created an electromagnetic interference that interrupted navigation and communications equipment.

Information Dissemination

Database trends must be analyzed, interpreted, and used to develop corrective actions. The resulting information can then be incorporated into the policy and procedure decisions made by various constituents of the aviation industry, especially the maintenance and engineering, flight and flight services, consumer relations, and insurance departments.

In addition, trend information leads to accurate and timely reports for senior management and may be used to document the company's procedural compliance and incident recurrence issues.

Safety Education

Companies should establish and implement their own internal safety education programs to help ensure compliance with regulations and federal mandates. Examples of typical programs that are available off the shelf although are not specific to the airline industry are those dealing with accident preparedness and bloodborne pathogens.

Airline flight safety departments should also provide employees with safety education from outside the company or in conjunction with other appropriate groups. Programs that are developed in conjunction with the FAA and the NTSB provide a good understanding of an airline carrier's operational requirements. These programs also foster investigators' awareness of the various areas of expertise that are available in the airline industry as well as in an airline's culture.

Safety departments produce training videos for firefighting and emergency response personnel to give them an understanding of an aircraft's specific procedures and to tailor response activities to various aircraft. By providing safety personnel who can give periodic speeches and by offering audiovisual presentations to emergency response organizations, a company promotes a better understanding of its operations and enhances teamwork and goodwill.

Airline safety departments can assist in the training of airport fire-fighting personnel by providing concise informational cards and videos specific to a certain aircraft. This information is especially helpful at fire stations in which personnel are not familiar with the aircraft's equipment, such as in rural areas and developing countries. If necessary, the information should be provided in more than one language.

Flight safety personnel also work with individuals who assist customers or employees in the event of an accident or incident. The safety department should oversee the development of airport disaster exercises to ensure compliance with regulations and emergency response. These exercises assist airport and corporate headquarters personnel in refining their response plans. Exercises dealing with emergency response should be conducted on a regular basis to develop a familiar and well-rehearsed plan.

ACCIDENT/INCIDENT PREPAREDNESS

Developing effective damage control plans can result in savings throughout the management of property and monetary losses. However, airline flight safety departments must develop a comprehensive plan of operation *before* a crisis arises. The plan should improve airline personnel's preparedness and designate specific responsibilities for responding to a minor or major accident or incident. The degree of preparedness and awareness of response activities must be the same regardless of the severity of the accident. In addition, employees' actions must become automatic so that they no longer need to refer to an instruction document or ask rudimentary questions such as, "Where do we go?" and "What do we do?"

Aircraft and Systems Engineering

Safety engineering and system safety techniques are critical to identifying loss or damage to equipment and facilities. Safety department personnel should have the educational backgrounds and abilities necessary to perform engineering system failure analysis of an aircraft, maintenance and flight procedures, and any new technologies introduced into the workplace. System engineering techniques may be used to detect obscure failures in aircraft equipment before purchasing that equipment. These techniques may also be used to conduct a hazard analysis to detect deficiencies in equipment following its placement into service and to provide recommended corrective action.

Cabin Safety

Airline flight safety departments are responsible for implementing flight, flight service, and field service policies and procedures for on-board cabin equipment and for cabin safety issues. These responsibilities also include designing cabin safety cards that meet federal regulations. In the past several years, public interest in regulatory and congressio-

nal action has focused attention on many areas never before considered. Because of the increased requirements for cabin safety improvements, the aviation industry is aware of the conditions that might injure or adversely affect passengers and employees while in flight.

Safety Audits

Various types of safety audits should be conducted to discover noncompliant activities in the areas of maintenance, cabin safety, flight, and flight training. The safety department should also audit corporations that provide materials and services to airlines. These audits must evaluate the equipment, staffing, work operations, and final products to ensure that the operations are in compliance with corporate as well as federal regulations. Flight safety department personnel should inspect the airfields at which the company intends to begin service for items such as crash and fire rescue teams, runway capabilities, air traffic control systems, and indications of previous accidents.

Aircraft Accident/Incident Investigation

A key component of the flight safety department's responsibility is investigating all actual and potential aircraft incidents and accidents. The department should be the leader of the company's accident investigation team unless the investigation is delegated to local management. The safety department, however, should review the results of all investigations. All aircraft accidents must be thoroughly investigated to determine the underlying causes so that corrective action can be recommended and implemented. The safety department should retain all records pertaining to an accident or incident and also act as the coordinator between the airline and the NTSB.

Investigation follow-up is another important responsibility of the safety department. The flight safety community is switching its focus from placing blame for accidents and incidents to preventing those accidents and incidents. A key reason for this new approach is the extensive efforts arising from every incident investigation to prevent a recurrence. Safety departments have taken a major step forward by increasing the involvement of responsible departments in the investigation and recommendation processes and by documenting investigations in reports. Because an airline's safety department is the corporate leader for investigating all major accidents, it should provide and maintain an investigation kit that contains supplies similar to those kept by the NTSB's Go Team members.

Interaction with the NTSB

Airline safety departments often cooperate and aid the NTSB in major accident investigations. An airline's flight safety department is responsible for notifying the NTSB or other agencies immediately after an aircraft accident or incident. In fact, most of the interaction between the airlines and the NTSB comes about because after an accident, the airlines must report certain items to the NTSB.

The following terms are used by the NTSB (and are generally accepted worldwide) during the investigative phase:

- *Aircraft accident.* An occurrence involving the operation of an aircraft with the intention of flight in which any person suffers death or serious injury or in which the aircraft receives substantial damage.
- *Aircraft incident.* An occurrence other than an accident that affects or could affect the safety of operations associated with the functioning of an aircraft.
- *Fatal injury.* Any injury that results in death within 30 days of the accident.
- *Serious injury.* Any injury that is required under 49 CFR 830 to be reported to the NTSB. These injuries include those that:
  - ○ require hospitalization for more than 48 hours
  - ○ cause the fracture of any bone (except fingers, toes, or nose)
  - ○ involve any internal organ
  - ○ cause second- or third-degree burns or any burns that affect more than 5% of body surface.
- *Substantial damage.* Damage or failure that adversely affects the structural strength, performance, or flight characteristics of an aircraft and that usually requires significant repair or replacement of the affected component. Single-engine failure or failure limited to that engine or damage to the landing gear, wheels, tires, flaps, brakes, or wingtips is not considered "substantial damage."

The Flight Safety Foundation

Independent organizations other than the NTSB also make safety recommendations. One such nongovernmental group is the Flight Safety Foundation (FSF). The Flight Safety Foundation has several characteristics that distinguish it from a trade organization. For example, the FSF was started by Jerome F. "Jerry" Lederer, who was an early pioneer in aviation safety. Also, the organization is an international nonprofit whose sole purpose is to provide impartial, independent, expert safety guidance and resources for the aviation and aerospace industry.

Some of the actions of the Flight Safety Foundation include:

- promoting and facilitating the global application of leading aviation safety assessments, standards, and practices
- developing global solutions to important aviation safety challenges
- disseminating interpretive and educational aviation safety materials worldwide and focusing principal efforts on community members with the most urgent needs

- being the constant and forceful "Just Culture" advocate for all global aviation community members
- providing opportunities for members of the global community to come together, interact, and learn from each other
- representing the global aviation community on safety matters with news media, industry and government groups, and the general public.

The Flight Safety Foundation sponsors events such as workshops, conferences, studies, and task forces. The FSF also investigates difficult safety issues, one example being the ICARUS committee, which studies human factors. One development of the Flight Safety Foundation is the Basic Aviation Risk Standard, which is a common standard and auditing process used by contracting companies to evaluate aviation operators to ensure regulatory compliance and their safety.

Trade Organizations

Several other organizations are involved in the field of aviation safety. For example, trade organizations advocate for various aviation parties. Some examples are the International Air Transport Association (IATA), Airlines for America (formerly known as the Air Transport Association), and the Aircraft Owners and Pilots Association (AOPA).

IATA, started in 1945, represents more than 200 international member airlines before regulators, decision makers, and governments. IATA's goal is to be the force for value creation and innovation driving a safe, secure, and profitable air transport industry that sustainably connects and enriches our world. IATA develops regulations on subjects ranging from safety and environmental matters to cargo transport.

Airlines for America is the primary trade organization of the leading U.S. airlines. Some member airlines include American Airlines, Delta Airlines, Federal Express Corporation, Southwest Airlines, and US Airways. Airlines for America advocates on behalf of its member airlines on a number of issues such as safety and operations, energy and the environment, customer service, and security.

AOPA, started in 1939, is an association that advocates for the field of general aviation. It also advocates for citizens' rights to fly private aircraft.

IOSA and ISAGO

One of the services that IATA both offers and requires of its members is the IATA Operational Safety Audit (IOSA) program. This audit assesses the operational management and control systems of an airline. Some safety-specific areas that an IOSA examines include safety management, safety risk management, safety assurance, and emergency response plans. An IOSA is designed to use standardized procedures that provide an efficient audit of an airline. The IOSA also ensures that the provisions of the IOSA Standards and Recommended Practices (ISARPs) are met. The program is overseen by the IOSA Oversight Council (IOC), which is part of the IATA governance structure. The audit itself is completed by a trained and IOSA-qualified auditor (IATA IOSA Program Manual [IPM] Operational Safety Audit).

A similar program to the IOSA is the IATA Safety Audit for Ground Operations (ISAGO). Ground operations may include baggage handling, aircraft towing, jet bridge operations, refueling, catering, and many other jobs. Ground operations are crucial components of modern aviation and present their own set of safety challenges. An ISAGO thus aims to improve safety and cut airline costs by drastically reducing ground accidents and injuries. One benefit to a company being audited is that the audit looks at all of the ground operations together instead of requiring different audits for specific ground operations. Some components of an ISAGO include the *IATA Ground Operational Manual (IGOM)*, which describes the proper procedures for and provides other information on all ground operations (ISAGO IATA Safety Audit for Ground Operations 2014).

CONCEPTS IN PRACTICE

Training

Crew resource management (CRM) training is an important step in reducing the risks associated with human factors in aviation safety. SKYbrary.aero defines CRM as the "effective use of all available resources for flight crew personnel to assure a safe and efficient operation, reducing error, avoiding stress and increasing efficiency."

CRM training was created in response to significant incidents such as the Tenerife Accident of 1977. Because of the availability of data from the cockpit voice recorders, accident investigators are able to determine that poor communication, decision making, and situational awareness are contributing factors in many such accidents. Ineffective communication, especially between a pilot and a copilot, also results in a higher risk of accidents. CRM training thus takes into account cultural factors, leadership, teamwork, assertiveness, time management, decision making, and dealing with interruptions. One incident that led to the adoption of CRM training was the crash of United Airlines Flight 173 on December 28, 1978, in Portland, Oregon. This accident occurred because the flight crew was preoccupied with troubleshooting a landing gear issue and did not successfully communicate to the pilot that the airplane was running low on fuel. The aircraft thus ran out of fuel in flight and crashed into a suburban neighborhood. CRM training thus acknowledges the cultural factors that could prevent a copilot from questioning the actions of the captain.

Safety Management System

How much emphasis is given to safety in aviation? Does safety have the same importance as quality? A Safety Management System (SMS) seeks to answer these questions. SMS is an approach to decision making in which safety is a priority for all employees in a business, regardless of their positions or departments. SMS involves all the participants in the aviation field because doing so increases the chances of a safe operation. SMS takes into account factors such as the culture in and the attitude of a business. The culture at all levels of the organization—from leadership positions to entry-level employees—is examined. Another goal of SMS is to integrate modern safety concepts into repeatable, proactive processes in a single system. SMS is designed to transition organizations from having a reactive nature, in which problems are only responded to, to having a proactive and predictive nature. Being proactive means that an organization actively seeks to identify hazardous conditions through analyzing its processes. Being predictive means that an organization analyzes system processes to identify potential or future problems.

The four components of SMS include safety policy, safety assurance, safety risk management, and safety promotion. Safety policy establishes senior management's commitment to continually improve safety and defines the methods, processes, and organizational structure needed to meet safety goals. Safety assurance evaluates the effectiveness of implemented risk control strategies and supports the identification of new hazards. Safety risk management determines, based on the assessment of acceptable risk, the need for, and the adequacy of, new or revised risk controls. Safety promotion includes training, communication, and other activities that create a positive safety culture within all levels of the workforce.

Flight Operational Quality Assurance

One safety program managed by the FAA is Flight Operational Quality Assurance (FOQA). The FOQA program's goal is to proactively identify potential hazard trends among airlines, air operators, airports, and air traffic control. Another goal is to examine, through data analysis, whether regulations are being followed. Airlines send their flight recorder data to the FAA, and the FAA analyzes these data. Note that the data are sent with all identifying information removed so that the analyzers do not know which airline it came from. This anonymity increases the incentive for airlines to send in their data, which ultimately leads to an in-depth, data-based analysis.

The FAA compares the data obtained through FOQA with the data obtained from other sources and its operational experiences to improve procedures and training. Air carriers who wish to implement a FOQA program should prepare an implementation and operations plan (I&O plan). This plan specifies how the carrier will collect FOQA data, how the carrier will take corrective action based on findings from the data, how the carrier will provide the FAA with the data, and how the carrier will inform the FAA of said corrective actions.

THE FUTURE

Next Generation
The field of commercial aviation is growing rapidly. The number of aircraft in the skies increases significantly every year, which in turn increases the strain on existing infrastructure and aviation safety practices. If no changes are made to accommodate this increase in air traffic, passengers will experience increasing delays at airports and a degrading quality of service. In response, the FAA has developed a program called Next Gen. Next Gen is an umbrella term that refers to changes in technology, changes in air management practices, and other changes needed to create a more efficient aviation network.

Many of the advances of Next Gen are in the area of technology. For example, Enhanced Flight Vision Systems (EFVS) is a technology that utilizes an infrared camera on the exterior of an aircraft and displays this video feed to the pilot. This video feed allows the pilot to see objects such as the runway in foggy or dark conditions, which is similar to wearing night vision goggles.

Traditional communication between an aircraft and air traffic controllers relies on analog, voice-based systems. Therefore, another technological improvement is the use of digital communications, which allow images to be sent between aircraft and ground control and allows pilots to see the weather picture that ground controllers see and see it at the same time that ground controllers see it. Known as Network Enabled Weather, this technology could create more efficient flight paths around weather systems, thus saving fuel.

Ground-based air traffic control equipment also needs upgrading. One new system is Airport Surface Detection Equipment, Model X (ASDE-X). This technology provides the air traffic controller with a digital view of all the airplanes currently taxiing or on the runway at a particular airport, which can help prevent congestion and runway incursions. Another seemingly simple technology that can help prevent runway incursions is known as runway status lights. This network of lights built into the runways and taxiways of an airport can alert pilots to stop before a runway incursion occurs.

Another interesting technology is the Automatic Dependent Surveillance Broadcast. This is a satellite global positioning system–based technology that is able to locate an aircraft's position more precisely than radar can. In areas of the world where there is little radar coverage, such as the Gulf of Mexico, these systems can help reduce the mandatory separation distances between aircraft, thus allowing a larger amount of air traffic over the Gulf.

Aircraft Telemetry
A relatively new technology involves the transfer of data from aircraft to ground locations via telemetry. Telemetry has been used in many different fields, and it also has potential and current uses in aviation. For example, aircraft manufacturers are able to build airplanes that can transmit performance data to ground centers that record the data and assist the pilots. Because data are able to be streamed in real time between an aircraft and the ground, telemetry prevents loss of data in the event that a test aircraft is destroyed. Telemetry can be of particular use to design engineers because large amounts of specific parameters can be measured. Telemetry is also used to test unmanned aircraft systems (UASs).

SUMMARY

- Air transportation is one of the safest methods of travel with the lowest incident rate per million people transported.
- The key to the growth of air transportation was the development and technological advancement of reciprocating engines, which were first used in 1903 to power aircraft in flight. The ever-increasing improvement in aviation technology has created faster, higher altitude flights over greater distances accompanied by improvements in aircraft construction and in-flight technology leading to unparalleled safety.
- The growth of the airline industry required both the assistance of the government as well as the need to regulate and guide the development of the industry. In the United States, the first aviation organization to incorporate extensive safety practices was the U.S. Air Mail Service. Air Mail Service's safety practices included abiding by strict criteria when selecting pilots, requiring regular pilot medical examinations, performing stringent aircraft inspections; using a 180-item checklist at the end of every flight; and overhauling engines and aircraft regularly.
- The first aircraft requirements began with the Air Commerce Act of 1926, which called for publicizing the causes of civil aviation accidents, establishing air traffic rules, performing periodic examinations of aircraft, and rating pilots.

- The Department of Commerce developed the Bureau of Air Commerce in 1934, which advanced air safety by encouraging airlines to create air traffic control centers. In 1938, the Civil Aeronautics Act was passed creating the Civil Aeronautics Authority (CAA), which included a three-member Air Safety Board that investigated aircraft accidents and recommended actions to prevent future accidents. The CAA was divided into two agencies, the CAA and the Civil Aeronautics Board (CAB) which became part of the Department of Commerce but remained an independent agency.
- The federal government, under the Federal Aviation Act, created the Federal Aviation Administration (FAA) following a number of aircraft incidents, which killed hundreds of passengers in 1958. The Federal Aviation Act created the Federal Aviation Agency (FAA).
- The National Transportation Safety Board (NTSB) is an independent agency within the U.S. federal government and is tasked with investigating all U.S. aviation accidents in addition to all serious accidents in other transportation areas including rail, highway, and marine transportation; hazardous materials; and the nation's pipelines.
- The International Civil Aviation Organization (ICAO) was established as a specialized agency of the United Nations in 1947. The ICAO has five main objectives: safety, security, environmental protection, air navigation capacity and efficiency, and economic development of air transport.
- Many major airlines and aviation corporations realize the importance of safety and have developed competent, professional departments devoted to aviation safety. These flight safety departments are responsible for creating policies, procedures, and programs that ensure a safe environment for aircrews and provide the world with the safest form of transportation
- The Flight Safety Foundation (FSF) has several characteristics that distinguish it from a trade organization including its status as an international nonprofit whose sole purpose is to provide impartial, independent, expert safety guidance and resources for the aviation and aerospace industry.
- The International Air Transport Association (IATA) represents more than 200 international member airlines before regulators, decision makers, and governments. The IATA's goal is to be the force for value creation and innovation driving a safe, secure, and profitable air transport industry that sustainably connects and enriches the world. The IATA develops regulations on subjects ranging from safety and environmental matters to cargo transport.
- Airlines for America (A4A) is the primary trade organization of the leading U.S. airlines. Some member airlines include American Airlines, Delta Airlines, Federal Express Corporation, Southwest Airlines, and US Airways. Airlines for America advocates on behalf of its member airlines on various issues such as safety and operations, energy and the environment, customer service, and security.
- Crew resource management (CRM) training is an important step in reducing the risks associated with human factors in aviation safety. CRM is the "effective use of all available resources for flight crew personnel to assure a safe and efficient operation, reducing error, avoiding stress, and increasing efficiency."
- Safety Management System (SMS) involves all the participants in the aviation field, which increases the chances of a safe operation. SMS takes into account factors such as the culture and attitude of a business. The culture at all levels of the organization—from leadership positions to entry-level employees—is examined. Another goal of SMS is to integrate modern safety concepts into repeatable, proactive processes in a single system.
- The Flight Operational Quality Assurance (FOQA) program's goal is to proactively identify potential hazard trends among airlines, air operators, airports, and air traffic control, and to examine through data analysis whether regulations are being followed.

REFERENCES

Airlines for America, 1301 Pennsylvania Ave. NW, Suite 1100, Washington DC 20004. airlines.org/about-us/history/.

Aircraft Owners and Pilots Association, 421 Aviation Way, Frederick, Maryland 21701.

    Mission and History of AOPA. www.aopa.org/About-AOPA/Governance/Mission-and-History-of-AOPA.aspx.

European Aviation Safety Agency. www.easa.europa.eu/frequently-asked-questions.php#easa-independant-agency.

Federal Aviation Administration.

    Become a Pilot. www.faa.gov/pilots/become/.

    Brief History. www.faa.gov/about/history/brief_history/.

    Flight Operational Quality Assurance (FOQA). www.faa.gov/about/initiatives/atos/air_carrier/foqa/.

    History of Aviation Safety Oversight in the United States. July 2008. tc.faa.gov/its/worldpac/techrpt/ar0839.pdf.

    NextGen. www.faa.gov/nextgen/.

    Safety Management System. www.faa.gov/about/initiatives/sms/.

Flight Safety Foundation, 801 N. Fairfax Street, Suite 400, Alexandria, VA 22314. flightsafety.org/about-the-foundation.

Hansen, M., Carolyn McAndrews, and Emily Berkeley. *History of Aviation Safety Oversight in the United States*. Washington DC: Air Traffic Organization, Operations and Planning Office of Aviation Research and Development, July 2008.

International Air Transport Association. www.iata.org/about/Pages/index.aspx.

> IATA Operational Safety Audit (IOSA). www.iata.org/whatwedo/safety/audit/iosa/Pages/index.aspx.

> IATA Safety Audit for Ground Operations (ISAGO). www.iata.org/whatwedo/safety/audit/isago/Pages/index.aspx.

International Civil Aviation Organization. www.icao.int/about-icao/Pages/default.aspx

> Vision and Mission. www.icao.int/about-icao/Pages/vision-and-mission.aspx.

International Society of Air Safety Investigators.

History. www.isasi.org/About/History.aspx.

Convention on International Civil Aviation. December 7, 1944. www.mcgill.ca/files/iasl/chicago1944a.pdf.

IOSA Program Manual. 6th ed. Effective May 2014.

ISAGO IATA Safety Audit for Ground Operations: ISAGO Benefits for Airlines. 2014

National Transportation Safety Board, 490 L'Enfant Plaza, SW, Washington DC 20594. www.ntsb.gov/about/index.html

SKYbrary. www.skybrary.aero/index.php/Crew_Resource_Management

The Statues at Large of the United States of America. libraryonline.erau.edu/online-full-text/books-online/aircommerceact1926.pdf.

U.S. Government Publishing Office. Electronic Code of Federal Regulations. ecfr.gov/cgi-bin/text-idx?tpl=/ecfrbrowse/Title49/49tab_02.tpl.

REVIEW QUESTIONS

1. The Air Commerce Act of 1926 was a milestone in the early development of aviation. The act was important in that it was responsible for providing two specific elements to early aviation. What are these two elements?
   a.
   b.

2. What are the three main parts of the FAA's Air Traffic Control operations?
   a.
   b.
   c.

3. The FAA's current responsibilities include a broad range of air transportation safety measures. List five.
   a.
   b.
   c.
   d.
   e.

4. What are the responsibilities of the National Transportation Safety Board (NTSB)?

5. What are the NTSB's four strategic goals regarding aircraft investigations?
   a.
   b.
   c.
   d.

6. What types of injuries are required to be reported to the NTSB under its serious injury category?

7. What is ICAO, and where does it obtain its authority to regulate international aviation programs and policies?

8. What are ICAO's five main objectives?
   a.
   b.
   c.
   d.
   e.

9. What are the responsibilities of commercial airlines flight departments in relation to safety?

10. In order to prevent and to react to potential/real incidents, the airlines develop Accident/Incident Preparedness Programs. List five departments and/or programs within an airline that are involved in these programs.
    a.
    b.
    c.
    d.
    e.

11. Name three aviation-related trade organizations.
    a.
    b.
    c.

12. List the four components of SMS and identify the purpose of each.
    a.
    b.
    c.
    d.

Oil and Gas Safety

J.B. Gregory, MEd
John F. Montgomery, PhD, CSP, CHMM

Historical Overview

Petroleum Operations Overview

Regulatory Safety in the Petroleum Industry
Introduction ▸ Federal Oversight ▸ Offshore Drilling Operations ▸ State-Run Occupational Safety and Health Programs • Trade Associations ▸ National Consensus Standards

Safety and Health Management Program
Management/Employee Commitment and Involvement ▸ Hazard Management Process ▸ Implementation

Hazard Management in Upstream Operations
Introduction ▸ Hazard Recognition ▸ Hazard Identification and Evaluation ▸ Hazard Control

Conclusion

Summary

References

Review Questions

This chapter provides a general overview of the topic of health and safety in the petroleum industry, with specific emphasis on the following areas:
- brief historical overview of the petroleum industry in the United States
- the operational components that make up the oil and gas industry in the United States
- the governmental organizations involved in the regulation of employee safety for the petroleum industry in the United States
- the trade associations and national consensus organizations involved in the regulation of employee safety for the petroleum industry in the United States
- the importance of the hazard management process in oil and gas extraction operations.

HISTORICAL OVERVIEW

The oil and gas industry has existed in the United States for almost 150 years (Figures 26–1 and 26–2). Starting with the drilling of the first commercial oil well in 1859 in Titusville, Pennsylvania, the oil boom began and moved rapidly across the United States (Wall 2014). Over the next 46 years, the United States experienced a growth in the number of crude oil wells that were drilled, which ranged from the East Coast to the West Coast and into Alaska. However, with the sudden increase in the number of drilling operations nationwide, it became quickly apparent that the ability to easily transport the produced crude oil to where it could be refined into more usable and commercially available products created a new set of industry challenges. In response to the growing need for reliable transportation and refining processes, John D. Rockefeller began moving toward the creation of his first refinery in Pennsylvania. In 1867, along with his two partners, Rockefeller founded the Standard Oil Company. Not long after, Standard Oil began developing pipeline and rail networks that would help simplify and connect the producing wells to the refining process (Wall 2014).

When looking at today's modern petroleum industry through the lens of history, it is easy to see that even though there have been tremendous advancements in knowledge and technology, having a drilled well that successfully produces oil, which can then be transported, refined, and sold, is still the industry's ultimate goal—just as it was at the Drake Well in 1859.

PETROLEUM OPERATIONS OVERVIEW

From an operations standpoint, the petroleum industry is divided into three distinct yet interrelated sectors: upstream (also called exploration and production, or E&P), midstream, and downstream operations (see Figure 26–3).

Figure 26–1. Early oil field drilling in West Virginia *(Photo courtesy of West Virginia Geological and Economic Survey; www.wvgs.wvnet.edu/www/geology/geoldvog.htm.)*

Figure 26–2. Early oil field drilling in Pennsylvania. *(Photos.com/Thinkstock)*

Even though each sector is a component that makes up the entire oil and gas industry, the nature of the work being performed and the operational challenges encountered vary greatly (Figures 26–4 through 26–8). From a safety and health perspective, the ways in which each sector's occupational hazards are identified and managed differ as well.

Unfortunately, the variance among the three operational sectors becomes very evident when analyzing the overall number of workplace fatalities occurring throughout the petroleum industry.

According to the U.S. Bureau of Labor Statistics in 2014, the number of fatalities that occurred during oil and gas

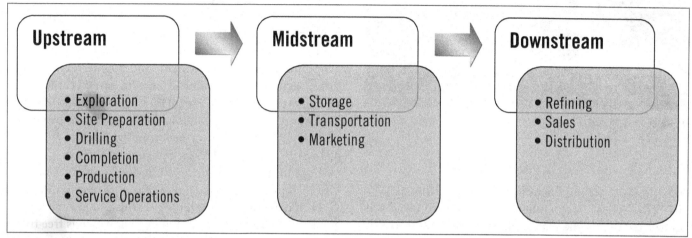

Figure 26–3. The three segments of the oil industry.

Figure 26–4. Upstream: oil field drilling. *(Ingram Publishing/ Thinkstock)*

Figure 26–5. Upstream: oil field pumping. *(Huyangshu/iStock/ Thinkstock)*

Figure 26–6. Midstream: oil and gas storage tanks. *(Jim Parkin/ Hemera/Thinkstock)*

Figure 26–7. Midstream: oil and gas storage tanks. *(Oskari Porkka/ iStock/Thinkstock)*

Figure 26-8. Downstream: oil refining plant. *(David Edwards/iStock/Thinkstock)*

extraction (upstream) operations rose 27% from 2011 to 2012 (BLS 2014). Unfortunately, this upward trend was not isolated to this one year. From 2003 to 2012, there was an increase of more than 67% in the number of workers' deaths. Of the 142 E&P fatalities that occurred in 2012, 68 were related to transportation incidents; 24 resulted when a worker came into contact with objects and equipment; 23 were the result of fire and explosions; 18 were attributed to slips, trips, and falls; and 9 were caused when workers were exposed to either harmful substances or environments. Compare these numbers to the seven total deaths that occurred in downstream operations during the same time period, and the disparity becomes apparent between upstream and downstream approaches and implementations of occupational safety and health programs (BLS 2011).

REGULATING SAFETY IN THE PETROLEUM INDUSTRY

Introduction
Despite the growth in the number of E&P fatalities, worker safety in the petroleum industry is highly regulated. At all levels of government—from local municipalities to state and federal agencies—there are multiple standards that employers must comply with regarding worker safety and the creation of a safe and healthful work environment. Beyond regulatory agencies, there are also petroleum trade associations that set forth recommended safety practices and guidelines for all three sectors of the oil and gas industry.

Federal Oversight
As mentioned in previous chapters, Congress created the Occupational Safety and Health Administration (OSHA) to ensure safe and healthful working conditions for most private-sector employees working in the United States, its territories, or federal jurisdictions. Over its more than 40-year history, OSHA has worked to meet its congressionally mandated goals by setting and enforcing workplace safety standards and by providing training, outreach, and compliance assistance to employers (OSHA 2014).

With regard to the petroleum industry, OSHA does not have a set of specific, or vertical, standards that apply only while conducting oil and gas operations. Therefore, when identifying OSHA regulations that govern the workers' safety in upstream, midstream, and downstream operations, regulatory applicability must be based on the work being performed. OSHA regulations governing workplace safety and health are contained in the following five parts of Title 29 of the *Code of Federal Regulations*: §§ 1910, 1915, 1917, 1918, and 1926 (see Figure 26-9).

Bear in mind that in the absence of industry-specific OSHA regulations, the employer is still required by the OSH Act, Section (5)(a)(1), to "...furnish to his employees employment and a place of employment which is free from recognized hazards that are causing or are likely to cause death or serious physical harm to his employees." This section of the law, which is also known as the General Duty Clause, places the burden of creating and maintaining a safe and healthful working environment upon the employer. Ultimately, if an employer fails to meet this legal requirement, then OSHA may issue a citation accordingly.

Offshore Drilling Operations
Even though this chapter is primarily focused on land-based oil and gas operations, it is important to understand that upstream operations that are conducted in the offshore environment are also regulated with regard to worker safety (Figure 26-10).

While OSHA's primary responsibility is to oversee worker safety and health in the United States (OSHA 2014), when upstream operations are conducted in an offshore environment, not only does the employer fall under the authority of OSHA, it also comes under the jurisdiction

1910	Occupational Safety and Health Standards
1915	Occupational Safety and Health Standards for Shipyard Employment
1917	Marine Terminals
1918	Safety and Health Regulations for Longshoring
1926	Safety and Health Regulations for Construction

Figure 26-9. *Code of Federal Regulations* standards related to safety

Figure 26-10. Offshore drilling rig. *(Svetlana Tebenkova/iStock/Thinkstock)*

of the Bureau of Safety and Environmental Enforcement (BSEE 2014) and the U.S. Coast Guard (OSHA 2014).

Established in 2011 within the Department of the Interior (DOI), BSEE is responsible for the development and enforcement of standards that focus on the safety and environmental responsibility of offshore drilling operations (BSEE 2014) (Figure 26–10). BSEE regulatory requirements that include employee safety for offshore oil and gas drilling activities are found in 30 CFR 250.

Because most offshore drilling operations occur within the boundaries of the Offshore Continental Shelf (OCS), employers conducting drilling operations within these waters are also mandated to comply with safety standards set forth by the U.S. Coast Guard (OSHA 2014). These standards are found by referencing 33 CFR Chapter I.

State-Run Occupational Safety and Health Programs

Section 18 of the OSH Act allows each state, if it chooses, to develop and operate its own state-run safety and health program as long as it has been submitted and approved by the federal OSHA program. If a state chooses to become a "state-plan state," it is required to maintain health and safety regulations that, at a minimum, meet the federal OSHA requirements. Therefore, this allows the state-plan state to have health and safety regulations that may be more stringent than OSHA's federal standards (OSHA 2014). As of 2014, there are 25 states and two territories (Puerto Rico and the Virgin Islands) that maintain oversight of their own occupational safety and health programs (OSHA 2014). So, when operations are occurring in a state-plan state, the employer needs to be aware of any additional health and safety requirements that need to be met to ensure compliance. Even though OSHA does not have a vertical set of standards for the oil and gas industry, there are some state-plan states that have standards to regulate certain oil and gas operations. If a state does not choose to become a state-plan state, it may regulate particular aspects of oil and gas safety-related operations. For example, when drilling in a state that has lands containing hydrogen sulfide gas, the drilling operations might fall under the control of an agency or commission that regulates all aspects of the petroleum sector within the state.

Trade Associations

The health and safety of petroleum workers are not important to just state and federal agencies; the oil and gas industry as a whole is also concerned with the health and safety of its work force. Trade associations serve a vital role throughout the petroleum industry. Not only do they serve to unite their members into one influencing voice, they also assist their membership in creating safer work environments for their employees. There are petroleum trade associations that set forth recommended practices for employee protection when performing specific oil and gas operations. Industry associations such as the American Petroleum Institute (API), the Independent Association of Drilling Contractors (IADC), and the Association of Energy Service Companies (AESC) research, develop, and publish safety- and health-related work practices that specifically address worker safety when performing operations such as oil, gas, and well drilling and servicing involving hydrogen sulfide (OSHA 2014). Once published, these recommendations—or "industry best practices"—may also be used by OSHA to support use of the General Duty Clause.

National Consensus Standards

Separate from the oil and gas trade associations, there are also nationally recognized standard-producing organizations that develop and adopt guidelines that address worker and workplace safety as well as set criteria for the design and testing of safety-related equipment. Examples of these national consensus organizations include the American National Standards Institute (ANSI), the National Fire Protection Association (NFPA), the Compressed Gas Association (CGA), and the American Conference of Governmental Industrial Hygienists (ACGIH). Similar to the recommended practices set by trade associations, national consensus standards can be considered industry best practices. In fact, there are a number of national consensus standards that have been incorporated by reference into the OSHA regulations. A list of these documents can be found by referencing 29 CFR 1910.6 (OSHA 2014) and 29 CFR 1926.6 (OSHA 2014).

SAFETY AND HEALTH MANAGEMENT PROGRAM

Management/Employee Commitment and Involvement

With the increasing number of worker deaths in the upstream sector of the petroleum industry, it is very important that both the employer and employees not only understand each other's responsibilities for creating a safe and healthy workplace, but also commit to continually strive to maintain such an environment.

It is not only the employer's responsibility to provide a safe work environment for employees. The person(s) managing the employer's health and safety program should be aware of the various federal and state standards that must be met and ensure compliance of the employer's health and safety program. Following is a list of employer responsibilities under OSHA (OSHA 2014):

- Provide a workplace free from serious recognized hazards, and comply with standards, rules, and regulations issued under the OSH Act.
- Examine workplace conditions to make sure they conform to applicable OSHA standards.
- Make sure employees have and use safe tools and equipment, and ensure the tools and equipment are properly maintained.
- Use color codes, posters, labels, or signs to warn employees of potential hazards.
- Establish or update operating procedures and communicate them to employees.
- Provide safety training in a language and vocabulary workers can understand.
- When hazardous chemicals are used in the workplace, develop and implement a written hazard communication program and train employees on the hazards they are exposed to and the proper precautions they should take to adequately protect themselves.
- Provide medical examinations and training when required by OSHA standards.
- Post at a prominent location within the workplace the OSHA poster (or the state-plan equivalent) informing employees of their rights and responsibilities.
- Within 8 hours, report to the nearest OSHA office any fatal accident or one that results in the hospitalization of three or more employees.
- Maintain records of work-related injuries and illnesses.
- Provide employees, former employees, and their repre-

sentatives access to the Log of Work-Related Injuries and Illnesses.
- Provide access to employee medical records and exposure records to employees or their authorized representatives.
- Give the OSHA compliance officer the names of authorized employee representatives who may be asked to accompany the compliance officer during an inspection.
- Not discriminate against employees who exercise their rights under the act.
- Post OSHA citations at or near the work area involved.
- Correct cited violations by the deadline set in the OSHA citation, and submit required abatement verification documentation.

In addition, all employees have a responsibility to follow the health and safety policies and procedures required by the employer as well as take a proactive approach to protecting themselves and their co-workers.

Hazard Management Process

An effective hazard management process is a systematic process of identifying, evaluating, and controlling workplace hazards. Figure 26–11 identifies the four elements of the hazard management process (OSHA 1998).

In petroleum operations, the hazard management process is generally documented on a job hazard analysis (JHA) form. The person(s) responsible for completing and signing the JHA often depends on the work being conducted.

Hazard Identification

Many oil and gas employers rely on federal, state, or local regulations to serve as the sole criteria by which they recognize/identify and control workplace hazards. While the practice of "meeting what's required by law" may help the company from a regulatory compliance perspective, it must be remembered that these standards merely set forth the minimum requirements. In other words, hazards could be present that are not identified in the standards but are just as dangerous and have the potential of causing the employee(s) harm or even death.

In all three sectors of the petroleum industry, health and safety hazards exist. Some of these hazards may have already been identified and are being controlled. However, many may not. Therefore, it is important to be able to properly identify the potential for a hazardous condition.

A hazard is defined as "any condition or activity that if left uncontrolled may result in an occupational injury or illness" (OSHA 2002). Due to the numerous hazards that can be found throughout oil and gas operations, it will help to identify worksite hazards based on categories. Figure 26–12 provides a broad framework that is often used to categorize safety and health hazards. Note that a hazard may fall into multiple categories (WHO 2001).

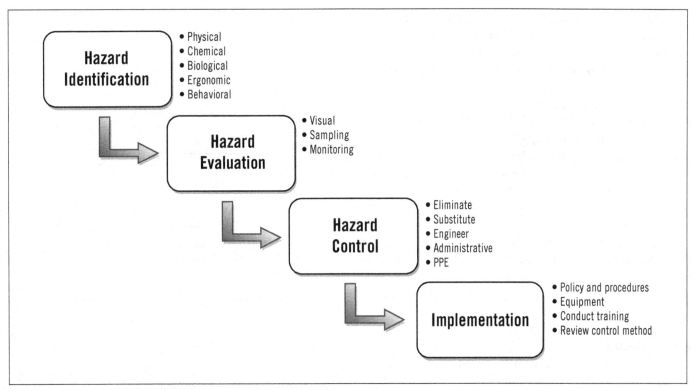

Figure 26–11. Four elements of the hazard management process.

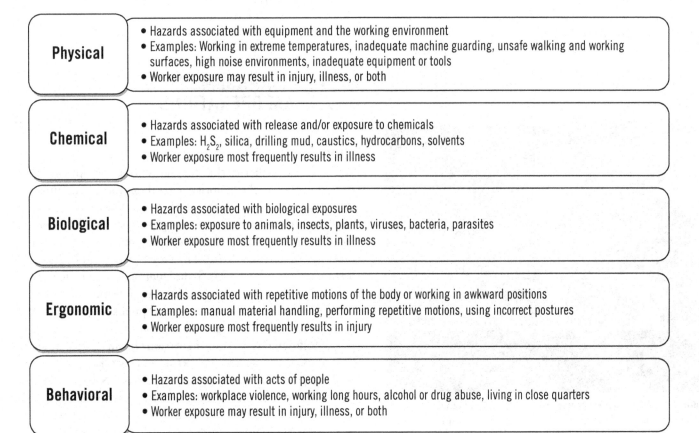

Figure 26–12. Safety and health hazard categories.

When conducting hazard recognition, the potential outcome of employee exposure should be identified, not just the existence of the hazard itself. Employee exposure to workplace hazards may result in injury, illness, or even death (Figure 26-13). Exposures that may result in an employee injury are considered safety related, whereas a hazard that results in employee illnesses is health related. Identifying the potential outcome of employee exposure helps to determine how the hazard should be evaluated and controlled. (For a more detailed explanation, see Chapter 9, Hazard Identification, in the *Administration & Programs* volume.)

Hazard Evaluation

When conducting a hazard evaluation, one of the first questions that should be asked is, "To what extent does the hazard exist?" There are three methods used to determine the degree to which a hazard is present: visual, sampling, and monitoring.

Hazard Control

Once the hazard has been identified and its presence evaluated, the next step is to determine how the hazard will be controlled. Figure 26-14 identifies the hierarchy of control from most to least effective (OSHA 2014).

Implementation

When implementing hazard controls there are four questions that need to be addressed:

1. Does a written policy and procedure need to be developed?
2. Is there any necessary equipment or devices that need to be provided?
3. Does employee training need to be conducted?
4. How will the implementation be reviewed for effectiveness?

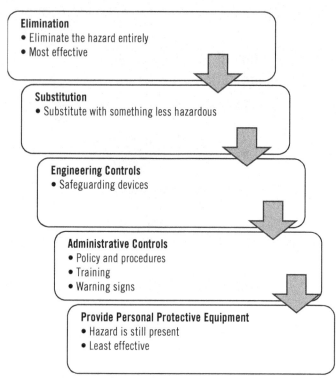

Figure 26-14. Hazard prevention and control.

Figure 26-13. Fire boat response crews battle the blazing remnants of the *Deepwater Horizon* offshore oil rig after an explosion in April 2010. Eleven workers were killed in the blast. *(U.S. Geological Survey/U.S. Coast Guard)*

HAZARD MANAGEMENT IN UPSTREAM OPERATIONS

Introduction

As mentioned earlier in this chapter, extraction operations are responsible for the majority of fatalities that occur annually in the petroleum industry (BLS 2014). The remainder of this chapter focuses on the extraction processes and the identification of general hazards most commonly associated with these operations (BLS 2014). The North American Industry Classification System (NAICS) designation is used by federal statistical agencies to classify business establishments for the purposes of collecting, analyzing, and publishing statistical data related to the U.S. business economy.

The following is a list of the various business groupings that compose the oil and gas extraction sector:
- Oil and gas extraction companies (operator) (NAICS 21111)
  - These companies are engaged in developing, operating, and recovering liquid hydrocarbons up to the point of shipment from the producing property.
- Oil and gas drilling companies (NAICS 213111)

○ These companies are hired contractually to either drill or perform workover of oil and gas wells.

- Support activities for oil and gas operations (service companies) (NAICS 213112)
  ○ These companies are engaged in providing various types of support services for upstream operations.
  ○ Examples of operations performed by these companies include:
    - exploration
    - site preparation
    - well surveying
    - well servicing
    - well cementing
    - well stimulation
    - well perforation
    - wellbore cleaning, treating, and swabbing.

Hazard Recognition

Except for administrative support operations, exploration and production activities are extremely mobile, and worksites and activities vary on a daily basis. Employees working in field operations are exposed to physical, chemical, biological, ergonomic, and behavior-based hazards every day. These hazardous conditions include:

- working long hours, generally 12 hours or more a day
- working outdoors and, with few exceptions, in all weather conditions
- working in environments that utilize heavy equipment
- working with high-pressure oil and gas systems
- working at locations that may contain explosive concentrations of hydrocarbons
- working at locations that contain hydrogen sulfide gas or respirable levels of silica
- working at elevated heights
- working around heavy power equipment
- working around high-voltage electrical systems.

Hazard Identification and Evaluation

Based on the work required for the exploration and production operation, there are several hazards that are present at the job site every day (OSHA 2013). These hazards can then be classified as being potentially harmful to workers' safety or health.

- safety/injury-related hazards:
  ○ explosions
  ○ fires
  ○ falls
  ○ struck by
  ○ caught in
  ○ caught between
  ○ electrocution hazards
- health/illness-related hazards:

  ○ chemical exposure
  ○ substance exposure
  ○ environmental exposure.

Hazard Control

In response to hazards, control measures need to be chosen based on the ability to ideally eliminate the hazard altogether. However, if this is not possible, the health and safety professional must try to identify a way to control the hazard via substitution or with engineering controls. Finally, if substitution or engineering methods are not possible, the health and safety professional should use administrative controls and personal protective equipment. Note: These last two control methods should be considered as a last resort and should be used only until a more effective method of hazard control can be implemented (CDC 2014).

CONCLUSION

Creating and maintaining a safe work environment goes beyond simply complying with regulatory requirements. Within the oil and gas industry, there are health and safety hazards present daily that are not specifically covered by federal standards. It is necessary to remember that federal and state standards merely set forth the minimum requirements for worker safety in the United States. Therefore, companies working in the oil and gas sector (especially those in extraction), along with their employees, need to commit to creating safer work environments and begin to actively manage the hazards associated with their work environment.

SUMMARY

- The first commercial oil well was drilled in 1859 in Titusville, Pennsylvania. Thus, the oil boom began—and moved rapidly across the United States. Over the next 46 years, the United States experienced a dramatic growth in the number of crude oil wells that were drilled; these wells ranged from the East Coast to the West Coast and into Alaska.
- The growth in the number of drilling operations nationwide made it necessary to develop a method to transport crude oil from its source to where it could be refined and, eventually, made commercially available to the public. In response to the growing need for reliable transportation and refining processes, John D. Rockefeller began moving toward the creation of his first refinery in Pennsylvania. In 1867, Rockefeller and his partners founded the Standard Oil Company. The company developed pipeline and rail networks that would help simplify and connect the producing wells to the refining process.

- The current petroleum industry is divided into three sectors: upstream (also called exploration and production [E&P]), midstream, and downstream operations. The operational challenges and the manner in which occupational hazards are identified and managed for each sector vary greatly.
- OSHA does not have a set of specific standards that apply only to oil and gas operations. When identifying OSHA regulations that govern the worker's safety in upstream, midstream, and downstream operations, regulatory applicability must be based on the work being performed.
- When upstream operations are conducted in an offshore environment, not only does the employer fall under the authority of OSHA, but it also comes under the jurisdiction of the Bureau of Safety and Environmental Enforcement and the U.S. Coast Guard.
- Trade associations serve a vital role throughout the petroleum industry. Trade associations such as the American Petroleum Institute (API), the Independent Association of Drilling Contractors (IADC), and the Association of Energy Service Companies (AESC) research, develop, and publish safety- and health-related work practices that specifically address worker safety when performing such operations as oil, gas, and well drilling and servicing involving hydrogen sulfide.
- National consensus organizations include the American National Standards Institute (ANSI), the National Fire Protection Association (NFPA), the Compressed Gas Association (CGA), and the American Conference of Governmental Industrial Hygienists (ACGIH). As with the recommended practices set forth by trade associations, national consensus standards are also considered industry best practices.
- The employer is primarily responsible for providing a safe work environment for its employees. Those managing the employer's health and safety program must be aware of the various federal and state standards that must be met and ensure compliance of the program.
- Identification and evaluation methods to control the hazards must be developed using control measures chosen based on the ability to, ideally, eliminate the hazard altogether. When this is not possible, the health and safety professional must identify a method to control the hazard via substitution or engineering controls. Finally, if substitution or engineering methods are not possible, the health and safety professional should use administrative controls and personal protective equipment.

REFERENCES

U.S. Bureau of Labor Statistics, Postal Square Building, 2 Massachusetts Avenue, Washington DC, 20210-0001.
Fatal Occupational Injuries. 2011. www.bls.gov/iif/ oshwc/cfoi/cftb0268.pdf.
Labor Statistics. 2014. www.bls.gov/iif/oshwc/cfoi/ cfch001pdf.
Labor Statistics. 2014. www.bls.gov/iif/oshwc/cfoi/ osar0018.htm.
NAICS, 2014. www.bls.gov/bls/naics.htm.

U.S. Department of Health and Human Services, Public Health Service, Centers for Disease Control and Prevention, 1600 Clifton Road, Atlanta, GA 30329-4027.
Emergency Controls. 2014. www.cdc.gov/niosh/ topics/engcontrols/.

U.S. Department of the Interior, Bureau of Safety and Environmental Enforcement (BSEE), 1849 C Street NW, Washington DC, 20240.
History Index. 2014. www.bsee.gov/About-BSEE/ BSEE-History/index/.

U.S. Department of Labor, Occupational Safety and Health Administration, 20 Constitution Avenue NW, Washington DC 20210.
Employer Responsibility. 2014. www.osha.gov/as/ opa/worker/employer-responsibility.html.
Gas and Oil Extraction. 2014. www.osha.gov/SLTC/ oilgaswelldrilling/otherresources.html.
Hazard Prevention and Control. 2014. www.osha. gov/SLTC/etools/safetyhealth/comp3.html.
Incorporated by Reform. 2014. www.osha. gov/pls/oshaweb/owadisp.show_document?p_ table=STANDARDS&p_id=9702.
Incorporated by Reform General. 2014. www.osha. gov/pls/oshaweb/owadisp.show_document?p_ table=STANDARDS&p_id.
Industrial Hygiene. 1998. osha.gov/Publications3143/ OSHA3143.html.
Instruction. 2014. www.osha.gov/OshDoc/ Directive_pdf/CPL_02-01-047.pdf.
Job Hazard Analysis. 2002. www.osha.gov/ Publications/osha3071.html.
Laws and Regulations, 2014. osha.gov.
List of State and Health Plans. 2014. www.osha.gov/ dcsp/osp/index.html.
Regional Notice. 2013. www.osha.gov/dep/leps/ RegionVI/reg6_fy2014_Oil-and-Gas_REP_FY14.pdf.
Regulatory Standard. 2014. www.osha.gov/ pls/oshaweb/owasrch.search_form?p_doc_ type=OSHACT&p_toc_level=0.

Wall, B. H. "Oil Industry." History Channel. 2014. history. com/topics/oil-industry.

World Health Organization. Regional Office for the East Mediterranean, Cairo, Egypt, 2001.
Occupational Health, A Manual for Primary Health Workers. 14 WHO-EM/OCH/85/E/L.

REVIEW QUESTIONS

1. Operationally, the petroleum industry is divided into three distinct yet interrelated sectors. List these sections and their components.
   a.
   b.
   c.

2. OSHA regulations governing workplace safety and health are contained in the five parts of Title 29 of the *Code of Federal Regulations*. List these parts and the section regulated.
   a.
   b.
   c.
   d.
   e.

3. When upstream operations are conducted in an offshore environment, not only does the employer fall under the authority of OSHA, but it also falls under the jurisdiction of what two other federal departments?
   a.
   b.

4. List four nationally recognized standard-producing organizations that have developed and have adopted guidelines for oil field workers and workplace safety.
   a.
   b.
   c.
   d.

5. List the three business groupings that make up the oil and gas extraction sector.
   a.
   b.
   c.

6. List at least four conditions or activities that could be hazardous for exploration and production workers.
   a.
   b.
   c.
   d.

Waste and Recycling Safety

27

Janice Comer Bradley, MS, CSP

Philip E. Hagan, JD, MBA, MPH, ARM, CIH, CET, CHMM, CHCM, CHSP, CEM

Introduction

Historical Waste Collection

Waste and Recycling Today
Landfills ▶ Transfer Stations ▶ Convenience Centers ▶ Materials Recovery Facilities

Safety Issues
Exposure to Potentially Hazardous Equipment ▶ Personal Protective Equipment ▶ Exposure to Extreme Temperatures ▶ Traffic ▶ Slips, Trips, and Falls

Ongoing Industry Efforts to Protect Workers

Summary

References

Review Questions

INTRODUCTION

The solid waste industry is historically among the top 10 most hazardous industries, based on the annual incidence rate of fatalities as documented by the U.S. Bureau of Labor Statistics. This industry is made up of three groups: (1) collection, (2) treatment and disposal, and (3) other waste remediation services. According to the Centers for Disease Control and Prevention (CDC), approximately 518,000 workers were employed in the solid waste industry in 2013. About 377,600 of those workers were in private industry, and about 72,500 of those employees were classified as refuse and recyclable materials collectors. Local government agencies employed another 49,000 collectors.

Transportation incidents are the leading causes of occupational fatalities in the waste and recycling industry. Contact with objects and equipment is the second leading cause of fatalities in the industry. Another significant source of worker fatalities occurs when workers are struck by passing vehicles while collecting garbage in residential neighborhoods.

Other safety problems stem from working outdoors in all kinds of weather, coupled with the risk of injury from heavy lifting. According to the Bureau of Labor Statistics, public-sector workers who collect refuse and recyclable material experience a days-away-from-work incidence rate approximately four times greater than the incidence rate for the same occupation in the private sector, largely due to musculoskeletal disorders. Workers across the industry can also be exposed to micro-organisms, chemicals such as hydrogen sulfide, diesel exhaust, and other airborne contaminants at landfills.

Due to the labor-intensive aspect of the work, the waste and recycling industry uses a large number of temporary workers. Industry efforts are currently focused on providing temporary workers with both a basic level of safety awareness training before venturing to a worksite, and site-specific training provided by onsite employers regarding hazards faced once on site.

This chapter will discuss the following topics:
- historical waste collection
- waste and recycling today
- safety issues related to landfills, transfer stations, and materials recovery facilities
- exposure to potentially hazardous equipment
- personal protective equipment
- exposure to extreme temperatures
- traffic
- slips, trips, and falls
- hazardous materials.

HISTORICAL WASTE COLLECTION

Throughout history, waste has been generated by humans. Yesterday's garbage dump was a pit or hill on the outskirts of town that played host to disease-carrying rodents, insects, and dangerous materials. During the Middle Ages, epidemics of disease were attributed to rats and fleas associated with mounds of trash that were disposed of at any convenient location. Following the onset of industrialization and the sustained urban growth of large population centers, the buildup of waste in urban environments caused a rapid deterioration in levels of sanitation and, subsequently, the general quality of life.

In the United States, both incineration and garbage dumps have been the traditional responses to the disposal of solid waste. Before the advent of modern landfills, open burning was often used to deal with waste, potentially exposing workers and the surrounding community to hazardous combustion by-products (Figure 27–1).

Early landfills were sources of contamination that quite often resulted in environmental degradation. Due to environmental safety concerns, municipalities have banned unregulated garbage dumps and burning because they lead to contamination of groundwater supplies, streams, and airways. Regulated landfills are typically a favored form of waste disposal and are used in many places around the world.

WASTE AND RECYCLING TODAY

The modern waste and recycling industry is an increasingly complex, technical operating environment. Traditionally a very labor-intensive environment, new methods are continually being devised for the responsible collection, transportation, containment, processing, treatment, and disposal of discarded or spent materials, defined by the Environmental

Figure 27–1. Open burning of solid waste produces air pollution and the potential for uncontrolled fire. *(ollirg/iStock)*

Figure 27-2. Municipal waste. *(bonottomario/iStock)*

Figure 27-3. A bulldozer moving garbage at a regulated landfill. *(likstudio/iStock)*

Protection Agency (EPA) as "solid waste." These methods both utilize tried-and-true technologies and procedures, which have been the mainstay of traditional operations, and also employ high-technology devices for the separation and movement of specific wastes as well as wastes in combinations and systems.

The need to increase efficiency in recycling and repurposing has also resulted in new technologies and automation. In many cases, the sheer enormity of the waste and recycling process has resulted in a multitude of scenarios in which workers and equipment are working in tandem. Labor-intensive environments, coupled with the use of large machinery (either stationary or mobile), by their nature tend to be areas where safety is a concern because these environments are a primary cause of fatalities in the waste and recycling industry.

Typically, garbage trucks collect waste from either residential or commercial providers. Most people see the end of household garbage when they leave it on the curbside for the garbage collection workers. After garbage is collected, it is transported to either a landfill, transfer station, or materials recovery/recycling facility.

Landfills

Properly licensed and constructed landfills are now the only sanctioned garbage disposal sites for municipalities in the United States. When garbage is taken away from a home or business, it is routed to a landfill, where it becomes part of the unending cycle of waste disposal.

Once the garbage truck arrives at a collection site or landfill, it is weighed and inspected for content. If regulated hazardous materials are found, the waste is typically returned to the originator if possible. If the garbage passes the inspection at the landfill, the garbage is unloaded onto a selected disposal location (the tipping face) and is then compacted by bulldozers and other machinery (Figure 27-2).

The location at a landfill where each day's garbage is placed is referred to as a *cell*. When a cell is filled, temporary coverings are used to seal the area—soil, foam spray, wood chips, and/or other similar materials. This covering keeps rain and wind from dispersing the garbage and controls insects, birds, and rodents from a vector control perspective. Once a landfill section is filled with waste, it is capped and the area is planted with grass or other similar covering (Figure 27-3).

Modern landfills are located in areas where clay deposits and/or other natural geographic features can act to buffer the environment from contamination. Landfills are required to use liners made of plastic, clay, or other nonporous materials to keep garbage by-products from leaking into the soil. Landfill operators employ a system of drainage pipes to route any resulting liquid waste (*leachate*) into nearby ponds or wells, where it is tested and treated to address identified pollutants. Groundwater around landfill sites is quality tested for pollutants for many years after a landfill has reached capacity. To ensure the safety of workers and the surrounding communities, the EPA has developed strict regulations governing the operation and closure of landfills to ensure the prevention of leachate and methane having an impact on the environment.

Newer types of landfills (bioreactors) use leachate and/or air to enhance biodegradation of waste products disposed of inside the landfill. In these types of landfills, leachate is used on site and does not require disposal. Another benefit of bioreactors is the production and collection of methane gas from the decomposition of organic waste. Methane has similar properties to natural gas and can be used as an energy source for fuel or by burning to generate steam and electricity.

The EPA requires the following for landfill design and operation:

- Location restrictions—ensure that landfills are built in suitable geological areas away from faults, wetlands, flood plains, or other restricted areas.
- Composite liner requirements—include a flexible membrane (geomembrane) overlaying 2 ft of compacted clay soil lining the bottom and sides of the landfill, protecting groundwater and the underlying soil from leachate releases.
- Leachate collection and removal systems—sit on top of the composite liner and remove leachate from the landfill for treatment and disposal.
- Operating practices—compact and cover waste frequently with several inches of soil to help reduce odor; control litter, insects, and rodents; and protect public health.
- Groundwater monitoring requirements—require testing groundwater wells to determine whether waste materials have escaped from the landfill.
- Closure and postclosure care requirements—include covering landfills and providing long-term care of closed landfills.
- Corrective action provisions—control and clean up landfill releases and achieve groundwater protection standards.
- Financial assurance—provides funding for environmental protection during and after landfill closure (i.e., closure and postclosure care).

Transfer Stations

Waste transfer stations are facilities where municipal solid waste is unloaded from collection vehicles and briefly held until it is reloaded onto larger, long-distance transport vehicles for shipment to landfills or other treatment or disposal facilities. Waste from customer communities is typically dumped on a wide-open cement space (tipping floor). The waste is then put into a transfer trailer and taken to a regulated disposal site.

By combining the loads of several individual waste collection trucks into a single shipment, the total number of vehicular trips traveling to and from the disposal site is reduced. Although waste transfer stations help reduce the impact of trucks traveling to and from the disposal site, they can cause an increase in traffic in the immediate area in which they are located. If not properly sited, designed, and operated, transfer stations can cause problems for residents living near them with impacts on roads, traffic congestion, and increased air pollution (dust, exhaust, and odors).

Although transfer stations vary significantly, they all serve the same basic purpose—consolidating waste from multiple collection vehicles into larger, high-volume transfer vehicles for more economical shipment to distant disposal sites. In its simplest form, a transfer station is a facility with a designated receiving area where waste collection vehicles discharge their loads. A significant safety issue in this scenario is the presence of moving equipment in close proximity to collection workers or those who work at the transfer station.

The waste is often compacted, and then loaded into larger vehicles (usually transfer trailers, but intermodal containers, railcars, and barges are also used) for long-haul shipment to a final disposal site—typically a landfill, waste-to-energy plant, or a composting facility. No long-term storage of waste occurs at a transfer station; waste is quickly consolidated and loaded into a larger vehicle and moved off site, usually in a matter of hours.

Automation of this process tends to result in a safer environment by separating personnel and heavy equipment. However, the high cost of automation often results in the use of manual labor, increasing the hazards for workers using the equipment. Many communities have installed full-service operations that provide public waste and recyclable drop-off accommodations on the same site as their transfer stations. Source reduction and recycling also play an integral role in a community's total waste management system. These two activities can significantly reduce the weight and volume of waste materials requiring disposal, which reduces transportation, landfill, and incinerator costs. *Source reduction* consists of reducing waste at the source by changing product design, manufacturing processes, and purchasing and sales practices to reduce the quantity or toxicity of materials before they reach the waste stream. Any time the waste volume is reduced at the source, the environment is made safer for workers.

The primary reason for using a transfer station is to reduce the cost of transporting waste to disposal facilities. Consolidating smaller loads from collection vehicles into larger transfer vehicles reduces disposal costs by enabling collection crews to spend less time traveling to and from distant disposal sites and more time collecting waste. This also reduces fuel consumption and vehicle maintenance costs; plus it produces less overall traffic, air emissions, and road wear.

A transfer station also provides an opportunity to screen waste prior to disposal. At many transfer stations, workers screen incoming wastes on conveyor systems, tipping floors, or in receiving pits. Waste screening has two components: (1) separating recyclables from the waste stream and (2) identifying any wastes that might be inappropriate for disposal and potentially harmful to the workers (e.g., hazardous wastes or materials, auto batteries, or infectious waste). Screening operations can result in potential adverse exposures for workers. Much of this screening and sorting process is performed by hand—workers are in close proximity to the potential hazards, and in the case of nonregulated hazardous waste, the potential for exposure could be significant.

According to the National Waste and Recycling Association, materials found in some "nonregulated hazardous" wastes and recyclables have the potential to cause injury to employees if the materials are not managed properly.

Examples of nonregulated hazardous materials include the following:
- discarded chemicals from home environments (e.g., cleaning and pool chemicals, bleach, paint)
- medical wastes (e.g., needles used in home care)
- commercial wastes that contain residual nonregulated hazardous materials (e.g., paint pigments, cement dust)
- industrial wastes that contain silica dust (e.g., foundry sand).

Employees who are permitted to handle "nonregulated hazardous wastes" must be provided with proper hazard recognition training and applicable personal protective equipment (see Chapter 7, Personal Protective Equipment, in this volume). Generally, this training should be included as a part of the employer's Hazard Communication program in accordance with 29 CFR 1910.1200.

Convenience Centers

Only a facility that receives some portion of its waste directly from collection vehicles, and then consolidates and reloads the waste onto larger vehicles for delivery to a final disposal facility, is considered a transfer station. Facilities serving only as citizen drop-off stations or community convenience centers are not considered waste transfer stations, but transfer stations often include convenience centers open to public use.

A *convenience center* is a designated area where residents manually discard waste and recyclables into Dumpsters or collection containers. These containers are periodically removed or emptied, and the waste is transported to the appropriate disposal site (or possibly to a transfer station first). The centers enable individual citizens to deliver waste directly to the transfer station facility for ultimate disposal. Some convenience centers offer programs to manage yard waste, bulky items, household hazardous waste, and recyclables.

It is important to ensure that citizens are protected during the drop-off procedures. In order to protect citizens, strict written guidelines should be put in place to keep residents separated from heavy, mobile equipment that is used to move the waste from the drop-off point to transportation vehicles. These guidelines should be strictly enforced and can be supported by processes as simple as painted lines on the floor or fences to indicate what areas should be avoided. Often, collection of the containers will occur after designated drop-off hours to minimize exposure to heavy equipment for residents who are dropping off waste.

Materials Recovery Facilities

Recyclable materials are prepared for shipment to markets in a special facility called a *materials recovery facility* (MRF), which is a special type of transfer station that separates, processes, and consolidates recyclable materials for shipment to one or more recovery facilities rather than to a landfill or other disposal site. Consequently, the concepts and practices discussed earlier in this chapter can be applied to MRFs as well. Aggressive community source reduction and recycling programs can substantially reduce the amount of waste destined for long-haul transfer and disposal. As indicated before, any reduction in volume results in a safer work environment.

MRFs are specialized facilities that receive, separate, and prepare recyclable materials for marketing to end-user manufacturers. Generally, there are two different types of materials recovery facilities: clean and dirty MRFs.

Clean MRF

A *clean MRF* accepts recyclable commingled materials that have already been separated from municipal solid waste generated by either residential or commercial sources. The most common types of MRFs are single stream or dual stream. In single-stream MRFs, recyclable material is mixed, while in dual-stream MRFs, source-separated recyclables are delivered in a mixed container stream. Material is sorted to specifications, baled, shredded, crushed, compacted, or otherwise prepared for shipment to market (Figure 27–4).

This process is often automated but can still be labor intensive. In all cases, it is important for employers to conduct hazard analysis evaluations to ensure workers are outfitted with proper personal protective equipment (see Chapter 7, Personal Protective Equipment).

Dirty MRF

A *dirty MRF* accepts a mixed solid waste stream that is separated by a combination of manual and mechanical sorting. The sorted recyclable materials may undergo further processing in order to be usable by end markets. Typically,

Figure 27–4. Recycled material packaged and ready for transport. *(paulprescott72/iStock)*

the remainder of the unsorted mixed waste stream is sent to a disposal facility, usually a landfill.

The dirty MRF process is necessarily labor intensive and can be especially challenging from a safety perspective. While some separation of some unacceptable materials may have occurred during initial collection, additional separation may occur when the materials are first unloaded at the MRF. Employees and machines at an MRF separate recyclable materials from the materials for disposal. Sorters may encounter sharp objects and hazardous materials during the separation. In addition, they may also encounter materials that should not be in the MRF (e.g., diabetic needles in plastic containers).

Operators of front loaders, forklifts, and skid-steer loaders can also be involved in materials handling. Generally, these operators are not in direct contact with the recyclables or unusable waste. However, these employees should remain alert for potential hazards when materials are unloaded or moved. A significant safety concern emerges when heavy mobile equipment is moving in the materials-handling area—workers are sorting materials while trying to avoid being struck by the moving equipment. Rules for separation of workers and mobile equipment should be strictly enforced (Figure 27–5).

When potentially hazardous materials are identified and removed from recyclables and unusable waste, they should be removed in a manner that does not endanger the worker performing the removal or other workers in the immediate work area. The material should be removed to a secure area where it will not create a hazard or be reintroduced into the recyclable materials or usable waste until it can be disposed of in an appropriate manner.

Employees who perform the collection and segregation of potentially hazardous materials may be classified as spill responders or as emergency responders and need to be trained in accordance with 29 CFR 1910.120, Hazardous Waste Operations and Emergency Response. These employees must be provided with the appropriate personal protective equipment (PPE) (see Chapter 7, Personal Protective Equipment, in this volume) and the specialized training (see Chapter 31, Safety and Health Training, in the *Administration & Programs* volume) that is required to safely handle and store these materials.

Wet materials recovery facility technologies are also used. This technology combines the mixed waste stream at a dirty MRF with water, which acts to stratify and clean the output streams. It also crushes and dissolves biodegradable organics to make them more suitable for anaerobic digestion. The use of water tends to make surfaces slippery, so either nonskid surfaces or special footwear should be used.

SAFETY ISSUES

Facility design, coupled with good operating practices, helps ensure landfills, transfer stations, and materials recovery facilities are safe places. These facilities should be designed and operated for the safety of employees, customers, and even persons illegally trespassing when the facility is closed. Most state regulations require security and access-control measures such as fences and gates that can be closed and locked after hours. Signs should be posted around the perimeter, with warnings about potential risks due to falls and contact with waste. Signs should be posted in multiple languages in jurisdictions with high percentages of non-English-speaking residents.

Federal Occupational Safety and Health Administration (OSHA) regulations require covered facilities to provide safe working conditions for all employees. Although regulations specific to landfills, waste transfer stations, and materials recovery facilities do not currently exist, general OSHA regulations apply—as they would to any other covered facility. State, tribal, and local workplace safety regulations, which can be more stringent than federal regulations, also might apply, depending on the jurisdiction. Some state, tribal, or local governments might require a facility's development permit to directly address employee and customer safety. State and tribal solid waste regulations, for instance, often require development of operating plans and contingency plans to address basic health and safety issues. Safety issues are the facility operator's responsibility. A facility must take steps to eliminate or reduce risk of injury from any identified hazards.

Exposure to Potentially Hazardous Equipment

Waste and recycling workers often complete their tasks in close proximity to a variety of hazards, including heavy equipment and machinery with moving parts such as conveyor belts, push blades, balers, and compactors. Facility operators

Figure 27–5. It is important to keep workers and visitors separated from operating mobile heavy equipment. *(padnpen/iStock)*

should develop an employee equipment orientation program and establish safety programs to minimize the risk of injury from both mobile and stationary equipment. Using locks or tags that prevent equipment from operating until the locks or tags are removed (lockout/tagout systems), for example, effectively minimizes hazards associated with energized machinery (see Chapter 5, Legal and Regulatory Issues for the Safety Manager, in the *Administration & Programs* volume). Facility operators must implement and strictly enforce rules requiring visiting children and pets to remain in the vehicle at all times. Posting signs and applying brightly colored paint or tape to hazards can alert customers to potential dangers.

Personal Protective Equipment

Employees coming in close contact with waste and heavy machinery should wear appropriate PPE. Common pieces of protective gear include hard hats, protective eye goggles, dust masks, steel-tipped boots, and protective gloves. If working in close proximity to loud machinery, hearing protection should be used as well. Check state and local codes and regulations to see if any additional personal protective equipment standards exist (see Chapter 7, Personal Protective Equipment). Ensure that all facility employees are using and properly maintaining appropriate equipment.

Exposure to Extreme Temperatures

Facilities located in areas of extreme weather must account for potential impacts to employees from prolonged exposure to heat or cold. Heat exhaustion and heatstroke are addressed with proper facility operations, including acclimation, good ventilation, access to water and shade, and periodic work breaks. Cold weather is addressed with proper clothing, protection from wind and precipitation, and access to warming areas. Extreme temperatures typically should not pose problems for customers because their exposure times are typically much less than those of facility workers.

Traffic

Controlled, safe traffic flows in and around the facility are critical to ensuring employee and customer safety. Ideally, a transfer station is designed so traffic from large waste-collecting vehicles is kept separate from self-haulers, who typically use cars and pickup trucks. Facility planners should consider the following issues when designing a facility:

- Direct traffic flow in a one-way loop through the main transfer building and around the entire site. Facilities with one-way traffic flow have buildings (and sometimes entire sites) with separate entrances and exits. The transfer trailers, in particular, are difficult to maneuver and require gentle slopes and sufficient turning radii. Ideally, these vehicles should not have to back up.
- Arrange buildings and roads on the site to eliminate or

minimize intersections, the need to back up vehicles, and sharp turns.
- Provide space for vehicles to queue when incoming traffic flow is greater than the facility's tipping area can accommodate. Sufficient queuing areas should be located after the scale house and before the tipping area. This is in addition to, and separate from, any queuing area required before the scale house to prevent traffic from backing up onto public roads.
- Provide easily understood and highly visible signs, pavement markings, and directions from transfer station staff to indicate proper traffic flow.
- Provide bright lighting, both artificial and natural, inside buildings. Using light-colored interior finishes that are easy to keep clean is also very helpful. When entering a building on a bright day, drivers' eyes need time to adjust to the building's darker interior. This adjustment period can be dangerous. Good interior lighting and light-colored surfaces can reduce the contrast and shorten adjustment time.
- Provide an area for self-haulers to unload separately from large trucks. Typically, self-haulers must manually unload the back of a pickup truck, car, or trailer. This process takes longer than the automated dumping of commercial waste collection vehicles and potentially exposes the driver to other traffic. It is often a good idea to provide staff to assist the public with safe unloading practices.
- Require facility staff to wear bright or conspicuous clothing. Personnel working in the tipping area especially must wear high-visibility clothing at all times.
- Install backup alarms on all moving facility equipment and train all vehicle operators in proper equipment operation safety. Backup alarms must be maintained in proper working condition at all times. Cameras and monitors can also be installed as an additional precaution.

Slips, Trips, and Falls

Accidental falls are another concern for facility employees and customers, especially in facilities with pits or direct dump designs where the drop at the edge of the tipping area might be 5 to 15 ft deep. Facilities with flat tipping areas offer greater safety in terms of reducing the height of falls, but they present their own hazards. These include standing and walking on floor surfaces that could be slippery from recently dumped waste material and being in close proximity to station operating equipment that removes waste after each load is dumped. Depending on the station design (pit or flat floor), a number of safety measures should be considered to reduce the risk of falls.

- For direct gravity loading of containers by citizens, a moderate grade separation will reduce the fall distance. For example, some facilities place rolloff boxes, used to collect waste below grade to facilitate easy loading of waste into the container, so that the top of the rolloff box is even with the surrounding ground. This approach, however, creates a fall hazard into an empty rolloff box. Alternatively, the rolloff box can be set about below grade, with the sides extending 3 ft or so feet above the floor. This height allows for relatively easy lifting over the box's edge yet is high enough to reduce the chance of accidental falls.

- For pit-type operations, the pit depth can be tapered to accommodate commercial unloading at the deep end and public unloading at the shallow end.
- Safety barriers, such as chains or ropes, can be placed around the pit edges at the end of the day or during cleaning periods to prevent falls.
- Wheel stops can be installed on the facility floor to prevent vehicles from backing into a pit or bin. Some curbs are removable to facilitate cleaning.
- Locating wheel stops a good distance from the edge of the unloading zone ensures that self-haul customers will not find themselves dangerously close to a ledge or the operating zone for station equipment.
- To prevent falls due to slipping, the floor should be cleaned regularly and designed with a skid-resistant surface. Designers need to provide sufficient slope in floors and pavements so that they drain readily and eliminate standing water. This is especially crucial in cold climate areas, where icing can cause an additional fall hazard. Because of transfer stations' large size and volume, and because of the constant flow of vehicles, it is impractical to design and operate the stations as heated facilities.
- Use of colored floor coatings (such as bright red or yellow) in special hazard zones (including the area immediately next to a pit) can provide customers with a strong visual cue.
- Designing unloading stalls with a generous width for self-haul customers maximizes the separation between adjacent unloading operations and reduces the likelihood of injury from activity in the next stall. For commercial customers, ensure that stall widths provide a similar safety cushion for the large vehicles. This is particularly necessary where self-haul and commercial stalls are located side by side.
- If backing movements are required, design the facility so vehicles can back in from the driver's side (i.e., left to right) to increase visibility.

ONGOING INDUSTRY EFFORTS TO PROTECT WORKERS

Several entities have the stated focus of enhancing safety efforts in the waste and recycling industry. The National Waste and Recycling Association (NWRA), which deals primarily with the private segment of the industry, provides educational resources on how to make the solid waste industry safer. A very successful marketing and education campaign—the "Slow Down to Get Around" campaign—is designed to protect garbage collectors out on their routes.

The Solid Waste Association of North America (SWANA), which represents primarily the municipal segment, has publications and training courses related to recycling and other aspects of the solid waste industry.

The mission of the Environmental Research and Education Foundation (EREF) is to develop and evaluate new approaches to manage municipal solid waste and to provide scholarships to students doing research in the waste and recycling area. As part of that mission, they provide educational offerings that enhance safety in that sector.

The National Institute for Occupational Safety and Health (NIOSH) has a National Occupational Research Agenda (NORA) Services Sector group that is working on safety issues in the North American Industry Classification System (NAICS) Administrative Support and Waste Management sector.

SUMMARY

- The solid waste industry is historically among the top 10 most hazardous industries and consists of three groups: (1) collection, (2) treatment and disposal, and (3) other waste remediation services.
- Transportation incidents, such as collisions and rollovers, are the leading causes of occupational fatalities in all three industry groups. Another leading cause of fatalities is when collection workers have been struck and killed by other motorists while out on their routes.
- Contact with objects and equipment is the second leading cause of fatalities in each of the industry groups. This category includes being struck by, struck against, or caught in between objects and equipment.
- Musculoskeletal injuries are the most prevalent injuries in the waste and recycling industry.
- Workers across the industry can also be exposed to micro-organisms, chemicals such as hydrogen sulfide, diesel exhaust, and other airborne contaminants in the course of their work.
- Due to the labor-intensive aspect of the work, the waste and recycling industry uses a large number of temporary workers. Industry efforts are currently focused on providing temporary workers with both a basic level of safety awareness training before venturing to a worksite, and site-specific training provided by onsite employers regarding hazards faced once on site.
- Traditionally a very labor-intensive environment, the modern waste and recycling industry is an increasingly complex, technical operating environment. Labor-intensive environments, coupled with the use of large machinery,

tend to be areas where safety is always a concern because these environments are a primary cause of fatalities in the waste and recycling industry.

- Typically, garbage trucks collect waste from either residential or commercial providers. After garbage is collected, it is transported to a landfill, transfer station, or materials recovery/recycling facility.
- Waste and recycling workers coming in close contact with waste and heavy machinery should wear appropriate personal protective equipment (PPE).
- Facilities located in areas of extreme weather must account for potential impacts to workers from prolonged exposure to heat or cold.
- Controlled, safe traffic flows in and around the facility are critical to ensuring worker and customer safety.
- Accidental falls are another concern for facility workers and customers.
- The National Waste and Recycling Association (NWRA), the Solid Waste Association of North America (SWANA), the Environmental Research and Education Foundation (EREF), and the National Institute for Occupational Safety and Health's (NIOSH) National Occupational Research Agenda (NORA) Services Sector group are some of the industry leaders that are working on safety issues in the waste and recycling sector.

REFERENCES

American National Standards Institute (ANSI), American National Standard for Mobile Refuse Collection and Compaction Equipment—Safety Requirements, ANSI Z245.1–2014.

Centers for Disease Control and Prevention, 1600 Clifton Rd., Atlanta, GA 30329-4027. cdc.gov/niosh/topics/solidwaste/.

Environmental Protection Agency, Office of Resource Conservation and Recovery, 1200 Pennsylvania Avenue NW, Washington DC 20460

40 CFR Part 258, Criteria for Municipal Solid Waste Landfills. epa.gov/solidwaste/nonhaz/municipal/landfill.htm.

40 CFR Part 261.2, Definition of Solid Waste. epa.gov/epawaste/hazard/dsw/index.htm.

Environmental Research and Education Foundation, 3301 Benson Drive, Suite 301, Raleigh, NC 27609.

Finstein, M. S., Y. Zadik, A. T. Marshall, and D. Brody. The ArrowBio Process for Mixed Municipal Solid Waste: Responses to "Requests for Information." Proceedings of the 1st UK Conference and Exhibition on Biodegradable and Residual Waste Management. Harrogate, England, February 2004. E. K. Papadimitriou and E. I. Stentiford eds. Leeds, UK: CalRecovery Europe Ltd, 407–13.

National Waste and Recycling Association, 4301 Connecticut Avenue NW, Suite 300, Washington DC 20008. Manual of Recommended Safety Practices.

Neal, H. A., and J. R. Schubel. Solid Waste Management and the Environment: The Mounting Garbage and Trash Crisis. January 1987.

Solid Waste Association of North America, 1100 Wayne Avenue, Suite 650, Silver Spring, MD 20910.

Standard Occupational Classification, 53-7080. Refuse and Recyclable Material Collectors. bls.gov/soc/2000/soc_v7i0.htm.

U.S. Bureau of Labor Statistics, Occupational Employment Statistics. bls.gov/oes/current/naics3_562000.htm.

U.S. Department of Labor, Occupational Safety and Health Administration, 200 Constitution Ave. NW, Washington DC 20210.

29 CFR 1910.120, Hazardous Waste Operations and Emergency Response.

29 CFR 1910.147, The Control of Hazardous Energy (Lockout/Tagout).

REVIEW QUESTIONS

1. After garbage is collected, what are three possible destinations for the garbage to end up?
   a.
   b.
   c.

2. List at least four of the EPA general guidelines for landfill design and operation.
   a.
   b.
   c.
   d.

3. What is the main purpose of a transfer station, and what is the primary safety concern?

4. Materials found in some "nonregulated hazardous" wastes and recyclables have the potential to cause injury to employees if the materials are not managed properly. List four examples of nonregulated hazardous wastes.
   a.
   b.
   c.
   d.

5. Define *clean MRF* and *dirty MRF*.
   a.
   b.
6. List two simple ways to alert customers to potential dangers associated with a waste and recycling facility.
   a.
   b.

7. How can a worker minimize hazards from extreme heat?
8. List four entities that are currently working to enhance safety efforts in the waste and recycling industry.
   a.
   b.
   c.
   d.

APPENDIX 1

Safety and Health Tables

References to Useful Handbooks of Engineering
Tables and Formulas

Table 1—Factors of Safety for Common Construction
Materials

Table 2—Safe Bearing Loads on Soils

Table 3—Specific Gravity of Gases and Liquids

Table 4—Approximate Specific Gravities and Densities

Table 5—Levels of Illumination

Table 6—Hot and Cold Work Environments

Table 7—Coefficients of Friction of Floors and Shoes

Table 8—Selected Common Abbreviations

Table 9—Table of Unit Prefixes

Table 10—Signs and Symbols

The safety and health professional requires information on many subjects. Specialized material must be sought out as required when the safety and health professional must deal with a specific subject in great detail and for a specific location. This latter information includes (1) applicable federal, state, provincial, and local code requirements that are beyond the scope of this Manual and that must be checked locally and (2) specific scientific and engineering information that generally is available in handbooks devoted to specific subject fields. For the reader's convenience in developing this additional information, a number of these handbooks are listed in the References section of this appendix and in Appendix 1, Sources of Help, in the *Administration & Programs* volume.

Engineers and others have long recognized that equipment must be stronger than is really necessary. Actual work conditions are usually worse than testing conditions, so a safety factor is often applied. By definition, this factor of safety is the ratio of the ultimate (breaking) strength of a member or piece of material to the actual working stress or to the maximum permissible (safe load) stress when in use. The magnitude of this factor depends on how great the cost of failure will be in terms of life or damage.

The selection of a safety factor is very important. The higher the number, the better, but a trade-off is always required. Cost, weight, supporting structure, speed or power requirement, size, and hazards involved are some of the common trade-offs considered.

The safety margin is the numerical value over 0 that results from strength divided by maximum stress. Typical factors of safety include the following:

	Factor of Safety
Stairs, landings	4
Standard railing	*
Scaffold, supporting member	4
Boiler—new	5
—Used	5.5
—10 years used	6
Unpressured vessel	5
Refrigerating system	5
Hydraulic pressure, piping, and hose	8
Cranes—hook	4-5
—Gears	8
—Structural steel	5
—Hot work	10
General hoisting equipment	5-8
Cast iron flywheel	10
Wood flywheel	20

*OSHA requires standard railings to withstand at least 200-lb (890-N) pressure applied in any direction at any point on the rail—§1910.23e(2)(v).

REFERENCES TO USEFUL HANDBOOKS OF ENGINEERING TABLES AND FORMULAS

Note: Because engineering and scientific handbooks are revised often, as frequently as once a year in some cases, no attempt has been made to list the latest edition in the following bibliography. In ordering, however, the most recent edition should be requested.

Alexander, J. M., and R. C. Brewer. *Manufacturing Properties of Materials*. Princeton, NJ 08540, VanNostrand Co., Inc.

Allegheny Ludlum Steel Corp., Pittsburgh, PA 15222. *Tool Steel Handbook*.

AM Best Company, Oldwick NJ 08858. *Best's Safety Directory*. 2 vols.

American National Metric Council, 1625 Massachusetts Avenue NW, Washington DC 20036. "Metrication for the Manager."

American Society for Testing and Materials, 1916 Race Street, Philadelphia, PA 19103. "Standard Metric Practice Guide," ASTM E380.

American Society of Heating, Refrigerating, and Air-Conditioning Engineers, 1791 Tullie Circle NE, Atlanta, GA, 30329. *Guide and Data Books: Applications, Handbook of Fundamentals, Systems and Equipment*.

Baumeister, T., III, and E. A. Avallone, eds. *Standard Handbook for Mechanical Engineers*. New York, NY 10036, McGraw-Hill Book Co.

Bennett, H., ed. *Concise Chemical and Technical Dictionary*. New York, NY 10003, Chemical Publishing Co., Inc.

Compressed Air Magazine Co, Phillipsburg, NJ 08865. *Compressed Air Data*.

Eshbach, O. W. *Handbook of Engineering Fundamentals*. New York, NY 10016, John Wiley & Sons.

Gardner, W., E. I. Cooke, and R. W. I. Cooke. *Handbook of Chemical Synonyms and Trade Names*. Cleveland, OH 44128, CRC Press.

Hudson, R. G. *The Engineer's Manual*. New York, NY 10016, John Wiley & Sons, Inc.

Illuminating Engineering Society of North America, 345 East 47th Street, New York, NY 10017. *IES Lighting Handbook (The Standard Lighting Guide)*.

Kidder, F. E., and H. Parker. *Kidder-Parker Architects' and Builders' Handbook*. New York, NY 10016, John Wiley & Sons, Inc.

Knowlton, A. E., ed. *Standard Handbook for Electrical Engineers*. New York, NY 10036, McGraw-Hill Book Co.

Kurtz, E. B., and T. M. Shoemaker. *The Linemans' and Cablemans' Handbook*. New York, NY 10036, McGraw-Hill Book Co.

LaLonde, W. S., Jr., and M. F. Janes. *Concrete Engineering Handbook*. New York, NY 10036, McGraw-Hill Book Co.

LeGrand, R, ed. *The New American Machinists' Handbook*. New York, NY 10036, McGraw-Hill Book Co.

Liebers, A. *The Engineer's Handbook Illustrated*. Los Angeles, CA 90047, Key Publishing Co.

Lindsey, F. R. *Pipefitters Handbook*. New York, NY 10016, The Industrial Press.

Mantell, C. L., ed. *Engineering Materials Handbook*. New York, NY 10036, McGraw-Hill Book Co.

Maynard, H. B. *Industrial Engineering Handbook*. New York, NY 10036, McGraw-Hill Book Co.

Morris, I. E. *Handbook of Structural Design*. New York, NY 10022, Reinhold Publishing Corp.

Morrow, L. C., ed. *Maintenance Engineering Handbook*. New York, NY 10036, McGraw-Hill Book Co.

National Association of Home Builders. *Construction Dictionary*.

National Fire Protection Association, 1 Batterymarch Park, Quincy, MA 02269. *Fire Protection Handbook*.

Oberg, E., and F. D. Jones. *Machinery's Handbook*. New York, NY 10016, The Industrial Press.

Pender, H., et al. *Electrical Engineers' Handbook*. New York, NY 10016, John Wiley & Sons, Inc.

Perry, J. H., and R. H. Perry. *Engineering Manual*. New York, NY 10036, McGraw-Hill Book Co.

Perry, R. H., et al., eds. *Chemical Engineers' Handbook*. New York, NY 10036, McGraw-Hill Book Co.

Rosaler, R., ed. *Standard Handbook of Plant Engineering* New York, NY 10036, McGraw-Hill Book Co.

Stanier, W. *Mechanical Power Transmission Handbook*. New York, NY 10036, McGraw-Hill Book Co.

Urquhart, L. C., ed. *Civil Engineering Handbook*. New York, NY 10036, McGraw-Hill Book Co.

Wilson, F. W., and P. D. Harvey, eds. *Tool Engineers Handbook*. New York, NY 10036, McGraw-Hill Book Co.

TABLE 1. Factors of Safety for Common Construction Materials

Material	Steady Load	Load Varying from Zero to Maximum in One Direction	Load Varying from Zero to Maximum in Both Directions	Suddenly Varying Loads and Shocks
Cast iron	6	10	15	20
Wrought iron	4	6	8	12
Steel	5	6	8	12
Wood	8	10	15	20
Brick	15	20	25	30
Stone	15	20	25	30

Reprinted with permission from *Machinery's Handbook*, 16th ed., The Industrial Press.

TABLE 2. Safe Bearing Loads on Soils

Nature of Soil	Safe Bearing Capacity (ton/ft²)
Solid ledge of hard rock such as granite, trap, etc.	25–100
Sound shale or other medium rock, requiring blasting for removal	10–15
Hardpan, cemented sand, and gravel; difficult to remove by picking	8–10
Soft rock, disintegrated ledge; in natural ledge, difficult to remove by picking	5–10
Compact sand and gravel, requiring picking for removal	4–6
Hard clay, requiring picking for removal	4–5
Gravel, coarse sand, in natural thick beds	4–5
Loose, medium, and coarse sand; fine compact sand	1.5–4
Medium clay, stiff but capable of being spaded	2–4
Fine loose sand	1–2
Soft clay	1

Reprinted with permission from *Standard Handbook for Mechanical Engineers*, rev. 9th ed., Edited by T. Baumeister III and E. A. Avallone. (Copyright 1987, McGraw-Hill Book Co.) Values approximate pressures allowed in major city building codes.

776   Appendix 1 Safety and Health Tables

TABLE 3. Specific Gravity of Gases and Liquids

Gas	Specific Gravity	Gas	Specific Gravity
Air	1.000	Hydrochloric acid	1.261
Acetylene	0.920	Hydrogen	0.069
Ethyl alcohol vapor	1.601	Mercury vapor	6.940
Ammonia	0.592	Marsh gas	0.555
Carbon dioxide	1.520	Methane	0.554
Carbon monoxide	0.967	Nitrogen	0.971
Chlorine	2.423	Nitric oxide	1.039
Ethane	1.049	Nitrous oxide	1.527
Ether vapor	2.586	Oxygen	1.106
Ethylene	0.967	Propane	1.554
Helium	0.138	Sulfur dioxide	2.250
Hydrofluoric acid	2.370	Water vapor	0.623

1 ft^3 of air at 32°F and atmospheric pressure weighs 0.0807 lb.

Liquid	Specific Gravity	Liquid	Specific Gravity
Acetic acid	1.06	Linseed oil	0.94
Alcohol, commercial	0.83	Mineral oil	0.92
Ammonia	0.77	Naphtha	0.76
Benzene	0.88	Olive oil	0.92
Bromine	2.97	Palm oil	0.97
Carbolic acid	0.96	Petroleum oil	0.82
Carbon disulfide	1.26	Phosphoric acid	1.78
Cottonseed oil	0.93	Rape oil	0.92
Fluoric acid	1.50	Vinegar	1.08
Gasoline	0.70	Water	1.00
Glycerin	1.26	Whale oil	0.92
Kerosene	0.80		

Reprinted with permission from *Machinery's Handbook,* 20th ed., The Industrial Press.
Gases at 32°F air = 1,000 liquids: water = 1,000
         1 ft^3 water at 39°F weighs 62.43 lb; 1 mL water at 4°C weighs 1 g

TABLE 4. Approximate Specific Gravities and Densities

Substance	Specific Gravity	Average density (lb/ft³)	Substance	Specific Gravity	Average density (lb/ft³)
Metals, Alloys, Ores			Sulfur	1.93–2.07	125
Aluminum, cast-hammered	2.55–2.80	165	Wool	1.32	82
Aluminum, bronze	7.7	481	**Timber, air-dry**		
Brass, cast-rolled	8.4–8.7	534	Apple	0.66–0.74	44
Bronze, 7.9 to 14% Sn	7.4–8.9	509	Ash, black	0.55	34
Bronze, phosphor	8.88	554	Ash, white	0.64–0.71	42
Copper, cast rolled	8.8–8.95	556	Birch, sweet, yellow	0.71–0.72	44
Copper ore, pyrites	4.1–4.3	262	Cedar, white, red	0.35	22
German silver	8.58	536	Cherry, wild red	0.43	27
Gold, cast-hammered	19.25–19.35	1205	Chestnut	0.48	30
Gold coin (U.S.)	17.18–17.2	1073	Cypress	0.45–0.48	29
Iridium	21.78–22.42	1383	Fir, douglas	0.48–0.55	32
Iron, gray cast	7.03–7.13	442	Fir, balsam	0.40	25
Iron, cast, pig	7.2	450	Elm, white	0.56	35
Iron, wrought	7.6–7.9	485	Hemlock	0.45–0.50	29
Iron, Spiegel-eisen	7.5	468	Hickory	0.74–0.80	48
Iron, ferrosilicon	6.7–7.3	437	Locust	0.67–0.77	45
Iron ore, hematite	5.2	325	Mahogany	0.56–0.85	44
Iron ore, limonite	3.6–4.0	237	Maple, sugar	0.68	43
Iron ore, magnetite	4.9–5.2	315	Maple, white	0.53	33
Iron slag	2.5–3.0	172	Oak, chestnut	0.74	46
Lead	11.34	710	Oak, live	0.87	54
Lead ore, galena	7.3–7.6	465	Oak, red, black	0.64–0.71	42
Manganese	7.42	475	Oak, white	0.77	48
Manganese ore, pyrolusite	3.7–4.6	259	Pine, Oregon	0.51	32
Mercury	13.546	847	Pine, red	0.48	30
Monel metal, rolled	8.97	555	Pine, white	0.43	27
Nickel	8.9	537	Pine, Southern	0.61–0.67	38–42
Platinum, cast-hammered	21.5	1330	Pine, Norway	0.55	34
Silver, cast-hammered	10.4–10.6	656	Poplar	0.43	27
Steel, cold-drawn	7.83	489	Redwood, California	0.42	26
Steel, machine	7.80	487	Spruce, white, red	0.45	28
Steel, tool	7.70–7.73	481	Teak, African	0.99	62
Tin, cast-hammered	7.2–7.5	459	Walnut, black	0.59	37
Tin ore, cassiterite	6.4–7.0	418	Willow	0.42–0.50	28
Tungsten	19.22	1200	**Various Liquids**		
Zinc, cast rolled	6.9–7.2	440	Alcohol, ethyl (100%)	0.789	49
Zinc, ore, blende	3.9–4.2	253	Alcohol, methyl (100%)	0.796	50
Various Solids			Acid, muriatic (HCl), 40%	1.20	75
Cereals, oats, bulk	0.41	26	Acid, nitric, 91%	1.50	94
Cereals, barley, bulk	0.62	39	Acid, sulfuric, 87%	1.80	112
Cereals, corn, rye, bulk	0.73	45	Chloroform	1.500	95
Cereals, wheat, bulk	0.77	48	Ether	0.736	46
Cork	0.22–0.26	15	Lye, soda, 66%	1.70	106
Cotton, flax, hemp	1.47–1.50	93	Oils, vegetable	0.91–0.94	58
Fats	0.90–0.97	58	Oils, mineral, lubricants	0.88–0.94	57
Flour, loose	0.40–0.50	28	Turpentine	0.861–0.867	54
Flour, pressed	0.70–0.80	47	Water, 4°C, max. density	1.0	62.428
Glass, common	2.40–2.80	162	Water 100°C	0.9584	59.830
Glass, plate or crown	2.45–2.72	161	Water, ice	0.88–0.92	56
Glass, crystal	2.90–3.00	184	Water, snow, fresh fallen	0.125	8
Glass, flint	3.2–4.7	247	Water, sea water	1.02–1.03	64
Hay and straw, bales	0.32	20	**Ashlar Masonry**		
Leather	0.86–1.02	59	Granite, syenite, gneiss	2.4–2.7	159
Paper	0.79–1.15	58	Limestone	2.1–2.8	153
Potatoes, piled	0.67	44	Marble	2.4–2.8	153
Rubber, caoutchouc	0.92–0.96	59	Sandstone	2.0–2.6	143
Rubber, goods	1.0–2.0	94	Bluestone	2.3–2.6	153
Salt, granulated, piled	0.77	48	**Rubble Masonry**		
Saltpeter	2.11	132	Granite, syenite, gneiss	2.3–2.6	153
Starch	1.53	96	Limestone	2.0–2.7	147

TABLE 4. *Continued.*

Substance	Specific Gravity	Average density (lb/ft³)	Substance	Specific Gravity	Average density (lb/ft³)
Sandstone	1.9–2.5	137	Borax	1.7–1.8	109
Bluestone	2.2–2.5	147	Chalk	1.8–2.8	143
Marble	2.3–2.7	156	Clay, marl	1.8–2.8	143
Dry Rubble Masonry			Dolomite	2.9	181
Granite, syenite, gneiss	1.9–2.3	130	Feldspar, orthoclase	2.5–2.7	162
Limestone, marble	1.9–2.1	125	Gneiss	2.7–2.9	175
Sandstone, bluestone	1.8–1.9	110	Granite	2.6–2.7	165
Brick Masonry			Greenstone, trap	2.8–3.2	187
Hard brick	1.8–2.3	128	Gypsum, alabaster	2.3–2.8	159
Medium brick	1.6–2.0	112	Hornblende	3.0	187
Soft brick	1.4–1.9	103	Limestone	2.1–2.86	155
Sand-lime brick	1.4–2.2	112	Marble	2.6–2.86	170
Concrete Masonry			Magnesite	3.0	187
Cement, stone, sand	2.2–2.4	144	Phosphate rock, apatite	3.2	200
Cement, slag, etc.	1.9–2.3	130	Porphyry	2.6–2.9	172
Cement, cinder, etc.	1.5–1.7	100	Pumice, natural	0.37–0.90	40
Various Building Mat'ls			Quartz, flint	2.5–2.8	165
Ashes, cinders	0.64–0.72	40–45	Sandstone	2.0–2.6	143
Cement, Portland, loose	1.5	94	Serpentine	2.7–2.8	171
Portland cement	3.1–3.2	196	Shale, slate	2.6–2.9	172
Lime, gypsum, loose	0.85–1.00	53–64	Soapstone, talc	2.6–2.8	169
Mortar, lime, set	1.4–1.9	103 / 94	Syenite	2.6–2.7	165
Mortar, Portland cement	2.08–2.25	135	**Stone, Quarried, Piled**		
Slags, bank slag	1.1–1.2	67–72	Basalt, granite, gneiss	1.5	96
Slags, bank screenings	1.5–1.9	98–117	Limestone, marble quartz	1.5	95
Sags, machine slag	1.5	96	Sandstone	1.3	82
Slags, slag sand	0.8–0.9	49–55	Shale	1.5	92
Earth, etc., Excavated			Greenstone, hornblende	1.7	107
Clay, dry	1.0	63	**Bituminous Substances**		
Clay, damp, plastic	1.76	110	Asphaltum	1.1–1.5	81
Clay and gravel, dry	1.6	100	Coal, anthracite	1.4–1.8	97
Earth, dry, loose	1.2	76	Coal, bituminous	1.2–1.5	84
Earth, dry, packed	1.5	95	Coal, lignite	1.1–1.4	78
Earth, moist, loose	1.3	78	Coal, peat, turf, dry	0.65–0.85	47
Earth, moist, packed	1.6	96	Coal, charcoal, pine	0.28–0.44	23
Earth, mud, flowing	1.7	108	Coal, charcoal, oak	0.47–0.57	33
Earth, mud, packed	1.8	115	Coal, coke	1.0–1.4	75
Riprap, limestone	1.3–1.4	80–85	Graphite	1.64–2.7	135
Riprap, sandstone	1.4	90	Paraffin	0.87–0.91	56
Riprap, shale	1.7	105	Petroleum	0.87	54
Sand, gravel, dry, loose	1.4–1.7	90–105	Petroleum, refined (kerosene)	0.78–0.82	50
Sand, gravel, dry, packed	1.6–1.9	100–120	Petroleum, benzene	0.73–0.75	46
Sand, gravel, wet	1.89–2.16	126	Petroleum, gasoline	0.70–0.75	45
Excavations in Water			Pitch	1.07–1.15	69
Sand or gravel	0.96	60	Tar, bituminous	1.20	75
Sand or gravel and clay	1.00	65	**Coal and Coke, Piled**		
Clay	1.28	80	Coal, anthracite	0.75–0.93	47–58
River mud	1.44	90	Coal, bituminous, lignite	0.64–0.87	40–54
Soil	1.12	70	Coal, peat, turf	0.32–0.42	20–26
Stone riprap	1.00	65	Coal, charcoal	0.16–0.23	10–14
Minerals			Coal, coke	0.37–0.51	23–32
Asbestos	2.1–2.8	153			
Barytes	4.50	281			
Basalt	2.7–3.2	184			
Bauxite	2.55	159			
Bluestone	2.5–2.6	159			

At room temperature with reference to water at 39°F
1lb/ft³ = 16.02 kg/m³

Reprinted with permission from *Standard Handbook for Mechanical Engineers,* rev. 9th ed. Edited by T. Baumeister III and E. A. Avallone. (Copyright 1987, McGraw-Hill Book Co.)

TABLE 5. Levels of Illumination

This material is adapted from ANSI/IES standard RP1, RP7. The values represent guides, and other factors—such as light source, colors in the work area, employee abilities, etc.—must also be considered. They are not regulatory minimum nor maximum standards. The reader is advised to consult RP7, Illumination Levels for Industry. Some currently recommended illuminances for industrial facility exteriors include:

Area/Activity	Lux	Footcandles	Area/Activity	Lux	Footcandles
Building (construction)			Rough difficult seeing	500	50
General construction	100	10	Medium	1000	100
Excavation work	20	2	Fine	5000	500
Building exteriors			Extra fine	10000	1000
Entrances			First manufacturing operations (first cut)		
Active (pedestrian and/or conveyance)	50	5	Marking, sheering, sawing	500	50
Dredging	20	2	Flight test and delivery area		
Loading and unloading platforms	200	20	On the horizontal plane	50	5
Freight car interiors	100	10	On the vertical plane	20	2
Lumber yards	10	1	Automotive Industry Facilities		
Parking areas			Elevators, steel furnace areas, locker rooms, exterior active storage areas	200	20
Main plant parking	20	2	Waste treatment facilities (interior), clay mold and kiln rooms	300	30
Secondary parking	10	1	Frame assembly, powerhouse, forgings, quick service dining, casting pouring and sorting	500	50
Ship yards			Control and dispatch rooms, kitchens, large casting core and molding areas (engines), machining operations (engine and parts)	750	75
Active	200	20	Chassis, body and component assembly	1000	100
Inactive	10	1	Parts inspection stations	1500	150
Warehousing	100-500	10-50	Final assembly, body finishing and assembly, difficult inspection, paint color comparison	2000	200
Aircraft Maintenance			Fine difficult inspection (casting cracks)	5000	500
Close up			Iron and Steel Industry		
Install plates, panels, fairings	750	75	Open hearth		
Seal plates	750		Stock yard	100	10
Paint (exterior or interior of aircraft where plates, panels, fairings, cowls, etc., must be in place before accomplishing)	750-1000	75	Charging floor	200	20
Docking			Mold yard	50	5
Position doors and control surfaces for docking	300	30	Hot top	300	30
Move aircraft into position in dock	500	50	Hot top storage	100	10
Maintenance, modification and repairs to airframe structures	750-1000	150	Rolling mills		
Specialty shops			Blooming, slabbing, hot strip hot sheet	300	30
Instruments, radio	1500-1000	150	Motor room, machine room	300	30
Electrical	1500	150	Inspection		
Hydraulic and pneumatic	1000	100	Black plate, bloom and billet chipping	1000	100
Components	1000	100	Tin plate and other bright surfaces	2000	200
Aircraft Manufacturing			Rubber Tires and Mechanical Rubber Goods		
Fabrication (preparation for assembly)			Rubber tire manufacturing		
Rough bench work and sheet metal operations such as shears, presses, punches, countersinking, spinning	500	50	Banbury	300	30
Drilling, riveting, screw fastening	750	75	Calendering		
Medium bench work and machining such as ordinary automatic machines, rough grinding, medium buffing and polishing	1000	100	General	300	30
			Letoff and windup	500	50
Fine bench work and machining such as ordinary automatic machines, rough grinding, medium buffing and polishing	5000	500	Stock cutting		
			General	300	30
Extra fine bench and machine work	10000	1000	Cutters and splicers	1000	100
Final assembly such as placing of motors, propellers, wing sections, landing gear	1000	100	Rubber goods—mechanical		
			Stock preparation		
General			Plasticating, milling, Banbury	300	30
Rough easy seeing	300	30	Calendering	500	50

TABLE 6. Hot and Cold Work Environments

Air Temperature and Relative Humidity versus Apparent Temperature

Relative Humidity (%)

Air Temperature (°F)	0	5	10	15	20	25	30	35	40	45	50	55	60	65	70	75	80	85	90	95	100
140	125																				
135	120	128																			
130	117	122	131																		
125	111	116	123	131	141																
120	107	111	116	123	130	139	148														
115	103	107	111	115	120	127	135	143	151												
110	99	102	105	108	112	117	123	130	137	143	150										
105	95	97	100	102	105	109	113	118	123	129	135	142	149								
100	91	93	95	97	99	101	104	107	110	115	120	126	132	138	144						
95	87	88	90	91	93	94	96	98	101	104	107	110	114	119	124	130	136				
90	83	84	85	86	87	88	90	91	93	95	96	98	100	102	106	109	113	117	122		
85	78	79	80	81	82	83	84	85	86	87	88	89	90	91	93	95	97	99	102	105	108
80	73	74	75	76	77	77	78	79	79	80	81	81	82	83	85	86	86	87	88	89	91
75	69	69	70	71	72	72	73	73	74	74	75	75	76	76	77	77	78	78	79	79	80
70	64	64	65	65	66	66	67	67	68	68	69	69	70	70	70	70	71	71	71	71	72

(Apparent Temperature)

Wind-Chill Chart

Estimated Wind Speed (MPH)	Actual Thermometer Reading (°F)												
	50	40	30	20	10	0	−10	−20	−30	−40	−50	−60	
	Equivalent Temperature (°F)												
Calm	50	40	30	20	10	0	−10	−20	−30	−40	−50	−60	
5	48	37	27	16	6	−5	−15	−26	−36	−47	−57	−68	
10	40	28	16	4	−9	−21	−33	−46	−58	−70	−83	−95	
15	36	22	9	−5	−18	−36	−45	−58	−72	−85	−99	−112	
20	32	18	4	−10	−25	−39	−53	−67	−82	−96	−110	−124	
25	30	16	0	−15	−29	−44	−59	−74	−88	−104	−118	−133	
30	28	13	−2	−18	−33	−48	−63	−79	−94	−109	−125	−140	
35	27	11	−4	−20	−35	−49	−67	−82	−98	−113	−129	−145	
40	26	10	−6	−21	−37	−53	−69	−85	−100	−116	−132	−148	

Wind speeds greater than 40 MPH have little additional effect	Little danger for properly clothed person	Increasing Danger	Great Danger (exposed flesh may freeze within 30 seconds)
		Danger from Freezing of Exposed Flesh	

The wind-chill chart indicates the importance of wearing proper attire to combat injury to exposed skin, even though the temperature may be comparatively mild. To convert Fahrenheit degrees to Celsius, subtract 32 from the Fahrenheit reading; then take five-ninths of that figure.

(Courtesy—American Petroleum Institute)

TABLE 7. Coefficients of Friction of Floors and Shoes

Coefficient of Friction	Floors	Floor Clean	Floor Soiled	Shoes: Soles
1.0	Soft rubber pad	0.8	0.6	Rubber-cork
0.8	End grain wood	0.75	0.55	U.S. Army–U.S. Air Force standard
0.7	Concrete, rough finish	0.7	0.5	Rubber-crepe
0.65	Working decorative, dry	0.6	0.4	Neoprene
0.5	Working decorative, soiled	0.5	0.3	Leather
0.4	Steel			
				Shoes: Heels
		0.7	0.5	Neoprene
		0.65	0.45	Nylon

Developed from Kroemer and Robinson, 1971; Kroemer, 1974.

TABLE 8. Selected Common Abbreviations

Selected Common Abbreviations

°A	Angstrom unit of length	L or l	liter
abs	absolute	L	lambert(s)
amb	ambient	lb	pound
amp	ampere	LP-gas	liquefied petroleum gas
app mol wt	apparent molecular weight	log	logarithm (common)
atm	atmospheric	ln	logarithm (natural)
at wt	atomic weight	m	meter
Be`	degrees Baume`	mA	milliampere
bp	boiling point	MAC	maximum allowable concentration
bbl	barrel	max	maximum
Btu	British thermal unit	mp	melting point
Btuh	Btu per hour	μ	micron
c	cycles per second (see Hz)	mks system	meter-kilogram-second system
cal	calorie	mph	miles per hour
cfh	cubic feet per hour	mg	milligram
cfm	cubic feet per minute	mL	milliliter
cfs	cubic feet per second	mm	millimeter
cg	centigram	mm (Hg)	mm of mercury
cm	centimeter	mμ	millimicron
cgs system	centimeter-gram-second system	mppcf	million particles per cu ft
conc	concentrated, concentration	mr	millirem
cc, cm³	cubic centimeter	mR	1/1000 roentgen
cu ft, ft³	cubic foot	min	minute or minimum
cu in.	cubic inch	mol wt, MW	molecular weight
° or deg	degree	muc	maximum use concentration
°C	degree Centigrade, degree Celsius	nm	nanometer
°F	degree Fahrenheit	N	newton
K	degree Kelvin	OD	outside diameter
R	degree Reaumur, degree Rankine	oz	ounce
dB	decibel	ppb	parts per billion
ET	effective temperature	pphm	parts per hundred million
ft	foot	ppm	parts per million
ft-c	foot-candle	psf	pounds per square foot
ft-lb	foot-pound	psi	pounds per square inch
fpm	feet per minute	psia	pounds per square inch absolute
fps	feet per second	psig	pounds per square inch gauge
fps system	foot-pound-second system	rem	roentgen equivalent man
fp	freezing point	rpm	revolution per minute
gal	gallon	sec	second
gr	grain	sp gr	specific gravity
g	gram	sp ht	specific heat
gpm	gallons per minute	sp wt	specific weight
Hz	hertz (cycles per second)	scf	standard cubic foot
hp	horsepower	STP	standard temperature and pressure
hr	hour	temp	temperature
ID	inside diameter	TLV	Threshold Limit Value
in.	inch	ton	NOT t or tn. Do not abbreviate.
kcal	kilocalorie	v	volt
kg	kilogram	W	watt
km	kilometer	wt	weight
liq	liquid	yr	year

Note: Symbols are always written in singular form. Unabbreviated units form plurals in the usual manner.

Appendix 1 Safety and Health Tables

TABLE 9. Table of Unit Prefixes

Multiples and Submultiples	Prefixes	Symbols
$1,000,000,000,000 = 10^{12}$	tera-	T
$1,000,000,000 = 10^{9}$	giga-	G
$1,000,000 = 10^{6}$	mega-	M
$1,000 = 10^{3}$	kilo-	k
$100 = 10^{2}$	hecto-	h
$10 = 10$	deka-	D
$0.1 = 10^{-1}$	deci-	d
$0.01 = 10^{-2}$	centi-	c
$0.001 = 10^{-3}$	milli-	m
$0.000001 = 10^{-6}$	micro-	μ
$0.000000001 = 10^{-9}$	nano-	n
$0.000000000001 = 10^{-12}$	pico	p

Data from National Bureau of Standards.

TABLE 10. Signs and Symbols

+	plus, addition, positive	$\sqrt{}$	square root
−	minus, subtraction, negative	$\sqrt[n]{}$	nth root
±	plus or minus, positive or negative	a^{n}	nth power of a
∓	minus or plus, negative or positive	$\log, \log_{10}$	common logarithm
÷, /, —	division	$\ln, \log_{e}$	natural logarithm
×, •, ( ) ( ), *	multiplication	e or ε	base of natural logs, 2.718
( ) [ ]	collection	π	pi, 3.146
=	is equal to	∠	angle
≠	is not equal to	⊥	perpendicular to
≡	is identical to	‖	parallel to
≅	equals approximately, congruent	n	any number
>	greater than	$|n|$	absolute value of n
≯	not greater than	$\bar{n}$	average value of n
≧	greater than or equal to	a^{-n}	reciprocal of nth power of a, of a, or $\{1/a^{n}\}$
<	less than	n°	n degrees (angle)
≮	not less than	n′	n minutes, n feet
≦	less than or equal to	n″	n seconds, n inches
∷	proportional to	f(x)	function of x
:	ratio	Δx	increment of x
~	similar to	dx	differential of x
∝	varies as, proportional	Σ	summation of
→	approaches	sin	sine
∞	infinity	cos	cosine
∴	therefore	tan	tangent

APPENDIX 2

Conversion of Units

Fundamental Units

Helpful Organizations

Conversion of Units
Fahrenheit-Celsius Conversion Table

All physical units of measurement can be reduced to three basic dimensions—mass, length, and time. Not only does reducing units to these basic dimensions simplify the solution of problems, but standardization of units makes comparison between operations (and between operations and standards) easier.

For example, air flows are usually measured in liters per minute, cubic meters per second, or cubic feet per minute. The total volume of air sampled can be easily converted to cubic meters or cubic feet. In another situation, the results of atmospheric pollution studies and stack sampling surveys are often reported as grains per cubic foot, grams per cubic foot, or pounds per cubic foot. The degree of contamination is usually reported as parts of contaminant per million parts of air.

If physical measurements are made or reported in different units, they must be converted to the standard units if any comparisons are to be meaningful.

To save time and space in reporting data, many units have standard abbreviations. Because the metric system (SI) is becoming more frequently used, conversion factors are given for the standard units of measurement.

FUNDAMENTAL UNITS

Conversion factors for various measurement units are listed in the tables in this section. To use a table to find the numerical value of the quantity desired, locate the unit to be converted in the first column. Then multiply this value by the number appearing at the intersection of the row and the column containing the desired unit. The answer will be the numerical value in the desired unit.

Various English system and metric system units are given for the reader's convenience. The accepted system of measurement, however, is the International System of Units (SI). The official conversion factors and an explanation of the system are given to 6- or 7-place accuracy in ASTM Standard E 380-76 (ANSI Z210.0–1976).

Briefly, the SI system being used throughout the world is a modern version of the MKSA (meter, kilogram, second, ampere) system. Its details are published and controlled by an international treaty organization, the International Bureau of Weights and Measures (BIPM), set up by the Metre Convention signed in Paris, France, on May 20, 1875. The United States and Canada are member states of this Convention, as implemented by the Metric Conversion Act of 1975 (Public Law 94-168).

HELPFUL ORGANIZATIONS

The following four groups in the United States and Canada are deeply involved in planning and implementing metric conversion:

American National Metric Council
5410 Grosvenor Lane
Bethesda, MD 20814

Metric Commission Canada
240 Sparks Street
Ottawa, Ontario, Canada K1A 0H5

U.S. Metric Association Inc.
Boulder, CO 80302

Office of Metric Programs
U.S. Department of Commerce
Washington DC 20230

CONVERSION OF UNITS

Fahrenheit-Celsius Conversion Table

Fahrenheit-Celsius Conversion—A simple way to convert a Fahrenheit temperature reading into a Celsius temperature reading or vice versa is to enter the accompanying table on the following page in the center or boldface column of figures. These figures refer to the temperature in either Fahrenheit or Celsius degrees. If it is desired to convert from Fahrenheit to Celsius degrees, consider the center column as a table of Fahrenheit temperatures and read the corresponding Celsius temperature in the column on the left. If it is desired to convert from Celsius to Fahrenheit degrees, consider the center column as a table of Celsius values and read the corresponding Fahrenheit temperature on the right.

To convert from "degrees Fahrenheit" to "degrees Celsius" (formerly called "degrees centigrade"), use the formula:

$$t_c = (t_f - 32)/1.8 \text{ or } {}^5/_9(t_f - 32)$$

Conversely,

$$t_f = 1.8t_c + 32 \text{ or } {}^5/_9 t_c + 32$$

For example, convert the boiling point of water in Fahrenheit to Celsius:

$$212°F - 32 = 180$$

$${}^5/_9(180) = 100°C$$

TABLE 1. Fahrenheit-Celsius Conversion

°C		°F	°C		°F	°C		°F	°C		°F
−273	**−459.4**	…	−17.8	**0**	32	9.4	**49**	120.2	35.6	**96**	204.8
−268	**−450**	…	−17.2	**1**	33.8	10.0	**50**	122.0	36.1	**97**	206.6
−262	**−440**	…	−16.1	**3**	37.4	10.6	**51**	123.8	36.7	**98**	208.4
−257	**−430**	…	15.6	**4**	39.2	11.1	**52**	125.6	37.2	**99**	210.2
−251	**−420**	…	15.0	**5**	41.0	11.7	**53**	127.4	38.3	**101**	213.8
−246	**−410**	…	14.4	**6**	42.8	12.2	**54**	129.2	38.9	**102**	215.6
−240	**−400**	…	−13.9	**7**	44.6	12.8	**55**	131.0	39.4	**103**	217.4
−234	**−390**	…	−13.3	**8**	46.4	13.3	**56**	132.8	40.0	**104**	219.2
−229	**−380**	…	−12.8	**9**	48.2	13.9	**57**	134.6	40.6	**105**	221.0
−223	**−370**	…	−12.2	**10**	50.0	14.4	**58**	136.4	41.1	**106**	222.8
−218	**−360**	…	−11.7	**11**	51.8	15.0	**59**	138.2	41.7	**107**	224.6
−212	**−350**	…	−11.1	**12**	53.6	15.6	**60**	140.0	42.2	**108**	226.4
−207	**−340**	…	−10.6	**13**	55.4	16.1	**61**	141.8	42.8	**109**	228.2
−201	**−330**	…	−10.0	**14**	57.2	16.7	**62**	143.6	43.3	**110**	230.0
−196	**−320**	…	−9.4	**15**	59.0	17.2	**63**	145.4	43.9	**111**	231.8
−190	**−310**	…	−8.9	**16**	60.8	17.8	**64**	147.2	44.4	**112**	233.6
−184	**−300**	…	−8.3	**17**	62.6	18.3	**65**	149.0	45.0	**113**	235.4
−179	**−290**	…	−7.8	**18**	64.4	18.9	**66**	150.8	45.6	**114**	237.2
−173	**−280**	…	−7.2	**19**	66.2	19.4	**67**	152.6	46.1	**115**	239.0
−169	**−273**	−459.4	−6.7	**20**	68.0	20.0	**68**	154.4	46.7	**116**	240.8
−168	**−270**	−454	−6.1	**21**	69.8	20.6	**69**	156.2	47.2	**117**	242.6
−162	**−260**	−436	−5.6	**22**	71.6	21.1	**70**	158.0	47.8	**118**	244.4
−157	**−250**	−418	−5.0	**23**	73.4	21.7	**71**	159.8	48.3	**119**	246.2
−151	**−240**	−400	−4.4	**24**	75.2	22.2	**72**	161.6	48.9	**120**	248.0
−146	**−230**	−382	−3.9	**25**	77.0	22.8	**73**	163.4	49.4	**121**	249.8
−140	**−220**	−364	−3.3	**26**	78.8	23.3	**74**	165.2	50.0	**122**	251.6
−134	**−210**	−346	−2.8	**27**	80.6	23.9	**75**	167.0	50.6	**123**	253.4
−129	**−200**	−328	−2.2	**28**	82.4	24.4	**76**	168.8	51.1	**124**	255.2
−123	**−190**	−310	−1.7	**29**	84.2	25.0	**77**	170.6	51.7	**125**	257.0
−118	**−180**	−292	−1.1	**30**	86.0	25.6	**78**	172.4	52.2	**126**	258.8
−112	**−170**	−274	−0.6	**31**	87.8	26.1	**79**	174.2	52.8	**127**	260.6
−107	**−160**	−256	0−	**32**	89.6	26.7	**80**	176.0	53.3	**128**	262.4
−101	**−150**	−238	0.6	**33**	91.4	27.2	**81**	177.8	53.9	**129**	264.2
−96	**−140**	−220	1.1	**34**	93.2	27.8	**82**	179.6	54.4	**130**	266.0
−90	**−130**	−202	1.7	**35**	95.0	28.3	**83**	181.4	55.0	**131**	267.8
−84	**−120**	−184	2.2	**37**	98.6	28.9	**84**	183.2	55.6	**132**	269.6
−79	**−110**	−166	3.3	**38**	100.4	29.4	**85**	185.0	56.1	**133**	271.4
−73	**−100**	−148	3.9	**39**	102.2	30.0	**86**	186.8	56.7	**134**	273.2
−68	**−90**	−130	4.4	**40**	104.0	30.6	**87**	188.6	57.2	**135**	275.0
−62	**−80**	−112	5.0	**41**	105.8	31.1	**88**	190.4	57.8	**136**	276.8
−57	**−70**	−94	5.6	**42**	107.6	31.7	**89**	192.2	58.3	**137**	278.6
−51	**−60**	−76	6.1	**43**	109.4	32.2	**90**	194.0	58.9	**138**	280.4
−46	**−50**	−58	6.7	**44**	111.2	32.8	**91**	195.8	59.4	**139**	282.2
−40	**−40**	−40	7.2	**45**	113.0	33.3	**92**	197.6	60.0	**140**	284.0
−34	**−30**	−22	7.8	**46**	114.8	33.9	**93**	199.4	60.6	**141**	285.8
−29	**−20**	−4	8.3	**47**	116.6	34.4	**94**	201.2	61.1	**142**	287.6
−23	**−10**	14	8.9	**48**	118.4	35.0	**95**	203.0	61.7	**143**	289.4

Appendix 2 Conversion of Units

TABLE 1. Fahrenheit-Celsius Conversion

°C		°F	°C		°F	°C		°F	°C		°F
62.2	144	291.2	85.0	185	365.0	176.7	350	662.0	404.4	760	1400
62.8	145	293.0	85.6	186	366.8	182.2	360	680.0	410	770	1418
63.3	146	294.8	86.1	187	368.6	187.8	370	698.0	415.6	780	1436
63.9	147	296.6	86.7	188	370.4	193.3	380	716.0	421.1	790	1454
64.4	148	298.4	87.2	189	372.2	198.9	390	734.0	426.7	800	1472
65.0	149	300.2	87.8	190	374.0	204.4	400	752.0	432.2	810	1490
65.6	150	302.0	88.3	191	375.8	210	410	770.0	437.8	820	1508
66.1	151	303.8	88.9	192	377.6	215.6	420	788	443.3	830	1526
66.7	152	305.6	89.4	193	379.4	221.1	430	806	448.9	840	1544
67.2	153	307.4	90.0	194	381.2	226.7	440	824	454.4	850	1562
67.8	154	309.2	90.6	195	383.0	232.2	450	842	460.0	860	1580
68.3	155	311.0	91.1	196	384.8	237.8	460	860	465.6	870	1598
68.9	156	312.8	91.7	197	386.6	243.3	470	878	471.1	880	1616
69.4	157	314.6	92.2	198	388.4	248.9	480	896	476.7	890	1634
70.0	158	316.4	92.8	199	390.2	254.4	490	914	482.2	900	1652
70.6	159	318.2	93.3	200	392.0	260.0	500	932	487.8	910	1670
71.1	160	320.0	93.9	201	393.8	265.6	510	950	493.3	920	1688
71.7	161	321.8	94.4	202	395.6	271.1	520	968	498.9	930	1706
72.2	162	323.6	95.0	203	397.4	276.7	530	986	504.4	940	1724
72.8	163	325.4	95.6	204	399.2	282.2	540	1004	510.0	950	1742
73.3	164	327.2	96.1	205	401.0	287.8	550	1022	515.6	960	1760
73.9	165	329.0	96.7	206	402.8	293.3	560	1040	521.1	970	1778
74.4	166	330.8	97.2	207	404.6	298.9	570	1058	526.7	980	1796
75.0	167	332.6	97.8	208	406.4	304.4	580	1076	532.2	990	1814
75.6	168	334.4	98.3	209	408.2	310.0	590	1094	537.8	1000	1832
76.1	169	336.2	98.9	210	410.0	315.6	600	1112	565.6	1050	1922
76.7	170	338.0	99.4	211	411.8	321.1	610	1130	593.3	1100	2012
77.2	171	339.8	100.0	212	413.6	326.7	620	1148	621.1	1150	2102
71.8	172	341.6	104.4	220	428.0	332.2	630	1166	648.9	1200	2192
78.3	173	343.4	110.0	230	446.0	337.8	640	1184	676.7	1250	2282
78.9	174	345.2	115.6	240	464.0	343.3	650	1202	704.4	1300	2372
79.4	175	347.0	121.1	250	482.0	348.9	660	1220	732.2	1350	2462
80.0	176	348.8	126.7	260	500.0	354.4	670	1238	760.0	1400	2552
80.6	177	350.6	132.2	270	518.0	360.0	680	1256	787.8	1450	2642
81.1	178	352.4	137.8	280	536.0	365.6	690	1274	815.6	1500	2732
81.7	179	354.2	143.3	290	554.0	371.1	700	1292	1093.9	2000	3632
82.2	180	356.0	148.9	300	572.0	376.7	710	1310	1648.9	3000	5432
82.8	181	357.8	154.4	310	590.0	382.2	720	1328	2760.0	5000	9032
83.3	182	359.6	160.0	320	608.0	387.8	730	1346			
83.9	183	361.4	165.6	330	626.0	393.3	740	1364			
84.4	184	363.2	171.1	340	644.0	398.9	750	1382			

Above 1000 in the center column, the table increases in increments of 50. To convert 1,462°F to Celsius, for instance, add to the Celsius equivalent of 1400°F $\frac{5}{9}$ of 62, or 34 degrees, which equals 792°C.

Appendix 2 Conversion of Units – Density

LENGTH

To Obtain → Multiply number of by ↓ ↘	meters (m)	centimeters (cm)	millimeters (mm)	microns (μ) or micrometers (μm)	angstrom units, (A)	inches (in.)	feet (ft)
meters	1	100	1,000	10^6	10^{10}	39.37	3.28
centimeters	0.01	1	10	10^4	10^8	0.394	0.0328
millimeters	0.001	0.1	1	10^3	10^7	0.0394	0.00328
microns	10^{-6}	10^{-4}	10^{-3}	1	10^4	3.94×10^{-5}	3.28×10^{-6}
angstroms	10^{-10}	10^{-8}	10^{-7}	10^{-4}	1	3.94×10^{-9}	3.28×10^{-10}
inches	0.0254	2.540	25.40	2.54×10^4	2.54×10^8	1	0.0833
feet	0.305	30.48	304.8	304,800	3.048×10^9	12	1

AREA

To Obtain → Multiply number of by ↓ ↘	square meters (m²)	square inches (in.²)	square feet (ft²)	square centimeters (cm²)	square millimeters (mm²)
square meters	1	1,550	10.76	10,000	10^6
square inches	6.452×10^{-3}	1	6.94×10^{-3}	6.452	645.2
square feet	0.0929	144	1	929.0	92,903
square centimeters	0.0001	0.155	0.001	1	100
square millimeters	10^{-6}	0.00155	0.00001	0.01	1

DENSITY

To Obtain → Multiply number of by ↓ ↘	grams per cubic centimeter (gm/cm³)	pounds per cubic foot (lb/ft³)	pounds per gallon (U.S.) (lb/gal)
grams per cubic centimeter	1	62.43	8.345
pounds per cubic foot	0.01602	1	0.1337
pounds per gallon (U.S.)	0.1198	7.481	1

1 grain/ft³ = 2.28 mg/m³

Appendix 2 Conversion of Units

FORCE

To Obtain → Multiply number of by ↓ ↘	dynes	newtons (N)	kilogram-force	pound-force (lbf)
dynes	1	1.0×10^{-5}	1.02×10^{-6}	2.248×10^{-6}
newtons	1.0×10^{5}	1	0.1020	0.2248
kilogram-force	9.807×10^{5}	9.807	1	2.205
pound-force	4.4748×10^{5}	4.448	0.4536	1

MASS

To Obtain → Multiply number of by ↓ ↘	grams (gm)	kilograms (kg)	grains (gr)	ounces (avoir) (oz)	pounds (avoir) (lb)
grams	1	0.001	15.432	0.03527	0.00220
kilograms	1,000	1	15,432	35.27	2.205
grains	0.0648	6.480×10^{-5}	1	2.286×10^{-3}	1.429×10^{-4}
ounces	28.35	0.02835	437.5	1	0.0625
pounds	453.59	0.4536	7,000	16	1

VOLUME

To Obtain → Multiply number of by ↓ ↘	cubic feet (ft³)	gallons (U.S. liquid)	liters	cubic centimeters (cm³)	cubic meters (m³)
cubic feet	1	7.481	28.32	28,320	0.0283
gallons (U.S. liquid)	0.1337	1	3.785	3,785	3.79×10^{-3}
liters	0.03531	0.2642	1	1,000	1×10^{-3}
cubic centimeters	3.351×10^{-5}	2.64×10^{-4}	0.001	1	10^{-6}
cubic meters	35.31	264.2	1,000	10^{6}	1

VELOCITY

To Obtain → Multiply number of by ↓ ↘	centimeters per second (cm/s)	meters per second (m/s)	kilometers per hour (km/hr)	feet per second (ft/s)	feet per minute (ft/min)	miles per hour (mph)
centimeters per second	1	0.01	0.036	0.0328	1.968	0.02237
meters per second	100	1	3.6	3.281	196.85	2.237
kilometers per hour	27.78	0.2778	1	0.9113	54.68	0.6214
feet per second	30.48	0.3048	18.29	1	60	0.6818
feet per minute	0.5080	0.00508	0.0183	0.0166	1	0.01136
miles per hour	44.70	0.4470	1.609	1.467	88	1

PRESSURE

To Obtain → Multiply number of by ↓ ↘	pounds per square inch, lb/sq in.² (psi)	atmospheres (atm)	inches (Hg) 32°F 0°C	millimeters (Hg) 32°F 0°C	kilopascals, kPa (kN/m²)	feet (H_2O) 60°F 15°C	inches (H_2O)	pounds per square foot (lb/ft²)
pounds per square inch	1	0.068	2.036	51.71	6.895	2.309	27.71	144
atmospheres	14.696	1	29.92	760.0	101.32	33.93	407.2	2,116
inches (Hg)	0.4912	0.033	1	25.40	3.386	1.134	13.61	70.73
millimeters (Hg)	0.01934	0.0013	0.039	1	0.1333	0.04464	0.5357	2.785
kilopascals	0.1450	9.87×10^{-3}	0.2953	7.502	1	0.3460*	4.019	20.89
feet (H_2O) 15°C	0.4332	0.0294	0.8819	22.40	2.989*	1	12.00	62.37
inches (H_2O)	0.03609	0.0024	0.073	1.867	0.2488	0.0833	1	5.197
pounds per square foot	0.0069	4.72×10^{-4}	0.014	0.359	0.04788	0.016	0.193	1

* at 4°C

APPENDIX Glossary 3

Abrasive blasting. A process for cleaning surfaces by means of high-pressure air or water with sand, alumina, nonfree silica, or organic materials.

Absorbent. A substance that takes in (absorbs) other material.

Accident. That occurrence in a sequence of events that produces unintended injury, death, or property damage. Refers to the event, not the result of the event. (See *Unintentional injury.*)

Accident causes. Hazards and those factors that, individually or in combination, directly cause accidents.

Accident prevention. The application of countermeasures designed to reduce accidents.

Accident rate. Accident experience in relation to a base unit of measure. For example:
- number of accidents per worker hours of exposure
- number of accidents per worker days worked
- number of accidents per miles traveled
- number of accidents per 100 employees.

Acclimation. The process of becoming adjusted to new climatic conditions (e.g., heat, cold, humidity, altitude, etc.).

Accuracy (instrument). Refers to the agreement of a reading or observation obtained from an instrument or a technique with the true value. Quite often used incorrectly as "precision." (See *Precision.*)

ACGIH. American Conference of Governmental Industrial Hygienists. Develops and publishes recommended occupational exposure limits for chemical substances and physical agents. (See *TLV.*)

Acoustic, Acoustical. Containing, producing, arising from, actuated by, related to, or associated with sound.

Action level. Term used by U.S. OSHA and NIOSH to express the level of toxicant that requires medical surveillance, usually one-half the permissible exposure limit. (See *PEL.*)

Acute effect. An adverse effect on a human or animal body, with several symptoms developing rapidly and coming quickly to a crisis.

Acute toxicity. The acute adverse effects resulting from a single dose of or exposure to a substance.

ADA. Americans with Disabilities Act. Legal standard requiring protections and accommodations be extended to the disabled, including employees and patrons of businesses.

Adjustable-barrier guard. See *Barrier guard.*

Administrative controls. Methods of controlling employee exposures by job rotation, varying tasks, work assignment, operational procedures, or time periods away from the hazard(s).

Adsorption. The condensation of gases, liquids, or dissolved substances on the surfaces of solids.

Affirmative action. Positive action taken to ensure nondiscriminatory treatment of all groups in employment regardless of sex, religion, age, handicap (disabilities), or national origin.

Agency or agent. The principal object, such as a tool, machine, or material, involved in an accident that inflicts injury, illness, or property damage.

Air. The mixture of gases that surrounds the earth; its major components are as follows: 78.08% nitrogen, 20.95% oxygen, 0.03% carbon dioxide, 0.93% argon, and varying amounts of water vapor. The mixture changes with altitude. (See *Standard air.*)

Air cleaner. A device designed to remove atmospheric airborne impurities, such as dusts, gases, vapors, fumes, and smoke.

Aircraft accident. An occurrence involving the operation of an aircraft with the intention of flight in which any person suffers death or serious injury or in which the aircraft receives substantial damage.

Aircraft fatal injury. Any injury that results in death within 30 days of the accident.

Aircraft incident. An occurrence other than an accident that affects or could affect the safety of operations associated with the functioning of an aircraft.

Aircraft serious injury. Any injury that is required to be reported to the NTSB under CFR Title 49, Part 830. These injuries include those that:
- require hospitalization for more than 48 hours
- cause the fracture of any bone (except fingers, toes, or nose)
- involve any internal organ
- cause second- or third-degree burns or any burns that affect more than 5% of body surface.

Aircraft substantial damage. Damage or failure that adversely affects the structural strength, performance, or flight characteristics of an aircraft and that usually requires significant repair or replacement of the affected component. Single-engine failure or failure limited to that engine or damage to the landing gear, wheels, tires, flaps, brakes, or wingtips is not considered "substantial damage."

Air-line respirator. A respirator that is connected to a compressed breathing air source by a hose.

Air monitoring. The sampling for and measuring of contaminants in a free or captive atmosphere.

Air-powered tools. Tools that use air under pressure to drive various rotating or percussion attachments.

Air-purifying respirators. Respirators that use filters or sorbents to remove harmful substances from the air.

Air-regulating valve. An adjustable valve used to regulate air pressure and flow rate, such as to the facepiece, helmet, or hood of an air-line respirator.

Air-supplied respirator. Respirator that provides a supply of breathable air from a clean air source.

Alloy. A mixture of metals.

Alpha-particle (alpha ray, alpha-radiation). A small electrically charged atomic particle of very high velocity thrown off by many radioactive materials, including uranium and radium. It is made up of two neutrons and two protons. Its electric charge is positive.

Aluminosis. A form of pneumoconiosis due to the presence of aluminum-bearing dust in the lungs, especially that of alum, bauxite, or clay.

Ambient noise. The all-encompassing noise associated with a given environment, being usually a composite of sounds from many sources.

Ampere. The standard unit for measuring the strength of an electrical current.

Anemometer. A device to measure air velocity.

Anneal. To treat by heat with subsequent cooling for drawing the temper of metals, that is, to soften and render them less brittle.

ANSI. American National Standards Institute. A nonprofit, voluntary membership organization that coordinates the U.S. Voluntary Consensus Standards System and approves American National Standards.

Anthropometric evaluation. A study of human body sizes and modes of action to better design tools and machines to human capabilities.

Anthropometry. The science of measuring the human body for differences in various characteristics.

Antirepeat. The part of the clutch/brake control system designed to limit the press to a single stroke if the actuating means is held or stuck on "operate." Antirepeat requires release of all actuating mechanisms before another stroke can be initiated. Antirepeat is also called "single-stroke reset" or "reset circuit."

Approved. Tested and/or listed as satisfactory; meeting predetermined requirements of some qualifying organization.

Arc welding. One form of electrical resistance welding using either uncoated or coated rods.

Arc-welding electrode. A component of the welding circuit through which current is conducted between the electrode holder and the arc.

Asbestos. A fibrous hydrated magnesium silicate.

Asbestosis. A disease of the lungs caused by the inhalation of fine airborne fibers of asbestos.

Asphyxia. Suffocation from lack of oxygen.

Asphyxiant. A vapor or gas that can cause unconsciousness or death by suffocation (lack of oxygen).

ASTM. American Society for Testing and Materials. Voluntary membership organization with members

from broad a spectrum of individuals, agencies, and industries concerned with materials.

Atmosphere-supply respirator. A respirator that provides breathing air from a source independent of the surrounding atmosphere.

Atmospheric pressure. The pressure exerted in all directions by the atmosphere. At sea level, mean atmospheric pressure is 29.92 in. Hg, 14.7 psi, or 407 in. w.g.

Atomic energy. Energy released in nuclear reactions. The energy is released when a neutron splits an atom nucleus into smaller pieces (fission) or when two nuclei are joined together under millions of degrees of heat (fusion).

Audible range. The normal frequency range for human hearing is approximately 20 Hz through 20,000 Hz. Above the range of 20,000 Hz, the term *ultrasonic* is used. Below 20 Hz, the term *subsonic* is used.

Audiogram. A record of hearing loss (i.e., hearing level measured at several different frequencies—usually 500 to 6,000 Hz). The audiogram may be presented graphically or numerically. Hearing level is shown as a function of frequency.

Audiometer. An instrument that measures a person's ability to hear a pure tone at various frequencies. (See *Frequency.*)

Aural insert. Usually called earplugs or inserts. The pliable material is inserted into the ear canal to reduce the amount of noise reaching the inner ear.

Auto-ignition temperature. The lowest temperature at which a flammable gas-air or vapor-air mixture will ignite from its own heat source or contact with a hot surface, without spark or flame.

Background noise. Noise coming from sources other than the particular noise source being monitored.

Baghouse. Term commonly used for the housing containing bag filters for recovery of airborne particulates from the exhausts of industrial operations.

Barrier guard. Physical protection for operators and other individuals from hazard points on machinery and equipment.

- *Fixed-barrier guard*—A nonmovable physical enclosure attached to the machine or equipment.
- *Interlocked-barrier guard*—An enclosure attached to the machinery or equipment frame and interlocked with the power switch so that the operating cycle cannot be started unless the guard is in its proper position.
- *Adjustable-barrier guard*—An enclosure attached to the frame of the machinery or equipment with front and side sections that can be adjusted.
- *Gate or movable-barrier guard*—A device designed to enclose the point of operation to exclude entry prior to equipment operation.

Base. A compound that reacts with an acid to form a salt. It is another term for alkali. It turns litmus paper blue.

Benign. Not malignant. A benign tumor is one that does not metastasize or invade tissue. Benign tumors may still be lethal, due to pressure on vital organs.

Biodegradable. Capable of being broken down into innocuous products by the action of living things.

Bioengineering. Designing equipment, machines, and other structures to fit the characteristics of people. (See *Ergonomics.*)

Biohazard. A biological hazard. Organisms or products of organisms that present a hazard to humans.

Biohazard area. Any area (a complete operating complex, a single facility, a room within a facility, etc.) in which work has been or is being performed with biohazardous agents or materials.

Biohazard control. Any set of equipment and procedures utilized to prevent or minimize the exposure of humans and their environment to biohazardous agents or materials.

Biomechanics. The study of the human body as a system operating under two sets of laws: the laws of Newtonian mechanics and the biological laws of life.

Black light. Ultraviolet (UV) light radiation between 3,000 and 4,000 angstroms (0.3 to 0.4 micrometers).

Boiler codes. Standards prescribing requirements for the design, construction, testing, and installation of boilers and unfired pressure vessels (e.g., American Society of Mechanical Engineers Boiler and Pressure Vessel Code).

Boiling point. The temperature at which a liquid changes to a vapor state, expressed in degrees.

Bonding. The interconnecting of two objects by means of an electrical conductor. Its purpose is to equalize the electrical potential between objects. (See *Grounding.*)

Brake. The mechanism used on a mechanical power press component to stop and/or hold the slides, either directly or through a gear train, when the clutch is disengaged.

Brake monitor. A sensor that has been designed, constructed, and arranged to monitor the effectiveness of the press braking system.

Braze. To solder with any alloy that is relatively infusible.

Breathing tube. A tube through which air or oxygen flows to the facepiece, helmet, or hood.

Breathing zone. The area encompassed by an imaginary globe of 2-foot radius surrounding the head.

Btu. British thermal unit.

Bubble tube. A device used to calibrate air-sampling pumps.

Buffer. Any substance in a fluid that tends to resist the change in pH when acid or alkali is added.

Building code. An assembly of regulations that set forth the standards to which buildings must be constructed.

Bulk density. Mass of powdered or granulated solid material per unit of volume.

Bulk plant. That portion of a property where flammable or combustible liquids are received by tank vessel, pipelines, tank car, or tank vehicle and are sorted or blended in bulk for the purpose of distributing such liquids by tank vessel, pipeline, tank car, tank vehicle, or container.

Bump cap. A hard-shell cap, without an interior suspension system, designed to protect the wearer's head in situations where the employee might bump into something.

Burns. See *Chemical burns, Thermal burns.*

Calender. An assembly of rollers for producing a desired finish on paper, rubber, artificial leather, plastics, or other sheet material.

Capture velocity. Air velocity at any point outside of an exhaust opening necessary to overcome opposing air currents and to capture the contaminated air by causing it to flow into the exhaust opening.

Carbon monoxide. A colorless, odorless toxic gas produced by any process that involves the incomplete combustion of carbon-containing substances.

Carcinogen. A substance or agent that can cause a growth of abnormal tissue or tumors in humans or animals.

Carcinogenic. Cancer producing.

Carpal tunnel. A passage in the wrist through which the median nerve and many tendons pass between the hand and the forearm.

Carpal tunnel syndrome. An affliction caused by compression of the median nerve in the carpal tunnel.

CAS number. Identifies a particular chemical by the Chemical Abstract Service, a service of the American Chemical Society that indexes and compiles abstracts of worldwide chemical literature called "Chemical Abstracts."

Casting. The pouring of a liquid material into a mold and permitting it to solidify to the desired shape.

Catalyst. A substance that changes the speed of a chemical reaction but undergoes no permanent change itself.

Catwalk. A narrow footway constructed usually for inspection or maintenance purposes.

Causal factor (of an accident). One or a combination of simultaneous or sequential circumstances directly or indirectly contributing to an accident. Modified to identify several kinds of causes such as direct, early, mediate, proximate, distal, etc.

Caustic. Something that strongly irritates, burns, corrodes, or destroys living tissue.

cc. Cubic centimeter; a volume measurement in the metric system, equal in capacity to 1 milliliter (mL)—approximately 20 drops. There are 16.4 cc in 1 in.^{3}.

Ceiling limit (C). In ACGIH terminology, the airborne concentration that should not be exceeded during any part of the working exposure. (See *TLV*.)

Celsius. The Celsius temperature scale is a designation of the scale previously known as the centigrade scale.

Centrifuge. An apparatus that uses centrifugal force to separate or remove particulate matter suspended in a liquid.

Ceramic. A term applied to pottery, brick, and the tile products molded from clay and subsequently calcined.

CERCLA. Comprehensive Environmental Response, Compensation, and Liability Act.

CEU. Continuing education unit. Needed by individuals for some educational programs.

CFR. *Code of Federal Regulations*. A collection of the regulations that have been promulgated under U.S. law.

CHCM. Certified Hazard Control Manager. A designation issued by the Board of Certified Hazard Control Management.

Chemical burns. Generally similar to those caused by heat. After emergency first aid, their treatment is the same as that for thermal burns.

Chemical cartridge. A changeable container filled with various chemical substances for removal of low concentrations of specific vapors, mists, gases, and fumes from the air passing through it.

Chemical cartridge respirator. A respirator that uses changeable cartridges containing various chemical substances to purify inhaled air of certain gases, vapors, mists, and fumes.

Chemical engineering. That branch of engineering concerned with the development and application of manufacturing processes in which chemical or certain physical changes of materials are involved.

Chemical reaction. A change in the arrangement of atoms or molecules to yield substances of different compositions and properties. Common types of reactions are combination, decomposition, double decomposition, replacement, and double replacement.

CHEMTREC. Chemical Transportation Emergency Center.

Circuit. A complete path over which electrical current may flow.

Circuit breaker. A device that automatically interrupts the flow of an electrical current when the current exceeds a specified level.

Citation. A written charge issued by regulatory representatives alleging specific conditions or actions that violate maritime, construction, environmental, mining, or general industry laws and standards.

Clean Air Act. U.S. law enacted to regulate/reduce air pollution. Administered by EPA.

Clean Water Act. U.S. law enacted to regulate/reduce water pollution. Administered by EPA.

Clutch. The coupling mechanism used on a mechanical power press component to couple the flywheel with the crankshaft to produce slide motion, either directly or through a gear train.

Coated electrode. A composite filler metal electrode consisting of a core of bare electrode or metal-cored electrode to which a covering (sufficient to provide a slag layer on the weld metal) has been applied—the covering may contain materials providing such functions as shielding from the atmosphere, deoxidation, and arc stabilization and can serve as a source of metallic additions to the weld.

Coated welding rods. Welding rods coated with various materials such as manganese, titanium, and a silicate, for the purpose of facilitating a solid welding bond on various kinds of iron and steel. (See *Coated electrode*.)

Code of Federal Regulations **(CFR).** The rules promulgated under U.S. law, published in the *Federal Register,* and actually in force at the end of a calendar year are incorporated into this code.

Codes. Rules and standards that have been adopted by a government agency as mandatory regulations having the force and effect of law. Also used to describe a body of standards.

Combustible. Able to catch fire and burn.

Combustible liquids. Combustible liquids are those having a flash point at or above 37.8°C (100°F) and below 93.3°C (200°F).

Common name. Any designation or identification such as code name, code number, trade name, brand name, or generic name used to identify something other than by its proper name.

Communicable. Refers to a disease whose causative agent is readily transferred from one person to another.

Competent person. One who is capable of identifying existing and predictable hazards in the surroundings or working conditions that are unsanitary, hazardous, or dangerous to employees, and who has the authorization to take prompt corrective measures to eliminate them. (See 29 CFR 1926.32.)

Compound. A substance composed of two or more elements joined according to the laws of chemical combination. Each compound has its own characteristic properties different from those of its constituent elements.

Compressed gas cylinder. A cylinder containing vapor or gas under higher than atmospheric pressure, sometimes to the point where it is liquified.

Concurrent. Acting in conjunction, and used to describe a situation wherein two or more controls exist in an operating condition at the same time.

Conductive hearing loss. Type of hearing loss; not caused by noise exposure, but due to any disorder in the middle or external ear that prevents sound from reaching the inner ear.

Confined space. Any area that has limited openings for entry and exit that would make escape difficult in an emergency, has a lack of ventilation, contains known and potential hazards, and is not intended nor designed for continuous human occupancy.

Connection. The part of the power press brake that transmits motion and force from the revolving crank or eccentric to the power press brake ram.

Consensus standard. A standard developed through a consensus process or general opinion among representatives of various interested or affected organizations and individuals.

Contact dermatitis. Dermatitis caused by skin contact with a substance—gaseous, liquid, or solid. May be due to primary irritation or an allergy.

Corrective lens. A lens ground to the wearer's individual prescription to improve vision.

Counter. A device for counting.

Cover guard. An enclosure that covers moving machine parts (excluding point of operation).

CPR. Cardiopulmonary resuscitation.

cps. Cycles per second (frequency). In electricity, it is called hertz.

CPSC. Consumer Product Safety Commission. U.S. agency with responsibility for regulating hazardous materials when they appear in consumer goods.

Critical pressure. The pressure under which a substance may exist as a gas in equilibrium with the liquid at the critical temperature.

Critical temperature. The temperature above which a gas cannot be liquefied by pressure alone.

Crucible. A heat-resistant barrel-shaped pot used to hold metal during melting in a furnace.

Cry-, cryo- (prefix). Very cold.

Cryogenics. The field of science dealing with the behavior of matter at very low temperatures.

CSP. Certified Safety Professional. A designation from the Board of Certified Safety Professionals.

Cubic centimeter (cc). A volumetric measurement that is equal to one milliliter (mL).

Cubic meter (m³). A measure of volume in the metric system.

Cumulative-trauma disorder (CTD). A disorder caused by one or more of the following: repetitive excessive motion of a body part, excessive force, or awkward body posture.

Current. Flow of electrons in an electrical circuit measured in amperes (amps). (See *Ampere.*)

Dampers. Adjustable sources of air-flow resistance used to regulate air-flow in a ventilation intake or exhaust system.

Dangerous to life or health, immediately (IDLH). Used to describe very hazardous atmospheres where employee exposure can cause serious injury or death within a short time or serious delayed effects.

dBA. Sound level in decibels read on the A-scale of a sound-level meter. The A-scale discriminates against very low frequencies (as does the human ear) and is therefore better for measuring general sound levels. (See also *Decibel.*)

Decibel (dB). A unit used to express sound power level (Lw). Sound power is the total acoustic output of a sound source in watts (W). By definition, sound power level, in decibels, is: $Lw = 10 \log W/W_o$, where W is the sound power of the source and W_o is the reference sound power.

Decontaminate. To make safe by eliminating poisonous or otherwise harmful substances, such as noxious chemicals or radioactive material.

Density. The mass (weight) per unit volume of a substance.

Dermatitis. Inflammation of the skin.

Dermatosis. A broader term than dermatitis; it includes any cutaneous abnormality, thus encompassing folliculitis, acne, pigmentary changes, and nodules and tumors.

Die. A (hard metal or plastic) form used to shape material to a particular contour or section. The complete (or portion of the) tooling component used for cutting, forming, or assembling material within its point of operation.

- *General-purpose dies.* The universal dies used to perform bending and forming operations on a variety of piece-parts or products.
- *Special-purpose dies.* Designed to perform work not normally done on general-purpose dies, or for performing a common bending or forming operation that eliminates piece-part whip-up or the need for the power press brake operator to handhold the piece-part component.

Die set. A tool holder held in alignment by guide posts and bushings and consisting of a lower shoe, an upper shoe or punch holder, and guide posts and bushings.

Differential pressure. The difference in static pressure between two locations.

Diffusion rate. A measure of the tendency of one gas or vapor to disperse into or mix with another gas or vapor.

Dike. A barrier constructed to control or confine solid or liquid substances and prevent their movement.

Direct costs (insured costs). Those costs that are paid by the organization for accidents. Usually include compensation insurance, medical, damage, etc. (See *Indirect costs.*)

Direct-reading instrumentation. Those instruments that give an immediate indication of the concentration of aerosols, gases, or vapors or magnitude of physical hazard by some means such as a dial or meter.

DOL. U.S. Department of Labor. Includes the Occupational Safety and Health Administration (OSHA) and Mine Safety and Health Administration (MSHA).

Dose. (1) Term used to express the amount of a chemical or of ionizing radiation energy absorbed in a unit volume or an organ or individual. Dose rate is the dose delivered per unit of time. (2) Term used to express amount of exposure to a chemical substance.

Dose equivalent, maximum permissible (MPD). The largest equivalent received within a specified period that is permitted by a regulatory agency or other authoritative group on the assumption that receipt of such dose equivalent creates no appreciable somatic or genetic injury. Different levels of MPD may be set for different groups within a population. (By popular usage, "dose, maximum permissible," is an accepted synonym.)

DOT. U.S. Department of Transportation.

DOT hazard class. DOT requires that hazardous materials offered for shipment be labeled with the proper DOT hazard class. These classes include corrosive, flammable liquid, organic peroxide, ORM-E, poison B, etc. The DOT hazard class may not adequately describe all the hazard properties of the material.

Double insulated. A method of encasing electric components of tools so that the operator cannot touch parts that could become energized during normal operation or in the event of tool failure.

Drop forge. To forge between dies by a drop hammer or drop press.

Dry chemical. A powdered fire-extinguishing agent usually composed of sodium bicarbonate, monoammonium phosphate, potassium bicarbonate, etc.

Duct. A conduit used for conveying air at low pressures.

Dust collector. An air-cleaning device to remove heavy particulate loadings from exhaust systems before discharge to the outdoors; usual range is loadings of 0.003 grain per cubic foot (gr/ft^3) (0.007 mg/m^3) and higher.

Dusts. Solid particles generated by handling, crushing, grinding, rapid impact, detonation, and decrepitation of organic or inorganic materials, such as rock, ore, metal, coal, wood, and grain. Dusts do not tend to flocculate, except under electrostatic forces; they do not diffuse in air but settle under the influence of gravity.

EAP. Employee assistance program.

Ear. The entire human hearing apparatus, consisting of three parts: external ear; middle ear or tympanic cavity, membrane, and eustachian tube; and the inner ear or labyrinth.

Effective temperature. An arbitrary index that combines into a single value the effects of temperature, humidity, and air movement on the human body's sensation of warmth and cold.

Ejector. A mechanism for removing work or material from between dies.

Electrical current. The flow of electricity measured in amperes.

Electrical precipitator. A device that removes particles from an air stream by applying a positive charge to the particles and collecting the charged particles on a negatively charged surface.

Element. Solid, liquid, or gaseous matter that cannot be further decomposed into simpler substances by chemical means.

Emergency plan. A plan of action for an anticipated, unwanted occurrence/disaster.
- *Shower.* A water shower for an employee when the employee has had chemical contamination that needs to be washed off quickly.
- *STOP (switch).* A switch or other device that, when activated, disengages the power source of and quickly stops the controlled mechanisms.

Emery. Aluminum oxide; natural and synthetic abrasive.

Emission. The release of some by-product or product from an operation or process.

Emission standards. The maximum amount of pollutant emissions permitted to be discharged into the water or air from a single polluting source.

Encapsulate. To cover or coat over with another substance.

Energy control program. A program consisting of an energy control procedure and employee training to ensure that a machine or equipment is isolated and inoperative before servicing or maintenance, thus protecting the employee from unexpected machine start-up or energizing.

Energy-isolating device. A mechanical device that physically prevents the release or transmission of energy. Some examples of energy-isolating devices include: a manually operated circuit breaker, a disconnect switch, a line valve, a block, and other similar devices. The following are not energy-isolating devices: push buttons, selector switches, and other circuit control devices.

Energy isolation. See *Energy control program* and *Energy-isolating device.*

Engineer. A person who can apply scientific principles creatively to design, operate, and maintain structures, machines, and apparatus.

Engineering controls. Methods of controlling employee exposures by modifying the source or the means of exposure or by reducing the quantity of hazards.

Environmental toxicity. Information obtained as a result of conducting environmental testing designed to study the effects on aquatic and plant life.

EOE. Equal opportunity employer.

EPA. U.S. Environmental Protection Agency.

EPA ID number. The number assigned to chemicals regulated by the U.S. Environmental Protection Agency.

Ergonomics. The study of human characteristics for the appropriate design of living and work environments.

Exhalation valve. A device that allows exhaled air to leave a respirator and prevents outside air from entering through the valve.

Exhaust ventilation. The removal of air (usually by mechanical means) from any space. The flow of air between two points is due to the occurrence of a pressure difference between the two points. This pressure difference will cause air to flow from the high-pressure zone to the low-pressure zone.

Explosion. A reaction that causes a sudden, almost instantaneous release of pressure, gas, and heat.

Explosive limit. See *Lower explosive limit* and *Upper explosive limit*.

Exposure. Contact with a chemical, biological, or radiological hazard. Also, the near proximity to an unprotected physical hazard.

Extinguishing medium. The fire-fighting substance used to stop combustion. It is usually referred to by its generic name, such as CO_2, foam, water, dry chemical, etc.

Extrusion. The forcing of raw material through a die or a form in either a heated or cold state, in a solid state, or in partial fluid.

Eyepiece. Gas-tight, transparent window(s) in a full facepiece through which the wearer may see.

Eye protection. "Safety" glasses, goggles, face shields, etc., used to protect against physical, chemical, and nonionizing radiation hazards.

Facepiece. That portion of a respirator that covers the wearer's nose and mouth (in a half-mask facepiece) or the nose, mouth, and eyes (in a full-face respirator).

Face velocity. Average air velocity into the exhaust system measured at the opening into the hood or booth.

Facilitator. A person who makes learning easier, assists interactions and the execution of tasks, and clarifies goals and processes.

Factor of safety. The ratio of ultimate strength of a material or structure to the specified stress allowable.

FDA. The U.S. Food and Drug Administration. Establishes requirements for the labeling of foods and drugs to protect consumers from misbranded, unwholesome, ineffective, and hazardous products. The FDA also regulates materials for food contact service and the conditions under which such materials are approved.

Federal Aviation Administration (FAA). Created by the Federal Aviation Act of 1958, the Federal Aviation Administration is an agency of the United States Department of Transportation and has national aviation authority for all aspects of flight in the United States including the authority to regulate and to oversee the United States Department of Transportation.

***Federal Register* (FR).** Official publication of U.S. government documents and other communications promulgated under the law, documents whose validity depends upon the publication. (See *Code of Federal Regulations*.)

Feeding. The process of placing or removing material within or from the point of operation.

- *Automatic feeding.* Feeding wherein the material or part being processed is placed within or removed from the point of operation by a method or means not requiring action by an operator on each stroke.
- *Hand feeding.* A type of manual feeding wherein the material is placed within, and processed parts removed from, the point of operation by use of a hand-feeding tool.
- *Semiautomatic feeding.* Feeding wherein the material or part being processed is placed within or removed from the point of operation by an auxiliary means controlled by the operator on each stroke.
- *Manual feeding.* Feeding wherein the material or part being processed is handled by the operator on each stroke of the press.
- *Push or slide feeding (hand operated).* A pusher or slide can be used to feed a blank under the upper die and withdraw it after the operation is performed. The pusher or slide may have a machined nest to fit the shape of the part. If the part neither drops through the die nor ejects by other means, it can be withdrawn by the pusher or slide.

Filter. (1) A device for separating components of a signal on the basis of its frequency. It allows components in one or more frequency bands to pass relatively unattenuated, and it attenuates greatly components in other frequency bands. (2) A fibrous medium used in respirators to remove solid or liquid particles from the air stream entering the respirator. (3) A sheet of material that is interposed between patient and the source of x-rays to absorb a selective part of the x-rays. (4) A fibrous or membranous medium used to collect dust, fume, or mist air samples.

Filter, HEPA. High-efficiency particulate air filter that is at least 99.97% efficient in removing thermally generated monodisperse dioctylphthalate smoke particles with a diameter of 0.3 m.

Fire brigade. An organized group trained in fire-fighting operations.

Fire doors. Doors tested and rated for resistance to various degrees of fire exposure and utilized to prevent the spread of fire through horizontal and vertical openings.

Fire resistant. See *Flameproof.*

First aid. The immediate care given to the injured or suddenly ill person.

Fission. The splitting of an atomic nucleus into two parts accompanied by the release of a large amount of radioactivity and heat.

Fixed-barrier guard. See *Barrier guard.*

Flameproof. Material incapable of burning. The term "fireproof" is incorrect. No material is immune to the effects of fire possessing sufficient intensity and duration. The term is commonly, although erroneously, used synonymously with "fire resistive."

Flame propagation. See *Propagation of flame.*

Flammable. Any substance that is easily ignited, burns intensely, or has a rapid rate of flame spread. "Flammable" and "inflammable" are identical in meaning; however, the prefix "in" indicates "negative" in many words and can cause confusion. *Flammable,* therefore, is the preferred term.

Flammable liquid. Any liquid having a flash point below 37.8°C (100°F).

Flammable range. The difference between the lower and upper flammable limits, expressed in terms of percentage of vapor or gas in air by volume; often referred to as the "explosive range." (See *Lower explosive limit* and *Upper explosive limit.*)

Flashback. Occurs when flame from a torch burns back into the tip, the torch, or the hose.

Flash blindness. Temporary visual disturbance resulting from viewing an intense light source.

Flash ignition. See *Flash point.*

Flash point. The lowest temperature at which a liquid gives off enough vapor to form an ignitable mixture with air and produce a flame when a source of ignition is present.

Floor load. (1) The weight that may be safely placed on a floor without danger of structural collapse. (2) The actual load (weight) placed on a floor.

Flowmeter. An instrument for measuring the rate of flow of a fluid or gas.

Fluid. A substance tending to flow or conform to the outline of its container. It may be liquid, vapor, gas, or semisolid (like raw rubber).

Fluorescent screen. A screen coated with a fluorescent substance that emits light when irradiated with x-rays or electromagnetic radiation.

Fly ash. Finely divided particles of ash entrained in flue gases arising from the combustion of solid fuel.

Fog. The visible presence of small water droplets suspended in air.

Footcandle. A unit of illumination.

Foot control. The foot-operated control mechanism designed to be used with a clutch or clutch/brake control system.

Foot-pound. A unit of work equal to the energy required to raise 1 pound a distance of 1 foot.

Force. That which changes the state of rest or motion in matter.

Frequency (in hertz, or Hz). Rate at which oscillations are produced. One hertz is equivalent to one cycle per second.

Friable. Readily crumbled or crumbling state.

Full-revolution clutch. A type of clutch that, when tripped, cannot be disengaged until the drive mechanism (usually a crankshaft) has completed a full revolution and the slide, a full stroke.

Fume. Airborne particulate formed by the evaporation of solid materials (e.g., metal fume emitted during welding). Usually less than 1 micron in diameter.

Fuse. A wire or strip of metal with known electrical resistance, usually set in a plug, placed in an electrical circuit as a safeguard. As the electrical current increases, the metal's resistance to flow causes it to heat until it reaches the point where the metal melts, breaking the current at the rated amperage.

Fusion. The joining of atomic nuclei to form a heavier nucleus, accomplished under conditions of extreme heat (millions of degrees). If two nuclei of light atoms fuse, the fusion is accompanied by the release of a great deal of energy. The energy of the sun is believed to be derived from the fusion of hydrogen atoms to form helium. In welding, the melting together of filler metal and base metal (substrate) or of base metal only.

Galvanizing. An old but still used method of providing corrosion protection for metals by dipping them in a bath of molten zinc.

Gas. A state of matter in which the material has very low density and viscosity; can expand and contract greatly in response to changes in temperature and pressure; easily diffuses; and is neither a solid nor a liquid.

Gas-metal arc welding (GMAW). An arc-welding process that produces coalescence of metals by heating them with an arc between a continuous filler metal (consumable) electrode and the work. Shielding of this process from surrounding air is required and is obtained from an externally supplied gas or gas mixture. Some variants of this process are called MIG or CO_2 welding.

Gas-tungsten arc welding (GTAW). An arc-welding process that produces coalescence of metals by heating them with an arc between a tungsten (nonconsumable) electrode and the work. Shielding is obtained from a gas or gas mixture. Pressure may or may not be used and filler metal may or may not be used. (This process has sometimes been called TIG welding.)

Gate or movable-barrier device. See *Barrier guard*.

Gauge pressure. Pressure measured with respect to atmospheric pressure.

General exhaust. A system for exhausting air from a general work area, accomplished mechanically by air-handling units that drain air from the space.

General ventilation. System of exchanging air in a general work area by either natural or mechanically induced fresh air movements to mix with the existing room air and escape by natural means.

Generic name. A nonproprietary name for a material or product.

GFCI. See *Ground-fault circuit interrupter*.

Glove box. A sealed enclosure in which all handling of items inside the box is carried out through long impervious gloves sealed to ports in the walls of the enclosure.

GMAW. See *Gas-metal arc welding*.

Grab sample. A sample that is taken within a very short time period during which atmospheric concentration is assumed to be constant throughout the sample.

Gram (g). A metric unit of weight equal to 0.035 ounce (avoir).

Gravity, specific. The ratio of the mass of a unit volume of a substance to the mass of the same volume of a standard substance at a standard temperature. Water at 4°C (39.2°F) is the standard substance usually referred to. For gases, dry air, at the same temperature and pressure as the gas, is often taken as the standard substance.

Gravity, standard. A gravitational force that will produce an acceleration equal to 9.8 m/sec^2 or 32.17 ft/sec^2. The actual force of gravity varies slightly with altitude and latitude. The standard was arbitrarily established as that at sea level and 45 degrees latitude.

Ground. A contact with the ground that becomes part of the electrical circuit.

Ground-fault circuit interrupter (GFCI). A device that measures the amount of current flowing to and from an electrical source. When a difference between the two is sensed, indicating a leakage of current, the device very quickly breaks the circuit.

Grounding. The procedure used to carry an electrical charge to ground through a conductive path. (See *Bonding*.)

GTAW. See *Gas-tungsten arc welding*.

Guard. A generic term applied to physical barriers, extraction, and presence-sensing devices designed to prevent contact with hazards. (See *Barrier guard*.)

Halogenated hydrocarbon. A chemical substance that has

carbon plus one or more of these elements: chlorine, fluorine, bromine, or iodine.

Hammer mill. A machine for reducing the size of stone or other bulk material by means of hammers usually placed on a rotating axle inside a steel cylinder.

Hand-feeding tool. Any hand-held tool designed for placing within or removing from the point-of-operation material or parts to be processed.

Hand protection. Coverings worn over the hands to protect against physical, chemical, biological, thermal, and electrical hazards.

Hard hat. A helmet so constructed as to help prevent head injuries from falling objects of limited size.

Hazard. An unsafe condition or activity that, if left uncontrolled, can contribute to an accident.

Hazard analysis. An analysis performed to identify and evaluate hazards for the purpose of their elimination or control.

Hazard control. A program to recognize, evaluate, eliminate, or control the existence of and exposure to hazards.

Hazardous material. Any substance or compound that has the capability of producing adverse effects on the health and safety of humans.

Health. Personal freedom from physical or mental defect, pain, injury, or disease.

Health hazard. A chemical, biological, or radiological material for which there is statistically significant scientific evidence that acute or chronic health effects may occur in exposed employees.

Hearing conservation. The prevention or minimizing of noise-induced hearing loss through the use of hearing protection devices; the control of noise through engineering and administrative methods, audiometric tests, and employee training.

Hearing level. The deviation in decibels of an individual's threshold from the zero reference of the audiometer.

Heat stress. Relative amount of thermal strain from the environment.

Heat stress index. Index that combines the environmental heat and metabolic heat into an expression of stress in terms of the requirement for evaporation of sweat.

Helmet. A device that shields the eyes, face, neck, and other parts of the head.

HEPA filter. See *Filter, HEPA*.

Hertz. Frequency of oscillation measured in cycles per second. 1 cps = 1 Hz.

High-frequency loss. Refers to a hearing deficit starting with frequencies of 2,000 Hz and higher.

Hold harmless. A written agreement in which a party absolves or is absolved by another for liability arising from a specified cause.

Holdout or restraint device. A mechanism, including attachments for the operator's hands, that when anchored and adjusted, prevents the operator's hands from entering the point of operation.

Hood. (1) Enclosure, part of a local exhaust system. (2) A device that completely covers the head, neck, and portions of the shoulders.

Horsepower. A unit of power equivalent to 33,000 foot-pounds per minute (746 W).

Hostage control device. A device designed, constructed, and arranged on a special-purpose mechanical and/or hydraulic power press brake to restrain and maintain the operator(s) at a control station located a safe distance from the point of operation or maintained by hand during the closing portion of the stroke. The use of the term *near the point of operation* means no closer than the distance referred to as the *safe distance*.

Hot. In addition to meaning "having a relatively high temperature," this is a colloquial term meaning "highly radioactive."

Human–equipment interface. Areas of physical or perceptual contact between man and equipment. The design characteristics of the human–equipment interface determine the quality of information. Poorly designed interfaces may lead to excessive fatigue or localized trauma.

Human factors. See *Ergonomics*.

Human factors engineering. See *Ergonomics* and the expressed application of engineering to human factors.

Humidify. To add water vapor to the atmosphere; to add water vapor or moisture to any material.

Humidity. (1) Absolute humidity is the weight of water vapor per unit volume, pounds per cubic foot, or grams per cubic centimeter. (2) Relative humidity is the ratio of the actual partial vapor pressure of the water vapor to the saturation pressure of pure water at the same temperature.

Hydrocarbons. Organic compounds composed solely of carbon and hydrogen.

ICC. U.S. Interstate Commerce Commission.

IDLH. See *Dangerous to life or health, immediately*.

Ignitable. Capable of being set afire.

Imminent danger. An impending or threatening hazard that could be expected to cause death or serious injury to persons in the immediate future unless corrective measures are taken.

Impervious. A material that does not allow another substance to pass through or penetrate it.

Impingement. In air sampling, refers to a process for the collection of particulate or gaseous matter in which the gas containing the contaminant is directed into the collecting solution and the particles or gas are retained by the liquid.

Inches of mercury column. A unit used in measuring pressures. One inch of mercury column equals a pressure of 1.66 kPa (0.491 lb/in.²).

Inches of water column. A unit used in measuring pressures. One inch of water column equals a pressure of 0.25 kPa (0.036 lb/in.²).

Incidence rate (as defined by U.S. OSHA). The number of injuries and/or illnesses or lost workdays per 100 full-time employees per year or 200,000 hours of exposure.

Incident. An unintentional event that may cause personal harm or other damage. In the United States, OSHA specifies that incidents of a certain severity be recorded. (See also *Near-miss incident* and *Accident*.)

Indirect costs. Losses ultimately measurable in a monetary sense resulting from an accident other than those costs that are insurable. (See *Direct costs [insured costs]*.)

Industrial hygiene. The science (or art) devoted to the anticipation, recognition, evaluation, and control of those environmental factors or stresses (i.e., chemical, physical, biological, and ergonomic) that may cause sickness, impaired health, or significant discomfort to employees or residents of the community.

Inert gas. A gas that does not normally combine chemically with other substances.

Inert gas welding. An electric welding operation utilizing an inert gas such as helium to shield the metal being welded from exposure to air, preventing oxidation.

Infrared radiation. Electromagnetic energy with wavelengths from 770 nm to 12,000 nm.

Ingestion. (1) The process of taking substances into the stomach, as food, drink, medicine, etc. (2) With regard to certain cells, the act of engulfing or taking up bacteria and other foreign matter.

Inhalation. The breathing in of a substance in the form of a gas, vapor, fume, mist, or dust.

Inhalation valve. A device that allows respirable air to enter the facepiece and prevents exhaled air from leaving the facepiece through the intake opening.

Injury. Physical harm or damage to the body resulting from an exchange of mechanical, chemical, thermal, or other environmental energy that exceeds the body's tolerance.

In-running nip (point). A rotating mechanism that can seize loose clothing, belts, hair, body parts, etc. It exists when two or more shafts or rolls rotate parallel to one another in opposite directions. (See *Nip point* and *Pinch point*.)

Insoluble. Incapable of being dissolved.

Inspection. Monitoring function conducted in an organization to locate and report existing and potential hazards having the capacity to cause accidents in the workplace.

Interlock. A device that interacts with another device or mechanism to govern succeeding operations. For example, an interlocked machine guard will prevent the machine from operating unless the guard is in its proper place. An interlock on an elevator door will prevent the car from moving unless the door is properly closed.

Interlocked-barrier guard. See *Barrier guard.*

International Civil Aviation Organization (ICAO). A "specialized" agency of the United Nations since 1947, ICAO has five main objectives: safety, security, environmental protection, air navigation capacity and efficiency, and economic development of air transport.

Ionizing radiation. Refers to (1) electrically charged or neutral particles or (2) electromagnetic radiation that will interact with gases, liquids, or solids to produce ions. There are five major types: alpha, beta, x (or x-ray), gamma, and neutrons.

Irradiation. The exposure of something to radiation.

Irritant. A substance that produces an irritating effect when it contacts skin, eyes, nose, or respiratory system.

Jigs and fixtures. Often used interchangeably; precisely, a "jig" holds work in position and guides the tools acting on the work, while a "fixture" holds but does not guide.

Job safety analysis. A method for studying a job in order to (1) identify hazards or potential accidents associated with each step or task and (2) develop solutions that will eliminate, nullify, or prevent such hazards or accidents. Sometimes called *job hazard analysis.*

Kilogram (kg). A unit of weight in the metric system equal to 2.2 lb.

Knockout. A mechanism for releasing material from either the upper or the lower die. Also known as *liftout.*

L (sometimes l). See *Liter.*

Laser. The acronym for light amplification by stimulated emission of radiation.

Laser light region. The portion of the electromagnetic spectrum that includes ultraviolet, visible, and infrared light.

Lathe. A machine tool used to perform cutting operations on wood or metal by the rotation of the workpiece against a blade.

LC. Lethal concentration. A concentration of a substance being tested that will kill a test animal.

LD. Lethal dose. An amount of a substance being tested that will kill a test animal.

Lead poisoning. Lead compounds can produce poisoning when they are swallowed or inhaled. Inorganic lead compounds commonly cause symptoms of lead colic and lead anemia. Organic lead compounds can attack the nervous system.

LEL. See *Lower explosive limit* and *Upper explosive limit.*

Lethal. Capable of causing death.

LFL. Lower flammable limit. (See *Lower explosive limit.*)

Liability. The state of being bound or obliged in law to do, pay, or make good on something. As to the law of torts, usually based on the law of negligence.

Liability, strict. The imposition of liability for damages resulting from any and all defective and hazardous products without requiring proof of negligence. Disclaimers are not valid; traditional warranty concepts, privity, and notice of injury are eliminated.

Liftout. See *Knockout.*

Liquefied petroleum gas. A compressed or liquefied gas usually composed of propane, some butane, and lesser quantities of other light hydrocarbons and impurities; obtained as a by-product in petroleum refining. Used chiefly as a fuel and in chemical synthesis.

Liquid. A state of matter in which the substance is a formless fluid that flows in accord with a law of gravity.

Liter. A metric measure of capacity—1 qt = 0.908 L (dry measure); 1 liter = 1.057 qt (liquid).

Load limit. The upper weight limit capable of safe support by a vehicle, floor, or roof structure.

Local exhaust. A system for capturing and exhausting contaminants from the air at the point where the contaminants are produced.

Local exhaust ventilation. A ventilation system that captures and removes contaminants at the point they are being produced before they escape into the workroom air.

Lockout/tagout. A program or procedure that prevents injury by eliminating unintentional operation or release of energy within machinery or processes during setup, start-up, or maintenance. (See *Energy control program.*)

Long-term sample. Sample taken over a sufficiently long period of time that the variations in exposure cycles are averaged.

Loss control. A program designed to minimize accident-based financial losses. The concept of total loss control is based on detailed analysis of both indirect and direct accident costs. Property damage as well as injurious and potentially injurious accidents are included in the analysis.

Loss prevention. A before-the-loss program designed to identify and correct hazards before they result in incidents that produce actual financial loss or injury.

Lost workday. The number of workdays (consecutive or not), beyond the day of injury or onset of illness, that an employee was away from work or limited to restricted work activity because of an occupational injury or illness.

Loudness. The intensity of an auditory sensation, in terms of which sounds may be ordered on a scale extending from soft to loud. Loudness depends primarily on the sound pressure of the stimulus, but it also depends on the frequency and waveform of the stimulus.

Lower explosive limit (LEL). The lower limit of flammability of a gas or vapor at ordinary ambient temperatures expressed in percent of the gas or vapor in the air by volume. (See *Upper explosive limit* and *Flammable range*.)

LP-gas. See *Liquefied petroleum gas*. Also used, LPG.

Lumen. The luminous flux on 1 square foot of a sphere, 1 foot in radius, with a light source of 1 candela at the center that radiates uniformly in all directions.

m^3. Cubic meter; a metric measure of volume, about 35.3 cubic feet or 1.3 cubic yards.

Maintenance personnel. Individuals who care for, inspect, and maintain mechanical equipment.

Makeup air. Clean, tempered outdoor air supplied to a work space to replace air removed by exhaust ventilation or some industrial process.

Manometer. Instrument for measuring pressure; essentially a U-tube partially filled with a liquid (usually water, mercury, or a light oil) and so constructed that the amount of displacement of the liquid indicates the pressure being exerted on the instrument.

Maser. Microwave amplification by stimulated emission of radiation.

Material Safety Data Sheet (MSDS). See *Safety Data Sheet*.

Maximum permissible concentration (MPC). These concentrations are set by the National Committee on Radiation Protection (NCRP). They are recommended maximum average concentrations of radionuclides to which a worker may be exposed, assuming that he or she works 8 hours a day, 5 days a week, and 50 weeks a year.

Maximum permissible dose (MPD). Currently, a permissible dose is defined as the dose of ionizing radiation that, in the light of present knowledge, is not expected to cause appreciable bodily injury to a person at any time during his or her lifetime. NRC and OSHA have established a maximum permissible dose of 5 rem per year for persons over age 18 and a lifetime dose of 5(N – 18), where N is a person's present age.

Maximum use concentration (muc). The product of the protection factor of the respiratory protection equipment and the permissible exposure limit (PEL).

Mechanical filter respirator. A respirator used to protect against airborne particulate matter like dusts, mists, metal fumes, and smoke. Mechanical filter respirators do not provide protection against gases, vapors, or oxygen-deficient atmospheres.

Mechanical ventilation. A powered device, such as a motor-driven fan or vacuum hose attachment, for exhausting contaminants from a workplace, vessel, or enclosure.

Mega. One million—for example, megacurie = 1 million curies.

Melting point. The temperature at which a solid substance changes to a liquid state.

Meter (m). A unit of length in the metric system. One meter is about 39.37 in.

MeV. Million electron volts.

mg. Milligram; a metric unit of weight. There are 1,000 milligrams in 1 gram (g) of a substance. One gram is equivalent to almost $^4/_{100}$ of an ounce.

mg/kg. Milligrams per kilogram.

mg/m^3. Milligrams per cubic meter.

Mica. A large group of silicates of varying compositions, but similar in physical properties. All have cleavage characteristics that allow them to be split into very thin sheets. Used in electrical insulation.

Microphone. An electroacoustic transducer that responds to sound waves and delivers essentially equivalent electric waves.

Milliampere. $^1/_{1000}$ of an ampere.

Milligram (mg). A unit of weight in the metric system. One thousand milligrams equal 1 gram. (See *mg*.)

Milligrams per cubic meter (mg/m^3). Unit used in the measurement of concentrations of dusts, gases, mists, and fumes in air.

Milliliter (mL). A metric unit used to measure volume. One milliliter equals 1 cubic centimeter or about $^1/_{16}$ cubic inch.

Millimeter of mercury (mm Hg). The unit of pressure equal to the pressure exerted by a column of liquid mercury 1 millimeter high at a standard temperature.

Mist. Suspended liquid droplets generated by condensation from the gaseous to the liquid state or by breaking up a liquid into a dispersed state, such as by spraying or atomizing.

Mixture. A combination of two or more substances that may be separated by mechanical means. The components may not be uniformly dispersed. (See also *Solution*.)

mL. See *Milliliter*.

mm Hg. Millimeters (mm) of mercury (Hg).

Monaural hearing. Refers to hearing with one ear only.

Monitoring. Testing to determine if the parameters being measured are within acceptable limits. This includes environmental and medical (biological) monitoring in the workplace.

MORT. Management Oversight and Risk Tree.

MPC. See *Maximum permissible concentration*.

MPD. See *Dose equivalent, maximum permissible*.

MPE. Maximum permissible exposure.

MPL. May be either maximum permissible level or limit, or dose. Refers to the tolerable dose rate of humans exposed to nuclear radiation.

MSHA. The Mine Safety and Health Administration of the U.S. Department of Labor; federal agency with safety and health regulatory and enforcement authority for the mining industry; established by the Mine Safety and Health Act.

muc. See *Maximum use concentration*.

Muff. A covering over the outside ear to reduce noise exposure.

National Transportation Safety Board (NTSB). An independent agency within the U.S. federal government responsible for the investigation of all U.S. aviation accidents; serious accidents in other transportation areas such as rail, highway, and marine; hazardous materials; and the nation's pipelines.

Nature of injury. The type of injury inflicted, such as: sprain, burn, contusion, laceration, etc.

Near-miss incident. For purposes of internal reporting, some employers choose to classify as "incidents" the near-miss incident; an injury requiring first aid; the newly discovered unsafe condition; fires of any size; or nontrivial incidents of damage to equipment, building, property, or product.

Negligence. The lack of required, expected, or reasonable conduct or care that a prudent person would ordinarily exhibit. There need not be a legal duty.

NEISS. National Electronic Injury Surveillance System. Collects data from 119 representative hospital emergency rooms on product-related injuries receiving emergency room treatment. A part of the U.S. Consumer Product Safety Commission.

Neutral wire. Wire carrying electrical current back to its source, thus completing a circuit.

NFPA. National Fire Protection Association. A voluntary organization whose aim is to promote and improve fire protection and prevention.

NIOSH. National Institute for Occupational Safety and Health. A branch of the U.S. Department of Labor, it conducts research on health and safety concerns, tests and certifies respirators, and trains occupational health and safety professionals.

Nip point. The point of intersection or contact between two or more surfaces when one or more are moving.

Noise. Any unwanted sound.

Noise-induced hearing loss. The slowly progressive inner ear hearing loss that results from exposure to continuous noise over a long period of time, as contrasted to acoustic trauma or physical injury to the ear.

Noise reduction rating (NRR). As applied to ear protection, the amount of sound intensity reduction afforded by the device, measured in dB.

Nonflammable. Not easily ignited, or if ignited, not burning with a flame (smolders).

Nonionizing radiation. Electromagnetic radiation that does not cause ionization. Includes ultraviolet, laser, infrared, microwave, and radiofrequency radiation.

Nonsparking tools. Tools made from beryllium-copper or aluminum-bronze that produce no sparks, or low-energy sparks, when used to strike other objects.

Nonvolatile matter. The portion of a material that does not evaporate at ordinary temperatures.

Not readily removable. Refers to using fastening procedures requiring effort and time to remove rather than quick-release fasteners such as wing nuts, and so forth.

NRC. (1) U.S. National Response Center; a notification center in the Coast Guard Building in Washington DC. (2) Nuclear Regulatory Commission.

Nuisance dust. Has a long history of little adverse effect on the lungs and does not produce significant organic disease or toxic effect when exposures are kept under reasonable control.

Offshore Continental Shelf (OCS). OCS consists of all submerged lands lying seaward of state coastal waters up to 3 miles offshore, which are under U.S. jurisdiction.

Ohm. The unit of electrical resistance.

Ohm's law. The current (I) through an electrical circuit is directly proportional to the applied electromotive force (voltage) (E) and the resistance (R) of the conductor. $I = E/R$.

Operator. Any individual performing production work on the mechanical equipment and controlling the production output.

Orifice. The opening that serves as an entrance and/or outlet. May apply to a body cavity, organ, or some types of equipment, especially the opening of a canal or a passage.

Orifice meter. A flowmeter employing as the measure of flow rate the difference between the pressures measured on the upstream and downstream sides of a restriction within a pipe or duct.

OSHA. U.S. Occupational Safety and Health Administration of the Department of Labor. Federal agency with safety and health regulatory and enforcement authorities for general U.S. industry and business.

Oxidation. Process of combining oxygen with some other substance; technically, a chemical change in which an atom loses one or more electrons whether or not oxygen is involved. Opposite of reduction.

Oxygen deficiency. An atmosphere containing a lower percentage of oxygen by volume than is contained in free air at sea level.

Particulate matter. A suspension of fine solid or liquid particles in air, such as dust, fog, fume, mist, smoke, or sprays. Particulate matter suspended in air is commonly known as an aerosol.

Partial-revolution clutch. A type of clutch that can be disengaged at any point before the drive mechanism (usually a crankshaft) has completed a full revolution and before the press slide has completed a full stroke.

PAW. See *Plasma arc welding*.

PEL. See *Permissible exposure limit*.

Permanent disability or permanent impairment. The partial or complete loss or impairment of any part or function of the body.

Permissible exposure limit (PEL). The legally enforced exposure limit for a substance established by the U.S. OSHA. The PEL indicates the permissible concentration of air contaminants to which nearly all workers may be repeatedly exposed 8 hours a day, 40 hours a week, over a working lifetime (30 years) without adverse health effects.

Personal protective equipment (PPE). Devices worn by the worker to protect against hazards in the environment.

Pesticides. General term for that group of chemicals used to control or kill such pests as rats, insects, fungi, bacteria, weeds, etc., that prey on man or agricultural products. Pesticides include insecticides, herbicides, fungicides, rodenticides, miticides, fumigants, and repellents.

PF. See *Protection factor*.

Physical hazards of chemicals. A chemical for which there is scientifically valid evidence that it is a combustible liquid or a compressed gas or is explosive, flammable, an organic peroxide, an oxidizer, pyrochloric, unstable (reactive), or water reactive.

Pinch point. Any point at which it is possible to be caught between the moving parts, stationary parts, or the material being processed. (See *Nip point* and *In-running nip [point]*.)

Plasma arc welding (PAW). An arc-welding process that produces coalescence of metals by heating them with a constricted arc between an electrode and the workpiece (transferred arc) or the electrode and the constricting nozzle (nontransferred arc). Shielding is obtained by the hot, ionized gas issuing from the orifice, which may be supplemented by an auxiliary source of shielding gas. Shielding gas can be an inert gas or a mixture of gases. Pressure may or may not be used, and filler metal may or may not be supplied.

Point of operation. The area in a process where material is positioned and work is performed during any process such as cutting, forming, or assembling.

Poison, Class A. A U.S. DOT hazard class for extremely dangerous poisons, that is, poisonous gases or liquids of such nature that a very small amount of the gas, or vapor of the liquid, mixed with air is dangerous to life. Some examples: phosgene, cyanogen, hydrocyanic acid, nitrogen peroxide.

Poison, Class B. A U.S. DOT hazard class for liquid, solid, paste, or semisolid substances—other than Class A poisons or irritating materials—that are known (or presumed on the basis of animal tests) to be so toxic to man as to afford a hazard to health during transportation. Some examples: arsenic, beryllium chloride, cyanide, mercuric oxide.

Pollution. Contamination of soil, water, or atmosphere beyond that which is natural.

Potential energy. Energy due to the position of one body with respect to another or to the relative parts of the same body.

Power. Time rate at which work is done; units are the watt (1 J/s) and the horsepower (33,000 ft-lb/min). One horsepower = 746 W.

PPE. See *Personal protective equipment*.

ppm. Parts per million part of air by volume of vapor or gas or other contaminant.

Precision. The degree of agreement of repeated measurements of the same property, expressed in terms of dispersion of test results about the mean result obtained by repetitive testing of a homogeneous sample under specified conditions.

Presence-sensing device. A device designed, constructed, and arranged to create a sensing field or area and to deactivate a moving component when an operator's hand or any other body part is detected within such field or area.

Press machine (or mechanical power press machine). The combination of the press component, tooling component, safeguarding component(s), and feeding components; a complete machine capable of processing the specific job requirement for which it is outfitted by the user (i.e., the production system).

Pressure. Force applied to, or distributed over, a surface; measured as force per unit area. (See *Atmospheric pressure, Gauge pressure, Standard air,* and *Static pressure*.)

Pressure vessel. A storage tank or vessel designed to operate at pressures greater than 15 psig (103 kPa).

Preventive maintenance. The systematic actions performed to maintain equipment in normal working condition and prevent failure.

Primary operation. Any machine operation with material to be subsequently processed.

Probe. A tube used for sampling or for measuring pressures at a distance from the actual collection or measuring apparatus. It is commonly used for reaching inside stacks or ducts.

Product liability. The liability a merchant or a manufacturer may incur as the result of some defect in the product sold or manufactured, or the liability a contractor might incur after job completion from improperly performed work.

Propagation of flame. The spread of flame through the entire volume of the flammable vapor-air mixture from a single source of ignition.

Protection factor (PF). With respiratory protective equipment, the ratio of the ambient airborne concentration of the contaminant to the concentration inside the facepiece.

Protective atmosphere. A gas envelope surrounding the part to be brazed, welded, or thermal sprayed, with the gas composition controlled with respect to chemical composition, dew point, pressure, flow rate, etc.

Protective coating. A thin layer of metal or organic material, as paint applied to a surface primarily to protect it from oxidation, weathering, and corrosion.

psi. Pounds per square inch. For technical accuracy, pressure must be expressed as psig (pounds per square inch gauge) or psia (pounds per square inch absolute; i.e., gauge pressure plus sea level atmospheric pressure, of psig plus about 14.7 pounds per square inch). (See also *mm Hg*.)

psig. Pounds per square inch gauge.

Quality assurance (Quality control). A management function to ensure that the products or goods are produced as intended.

Radar (radio detection and ranging). A radio-detecting instrument able to measure distance to an object, among other characteristics.

Radiation (nuclear). The emission of atomic particles or electromagnetic radiation from the nucleus of an atom.

Radiation (thermal). The transmission of energy by means of electromagnetic waves longer than visible light. Radiant energy of any wavelength may, when absorbed, become thermal energy and result in the increase in the temperature of the absorbing body.

Radiation protection guide (RPG). The radiation dose that should not be exceeded without careful consideration of the reasons for doing so; every effort should be made to encourage the maintenance of radiation doses as far below this guide as practicable.

Radiation source. An apparatus or a material emitting or capable of emitting ionizing radiation.

Radiator. That which is capable of emitting energy in wave form.

Radioactive. The property of an isotope or element that is characterized by spontaneous decay and emission of radiation.

Ram (slide). The powered movable portion of a power press brake structure, with die attachment surface, that imparts the pressing load through dies and the piece-part and against the stationary portion of the press brake bed.

Rated line voltage. The range of potentials in volts of the supply line.

Reaction. A chemical transformation or change; the interaction of two or more substances to form new substances.

Relative humidity. See *Humidity*.

Reliability. The degree to which an instrument, component, or system retains its performance characteristics over a period of time.

Repeat. An unintended or unexpected successive stroke resulting from a malfunction.

Resistance. (1) In electricity, any condition that retards current (or the flow of electrons); it is measured in ohms. (2) Opposition to the flow of air, as through a canister, cartridge, particulate filter, or orifice. (3) A property of conductors, depending on their dimensions, material, and temperature, that determines the current produced by a given difference in electrical potential.

Respirable size particulates. Particles in the size range that permits them to penetrate deep into the lungs upon inhalation.

Respirator. A device to protect the wearer from inhalation of harmful contaminants.

Respiratory protection. Devices that will protect the wearer's respiratory system from overexposure due to inhaling airborne contaminants.

Respiratory system. Consists of (in descending order) the nose, mouth, nasal passages, nasal pharynx, pharynx, larynx, trachea, bronchi, bronchiole, air sacs (alveoli) of the lungs, and muscles of respiration.

Restraint device. A mechanism, including attachments for the operator's hands, that when anchored and adjusted, inhibits the operator's hands from entering the point of operation.

Risk. (1) An insurance term for insured value and another name for the insured or prospective insured. (2) A term applied to the individual or combined assessments of "probability of loss" and potential amount of loss.

Route of entry. The path by which chemicals can enter the body, primarily inhalation, ingestion, skin absorption, and injection.

RPG. See *Radiation protection guide*.

Safe. A condition of relative freedom from danger.

Safe distance. A minimum distance between the operator's hand (or hands) and the point of operation.

Safeguarding. Term used to cover all methods of protection against injury or illness. Two basic categories exist under this umbrella term: the guard and the device. A guard is a physical barrier that absolutely prevents access to a point-of-operation die hazard when it is in place and while it remains in place during a production run of successful cycles. A device is a safeguarding means that controls access to the point of operation.

Safety. The control of recognized hazards to attain an acceptable level of risk.

Safety belt. (1) A life belt worn by linesmen, window washers, etc., attached to a secure object (window sill, etc.) to prevent falling. (2) A seat or torso belt securing a passenger in an automobile or airplane to provide body protection during a collision, sudden stop, air turbulence, etc.

Safety block. A prop that, when inserted between the upper and lower dies or between the bolster plate and the face of the slide, prevents the slide from falling of its own dead weight.

Safety can. An approved container, of not more than 19-L (5-gal) capacity, having a spring-closing lid and spout cover, and so designed that it will safely relieve internal pressure when subjected to fire exposure.

Safety Data Sheet (SDS). A document prepared by a chemical manufacturer describing the composition, properties, and hazards of a chemical along with recommended safeguards for handling, storage, and use. Formerly known as a Material Safety Data Sheet (MSDS).

Safety factor. See *Factor of safety.*

Safety program. Activities designed to assist employees in the recognition, understanding, and control of hazards in the workplace.

Safety shoes. Term commonly used to describe protective footwear meeting ANSI Z41 requirements.

Salamander. A small furnace usually cylindrical in shape, without grates, used for heating. Also a term used to refer to open barrels on a construction site used to provide a heat source.

Salt. A product of the reaction between an acid and a base.

Sampling. A process consisting of the withdrawal or isolation of a fractional part of a whole.

Sanitize. To reduce the microbial flora in or on articles, such as eating utensils, to levels judged safe by public health authorities.

SARA. Superfund Amendments and Reauthorization Act.

SCBA. See *Self-contained breathing apparatus.*

Secondary operation. Press machine operations in which a pre-worked part is further processed.

Self-contained breathing apparatus (SCBA). A respiratory protection device that consists of a supply or a means of respirable air, oxygen, or oxygen-generating material, carried by the wearer.

Self-ignition. See *Auto-ignition temperature.*

Self-insurance. Term used to describe the assumption of one's own financial risk.

Sensible. Capable of being perceived by the sense organs.

Serious violation. Any violation in which there is a substantial probability that death or serious physical harm could result from the violative condition (OSH Act).

Shakeout. In the foundry industry, the separation of the solid, but still not cold, casting from its molding sand.

Shielded-metal arc welding (SMAW). An arc-welding process that produces coalescence of metals by heating them with an arc between a covered metal electrode and the work. Shielding is obtained from decomposition of the electrode covering. Pressure is not used, and filler metal is obtained from the electrode.

Shock. The physical effects of trauma to the body.

Short-term exposure limit (STEL). See *TLV; Standard Industrial Classification.*

Shut height. The distance between the bed and the ram when the ram is at the bottom of its stroke.

SIC. See *Standard Industrial Classification.*

Silicon. A nonmetallic element being, next to oxygen, the chief elementary constituent of the earth's crust.

Single-stroke capability. An arrangement wherein the operating means (lever, pedal, switch, or buttons), when held depressed, normally do not result in more than a single stroke of the slide. Release and reapplication of the operating means are required to obtain a successive stroke. Single-stroke capability is provided by antirepeat or by a single-stroke mechanism.

Slide. The main reciprocating press component member. A slide may be called a ram, plunger, head, or platen.

Sludge. In general, any muddy or slushy mass.

Slurry. A thick, creamy liquid resulting from the mixing and grinding of limestone, clay, and other raw materials with water.

SMAW. See *Shielded-metal arc welding.*

Smelting. One step in the procurement of metals from ore— hence to reduce, to refine, to flux, or to scorify.

Smog. Irritating hazard resulting from the sun's effect on certain pollutants in the air.

Smoke. An air suspension (aerosol) of particles, originating from combustion or sublimation.

Solder. A material used for joining metal surfaces together by filling a joint or covering a junction.

Solution. Mixture in which the components lose their identities and are uniformly dispersed. All solutions are composed of a solvent (water or other fluid) and the substance dissolved, called the "solute." A true solution is homogeneous, as salt is in water.

Solvent. A substance that dissolves another substance.

Soot. Agglomerations of particles of carbon impregnated with tar, formed in the incomplete combustion of carbonaceous material.

Sorbent(s). (1) A material that removes toxic gases and vapors from air inhaled through a canister or cartridge. (2) Material used to collect gases and vapors during air sampling. (3) Nonreactive materials used to clean up chemical spills. Examples: clay and vermiculite.

Sound. An oscillation in pressure, stress, particle displacement, particle velocity, etc., that is propagated in an elastic material, in a medium with internal forces (e.g., elastic, viscous), or by the superposition of such propagated oscillations.

Sound level. A weighted sound pressure level obtained by the use of metering instruments using weighting scales specified in ANSI S1.4.

Sound-level meter and octave-band analyzer. Instruments for measuring sound pressure levels in decibels referenced to 0.0002 microbar.

Sound pressure level (SPL). The level, in decibels, of a sound is 20 times the logarithm to the base 10 of the ratio of the pressure of this sound to the reference pressure. The reference pressure must be explicitly stated.

Specific gravity. The weight of a material compared to the weight of an equal volume of water; an expression of the density (or heaviness) of the material.

Specific weight. The weight per unit volume of a substance; same as density.

SPL. See *Sound pressure level.*

Spontaneously combustible. A material that ignites as a result of retained heat from processing, or that will oxidize to generate heat and ignite, or that absorbs moisture to generate heat and ignite.

Spot welding. One form of electrical-resistance welding in which the current and pressure are restricted to the spots of metal surfaces directly in contact.

Spray coating painting. The result of the application of a spray in painting as a substitute for brush painting or dipping.

Stamping. Many different usages in industry, but a common one is the cutting or forming of sheet metals with a power press.

Standard. A written guide that may or may not be a legal requirement.

Standard air. Air at standard temperature and pressure. The most common values are 21.1°C (70°F) and 101.3 kPa (29.92 in. Hg).

Standard conditions. In industrial ventilation, 21.1°C

(70°F), 50% relative humidity, and 101.3 kPa (29.92 in. of mercury) atmosphere pressure.

Standard Industrial Classification (SIC). A U.S. government classification system for places of employment according to business activity.

Standard man. A theoretical physically fit man of standard (average) height, weight dimensions, and other parameters (blood composition, percentage of water, mass of salivary glands, to name a few).

Standard temperature and pressure. See *Standard air.*

Static pressure. The potential pressure exerted in all directions by a fluid at rest.

STEL. See *Short-term exposure limit* and *TLV.*

Sterilization. The process of making sterile; the killing of all forms of life.

Stop control. An operator control designed to immediately deactivate the clutch control and activate the brake to stop slide motion.

Stress. (1) A physical, chemical, or emotional factor that causes bodily or mental tension and may be a factor in disease causation or fatigue. (2) An applied force or system of forces that tends to strain or deform a body.

Stressor. Any agent or thing causing a condition of stress.

Strict liability. See *Liability, strict.*

Stroking selector. The part of the clutch/brake control that determines the type of stroking or vertical movement when the operating means is actuated. Stroking selectors are normally furnished on hydraulic and special-purpose mechanical power press brakes. The stroking selector generally includes positions for OFF (clutch control), INCH, SINGLE STROKE, and CONTINUOUS (when CONTINUOUS is furnished).

Superfund. See *CERCLA.*

Supplied-air respirators. Air-line respirators or self-contained breathing apparatus.

Supplied-air suit. A one- or two-piece suit that is impermeable to most particulate and gaseous contaminants and is provided with an adequate supply of respirable air.

Suspect carcinogen. A material that is believed to be capable of causing cancer but for which there is limited scientific evidence.

Sweating. (1) Visible perspiration. (2) The process of uniting metal parts by heating solder so that it runs between the parts.

Synthetic. (From Greek word *synthetikos*—that which is put together.) "Man-made 'synthetic' should not be thought of as a substitute for the natural," states the *Encyclopedia of the Chemical Process Industries*; it adds, "Synthetic chemicals are frequently more pure and uniform than those obtained naturally."

Systemic toxicity. Adverse effects caused by a substance that affects the body in a general rather than local manner.

Temporary total disability. An injury that does not result in death or permanent disability, but renders the injured person unable to perform regular duties or activities on one or more calendar days after the day of injury. (This is a definition established by U.S. OSHA.)

Tenosynovitis. Inflammation of the connective tissue sheath of a tendon.

Teratogen. A substance or agent to which exposure of a pregnant female can result in malformations in the fetus. An example is thalidomide.

Thermal burns. Result of the application of too much heat to the skin. First-degree burns show redness of the unbroken skin; second-degree, skin blisters and some breaking of the skin; third-degree, skin blisters and destruction of the skin and underlying tissues, which can include charring and blackening.

Thermal pollution. Discharge of heat into bodies of water to the point that increased warmth activates all sewage, depletes the oxygen the water needs to cleanse itself, and eventually destroys some of the fish and other organisms in the water.

Three segments of the oil industry. From an operations standpoint, the petroleum industry is divided into three distinct yet interrelated sectors: upstream (exploration and production), midstream (storage, transportation, and marketing) and downstream operations (refining, sales, and distribution).

Threshold. The level where the first effects occur; also the point at which a person just begins to notice the tone is becoming audible.

Threshold Limit Value. See *TLV*.

Time-weighted average concentration (TWA). Refers to concentrations of airborne toxic materials that have been weighted for a certain time duration, usually 8 hours (ACGIH).

TLV. Threshold Limit Value. Term used by ACGIH to express the airborne concentration of a material to which nearly all persons can be exposed day after day, without adverse effects. ACGIH expresses TLVs in three ways:

- *TLV-C*. The ceiling limit—the concentration that should not be exceeded even instantaneously.
- *TLV-STEL*. The short-term exposure limit, or maximum concentration for a continuous 15-minute exposure period (maximum of four such periods per day, with at least 60 minutes between exposure periods, and provided that the daily TLV-TWA is not exceeded).
- *TLV-TWA*. The allowable time-weighted average concentration for a normal 8-hour workday or 40-hour workweek.

Tort. A civil wrong, other than breach of contract, for which the law allows compensation by payment of money damages.

Toxicity. The sum of adverse effects resulting from exposure to a material, generally by the mouth, skin, or respiratory tract.

Toxic substance. Any substance that can cause acute or chronic injury to the human body, or that is suspected of being able to cause diseases or injury under some conditions.

Toxin. A poisonous substance that is derived from an organism.

Trade name. The commercial name or trademark by which a chemical is known.

Trade secret. Any confidential formula, pattern, process, device, information, or compilation of information (including chemical name or other unique chemical identifier) that is used in an employer's business and that gives the employer an opportunity to obtain an advantage over competitors who do not know or use it.

Trauma. An injury or wound brought about by an outside force.

Trip (or tripping). Activation of the drive mechanism to run a machine.

TSCA. Toxic Substances Control Act. U.S. environmental legislation, administered by EPA, for regulating the manufacture, handling, and use of materials classified as "toxic substances."

TWA. Time-weighted average exposure. (See *TLV*.)

Two-hand control device. Actuating control that requires concurrent use of both hands of the operator.

Two-hand trip. A clutch- or clutch/brake-actuating method requiring the momentary concurrent use of both hands of each operator.

Type-A movable-barrier device. A self-powered movable barrier that, in normal operation, is designed to (1) close off access to the point of operation in response to operation of the press-tripping control; (2) prevent engagement of the clutch prior to closing of the barrier; (3) hold itself in the closed position; and (4) remain in the closed position until the slide has stopped at the top of the stroke.

Type-B movable-barrier device. A self-powered movable barrier that, in normal single-stroke operation, is designed to (1) close off access to the point of operation in response to operation of the press-tripping control; (2) prevent engagement of the clutch prior to closing of the barrier; (3) hold itself in the closed position during the downward portion of the stroke while the slide is in motion, but be permitted to open during the downward portion of the stroke if the slide is stopped due to clutch control action; and (4) open during the upstroke of the slide in normal single-stroke operations.

UEL. See *Upper explosive limit* and *Lower explosive limit.*

Ultraviolet. Those wavelengths of the electromagnetic spectrum that are shorter than those of visible light and longer than x-rays, 10^{-5} cm to 10^{-6} cm wavelength.

Unintentional injury. The preferred term for accidental injury in the public health community. It refers to the result of an accident.

Upper explosive limit (UEL). The highest concentration (expressed in percent vapor or gas in the air by volume) of a substance that will burn or explode when an ignition source is present. (See *Lower explosive limit* and *Flammable range.*)

USC. United States Code. The official compilation of federal statutes.

USDA. U.S. Department of Agriculture.

Vapors. The gaseous form of substances that are normally in the solid or liquid state (at room temperature and pressure).

Ventilation. Circulating fresh air to replace contaminated air.
- *Dilution.* Air flow designed to dilute contaminants to acceptable levels.
- *Mechanical.* Air movement caused by a fan or other air-moving device.
- *Natural.* Air movement caused by wind, temperature difference, or other nonmechanical factors.

Vibration. An oscillating motion about an equilibrium position produced by a distributing force.

Volatile. Percentage of volatility by volume; the percentage of a liquid or solid (by volume) that will evaporate at an ambient temperature of 70°F (21.1°C) (unless some other temperature is stated). Examples: butane, gasoline, and paint thinner (mineral spirits) are 100% volatile; their individual evaporation rates vary, but over a period of time each will evaporate completely.

Volt. The practical unit of electromotive force or difference in potential between two points in an electrical field.

Warranty. A promise that a proposition of fact is true, and if not true, a consideration is available.
- *Expressed warranty.* A written warranty.
- *Implied warranty.* A generally nonwritten warranty but expressed to the other party in the action.

Watt (W). A unit of electrical power, equal to 1 joule per second.

Weight. The force with which a body is attracted toward the earth.

Weld (welding). A localized coalescence of metals or nonmetals produced either by heating the materials to suitable temperatures, with or without the application of pressure, or by the application of pressure alone, and with or without the use of filler material.

Welding. The several types of welding are electric arc welding, oxyacetylene welding, spot welding, and inert or shielded gas welding utilizing helium or argon. The hazards involved in welding stem from (1) the fumes from the weld metal such as lead or cadmium metal, (2) the gases created by the process, or (3) the fumes or gases arising from the flux.

Welding rod. A rod or heavy wire that is melted and fused into metals in arc welding.

Wellness. The practice of a healthy lifestyle.

Wind load (force or pressure). The pressure exerted on a building or structure from moving air.

Work. When a force acts against resistance to produce motion in a body, the force is said to work. Work is measured by the product of the force acting and the distance moved through against resistance. The units of measurement are the erg (the joule is 1×10^7 erg) and the foot-pound.

Workers' compensation. An insurance system under law, financed by employers, that provides payment to injured and diseased employees or relatives for job-related injuries and illnesses.

Work hours. The total number of hours worked by all employees.

Work injuries. Injuries (including occupational illnesses) that arise out of or in the course of gainful employment regardless of where the accident occurs. Excluded are work injuries to private household workers and injuries occurring in connection with farm chores, which are classified as home injuries.

Work stress. Biomechanically, any external force acting on the body during the performance of a task. Application of work stress to the human body is the inevitable consequence of performance of any task, and is, therefore, synonymous with "stressful work conditions" only when excessive. Work stress analysis is an integral part of task design.

Zero energy state. See *Zero mechanical energy.*

Zero mechanical energy (ZME). An old term, now called "energy isolation"; indicates a piece of equipment without any source of power that could harm someone.

Index

A

Abandonment, of tanks, 341
Aboveground rescue systems, 197
Aboveground tanks, 38–39, 336, 337–339
Abrasive blasting respirators, 205
Abrasive wheels and disks
    dressing, 615–616
    guard devices for, 614
    handling, 612
    inspecting, 612
    mounting, 612–613
    operating, 613–614
    for portable grinders, 539
    safe speeds for, 614–615
    storing, 612
    work rests for, 615
Acceptable risk, 6, 9
Acceptance certifications, 96
Accessories
    hand tools, 113–114, 363
    hand trucks, 364–365
    jacks, 363–364
    for materials handling, 363–365
Accidental losses, 77
Accident prevention signs, colors of, 47–49
Accident reports, for contract close-out, 97
Accidents
    aviation, 737, 739–740
    frequency and severity rates, 88–89
    heavy-equipment, 502
    investigation and reporting of, 67, 95
    preparedness programs, 744–746
    robotic equipment and, 691–693
    speed as factor in, 484
Accountability, in safety programs, 63, 721
Acetylene, 585–587, 589
Acid hoods, 188
Acoustic trauma, 190
Acrolein, 650
Action limit (AL), 358
Actions, guarding, 167
Active fall arrest systems, 193, 195–197
Adjustable-barrier guards, 626, 627–628
Adjustable wrenches, 530–531
Administrative controls, in hazard prevention, 11, 180
Advancing Process Safety (AFPM), 719
Aerial baskets, 413–415
Aerial conveyors, 423–424
A-frame derricks, 404
Age of facility, color considerations and, 46
Airborne hazards, 200–201
Air cleaners, 582
Air Commerce Act of 1926, 736, 738–739
Air conditioning. *See* Heating, ventilation, and air conditioning (HVAC)
Aircraft Owners and Pilots Association (AOPA), 745
Air hammers, 664, 665
Air heaters, 286–287
Air hoists, 391
Air hoses, 541, 653
Air jets, 631–632
Air-line respirators, 204–206
Airlines for America, 745
Air Mail Service, U.S., 736
Air pollution, 29, 32
Airport Surface Detection Equipment, Model X (ASDE-X), 747

Air-powered tools, 541–544
Air preheaters, 133–134
Air-purifying respirators, 202, 206–207, 210
Air quality, 111–112
Air-supplied hoods, 205
Air-supplied suits, 205–206
Air-supplying respirators, 202–204, 210–211
Air traffic control, 737, 738, 747
Air transportation safety. *See* Aviation safety
Air valves, 636
Airworthiness Certifications, 738
Aisles
    in facility layout, 39, 50
    in foundries, 653
    maintenance of, 105
    in warehouses and storerooms, 380
Alarm systems, for fire protection, 294–295
Alcohol-testing programs, 68
Alignment, of electric motors, 240
"Alligator cracks," in crucibles, 659
Alligator shears, 645–646
Alloy steel chains, 466–469
Alternating current (AC), 223–225
Aluminosis, 575
Aluminum, 650
American Association of State Highway and Transportation Officials, 510
American Chemistry Council (ACC), 718, 719
American Conference of Governmental Industrial Hygienists (ACGIH), 111, 188, 326, 491, 574
American Fuel and Petrochemical Manufacturers (AFPM), 719
American Institute of Architects (AIA), 83
American Institute of Chemical Engineers (AIChE), 698, 718
American Iron and Steel Institute (AISI), 454
American National Standards Institute (ANSI)
    ANSI A10.28, Safety Requirements for Work Platforms Suspended from Cranes or Derricks for Construction and Demolition Operations, 409
    ANSI A13.1, Scheme for the Identification of Piping Systems, 48–49, 122, 144, 322, 370
    ANSI A14.3, Fixed Ladders, 393
    ANSI A103–1984, powder-activated tools, 546
    ANSI A1264.1, Safety Requirements for Workplace Floor and Wall Openings, Stairs, and Railing Systems, 29, 51, 52, 107, 606, 653
    ANSI B7.1, Safety Requirements for the Use, Care, and Protection of Abrasive Wheels, 539, 544, 611, 613
    ANSI B11, Machine Tools Safety Package, 617, 685
    ANSI B11.1–2001, Safety Requirements for the Construction, Care, and Use of Mechanical Power Presses, 625, 626, 630
    ANSI B11.1 through B11.21, Machine Safety Guarding Standards, 160–161, 171
    ANSI B11.3–2002, Safety Requirements for the Construction, Care, and Use of Power Press Brakes, 635, 646, 648, 649
    ANSI B11.4–2003, Safety Requirements for Shears, 644, 645
    ANSI B11.19, Performance Requirements for Safeguarding, 685–686
    ANSI B11.20, Integrated Manufacturing Systems, 685
    ANSI B11.21, Machine Tools Using Lasers for Processing Material, 685
    ANSI B11.22, Turning Centers and Automatic, Numerically Controlled Turning Machines, 685
    ANSI B11.23, Machining Centers and Automatic, Numerically Controlled Milling, Drilling, and Boring Machines, 685
    ANSI B11.24, Transfer Machines, 685
    ANSI B56.5, Safety Standard for Guided Industrial Vehicles and

811

Automated Functions of Manned Industrial Vehicles, 161, 690

ANSI B65.1, Safety Standards for Printing Press Systems, 161

ANSI B74.2, Specifications for Shapes and Sizes of Grinding Wheels, and Shapes, Sizes, and Identification of Mounted Wheels, 539

ANSI B151.27, Safety Requirements for the Integration, Care, and Use of Robots Used with Horizontal and Vertical Injection Molding Machines, 161

ANSI C2, National Electrical Safety Code, 229, 408, 603

ANSI S2.73, Mechanical Vibration and Shock, 547

ANSI SNT 101–2002, power nailers and staplers, 543–544

ANSI Z21.22, Relief Valves and Automatic Gas Shut-Off Devices for Hot Water Supply Systems, 135

ANSI Z41, Protective Footwear, 373

ANSI Z49.1, Safety in Welding and Cutting, 578, 593

ANSI Z136.1, Safe Use of Lasers, 188

ANSI Z241.1, Safety Requirements for Sand Preparation, Molding, and Coremaking in the Sand Foundry Industry, 652

ANSI Z244.1, Safety Requirements for Conveyors and Related Equipment, 422, 687

ANSI Z535.1, Safety Color Code for Marking Physical Hazards, 39, 48

ANSI Z535.4, Product Safety Signs and Labels, 233

ANSI/AIHA Z10–2005, Occupational Health and Safety Management Systems, 4, 9–10, 20, 711

ANSI/ASME A17.1, Safety Code for Elevators and Escalators, 426–432, 434, 437, 440

ANSI/ASME A17.2, Guide for Inspection of Elevators, Escalators, and Moving Walks, 432–435

ANSI/ASME A17.4, Guide for Emergency Personnel, 436

ANSI/ASME A90.1, Safety Standard for Belt Manlifts, 440

ANSI/ASME B15.1, Safety Standard for Mechanical Power Transmission Apparatus, 420, 554

ANSI/ASME B20.1A, Safety Standards for Conveyors and Related Equipment, 417, 422, 688

ANSI/ASME B30 Series, Safety Requirements for Cranes, Derricks, Hoists, Hooks, Jacks and Slings, 392, 395–397

ANSI/ASME B30.2, Overhead and Gantry Cranes, 391, 392, 394, 399, 401

ANSI/ASME B30.5, Mobile and Locomotive Cranes, 366, 506

ANSI/ASME B30.9, Slings, 466

ANSI/ASME B31.3, Series, Pressure Piping code, 136

ANSI/ASME B56.1–7, Safety Standards for Powered Industrial Trucks, 366, 480, 481, 483, 488

ANSI/ASSE A10.11, Safety Requirements for Personnel and Debris Nets, 194

ANSI/ASSE Z359, Fall Protection Code, 198

ANSI/AWS A3.0, Welding Terms and Definitions, 574, 596

ANSI/AWS F4.1, Safe Practices for the Preparation of Containers and Piping for Welding and Cutting, 579–580

ANSI/AWS F6.1, Method for Sound Level Measurement of Manual Arc Welding and Cutting Processes, 583

ANSI/CEMA 102, Terms and Conveyor Definitions, 417

ANSI/CGA V-1–1987, Compressed Gas Cylinder Valve Outlet and Inlet Connections, 589

ANSI/IES RP-7, Practice for Industrial Lighting, 42, 43, 109, 380

ANSI/ISEA 107–2010, High-Visibility Safety Apparel, 217

ANSI/ISEA Z87.1, Occupational and Educational Eye and Face Protection, 185, 186, 189, 543, 546, 556, 583

ANSI/ISEA Z89.1, Industrial Head Protection, 182–183, 546, 583

ANSI/ISEA Z308.1, Minimum Requirements for Workplace First Aid Kits, 546

ANSI/NEMA EW1, Electric Arc Welding Power Sources, 593

ANSI/RIA R15.06, Safety Requirements for Industrial Robots and Robot Systems, 161, 684, 688, 693

on hazard identification, 684

on risk reduction, 154

American Optometric Association (AOA), 186, 187

American Petroleum Institute (API)

API 2015, Safe Entry and Cleaning of Petroleum Storage Tanks, 340, 369

on auditing, 700

on hazard analysis updates, 699

on operating manuals, 699–700

on pressure vessels, 140

on pre-start-up safety reviews, 699

on process safety management, 718, 719

RP 752, Management of Hazards Associated with Location of Process Plant Buildings, 719

RP 754, Process Safety Performance Indicators for the Refining and Petrochemical Industries, 719

RP 2000, Venting Atmospheric and Low-Pressure Storage Tanks, 339

RP 2003, Protection against Ignitions Arising Out of Static, Lightning, and Stray Currents, 326

on worker safety, 755

American Railway Engineering Association (AREA), 34, 35

American Society for Non-destructive Testing (ASNT), 143–144

American Society for Testing and Materials (ASTM)

ASTM A391–1975, Specifications for Alloy Steel Chains, 466

ASTM B783–2004, Specifications for Materials for Ferrous Powder Metallurgy Structural Parts, 471

ASTM D323–2008, Standard Test Method for Vapor Pressure of Petroleum Products (Reid Method), 321

ASTM F496, Specifications for In-Service Care of Insulating Gloves and Sleeves, 260

ASTM F2413–2011, Standard Specification for Performance Requirements for Protective Toe Cap Footwear, 214, 215

ASTM F2704–2010, Standard Specification for Air-Fed Protective Ensembles, 205

American Society of Mechanical Engineers (ASME)

ANSI/ASME A17.1, Safety Code for Elevators and Escalators, 426–432, 434, 437, 440

ANSI/ASME A17.2, Guide for Inspection of Elevators, Escalators, and Moving Walks, 432–435

ANSI/ASME A17.4, Guide for Emergency Personnel, 436

ANSI/ASME A90.1, Safety Standard for Belt Manlifts, 440

ANSI/ASME B15.1, Safety Standard for Mechanical Power Transmission Apparatus, 420, 554

ANSI/ASME B20.1A, Safety Standards for Conveyors and Related Equipment, 417, 422, 688

ANSI/ASME B30 Series, Safety Requirements for Cranes, Derricks, Hoists, Hooks, Jacks and Slings, 392, 395–397

ANSI/ASME B30.2, Overhead and Gantry Cranes, 391, 392, 394, 399, 401

ANSI/ASME B30.5, Mobile and Locomotive Cranes, 366, 506

ANSI/ASME B30.9, Slings, 466

ANSI/ASME B56.1–7, Safety Standards for Powered Industrial Trucks, 366, 480, 481, 483, 488

ASME/ANSI, A13.1–2007, Scheme for the Identification of Piping Systems, 48–49, 122, 144, 322, 370

ASME CSD-1, Controls and Safety Devices for Automatically Fired Boilers, 131, 133, 138, 139

Boiler and Pressure Vessel Code, 37, 130, 131, 133, 134, 139–140, 142, 145

Division 1, 139–140

Division 2, 140

Americans with Disabilities Act of 1990 (ADA), 31–32

American Welding Society (AWS)

ANSI/AWS A3.0, Welding Terms and Definitions, 574, 596

ANSI/AWS F4.1, Safe Practices for the Preparation of Containers and Piping for Welding and Cutting, 579–580

ANSI/AWS F6.1, Method for Sound Level Measurement of Manual Arc Welding and Cutting Processes, 583

AWS A6.0, Safe Practices for Welding and Cutting Containers That Have Held Combustibles, 333

training and qualification standards of, 584

Welding Handbook, 590, 591

American Wire Gauge (AWG) system, 230

Ammeters, 256, 257

Ammonium phosphate–based dry chemical extinguishers, 301–302

Amperage, defined, 222

Anchorage, roof, 107

Anchor/anchorage points, 195

Anchoring, of tanks, 336

Annealing process, 659

ANSI. *See* American National Standards Institute

Anthracosis, 575

Antilaser eyeshields, 188

Antirepeat, for power presses, 630, 636

API. *See* American Petroleum Institute

Appliances. *See* Tools and appliances

Approach boundaries, 259

Approach zone, 173, 174

Apron conveyors, 420–421
Aprons, 215
Arc blasts, 222
Arc-fault circuit interrupters (AFCIs), 236
Arc flashes, 222
Architects, 74
Arc welding and cutting, 593–596
Arm protection, 211–214
Army Corps of Engineers, U.S., 424
Asbestos, 578
Aseptic environments, 683
Ash disposal equipment, 134
As low as reasonably practicable (ALARP), 6
ASME. *See* American Society of Mechanical Engineers
Asphalt flooring, 52–53, 104, 105
Asphyxiation, 378, 575
Aspirating smoke detectors, 292
Assigned protection factors (APFs), 202, 207
Association for Manufacturing Technology (AMT), 602
Association of American Railroads (AAR), 327–329
Association of Energy Service Companies (AESC), 755
ASTM. *See* American Society for Testing and Materials
Atmosphere testing, 33
Attorney–client privilege, 731
Audiometric testing, 190, 191
Audits
　　for aviation safety, 744
　　of chemical processes, 700–701
　　confidentiality and, 731
　　of contractor performance, 91
　　in process safety management, 700–701, 728–731
　　of safety practices, 73
　　self-audits, 173
Aural insert hearing protectors, 191, 192
Autoclaves, 146–147
Auto-ignition temperature, 320–321
Automated guided vehicles (AGVs), 479, 480, 483, 684, 685, 688–689
Automated systems, 679–701
　　barriers and interlocked barriers in, 685–687
　　for chemical processes, 695–701
　　in food processing, 683–684
　　hazard identification and controls in, 682–685
　　manufacturing philosophies, 680–682
　　materials handling in, 688–690
　　robotic equipment, 173–175, 684–685, 690–695
　　training for, 687–688
　　at waste transfer stations, 766
Automatic Dependent Surveillance Broadcast, 747
Automatic external defibrillators (AEDs), 222–223, 228
Automatic feeds, 624, 630, 638
Automatic fire detection systems, 290–293
Automatic safeguarding devices, 162–164
Automatic single-stroke control, 638
Auxiliary fire alarm systems, 295
Auxiliary mechanisms, of power presses, 629–630
Aviation safety, 735–748
　　accident/incident preparedness, 744–746
　　aircraft and systems engineering, 744
　　air traffic control and, 737, 738, 747
　　audits for, 744
　　cabin safety, 744
　　concepts in practice, 746–747
　　data collection and analysis on, 743
　　future of, 747
　　governmental organizations and, 736–740
　　history of, 736
　　information dissemination regarding, 743
　　international, 740–742
　　investigations and, 739–740, 744
　　stakeholders in, 742–744
　　terminology, 745
　　trade associations and, 745
　　training on, 738, 743–744, 746
Aviation Safety Information Analysis and Sharing (ASIAS) system, 737
Awls, 528
AWS. *See* American Welding Society
Axes, 526–527

B

Back belts, 356–357
Back injuries, 356–357
Backrests, 480, 482
Backward movement, of heavy equipment, 116, 504
Baffle furnaces, 651
Bagged material, 367, 373–374
Balagna v. Shawnee County (1983), 74
Ballistic nylon patches, for leg protection, 546
Ball peen hammers, 524
Band brakes, 644
Band saws, 536, 553, 565–566, 608
Barrels and kegs
　　handling and lifting, 361
　　of hazardous liquids, 370
　　opening, 383
　　storage of, 367
Barrier guards, 162, 164, 232–233, 626, 627–628, 685–687
Bar stock, storage of, 367
Baseline safety reviews, 70
Basket hitches, 462–463, 473
Baskets, aerial, 413–415
Batteries, for powered industrial trucks, 489–491
Battery-charging rooms, 40
Battery-powered locomotives, 37
Beam photoelectric detectors, 291–292
Beams, maintenance of, 103
Bearings, for power presses, 642
Behavior modification vs. workplace redesign, 12–13
Bell warning system, for blind level crossings, 34, 35
Belowground tanks, rescue systems for, 197
Belt conveyors, 419–420
Belt-driven elevators, 427
Belted machines, inspection and maintenance of, 435
Belt Manlift Inspection Report, 442
Belts, body, 195
Belts, for back, 356–357
Bending actions, 169
Bending equipment, 671
Beneficial occupancy, 73
Benign pneumoconioses, 575
Beryllium, 576, 650
Bid packages, 71
Biodegradation, of waste products, 765
Biological agents, 200
Bioreactors, 765
Black boxes, on airplanes, 740
Blades, for saws, 535–536, 560–561
Blankets, fire, 303
Blank feeds, 624
Blast gates, 657
Blind level crossings, bell warning system for, 34, 35
Block and tackle, 415–416
Blowdown pipes and valves, 134
BOAC National Building Code, 110
Body belts, 195
Boiler rooms, 137–138
Boilers, 130–139
　　annual maintenance for, 137
　　cleaning, 137
　　codes and standards for, 130–131
　　design and construction of, 19, 133–136
　　detecting cracks and measuring thickness in, 142–144
　　emergency procedures, 138–139
　　hydrostatic tests in, 142
　　inspection of, 131–132, 136–137
　　operator training, 138
　　overview, 130, 132–133
　　precautions for entering, 137
　　rooms housing, 137–138
　　safety of high-temperature water in, 139
Bolster plates, for power presses, 627, 639–641
Bolt headers, 672
Bonding
　　defined, 223
　　in fluid transfer, 323–326
Booms, crane, 410, 508–509
Boots, 214–215

Boring machines, 605–606
Boring mills, 606
Bott and botting-up, 657
Boundaries
    controls and, 686
    between restricted and nonrestricted areas, 684–685
Boxes
    lifting, 361
    opening, 383
    toolboxes, 520–521
Box ovens, 341–342
Box wrenches, 529–530
Brad awls, 528
Braided slings, 463, 464
Brakes
    on dumbwaiters, 438
    on electrical woodworking equipment, 554, 555
    on escalators, 438
    on hand elevators, 437–438
    inspection and maintenance of, 644
    on man-lifts, 441–442
    monitoring, in single-stroke control, 638
    for power presses, 635, 644
Breast derricks, 405
Bricklayer's tools, 522, 524
Bridge plates, powered industrial trucks on, 484–485
Bridges
    crane, 402
    for loading or unloading flammable and combustible liquids, 328
    pedestrian, 30
Broaches, 611
Bucket conveyors, 422, 423
Buffers
    elevator, 427, 435
    grinding wheels and, 539
Buffing wheels, 617
Builder's risk insurance, 78–79
Building design. See Design
Building information modeling (BIM), 69
Building layout, 27–55
    codes and standards for, 28–29
    color considerations in, 44–49
    design considerations, 28–29
    facility railways, 34–38
    internal facility layout, 38–41
    lighting, 30, 32, 34–35, 41–44
    outside facilities and, 30–34
    security systems, 44
    site selection and, 29–30
    of structures, 49–55
Building-related illnesses (BRIs), 111
Buildings, location of, 38–39
Built-in safeguards, 162
Bulb detection systems, 291
Bulk storage, of flammable and combustible liquids, 334–340
Bulldozers, 505, 509–512, 672, 765
Bump caps, 184
Bureau of Labor Statistics (BLS), 88, 355, 356, 753–754, 764
Bureau of Safety and Environmental Enforcement (BSEE), 755
Burlap sacking, 368, 382
Burning, of waste products, 764
Burns
    cold-contact, 377
    electrical, 227
    hazards causing, 683–684
Bushings, 231

C
Cabinets
    for storage of flammable and combustible liquids, 332–333
    for tools, 521
Cabin safety, on airplanes, 744
Cable-laid slings, 463
Cables, in welding, 592, 593–594
Cadmium, 576–577
Calcium carbide, 586
Calculation of Daylight Availability (IES), 41
Calipers, wire ropes and, 456

Canadian Centre for Occupational Health and Safety (CCOHS), 707, 710
Canadian Standard Association (CSA), 195
Canal caps, 191, 192
Canisters, respirator, 202
Canopies, maintenance of, 108
Capacity ratings. See also Loads
    cranes, 392, 408–409
    for elevators, 427–428, 432
    floor load, 54, 105
    of powered industrial trucks, 481, 488
    of slings, 464–469
Capstan work, 448
Carbon dioxide, in welding, 575
Carbon dioxide extinguishers, 302
Carbon dust, 650
Carbon monoxide
    in foundries, 650, 656–657
    from liquefied petroleum gas, 491
    in welding, 575
Carboys, 370–371
Cardboard cartons, 366–367
Cardiopulmonary resuscitation (CPR), 228
Car movers, for facility railways, 37
Cars, elevator, 431–433
Carton cutters, 528
Cartons
    fiberboard/cardboard, 366–367
    lifting, 361
    opening, 383
    storage of, 366–367
Cartridge fuses, 235
Cartridge-operated dry chemical extinguishers, 301
Cartridges, respirator, 202, 206
Carvlho v. Toll Brothers (1995), 74
Cast-iron plates, 53
Catastrophic losses, 77
Catwalks, on forging hammers, 665
Ceilings, maintenance of, 103–104
Cells, of landfills, 765
Center for Chemical Process Safety (CCPS), 718–719, 721
Ceramic glazed tile, 53
Certifications
    aviation, 737–738
    of safety equipment, 181
Chain conveyors, 421
Chain-driven elevators, 427
Chain hoists, 391
Chain reaction, in fire, 269
Chain saws, 544–545
Chain slings, 466–470
    alloy steel, 466–469
    double-chain, 465
    hooks and attachments for, 467
    inspection of, 467, 469
    maximum allowable wear of, 470
    properties and working load of, 466–469
    safe practices for, 469–470
    types of, 466
Change analysis, 19
Change management, 19–20, 700, 724
Chapanis, Alphonse, 13
Charging, of crucibles and cupolas, 656, 659
Charging floors, 656
Charter for Sustainable Development (ICC), 719
Checklists
    crane inspection, 399
    ergonomic factors of hand tools, 518
    forging hammer maintenance, 668, 669
    general safety design, 14–19
    grinder operation, 613
    for operators, 499
    powered industrial truck inspections, 488, 489–490, 499
    pressure vessel operation, 138, 144
    pre-start-up reviews, 699
    process, 697–698, 724
    rope, 448
    self-evaluation process safety audits, 731

tool inspection, 520
Chemical-cartridge respirators, 206–207
Chemical extinguishers, 301–303
Chemical goggles, 183, 188
Chemical hazards, 683
Chemical processes, 695–701
    auditing, 700–701
    documentation of, 695–696
    hazard and risk analyses for, 696–698
    management of change in, 700
    pre-start-up reviews for, 699
    procedure manuals for, 695, 699–700
    safety information for, 695–696, 722
Chemical Process Operator Certification Training (SOCMA), 719
Chemical Process Safety: Fundamentals with Applications (Crowl & Louvar), 696, 698
Chemicals, use in grounds maintenance, 116–117
Chemical Safety and Hazard Investigation Board, 719
Chemistry of fire, 268–269
Chests, tool, 521
Child labor laws, 492
Chimneys, 107, 134
Chin straps, 183
Chipping hammers, 542, 582–583
Chips, 603, 663
Chisels
    for cutting metal, 524–525
    for cutting wood, 526
    redressing, 522
Chlamydia psittaci, 200
Chlorinated hydrocarbons, 578
Choker hitches, 462
Chromium, 577, 650
Chute conveyors, 424–425
Chute feeds, 630–631
Circuit breakers, 235–236
Circuit testers, receptacle, 240, 245
Circular saws, 556–561
    blades for, 535–536, 560–561
    for cold metal cutting, 608
    cords and, 536
    crosscutting, 560
    guard devices for, 557, 559
    hazards of, 556, 559
    kickbacks and ripping, 556–558
    operating methods, 559–560
    rabbeting and dadoing, 557, 559
    safe operation of, 535–536, 553
Circumaural hearing protectors, 191, 192
Civil Aeronautics Act of 1938, 737
Civil Aeronautics Authority (CAA), 737
Civil Aeronautics Board (CAB), 737
Clamp-on ammeters, 256, 257
Clamp shuttle extractors, 633
Class A fire extinguishers, 283, 297–302
Class A fires, 269
Class B fire extinguishers, 297–302
Class B fires, 269
Class C fire extinguishers, 283, 297, 298, 300, 301
Class C fires, 269
Class D fire extinguishers, 297, 298, 300
Class D fires, 269–270
Classification
    of fire extinguishers, 297
    of fires, 269–270
    of hazardous locations, 247–252
    of wire ropes, 454
Classified locations. *See* Hazardous (classified) locations
Class I flammable liquids, 320, 332, 333, 336, 337
Class I hazardous locations, 247, 249–250, 481
Class II flammable liquids, 320, 332, 333, 335–337, 339
Class II hazardous locations, 247, 250–252, 481
Class III flammable liquids, 320, 332, 333, 335–337, 339
Class III hazardous locations, 247, 252, 481
Class K fire extinguishers, 297, 298, 300, 302–303
Class K fires, 270
Clean Air Act Amendments of 1990 (CAAA), 719, 720–721
Cleaning and sanitizing

of boilers and unfired pressure vessels, 137, 142
    containers, 580
    core-blowing machines, 661–662
    fall arrest equipment, 197–198
    indoor air quality and, 112
    metal parts, 342
    respirators, 210
    rubber boots, 215
    sanders, 540
    tanks and containers, 333–334, 340–341, 369
    wire ropes, 458
    work clothing, 217
Cleaning compounds, 578
Cleaning in place (CIP), 683
Clean materials recovery facilities, 767
Clean Water Act of 1972, 331, 370
Clearances
    for automated guided vehicles, 684
    for construction sites, 411, 509
    for cranes, 411
    for facility railways, 34–35
Clients, roles and responsibilities of, 74
Climate, in site selection, 29
Closed-circuit SCBAs, 203
Close-out meetings, 73–74
Clothing
    cold-weather, 216–217
    flame-retardant, 216, 355
    heat stress, 216
    high-visibility, 217
    impervious, 213
    protective, 216–217
    for welding, 583, 584
Clutches, for power presses, 627, 629, 643–644
Coal and coke, handling of, 654
Coast Guard, U.S., 755
Code letter ratings, for motors, 239
Code of Federal Regulations, 154, 166, 194, 754
Code of Practice on the Prevention of Major Accident Hazards (ILO), 719
Codes and standards. *See also specific codes and standards*
    boilers and unfired pressure vessels, 130–131
    building layout, 28–29
    color-coding, 46, 47–49, 122, 123, 206
    contractor compliance with, 91
    for electrical equipment, 229
    fall protection, 198
    for flammable and combustible liquids, 334
    for lockout/tagout, 687
    for metalworking machinery, 617
    for petroleum industry, 754–756
    for process safety management, 729
    for woodworking machinery, 556
Coil feeds, 624
Cold-coin dies, 672
Cold-contact burns, 377
Cold forming of metals, 622–649
    defined, 622
    kick presses and, 633–635
    metal shears and, 644–646
    power presses and, 622–644 (*See also* Power presses)
    press brakes, 646–649
Cold-heading machines, 672
Cold-trim dies and punches, 671
Cold water vs. high-temperature water, 139
Cold-weather clothing, 216–217
Collapsible cradles, 197
Collision accidents, 692
Color
    in building layout, 44–49
    human response to, 46–48
    for marking physical hazards, 39, 47–48
Color-coding, 46, 47–49, 122, 123, 184, 206
Columns, maintenance of, 102, 103
Combination supplied-air SCBA respirators, 206
Combination wrenches, 530
Combustible-gas indicators, 326–327
Combustible liquids, 319–344

bulk storage of, 334–340
cleaning tanks and containers for, 333–334, 340–341
common uses of, 341–343
definitions and terminology, 320–321
general safety measures for, 321–327
loading and unloading, 327–331
nonbulk storage of, 331–333
regulatory issues for, 334
storage of, 38
in welding, 586
Combustible materials
collection and storage of, 288
dust, 373
welding near, 579
Combustion: Fossil Power Systems (Singer), 133
Combustion gas detectors, 292–293
Commerce Department, U.S., 736, 737, 739
Commercial airlines, 742–743
Commitment, to safety and health programs, 756
Communication
in aviation safety, 747
fire detection and, 293
on indoor environmental quality issues, 111
privileged, 731
as safety component, 105, 144
Competent persons, defined, 64
Competitive states, 79
Completed operations liability insurance, 78
"Comply-with-all-laws" clauses, 84–85
Component failure, in stroking control systems, 638
Composite ropes, 449, 451
Comprehensive general liability insurance, 77–79
Compressed air hoses, 653
Compressed air locomotives, 37
Compressed gas cylinders. *See* Gas cylinders
Computer-integrated manufacturing (CIM), 695
Computerized maintenance management systems (CMMSs), 118, 681–682, 687
Computerized predictive maintenance (CPM), 118–120
Computer rooms, fire protection considerations for, 283
Computers, 705–711
as networking tools, 709–710
as program management tools, 711
as reference tools, 706–708
as safety culture tools, 710–711
Conceptualization, of projects, 69
Concrete construction, for fire prevention, 279
Concrete flooring, 53, 105, 106
Concrete parts, maintenance of, 103
Conductive clothing, 217
Conductive footwear, 214
Conductors, electric, 229, 230, 234–235
Confidentiality considerations, 729, 731
Confined spaces
in building layout, 32–33
design checklist for, 14–15
in facilities maintenance, 108
hazards in, 141
rescue systems for, 197
silos, 372
tanks, 369
Confining fire, 280
Connections
for sprinklers and standpipes, 283
for tank cars and trucks, 330–331
Connectors, 196
Consolidated insurance programs, 81
Constructability reviews, 70–71
Construction. *See also* Facilities construction
of boilers, 133–136
of cranes, 391–393
features for two-hand systems, 636–638
fire prevention materials in, 278–280
of guard devices, 159, 160
of man-lifts, 440
of tanks, 334, 335
Construction managers (CMs), 60, 61, 75
Construction Site Safety: A Guide for Managing Contractors

(Hislop), 60
Consumer Product Safety Commission, 115
Contact lenses, 185–187
Containers
cleaning, 580
for cryogenic liquids, 378, 379
for flammable liquids, 322
for gasoline, 545
for hazardous liquids and solids, 370–371, 373–374
storage of, 366–367
unpacking, 382–383
welding of, 579–580
Continuous control systems, 638
Continuous-conveyor ovens, 341, 342
Continuous-flow air-line respirators, 204
Contract close-outs, 73–74, 95–97
Contracting phase, of facilities construction, 71–72
Contractor-controlled insurance programs (CCIPs), 81
Contractors
bidding process for, 71
equipment and tool inspections, 72
general, 60, 75–76
process safety management for, 726
progress reports issued by, 73
request for proposal and pre-bid meeting, 90–92
risks of using, 61
safety responsibilities of, 60, 72, 73
screening of, 87
selection of, 71, 86–93
subcontractors, 60, 76
technical evaluation of, 87–90, 92–93
Contracts
in facilities construction, 82–86
safety clauses in, 84–86
termination/completion criteria, 92
Control components, 635
Control panels, 635
Control reliability, in single-stroke control, 638
Controls
for boilers, 133
boundaries and, 686
electrical, 237–238
for machines and equipment, 166
Convenience centers, for waste disposal, 767
Convention on International Civil Aviation, 740
Conveyor Equipment Manufacturers Association (CEMA), 417, 422
Conveyors, 417–425
aerial, 423–424
in automated systems and, 688
belt, 419–420
bucket, 422, 423
chain, 421
defined, 417
as fire hazards, 425
in foundries, 655
general precautions, 417–419
gravity, 424
live roll, 425
maintenance, 419
monorails and, 421
operation of, 418–419
pneumatic, 422–423
portable, 424
for power press operations, 633
roller or wheel, 425
screw, 421–422
shackle, 421
slat and apron, 420–421
training and, 420
vertical, 425
Cooling, of fire, 268–269
Coordinating engineers, 69
Coordination meetings, 72–73
Copper, 577
Cords
circular saws and, 536
electrical, 535
extension, 113, 240–241, 265

Cored-wire welding, 596
Cores and core-blowing machines, in foundries, 661–662
Cork tile, 53
Corporate safety philosophy, 88
Corridors, in facility layout, 50
Corrosion, in wire ropes, 458
Costs
    as criteria for contractor selection, 93
    estimating, 69–70
    of insurance, 76
Counterbalance, for power presses, 638, 643
Counterweights
    of cranes, 410
    ropes and, 432
    for saw tables, 562
Coupling equipment, 511
Covers, in open areas, 35
Crabs, 415
Cracks, 102, 103, 142–144
Craftspeople, 76
Crane Manufacturers Association of America, 399–400
Crane runways, 36, 123–124
Cranes, 391–417
    aerial baskets, 413–415
    block and tackle, 415–416
    crabs and winches, 415
    crane and hoist signals, 395–398
    crawler and wheel-mounted, 409, 411
    derricks, 404–405
    design and construction of, 391–393
    electric wires and, 411–413
    electromagnets and hook-on devices, 402
    fatalities involving, 408, 409
    in foundries, 655
    guards and limit devices for, 393–394
    inspection of, 399, 407–408, 412–413
    jib, 403–404
    load capacity, 392, 408–409
    locomotive, 409–410
    maintenance and safety rules for, 401
    mobile, 408–413, 508–509
    monorails systems for, 403
    operating rules for, 399–400, 407–408, 410–411
    operators for, 398, 407
    overhead, 391, 401–402
    overhead-crane runways, 36
    portable floor cranes and hoists, 416–417
    ropes and sheaves, 394
    storage bridge and gantry, 391, 392, 403
    tiering hoists and stackers, 417
    tower, 405–408
    travel of, 411
    types of, 392, 393
Crane trucks, 482
Crawler cranes, 409, 411
Crew resource management (CRM) training, 746
Crews. See Employees; Operators; Personnel; Workers
Crosscutting, 560, 563, 564
Crowbars, 363, 528
Crowl, Daniel A., 696, 698
Crucibles, 659
Crude oil, 752
Crunch zone, 504
Crushing accidents, 692
Cryogenic liquids, 374–380
    characteristics of, 375–376
    containers for, 378, 379
    housekeeping practices and, 378
    safety practices for, 376
    special precautions for, 376–378
    storage and handling of, 378–380
    training on, 378
    transfer lines for, 380
Cupolas
    blast gates and, 657
    carbon monoxide in, 656–657
    charging and charging floors of, 656
    dropping bottom doors of, 657–658

    hazards of, 656
    lining repairs for, 658
    safety tuyeres in, 657
    tapping out, 657
Current. See also Electricity
    alternating, 223–225
    defined, 223
    direct, 223, 225, 595
    inrush, 238–239
    leaks, 236, 237
Current-carrying conductors, 234–235
Curtiss JN-4H *Jenny* aircraft, 736
Custom-molded aural inserts, 192
Cutoff saws, 561–563
Cuts, clothing for protection from, 216
Cutters, 526, 528, 531–532, 608–609
Cutting. See Welding and cutting
Cutting actions, 169
Cuttings, removal of, 603
Cutting tools, 524–528, 531–532
Cylinders, gas. See Gas cylinders

D
Dadoing and rabbeting, 557, 559
Daily-fit tests, for respirators, 208
Daily participation, in discussion groups, 709
Databases, 708
Data collection, 711, 743
Daylight lighting, 41
Debris nets, 194–195
Decontamination considerations, 213–214
Deductibles, 79
Deepwater Horizon explosion (2010), 758
Defibrillation, 223
Deluge systems, 305–306
Demand respirators, 203
Deming, W. Edwards, 5
Demographics, color considerations and, 46
Derailers, 35
Derricks, 404–405
Descent devices, 197
Design
    of boilers and unfired pressure vessels, 19, 133–136, 139–140
    building layout and, 28–29
    checklist for general safety, 14–19
    of cranes, 391–393
    of dies, 666–667, 670
    for fire prevention, 276–277, 280–283, 310, 311–312
    of landfills, 766
    for power presses, 622–623
    process safety management and, 723, 727
    risk assessments in, 699
    risk elimination and reduction in, 10
    for safety, 13, 28–29
    of wire ropes for hoisting, 455
Designated safety representatives, 63–64
Design engineers, 74
Design-in safety, 681, 682, 689
Design phase, of facilities construction, 69–71
Design reviews, 70–71
Design safety reports, 23
Detection systems. See Fire detection systems
Devices, defined, 154
Dewar containers, 378, 379
Diagnostic technology, for maintenance, 119–120, 687
Dial feeds, 631
Die-enclosure guards, 627, 631
Die keys, 666
Dies
    cleaning, 625
    cold-coin, 672
    cold-trim, 671
    design of, 666–667, 670
    forging hammers and, 666–668
    for forging presses, 671
    for forging upsetters, 670
    guards for, 627, 631
    hot-trim, 671

pneumatic cushions and springs, 644
rolling-bolster carriages, 644
setup and removal of, 640–641, 667–668, 670, 671
tap and die work, 525
transferring, 639–640
Diesel locomotives, 36–37
Die setters, 626, 627
Die shoes, 640
Differential chain hoists, 391
Diffusion, rate of, 321
Digital multimeters, 256, 257
Dip tanks, 341
Direct current (DC), 223, 225, 595
Direct glare, 42
Dirty materials recovery facilities, 767–768
Disability glare, 42
Disabled workers. *See* Workers with disabilities
Disciplinary programs, 64–65
Discomfort glare, 42
Disconnecting reflectors, 109–110
Discussion groups, 708, 709–710
Displays, for machines and equipment, 166
Disposable clothing, 217
Disposable respirators, 202, 210
Disposal. *See* Waste collection and disposal
Distribution, of fire extinguishers, 299–300
Distribution piping, 588–589
Dock boards, 35, 381
Docks, loading, 34, 108
Documentation
of chemical processes, 695–696
of end of construction, 73
for lockout/tagout procedures, 171
of maintenance procedures and repairs, 686
of operating procedure changes, 700
of results, 8
of safety programs, 96–97
of training procedures, 694, 726
Doors
in bottom of cupola, 657–658
elevator car, 432
elevator hoistway, 429–430, 437
fire, 281
DOT. *See* Transportation Department, U.S.
Double-insulated tools, 246, 534
Dough mixers, 660
Downstream operations, in petroleum industry, 752, 754
Drag brakes, 644
Drainage
of driveways, 108
of floors, 52
at landfills, 765
Drain connections, inspection of, 106
Dressing, of abrasive wheels and disks, 615–616
Drilling, oil, 752, 754–755, 758
Drills
drill press machines, 605–606
electric, 535
redressing, 522
Drills, evacuation, 293
Drivers. *See* Operators
Driveways, maintenance of, 108
Driving conditions, on haul roads, 503
Droplines, 196
Drug-testing programs, 68
Drums
handling and lifting, 361, 362
of hazardous liquids, 368–370
opening, 383
welding of, 579–580
wire ropes and, 458–460
Dry chemical extinguishers, 301–302
Dry friction clutches, 643
Drying ovens, 341–342
Dry-pipe systems, 305, 306
Dry powder extinguishers, 302
Dual-chamber ionization detectors, 292
Dual-stream materials recovery facilities, 767

Dumbwaiters, 438
Dumpers, mobile, 354
Dumping procedures, 504–505
Dust
combustible, 247, 373
electric motors and, 238
explosions and, 289
exposure to, 556
formation of, 200
in foundries, 650, 662, 663
in magnesium grinding, 663
from sanders, 539, 540
in silos, 373
Dust masks, 202

E

Earmuffs, 191
Earplugs, 191, 192
Ear protection. *See* Hearing protection
Earth-moving equipment, 509–512
Economizers, 133
Eddy-current tests, 673
Edgers, 113–114
Edison, Thomas, 235–236
Education. *See* Training
Egress, means of, 16
Ejecting mechanisms, for power presses, 624, 631–633
Electrical controls
inspection and maintenance of, 642
on metalworking machinery, 603
on power presses, 635–639, 642
Electrical cords, 535
Electrical discharge machining (EDM), 609–610
Electrical equipment, 229–241
control equipment, 237–238
extension cords for, 240–241, 265
in facility layout, 40
flammable vapors and, 326
for grounds maintenance, 113–114
hand tools, 113–114
high-voltage, 226, 241
hoists, 390–391
industrial fires and, 285–286
inspecting, 122
installing, 229
lighting, 41–42
locomotives, 37
motors in, 238–240
portable power tools, 534–541
protective devices for, 234–237
safeguarding, 230–233
safe operation of, 552, 554
selecting, 229
specialized processes, 241
switches on, 233–234
testing devices, 256–257
trimmers and edgers, 113–114
warning signs and field markings for, 233
wiring, 229–230
Electrical hazard footwear, 214–215
Electrical safety, 221–263
definitions and terminology, 222–226
design checklist for, 15–16
employee training for, 260–261
equipment for, 229–241, 255–258
grounding, 224–225, 242–246
hazardous locations, 246–255
injuries and, 226–228
inspections and, 261–262
in loading and unloading of flammable and combustible liquids, 329–330
maintenance and, 255–258
qualified persons and, 225
work practices and protective equipment for, 258–260
Electrical shock
boilers and, 137
defined, 224
power tools and, 534

in welding, 594–595
Electric clutches, 643
Electric-drive elevators, 426–427
Electricity. *See also* Current
  dust explosions and, 373
  injuries resulting from, 228
  static, 323–326, 373, 377, 419
Electric trucks, inspection and maintenance of, 489–490
Electric wires, cranes and, 411–413
Electrocution, 224
Electrode holders, in arc welding, 594
Electromagnetic fields (EMF), 226, 686
Electromagnetic radiation suits, 217
Electromagnetic testing (ET), 143, 673
Electromagnets, 402
Electronically controlled vehicles (ECVs), 479
Electronic information. *See* Internet
Electrostatic dispensating (ESD) footwear, 373
Elevator Code (ANSI/ASME A17.1), 426–432, 434, 437, 440
Elevators, 426–438
  belt- and chain-drive, 427
  capacity ratings for, 427–428, 432
  cars, 431–433
  disabled persons and, 437
  electric-drive, 426–427
  emergency procedures for, 436–437
  fire prevention in, 428
  freight, 430, 431
  hand, 437–438
  hoisting ropes, 394, 426–427, 432–433
  hoistway doors and landings, 429–431, 434, 437
  hoistways, pits, and machine rooms, 428–429
  hydraulic-drive, 427
  inspection and maintenance of, 433–435
  interlocks, 430–431
  operating controls, 433
  operation of, 435–436
  powered industrial trucks in, 484–485
  for power press operations, 633
  sidewalk, 437
E-mail lists, 708, 709–710
Emergency drills, for fire protection, 293
Emergency lighting, 42
Emergency medical services, 33, 68
Emergency Planning and Community Right to Know Act of 1986 (EPCRA), 720
Emergency procedures
  for boilers, 138–139
  on construction sites, 68–69
  for elevators, 436–437
  for pesticides, 118
  in process safety management, 728
Emergency safety systems, design checklist for, 16
Emergency stop switches
  in broaches, 611
  conveyors and, 421
  on electrical woodworking equipment, 552
  in elevators, 427, 428, 433
  for robotic equipment, 693
Emission standards, 29, 32
Employees. *See also* Personnel; Workers
  contractor determination of hours worked, 88–89
  maintenance, 120–125
  process safety management and, 723
  training for (*See* Training)
Enclosure hearing protectors, 191, 192
Enclosures
  in building layout, 30, 50
  defined, 154
  elevator cars, 431
  maintenance and servicing, 170
  as safeguards, 164, 627, 631
End-of-service-life indicators (ESLIs), 201
End of work/project documentation, 73
Energy isolation, 170–172, 687
Energy sources, hazardous, 170–173
Enforcement
  of personal protective equipment use, 181

of process safety management, 729–730
of safety standards, 94–95
Engineered safety features (ESFs), 11
Engineering controls, in hazard prevention, 180, 351–352, 759
Engineering phase, of facilities construction, 69–71
Engineers
  coordinating, 69
  design, 74
  field, 65, 70–72, 74–75, 93–95
  project, 74–75
Engine lathes, 604
Enhanced Flight Vision Systems (EFVS), 747
Entrances
  in building layout, 30
  to foundries, 653
  for parking lots, 32
Entry, to unfired pressure vessels, 141–142
Envelope (space), maximum, 684
Environment, safety, and health (ES&H) requirements, 96
Environmental considerations
  design checklist for, 16–17
  indoor air quality, 111–112
  in loading and unloading of flammable and combustible liquids, 329
  pollution, 29, 32, 111
Environmental Protection Agency (EPA)
  Emergency Management program, 720
  emission standards, 29
  on flammable and combustible liquids, 331
  on landfills, 765–766
  on noise reduction ratings, 191
  on pesticides, 112–113, 117
  on process safety management, 720–721
  on risk management plans, 721
  tank storage requirements, 369
  on waste collection and disposal, 109, 764–765
  Worker Protection Standard, 116, 117
Environmental Research and Education Foundation (EREF), 770
EPA. *See* Environmental Protection Agency
Equipment. *See also* Electrical equipment; Haulage and off-road equipment; Personal protective equipment (PPE)
  color considerations, 46
  controls and displays for, 166
  electrical, 229–241 (*See also* Electrical equipment)
  explosion-proof, 255
  fall arrest systems, 197–198
  grounding, 224–225, 243–245
  inspection of, 72, 94, 122, 197
  layout of, 39
  maintenance of, 124–125, 197
  matching to operator, 163, 166
  process safety information for, 723
  restoring to service, 173
  shop equipment, 124–125
  for waste collection and disposal, 768–769
Ergonomics
  design checklist for, 17
  in safety through design, 13
  tool selection checklist for, 518
  in workstation design, 17, 54–55, 166
Error-provocative situations, 13
Escalators, 438–440
Escape-only SCBAs, 204
European Aviation Safety Agency (EASA), 741–742
Eutectic salt line thermal detectors, 291
Evacuation and evacuation drills, 282, 293
Evaluation
  of fire risk, 271–275
  of hazards, 200–201, 758
  of safety programs, 71–72, 89
  of training programs, 726
Evaporating pans, 147
Event tree analysis (ETA), 698, 723
Exhaust hoods, 374, 419
Exits
  in boiler rooms, 138
  in building layout, 51–52
  for fire protection, 281

to foundries, 653
    maintenance of, 110–111
    for parking lots, 32
Experience modification rates (EMRs), 79–80, 87–88
Exploration and production (E&P) operations, 752, 754, 759
Explosion-proof equipment, 255
Explosions
    boiler hazards, 132
    dust and, 373
    in foundries, 650, 656
    furnace, 130
    nanomaterials and, 352
    wood sanding and, 540
    worldwide industry incidents, 719
Explosive atmospheres, 289
Explosives, storage of, 374
Exposure. *See also specific materials*
    to dust, 556
    frequency and duration of, 7
    to nanomaterials, 351
    permissible limits to, 190–191, 491, 580–581, 657
    in welding, 580–583
Exposure protection, in fire protection site planning, 277–278
Extension cords, 113, 240–241, 265
Exterior site planning, for fire protection, 277–278
Exterior walls, maintenance of, 103
Extinguishing agents, 269. *See also* Fire extinguishers; Fire suppression and extinguishment
Extinguishing system attachments, 293
Extracting, tools for, 629
Extrusion lines, 683
Eye injuries, from electricity, 228
Eye protection. *See also* Goggles and glasses
    comfort and fit of, 187
    contact lenses and, 185–187
    for kick presses, 633
    with landscaping tools, 113
    from laser beams, 188
    for materials handling, 356
    power tools and, 546
    selection of, 185
    for welding, 188–190, 583
    for woodworking, 556

F

Face protection, 185, 187–188, 556
Facilities
    fire protection programs for, 309–312
    foundry structures in, 652–653
    layout for, 38–41
    process safety management for, 720, 721
    railways for, 34–38
    for shipping and receiving, 30, 380–381
    for waste collection and disposal, 768, 769
Facilities construction, 59–98
    contract close-outs in, 73–74, 95–97
    contractor selection in, 71, 86–93
    contracts in, 82–86
    cost and schedule estimates for, 69–70
    engineering and design phase, 69–71
    field engineers in, 65, 70–72, 74–75, 93–95
    for fire prevention, 276–283, 310, 311–312
    insurance for, 76–82
    integrating safety into, 69–74
    procurement–contracting phase, 71–72
    roles and responsibilities in, 74–76
    safety programs for, 60–69
    work–construction phase, 72–73
Facilities maintenance, 101–126
    canopies, 108
    computerized predictive maintenance, 118–120
    crews for, 120–125
    for fire prevention, 283–289, 312
    fixed ladders, 107–108
    floors, 104–106
    foundations, 102
    grounds maintenance, 112–118
    heating equipment, 111

indoor environmental quality issues in, 111–112
    lighting systems, 109–110
    platforms and loading docks, 108
    preventive and predictive, 102, 112, 121, 124
    roofs, 106–107
    sidewalks and driveways, 108
    stacks and chimneys, 107
    stairs and exits, 110–111
    structural members, 102–103
    tanks and towers, 107
    underground utilities, 108–109
    walls, 102, 103–104
Factory Mutual (FM), 322, 491, 589, 591
Failure mode and effect analysis (FMEA), 7, 24, 680, 698, 723–724
Failure modes, effects, and criticality analysis (FMECA), 698
Fair Labor Standards Act of 1938 (FLSA), 492
Fall arresters, 196
Fall arresting systems (FASs), 196–197
Fall arrest systems, 192–198
    active, 193, 195–197
    cleaning equipment, 197–198
    defined, 192–193
    elements of program for, 193
    inspection and maintenance of, 197, 198
    passive, 193, 194–195
    rescue systems, 197
    selecting, 193–194
    storage of, 198
Fall avoidance, design checklist for, 17
Fall hazards
    electrical shock and, 228
    floor maintenance and, 104–105
    in materials handling, 354, 357
    power tools and, 546
    in waste collection and disposal, 769–770
Fall protection standards, 198
Fastenings, rope, 433
Fatalities
    crane-related, 408, 409
    in petroleum industry, 753–754
    robotic equipment and, 693
    in waste industry, 764
Fatigue, 166, 633–634
Fault tree analysis (FTA), 680, 698, 723
Feather boards, 559, 569
Federal Aviation Administration (FAA), 737–738, 743, 747
Federal Insecticide, Fungicide, and Rodenticide Act of 1996 (FIFRA), 116, 117
Federal Mine Safety and Health Act of 1977, 199
Federal Railroad Administration (FRA), 35
Feeding mechanisms, for power presses, 623–625, 629–632
Feet. *See* Foot protection
Fencing, 30, 154
Fiberboard cartons, 366–367
Fiber cores (FCs), 455
Fiber ropes, 448–454
    care of, 453–454
    inspection of, 451–453
    slings, 463–466
    specifications for, 450–452
    types of, 448–449
    working load of, 449–451
Fibers, combustible, 247
Field engineers, 65, 70–72, 74–75, 93–95
Field markings, for electrical equipment, 233
Files
    for cutting metal, 525–526
    redressing, 522
    work safety, 95
Filtering facepiece respirators, 202
Filter lenses, 189, 190
Filters, respirator, 202, 207
Finger cots/stalls, 212
Fire alarm systems, 294–295
Fire blankets, 303
Fire brigades, 293–294
Fire department access, in site planning, 277
Fire detection systems, 289–296

alarm systems, 294–295
    automatic, 290–293
    communications and, 293
    in facility fire protection programs, 310–311
    human observers as, 290
    location of, 296
    maintenance of, 296
    spacing of, 296
Fire doors, 281, 428, 429
Fired pressure vessels. *See* Boilers
Fire extinguishers, 296–303
    classification of, 297
    distribution of, 299–300
    equipment and accessories, 303
    location of, 297–299
    maintenance and inspection of, 303
    portable, 296–303
    principles of use, 296–297
    requirements for, 29
    symbols and markings on, 298
    types of, 300–303
Fire hazard analysis, 270, 312–314
Fire hoses, 308–309
Fire hydrants, 307–308
Fire loads, 272–273
Fire prevention
    construction of facilities for, 276–283, 310, 311–312
    detection and response systems, 289–296, 310–311
    in elevators, 428
    in facility fire protection programs, 310
    gasoline and, 545
    inspections for, 284
    in magnesium grinding, 663
    maintaining facilities for, 283–289, 312
    nanomaterials and, 352
    in tanks, 340
Fire protection, 267–315
    for boilers, 132
    building design, methods and concepts in, 276–277, 280–283, 310
    chemistry of fire, 268–269
    classification of fires, 269–270
    construction of facilities for, 276–283, 310, 311–312
    design checklist for, 17
    detection and response systems, 289–296, 310–311
    escalator floors and, 439–440
    facility fire protection programs, 309–312
    fire risk and, 270–275
    fire suppression and extinguishment, 296–309
    in foundries, 652, 663
    liquefied petroleum gas and, 492
    in loading and unloading of flammable and combustible liquids, 327–328
    maintaining facilities for, 283–289, 312
    planning for, 276–278
    in welding, 578–579
Fire pumps, 307
Fire risk, 270–275
    assessment process, 270
    conveyors and, 425
    evaluating, 271–275
    hazard analysis for, 270, 312–314
    identification of hazardous materials and, 271, 272–274
    loss scenarios and, 271–275
    sanding and, 540
Fires
    chemistry of, 268–269
    classification of, 269–270
    confining, 280
    cooling of, 268–269
    emergency preparedness plan for, 68–69
    industrial fires, factors contributing to, 285–289
    recovering from, 312
    in tanks, 340
    worldwide industry incidents, 719
Fire suppression and extinguishment, 296–309
    agents for, 269
    in facility fire protection programs, 311–312

fire extinguishers, 296–303
    fire hydrants, 307–308
    hoses, 308–309
    special systems and agents, 309
    sprinkler and water-spray systems, 303–307
Firetube boilers, 133
Fire watchers, 579
First-aid kits, 546–547
First-line supervisors, responsibilities in safety programs, 63
Fit tests, for respirators, 207–208
Fixed-barrier guards, 627
Fixed die-enclosure guards, 627
Fixed stairs and ladders
    design checklist for, 19
    in facility layout, 50–51
    maintenance of, 107–108
    safety requirements for, 393
Fixed-temperature fire detectors, 291
Flame, propagation of, 321
Flame detectors, 292
Flame fires, 268
Flame-retardant work clothes, 216, 355
Flammability hazards, 271, 273
Flammable gas precautions, 377–378
Flammable liquids, 319–344
    auto-ignition temperature of, 320–321
    bulk storage of, 334–340
    cleaning tanks and containers for, 333–334, 340–341
    common uses of, 341–343
    definitions and terminology, 320–321
    disposal of, 334
    general safety measures for, 321–327
    loading and unloading, 327–331
    nonbulk storage of, 331–333
    process safety management for, 720
    regulatory issues for, 334
    space requirements for, 29
    storage of, 38
    upper/lower flammable limit of, 321
    utility tractors near, 116
    in welding, 586
Flammable ranges, 321
Flashback, in hoses for welding gases, 591
Flashover, 222
Flash point, 320
Flasks, 662
Fleet angle, wire ropes and, 460
Flexible-shaft tools, 544
Flicker, 43
Flight data recorders, 740
Flight Operational Quality Assurance (FOQA), 746–747
Flight safety departments, 742–743
Flight Safety Foundation (FSF), 745
Float switches, 136
Floor cranes, 416–417
Floor openings, 19, 29, 441
Floors
    automated guided vehicles and, 688, 689
    in boiler rooms, 137
    charging, 656
    design checklist for, 19
    elevator cars, 432
    in foundries, 652, 653
    load capacity and distribution, 54, 105, 652
    maintenance of, 104–106
    materials for, 52–54
    open-sided, 49
    overloading of, 105–106
    in restricted areas, 684
    slipping hazards and, 104–105, 770
    in warehouses and shipping rooms, 380
    welding and, 579
Flow sheets, 39
Fluid transfer, bonding and grounding during, 323–326
Fluorides, 577, 650
Flux-cored arc welding (FCAW), 596
Flyings, combustible, 247
Flywheel brakes, 635

Flywheels, on forging hammers, 665–666
FM Global, 285
Follow feeds, 625
Follow up, on actions taken, 8
Food processing, safety precautions in, 683–684
Foot control, for power presses, 629–630, 639
Footings, maintenance of, 102
Foot protection
    conductive, 214
    for electrical hazards, 214–215
    electrostatic dispensating, 373
    for foundries, 215
    for materials handling, 356, 363
    sole puncture resistant, 215
    static dissipative, 215, 377
    for welding, 546
Footwalks, on cranes, 402
Foreign objects, in industrial fires, 286
Forging hammers, 664–669
    dies and die keys for, 666–668
    flywheels and pulleys on, 665–666
    gravity-drop, 664–665
    guard devices for, 664–666
    hand tools for, 666
    hazards of, 664
    inspection and maintenance of, 668–669
    key-driving rams for, 665, 666
    lead casts and, 669
    open-frame, 664
    personal protective equipment for, 668
    safe operating practices for, 668
    safety props for, 666
    scale guards for, 665
    steam and air, 664, 665
    treadles and pedals on, 665
Forging presses, 670–671
Forging upsetters, 669–670
Fork trucks (forklifts)
    in hazardous locations, 481
    high-lift, 478
    performance tests for, 494, 495
Formable aural inserts, 192
Formal databases, 708
Forward movement, of heavy equipment, 504
Foundations, maintenance of, 102
Foundries. *See also* Hot working of metals
    chips in, 663
    cleaning and finishing products, 663–664
    compressed air hoses in, 653
    conveyors in, 655
    cores and core-blowing machines in, 661–662
    cupolas in, 656–658
    dough mixers, 660
    dust in, 650, 662, 663
    explosions in, 650, 656
    facility structures in, 652–653
    fire protection in, 652, 663
    flasks in, 662
    floor loading in, 652
    footwear for, 215
    hazards of, 650–652, 654
    hoists and cranes in, 655
    housekeeping practices in, 652, 663
    inspection and maintenance of, 652
    ladles in, 654–655
    lighting in, 652
    magnesium grinding in, 663
    materials handling in, 653–656
    medical programs in, 651–652
    molds and molding machines in, 655, 661
    noise control in, 652
    personnel facilities in, 652
    power presses in, 663–664
    production equipment in, 660–662
    sand, coal, and coke in, 654
    sandblast rooms in, 662
    sand cutters in, 660–661
    sand mills in, 660

    scrap breakers in, 655
    shake-out machines in, 662
    sifters in, 661
    slag disposal in, 656
    storage in, 655–656
    tumbling barrels in, 662
    ventilation in, 652
    welding in, 663
    work environments in, 652–653
Four-wheeled trucks, 365
Frames, for power presses, 642
Freight elevators, 430, 431
Frequency, defined, 223
Friction
    electric motors and, 240
    industrial fires and, 286
Fuel, removing from fire, 269
Fuel-fired locomotives, 36
Full-body harnesses, 195
Full-face respirators, 206
Full-revolution clutch power presses, 629, 630, 643–644
Fumes, 200, 532, 576, 582
Furnaces
    crucibles and, 659
    cupolas, 656–658
    explosions in, 130
    industrial fires and, 287
    precautions for entering, 137
Furnishings, for fire prevention, 278–280
Fuses, 235, 258
Fusible plugs, 134–135

G

Galleries, in foundries, 653
Gang boxes, 521
Gangways, 328, 653
Gantry cranes, 391, 392, 403
Garbage. *See* Waste collection and disposal
Gas cylinders, 586–592
    distribution piping for, 588–589
    handling, 586–587
    hoses and hose connections for, 591
    manifolds and, 588
    portable outlet headers for, 589
    powder-cutting processes and, 592
    pressure regulators for, 589–590
    safe practices for use of, 587–588
    storing, 367, 374, 587
    torches and, 591–592
Gases
    defined, 200
    explosions and, 289
    flammable, 247
    hazardous, handling and storage of, 374
    in welding, 574–575
    for welding and cutting, 585–588
Gas-fired ovens, 660
Gas industry. *See* Petroleum industry
Gas lines, industrial, 123
Gas masks, 207
Gas-metal arc welding (GMAW), 574, 595–596
Gasoline, containers for, 545
Gasoline-operated trucks, 491
Gasoline-powered tools and equipment, 114–116, 544–545
Gas-removing respirators, 202, 206–207
Gas-tungsten arc welding (GTAW), 574, 595
Gates, elevator, 430, 432
Gauging, for tanks, 336
Gear cutters, 608–609
General contractors, 60, 75–76
General lighting, 41–42
General-purpose press brakes, 647
General safety design checklist, 14–19
Generators, 44
Gin pole derricks, 405
Girders, maintenance of, 103
Glare, 42
Glass

in fire prevention, 280
handling and lifting, 361–362
unpacking, 382
Glasses. *See* Goggles and glasses
Global Air Navigation Plan (GANP), 741
Global Aviation Safety Plan (GASP), 741
Global Harmonized System, of hazardous materials identification, 271
Gloves, 212–214, 260, 354–356
Glue, 383
Goggles and glasses. *See also* Eye protection
chemical, 183, 188
comfort and fit of, 187
laser-protective, 188
selection of, 185
for welding, 583, 584
in woodworking, 556
G-1 powdered agent extinguishers, 302
Governor ropes, 433
Graders, 509–512
Grades, driving on, 487–488
Gravity conveyors, 424
Gravity-drop hammers, 664–665
Gravity feeds, 624–625
Grinding machines, 611–617
abrasive disks and wheels, 539, 612–616
air-powered, 541
checklist for safe operation, 613
grindstones, 616–617
guard adjustments for, 614
hazards of, 611–612
operation of, 613–614
polishing and buffing wheels, 617
rotary die, 538–539
safe speeds for, 614–615, 617
surface and internal, 616
wire brush wheels, 539, 617
work rests for, 615
Grinding wheels, 539
Grindstones, 616–617
Grips, for lifting, 360–361
Ground-fault circuit interrupters (GFCIs), 113, 236–237, 240, 246, 519, 534–535
Grounding
of cranes, 411
defined, 224–225
of elevators, 433
equipment, 243–245
in fluid transfer, 323–326
of heavy equipment, 506–507
maintenance of, 245
of portable power tools, 519, 534–535
of stationary equipment, 377
systems for, 40, 242–243
of two-hand control systems, 639
of welding equipment, 594
Grounding bushings, 231
Grounds, electrical, 224
Grounds maintenance, 112–118
chemicals and pesticides used in, 116–118
electric-powered hand tools for, 113–114
gasoline-powered equipment for, 114–116
landscaping, 32, 113
snow shoveling, 116
Groundwater monitoring requirements, at landfills, 765, 766
Group lockout, 173
Guard devices. *See also* Safeguarding; Shields
automatic or semiautomatic, 162–164
for band saws, 536
barrier guards, 162, 164, 232–233, 626, 627–628, 685–687
for boring mills, 606, 607
built-in, 162
construction of, 159, 160
for conveyors, 419
for cranes, 393–394
defined, 154
design checklist for, 18
enclosures, 164, 627, 631

for forging hammers, 664–666
for forging presses, 671
in foundries, 655
for gear cutters, 608–609
for grinding machines, 614, 616, 617
hand removal/restraint devices, 164
for hand trucks, 483
for in-running rolls, 157, 158–159
inspection and maintenance of, 644
interlocking, 162, 627, 685–687
for kick presses, 633, 634
maintenance and, 507
maintenance and servicing, 170
for man-lifts, 441
mechanic check of, 124
for milling machines, 607–608
miscellaneous, 165
noise control and, 175
openings used for safeguarding, 156–159
for planers, 567, 568–569, 610
for powered industrial trucks, 480, 481
for power presses, 626–628, 644
for power transmissions, 166–170
for press brakes, 648, 649
for sanders, 569
for saws, 557, 559, 562, 563, 566
for shapers, 568
standard materials and dimensions for, 167
test data for, 157, 158
two-hand trip, 165
for woodworking equipment, 554
Guarding. *See* Safeguarding
Guidelines for Hazard Evaluation Procedures (American Institute of Chemical Engineers), 698
Guide posts, for power presses, 623
Gutters, inspection of, 106
Guy derricks, 404–406
Gypsum, in fire prevention, 280

H
Hacksaws, 525
Haddon, William, 13–14
Hair protection, 184–185
Half-face respirators, 206
Hammers, types and uses of, 523–524, 664–669
Hand-arm vibration syndrome (HAVS), 518, 542, 547
Hand-drilling hammers, 524
Hand elevators, 437–438
Hand-fed material, 624
Hand-feeding tools, 624
Handholds, on man-lifts, 440
Handicapped workers. *See* Workers with disabilities
Hand leathers, 213
Handles, for tools, 522–523
Hand protection, 211–214, 356
Handrails, 29, 35, 49, 51
Hand removal devices, 164
Hand snips, 526
Hand tools, 523–532. *See also specific tools*
carrying, 521
central control of, 520
for cutting, 524–528, 531–532
for forging hammers, 666
for ground maintenance, 113–114
hammers, 523–524
maintenance and repair, 521–523
for materials handling, 363, 528–529
for power presses, 625–626
preventing incidents with, 518–521
punches, 522, 524
safety practices for, 113–114, 519–520
screwdrivers, 522, 523
selection of, 518–519
soldering irons, 532
spark-resistant, 532
storage of, 520–521
wrenches, 529–532
Hand trucks, 364–365, 479, 483

Hard hats, 182
Hardware connectors, 196
Harnesses, 141–142, 195, 197
Hatch covers, for sidewalk elevators, 437
Hatchets, 522, 527
Haulage and off-road equipment, 501–512
    accident prevention, 502
    for clearing work, 511–512
    driver qualifications and training, 503–504
    dumping procedures, 504–505
    general safety requirements for, 502–506
    graders, bulldozers, and scrapers, 509–512
    grounding systems for, 506–507
    haul roads, 503
    maintenance practices for, 507, 510
    mobile cranes, 508–509
    operating practices for, 503–504, 509, 510–512
    operating vehicles near workers, 504
    power shovels, 506–508
    protective frames for, 505
    for towing, 506, 511
    for transportation of workers, 505–506
Haul roads, 503
Hazard analysis
    for chemical processes, 696–698
    for fire risk, 270, 312–314
    flowchart for, 155
    process hazard analysis, 680
    in process safety management, 723–724
    respiratory, 200–201
    risk assessments and, 6–9
Hazard and operability studies (HAZOPs), 680, 698, 723
Hazard management process, 756–759
Hazardous (classified) locations
    Class I locations, 247, 249–250
    Class II locations, 247, 250–252
    Class III locations, 247, 252
    dumps, excavations, and slopes, 512
    electrical, 246–255
    establishing limits of, 252–254
    explosion-proof equipment, 255
    lighting in, 43
    planning electrical installations, 253
    powered industrial trucks in, 481–482
    power lines, 411, 509
    reducing hazards, 253
    sample installations, 253–255
    shock hazards and, 534
    welding near, 579
    work sites with vehicles, 504
Hazardous energy sources, control of, 170–173
Hazardous materials
    categories for control of, 180
    design checklist for, 17–18
    in foundries, 650–651
    gases, 374
    handling and storage of, 368–374
    identification for fire risk, 271, 272–274
    liquids, 368–372
    nonregulated, 766–767
    process safety management for, 718–719
    release incidents of, 718
    solids, 372–374
Hazardous Materials Transportation Uniform Safety Act of 1990, 327
Hazard rating system, identification of materials by, 271, 272–274
Hazards. See also Health hazards; Injuries
    airborne, 200–201
    alligator shears, 645–646
    burn, 683–684
    categorization of, 756, 757
    chemical, 683
    in confined spaces, 141
    defined, 6, 756
    discharge gases, 610
    dust, 556
    emergency procedures for, 68–69
    employee awareness of, 651
    evaluation of, 200–201, 758

flammability, 271, 273
forging hammers, 664
in foundries, 650–652, 654
health (See Health hazards)
identification and control of, 7, 70, 200, 682–686, 696–698, 756, 758–759
instability, 271, 274
jointer/planers, 566–567
metalworking machinery, 604–607, 611–612
ovens, 659–660
in petroleum industry, 753–754, 756–759
pinch, 684
planers and shapers, 610
of pneumatic-impact tools, 542
of portable power tools, 519–520, 532–533
power presses, 623, 624
press brakes, 649
radiation, 684
respiratory, 200–201
of robotic equipment, 173–174, 691–695
saws, 544–545, 556, 559, 565
signs and warning labels, 422
types of, 683–684
in waste collection and disposal, 764, 766–770
welding, 574–580
in wood sanding, 540
Hazard surveys, 696
Head protection, 182–185, 189
Health hazards
    fire risk and, 271, 272
    of flammable and combustible liquids, 326
    in foundries, 650–652
    in welding, 574–578
Health programs, in petroleum industry, 755, 756–758
Hearing conservation programs, 190–191, 556, 664
Hearing protection, 190–192, 546, 555–556, 605, 664
Heat exhaustion, 769
Heating, ventilation, and air conditioning (HVAC), 40, 111, 112
Heating equipment
    extension cords for, 240
    facility layout and, 40
    maintenance of, 111
Heat-protective clothing, 215–216
Heat stress clothing, 216
Heatstroke, 769
Heavy equipment. See Haulage and off-road equipment
Heavy timber construction, for fire prevention, 278
Hedge trimmers, 113
Helmets, 182–184, 189, 192, 546, 583
Hidden provisions, in contracts, 85
Hierarchy of controls, 6, 9–11, 758
High-bay storage, 366
High-pressure boilers, 133
High-pressure systems, 147–148
High-temperature water (HTW), 133, 139
High-visibility clothing, 217
High-voltage equipment, 226, 241
Histoplasma capsulatum, 200
Hitches, sling, 462–463, 473
Hoisting and conveying equipment, 389–443
    conveyors, 417–425 (See also Conveyors)
    cranes, 391–417 (See also Cranes)
    design factors for wire rope used in, 455
    dumbwaiters, 438
    elevators, 426–438 (See also Elevators)
    escalators, 438–440
    in foundries, 655
    hoisting apparatus, 390–391
    man-lifts, 440–443
    moving walks, 440
Hoistways
    dumbwaiter, 438
    elevator, 428, 429–431, 434, 437
Holding time, in single-stroke control, 638
Holdout devices, for power presses, 628
Hoods
    acid, 188
    air-supplied, 205

on baffle furnaces, 651
for saws, 557, 562
Hook-on devices, 402
Hooks
inspection of, 399
for materials handling, 363
slings and, 464, 467
use of, 528
Horizontal boring mills, 606
Horizontal lifelines, 195–196
Hoses
air, 541, 653
fire, 308–309
for welding, 591
Hot saws, 672
Hot-trim dies and punches, 671
Hot-water boilers, 133
Hot work
in fire prevention, 284–285
permits for, 284, 578–579, 725
Hot working of metals, 649–674. *See also* Foundries
bolt headers and riveting machines in, 672
bulldozers in, 672
cold-heading machines in, 672
crucibles for, 659
cupolas in, 656–658
forging hammers for, 664–669
forging presses for, 670–671
forging upsetters for, 669–670
hot saws in, 672
nondestructive testing and, 672–674
ovens for, 659–660
padding, bending, and straightening equipment for, 671
protective clothing for, 215–216
Housekeeping practices
in construction sites, 67–68
cryogenic liquids and, 378
for facilities maintenance, 104–105
in foundries, 652, 663
industrial fires and, 288–289
woodworking and, 554–555
Howard, John, 4
Hue. *See* Color
Human error analysis, 698
Human observers, as fire detection systems, 290
Hydrants, fire, 307–308
Hydraulic dock levelers, 485
Hydraulic-drive elevators, 427
Hydraulic power tools, 544
Hydraulic press brakes, 646, 647
Hydrogen peroxide, 683
Hydrostatic testing, 142

I

IATA Ground Operational Manual (IGOM), 746
IATA Operational Safety Audit (IOSA) program, 745–746
IATA Safety Audit for Ground Operations (ISAGO), 746
Identification
of hazards and hazardous materials, 271, 272–274, 756, 758–759
in helmets, 182
of piping, 122
of respiratory hazards, 200
Illuminating Engineering Society (IES)
ANSI/IES RP-7, Practice for Industrial Lighting, 42, 43, 109, 380
Calculation of Daylight Availability, 41
Illumination. *See* Lighting
Immediately dangerous to life and health (IDLH) atmospheres, 201, 203, 207
Impact, clothing for protection from, 216
Impact accidents, 692
Impact wrenches, 542–543
Impedance, defined, 225
Impervious clothing, 213
Inch control, for power presses, 638
Incident rate calculation, 88
Incident reports and investigations, 67, 95, 175, 729
Inclined ladders, 51

Indemnity agreements, 60, 78
Independent Association of Drilling Contractors (IADC), 755
Independent contractor's protective liability insurance, 77–78
Independent wire rope cores (IWRCs), 454, 455, 463
Indoor air quality (IAQ), 111–112
Indoor environmental quality (IEQ), 111–112
Industrial by-product disposal, 29
Industrial fires, factors contributing to, 285–289
Industrial gas lines, 123
Industrial robots. *See* Robotic equipment
Industrial Ventilation Manual (ACGIH), 111
Inert gas displacement, 334
Inert gas precautions, 377
Informal knowledge, 708
Information, electronic. *See* Internet
Infrared detectors, 292
Infrared imaging, 120
Infrared radiation, in welding, 189, 190, 582
Infrared thermography, 256
Injuries. *See also* Hazards
to back, 356–357
cold-contact burns, 377
contractual notice of, 85
crane-related, 408, 409
in die setup and removal, 641
electrical, 226–228
emergency procedures for, 68–69
frequency and severity rates, 88–89
jointers/planers, 566–567
kick presses, 633–634
manual lifting and, 356
materials handling and, 354–357
metalworking machinery and, 602
personal protection, 354–355
reporting requirements, 175
ropes and, 450
from saws, 561–562, 565
shapers, 567
tool-related, 518–521, 532
in waste industry, 764, 766–767
In-running nip points, 168
In-running rolls, 157, 158–159
Inrush current, 238–239
Insecticides, 117
Inside storage, 40–41, 332–333
Inspections
abrasive wheels and disks, 612
aerial baskets, 414–415
boilers and unfired pressure vessels, 131–132, 136–137, 140–142
during construction, 65, 73
cranes, 399, 407–408, 412–413
electrical, 261–262
elevators, 433–435
equipment, 72, 94, 122, 197
by field engineers, 94
fire extinguishers, 303
for fire prevention, 284
forging hammers, 668–669
forging upsetters, 670
of foundries, 652
for haulage and off-road equipment, 503
hoisting ropes, 394
hoists, 391
in loading and unloading of flammable and combustible liquids, 330
man-lifts, 442
ovens, 660
pallets, 488
powered industrial trucks, 488–492
power presses, 641–643
in process safety management, 728
respirators, 210–211
roofs, 106
ropes, 451–453, 455–458
routines for, 435
of safety programs, 65
slings, 466, 467, 469, 471–473
tools, 72, 520, 521, 533–534
waste products, 765

woodworking machinery, 555
Instability hazards, 271, 274
Installation, of electrical controls on power presses, 635–636
Institute of Makers of Explosives, 38
Instrumentation, for boilers, 133
Insulation, electric, 230
Insulation-resistance testers, 533
Insurance, 76–82
    comprehensive general liability coverage, 77–79
    costs of, 76
    deductibles, 79
    defined, 77
    premiums, 76
    workers' compensation, 79–81
    wrap-up, 81–82
Integrated pest management (IPM), 117
Interior building functions, in fire prevention, 277
Interior furnishings, for fire prevention, 278–280
Interior walls, maintenance of, 103–104
Interlocks
    barrier guards and, 162, 627, 685–687
    for electrical equipment, 231–232
    elevator, 430–431
    for two-hand control systems, 638–639
Intermittent-start equipment, electrical, 258
Internal-combustion engines, 342
Internal grinders, 616
Internal inspection, of boilers, 136–137, 140–142
International Air Transport Association (IATA), 745–746
International aviation safety, 740–742
International Building Code, 110, 276, 280
International Chamber of Commerce (ICC), 719
International Civil Aviation Organization (ICAO), 740–741
International Electrotechnical Commission (IEC), 238
International Federation of Robotics (IFR), 690
International Labour Organization (ILO), 719
International Protection (IP) ratings, 238, 239
International Safety Equipment Association (ISEA)
    ANSI/ISEA 107-2010, High-Visibility Safety Apparel, 217
    ANSI/ISEA Z87.1, Occupational and Educational Eye and Face Protection, 185, 186, 189, 543, 546, 556, 583
    ANSI/ISEA Z89.1, Industrial Head Protection, 182–183, 546, 583
    ANSI/ISEA Z308.1, Minimum Requirements for Workplace First Aid Kits, 546
Internet, 705–711
    as networking tool, 709–710
    as program management tool, 711
    as reference tool, 706–708
    as safety culture tool, 710–711
Interrupted stroke protection, in single-stroke control, 638
Investigations, 67, 95, 436, 729, 739–740, 744
Ionization detectors, 292
IOSA Standards and Recommended Practices (ISARPs), 746
Iron-oxide fumes and dust, 650
Irregularly shaped objects, handling and lifting, 362

J
Jackhammers, 542
Jacks, 363–364
Japanning ovens, 341–342
Jenny aircraft, 736
Jet turbines, 741
Jib cranes, 403–404
Jigsaws, 538, 566
Job hazard analysis (JHA), 756
Job safety analysis (JSA), 63, 67, 71–72, 75–76, 259, 355
Job site
    contractor safety program requirements, 91
    monitoring by field engineers, 94
Jointers/planers, 553, 566–567
Joint ventures, 78
Joists, maintenance of, 103
Jurisdictional laws, 85
Just-in-time (JIT) production, 681

K
Kegs. *See* Barrels and kegs
Key-driving rams, 665, 666

Keywords, in Internet searching, 708
Kickback, of tools, 537, 544–545, 556–557, 563–564
Kick presses, 633–635
Knee pads, 216
Knife switches, 234
Knives
    of shapers, 567–568
    use of, 527–528
Knockout pins, for power presses, 633
Knots, wire ropes and, 461–462
Knowledge, informal, 708

L
Labeling
    of escape-only pressure-demand SCBA, 204
    of flammable liquid storage cabinets, 333
    of pesticides, 117
Labor-saving devices, for lighting maintenance, 109–110
Lacquer method, for crack detection, 144
Ladders
    in boiler rooms, 138
    design checklist for, 19
    fixed, 19, 50–51, 107–108, 393
    inclined, 51
    for lighting maintenance, 110
    maintenance of, 107–108
    on man-lifts, 441–442
    in unfired pressure vessels, 141
    in warehouses and storerooms, 381
Ladles, 654–655
Lamps. *See* Lighting
Landfills, 32, 764, 765–766
Landings
    elevator, 429, 434
    on man-lifts, 440
Landscaping, 32, 113. *See also* Grounds maintenance
Lanyards, 195
Laser beam protection, 188
Laser monitors, 687
Lasers, in welding, 596
Laser safety goggles, 188
Laser shaft alignment, 119
Lathes, 553, 569–570, 604–605
Lawn mowers, 114–115, 511
Lawn trimmers, 113–114
Lawsuits, for construction injuries, 75
Layout. *See* Building layout
Leachate, 765, 766
Lead, 577, 650
Lead casts, 669
Leaded clothing, 217
Leaks
    of current, 236, 237
    in gas cylinders, 588, 590
    in pressure vessels, 142–143, 147–148
    roof, 107
Leather clothing, 215
Lederer, Jerome F., 745
Leggings, 215
Lenses, for eye protection, 185, 189–190, 583–584
Liability, of project engineers, 74–75
Liability insurance, 77–79
Liberty Mutual Insurance Company, 157
Licensure requirements, 91
Lifelines, 195–196
Lifting
    grips for, 360–361
    guidelines for, 357–363
    heavy objects, 362–363
    limits to, 358–359
    load held close to body, 360
    machines, 362–363
    manual, 356
    rules for, 359
    team, 361
    techniques for specific shapes, 361–362
Lifts, scissor, 110
Lift trucks. *See also* Fork trucks (forklifts)

in hazardous locations, 481
    maneuvering, 487
    parking for, 39–40
    styles of, 478
Light beam sensors, 232
Lighting
    in boilers rooms, 137–138
    color effects and, 44–46
    daylight, 41
    electric, 41–42
    elevator, 432
    emergency, 42
    extension, 240–241
    in foundries, 652
    glare, 42
    in hazardous (classified) locations, 43
    maintenance of, 109–110
    management of, 42–43
    on man-lifts, 440
    outside, 30, 32, 34–35
    protective, 44
    safety, 43
    in warehouses and storerooms, 380–381
    in waste disposal facilities, 769
    in wet locations, 43–44
    for woodworking areas, 555
Light-reflectance values (LRVs), 45–46
Limit devices
    for cranes, 393–394
    in elevators, 427, 435
    inspection of, 435
Limited-combustible construction, for fire prevention, 278–279
Line thermal detectors, 291
Linings, of cupolas, 658
Link fuses, 235
Linoleum floors, 104, 105
Liquefied gases, 374–380
Liquefied natural gas (LNG), 375
Liquefied petroleum gas (LPG)
    industrial fires and, 287
    storage and handling of, 38, 491–492
    for trucks, 491
    uses of, 343
Liquids. *See also* Combustible liquids; Flammable liquids
    cryogenic, 374–380
    hazardous, handling and storage of, 368–372
    volatility of, 321
List-servs, 708, 709–710
Lith-X dry powder extinguishers, 302
Live roll conveyors, 425
Loading
    of flammable and combustible liquids, 327–331
    powered industrial trucks, 485–486
    roofs, 106–107
    tracks at areas for, 35–36
Loading docks, 34, 108
Loads. *See also* Capacity ratings
    automated guided vehicles and, 688–689
    of cranes, 392, 408–409
    dynamic and working loads, 451
    elevator, 432
    floor load capacity and distribution, 54, 105, 652
    in foundries, 652
    ropes and, 449–451
    of slings, 464–469
    stability of, 486
Load transfer stations, 684
Localized general lighting, 42
Local (protected-premises) fire alarm systems, 295
Location. *See also* Hazardous (classified) locations
    of buildings and structures, 38–39
    color preferences and, 46
    of fire detection systems, 296
    of fire extinguishers, 297–299
    safeguarding by, 154, 162–163
    in site selection, 29
Locker rooms, 288–289
Lockers, for storage of flammable and combustible liquids, 333

Lockout/tagout
    in control of hazardous energy sources, 170–173
    design checklist for, 18
    documentation of, 171
    for electrical systems, 257–258
    for electrical woodworking equipment, 552, 554, 555
    equipment in, 173
    group, 173
    maintenance workers and, 122
    standards for, 687
    training in, 173
Locomotive cranes, 409–410
Locomotives, 36–37
Logs, on unfired pressure vessel inspections, 141
Long objects, handling and lifting, 362
Loose-fitting respirators, 202
Loss scenarios, 271–275
Louvar, Joseph F., 696, 698
Lower explosive level (LEL), 246, 326, 327
Lower flammable limit (LFL), 321, 335, 340
Low-pressure boilers, 133
Low voltage, defined, 226
Low-voltage ionization detectors, 292
Lubrication
    of conveyor ropes, 423
    of power presses, 644
    in preventive maintenance, 124
    of wire ropes, 458
Lumber, storage of, 367
Lurking, 709
Lymphocyte transformation test (LTT), 576

M
Machine guarding, design checklist for, 18
Machinery. *See also* Metalworking machinery; Woodworking machinery
    controls and displays for, 166
    elevators, 428
    escalators, 438–439
    handling and lifting, 362–363
    matching to operator, 163, 166
    for shipping and receiving, 381
Machinery guards, 167, 381. *See also* Guard devices
Magazine feeds, 625
Magazines, storage of explosives in, 374
Magnesite, 53
Magnesium, 577, 650
Magnesium grinding, 663
Magnetic air valves, 636
Magnetic circuit breakers, 236
Magnetic particle testing (MT), 143, 672
Magneto-inductive testing, 673
Maintenance. *See also* Facilities maintenance
    computerized management of, 118, 681–682, 687
    computerized predictive, 118–120
    conveyors, 419
    cranes, 401
    diagnostic technology for, 119–120, 687
    earth-moving equipment, 510
    electrical equipment, 255–258
    of elevators, 433–435
    fall arrest equipment, 197, 198
    of fiber ropes, 453–454
    fire detection systems, 296
    fire extinguishers, 303
    fire hoses, 308
    for fire prevention, 283–289, 312
    forging hammers, 668–669
    forging presses, 671
    forging upsetters, 670
    of foundries, 652
    grounding, 245
    guard devices, 170
    hand and power tools, 521–523
    helmets, 183–184
    housekeeping practices and, 104–105
    kick presses, 635
    motors, 239–240

pallets, 488
personal protective equipment, 181
personnel, 120–125
powered industrial trucks, 488–492
power presses, 643–644
power shovels, 507
pressure vessels, 137, 146–147
preventive and predictive, 102, 112, 121, 124, 682, 686
in process safety management, 727–728
respirators, 209–210
robotic equipment, 694–695
safety and, 686–687
saw blades, 560–561
shop equipment, 124–125
of wire ropes, 455
Management
of change processes, 19–20, 700, 724
of electrical safety programs, 261
of lighting, 42–43
of risk, 721–722
of safety and health programs, 61, 62–63, 756
Managers, construction, 60, 61, 75
Maneuvering, of powered industrial trucks, 487
Manganese, 577, 650
Manifolds, 588, 609
Manila ropes, 448, 450, 454
Man-lifts, 440–443
Manual feeding methods, 624
Manual handling, accessories for, 363–365
Manual lifting, 356
Manual materials handling (MMH), 359, 360
Manual of Standards and Recommended Practices (AAR), 327
Manual rates, 80
Manuals
for automated systems, 686
for chemical processes, 695, 699–700
Manufacturers
recommendations for inspection and maintenance, 686
responsibility for guarding and safeguarding press brakes, 649
Manufacturing philosophies, 680–682
Marble floors, 104
Marking
guidelines for, 39, 47–48
tools for, 525
Masonry, 280, 544
Masts, 482
Material Safety Data Sheets (MSDSs). *See* Safety Data Sheets (SDSs)
Materials handling and storage, 353–384. *See also* Waste collection
and disposal
accessories for, 363–365
aids for, 166
in automated systems, 688–690
cryogenic liquids, 374–380
fall protection in, 357
in foundries, 653–656
of hazardous materials, 368–374
injury prevention in, 354–357
lifting guidelines for, 357–363
manual handling accessories, 363–365
personal protective equipment for, 354–355, 363
personnel selection for, 359–361
planning for, 365–366
power presses and, 623
of raw and finished products, 40–41
rigid containers, 366–367
shipping and receiving, 380–384
tools for, 363–364, 528–529
training for, 358–359
uncrated stock, 367–368
in woodworking, 555
Materials recovery facilities (MRFs), 767–768
Matrices, for risk assessments, 8–9
Maximum permissible exposure (MPE), 556
Maximum permissible limit (MPL), 358
Means of egress, design checklist for, 16
Mechanical filter respirators, 207
Mechanical integrity, 728
Mechanical parts accidents, 692

Mechanical press brakes, 647
Mechanical ventilation, 581
Mechanical/visual warnings, 685
Medical programs, in foundries, 651–652
Medical surveillance, respirator use and, 211
Mercury, 577–578
Metal active gas (MAG) process, 595–596
Metal fume fever, 576
Metal grates, 53
Metal inert gas (MIG) process, 595
Metal mesh slings, 472–473
Metals. *See also* Cold forming of metals; Hot working of metals
hot slag, 579
nondestructive testing of, 672–674
powdered, 373
scrap, handling and lifting, 362
sheet, 361, 367–368
storage of, 367–368
tools for cutting, 524–526
Metal saws, 608
Metal shears, 644–646
Metalworking machinery, 601–618
boring machines, 605–606
codes and standards for, 617
electrical controls on, 603
general safety rules for, 602–603
grinding machines, 611–617 (*See also* Grinding machines)
injuries caused by, 602
milling machines, 606–610
personal protective equipment for, 603
planing machines, 610–611
turning machines, 604–605
Metatarsal footwear, 214
Meters, for trace metal detection, 651
Methane gas, 765
Methodologies, of process safety management, 680
Met-L-Kyl extinguishers, 302
Met-L-X dry powder extinguishers, 302
Micro switches, 234
Midstream operations, in petroleum industry, 752–754
Milling machines, 606–610
Mine Safety and Health Administration (MSHA), 204, 205, 503
Minors, labor laws for, 492
Mists, 200
Miter saws, 537–538
Mittens, 212
Mixing rooms, 332
Mixtures, flammable and combustible, 322
Mobile cranes, 408–413, 508–509
Mobile dumpers, 354
Mobile tool cabinets, 521
Moisture, in electric motors, 240
Molded aural inserts, 192
Molds and molding machines, in foundries, 655, 661
Molybdenum, 578
Monitor nozzles, 309
Monitors, laser, 687
Monopolistic states, 79
Monorail systems, 403, 421
Motions, guarding, 167–168
Motive power, for facility railways, 36–37
Motorized hand trucks, 479, 483
Motors, 238–240, 642
Mouthpiece-style respirators, 206
Movable-barrier devices, 628
Moving walks, 440
Mowing and mowers, 114–115, 511
Multimeters, 256, 257
Multiple-operator machines, control system in, 638
Municipal waste, 765, 766
Musculoskeletal disorders (MSDs), 355, 764

N

Nailers, 543–544
Nail hammers, 524
Nails, 382–383
Nanomaterials, 349–352
clearing for use at workplace, 350–351

defined, 350
engineering controls for, 351–352
explosion and fire risks, 352
monitoring results of use, 351
personal protective equipment for, 351–352
preparing for workplace exposure to, 351
regulatory issues, 352
Nanoscience, defined, 350
Nanotechnology, defined, 350
National Association of Chain Manufacturers, 467
National Board of Boiler and Pressure Vessel Inspectors Code (NBIC), 130–131, 134, 135, 142, 145
National consensus standards, 755–756. *See also* Codes and standards
National Council on Compensation Insurance (NCCI), 79, 80
National Electrical Code (NEC)
on cranes and hoists, 394
on double-insulated tools, 534
electrical fixtures compatible with room contents, 332
elevators and, 428, 433
explosion-proof equipment, 255
extension cords, 240
field marking, 233
on flammable gas precautions, 377
flammable materials, 43
grounding requirements, 44, 242, 243, 245, 246, 552
on hazardous locations, 373, 481
lighting conditions, 110
on magnetic particle inspection, 672
protective devices, 236
on welding equipment, 593, 594
wiring and installations, 29, 40, 229, 230–231, 247, 326, 602–603, 635
National Electrical Manufacturers Association (NEMA), 244, 593
National Elevator Industry, Minimum Passenger Elevator Requirements for the Handicapped, 437
National Fire Protection Association (NFPA)
alarm systems and, 295
classification of fire, 269
Fire Protection Handbook, 270, 309
National Fire Codes, 29, 280
NFPA 1, Uniform Fire Code, 276
NFPA 10, Portable Fire Extinguishers, 285, 297, 298, 303
NFPA 11, foam and foam systems, 309
NFPA 12, carbon dioxide extinguishing systems, 309
NFPA 13, Installation of Sprinkler Systems, 304, 305, 309, 365, 380
NFPA 14, Installation of Standpipe and Hose Systems, 308
NFPA 15, Water Spray Fixed Systems for Fire Protection, 307, 309
NFPA 16, foam and foam systems, 309
NFPA 17, dry chemical extinguishing systems, 309
NFPA 17A, wet chemical extinguishing systems, 309
NFPA 22, Water Tanks for Private Fire Protection, 304
NFPA 24, Installation of Private Fire Service Mains and Their Appurtenances, 307
NFPA 25, Inspection, Testing, and Maintenance of Water-Based Fire Protection Systems, 304–305
NFPA 30, Flammable and Combustible Liquids Code, 38, 39, 320, 326, 332, 333, 335–337, 339, 369, 491, 656
NFPA 33, Spray Application Using Flammable or Combustible Materials, 342
NFPA 34, Dipping, Coating and Printing Processes Using Flammable or Combustible Liquids, 341
NFPA 37, Installation and Use of Stationary Combustion Engines and Gas Turbines, 342
NFPA 50B, Liquefied Hydrogen Systems at Consumer Sites, 377
NFPA 51, Oxygen-Fuel Gas Systems for Welding, Cutting, and Allied Processes, 587–589
NFPA 51B, Fire Prevention during Welding, Cutting, and Other Hot Work, 284, 578
NFPA 58, Liquefied Petroleum Gas Code, 491, 492
NFPA 59A, Production, Storage, and Handling of Liquefied Natural Gas, 377
NFPA 61A–D, fire and dust explosions, 373
NFPA 69, explosion prevention systems, 309
NFPA 70 (See National Electrical Code (NEC))
NFPA 70E, Electrical Safety Requirements for Employee Workplaces, 161, 225, 228, 258, 688
NFPA 72, National Fire Alarm and Signalling Code, 296

NFPA 77, Recommended Practice on Static Electricity, 323, 326
NFPA 79, Electrical Standard for Industrial Machinery, 161, 603, 635, 688
NFPA 80A, Recommended Practice for Protection of Buildings from Exterior Fire Exposures, 38, 278
NFPA 80, Fire Doors and Other Opening Protectives, 281
NFPA 85A, 85B, 85D, 85E, boiler-furnace standards, 131, 133
NFPA 86, Ovens and Furnaces, 342
NFPA 91, Exhaust Systems for Air Conveying of Vapors, Gases, Mists, and Noncombustible Particulate Solids, 332, 341
NFPA 101, Life Safety Code, 30, 42, 44, 50, 52, 276, 418, 440
NFPA 231C, Rack Storage of Materials, 366
NFPA 306, Control of Gas Hazards on Vessels, 580
NFPA 321, Classification of Flammable and Combustible Liquids, 320
NFPA 326, Safeguarding of Tanks and Containers for Entry, Cleaning, or Repair, 333, 340, 579
NFPA 327, Cleaning and Safeguarding Small Tanks and Containers, 369, 579
NFPA 495, Explosive Materials Code, 374
NFPA 505, Powered Industrial Trucks, 481, 490, 492, 684
NFPA 704, Standard System for Identification of Hazards of Materials for Emergency Response, 271, 369
NFPA 5000, Building Construction and Safety Code, 276
Principles of Fire Protection, 276
National Institute for Occupational Safety and Health (NIOSH)
on fatal accidents, 693
on hearing protectors, 191
on nanomaterials, 351, 352
National Occupational Research Agenda Services Sector group, 770
NIOSH 76–128, fall prevention, 357
NIOSH 94–127, back belts, 357
on personal protective equipment and industrial hazard-measuring instruments, 199, 201
respirator testing and certification, 203–205, 207, 210
Work Practices Guide to Manual Lifting, 356
Workshop on Prevention through Design (PtD), 4
National Institute of Standards and Technology (NIST), 375
National Occupational Research Agenda (NORA), 770
National Safety Council (NSC)
Aviation Ground Operation Safety Handbook, 478
on back belts, 357
Fundamentals of Industrial Hygiene, 185, 188, 191, 650, 652
Hearing Conservation in the Workplace: A Practical Guide, 605
injury frequency and severity as reported by, 88
NSC Data Sheet 12304-0445, Soldering and Brazing, 532
NSC Industrial Data Sheet 359, Electromagnets Used with Crane Hoists, 402
NSC Industrial Data Sheet 12304-0595, Floor Mats and Runners, 429
NSC Industrial Data Sheet 12304-0635, Lead-Acid Storage Batteries, 491
NSC Industrial Data Sheet 12304-D717, Robots, 685
NSC Occupational Safety and Health Data Sheet 622, Tractor Operation and Roll-Over Protective Structures, 505
NSC Occupational Safety and Health Data Sheet 743, Mobile Cranes and Power Lines, 412
NSC Occupational Safety and Health Data Sheet, 12304-0236, Powder-Activated Hand Tools, 545
NSC Occupational Safety and Health Data Sheet 12304-0662, Ultrasonic Nondestructive Testing for Metals, 672
Occupational Vibration: Preventing Injuries and Illness, 542
National Transportation Safety Board (NTSB), 737, 738–740, 743, 744–745
National Waste and Recycling Association (NWRA), 766, 770
Natural fiber ropes, 448, 450–451
Natural rubber latex (NRL) products, 212
Natural ventilation, 581
NEC. *See* National Electrical Code
Negative-pressure respirators, 202, 203
Negative-pressure tests, for respirators, 208, 209
Nets, in fall arrest systems, 194–195
Network Enabled Weather technology, 747
Networking tools, 709–710
Newton, Isaac, 46
Next Gen program, 747
NFPA. *See* National Fire Protection Association

Nickel, 578
NIOSH. *See* National Institute for Occupational Safety and Health
Nip points or bites, 154, 158–159, 168
Nitrogen oxides, 574–575
Noise and noise control
color considerations and, 46
design checklist for, 18–19
in foundries, 652
guards and, 175
for planers, 568
for pneumatic-impact tools, 541
of screw machines, 605
in welding, 582
in woodworking, 555–556
Noise exposure, permissible levels of, 190–191
Noise-induced hearing loss (NIHL), 190
Noise reduction ratings (NRRs), 191
Nonbulk storage, of flammable and combustible liquids, 331–333
Noncombustible construction, for fire prevention, 278–279
Nondestructive testing
electromagnetic, 143, 673
magnetic particle, 143, 672
of metals, 672–674
penetrant, 143, 672–673
of pressure vessels, 142–144
radiography, 143, 673–674
triboelectric method, 673
ultrasonic, 119, 673
Nonionizing radiation, 582
Nonpulmonary gases, in welding, 575
North American Industry Classification System (NAICS), 758, 770
Notice of injury clauses, 85
Nozzles, hose, 308–309
NSC. *See* National Safety Council
N-series filters, 207
Nylon ropes, 448–449, 451–452

O

Occupational acoustic trauma, 190
Occupational noise-induced hearing loss, 190
Occupational Safety and Health Act of 1970 (OSH Act)
confined-space hazards in, 109
employer responsibilities under, 756
on management of change processes, 20
Personal Protective Equipment Standard, 199
Process Safety Management standard, 92, 718, 720, 728, 729–730
on state-run occupational safety and health programs, 755
Occupational Safety and Health Administration (OSHA)
on anchorage points, 195
on carbon monoxide exposure, 657
29 CFR 1910, Subpart O, Machinery and Machine Guarding, 628
29 CFR 1910.6, national consensus standards, 756
29 CFR 1910.22, marking guidelines, 39
29 CFR 1910.27, Fixed Ladders, 108
29 CFR 1910.95, Occupational Noise Exposure, 190
29 CFR 1910.119, Process Safety Management of Highly Hazardous Chemicals, 720, 723, 729
29 CFR 1910.120, Hazardous Waste Operations and Emergency Response, 768
29 CFR 1910.123–126, dip tanks, 341
29 CFR 1910.132, personal protective equipment, 182, 214, 260, 261
29 CFR 1910.133, eye and face protection, 261, 535
29 CFR 1910.134, respirators, 199, 211, 519
29 CFR 1910.135, head protection, 261
29 CFR 1910.136, foot protection, 214, 261
29 CFR 1910.137, electrical protective equipment, 260, 261
29 CFR 1910.138, hand protection, 261
29 CFR 1910.139, biological agents, 200
29 CFR 1910.145, Specifications for Accident Prevention Signs and Tags, 233
29 CFR 1910.146, Permit-Required Confined Spaces, 33, 141, 327, 369
29 CFR 1910.147, lockout standard, 141, 170, 419, 609, 642, 687, 694
29 CFR 1910.177, Servicing Multi-Piece and Single-Piece Rim

Wheels, 510
29 CFR 1910.178, powered industrial truck training, 478, 480, 498
29 CFR 1910.179, charging floors, 656
29 CFR 1910.184, synthetic slings, 449, 463, 467, 472
29 CFR 1910.211 through 1910.222, Machinery and Machine Guarding, 156, 161, 171
29 CFR 1910.213, Woodworking Machine Requirements, 556
29 CFR 1910.217, injury reporting requirements, 175
29 CFR 1910.243, Explosive Actuated Fastening Tools, 545
29 CFR 1910.269, General Industry Standard, 504
29 CFR 1910.331–333, safety programs, 261
29 CFR 1910.1000, air contaminants standards, 539
29 CFR 1910.1200, Hazard Communication Standard, 116, 117, 371, 540, 656, 683, 720, 725–726, 767
29 CFR 1926, Subpart M, Fall Protection, 546
29 CFR 1926, Subpart P, Excavations, 109
29 CFR 1926.6, national consensus standards, 756
29 CFR 1926.404, grounding, 519
29 CFR 1926.550, construction site clearances, 411, 509
29 CFR 1926.600, construction site rules, 510
29 CFR 1926.601, heavy equipment, 503, 504
29 CFR 1926.602, operator-restraint systems, 503
29 CFR 3133, Process Safety Management Guidelines for Compliance, 680
Combustible Dust National Emphasis Program, 539
on compressed air cleaning, 603
contractor citation history, 89
on cranes, 391, 401–402, 411
electrical equipment regulations, 229
employer responsibilities under, 756
on facility safety regulations, 768
on fiber ropes, 449, 452
on flammable and combustible liquids, 320
frequently asked questions, 498
General Duty Clause, 754, 755
on gloves, 212
on hardware connectors, 196
on hazards assessment, 180
Instruction CPL 2-2.45, Systems Safety Evaluation of Operations with Catastrophic Potential, 720
on pesticides, 118
Petrochemical Industries Compliance Directive, 720
petroleum industry and, 754–755
on portable power tools, 533
record retention requirements, 97
on risk reduction, 154
robotic accidents, categorization of, 691–692
slings, 449, 463, 464, 466, 467, 472, 473
website of, 708
on wire ropes, 455, 458, 460
Occurrence probabilities, 7–8
Office of Accident Investigation and Prevention, 737
Off-road equipment. *See* Haulage and off-road equipment
Offshore Continental Shelf (OCS), 755
Offshore drilling operations, 754–755, 758
Offsite limits, 79
Ohm's law, 222
Oil. *See also* Petroleum industry
analysis of, 119, 120
crucibles and, 659
electric motors and, 240
Oil burners, 342
Oil house layouts, 335
Oil-wet clutches, 643
Online training, 710
On-property transportation, for flammable and combustible liquids, 332
Open back stationary (OBS) press, 629
Open-circuit SCBAs, 203–204
Open-end wrenches, 529
Open flames, in industrial fires, 286–287
Open-frame forging hammers, 664
Open-sided floors, 49
Open states, 79
Operating control stations, 646
Operating envelope (space), 685
Operating modes, of press brakes, 648

Operating procedures manuals, 699–700
Operations and premises liability insurance, 77
Operators
  backward and forward driving by, 504
  boilers and unfired pressure vessels, 138, 144
  chain saws, 545
  cranes, 398, 407
  daily checklist for, 499
  for elevators, 435–436
  fatigue of, 166
  injury reporting requirements, 175
  maintenance practices for, 507
  matching machine or equipment to, 163, 166
  performance testing of, 494, 495
  powder-activated tools, 545, 546
  powered industrial trucks, 487, 492–494, 500
  power presses, 626
  press brakes, 646
  qualifications and training of, 503–504
  robotic equipment, 693–694
  safety practices of, 503–504, 509, 510–512
  seat belt use by, 503
  of woodworking machinery, 555
Ordinary construction, for fire prevention, 279
Orientations, safety, 66–67, 91
OSHA. See Occupational Safety and Health Administration
Outdoor Power Equipment Institute, 116
Outriggers, of cranes, 410
Outside facilities, 30–34
Ovens
  drying, 341–342
  gas-fired, 660
  hazards of, 659–660
  inspection of, 660
  Japanning, 341–342
  ventilation for, 660
Over-current devices, 234
Overhead-crane runways, 36
Overhead cranes, 391, 401–402
Overhead protection, in powered industrial trucks, 482
Overhead swing saws, 561–563
Overloading
  of electric motors, 238
  of elevators, 436
  of floors, 105–106
  of roofs, 106
Overload protective devices, in conveyors, 418–419
Overturning, of utility tractors, 115
Owner-controlled insurance programs (OCIPs), 81
Oxidation-reduction reactions, 268
Oxides of nitrogen, 574–575
Oxyfuel welding and cutting, 585–592
Oxygen
  levels for combustion, 321
  limiting in fire, 269
  special precautions for, 377
  in welding, 585–589
Oxygen deficiency, 108, 200, 326
Ozone, 574

P

Packing materials, 368
Padding equipment, 671
Page, Charles G., 235
Painting, color considerations and, 44–49
Pallet lifts, 480
Pallets, 485, 488
Palmar grip, 360
Pan shuttle extractors, 633
Paper, storage of, 367
Parapets, maintenance of, 103
Parking, for facility equipment, 39–40
Parking lots, 30, 31–32
Parquet flooring, 53
Partial-revolution clutch power presses, 629, 630
Participation, in discussion groups, 709
Particles, protection against, 546
Particulate matter, in welding, 575–576

Particulate-removing respirators, 202, 206, 207
Passengers, on escalators, 439
Passive fall arrest systems, 193
Paving brick, 53
Payment, for personal protective equipment, 182
Payroll, 80
Pedals, on forging hammers, 665
Pedestrian bridges, 30
Pedestrian safety, powered industrial trucks and, 487
Pendant switches, 234
Penetrant testing (PT), 143, 672–673
Percussion tools, 544
Permissible exposure limits (PELs)
  for carbon monoxide, 491, 657
  noise and, 190–191
  in welding and cutting, 580–581
Permits
  construction, 73, 91
  electrical, 266
  hot-work, 284, 578–579, 725
  safe work, 725
Personal protective equipment (PPE), 179–218
  for battery handling, 489–490
  for boiler-related work, 134
  for chemical hazards, 683
  clothing, 215–217, 355, 583, 584
  for cryogenic liquids, 377
  decontamination considerations, 213–214
  effectiveness of, 11
  for electrical safety, 258–260
  for eyes (See Eye protection)
  face protection, 185, 187–188, 556
  fall arrest systems, 192–198
  footwear (See Foot protection)
  forging hammers and, 668
  hand and arm protection, 211–214, 356
  for hazardous liquids, 369
  head protection, 182–185
  hearing protection, 190–192, 546, 555–556, 605, 664
  for maintenance workers, 122
  for materials handling, 354–355, 363
  for metalworking machinery, 603
  for nanomaterial exposure, 351–352
  for pesticides and chemicals and, 116–117
  policy statements regarding, 180
  for power tools, 546–547
  program for introducing, 180–182
  respirators, 198–211 (See also Respiratory protection)
  selection of, 180–181
  static electricity and, 373
  supervisor and management responsibilities for, 63
  training on, 181
  for waste collection and disposal, 767–769
  for welding, 183, 188–190, 583–584
  for woodworking, 556
Personnel. See also Employees; Operators; Workers
  designated safety representative, 63–64
  facilities for, 28, 652
  keeping up-to-date, 125
  maintenance, 120–125
  for materials handling, 359–361
  procurement, 70–71
  safety in loading and unloading flammable and combustible liquids, 328–329
  selection of, 398, 435–436, 492–493
Personnel nets, 194
Pesticides, 112–113, 116–118
Petroleum industry, 751–760
  codes and standards for, 754–756
  hazards and hazard management in, 753–754, 756–759
  health and safety management programs for, 756–758
  historical overview, 752
  offshore drilling operations, 754–755, 758
  operational sectors of, 752–754
  safety in, 754–756
  trade associations in, 755
Petroleum liquids, electrical charges in, 323, 324
Philosophies, of manufacturing, 680–682

Phosgene, 575
Phosphine, 575
Phosphorus, 650
Photoelectric smoke detectors, 291–292
Pile extractors, 410
Pilots, certification for, 738. *See also* Aviation safety
Pinch hazards, 684
Pinch points, 154, 623, 624, 633
Pipelines and piping
    for aboveground storage tanks, 337
    blowdown, 134
    on boilers, 134, 136
    color-coding of, 48–49, 122, 123
    for gas cylinders, 588–589
    handling, 123
    handling and lifting, 362
    hazardous liquids in, 369–370
    identification of, 122
    isolation of, 123
    at landfills, 765
    maintenance of, 109, 122–123
    storage of, 367
Pipe wrenches, 531
Pitch, 383
Pits
    elevator, 428
    foundation, 102
    in foundries, 653
    maintenance of, 102
PITs. *See* Powered industrial trucks
Planers, 553, 566–569, 610–611
Planning
    for fire protection, 276–278
    for materials storage, 365–366
    for safety, 680–681
Plasma arc welding, 595
Plastics, in fire prevention, 280
Plate glass, lifting, 361–362
Platforms, 49–50, 108, 110, 402
Pliers, 531
Plug doors, 36
Plug fuses, 235
Plunge cutting, 537
Plunger feeds, 631, 632
Pneumatically powered cylinders, 632–633
Pneumatic conveyors, 422–423
Pneumatic devices, 362
Pneumatic die cushions and springs, 644
Pneumatic-impact tools, 541–542
Point of operation. *See also* Guard devices
    defined, 154
    for power presses, 626–629
    protective devices, 156–159, 628–629
    safeguards, 160–166, 554, 626–629
Polarity plugs, 246
Policy statements, 62, 180
Polishing wheels, 617
Pollution, 29, 32, 111
Polyester ropes, 449, 450
Polyolefin ropes, 449
Polypropylene ropes, 450
Portable containers, for hazardous liquids and solids, 370–371, 373–374
Portable conveyors, 424
Portable fire extinguishers, 296–303
Portable furnaces, 287
Portable outlet headers, 589
Portable power tools, 532–546. *See also specific tools*
    air-powered, 541–544
    carrying, 521
    central control of, 520
    checklist for inspection of, 520
    electrical, 534–541
    flexible shaft, 544
    gasoline-powered, 544–545
    hazards and safety precautions for, 519–520, 532–533
    hydraulic, 544
    maintenance, inspection, and repair, 521–523, 533–534

percussion, 544
    personal protective equipment for, 546–547
    powder-actuated, 545–546
    preventing incidents with, 518–521
    selection of, 518–519, 533
    storage of, 520–521
Portable toolboxes, 520–521
Positioning device systems, 195
Positive-pressure blowers, in foundries, 657
Positive-pressure respirators, 202, 203–204
Positive-pressure tests, for respirators, 208, 209
Potassium bicarbonate–based dry chemical extinguishers, 301
Powder-Actuated Tool Manufacturer's Institute, 545
Powder-actuated tools, 545–546
Powder-cutting processes, 592
Powdered metals, 373
Powered industrial trucks (PITs), 477–496
    batteries for, 489–491
    care of, 486
    checklists for, 488, 489–490, 499
    classifications of, 478
    electric, 489–491
    gasoline-operated, 491
    general operating guidelines, 484–488
    in hazardous locations, 481–482
    identification of, 481
    inspection and maintenance, 488–492
    load capacity of, 481, 488
    loading and unloading, 485–486
    maneuvering, 487
    operators of, 487, 492–494, 500
    rider-controlled, 478–479
    safeguards for, 479–483
    safety requirements for, 480–481, 487
    training for, 478, 493–494, 500
    types of, 478–479
Power elevators, 426–437. *See also* Elevators
Power-feed ripsaws, 564–565
Power-feed (thickness) planers, 568–569
Power lawn mowers, 114
Power lift tables, 354
Power nailers, 543–544
Power press brakes. *See* Press brakes (bending brakes)
Power presses, 622–644
    auxiliary mechanisms, 629–630
    cleaning dies and clearing jams in, 625
    electrical controls on, 635–639, 642
    elevators and conveyors for, 633
    feeding and ejecting mechanisms, 623–625, 629, 630–633
    foot control and shielding, 629–630
    in foundries, 663–664
    hand tools for, 625–626
    inclining, 624
    inspection and maintenance of, 641–644
    material-handling hazards, 623
    open back stationary press, 629
    part and scrap ejection, 624
    primary operations (blanking), 623–624, 630
    safeguarding, 626–629, 644
    secondary operations, 624–626, 630
    setup and die removal, 639–641
    single-stroke attachments for, 630
    stock material for, 623–624
Power shovels, 506–508
Power sources, inspection and maintenance of, 642
Power squaring shears, 644–645
Power staplers, 543–544
Power supply, in welding, 592, 593
Power tools. *See* Portable power tools
Power transmissions, 154, 166–170
PPE. *See* Personal protective equipment
Pre-action systems, 305
Pre-bid meetings, 71, 90–92
Preconstruction meetings, 71
Predictive maintenance. *See* Preventive and predictive maintenance (PPM)
Pre-job planning and safety analysis outline, 20, 24–25
Premiums, insurance, 76

Pre-Operational Equipment Safety Inspections (POIs), 503
Presence-sensing devices, 629
Press brakes (bending brakes), 646–649
    general-purpose, 647
    hydraulic, 646, 647
    mechanical, 647
    in production system, 646
    special-purpose, 647–648
    tools for, 648–649
Pressure-demand respirators, 203–204
Pressure gauges, 148
Pressure parts, 130
Pressure regulators, 589–590
Pressure relief
    for cryogenic containers, 375, 377
    in flammable and combustible liquids tank cars, 330
Pressure vessels. *See* Boilers; Unfired pressure vessels
Pressurized liquid cylinders, 379
Pre-start-up reviews, 699, 727
Prevent Blindness America, 182, 185–186
Prevention
    of dangerous mixtures, 322
    of fire (*See* Fire prevention)
    of indoor air quality problems, 112
    of spill of flammable and combustible liquids, 331
Prevention through Design (PtD), 4
Preventive and predictive maintenance (PPM), 102, 112, 121, 124, 682, 686
Pre-work meetings, 67, 72, 93–94
Pre-work planning, 71
Price-related criteria, in contractor selection, 93
Primary operations, of power presses, 623–624, 630
Privileged communication, 731
Probability, defined, 6
Probability levels, in fire risk, 274
Process checklists, 697–698
Process hazard analysis (PHA), 680
Process safety management (PSM), 717–732
    accountability in, 721
    auditing in, 700–701, 728–731
    change management in, 724
    confidentiality and trade secrets in, 729
    for contractors, 726
    design and quality assurances in, 727
    elements of, 680, 722–729
    emergency planning, preparedness, and response in, 728
    employee involvement in, 723
    enforcement of, 729–730
    government involvement in, 720–721
    hazard analysis in, 723–724
    historical perspective on, 718
    incident investigations and, 729
    industry and labor involvement with, 718–719
    information needed in, 722–723
    maintenance and mechanical integrity in, 727–728
    methodologies of, 680
    operating procedures in, 724–725
    pre-start-up reviews and, 727
    requirements for, 721
    risk management and, 721–722
    safe work practices and permits in, 725
    standards and regulations for, 729
    training on, 725–726
Procurement personnel, 70–71
Procurement phase, of facilities construction, 71–72
Products of combustion (ionization) detectors, 292
Professional interpretations, 708
Program quality verification (PQV), 729–730
Progress meetings, 72–73, 95
Progress reports, 73
Project conceptualization, 69
Project engineers, 74–75
Project safety managers, 64
Project safety plans, 62
Proof tests, 142
Propagation of flame, 321
Protected ordinary construction, 279
Protected-premises (local) fire alarm systems, 295

Protective devices. *See also* Personal protective equipment (PPE)
    for electrical equipment, 234–237
    point of operation, 156–159
Protective lighting, 44
PSM. *See* Process safety management
Psychological factors, color considerations and, 46
Pull-back devices, 628, 629
Pullers, 532
Pulleys, on forging hammers, 665–666
Pulmonary gases, in welding, 574–575
Pulmonary irritants, in welding, 576–578
Pumps
    fire, 307
    for flammable-liquid transfer, 335–336
Punches, 522, 524, 671
Punching actions, 169
Purchasing documents, including safety specifications in, 20
Purging, of unfired pressure vessels, 142
Purpose, in control of hazardous energy sources, 171
Push-button switches, 234
Push feeds, 625, 631, 632

Q

Qualified Operator's Card, for powder-actuated tools, 545, 546
Qualitative fit tests, for respirators, 208
Quality assurances, 727
Quality management, correlation with safety through design, 5–6
Quantitative fit tests, for respirators, 208
Quarter-face respirators, 206

R

Rabbeting and dadoing, 557, 559
Radar-frequency tests, 673
Radial saws, 553, 563–564
Radiation
    hazards of, 684
    nonionizing, 582
    in welding, 188–190, 574
Radiation hazard symbol, colors of, 48
Radio frequency interference (RFI), 686
Radiography, 143, 673–674
Railcars, 36, 37, 384
Railings, 29, 35, 49, 51, 656
Railroad tracks, powered industrial trucks on, 484–485
Railway crossings, warning systems for, 34–35
Railways, facility, 34–38
Rakes, 528–529
Ramps, 40, 49–50, 380
Rams (slides), 642–643, 665
Rate-compensated thermal detectors, 291
Rate of diffusion, 321
Rate-of-rise thermal detectors, 291
Rebreather devices, 203
Receiving facilities, 30
Receptacle testers, 240, 245
Reciprocating motions, hazards created by, 167–168
Reciprocating saws, 537
Recognition clubs, 182
Records and record keeping
    for contractor close-out, 97
    by contractors, 92
    on unfired pressure vessel inspections, 141
Recycling, 764, 765, 767–768. *See also* Waste collection and disposal
Redesign, risk elimination and reduction in, 10
Redox reactions, 268
Redressing tools, 521–522
Reducing valves, 146
Reels, storage of, 367
Reference gauges, 148
Reference library, Internet as, 710
References, in contractor evaluations, 89–90
Reflected beam photoelectric detectors, 292
Reflected glare, 42
Reflection method, of nondestructive testing, 673
Regenerative-style air heaters, 133–134
Regulating valves, 146
Regulations. *See* Codes and standards
Regulators, in gas cylinders, 589–590

Reinforcement programs, 64–65
Remediation proposals, 8
Repairs
    aircraft, 742
    cupolas, 658
    floors, 106
    hand and power tools, 521–523
    portable power tools, 533–534
    robotic equipment, 694–695
    roofs, 107
Reports
    on aviation safety, 743
    of construction progress, 73
    design safety, 23
    to DOT, 329
    incident, 67, 95, 175, 729
Request for bid (RFB), 90–92
Rescue systems, 197
Rescue teams, 33, 34
Residual risk, 6, 8
Resins, 650–651
Resistance, defined, 225
Resistance welding, 592–593
Resonant-frequency method, of nondestructive testing, 673
Resource Conservation and Recovery Act of 1976 (RCRA), 109, 334
Respiratory protection, 198–211
    cleaning and sanitizing, 210
    in confined spaces, 33, 108
    for cryogenic gases, 378
    effectiveness of, 202
    fitting, 207–208
    in foundries, 657
    hazard identification and evaluation for, 200–201
    inspection of, 210–211
    maintenance of, 209–210
    medical surveillance and, 211
    for pesticide use, 117, 118
    programs for, 198–199
    selecting, 199, 207
    storage of, 209
    training on, 211
    types of, 201, 202–207
    in welding, 583
Response systems, for fire protection, 289–296
Responsible Care Program, 719
Restraint devices
    for powered industrial trucks, 479–480
    for power presses, 628
    as safeguard, 164
Restraint systems, 197, 503
Restricted areas
    boundaries between nonrestricted and, 684–685
    landfills and, 766
    for robotic equipment, 693
Restricted envelope (space), 684–685
Restricted-use pesticides (RUPs), 116
Retracting lifeline devices, 195
Reviews, of design and constructability, 70–71
Revolving die feeds, 631
Rider-controlled trucks, 478–479
Riding mowers, 115
Rigging, 462–463
Rigid containers, 366–367
Ring knives, 528
Ripping, with saws, 556–558, 563–564
Ripsaws, 564–565
Risers, 51
Risk
    acceptable, 6, 9
    defining, 6, 8
    of fire, 270–275
    ranking, 8
    residual, 6, 8
    ropes and, 450
Risk assessments
    of chemical processes, 696–698
    defined, 6
    in design, 699

    flowchart for, 155
    matrices for, 8–9
    for nanomaterials, 351
    overview, 6–7
    process for, 7–8
    for robotic equipment, 692
    timing of, 698–699
Risk management, 721–722
Risk matrix, 275
Riveting hammers, 524
Riveting machines, 672
Roadways
    in building layout, 30–31
    for heavy equipment, 503, 510–511
    safe procedures on, 510–511
Robotic equipment, 690–695
    accidents involving, 691–693
    classification of, 690
    computer-integrated controls for, 695
    definitions and terminology, 684–685, 690
    functions of, 690–691
    hazards of, 173–174, 691–695
    maintenance and repair personnel for, 694–695
    operators of, 693–694
    prevalence of, 690
    risk assessments for, 692
    safeguarding, 173–175, 692
    teachers of, 692–693
Robotic Industries Association (RIA), 690, 692
Robot movement zone, 173–174
Rockefeller, John D., 752
Rolled paper and reels, storage of, 367
Roller conveyors, 425
Rollers, 363
Rolling-bolster die carriages, 644
Rollover protective structures (ROPS), 502, 503, 505, 511
Roofs, maintenance of, 106–107
Ropes, 448–474
    checklist for selection of, 448
    conveyor hauling, 423
    elevator hoisting, 394, 426–427, 432–433
    fastenings, 433
    fiber, 448–454 (See also Fiber ropes)
    inspection of, 451–453, 455–458
    rigging, 462–463
    risks and, 450
    slings, 463–473 (See also Slings)
    synthetic, 197
    wire, 454–462 (See also Wire ropes)
Rotary die grinders, 538–539
Rotary limit switches, 635
Rotating equipment, electrical, 258
Rotating motions, hazards created by, 167–168
Round objects, handling and lifting, 362
Routers, 540–541
Rubber boots, 215
Rubber flooring, 53, 104
Rubbish, disposal of, 288
Runways
    in boiler rooms, 138
    crane, 36, 123–124
    design specifications, 49–50
Rupture disks, 145

S
Saber saws, 538
Sacks
    lifting, 361
    storage of, 368
    unpacking, 382
Safeguarding, 153–176. See also Guard devices
    aerial baskets, 415
    of core-blowing machines, 661
    definitions and terminology, 154
    electrical equipment, 230–233
    exclusions for two-hand systems, 636
    of forging hammers, 664–666
    of hazardous energy sources, 170–173

by location, 154, 162–163
maintenance and servicing, 170
matching machines to operators, 163, 166
materials for, 169–170
for metal shears, 644–645
openings used for, 156–159
point-of-operation, 156–166, 626–629
powered industrial trucks, 479–483
for power presses, 626–629, 644
for power transmissions, 166–170
for press brakes, 648, 649
robotic equipment, 173–175, 692
substitution and, 163
types of, 161–165
Safety
aviation, 735–748 (*See also* Aviation safety)
awareness, methods for maintaining, 710–711
communication as component of, 105, 144
cranes and, 401, 410–411
defined, 6
design for, 13, 28–29
design-in, 681, 682, 689
electrical, 221–263 (*See also* Electrical safety)
elevators and, 433, 434–435, 437–438
in facilities construction, 60–74
facility railways, 38
fire protection, 267–315 (*See also* Fire protection)
for flammable and combustible materials, 321–327
hand tools and, 113–114, 519–520
for haulage and off-road equipment, 502–506
lighting, 43
maintenance and, 686–687
manufacturing philosophies and, 680–682
metalworking machinery and, 602–603
for mowing, 114–115, 511
in petroleum industry, 754–756
portable power tools and, 519–520, 532–533
powered industrial trucks and, 480–481, 487
slings and, 466, 469–470, 472–473
as technical evaluation criterion, 87–90, 92
tool-related, 519–520
up-front planning for, 680–681
in waste collection and disposal, 768–770
in welding, 578–580, 584–585
woodworking machinery and, 552–556
Safety Assessment of Foreign Aircraft (SAFA), 741
Safety blocks, for power presses, 623, 636
Safety clauses, in contracts, 84–86
Safety Data Sheets (SDSs)
for flammable and combustible liquids, 322, 333
in foundries, 651
for hazardous materials, 271, 722
materials handling and, 356
for pesticides, 118
in welding, 581
Safety design processes, defined, 6
Safety devices
for automated guided vehicles, 689–690
on dumbwaiters, 438
on escalators, 438
on hand elevators, 437–438
incorporation of, 10
on man-lifts, 441–442
for pressure vessels, 145–146
Safety Equipment Institute (SEI), 181
Safety goggles. *See* Goggles and glasses
Safety harnesses, 141–142
Safety hazards. *See* Hazards
Safety management, 61, 62–63
Safety Management System (SMS), 746
Safety orientations, 66–67, 91
Safety professionals, role of, 12–14
Safety programs, 60–69
benefits of, 60
designated safety representatives for, 63–64
documentation of, 96–97
electrical, 261, 262
emergency procedures in, 68–69

establishing, 60–61
evaluation of, 71–72, 89
housekeeping practices and, 67–68
incident reporting and investigations, 67
inspections in, 65
Internet as management tool for, 711
managing, 61, 62–63, 756
in petroleum industry, 756–758
policy statements for, 62
reinforcement and discipline in, 64–65
responsibility and accountability in, 63
state-run, 755
substance abuse programs and, 68
training in, 65–67
Safety props, for forging hammers, 666
Safety reviews, 698
Safety through design, 4–6
Safety training. *See* Training
Safety valves, on boilers and unfired pressure vessels, 133, 134–135, 145, 146
Safe work permits, 725
Sand, handling of, 654
Sandblast rooms, 662
Sand cutters, 660–661
Sanders, 539–540, 553, 569
Sand mills, 660
Sanitizing. *See* Cleaning and sanitizing
Saws, 556–566
band, 536, 553, 565–566, 608
blades for, 535–536, 560–561
chain, 544–545
circular, 535–536, 553, 556–561, 608
cutoff, 561–563
hacksaws, 525
hot, 672
jigsaws, 538, 566
kickback of, 537, 544–545, 556–557, 563–564
metal, 608
miter, 537–538
radial, 553, 563–564
reciprocating, 537
ripsaws, 564–565
saber, 538
safe operation of, 536, 553
swing, 561–563, 608
table, 553, 557, 560
woodworking, 526
work space for, 554
Scale, for forging presses, 671
Scale guards, 665
Schedule estimates, 69–70
Scissor lifts, 110, 585
Scope, in control of hazardous energy sources, 171
Scrap breakers, in foundries, 655
Scrapers, 502, 505, 509–512
Scrap metals, handling and lifting, 362
Scratch wheels, 617
Screening
of contractors, 87
for drugs and alcohol, 68
of waste products, 766
Screw conveyors, 421–422
Screwdrivers, 522, 523
Screw-geared chain hoists, 391
Screw machines, 604–605
Search engines, 707
Search strategies, for Internet, 707–708
Seat belt use, 479, 503
Secondary operations, of power presses, 624–626, 630
Secondhand pressure vessels, 140
Security systems, 44
Selective participation, in discussion groups, 709
Self-audits, 173
Self-contained breathing apparatus (SCBA), 201, 203–204, 330, 378
Self-evaluation privilege, 731
Self-insurance states, 79
Semiautomatic feeds, 164, 630–631
Sensor systems, 293

Separate units, for fire prevention, 280–281
Sequence of lockout, in control of hazardous energy sources, 171
Servicing. *See* Maintenance
Setting hammers, 524
Settlement, 102, 103
Severity, of harm or danger, 6
Severity categories, in fire risk, 274
Sewers, 108
Shackle conveyors, 421
Shaft alignment, 119
Shakeout machines, 662
Shapers, 553, 567–570, 610
Shavings, removal of, 603
Shearing actions, 169
Shear points, 154
Shears, 644–646, 655
Sheaves
    of fiber ropes, 454
    hoisting ropes and, 394
    wire ropes and, 458–460
Sheet metal, 361, 367–368
Shielded-metal arc welding (SMAW), 582, 594, 596
Shielding, for foot controls, 629–630
Shields. *See also* Guard devices
    on drill presses, 605, 606
    face, 187–188
    on lathes, 604
    on milling machines, 607
Shipping and receiving, 30, 380–384
"Ship's ladders," 51
Shock, electrical. *See* Electrical shock
Shock absorbers, 196
Shoes, 214–215, 354, 356
Shop equipment maintenance, 124–125
Shovels, 506–508, 528
Sick building syndrome, 111
Siderosis, 575
Sidewalk elevators, 437
Sidewalk maintenance, 108
Sidewalks, 31
Sifters, 661
Signage
    accident prevention, 47–49
    color-coding and, 47–49
    DANGER, 229
    on dumbwaiters, 438
    for electrical equipment, 233
    in elevators, 436, 437
    on escalators, 439
    for floor load capacity, 105, 106
    for man-lifts, 442
    road construction and, 510
    traffic, 31
    warning, 422
Signals and signaling systems
    crane and hoist, 395–398
    for dumping, 504
    for elevators, 433
    for flaggers, 511
    in pressure vessel maintenance, 144
Silica, 651
Silicones, 651
Silos, 372–373
Single-chamber ionization detectors, 292
Single-stream materials recovery facilities, 767
Single-stroke attachments, for power presses, 630
Single-stroke control, for power presses, 638
Single-stroke mode, for press brakes, 648
Single-use respirators, 202
Sisal ropes, 448, 451
Site orientations, 66–67, 91
Site planning, for fire protection, 277–278
Site selection, 29–30
"Skelping," in crucibles, 659
Skin injuries, from electricity, 228
Skylights, 41, 107
Slag, 579, 656
Slag wool, 657

Slat conveyors, 420–421
Sledgehammers, 524
Slide feeds, 631
Slides (rams), 642–643
Sliding-die feeds, 631, 632
Slings, 463–473
    braided, 463, 464
    chain, 465, 466–470
    fiber and wire rope, 463–466
    hitches, 462–463, 473
    inspection of, 466, 467, 469, 471–473
    metal mesh slings, 472–473
    methods of attachment, 464, 467
    safe operating practices for, 466, 469–470, 472–473
    synthetic web slings, 470–472
    working load of, 464–469
Slipping hazards
    floor maintenance and, 104–105, 770
    in waste collection and disposal, 769–770
Slotters, 611
"Slow Down to Get Around" campaign, 770
Slow-moving vehicle (SMV) emblems, 511
Smith forging hammers, 664
Smoke
    controlling, 282–283
    defined, 200
    detectors for, 291–292, 296
Smoking, 286, 322, 376–377
Snaphooks, 196
Snap switches, 234
Snips, 526
Snow shoveling, 116
Snow throwers, 116, 232
Society of Automotive Engineers (SAE), 503
Sockets, for impact wrenches, 542
Socket wrenches, 529–530
Sodium bicarbonate–based dry chemical extinguishers, 301
Soldering irons, 532
Sole puncture resistant footwear, 215
Solids, hazardous, handling and storage of, 372–374
Solid Waste Association of North America (SWANA), 770
Solvents, 650, 687
Source reduction, for waste, 766
Space requirements
    for aboveground storage tanks, 337, 338–339
    in site selection, 29–30
    for woodworking machinery, 554–555
Spacing, of fire detectors, 296
Spark-resistant tools, 125, 326, 532
Special-purpose press brakes, 647–648
Speed
    conditions governing, 510
    as factor in accidents, 484
    for grinding machines, 614–615, 617
Spill containment caddies, 362
Spill prevention, control, and countermeasure (SPCC) regulations, 331
Spinning lathes, 605
Splices, wire ropes and, 460–461
Spontaneous ignition, in industrial fires, 287–288
Spray booths, 342–343
Spray nozzles, 309
Spring buffers, in elevators, 427, 435
Sprinklers
    alarms in, 306
    automatic, 304–306
    causes of failure in, 306–307
    clearance under, 365
    connections for, 283
    temperature rating of, 306
    types of, 306
    in warehouses and storerooms, 380
    water supply and storage for, 303–304
Spur-geared chain hoists, 391
Stacking of containers, 365–366
Stacks, maintenance of, 107
Staff. *See* Employees; Operators; Personnel; Workers
Stairways
    in boiler rooms, 138

design checklist for, 19
in facility layout, 50–51
fixed, 19, 50–51
in foundries, 653
maintenance of, 110–111
Stakeholders, in aviation safety, 742–744
Stamping and marking tools, 525
Standard Handbook for Mechanical Engineers (Avallone & Baumeister), 133
Standard Oil Company, 752
Standard operating procedures, 104
Standards. *See* Codes and standards
Standpipes, 283
Stannosis, 576
Staplers, 543–544
Star drills, redressing, 522
Statement of work, 71
State-run occupational safety and health programs, 755
Static dissipative footwear, 215, 377
Static electricity
conveyors and, 419
dust and, 373
flammable and combustible liquids and, 323–326
liquid oxygen and, 377
Steam boilers, 133
Steam hammers, 664, 665
Steaming, for cleaning tanks and containers, 333–334
Steam-jacketed vessels, 147
Steam locomotives, 37
Steam pressure indicators and controls, 135–136
Steam vs. high-temperature water, 139
Steel chutes, 633
Steel construction, for fire prevention, 278
Steel parts, maintenance of, 103
Step-down transformers, 225, 226
Stiff-leg derricks, 404, 405
Stock, uncrated, 367–368
Stock picking, 381
Stone caps, maintenance of, 103
Stop control, for power presses, 637
Stop work, 74
Storage. *See also* Materials handling and storage
of abrasive wheels and disks, 612
of combustible materials, 288
of crucibles, 659
of cryogenic liquids, 378–380
of fall arrest systems, 198
of flammable and combustible liquids, 38, 331–340
floor loads and, 106
in foundries, 655–656
of gas cylinders, 367, 374
inside, 40–41, 332–333
of liquefied petroleum gas, 38, 491–492
of pesticides, 118
planning for, 365–366
of respirators, 209
of rigid containers, 366–367
of safety records, 97
for tools, 520–521
of uncrated stock, 367–368
Storage bridge cranes, 403
Storage of fiber ropes, 453–454
Stored-pressure extinguishers, 301, 302
Straddle trucks, 479, 482
Straightening equipment, 671
Straight-line pull cutoff saws, 561–563
Strain, kick presses and, 633–634
Strands, in wire ropes, 454
Strapping, application and removal of, 381–382
Strategies for Today's Environmental Partnership (STEP), 719
Stretch, of ropes, 451
Strip feeds, 624
Strippers, for power presses, 627
Stroboscopic effect, 43
Stroking control systems, 638, 647–648
Stroking selectors, 637–638
Structural members, maintenance of, 102–103
Structures. *See also* Facilities

layout of, 49–55
location of, 38–39
roof-mounted, 107
Subcontractors, 60, 76
Subcontracts, 84
Subject-specific indexes, 707
Substance abuse programs, 68
Substitution
of less hazardous methods or materials, 10, 759
as safeguard, 163
Suffocation, 575
Sulfur dioxide, 651
Superaural hearing protectors, 191, 192
Superheaters, 133
Supervising-station fire alarm systems, 295
Supervisors
competence of, 64
responsibilities in safety programs, 63
for robotics operators, 694
in welding, 585
Supplementary lighting, 42
Supplied-air respirators (SARs), 201, 204–206
Surface fires, 268
Surface grinders, 616
Surveys, hazard, 697
Swain, Alan D., 13
Sweatbands, 187
Swing saws, 561–563, 608
Switch crews, in overhead-crane area, 36
Switches
on boilers, 136
electrical, 233–234
for elevator cars, 433
in facility layout, 35
warnings of, 36
Synthetic fiber ropes, 448–452
Synthetic harnesses, 197
Synthetic Organic Chemical Manufacturers Association (SOCMA), 719
Synthetic web slings, 470–472
System grounding, 40, 242–243

T
Table saws, 553, 557, 560
Tackle blocks, 415–416
Tagout. *See* Lockout/tagout
Tank cars and trucks
connections for, 330–331
hazardous liquids in, 371–372
in loading and unloading of flammable and combustible liquids, 327, 328–330
placards and shipping papers for, 331
pressure relief in, 330
removing covers of, 330
spill prevention, control, and countermeasures for, 331
static electricity in, 323–325
Tanks
abandonment of, 341
aboveground, 38–39, 336, 337–339
anchoring of, 336
cleaning, 333–334, 340–341, 369
construction of, 334, 335
facility layout of, 38–39
fires in, 340
gauging for, 336
of hazardous liquids, 369
maintenance of, 107
pumps for, 335–336
spill control for, 339–340
underground, 336–337, 369
ventilation for, 335, 337
welding of, 579–580
Tap and die work, 525
Tapping out, in foundries, 657
Teachers, of robotic equipment, 692–693
Team lifting, 361
Technical evaluation, of contractors, 87–90, 92–93
Telemetry, 747

Temperature
  alloy steel chain and, 467
  of auto-ignition of flammable liquids, 320–321
Temperature rating, of sprinklers, 306
Tensile strength, of steel chains, 467
Terminal Radar Approach Control Facilities (TRACON), 737
Terrain, in site selection, 29
Terrazzo flooring, 53, 104, 105
Test data, for guard devices, 157, 158
Testing equipment
  combustible gas indicators, 326–327
  electrical, 256–257
  insulation-resistance testers, 533
Thermal circuit breakers, 236
Thermal fire detectors, 291
Thermographs, 120, 256
Thickness measurements, in pressure vessels, 142–144
Threading, on gas regulators, 589
Three-wire adapters, 245–246
Threshold Limit Values (TLVs), 491, 556, 575, 581
Through-transmission method, of nondestructive testing, 673
Tiering hoists, 417
Tight building syndrome, 111
Tight-fitting respirators, 202
Tilting spouts, 657
Timing, of risk assessments, 698–699
Titanium, 578
Toe protection, 214–215, 356. *See also* Foot protection
Tongs, 531, 670
Toolboxes, 520–521
Toolbox talks, 67
Tool chests, 521
Tools and appliances. *See also* Hand tools; Portable power tools; *specific tools*
  air-powered, 541–544
  for bricklayers, 522, 524
  carrying, 521
  central control of, 520
  changing, 519
  checklist for inspection of, 520
  for cutting, 524–528, 531–532
  double-insulated, 246, 534
  electrical, 534–541
  ergonomic analysis of, 518
  extension cords for portable, 240
  for facility railways, 37
  for forging hammers, 666
  for grounds maintenance, 113–114
  handles of, 522–523
  identification of, 625–626
  inspections of, 72, 520, 521, 533–534
  machine tool classifications, 602
  for materials handling, 363–364, 528–529
  mechanic check of, 124
  percussion, 544
  powder-actuated, 545–546
  for power presses, 625–626
  for press brakes, 648–649
  safety practices for, 519–520
  selection of, 518–519, 533
  spark-resistant, 125, 326, 532
Torches, 287, 591–592
Torque wrenches, 530
Total quality management (TQM), 681
Tow conveyors, 421
Tower cranes, 405–408
Towers, maintenance of, 107
Towing, 506, 511
Toxic materials. *See* Hazardous materials
Tracks, in facility railways, 35
Traction-drive elevators, 426
Tractors and trailers, 115–116, 482
Trade associations, 745, 755
Trade secrets, 729
Traffic
  air traffic control, 737, 738, 747
  in fire protection site planning, 277
  in parking lots, 32

signs and signals for, 31
  in waste collection and disposal, 769
Training
  for automated systems, 687–688
  on aviation safety, 738, 743–744, 746
  for boiler and unfired pressure vessel operators, 138, 144
  on confined spaces, 33
  on construction sites, 65–67, 72, 91
  conveyors and, 420
  of crane operators, 398
  on cryogenic liquids, 378
  documentation of, 694, 726
  electrical safety, 260–261
  for fire prevention, 285
  for heavy-equipment operators, 503–504
  in lockout/tagout procedures, 173
  for maintenance and repair personnel, 120, 687
  in manufacturer-recommended procedures, 506
  for materials handling, 358–359
  online, 710
  on personal protective equipment, 181
  for powder-activated tools, 545
  for powered industrial trucks, 478, 493–494, 500
  on process safety management, 725–726
  on respiratory protection, 211
  of robotics operators, 694
  for welders, 584–585
Transfer lines, for cryogenic liquids, 380
Transfer stations, for waste disposal, 766–767
Transformers, 225, 226
Transit limits, 79
Transportation. *See also* Tank cars and trucks
  in fire protection site planning, 277
  of flammable and combustible liquids, 332
  personnel safe practices, 38
  of tools, 521
  of waste products, 766
  of workers, 505–506
Transportation Department, U.S. (DOT)
  49 CFR 170–179, Hazardous Materials Regulations, 320, 327
  49 CFR 174.67, Tank Car Unloading, 330
  compressed gas cylinder regulations, 586
  Manual on Uniform Traffic Control Devices for Streets and Highways, 31, 32, 503, 510
  National Transportation Safety Board and, 739
  reporting using Form F 5800.1, 329
  specifications for liquid cylinders, 379
Transverse motions, hazards created by, 167–168
Trapping accidents, 692
Trash. *See* Waste collection and disposal
Travel, of cranes, 411
Treadles, on forging hammers, 665
Treads, 50, 51
Trenches, maintenance of, 108–109
Trend analysis, 743
Trent, Linda, 44
Trestles, 31, 35
Triboelectric method of nondestructive testing, 673
Tripping hazards, in waste collection and disposal, 769–770
Trolley conveyors, 421
Trucks. *See* Hand trucks; Lift trucks; Powered industrial trucks (PITs); Tank cars and trucks
Trunnions, of foundry flasks, 662
Tuberculosis (TB), 200
Tumbling barrels, 662
Tunnels, maintenance of, 108–109
Turning machines, for metalworking, 604–605
Turnover bar, for power presses, 642
Turret lathes, 604–605
Tuyeres, 657
TWA Flight 2 crash (1956), 737
Twinned extinguishers, 303
Two-hand control systems
  circuits and grounding in, 639
  construction features of, 637–638
  devices for power presses, 628, 629
  foot operation of, 639
  installation of, 635–636

interlocks for, 638–639
minimum safety distance between hand controls and nearest hazard, 637
safeguarding exclusions for, 636
Two-hand tripping systems
advantages and limitations of, 165
construction features of, 636–637
devices for power presses, 628
installation of, 635–636
minimum safety distance between hand controls and nearest hazard, 636
safeguarding exclusions for, 636
Two-wheeled trucks, 364–365
Type-A movable-barrier devices, 628
Type-B movable-barrier devices, 628

U

U-groove, in elevators, 426
Ultrasonic examination (UT), 143
Ultrasonic testing, 119, 673
Ultraviolet detectors, 292
Ultraviolet radiation, in welding, 188–190, 574, 582
Uncrated stock, 367–368
Underground pipelines, maintenance of, 109
Underground storage tanks (USTs), 336–337, 369
Underground utilities, maintenance of, 108–109
Underslung cutoff saws, 563
Underwriters Laboratories (UL)
containers for handling flammable liquids, 322
Electro-Sensitive Protective Equipment, 161
fire, security, and emergency facility condition indicators at, 295
on gas distribution systems, 589, 591
on safety cans, 491
Standard for Safety No. 558 and 583, powered industrial trucks, 481, 491, 684
tools approved by, 113
Unfired pressure vessels, 139–147
autoclaves, 146–147
cleaning and purging, 142
codes and standards for, 130–131
design of, 19, 139–140
detecting cracks and measuring thickness in, 142–144
entering, 141–142
evaporating pans, 147
hydrostatic tests in, 142
inspections of, 131–132, 140–142
operator training, 144
overview, 130
safety devices for, 145–146
steam-jacketed vessels, 147
Uniform Building Code, 110
Universal Safety Oversight Audit Program (USOAP), 741
Unloading
of flammable and combustible liquids, 327–331
powered industrial trucks, 485–486
tracks at areas for, 35–36
Unmanned aircraft systems (UASs), 747
Unwanted energy release concept, 13–14
Up-front planning for safety, 680–681
Upper explosive level (UEL), 246
Upper flammable limit (UFL), 321
Upstream operations, in petroleum industry, 752–754, 758–759
U.S. Air Mail Service, 736
U.S. Army Corps of Engineers, 424
U.S. Coast Guard, 755
Utilities, maintenance of, 108–109
Utility tractors, 115–116
Utility trenches and tunnels, 108–109

V

Vacuum, in pressure devices, 368
Vacuum breakers, 145
Valves
on aboveground storage tanks, 337
blowdown, 134
on gas cylinders, 589–590
magnetic air, 636
safety, 133, 134–135, 145, 146

Vanadium, 578
Vapor pressure, 321
Vapor-removing respirators, 202, 206–207
Vapors
defined, 200
explosions and, 289
flammable, 247
in welding, 574
Vehicle-mounted extinguishers, 303
Vehicle restraints, 382
Vehicles. *See* Haulage and off-road equipment; Lift trucks; Tank cars and trucks
Vendors, manuals from, 686
Ventilation. *See also* Heating, ventilation, and air conditioning (HVAC)
air cleaners, 582
conveyors and, 419
design checklist for, 19
facility layout and, 40
for fire protection, 282
in foundries, 652
liquefied petroleum gas trucks and, 492
mechanical, 581
natural, 581
for ovens, 660
for tanks, 335, 337
in unfired pressure vessels, 141, 142
in welding, 581–582
Ventricular fibrillation, 227
Vents, in pressure vessels, 145–146
Vertical boring mills, 606
Vertical conveyors, 425
Vertical shear, in guard tests, 158
V-groove, in elevators, 426
Vibration
analysis of, 120
of electric motors, 240
of foundry sifters, 661
hand-arm vibration syndrome, 518, 542, 547
protection against, 518, 542
Vinyl-based tile, 53, 105
Violence, emergency procedures for, 68–69
Visual/mechanical warnings, 685
Visual testing (VT), 143
Volatile organic compound (VOC) restrictions, 687
Volatility, of liquids, 321
Voltage
in arc welding, 593
defined, 222, 226
in two-hand control systems, 639
Voltage detectors, 256, 257

W

Waivers of subrogation, 78
Walking surfaces, design checklist for, 19
Walkways
in building layout, 30, 31, 51
conveyors and, 417–418
cover in open areas, 35
on cranes, 402
sidewalk elevators and, 437
Wall openings, 19, 29
Walls
design checklist for, 19
foundation, 102
maintenance of, 102, 103–104
Warning devices and methods
in ash removal areas, 134
cranes and, 403, 409
effectiveness of, 10–11
for facility railways, 34–35
on powered industrial trucks, 481
visual/mechanical, 685
Warning signs. *See* Signage
Washington State law HB 2171, for crane inspection, 399
Waste collection and disposal, 763–771
in building layout, 30, 32
convenience centers for, 767

equipment for, 768–769
facility maintenance, 109
of flammable liquids, 334
hazards of, 764, 766–770
history of, 764
industrial fires and, 288
landfills for, 32, 764, 765–766
modern technology for, 764–765
personal protective equipment for, 767–769
pesticides, 118
recycling and, 764, 765, 767–768
safety issues in, 768–770
traffic considerations in, 769
transfer stations for, 766–767
weather considerations in, 769
worker protection in, 764, 770
Waste transfer stations, 766–767
Water, supply and storage of, 277, 303–304
Water-flow alarms, in sprinklers, 306
Water-level indicators and controls, 135–136
Water seals, 145
Water solution extinguishers, 300
Water-spray systems, 307
Water treatment, in boilers, 134
Watertube boilers, 133
Watts, defined, 226
Wear particle analysis, 119–120
Weather considerations, in waste collection and disposal, 769
Web directories, 707
Weed trimmers, 113–114
Welding and cutting, 573–597
    arc welding and cutting, 593–596
    controlling hazardous exposures, 580–583
    definitions and terminology, 574
    eye protection for, 188–190
    in foundries, 663
    head protection for, 183, 189
    health hazards of, 574–578
    industrial fires and, 287
    lasers in, 596
    oxyfuel welding and cutting, 585–592
    personal protective equipment for, 183, 188–190, 583–584
    resistance welding, 592–593
    safety hazards in, 578–580
    training for, 584–585
Wells, oil, 752
Wet chemical extinguishers, 302–303
Wet locations
    electrical tool use in, 534
    lighting in, 43–44
Wet-pipe systems, 305
Wharves, designing, 34
"What if?" analysis, 680, 698, 724
Wheelbarrows, 364
Wheel checks, 381
Wheel conveyors, 425
Wheeled extinguishers, 303
Wheel-mounted cranes, 409, 411
Wheels, grinding, 539
Winches, 197, 415
Winding-drum elevators, 426
Windows, 41, 103
Wire brushes and wheels, 539, 617
Wire gauge, 113
Wire ropes, 454–462
    breaking strength of, 455
    care of, 458
    classifications of, 454
    corrosion and, 458
    design factors for hoisting, 455
    fittings for, 460–462
    inspections and replacement, 455–458
    measuring, 456, 458
    service requirements for, 455
    sheave and drum recommendations for, 458–460
    slings, 463–466

spacing of clips for, 461
types of, 454–455
wear and damage of, 456–457
Wire Rope Users Manual (AISI), 454
Wire screen enclosures, 50
Wire strand cores (WSCs), 455
Wiring systems, for electrical equipment, 229–230
Wood
    blades for cutting, 561
    in fire prevention, 280
    flooring, 53–54, 104, 106
    tools for cutting, 526–527
Wood-frame construction, for fire prevention, 279
Wood lathes, 553, 569–570
Wood parts, maintenance of, 103
Wood shapers, 553
Woodworking machinery, 551–570
    codes and standards for, 556
    electrical, 552, 554
    general safety principles for, 552–556
    heights and work space for, 554–555
    inspection of, 555
    jointer-planers, 553, 566–567
    lathes, 553, 569–570
    materials handling and, 555
    personal protective equipment for, 556
    power-feed (thickness) planers, 568–569
    sanders, 539–540, 553, 569
    saws, 556–566 (*See also* Saws)
    shapers, 553, 567–570
Work clothing, protective, 215–217
Work environments
    in foundries, 652–653
    for powered industrial trucks, 479
Worker Protection Standard (EPA), 116, 117
Workers. *See also* Employees; Personnel; Workers with disabilities
    operating heavy equipment near, 504
    responsibilities in safety programs, 63
    transportation of, 505–506
    in waste industry, 764, 770
Workers' compensation insurance, 79–81
Workers with disabilities
    design checklist for, 14
    elevator requirements for, 437
    elevators and, 437
    parking lot design for, 31–32
Work height, for machines and equipment, 166, 554
Working surfaces, design checklist for, 19
Work methods design, 17
Work permits, 73, 91, 266, 284, 578–579
Work phase, of facilities construction, 72–73
Workplace
    behavior modification vs. redesign in, 12–13
    color and light in, 44–45
    design considerations, 28
    drug-testing programs in, 68
    first-aid kits at, 546–547
    inspections during construction, 65
    machine and equipment layout in, 166
    nanomaterials in, 350–351
    for woodworking equipment, 554–555
Work positioning systems, 197
Work procedures, evaluation of, 62–63
Work-product privilege, 731
Work release meetings, 93–94
Work rests, 615
Work safety files, 95
Workstation design, 17, 54–55, 166, 519. *See also* Ergonomics
Wrap-up insurance, 81–82
Wrenches, 529–532, 542–543
Wristlets, 213

Z
Zero mechanical state, 170
Zinc, 578
Zircon, 651